THE WATER ENCYCLOPEDIA

Second Edition

Frits van der Leeden
Fred L. Troise
David Keith Todd

LEWIS PUBLISHERS

Library of Congress Cataloging-in-Publication Data

van der Leeden, Frits.
The water encyclopedia / Frits van der Leeden, Fred L. Troise, David Keith Todd.—2nd ed.
p. cm.
Includes bibliographical references.
First ed. by: David Keith Todd.
1. Water-supply—Tables. 2. Hydrology—Tables.
I. Troise, Fred L. (Fred Louis) II. Todd, David Keith.
III. Title.
TD351.V36 1990
553.7'P021—dc20 89-14011
ISBN 0-87371-120-3

LEWIS PUBLISHERS, INC.
121 South Main Street, Chelsea, Michigan 48118

PRINTED IN THE UNITED STATES OF AMERICA

Frits van der Leeden is a consultant in ground-water hydrology and a Senior Vice President of Geraghty & Miller, Inc., a firm specializing in ground-water development and assessment and remediation of soil and water contamination problems.

Mr. van der Leeden has provided consulting services to industry and governmental and international agencies, including the U.S. Environmental Protection Agency, the Organization of American States, and the United Nations Development Programme, and has carried out water resources studies in many parts of the world. He is a member of the American Water Resources Association, the International Water Resources Association, and the Association of Ground Water Scientists and Engineers.

Mr. van der Leeden holds a BS in Geology from the College of the City of New York, and an MS in Geology from New York University. He is a registered professional geologist in several states and the author of numerous technical reports and publications, including the *Geraghty & Miller Ground-Water Bibliography,* the *Water Atlas of the United States* (co-author), and *Water Resources of the World.* He is a contributing editor to the Water Information Center, Inc.

Fred Louis Troise is a hydrogeologist and the Senior Vice President of Marketing of Geraghty & Miller, Inc. He also holds the position of Vice President of Water Information Center, Inc., a publishing subsidiary of Geraghty & Miller, which he manages for the parent company.

Mr. Troise has acted as a consultant to industry and local government on ground-water development projects in the United States and Puerto Rico, and has contributed to national, federally sponsored studies on ground-water contamination. His educational background includes a BS in Geology from Brooklyn College of the City University of New York, and graduate work in Geology at New York University. He serves on the executive committee of the northeast section of the American Institute of Professional Geologists, and is a registered Geologist in several states.

As a publisher, author, and editor, Mr. Troise has developed and produced many reference books, training guides, and periodicals on the subject of water. He is editor of *The Groundwater Newsletter,* a major subscriber-supported newsletter in the field, and managing editor of the *Water Newsletter* and *The International Water Report.* He is the Series Editor for the new *Geraghty & Miller Ground-Water Series* and co-author of the *Water Atlas of the United States,* and he contributed to and published David Keith Todd's first edition of *The Water Encyclopedia.*

David Keith Todd is President of David Keith Todd Consulting Engineers, Inc., located in Berkeley, California. The firm, established in 1978, specializes in the planning, development, management, and protection of water resources, with particular emphasis on ground water.

Dr. Todd also holds the position of Professor of Civil Engineering, Emeritus at the University of California, Berkeley. While on campus he taught all of the hydrology courses and was in charge of the graduate program in Water Resources Engineering. As a result of his extensive experience in teaching, research, and consulting, he has been associated with a wide variety of water problems, thereby earning an international reputation, particularly relating to underground water.

His educational background includes a BS in Civil Engineering from Purdue University, an MS in Meteorology from New York University, and a PhD in Civil Engineering from the University of California, Berkeley. He is a registered Civil Engineer in the states of California and Indiana. Best known among his many publications is his text *Groundwater Hydrology,* which has been translated into six languages. Dr. Todd served as the editor of the first edition of *The Water Encyclopedia.*

ACKNOWLEDGMENTS

The authors wish to express their thanks and appreciation to the many people who have, in one way or another, helped to make the second edition become a reality. More than 100 organizations, publishing firms, and individual authors gracefully gave permission to reproduce tables and charts or other source materials.

Although it is not possible to list all contributors, a special 'thank you' is in order for the following people: Vivian Collier, managing editor at Lewis Publishers, who cheerfully reviewed and edited the hundreds of tables and charts and carefully guided the composition and production of the book; Marie-Claire Troise who took on the exacting and exhausting task of proofreading the bulk of this second edition, duplicating her great efforts of 20 years ago on the first edition; Jeanette Manheim of Geraghty & Miller who handled all correspondence and requests for copyright permission; Mary Schwarz of the Water Information Center who computerized the contents of the various chapters during the compilation stage of the book; Jacqueline van der Leeden who, at a critical time, offered her assistance in proofreading and helped us meet our last deadlines; and finally, Brian Lewis of Lewis Publishers and David Miller of Geraghty & Miller who, over a period of three years, provided the steady support necessary to finish this project.

PREFACE

The second edition of *The Water Encyclopedia* is a completely revised and expanded version of the original edition published in 1970. Every entry and feature of the previous edition has been reexamined and, where necessary, replaced with new, more useful material, yet this edition preserves the best features of the first edition. The *Encyclopedia* is meant to serve the scientist as well as the general public as an important source of information about water and all its related aspects. Professional hydrologists, engineers, irrigation specialists, hydrogeologists, and persons interested in the hydrologic environment—all will find this encyclopedia a reliable guide to information about our world's most important and priceless commodity.

The second edition of the *Encyclopedia* has been greatly expanded and now contains more than 600 tables and 100 illustrations. As in the first edition, the subject matter has been broadly interpreted to include climates, hydrology, surface and ground water, water use, water quality, water management, water resources agencies, and legislation. The book ends with a chapter on constants and conversion factors. An entire new chapter, Environmental Problems, has been added, reflecting the greater awareness and concern by the people in this country and elsewhere about contamination of soil and water and environmental quality trends. Among the extensive body of material included in this chapter is information on pollution sources and pathways; contamination of surface water and ground water; use of pesticides and fertilizers; domestic, municipal and industrial waste disposal; water treatment and reuse; hazardous waste sites; offshore disposal; air pollution; acid rain; and the projected rise of sea levels due to the greenhouse effect. The environmental section concludes with a greatly expanded section on waterborne diseases and toxic effects of chemicals in drinking water.

The information is presented in tabular form, except in a few places where text is provided to explain a significant process or activities of a particular organization. Time-dependent data are identified so that the reader can judge the relation of information to the current situation or to his or her particular purpose. To facilitate the rapid location of specific data, particular attention has been given to making the index as complete as possible.

The second edition of *The Water Encyclopedia* is the result of much patient labor by the editors, specifically by Frits van der Leeden, who assumed the role of lead editor and who was instrumental in transforming a good idea into reality. Many governmental and private organizations, publishers, and individuals supplied material and gave permission to reproduce data, and these sources are acknowledged throughout the book.

This new edition is offered to the user in the same spirit as was the earlier edition—to make a small contribution to the goals for wide dissemination of water data expressed by the International Hydrological Decade program.

The Editors

CONTENTS

THE WATER ENCYCLOPEDIA

Second Edition

CHAPTER 1

Climate and Precipitation

PAGE

SECTION A. CLIMATIC DATA—UNITED STATES

TABLE 1–1. NORMAL DAILY MEAN TEMPERATURE—SELECTED CITIES OF THE UNITED STATES

[In Fahrenheit degrees. Airport data except as noted. Based on standard 30–year period 1951 through 1980]

ST.	STATION	JAN	FEB	MAR	APR	MAY	JUN	JUL	AUG	SEP	OCT	NOV	DEC	ANNUAL
AL	BIRMINGHAM	42.9	46.3	53.7	62.8	70.0	77.0	80.1	79.5	74.1	62.6	52.1	45.6	62.2
	MOBILE	50.8	53.6	60.1	68.0	74.9	80.5	82.2	81.8	78.2	68.5	58.6	53.1	67.5
AK	ANCHORAGE	13.0	17.9	23.7	35.4	46.3	54.4	58.1	56.2	48.2	34.6	21.7	13.8	35.3
	BARROW	-14.4	-19.6	-15.9	-1.7	18.8	33.3	38.9	38.0	30.6	14.2	-1.0	-12.9	9.0
	FAIRBANKS	-12.8	-4.0	8.5	30.2	48.2	59.3	61.5	56.6	44.9	25.0	3.9	-10.1	25.9
	JUNEAU	21.8	27.8	31.2	39.1	46.5	52.7	55.7	54.6	49.2	41.8	32.7	26.8	40.0
AZ	PHOENIX	52.3	56.1	60.6	68.0	77.0	86.5	92.3	89.9	84.6	73.4	60.6	53.3	71.2
	TUCSON	51.1	53.8	57.8	65.0	73.2	82.9	86.2	84.0	80.4	70.4	58.7	52.0	68.0
AR	FORT SMITH	37.5	42.4	50.5	61.4	69.6	77.5	82.1	80.9	73.9	62.5	49.8	41.2	60.8
	LITTLE ROCK	39.9	44.1	52.2	62.4	70.5	78.5	82.1	81.0	74.3	63.1	51.2	43.2	61.9
CA	BAKERSFIELD	48.2	53.2	57.1	62.7	70.6	78.3	84.5	82.4	77.3	68.0	56.2	48.2	65.6
	FRESNO	45.5	50.5	54.3	60.1	67.7	75.0	81.0	78.9	74.1	64.8	53.2	45.3	62.5
	LOS ANGELES[1]	57.2	58.9	59.5	62.0	64.9	69.2	74.1	74.7	73.4	68.9	62.7	58.3	65.3
	SACRAMENTO	45.3	50.3	53.2	58.2	64.9	71.2	75.6	74.7	71.7	63.9	53.0	45.6	60.6
	SAN DIEGO	56.8	58.4	59.0	61.2	63.4	66.3	70.3	72.2	71.0	67.5	61.8	57.4	63.8
	SAN FRANCISCO[1]	51.2	53.9	54.3	55.2	56.6	58.4	58.5	59.6	62.4	61.6	57.2	52.0	56.7
CO	DENVER	29.5	33.6	38.0	47.4	57.2	67.0	73.3	71.4	62.6	51.9	38.7	32.6	50.3
	GRAND JUNCTION	25.5	33.5	41.9	51.7	62.1	72.3	78.9	75.9	67.1	54.9	39.6	28.3	52.7
CT	HARTFORD	25.2	27.6	36.8	48.8	59.4	68.6	73.4	71.3	63.3	52.4	41.6	29.1	49.8
DE	WILMINGTON	31.2	33.2	41.8	52.4	62.2	71.2	76.0	74.8	67.9	56.3	45.6	35.5	54.0
DC	WASHINGTON	35.2	37.5	45.8	56.7	66.0	74.5	78.9	77.6	71.1	59.3	48.7	38.9	57.5
FL	JACKSONVILLE	53.2	55.1	61.3	67.7	74.1	79.0	81.3	81.0	78.2	69.5	60.8	54.8	68.0
	MIAMI	67.1	67.8	71.7	75.3	78.5	81.0	82.4	82.8	81.8	77.9	72.8	68.5	75.6
	TAMPA	59.8	60.8	66.2	71.6	77.1	80.9	82.2	82.2	80.9	74.5	66.7	61.3	72.0
GA	ATLANTA	41.9	44.9	52.5	61.8	69.3	75.8	78.6	78.2	73.0	62.2	52.0	44.5	61.2
	SAVANNAH	49.2	51.6	58.4	66.0	73.3	78.6	81.2	80.8	76.6	66.9	57.5	51.0	65.9
HI	HONOLULU	72.6	72.9	74.4	75.7	77.5	79.1	80.1	81.0	80.6	79.5	76.6	74.0	77.0
	LIHUE	71.3	71.4	72.0	73.4	75.5	77.4	78.4	79.1	78.9	77.3	75.3	72.9	75.2
ID	BOISE	29.9	36.1	41.4	48.6	57.4	65.8	74.6	72.0	63.2	51.9	39.7	32.0	51.1
	POCATELLO	23.8	29.5	35.5	44.6	54.0	62.5	71.2	68.9	59.2	48.1	35.2	26.6	46.6
IL	CHICAGO	21.4	26.0	36.0	48.8	59.1	68.6	73.0	71.9	64.7	53.5	39.8	27.7	49.2
IN	EVANSVILLE	30.6	34.9	44.6	56.5	65.6	74.6	78.1	76.2	69.4	57.3	45.1	35.6	55.7
	INDIANAPOLIS	26.0	29.9	40.0	52.4	62.5	71.6	75.1	73.2	66.6	54.8	41.8	31.5	52.1
IA	DES MOINES	18.6	24.5	35.1	50.5	62.1	71.6	76.3	73.9	65.1	54.2	38.6	25.7	49.7
	SIOUX CITY	16.2	23.2	33.8	49.7	61.5	70.9	75.6	73.3	64.5	52.5	36.4	23.3	48.4
KS	TOPEKA	26.1	32.5	41.8	54.6	64.6	73.9	78.6	77.0	68.5	57.0	42.7	31.8	54.1
	WICHITA	29.6	35.1	44.1	56.3	65.9	76.1	81.4	79.7	70.6	59.1	44.3	34.4	56.4
KY	LEXINGTON	31.5	34.6	43.9	55.1	64.2	72.2	76.0	74.9	68.7	56.8	44.9	36.1	54.9
LA	NEW ORLEANS	52.4	54.7	61.4	68.7	74.9	80.3	82.1	81.7	78.5	69.2	60.0	54.6	68.2
	SHREVEPORT	46.0	49.8	57.0	65.7	73.0	79.8	82.9	82.4	77.1	66.7	55.7	48.7	65.4
ME	PORTLAND	21.5	23.0	32.1	42.8	52.8	62.2	68.1	66.6	58.6	48.4	38.4	25.8	45.0
MD	BALTIMORE	32.7	34.7	43.3	54.0	63.4	72.2	76.8	75.6	68.9	56.9	46.3	36.5	55.1
MA	BOSTON	29.6	30.7	38.4	48.7	58.5	68.0	73.5	71.9	64.6	54.8	45.2	33.7	51.5
MI	DETROIT	23.4	25.8	35.0	47.4	58.1	67.7	71.9	70.5	63.3	51.9	39.5	28.5	48.6
	GRAND RAPIDS	22.0	23.7	33.1	46.3	57.5	67.1	71.4	69.6	62.1	50.9	38.5	27.3	47.5
	SAULT STE. MARIE	13.3	14.3	23.9	38.1	49.7	58.4	63.5	62.9	54.8	45.3	32.8	19.7	39.7
MN	DULUTH	6.3	12.0	22.9	38.3	50.3	59.4	65.3	63.2	54.0	44.2	28.2	13.8	38.2
	MINNEAPOLIS-ST. PAUL	11.2	17.5	29.2	46.0	58.5	68.1	73.1	70.6	60.6	49.6	33.2	19.2	44.7
MS	JACKSON	45.7	49.1	56.3	65.1	72.5	79.2	81.9	81.2	76.4	65.0	54.9	48.6	64.6
MO	KANSAS CITY	28.4	34.4	43.6	56.9	66.9	75.9	80.9	79.3	71.0	59.6	44.6	34.2	56.3
	ST. LOUIS	28.8	33.8	43.2	56.1	65.6	74.8	78.9	77.0	69.7	57.9	44.6	34.2	55.4
	SPRINGFIELD	31.5	36.2	44.6	56.2	64.8	73.4	78.0	77.0	69.5	58.2	45.1	36.2	55.9
MT	BILLINGS	20.9	28.4	33.8	44.6	54.9	64.0	72.3	70.3	59.4	49.3	35.0	27.1	46.7
	GREAT FALLS	18.7	26.7	31.4	42.7	53.2	61.9	69.3	67.5	57.4	47.9	34.0	25.7	44.7
	HELENA	18.1	26.0	31.6	42.3	52.2	60.1	67.9	65.9	55.6	45.1	31.4	23.5	43.3
NE	NORTH PLATTE	21.3	27.3	34.7	47.6	58.2	68.3	74.2	72.4	61.9	50.2	35.0	25.9	48.1
	OMAHA	20.2	27.2	37.3	52.2	63.3	73.0	77.7	75.2	65.8	54.5	39.5	27.2	51.1
	SCOTTSBLUFF	24.2	30.1	35.6	46.4	57.2	67.5	74.2	71.6	61.6	50.2	35.9	27.8	48.5
NV	ELKO	25.0	31.0	36.0	43.4	52.4	61.2	70.1	67.6	58.4	47.5	35.3	26.1	46.2
	LAS VEGAS	44.6	50.1	55.3	63.5	73.3	83.6	90.3	88.0	80.1	67.6	53.6	45.4	66.3
	RENO	32.2	37.4	40.6	46.4	54.6	62.4	69.5	66.9	60.2	50.3	39.7	32.5	49.4
NH	CONCORD	19.9	22.2	32.1	44.1	55.2	64.7	69.5	67.3	59.1	48.2	37.3	24.5	45.3
	MT. WASHINGTON	5.1	4.8	12.0	22.4	34.4	44.6	48.7	47.1	40.6	30.5	20.3	9.1	26.6
NJ	ATLANTIC CITY[1]	34.1	35.4	41.7	51.0	59.5	68.4	74.1	73.8	67.8	57.6	47.7	38.4	54.1
	NEWARK	31.3	32.8	41.2	52.1	62.3	71.5	76.8	75.5	68.2	57.2	46.5	35.5	54.2
NM	ALBUQUERQUE	34.8	39.4	46.2	55.1	64.3	74.5	78.8	76.1	69.0	57.4	44.0	35.6	56.2
NY	ALBANY	21.1	23.4	33.6	46.6	57.5	66.7	71.4	69.2	61.2	50.5	39.3	26.5	47.3

TABLE 1-1. NORMAL DAILY MEAN TEMPERATURE— SELECTED CITIES OF THE UNITED STATES (continued)

[In Fahrenheit degrees. Airport data except as noted. Based on standard 30-year period 1951 through 1980]

ST.	STATION	JAN	FEB	MAR	APR	MAY	JUN	JUL	AUG	SEP	OCT	NOV	DEC	ANNUAL
	BUFFALO	23.5	24.5	33.0	45.4	56.1	66.0	70.7	68.9	62.1	51.5	40.3	28.8	47.6
	NEW YORK[1]	31.8	33.4	41.4	52.4	62.5	71.4	76.7	75.4	68.3	57.7	47.2	36.2	54.5
	SYRACUSE	22.8	24.0	33.3	46.1	57.0	66.3	70.9	69.3	62.1	51.3	40.6	28.3	47.7
NC	ASHEVILLE	36.8	39.1	46.4	55.7	63.3	69.8	73.2	72.6	66.9	56.0	46.4	39.3	55.5
	RALEIGH	39.6	41.6	49.3	59.5	67.2	73.9	77.7	77.0	71.0	59.7	50.0	42.0	59.0
	WILMINGTON	45.6	47.4	54.1	63.1	70.7	76.6	80.3	79.7	74.8	64.5	55.4	48.2	63.4
ND	BISMARCK	6.7	14.5	26.0	42.5	54.9	64.3	70.4	68.8	57.3	46.1	28.6	15.4	41.3
	WILLISTON	6.6	14.6	25.3	41.8	54.5	63.8	70.0	68.1	56.6	45.1	27.6	14.9	40.8
OH	CINCINNATI	28.9	32.1	41.8	53.5	63.0	71.4	75.4	74.1	67.5	55.3	43.4	33.8	53.4
	CLEVELAND	25.5	27.4	36.6	48.1	58.2	67.6	71.6	70.4	64.1	53.2	41.8	31.1	49.6
	COLUMBUS	27.1	29.8	40.0	51.4	61.4	70.2	73.8	72.4	65.8	53.9	42.1	32.1	51.7
	TOLEDO	23.1	25.8	35.4	47.8	58.6	68.0	71.8	70.1	63.2	51.7	39.3	28.1	48.6
OK	OKLAHOMA CITY	35.9	40.8	49.1	60.2	68.4	77.0	82.1	81.1	73.3	62.3	48.8	39.9	59.9
	TULSA	35.2	40.7	49.3	60.9	69.1	77.7	83.2	81.7	73.8	62.6	49.2	39.8	60.3
OR	EUGENE	40.1	43.5	45.8	49.6	55.1	61.1	66.8	66.2	62.1	53.3	45.3	41.3	52.5
	MEDFORD	37.6	42.4	45.5	50.3	57.5	65.2	72.5	71.1	65.1	54.2	43.6	37.7	53.6
	PORTLAND	38.9	43.2	45.9	50.4	56.7	62.5	67.7	67.3	62.7	54.3	45.5	40.9	53.0
PC	GUAM	77.2	77.3	77.8	79.1	79.7	80.0	79.5	79.1	79.0	79.1	79.2	78.4	78.8
	JOHNSTON ISLAND	76.7	76.6	76.9	77.7	79.0	80.0	80.6	81.1	81.0	80.7	79.2	77.6	78.9
	KWAJALEIN, MARSHALL IS.	81.3	81.7	82.2	81.8	81.6	81.7	81.7	81.9	82.0	81.9	81.8	81.6	81.8
	PAGO PAGO, AMER SAMOA	80.7	80.8	81.0	80.5	79.9	79.7	78.9	78.7	79.4	80.0	80.3	80.6	80.0
	TRUK, E. CAROLINE IS.	81.2	81.2	81.4	81.5	81.5	81.5	81.0	81.1	81.2	81.3	81.5	81.4	81.3
	WAKE ISLAND	76.9	76.6	77.7	78.3	80.0	82.0	82.5	82.6	82.7	81.9	80.4	78.4	80.0
	YAP, W. CAROLINE IS.	80.3	80.5	81.0	81.7	81.7	81.4	81.0	80.8	81.0	81.2	81.3	80.8	81.1
PA	ERIE	24.5	25.0	33.5	44.9	55.0	64.6	69.1	68.2	62.1	51.7	40.7	30.0	47.5
	HARRISBURG	29.4	31.5	40.6	52.2	62.0	71.2	75.8	74.3	66.9	55.0	43.9	33.4	53.0
	PHILADELPHIA	31.2	33.1	41.8	52.9	62.8	71.6	76.5	75.3	68.2	56.5	45.8	35.5	54.3
	PITTSBURGH	26.7	28.8	38.5	50.1	59.7	68.1	72.0	70.6	64.1	52.5	41.6	31.4	50.3
PR	SAN JUAN	76.5	76.6	77.5	78.8	80.3	81.7	82.0	82.2	81.9	81.4	79.6	77.7	79.7
RI	PROVIDENCE	28.2	29.3	37.4	47.9	57.6	66.8	72.5	71.1	63.5	53.2	43.4	32.3	50.3
	CHARLESTON	47.9	49.8	56.7	64.3	72.2	77.6	80.5	80.0	75.7	65.8	56.7	50.0	64.8
SC	COLUMBIA	44.7	47.1	54.5	63.8	71.5	77.7	81.0	80.2	74.8	63.4	53.9	46.7	63.3
	GREENVILLE-SPARTANBURG	41.1	43.6	51.0	60.5	68.3	74.8	78.2	77.5	71.7	60.7	50.9	43.4	60.1
SD	RAPID CITY	20.8	26.0	32.6	44.6	55.6	65.2	72.6	71.4	60.9	49.7	34.9	26.1	46.7
	SIOUX FALLS	12.4	19.1	30.4	46.3	58.2	68.4	74.0	71.8	61.0	49.4	33.0	19.7	45.3
TN	BRISTOL-JOHNSON CITY-KINGSPORT	35.0	37.9	46.3	56.1	64.4	71.5	74.9	74.3	68.8	57.1	46.2	38.2	55.9
	KNOXVILLE	38.2	41.5	49.7	59.6	67.4	74.3	77.6	77.0	71.5	59.5	48.8	41.1	58.9
	MEMPHIS	39.6	43.5	51.7	62.6	71.0	78.7	82.1	80.6	74.2	62.9	51.3	43.3	61.8
	NASHVILLE	37.1	40.4	49.0	59.6	68.1	75.8	79.4	78.4	72.3	60.2	48.6	40.9	59.2
TX	AUSTIN	49.1	53.2	60.5	68.7	74.9	81.6	84.7	84.5	79.2	69.8	58.7	52.1	68.1
	CORPUS CHRISTI	56.3	59.3	65.9	73.0	78.1	82.7	84.9	85.0	81.5	74.0	65.0	59.1	72.1
	DALLAS-FORT WORTH	44.0	48.5	56.1	65.9	73.7	82.0	86.3	85.5	78.6	67.9	55.6	47.8	66.0
	EL PASO	44.2	48.4	55.1	63.6	71.9	80.8	82.5	80.3	74.1	63.6	51.4	44.4	63.4
	HOUSTON	51.4	54.5	61.0	68.7	74.9	80.6	83.1	82.6	78.4	69.7	60.1	54.0	68.3
	LUBBOCK	38.8	42.6	50.2	60.3	69.0	77.6	79.8	77.9	71.2	61.0	48.5	41.5	59.9
UT	SALT LAKE CITY	28.6	34.1	40.7	49.2	58.8	68.3	77.5	74.9	65.0	53.0	39.7	30.3	51.7
VT	BURLINGTON	16.6	18.1	29.2	42.7	55.2	64.9	69.6	67.1	58.8	47.9	36.6	22.6	44.1
VA	LYNCHBURG	35.1	37.4	45.7	56.4	64.5	71.6	75.7	74.8	68.4	57.1	47.0	38.3	56.0
	NORFOLK	39.9	41.1	48.5	58.2	66.4	74.3	78.4	77.7	72.2	61.3	51.9	43.5	59.5
	RICHMOND	36.6	38.9	47.2	57.9	66.1	73.5	77.8	76.8	70.2	58.6	48.9	39.9	57.7
WA	SEATTLE[1]	40.6	44.2	45.7	50.4	56.3	61.0	65.3	65.0	61.1	53.8	46.2	42.6	52.7
WA	SPOKANE	25.7	32.4	37.6	45.8	54.3	61.7	69.7	68.1	59.4	47.6	34.9	29.0	47.2
	WALLA WALLA	34.3	40.7	45.8	52.3	60.1	67.6	75.5	73.6	65.5	54.3	42.5	37.2	54.1
WV	CHARLESTON	32.9	35.6	44.8	55.3	63.9	71.0	74.5	73.7	67.6	55.9	45.3	36.9	54.8
WI	GREEN BAY	14.0	17.8	28.6	43.7	55.1	64.7	69.5	67.5	58.9	48.4	34.2	20.8	43.6
	MADISON	15.6	20.5	31.2	45.8	57.0	66.3	70.6	68.5	60.1	49.5	35.1	22.4	45.2
	MILWAUKEE	18.7	23.0	32.1	44.6	54.8	64.9	70.5	69.3	61.9	50.9	37.3	25.1	46.1
WY	CASPER	22.2	26.9	31.9	42.1	52.6	62.9	70.9	68.8	58.4	47.2	32.9	25.7	45.2
	CHEYENNE	26.1	29.3	32.1	41.8	52.2	62.0	68.9	66.8	57.9	47.5	34.8	29.3	45.7

[1] City office data.

Source: U.S. National Oceanic and Atmospheric Administration, Comparative Climatic Data for the United States through 1986

TABLE 1–2. NORMAL DAILY MINIMUM TEMPERATURE— SELECTED CITIES OF THE UNITED STATES

[In Fahrenheit degrees. Airport data except as noted. Based on standard 30–year period 1951 through 1980]

ST.	STATION	JAN	FEB	MAR	APR	MAY	JUN	JUL	AUG	SEP	OCT	NOV	DEC	ANNUAL
AL	BIRMINGHAM	33.0	35.2	42.1	50.4	58.3	65.9	69.8	69.1	63.6	50.4	40.5	35.2	51.1
	MOBILE	40.9	43.2	49.8	57.7	64.8	70.8	73.2	72.9	69.3	57.5	47.9	42.9	57.6
AK	ANCHORAGE	6.0	10.3	15.7	28.2	38.3	47.0	51.1	49.2	41.1	28.4	15.4	7.1	28.2
	BARROW	-20.8	-25.5	-22.1	-8.8	13.9	29.3	33.2	33.5	27.3	9.5	-6.6	-18.8	3.7
	FAIRBANKS	-21.6	-15.4	-4.8	19.5	37.2	48.5	51.2	46.5	35.4	17.5	-4.6	-18.4	15.9
	JUNEAU	16.1	21.9	25.0	31.3	38.1	44.2	47.4	46.6	42.3	36.5	28.0	22.1	33.3
AZ	PHOENIX	39.4	42.5	46.7	53.0	61.5	70.6	79.5	77.5	70.9	59.1	46.9	40.2	57.3
	TUCSON	38.1	40.0	43.8	49.7	57.5	67.4	73.8	72.0	67.3	56.7	45.2	39.0	54.2
AR	FORT SMITH	26.6	30.9	38.5	49.1	58.2	66.3	70.5	68.9	62.1	49.0	37.7	30.2	49.0
	LITTLE ROCK	29.9	33.6	41.2	50.9	59.2	67.5	71.4	69.6	63.0	50.4	40.0	33.2	50.8
CA	BAKERSFIELD	38.9	42.6	45.5	50.1	57.2	64.3	70.1	68.5	63.8	54.9	44.9	38.7	53.3
	FRESNO	36.8	39.7	42.0	46.5	52.7	58.9	64.1	62.2	57.8	49.7	41.1	36.3	49.0
	LOS ANGELES[1]	47.7	49.2	50.2	53.0	56.6	60.4	64.3	65.3	63.7	59.2	52.7	48.4	55.9
	SACRAMENTO	37.9	41.2	42.4	45.3	50.1	55.1	57.9	57.6	55.8	50.0	42.8	37.9	47.8
	SAN DIEGO	48.4	50.3	52.1	54.5	58.2	61.2	64.9	66.8	65.1	60.3	53.6	48.7	57.0
	SAN FRANCISCO[1]	46.2	48.4	48.6	49.2	50.7	52.5	53.1	54.2	55.8	54.8	51.5	47.2	51.0
CO	DENVER	15.9	20.2	24.7	33.7	43.6	52.4	58.7	57.0	47.7	36.9	25.1	18.9	36.2
	GRAND JUNCTION	15.2	22.4	29.7	38.2	48.0	56.6	63.8	61.5	52.2	41.1	28.2	17.9	39.6
CT	HARTFORD	16.7	18.8	28.0	37.6	47.3	57.0	61.9	60.0	51.7	40.9	32.5	20.9	39.5
DE	WILMINGTON	23.2	24.6	32.6	41.8	51.7	61.2	66.3	65.4	58.0	45.9	36.4	27.3	44.5
DC	WASHINGTON	27.5	29.0	36.6	46.2	56.1	65.0	69.9	68.7	62.0	49.7	39.9	31.2	48.5
FL	JACKSONVILLE	41.7	43.3	49.3	55.7	63.0	69.1	71.8	71.8	69.4	59.2	49.2	43.2	57.2
	MIAMI	59.2	59.7	64.1	68.2	71.9	74.6	76.2	76.5	75.7	71.6	65.8	60.8	68.7
	TAMPA	49.5	50.4	56.1	61.1	67.2	72.3	74.2	74.2	72.8	65.1	56.4	50.9	62.5
GA	ATLANTA	32.6	34.5	41.7	50.4	58.7	65.9	69.2	68.7	63.6	51.4	41.3	34.8	51.1
	SAVANNAH	37.9	40.0	46.8	54.1	62.3	68.5	71.5	71.4	67.6	55.9	45.5	39.4	55.1
HI	HONOLULU	65.3	65.3	67.3	68.7	70.2	71.9	73.1	73.6	72.9	72.2	69.2	66.5	69.7
	LIHUE	64.6	64.7	65.9	67.6	69.7	71.8	72.9	73.5	72.9	71.4	69.7	66.9	69.3
ID	BOISE	22.6	27.9	30.9	36.4	44.0	51.8	58.5	56.7	48.7	39.1	30.5	24.6	39.3
	POCATELLO	15.1	20.4	25.2	32.3	40.3	47.3	53.8	51.7	42.7	33.3	24.8	17.9	33.7
IL	CHICAGO	13.6	18.1	27.6	38.8	48.1	57.7	62.7	61.7	53.9	42.9	31.4	20.3	39.7
	PEORIA	13.3	18.4	28.1	40.6	50.6	60.2	64.6	62.7	54.5	42.9	30.9	20.2	40.6
IN	EVANSVILLE	21.9	25.5	34.6	45.4	54.3	63.2	67.4	65.0	57.4	44.5	35.0	27.2	45.1
	INDIANAPOLIS	17.8	21.1	30.7	41.7	51.5	60.9	64.9	62.7	55.3	43.4	32.8	23.7	42.2
IA	DES MOINES	10.1	15.8	26.0	39.9	51.6	61.4	66.3	63.7	54.4	43.3	29.5	17.6	40.0
	SIOUX CITY	6.3	13.3	24.0	37.9	49.8	59.7	64.6	62.3	53.6	40.0	25.8	13.9	37.6
KS	TOPEKA	15.7	21.9	30.4	42.7	53.2	63.1	67.5	65.6	56.2	44.1	31.5	21.8	42.8
	WICHITA	19.4	24.1	32.4	44.5	54.6	64.7	69.8	67.9	59.2	46.9	33.5	24.2	45.1
KY	LEXINGTON	23.1	25.4	34.1	44.3	53.6	61.8	65.9	64.8	58.1	45.9	35.7	27.8	45.0
LA	NEW ORLEANS	43.0	44.8	51.6	58.8	65.3	70.9	73.5	73.1	70.1	59.0	49.9	44.8	58.7
	SHREVEPORT	36.2	39.0	45.8	54.6	62.4	69.4	72.5	71.5	66.5	54.5	44.5	38.2	54.6
ME	PORTLAND	11.9	12.9	23.7	33.0	42.1	51.4	57.3	55.8	47.7	37.9	29.6	16.7	35.0
MD	BALTIMORE	24.3	25.7	33.4	42.9	52.5	61.5	66.5	65.7	58.6	46.1	36.6	27.9	45.1
MA	BOSTON	22.8	23.7	31.8	40.8	50.0	59.3	65.1	63.9	56.9	47.1	38.7	27.1	43.9
MI	DETROIT	16.1	18.0	26.5	36.9	46.7	56.3	60.7	59.4	52.2	41.2	31.4	21.6	38.9
	GRAND RAPIDS	14.9	15.6	24.5	35.6	45.5	55.3	59.8	58.1	50.8	40.4	30.9	20.7	37.7
	SAULT STE. MARIE	5.4	5.3	15.4	29.0	38.3	46.7	51.9	52.4	45.3	36.9	26.4	12.7	30.5
MN	DULUTH	-2.9	2.2	13.9	28.9	39.3	48.2	54.3	52.8	44.3	35.4	21.2	5.8	28.6
	MINNEAPOLIS-ST. PAUL	2.4	8.5	20.8	36.0	47.6	57.7	62.7	60.3	50.2	39.4	25.3	11.7	35.2
MS	JACKSON	34.9	37.2	44.2	52.9	60.8	67.9	71.3	70.2	65.1	51.4	42.3	37.1	52.9
MO	KANSAS CITY	19.4	25.2	33.6	46.4	57.0	66.4	71.3	69.4	60.5	48.6	34.4	25.6	46.5
	ST. LOUIS	19.9	24.5	33.0	45.1	54.7	64.3	68.8	66.6	58.6	46.7	35.1	25.7	45.3
	SPRINGFIELD	20.8	25.3	33.0	44.0	53.1	61.9	66.2	64.7	57.3	45.5	33.9	25.9	44.3
MT	BILLINGS	11.8	18.8	23.6	33.2	43.3	51.6	58.0	56.2	46.5	37.5	25.5	18.2	35.4
	GREAT FALLS	9.2	16.8	21.1	31.3	41.1	49.4	54.4	53.0	44.2	36.2	24.5	16.6	33.2
	HELENA	8.1	15.7	20.6	29.8	39.5	47.0	52.2	50.3	40.8	31.5	20.4	13.5	30.8
NE	NORTH PLATTE	8.3	14.1	21.5	33.6	44.8	55.0	60.6	58.3	46.5	33.7	20.5	12.5	34.1
	OMAHA	10.2	17.1	26.9	40.3	51.8	61.7	66.8	64.2	54.0	42.0	28.6	17.4	40.1
	SCOTTSBLUFF	11.2	16.7	22.3	32.5	43.6	53.3	59.2	56.5	45.6	34.3	22.0	14.8	34.3
NV	ELKO	13.2	19.4	23.0	28.6	36.1	43.3	49.8	47.3	38.0	28.7	21.2	13.9	30.2
	LAS VEGAS	33.0	37.7	42.3	49.8	59.0	68.6	75.9	73.9	65.6	53.5	41.2	33.6	52.8
	RENO	19.5	23.5	25.4	29.4	36.9	43.0	47.7	45.2	38.9	30.5	23.8	18.9	31.9
NH	CONCORD	9.0	11.0	22.2	31.6	41.4	51.6	56.4	54.5	46.2	35.5	27.3	14.5	33.4
	MT. WASHINGTON	-3.3	-3.4	4.7	15.9	28.1	38.4	42.9	41.5	34.9	24.5	13.8	1.2	19.9
NJ	ATLANTIC CITY[1]	27.9	28.9	35.6	44.2	53.3	62.3	68.2	68.0	61.8	50.7	41.1	31.9	47.8
	NEWARK	24.2	25.3	33.3	42.9	53.0	62.4	67.9	67.0	59.4	48.3	39.0	28.6	45.9
NM	ALBUQUERQUE	22.3	25.9	31.7	39.5	48.6	58.4	64.7	62.8	54.9	43.1	30.7	23.2	42.1
	ROSWELL	27.4	31.4	37.9	46.8	55.6	64.8	69.0	67.0	59.6	47.5	35.0	28.2	47.5

TABLE 1–2. NORMAL DAILY MINIMUM TEMPERATURE— SELECTED CITIES OF THE UNITED STATES (continued)

[In Fahrenheit degrees. Airport data except as noted. Based on standard 30–year period 1951 through 1980]

ST.	STATION	JAN	FEB	MAR	APR	MAY	JUN	JUL	AUG	SEP	OCT	NOV	DEC	ANNUAL
NY	ALBANY	11.9	14.0	24.6	35.5	45.4	55.0	59.6	57.6	49.6	39.4	30.8	18.2	36.8
	BUFFALO	17.0	17.5	25.6	36.3	46.3	56.4	61.2	59.6	52.7	42.7	33.6	22.5	39.3
	NEW YORK[1]	25.6	26.6	34.1	43.8	53.3	62.7	68.2	67.1	60.1	49.9	40.8	30.3	46.9
	SYRACUSE	15.0	15.8	25.2	36.0	46.0	55.4	60.3	58.9	51.8	41.7	33.3	21.3	38.4
NC	ASHEVILLE	26.0	27.6	34.4	42.7	51.0	58.2	62.4	61.6	55.8	43.3	34.2	28.2	43.8
	RALEIGH	29.1	30.3	37.7	46.5	55.3	62.6	67.1	66.8	60.4	47.7	38.1	31.2	47.7
	WILMINGTON	35.3	36.6	43.3	51.8	60.4	67.1	71.3	70.8	65.7	53.7	43.9	37.2	53.1
ND	BISMARCK	-4.2	3.7	15.6	30.8	42.0	51.8	56.4	54.2	43.2	32.8	17.7	4.8	29.1
	WILLISTON	-4.3	3.5	14.3	29.6	41.5	51.1	56.0	53.7	43.0	32.0	17.0	4.3	28.5
OH	CINCINNATI	20.4	23.0	32.0	42.4	51.7	60.5	64.9	63.3	56.3	43.9	34.1	25.7	43.2
	CLEVELAND	18.5	19.9	28.4	38.3	47.9	57.2	61.4	60.5	54.0	43.6	34.3	24.6	40.7
	COLUMBUS	19.4	21.5	30.6	40.5	50.2	59.0	63.2	61.7	54.6	42.8	33.5	24.7	41.8
	TOLEDO	15.5	17.5	26.1	36.5	46.6	56.0	60.2	58.4	51.2	40.1	30.6	20.6	38.3
OK	OKLAHOMA CITY	25.2	29.4	37.1	48.6	57.7	66.3	70.6	69.4	61.9	50.2	37.6	29.1	48.6
	TULSA	24.8	29.5	37.7	49.5	58.5	67.5	72.4	70.3	62.5	50.3	38.1	29.3	49.2
OR	EUGENE	33.8	35.5	36.5	38.7	42.9	48.0	51.0	51.1	47.7	42.0	37.8	35.3	41.7
	MEDFORD	30.2	31.9	33.9	36.8	42.7	49.3	54.2	53.4	47.4	39.6	34.5	31.2	40.4
	PORTLAND	33.5	36.0	37.4	40.6	46.4	52.2	55.8	55.8	51.1	44.6	38.6	35.4	44.0
PC	GUAM	71.0	71.1	71.2	72.4	72.9	73.1	72.5	72.2	72.2	72.4	73.1	72.6	72.2
	JOHNSTON ISLAND	72.8	72.7	72.9	73.9	75.3	76.3	76.9	77.4	77.2	76.9	75.5	73.9	75.1
	KWAJALEIN, MARSHALL IS.	76.8	76.9	77.4	77.1	76.9	76.9	76.9	77.0	77.1	77.0	77.1	77.2	77.0
	PAGO PAGO, AMER SAMOA	74.9	74.8	75.1	74.6	74.5	75.0	74.3	73.9	74.2	75.0	75.1	75.1	74.7
	TRUK, E. CAROLINE IS.	76.9	76.9	77.0	76.8	76.4	76.0	75.2	75.1	75.3	75.5	76.1	76.8	76.2
	WAKE ISLAND	72.1	71.5	72.5	73.0	74.7	76.4	76.9	77.0	77.3	76.7	75.6	73.7	74.8
	YAP, W. CAROLINE IS.	75.0	75.2	75.4	75.9	75.8	75.4	74.9	74.7	74.8	74.9	75.4	75.4	75.2
PA	ERIE	18.0	17.7	25.8	36.1	45.4	55.2	59.9	59.4	53.1	43.2	34.3	24.2	39.4
	HARRISBURG	22.1	23.5	31.5	41.5	51.0	60.5	65.3	64.2	56.6	44.6	35.4	26.2	43.5
	PHILADELPHIA	23.8	25.0	33.1	42.6	52.5	61.5	66.8	66.0	58.6	46.5	37.1	28.0	45.1
	PITTSBURGH	19.2	20.7	29.4	39.4	48.5	57.1	61.3	60.1	53.3	42.1	33.3	24.3	40.7
PR	SAN JUAN	70.3	70.0	70.8	72.3	73.9	75.3	76.1	76.1	75.5	74.9	73.4	71.8	73.4
RI	PROVIDENCE	20.0	20.9	29.2	38.3	47.6	57.0	63.3	61.9	53.8	43.1	34.8	24.1	41.2
SC	CHARLESTON	36.9	38.4	45.3	52.5	61.4	68.0	71.6	71.2	66.7	54.7	44.6	38.5	54.2
	COLUMBIA	33.2	34.6	41.9	50.5	59.1	66.1	70.1	69.4	63.9	50.3	40.6	34.7	51.2
	GREENVILLE-SPARTANBURG	31.2	32.6	39.4	48.3	56.9	64.2	68.2	67.4	61.7	49.1	39.6	33.2	49.3
SD	RAPID CITY	9.2	14.6	21.0	32.1	43.0	52.5	58.7	57.0	46.4	36.1	23.0	14.8	34.0
	SIOUX FALLS	1.9	8.9	20.6	34.6	45.7	56.3	61.8	59.7	48.5	36.7	22.3	10.1	33.9
TN	BRISTOL-JOHNSON CITY-KINGSPORT	25.5	27.4	34.9	43.8	52.5	60.1	64.3	63.4	57.2	44.7	35.1	28.3	44.8
	KNOXVILLE	29.5	31.7	39.3	48.2	56.5	64.0	68.0	67.1	61.2	48.1	38.4	31.9	48.7
	MEMPHIS	30.9	34.1	41.9	52.2	60.9	68.9	72.6	70.8	64.1	51.3	41.1	34.3	51.9
	NASHVILLE	27.8	30.1	38.3	48.1	56.9	64.8	69.0	67.8	61.3	48.0	38.0	31.3	48.5
TX	AUSTIN	38.8	42.2	49.3	58.3	65.1	71.5	73.9	73.7	69.1	58.7	48.1	41.4	57.5
	CORPUS CHRISTI	46.1	48.7	55.7	63.9	69.5	74.1	75.6	75.8	72.8	64.1	54.9	48.8	62.5
	DALLAS-FORT WORTH	33.9	37.8	44.9	55.0	62.9	70.8	74.7	73.7	67.5	56.3	44.9	37.4	55.0
	EL PASO	30.4	34.1	40.5	48.5	56.6	65.7	69.6	67.5	60.6	48.7	37.0	30.6	49.2
	HOUSTON	40.8	43.2	49.8	58.3	64.7	70.2	72.5	72.1	68.1	57.5	48.6	42.7	57.4
	LUBBOCK	24.3	27.9	35.2	45.8	55.2	64.3	67.6	65.7	58.7	47.3	34.8	27.4	46.2
UT	SALT LAKE CITY	19.7	24.4	29.9	37.2	45.2	53.3	61.8	59.7	50.0	39.3	29.2	21.6	39.3
VT	BURLINGTON	7.7	8.8	20.8	32.7	44.0	54.0	58.6	56.6	48.7	38.7	29.6	14.9	34.6
VA	LYNCHBURG	25.9	27.6	35.1	44.6	53.2	60.6	65.1	64.5	57.9	46.1	36.7	29.1	45.5
	NORFOLK	31.7	32.3	39.4	48.1	57.2	65.3	69.9	69.6	64.2	52.8	43.0	35.0	50.7
	RICHMOND	26.5	28.1	35.8	45.1	54.2	62.2	67.2	66.4	59.3	46.7	37.3	29.6	46.5
WA	SEATTLE[1]	35.9	38.2	38.8	42.4	47.7	53.0	56.0	56.3	52.9	47.1	41.1	38.1	45.6
	SPOKANE	20.0	25.7	29.0	34.9	42.5	49.3	55.3	54.3	46.5	36.7	28.5	23.7	37.2
	WALLA WALLA	28.4	33.9	37.3	42.4	49.3	55.8	62.2	61.2	53.6	44.5	35.9	31.4	44.7
WV	CHARLESTON	23.9	25.8	34.1	43.3	51.8	59.4	63.8	63.1	56.4	44.0	35.0	27.8	44.0
WI	GREEN BAY	5.4	8.7	20.1	33.6	43.5	53.1	58.1	56.3	47.9	38.2	26.3	13.0	33.7
	MADISON	6.7	11.0	21.5	34.1	44.2	53.8	58.3	56.3	47.8	37.8	26.0	14.1	34.3
	MILWAUKEE	11.3	15.8	24.9	35.6	44.7	54.7	61.1	60.2	52.5	41.9	29.9	18.2	37.6
WY	CASPER	11.9	16.3	20.2	29.3	38.9	47.6	54.7	52.8	42.5	33.2	21.9	15.7	32.1
	CHEYENNE	14.8	17.9	20.6	29.6	39.7	48.5	54.6	52.8	43.7	34.0	23.1	18.2	33.1

[1] City office data.

Source: U.S. National Oceanic and Atmospheric Administration, Comparative Climatic Data for the United States through 1986

TABLE 1–3. NORMAL DAILY MAXIMUM TEMPERATURE— SELECTED CITIES OF THE UNITED STATES

[In Fahrenheit degrees. Airport data except as noted. Based on standard 30–year period 1951 through 1980]

ST.	STATION	JAN	FEB	MAR	APR	MAY	JUN	JUL	AUG	SEP	OCT	NOV	DEC	ANNUAL
AL	BIRMINGHAM	52.7	57.3	65.2	75.2	81.6	87.9	90.3	89.7	84.6	74.8	63.7	55.9	73.2
	MOBILE	60.6	63.9	70.3	78.3	84.9	90.2	91.2	90.7	87.0	79.4	69.3	63.1	77.4
AK	ANCHORAGE	20.0	25.5	31.7	42.6	54.2	61.8	65.1	63.2	55.2	40.8	27.9	20.4	42.4
	BARROW	-8.0	-13.8	-9.7	5.4	23.6	37.4	44.6	42.4	33.8	18.9	4.6	-7.0	14.4
	FAIRBANKS	-3.9	7.3	21.7	40.8	59.2	70.1	71.8	66.5	54.4	32.6	12.4	-1.7	35.9
	JUNEAU	27.4	33.7	37.4	46.8	54.7	61.1	64.0	62.6	55.9	47.0	37.5	31.5	46.6
AZ	PHOENIX	65.2	69.7	74.5	83.1	92.4	102.3	105.0	102.3	98.2	87.7	74.3	66.4	85.1
	TUCSON	64.1	67.4	71.8	80.1	88.8	98.5	98.5	95.9	93.5	84.1	72.2	65.0	81.7
AR	FORT SMITH	48.4	53.8	62.5	73.7	81.0	88.5	93.6	92.9	85.7	75.9	61.9	52.1	72.5
	LITTLE ROCK	49.8	54.5	63.2	73.8	81.7	89.5	92.7	92.3	85.6	75.8	62.4	53.2	72.9
CA	BAKERSFIELD	57.4	63.7	68.6	75.1	83.9	92.2	98.8	96.4	90.8	81.0	67.4	57.6	77.7
	FRESNO	54.2	61.2	66.5	73.7	82.7	91.1	97.9	95.5	90.3	79.9	65.2	54.4	76.1
	LOS ANGELES[1]	66.6	68.5	68.7	70.9	73.2	77.9	83.8	84.1	83.0	78.5	72.7	68.1	74.7
	SACRAMENTO	52.6	59.4	64.1	71.0	79.7	87.4	93.3	91.7	87.6	77.7	63.2	53.2	73.4
	SAN DIEGO	65.2	66.4	65.9	67.8	68.6	71.3	75.6	77.6	76.8	74.6	69.9	66.1	70.5
	SAN FRANCISCO[1]	56.1	59.4	60.0	61.1	62.5	64.3	64.0	65.0	68.9	68.3	62.9	56.9	62.5
CO	DENVER	43.1	46.9	51.2	61.0	70.7	81.6	88.0	85.8	77.5	66.8	52.4	46.1	64.3
	GRAND JUNCTION	35.7	44.5	54.1	65.2	76.2	87.9	94.0	90.3	81.9	68.7	51.0	38.7	65.7
CT	HARTFORD	33.6	36.3	45.5	60.0	71.4	80.1	84.8	82.6	74.8	63.9	50.6	37.3	60.1
DE	WILMINGTON	39.2	41.8	50.9	63.0	72.7	81.2	85.6	84.1	77.8	66.7	54.8	43.6	63.5
DC	WASHINGTON	42.9	45.9	55.0	67.1	75.9	84.0	87.9	86.4	80.1	68.9	57.4	46.6	66.5
FL	JACKSONVILLE	64.6	66.8	73.3	79.7	85.2	88.9	90.7	90.2	86.9	79.7	72.4	66.3	78.7
	MIAMI	75.0	75.8	79.3	82.4	85.1	87.3	88.7	89.2	87.8	84.2	79.8	76.2	82.6
	TAMPA	70.0	71.0	76.2	81.9	87.1	89.5	90.0	90.3	88.9	83.7	76.9	71.6	81.4
GA	ATLANTA	51.2	55.3	63.2	73.2	79.8	85.6	87.9	87.6	82.3	72.9	62.6	54.1	71.3
	SAVANNAH	60.3	63.1	69.9	77.8	84.2	88.6	90.8	90.1	85.6	77.8	69.5	62.5	76.7
HI	HONOLULU	79.9	80.4	81.4	82.7	84.8	86.2	87.1	88.3	88.2	86.7	83.9	81.4	84.2
	LIHUE	77.8	78.0	78.1	79.2	81.2	83.0	83.8	84.6	84.7	83.2	80.8	78.8	81.1
ID	BOISE	37.1	44.3	51.8	60.8	70.8	79.8	90.6	87.3	77.6	64.6	49.0	39.3	62.8
	POCATELLO	32.4	38.6	45.8	56.8	67.7	77.6	88.6	86.0	75.7	62.8	45.6	35.3	59.4
IL	CHICAGO	29.2	33.9	44.3	58.8	70.0	79.4	83.3	82.1	75.5	64.1	48.2	35.0	58.7
	PEORIA	29.7	35.2	46.5	61.9	72.5	82.1	85.5	83.4	76.7	64.8	48.5	35.4	60.2
IN	EVANSVILLE	39.3	44.2	54.5	67.6	76.8	85.9	88.8	87.3	81.4	70.0	55.1	44.0	66.2
	INDIANAPOLIS	34.2	38.5	49.3	63.1	73.4	82.3	85.2	83.7	77.9	66.1	50.8	39.2	62.0
IA	DES MOINES	27.0	33.2	44.2	61.0	72.6	81.8	86.2	84.0	75.7	65.0	47.6	33.7	59.3
	SIOUX CITY	26.0	33.0	43.5	61.5	73.1	82.0	86.5	84.2	75.4	64.9	46.9	32.7	59.1
KS	TOPEKA	36.3	43.0	53.2	66.5	76.0	84.6	89.6	88.5	80.7	69.9	53.8	41.8	65.3
	WICHITA	39.8	46.1	55.8	68.1	77.1	87.4	92.9	91.5	82.0	71.2	55.1	44.6	67.6
KY	LEXINGTON	39.8	43.7	53.7	65.8	74.9	82.6	85.9	85.0	79.3	67.6	54.1	44.4	64.7
LA	NEW ORLEANS	61.8	64.6	71.2	78.6	84.5	89.5	90.7	90.2	86.8	79.4	70.1	64.4	77.7
	SHREVEPORT	55.8	60.6	68.1	76.7	83.5	90.1	93.3	93.2	87.7	78.9	66.8	59.2	76.2
ME	PORTLAND	31.0	33.1	40.5	52.5	63.4	72.8	78.9	77.5	69.6	59.0	47.1	34.9	55.0
MD	BALTIMORE	41.0	43.7	53.1	65.1	74.2	82.9	87.1	85.5	79.1	67.7	55.9	45.1	65.0
MA	BOSTON	36.4	37.7	45.0	56.6	67.0	76.6	81.8	79.8	72.3	62.5	51.6	40.3	59.0
MI	DETROIT	30.6	33.5	43.4	57.7	69.4	79.0	83.1	81.5	74.4	62.5	47.6	35.4	58.2
	GRAND RAPIDS	29.0	31.7	41.6	56.9	69.4	78.9	83.0	81.1	73.4	61.4	46.0	33.8	57.2
	SAULT STE. MARIE	21.2	23.1	32.3	47.1	61.0	70.1	75.1	73.4	64.2	53.6	39.0	26.6	48.9
MN	DULUTH	15.5	21.7	31.9	47.6	61.3	70.5	76.4	73.6	63.6	53.0	35.2	21.8	47.7
	MINNEAPOLIS-ST. PAUL	19.9	26.4	37.5	56.0	69.4	78.5	83.4	80.9	71.0	59.7	41.1	26.7	54.2
MS	JACKSON	56.5	60.9	68.4	77.3	84.1	90.5	92.5	92.1	87.6	78.6	67.5	60.0	76.3
MO	KANSAS CITY	37.4	43.5	53.6	67.4	76.8	85.3	90.5	89.1	81.4	70.6	54.8	42.8	66.1
	ST. LOUIS	37.6	43.1	53.4	67.1	76.4	85.2	89.0	87.4	80.7	69.1	54.0	42.6	65.5
	SPRINGFIELD	42.2	47.1	56.1	68.3	76.5	84.9	89.8	89.3	81.6	70.8	56.2	46.4	67.4
MT	BILLINGS	29.9	37.9	44.0	55.9	66.4	76.3	86.6	84.3	72.3	61.0	44.4	36.0	57.9
	GREAT FALLS	28.2	36.5	41.7	54.0	65.3	74.3	84.2	82.0	70.5	59.5	43.5	34.7	56.2
	HELENA	28.1	36.2	42.5	54.7	64.9	73.1	83.6	81.3	70.3	58.6	42.3	33.3	55.7
NE	NORTH PLATTE	34.2	40.5	47.8	61.5	71.5	81.6	87.8	86.4	77.3	66.7	49.4	39.3	62.0
	OMAHA EPPLEY	30.2	37.3	47.7	64.0	74.7	84.2	88.5	86.2	77.5	67.0	50.3	36.9	62.0
	SCOTTSBLUFF	37.2	43.4	48.8	60.3	70.8	81.7	89.2	86.7	77.5	66.0	49.8	40.8	62.7
NV	ELKO	36.6	42.6	48.9	58.2	68.5	79.2	90.4	87.8	78.8	66.3	49.4	38.3	62.1
	LAS VEGAS	56.0	62.4	68.3	77.2	87.4	98.6	104.5	101.9	94.7	81.5	66.0	57.1	79.6
	RENO	44.8	51.1	55.8	63.3	72.3	81.8	91.3	88.7	81.4	70.0	55.6	46.2	66.9
NH	CONCORD	30.8	33.2	41.9	56.5	68.9	77.7	82.6	80.1	71.9	61.0	47.2	34.4	57.2
	MT. WASHINGTON	13.4	13.1	19.1	28.9	40.7	50.7	54.4	52.7	46.3	36.5	26.9	17.0	33.3
NJ	ATLANTIC CITY[1]	40.3	41.8	47.7	57.7	65.7	74.5	79.9	79.5	73.8	64.4	54.3	44.8	60.4
	NEWARK	38.2	40.3	49.1	61.3	71.6	80.6	85.6	84.0	76.9	66.0	54.0	42.3	62.5
NM	ALBUQUERQUE	47.2	52.9	60.7	70.6	79.9	90.6	92.8	89.4	83.0	71.7	57.2	48.0	70.3

TABLE 1–3. NORMAL DAILY MAXIMUM TEMPERATURE— SELECTED CITIES OF THE UNITED STATES (continued)

[In Fahrenheit degrees. Airport data except as noted. Based on standard 30–year period 1951 through 1980]

ST.	STATION	JAN	FEB	MAR	APR	MAY	JUN	JUL	AUG	SEP	OCT	NOV	DEC	ANNUAL
	ROSWELL	55.4	60.4	67.7	76.9	85.0	93.1	93.7	91.3	84.9	75.8	63.1	56.7	75.3
NY	ALBANY	30.2	32.7	42.5	57.6	69.5	78.3	83.2	80.7	72.8	61.5	47.8	34.6	57.6
	BUFFALO	30.0	31.4	40.4	54.4	65.9	75.6	80.2	78.2	71.4	60.2	47.0	35.0	55.8
	NEW YORK[1]	38.0	40.1	48.6	61.1	71.5	80.1	85.3	83.7	76.4	65.6	53.6	42.1	62.2
	SYRACUSE	30.6	32.2	41.4	56.2	67.9	77.2	81.6	79.6	72.3	60.9	47.9	35.3	56.9
NC	ASHEVILLE	47.5	50.6	58.4	68.6	75.6	81.4	84.0	83.5	77.9	68.7	58.6	50.3	67.1
	RALEIGH	50.1	52.8	61.0	72.3	79.0	85.2	88.2	87.1	81.6	71.6	61.8	52.7	70.3
	WILMINGTON	55.9	58.1	64.8	74.3	80.9	86.1	89.3	88.6	83.9	75.2	66.8	59.1	73.6
ND	BISMARCK	17.5	25.2	36.4	54.2	67.7	76.8	84.4	83.3	71.4	59.3	39.4	25.9	53.5
	WILLISTON	17.5	25.6	36.3	54.0	67.5	76.5	84.0	82.5	70.2	58.2	38.1	25.5	53.0
OH	CINCINNATI	37.3	41.2	51.5	64.5	74.2	82.3	85.8	84.8	78.7	66.7	52.6	41.9	63.5
	CLEVELAND	32.5	34.8	44.8	57.9	68.5	78.0	81.7	80.3	74.2	62.7	49.3	37.5	58.5
	COLUMBUS	34.7	38.1	49.3	62.3	72.6	81.3	84.4	83.0	76.9	65.0	50.7	39.4	61.5
	TOLEDO	30.7	34.0	44.6	59.1	70.5	79.9	83.4	81.8	75.1	63.3	47.9	35.5	58.8
OK	OKLAHOMA CITY	46.6	52.2	61.0	71.7	79.0	87.6	93.5	92.8	84.7	74.3	59.9	50.7	71.2
	TULSA	45.6	51.9	60.8	72.4	79.7	87.9	93.9	93.0	85.0	74.9	60.2	50.3	71.3
OR	EUGENE	46.3	51.4	55.0	60.5	67.2	74.2	82.6	81.3	76.4	64.6	52.8	47.3	63.3
	MEDFORD	45.0	52.9	57.1	63.8	72.2	81.0	90.7	88.8	82.8	68.7	52.6	44.2	66.7
	PORTLAND	44.3	50.4	54.5	60.2	66.9	72.7	79.5	78.6	74.2	63.9	52.3	46.4	62.0
PC	GUAM	83.4	83.4	84.4	85.7	86.5	86.9	86.4	86.0	85.8	85.7	85.3	84.2	85.3
	JOHNSTON ISLAND	80.5	80.4	80.8	81.4	82.6	83.7	84.3	84.7	84.8	84.4	82.8	81.3	82.6
	KWAJALEIN, MARSHALL IS.	85.7	86.4	87.0	86.5	86.3	86.4	86.4	86.8	86.9	86.7	86.5	86.0	86.5
	PAGO PAGO, AMER SAMOA	86.5	86.7	86.9	86.4	85.2	84.4	83.4	83.4	84.5	85.0	85.5	86.1	85.3
	TRUK, E. CAROLINE IS.	85.4	85.5	85.8	86.2	86.6	86.9	86.7	87.0	87.1	87.0	86.8	86.0	86.4
	WAKE ISLAND	81.7	81.7	82.8	83.5	85.3	87.5	88.0	88.1	88.0	87.1	85.1	83.1	85.2
	YAP, W. CAROLINE IS.	85.5	85.8	86.5	87.4	87.5	87.4	87.1	86.9	87.2	87.4	87.2	86.2	86.8
PA	ERIE	30.9	32.2	41.1	53.7	64.6	74.0	78.2	77.0	71.0	60.1	47.1	35.7	55.5
	HARRISBURG	36.7	39.5	49.6	62.9	73.0	81.8	86.2	84.4	77.2	65.4	52.4	40.6	62.5
	PHILADELPHIA	38.6	41.1	50.5	63.2	73.0	81.7	86.1	84.6	77.8	66.5	54.5	43.0	63.4
	PITTSBURGH	34.1	36.8	47.6	60.7	70.8	79.1	82.7	81.1	74.8	62.9	49.8	38.4	59.9
PR	SAN JUAN	82.7	83.2	84.2	85.2	86.7	88.0	87.9	88.2	88.2	87.9	85.7	83.6	86.0
RI	PROVIDENCE	36.4	37.7	45.5	57.5	67.6	76.6	81.7	80.3	73.1	63.2	51.9	40.5	59.3
SC	CHARLESTON	58.8	61.2	68.0	76.0	82.9	87.0	89.4	88.8	84.6	76.8	68.7	61.4	75.3
	COLUMBIA	56.2	59.5	67.1	77.0	83.8	89.2	91.9	91.0	85.5	76.5	67.1	58.8	75.3
	GREENVILLE-SPARTANBURG	51.0	54.5	62.5	72.6	79.7	85.4	88.2	87.5	81.7	72.2	62.1	53.5	70.9
SD	RAPID CITY	32.4	37.4	44.2	57.0	68.1	77.9	86.5	85.7	75.4	63.2	46.7	37.4	59.3
	SIOUX FALLS	22.9	29.3	40.1	58.1	70.5	80.3	86.2	83.9	73.5	62.1	43.7	29.3	56.7
TN	BRISTOL-JOHNSON CITY-KINGSPORT	44.5	48.4	57.6	68.3	76.3	82.9	85.5	85.2	80.4	69.5	57.3	48.1	67.0
	KNOXVILLE	46.9	51.2	60.1	71.0	78.3	84.6	87.2	86.9	81.7	70.9	59.1	50.3	69.0
	MEMPHIS	48.3	53.0	61.4	72.9	81.0	88.4	91.5	90.3	84.3	74.5	61.4	52.3	71.6
	NASHVILLE	46.3	50.7	59.6	71.2	79.2	86.7	89.8	89.0	83.2	72.3	59.2	50.4	69.8
TX	AUSTIN	59.4	64.1	71.7	79.0	84.7	91.6	95.4	95.3	89.3	80.8	69.2	62.8	78.6
	CORPUS CHRISTI	66.5	69.9	76.1	82.1	86.7	91.2	94.2	94.1	90.1	83.9	75.1	69.3	81.6
	DALLAS-FORT WORTH	54.0	59.1	67.2	76.8	84.4	93.2	97.8	97.3	89.7	79.5	66.2	58.1	76.9
	EL PASO	57.9	62.7	69.6	78.7	87.1	95.9	95.3	93.0	87.5	78.5	65.7	58.2	77.5
	HOUSTON	61.9	65.7	72.1	79.0	85.1	90.9	93.6	93.1	88.7	81.9	71.6	65.2	79.1
	LUBBOCK	53.3	57.3	65.1	74.8	82.8	90.8	91.9	90.1	83.6	74.7	62.1	55.5	73.5
UT	SALT LAKE CITY	37.4	43.7	51.5	61.1	72.4	83.3	93.2	90.0	80.0	66.7	50.2	38.9	64.0
VT	BURLINGTON	25.4	27.3	37.7	52.6	66.4	75.9	80.5	77.6	68.8	57.0	43.6	30.3	53.6
VA	NORFOLK	48.1	49.9	57.5	68.2	75.7	83.2	86.9	85.7	80.2	69.8	60.8	51.9	68.2
	RICHMOND	46.7	49.6	58.5	70.6	77.9	84.8	88.4	87.1	81.0	70.5	60.5	50.2	68.8
WA	SEATTLE[1]	45.3	50.1	52.6	58.3	64.8	69.0	74.6	73.6	69.2	60.4	51.3	46.9	59.7
	SPOKANE	31.3	39.0	46.2	56.7	66.1	74.0	84.0	81.7	72.4	58.3	41.4	34.2	57.1
	WALLA WALLA	40.1	47.3	54.3	62.2	70.9	79.3	88.8	86.0	77.3	64.0	49.0	43.0	63.5
WV	CHARLESTON	41.8	45.4	55.4	67.3	76.0	82.5	85.2	84.2	78.7	67.7	55.6	45.9	65.5
WI	GREEN BAY	22.5	26.9	37.0	53.7	66.6	76.2	80.9	78.7	69.8	58.5	42.0	28.5	53.4
	MADISON	24.5	30.0	40.8	57.5	69.8	78.8	82.8	80.6	72.3	61.1	44.1	30.6	56.1
	MILWAUKEE	26.0	30.1	39.2	53.5	64.8	75.0	79.8	78.4	71.2	59.9	44.7	32.0	54.6
WY	CASPER	32.5	37.4	43.4	54.9	66.2	78.1	87.1	84.8	74.2	61.0	43.9	35.6	58.3
	CHEYENNE	37.3	40.7	43.6	54.0	64.6	75.4	83.1	80.8	72.1	61.0	46.5	40.4	58.3

[1] City office data.

Source: U.S. National Oceanic and Atmospheric Administration, Comparative Climatic Data for the United States through 1986

TABLE 1–4. AVERAGE NUMBER OF DAYS WITH FREEZING TEMPERATURES— SELECTED CITIES OF THE UNITED STATES

[Minimum temperature 32 degrees Fahrenheit or less]

ST.	STATION	YR.	JAN	FEB	MAR	APR	MAY	JUN	JUL	AUG	SEP	OCT	NOV	DEC	ANNUAL
AL	BIRMINGHAM	23	18	14	6	1	0	0	0	0	0	*	7	14	61
	MOBILE	24	9	6	1	0	0	0	0	0	0	0	1	6	23
AK	ANCHORAGE	22	31	27	28	22	3	0	0	*	3	20	28	30	191
	BARROW	66	31	28	31	30	31	24	14	16	26	31	30	31	323
	FAIRBANKS	23	31	28	31	27	6	0	0	1	8	28	30	31	222
	JUNEAU	42	26	23	24	16	4	*	0	*	2	8	19	24	145
AZ	PHOENIX	26	4	2	*	0	0	0	0	0	0	0	*	2	8
	TUCSON	46	6	4	1	*	0	0	0	0	0	*	1	5	18
AR	FORT SMITH	22	24	18	8	1	0	0	0	0	0	1	9	20	80
	LITTLE ROCK	26	21	15	4	*	0	0	0	0	0	*	5	16	62
CA	BAKERSFIELD	23	4	1	*	0	0	0	0	0	0	*	*	4	9
	FRESNO	23	7	3	1	*	0	0	0	0	0	*	2	8	21
	LOS ANGELES[1]	46	*	0	0	0	0	0	0	0	0	0	0	*	*
	SACRAMENTO	36	6	2	1	*	0	0	0	0	0	0	1	6	16
	SAN DIEGO	26	*	0	0	0	0	0	0	0	0	0	0	0	*
	SAN FRANCISCO[1]	50	*	0	0	0	0	0	0	0	0	0	0	*	*
CO	DENVER	26	30	26	25	12	2	0	0	0	1	9	25	29	158
	GRAND JUNCTION	23	30	25	17	7	1	0	0	0	*	3	20	29	132
CT	HARTFORD	27	29	26	22	9	1	0	0	0	*	6	16	27	136
DE	WILMINGTON	39	26	22	15	3	*	0	0	0	0	2	11	22	101
DC	WASHINGTON	26	23	19	8	1	0	0	0	0	0	*	4	16	71
FL	JACKSONVILLE	45	6	4	1	0	0	0	0	0	0	0	1	4	15
	MIAMI	22	*	0	*	0	0	0	0	0	0	0	0	0	*
	TAMPA	23	2	1	*	0	0	0	0	0	0	0	*	1	4
GA	ATLANTA	26	18	14	6	1	0	0	0	0	0	*	5	14	57
	SAVANNAH	22	11	8	2	0	0	0	0	0	0	*	3	8	32
HI	HONOLULU	17	0	0	0	0	0	0	0	0	0	0	0	0	0
	LIHUE	37	0	0	0	0	0	0	0	0	0	0	0	0	0
ID	BOISE	47	26	21	18	8	2	*	0	0	*	6	18	25	124
	POCATELLO	23	28	25	26	17	5	*	0	*	3	15	23	28	170
IL	CHICAGO	28	29	25	21	8	1	0	0	0	*	5	17	26	133
	PEORIA	27	30	25	20	6	*	0	0	0	*	4	17	27	129
IN	EVANSVILLE	25	26	21	13	2	*	0	0	0	0	3	12	22	99
	INDIANAPOLIS	27	28	24	17	5	*	0	0	0	0	4	15	25	119
IA	DES MOINES	25	30	25	21	6	*	0	0	0	*	4	18	29	135
	SIOUX CITY	27	31	27	23	8	1	0	0	0	1	7	22	30	149
KS	TOPEKA	22	29	23	17	4	*	0	0	0	*	4	17	27	120
	WICHITA	33	29	23	15	3	*	0	0	0	*	1	15	27	112
KY	LEXINGTON	23	25	21	14	3	*	0	0	0	0	2	11	20	96
LA	NEW ORLEANS	40	5	3	1	*	0	0	0	0	0	0	1	4	14
	SHREVEPORT	34	13	8	3	*	0	0	0	0	0	*	3	10	37
ME	PORTLAND	46	30	27	26	14	3	0	0	0	1	9	20	29	158
MD	BALTIMORE	36	26	21	14	3	*	0	0	0	0	2	11	21	98
MA	BOSTON	22	26	23	17	3	0	0	0	0	0	1	7	22	98
MI	DETROIT	28	30	26	23	10	1	0	0	0	*	5	17	26	137
	GRAND RAPIDS	23	30	26	24	12	2	0	0	0	*	6	18	28	145
	SAULT STE. MARIE	45	31	28	29	21	7	1	0	*	2	9	22	30	181
MN	DULUTH	25	31	28	29	20	5	*	0	*	3	11	25	31	184
	MINNEAPOLIS-ST. PAUL	27	31	27	26	11	1	0	0	0	1	7	23	30	156
MS	JACKSON	23	16	12	4	*	0	0	0	0	0	*	6	12	51
MO	KANSAS CITY	18	27	21	14	2	0	0	0	0	0	1	11	23	98
	ST. LOUIS	26	27	22	14	3	*	0	0	0	0	2	12	24	103
	SPRINGFIELD	26	27	21	14	3	*	0	0	0	*	2	13	23	103
MT	BILLINGS	27	28	24	24	13	2	*	0	0	2	8	22	28	149
	GREAT FALLS	25	27	24	25	16	3	*	0	0	3	11	21	27	158
	HELENA	23	29	27	27	19	4	*	0	0	4	18	27	30	184
NE	NORTH PLATTE	22	31	28	27	13	3	*	0	0	3	14	28	31	177
	OMAHA EPPLEY	22	30	26	22	6	*	0	0	0	1	5	20	29	140
	SCOTTSBLUFF	22	30	27	26	13	2	*	0	0	2	12	27	30	168
NV	ELKO	22	29	25	26	20	7	1	*	*	6	20	24	29	186
	LAS VEGAS	26	13	5	1	*	0	0	0	0	0	*	2	11	33
	RENO	23	28	25	25	19	7	1	0	*	5	19	24	28	180
NH	CONCORD	21	30	27	26	18	6	*	0	*	2	14	21	29	174
	MT. WASHINGTON	54	31	28	31	28	20	7	2	3	12	23	28	31	243
NJ	ATLANTIC CITY[1]	27	23	18	10	1	0	0	0	0	0	*	4	16	74
	NEWARK	21	25	21	12	2	0	0	0	0	0	1	6	19	85
NM	ALBUQUERQUE	26	29	23	16	5	*	0	0	0	0	2	16	28	120
	ROSWELL	14	25	18	7	2	0	0	0	0	0	1	11	25	89
NY	ALBANY	21	30	26	24	13	2	0	0	0	1	9	18	28	150
	BUFFALO	26	29	26	24	11	1	0	0	0	*	3	14	26	134

TABLE 1–4. AVERAGE NUMBER OF DAYS WITH FREEZING TEMPERATURES—SELECTED CITIES OF THE UNITED STATES (continued)

[Minimum temperature 32 degrees Fahrenheit or less]

ST.	STATION	YR.	JAN	FEB	MAR	APR	MAY	JUN	JUL	AUG	SEP	OCT	NOV	DEC	ANNUAL
	NEW YORK[1]	73	23	20	12	2	*	0	0	0	0	*	5	18	80
	SYRACUSE	23	29	25	24	12	1	0	0	0	*	5	15	26	137
NC	ASHEVILLE	22	24	21	14	4	*	0	0	0	*	4	13	21	101
	RALEIGH	22	22	18	10	2	*	0	0	0	0	1	10	17	80
	WILMINGTON	23	15	12	4	*	0	0	0	0	0	*	3	11	44
ND	BISMARCK	27	31	28	29	17	4	*	0	0	3	14	28	31	185
	WILLISTON	25	31	28	29	18	4	*	0	0	3	15	28	31	188
OH	CINCINNATI	24	27	23	16	5	*	0	0	0	0	3	13	22	109
	CLEVELAND	26	28	24	21	9	1	*	0	0	0	3	13	25	124
	COLUMBUS	27	27	24	19	7	1	0	0	0	*	4	14	24	119
	TOLEDO	31	30	26	23	11	2	*	0	0	*	6	18	27	143
OK	OKLAHOMA CITY	21	24	17	8	1	0	0	0	0	0	1	9	21	80
	TULSA	26	25	18	8	1	0	0	0	0	0	*	8	20	80
OR	EUGENE	44	15	9	7	3	*	*	0	0	*	2	8	11	56
	MEDFORD	25	20	15	12	7	1	0	0	0	*	4	10	16	86
	PORTLAND	46	13	8	5	1	*	0	0	0	0	1	5	9	43
PC	GUAM	29	0	0	0	0	0	0	0	0	0	0	0	0	0
	JOHNSTON ISLAND	28	0	0	0	0	0	0	0	0	0	0	0	0	0
	KWAJALEIN, MARSHALL IS.	34	0	0	0	0	0	0	0	0	0	0	0	0	0
	PAGO PAGO, AMER SAMOA	27	0	0	0	0	0	0	0	0	0	0	0	0	0
	TRUK, E. CAROLINE IS.	35	0	0	0	0	0	0	0	0	0	0	0	0	0
	WAKE ISLAND	39	0	0	0	0	0	0	0	0	0	0	0	0	0
	YAP, W. CAROLINE IS.	38	0	0	0	0	0	0	0	0	0	0	0	0	0
PA	ERIE	21	28	25	23	12	1	*	0	0	0	2	12	25	129
	HARRISBURG	48	26	23	17	4	*	0	0	0	*	2	11	23	106
	PHILADELPHIA	27	26	23	15	3	*	0	0	0	0	2	9	21	98
	PITTSBURGH	27	28	24	20	8	1	0	0	0	0	4	14	25	124
PR	SAN JUAN	31	0	0	0	0	0	0	0	0	0	0	0	0	0
RI	PROVIDENCE	23	28	24	20	6	*	0	0	0	0	4	13	25	119
SC	CHARLESTON	44	11	8	3	*	0	0	0	0	0	*	3	9	36
	COLUMBIA	20	18	14	7	1	0	0	0	0	0	1	8	14	63
	GREENVILLE-SPARTANBURG	24	20	16	7	1	0	0	0	0	0	1	7	15	67
SD	RAPID CITY	44	30	27	27	15	3	*	0	0	2	10	25	30	168
	SIOUX FALLS	23	31	27	26	12	2	0	0	0	1	11	25	31	167
TN	BRISTOL-JOHNSON CITY-KINGSPORT	25	24	20	14	4	*	0	0	0	0	3	12	21	97
	KNOXVILLE	26	21	17	9	2	*	0	0	0	0	1	8	17	74
	MEMPHIS	45	18	13	5	*	0	0	0	0	0	*	6	15	58
	NASHVILLE	21	23	17	9	2	0	0	0	0	0	1	8	17	77
TX	AUSTIN	25	9	5	1	0	0	0	0	0	0	0	1	6	22
	CORPUS CHRISTI	22	3	1	*	0	0	0	0	0	0	0	*	1	7
	DALLAS-FORT WORTH	23	16	10	3	*	0	0	0	0	0	*	3	10	41
	EL PASO	26	19	12	5	1	*	0	0	0	0	*	7	19	63
	HOUSTON	17	8	5	1	*	0	0	0	0	0	0	2	5	22
	LUBBOCK	39	26	19	11	2	*	0	0	0	0	1	12	24	95
UT	SALT LAKE CITY	27	27	22	17	7	1	0	0	0	*	5	18	27	125
VT	BURLINGTON	22	30	26	26	16	3	0	0	0	1	9	19	28	157
VA	LYNCHBURG	23	24	21	12	3	*	0	0	0	0	2	10	19	92
	NORFOLK	38	17	14	6	*	0	0	0	0	0	*	3	13	55
	RICHMOND	57	22	19	10	2	*	0	0	0	0	2	10	20	85
WA	SEATTLE[1]	36	6	3	1	0	0	0	0	0	0	*	2	4	17
	SPOKANE	27	27	22	21	11	2	0	0	0	1	9	20	26	140
	WALLA WALLA	72	19	12	5	1	*	0	0	0	*	1	9	18	65
WV	CHARLESTON	39	23	20	15	5	*	0	0	0	0	3	13	21	100
WI	GREEN BAY	25	31	27	27	14	3	0	0	0	1	8	22	29	161
	MADISON	27	30	27	26	14	3	*	0	0	1	9	22	29	163
	MILWAUKEE	26	30	26	24	10	1	0	0	0	*	4	19	28	142
WY	CASPER	22	29	26	27	19	8	*	*	0	4	16	25	29	184
	CHEYENNE	27	29	26	28	18	4	0	0	0	2	12	25	29	172

THE ANNUAL VALUE IS THE TOTAL OF THE UNROUNDED MONTHLY VALUES.
IT MAY NOT AGREE WITH THE SUM OF THE ROUNDED MONTHLY VALUES.

*Average frequency of occurrence greater than zero, but smaller than one half.

[1] City office data.

Source: U.S. National Oceanic and Atmospheric Administration, Comparative Climatic Data for the United States through 1986

TABLE 1–5. NORMAL HEATING DEGREE DAYS— SELECTED CITIES OF THE UNITED STATES

[Airport data except as noted. Based on standard 30-year period 1951 through 1980]

ST.	STATION	JUL	AUG	SEP	OCT	NOV	DEC	JAN	FEB	MAR	APR	MAY	JUN	ANNUAL
AL	BIRMINGHAM	0	0	7	137	387	601	685	532	368	110	36	0	2863
	MOBILE	0	0	0	50	218	382	469	342	191	43	0	0	1695
AK	ANCHORAGE	214	273	504	942	1299	1587	1612	1319	1280	888	580	318	10816
	BARROW	809	837	1032	1575	1980	2415	2461	2369	2508	2001	1432	951	20370
	FAIRBANKS	141	270	603	1240	1833	2328	2412	1932	1752	1044	521	198	14274
	JUNEAU	288	322	474	719	969	1184	1339	1042	1048	777	574	369	9105
AZ	PHOENIX	0	0	0	13	159	368	394	269	187	52	0	0	1442
	TUCSON	0	0	0	30	204	403	431	326	246	86	8	0	1734
AR	FORT SMITH	0	0	13	143	456	738	853	633	461	147	33	0	3477
	LITTLE ROCK	0	0	8	132	414	676	778	585	417	124	18	0	3152
CA	BAKERSFIELD	0	0	0	50	268	521	521	335	255	137	35	6	2128
	FRESNO	0	0	0	88	354	611	605	406	336	187	52	8	2647
	LOS ANGELES[1]	0	0	0	27	108	218	252	191	190	129	62	27	1204
	SACRAMENTO	0	0	7	82	360	601	611	412	366	229	83	21	2772
	SAN DIEGO	5	0	7	32	118	240	258	196	193	124	71	40	1284
	SAN FRANCISCO[1]	202	173	109	124	234	403	428	311	332	294	260	201	3071
CO	DENVER	0	0	135	414	789	1004	1101	879	837	528	253	74	6014
	GRAND JUNCTION	0	0	65	325	762	1138	1225	882	716	403	148	19	5683
CT	HARTFORD	0	8	102	391	702	1113	1234	1047	874	486	197	20	6174
DE	WILMINGTON	0	0	36	282	582	915	1048	890	719	378	130	6	4986
DC	WASHINGTON	0	0	13	197	489	809	924	770	595	257	68	0	4122
FL	JACKSONVILLE	0	0	0	21	164	332	396	302	166	21	0	0	1402
	MIAMI	0	0	0	0	5	42	76	62	14	0	0	0	199
	TAMPA	0	0	0	0	65	173	228	186	87	0	0	0	739
GA	ATLANTA	0	0	7	130	394	636	716	563	400	133	37	5	3021
	SAVANNAH	0	0	0	58	240	444	507	387	243	42	0	0	1921
HI	HONOLULU	0	0	0	0	0	0	0	0	0	0	0	0	0
	LIHUE	0	0	0	0	0	0	0	0	0	0	0	0	0
ID	BOISE	0	23	134	406	759	1023	1088	809	732	492	253	83	5802
	POCATELLO	0	32	209	524	894	1190	1277	994	915	612	348	128	7123
IL	CHICAGO	0	9	75	368	756	1156	1352	1092	899	486	224	38	6455
	PEORIA	0	5	64	361	756	1153	1349	1070	859	411	176	22	6226
IN	EVANSVILLE	0	0	40	259	597	911	1066	843	640	267	100	6	4729
	INDIANAPOLIS	0	0	63	330	696	1039	1209	983	775	382	158	15	5650
IA	DES MOINES	0	0	80	357	792	1218	1438	1134	927	435	156	17	6554
	SIOUX CITY	0	8	94	398	858	1293	1513	1170	967	463	159	24	6947
KS	TOPEKA	5	0	53	276	669	1029	1206	910	719	321	117	14	5319
	WICHITA	0	0	37	219	621	949	1097	837	656	275	89	7	4787
KY	LEXINGTON	0	0	47	280	603	896	1039	851	661	306	121	10	4814
LA	NEW ORLEANS	0	0	0	31	186	336	423	318	171	25	0	0	1490
	SHREVEPORT	0	0	0	76	293	505	597	438	282	69	9	0	2269
ME	PORTLAND	22	54	201	515	798	1215	1349	1176	1020	666	378	107	7501
MD	BALTIMORE	0	0	29	261	561	884	1001	848	673	334	115	0	4706
MA	BOSTON	0	6	80	329	594	970	1097	960	825	489	218	25	5593
MI	DETROIT	5	12	106	414	765	1132	1290	1098	930	528	247	36	6563
	GRAND RAPIDS	12	23	130	443	795	1169	1333	1156	989	561	262	54	6927
	SAULT STE. MARIE	101	123	306	611	966	1404	1603	1420	1274	807	480	210	9305
MN	MINNEAPOLIS-ST. PAUL	12	16	160	488	954	1420	1668	1330	1110	570	238	41	8007
MS	JACKSON	0	0	0	98	316	513	611	462	303	77	9	0	2389
MO	KANSAS CITY	0	0	25	213	612	955	1135	857	670	260	78	7	4812
	ST. LOUIS	0	0	40	258	612	955	1122	874	676	279	110	12	4938
	SPRINGFIELD	0	0	43	242	597	893	1039	806	640	279	107	14	4660
MT	BILLINGS	9	27	214	487	900	1175	1367	1025	967	612	318	111	7212
	GREAT FALLS	20	66	268	536	930	1218	1435	1072	1042	669	369	141	7766
	HELENA	41	77	308	617	1008	1287	1454	1092	1035	681	397	179	8176
NE	NORTH PLATTE	8	9	151	463	900	1212	1355	1056	939	522	235	59	6909
	OMAHA	0	0	73	342	765	1172	1389	1058	859	390	130	16	6194
	SCOTTSBLUFF	6	5	172	459	873	1153	1265	977	911	558	254	69	6702
NV	ELKO	19	58	230	543	891	1206	1240	952	899	648	396	166	7248
	LAS VEGAS	0	0	0	63	346	608	632	417	313	131	22	0	2532
	RENO	16	59	171	456	759	1008	1017	773	756	558	333	124	6030
NH	CONCORD	20	39	191	521	831	1256	1398	1198	1020	627	314	67	7482
	MT. WASHINGTON	505	555	732	1070	1341	1733	1857	1686	1643	1278	949	612	13961
NJ	ATLANTIC CITY[1]	0	0	32	246	519	825	958	829	722	420	189	23	4763
	NEWARK	0	0	36	254	555	915	1045	902	738	387	140	0	4972
NM	ALBUQUERQUE	0	0	12	242	630	911	936	717	583	302	81	0	4414
	ROSWELL	0	0	10	143	477	698	732	535	386	134	11	0	3126
NY	ALBANY	7	15	149	450	771	1194	1361	1165	973	552	252	38	6927

TABLE 1–5. NORMAL HEATING DEGREE DAYS— SELECTED CITIES OF THE UNITED STATES (continued)

[Airport data except as noted. Based on standard 30-year period 1951 through 1980]

ST.	STATION	JUL	AUG	SEP	OCT	NOV	DEC	JAN	FEB	MAR	APR	MAY	JUN	ANNUAL
	BUFFALO	9	25	130	423	741	1122	1287	1134	992	588	294	53	6798
	NEW YORK[1]	0	0	36	240	534	893	1029	885	732	378	134	7	4868
	SYRACUSE	12	25	133	425	732	1138	1308	1148	983	567	269	47	6787
NC	ASHEVILLE	0	0	57	286	558	797	874	725	577	283	114	23	4294
	RALEIGH	0	0	9	187	450	713	787	655	496	181	53	0	3531
	WILMINGTON	0	0	0	94	295	521	607	498	350	94	10	0	2469
ND	BISMARCK	18	57	255	586	1092	1538	1807	1414	1209	675	324	100	9075
	WILLISTON	23	58	277	617	1122	1553	1810	1411	1231	696	335	108	9241
OH	CINCINNATI	0	0	52	316	648	967	1119	921	719	350	143	12	5247
	CLEVELAND	8	11	99	371	696	1051	1225	1053	880	507	244	33	6178
	COLUMBUS	0	5	78	355	687	1020	1175	986	775	408	178	19	5686
	TOLEDO	0	16	113	419	771	1144	1299	1098	918	516	237	39	6570
OK	OKLAHOMA CITY	0	0	15	145	486	778	902	678	506	184	41	0	3735
	TULSA	0	0	18	146	474	781	924	680	500	168	40	0	3731
OR	EUGENE	44	57	126	363	591	735	772	602	595	462	307	145	4799
	MEDFORD	6	19	92	335	642	846	849	633	605	441	245	85	4798
	PORTLAND	35	51	111	332	585	747	809	610	592	438	263	118	4691
PC	GUAM	0	0	0	0	0	0	0	0	0	0	0	0	0
	JOHNSTON ISLAND	0	0	0	0	0	0	0	0	0	0	0	0	0
	KWAJALEIN, MARSHALL IS.	0	0	0	0	0	0	0	0	0	0	0	0	0
	PAGO PAGO, AMER SAMOA	0	0	0	0	0	0	0	0	0	0	0	0	0
	TRUK, E. CAROLINE IS.	0	0	0	0	0	0	0	0	0	0	0	0	0
	WAKE ISLAND	0	0	0	0	0	0	0	0	0	0	0	0	0
	YAP, W. CAROLINE IS.	0	0	0	0	0	0	0	0	0	0	0	0	0
PA	ERIE	17	28	130	420	729	1085	1256	1120	977	603	323	80	6768
	HARRISBURG	0	0	58	320	633	980	1104	938	756	384	150	12	5335
	PHILADELPHIA	0	0	33	273	576	915	1048	893	719	363	127	0	4947
	PITTSBURGH	0	13	101	393	702	1042	1187	1014	822	447	201	28	5950
PR	SAN JUAN	0	0	0	0	0	0	0	0	0	0	0	0	0
RI	PROVIDENCE	0	6	94	366	648	1014	1141	1000	856	513	239	31	5908
SC	CHARLESTON	0	0	0	76	262	471	543	434	286	69	6	0	2147
	COLUMBIA	0	0	0	123	339	567	637	508	346	87	22	0	2629
	GREENVILLE-SPARTANBURG	0	0	7	162	423	670	741	599	442	154	41	0	3239
SD	RAPID CITY	21	24	188	482	903	1206	1370	1092	1004	612	298	101	7301
	SIOUX FALLS	14	15	161	489	960	1404	1631	1285	1073	561	240	52	7885
TN	BRISTOL-JOHNSON CITY-KINGSPORT	0	0	35	263	564	831	930	759	580	273	111	10	4356
	KNOXVILLE	0	0	14	201	486	741	831	658	483	181	63	0	3658
	MEMPHIS	0	0	9	137	415	673	787	602	433	126	25	0	3207
	NASHVILLE	0	0	19	193	492	747	865	689	510	186	55	0	3756
TX	AUSTIN	0	0	0	37	221	406	505	347	203	41	0	0	1760
	CORPUS CHRISTI	0	0	0	11	116	220	310	209	97	7	0	0	970
	DALLAS-FORT WORTH	0	0	0	56	300	533	651	469	313	85	0	0	2407
	EL PASO	0	0	0	96	408	639	645	465	318	93	0	0	2664
	HOUSTON	0	0	0	36	201	349	442	314	175	32	0	0	1549
	LUBBOCK	0	0	15	157	495	729	812	627	470	178	33	0	3516
UT	SALT LAKE CITY	0	0	97	377	759	1076	1128	865	753	474	220	53	5802
VT	BURLINGTON	23	50	202	530	852	1314	1500	1313	1110	669	326	64	7953
VA	LYNCHBURG	0	0	32	258	540	828	927	773	598	263	97	7	4323
	NORFOLK	0	0	9	146	393	667	778	669	512	219	53	0	3446
	RICHMOND	0	0	24	221	483	778	880	731	552	226	65	0	3960
WA	SEATTLE[1]	64	74	142	347	564	694	756	582	598	438	274	148	4681
	SPOKANE	17	63	209	539	903	1116	1218	913	849	576	339	140	6882
	WALLA WALLA	0	16	85	332	675	862	952	680	595	381	175	54	4807
WV	CHARLESTON	0	0	51	301	591	871	995	823	626	298	125	16	4697
WI	GREEN BAY	17	39	192	515	924	1370	1581	1322	1128	639	325	91	8143
	MADISON	12	29	161	490	897	1321	1531	1246	1048	576	273	58	7642
	MILWAUKEE	11	25	117	444	831	1237	1435	1176	1020	612	334	84	7326
WY	CASPER	16	31	240	552	963	1218	1327	1067	1026	687	384	131	7642
	CHEYENNE	24	37	235	543	906	1107	1206	1000	1020	696	397	139	7310

[1] City office data.

Degree days data are used to estimate amounts of energy required to maintain comfortable indoor temperature levels. Each degree that a day's mean temperature is below 65 degrees Fahrenheit is counted as one heating degree day.

Source: U.S. National Oceanic and Atmospheric Administration, Comparative Climatic Data for the United States through 1986

TABLE 1-6. NORMAL COOLING DEGREE DAYS— SELECTED CITIES OF THE UNITED STATES

[Airport data except as noted. Based on standard 30-year period 1951 through 1980]

ST.	STATION	JAN	FEB	MAR	APR	MAY	JUN	JUL	AUG	SEP	OCT	NOV	DEC	ANNUAL
AL	BIRMINGHAM	0	8	18	44	191	360	468	450	280	62	0	0	1881
	MOBILE	29	23	39	133	307	465	533	521	396	158	26	13	2643
AK	ANCHORAGE	0	0	0	0	0	0	0	0	0	0	0	0	0
	BARROW	0	0	0	0	0	0	0	0	0	0	0	0	0
	FAIRBANKS	0	0	0	0	0	27	33	10	0	0	0	0	70
	JUNEAU	0	0	0	0	0	0	0	0	0	0	0	0	0
AZ	PHOENIX	0	20	51	142	376	645	846	772	588	273	27	6	3746
	TUCSON	0	12	22	86	262	537	657	589	462	198	15	0	2840
AR	FORT SMITH	0	0	11	39	175	375	530	493	280	66	0	0	1969
	LITTLE ROCK	0	0	20	46	188	405	530	496	287	73	0	0	2045
CA	BAKERSFIELD	0	0	10	68	208	405	605	539	369	143	0	0	2347
	FRESNO	0	0	0	40	135	308	496	431	277	82	0	0	1769
	LOS ANGELES[1]	10	21	20	39	59	153	282	301	256	148	39	11	1339
	SACRAMENTO	0	0	0	25	80	207	329	301	208	48	0	0	1198
	SAN DIEGO	0	11	7	10	21	79	170	226	187	109	22	0	842
	SAN FRANCISCO[1]	0	0	0	0	0	0	0	6	31	19	0	0	56
CO	DENVER	0	0	0	0	11	134	261	203	63	8	0	0	680
	GRAND JUNCTION	0	0	0	0	58	238	431	338	128	12	0	0	1205
CT	HARTFORD	0	0	0	0	24	128	260	203	51	0	0	0	666
DE	WILMINGTON	0	0	0	0	43	192	341	304	123	12	0	0	1015
DC	WASHINGTON	0	0	0	8	99	285	431	391	196	20	0	0	1430
FL	JACKSONVILLE	30	25	51	102	282	420	505	496	396	160	38	15	2520
	MIAMI	141	140	222	309	419	480	539	552	504	400	239	150	4095
	TAMPA	66	68	124	202	375	477	533	533	477	295	116	58	3324
GA	ATLANTA	0	0	12	37	170	329	422	409	247	44	0	0	1670
	SAVANNAH	17	12	38	72	261	408	502	490	348	117	15	10	2290
HI	HONOLULU	236	221	291	321	388	423	468	496	468	450	348	279	4389
	LIHUE	200	184	220	252	326	372	415	437	417	381	309	245	3758
ID	BOISE	0	0	0	0	17	107	298	240	80	0	0	0	742
	POCATELLO	0	0	0	0	7	53	197	153	35	0	0	0	445
IL	CHICAGO	0	0	0	0	41	146	252	223	66	12	0	0	740
	PEORIA	0	0	0	0	71	208	314	256	82	17	0	0	948
IN	EVANSVILLE	0	0	8	12	119	294	406	347	172	20	0	0	1378
	INDIANAPOLIS	0	0	0	0	80	213	313	257	111	14	0	0	988
IA	DES MOINES	0	0	0	0	66	215	354	279	83	22	0	0	1019
	SIOUX CITY	0	0	0	0	51	201	333	266	79	10	0	0	940
KS	TOPEKA	0	0	0	9	105	281	427	372	158	28	0	0	1380
	WICHITA	0	0	8	14	117	340	508	456	205	36	0	0	1684
KY	LEXINGTON	0	0	7	9	96	226	341	307	158	26	0	0	1170
LA	NEW ORLEANS	32	30	59	136	307	459	530	518	405	161	36	13	2686
	SHREVEPORT	8	12	34	90	257	444	555	539	363	128	14	0	2444
ME	PORTLAND	0	0	0	0	0	23	118	104	9	0	0	0	254
MD	BALTIMORE	0	0	0	0	66	221	366	329	146	10	0	0	1138
MA	BOSTON	0	0	0	0	17	115	266	220	68	13	0	0	699
MI	DETROIT	0	0	0	0	33	117	219	183	55	8	0	0	615
	GRAND RAPIDS	0	0	0	0	29	117	210	165	43	6	0	0	570
	SAULT STE. MARIE	0	0	0	0	6	12	55	58	0	0	0	0	131
MN	DULUTH	0	0	0	0	0	11	80	59	0	0	0	0	150
	MINNEAPOLIS-ST. PAUL	0	0	0	0	36	134	263	190	28	11	0	0	662
MS	JACKSON	13	17	34	80	241	426	524	502	342	98	13	0	2290
MO	KANSAS CITY	0	0	7	17	137	334	493	443	205	45	0	0	1681
	ST. LOUIS	0	0	0	12	128	306	431	372	181	38	0	0	1468
	SPRINGFIELD	0	0	7	15	101	266	403	372	178	32	0	0	1374
MT	BILLINGS	0	0	0	0	0	81	235	191	46	0	0	0	553
	GREAT FALLS	0	0	0	0	0	48	153	144	40	6	0	0	391
	HELENA	0	0	0	0	0	32	131	105	26	0	0	0	294
NE	NORTH PLATTE	0	0	0	0	24	158	294	239	58	0	0	0	773
	OMAHA	0	0	0	6	77	256	394	320	97	16	0	0	1166
	SCOTTSBLUFF	0	0	0	0	13	144	291	210	70	0	0	0	728
NV	ELKO	0	0	0	0	6	52	178	138	32	0	0	0	406
	LAS VEGAS	0	0	12	86	279	558	784	713	453	144	0	0	3029
	RENO	0	0	0	0	11	46	156	117	27	0	0	0	357
NH	CONCORD	0	0	0	0	10	58	160	111	14	0	0	0	353
	MT. WASHINGTON	0	0	0	0	0	0	0	0	0	0	0	0	0
NJ	ATLANTIC CITY[1]	0	0	0	0	19	125	282	273	116	17	0	0	832
	NEWARK	0	0	0	0	56	199	366	326	132	12	0	0	1091
NM	ALBUQUERQUE	0	0	0	0	59	285	428	344	132	6	0	0	1254
	ROSWELL	0	0	8	41	176	420	508	440	229	41	0	0	1863
NY	ALBANY	0	0	0	0	19	89	206	145	35	0	0	0	494

TABLE 1–6. NORMAL COOLING DEGREE DAYS— SELECTED CITIES OF THE UNITED STATES (continued)

[Airport data except as noted. Based on standard 30-year period 1951 through 1980]

ST.	STATION	JAN	FEB	MAR	APR	MAY	JUN	JUL	AUG	SEP	OCT	NOV	DEC	ANNUAL
	BUFFALO	0	0	0	0	18	83	186	146	43	0	0	0	476
	NEW YORK[1]	0	0	0	0	56	199	363	322	135	14	0	0	1089
	SYRACUSE	0	0	0	0	21	86	195	158	46	0	0	0	506
NC	ASHEVILLE	0	0	0	0	61	167	254	239	114	7	0	0	842
	RALEIGH	0	0	9	16	121	270	394	372	189	23	0	0	1394
	WILMINGTON	6	5	12	37	187	348	474	456	294	78	7	0	1904
ND	BISMARCK	0	0	0	0	10	79	186	174	24	0	0	0	473
	WILLISTON	0	0	0	0	10	72	178	155	25	0	0	0	440
OH	CINCINNATI	0	0	0	5	81	204	322	282	127	16	0	0	1037
	CLEVELAND	0	0	0	0	33	111	213	178	72	5	0	0	612
	COLUMBUS	0	0	0	0	66	175	273	235	102	11	0	0	862
	TOLEDO	0	0	0	0	38	129	215	174	59	7	0	0	622
OK	OKLAHOMA CITY	0	0	13	40	147	360	530	499	264	61	0	0	1914
	TULSA	0	0	14	45	167	381	564	518	282	72	0	0	2043
OR	EUGENE	0	0	0	0	0	28	100	94	39	0	0	0	261
	MEDFORD	0	0	0	0	12	91	239	208	95	0	0	0	645
	PORTLAND	0	0	0	0	6	43	119	122	42	0	0	0	332
PC	GUAM	378	344	397	423	456	450	450	437	420	437	426	415	5033
	JOHNSTON ISLAND	363	325	369	381	434	450	484	499	480	487	426	391	5089
	KWAJALEIN, MARSHALL IS.	505	468	533	504	515	501	518	524	510	524	504	515	6121
	PAGO PAGO, AMER SAMOA	487	442	496	465	462	441	431	425	432	465	459	484	5489
	TRUK, E. CAROLINE IS.	502	454	508	495	512	495	496	499	486	505	495	508	5955
	WAKE ISLAND	369	325	394	399	465	510	543	546	531	524	462	415	5483
	YAP, W. CAROLINE IS.	474	434	496	501	518	492	496	490	480	502	489	490	5862
PA	ERIE	0	0	0	0	13	68	144	127	43	7	0	0	402
	HARRISBURG	0	0	0	0	57	198	335	291	115	10	0	0	1006
	PHILADELPHIA	0	0	0	0	59	202	357	319	129	9	0	0	1075
	PITTSBURGH	0	0	0	0	37	121	222	186	74	5	0	0	645
PR	SAN JUAN	357	325	388	414	474	501	527	533	507	508	438	394	5366
RI	PROVIDENCE	0	0	0	0	10	85	235	195	49	0	0	0	574
SC	CHARLESTON	13	9	29	48	229	378	481	465	321	101	13	6	2093
	COLUMBIA	8	6	20	51	223	381	496	471	297	74	6	0	2033
	GREENVILLE-SPARTANBURG	0	0	8	19	143	297	409	388	208	29	0	0	1501
SD	RAPID CITY	0	0	0	0	7	107	257	223	65	8	0	0	667
	SIOUX FALLS	0	0	0	0	29	154	293	226	41	6	0	0	749
TN	BRISTOL-JOHNSON CITY-KINGSPORT	0	0	0	6	93	205	307	288	149	18	0	0	1066
	KNOXVILLE	0	0	8	19	137	283	391	372	209	30	0	0	1449
	MEMPHIS	0	0	20	54	211	411	530	484	285	72	0	0	2067
	NASHVILLE	0	0	14	24	151	328	446	415	238	45	0	0	1661
TX	AUSTIN	12	16	63	152	307	498	611	605	426	186	32	6	2914
	CORPUS CHRISTI	40	50	125	247	406	531	617	620	495	290	116	37	3574
	DALLAS-FORT WORTH	0	7	37	112	275	510	660	636	408	146	18	0	2809
	EL PASO	0	0	11	51	218	474	543	474	273	52	0	0	2096
	HOUSTON	20	20	51	143	307	468	561	546	402	181	54	8	2761
	LUBBOCK	0	0	11	37	157	378	459	400	201	33	0	0	1676
UT	SALT LAKE CITY	0	0	0	0	28	152	388	311	97	5	0	0	981
VT	BURLINGTON	0	0	0	0	22	61	165	115	16	0	0	0	379
VA	LYNCHBURG	0	0	0	5	81	205	332	304	134	13	0	0	1074
	NORFOLK	0	0	0	15	96	282	415	394	225	31	0	0	1458
	RICHMOND	0	0	0	13	99	258	397	366	180	23	0	0	1336
WA	SEATTLE[1]	0	0	0	0	0	28	73	74	25	0	0	0	200
	SPOKANE	0	0	0	0	8	41	162	159	41	0	0	0	411
	WALLA WALLA	0	0	0	0	23	132	326	282	100	0	0	0	863
WV	CHARLESTON	0	0	0	7	91	196	295	270	129	19	0	0	1007
WI	GREEN BAY	0	0	0	0	18	82	156	116	9	0	0	0	381
	MADISON	0	0	0	0	25	97	185	137	14	9	0	0	467
	MILWAUKEE	0	0	0	0	18	81	182	158	24	7	0	0	470
WY	CASPER	0	0	0	0	0	68	199	148	42	0	0	0	457
	CHEYENNE	0	0	0	0	0	49	145	93	22	0	0	0	309

[1] City office data.

Degree days data are used to estimate amounts of energy required to maintain comfortable indoor temperature levels. Each degree that a day's mean temperature is above 65 degrees Fahrenheit is counted as one cooling degree day.

Source: U.S. National Oceanic and Atmospheric Administration, Comparative Climatic Data for the United States through 1986

TABLE 1–7. NORMAL MONTHLY PRECIPITATION— SELECTED CITIES OF THE UNITED STATES

[In inches. Includes liquid water equivalent of snowfall. Airport data except as noted. Based on standard 30-year period 1951 through 1980]

ST.	STATION	JAN	FEB	MAR	APR	MAY	JUN	JUL	AUG	SEP	OCT	NOV	DEC	ANNUAL
AL	BIRMINGHAM	5.23	4.72	6.62	5.00	4.53	3.61	5.39	3.85	4.34	2.64	3.64	4.95	54.52
	MOBILE	4.59	4.91	6.48	5.35	5.46	5.07	7.74	6.75	6.56	2.62	3.67	5.44	64.64
AK	ANCHORAGE	0.80	0.93	0.69	0.66	0.57	1.08	1.97	2.11	2.45	1.73	1.11	1.10	15.20
	BARROW	0.21	0.17	0.17	0.21	0.16	0.37	0.86	0.98	0.59	0.55	0.30	0.18	4.75
	FAIRBANKS	0.53	0.42	0.40	0.27	0.57	1.32	1.77	1.86	1.09	0.74	0.67	0.73	10.37
	JUNEAU	3.69	3.74	3.34	2.92	3.41	2.98	4.13	5.02	6.40	7.71	5.15	4.66	53.15
AZ	PHOENIX	0.73	0.59	0.81	0.27	0.14	0.17	0.74	1.02	0.64	0.63	0.54	0.83	7.11
	TUCSON	0.83	0.63	0.68	0.32	0.14	0.22	2.42	2.13	1.33	0.88	0.62	0.94	11.14
AR	FORT SMITH	1.86	2.53	3.88	4.20	4.79	3.67	3.15	3.02	3.22	3.24	3.50	2.85	39.91
	LITTLE ROCK	3.91	3.83	4.69	5.41	5.29	3.67	3.63	3.07	4.26	2.84	4.37	4.23	49.20
CA	BAKERSFIELD	0.98	1.07	0.87	0.70	0.24	0.07	0.01	0.05	0.13	0.30	0.65	0.65	5.72
	FRESNO	2.05	1.85	1.61	1.15	0.31	0.08	0.01	0.02	0.16	0.43	1.24	1.61	10.52
	LOS ANGELES[1]	3.69	2.96	2.35	1.17	0.23	0.03	0.00	0.12	0.27	0.21	1.85	1.97	14.85
	SACRAMENTO	4.03	2.88	2.06	1.31	0.33	0.11	0.05	0.07	0.27	0.86	2.23	2.90	17.10
	SAN DIEGO	2.11	1.43	1.60	0.78	0.24	0.06	0.01	0.11	0.19	0.33	1.10	1.36	9.32
	SAN FRANCISCO[1]	4.48	2.83	2.58	1.48	0.35	0.15	0.04	0.08	0.24	1.09	2.49	3.52	19.33
CO	DENVER	0.51	0.69	1.21	1.81	2.47	1.58	1.93	1.53	1.23	0.98	0.82	0.55	15.31
	GRAND JUNCTION	0.64	0.54	0.75	0.71	0.76	0.44	0.47	0.91	0.70	0.87	0.63	0.58	8.00
CT	HARTFORD	3.53	3.19	4.15	4.02	3.37	3.38	3.09	4.00	3.94	3.51	4.05	4.16	44.39
DE	WILMINGTON	3.11	2.99	3.87	3.39	3.23	3.51	3.90	4.03	3.59	2.89	3.33	3.54	41.38
DC	WASHINGTON	2.76	2.62	3.46	2.93	3.48	3.35	3.88	4.40	3.22	2.90	2.82	3.18	39.00
FL	JACKSONVILLE	3.07	3.48	3.72	3.32	4.91	5.37	6.54	7.15	7.26	3.41	1.94	2.59	52.76
	MIAMI	2.08	2.05	1.89	3.07	6.53	9.15	5.98	7.02	8.07	7.14	2.71	1.86	57.55
	TAMPA	2.17	3.04	3.46	1.82	3.38	5.29	7.35	7.64	6.23	2.34	1.87	2.14	46.73
GA	ATLANTA	4.91	4.43	5.91	4.43	4.02	3.41	4.73	3.41	3.17	2.53	3.43	4.23	48.61
	SAVANNAH	3.09	3.17	3.83	3.16	4.62	5.69	7.37	6.65	5.19	2.27	1.89	2.77	49.70
HI	HONOLULU	3.79	2.72	3.48	1.49	1.21	0.49	0.54	0.60	0.62	1.88	3.22	3.43	23.47
	LIHUE	6.24	3.68	4.52	3.29	2.99	1.64	2.03	1.85	2.25	4.52	5.55	5.46	44.02
ID	BOISE	1.64	1.07	1.03	1.19	1.21	0.95	0.26	0.40	0.58	0.75	1.29	1.34	11.71
	POCATELLO	1.13	0.86	0.94	1.16	1.20	1.06	0.47	0.60	0.65	0.92	0.91	0.96	10.86
IL	CHICAGO	1.60	1.31	2.59	3.66	3.15	4.08	3.63	3.53	3.35	2.28	2.06	2.10	33.34
	PEORIA	1.60	1.41	2.86	3.81	3.84	3.88	3.99	3.39	3.63	2.51	1.96	2.01	34.89
IN	EVANSVILLE	2.99	3.02	4.58	4.08	4.37	3.50	3.98	3.07	2.67	2.48	3.36	3.45	41.55
	INDIANAPOLIS	2.65	2.46	3.61	3.68	3.66	3.99	4.32	3.46	2.74	2.51	3.04	3.00	39.12
IA	DES MOINES	1.01	1.12	2.20	3.21	3.96	4.18	3.22	4.11	3.09	2.16	1.52	1.05	30.83
	SIOUX CITY	0.60	0.94	1.71	2.29	3.43	3.99	3.36	3.14	2.51	1.73	0.93	0.74	25.37
KS	TOPEKA	0.88	1.05	2.18	3.08	3.99	5.14	4.04	3.69	3.45	2.82	1.75	1.31	33.38
	WICHITA	0.68	0.85	2.01	2.30	3.91	4.06	3.62	2.80	3.45	2.47	1.47	0.99	28.61
KY	LEXINGTON	3.57	3.26	4.83	4.01	4.23	4.25	4.95	3.96	3.28	2.26	3.30	3.78	45.68
LA	NEW ORLEANS	4.97	5.23	4.73	4.50	5.07	4.63	6.73	6.02	5.87	2.66	4.06	5.27	59.74
	SHREVEPORT	4.02	3.46	3.77	4.71	4.70	3.54	3.56	2.52	3.29	2.63	3.77	3.87	43.84
ME	PORTLAND	3.78	3.57	3.98	3.90	3.27	3.06	2.83	2.82	3.27	3.83	4.70	4.51	43.52
MD	BALTIMORE	3.00	2.98	3.72	3.35	3.44	3.76	3.89	4.62	3.46	3.11	3.11	3.40	41.84
MA	BOSTON	3.99	3.70	4.13	3.73	3.52	2.92	2.68	3.68	3.41	3.36	4.21	4.48	43.81
MI	DETROIT	1.86	1.69	2.54	3.15	2.77	3.43	3.10	3.21	2.25	2.12	2.33	2.52	30.97
	GRAND RAPIDS	1.91	1.53	2.48	3.56	3.03	3.86	3.02	3.45	3.14	2.89	2.93	2.55	34.35
	SAULT STE. MARIE	2.20	1.69	2.03	2.38	2.90	3.26	3.00	3.46	3.90	2.89	3.20	2.57	33.48
MN	DULUTH	1.20	0.90	1.78	2.16	3.15	3.96	3.96	4.12	3.26	2.21	1.69	1.29	29.68
	MINNEAPOLIS-ST. PAUL	0.82	0.85	1.71	2.05	3.20	4.07	3.51	3.64	2.50	1.85	1.29	0.87	26.36
MS	JACKSON	5.00	4.48	5.86	5.85	4.83	2.94	4.40	3.71	3.55	2.62	4.18	5.40	52.82
MO	KANSAS CITY	0.98	1.05	2.12	2.68	3.42	4.13	3.49	3.16	3.33	2.54	1.23	1.14	29.27
	ST. LOUIS	1.72	2.14	3.28	3.55	3.54	3.73	3.63	2.55	2.70	2.32	2.53	2.22	33.91
	SPRINGFIELD	1.60	2.13	3.44	4.03	4.32	4.66	3.58	2.83	4.24	3.20	2.89	2.55	39.47
MT	BILLINGS	0.97	0.71	1.05	1.93	2.39	2.07	0.85	1.05	1.26	1.16	0.85	0.80	15.09
	GREAT FALLS	1.00	0.75	0.93	1.49	2.52	2.75	1.10	1.31	1.03	0.82	0.74	0.80	15.24
	HELENA	0.66	0.44	0.69	1.01	1.72	2.01	1.04	1.18	0.83	0.65	0.54	0.60	11.37
NE	NORTH PLATTE	0.40	0.55	1.12	1.85	3.36	3.72	2.98	1.92	1.67	0.91	0.56	0.43	19.47
	OMAHA	0.77	0.91	1.91	2.94	4.33	4.08	3.62	4.10	3.50	2.09	1.32	0.77	30.34
	SCOTTSBLUFF	0.44	0.37	0.97	1.43	2.66	2.93	1.96	0.97	1.08	0.75	0.52	0.51	14.59
NV	ELKO	1.16	0.81	0.85	0.79	1.03	0.91	0.33	0.58	0.47	0.56	0.83	0.98	9.30
	LAS VEGAS	0.50	0.46	0.41	0.22	0.20	0.09	0.45	0.54	0.32	0.25	0.43	0.32	4.19
	RENO	1.24	0.95	0.74	0.46	0.74	0.34	0.30	0.27	0.30	0.34	0.60	1.21	7.49
NH	CONCORD	2.78	2.47	2.93	3.01	2.93	2.91	2.93	3.26	3.12	3.10	3.66	3.43	36.53
	MT. WASHINGTON	7.31	8.01	8.19	7.03	6.46	7.06	6.90	7.60	7.15	6.73	8.54	8.94	89.92
NJ	ATLANTIC CITY[1]	3.25	3.22	3.71	3.12	2.91	2.88	3.89	4.54	2.73	2.76	3.54	3.51	40.06
	NEWARK	3.13	3.05	4.15	3.57	3.59	2.94	3.85	4.30	3.66	3.09	3.59	3.42	42.34
NM	ALBUQUERQUE	0.41	0.40	0.52	0.40	0.46	0.51	1.30	1.51	0.85	0.86	0.38	0.52	8.12
	ROSWELL	0.24	0.28	0.27	0.37	0.77	0.91	1.38	2.17	1.72	0.99	0.33	0.27	9.70

TABLE 1–7. NORMAL MONTHLY PRECIPITATION— SELECTED CITIES OF THE UNITED STATES (continued)

[In inches. Includes liquid water equivalent of snowfall. Airport data except as noted. Based on standard 30-year period 1951 through 1980]

ST.	STATION	JAN	FEB	MAR	APR	MAY	JUN	JUL	AUG	SEP	OCT	NOV	DEC	ANNUAL
NY	ALBANY	2.39	2.26	3.01	2.94	3.31	3.29	3.00	3.34	3.23	2.93	3.04	3.00	35.74
	BUFFALO	3.02	2.40	2.97	3.06	2.89	2.72	2.96	4.16	3.37	2.93	3.62	3.42	37.52
	NEW YORK[1]	3.21	3.13	4.22	3.75	3.76	3.23	3.77	4.03	3.66	3.41	4.14	3.81	44.12
	SYRACUSE	2.61	2.65	3.11	3.34	3.16	3.63	3.76	3.77	3.29	3.14	3.45	3.20	39.11
NC	ASHEVILLE	3.48	3.60	5.13	3.84	4.19	4.20	4.43	4.79	3.96	3.29	3.29	3.51	47.71
	RALEIGH	3.55	3.43	3.69	2.91	3.67	3.66	4.38	4.44	3.29	2.73	2.87	3.14	41.76
	WILMINGTON	3.64	3.44	4.04	2.98	4.22	5.65	7.44	6.64	5.71	2.97	3.19	3.43	53.35
ND	BISMARCK	0.51	0.45	0.70	1.51	2.23	3.01	2.05	1.69	1.38	0.81	0.51	0.51	15.36
	WILLISTON	0.55	0.50	0.57	1.29	1.85	2.68	1.83	1.42	1.37	0.74	0.50	0.55	13.85
OH	CINCINNATI	3.13	2.73	3.95	3.58	3.84	4.09	4.28	2.97	2.91	2.54	3.12	3.00	40.14
	CLEVELAND	2.47	2.20	2.99	3.32	3.30	3.49	3.37	3.38	2.92	2.45	2.76	2.75	35.40
	COLUMBUS	2.75	2.18	3.23	3.41	3.76	4.01	4.01	3.70	2.76	1.91	2.64	2.61	36.97
	TOLEDO	1.99	1.80	2.64	3.04	2.90	3.49	3.26	3.19	2.53	1.94	2.41	2.59	31.78
OK	OKLAHOMA CITY	0.96	1.29	2.07	2.91	5.50	3.87	3.04	2.40	3.41	2.71	1.53	1.20	30.89
	TULSA	1.35	1.74	3.14	4.15	5.14	4.57	3.51	3.01	4.37	3.41	2.56	1.82	38.77
OR	EUGENE	8.39	5.12	5.11	2.76	1.97	1.24	0.27	0.95	1.45	3.47	6.82	8.49	46.04
	MEDFORD	3.42	2.12	1.85	1.07	1.19	0.67	0.25	0.46	0.75	1.68	2.89	3.49	19.84
	PORTLAND	6.16	3.93	3.61	2.31	2.08	1.47	0.46	1.13	1.61	3.05	5.17	6.41	37.39
PC	GUAM	5.43	4.76	4.19	4.14	6.41	5.53	10.31	13.92	14.25	13.87	8.98	6.07	97.86
	JOHNSTON ISLAND	2.35	1.88	2.72	2.41	1.81	0.92	1.05	2.11	2.09	3.02	3.06	3.10	26.52
	KWAJALEIN, MARSHALL IS.	4.91	2.97	5.17	7.60	11.24	10.11	10.30	10.30	10.94	12.24	10.97	7.96	104.71
	PAGO PAGO, AMER SAMOA	12.78	12.53	11.38	11.25	10.72	8.56	6.51	7.08	6.69	11.05	11.20	14.21	123.96
	TRUK, E. CAROLINE IS.	8.36	6.67	9.11	12.76	15.64	12.37	14.32	14.04	13.23	14.68	12.07	12.59	145.84
	WAKE ISLAND	1.17	1.20	1.94	2.09	1.87	2.34	3.84	5.46	5.66	4.76	2.77	1.76	34.86
	YAP, W. CAROLINE IS.	7.92	5.54	6.28	6.56	9.96	11.39	14.24	14.64	13.18	12.56	9.84	10.07	122.18
PA	ERIE	2.49	2.12	2.91	3.49	3.28	3.72	3.28	3.85	3.89	3.37	3.74	3.25	39.39
	HARRISBURG	2.96	2.73	3.50	3.19	3.67	3.63	3.32	3.29	3.60	2.73	3.24	3.23	39.09
	PHILADELPHIA	3.18	2.81	3.86	3.47	3.18	3.92	3.88	4.10	3.42	2.83	3.32	3.45	41.42
	PITTSBURGH	2.86	2.40	3.58	3.28	3.54	3.30	3.83	3.31	2.80	2.49	2.34	2.57	36.30
PR	SAN JUAN	3.01	2.02	2.31	3.62	5.64	4.66	4.87	5.93	5.99	5.89	5.59	4.46	53.99
RI	PROVIDENCE	4.06	3.72	4.29	3.95	3.48	2.79	3.01	4.04	3.54	3.75	4.22	4.47	45.32
SC	CHARLESTON	3.33	3.37	4.38	2.58	4.41	6.54	7.33	6.50	4.94	2.92	2.18	3.11	51.59
	COLUMBIA	4.38	3.99	5.16	3.59	3.85	4.45	5.35	5.56	4.23	2.55	2.51	3.50	49.12
	GREENVILLE-SPARTANBURG	4.21	4.39	5.87	4.35	4.22	4.77	4.08	3.66	4.35	3.49	3.21	3.93	50.53
SD	RAPID CITY	0.42	0.62	1.02	1.96	2.63	3.26	2.12	1.44	1.03	0.81	0.51	0.45	16.27
	SIOUX FALLS	0.50	0.93	1.58	2.36	3.21	3.70	2.71	3.13	2.79	1.57	0.92	0.72	24.12
TN	BRISTOL-JOHNSON CITY-KINGSPORT	3.56	3.43	4.29	3.46	3.61	3.46	4.19	3.23	3.00	2.50	2.98	3.53	41.24
	KNOXVILLE	4.65	4.18	5.49	3.87	3.71	3.95	4.33	3.02	2.99	2.73	3.78	4.59	47.29
	MEMPHIS	4.61	4.33	5.44	5.77	5.06	3.58	4.03	3.74	3.62	2.37	4.17	4.85	51.57
	NASHVILLE	4.49	4.03	5.58	4.47	4.56	3.70	3.82	3.40	3.71	2.58	3.52	4.63	48.49
TX	AUSTIN	1.60	2.49	1.68	3.11	4.19	3.06	1.89	2.24	3.60	3.38	2.20	2.06	31.50
	CORPUS CHRISTI	1.63	1.55	0.84	1.99	3.05	3.36	1.96	3.51	6.15	3.19	1.55	1.40	30.18
	DALLAS-FORT WORTH	1.65	1.93	2.42	3.63	4.27	2.59	2.00	1.76	3.31	2.47	1.76	1.67	29.46
	EL PASO	0.38	0.45	0.32	0.19	0.24	0.56	1.60	1.21	1.42	0.73	0.33	0.39	7.82
	HOUSTON	3.21	3.25	2.68	4.24	4.69	4.06	3.33	3.66	4.93	3.67	3.38	3.66	44.76
	LUBBOCK	0.38	0.57	0.90	1.08	2.59	2.81	2.34	2.20	2.06	1.81	0.59	0.43	17.76
UT	SALT LAKE CITY	1.35	1.33	1.72	2.21	1.47	0.97	0.72	0.92	0.89	1.14	1.22	1.37	15.31
VT	BURLINGTON	1.85	1.73	2.20	2.77	2.96	3.64	3.43	3.87	3.20	2.81	2.80	2.43	33.69
VA	LYNCHBURG	3.06	2.93	3.69	2.90	3.65	3.47	3.85	3.69	3.23	3.36	2.92	3.16	39.91
	NORFOLK	3.72	3.28	3.86	2.87	3.75	3.45	5.15	5.33	4.35	3.41	2.88	3.17	45.22
	RICHMOND	3.23	3.13	3.57	2.90	3.55	3.60	5.14	5.01	3.52	3.74	3.29	3.39	44.07
WA	SEATTLE[1]	5.94	4.20	3.70	2.46	1.66	1.53	0.89	1.38	2.03	3.40	5.36	6.29	38.84
	SPOKANE	2.47	1.61	1.36	1.08	1.38	1.23	0.50	0.74	0.71	1.08	2.06	2.49	16.71
	WALLA WALLA	2.12	1.40	1.41	1.35	1.40	0.93	0.35	0.71	0.83	1.40	1.87	2.19	15.96
WV	CHARLESTON	3.48	3.11	4.00	3.52	3.68	3.32	5.36	4.15	3.01	2.63	2.90	3.27	42.43
WI	GREEN BAY	1.19	1.05	1.90	2.70	3.13	3.17	3.25	3.16	3.17	2.10	1.76	1.42	28.00
	MADISON	1.11	1.02	2.15	3.10	3.34	3.89	3.75	3.82	3.06	2.24	1.83	1.53	30.84
	MILWAUKEE	1.64	1.33	2.58	3.37	2.66	3.59	3.54	3.09	2.88	2.25	1.98	2.03	30.94
WY	CASPER	0.50	0.56	0.99	1.51	2.13	1.24	1.06	0.63	0.76	0.88	0.66	0.51	11.43
	CHEYENNE	0.41	0.40	0.97	1.24	2.39	2.00	1.87	1.39	1.06	0.68	0.53	0.37	13.31

[1] City office data.

Source: U.S. National Oceanic and Atmospheric Administration, Comparative Climatic Data for the United States through 1986

TABLE 1–8. AVERAGE NUMBER OF DAYS WITH PRECIPITATION OF 0.01 INCH OR MORE— SELECTED CITIES OF THE UNITED STATES

[Airport data except as noted. Includes liquid water equivalent of frozen precipitation]

ST.	STATION	YRS.	JAN	FEB	MAR	APR	MAY	JUN	JUL	AUG	SEP	OCT	NOV	DEC	ANNUAL
AL	BIRMINGHAM	43	11	10	11	9	10	9	12	10	8	6	9	11	117
	MOBILE	45	11	10	11	7	8	11	16	14	10	6	8	10	122
AK	ANCHORAGE	22	7	8	8	7	7	8	12	13	14	12	9	11	115
	BARROW	66	5	4	4	4	4	5	9	11	10	11	6	5	78
	FAIRBANKS	35	7	7	6	5	7	10	13	13	10	11	10	9	106
	JUNEAU	42	18	17	18	17	17	16	17	18	20	24	19	21	220
AZ	PHOENIX	47	4	4	4	2	1	1	4	5	3	3	2	4	36
	TUCSON	46	4	4	4	2	1	2	11	9	5	3	3	4	53
AR	FORT SMITH	41	7	8	9	10	10	8	8	7	8	7	7	7	96
	LITTLE ROCK	44	10	9	10	10	10	8	8	7	7	7	8	9	104
CA	BAKERSFIELD	49	6	6	7	4	2	*	*	*	1	2	4	5	37
	FRESNO	37	8	7	7	4	2	1	*	*	1	2	6	7	45
	LOS ANGELES[1]	46	6	5	6	4	1	1	*	1	1	2	4	5	36
	SACRAMENTO	47	10	9	8	5	3	1	*	*	1	3	7	9	58
	SAN DIEGO	46	7	6	7	5	2	1	*	1	1	2	5	6	42
	SAN FRANCISCO[1]	50	11	10	11	6	3	1	1	1	2	4	8	10	68
CO	DENVER	52	6	6	9	9	11	9	9	9	6	5	5	5	88
	GRAND JUNCTION	40	7	6	8	6	6	4	5	6	6	6	6	6	72
CT	HARTFORD	32	11	10	11	11	12	11	10	10	9	8	11	12	127
DE	WILMINGTON	39	11	10	11	11	11	10	9	9	8	8	10	10	117
DC	WASHINGTON	45	10	9	11	10	11	9	10	9	8	7	8	9	112
FL	JACKSONVILLE	45	8	8	8	6	8	12	15	15	13	9	6	8	116
	MIAMI	44	7	6	6	6	10	15	16	17	17	14	8	7	129
	TAMPA	40	6	7	7	5	6	12	16	17	13	7	5	6	107
GA	ATLANTA	52	11	10	11	9	9	10	12	10	8	6	8	10	115
	SAVANNAH	36	9	9	9	7	9	11	14	13	10	6	6	8	111
HI	HONOLULU	37	10	9	9	9	7	6	8	6	7	9	9	10	100
	LIHUE	36	15	14	17	18	16	17	19	18	16	18	18	17	201
ID	BOISE	47	12	11	10	8	8	6	2	3	4	6	10	12	92
	POCATELLO	37	12	10	10	8	9	7	4	5	5	5	9	11	96
IL	CHICAGO	28	11	10	13	13	11	10	10	9	10	9	10	11	127
	PEORIA	47	9	8	11	12	12	10	9	8	9	8	9	10	114
IN	EVANSVILLE	46	10	9	12	12	11	10	9	8	7	7	10	10	116
	INDIANAPOLIS	47	12	10	13	12	12	10	9	9	8	8	10	12	125
IA	DES MOINES	47	7	7	10	11	11	11	9	9	9	8	7	8	107
	SIOUX CITY	46	7	7	9	10	11	11	9	9	8	7	5	7	99
KS	TOPEKA	40	6	6	9	10	12	10	9	8	8	7	6	6	97
	WICHITA	33	5	5	8	8	11	9	7	7	8	6	5	6	86
KY	LEXINGTON	42	12	11	13	12	12	11	11	9	8	8	11	11	130
LA	NEW ORLEANS	38	10	9	9	7	8	10	15	13	10	6	7	10	114
	SHREVEPORT	34	9	8	9	9	9	8	8	7	7	7	8	9	97
ME	PORTLAND	46	11	10	11	12	13	11	10	9	8	9	12	12	128
MD	BALTIMORE	36	10	9	11	11	11	9	9	10	7	7	9	9	113
MA	BOSTON	35	12	11	12	11	12	11	9	10	9	9	11	12	127
MI	DETROIT	28	13	11	13	13	11	10	9	9	10	9	11	14	134
	GRAND RAPIDS	23	16	12	13	13	11	10	9	9	10	11	13	17	145
	SAULT STE.MARIE	45	19	15	13	11	11	12	10	11	13	13	17	20	166
MN	DULUTH	45	12	10	11	11	12	13	11	11	12	10	11	12	135
	MINNEAPOLIS-ST. PAUL	48	9	8	10	10	11	12	10	10	10	8	8	9	115
MS	JACKSON	23	11	9	10	9	9	8	10	10	8	6	8	10	109
MO	KANSAS CITY	45	6	7	9	10	11	10	8	8	8	7	6	7	98
	ST. LOUIS	29	8	8	11	11	11	10	9	8	8	8	9	9	111
	SPRINGFIELD	41	8	8	10	11	11	10	8	9	8	8	8	9	108
MT	BILLINGS	52	8	8	9	9	11	11	7	7	7	6	6	7	96
	GREAT FALLS	49	9	8	9	9	11	12	7	8	7	6	7	8	101
	HELENA	46	8	7	9	8	11	11	7	8	7	6	7	8	96
NE	NORTH PLATTE	34	5	5	7	8	11	9	10	8	7	5	5	4	84
	OMAHA	50	6	7	9	10	12	11	9	9	9	7	5	6	98
	SCOTTSBLUFF	43	6	5	7	9	12	11	9	7	7	5	5	5	86
NV	ELKO	56	9	9	9	8	8	6	4	4	4	5	7	9	79
	LAS VEGAS	38	3	3	3	2	1	1	3	3	2	2	2	3	26
	RENO	44	6	6	6	4	4	3	2	2	2	3	5	6	51
NH	CONCORD	45	11	10	11	12	12	11	10	10	9	9	11	11	125
	MT. WASHINGTON	54	19	18	19	18	17	16	17	16	15	15	19	20	209
NJ	ATLANTIC CITY[1]	27	10	10	10	10	10	9	8	9	8	8	10	10	111
	NEWARK	45	11	10	11	11	12	10	10	9	8	8	10	11	122
NM	ALBUQUERQUE	47	4	4	5	3	4	4	9	9	6	5	3	4	60
	ROSWELL	14	5	3	3	3	4	5	6	8	7	5	3	4	55
NY	ALBANY	40	12	11	12	12	13	11	10	10	10	9	12	12	135
	BUFFALO	43	20	17	16	14	12	10	10	11	11	11	16	20	169

TABLE 1–8. AVERAGE NUMBER OF DAYS WITH PRECIPITATION OF 0.01 INCH OR MORE—SELECTED CITIES OF THE UNITED STATES (continued)

[Airport data except as noted. Includes liquid water equivalent of frozen precipitation]

ST.	STATION	YRS.	JAN	FEB	MAR	APR	MAY	JUN	JUL	AUG	SEP	OCT	NOV	DEC	ANNUAL
	NEW YORK[1]	117	11	10	11	11	11	10	10	10	8	8	9	10	121
	SYRACUSE	37	19	16	17	14	13	11	11	11	11	12	16	19	170
NC	ASHEVILLE	22	10	9	11	9	12	11	12	12	9	8	9	10	124
	RALEIGH	42	10	10	10	9	10	9	11	10	7	7	8	9	111
	WILMINGTON	35	10	10	10	8	10	10	13	12	9	7	8	9	116
ND	BISMARCK	47	8	7	8	8	10	12	9	9	7	6	6	8	97
	WILLISTON	25	8	7	8	8	10	10	9	7	7	5	6	8	93
OH	CINCINNATI	39	12	11	13	13	12	11	10	9	8	8	11	12	129
	CLEVELAND	45	16	14	16	14	13	11	10	10	10	11	14	16	156
	COLUMBUS	47	13	12	14	13	13	11	11	9	8	9	11	13	137
	TOLEDO	31	14	11	13	13	12	10	9	9	10	9	12	14	137
OK	OKLAHOMA CITY	47	5	6	7	8	10	9	6	6	7	6	5	5	82
	TULSA	48	6	7	8	9	11	9	6	7	7	7	6	7	90
OR	EUGENE	44	18	15	17	13	10	7	2	4	6	11	16	18	138
	MEDFORD	57	14	12	12	9	8	5	1	2	4	8	12	14	102
	PORTLAND	46	18	16	17	14	12	9	4	5	8	13	18	19	153
PC	GUAM	30	20	18	19	20	20	24	26	25	26	26	25	23	272
	JOHNSTON ISLAND	28	11	12	15	14	13	12	12	13	14	16	15	16	162
	KWAJALEIN, MARSHALL IS.	34	15	13	15	18	21	23	23	23	23	23	23	19	238
	PAGO PAGO, AMER SAMOA	20	24	22	23	22	21	19	18	18	18	22	20	23	250
	TRUK, E. CAROLINE IS.	35	19	16	19	20	25	24	24	24	23	24	24	23	265
	WAKE ISLAND	39	11	10	12	14	15	16	19	19	19	19	15	13	181
	YAP, W. CAROLINE IS.	38	20	17	18	17	21	24	25	24	23	24	23	22	259
PA	ERIE	33	18	15	15	14	12	10	10	11	11	13	16	19	164
	HARRISBURG	8	11	11	12	12	13	11	9	10	7	10	11	10	128
	PHILADELPHIA	46	11	9	11	11	11	10	9	9	8	8	10	10	117
	PITTSBURGH	34	17	14	16	14	12	12	11	10	9	11	13	16	154
PR	SAN JUAN	31	16	13	12	13	17	16	19	18	17	17	18	19	196
RI	PROVIDENCE	33	11	10	12	11	11	11	9	10	8	8	11	12	125
SC	CHARLESTON	44	10	9	10	7	9	11	14	12	9	6	7	8	113
	COLUMBIA	39	10	10	10	8	9	9	12	11	7	6	7	9	109
	GREENVILLE-SPARTANBURG	24	11	9	11	9	11	10	12	10	8	7	9	10	118
SD	RAPID CITY	44	7	7	9	9	12	13	9	8	7	5	6	6	97
	SIOUX FALLS	41	6	6	9	9	10	11	9	9	8	6	6	6	97
TN	BRISTOL-JOHNSON CITY-KINGSPORT	41	14	12	13	11	12	11	12	11	8	8	11	11	133
	KNOXVILLE	44	12	11	13	11	11	10	11	10	8	8	10	11	127
	MEMPHIS	36	10	9	11	10	9	8	9	8	7	6	9	10	106
	NASHVILLE	45	11	11	12	11	11	9	10	9	8	7	10	11	119
TX	AUSTIN	45	8	8	7	8	9	6	5	5	7	7	7	7	83
	CORPUS CHRISTI	47	8	7	6	5	6	6	5	6	9	7	6	7	77
	DALLAS-FORT WORTH	33	7	6	7	8	9	6	5	5	7	6	6	6	78
	EL PASO	47	4	3	2	2	2	3	8	8	5	4	3	4	48
	HOUSTON	17	10	8	10	7	8	8	9	10	10	8	9	9	106
	LUBBOCK	40	4	4	4	4	7	7	7	7	6	5	4	4	62
UT	SALT LAKE CITY	58	10	9	10	10	8	5	5	6	5	6	8	9	91
VT	BURLINGTON	43	14	12	13	12	14	13	12	12	12	12	14	15	154
VA	LYNCHBURG	42	11	10	11	10	11	10	11	10	8	8	9	10	119
	NORFOLK	38	10	10	11	10	10	9	11	10	8	8	8	9	114
	RICHMOND	49	10	9	11	9	11	9	11	10	8	7	8	9	113
WA	SEATTLE[1]	36	19	16	17	14	11	9	5	6	9	10	18	19	152
	SPOKANE	39	14	12	11	9	9	8	4	5	6	8	13	15	114
	WALLA WALLA	72	13	12	12	9	8	7	3	3	6	8	12	14	107
WV	CHARLESTON	39	15	14	15	14	13	11	13	11	9	10	12	14	152
WI	GREEN BAY	37	10	8	11	11	11	11	10	10	10	9	9	11	122
	MADISON	38	10	8	11	11	11	11	10	9	9	9	9	10	119
	MILWAUKEE	46	11	10	12	12	12	11	10	9	9	9	10	11	125
WY	CASPER	36	7	8	9	10	11	9	8	5	6	7	7	7	95
	CHEYENNE	51	6	6	9	10	12	11	11	10	7	6	6	5	98

THE ANNUAL VALUE IS THE TOTAL OF THE UNROUNDED MONTHLY VALUES.
IT MAY NOT AGREE WITH THE SUM OF THE ROUNDED MONTHLY VALUES.

[1] City office data.

Source: U.S. National Oceanic and Atmospheric Administration, Comparative Climatic Data for the United States through 1986

TABLE 1-9. AVERAGE TOTAL SNOWFALL— SELECTED CITIES OF THE UNITED STATES

[In inches. Includes ice pellets. Airport data except as noted]

ST.	STATION	YRS.	JAN	FEB	MAR	APR	MAY	JUN	JUL	AUG	SEP	OCT	NOV	DEC	ANNUAL
AL	BIRMINGHAM	43	0.6	0.2	0.1	T	0.0	0.0	0.0	0.0	0.0	T	0.0	0.4	1.3
	MOBILE	45	0.1	0.1	0.0	0.0	0.0	0.0	0.0	0.0	0.0	0.0	T	0.1	0.3
AK	ANCHORAGE	43	10.0	11.5	9.4	5.5	0.5	0.0	0.0	0.0	0.2	7.1	9.8	14.2	68.2
	BARROW	66	2.2	2.1	1.8	2.1	1.8	0.6	0.6	0.7	3.2	6.8	3.5	2.5	27.9
	FAIRBANKS	35	9.6	8.7	6.6	3.7	0.5	T	0.0	T	1.0	10.8	12.6	12.7	66.2
	JUNEAU	42	26.3	19.4	16.6	4.1	0.1	T	0.0	0.0	T	1.1	11.3	23.0	101.9
AZ	PHOENIX	49	T	0.0	T	T	0.0	0.0	0.0	0.0	0.0	0.0	0.0	0.0	T
	TUCSON	46	0.2	0.2	0.3	0.1	0.0	0.0	0.0	0.0	0.0	T	0.1	0.3	1.2
AR	FORT SMITH	41	2.6	2.1	0.6	T	0.0	0.0	0.0	0.0	0.0	0.0	0.5	0.9	6.7
	LITTLE ROCK	44	2.3	1.6	0.6	T	0.0	0.0	0.0	0.0	0.0	0.0	0.2	0.8	5.5
CA	BAKERSFIELD	49	T	T	0.0	0.0	0.0	0.0	0.0	0.0	0.0	0.0	0.0	T	T
	FRESNO	37	0.1	T	T	0.0	0.0	0.0	0.0	0.0	0.0	T	0.0	0.0	0.1
	LOS ANGELES[1]	44	0.0	T	0.0	0.0	0.0	0.0	0.0	0.0	0.0	0.0	0.0	T	T
	SACRAMENTO	38	T	0.1	T	0.0	0.0	0.0	0.0	0.0	0.0	0.0	0.0	T	0.1
	SAN DIEGO	46	T	0.0	T	0.0	0.0	0.0	0.0	0.0	0.0	0.0	T	T	T
	SAN FRANCISCO[1]	36	T	T	T	0.0	0.0	0.0	0.0	0.0	0.0	0.0	0.0	T	T
CO	DENVER	52	7.7	7.4	12.8	9.2	1.8	0.0	0.0	0.0	1.7	3.7	8.3	7.0	59.6
	GRAND JUNCTION	40	7.2	3.9	4.0	1.1	0.2	0.0	0.0	0.0	0.1	0.6	3.0	5.4	25.5
CT	HARTFORD	32	11.8	12.1	10.2	1.8	0.0	0.0	0.0	0.0	0.0	0.1	1.9	10.9	48.8
DE	WILMINGTON	39	6.5	6.3	3.5	0.2	T	0.0	0.0	0.0	0.0	0.1	0.9	3.4	20.9
DC	WASHINGTON	43	5.1	5.7	2.2	0.0	T	0.0	0.0	0.0	0.0	0.0	0.6	3.2	16.8
FL	JACKSONVILLE	45	T	0.0	0.0	0.0	0.0	0.0	0.0	0.0	0.0	0.0	0.0	T	T
	MIAMI	44	0.0	0.0	0.0	0.0	0.0	0.0	0.0	0.0	0.0	0.0	0.0	0.0	0.0
	TAMPA	40	0.0	T	T	0.0	0.0	0.0	0.0	0.0	0.0	0.0	0.0	0.0	T
GA	ATLANTA	52	0.8	0.6	0.4	T	0.0	0.0	0.0	0.0	0.0	0.0	0.0	0.2	2.0
	SAVANNAH	36	0.1	0.2	0.0	0.0	0.0	0.0	0.0	0.0	0.0	0.0	0.0	T	0.3
HI	HONOLULU	40	0.0	0.0	0.0	0.0	0.0	0.0	0.0	0.0	0.0	0.0	0.0	0.0	0.0
	LIHUE	37	0.0	0.0	0.0	0.0	0.0	0.0	0.0	0.0	0.0	0.0	0.0	0.0	0.0
ID	BOISE	47	7.0	3.7	1.8	0.7	0.1	T	T	0.0	0.0	0.1	2.3	5.9	21.6
	POCATELLO	37	9.9	6.2	6.0	4.7	0.5	0.0	0.0	0.0	0.1	1.9	5.1	8.7	43.1
IL	CHICAGO	28	11.2	8.1	7.2	2.0	0.1	0.0	0.0	0.0	T	0.3	2.1	8.7	39.7
	PEORIA	43	6.5	5.2	4.4	1.0	0.0	0.0	0.0	0.0	0.0	0.0	2.2	5.9	25.2
IN	EVANSVILLE	46	4.5	3.7	2.7	0.3	0.0	0.0	0.0	0.0	0.0	T	0.7	2.3	14.2
	INDIANAPOLIS	55	6.2	5.9	3.6	0.5	T	0.0	0.0	0.0	0.0	0.0	1.9	4.9	23.0
IA	DES MOINES	47	8.5	7.2	6.9	2.1	0.0	0.0	0.0	0.0	T	0.3	2.9	6.8	34.7
	SIOUX CITY	46	6.4	5.6	7.9	1.5	0.1	0.0	0.0	0.0	0.0	0.6	3.5	6.2	31.8
KS	TOPEKA	40	5.9	4.7	3.9	0.6	0.0	0.0	0.0	0.0	0.0	0.0	1.2	5.0	21.3
	WICHITA	33	4.4	4.5	2.5	0.3	0.0	0.0	0.0	0.0	0.0	0.0	1.1	3.2	16.0
KY	LEXINGTON	42	6.1	5.0	2.7	0.2	0.0	0.0	0.0	0.0	0.0	0.0	0.6	1.7	16.3
LA	NEW ORLEANS	40	0.0	0.1	T	0.0	0.0	0.0	0.0	0.0	0.0	0.0	T	0.1	0.2
	SHREVEPORT	34	0.9	0.5	0.2	0.0	0.0	0.0	0.0	0.0	0.0	0.0	0.0	0.3	1.9
ME	PORTLAND	46	19.1	17.7	12.9	2.9	0.2	0.0	0.0	0.0	T	0.3	3.1	15.2	71.4
MD	BALTIMORE	36	5.5	7.1	4.1	0.1	T	0.0	0.0	0.0	0.0	0.0	0.9	3.6	21.3
MA	BOSTON	51	12.2	11.5	7.6	0.9	0.0	0.0	0.0	0.0	0.0	0.0	1.2	7.6	41.0
MI	DETROIT	28	9.9	8.9	6.8	1.8	T	0.0	0.0	0.0	0.0	0.1	3.4	10.5	41.4
	GRAND RAPIDS	23	21.4	11.2	11.0	3.3	0.0	0.0	0.0	0.0	T	0.5	7.5	17.7	72.6
	SAULT STE. MARIE	45	28.2	19.3	15.1	5.4	0.6	T	0.0	T	0.1	2.2	14.7	29.6	115.2
MN	DULUTH	43	17.0	11.7	13.9	6.5	0.8	0.0	0.0	T	0.0	1.2	11.0	15.5	77.6
	MINNEAPOLIS-ST. PAUL	48	10.0	8.4	11.1	3.1	0.2	0.0	0.0	0.0	0.0	0.4	6.9	9.5	49.6
MS	JACKSON	23	0.7	0.2	0.2	T	0.0	0.0	0.0	0.0	0.0	0.0	0.0	0.0	1.1
MO	KANSAS CITY	45	5.8	4.2	3.7	0.7	T	0.0	0.0	0.0	0.0	0.0	0.9	4.4	19.7
	ST. LOUIS	50	5.3	4.5	4.2	0.4	T	0.0	0.0	0.0	0.0	T	1.4	3.6	19.4
	SPRINGFIELD	41	4.4	4.3	3.3	0.4	0.0	0.0	0.0	0.0	0.0	0.0	1.8	2.6	16.8
MT	BILLINGS	52	9.3	8.1	10.2	7.5	1.6	0.0	0.0	0.0	1.3	3.4	7.1	8.7	57.2
	GREAT FALLS	49	10.0	8.5	10.0	7.3	1.6	0.3	T	T	1.5	3.2	7.6	8.8	58.8
	HELENA	46	8.9	6.2	7.2	5.0	1.5	0.1	T	0.0	1.6	2.2	6.6	8.6	47.9
NE	NORTH PLATTE	34	5.1	5.0	7.2	3.1	0.2	0.0	0.0	0.0	0.1	1.3	3.9	5.1	31.0
	OMAHA	51	7.7	6.9	6.6	0.9	0.1	0.0	0.0	0.0	T	0.3	2.4	5.7	30.6
	SCOTTSBLUFF	43	6.5	4.9	8.7	4.9	1.0	0.0	0.0	0.0	0.4	2.3	5.2	6.3	40.2
NV	ELKO	55	9.6	6.1	5.6	2.6	0.8	T	0.0	0.0	0.1	0.8	4.6	8.1	38.3
	LAS VEGAS	38	1.1	0.0	0.0	T	0.0	0.0	0.0	0.0	0.0	T	0.1	0.1	1.3
	RENO	44	5.9	4.7	4.9	1.4	1.0	0.0	0.0	0.0	0.0	0.4	2.3	4.3	24.9
NH	CONCORD	45	17.6	14.9	11.0	2.3	0.1	0.0	0.0	0.0	0.0	0.1	4.1	14.1	64.2
	MT. WASHINGTON	54	39.0	40.5	41.8	29.2	10.7	1.1	0.0	0.2	1.7	11.5	30.7	42.5	248.9
NJ	ATLANTIC CITY	42	5.2	5.5	2.8	0.3	T	0.0	0.0	0.0	0.0	T	0.3	2.2	16.3
	NEWARK	45	7.4	8.4	4.8	0.8	T	0.0	0.0	0.0	0.0	0.0	0.4	6.0	27.8
NM	ALBUQUERQUE	47	2.5	2.2	1.9	0.5	0.0	0.0	0.0	0.0	T	0.1	1.1	2.6	10.9
	ROSWELL	39	2.9	2.5	1.3	0.3	T	0.0	0.0	0.0	0.0	0.1	1.5	2.8	11.4
NY	ALBANY	40	15.8	14.3	11.6	3.0	0.1	0.0	0.0	0.0	0.0	0.1	4.5	15.8	65.2

TABLE 1–9. AVERAGE TOTAL SNOWFALL— SELECTED CITIES OF THE UNITED STATES (continued)

[In inches. Includes ice pellets. Airport data except as noted]

ST.	STATION	YRS.	JAN	FEB	MAR	APR	MAY	JUN	JUL	AUG	SEP	OCT	NOV	DEC	ANNUAL
	BUFFALO	43	24.7	17.9	11.5	3.2	0.1	T	0.0	0.0	T	0.2	12.2	23.2	93.0
	NEW YORK[1]	118	7.6	8.7	5.1	0.9	T	0.0	0.0	0.0	0.0	0.0	0.9	5.6	28.8
	SYRACUSE	37	28.7	25.1	17.0	3.6	0.1	0.0	0.0	0.0	0.0	0.5	9.3	25.5	109.8
NC	ASHEVILLE	22	4.7	5.5	3.0	0.3	T	0.0	0.0	0.0	0.0	T	0.9	2.1	16.5
	RALEIGH	42	2.4	2.5	1.5	0.0	0.0	0.0	0.0	0.0	0.0	0.0	0.1	0.8	7.3
	WILMINGTON	35	0.3	0.6	0.5	0.0	0.0	0.0	0.0	0.0	0.0	0.0	T	0.3	1.7
ND	BISMARCK	47	7.2	6.3	8.4	4.0	0.9	T	0.0	0.0	0.3	1.3	6.0	6.7	41.1
	WILLISTON	25	6.8	5.2	7.0	5.7	0.8	0.0	0.0	0.0	0.4	1.7	4.8	7.4	39.8
OH	CINCINNATI	39	7.5	5.6	4.3	0.5	T	0.0	0.0	0.0	0.0	0.0	2.3	3.7	23.9
	CLEVELAND	45	12.5	11.6	9.9	2.2	0.1	0.0	0.0	0.0	T	0.7	5.2	11.5	53.7
	COLUMBUS	39	8.8	6.4	4.5	0.7	T	0.0	0.0	0.0	T	0.0	2.5	5.6	28.5
	TOLEDO	31	9.6	8.3	6.1	1.7	T	0.0	0.0	0.0	T	0.0	3.3	8.9	37.9
OK	OKLAHOMA CITY	47	2.8	2.7	1.5	0.0	0.0	0.0	0.0	0.0	0.0	T	0.5	1.6	9.1
	TULSA	48	3.2	2.5	1.4	0.0	0.0	0.0	0.0	0.0	0.0	T	0.4	1.5	9.0
OR	EUGENE	44	4.1	0.4	0.6	T	T	T	0.0	0.0	T	T	0.3	1.2	6.6
	MEDFORD	57	3.5	1.3	0.8	0.2	T	0.0	0.0	0.0	0.0	0.0	0.5	1.4	7.7
	PORTLAND	46	3.7	0.8	0.4	T	0.0	T	0.0	0.0	T	0.0	0.5	1.4	6.8
PC	GUAM	30	0.0	0.0	0.0	0.0	0.0	0.0	0.0	0.0	0.0	0.0	0.0	0.0	0.0
	JOHNSTON ISLAND	28	0.0	0.0	0.0	0.0	0.0	0.0	0.0	0.0	0.0	0.0	0.0	0.0	0.0
	KWAJALEIN, MARSHALL IS.	42	0.0	0.0	0.0	0.0	0.0	0.0	0.0	0.0	0.0	0.0	0.0	0.0	0.0
	PAGO PAGO, AMER SAMOA	27	0.0	0.0	0.0	0.0	0.0	0.0	0.0	0.0	0.0	0.0	0.0	0.0	0.0
	TRUK, E. CAROLINE IS.	37	0.0	0.0	0.0	0.0	0.0	0.0	0.0	0.0	0.0	0.0	0.0	0.0	0.0
	WAKE ISLAND	39	0.0	0.0	0.0	0.0	0.0	0.0	0.0	0.0	0.0	0.0	0.0	0.0	0.0
	YAP, W. CAROLINE IS.	38	0.0	0.0	0.0	0.0	0.0	0.0	0.0	0.0	0.0	0.0	0.0	0.0	0.0
PA	ERIE	32	23.0	15.6	10.1	2.6	0.0	0.0	0.0	0.0	0.0	0.4	10.6	21.4	83.7
	HARRISBURG	47	9.5	9.9	6.3	0.5	T	0.0	0.0	0.0	0.0	0.1	1.9	7.0	35.2
	PHILADELPHIA	44	6.4	6.8	3.9	0.3	T	0.0	0.0	0.0	0.0	0.0	0.6	3.6	21.6
	PITTSBURGH	34	12.2	10.1	8.3	1.6	0.1	0.0	0.0	0.0	0.0	0.2	3.6	8.3	44.4
PR	SAN JUAN	31	0.0	0.0	0.0	0.0	0.0	0.0	0.0	0.0	0.0	0.0	0.0	0.0	0.0
RI	PROVIDENCE	33	9.6	10.1	7.8	0.7	0.2	0.0	0.0	0.0	0.0	0.1	0.7	7.0	36.2
SC	CHARLESTON	44	0.1	0.3	0.1	T	0.0	0.0	0.0	0.0	0.0	0.0	T	0.1	0.6
	COLUMBIA	39	0.4	0.9	0.3	0.0	0.0	0.0	0.0	0.0	0.0	0.0	T	0.3	1.9
	GREENVILLE-SPARTANBURG	24	2.1	2.3	1.2	0.0	0.0	0.0	0.0	0.0	0.0	0.0	0.1	0.7	6.4
SD	RAPID CITY	44	5.1	6.2	9.0	6.1	0.8	0.1	0.0	0.0	0.2	1.5	5.2	4.7	38.9
	SIOUX FALLS	41	6.6	8.2	10.1	2.2	0.0	0.0	0.0	0.0	0.0	0.5	5.0	7.5	40.1
TN	BRISTOL-JOHNSON CITY-KINGSPORT	49	5.3	4.6	2.3	0.2	T	0.0	0.0	0.0	0.0	T	1.1	2.7	16.2
	KNOXVILLE	44	4.1	4.0	1.6	0.2	T	0.0	0.0	0.0	0.0	T	0.7	1.7	12.3
	MEMPHIS	36	2.5	1.5	1.0	T	0.0	0.0	0.0	0.0	0.0	0.0	0.1	0.7	5.8
	NASHVILLE	45	4.2	3.2	1.5	0.0	0.0	0.0	0.0	0.0	0.0	0.0	0.5	1.7	11.1
TX	AUSTIN	45	0.6	0.4	0.0	0.0	0.0	0.0	0.0	0.0	0.0	0.0	0.1	T	1.1
	CORPUS CHRISTI	47	0.1	0.0	T	0.0	0.0	0.0	0.0	0.0	0.0	0.0	T	T	0.1
	DALLAS-FORT WORTH	33	1.4	1.1	0.2	0.0	0.0	0.0	0.0	0.0	0.0	0.0	0.2	0.3	3.2
	EL PASO	47	1.3	0.8	0.4	0.4	0.0	0.0	0.0	0.0	0.0	0.0	1.0	1.3	5.2
	HOUSTON	52	0.2	0.2	0.0	0.0	0.0	0.0	0.0	0.0	0.0	0.0	T	T	0.4
	LUBBOCK	38	2.5	3.2	1.6	0.2	0.0	0.0	0.0	0.0	0.0	0.2	1.2	1.8	10.7
UT	SALT LAKE CITY	58	13.0	9.4	10.1	5.3	0.7	T	0.0	0.0	0.1	1.3	6.5	12.2	58.6
VT	BURLINGTON	43	19.0	16.6	12.4	3.7	0.2	0.0	0.0	0.0	T	0.2	7.0	19.2	78.3
VA	LYNCHBURG	42	5.3	5.5	3.5	0.3	0.0	0.0	0.0	0.0	0.0	0.1	0.8	2.9	18.4
	NORFOLK	38	3.0	2.7	1.1	0.0	0.0	0.0	0.0	0.0	0.0	0.0	0.0	1.0	7.8
	RICHMOND	49	5.0	4.2	2.7	0.1	0.0	0.0	0.0	0.0	0.0	T	0.4	2.0	14.4
WA	SEATTLE[1]	36	3.4	0.8	0.6	0.0	T	0.0	0.0	0.0	0.0	T	0.8	1.9	7.5
	SPOKANE	39	16.9	7.6	4.2	0.7	0.1	T	0.0	0.0	0.0	0.5	6.3	15.2	51.5
	WALLA WALLA	72	7.5	3.9	1.2	0.2	T	0.0	0.0	0.0	0.0	0.1	2.0	5.0	19.9
WV	CHARLESTON	39	10.7	9.1	4.7	0.4	0.0	0.0	0.0	0.0	0.0	0.1	2.3	4.8	32.1
WI	GREEN BAY	37	10.8	8.5	9.0	2.2	0.1	0.0	0.0	0.0	T	0.1	4.6	10.6	45.9
	MADISON	38	9.7	7.1	8.6	2.3	0.0	0.0	0.0	0.0	T	0.1	3.5	10.6	41.9
	MILWAUKEE	46	13.1	9.5	9.0	1.9	0.0	0.0	0.0	0.0	T	0.1	2.9	10.5	47.0
WY	CASPER	36	10.1	9.8	14.8	13.6	4.2	0.2	0.0	T	1.3	5.3	10.8	11.1	81.2
	CHEYENNE	51	6.3	5.5	11.9	9.3	3.7	0.3	0.0	0.0	0.8	3.6	7.0	5.9	54.3

[1] City office data

T=Trace, less than 0.05 inch of snowfall.

Source: U.S. National Oceanic and Atmospheric Administration, Comparative Climatic Data for the United States through 1986

FIGURE 1–1. AVERAGE ANNUAL PRECIPITATION IN THE UNITED STATES

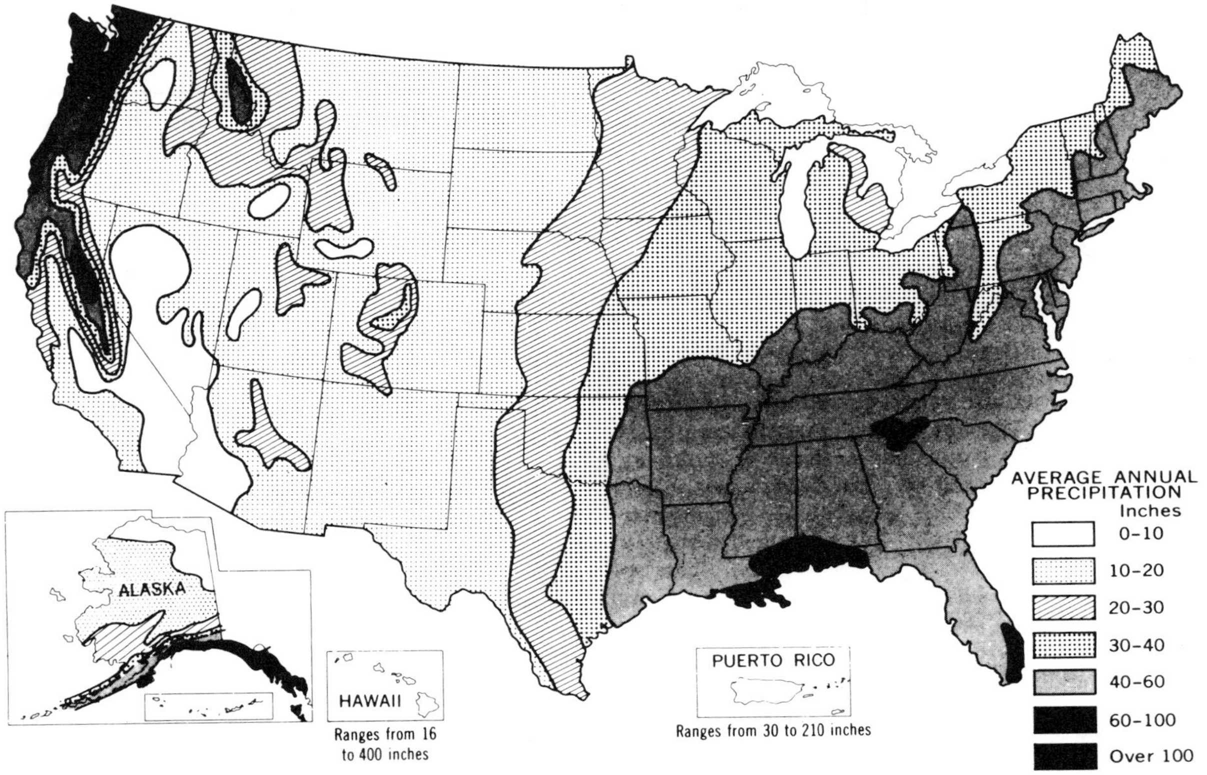

Source: U.S. Water Resources Council, 1968, The Nation's Water Resources

FIGURE 1–2. AVERAGE ANNUAL PRECIPITATION IN CANADA

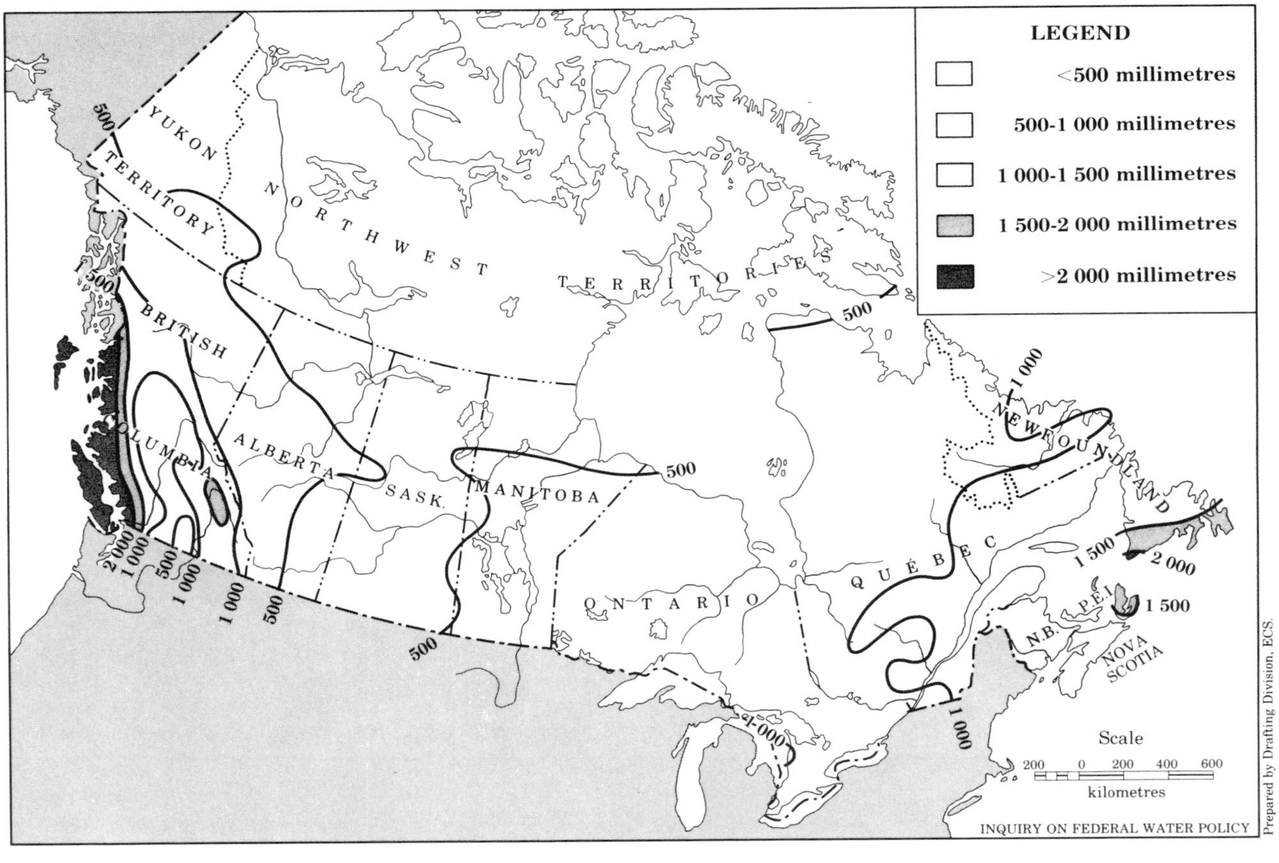

Source: Pearse, P.H., and others, 1985, Currents of Change, Final Report, Inquiry on Federal Water Policy, Ottawa

SECTION B. CLIMATIC DATA—WORLD

TABLE 1–10. TEMPERATURE AND PRECIPITATION DATA FOR REPRESENTATIVE WORLD-WIDE STATIONS

COUNTRY AND STATION	LATITUDE	LONGITUDE	ELEVATION	Temperature: Length of record	Average Daily JAN. Maximum	Average Daily JAN. Minimum	Average Daily APR. Maximum	Average Daily APR. Minimum	Average Daily JULY Maximum	Average Daily JULY Minimum	Average Daily OCT. Maximum	Average Daily OCT. Minimum	Extreme Maximum	Extreme Minimum	Precipitation: Length of record	January	February	March	April	May	June	July	August	September	October	November	December	YEAR
	° '	° '	FEET	YEAR	°F	°F	°F	°F	°F	°F	°F	°F	°F	°F	YEAR	IN.	IN.	IN.	IN.	IN.	IN.	IN.	IN.	IN.	IN.	IN.	IN.	IN.
NORTH AMERICA																												
United States:																												
Albuquerque, N. Mex.	35 03N	106 37W	5,311	30	46	24	69	42	91	66	71	45	104	-16	30	0.4	0.4	0.5	0.5	0.8	0.6	1.2	1.3	1.0	0.8	0.4	0.5	8.4
Asheville, N. C.	35 26N	82 32W	2,140	30	48	28	67	42	84	61	68	45	99	-7	30	4.2	4.0	4.8	4.0	3.7	3.5	5.9	4.9	3.6	3.1	2.8	3.6	48.1
Atlanta, Ga.	33 39N	84 26W	1,010	30	52	37	70	50	87	71	72	52	103	-9	30	4.4	4.5	5.4	4.5	3.2	3.8	4.7	3.6	3.3	2.4	3.0	4.4	47.2
Austin, Tex.	30 18N	97 42W	597	30	60	41	78	57	95	74	82	60	109	-2	30	2.4	2.6	2.1	3.6	3.7	3.2	2.2	1.9	3.4	2.8	2.1	2.5	32.5
Birmingham, Ala.	33 34N	86 45W	620	30	57	36	76	50	93	71	79	52	107	-10	30	5.0	5.3	6.0	4.5	3.4	4.0	5.2	4.9	3.3	3.0	3.5	5.0	53.1
Bismark, N. Dak.	46 46N	100 45W	1,647	30	20	0	55	32	86	58	59	34	114	-45	30	0.4	0.4	0.8	1.2	2.0	3.4	2.2	1.7	1.2	0.9	0.6	0.4	15.2
Boise, Idaho	43 34N	116 13W	2,838	30	36	22	63	37	91	59	65	38	112	-28	30	1.3	1.3	1.3	1.2	1.3	0.9	0.2	0.2	0.4	0.8	1.2	1.3	11.4
Brownsville, Tex.	25 54N	97 26W	16	30	71	52	82	66	93	76	85	67	104	12	30	1.4	1.5	1.0	1.6	2.4	3.0	1.7	2.8	5.0	3.5	1.3	1.7	26.9
Buffalo, N.Y.	42 56N	78 44W	705	30	31	18	53	34	80	59	60	41	99	-21	30	2.8	2.7	3.2	3.0	3.0	2.5	2.6	3.1	3.1	3.0	3.6	3.0	35.6
Cheyenne, Wyo.	41 09N	104 49W	6,126	30	37	14	56	30	85	55	63	32	100	-38	30	0.5	0.6	1.2	1.9	2.5	2.1	1.8	1.4	1.1	0.8	0.6	0.5	15.0
Chicago, Ill.	41 47N	87 45W	607	30	33	19	57	41	84	67	63	47	105	-23	30	1.9	1.6	2.7	3.0	3.7	4.1	3.4	3.2	2.7	2.8	2.2	1.9	33.2
Des Moines, Iowa	41 32N	93 39W	938	30	29	11	59	38	87	65	66	43	110	-30	30	1.3	1.1	2.1	2.5	4.1	4.7	3.1	3.7	2.9	2.1	1.8	1.1	30.5
Dodge City, Kans.	37 46N	99 58W	2,582	30	42	20	66	41	93	68	71	46	109	-26	30	0.6	0.7	1.2	1.8	3.2	3.0	2.3	2.4	1.5	1.4	0.6	0.5	19.2
El Paso, Tex.	31 48N	106 24W	3,918	30	56	30	78	49	95	69	79	50	109	-8	30	0.5	0.4	0.4	0.3	0.4	0.7	1.3	1.2	1.1	0.9	0.3	0.5	8.0
Indianapolis, Ind.	39 44N	86 17W	792	30	37	21	61	40	86	64	67	44	107	-25	30	3.1	2.3	3.4	3.7	4.0	4.6	3.5	3.0	3.2	2.6	3.1	2.7	39.2
Jacksonville, Fla.	30 25N	81 39W	20	30	67	45	80	58	92	73	80	62	105	10	30	2.5	2.9	3.5	3.6	3.5	6.3	7.7	6.9	7.6	5.2	1.7	2.2	53.6
Kansas City, Mo.	39 07N	94 36W	742	30	40	23	66	46	92	71	72	49	113	-22	30	1.4	1.2	2.5	3.6	4.4	4.6	3.2	3.8	3.3	2.9	1.8	1.5	34.2
Las Vegas, Nev.	36 05N	115 10W	2,162	30	54	32	78	51	104	76	80	53	117	8	30	0.5	0.4	0.4	0.2	0.1	*	0.5	0.5	0.3	0.2	0.3	0.4	3.8
Los Angeles, Calif.	33 56N	118 23W	97	30	64	45	67	52	76	62	73	57	110	23	30	2.7	2.9	1.8	1.1	0.1	0.1	*	*	0.2	0.4	1.1	2.4	12.8
Louisville, Ky.	38 11N	85 44W	477	30	44	27	66	43	89	67	70	46	107	-20	30	4.1	3.3	4.6	3.8	3.9	4.0	3.4	3.0	2.6	2.3	3.2	3.2	41.4
Miami, Fla.	25 48N	80 16W	7	30	76	58	83	66	89	75	85	71	100	28	30	2.0	1.9	2.3	3.9	6.4	7.4	6.8	7.0	9.5	8.2	2.8	1.7	59.9
Minneapolis, Minn.	44 53N	93 13W	834	30	22	2	56	33	84	61	61	37	108	-34	30	0.7	0.8	1.5	1.9	3.2	4.0	3.3	3.2	2.4	1.6	1.4	0.9	24.9
Missoula, Mont.	46 55N	114 05W	3,190	30	28	10	57	31	85	49	58	30	105	-33	30	0.9	0.9	0.7	1.0	1.9	1.9	0.9	0.7	1.0	1.0	0.9	1.1	12.9
Nashville, Tenn.	36 07N	86 41W	590	30	49	31	71	48	91	70	74	49	107	-15	30	5.5	4.5	5.2	3.7	3.7	3.3	3.7	2.9	2.9	2.3	3.3	4.2	45.2
New Orleans, La.	29 59N	90 15W	3	30	64	45	78	58	91	73	80	61	102	7	30	3.8	4.0	5.3	4.6	4.4	4.4	6.7	5.3	5.0	2.8	3.3	4.1	53.7
New York, N.Y.	40 47N	73 58W	132	30	40	27	60	43	85	68	66	50	106	-15	30	3.3	2.8	4.0	3.4	3.7	3.3	3.7	4.4	3.9	3.1	3.4	3.3	42.3
Oklahoma City, Okla.	35 24N	97 36W	1,285	30	46	28	71	49	93	72	74	52	113	-17	30	1.3	1.4	2.0	3.1	5.2	4.5	2.4	2.5	3.0	2.5	1.6	1.4	30.9
Phoenix, Ariz.	33 26N	112 01W	1,117	30	64	35	84	50	105	75	87	55	118	16	30	0.7	0.9	0.7	0.3	0.1	0.1	0.8	1.1	0.7	0.5	0.5	0.9	7.3
Pittsburgh, Pa.	40 27N	80 00W	747	30	40	25	63	42	85	65	65	45	103	-20	30	2.8	2.3	3.5	3.4	3.8	4.0	3.6	3.5	2.7	2.5	2.3	2.5	36.9
Portland, Maine	43 39N	70 19W	47	30	32	12	53	32	80	57	60	37	103	-39	30	4.4	3.8	4.3	3.7	3.4	3.2	2.9	2.4	3.5	3.2	4.2	3.9	42.9
Portland, Oreg.	45 36N	122 36W	21	30	44	33	62	42	79	56	63	45	107	-3	30	5.4	4.2	3.8	2.1	2.0	1.7	0.4	0.7	1.6	3.6	5.3	6.4	37.2
Reno, Nev.	39 30N	119 47W	4,404	30	45	16	65	31	89	46	69	29	106	-19	30	1.2	1.0	0.7	0.5	0.5	0.4	0.3	0.2	0.2	0.5	0.6	1.1	7.2

TABLE 1–10. TEMPERATURE AND PRECIPITATION DATA FOR REPRESENTATIVE WORLD-WIDE STATIONS (continued)

COUNTRY AND STATION	LATI-TUDE	LONGI-TUDE	ELE-VATION	TEMPERATURE: Length of record	Average Daily JAN. Maximum	JAN. Minimum	APR. Maximum	APR. Minimum	JULY Maximum	JULY Minimum	OCT. Maximum	OCT. Minimum	Extreme Maximum	Extreme Minimum	AVERAGE PRECIPITATION: Length of record	January	February	March	April	May	June	July	August	September	October	November	December	YEAR
	° '	° '	FEET	YEAR	°F	°F	°F	°F	°F	°F	°F	°F	°F	°F	YEAR	IN.	IN.	IN.	IN.	IN.	IN.	IN.	IN.	IN.	IN.	IN.	IN.	IN.
Salt Lake City, Utah	40 46N	111 58W	4,220	30	37	18	63	36	94	60	65	38	107	-30	30	1.4	1.2	1.6	1.8	1.4	1.0	0.6	0.9	0.5	1.2	1.3	1.2	14.1
San Francisco, Calif.	37 37N	122 23W	8	30	55	42	64	47	72	54	71	51	106	20	30	4.0	3.5	2.7	1.3	0.5	0.1	*	*	0.2	0.7	1.6	4.1	18.7
Sault Ste. Marie, Mich.	46 28N	84 22W	721	30	23	8	46	30	76	54	55	38	98	-37	30	2.1	1.5	1.8	2.2	2.8	3.3	2.5	2.9	3.8	2.8	3.3	2.3	31.3
Seattle, Wash.	47 27N	122 18W	400	30	44	33	58	40	76	54	60	44	100	0	30	5.7	4.2	3.8	2.4	1.7	1.6	0.8	1.0	2.1	4.0	5.4	6.3	39.0
Sheridan, Wyo.	44 46N	106 58W	3,964	30	34	9	56	31	87	56	62	33	106	-41	30	0.6	0.7	1.4	2.2	2.6	2.6	1.2	0.9	1.2	1.1	0.8	0.6	15.9
Spokane, Wash.	47 38N	117 32W	2,356	30	31	19	59	36	86	55	60	38	108	-30	30	2.4	1.9	1.5	0.9	1.2	1.5	0.4	0.4	0.8	1.6	2.2	2.4	17.2
Washington, D.C.	38 51N	77 03W	14	30	44	30	66	46	87	69	68	50	106	-15	30	3.0	2.5	3.2	3.2	4.1	3.2	4.2	4.9	3.8	3.1	2.8	2.8	40.8
Wilmington, N.C.	34 16N	77 55W	28	30	58	37	74	51	89	71	76	55	104	5	30	2.9	3.4	4.0	2.9	3.5	4.3	7.7	6.9	6.3	3.0	3.1	3.4	51.4
United States, Alaska:																												
Anchorage	61 13N	149 52W	85	30	21	4	44	28	65	50	42	28	86	-38	30	0.8	0.7	0.5	0.4	0.5	1.0	1.9	2.6	2.5	1.9	1.0	0.9	14.7
Annette	55 02N	131 34W	110	30	38	30	50	37	63	51	51	42	90	-4	30	11.4	8.5	9.6	9.1	7.1	5.7	6.0	7.5	9.9	16.9	14.7	12.1	118.5
Barrow	71 18N	156 47W	31	30	-9	-23	7	-7	45	33	21	12	78	-56	30	0.2	0.2	0.1	0.1	0.1	0.4	0.8	0.9	0.6	0.5	0.2	0.2	4.3
Bethel	60 47N	161 48W	125	30	11	-4	34	18	62	48	38	25	90	-52	30	1.1	1.1	1.0	0.6	1.0	1.2	2.0	4.2	2.6	1.5	1.1	1.0	18.4
Cold Bay	55 12N	162 43W	96	30	33	23	38	28	54	45	45	36	78	-9	30	2.3	3.2	1.8	1.5	2.3	2.0	1.8	4.3	4.3	4.6	3.8	2.6	34.5
Fairbanks	64 49N	147 52W	436	30	-1	-21	42	17	72	48	35	17	99	-66	30	0.9	0.5	0.4	0.3	0.7	1.4	1.8	2.2	1.1	0.9	0.6	0.5	11.3
Juneau	58 22N	134 35W	12	30	30	20	45	31	63	48	47	37	89	-21	30	4.0	3.1	3.3	2.9	3.2	3.4	4.5	5.0	6.7	8.3	6.1	4.2	54.7
King Salmon	58 41N	156 39W	49	30	21	6	41	25	63	47	43	29	88	-40	30	1.1	1.0	1.0	0.6	1.0	1.4	2.1	3.4	3.1	2.2	1.5	1.0	19.4
Nome	64 30N	165 26W	13	30	12	-3	28	14	55	44	35	24	84	-47	30	1.0	0.9	0.9	0.8	0.7	0.9	2.3	3.8	2.7	1.7	1.2	1.0	17.9
St. Paul Island	57 09N	170 13W	22	30	30	21	33	24	49	42	41	33	64	-26	30	1.8	1.2	1.1	1.0	1.3	1.2	2.3	3.3	3.1	3.2	2.5	1.8	23.8
Shemya	52 43N	174 06E	122	30	34	29	38	33	49	44	42	38	63	16	30	2.5	2.3	2.6	2.1	2.4	1.3	2.2	2.1	2.3	2.8	2.7	2.1	27.4
Yakutat	59 31N	139 40W	28	30	34	20	45	29	61	48	49	35	86	-24	30	10.9	8.2	8.7	7.2	8.0	5.1	8.4	10.9	16.6	19.6	16.1	12.3	132.0
Canada:																												
Aklavik, N.W.T.	68 14N	135 00W	30	22	-10	-26	19	-2	66	47	25	15	93	-62	22	0.5	0.5	0.4	0.5	0.5	0.8	1.4	1.4	0.9	0.9	0.8	0.4	9.0
Alert, N.W.T.	82 31N	62 20W	95	9	-19	-29	-8	-18	44	36	2	-7	67	-53	10	0.2	0.3	0.3	0.3	0.5	0.6	0.5	1.1	1.0	0.9	0.2	0.4	6.3
Calgary, Alta.	51 06N	114 01W	3,540	55	24	2	53	27	76	47	54	29	97	-49	55	0.5	0.5	0.8	1.0	2.3	3.1	2.5	2.3	1.5	0.7	0.7	0.6	16.7
Charlottetown, P.E.I.	46 17N	63 08W	181	65	26	10	43	30	73	58	54	41	98	-27	65	3.8	3.0	3.2	2.8	2.7	2.6	3.0	3.4	3.4	4.1	3.8	4.0	39.8
Chatham, N.B.	47 00N	65 27W	109	50	23	2	47	28	77	56	55	37	102	-43	50	3.4	2.7	3.3	3.0	3.2	3.6	3.9	4.0	3.1	4.0	3.4	3.2	40.8
Churchhill, Man.	58 45N	94 04W	94	30	-11	-27	24	4	64	43	34	20	96	-57	30	0.5	0.6	0.9	0.9	0.9	1.9	2.2	2.7	2.3	1.4	1.0	0.7	16.0
Edmonton, Alta.	53 34N	113 31W	2.219	71	16	-3	52	28	74	50	51	30	99	-57	71	0.9	0.7	0.7	1.0	1.9	3.2	3.3	2.4	1.3	0.8	0.9	0.9	18.0
Fort Nelson, B.C.	58 50N	122 35W	1.253	12	1	-15	47	25	74	51	43	25	98	-61	13	0.9	1.2	0.7	0.8	1.4	2.5	2.4	1.5	1.3	1.0	1.4	1.2	16.3
Fort Simpson, N.W.T.	61 45N	121 14W	554	42	-10	-27	38	14	74	50	36	21	97	-70	42	0.7	0.7	0.5	0.7	1.4	1.5	2.0	1.5	1.3	1.1	0.9	0.8	13.1
Frobisher Bay, N.W.T.	63 45N	68 33W	110	18	-9	-23	16	-1	53	39	29	18	76	-49	10	0.7	0.9	0.8	0.8	0.7	0.9	1.5	2.0	1.8	1.1	1.1	1.0	13.3
Gander, Nfld.	48 57N	54 34W	496	14	27	13	40	27	71	52	51	37	96	-17	14	2.6	3.3	2.8	2.6	2.6	2.8	3.6	3.6	3.7	4.1	4.2	3.7	39.6
Halifax, N.S.	44 39N	63 34W	83	75	32	15	47	31	74	55	57	41	99	-21	71	5.4	4.4	4.9	4.5	4.1	4.0	3.8	4.4	4.1	5.4	5.3	5.4	55.7
Kapuskasing, Ont.	49 25N	82 28W	743	19	10	-14	43	19	75	50	47	31	101	-53	19	2.0	1.1	1.6	1.8	2.1	2.3	3.4	2.9	3.5	2.5	2.4	1.9	27.5
Knob Lake, Que.	54 48N	66 49W	1,712	30	-3	-21	30	12	64	46	37	25	88	-59	10	1.9	1.9	1.4	1.6	1.7	3.3	3.3	4.4	3.4	2.9	2.4	1.5	29.7

Montreal, Que.	45 30N	73 34W	187	67	21	6	50	33	78	61	54	40	97	-35	77	3.8	3.0	3.5	2.6	3.1	3.4	3.7	3.5	3.7	3.4	3.5	3.6	40.8
North Bay, Ont.	46 21N	79 25W	1,216	17	22	2	48	28	78	56	49	36	99	-46	23	2.0	1.5	1.8	2.2	2.5	3.2	3.2	2.7	3.7	3.2	2.7	2.1	30.8
Ottawa, Ont.	45 19N	75 40W	374	65	21	3	51	31	81	58	54	37	102	-38	65	2.9	2.2	2.8	2.7	2.5	3.5	3.4	2.6	3.2	2.9	3.0	2.6	34.3
Penticton, B.C.	49 28N	119 36W	1,129	32	32	21	61	35	84	53	59	38	105	-16	32	1.0	0.7	0.7	0.7	1.1	1.2	0.8	0.8	1.0	0.8	0.9	1.1	10.8
Port Arthur, Ont.	48 22N	89 19W	644	62	17	-4	44	26	74	52	50	34	104	-42	59	0.9	0.8	1.0	1.5	2.1	2.8	3.6	2.8	3.4	2.5	1.5	0.9	23.8
Prince George, B.C.	53 53N	122 41W	2,218	27	23	3	54	27	75	44	52	30	102	-58	27	1.8	1.2	1.4	0.8	1.3	2.1	1.6	1.9	2.0	2.0	1.9	1.9	19.9
Prince Rupert, B.C.	54 17N	130 23W	170	26	39	30	50	37	62	49	53	42	90	-3	26	9.8	7.6	8.4	6.7	5.3	4.1	4.8	5.1	7.7	12.2	12.3	11.3	95.3
Quebec, Que.	46 48N	71 23W	239	72	18	2	44	29	76	57	51	37	97	-34	72	3.5	2.7	3.0	2.4	3.1	3.7	4.0	4.0	3.6	3.4	3.2	3.2	39.8
Regina, Sask.	50 26N	104 40W	1,884	55	10	-11	50	26	79	51	52	27	110	-56	49	0.5	0.3	0.7	0.7	1.8	3.3	2.4	1.8	1.3	0.9	0.6	0.4	14.7
Resolute, N.W.T.	74 43N	94 59W	220	13	-20	-33	-1	-16	45	35	11	0	61	-61	7	0.1	0.1	0.2	0.2	0.5	0.8	0.9	1.1	0.8	0.5	0.2	0.1	5.5
St. John, N.B.	45 17N	66 04W	119	61	28	11	43	32	69	54	54	41	93	-24	61	4.1	3.1	3.7	3.2	3.1	3.2	3.1	3.6	3.7	4.1	3.9	3.8	42.6
St. Johns, Nfld.	47 32N	52 44W	211	68	30	18	41	29	69	51	53	40	93	-21	58	5.3	5.1	4.6	3.8	3.9	3.1	3.1	4.0	3.7	4.8	5.7	6.0	53.1
Saskatoon, Sask.	52 08N	106 38W	1,690	38	9	-11	49	26	77	52	51	27	104	-55	38	0.9	0.5	0.7	0.7	1.4	2.6	2.4	1.9	1.5	0.9	0.5	0.6	14.6
The Pas, Man.	53 49N	101 15W	890	27	1	-18	45	21	75	54	45	26	100	-54	27	0.6	0.5	0.7	0.8	1.4	2.2	2.2	2.1	2.0	1.2	1.0	0.8	15.5
Toronto, Ont.	43 40N	79 24W	379	105	30	16	50	34	79	59	56	40	105	-26	105	2.7	2.4	2.6	2.5	2.9	2.7	3.0	2.7	2.9	2.4	2.8	2.6	32.2
Vancouver, B.C.	49 17N	123 05W	127	43	41	32	58	40	74	54	57	44	92	2	41	8.6	5.8	5.0	3.3	2.8	2.5	1.2	1.7	3.6	5.8	8.3	8.8	57.4
Whitehorse, Y.T.	60 43N	135 04W	2,303	10	13	-3	41	22	67	45	41	28	91	-62	10	0.6	0.5	0.6	0.4	0.6	1.0	1.6	1.5	1.3	0.7	1.0	0.8	10.6
Winnipeg, Man.	49 54N	97 14W	783	66	7	-13	48	27	79	55	51	31	108	-54	66	0.9	0.9	1.2	1.4	2.3	3.1	3.1	2.5	2.3	1.5	1.1	0.9	21.2
Yellow Knife, N.W.T.	62 28N	114 27W	674	13	-8	-23	29	9	69	52	36	26	90	-60	13	0.8	0.6	0.7	0.4	0.7	0.6	1.5	1.4	1.0	1.3	1.0	0.8	10.8
Greenland:																												
Angmagssalik	65 36N	37 33W	95	30	23	10	35	16	54	37	35	25	77	-26	38	2.9	2.4	2.6	2.1	2.0	1.8	1.5	2.1	3.3	4.7	3.0	2.7	31.1
Denmarkshaven	76 46N	19 00W	7	2	-1	-15	6	-13	47	34	13	2	63	-42	2	1.2	0.7	0.7	0.1	0.2	0.2	0.5	0.6	0.3	0.3	1.0	0.7	6.0
Eismitte	70 53N	40 42W	9,843	1	-33	-53	-14	-37	19	1	-23	-42	27	-85	1	0.6	0.2	0.3	0.2	0.1	0.1	0.1	0.4	0.3	0.5	0.5	1.0	4.3
Godthaab	64 10N	51 43W	66	40	19	10	31	20	52	38	35	26	76	-20	45	1.4	1.7	1.6	1.2	1.7	1.4	2.2	3.1	3.3	2.5	1.9	1.5	23.5
Ivigtut	61 12N	48 10W	98	48	24	12	38	24	57	42	40	29	86	-20	50	3.3	2.6	3.4	2.5	3.5	3.2	3.1	3.7	5.9	5.7	4.6	3.1	44.6
Jacobshavn	69 13N	51 02W	104	32	8	-7	24	6	51	40	31	20	71	-46	52	0.4	0.4	0.5	0.5	0.6	0.8	1.2	1.4	1.3	0.9	0.7	0.5	9.2
Nord	81 36N	16 40W	118	8	-15	-28	-5	-18	44	35	3	-6	61	-60	8	0.8	0.8	0.5	0.3	0.1	0.3	1.0	1.4	1.2	0.6	1.4	0.5	8.9
Scoresbysund	70 29N	21 58W	56	12	12	-3	22	6	49	36	25	15	63	-42	12	1.8	1.4	0.9	1.4	0.4	0.8	1.5	0.7	1.7	1.4	1.1	1.9	15.0
Thule	76 31N	68 44W	251	12	-4	-17	10	-7	46	38	19	8	63	-44	12	0.4	0.3	0.2	0.2	0.3	0.2	0.7	0.6	0.6	0.7	0.5	0.2	4.9
Upernivik	72 47N	56 07W	59	40	-1	-13	15	-1	48	35	29	21	69	-44	50	0.4	0.5	0.7	0.6	0.6	0.5	0.9	1.1	1.1	1.1	1.1	0.6	9.2
Mexico:																												
Acapulco	16 50N	99 56W	10	8	85	70	87	71	89	75	88	74	97	60	40	0.3	*	0.0	*	1.4	12.8	9.1	9.3	13.9	6.7	1.2	0.4	55.1
Chihuahua	28 42N	105 57W	4,429	9	65	36	81	51	89	66	79	51	102	12	22	0.2	0.4	0.3	0.2	0.2	1.7	3.6	3.7	3.3	0.9	0.5	0.4	15.4
Guadalajara	20 41N	103 20W	5,194	26	73	45	85	53	79	60	78	56	101	26	33	0.4	0.2	0.2	0.2	1.1	8.8	9.4	8.5	7.2	2.2	0.8	0.7	39.7
Guaymas	27 57N	110 55W	58	9	74	57	84	65	96	82	91	75	117	41	41	0.5	0.2	0.2	0.1	*	0.1	1.7	2.7	2.1	0.7	0.3	0.8	9.4
La Paz	24 07N	110 17W	85	9	74	54	86	58	96	73	90	68	108	31	12	0.2	0.1	0.0	0.0	0.0	0.2	0.4	1.2	1.4	0.6	0.5	1.1	5.7
Lerdo	25 30N	103 32W	3,740	10	72	45	86	57	90	68	82	58	105	23	14	0.4	0.1	0.2	0.3	0.8	1.5	1.5	1.3	2.0	0.8	0.8	0.5	10.2
Manzanillo	19 04N	104 20W	26	17	86	68	87	67	93	76	91	76	103	54	17	0.1	0.2	*	0.0	0.1	4.7	5.7	6.4	14.5	5.1	0.9	1.8	39.5
Mazatlan	23 11N	106 25W	256	10	71	61	76	65	86	77	85	76	93	52	46	0.8	0.5	0.2	0.1	0.1	1.5	5.9	8.3	8.0	2.6	0.9	1.3	30.2
Merida	20 58N	89 38W	72	22	83	62	92	69	92	73	87	71	106	51	40	1.2	0.9	0.7	0.8	3.2	5.6	5.2	5.6	6.8	3.8	1.3	1.3	36.5
Mexico City	19 26N	99 04W	7,340	42	66	42	78	52	74	54	70	50	92	24	48	0.2	0.3	0.5	0.7	1.9	4.1	4.5	4.3	4.1	1.6	0.5	0.3	23.0
Monterrey	25 40N	100 18W	1,732	11	68	48	84	62	90	71	80	64	107	25	33	0.6	0.7	0.8	1.3	1.3	3.0	2.3	2.4	5.2	3.0	1.5	0.8	22.9
Salina Cruz	16 12N	95 12W	184	10	85	72	88	76	89	76	87	75	98	62	22	*	0.4	0.6	0.5	3.3	11.6	4.5	5.5	7.1	4.0	0.9	0.1	38.5
Tampico	22 16N	97 51W	78	12	75	59	83	69	89	75	85	71	104	34	12	1.5	1.2	1.0	1.5	1.9	8.7	4.9	4.8	10.8	5.0	2.0	1.6	44.9
Vera Cruz	19 12N	96 08W	52	10	77	66	83	72	87	74	85	73	98	53	40	0.9	0.6	0.6	0.8	2.6	10.4	4.1	11.1	13.9	6.9	3.0	1.0	65.7

TABLE 1–10. TEMPERATURE AND PRECIPITATION DATA FOR REPRESENTATIVE WORLD-WIDE STATIONS (continued)

COUNTRY AND STATION	LATITUDE	LONGITUDE	ELEVATION	TEMPERATURE: Length of record	Average Daily JAN. Maximum	Average Daily JAN. Minimum	Average Daily APR. Maximum	Average Daily APR. Minimum	Average Daily JULY Maximum	Average Daily JULY Minimum	Average Daily OCT. Maximum	Average Daily OCT. Minimum	Extreme Maximum	Extreme Minimum	AVERAGE PRECIPITATION: Length of record	January	February	March	April	May	June	July	August	September	October	November	December	YEAR
	° ′	° ′	FEET	YEAR	°F	°F	°F	°F	°F	°F	°F	°F	°F	°F	YEAR	IN.	IN.	IN.	IN.	IN.	IN.	IN.	IN.	IN.	IN.	IN.	IN.	IN.
CENTRAL AMERICA																												
Belize:																												
Belize	17 31N	88 11W	17	27	81	67	86	74	87	75	86	72	97	49	33	5.4	2.4	1.5	2.2	4.3	7.7	6.4	6.7	9.6	12.0	8.9	7.3	74.4
Canal Zone:																												
Balboa Heights	08 57N	79 33W	118	34	88	71	90	74	87	74	85	73	97	63	46	1.0	0.4	0.7	2.9	8.0	8.4	7.1	7.9	8.2	10.1	10.2	4.8	69.7
Cristobal	09 21N	79 54W	35	36	84	76	86	77	85	76	86	75	97	66	73	3.4	1.5	1.5	4.1	12.5	13.9	15.6	15.3	12.7	15.8	22.3	11.7	130.3
Costa Rica:																												
San Jose	09 56N	84 08W	3,760	8	75	58	79	62	77	62	77	60	92	49	34	0.6	0.2	0.8	1.8	9.0	9.5	8.3	9.5	12.0	11.8	5.7	1.6	70.8
El Salvador:																												
San Salvador	13 42N	89 13W	2,238	39	90	60	93	65	89	65	87	65	105	45	39	0.3	0.2	0.4	1.7	7.7	12.9	11.5	11.7	12.1	9.5	1.6	0.4	70.0
Guatemala:																												
Guatemala City	14 37N	90 31W	4,855	6	73	53	82	58	78	60	76	60	90	41	29	0.3	0.1	0.5	1.2	6.0	10.8	8.0	7.8	9.1	6.8	0.9	0.3	51.8
Honduras:																												
Tela	15 46N	87 27W	41	4	82	67	87	72	88	73	86	71	96	58	20	8.9	5.1	2.6	3.3	4.3	5.0	6.4	9.4	7.7	13.5	15.9	14.0	96.1
WEST INDIES																												
Bridgetown, Barbados	13 08N	59 36W	181	35	83	70	86	72	86	74	86	73	95	61	22	2.6	1.1	1.3	1.4	2.3	4.4	5.8	5.8	6.7	7.0	8.1	3.8	50.3
Camp Jacob, Guadaloupe	16 01N	61 42W	1,750	19	77	64	79	65	81	68	81	68	92	54	21	9.2	6.1	8.1	7.3	11.5	14.1	17.6	15.3	16.4	12.4	12.3	10.1	140.4
Fort-de-France, Martinique	14 37N	61 05W	13	22	83	69	86	71	86	74	87	73	96	56	31	4.7	4.3	2.9	3.9	4.7	7.4	9.4	10.3	9.3	9.7	7.9	5.9	80.4
Hamilton, Bermuda	32 17N	64 46W	151	59	68	58	71	59	85	73	79	69	99	40	62	4.4	4.7	4.8	4.1	4.6	4.4	4.5	5.4	5.2	5.8	5.0	4.7	57.6
Havana, Cuba	23 08N	82 21W	80	25	79	65	84	69	89	75	85	73	104	43	72	2.8	1.8	1.8	2.3	4.7	6.5	4.9	5.3	5.9	6.8	3.1	2.3	48.2
Kingston, Jamaica	17 58N	76 48W	110	33	86	67	87	70	90	73	88	73	97	56	59	0.9	0.6	0.9	1.2	4.0	3.5	1.5	3.6	3.9	7.1	2.9	1.4	31.5
La Guerite, St. Christopher (St. Kitts)	17 20N	62 45W	157	19	80	71	83	73	86	76	85	75	91	61	21	4.1	2.0	2.3	2.3	3.8	3.6	4.4	5.2	6.0	5.4	7.3	4.5	50.9
Nassau, Bahamas	25 05N	77 21W	12	35	77	65	81	69	88	75	85	73	94	41	57	1.4	1.5	1.4	2.5	4.6	6.4	5.8	5.3	6.9	6.5	2.8	1.3	46.4
Port-au-Prince, Haiti	18 33N	72 20W	121	42	87	68	89	71	94	74	90	72	101	58	70	1.3	2.3	3.4	6.3	9.1	4.0	2.9	5.7	6.9	6.7	3.4	1.3	53.3
Saint Clair, Trinidad	10 40N	61 31W	67	49	87	69	90	69	88	71	89	71	101	52	97	2.7	1.6	1.8	2.1	3.7	7.6	8.6	9.7	7.6	6.7	7.2	4.9	64.2
Saint Thomas, Virgin Is.	18 20N	64 58W	11	9	82	71	85	74	88	77	87	76	92	63	9	2.5	1.9	1.7	2.2	4.6	3.2	3.2	4.1	6.9	5.6	3.9	3.9	43.7
San Juan, Puerto Rico	18 26N	66 00W	13	30	81	67	84	69	87	74	87	73	94	60	30	4.7	2.9	2.2	3.7	7.1	5.7	6.3	7.1	6.8	5.8	6.5	5.4	64.2
Santo Domingo, Dom. Rep	18 29N	69 54W	57	26	84	66	85	69	88	72	87	72	98	59	25	2.4	1.4	1.9	3.9	6.8	6.2	6.4	6.3	7.3	6.0	4.8	2.4	55.8
SOUTH AMERICA																												
Argentina:																												
Bahia Blanca	38 43S	62 16W	95	33	88	62	71	51	57	39	71	48	109	18	46	1.7	2.2	2.5	2.3	1.2	0.9	1.0	1.0	1.6	2.2	2.1	1.9	20.6
Buenos Aires	34 35S	58 29W	89	23	85	63	72	53	57	42	69	50	104	22	70	3.1	2.8	4.3	3.5	3.0	2.4	2.2	2.4	3.1	3.4	3.3	3.9	37.4
Cipolletti	38 57S	67 59W	889	9	89	56	72	40	55	29	72	43	107	9	24	0.4	0.4	0.7	0.4	0.6	0.6	0.5	0.3	0.6	0.9	0.5	0.5	6.4
Corrientes	27 28S	58 50W	177	39	93	71	81	63	71	53	82	60	112	30	40	4.7	4.5	5.3	5.6	3.3	1.9	1.7	1.5	2.8	4.7	5.2	5.2	46.4
La Quiaca	22 06S	65 36W	11,345	23	70	41	69	32	60	16	71	32	95	0	25	3.5	2.6	1.8	0.3	*	0.0	*	*	0.1	0.3	1.0	2.7	12.3

Mendoza	32 53S	68 49W	2,625	23	90	60	73	47	59	35	76	50	109	15	46	0.9	1.2	1.1	0.5	0.4	0.3	0.2	0.3	0.5	0.7	0.7	0.7	7.5
Parana	31 44S	60 31W	210	12	91	67	77	58	62	45	75	54	113	21	23	3.1	3.1	3.9	4.9	2.6	1.2	1.2	1.6	2.4	2.8	3.7	4.5	35.0
Puerto Madryn	42 47S	65 01W	26	50	81	57	70	46	55	36	68	45	104	10	50	0.4	0.6	0.7	0.5	0.9	0.6	0.6	0.4	0.6	0.7	0.4	0.6	7.0
Santa Cruz	50 01S	68 32W	39	12	70	48	57	39	41	28	58	39	94	1	20	0.6	0.3	0.3	0.6	0.4	0.5	0.4	0.5	0.3	0.3	0.4	0.7	5.3
Santiago del Estero	27 46S	64 18W	653	28	97	69	82	59	70	44	87	59	116	19	20	3.4	3.0	3.0	1.3	0.6	0.3	0.2	0.2	0.5	1.4	2.5	4.1	20.4
Ushuaia	54 50S	68 20W	26	16	57	41	48	33	39	25	52	35	85	6	21	2.0	2.6	1.9	2.1	1.5	1.2	1.2	1.1	1.3	1.6	1.5	1.9	19.9
Bolivia:																												
Concepcion	16 15S	62 03W	1,607	5	85	66	86	62	81	54	88	62	101	32	16	7.2	4.7	4.4	1.8	2.0	1.5	1.1	0.9	1.2	2.9	5.0	5.9	38.6
La Paz	16 30S	68 08W	12,001	31	63	43	65	40	62	33	66	40	80	26	50	4.5	4.2	2.6	1.3	0.5	0.3	0.4	0.5	1.1	1.6	1.9	3.7	22.6
Sucre	19 03S	65 17W	9,344	5	63	48	63	45	61	37	65	46	88	25	52	7.3	4.9	3.7	1.6	0.2	0.1	0.2	0.3	1.0	1.6	2.6	4.3	27.8
Brazil:																												
Barra do Corda	05 35S	45 28W	266	9	89	71	89	71	92	64	94	72	103	45	9	6.7	8.7	8.0	6.1	2.3	1.0	0.7	0.7	1.0	2.5	3.9	5.7	47.2
Bela Vista	22 06S	56 22W	525	13	91	67	85	61	77	49	87	61	108	20	20	6.6	4.9	4.4	4.3	5.0	2.8	1.3	1.8	2.9	5.4	5.8	7.0	52.2
Belem	01 27S	48 29W	42	16	87	72	87	73	88	71	89	71	98	61	20	12.5	14.1	14.1	12.6	10.2	6.7	5.9	4.4	3.5	3.3	2.6	6.1	96.0
Brasilia	15 51S	47 56W	3,481	3	80	65	82	62	78	51	82	64	93	46	3	9.0	7.8	4.8	3.4	1.4	*	0.0	*	1.3	4.9	9.7	11.7	54.0
Conceicao do Araguaia	08 15S	49 12W	53	5	88	70	91	68	95	63	93	68	102	55	5	14.9	12.1	10.8	4.1	1.9	0.4	*	0.5	1.5	6.6	4.9	8.6	66.2
Corumba	19 00S	57 39W	381	8	94	73	92	73	84	64	93	70	106	33	11	7.3	5.9	5.1	4.6	2.9	1.9	0.3	1.2	2.6	4.0	5.6	7.1	48.5
Florianopolis	27 35S	48 33W	96	17	83	72	74	64	68	57	73	63	102	32	25	7.6	5.6	6.3	4.1	3.6	3.5	2.2	3.7	4.3	5.1	3.5	4.3	53.1
Goias	15 58S	50 04W	1,706	11	86	63	91	63	89	56	94	63	104	41	11	12.5	9.9	10.2	4.6	0.4	0.3	0.0	0.3	2.3	5.3	9.4	9.5	64.8
Guarapuava	25 16S	51 30W	3,592	10	79	61	73	55	66	47	74	53	94	23	5	8.7	5.8	5.4	4.5	4.6	6.5	2.7	3.6	4.6	6.9	6.6	6.1	65.8
Manaus	03 08S	60 01W	144	11	88	75	87	75	89	75	92	76	101	63	25	9.8	9.1	10.3	8.7	6.7	3.3	2.3	1.5	1.8	4.2	5.6	8.0	71.3
Natal	05 46S	35 12W	52	18	87	76	86	73	82	69	85	75	100	61	18	1.9	4.8	7.0	9.2	7.1	8.7	7.7	3.8	1.4	0.8	0.7	1.1	54.2
Parana	12 26S	48 06W	853	19	90	58	90	58	91	48	94	58	105	37	19	11.3	9.3	9.4	4.0	0.5	*	0.1	0.2	1.1	5.0	9.1	12.2	62.3
Porto Alegre	30 02S	51 13W	33	22	87	67	78	60	66	49	74	57	105	25	22	3.5	3.2	3.9	4.1	4.5	5.1	4.5	5.0	5.2	3.4	3.1	3.5	49.1
Quixeramobim	05 12S	39 18W	653	9	92	79	86	76	88	74	93	77	100	63	13	0.7	5.0	6.6	5.0	7.0	1.7	0.7	0.6	0.4	0.6	0.7	0.6	29.6
Recife	08 04S	34 53W	97	27	86	77	85	75	80	71	84	75	94	50	56	2.1	3.3	6.3	8.7	10.5	10.9	10.0	6.0	2.5	1.0	1.0	1.1	63.4
Rio de Janeiro	22 55S	43 12W	201	38	84	73	80	69	75	63	77	66	102	46	84	4.9	4.8	5.1	4.2	3.1	2.1	1.6	1.7	2.6	3.1	4.1	5.4	42.6
Salvador (Bahia)	13 00S	38 30W	154	25	86	74	84	74	79	69	83	71	100	50	20	2.6	5.3	6.1	11.2	10.8	9.4	7.2	4.8	3.3	4.0	4.5	5.6	74.8
Santarem	02 30S	54 42W	66	22	86	73	85	73	87	71	91	73	99	65	22	6.8	10.9	13.2	12.9	11.3	6.9	4.1	1.7	1.5	1.9	2.3	4.1	77.9
Sao Paulo	23 37S	46 39W	2,628	44	77	63	73	59	66	53	68	57	100	32	24	8.8	7.8	6.0	2.2	3.0	2.4	1.5	2.1	3.5	4.6	6.0	9.4	57.3
Sena Madureira	09 04S	68 39W	443	12	92	69	91	68	91	63	93	69	100	41	17	11.2	11.3	10.2	9.4	4.1	2.2	1.1	1.5	4.0	7.0	7.5	11.7	81.2
Uaupes	00 08S	67 05W	272	15	88	72	88	72	85	70	89	71	100	52	10	10.3	7.7	10.0	10.6	12.0	9.2	8.8	7.2	5.1	6.9	7.2	10.4	105.4
Uruguaiana	29 46S	57 07W	246	15	91	69	78	59	66	48	77	55	108	27	12	3.6	3.6	5.6	5.1	3.7	4.2	3.2	2.8	3.6	4.1	2.9	4.1	46.6
Chile:																												
Ancud	41 47S	73 52W	184	30	62	51	57	47	50	42	55	45	82	30	46	3.1	3.7	5.3	7.4	9.9	11.0	10.3	9.4	6.5	4.2	4.7	4.6	80.1
Antofagasta	23 42S	70 24W	308	22	76	63	70	58	63	51	66	55	86	37	32	0.0	0.0	0.0	*	*	0.1	0.2	0.1	*	0.1	*	0.0	0.5
Arica	18 28S	70 20W	95	15	78	64	74	60	66	54	69	58	93	39	25	*	0.0	0.0	0.0	0.0	0.0	0.0	*	0.0	0.0	0.0	*	*
Cabo Raper	46 50S	75 38W	131	8	58	46	54	44	47	38	51	40	72	28	10	7.8	5.8	7.1	7.7	7.5	7.9	9.5	7.5	5.6	7.0	6.7	7.0	87.1
Los Evangelistas	52 23S	75 07W	190	16	50	44	48	41	43	36	45	39	66	19	27	11.7	10.0	11.3	11.4	9.6	9.4	9.4	8.6	9.2	8.8	9.9	10.1	119.4
Potrerillos	26 30S	69 27W	9,350	7	65	49	63	47	57	40	61	44	75	20	7	*	*	0.3	*	0.7	*	0.5	0.3	0.2	0.2	0.0	*	2.2
Puerto Aisen	42 24S	72 42W	33	8	63	50	55	43	45	37	55	42	93	18	11	7.8	7.8	8.3	7.5	14.7	10.4	11.1	11.1	6.5	7.8	7.0	7.9	107.9
Punta Arenas	53 10S	70 54W	26	15	58	45	50	39	40	31	51	38	86	11	15	1.5	0.9	1.3	1.4	1.3	1.6	1.1	1.2	0.9	1.1	0.7	1.4	14.4
Santiago	33 27S	70 42W	1,706	14	85	53	74	45	59	37	72	45	99	24	58	0.1	0.1	0.2	0.5	2.5	3.3	3.0	2.2	1.2	0.6	0.3	0.2	14.2
Valdivia	39 48S	73 14W	16	29	73	52	62	46	52	41	63	44	97	19	60	2.6	2.9	5.2	9.2	14.2	17.7	15.5	12.9	8.2	5.0	4.9	4.1	102.4
Valparaiso	33 01S	71 38W	135	30	72	56	67	52	60	47	65	50	94	32	41	0.1	*	0.3	0.6	4.1	5.9	3.9	2.9	1.3	0.4	0.2	0.2	19.9

TABLE 1–10. TEMPERATURE AND PRECIPITATION DATA FOR REPRESENTATIVE WORLD-WIDE STATIONS (continued)

COUNTRY AND STATION	LATITUDE	LONGITUDE	ELEVATION	TEMPERATURE Length of record	Average Daily JAN. Maximum	JAN. Minimum	APR. Maximum	APR. Minimum	JULY Maximum	JULY Minimum	OCT. Maximum	OCT. Minimum	Extreme Maximum	Extreme Minimum	AVERAGE PRECIPITATION Length of record	January	February	March	April	May	June	July	August	September	October	November	December	YEAR
	° ′	° ′	FEET	YEAR	°F	°F	°F	°F	°F	°F	°F	°F	°F	°F	YEAR	IN.	IN.	IN.	IN.	IN.	IN.	IN.	IN.	IN.	IN.	IN.	IN.	IN.
Colombia:																												
Andagoya	05 06N	76 40W	197	8	90	75	90	75	89	74	90	74	97	62	15	25.0	21.4	19.5	26.1	25.5	25.8	23.3	25.3	24.6	22.7	22.4	19.5	281.1
Bogota	04 42N	74 08W	8,355	10	67	48	67	51	64	50	66	50	75	30	49	2.3	2.6	4.0	5.8	4.5	2.4	2.0	2.2	2.4	6.3	4.7	2.6	41.8
Cartagena	10 28N	75 30W	39	6	84	73	87	76	88	78	87	77	98	61	10	0.4	0.0	0.4	0.9	3.4	3.4	3.0	0.6	0.5	10.8	8.9	4.5	36.8
Ipiales	00 50N	77 42W	9,680	9	61	50	60	49	57	42	62	49	77	32	13	3.1	2.3	3.5	3.5	2.8	1.9	1.3	1.1	1.4	3.1	3.3	2.6	29.9
Tumaco	01 49N	78 45W	7	10	82	75	84	76	82	75	82	75	90	64	10	16.9	11.7	9.6	14.6	17.4	12.0	7.7	7.3	7.3	5.9	4.9	7.0	122.3
Ecuador:																												
Cuenca	02 53S	78 39W	8,301	7	69	50	69	50	65	47	70	49	81	29	10	2.0	1.8	3.2	4.3	4.3	1.7	0.9	1.1	1.6	3.1	1.8	2.5	28.3
Guayaquil	02 10S	79 53W	20	5	87	72	88	72	84	67	86	68	98	52	10	8.3	11.4	11.5	8.1	2.1	0.4	0.2	*	*	*	0.1	1.1	43.2
Quito	00 08S	78 29W	9,222	54	67	46	69	47	71	44	71	46	86	25	33	3.9	4.4	5.6	6.9	5.4	1.7	0.8	1.2	2.7	4.4	3.8	3.1	43.9
French Guiana:																												
Cayenne	04 56N	52 27W	20	38	84	74	86	75	88	73	91	74	97	65	51	14.4	12.3	15.8	18.9	21.7	15.5	6.9	2.8	1.2	1.3	4.6	10.7	126.1
Guyana:																												
Georgetown	06 50N	58 12W	6	54	84	74	85	76	85	75	87	76	93	68	35	8.0	4.5	6.9	5.5	11.4	11.9	10.0	6.9	3.2	3.0	6.1	11.3	88.7
Lethem	03 24N	59 38W	270	3	91	73	91	74	87	73	92	76	97	63	9	1.2	1.4	1.3	5.7	11.5	11.9	14.8	9.4	3.4	2.3	4.3	1.3	68.5
Paraguay:																												
Asuncion	25 17S	57 30W	456	15	95	71	84	65	74	53	86	62	110	29	30	5.5	5.1	4.3	5.2	4.6	2.7	2.2	1.5	3.1	5.5	5.9	6.2	51.8
Bahia Negra	20 14S	58 10W	318	20	92	74	87	68	79	61	90	69	106	35	20	5.4	5.3	4.9	2.9	2.3	1.6	1.5	0.6	2.3	4.2	5.3	4.3	40.6
Peru																												
Arequipa	16 21S	71 34W	8,460	13	67	49	67	48	67	47	68	47	82	25	37	1.3	1.8	0.7	0.2	*	*	*	*	0.0	*	*	0.4	4.4
Cajamarca	07 09S	78 30W	8,662	9	71	48	70	47	70	41	71	47	79	25	9	3.6	4.2	4.6	3.4	1.7	9.5	9.2	9.3	2.3	2.3	1.9	3.2	28.2
Cusco	13 33S	71 59W	10,866	13	68	45	71	40	70	31	72	43	86	16	12	6.4	5.9	4.3	2.0	0.6	0.2	0.2	0.4	1.0	2.6	3.0	5.4	32.0
Iquitos	03 45S	73 13W	384	5	90	71	87	71	88	68	90	70	100	54	5	9.1	10.4	9.4	13.6	10.7	5.7	6.4	5.2	10.5	7.3	9.1	10.3	107.7
Lima	12 05S	77 03W	394	15	82	66	80	63	67	57	71	58	93	49	15	0.1	*	*	*	0.2	0.2	0.3	0.3	0.3	0.1	0.1	*	1.6
Mollendo	17 00S	72 07W	80	10	79	66	76	63	67	57	70	59	90	50	10	*	0.1	*	*	0.1	0.1	*	0.2	0.2	0.1	0.1	*	0.9
Surinam:																												
Paramaribo	05 49N	55 09W	12	35	85	72	86	73	87	73	91	73	99	62	75	8.4	6.5	7.9	9.0	12.2	11.9	9.1	6.2	3.1	3.0	4.9	8.8	91.0
Uruguay:																												
Artigas	30 24S	56 23W	384	13	91	65	77	55	65	45	75	54	107	24	50	4.3	3.9	4.7	5.1	4.1	4.1	2.8	3.0	4.0	4.7	3.8	4.1	48.6
Montevideo	34 52S	56 12W	72	56	83	62	71	53	58	43	68	49	109	25	56	2.9	2.6	3.9	3.9	3.3	3.2	2.9	3.1	3.0	2.6	2.9	3.1	37.4
Venezuela:																												
Caracas	10 30N	66 56W	3,418	30	75	56	81	60	78	61	79	61	91	45	46	0.9	0.4	0.6	1.3	3.1	4.0	4.3	4.3	4.2	4.3	3.7	1.8	32.9
Ciudad Bolivar	08 07N	63 32W	197	10	90	72	93	75	90	75	93	75	100	64	10	1.4	0.8	0.7	1.0	3.8	5.5	6.3	7.1	3.6	4.0	2.8	1.3	38.3
Maracaibo	10 39N	71 36W	20	12	90	73	92	76	94	76	92	76	102	66	36	0.1	*	0.3	0.8	2.7	2.2	1.8	2.2	2.8	5.9	3.3	0.6	22.7

Merida	08 36N	71 10W	5,293	14	73	56	75	60	76	59	75	60	90	48	14	2.5	1.5	3.6	6.7	9.8	7.3	4.7	5.7	6.7	9.5	8.2	3.4	69.7
Santa Elena	04 36N	61 07W	2,976	10	82	61	82	63	81	61	84	61	95	48	10	3.2	3.2	3.2	5.7	9.6	9.5	9.1	7.6	5.3	4.9	4.9	4.5	70.7
PACIFIC ISLANDS																												
Easter Is. (Isla de Pascua)	27 10S	109 26W	98	4	77	64	78	63	70	58	73	58	88	46	10	4.8	3.7	4.6	4.2	4.6	4.3	3.5	3.0	2.7	3.7	4.6	4.9	48.6
Mas a Tierra (Juan																												
Fernandez)	33 37S	78 52W	20	25	72	60	68	57	60	50	61	51	86	39	29	0.8	1.2	1.6	3.4	5.9	6.4	5.8	4.4	2.9	1.9	1.6	1.0	36.9
Seymour Is. (Galapagos Is.)	00 28S	90 18W	36	3	86	72	87	75	81	69	81	67	93	58	3	0.8	1.4	1.1	0.7	*	*	*	*	*	*	*	*	4.0
ATLANTIC ISLANDS																												
Fernando de Noronha	03 50S	32 25W	148	32	84	75	82	75	81	73	82	75	93	63	32	1.7	4.7	7.4	10.5	10.5	7.3	5.4	1.9	0.7	0.3	0.4	0.5	51.3
Cumberland Bay,																												
South Georgia	54 16S	36 30W	8	23	48	35	42	29	34	23	41	28	84	-3	24	3.3	4.3	5.3	5.4	5.2	4.9	5.5	5.3	3.5	2.6	3.4	3.0	51.7
Laurie Is., South Orkneys	60 44S	44 44W	13	48	35	29	31	21	20	4	30	19	54	-40	46	1.4	1.5	1.9	1.6	1.2	1.0	1.3	1.3	1.1	1.1	1.3	1.0	15.7
Stanley, Falkland Isles	51 42S	57 51W	6	25	56	42	49	37	40	31	48	35	76	12	41	2.8	2.3	2.5	2.6	2.6	2.1	2.0	2.0	1.5	1.6	2.0	2.8	26.8
EUROPE																												
Albania:																												
Durres	41 19N	19 28E	23	10	51	42	63	55	83	74	68	58	95	21	10	3.0	3.3	3.9	2.2	1.6	1.9	0.5	1.9	1.7	7.1	8.5	7.3	42.9
Andorra:																												
Les Escaldes	42 30N	01 31E	3,543	5	43	29	59	39	78	55	61	42	91	0	9	1.5	1.7	2.9	2.4	4.7	3.1	2.2	3.4	3.1	3.5	3.3	2.5	34.3
Austria:																												
Innsbruck	47 16N	11 24E	1,909	34	34	20	60	39	78	55	58	40	97	-16	35	2.1	1.8	1.5	2.2	2.9	4.1	5.1	4.5	3.1	2.4	2.2	1.9	33.8
Vienna	48 15N	16 22E	664	50	34	26	57	41	75	59	55	44	98	-14	100	1.5	1.4	1.8	2.0	2.8	2.7	3.0	2.7	2.0	2.0	1.9	1.8	25.6
Bulgaria:																												
Sofia	42 42N	23 20E	1,805	30	34	22	62	41	82	57	63	42	99	-17	27	1.3	1.1	1.7	2.3	3.3	3.2	2.4	2.0	2.3	2.1	1.9	1.4	25.0
Varna	43 12N	27 55E	115	30	40	30	59	43	84	63	67	50	107	-12	20	1.5	0.9	1.2	1.2	1.8	2.6	1.9	1.2	1.5	1.9	1.9	2.0	19.6
Cyprus:																												
Nicosia	35 09N	33 17E	716	40	58	42	74	50	97	69	81	58	116	23	64	2.9	2.0	1.3	0.8	1.1	0.4	*	*	0.2	0.9	1.7	3.0	14.6
Czechoslovakia:																												
Prague	50 05N	14 25E	662	40	34	25	55	40	74	58	54	44	98	-16	70	0.9	0.8	1.1	1.5	2.4	2.8	2.6	2.2	1.7	1.2	1.2	0.9	19.3
Prerov	49 27N	17 27E	702	20	34	25	57	38	77	55	56	40	100	-23	21	1.3	1.1	1.1	2.0	2.4	2.9	3.5	3.2	2.0	2.4	1.5	1.4	24.8
Denmark:																												
Copenhagen	55 41N	12 33E	43	30	36	29	50	37	72	55	53	42	91	-3	30	1.6	1.3	1.2	1.7	1.7	2.1	2.2	3.2	1.9	2.1	2.2	2.1	23.3
Aarhus	56 08N	10 12E	161	21	35	27	51	37	70	54	53	42	87	-12	21	2.3	1.5	1.4	1.8	1.2	2.2	2.5	3.3	3.2	2.6	2.5	2.1	26.6
Finland:																												
Helsinki	60 10N	24 57E	30	20	27	17	43	31	71	57	45	37	89	-23	50	2.2	1.7	1.7	1.7	1.9	2.0	2.3	3.3	2.8	2.9	2.7	2.4	27.6
Kuusamo	65 57N	29 12E	843	20	17	2	35	18	68	50	36	27	90	-40	20	1.1	1.1	1.1	1.1	1.4	2.3	2.8	3.0	2.1	2.1	1.6	1.1	20.8
Vaasa	63 05N	21 36E	13	18	26	16	41	28	69	55	44	36	89	-29	19	1.1	0.8	0.8	1.0	1.4	1.8	2.4	2.5	2.7	2.3	1.7	1.1	19.6
France:																												
Ajaccio (Corsica)	41 52N	08 35E	243	46	56	40	66	48	85	64	72	55	103	23	86	3.0	2.3	2.6	2.2	1.6	0.9	2.8	0.7	1.7	3.8	4.4	3.1	29.1
Bordeaux	44 50N	00 43W	157	51	48	35	63	44	80	58	66	47	102	9	47	2.7	2.8	2.9	2.6	2.5	2.3	2.0	1.9	2.2	3.0	3.9	3.9	32.7
Brest	48 19N	04 47W	56	56	49	40	57	44	70	56	61	49	95	7	56	3.5	3.0	2.5	2.5	1.9	2.0	2.0	2.2	2.3	3.6	4.2	4.4	34.1
Cherbourg	49 39N	01 38W	30	47	47	40	54	43	67	57	59	50	91	14	47	3.3	2.9	2.7	2.0	1.9	1.8	1.9	3.0	2.9	4.6	5.1	5.2	37.3
Lille	50 35N	03 05W	141	40	42	33	58	40	75	55	59	45	96	0	40	2.5	1.9	2.5	2.0	2.4	2.2	2.8	2.3	2.6	3.0	3.0	3.2	30.3
Lyon	45 42N	04 47E	938	70	41	30	61	42	80	58	61	45	105	-13	70	1.4	1.4	1.8	2.1	2.8	2.9	2.8	2.9	3.1	3.1	2.6	1.9	28.8
Marseille	43 18N	05 23E	246	72	53	38	59	41	78	58	76	57	101	9	102	1.9	1.5	1.8	2.0	1.9	1.0	0.6	0.9	2.6	3.7	3.1	2.2	23.2
Paris	48 49N	02 29E	164	66	42	32	60	41	76	55	59	44	105	1	118	1.5	1.3	1.5	1.7	2.0	2.1	2.1	2.0	2.0	2.2	2.0	1.9	22.3
Strasbourg	48 35N	07 46E	465	20	40	31	59	41	78	57	58	43	101	-8	20	1.6	1.4	1.7	2.6	2.6	3.1	3.4	3.4	3.1	2.7	2.0	1.9	29.5
Toulouse	43 33N	01 23E	538	47	47	35	62	43	82	59	66	48	111	1	47	1.9	1.7	2.3	2.7	2.9	2.4	1.5	2.1	2.3	2.2	2.4	2.3	26.7

TABLE 1–10. TEMPERATURE AND PRECIPITATION DATA FOR REPRESENTATIVE WORLD-WIDE STATIONS (continued)

COUNTRY AND STATION	LATITUDE	LONGITUDE	ELEVATION	TEMPERATURE: Length of record	Average Daily JAN. Maximum	JAN. Minimum	APR. Maximum	APR. Minimum	JULY Maximum	JULY Minimum	OCT. Maximum	OCT. Minimum	Extreme Maximum	Extreme Minimum	AVERAGE PRECIPITATION: Length of record	January	February	March	April	May	June	July	August	September	October	November	December	YEAR
	° '	° '	FEET	YEAR	°F	°F	°F	°F	°F	°F	°F	°F	°F	°F	YEAR	IN.	IN.	IN.	IN.	IN.	IN.	IN.	IN.	IN.	IN.	IN.	IN.	IN.
Germany:																												
Berlin	52 27N	13 18E	187	50	35	26	55	38	74	55	55	41	96	-15	40	1.9	1.3	1.5	1.7	1.9	2.3	3.1	2.2	1.9	1.7	1.7	1.9	23.1
Bremen	53 05N	08 47E	52	50	37	30	53	38	71	55	54	43	94	-7	80	1.9	1.6	1.8	1.5	2.1	2.6	3.2	2.8	2.1	2.2	2.0	2.2	26.0
Frankfurt A/M	50 07N	08 40E	338	50	37	29	58	41	75	56	56	43	100	-7	80	1.7	1.3	1.6	1.5	2.0	2.5	2.8	2.6	1.9	2.2	2.0	2.0	24.1
Hamburg	53 33N	09 58E	66	50	35	28	51	39	69	56	53	44	92	-4	80	2.1	1.9	2.0	1.8	2.1	2.7	3.4	3.2	2.5	2.6	2.1	2.5	28.9
Munich	48 09N	11 34E	1,739	50	33	23	54	37	72	54	53	40	92	-14	80	1.7	1.4	1.9	2.7	3.7	4.6	4.7	4.2	3.2	2.2	1.9	1.9	34.1
Munster	51 58N	07 38E	207	50	39	29	56	38	73	54	56	42	96	-17	40	2.6	1.9	2.2	2.0	2.2	2.7	3.3	3.1	2.5	2.7	2.4	2.9	30.5
Nurnberg	49 27N	11 03E	1,050	50	35	26	56	38	74	55	55	41	99	-18	80	1.5	1.2	1.3	1.7	2.2	2.5	3.1	3.1	2.1	2.1	1.9	1.7	24.4
Gibraltar:																												
Windmill Hill	36 06N	05 21W	400	12	58	50	64	55	77	66	70	61	97	35	12	4.6	3.4	3.7	2.5	1.4	0.2	*	0.1	0.8	3.5	4.1	5.4	29.7
Greece:																												
Athens	37 58N	23 43E	351	72	54	42	67	52	90	72	74	60	109	20	80	2.2	1.6	1.4	0.8	0.8	0.6	0.2	0.4	0.6	1.7	2.8	2.8	15.8
Iraklion (Crete)	35 20N	25 08E	98	21	60	48	70	54	85	72	77	62	114	32	22	3.7	3.0	1.6	0.9	0.7	0.1	*	0.1	0.7	1.7	2.7	4.0	19.2
Rhodes	36 26N	28 15E	289	10	59	51	67	59	83	74	76	68	104	30	6	5.7	3.9	2.6	1.7	0.5	0.3	0.0	*	0.4	1.7	5.2	6.7	28.5
Thessaloniki	40 37N	22 57E	78	9	49	37	66	49	90	70	73	56	107	15	26	1.5	1.5	1.6	1.9	2.0	1.2	1.0	0.7	1.2	2.4	2.1	1.9	19.0
Hungary:																												
Budapest	47 31N	19 02E	394	50	35	26	62	44	82	61	61	45	103	-10	50	1.5	1.5	1.7	2.0	2.7	2.6	2.0	1.9	1.8	2.1	2.4	2.0	24.2
Debrecen	47 36N	21 39E	430	50	33	21	61	39	81	57	60	41	102	-22	80	1.2	1.1	1.4	1.8	2.4	2.8	2.5	2.3	1.8	2.2	2.0	1.6	23.1
Iceland:																												
Akureyri	65 41N	18 05W	16	23	34	26	40	30	57	47	43	34	83	-8	26	1.7	1.5	1.7	1.3	0.6	0.9	1.3	1.6	1.9	2.3	1.9	1.9	18.6
Reykjavik	64 09N	21 56W	92	25	36	28	43	33	58	48	44	36	74	4	30	4.0	3.1	3.0	2.1	1.6	1.7	2.0	2.6	3.1	3.4	3.6	3.7	33.9
Ireland:																												
Cork	51 54N	08 29W	56	27	48	38	55	41	68	53	58	44	85	15	35	4.9	3.6	3.3	2.6	2.9	2.0	2.9	3.1	2.9	3.9	4.5	4.7	41.3
Dublin	53 22N	06 21W	155	30	47	35	54	38	67	51	57	43	86	8	35	2.7	2.2	2.0	1.9	2.3	2.0	2.8	3.0	2.8	2.7	2.7	2.6	29.7
Shannon Airport	52 41N	08 55W	8	9	46	36	55	41	66	53	58	45	87	12	12	3.8	3.0	2.0	2.2	2.4	2.1	3.1	3.0	3.0	3.4	4.2	4.3	36.5
Italy:																												
Ancona	43 37N	13 32E	52	30	46	36	62	50	83	68	67	55	102	18	30	2.6	1.7	1.6	2.3	2.1	1.9	1.5	1.5	3.5	3.7	2.5	3.0	28.0
Cagliari (Sardinia)	39 15N	09 03E	3	30	56	43	66	50	86	67	72	58	102	25	25	2.2	1.5	1.5	1.2	1.5	0.5	0.1	0.4	1.0	3.0	1.8	2.3	17.0
Genoa	44 24N	08 55E	318	10	50	41	65	53	82	70	73	58	100	18	10	3.9	4.0	3.3	3.4	4.6	1.4	1.6	2.3	4.7	6.1	7.2	4.1	46.6
Naples	40 51N	14 15E	82	30	54	40	65	52	84	70	71	60	101	24	30	3.7	3.2	3.0	2.6	1.8	1.8	0.6	0.7	2.8	5.1	4.5	5.4	35.2
Palermo (Sicily)	38 07N	13 19E	354	10	58	47	67	53	86	71	75	62	113	31	30	3.8	3.4	2.4	1.9	1.1	0.6	0.2	0.6	2.0	3.7	4.1	4.5	28.3
Rome	41 48N	12 36E	377	10	54	39	68	46	88	64	73	53	104	20	30	3.3	2.9	2.0	2.0	1.9	0.7	0.4	0.7	2.8	4.3	4.4	4.1	29.5
Taranto	40 28N	17 17E	56	10	55	43	59	50	89	70	73	58	108	26	10	1.6	0.9	1.3	0.8	1.0	0.6	0.4	0.7	1.0	2.2	1.8	1.9	14.2
Venice	45 26N	12 23E	82	10	43	33	63	49	82	67	65	52	97	14	30	2.0	2.1	2.4	2.8	3.2	3.3	2.6	2.6	2.6	3.7	3.5	2.6	33.4
Luxembourg:																												
Luxembourg	49 37N	06 03E	1,096	7	36	29	58	40	74	55	56	43	99	-10	100	2.3	2.0	1.9	2.1	2.4	2.5	2.8	2.6	2.4	2.7	2.7	2.8	29.2
Malta:																												
Valletta	35 54N	14 31E	233	90	59	51	66	56	84	72	76	66	105	34	90	3.3	2.3	1.5	0.8	0.4	0.1	*	0.2	1.3	2.7	3.6	3.9	20.3

Monaco:																												
Monaco	43 44N	07 25E	180	60	54	46	61	53	77	70	67	60	93	27	60	2.4	2.3	3.1	2.2	2.1	1.4	0.7	1.1	2.3	4.7	4.3	3.5	30.1
Netherlands:																												
Amsterdam	52 23N	04 55E	5	29	40	34	52	43	69	59	56	48	95	3	29	2.0	1.4	1.3	1.6	1.8	1.8	2.6	2.7	2.8	2.8	2.6	2.2	25.6
Norway:																												
Bergen	60 24N	05 19E	141	49	43	27	55	34	72	51	57	38	89	3	75	7.9	6.0	5.4	4.4	3.9	4.2	5.2	7.3	9.2	9.2	8.0	8.1	78.8
Kristiansand	58 10N	07 59E	175	11	32	25	50	35	71	53	53	39	90	-14	56	5.0	3.6	3.6	2.7	2.5	2.8	3.5	5.3	4.7	6.2	5.7	6.4	52.0
Oslo	59 56N	10 44E	308	44	30	20	50	34	73	56	49	37	93	-21	56	1.7	1.3	1.4	1.6	1.8	2.4	2.9	3.8	2.5	2.9	2.3	2.3	26.9
Tromso	69 39N	18 57E	335	47	30	22	37	27	59	48	40	33	83	-1	75	4.1	3.8	3.3	2.4	2.1	2.1	2.3	2.9	4.7	4.5	4.0	3.9	40.1
Trondheim	63 25N	10 27E	417	44	31	22	45	32	66	51	46	36	95	-22	65	3.1	2.7	2.6	2.0	1.7	1.9	2.4	3.0	3.4	3.7	2.8	2.8	32.1
Vardo	70 22N	31 06E	43	40	27	19	34	26	53	44	38	32	80	-11	56	2.5	2.5	2.3	1.5	1.3	1.3	1.5	1.7	1.9	2.5	2.1	2.4	23.5
Poland:																												
Danzig	54 24N	18 40E	36	36	33	25	49	37	70	56	53	42	94	-16	35	1.2	1.0	1.3	1.5	1.8	2.3	2.8	2.6	2.1	1.8	1.8	1.5	21.7
Krakow	50 04N	19 57E	723	35	32	22	55	38	76	57	56	41	97	-28	35	1.1	1.3	1.4	1.8	2.8	4.0	4.5	3.8	2.7	2.2	1.7	1.3	28.6
Warsaw	52 13N	21 02E	294	25	30	21	54	38	75	56	54	41	98	-22	113	1.2	1.1	1.3	1.5	1.9	2.6	3.0	3.0	1.9	1.7	1.4	1.4	22.0
Wroclaw (Breslau)	51 07N	17 05E	482	50	35	25	55	39	74	57	55	42	98	-26	40	1.5	1.1	1.5	1.7	2.4	2.4	3.4	2.7	1.8	1.7	1.5	1.5	23.2
Portugal:																												
Braganca	41 49N	06 47W	2,395	11	46	31	59	39	80	54	62	42	103	10	11	11.9	6.9	7.7	3.7	3.0	1.6	0.5	0.6	1.5	3.0	6.3	7.1	53.8
Lagos	37 06N	08 38W	46	21	61	47	67	52	83	64	73	58	107	28	17	3.2	2.6	2.8	1.4	0.8	0.2	*	*	0.4	1.5	2.6	2.8	18.3
Lisbon	38 43N	09 08W	313	75	56	46	64	52	79	63	69	57	103	29	75	3.3	3.2	3.1	2.4	1.7	0.7	0.2	0.2	1.4	3.1	4.2	3.6	27.0
Romania:																												
Bucharest	44 25N	26 06E	269	41	33	20	63	41	86	61	65	44	105	-18	41	1.5	1.1	1.7	1.6	2.5	3.8	2.3	1.8	1.5	1.6	1.9	1.5	22.8
Cluj	46 47N	23 40E	1,286	15	31	18	58	38	79	56	60	41	100	-26	16	1.3	1.2	1.0	2.1	3.3	3.3	2.6	3.3	2.0	1.7	1.0	1.2	24.0
Constanta	44 11N	28 39E	13	20	37	25	55	42	79	63	62	49	101	-13	39	1.2	1.2	1.1	1.1	1.3	1.7	1.3	1.1	1.1	1.4	1.2	1.4	15.1
Spain:																												
Almeria	36 51N	02 28W	213	20	61	47	69	54	85	69	76	62	108	34	20	0.9	1.0	0.7	0.9	0.7	0.2	*	0.1	0.6	0.9	1.5	1.1	8.6
Barcelona	41 24N	02 09E	312	20	56	42	64	51	81	69	71	58	98	24	30	1.2	2.1	1.9	1.8	1.8	1.3	1.2	1.7	2.6	3.4	2.7	1.8	23.5
Burgos	42 20N	03 42W	2,825	29	42	30	57	38	77	53	61	43	99	0	29	1.5	1.5	2.1	1.9	2.4	1.7	0.8	0.7	1.4	2.0	2.2	2.0	20.2
Madrid	40 25N	03 41W	2,188	30	47	33	64	44	87	62	66	48	102	14	30	1.1	1.7	1.7	1.7	1.5	1.2	0.4	0.3	1.2	1.9	2.2	1.6	16.5
Sevilla	37 29N	05 59W	98	26	59	41	73	51	96	67	78	57	117	27	26	2.2	2.9	3.3	2.3	1.3	0.9	0.1	0.1	1.1	2.6	3.7	2.8	23.3
Valencia	39 28N	00 23W	79	26	58	41	67	51	83	68	73	57	107	20	29	0.9	1.5	0.9	1.2	1.1	1.3	0.4	0.5	2.2	1.6	2.5	1.3	15.4
Sweden:																												
Abisko	68 21N	18 49E	1,273	11	20	6	33	19	61	45	35	24	82	-30	11	0.7	0.6	0.5	0.5	0.7	1.8	1.6	1.8	1.2	1.0	0.6	0.6	11.7
Goteberg	57 42N	11 58E	55	39	35	27	48	36	69	56	51	42	88	-13	61	2.5	2.0	2.0	1.7	1.9	2.2	2.8	3.7	3.1	3.1	2.7	2.8	30.5
Haparanda	65 50N	24 09E	30	20	22	10	38	23	71	53	39	30	89	-34	20	2.2	1.6	1.2	1.5	1.4	1.7	2.1	2.8	2.6	2.8	2.5	2.0	24.4
Karlstad	59 23N	13 30E	164	30	30	20	49	32	73	56	49	38	93	-21	30	1.9	1.2	1.2	1.4	1.9	1.9	2.6	3.1	2.9	2.4	2.4	1.9	24.8
Sarna	61 41N	13 07E	1,504	20	19	4	42	23	69	46	42	28	91	-51	20	1.6	0.8	0.9	1.2	1.6	2.8	3.6	3.3	2.6	2.3	1.8	1.8	24.3
Stockholm	59 21N	18 04E	146	30	31	23	45	32	70	55	48	39	97	-26	30	1.5	1.1	1.1	1.5	1.6	1.9	2.8	3.1	2.1	2.1	1.9	1.9	22.4
Visby (Gotland)	57 39N	18 18E	36	30	35	28	44	33	67	55	50	41	88	1	30	1.7	1.1	1.2	1.4	1.1	1.4	2.0	2.7	1.7	1.9	2.1	2.0	20.3
Switzerland:																												
Berne	46 57N	07 26E	1,877	30	35	26	56	39	74	56	55	42	96	-9	77	1.9	2.0	2.6	3.0	3.7	4.4	4.4	4.3	3.5	3.5	2.7	2.5	38.5
Geneva	46 12N	06 09E	1,329	30	39	29	58	41	77	58	58	44	101	-1	125	1.9	1.8	2.2	2.5	3.0	3.1	2.9	3.6	3.6	3.8	3.1	2.4	33.9
Zurich	47 23N	08 33E	1,617	23	38	28	57	39	76	55	57	42	98	-12	23	2.3	1.9	2.9	3.4	4.0	4.9	5.0	4.6	3.3	3.2	2.5	2.9	40.9
Turkey:																												
Edirne	41 39N	26 34E	154	18	41	28	66	44	88	63	70	49	107	-8	18	2.2	1.9	1.7	1.9	1.7	2.1	1.5	1.1	1.1	2.1	2.9	3.0	23.2
Istanbul	40 58N	28 50E	59	18	45	36	61	45	81	65	67	54	100	17	18	3.7	2.3	2.6	1.9	1.4	1.3	1.7	1.5	2.3	3.8	4.1	4.9	31.5
United Kingdom:																												
Belfast	54 35N	05 56W	57	7	42	34	53	38	65	52	55	44	82	14	30	4.2	2.8	2.3	2.4	2.3	2.5	3.5	3.5	3.4	3.8	3.6	3.9	38.2
Birmingham	52 29N	01 56W	535	30	42	35	53	40	69	54	55	45	92	11	30	2.9	2.1	1.7	2.2	2.5	1.8	2.8	2.7	2.3	2.9	3.2	2.6	29.7
Cardiff	51 28N	03 10W	203	30	45	36	55	41	69	54	57	45	91	2	30	4.6	3.0	2.3	2.5	3.0	2.2	3.4	3.9	3.6	4.5	4.6	4.3	41.9

TABLE 1–10. TEMPERATURE AND PRECIPITATION DATA FOR REPRESENTATIVE WORLD-WIDE STATIONS (continued)

COUNTRY AND STATION	LATITUDE (° ′)	LONGITUDE (° ′)	ELEVATION (FEET)	Temp. Length of record (YEAR)	JAN. Maximum (°F)	JAN. Minimum (°F)	APR. Maximum (°F)	APR. Minimum (°F)	JULY Maximum (°F)	JULY Minimum (°F)	OCT. Maximum (°F)	OCT. Minimum (°F)	Extreme Maximum (°F)	Extreme Minimum (°F)	Precip. Length of record (YEAR)	January (IN.)	February (IN.)	March (IN.)	April (IN.)	May (IN.)	June (IN.)	July (IN.)	August (IN.)	September (IN.)	October (IN.)	November (IN.)	December (IN.)	YEAR (IN.)
Dublin	53 22N	06 21W	155	30	47	35	54	38	67	51	57	43	86	8	35	2.7	2.2	2.0	1.9	2.3	2.0	2.8	3.0	2.8	2.7	2.7	2.6	29.7
Edinburgh	55 55N	03 11W	441	30	43	35	50	39	65	52	53	44	83	15	30	2.5	1.6	1.6	1.6	2.2	1.9	3.1	3.1	2.6	2.9	2.4	2.1	27.6
London	51 29N	00 00	149	30	44	35	56	40	73	55	58	44	99	9	30	2.0	1.5	1.4	1.8	1.8	1.6	2.0	2.2	1.8	2.3	2.5	2.0	22.9
Liverpool	53 24N	03 04W	198	30	44	36	52	41	66	55	55	46	87	15	30	2.7	1.9	1.5	1.6	2.2	2.0	2.8	3.1	2.6	3.0	3.0	2.5	28.9
Perth	56 24N	03 27W	77	30	43	32	53	38	68	51	55	41	89	0	30	3.1	2.2	1.9	1.7	2.3	2.0	3.1	2.9	2.8	3.3	2.7	2.7	30.7
Plymouth	50 21N	04 07W	87	30	47	40	54	43	66	55	58	49	88	16	30	4.3	3.0	2.6	2.3	2.5	2.0	2.6	2.9	2.8	3.8	4.6	4.4	37.8
Wick	58 26N	03 05W	119	30	42	35	48	38	59	50	52	43	80	8	30	2.9	2.1	1.8	2.1	1.8	2.0	2.6	2.6	2.9	3.2	3.1	2.9	30.0
U.S.S.R.:																												
Arkhangelsk	64 33N	40 32E	22	23	9	2	36	23	64	51	36	30	91	-49	25	1.2	1.1	1.1	0.7	1.3	1.9	2.6	2.7	2.2	1.9	1.6	1.3	19.8
Astrakhan	46 21N	48 02E	45	10	23	14	57	40	85	69	56	40	99	-22	25	0.5	0.5	0.4	0.6	0.6	0.7	0.5	0.4	0.6	0.4	0.6	0.6	6.4
Dnepropetrovsk	48 27N	35 04E	259	18	25	16	53	39	80	62	56	40	101	-25	17	1.4	1.1	1.2	1.4	1.8	3.0	1.9	1.6	1.0	1.8	1.6	1.6	19.4
Kaunas	54 54N	23 53E	118	19	26	18	49	34	72	53	50	38	96	-23	19	1.6	1.3	1.3	1.8	2.0	3.2	3.3	3.5	1.9	1.9	1.6	1.6	25.0
Kirov	58 36N	49 41E	594	20	6	-2	41	27	72	55	37	29	92	-43	29	1.2	1.0	0.9	0.9	1.9	2.5	2.1	2.9	2.3	2.0	1.6	1.3	20.6
Kursk	51 45N	36 12E	773	15	19	11	47	35	74	58	48	36	91	-23	20	1.5	1.3	1.2	1.5	2.2	2.5	3.2	2.3	1.6	1.8	1.5	1.7	22.3
Leningrad	59 56N	30 16E	16	26	23	12	45	31	71	57	45	37	91	-36	95	1.0	0.9	0.9	1.0	1.6	2.0	2.5	2.8	2.1	1.8	1.4	1.2	19.2
Lvov	49 50N	24 01E	978	9	31	22	53	38	77	59	55	43	97	-29	35	1.3	1.5	1.8	2.0	2.8	3.7	4.1	3.1	2.4	2.1	0.8	1.6	28.2
Minsk	53 54N	27 33E	738	12	22	13	47	33	70	54	47	36	92	-27	20	1.4	1.5	1.3	1.5	2.0	2.8	3.0	3.1	1.6	1.5	1.5	1.7	22.9
Moscow	55 46N	37 40E	505	15	21	9	47	31	76	55	46	34	96	-27	11	1.5	1.4	1.1	1.9	2.2	2.9	3.0	2.9	1.9	2.7	1.7	1.6	24.8
Odessa	46 29N	30 44E	214	20	28	22	52	41	79	65	57	47	99	-13	15	1.0	0.7	0.7	1.1	1.1	1.9	1.6	1.4	1.1	1.4	1.1	1.1	14.3
Riga	56 57N	24 06E	67	30	29	20	48	35	72	56	49	39	93	-20	57	1.3	1.0	1.1	1.2	1.7	2.4	3.0	3.0	2.1	2.0	1.9	1.5	22.2
Saratov	51 32N	46 03E	197	14	15	7	50	35	82	64	48	36	102	-27	15	1.0	1.0	0.8	1.0	1.3	1.8	1.2	1.3	1.1	1.4	1.4	1.2	14.5
Sevastopol	44 37N	33 31E	75	20	39	30	55	42	79	65	63	50	97	-4	30	1.1	1.1	1.1	0.9	0.6	1.1	0.8	0.6	1.1	1.5	1.2	1.1	12.2
Stalingrad	48 42N	44 31E	136	8	15	4	52	36	84	65	53	37	106	-30	12	0.9	1.0	0.6	0.6	1.0	1.9	0.9	0.8	0.7	1.0	1.5	1.3	12.2
Stavropol	45 02N	41 58E	1,886	18	26	17	50	37	76	60	55	42	95	-22	41	1.4	1.1	1.5	2.4	3.0	4.1	3.0	2.0	2.5	2.3	1.8	1.8	26.9
Tallin	59 26N	24 48E	146	15	27	18	42	31	70	55	47	38	89	-19	63	1.1	1.0	0.9	1.1	1.7	1.9	2.1	2.7	2.3	2.1	1.9	1.5	20.2
Tbilisi	41 43N	44 48E	1,325	10	39	26	61	44	83	65	64	48	95	6	10	0.7	0.8	1.3	1.6	3.6	3.1	2.2	1.7	1.9	1.3	2.0	1.2	21.4
Ust'Shchugor	64 16N	57 34E	279	15	4	-14	35	17	65	49	33	23	90	-67	15	1.1	0.8	0.8	0.7	1.4	2.2	3.0	3.2	2.4	2.2	1.5	1.3	20.6
Ufy	54 43N	55 56E	571	20	6	-3	44	30	75	58	41	31	99	-42	23	1.6	1.3	1.2	0.9	1.6	2.4	2.6	2.2	1.8	2.3	2.2	2.3	22.5
Yugoslavia:																												
Belgrade	44 48N	20 28E	453	16	37	27	64	45	84	61	65	47	107	-14	16	1.6	1.3	1.6	2.2	2.6	2.8	1.9	2.5	1.7	2.7	1.8	1.9	24.6
Skopje	41 59N	21 28E	787	10	40	26	67	42	88	60	65	43	105	-11	10	1.5	1.2	1.3	1.5	1.9	1.9	1.3	1.1	1.1	2.6	2.3	1.8	19.5
Split	43 31N	16 26E	420	14	51	29	65	50	87	68	69	55	100	17	51	3.1	2.5	3.2	3.0	2.5	2.1	1.2	1.6	2.9	4.4	4.2	4.4	35.1
OCEAN ISLANDS																												
Bjornoya, Bear Island	74 31N	19 01E	49	10	26	17	27	16	44	36	36	29	71	-25	25	1.6	1.3	1.3	0.9	0.8	0.7	0.8	1.2	1.8	1.7	1.4	1.6	15.1
Gronfjorden, Spitzbergen	78 02N	14 15E	23	19	10	-4	15	-3	46	38	25	17	60	-57	15	1.4	1.3	1.1	0.9	0.5	0.4	0.6	0.9	1.0	1.2	0.9	1.5	11.7
Horta, Azores	38 32N	28 38W	200	30	62	54	64	55	76	65	71	62	88	38	30	4.5	4.1	4.2	3.0	2.9	2.0	1.5	1.9	3.2	4.4	4.1	4.5	40.3

Jan Mayen	71 01N	08 28W	131	5	31	21	31	22	46	38	39	29	60	-18	29	2.1	1.7	1.6	1.4	0.9	0.9	1.4	1.8	2.5	2.5	2.2	2.2	21.2
Lerwick, Shetland Island	60 08N	01 11W	269	30	42	35	46	37	58	49	50	42	71	17	30	4.5	3.4	2.9	2.7	2.2	2.2	2.7	2.9	3.7	4.3	4.5	4.5	40.5
Matochikin Shar,																												
Novaya Zemlya	73 16N	56 24E	61	9	8	-6	13	-1	47	36	30	21	68	-41	9	0.6	0.6	0.6	0.4	0.3	0.4	1.4	1.5	1.5	0.6	0.6	0.4	8.9
Ponta Delgada, Azores	37 45N	25 40W	118	30	62	54	64	55	76	64	71	61	85	37	30	4.0	3.5	3.5	2.5	2.3	1.4	1.0	1.2	2.9	3.6	3.7	3.0	32.6
Stornoway, Hebrides	58 11N	06 21W	34	30	44	37	49	39	61	51	53	44	78	11	15	6.4	3.2	3.2	3.1	2.5	2.4	3.0	4.3	4.7	6.2	4.6	5.5	49.1
Thorshavn, Faeroes	62 02N	06 45W	82	50	42	33	45	36	56	47	58	40	70	8	50	6.6	5.2	4.8	3.6	3.4	2.5	3.1	3.5	4.7	5.9	6.3	6.6	56.2
AFRICA																												
Algeria:																												
Adrar	27 52N	00 17W	948	15	69	39	92	60	115	82	92	63	124	25	15	*	*	0.1	*	*	*	*	*	*	0.2	0.2	*	0.6
Algiers	36 46N	03 03E	194	25	59	49	68	55	83	70	74	63	107	32	25	4.4	3.3	2.9	1.6	1.8	0.6	*	0.2	1.6	3.1	5.1	5.4	30.0
Annaba	36 54N	07 46E	66	26	59	46	67	52	85	69	75	61	115	32	26	5.6	4.1	2.9	2.2	1.5	0.6	0.1	0.3	1.2	3.0	4.3	5.2	31.0
Bordj Omar Driss	28 06N	06 42E	1,224	15	67	38	90	59	110	78	92	63	124	19	15	0.3	0.1	0.1	0.2	*	*	0.0	*	*	*	0.2	0.2	1.1
El Golea	30 35N	02 53E	1,247	15	63	37	84	56	107	79	87	60	120	23	15	0.1	0.3	0.5	*	*	*	*	*	*	0.3	0.4	0.3	1.9
Tamanrasset	22 42N	05 31E	4,593	15	67	39	86	56	95	71	85	59	102	20	15	0.2	*	*	0.2	0.4	0.1	0.1	0.4	0.1	*	*	*	1.5
Touggourt	33 07N	06 04E	226	26	62	38	83	55	107	77	84	59	122	26	26	0.2	0.4	0.5	0.2	0.2	0.2	*	*	0.1	0.3	0.5	0.3	2.9
Angola:																												
Cangamba	13 41S	19 52E	4,331	6	84	62	89	58	82	46	87	59	109	20	7	8.9	7.4	6.8	1.8	0.1	0.0	0.0	0.2	0.2	1.6	5.1	8.5	40.6
Huambo	12 48S	15 45E	5,577	14	78	58	78	57	77	47	81	58	90	36	14	8.7	7.8	9.8	5.7	0.4	0.0	*	*	0.6	5.5	9.6	8.9	57.0
Luanda	08 49S	13 13E	194	27	83	74	85	75	74	65	79	71	98	58	59	1.0	1.4	3.0	4.6	0.5	*	*	*	0.1	0.2	1.1	0.8	12.7
Moçâmedes	15 12S	12 09E	10	15	79	65	82	66	68	56	74	61	102	44	21	0.3	0.4	0.7	0.5	*	*	*	*	*	*	0.1	0.1	2.1
Benin:																												
Cotonou	06 21N	02 26E	23	5	80	74	83	78	78	74	80	75	95	65	10	1.3	1.3	4.6	4.9	10.0	14.4	3.5	1.5	2.6	5.3	2.3	0.5	52.4
Botswana:																												
Francistown	21 13S	27 30E	3,294	20	88	65	83	56	75	41	90	61	107	24	28	4.2	3.1	2.8	0.7	0.2	0.1	*	*	*	0.9	2.3	3.4	17.7
Maun	19 59S	23 25E	3,091	20	90	66	87	58	77	42	95	64	110	24	20	4.3	3.8	3.5	1.1	0.2	*	0.0	0.0	*	0.5	1.9	2.8	18.2
Tsabong	26 03S	22 27E	3,156	10	94	65	83	51	71	34	88	54	107	15	14	2.0	1.9	1.9	1.3	0.4	0.4	0.1	*	0.2	0.7	1.1	1.5	11.5
Burkina Faso:																												
Bobo Dioulasso	11 10N	04 15W	1,411	11	92	58	99	71	87	69	90	70	115	46	10	0.1	0.2	1.1	2.1	4.6	4.8	9.8	12.0	8.5	2.5	0.7	0.0	46.4
Ouagadougou	12 22N	01 31W	991	10	92	60	103	79	91	74	95	74	118	48	15	*	0.1	0.5	0.6	3.3	4.8	8.0	10.9	5.7	1.3	*	0.0	35.2
Cameroon:																												
Ngaoundere	07 17N	13 19E	3,601	9	87	55	87	64	82	63	82	61	102	46	10	*	*	1.1	5.5	7.0	8.4	10.6	9.6	9.2	5.3	0.5	*	57.2
Yaounde	03 53N	11 32E	2,526	11	85	67	85	66	80	66	81	65	96	57	11	0.9	2.6	5.8	6.7	7.7	6.0	2.9	3.1	8.4	11.6	4.6	0.9	61.2
Central African Republic:																												
Bangui	04 22N	18 34E	1,270	5	90	68	91	71	85	69	87	69	101	57	5	1.0	1.7	5.0	5.3	7.4	4.5	8.9	8.1	5.9	7.9	4.9	0.2	60.8
Ndele	08 24N	20 39E	1,939	3	99	67	98	73	86	69	90	68	109	58	3	0.2	1.3	0.6	1.7	8.4	6.1	8.3	10.1	10.7	7.8	0.6	0.0	55.8
Chad:																												
Am Timan	11 02N	20 17E	1,430	3	98	56	105	68	89	70	96	67	113	43	3	0.0	0.0	0.1	1.2	4.3	5.0	7.3	12.3	5.8	1.2	0.0	0.0	37.2
Fort Lamy	12 07N	15 02E	968	5	93	57	107	74	92	72	97	70	114	47	5	0.0	0.0	0.0	0.1	1.2	2.6	6.7	12.6	4.7	1.4	0.0	0.0	29.3
Largeau (Faya)	18 00N	19 10E	837	5	84	54	104	69	109	76	103	72	121	37	5	0.0	0.0	0.0	0.0	*	0.0	*	0.7	*	0.0	0.0	0.0	0.7
Congo:																												
Brazzaville	04 15S	15 15E	1,043	15	88	69	91	71	82	63	89	70	98	54	18	6.3	4.9	7.4	7.0	4.3	0.6	*	*	2.2	5.4	11.5	8.4	58.0
Ouesso	01 37N	16 04E	1,132	4	88	69	91	71	85	69	87	69	106	60	4	2.4	3.6	6.4	3.2	5.8	4.6	2.9	3.7	7.9	10.0	5.7	2.4	58.6
Pointe Noire (Loango)	04 39S	11 48E	164	7	85	73	87	74	78	66	83	72	93	59	7	5.4	6.7	6.4	8.0	3.9	0.0	0.0	0.0	0.4	4.1	6.6	6.6	48.1

TABLE 1–10. TEMPERATURE AND PRECIPITATION DATA FOR REPRESENTATIVE WORLD-WIDE STATIONS (continued)

COUNTRY AND STATION	LATITUDE	LONGITUDE	ELEVATION	TEMPERATURE: Length of record	Average Daily JAN. Maximum	Average Daily JAN. Minimum	Average Daily APR. Maximum	Average Daily APR. Minimum	Average Daily JULY Maximum	Average Daily JULY Minimum	Average Daily OCT. Maximum	Average Daily OCT. Minimum	Extreme Maximum	Extreme Minimum	AVERAGE PRECIPITATION: Length of record	January	February	March	April	May	June	July	August	September	October	November	December	YEAR
	° ′	° ′	FEET	YEAR	°F	°F	°F	°F	°F	°F	°F	°F	°F	°F	YEAR	IN.	IN.	IN.	IN.	IN.	IN.	IN.	IN.	IN.	IN.	IN.	IN.	IN.
Djibouti:																												
Djibouti	11 36N	43 09E	23	16	84	73	90	79	106	87	92	80	117	63	46	0.4	0.5	1.0	0.5	0.2	*	0.1	0.3	0.3	0.4	0.9	0.5	46.0
Egypt:																												
Alexandria	31 12N	29 53E	105	45	65	51	74	59	85	73	83	68	111	37	61	1.9	0.9	0.4	0.1	*	*	*	*	*	0.2	1.3	2.2	7.0
Aswan	24 02N	32 53E	366	46	74	50	96	66	106	79	98	71	124	35	11	*	*	*	*	*	*	0.0	0.0	0.0	*	*	*	*
Cairo	29 52N	31 20E	381	42	65	47	83	57	96	70	86	65	117	34	42	0.2	0.2	0.2	0.1	0.1	*	0.0	0.0	*	*	0.1	0.2	1.1
Ethiopia:																												
Addis Ababa	09 20N	38 45E	8,038	15	75	43	77	50	69	50	75	45	94	32	37	0.5	1.5	2.6	3.4	3.4	5.4	11.0	11.8	7.5	0.8	0.6	0.2	48.7
Asmara	15 17N	38 55E	7,628	9	74	44	78	51	71	53	72	53	88	31	17	*	*	0.4	1.5	1.5	1.3	6.7	5.0	1.3	0.3	0.4	*	18.4
Diredawa	09 02N	41 45E	3,937	8	81	58	91	69	90	68	89	67	100	49	8	0.8	0.8	3.3	3.0	2.8	1.5	4.3	3.8	2.2	0.5	0.3	0.8	24.1
Gambela	08 15N	34 35E	1,345	26	98	64	98	71	87	69	92	67	111	48	30	0.2	0.4	1.4	3.2	5.9	6.7	8.5	9.5	7.3	3.5	1.8	0.4	48.8
Gabon:																												
Libreville	00 23N	09 26E	115	11	87	73	89	73	83	68	86	71	99	62	21	9.8	9.3	13.2	13.4	9.6	0.5	0.1	0.7	4.1	13.6	14.7	9.8	98.8
Mayoumba	03 25S	10 38E	200	8	84	73	86	73	78	68	82	72	91	60	8	6.5	9.3	6.2	10.2	2.3	0.1	0.0	0.2	2.6	9.3	10.7	4.6	62.0
Gambia:																												
Banjul	13 21N	16 40W	90	9	88	59	91	65	86	74	89	72	106	45	9	0.1	0.1	*	*	0.4	2.3	11.1	19.7	12.2	4.3	0.7	0.1	51.0
Ghana:																												
Accra	05 33N	00 12W	88	17	87	73	88	76	81	73	85	74	100	59	65	0.6	1.3	2.2	3.2	5.6	7.0	1.8	0.6	1.4	2.5	1.4	0.9	28.5
Kumasi	06 40N	01 37W	942	10	88	66	89	71	82	70	86	70	100	51	10	0.8	2.3	5.7	5.1	7.5	7.9	4.3	3.1	6.8	7.1	3.7	0.8	55.2
Guinea:																												
Conakry	09 31N	13 43W	23	7	88	72	90	73	83	72	87	73	96	63	10	0.1	0.1	0.4	0.9	6.2	22.0	51.1	41.5	26.9	14.6	4.8	0.4	169.0
Kouroussa	10 39N	09 53W	1,217	9	93	60	99	73	87	69	90	69	109	39	10	0.4	0.3	0.9	2.8	5.3	9.7	11.7	13.6	13.4	6.6	1.3	0.4	66.4
Guinea-Bissau:																												
Bolama	11 34N	15 26W	62	31	88	67	91	73	84	74	87	74	106	59	37	*	*	*	*	0.8	7.8	23.1	27.6	16.9	8.0	1.6	0.1	85.9
Ifni (now in Morocco):																												
Sidi Ifni	29 27N	10 11W	148	14	66	52	71	59	75	64	75	62	124	40	14	1.0	0.6	0.5	0.6	0.1	0.1	*	*	0.4	0.1	0.9	1.8	6.1
Ivory Coast:																												
Abidjan	05 19N	04 01W	65	13	88	73	90	75	83	73	85	74	96	59	10	1.6	2.1	3.9	4.9	14.2	19.5	8.4	2.1	2.8	6.6	7.9	3.1	77.1
Bouake	07 42N	05 00W	1,194	12	91	68	92	70	85	68	89	68	104	57	10	0.4	1.5	4.1	5.8	5.3	6.0	3.1	4.6	8.2	5.2	1.5	1.0	46.7
Kenya:																												
Mombasa	04 03S	39 39E	52	45	87	75	86	76	81	71	84	74	96	61	54	1.0	0.7	2.5	7.7	12.6	4.7	3.5	2.5	2.5	3.4	3.8	2.4	47.3
Nairobi	01 16S	36 48E	5,971	15	77	54	75	58	69	51	76	55	87	41	17	1.5	2.5	4.9	8.3	6.2	1.8	0.6	0.9	1.2	2.1	4.3	3.4	37.7
Liberia:																												
Monrovia	06 18N	10 48W	75	6	89	71	90	72	80	72	86	72	97	62	4	0.2	0.1	4.4	11.7	13.4	36.1	24.2	18.6	29.9	25.2	8.2	2.9	174.9
Libya:																												
Benghazi	32 06N	20 04E	82	46	63	50	74	58	84	71	80	66	109	37	46	2.6	1.6	0.8	0.2	0.1	*	*	*	0.1	0.7	1.8	2.6	10.5
Kufra	24 12N	23 21E	1,276	7	69	43	90	62	101	75	90	64	122	26	7	*	0.0	0.0	0.0	*	0.0	0.0	0.0	0.0	0.0	0.0	*	*
Sabhah	27 01N	14 26E	1,457	3	64	41	89	60	102	74	91	64	120	24	10	*	*	*	*	0.1	0.1	0.0	0.0	0.0	*	*	*	0.3
Tripoli	32 54N	13 11E	72	47	61	47	72	57	85	71	80	65	114	33	56	3.2	1.8	1.1	0.4	0.2	0.1	*	*	0.4	1.6	2.6	3.7	15.1
Malagasy Republic:																												
Diego Suarez	12 17S	49 17E	100	11	88	75	88	75	84	69	86	72	98	63	31	10.6	9.5	7.6	2.2	0.3	0.2	0.2	0.3	0.3	0.7	1.1	5.8	38.7

Tananarive	18 55S	47 33E	4,500	44	79	61	76	58	68	48	80	54	95	34	62	11.8	11.0	7.0	2.1	0.7	0.3	0.3	0.4	0.7	2.4	5.3	11.3	53.4
Tulear	23 20S	43 41E	20	27	92	72	89	64	81	58	86	65	108	43	15	3.1	3.2	1.4	0.3	0.7	0.4	0.1	0.2	0.3	0.7	1.4	1.7	13.5
Malawi:																												
Karonga	09 57S	33 56E	1,596	8	86	71	85	70	81	59	91	66	99	51	8	7.1	7.0	10.8	6.2	1.7	0.1	*	*	0.0	0.3	0.3	4.7	38.3
Zomba	15 23S	35 19E	3,141	27	80	65	78	62	72	53	85	64	95	41	29	12.1	9.9	10.1	2.7	0.7	0.4	0.3	0.3	0.2	1.0	4.3	10.9	52.9
Mali:																												
Araouane	18 54N	03 33W	935	8	81	48	110	67	111	79	103	70	130	37	10	*	*	0.0	0.0	0.0	0.2	0.2	0.5	0.6	0.1	0.1	*	1.7
Bamako	12 39N	07 58W	1,116	11	91	61	103	76	89	71	93	71	117	47	10	*	*	0.1	0.6	2.9	5.4	11.0	13.7	8.1	1.7	0.6	*	44.1
Gao	16 16N	00 03W	902	15	83	58	105	77	97	80	100	78	116	44	19	*	0.0	*	0.1	0.4	1.0	2.9	5.4	1.5	0.2	*	0.0	11.5
Mauritania:																												
Atar	20 31N	13 04W	761	7	84	54	97	67	106	81	98	72	117	39	10	*	0.0	*	*	*	0.1	0.3	1.2	1.1	0.1	*	*	2.8
Nema	16 36N	07 16W	883	9	86	62	105	79	99	78	101	79	120	47	10	0.1	*	*	*	0.7	1.1	2.3	4.7	2.1	0.7	*	0.1	11.6
Nouakchott	18 07N	15 36W	69	5	85	57	90	64	89	74	91	71	115	44	10	*	0.1	*	*	*	0.1	0.5	4.1	0.9	0.4	0.1	*	6.2
Morocco:																												
Casablanca	33 35N	07 39W	164	48	63	45	69	52	79	65	76	58	110	31	40	2.1	1.9	2.2	1.4	0.9	0.2	0.0	*	0.3	1.5	2.6	2.8	15.9
Marrakech	31 36N	08 01W	1,509	35	65	40	79	52	101	67	83	57	120	27	31	1.0	1.1	1.3	1.2	0.6	0.3	0.1	0.1	0.4	0.9	1.2	1.2	9.4
Rabat	34 00N	06 50W	213	35	63	46	71	52	82	63	77	58	118	32	29	2.6	2.5	2.6	1.7	1.1	0.3	*	*	0.4	1.9	3.3	3.4	19.8
Tangier	35 48N	05 49W	239	35	60	47	65	51	80	64	72	59	106	28	35	4.5	4.2	4.8	3.5	1.7	0.6	*	*	0.9	3.9	5.8	5.4	35.3
Mozambique:																												
Beira	19 50S	34 51E	28	37	89	75	86	71	77	61	87	71	109	48	39	10.9	8.4	10.1	4.2	2.2	1.3	1.2	1.1	0.8	5.2	5.3	9.2	59.9
Chicoa	15 36S	32 21E	899	8	96	65	93	63	86	55	101	68	117	32	8	7.8	5.7	4.4	0.6	*	*	*	*	*	1.1	2.6	5.2	27.4
Maputo	25 58S	32 36E	194	42	86	71	83	66	76	55	82	64	114	45	42	5.1	4.9	4.9	2.1	1.1	0.8	0.5	0.5	1.1	1.9	3.2	3.8	29.9
Namibia:																												
Keetmanshoop	26 35S	18 08E	3,295	17	95	65	85	57	70	42	87	55	108	26	45	0.8	1.1	1.4	0.6	0.2	*	*	*	0.1	0.2	0.3	0.4	5.2
Windhoek	22 34S	17 06E	5,669	30	85	63	77	55	68	43	84	59	97	25	60	3.0	2.9	3.1	1.6	0.3	*	*	*	0.1	0.4	0.9	1.9	14.3
Niger:																												
Agades	16 59N	07 59E	1,706	8	86	50	105	70	104	75	101	68	115	40	10	0.0	0.0	*	*	0.2	0.3	1.9	3.7	0.7	0.0	0.0	0.0	6.8
Bilma	18 41N	12 55E	1,171	9	81	45	101	63	108	75	101	62	116	29	10	0.0	0.0	0.0	*	*	0.0	0.1	0.5	0.3	0.0	0.0	0.0	0.9
Niamey	13 31N	02 06E	709	10	93	58	108	77	94	74	101	74	114	47	10	*	*	0.2	0.3	1.3	3.2	5.2	7.4	3.7	0.5	*	0.0	21.6
Nigeria:																												
Enugu	06 27N	07 29E	763	11	90	72	91	74	83	71	87	71	99	55	33	0.7	1.1	2.6	5.9	10.4	11.4	7.6	6.7	12.8	9.8	2.1	0.5	71.5
Kaduna	10 35N	06 26E	2,113	18	89	59	95	72	83	68	89	66	105	46	34	*	0.1	0.5	2.5	5.9	7.1	8.5	11.9	10.6	2.9	0.1	*	50.1
Lagos	06 27N	03 24E	10	32	88	74	89	77	83	74	85	74	104	60	47	1.1	1.8	4.0	5.9	10.6	18.1	11.0	2.5	5.5	8.1	2.7	1.0	72.3
Maiduguri	11 51N	13 05E	1,162	15	90	54	104	72	90	73	96	68	112	43	40	*	*	*	0.3	1.6	2.7	7.1	8.7	4.2	0.7	*	0.0	25.3
Senegal:																												
Dakar	14 42N	17 29W	131	25	79	64	81	65	88	76	89	76	109	53	26	*	*	*	*	*	0.7	3.5	10.0	5.2	1.5	0.1	0.3	21.3
Kaolack	14 08N	16 04W	20	9	93	60	103	68	91	75	93	74	114	48	10	*	0.0	*	*	0.3	2.6	6.9	10.7	7.0	2.7	0.1	*	30.3
Sierra Leone:																												
Freetown/Lungi	08 37N	13 12W	92	8	87	73	88	76	82	73	85	72	98	62	8	0.4	0.2	1.2	3.1	9.5	14.3	29.2	36.5	22.3	14.2	5.5	1.2	137.6
Somalia:																												
Berbera	10 26N	45 02E	45	30	84	68	89	77	107	88	92	76	117	58	30	0.3	0.1	0.2	0.5	0.3	*	*	0.1	*	0.1	0.2	0.2	2.0
Mogadiscio	02 02N	45 21E	39	13	86	73	90	78	83	73	86	76	97	59	21	*	*	*	2.3	2.3	3.8	2.5	1.9	1.0	0.9	1.6	0.5	16.9
South Africa, Republic of:																												
Cape Town	33 54S	18 32E	56	19	78	60	72	53	63	45	70	52	103	28	18	0.6	0.3	0.7	1.9	3.1	3.3	3.5	2.6	1.7	1.2	0.7	0.4	20.0
Durban	29 50S	31 02E	16	15	81	69	78	64	72	52	75	62	107	39	78	4.3	4.8	5.1	3.0	2.0	1.3	1.1	1.5	2.8	4.3	4.8	4.7	39.7
Kimberley	28 48S	24 46E	3,927	19	91	64	77	52	65	36	83	54	103	20	57	2.4	2.5	3.1	1.5	0.7	0.2	0.2	0.3	0.6	1.0	1.6	2.0	16.1
Port Elizabeth	33 59S	25 36E	190	14	78	61	73	55	67	45	70	54	104	31	84	1.2	1.3	1.9	1.8	2.4	1.8	1.9	2.0	2.3	2.2	2.2	1.7	22.7
Port Nolloth	29 14S	16 52E	23	20	67	53	66	50	62	45	64	49	107	31	64	0.1	0.1	0.2	0.2	0.3	0.3	0.3	0.3	0.2	0.1	0.1	0.1	2.3
Pretoria	25 45S	28 14E	4,491	13	81	60	75	50	66	37	80	55	96	24	12	5.0	4.3	4.5	1.7	0.9	0.6	0.3	0.2	0.8	2.2	5.2	5.2	30.9
Walvis Bay	22 56S	14 30E	24	20	73	59	75	55	70	47	67	51	104	25	20	*	0.2	0.3	0.1	0.1	*	*	0.1	*	*	*	*	0.9

TABLE 1–10. TEMPERATURE AND PRECIPITATION DATA FOR REPRESENTATIVE WORLD-WIDE STATIONS (continued)

COUNTRY AND STATION	LATI-TUDE	LONGI-TUDE	ELE-VATION	TEMPERATURE: Length of record	Average Daily JAN. Maximum	JAN. Minimum	APR. Maximum	APR. Minimum	JULY Maximum	JULY Minimum	OCT. Maximum	OCT. Minimum	Extreme Maximum	Extreme Minimum	AVERAGE PRECIPITATION: Length of record	January	February	March	April	May	June	July	August	September	October	November	December	YEAR
	° ′	° ′	FEET	YEAR	°F	°F	°F	°F	°F	°F	°F	°F	°F	°F	YEAR	IN.	IN.	IN.	IN.	IN.	IN.	IN.	IN.	IN.	IN.	IN.	IN.	IN.
Sudan:																												
El Fasher	13 38N	25 21E	2,395	17	88	50	102	64	96	70	99	64	113	33	17	*	0.0	*	*	0.3	0.7	4.5	5.3	1.2	0.2	0.0	0.0	12.2
Khartoum	15 37N	32 33E	1,279	46	90	59	105	72	101	77	104	75	118	41	46	*	*	*	*	0.1	0.3	2.1	2.8	0.7	0.2	*	0.0	6.2
Port Sudan	19 37N	37 13E	18	30	81	68	89	71	106	83	93	76	117	50	40	0.2	0.1	*	*	*	*	0.3	0.1	*	0.4	1.7	0.9	3.7
Wadi Halfa	21 55N	31 20E	410	39	75	46	98	62	106	74	98	67	127	28	39	*	*	*	*	*	0.0	*	*	*	*	*	0.0	*
Wau	07 42N	28 03E	1,443	38	96	64	99	72	89	69	93	69	115	50	38	*	0.2	0.9	2.6	5.3	6.5	7.5	8.2	6.6	4.9	0.6	*	43.3
Tanzania:																												
Dares Salaam	06 50S	39 18E	47	44	83	77	86	73	83	66	85	69	96	59	49	2.6	2.6	5.1	11.4	7.4	1.3	1.2	1.0	1.2	1.6	2.9	3.6	41.9
Iringa	07 47S	35 42E	5,330	14	76	59	75	59	72	52	80	57	90	42	24	6.8	5.1	7.1	3.5	0.5	*	*	*	0.1	0.2	1.5	4.5	29.3
Kigoma	04 53S	29 38E	2,903	26	80	67	81	67	83	63	84	69	100	53	18	4.8	5.0	5.9	5.1	1.7	0.2	0.1	0.2	0.7	1.9	5.6	5.3	36.5
Togo:																												
Lome	06 10N	01 15E	72	5	85	72	86	74	80	71	83	72	94	58	15	0.6	0.9	1.9	4.6	5.7	8.8	2.8	0.4	1.4	2.4	1.1	0.4	31.0
Tunisia:																												
Gabes	33 53N	10 07E	7	50	61	43	74	54	89	71	81	62	122	27	50	0.9	0.7	0.8	0.4	0.3	*	*	0.1	0.5	1.2	1.2	0.6	6.7
Tunis	36 47N	10 12E	217	50	58	43	70	51	90	68	77	59	118	30	50	2.5	2.0	1.6	1.4	0.7	0.3	0.1	0.3	1.3	2.0	1.9	2.4	16.5
Uganda																												
Kampala	00 20N	32 36E	4,304	15	83	65	79	64	77	62	81	63	97	53	15	1.8	2.4	5.1	6.9	5.8	2.9	1.8	3.4	3.6	3.8	4.8	3.9	46.2
Lira	02 15N	32 54E	3,560	14	91	61	86	64	81	61	86	61	100	50	14	0.7	1.0	3.5	6.9	7.9	4.9	6.4	10.0	8.3	6.1	3.2	1.8	60.7
Western Sahara:																												
Semara	26 46N	11 31W	1,509	6	73	47	88	58	99	66	88	61	121	37	6	0.1	*	0.0	*	*	0.0	0.0	*	1.0	*	0.4	0.0	1.5
Villa Cisneros	23 42N	15 52W	35	12	71	56	74	60	78	65	80	65	107	48	14	*	*	*	*	0.1	0.0	*	0.2	1.4	0.1	0.2	1.0	3.0
Zaire:																												
Kalemie	05 54S	29 12E	2,493	5	85	66	83	67	82	58	87	67	92	50	20	4.2	4.7	6.3	8.4	3.3	0.3	0.1	0.3	0.8	2.8	7.9	6.3	45.4
Kinshasa	04 20S	15 18E	1,066	8	87	70	89	71	81	64	88	70	97	58	12	5.3	5.7	7.7	7.7	6.2	0.3	0.1	0.1	1.2	4.7	8.7	5.6	53.3
Kisangani	00 26N	25 14E	1,370	8	88	69	88	70	84	67	86	68	97	61	14	2.1	3.3	7.0	6.2	5.4	4.5	5.2	6.5	7.2	8.6	7.8	3.3	67.1
Luluabourg	05 54S	22 25E	2,198	3	85	68	86	68	85	63	85	68	94	57	14	5.4	5.6	7.7	7.6	3.3	0.8	0.5	2.3	4.6	6.5	9.1	8.9	62.3
Zambia:																												
Balovale	13 34S	23 06E	3,577	8	82	65	84	61	81	47	91	64	108	38	9	8.5	6.9	5.8	1.2	*	0.0	0.0	*	0.3	2.3	4.4	8.9	38.3
Kasama	10 12S	31 11E	4,544	10	79	61	79	60	76	50	87	62	95	39	10	10.7	9.9	10.9	2.8	0.5	*	*	*	*	0.8	6.4	9.5	51.5
Lusaka	15 25S	28 19E	4,191	10	78	63	79	59	73	49	88	64	100	39	10	9.1	7.5	5.6	0.7	0.1	*	*	0.0	*	0.4	3.6	5.9	32.9
Zimbabwe:																												
Bulawayo	20 09S	28 37E	4,405	15	81	61	79	56	70	45	85	59	99	28	50	5.6	4.3	3.3	0.7	0.4	0.1	*	*	0.2	0.8	3.2	4.8	23.4
Salisbury	17 50S	31 08E	4,831	15	78	60	78	55	70	44	83	58	95	32	50	7.7	7.0	4.6	1.1	0.5	0.1	*	0.1	0.2	1.1	3.8	6.4	32.6
ATLANTIC ISLANDS:																												
Funchal, Madeira Island	32 38N	16 55W	82	30	66	56	67	58	75	66	74	65	103	40	30	2.5	2.9	3.1	1.3	0.7	0.2	*	*	1.0	3.0	3.5	3.3	21.5
Georgetown, Ascension Island	07 56S	14 25W	55	29	85	73	88	75	84	72	83	71	95	65	45	0.2	0.4	0.7	1.1	0.5	0.5	0.5	0.4	0.3	0.3	0.2	0.1	5.2
Hutts Gate, St. Helena	15 57S	05 40W	2,062	30	68	60	69	61	62	55	61	54	82	50	30	2.1	3.1	4.2	3.1	2.8	3.2	4.3	2.6	2.2	1.7	1.2	1.6	32.1
Las Palmas, Canary Islands	28 11N	15 28W	20	45	70	58	71	61	77	67	79	67	99	46	48	1.4	0.9	0.9	0.5	0.2	*	*	*	0.2	1.1	2.1	1.6	8.6

Porto da Praia, Cape Verde Is.	14 54N	23 31W	112	25	77	68	79	69	83	75	85	76	94	56	25	0.1	*	*	*	0.0	*	0.2	3.8	4.5	1.2	0.3	0.1	10.2
Santa Isabel, Fernando Po	03 46N	08 46E	—	2	87	67	89	70	84	69	86	70	102	61	16	1.3	2.5	4.2	7.2	9.4	11.1	7.4	6.6	9.6	10.4	3.5	1.7	74.9
Sao Tome, Sao Tome	00 20N	06 43E	16	10	86	73	86	73	82	69	84	71	91	56	10	3.2	4.2	5.9	5.0	5.3	1.1	*	*	0.9	4.3	4.6	3.5	38.0
Tristan da Cunha	37 03S	12 19W	75	5	66	59	64	57	57	50	59	51	75	38	5	3.5	3.5	6.4	4.7	7.1	5.9	6.1	6.9	7.9	5.8	4.3	4.0	66.1
INDIAN OCEAN ISLANDS:																												
Agalega Island	10 26S	56 40E	10	3	86	77	87	77	83	75	84	75	91	69	2	5.9	10.1	4.9	6.9	13.2	8.9	8.7	3.2	1.8	4.2	7.0	10.0	84.7
Cocos (Keeling) Island	12 05S	96 53E	15	36	86	77	85	78	82	76	84	76	94	68	38	5.4	7.7	8.5	10.4	7.9	9.0	8.7	4.8	3.7	3.3	4.2	4.6	78.2
Heard Island	53 01S	73 23E	16	5	41	35	39	33	34	27	35	28	58	13	5	5.8	5.8	5.7	6.1	5.8	3.9	3.6	2.2	2.5	3.7	4.0	5.1	54.3
Hellburg, Reunion Island	21 04S	55 22E	3,070	5	74	59	73	56	65	48	69	51	84	40	11	22.4	8.0	16.4	7.2	5.3	4.4	3.1	3.0	2.0	2.3	3.5	12.9	90.5
Port Victoria, Seychelles	04 37S	55 27E	15	60	83	76	86	77	81	75	83	75	92	67	64	15.2	10.5	9.2	7.2	6.7	4.0	3.3	2.7	5.1	6.1	9.1	13.4	92.5
Royal Alfred Observatory, Mauritius	20 06S	57 32E	181	40	86	73	82	70	75	62	80	64	95	50	43	8.5	7.8	8.7	5.0	3.8	2.6	2.3	2.5	1.4	1.6	1.8	4.6	50.6
ASIA–FAR EAST																												
China																												
Canton	23 10N	113 20E	59	26	65	49	77	65	91	77	85	67	101	31	36	0.9	1.9	4.2	6.8	10.6	10.6	8.1	8.5	6.5	3.4	1.2	0.9	63.6
Chanasha	28 15N	112 58E	161	14	45	35	70	56	94	78	75	59	109	16	26	1.9	3.7	5.3	5.7	8.2	8.7	4.4	4.3	2.7	3.0	2.7	1.5	52.1
Chungking	29 30N	106 33E	855	27	51	42	73	59	93	76	71	61	111	28	60	0.7	0.8	1.5	3.8	5.7	7.1	5.6	4.7	5.8	4.3	1.9	0.8	42.9
Hankow	30 35N	114 17E	75	29	46	34	69	55	93	78	74	60	108	9	55	1.8	1.9	3.6	5.8	7.0	9.0	7.0	4.1	3.0	3.1	1.9	1.2	49.4
Harbin	45 45N	126 38E	476	35	7	-14	54	31	84	65	54	31	102	-43	38	0.2	0.2	0.4	0.9	1.7	3.7	6.6	4.7	2.3	1.2	0.5	0.2	22.6
Kashgar	39 24N	76 07E	4,296	27	33	12	71	48	92	68	71	43	106	-15	18	0.6	0.1	0.5	0.2	0.3	0.2	0.4	0.3	0.1	0.1	0.2	0.3	3.2
Kunming	25 02N	102 43E	6,211	32	61	37	76	51	77	62	70	53	91	22	31	0.4	0.5	0.7	0.8	4.3	6.3	8.8	8.6	5.0	3.0	1.7	0.4	40.5
Lanchow	36 06N	103 55E	5,105	8	33	7	65	40	84	61	62	39	100	-3	4	0.2	0.2	0.2	0.5	0.8	0.7	3.3	5.1	2.2	0.6	0.0	0.3	14.1
Mukden	41 47N	123 24E	138	40	20	-2	60	36	87	69	62	39	103	-28	42	0.2	0.2	0.7	1.2	2.6	3.8	7.0	6.3	2.9	1.7	0.9	0.4	28.2
Shanghai	31 12N	121 26E	16	56	47	32	67	49	91	75	75	56	104	10	81	1.9	2.4	3.3	3.6	3.8	7.0	5.8	5.5	5.2	2.9	2.1	1.5	45.0
Tientsin	39 10N	117 10E	13	24	33	16	68	45	90	73	68	48	109	-3	25	0.2	0.1	0.4	0.5	1.1	2.4	7.6	6.0	1.7	0.6	0.4	0.2	21.0
Urumchi	43 45N	87 40E	2,972	6	13	-7	60	36	82	58	50	31	112	-30	6	0.6	0.3	0.5	1.5	1.1	1.5	0.7	1.0	0.6	1.7	1.6	0.4	11.5
Hong Kong:	22 18N	114 10E	109	50	64	56	75	67	87	78	81	73	97	32	50	1.3	1.8	2.9	5.4	11.5	15.5	15.0	14.2	10.1	4.5	1.7	1.2	85.1
Japan:																												
Kushiro	43 02N	144 12E	315	41	30	8	44	31	66	55	58	40	87	-19	41	1.8	1.4	2.8	3.6	3.8	4.1	4.4	4.9	6.6	4.0	3.1	2.0	42.9
Miyako	39 38N	141 59E	98	30	43	23	58	37	77	62	66	46	99	1	30	2.9	3.0	3.2	3.5	4.5	5.0	5.0	7.2	9.5	6.8	3.0	2.6	56.2
Nagasaki	32 44N	129 53E	436	59	49	36	66	50	85	73	72	58	98	22	59	2.8	3.3	4.9	7.3	6.7	12.3	10.1	6.9	9.8	4.5	3.7	3.2	75.5
Osaka	34 47N	135 26E	49	60	47	32	65	47	87	73	72	55	102	19	60	1.7	2.3	3.8	5.2	4.9	7.4	5.9	4.4	7.0	5.1	3.0	1.9	52.6
Tokyo	35 41N	139 46E	19	60	47	29	63	46	83	70	69	55	101	17	60	1.9	2.9	4.2	5.3	5.8	6.5	5.6	6.0	9.2	8.2	3.8	2.2	61.6
Korea:																												
Pusan	35 10N	129 07E	6	36	43	29	62	47	81	71	70	54	97	7	36	1.7	1.4	2.7	5.5	5.2	7.9	11.6	5.1	6.8	2.9	1.6	1.2	53.6
Pyongyang	39 01N	125 49E	94	43	27	8	61	38	84	69	65	43	100	-19	43	0.6	0.4	1.0	1.8	2.6	3.0	9.3	9.0	4.4	1.8	1.6	0.8	36.4
Seoul	37 31N	126 55E	34	22	32	15	62	41	84	70	67	45	99	-12	22	1.2	0.8	1.5	3.0	3.2	5.1	14.8	10.5	4.7	1.6	1.8	1.0	49.2
Mongolia:																												
Ulan Bator	47 54N	106 56E	4,287	13	-2	-27	45	18	71	50	44	17	97	-48	15	*	*	0.1	0.2	0.3	1.0	2.9	1.9	0.8	0.2	0.2	0.1	7.7
Taiwan:																												
Tainan	22 57N	120 12E	53	13	72	55	82	67	89	77	86	70	95	39	13	0.7	0.7	1.1	3.2	6.3	15.6	16.0	15.8	8.4	1.2	0.9	0.6	70.5
Taipei	25 04N	121 32E	21	12	66	53	77	64	92	76	80	68	101	32	12	3.8	5.3	4.3	5.3	6.9	8.8	8.8	8.7	8.2	5.5	4.2	2.9	72.7
Union of Soviet Socialist Republics:																												
Alma-Ata	43 16N	76 53E	2,543	19	23	7	56	38	81	60	55	35	100	-30	27	1.3	0.9	2.2	4.0	3.7	2.6	1.4	1.2	1.0	2.0	1.9	1.3	23.5
Chita	52 02N	113 30E	2,218	10	-10	-27	42	19	75	51	38	18	99	-52	24	0.1	0.1	0.1	0.4	1.1	1.8	3.3	3.3	1.2	0.5	0.2	0.2	12.3
Dubinka	69 07N	87 00E	141	5	-23	-31	6	-10	59	47	19	11	84	-62	5	0.3	0.4	0.2	0.3	0.6	1.9	1.5	2.1	1.8	0.9	0.4	0.3	10.7

TABLE 1–10. TEMPERATURE AND PRECIPITATION DATA FOR REPRESENTATIVE WORLD-WIDE STATIONS (continued)

COUNTRY AND STATION	LATITUDE	LONGITUDE	ELEVATION	TEMPERATURE: Length of record	Average Daily JAN. Maximum	Average Daily JAN. Minimum	Average Daily APR. Maximum	Average Daily APR. Minimum	Average Daily JULY Maximum	Average Daily JULY Minimum	Average Daily OCT. Maximum	Average Daily OCT. Minimum	Extreme Maximum	Extreme Minimum	AVERAGE PRECIPITATION: Length of record	January	February	March	April	May	June	July	August	September	October	November	December	YEAR
	° ′	° ′	FEET	YEAR	°F	°F	°F	°F	°F	°F	°F	°F	°F	°F	YEAR	IN.	IN.	IN.	IN.	IN.	IN.	IN.	IN.	IN.	IN.	IN.	IN.	IN.
Irkutsk	52 16N	104 19E	1,532	10	3	-15	42	20	70	50	41	21	98	-58	38	0.5	0.4	0.3	0.6	1.3	2.2	3.1	2.8	1.7	0.7	0.6	0.6	14.9
Kazalinsk	45 46N	62 06E	207	10	16	5	58	27	90	65	57	35	108	-27	19	0.4	0.4	0.5	0.5	0.6	0.2	0.2	0.3	0.3	0.4	0.5	0.6	4.9
Khabarovsk	48 28N	135 03E	165	7	-2	-13	41	28	75	63	48	34	91	-46	8	0.3	0.2	0.3	0.7	2.0	3.5	4.1	3.3	3.0	0.7	0.6	0.5	19.2
Kirensk	57 47N	108 07E	938	18	-14	-28	38	15	74	51	10	-4	95	-71	19	0.8	0.5	0.5	0.5	1.0	1.8	2.1	2.1	1.7	1.0	1.0	1.0	14.0
Krasnoyarsk	56 01N	92 52E	498	10	3	-10	34	23	67	55	34	26	103	-47	8	0.1	0.2	0.1	0.2	1.0	1.4	1.2	2.1	1.7	0.9	0.5	0.4	9.8
Markovo	64 45N	170 50E	85	15	-19	-29	5	-8	59	47	16	9	84	-72	16	0.2	0.2	0.3	0.1	0.3	0.8	1.0	1.9	1.1	0.4	0.4	0.3	7.0
Narym	58 50N	81 39E	197	13	-7	-18	35	19	71	56	35	25	94	-61	14	0.8	0.5	0.8	0.5	1.3	2.6	2.4	2.7	1.7	1.4	1.1	0.9	16.8
Okhotsk	59 21N	143 17E	18	19	-6	-17	29	10	57	48	33	21	78	-50	25	0.1	0.1	0.2	0.4	0.9	1.6	2.2	2.6	2.4	1.0	0.2	0.1	11.8
Omsk	54 58N	73 20E	279	19	-1	-14	39	21	74	56	40	27	102	-56	22	0.6	0.3	0.3	0.5	1.2	2.0	2.0	2.0	1.1	1.0	0.7	0.8	12.5
Petropavlovsk	52 53N	158 42E	286	7	23	11	35	25	56	47	46	34	84	-29	35	3.0	2.2	3.4	2.5	2.2	2.0	3.1	3.2	3.8	3.9	3.6	3.0	35.9
Salehkard	66 31N	66 35E	60	18	-13	-21	18	4	61	49	26	20	85	-65	27	0.3	0.3	0.3	0.3	0.7	1.3	1.9	2.0	1.5	0.7	0.5	0.4	10.2
Semipalatinsk	50 24N	80 13E	709	10	8	-7	45	26	81	57	46	30	101	-47	10	0.9	0.5	0.5	0.6	1.2	1.5	1.1	1.3	0.7	1.2	1.1	1.0	11.6
Sverdlovsk	56 49N	60 38E	894	21	6	-5	42	26	70	54	37	28	94	-45	29	0.5	0.4	0.5	0.7	1.9	2.7	2.6	2.7	1.6	1.2	1.1	0.8	16.7
Tashkent	41 20N	69 18E	1,569	19	37	21	65	47	92	64	65	41	106	-19	19	2.1	1.1	2.6	2.3	1.4	0.5	0.2	0.1	0.1	1.2	1.5	1.6	14.7
Verkhoyansk	67 34N	133 51E	328	24	-54	-63	19	-10	66	47	12	-3	98	-90	44	0.2	0.2	0.1	0.2	0.3	0.9	1.1	1.0	0.5	0.3	0.3	0.2	5.3
Vladivostok	43 07N	131 55E	94	14	13	0	46	34	71	60	55	41	92	-22	53	0.3	0.4	0.7	1.2	2.1	2.9	3.3	4.7	4.3	1.9	1.2	0.6	23.6
Yakutsk	62 01N	129 43E	535	19	-45	-53	27	6	73	54	23	11	97	-84	22	0.3	0.2	0.1	0.3	0.4	1.1	1.6	1.3	1.1	0.5	0.4	0.3	7.4
ASIA–SOUTHEAST																												
Brunei:																												
Brunei	04 55N	114 55E	10	5	85	76	87	77	87	76	86	77	99	70	12	14.6	7.6	7.8	9.8	10.9	9.5	9.0	7.3	11.8	14.5	15.2	13.0	131.0
Burma:																												
Mandalay	21 59N	96 06E	252	20	82	55	101	77	93	78	89	73	111	44	20	0.1	0.1	0.2	1.2	5.8	6.3	2.7	4.1	5.4	4.3	2.0	0.4	32.6
Moulmein	16 26N	97 39E	150	43	89	65	95	77	83	74	88	75	103	52	60	0.2	0.2	0.5	3.0	19.9	37.1	47.5	44.2	27.1	8.5	1.7	0.3	190.2
Cambodia:																												
Phnom Penh	11 33N	104 51E	39	37	88	71	95	76	90	76	87	76	105	55	49	0.3	0.4	1.4	3.1	5.7	5.8	6.0	6.1	8.9	9.9	5.5	1.7	54.8
Indonesia:																												
Jakarta	06 11S	106 50E	26	80	84	74	87	75	87	73	87	74	98	66	78	11.8	11.8	8.3	5.8	4.5	3.8	2.5	1.7	2.6	4.4	5.6	8.0	70.8
Manokwari	00 53S	134 03E	10	5	86	73	86	74	86	74	87	74	93	68	40	12.0	9.4	13.2	11.1	7.8	7.2	5.4	5.6	4.9	4.7	6.5	10.3	98.1
Mapanget	01 32N	124 55E	264	21	85	73	86	73	87	73	89	72	97	65	63	18.6	13.8	12.2	8.0	6.4	6.5	4.8	4.0	3.3	4.9	8.9	14.7	106.1
Penfui	10 10S	123 39E	335	21	87	75	89	72	88	70	92	72	101	58	63	15.2	13.7	9.2	2.6	1.2	0.4	0.2	0.0	0.0	0.7	3.3	9.1	55.7
Pontianak	00 00N	109 20E	13	20	87	74	89	75	89	74	89	75	96	68	63	10.8	8.2	9.5	10.9	11.1	8.7	6.5	8.0	9.0	14.4	15.3	12.7	125.1
Tabing	00 52S	100 21E	19	21	87	74	87	75	87	74	86	74	94	68	63	13.9	10.1	12.2	14.5	12.8	11.7	10.5	13.7	16.2	20.1	20.5	19.2	175.4
Tarakan	03 19N	117 33E	20	19	85	73	86	75	87	74	87	74	94	67	31	10.9	10.2	14.0	13.9	13.5	12.6	10.3	12.4	11.6	14.3	15.2	13.4	152.3
Laos:																												
Vientiane	17 58N	102 34E	559	13	83	58	95	73	89	75	88	71	108	32	27	0.2	0.6	1.5	3.9	10.5	11.9	10.5	11.5	11.9	4.3	0.6	0.1	67.5
Malaysia:																												
Kuala Lumpur	03 06N	101 42E	111	19	90	72	91	74	90	72	89	73	99	64	19	6.2	7.9	10.2	11.5	8.8	5.1	3.9	6.4	8.6	9.8	10.2	7.5	96.1

North Borneo:																										
Sandakan	05 54N	118 03E	38	45	85	74	89	76	89	75	88	75	99	70	46	19.0	10.9	8.6	4.5	6.2	7.4	6.7	7.9	9.3	10.2	14.5	18.5	123.7
Philippine Islands:																										
Davao	07 07N	125 38E	88	15	87	72	91	73	88	73	89	73	97	65	34	4.8	4.5	5.2	5.8	9.2	9.1	6.5	6.5	6.7	7.9	5.3	6.1	77.6
Manila	14 31N	121 00E	49	61	86	69	93	73	88	75	88	74	101	58	75	0.9	0.5	0.7	1.3	5.1	10.0	17.0	16.6	14.0	7.6	5.7	2.6	82.0
Sarawak:																										
Kuching	01 29N	110 20E	85	5	85	72	90	73	90	72	89	73	98	64	19	24.0	20.1	12.9	11.0	10.3	7.1	7.7	9.2	8.6	10.5	14.1	18.2	153.7
Singapore:																										
Singapore	01 18N	103 50E	33	39	86	73	88	75	88	75	87	74	97	66	64	9.9	6.8	7.6	7.4	6.8	6.8	6.7	7.7	7.0	8.2	10.0	10.1	95.0
Thailand:																										
Bangkok	13 44N	100 30E	53	10	89	67	95	78	90	76	88	76	104	50	10	0.2	1.1	1.1	2.3	5.2	6.0	6.9	9.2	14.0	9.9	1.8	0.1	57.8
Viet Nam																										
Hanoi	21 03N	105 52E	20	12	68	58	80	70	92	79	84	72	108	41	12	0.8	1.2	2.5	3.6	4.1	11.2	11.9	15.2	10.0	3.5	2.6	2.8	69.4
Saigon	10 49N	106 39E	33	31	89	70	95	76	88	75	88	74	104	57	33	0.6	0.1	0.5	1.7	8.7	13.0	12.4	10.6	13.2	10.6	4.5	2.2	78.1
ASIA–MIDDLE EAST																										
Aden:																										
Riyan	14 39N	49 19E	83	13	82	67	88	74	92	77	88	72	111	57	13	0.3	0.1	0.6	0.2	*	0.1	0.1	0.1	*	*	0.7	0.3	2.5
Afghanistan:																										
Kabul	34 30N	69 13E	5,955	9	36	18	66	43	92	61	73	42	104	-6	45	1.3	1.5	3.6	3.3	0.9	0.2	0.1	0.1	*	0.4	0.6	0.6	12.6
Kandhar	31 36N	65 40E	3,462	7	56	31	83	50	102	66	85	44	111	14	7	3.1	1.7	0.8	0.3	0.2	*	0.1	*	0.0	*	*	0.8	7.0
Bangladesh:																										
Dacca	23 46N	90 23E	24	60	77	56	92	74	88	79	88	75	108	43	61	0.3	1.2	2.4	5.4	9.6	12.4	13.0	13.3	9.8	5.3	1.0	0.2	73.9
India:																										
Ahmadabad	23 03N	72 37E	180	45	85	58	104	75	93	79	97	73	118	36	45	*	0.1	0.1	*	0.4	3.7	12.2	8.1	4.2	0.4	0.1	*	29.3
Bangalore	12 57N	77 40E	2,937	60	80	57	93	69	81	66	82	65	102	46	60	0.2	0.3	0.4	1.6	4.2	2.9	3.9	5.0	6.7	5.9	2.7	0.4	34.2
Bombay	19 06N	72 51E	27	60	88	62	93	74	88	75	93	73	110	46	60	0.1	0.1	0.1	*	0.7	19.1	24.3	13.4	10.4	2.5	0.5	0.1	71.2
Calcutta	22 32N	88 20E	21	60	80	55	97	76	90	79	89	74	111	44	60	0.4	1.2	1.4	1.7	5.5	11.7	12.8	12.9	9.9	4.5	0.8	0.2	63.0
Cherrapunji	25 15N	91 44E	4,309	35	60	46	71	59	72	65	72	61	87	33	35	0.7	2.1	7.3	26.2	50.4	106.1	96.3	70.1	43.3	19.4	2.7	0.5	425.1
Hyderabad	17 27N	78 28E	1,741	50	85	59	101	75	87	73	88	68	112	43	45	0.3	0.4	0.5	1.2	1.1	4.4	6.0	5.3	6.5	2.5	1.1	0.3	29.6
Jalpaiguri	26 32N	88 43E	272	50	74	50	90	68	89	77	87	70	104	36	55	0.3	0.7	1.3	3.7	11.8	25.9	32.2	25.3	21.2	5.6	0.5	0.2	128.7
Lucknow	26 45N	80 52E	400	60	74	47	101	71	92	80	91	67	119	34	60	0.8	0.7	0.3	0.3	0.8	4.5	12.0	11.5	7.4	1.3	0.2	0.3	40.1
Madras	13 04N	80 15E	51	60	85	67	95	78	96	79	90	75	113	57	60	1.4	0.4	0.3	0.6	1.0	1.9	3.6	4.6	4.7	12.0	14.0	5.5	50.0
Mormugao	15 22N	73 49E	157	10	86	70	88	79	83	75	86	75	98	59	30	*	*	*	0.7	2.6	29.6	31.2	15.9	9.5	3.8	1.3	0.2	94.8
New Delhi	28 35N	77 12E	695	10	71	43	97	68	95	80	93	64	115	31	75	0.9	0.7	0.5	0.3	0.5	2.9	7.1	6.8	4.6	0.4	0.1	0.4	25.2
Silchar	24 49N	92 48E	95	60	78	52	88	69	90	77	88	72	103	41	53	0.8	2.1	7.9	14.3	15.6	21.7	19.7	19.7	14.4	6.5	1.4	0.4	124.5
Indian Ocean Islands:																										
Port Blair, Andaman Is.	11 40N	92 43E	261	60	84	72	89	75	84	75	84	74	97	62	60	1.8	1.1	1.1	2.4	15.1	21.7	15.4	16.3	17.4	12.5	10.5	7.9	123.2
Amini Divi, Laccadive Is.	11 07N	72 44E	13	29	86	74	92	80	86	77	86	77	99	65	30	0.7	*	*	1.5	3.7	14.3	12.0	7.7	6.3	5.8	2.6	1.3	56.0
Minicoy, Maldive Is.	08 18N	73 00E	9	20	85	73	87	80	85	76	85	76	98	63	50	1.8	0.7	0.9	2.3	7.0	11.6	8.9	7.8	6.3	7.3	5.5	3.4	63.5
Car Nicobar, Nicobar Is.	09 09N	92 49E	47	13	86	77	90	77	86	77	85	75	95	66	30	3.9	1.2	2.1	3.5	12.5	12.4	9.3	10.2	12.9	11.6	11.4	7.8	98.8
Iran:																										
Abadan	30 21N	48 13E	10	12	64	44	90	62	112	81	98	63	127	24	10	1.5	1.7	0.6	0.8	0.1	0.0	0.0	0.0	0.0	0.1	1.0	1.8	7.6
Isfahan	32 37N	51 41E	5,238	45	47	25	72	46	98	67	78	47	108	-4	45	0.7	0.6	0.8	0.6	0.3	*	0.1	*	*	0.1	0.4	0.7	4.4
Kermanshah	34 19N	47 07E	4,331	15	45	23	68	38	99	56	79	38	108	-13	15	2.6	2.3	2.8	2.2	1.6	*	*	*	*	0.4	2.0	2.4	16.4
Rezaiyeh	37 32N	45 05E	4,364	3	32	17	67	45	91	64	67	47	99	-11	3	1.9	2.3	2.0	1.7	1.2	0.5	*	0.1	0.2	1.5	0.8	1.6	13.8

TABLE 1–10. TEMPERATURE AND PRECIPITATION DATA FOR REPRESENTATIVE WORLD-WIDE STATIONS (continued)

COUNTRY AND STATION	LATI-TUDE	LONGI-TUDE	ELE-VATION	TEMPERATURE: Length of record	Average Daily JAN. Maximum	JAN. Minimum	APR. Maximum	APR. Minimum	JULY Maximum	JULY Minimum	OCT. Maximum	OCT. Minimum	Extreme Maximum	Extreme Minimum	AVERAGE PRECIPITATION: Length of record	January	February	March	April	May	June	July	August	September	October	November	December	YEAR
	° '	° '	FEET	YEAR	°F	°F	°F	°F	°F	°F	°F	°F	°F	°F	YEAR	IN.	IN.	IN.	IN.	IN.	IN.	IN.	IN.	IN.	IN.	IN.	IN.	IN.
Tehran	35 41N	51 19E	3,937	24	45	27	71	49	99	72	76	53	109	-5	33	1.8	1.5	1.8	1.4	0.5	0.1	0.1	0.1	0.1	0.3	0.8	1.2	9.7
Iraq:																												
Baghdad	33 20N	44 24E	111	15	60	39	85	57	110	76	92	61	121	18	15	0.9	1.0	1.1	0.5	0.1	*	*	*	*	0.1	0.8	1.0	5.5
Basra	30 34N	47 47E	8	10	64	45	85	63	104	81	94	64	123	24	10	1.4	1.1	1.2	1.2	0.2	0.0	*	*	*	*	1.4	0.8	7.3
Mosul	36 19N	43 09E	730	26	54	35	77	49	109	72	88	51	124	12	29	2.8	3.1	2.1	1.9	0.7	*	*	*	*	0.2	1.9	2.4	15.2
Israel:																												
Haifa	32 48N	35 02E	23	16	65	49	77	58	88	75	85	68	112	27	30	6.9	4.3	1.6	1.0	0.2	*	*	*	0.1	1.0	3.7	7.3	26.2
Jerusalem	31 47N	35 13E	2,654	19	55	41	73	50	87	63	81	59	107	26	50	5.1	4.7	2.9	0.9	0.1	*	0.0	0.0	*	0.3	2.2	3.5	19.7
Tel Aviv	32 06N	34 46E	33	10	64	50	70	57	82	72	79	65	102	34	10	4.9	2.7	2.0	0.7	0.1	0.0	0.0	0.0	0.1	0.4	4.1	6.1	21.1
Jammu/Kashmir:																												
Srinagar	33 58N	74 46E	5,458	50	41	24	67	45	88	64	74	41	106	-4	50	2.9	2.8	3.6	3.7	2.4	1.4	2.3	2.4	1.5	1.2	0.4	1.3	25.9
Jordan:																												
Amman	31 58N	35 59E	2,547	25	54	39	73	49	89	65	81	57	109	21	25	2.7	2.9	1.2	0.6	0.2	0.0	0.0	0.0	*	0.2	1.3	1.8	10.9
Kuwait:																												
Kuwait	29 21N	48 00E	16	14	61	49	83	68	103	86	91	73	119	33	10	0.9	0.9	1.1	0.2	*	0.0	0.0	0.0	0.0	0.1	0.6	1.1	5.1
Lebanon:																												
Beirut	33 54N	35 28E	111	62	62	51	72	58	87	73	81	69	107	30	71	7.5	6.2	3.7	2.2	0.7	0.1	*	*	0.2	2.0	5.2	7.3	35.1
Nepal:																												
Katmandu	27 42N	85 22E	4,423	27	65	36	84	53	84	69	80	56	99	27	9	0.6	1.6	0.9	2.3	4.8	9.7	14.7	13.6	6.1	1.5	0.3	0.1	56.2
Oman and Muscat:																												
Muscat	23 37N	58 35E	15	23	77	66	90	78	97	87	93	80	116	51	38	1.1	0.7	0.4	0.4	*	0.1	*	*	0.0	0.1	0.4	0.7	3.9
Pakistan:																												
Karachi	24 48N	66 59E	13	43	77	55	90	73	91	81	91	72	118	39	59	0.5	0.4	0.3	0.1	0.1	0.7	3.2	1.6	0.5	0.1	0.1	0.2	7.8
Multan	30 11N	71 25E	400	60	68	42	95	68	102	86	94	64	122	29	60	0.4	0.4	0.4	0.3	0.3	0.6	2.0	1.8	0.5	0.1	0.1	0.2	7.1
Rawalpindi	33 35N	73 03E	1,676	60	62	38	86	59	98	77	89	57	118	25	60	2.5	2.5	2.7	1.9	1.3	2.3	8.1	9.2	3.9	0.6	0.3	1.2	36.5
Saudi Arabia:																												
Dhahran	26 16N	50 10E	78	10	69	54	90	69	107	86	95	73	120	40	10	1.1	0.6	0.4	0.2	0.1	0.0	0.0	0.0	0.0	0.0	0.2	0.9	3.5
Jidda	21 28N	39 10E	20	5	84	66	91	70	99	79	95	73	117	49	5	0.2	*	*	*	*	0.0	*	*	*	*	1.0	1.2	2.5
Riyadh	24 39N	46 42E	1,938	3	70	46	89	64	107	78	94	61	113	19	3	0.1	0.8	0.9	1.0	0.4	*	0.0	*	0.0	0.0	*	*	3.2
Sri Lanka:																												
Colombo	06 54N	79 52E	22	25	86	72	88	76	85	77	85	75	99	59	40	3.5	2.7	5.8	9.1	14.6	8.8	5.3	4.3	6.3	13.7	12.4	5.8	92.3
Syria:																												
Deir Ez Zor	35 21N	40 09E	699	5	53	35	80	52	105	78	86	56	114	16	8	1.6	0.8	0.3	0.8	0.1	*	0.0	0.0	0.0	0.2	1.5	0.9	6.2
Damascus	33 30N	36 20E	2,362	13	53	36	75	49	96	64	81	54	113	21	7	1.7	1.7	0.3	0.5	0.1	*	*	0.0	0.7	0.4	1.6	1.6	8.6

Aleppo	36 14N	37 08E	1,280	8	50	34	75	48	97	69	81	54	117	9	10	3.5	2.5	1.5	1.1	0.3	0.1	0.0	*	*	1.0	2.2	3.3	15.5
Turkey:																												
Adana	36 59N	35 18E	82	21	57	39	74	51	93	71	84	58	109	19	31	4.3	4.0	2.5	1.6	2.0	0.7	0.2	0.2	0.7	1.9	2.4	3.8	24.3
Ankara	39 57N	32 53E	2,825	26	39	24	63	40	86	59	69	44	104	-13	24	1.3	1.2	1.3	1.3	1.9	1.0	0.5	0.4	0.7	0.9	1.2	1.9	13.6
Erzurum	39 54N	41 16E	6,402	16	24	8	50	32	78	53	59	37	93	-22	16	1.4	1.6	2.0	2.5	3.1	2.1	1.3	0.9	1.1	2.3	1.8	1.1	21.2
Izmir	38 27N	27 15E	92	39	55	39	70	49	92	69	76	55	108	12	58	4.4	3.3	3.0	1.7	1.3	0.6	0.2	0.2	0.8	2.1	3.3	4.8	25.5
Samsun	41 17N	36 19E	131	24	50	38	59	45	79	65	69	56	103	20	27	2.9	2.6	2.7	2.3	1.8	1.5	1.5	1.3	2.4	3.2	3.5	2.4	29.1
United Arab Emirates:																												
Sharjah	25 20N	55 24E	18	11	74	54	86	65	100	82	92	71	118	37	12	0.9	0.9	0.4	0.2	0.0	0.0	0.0	0.0	0.0	0.0	0.4	1.4	4.2
Yemen:																												
Kamaran I.	15 20N	42 37E	20	26	82	74	89	79	98	85	93	82	105	66	21	0.2	0.2	0.1	0.1	0.1	*	0.5	0.7	0.1	0.1	0.4	0.9	3.4

AUSTRALIA & PACIFIC ISLANDS

Australia:																												
Adelaide	34 57S	138 32E	20	86	86	61	73	55	59	45	73	51	118	32	104	0.8	0.7	1.0	1.8	2.7	3.0	2.6	2.6	2.1	1.7	1.1	1.0	21.1
Alice Springs	23 48S	133 53E	1,791	62	97	70	81	54	67	39	88	58	111	19	30	1.7	1.3	1.1	0.4	0.6	0.5	0.3	0.3	0.3	0.7	1.2	1.5	9.9
Bourke	30 05S	145 58E	361	63	99	70	82	55	65	40	85	56	125	25	72	1.4	1.5	1.1	1.1	1.0	1.1	0.9	0.8	0.8	0.9	1.2	1.4	13.2
Brisbane	27 25S	153 05E	17	53	85	69	79	61	68	49	80	60	110	35	91	6.4	6.3	5.7	3.7	2.8	2.6	2.2	1.9	1.9	2.5	3.7	5.0	44.7
Broome	17 57S	122 13E	56	41	92	79	93	72	82	58	91	72	113	40	50	6.3	5.8	3.9	1.2	0.6	0.9	0.2	0.1	*	*	0.6	3.3	22.9
Burketown	17 45S	139 33E	30	31	93	77	91	69	82	55	93	70	110	40	53	8.2	6.3	5.2	1.0	0.2	0.3	*	*	*	0.4	1.5	4.4	27.5
Canberra	35 18S	149 11E	1,886	23	82	55	67	44	52	33	68	43	109	14	25	1.9	1.7	2.2	1.6	1.8	2.1	1.8	2.2	1.6	2.2	1.9	2.0	23.0
Carnarvon	24 53S	113 40E	13	43	88	72	84	66	71	51	78	61	118	37	57	0.4	0.7	0.7	0.6	1.5	2.4	1.6	0.7	0.2	0.1	*	0.2	9.1
Cloncurry	20 40S	140 30E	622	32	99	77	90	67	77	51	95	68	127	35	59	4.4	4.2	2.4	0.7	0.5	0.6	0.3	0.1	0.3	0.5	1.3	2.7	18.0
Esperance	33 50S	121 55E	14	44	77	60	72	54	62	45	68	50	117	31	60	0.7	0.7	1.2	1.8	3.3	4.1	4.0	3.8	2.7	2.2	1.0	0.9	26.4
Laverton	28 40S	122 23E	1,510	30	96	69	81	57	64	41	82	55	115	25	30	0.8	0.8	1.6	0.8	0.9	0.7	0.6	0.5	0.2	0.3	0.8	0.8	8.8
Melbourne	37 49S	144 58E	115	88	78	57	68	51	56	42	67	48	114	27	88	1.9	1.8	2.2	2.3	2.1	2.1	1.9	1.9	2.3	2.6	2.3	2.3	25.7
Mundiwindi	23 52S	120 10E	1,840	15	101	64	87	61	70	41	89	58	112	22	15	1.0	1.9	2.0	0.8	0.6	0.9	0.1	0.3	0.3	0.5	0.5	1.2	10.1
Perth	31 56S	115 58E	64	44	85	63	76	57	63	48	70	53	112	31	63	0.3	0.4	0.8	1.7	5.1	7.1	6.7	5.7	3.4	2.2	0.8	0.5	34.7
Port Darwin	12 25S	130 52E	104	58	90	77	92	76	87	67	93	77	105	55	70	15.2	12.3	10.0	3.8	0.6	0.1	*	0.1	0.5	2.0	4.7	9.4	58.7
Sydney	33 52S	151 02E	62	87	78	65	71	58	60	46	71	56	114	35	87	3.5	4.0	5.0	5.3	5.0	4.6	4.6	3.0	2.9	2.8	2.9	2.9	46.5
Thursday Island	10 35S	142 13E	200	31	87	77	86	77	82	73	86	76	98	64	49	18.2	15.8	13.9	8.0	1.6	0.5	0.4	0.2	0.1	0.3	1.5	7.0	67.5
Townsville	19 15S	146 46E	18	31	87	76	84	70	75	59	83	71	110	39	67	10.9	11.2	7.2	3.3	1.3	1.4	0.6	0.5	0.7	1.3	1.9	5.4	45.7
William Creek	28 55S	136 21E	247	39	96	69	80	55	65	41	84	56	119	25	30	0.5	0.6	0.3	0.3	0.3	0.5	0.2	0.3	0.3	0.5	0.5	0.7	5.0
Windorah	25 26S	142 36E	390	29	101	74	86	59	70	43	91	61	116	26	50	1.4	1.6	1.6	0.9	0.8	0.8	0.5	0.4	0.5	0.6	0.9	1.4	11.4
Tasmania:																												
Hobart	42 53S	147 20E	177	70	71	53	63	48	52	40	63	46	105	28	100	1.9	1.5	1.8	1.9	1.8	2.2	2.1	1.9	2.1	2.3	2.4	2.1	24.0
New Zealand																												
Auckland	37 00S	174 47E	23	36	73	60	67	56	56	46	63	52	90	33	92	3.1	3.7	3.2	3.8	5.0	5.4	5.7	4.6	4.0	4.0	3.5	3.1	49.1
Christchurch	43 29S	172 32E	118	52	70	53	62	45	50	35	62	44	96	21	64	2.2	1.7	1.9	1.9	2.6	2.6	2.7	1.9	1.8	1.7	1.9	2.2	25.1
Dunedin	45 55S	170 12E	4	77	66	50	59	45	48	37	59	42	94	23	77	3.4	2.8	3.0	2.8	3.2	3.2	3.1	3.0	2.7	3.0	3.2	3.5	36.9
Wellington	41 17S	174 46E	415	66	69	56	63	51	53	42	60	48	88	29	79	3.2	3.2	3.2	3.8	4.6	4.6	5.4	4.6	3.8	4.0	3.5	3.5	47.4

PACIFIC ISLANDS:

TABLE 1–10. TEMPERATURE AND PRECIPITATION DATA FOR REPRESENTATIVE WORLD-WIDE STATIONS (continued)

COUNTRY AND STATION	LATI-TUDE	LONGI-TUDE	ELE-VATION	TEMPERATURE: Length of record	Average Daily JAN. Maximum	JAN. Minimum	APR. Maximum	APR. Minimum	JULY Maximum	JULY Minimum	OCT. Maximum	OCT. Minimum	Extreme Maximum	Extreme Minimum	AVERAGE PRECIPITATION: Length of record	January	February	March	April	May	June	July	August	September	October	November	December	YEAR
	° ′	° ′	FEET	YEAR	°F	°F	°F	°F	°F	°F	°F	°F	°F	°F	YEAR	IN.	IN.	IN.	IN.	IN.	IN.	IN.	IN.	IN.	IN.	IN.	IN.	IN.
Canton, Phoenix Is.	02 46S	171 43W	9	12	88	78	89	78	89	78	90	78	98	70	30	2.6	2.2	2.5	3.6	4.3	2.6	2.6	2.5	1.2	1.1	1.6	2.6	29.4
Guam, Marianas Is.	13 33N	144 50E	361	30	84	72	86	73	87	72	86	73	95	54	30	4.6	3.5	2.6	3.0	4.2	5.9	9.0	12.8	13.4	13.1	10.3	6.1	88.5
Honolulu, Hawaii	21 20N	157 55W	7	30	79	66	80	68	85	73	84	72	93	56	30	3.8	3.3	2.9	1.3	1.0	0.3	0.4	0.9	1.0	1.8	2.2	3.0	21.9
Iwo Jima, Bonin Is.	24 47N	141 19E	353	15	71	64	77	69	86	78	84	76	95	46	17	3.2	2.5	2.1	3.7	4.9	4.0	6.4	6.5	4.6	5.9	4.8	4.3	52.8
Madang, New Guinea	05 12S	145 47E	19	12	87	75	88	74	88	74	88	75	98	62	20	12.1	11.9	14.9	16.9	15.1	10.8	7.6	4.8	5.3	10.0	13.3	14.5	137.2
Midway Is.	28 13N	177 23W	29	21	69	62	71	64	81	74	79	72	92	46	20	4.6	3.7	3.1	2.5	1.9	1.3	2.9	3.9	3.7	3.7	3.6	4.2	40.7
Naha, Okinawa	26 12N	127 39E	96	30	67	56	76	64	89	77	81	69	96	41	30	5.3	5.4	6.1	6.1	8.9	10.0	7.1	10.0	7.1	6.6	5.9	4.3	82.8
Noumea, New Caledonia	22 16S	166 27E	246	24	86	72	83	70	76	62	80	65	99	52	52	3.7	5.1	5.7	5.2	4.4	3.7	3.6	2.6	2.5	2.0	2.4	2.6	43.5
Pago Pago, Samoa	14 19S	170 43W	29	2	87	75	87	76	83	74	85	75	98	67	41	24.5	20.5	19.2	16.5	15.4	12.3	10.0	8.2	13.1	14.9	19.2	19.8	193.6
Ponape, Caroline Is.	06 58N	158 13E	123	30	86	75	86	75	87	73	87	72	96	67	30	11.1	9.7	14.6	20.0	20.3	16.7	16.2	16.3	15.8	16.0	16.9	18.3	191.9
Port Moresby, New Guinea	09 29S	147 09E	126	20	89	76	87	75	83	73	86	75	98	64	38	7.0	7.6	6.7	4.2	2.5	1.3	1.1	0.7	1.0	1.4	1.9	4.4	39.8
Rabaul, New Guinea	04 13S	152 11E	28	19	90	73	90	73	89	73	92	73	100	65	24	14.8	10.4	10.2	10.0	5.2	3.3	5.4	3.7	3.5	5.1	7.1	10.1	88.8
Suva, Fiji Is.	18 08S	178 26E	20	43	86	74	84	73	79	68	81	70	98	55	43	11.4	10.7	14.5	12.2	10.1	6.7	4.9	8.3	7.7	8.3	9.8	12.5	117.1
Tahiti, Society Is.	17 33S	149 36W	7	23	89	72	89	72	86	68	87	70	93	61	27	13.2	11.5	6.5	6.8	4.9	3.2	2.6	1.9	2.3	3.4	6.5	11.9	74.7
Tulagi, Solomon Is.	09 05S	160 10E	8	20	88	76	88	76	86	76	87	76	96	68	37	14.3	15.8	15.0	10.0	8.1	6.8	7.6	8.7	8.0	8.7	10.0	10.4	123.4
Wake Is.	19 17N	166 39E	11	30	82	73	83	74	87	77	86	77	92	64	30	1.1	1.4	1.5	1.9	2.0	1.9	4.6	7.1	5.2	5.3	3.1	1.8	36.9
Yap, Caroline Is.	9 31N	138 08E	62	30	85	76	87	77	88	75	88	75	97	69	30	7.9	4.6	5.4	6.4	9.5	10.7	13.8	14.7	14.0	13.2	11.2	10.2	121.6
ANTARCTICA																												
Byrd Station	80 01S	119 32W	5,095	6	10	-2	-11	-30	-25	-45	-15	-33	31	-82	6	0.4	0.4	0.2	0.3	0.4	0.5	0.7	0.7	0.3	0.7	0.0	0.3	4.9
Ellsworth	77 44S	41 07W	139	6	22	12	-10	-25	-21	-35	-2	-15	36	-70	6	0.3	0.2	0.3	0.6	0.2	0.2	0.2	0.2	0.3	0.4	0.5	0.2	3.6
McMurdo Station	77 53S	166 48W	8	10	30	21	-1	-13	-9	-24	2	-12	42	-59	10	0.5	0.7	0.4	0.4	0.4	0.3	0.2	0.3	0.4	0.2	0.2	0.3	4.3
South Pole Station	89 59S	000 00W	9,186	5	-16	-23	-66	-79	-67	-81	-55	-64	6	-107	5	*	0.1	0.0	0.0	0.0	0.0	0.0	0.0	0.0	*	0.0	*	0.1
Wilkes	66 16S	110 31E	31	7	34	28	17	9	8	-3	16	6	46	-35	7	0.5	0.4	1.7	1.1	1.4	1.2	1.3	0.8	1.5	1.2	0.8	0.3	12.2

NOTES

1. "Length of Record" refers to average daily maximum and minimum temperatures and precipitation. A standard period of the 30 years from 1931-1960 had been used for locations in the United States and some other countries. The length of record of extreme maximum and minimum temperatures includes all available years of data for a given location and is usually for a longer period.

2. * = Less than 0.05"

3. Except for Antarctica, amounts of solid precipitation such as snow or hail have been converted to their water equivalent. Because of the frequent occurrence of blowing snow, it has not been possible to determine the precise amount of precipitation actually falling in Antarctica. The values shown are the average amounts of solid snow accumulating in a given period as determined by snow markers. The liquid content of the accumulation is undetermined.

Source: Environmental Science Services Administration, Climates of the World, 1969. Geographic names revised by editors in accordance with 1987 usage

FIGURE 1–3. GENERAL PATTERN OF ANNUAL WORLD PRECIPITATION

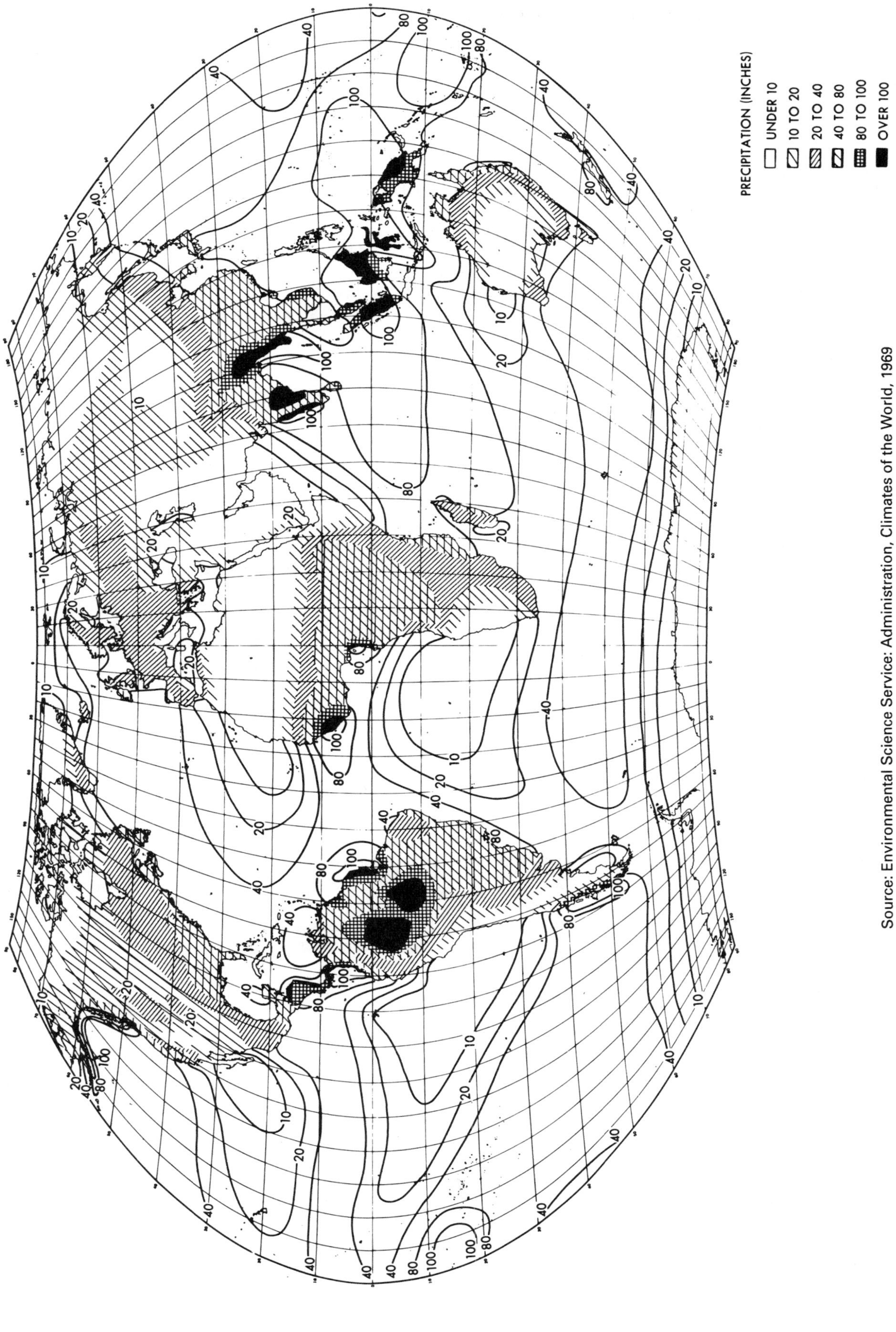

Source: Environmental Science Service: Administration, Climates of the World, 1969

SECTION C. WEATHER EXTREMES

TABLE 1–11. WORLD-WIDE EXTREMES OF TEMPERATURE AND PRECIPITATION

TEMPERATURE

A. Highest

World, 58°C (136°F), El Azizia, Libya, 13 September 1922
Western Hemisphere, 57°C (134°F), Death Valley, California, 10 July 1913
Asia, 54°C (129°F), Tirat Tsvi, Israel, 21 June 1942
Australia, 53°C (128°F), Cloncurry, Queensland, 16 January 1889
Europe, 50°C (122°F), Seville, Spain, 4 August 1881
South America, 49°C (120°F), Rivadavia, Argentina, 11 December 1905
Canada, 45°C (113°F), Midale and Yellow Grass, Saskatchewan, 5 July 1937
Vanda Station, Antarctica, had a 15°C (59°F) maximum, 5 January 1974 (possibly Antarctica's highest)
South Pole, -14°C (7.5°F), 27 December 1978
Persian Gulf had a 36°C (96°F) sea-surface, 5 August 1924

B. Lowest

World, -89°C (-129°F), Vostok, Antarctica, 21 July 1983
Northern Hemisphere, -68°C (-90°F), Verkhoyansk, U.S.S.R., 5 and 7 February 1892 and Oimekon, U.S.S.R., 6 February 1933
Greenland, -66°C (-87°F), Northice, 9 January 1954
North America, excluding Greenland, -63°C (-81°F), Snag, Yukon Territory, 3 February 1947
U.S., -62°C (-80°F), Prospect Creek, Endicott Mts., Alaska, 23 January 1971
U.S., excluding Alaska, -56.5°C (-70°F), Rogers Pass, Montana, 20 January 1954
Europe, -55°C (-67°F), Ust 'Shchugor, U.S.S.R., January (date not known, lowest in 15-year period)
South America, -33°C (-27°F), Sarmiento, Argentina, 1 June 1907
Africa, -24°C (-11°F), Ifrane, Morocco, 11 February 1935
Australia, -22°C (-8°F), Charlotte Pass, New South Wales, 14 June 1945 and 22 July 1947

PRECIPITATION

A. Greatest Rainfall

World, 1-minute, 3.1 cm (1.23"), Unionville, Maryland, 4 July 1956
World, 20-minute, 20.5 cm (8.10"), Curtea-de-Arges, Romania, 7 July 1889
World, 42-minute, 30.5 cm (12"), Holt, Missouri, 22 June 1947
World, 12-hour, 135 cm (53"), Belouve, La Réunion I., 28–29 February 1964
World, 24-hour, 188 cm (74"), Cilaos, La Réunion I., 15–16 March 1952
Northern Hemisphere, 24-hour, 125 cm (49"), Paishih, Taiwan, 10–11 September 1963
Australia, 24-hour, 91 cm (36"), Crohamhurst, Queensland, 3 February 1893 (updated to 114 cm or 44", Bellenden Ker, Queensland, 4 January 1979)
Alvin, Texas, had a 24-hour rainfall of 109 cm (43"), 25–26 July 1979 (possibly the world's greatest on flat terrain)
Canada, 24-hour, 49 cm (19"), Ucluelet Brynnor Mines, British Columbia, 6 October 1967
World, 5-day, 386 cm (152"), Cilaos, La Réunion I., 13–18 March 1952
World, 1-month, 930 cm (366"), Cherrapunji, India July 1861
World, 12-month, 2647 cm (1042"), Cherrapunji, India, August 1860 to August 1861
U.S., 12 month, 1878 cm (739"), Kukui, Maui, Hawaii, December 1981 to December 1982

B. Greatest Average Yearly Precipitation

World, 1168 cm (460") during a 32-year period, Mt. Waialeale, Kauai, Hawaii
Asia, 1143 cm (450") during a 74-year period, Cherrapunji, India
Africa, 1029 cm (405") during a 32-year period, Debundscha, Cameroon
South America, 899 cm (354") during a 10–16 year period, Quibdó, Colombia
North America, 650 cm (256") during a 14-year period Henderson Lake, British Columbia
Europe, 465 cm (183") during a 22-year period, Crkvice, Yugoslavia
Australia, 455 cm (179") during a 31-year period, Tully, Queensland (updated to 425 cm (167") during a 59-year period)
Bahia Felix, Chile, averages 325 days/year with rain
Canada, highest frequency of days with precipitation, 242 per year average, Langara, Queen Charlotte Islands, British Columbia

TABLE 1–11. WORLD-WIDE EXTREMES OF TEMPERATURE AND PRECIPITATION (continued)

C. Least Precipitation

Arica, Chile, had no rain for more than 14 consecutive years, October 1903 to January 1918
U.S., longest dry period, 767 days from 3 October 1912 to 8 November 1914, Bagdad, California
Canada, least precipitation during a calendar year, 1.27 cm (0.05"), Arctic Bay, Northwest Territories, 1949
Canada, lowest frequency of days with precipitation, 8 per year average, Rea Point, Northwest Territories

D. Lowest Average Yearly Precipitation

World, 0.08 cm (0.03") during a 59-year period, Arica, Chile
Africa, <0.25 cm (<0.1") during a 39-year period, Wadi Halfa, Sudan
North America, 3.0 cm (1.2") during a 14-year period, Bataques, Mexico
U.S., 4.1 cm (1.63") during a 42-year period, Death Valley, California
Asia, 4.6 cm (1.8") during a 50-year period, Aden, South Yemen
Australia, 10 cm (4.05") during a 34-year period, Mulka, South Australia (updated to during a 42-year period, Troudaninna, South Australia)
Europe, 16 cm (6.4") during a 25-year period, Astrakhan, U.S.S.R.

E. Hailstones

U.S., largest hailstone, 44.5 cm (17.5") circumference, Coffeyville, Kansas, 3 September 1979
Canada, heaviest hailstone, 290 gm (10.23 oz), Cedoux, Saskatchewan, 27 August 1973
Canada, highest frequency of days with hail, 7 per year average, Edson and Red Deer, Alberta

F. Greatest Snowfall

North America, 24-hour, 192.5 cm (76"), Silver Lake, Colorado, 14–15 April 1921
Alaska, 24-hour, 157.5 cm (62"), Thompson Pass, 29 December 1955
Bessans, France, had a snowfall of 172 cm (68") in 19 hours, 5–6 April 1969
Canada, climatological day, 118 cm (46"), Lakelse Lake, British Columbia, 17 January 1974
North America, one storm, 480 cm (189"), Mt. Shasta Ski Bowl, California, 13–19 February 1959
Alaska, one storm, 445.5 cm (175"), Thompson Pass, 26–31 December 1955
North America, one season, 2850 cm (1122"), Rainier Paradise Ranger Station, Washington, 1971–1972
Alaska, one season, 2475 cm (974.5"), Thompson Pass, 1952–1953
Canada, one season, 2446.5 cm (964"), Revelstoke Mt. Copeland, British Columbia, 1971–1972
Canada, highest frequency of days with snow, 142 per year average, Old Glory Mountain, British Columbia
North America, greatest depth of snow on the ground, 1145.5 cm (451"), Tamarack, California, 11 March 1911
Canada, greatest depth of snow on the ground, 775 cm (305"), Loch Lomond, British Columbia

OTHER ELEMENTS

A. Thunderstorms

Kampala, Uganda, averages 242 days/year with thunderstorms, during a 10-year period
Bogor, Indonesia, averaged 322 days/year with thunderstorms from 1916 to 1920
Canada, highest frequency of days with thunderstorms, 34 per year average, Windsor, Ontario

B. Fog Frequency

U.S. West Coast, highest average, 2552 hours per year during a 10-year period or more, Cape Disappointment, Washington
U.S. East Coast, highest average, 1580 hours per year during a 10-year period or more, Moose Peak Lighthouse, Mistake Island, Maine
Canada, highest average, 158 days per year, Cape Race, Newfoundland

Source: Riordan, P. and Bourget, P.G., 1985, World Weather Extremes, U.S. Army Corps of Engineers, Engineer Topographic Laboratories, Fort Belvoir, VA 22060

TABLE 1–12. WORLD'S GREATEST OBSERVED POINT RAINFALLS

Duration	Inches	Location	Date
1 min.	1.23	Unionville, Md.	July 4, 1956
8 min.	4.96	Füssen, Bavaria	May 25, 1920
15 min.	7.80	Plumb Point, Jamaica	May 12, 1916
20 min.	8.10	Curtea-de-Arges, Rumania	July 7, 1889
42 min.	12.00	Holt, Mo.	June 22, 1947
2 hr. 10 min.	19.00	Rockport, W. Va.	July 18, 1889
2 hr. 45 min.	22.00	D'Hanis, Tex. (17 mi. NNW)	May 31, 1935
4 hr. 30 min.	30.8+	Smethport, Pa.	July 18, 1942
9 hr.	42.79	Belouve, La Réunion	Feb. 28, 1964
12 hr.	52.76	Belouve, La Réunion	Feb. 28–29, 1964
18 hr. 30 min.	66.49	Belouve, La Réunion	Feb. 28–29, 1964
24 hr.	73.62	Cilaos, La Réunion	Mar. 15–16, 1952
2 days	98.42	Cilaos, La Réunion	Mar. 15–17, 1952
3 days	127.56	Cilaos, La Réunion	Mar. 15–18, 1952
4 days	137.95	Cilaos, La Réunion	Mar. 14–18, 1952
5 days	151.73	Cilaos, La Réunion	Mar. 13–18, 1952
6 days	159.65	Cilaos, La Réunion	Mar. 13–19, 1952
7 days	161.81	Cilaos, La Réunion	Mar. 12–19, 1952
8 days	162.59	Cilaos, La Réunion	Mar. 11–19, 1952
15 days	188.88	Cherrapunji, India	June 24-July 8, 1931
31 days	366.14	Cherrapunji, India	July 1861
2 mo.	502.63	Cherrapunji, India	June-July, 1861
3 mo.	644.44	Cherrapunji, India	May-July, 1861
4 mo.	737.70	Cherrapunji, India	Apr-July, 1861
5 mo.	803.62	Cherrapunji, India	Apr.-Aug. 1861
6 mo.	884.03	Cherrapunji, India	Apr.-Sept., 1861
11 mo.	905.12	Cherrapunji, India	Jan.-Nov., 1861
1 yr.	1041.78	Cherrapunji, India	Aug. 1860-July 1861
2 yr.	1605.05	Cherrapunji, India	1860–1861

Source: ECAFE, United Nations, 1967

SECTION D. PRECIPITATION DATA

TABLE 1-13. MAXIMUM OBSERVED STORM RAINFALL IN THE UNITED STATES

[Average rainfall in inches]

Area, sq mi	Duration, hr 6	12	18	24	36	48	72
10	24.7*a*	29.8*b*	36.3*c*	38.7*c*	41.8*c*	43.1*c*	45.2*c*
100	19.6*b*	26.3*c*	32.5*c*	35.2*c*	37.9*c*	38.9*c*	40.6*c*
200	17.9*b*	25.6*c*	31.4*c*	34.2*c*	36.7*c*	37.7*c*	39.2*c*
500	15.4*b*	24.6*c*	29.7*c*	32.7*c*	35.0*c*	36.0*c*	37.3*c*
1000	13.4*b*	22.6*c*	27.4*c*	30.2*c*	32.9*c*	33.7*c*	34.9*c*
2000	11.2*b*	17.7*c*	22.5*c*	24.8*c*	27.3*c*	28.4*c*	29.7*c*
5000	8.1*bd*	11.1*b*	14.1*b*	15.5*c*	18.7*e*	20.7*e*	24.4*e*
10,000	5.7*d*	7.9*f*	10.1*g*	12.1*g*	15.1*e*	17.4*e*	21.3*e*
20,000	4.0*d*	6.0*f*	7.9*g*	9.6*g*	11.6*e*	13.8*e*	17.6*e*
50,000	2.5*gh*	4.2*i*	5.3*g*	6.3*g*	7.9*g*	8.9*g*	11.5*j*
100,000	1.7*h*	2.5*hk*	3.5*g*	4.3*g*	5.6*g*	6.6*j*	8.9*j*

Storm	Date	Location of Center	Storm	Date	Location of Center
a	July 17–18, 1942	Smethport, PA	*g*	Mar. 13–15, 1929	Elba, AL
b	Sept. 8–10, 1921	Thrall, TX	*h*	May 22–26, 1908	Chattanooga, OK
c	Sept. 3–7, 1950	Yankeetown, FL	*i*	Apr. 15–18, 1900	Eutaw, AL
d	June 27–July 4, 1936	Bebe, TX	*j*	July 5–10, 1916	Bonifay, FL
e	June 27–July 1, 1899	Hearne, TX	*k*	Nov. 19–22, 1934	Millry, AL
f	Apr. 12–16, 1927	Jefferson Parish, LA			

Source: Corps of Engineers, U.S. Army, 1960

TABLE 1-14. AVERAGE ANNUAL HOURS OF PRECIPITATION FOR SELECTED CITIES IN THE UNITED STATES

[By range in precipitation; compiled from National Climatic Data Center statistics, period of record 1951–1960]

	Average Annual Hours						
	Precipitation in Inches						
City	Trace	.01–.09	.10–.25	.25–.49	.50 or greater	Total Hours Measurable Precipitation	Percent of Total Annual Hours[1]
Akron, OH	1422	614	71	18	1	704	8.0
Albany, NY	880	595	76	17	2	690	7.8
Albuquerque, NM	314	132	14	1	~	147	1.7
Amarillo, TX	473	204	38	11	2	255	2.9
Anchorage, AK	1226	559	17	<1	~	576	6.6
Atlanta, GA	566	381	95	28	13	517	5.9
Austin, TX	517	260	49	21	7	337	3.8
Baltimore, MD	599	493	91	21	3	608	6.9
Baton Rouge, LA	500	286	83	33	15	417	4.8
Birmingham, AL	500	360	97	32	14	503	5.7
Bismarck, ND	1038	346	26	2	<1	374	4.3
Boise, ID	609	370	17	<1	~	387	4.4
Boston, MA	891	652	125	26	2	805	9.2
Brownsville, TX	397	212	35	16	4	267	3.1
Buffalo, NY	1349	759	81	15	<1	855	9.8

TABLE 1-14. AVERAGE ANNUAL HOURS OF PRECIPITATION FOR SELECTED CITIES IN THE UNITED STATES (continued)

[By range in precipitation; compiled from National Climatic Data Center statistics, period of record 1951–1960]

	Average Annual Hours						
	Precipitation in Inches						
City	Trace	.01–.09	.10–.25	.25–.49	.50 or greater	Total Hours Measurable Precipitation	Percent of Total Annual Hours[1]
Burbank, CA	252	146	35	11	~	192	2.2
Casper, WY	726	336	19	<1	~	355	4.1
Charleston, SC	564	394	95	45	18	552	6.3
Charlotte, NC	498	411	94	26	5	536	6.1
Chattanooga, TN	540	479	109	32	9	629	7.2
Chicago (Midway) IL	882	470	65	18	1	554	6.3
Cincinnati, OH	917	488	83	20	3	594	6.8
Cleveland, OH	1237	638	77	17	1	733	8.4
Columbia, SC	490	420	86	24	16	546	6.3
Columbus, OH	978	549	68	19	2	638	7.3
Corpus Christi, TX	410	207	40	20	20	287	3.3
Dallas, TX	445	239	58	21	6	324	3.7
Dayton, OH	1010	506	72	17	5	600	6.9
Denver, CO	624	347	31	2	<1	380	4.3
Des Moines, IA	686	403	57	17	4	481	5.5
Detroit, MI	1037	534	65	18	1	618	7.1
Duluth, MN	1482	554	54	12	1	621	7.1
El Paso, TX	233	111	19	3	~	133	1.5
Evansville, IN	678	406	90	24	4	524	6.2
Fairbank, AK	1368	382	6	<1	~	388	4.4
Fargo, ND	1057	309	34	6	~	349	4.0
Fort Wayne, IN	923	520	73	20	3	616	7.0
Fresno, CA	272	191	30	<1	~	221	2.5
Galveston, TX	380	242	57	24	7	330	3.8
Grand Rapids, MI	1345	594	58	16	4	672	7.7
Great Falls, MT	853	421	29	<1	~	450	5.1
Greensboro, NC	627	451	84	30	7	572	6.5
Harrisburg, PA	742	551	87	19	1	658	7.5
Honolulu, HA	1205	241	35	18	<1	294	3.4
Huron, SD	922	331	30	4	<1	365	4.2
Indianapolis, IN	903	499	86	22	8	980	11.2
Jackson, MS	491	328	83	36	15	462	5.3
Jacksonville, FL	450	314	80	32	20	446	5.1
Kansas City, MO	638	352	66	21	5	444	5.1
Knoxville, TN	648	476	108	28	4	616	7.0
Lake Charles, LA	453	259	76	36	22	393	4.5
Las Vegas, NV	137	76	3	<1	~	79	0.9
Little Rock, AR	473	358	96	32	17	503	5.7
Los Angeles, CA	216	139	24	7	<1	170	1.9
Louisville, KY	732	463	89	23	3	578	6.6
Madison, WI	902	437	62	15	2	516	5.9
Medford, OR	780	465	49	<1	~	514	5.9
Memphis, TN	499	368	100	35	16	519	5.9
Miami, FL	390	292	78	37	27	434	5.0
Midland-Odessa, TX	454	176	18	5	5	204	2.3

TABLE 1-14. AVERAGE ANNUAL HOURS OF PRECIPITATION FOR SELECTED CITIES IN THE UNITED STATES (continued)

[By range in precipitation; compiled from National Climatic Data Center statistics, period of record 1951–1960]

	Average Annual Hours						
	Precipitation in Inches						
City	Trace	.01–.09	.10–.25	.25–.49	.50 or greater	Total Hours Measurable Precipitation	Percent of Total Annual Hours[1]
Milwaukee, WI	958	480	59	18	<1	557	6.4
Minneapolis, MN	961	421	46	13	1	481	5.5
Moline, IL	788	415	67	21	2	505	5.8
Montgomery, AL	455	323	83	30	13	449	5.1
Nashville, TN	660	409	101	31	10	551	6.3
Newark, NJ	690	539	102	23	3	667	7.6
New Orleans, LA	464	309	82	38	23	452	5.2
New York (La Guardia), NY	651	568	104	30	5	707	8.1
Norfolk, VA	614	469	91	31	12	603	6.9
Oakland, CA	408	280	47	2	<1	329	3.8
Oklahoma City, OK	531	263	54	23	10	350	4.0
Omaha, NB	713	350	68	16	19	453	5.2
Philadelphia, PA	662	497	142	25	2	666	7.6
Phoenix, AZ	242	95	14	<1	<1	109	1.2
Portland, ME	781	666	103	18	1	788	9.0
Portland, OR	993	928	79	2	<1	1009	11.5
Providence, RI	757	592	104	30	2	728	8.3
Rapid City, SD	829	362	23	2	<1	387	4.4
Reno, NV	346	187	12	<1	~	199	2.3
Richmond, VA	591	460	86	28	8	582	6.6
Rochester, NY	1304	679	60	13	<1	752	8.6
Sacramento, CA	327	280	48	1	<1	329	3.8
St. Louis, MO	635	381	65	18	3	467	5.3
Salt Lake City, UT	607	346	27	4	2	379	4.3
San Antonio, TX	572	244	45	12	5	306	3.5
San Diego, CA	221	149	23	1	<1	173	2.0
San Francisco, CA	564	314	52	5	<1	371	4.2
Savannah, GA	457	329	81	34	15	459	5.2
Sioux City, IA	723	360	47	10	1	418	4.8
Spokane, WA	797	545	25	<1	~	570	6.5
Springfield, IL	715	405	64	19	1	489	5.6
Springfield, MO	680	378	73	23	8	482	5.5
Syracuse, NY	1378	828	71	15	1	915	10.4
Topeka, KS	658	333	59	21	6	419	4.8
Tucson, AZ	249	139	23	5	<1	167	1.9
Tulsa, OK	478	311	62	24	9	406	4.6
Washington, D.C.	684	450	82	24	5	561	6.4
Wichita, KS	577	275	51	17	8	351	4.0
Wichita Falls, TX	375	238	48	16	5	307	3.5
Youngstown, OH	1362	656	75	20	2	753	8.6

[1] There are 8760 hours in a non-leap year

Source: Doesken, N.J., and Eckrich, W.P., 1988, Colorado Climate Center, Colorado State University, Ft. Collins, CO; amended

TABLE 1–15. VELOCITY OF FALL, NUMBER OF DROPS, AND KINETIC ENERGY FOR RAINFALL OF VARIOUS INTENSITIES

	Intensity (In. per hr.)	Median diameter (Mm.)	Velocity of fall (Ft. per sec.)	Drops per square foot (No. per sec.)	Kinetic energy (Ft.-lbs. per sq. ft. per hr.)
Fog	0.005	0.01	0.01	6,264,000	4.043x10-8
Mist	.002	.1	.7	2,510	7.937x10-5
Drizzle	.01	.96	13.5	14	.148
Light rain	.04	1.24	15.7	26	.797
Moderate rain	.15	1.60	18.7	46	4.241
Heavy rain	.60	2.05	22.0	46	23.47
Excessive rain	1.60	2.40	24.0	76	74.48
Cloudburst	4.00	2.85	25.9	113	216.9
Do.	4.00	4.00	29.2	41	275.8
Do.	4.00	6.00	30.5	12	300.7

Source: Lull, H.W., 1959, Soil Compaction on Forest and Range Lands, U.S. Dept. of Agriculture, Forestry Service, Misc. Publ. No. 768

TABLE 1–16. SUMMARY OF CLOUD-SEEDING EXPERIMENTS, 1954–1975

[Randomized experiments on isolated cumulus clouds or cloud clusters]

Project	Dates	Main seeding agent(s) and delivery techniques	Results	Significance level
Central USA	1954	Dry ice, 5 or 15 kg km^{-1}, on penetrations above 0°C level	Greater probability of radar echoes	0.39
Caribbean	1954	Waterspray (1 m^3km^{-1}) into clouds	a. Greater frequency of radar echoes b. Time to first echo reduced	0.02 0.01
Australia	1962–65	AgI • NaI • acetone burners in updraft, 0.2 or 20 g per cloud	More rain at cloud base if tops colder than 10°C and 20 g treatment applied	0.02
Sierra Cumulus (pair seeding)	1966–68	Dry ice, AgI flares in updrafts	Increase probability of precipitation reaching ground	0.001
Flagstaff, Arizona	1967	AgI • NaI(?) • acetone burners in updrafts 120 or 240 g per cloud	a. Increase in height radar tops b. Increase of shower duration by 10 min. c. Increase in radar estimated rainfall	0.04 0.08 0.19
Rhodesia	1968–69	AgI • NaI solution burned in updrafts in cloud 4 km above sea level (~-6°C level) 500 g hr^{-1}	a. Increase rainfall at cloud base b. Increase shower duration	<0.01 <0.01
Florida	1968, 1970	AgI: About 2050 g pyrotechnics per cloud from cloud top	a. Increase in cloud height b. Increase in radar estimated rainfall	0.01 0.005
Cloud Catcher	1969–70	Salt, ~50 kg, or one to 6 120-g AgI flares in updrafts below cloud base (random choice, 1/3 cases no seed)	a. Radar echoes closer to cloud base b. Increase in echo tops for AgI seed cases c. Increases in radar estimated rainfall	0.01 0.10 0.06 (salt) 0.01 (AgI)
Rhodesia	1973–75	AgI pyrotechnics fired in updrafts ~250 m below cloud top	a. No effect if top temp. > - 10°C b. Rainfall increase if top temp. < - 10°C	— 0.07*
Ukraine	1973–	AgI • NH_4I in acetone, 500 g hr^{-1} in updrafts	Radar data through 1975 indicate rainfall increased over amounts predicted by regression equation	0.05

* Best result, found by measuring rain beginning 50 min after seeding

Source: Williams, M.C. and Elliott, R.D., 1985, Weather Modification, *in* Facets of Hydrology II, edited by J.C. Rodda. Copyright John Wiley & Sons Ltd.

SECTION E. SNOW AND SNOW MELT

TABLE 1–17. PHYSICAL PROPERTIES OF SNOW AND ICE

	Density, g/cm³	Porosity, per cent	Air permeability, g/cm²/sec	Grain size, mm
New Snow	0.01–0.3	99–67	> 400–40	0.01–5
Old Snow	0.2–0.6	78–35	100–20	0.5–3
Firn*	0.4–0.84	56–8	40–0	0.5–5
Glacier Ice	0.84–0.917	8–0	0	1–>100

* Firn is snow which has been modified into a dense compact material by deformation, refreezing, recrystallization, and other processes.
Source: Meier, *in* Chow, Handbook of Applied Hydrology, McGraw-Hill, Copyright 1964

TABLE 1–18. HEAT SUPPLIED TO MELTING SNOW BY DIFFERENT PROCESSES

Heat supply	Extreme conditions	Approximate heat supplied*
Convection from turbulent air	70° dry bulb, 20-mile wind	600
Condensation of atmospheric moisture	60° dew point, 20-mile wind	600
Absorption of solar radiation	Very moist air, cloudy at night	200
Warm rain	4 inches, 50° wet bulb	100
Conduction from soil	New snow	20

* Calories per square centimeter per day.
Source: Wilson, Trans. Amer. Geophysical Union, 1941

TABLE 1–19. RELATION OF SNOW MELT TO SNOW EVAPORATION

Air temperature	Relative humidity	Snow evaporated	Heat transfer from air to snow	Heat required in evaporation process*	Heat available to melt snow	Melted snow	Melt/evaporation
°C	%	g cm^{-2} day^{-1}	cal cm^{-2} day^{-1}	cal cm^{-2} day^{-1}	cal cm^{-2} day^{-1}	g cm^{-2} day^{-1}	
5	20	2.02	900	1370	0	0	
10	20	1.69	1790	1150	640	8.0	4.7
15	20	1.25	2690	850	1840	23.0	18.4
20	20	.67	3590	460	3130	39.1	58.4

* Heat required in evaporation process is equal to heat transfer from air to snow plus heat obtained by a lowering of snow-surface temperature.
Source: Corps of Engineers, U.S. Army

TABLE 1–20. MELTING CONSTANT FOR SNOW

[The melting constant is the depth of water in inches melted per degree day. A degree-day is a unit of heat resulting from a day with a mean temperature one degree Fahrenheit above 32°F.]

Location	Descriptive notes	Melting period	Melting constant
Albany, N.Y.	Tests of small cylinders	8–12 hours	0.04–0.06
Donner Summit, Calif.	Observations in 1917	Apr. 1–May 6	0.071
Gooseberry Creek, Utah	Field measurements	Apr. 23–May 9	0.091
Gooseberry Creek, Utah	Tests of cores	6–9 hours	0.05–0.07
Finland	All basins, 1934–1937	April	0.108
Soda Springs, Calif.	Average, 1936–1941	April	0.051
New England floods	Studies by Boston Soc. C.E.	1–14 days	0.01–0.04
N.Y. and Pa. basins	Flood runoff studies	March or April	0.04–0.07
La Grange Brook, N.Y.	Basin area, 36 acres	Mar. 28–Apr. 6	0.09
New England floods, 1936	Geol. Survey, average values	Mar. 9–22	0.03–0.05
Pemigewasset Basin, N.H.	Flood of March 1936	Mar. 17–20	0.16
Crater Lake, Ore.	Small test plots	Mar. 3–June 9	0.153
Crater Lake, Ore.	Small test plots	May 26–June 2	0.658

Source: Houk, Irrigation Engineering, v.1, John Wiley & Sons, Copyright 1951

TABLE 1–21. SNOW SURVEY REPORTS IN THE WESTERN UNITED STATES

[Snow surveys are conducted cooperatively by Federal, state, and private agencies. Coordination is provided by the U.S. Soil Conservation Service (SCS) in all western states except California where it is provided by the California Department of Water Resources (DWR). Data are collected manually and by automated remote sites (SNOTEL) that utilize a meteor burst communications system.]

REPORT	FREQUENCY OF ISSUE	LOCATION	COOPERATION
RIVER BASINS:			
Colorado - none (by SCS) for basin only - refer to Utah state			
Columbia - none (by SCS) for basin only - refer to Idaho state			
Upper Missouri - none (by SCS) for basin only - refer to Montana state			
West-wide	Monthly (Jan.–May) Oct. 1	NWS 220 N.W. 8th Room 121 Portland, OR 97209	All interested agencies and organizations
STATES:			
Alaska	Monthly (Feb.–May) Annual Sum. (Mar.)	USDA SCS 201 E. 9th Ave. Suite 300 Anchorage, AK 99501	All interested agencies and organizations
Arizona	Semimonthly (Jan. 1–Apr. 1) Oct. 1 Annual Sum. (Mar.)	USDA SCS 201 E. Indianola Ave. Suite 200 Phoenix, AZ 85012	All interested agencies and organizations

TABLE 1–21. SNOW SURVEY REPORTS IN THE WESTERN UNITED STATES (continued)

[Snow surveys are conducted cooperatively by Federal, state, and private agencies. Coordination is provided by the U.S. Soil Conservation Service (SCS) in all western states except California where it is provided by the California Department of Water Resources (DWR). Data are collected manually and by automated remote sites (SNOTEL) that utilize a meteor burst communications system.]

REPORT	FREQUENCY OF ISSUE	LOCATION	COOPERATION
California	Monthly (Feb.–May) Weekly (Feb.–May) Annual Sum. (Jan.)	State of California DWR P.O. Box 942836 Sacramento, CA 94236	All interested agencies and organizations
Colorado	Monthly (Jan.–June) Oct. 1, mid-May Annual Sum. (Mar.)	USDA SCS Diamond Hill, Bldg. A 2490 W. 26th Avenue Denver, CO 80211	All interested agencies and organizations
Idaho	Monthly (Jan.–June) Annual Sum. (Mar.)	USDA SCS Room 345 304 N. 8th St. Boise, ID 83702	State of Idaho DWR Boise, ID and all interested agencies and organizations
Montana	Monthly (Jan.–June) Oct. 1 Annual Sum. (Mar.)	USDA SCS Federal Bldg. Room 443 10 E. Babcock St. Bozeman, MT 59715	All interested agencies and organizations
Nevada	Monthly (Jan.–May) Annual Sum. (Mar.)	USDA SCS 1201 Terminal Way Second Floor Reno, NV 89502	Department of Conservation and Natural Resources Carson City, NV and all interested agencies and organizations
New Mexico	Monthly (Jan.–May) Annual Sum. (Mar.)	USDA SCS 517 Gold Ave., S.W. Room 3301 Albuquerque, NM 87102	All interested agencies and organizations
Oregon	Monthly (Jan.–June) Oct. 1 Annual Sum. (Mar.)	USDA SCS 1220 S.W. 3rd Ave. 16th Floor Portland, OR 97204	All interested agencies and organizations
Utah	Monthly (Jan.–June) Weekly (Dec.–June) Annual Sum. (Mar.)	USDA SCS 125 S. State St. P.O. Box 11350 Salt Lake City, UT 84147	Utah State Department of Natural Resources and all interested agencies and organizations
Washington	Monthly (Jan.–June) Annual Sum. (Mar.)	USDA SCS Room 360 U.S. Court House Spokane, WA 99201	All interested agencies and organizations
Wyoming	Monthly (Jan.–June) Annual Sum. (Mar.)	USDA SCS 100 E. `B' St. Room 3124 Casper, WY 82601	All interested agencies and organizations

Source: U. S. Department of Agriculture Soil Conservation Service, 1988

FIGURE 1–4. TYPICAL WINTER SNOWPACK DISTRIBUTION IN THE CONTERMINOUS UNITED STATES

The snow measurements were made at established stations that generally are at low elevation; in the mountains, snowpack usually is much thicker than that at lower elevations. The area of snow cover is based on data from images acquired by Geostationary Observational Environmental Satellite.

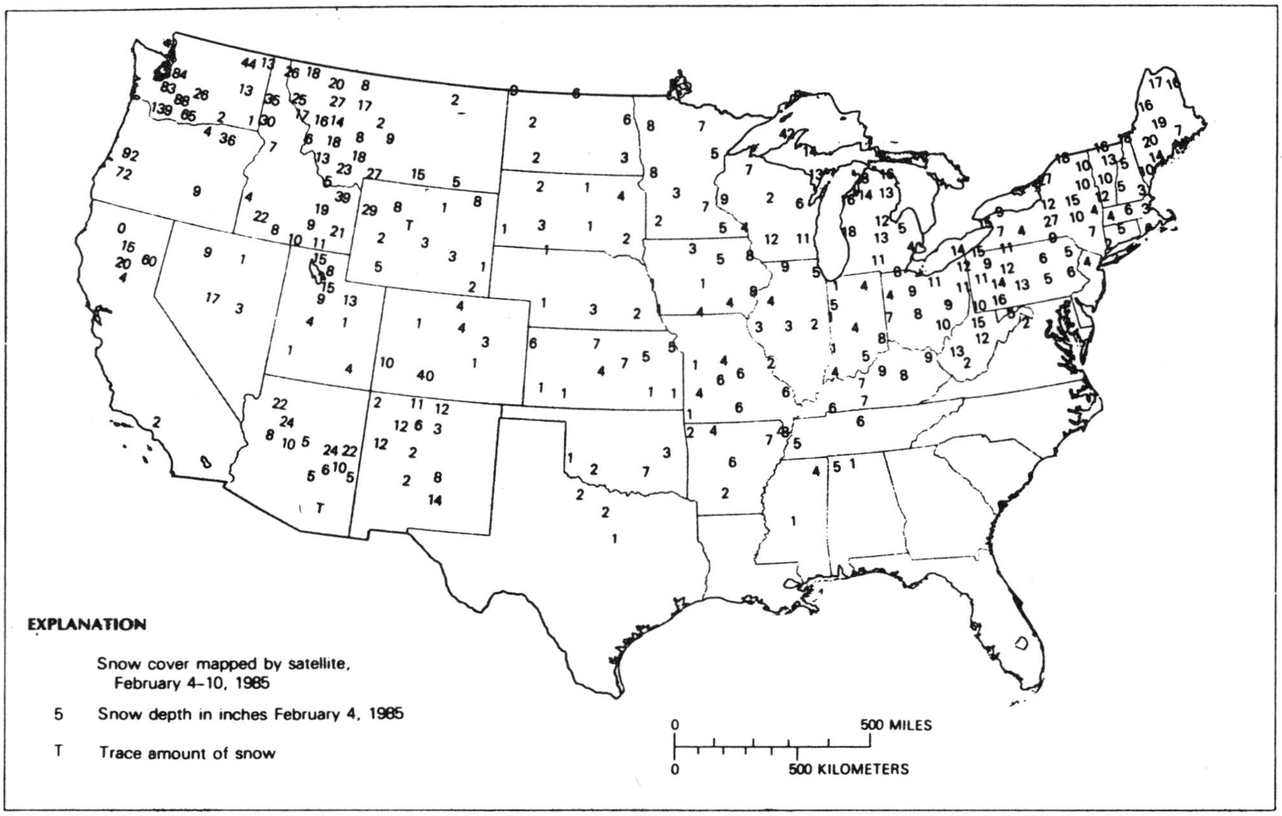

Source: U.S. Geological Survey, National Water Summary 1985—Hydrologic Events and Surface-Water Resources, U.S. Geological Survey Water-Supply Paper 2300

CHAPTER 2

Hydrologic Elements

PAGE

SECTION A. HYDROLOGIC CYCLE

FIGURE 2–1. THE HYDROLOGIC CYCLE - A DESCRIPTIVE REPRESENTATION

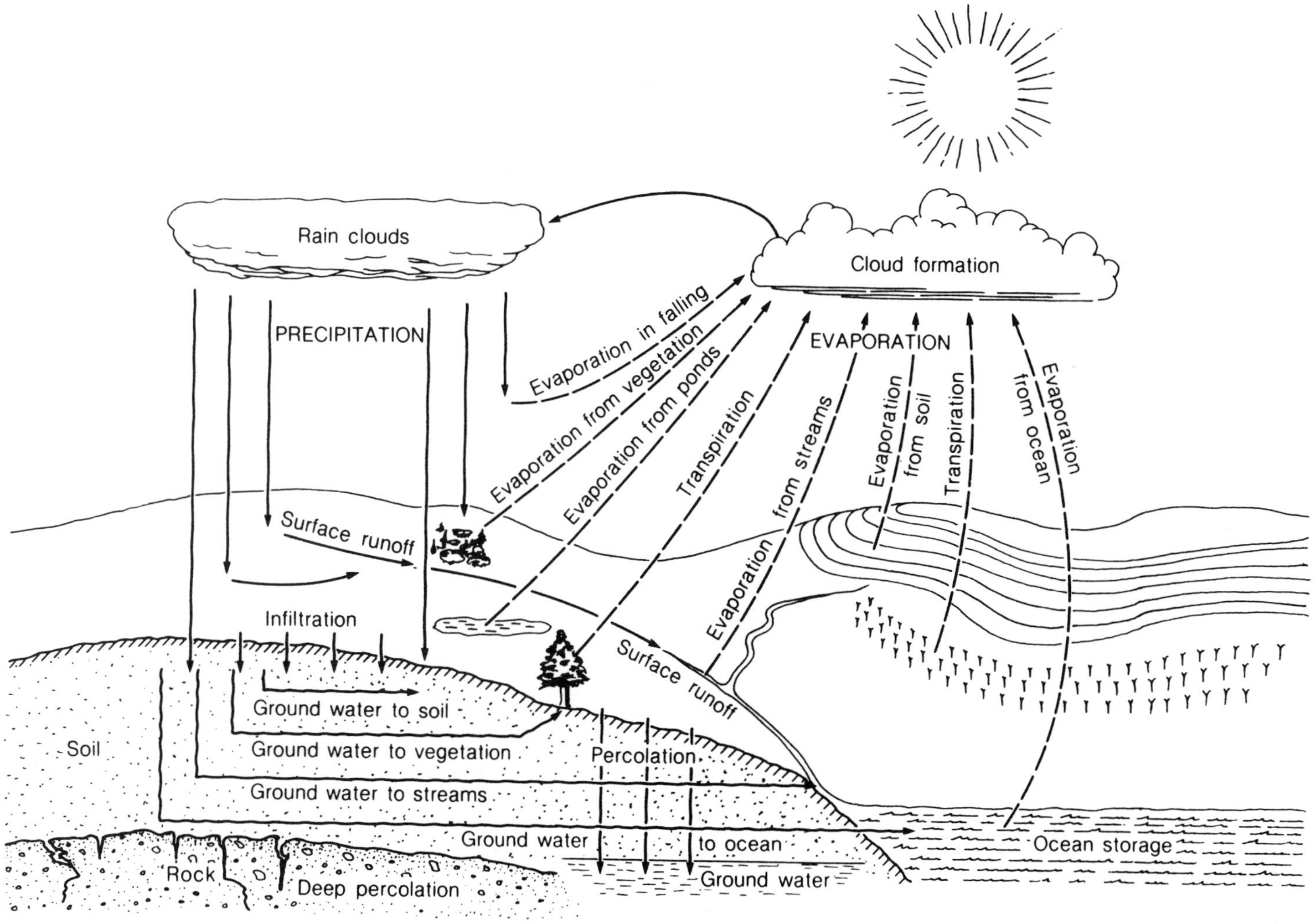

Source: Council on Environmental Quality, 1981, Washington, D.C.

TABLE 2–1. HYDROLOGIC EFFECTS OF URBANIZATION

[A selected sequence of changes in land and water use associated with urbanization]

Change in Land or Water Use	*Possible Hydrologic Effect*
Transition from pre-urban to early-urban stage:	
Removal of trees or vegetation	Decrease in transpiration and increase in storm flow. Increased sedimentation of streams.
Construction of scattered city-type houses and limited water and sewage facilities	
Drilling of wells	Some lowering of water table.
Construction of septic tanks and sanitary drains	Some increase in soil moisture and perhaps a rise in water table. Perhaps some waterlogging of land and contamination of nearby wells or streams from overloaded sanitary drain system.
Transition from early-urban to middle-urban stage:	
Bulldozing of land for mass housing, some topsoil removed, farm ponds filled in	Accelerated land erosion and stream sedimentation and aggradation. Increased flood flows. Elimination of smallest streams.
Mass construction of houses, paving of streets, building of culverts	Decreased infiltration, resulting in increased flood flows and lowered groundwater levels. Occasional flooding at channel constrictions (culverts) on remaining small streams. Occasional overtopping or undermining of banks of artificial channels on small streams.
Discontinued use and abandonment of some shallow wells	Rise in water table.
Diversion of nearby streams for public water supply	Decrease in runoff between points of diversion and disposal.
Untreated or inadequately treated sewage discharged into streams or disposal wells	Pollution of stream or wells. Death of fish and other aquatic life. Inferior quality of water available for supply and recreation at downstream populated areas.
Transition from middle-urban to late-urban stage:	
Urbanization of area completed by addition of more houses and streets and of public, commercial, and industrial buildings	Reduced infiltration and lowered water table. Streets and gutters act as storm drains, creating higher flood peaks and lower base flow of local streams.
Larger quantities of untreated waste discharged into local streams	Increased pollution of streams and concurrent increased loss of aquatic life. Additional degradation of water available to downstream users.
Abandonment of remaining shallow wells because of pollution	Rise in water table.
Increase in population requires establishment of new water-supply and distribution systems, construction of distant reservoirs diverting water from upstream sources within or outside basin	Increase in local streamflow if supply is from outside basin.
Channels of streams restricted at least in part to artificial channels and tunnels	Increased flood damage (higher stage for a given flow). Changes in channel geometry and sediment load. Aggradation.
Construction of sanitary drainage system and treatment plant for sewage	Removal of additional water from the area, further reducing infiltration and recharge of aquifer.
Improvement of storm drainage system	A definite effect is alleviation or elimination of flooding of basements, streets, and yards, with consequent reduction in damages, particularly with respect to frequency of flooding.
Drilling of deeper, large-capacity industrial wells	Lowered water-pressure surface of artesian aquifer; perhaps some local overdrafts (withdrawal from storage) and land subsidence. Overdraft of aquifer may result in salt-water encroachment in coastal areas and in pollution or contamination by inferior or brackish waters.
Increased use of water for air conditioning	Overloading of sewers and other drainage facilities. Possibly some recharge to water table, due to leakage of disposal lines.
Drilling of recharge wells	Raising of water-pressure surface.
Waste-water reclamation and utilization	Recharge to groundwater aquifers. More efficient use of water resources.

Source: U.S. Geological Survey

SECTION B. WATER RESOURCES — UNITED STATES

TABLE 2–2. DISTRIBUTION OF WATER IN THE CONTINENTAL UNITED STATES

	Volume		Annual Circulation	Replacement Period
	X 10^9 m^3	%	(X 10^9 m^3/year)	(in years)
Liquid water				
Groundwater				
Shallow (<800 m deep)	63,000	43.2	310	>200
Deep (>800 m deep)	63,000	43.2	6.2	>10,000
Freshwater lakes	19,000	13.0	190	100
Soil moisture (1-m root zone)	630	0.43	3,100	0.2
Salt lakes	58	0.04	5.7	>10
Average in stream channels	50	0.03	1,900	<0.03
Water vapor in atmosphere	190	0.13	6,200	>0.03
Frozen water, glaciers	67	0.05	1.6	>40

Source: Ad Hoc Panel on Hydrology, Scientific Hydrology, Washington, D.C.: Federal Council for Science and Technology, 1962

FIGURE 2–2. HYDROLOGIC CYCLE SHOWING THE GROSS WATER BUDGET OF THE CONTERMINOUS UNITED STATES

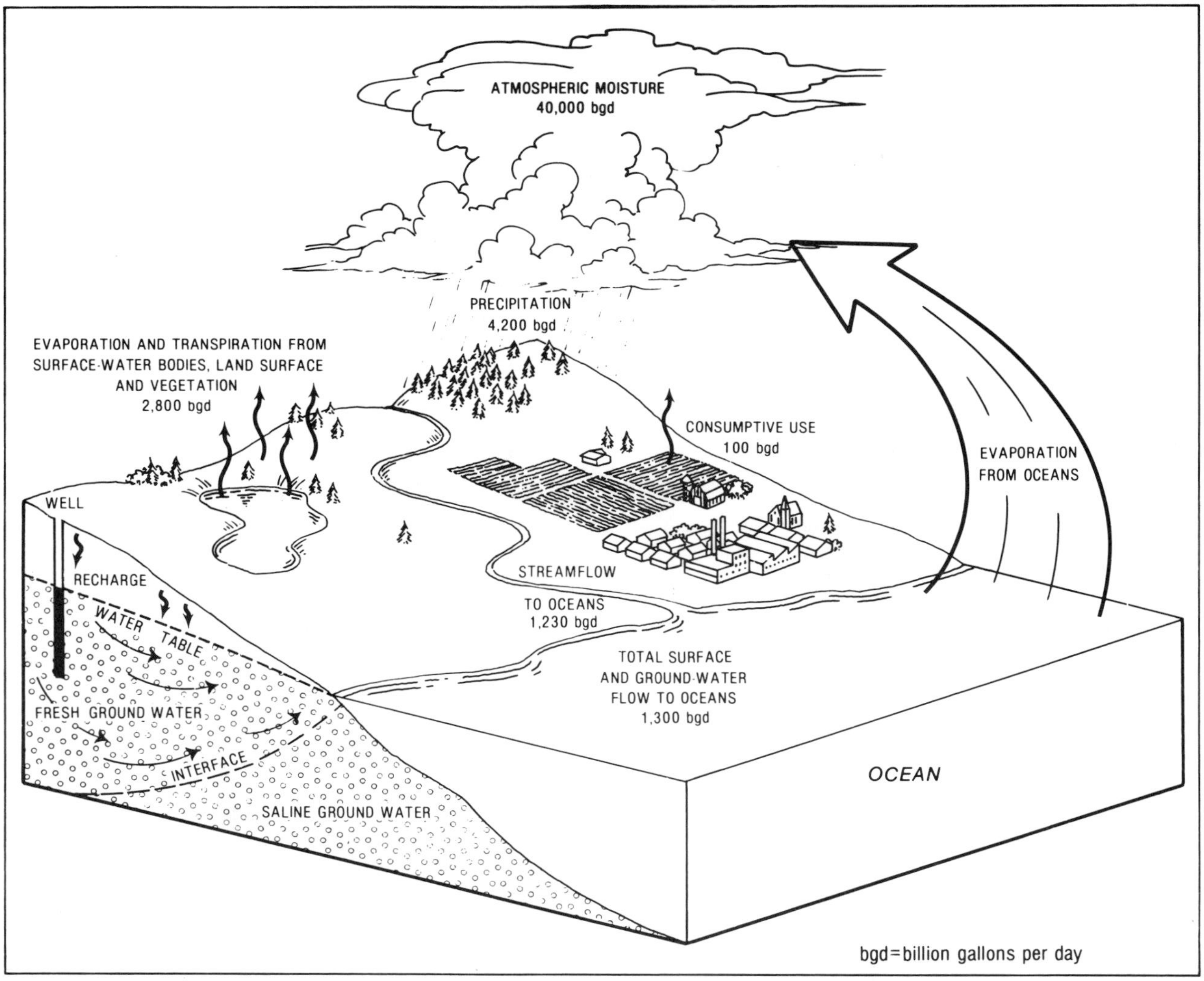

Source: U.S. Geological Survey, 1984, National Water Summary 1983 - Hydrologic Events and Issues, Water-Supply Paper 2250

TABLE 2–3. SOME PURPOSES OF WATER-RESOURCES DEVELOPMENT

Purpose	Description	Type of Works and Measures
Flood control	Flood-damage abatement or reduction, protection of economic development, conservation storage, river regulation, recharging of groundwater, water supply, development of power, protection of life	Dams, storage reservoirs, levees, floodwalls, channel improvements, floodways, pumping stations, flood-plain zoning, flood forecasting
Irrigation	Agricultural production	Dams, reservoirs, wells, canals, pumps and pumping plants, weed-control and desilting works, distribution systems, drainage facilities, farmland grading
Hydroelectricity	Provision of power for economic development and improved living standards	Dams, reservoirs, penstocks, power plants, transmission lines
Navigation	Transportation of goods and passengers	Dams, reservoirs, canals, locks, open-channel improvements, harbor improvements
Domestic and industrial water supply	Provision of water for domestic, industrial, commercial, municipal, and other uses	Dams, reservoirs, wells, conduits, pumping plants, treatment plants, saline-water conversion, distribution systems
Watershed management	Conservation and improvement of the soil, sediment abatement, runoff retardation, forests and grassland improvement, and protection of water supply	Soil-conservation practices, forest and range management practices, headwater-control structures, debris-detention dams, small reservoirs, and farm ponds
Recreational use of water	Increased well-being and health of the people	Reservoirs, facilities for recreational use, works for pollution control, preservation of scenic and wilderness areas
Fish and wildlife	Improvement of habitat for fish and wildlife, reduction or prevention of fish or wildlife losses associated with man's works, enhancement of sports opportunities, provision for expansion of commercial fishing	Wildlife refuges, fish hatcheries, fish ladders and screens, reservoir storage, regulation of streamflows, stocking of streams and reservoirs with fish, pollution control, and land management
Pollution abatement	Protection or improvement of water supplies for municipal, domestic, industrial and agricultural uses and for aquatic life and recreation	Treatment facilities, reservoir storage for augmenting low flows, sewage-collection systems, legal control measures
Insect control	Public health, protection of recreational values, protection of forests and crops	Proper design and operation of reservoirs and associated works, drainage, and extermination measures
Drainage	Agricultural production, urban development and protection of public health	Ditches, tile drains, levees, pumping stations, soil treatment
Sediment control	Reduction or control of silt load in streams and protection of reservoirs	Soil conservation, sound forest practices, proper highway and railroad construction, desilting works, channel and revetment works, bank stabilization, special dam construction and reservoir operations
Salinity control	Abatement or prevention of salt-water contamination of agricultural, industrial, and municipal water supplies	Reservoirs for augmenting low streamflow, barriers, groundwater recharge, coastal jetties

Source: Chow, V.T., 1979, Water as a World Resource, Water International, Vol.4, No.1, p.6

SECTION C. WORLD WATER BALANCE

TABLE 2–4. ESTIMATED WORLD WATER SUPPLY AND BUDGET

Water Item	Volume (Thousands)		Percent of Total Water
	Cubic Miles	Cubic Kilometers	
Water in land areas:			
Fresh-water lakes	30	125	0.009
Saline lakes and inland seas	25	104	.008
Rivers (average instantaneous volume)	.3	1.25	.0001
Soil moisture and vadose water	16	67	.005
Ground water to depth of 4,000 m (about 13,100 ft)	2,000	8,350	.61
Icecaps and glaciers	7,000	29,200	2.14
Total in land area (rounded)	9,100	37,800	2.8
Atmosphere	3.1	13	.001
World ocean	317,000	1,320,000	97.3
Total, all items (rounded)	326,000	1,360,000	100
Annual evaporation:[1]			
From world ocean	85	350	0.025
From land areas	17	70	.005
Total	102	420	0.031
Annual precipitation:			
On world ocean	78	320	0.024
On land areas	24	100	.007
Total	102	420	0.031
Annual runoff to oceans from rivers and icecaps	9	38	0.003
Ground-water outflow to oceans[2]	.4	1.6	.0001
Total	9.4	39.6	0.0031

[1] Evaporation (420,000 km^3) is a measure of total water participating annually in the hydrologic cycle.

[2] Arbitrarily set equal to about 5 percent of surface runoff.

Source: Nace, U.S. Geological Survey, 1967

TABLE 2–5. WORLD WATER BALANCE, BY CONTINENT

Water Balance Elements	Europe[1]	Asia	Africa	North America[2]	South America	Australia[3]	Total Land Area[4]
Area, millions of km^2	9.8	45.0	30.3	20.7	17.8	8.7	132.3
			in mm				
Precipitation, P	734	726	686	670	1,648	736	834
Total river runoff, R	319	293	139	287	583	226	294
Groundwater runoff, U	109	76	48	84	210	54	90
Surface water runoff, S	210	217	91	203	373	172	204
Total infiltration and soil moisture, W	524	509	595	467	1,275	564	630
Evaporation, E	415	433	547	383	1,065	510	540
			in km^3				
Precipitation	7,165	32,690	20,780	13,910	29,355	6,405	110,303
Total river runoff	3,110	13,190	4,225	5,960	10,380	1,965	38,830
Groundwater runoff	1,065	3,410	1,465	1,740	3,740	465	11,885
Surface water runoff	2,045	9,780	2,760	4,220	6,640	1,500	26,945
Total infiltration and soil moisture	5,120	22,910	18,020	9,690	22,715	4,905	83,360
Evaporation	4,055	19,500	16,555	7,950	18,975	4,440	71,475
			relative values				
Groundwater runoff as percent of total runoff	34	26	35	32	36	24	31
Coefficient of groundwater discharge into rivers	0.21	0.15	0.08	0.18	0.16	0.10	0.14
Coefficient of runoff	0.43	0.40	0.23	0.31	0.35	0.31	0.36

[1] Including Iceland.
[2] Excluding the Canadian archipelago and including Central America.
[3] Including Tasmania, New Guinea and New Zealand, only within the limits of the continent: P - 440 mm, R - 47 mm, U - 7 mm;, S - 40 mm, W - 400 mm, E - 393 mm.
[4] Excluding Greenland, Canadian archipelago and Antarctica.
Source: Lvovitch, M.I., EOS, Vol. 54, No. 1, Jan. 1973, Copyright by American Geophysical Union

TABLE 2–6. WORLD-WIDE PER CAPITA WATER RESOURCES, BY CONTINENT

	Population in Millions (1969)	Annual Volume of River Runoff, km^3		Runoff Volume per Capita, m^3/year	
		Total	Stable Runoff	Total	Stable Runoff
Europe	642	3,110	1,325	4,850	2,100
Asia	2,047	13,190	4,005	6,440	1,960
Africa	345	4,225	1,905	12,250	5,500
North America	312	5,960	2,380	19,100	7,640
South America	185	10,380	3,900	56,100	21,100
Australia[1]	18	1,965	495	10,900	2,750
Total land area	3,549	38,830	14,010	10,940	3,950

[1] Including New Guinea and New Zealand.
Source: Lvovitch, M.I., EOS, Vol. 54, No. 1, Jan. 1973, Copyright by American Geophysical Union

FIGURE 2–3. WATER AVAILABILITY ON EARTH

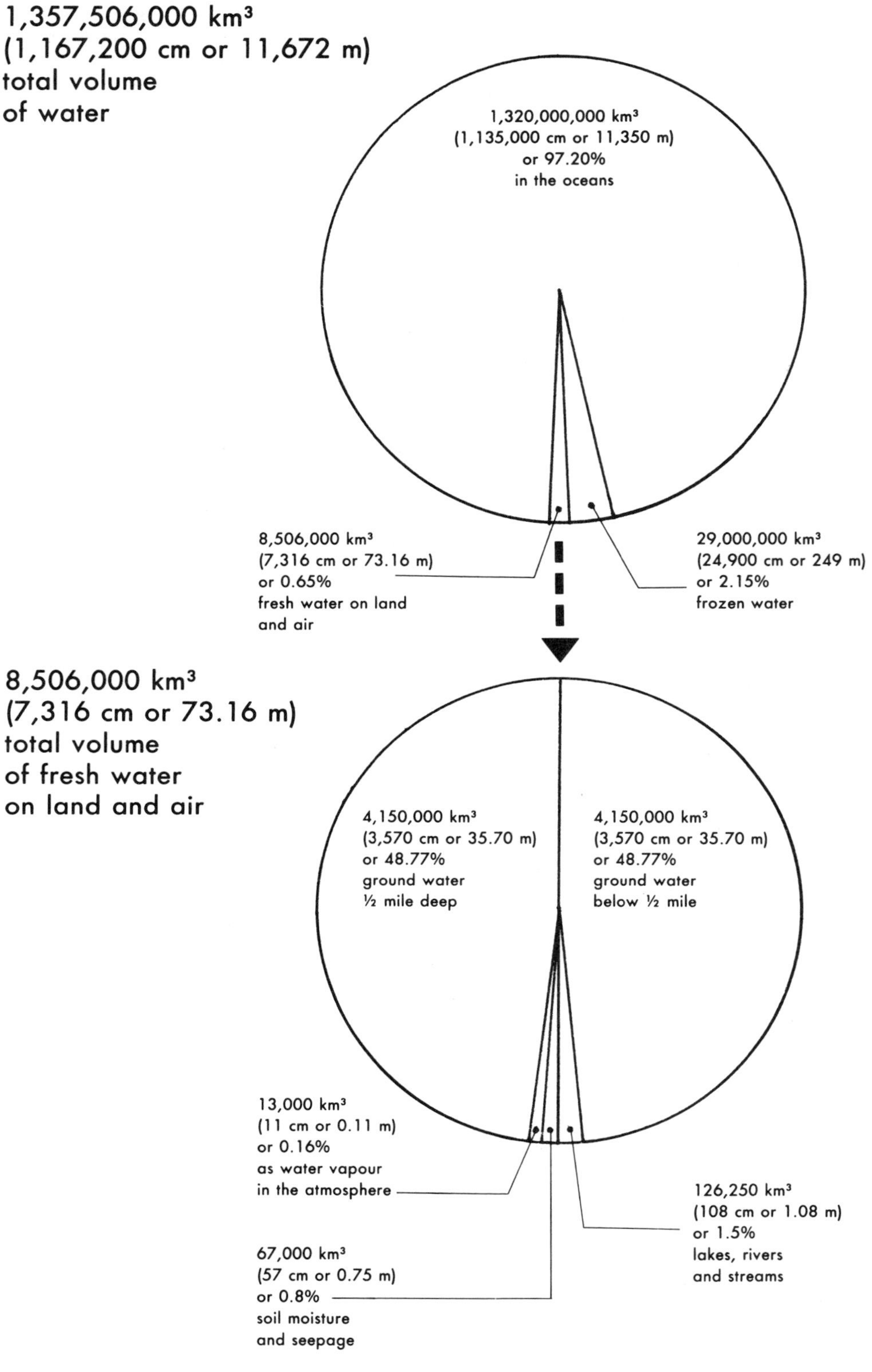

note: figures in brackets indicate the height that the relevant quantities of water would reach if they were placed on the whole non-frozen land area of the earth which is 116,400,000 km²

Source: Doxiadis, C.A., 1967, Water and Environment, International Conference on Water for Peace, Washington, D.C.

TABLE 2–7. EXPECTED CHANGES IN THE WORLD WATER BALANCE BY THE YEAR 2000

Elements of the Balance	Water Balance,[1] km³ 1974	2000	Nature of Transformation
Precipitation	110 300	110 300	Conversion of 700 km³ of surface runoff into soil moisture resources from plowland, and a 600 km³ increase of evapotranspiration in forests and from the surface of reservoirs
Total runoff	38 800[2]	37 500[3]	
Stable runoff	14 000	22 500	An increase of stable runoff by 8 500 km³ in the following ways:
Groundwater runoff and replenishable groundwater supply	12 000	17 000	Storage of 5 000 km³ of groundwater
Runoff regulated by lakes and reservoirs	2 000	5 500	Regulation of 3 500 km³ of floodwaters by reservoirs
Surface (flood) runoff	26 800	20 500	Use of 6 300 km³ of surface runoff, including 1 300 km³ of moisture retained in soil and larger evapotranspiration and 5 000 km³ for storage of groundwater
Total infiltration and soil moisture	83 500	89 800	Increase of 6 300 km³ due to additional wetting of unirrigated land and increased evapotranspiration—1 300 km³, and also by virtue of groundwater storage—5 000 km³
Evapotranspiration	71 500	72 800	Increase of 1 300 km³ because of higher yields and increased evaporation from reservoirs

[1] Rounded off.
[2] Not including runoff (ice) from polar glaciers.
[3] Not including runoff (ice) from polar glaciers and not taking into account consumptive use of water for human needs.
Source: Lvovitch, M.I., 1979, World Water Resources and the Future, Copyright American Geophysical Union

SECTION D. HYDROLOGIC DATA

FIGURE 2–4. LOCATIONS OF NASQAN AND NATIONAL HYDROLOGIC BENCH-MARK STATIONS IN THE UNITED STATES IN 1985

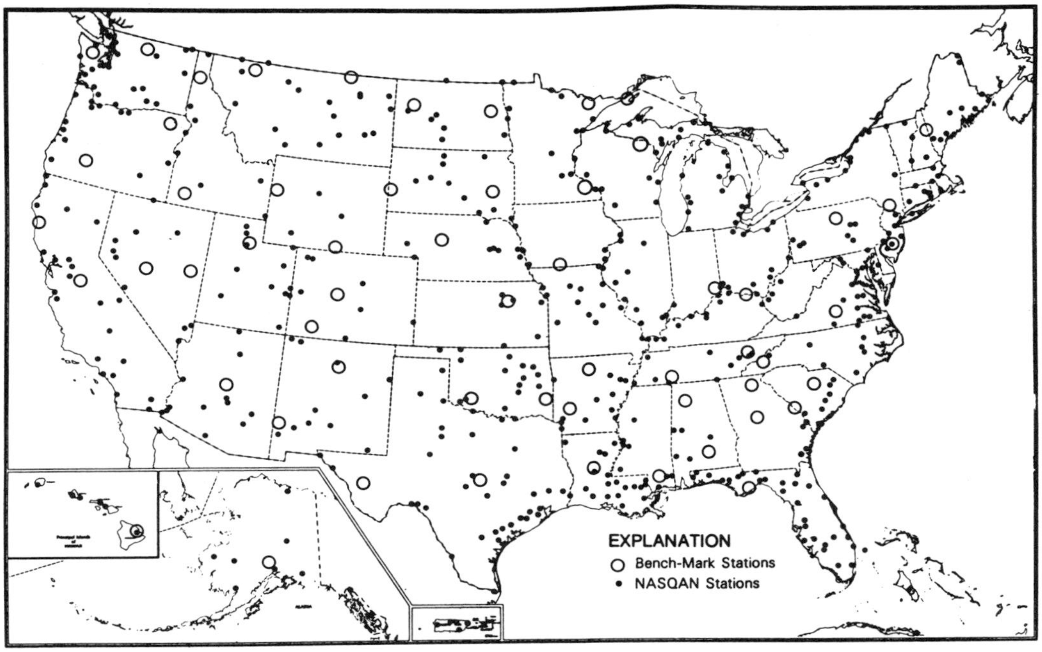

Source: Cardin, C.W. and others, 1986, Water Resources Division in the 1980's, U.S. Geological Survey Circular 1005

The U.S. Geological Survey operates two special-purpose programs to monitor surface-water quality—the National Stream Quality Accounting Network (NASQAN) and the National Hydrologic Bench-Mark Program. These programs are designed to answer the question, "What is the quality of water in the major river systems of the United States?" and "Is the quality changing?" Neither of these networks is designed to determine why the water quality might be changing. However, research is underway to improve knowledge of the processes that influence the natural and changing quality of water.

National Stream-Quality Accounting Network

Purpose:
- To obtain information on the quality and quantity of surface water moving within and from the United States through a systematic and uniform process of data collection, summarization, analysis and reporting so that the data may be used to—
  - Describe the variability of water-quality in the nation's streams.
  - Detect changes or trends in water-quality characteristics.
  - Provide a nationally consistent data base useful for water-quality assessment and hydrologic research.

Activities:
- Monitor a comprehensive set of physical and chemical characteristics of surface water at stations throughout the country to detect changes in stream quality and to determine long-term trends.
- Assess water quality from a regional and national perspective, through interpretive projects in the Districts of the Water Resources Division.

Accomplishments:
- Established more than 550 NASQAN stations in 1973–1985. NASQAN stations are generally located at or near the downstream end of hydrologic accounting units or at representative sites along coastal areas and the Great Lakes.
- Established within NASQAN a radiochemical surveillance subnetwork of 52 sampling sites.
- Developed and refined a statistical procedure for determining trends in water-quality constituents.
- Developed a computer-based station information file for use in preparing information reports, answering requests for information, and program operation and planning needs.

National Hydrologic Bench-Mark Program

Purpose:
- Provide consistent hydrologic data (including water quality and related factors) in representative undeveloped watersheds nationwide, and provide analyses on a continuing basis to compare and contrast conditions observed in basins more obviously affected by human activity.

Activities:
- Monitor streamflow and physical and chemical characteristics at selected "natural" sites throughout the country.
- Provide summary, analysis, and interpretation of the data.

Accomplishments:
- Established 58 monitoring stations.
- Utilized data from the program in a search for evidence of effects of acid precipitation on stream waters.
- Published reports that describe the environment and the natural water quality of remote watersheds not "influenced" by human activities. Compared these data with data on the water quality of major streams that drain the same regions.
- Started studies to summarize all streamflow quantity and quality data and to analyze trends in quantity and quality characteristics.

TABLE 2–8. CHARACTERISTICS MEASURED AT NASQAN STATIONS

[Frequencies: C, continuous; D, daily; M, monthly; Q, quarterly]

Characteristic	Frequency
Field determinations:	
Water temperature	[1]C, D, or M
Specific conductance	[1]C, D, or M
pH	M
Discharge	C
Coliform, fecal	M
Streptococci, fecal	M
Common constituents (dissolved):	[2]M or Q
(Bicarbonate, carbonate, total hardness, noncarbonate hardness, calcium, magnesium, fluoride, sodium, potassium, dissolved solids, silica, turbidity, chloride, and sulfate).	
Major nutrients:	
Phosphorus, total as P	M
Nitrite plus nitrate, total as N	M
Nitrogen, total Kjeldahl as N	M
Trace elements (total and dissolved):	Q
(Arsenic, cadmium, chromium, cobalt, copper, iron, lead, manganese, mercury, selenium, and zinc).	
Organics and biological:	
Organic carbon, total	Q
Phytoplankton, total, cells/ml	M
Phytoplankton, identification of 3 codominants	M
Phytoplankton, 3 codominants, percent of total	M
Periphyton, biomass, dry weight g/m^2	Q
Periphyton, biomass, ash weight g/m^2	Q
Periphyton, chlorophyll *a*	Q
Periphyton, chlorophyll *b*	Q
Suspended sediment:	
Suspended sediment concentration	M
Percent finer than 0.062-mm sieve diameter	M

[1] Continuous or daily depending upon whether the station is equipped with a monitor or whether daily observations are made. Monthly measurements made at stations where a long-term record is available.
[2] Quarterly or monthly, depending upon whether relationships have been established between conductance and concentrations of various common constituents.
Source: U.S. Geological Survey Annual Report, Fiscal Year 1976

TABLE 2–9. HYDROLOGIC AND RELATED DATA COLLECTION NETWORKS IN THE UNITED STATES

Type of Network	Number of Stations
Automatic meteorological observing stations (full parameter); temperature, dew point, wind, pressure, precipitation)	92
National weather service synoptic and basic observation stations (high quality observations for basic weather program)	67
Cooperative station services (observations by lay persons):	
Temperature and precipitation	5,568
Precipitation only—daily	3,200
Precipitation only—storage	32
Hourly precipitation stations equipped with recording precipitation gages	3,205
Cooperative stations equipped with both recording and nonrecording precipitation gages	1,995
Crop reporting stations	566
River and/or rainfall reporting stations	
River stage reports only	998
Rainfall reports only	3,656
River stage and rainfall reports	1,069
Evaporation storage	448
Reference Climatological Stations	21
Automated Hydrologic Observing System (AHOS) - river and rainfall data for flood forecasting	
AHOS/T[1]	506
AHOS/S[2]	75
Special reporting stations	293
Cooperative station data published	
Temperature and precipitation	8,256
Precipitation only	3,055
Evaporation	431
Soil temperature	308
Miscellaneous (snow density, special meteorological, etc.)	473

[1] Data transmitted by telephone.
[2] Data transmitted by satellite.
Source: National Weather Service, 1985, Operations of the National Weather Service

TABLE 2–10. HYDROLOGIC DATA COLLECTION STATIONS OPERATED BY THE U.S. GEOLOGICAL SURVEY IN 1987

Type of Station	Number of Stations by Source of Funding							Total Stations
	Single Program Support			Combined Support				
	Federal Program (Federal)	Federal-State Coopera-tive Program (COOP)	Reimburse-ment from Other Federal Agencies (OFA)	Federal, COOP	Federal, OFA	COOP, OFA	Federal, COOP, OFA	
SURFACE WATER								
Discharge								
Continuous record	481	3,158	1,575	1,130	146	472	38	7,000
Partial record	99	2,880	273	268	91	12	1	3,624
Stage Only - Streams								
Continuous record	13	86	245	93	3	8	0	448
Partial record	1	166	28	33	0	1	0	229
Stage Only - Lakes and Reservoirs								
Continuous record	12	274	277	210	1	4	1	779
Partial record	11	177	75	87	0	0	0	350
Quality								
Continuous record	66	243	217	62	9	2	0	599
Scheduled, long-term operation	389	1,109	330	312	24	19	2	2,185
Short-term or project stations	26	512	116	47	15	0	0	716
GROUND WATER								
Water Levels								
Continuous record	90	1,413	200	656	14	0	0	2,373
Scheduled, long-term operation	656	17,089	1,098	3,791	0	0	0	22,634
Short-term or project stations	1,157	4,202	1,061	3,219	315	0	0	9,954
Quality								
Scheduled, long-term operation	49	3,053	158	640	0	0	0	3,900
Short-term or project stations	560	3,475	730	429	26	0	0	5,220
SEDIMENT								
Daily Sampling	29	65	72	5	1	2	0	174
Periodic Sampling	460	225	161	17	10	5	0	878
PRECIPITATION								
Quantity	45	460	319	79	6	0	0	909
Quality	32	32	8	3	3	0	0	78

Source: Condes de la Torre, Alberto, 1987, U.S. Geological Survey Open File Report 87–663

TABLE 2–11. NUMBER AND TYPE OF WATER RESOURCES REPORTS ISSUED BY THE U.S. GEOLOGICAL SURVEY, 1975–1986

Type of Report	Number of Approved Reports											
	1975	1976	1977	1978	1979	1980	1981	1982	1983	1984	1985	1986
Water Supply Paper	25	18	17	7	13	32	28	13	30	22	22	22
Professional Paper	20	13	13	24	9	9	13	12	16	13	14	12
Bulletin	0	0	0	1	1	1	0	1	1	3	1	2
Circular	2	10	6	6	4	8	12	3	10	1	5	4
Techniques of Water Resources Investigations	1	4	0	4	0	2	3	3	3	2	1	2
Map Series	43	34	21	30	29	35	13	13	20	18	21	20
Water Resources Investigations Reports	61	134	136	144	151	222	200	222	283	398	344	270
Open-File Book and Map Reports	113	107	68	180	153	295	265	275	192	220	208	158
Water Resources Annual Data Reports	78	68	65	71	72	81	78	76	74	64	77	64
Journal of Research Articles[1]	27	28	19	13	...	...	...	...	...	...	...	...
Administrative Releases	0	26	23	12	12	16	2	10	4	5	4	7
Outside Publications (Includes State Publications, Journal Articles, Abstracts, and Other Related Items)	597	621	648	700	444	501	495	599	582	702	858	835
Flood-Prone Area Mapping Project:												
Maps[2]	1,210	451	171	103	285	222	129	173	...	...	...	...
Pamphlet[3]	330	57	...	...	...	...	...	...	...	...	...	...
Total (exclusive of flood-prone maps and pamphlets)	967	1,063	1,016	1,192	888	1,202	1,109	1,227	1,215	1,448	1,555	1,396

[1] The journal was discontinued in 1979.
[2] The flood-prone area mapping project was completed in 1982.
[3] Completed in 1976.
Source: Cardin, C.W. and others, 1986, Water Resources Division in the 1980's, U.S. Geological Survey Circular 1005

TABLE 2–12. GEOGRAPHICAL DISTRIBUTION OF GLOBAL ENVIRONMENTAL MONITORING SYSTEM (GEMS) HYDROLOGY STATIONS AND DATA COLLECTED AS OF END OF 1984

Area (WHO Region)	Number of Countries	Rivers		Lakes and Reservoirs		Ground Water		Total Stations	Total Number of Measurements
		Stations	Number of Measurements	Stations	Number of Measurements	Stations	Number of Measurements		
Africa	5	22	815	14	3 890	3	140	39	4 845
Americas	13	72	45 902	15	10 187	20	1 862	107	57 951
Eastern Mediterranean	10	28	16 182	12	7 450	29	1 721	69	25 353
Europe	17	100	67 652	13	15 124	11	2 679	124	85 455
South-East Asia	5	40	30 335	3	1 349	20	8 393	63	40 077
Western Pacific	10	35	44 840	9	8 241	3	1 767	47	54 848
Total	60	297 (66.1%)	205 726 (76.6%)	66 (14.7%)	46 241 (17.2%)	86 (19.2%)	16 562 (6.2%)	449	268 529

Four United Nations agencies coordinate the GEMS/Water project: the United Nations Environment Programme (UNEP), WHO, the World Meteorological Organization (WMO) and the United Nations Educational, Scientific and Cultural Organization (UNESCO). The specific objectives of the GEMS/Water project are: *(i)* monitoring the incidence and long-term trends of pollutants in water; *(ii)* assisting participating countries in establishing their own national monitoring network; (iii) improving the validity and comparability of water-quality data within and between participating countries; *(iv)* providing advance warning of any serious deterioration; and *(v)* prompting governments to initiate, individually and/or collectively, corrective action for the protection, restoration and improvement of the environment.

Source: Barabas, Silvio, 1986, Monitoring Natural Waters for Drinking-Water Quality, World Health Statistics Quarterly, Vol. 39,No.1

SECTION E. INTERCEPTION

TABLE 2–13. INTERCEPTION BY TREES

[Interception includes stemflow and is expressed as a percentage of annual precipitation]

Type or Species	Age or Size	Place in Succession	Locality	Interception, Percent
Hemlock	Mature	Climax	Connecticut	48
Douglas fir	25 yr	Climax	Washington	43
Hemlock	Mature	Climax	New Hampshire	38
Spruce-fir	Mature	Climax	Maine	37
Hemlock	Mature	Climax	Adirondacks, New York	34
Douglas fir	Mature	Climax	Washington	34
Hemlock	Mature	Climax	Ithaca, New York	31
Spruce—fir—paper birch	Mature	Climax	Maine	26
White pine—hemlock	Mature	Climax	Massachusetts	24
Western white pine—western hemlock	Overmature	Climax	Idaho	21
Maple—beech	Mature	Climax	New York	43
Mixed	Mature	Climax	New York	40
Maple—hemlock	Mature, cutover	Climax	Wisconsin	25
Beech—birch—maple	Mature	Climax	Ontario	21
Ponderosa pine	Mature	Preclimax	Arizona	40
Lodgepole pine	Mature	Preclimax	Colorado	32
Ponderosa pine	Mature	Preclimax	Idaho	27
Jeffrey pine	Mature	Preclimax	Southern California	26
Lodgepole pine	32 yr	Preclimax	Colorado	23
Ponderosa pine	Mature	Preclimax	Idaho	22
Ponderosa pine	Young	Pioneer	Colorado	18
Calif. scrub oak	6 ft	....	Southern California	31
Mixed brush	Mature	Preclimax	North Fork, California	19
White pine—red pine	40 yr	Preclimax	Ontario	37
Jack pine	50 yr	Pioneer	Wisconsin	21
Shortleaf pine	45 yr	Pioneer	North Carolina	16
Quaking aspen	32 yr	Pioneer	Colorado	16
Chaparral, mixed	6 ft		Southern California	17
Maple—hemlock	Mature (after leaf fall)	Climax (understocked)	Wisconsin	16
Hemlock	Mature	Climax	New York	13
Oak-pine	Open, second growth	Preclimax	New Jersey	13
Ponderosa—lodgepole pine	25 ft	Preclimax	Idaho	8
Beech—maple	Mature	Climax	New York	6
Chamise	6 ft	Pioneer	Southern California	3

Source: Compilation of data from various references, Kittredge, Forest Influences, McGraw-Hill, Copyright 1948

TABLE 2–14. INTERCEPTION BY VARIOUS FOREST TYPES

Forest Type	Gross interception		Stemflow		Net interception		Net Snow Interception, %
	With Leaves %	Without Leaves, %	With Leaves, %	Without Leaves, %	With Leaves, %	Without Leaves, %	
Northern hardwood	20	17	5	10	15	7	10
Aspen—birch	15	12	5	8	10	4	7
Spruce—spruce-fir	35	...	3	...	32	...	35
White pine	30	...	4	...	26	...	25
Hemlock	30	...	2	...	28	...	25
Red pine	32	...	3	...	29	...	30

Source: U.S. Forest Service

TABLE 2–15. INTERCEPTION BY VARIOUS CROPS

Description	Alfalfa	Corn	Soybean	Oats
During growing season:				
Rainfall, in.	10.81	7.12	6.25	6.77
Canopy penetration, in.	6.18	4.84	4.06	6.30
Stemflow, in.	0.76	1.18	1.28	
Interception, in.	3.87	1.10	0.91	0.47
Interception, %	35.8	15.5	14.6	6.9
During low-vegetation development, %	21.9	3.4	9.1	3.1

Source: U.S. Dept. of Agriculture

SECTION F. INFILTRATION

TABLE 2–16. SEEPAGE RATES FOR CANALS

[Values are average maximum rates through the wetted area]

Canal Soil Material	Seepage, Feet per Day	Canal Soil Material	Seepage, Feet per Day
Sandy loam	8.2	Loam and adobe	1.4
Gravelly loam	5.3	Loam	1.1
Fine sandy loam and adobe	3.8	Silty clay	0.9
Sand and sandy loam	3.4	Sand and silty clay	0.4
Loam and sandy loam	3.3	Sand and clay	0.1
Adobe	3.0	Loam and gravelly loam	0.1
Fine sandy loam	2.1		

Source: Rohwer and Stout, Colo. Agric. Exp. Sta. Bull., 1948

TABLE 2–17. INFILTRATION RATE AND LAND USE

[Rank of land uses in order of infiltration rate; first use listed has lowest rate]

1. Fallow
2. Row crops, poor rotation[1]
3. Row crops, good rotation[2]
4. Pasture, poor
5. Legumes after row crops
6. Small grains, poor rotation
7. Small grains, good rotation
8. Pasture, fair
9. Woods, poor
10. Pasture, good
11. Woods, fair
12. Meadows
13. Woods, good

[1] One-fourth or less in hay or sod.
[2] More than one-fourth of rotation in hay or sod.
Source: U.S. Soil Conservation Service

FIGURE 2–5. TOTAL ANNUAL INFILTRATION AND SOIL MOISTURE IN THE WORLD (in mm)

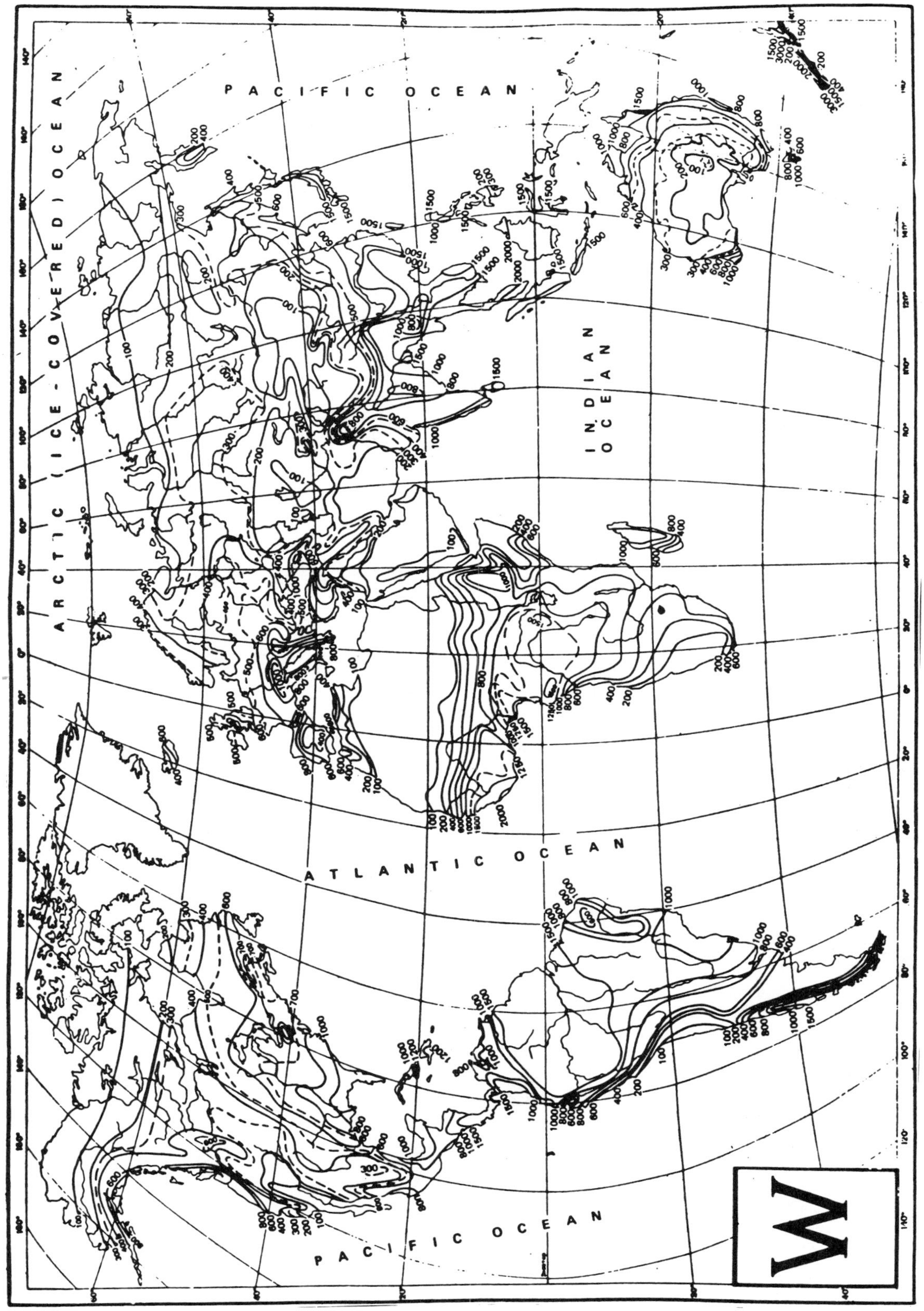

Source: Lvovitch, M.I., EOS, Vol. 54, No. 1, 1973. Copyright by American Geophysical Union

SECTION G. RUNOFF

TABLE 2–18. RUNOFF IN THE UNITED STATES

[Annual natural runoff in billions of gallons per day; regions are shown in Fig. 2–6]

Region	Mean	50%[1]	90%[1]	95%[1]
North Atlantic[2]	163	163	123	112
South Atlantic-Gulf	197	188	131	116
Great Lakes[2,3]	63.2	61.4	46.3	42.4
Ohio[4]	125	125	80.0	67.5
Tennessee	41.5	41.5	28.2	24.4
Upper Mississippi[4]	64.6	64.6	36.4	28.5
Lower Mississippi[4]	48.4	48.4	29.7	24.6
Souris-Red-Rainy[2]	6.17	5.95	2.60	1.91
Missouri[2]	54.1	53.7	29.9	23.9
Arkansas-White-Red	95.8	93.4	44.3	33.4
Texas-Gulf	39.1	37.5	15.8	11.4
Rio Grande[5]	4.9	4.9	2.6	2.1
Upper Colorado[6]	13.45	13.45	8.82	7.50
Lower Colorado[4,5]	3.19	2.51	1.07	0.85
Great Basin[4]	5.89	5.82	3.12	2.46
Columbia-North Pacific[2]	210	210	154	138
California[5]	65.1	64.1	32.8	25.6
Conterminous United States[7]	1,201			
Alaska[2]	580	8	8	8
Hawaii	13.3	8	8	8
United States[7]	1,794			

[1] Flow exceeded in indicated percent of years.
[2] Does not include runoff from Canada.
[3] Does not include net precipitation on the Lakes.
[4] Does not include runoff from upstream regions.
[5] Does not include runoff from Mexico.
[6] Virgin flow. Mean annual natural runoff estimated to be 13.7 bgd.
[7] Rounded.
[8] Not available.
Source: U.S. Water Resources Council, 1968

FIGURE 2–6. WATER RESOURCES REGIONS OF THE UNITED STATES

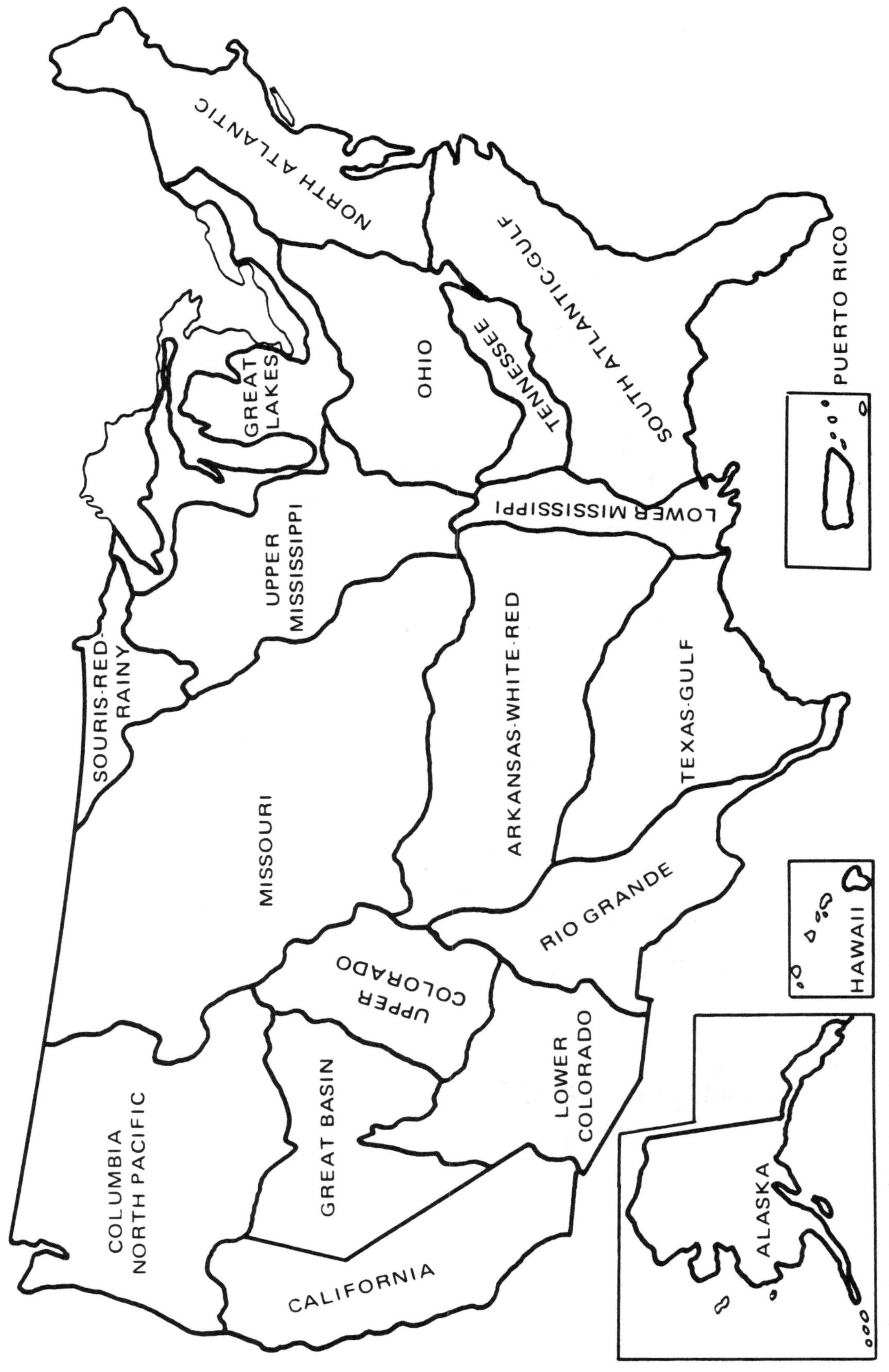

Source: U.S. Water Resources Council, 1968

FIGURE 2–7. AVERAGE ANNUAL RUNOFF IN THE UNITED STATES.

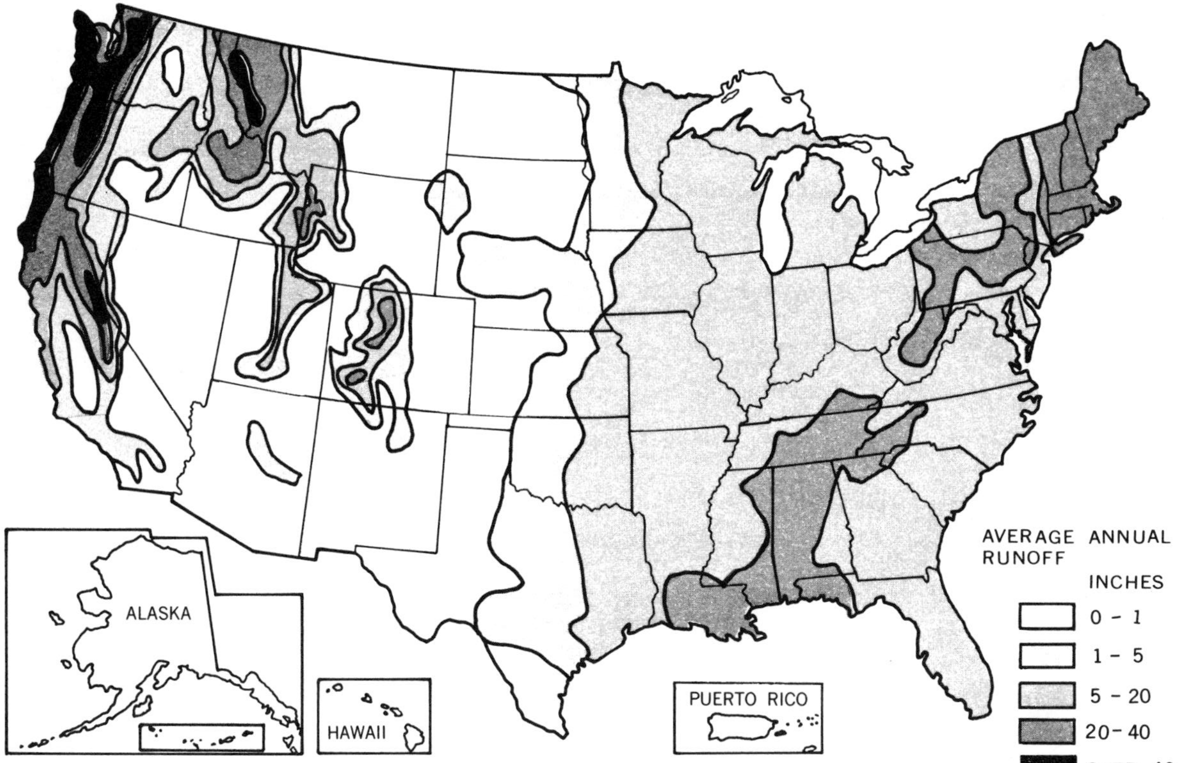

Regional data not available

Regional data not available

Source: U.S. Water Resources Council, 1968, The Nation's Water Resources

FIGURE 2–8. DRAINAGE REGIONS OF CANADA

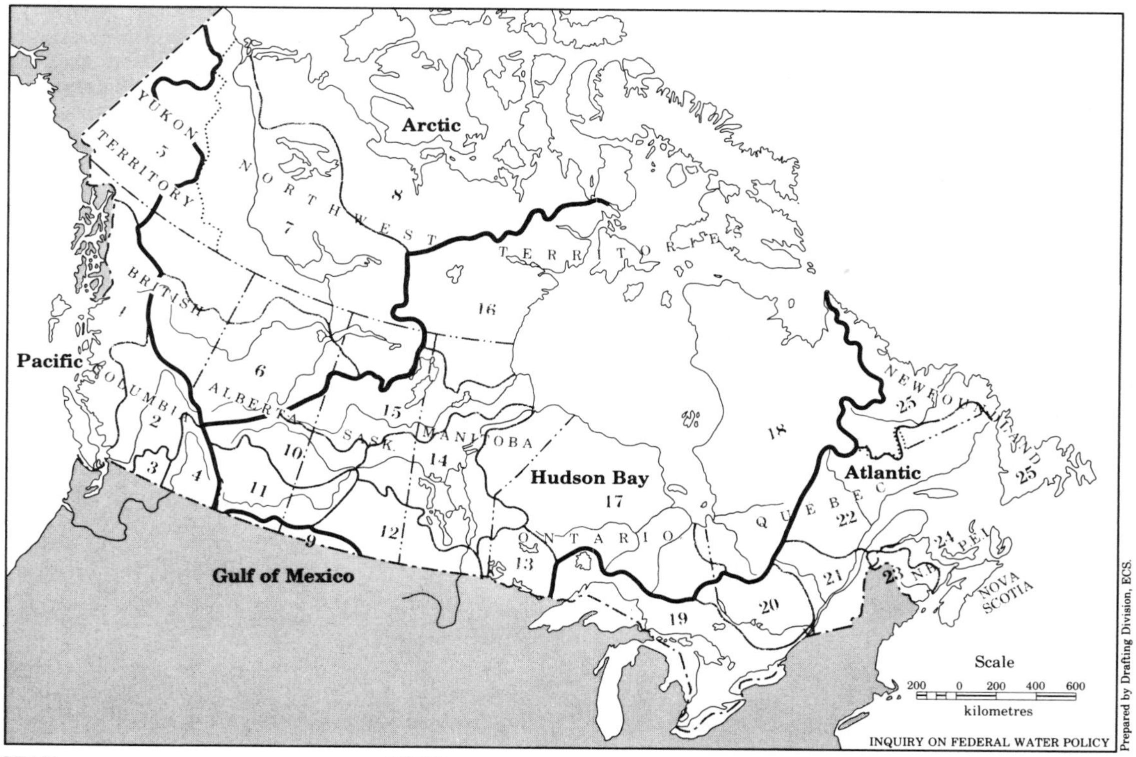

OCEAN BASIN REGION		RIVER BASIN REGION	AREA IN 000s km²	POPULATION IN 000s 1981
	1	Pacific Coastal	352	616
	2	Fraser-Lower Mainland	234	1 722
Pacific	3	Okanagan-Similkameen[a]	14	189
	4	Columbia[a]	90	161
	5	Yukon[a]	328	23
	6	Peace-Athabasca	487	286
Arctic	7	Lower Mackenzie	1 300	43
	8	Arctic Coast-Islands	2 025	13
Gulf of Mexico	9	Missouri[a]	26	14
	10	North Saskatchewan	146	1 084
	11	South Saskatchewan[a]	170	1 282
	12	Assiniboine-Red[a]	190	1 300
	13	Winnipeg[a]	107	77
Hudson Bay	14	Lower Saskatchewan-Nelson	363	224
	15	Churchill	298	68
	16	Keewatin	689	5
	17	Northern Ontario	694	157
	18	Northern Quebec	950	109
	19	Great Lakes[a]	319	7 579
	20	Ottawa	146	1 270
	21	St. Lawrence[a]	116	5 193
Atlantic	22	North Shore-Gaspé	403	653
	23	St. John-St. Croix[a]	37	393
	24	Maritime Coastal	114	1 314
	25	Newfoundland-Labrador	376	568
CANADA			9 974	24 343

[a]Canadian portion only; area and population on U.S. side of international basin regions are excluded from totals.
Source: Pearse, P.H., 1985, Currents of Change, Final Report Inquiry on Federal Water Policy, Ottawa, Canada

TABLE 2–19. RUNOFF DISTRIBUTION IN THE UNITED STATES

Range in Runoff (Inches per Year	Area (Square Miles)	Percent of Total Area	Percent of Total Runoff
0 to 0.25	306,000	10.1	0.1
0.25 to 0.5	380,000	12.6	.5
0.5 to 1.0	266,000	8.8	.8
1.0 to 2.5	413,000	13.7	2.8
2.5 to 5	247,000	8.2	3.6
5 to 10	258,000	8.5	7.4
10 to 20	830,000	27.4	44.8
20 to 40	290,000	9.6	32.4
40 to 80	30,000	1.0	6.9
Over 80	2,000	.1	.7
Total	3,022,000	100	100

Source: House of Representatives, U.S. Congress

TABLE 2–20. SEASONAL VARIATION OF NATURAL RUNOFF BY REGIONS OF THE UNITED STATES

[For regions see Fig. 2–6]

Region	Months of High Flow	Months of Low Flow
North Atlantic	March, April	August, September
South Atlantic-Gulf	February, March	September, October
Great Lakes	April	January, August, September
Ohio	March	September, October
Tennessee	March	October
Upper Mississippi	March, April	January, September, October
Lower Mississippi	March	October
Souris-Red-Rainy	April	January, February
Missouri	March, June	January
Arkansas-White-Red	April, May, June	January, September
Texas-Gulf	March, May	August, October
Rio Grande	May	June
Upper Colorado	June	January, February
Lower Colorado	March, April	June, November
Great Basin	June	September, January
Columbia-North Pacific	February, April, May	January, February, August, September
California	April, May	September, October December

Source: U.S. Water Resources Council, 1968

TABLE 2–21. PROJECTIONS OF AVERAGE WATER AVAILABILITY IN THE UNITED STATES[1]

[Values in billion gallons per day; for regions, see Fig. 2–6]

Region	1965	1980	2000	2020
North Atlantic	163	163	163	163
South Atlantic-Gulf	197	197	197	197
Great Lakes[2]	80.3	80.3	80.3	80.3
Ohio	125	125	125	125
Tennessee	41.5	41.5	41.5	41.5
Upper Mississippi[3]	66.7	66.7	66.7	66.7
Lower Mississippi[4]	408	401	395	390
Souris-Red-Rainy[5]	6.2	6.4	6.8	6.8
Missouri[6]	54.5	54.6	54.8	54.9
Arkansas-White-Red[6]	95.8	95.9	96.0	96.0
Texas-Gulf[7]	39.1	39.2	39.2	39.2
Rio Grande[8]	5.2	5.3	5.3	5.3
Upper Colorado[9]	13.5	13.5	13.5	13.5
Lower Colorado[10]	14.1	12.6	11.9	11.6
Great Basin[11]	6.9	7.0	7.1	7.2
Columbia-North Pacific[12]	258	258	258	258
California[13]	69.7	69.4	69.3	69.2
Alaska[12]	710	710	710	710
Hawaii	13.3	13.3	13.3	13.3

[1] Natural runoff adjusted for imports and upstream runoff where appropriate, values rounded.
[2] Includes net precipitation on U.S. portion of Great Lakes.
[3] Includes import from Great Lakes Region.
[4] Includes net upstream runoff and imports.
[5] Includes import from Missouri Region.
[6] Includes imports from Upper Colorado Region.
[7] Includes import from Arkansas-White-Red Region.
[8] Includes imports from Upper Colorado Region and Mexican Treaty deliveries.
[9] Virgin flow at Lee Ferry Compact point.
[10] Includes net upstream runoff.
[11] Includes imports from Upper Colorado Region and natural runoff from California Region.
[12] Includes natural runoff from Canada.
[13] Includes imports from Lower Colorado Region.
Source: U.S. Water Resources Council, 1968

TABLE 2–22. VALUES OF RUNOFF COEFFICIENT IN THE RATIONAL FORMULA

[Formula is applicable to drainage areas less than about 5000 acres and has the form Q = CiA where Q is peak discharge in cfs, C is a dimensionless runoff coefficient, i is rainfall intensity for the time of concentration (see Table 2–24) in inches per hour, and A is drainage area in acres.]

Type of Drainage Area	Runoff Coefficient, C	Type of Drainage Area	Runoff Coefficient, C
Lawns:		Industrial:	
Sandy soil, flat, 2%	0.05–0.10	Light areas	0.50–0.80
Sandy soil, average, 2–7%	0.10–0.15	Heavy areas	0.60–0.90
Sandy soil, steep, 7%	0.15–0.20	Parks, cemeteries	0.10–0.25
Heavy soil, flat, 2%	0.13–0.17	Playgrounds	0.20–0.35
Heavy soil, average, 2–7%	0.18–0.22	Railroad yard areas	0.20–0.40
Heavy soil, steep, 7%	0.25–0.35	Unimproved areas	0.10–0.30
Business:		Streets:	
Downtown areas	0.70–0.95	Asphaltic	0.70–0.95
Neighborhood areas	0.50–0.70	Concrete	0.80–0.95
Residential:		Brick	0.70–0.85
Single-family areas	0.30–0.50	Drives and walks	0.75–0.85
Multi units, detached	0.40–0.60	Roofs	0.75–0.95
Multi units, attached	0.60–0.75		
Suburban	0.25–0.40		
Apartment dwelling areas	0.50–0.70		

Source: Amer. Soc. Civil Engrs.

TABLE 2–23. WATERSHED CHARACTERISTICS FOR DETERMINING RUNOFF COEFFICIENT IN THE RATIONAL FORMULA

[For each watershed characteristic in left column select appropriate descriptive box; add four numerical values given in parentheses to obtain runoff coefficient as a percentage.]

Designation of Watershed Characteristics	Runoff-Producing Characteristics			
	100 Extreme	75 High	50 Normal	25 Low
Relief	(40) Steep, rugged terrain, with average slopes generally above 30%	(30) Hilly, with average slopes of 10 to 30%	(20) Rolling, with average slopes of 5 to 10%	(10) Relatively flat land, with average slopes of 0 to 5%
Soil infiltration	(20) No effective soil cover, either rock or thin soil mantle of negligible infiltration capacity	(15) Slow to take up water; clay or other soil of low infiltration capacity, such as heavy gumbo	(10) Normal; deep loam with infiltration about equal to that of typical prairie soil	(5) High; deep sand or other soil that takes up water readily and rapidly
Vegetal cover	(20) No effective plant cover; bare or very sparse cover	(15) Poor to fair; clean-cultivated crops or poor natural cover; less than 10% of drainage area under good cover	(10) Fair to good; about 50% of drainage area in good grassland, woodland, or equivalent cover; not more than 50% of area in clean-cultivated crops	(5) Good to excellent; about 90% of drainage area in good grassland, woodland, or equivalent cover
Surface storage	(20) Negligible; surface depressions few and shallow; drainage-ways steep and small; no ponds or marshes	(15) Low; well-defined system of small drainage-ways; no ponds or marshes	(10) Normal; considerable surface-depression storage; drainage system similar to that of typical prairie lands; lakes, ponds and marshes less than 2% of drainage area	(5) High; surface-depression storage high; drainage system not sharply defined; large flood-plain storage or a large number of lakes, ponds or marshes

Source: U.S. Soil Conservation Service

TABLE 2-24. TIME OF CONCENTRATION OF A WATERSHED

[Values are time in minutes for water to travel from the most distant point in a watershed to the watershed outlet. Length is distance along the main stream from the watershed outlet to the most distant ridge; height is difference in elevation between the watershed outlet and the most distant ridge.]

Length, Feet	Height, Feet										
	5	10	20	40	60	100	200	400	600	800	1000
200	2	2	1	1	1	1					
400	5	4	3	2	2	1	1				
600	8	6	4	3	3	2	2	1	1		
800	10	8	6	5	4	3	2	2	2	1	
1,000	14	10	8	6	5	4	3	2	2	2	2
2,000	28	21	17	13	10	9	7	5	4	4	3
3,000	48	35	27	20	17	13	10	8	7	6	5
4,000	65	50	36	28	24	19	14	11	9	8	7
5,000	100	65	48	35	28	22	19	14	12	11	9
7,000	150	100	70	52	42	34	27	20	17	15	13
10,000	240	180	120	85	70	54	40	28	25	22	20
15,000	330	270	210	150	120	90	60	46	44	34	30
20,000	430	350	280	210	180	140	95	65	54	48	43
30,000	600	500	450	320	280	230	170	120	100	90	75
40,000	870	660	510	420	360	300	240	180	150	130	120
50,000	1080	840	600	480	430	360	300	230	190	160	150

Source: Kirpich, Civil Engineering, 1940

TABLE 2-25. RUNOFF FOR NATIONAL FOREST AND NON-NATIONAL FOREST AREAS IN SELECTED WESTERN DRAINAGE BASINS

Drainage Basin or Area	Area		AVERAGE ANNUAL WATER PRODUCTION				
				NF		Outside NF	
	NF, Percent	Outside NF, Percent	Whole Area, Inches	Inches	Percent of Total Volume	Inches	Percent of Total Volume
Columbia (in U.S.)	37	63	10.4	16.7	59	6.7	41
Colorado (in U.S.)	19	81	2.5	7.2	56	1.3	44
Rio Grande above El Paso	25	75	1.7	3.8	58	0.9	42
Central Valley (California only)	32	68	11.8	23.5	63	6.4	37
Rogue-Umpqua Area	40	60	35.5	37.0	42	34.2	58
Northwest Washington (State less Columbia)	32	68	39.3	51.4	41	33.7	59
Southern California Coast (Los Angeles watershed to Mexican border)	25	75	3.7	6.3	43	2.8	57
North Platte and South Platte	11	89	1.7	6.2	41	1.2	59
Missouri above Fort Randall Dam	9	91	1.7	6.9	37	1.2	63
Arkansas above Dodge City	9	91	1.2	4.7	38	0.7	62

Source: U.S. Geological Survey

TABLE 2–26. TOTAL IMPERVIOUS AREA FOR SPECIFIC LAND-USE CATEGORIES

Land-Use Category	Typical Values of Total Impervious Area (Percent)		
	Low	Intermediate	High
Single-family residential[1]	16	27	45
Multifamily residential[2]	50	60	70
Commercial[3]	80	88	95
Industrial[4]	50	75	90
Public facilities[5]	50	60	75
Parks and undeveloped land[6]	0	1	3

[1] Single-family residential—Single-family dwellings predominate.
[2] Multifamily residential—Multiple-family units predominate. These include duplexes, apartment buildings; and condominiums.
[3] Commercial—Zones consisting of various types of business.
[4] Industrial—Manufacturing complexes, railroad yards, and large utilities.
[5] Public facilities—School, hospitals, churches, airports, and other public buildings.
[6] Parks and undeveloped land—Parks, forests, and open undeveloped land.
Source: Conger, D.H., 1986, Estimating Magnitude and Frequency of Floods for Wisconsin Urban Streams, U.S. Geological Survey Water-Resources Investigations Report 86–4005

TABLE 2–27. WORLD-WIDE STABLE RUNOFF, BY CONTINENT

	Stable Runoff[1], km ^{3}					
	Of Underground Origin	Regulated by Lakes	Regulated by Water Reservoirs	Total	Total River Runoff[2]	Total Stable Runoff as Percent of Total River Runoff
Europe	1,065	60	200	1,325	3,110	43
Asia	3,410	35	560	4,005	13,190	30
Africa	1,465	40	400	1,905	4,225	45
North America	1,740	150	490	2,380	5,960	40
South America	3,740	–	160	3,900	10,380	38
Australia[3]	465	–	30	495	1,965	25
Total land area except polar zones	11,885	285	1,840	14,010	38,830	36

[1] Excluding flood flows.
[2] Including flood flow.
[3] Including Tasmania, New Guinea and New Zealand.
Source: Lvovitch, M.I., EOS, Vol. 54, No. 1, Jan. 1973, Copyright by American Geophysical Union

FIGURE 2–9. ANNUAL TOTAL RIVER RUNOFF IN THE WORLD

[Includes groundwater discharge to rivers; in mm]

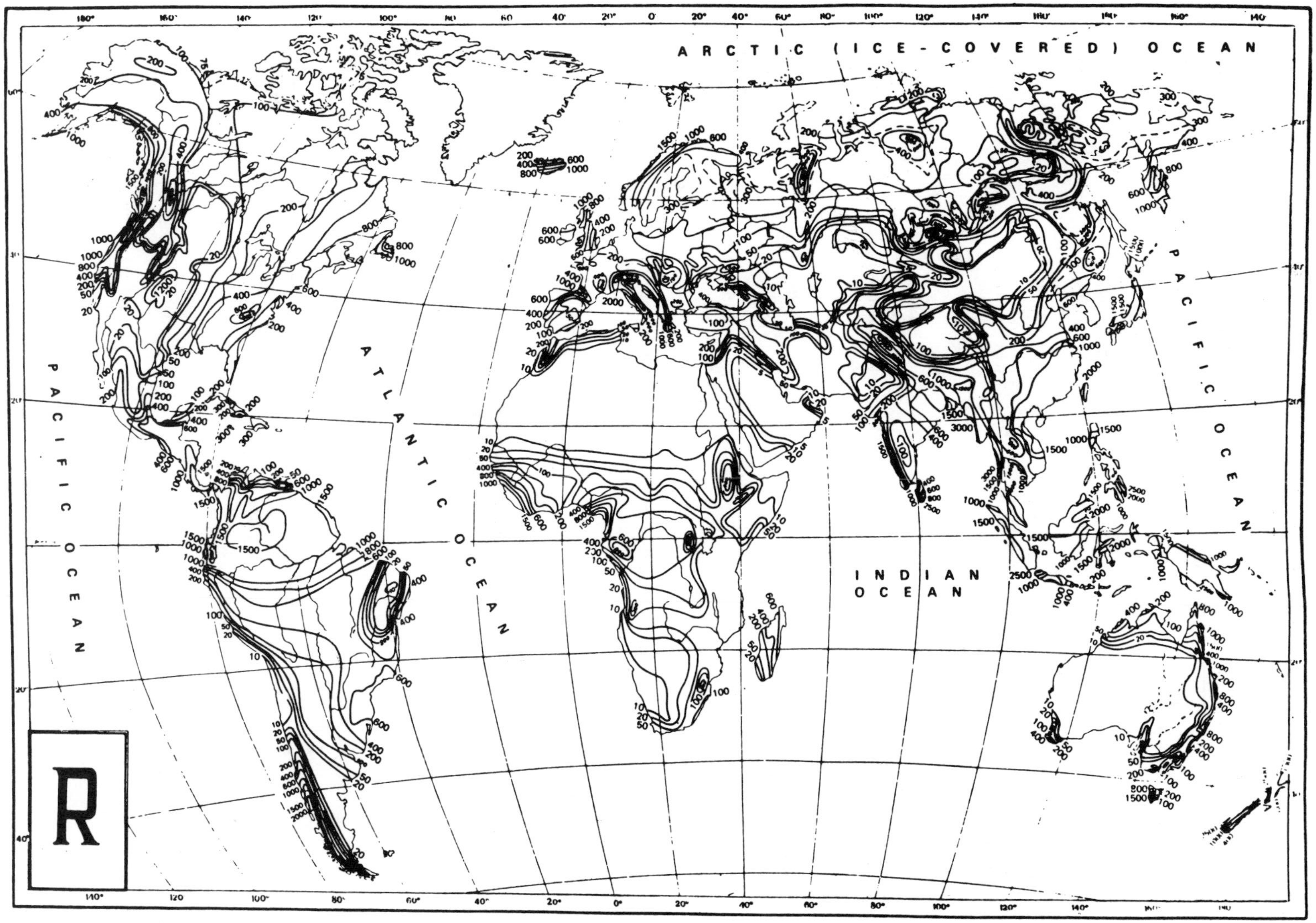

Source: Lvovitch, M.I., EOS, Vol. 54, No. 1, 1973. Copyright by American Geophysical Union

SECTION H. EROSION AND SEDIMENTATION

TABLE 2–28. LOSSES OF LAND BY RIVERBANK EROSION IN THE UNITED STATES

[Estimated average annual losses as of 1966; for regions see Fig. 2–6]

Region	Areas of Significant Erosion	Average Land Losses
	River-miles	*Acres/year*
North Atlantic	Not significant	Not significant
South Atlantic-Gulf	5,100	350
Great Lakes	Not significant	Not significant
Ohio	Not significant	Not significant
Tennessee	Not significant	Not significant
Upper Mississippi	Not available	Not available
Lower Mississippi	1,044	4,705
Souris-Red-Rainy	Not available	Not available
Missouri	1,692	5,000
Arkansas-White-Red	2,300	7,300
Texas-Gulf	1,698	1,045
Rio Grande	250	150
Upper Colorado	Not available	Not available
Lower Colorado	Not available	Not available
Great Basin	265	150
Columbia-North Pacific	13,500	1,300
California	2,600	3,837
Alaska	80,000	5,000
Hawaii	Not significant	Not significant
Puerto Rico	Not available	Not available

Source: U.S. Water Resources Council, 1968

TABLE 2–29. LOSSES OF LAND BY EROSION OF BEACHES AND ESTUARY SHORES IN THE UNITED STATES

[Estimated average annual losses as of 1966; for regions see Fig. 2–6]

Region	Average Annual Land Losses
	Acres/mile
North Atlantic	0.12
South Atlantic-Gulf	0.11
Great Lakes	0.07
Lower Mississippi	–
Texas-Gulf	0.27
Columbia-North Pacific	0.02
California	0.13
Alaska	0.06
Hawaii	0.24
Puerto Rico	–

Source: U.S. Water Resources Council, 1968

TABLE 2–30. EROSION PROBLEMS IN THE PUBLIC DOMAIN OF THE UNITED STATES

[Values in millions of acres; conterminous United States only; for regions see Fig. 2–6]

Region	Extent of Erosion			
	Slight	Moderate	Critical	Total
Missouri	7.5	9.0	5.2	21.7
Arkansas-White-Red	.1	.4	.2	.7
Rio Grande	1.3	6.4	3.6	11.3
Upper Colorado	4.1	14.4	8.3	26.8
Lower Colorado	2.4	13.8	7.9	24.1
Great Basin	15.4	20.1	11.6	47.1
Columbia-North Pacific	14.6	8.5	5.0	28.1
California	6.8	5.4	3.2	15.4
Total	52.2	78.0	45.0	175.2

Source: U.S. Water Resources Council, 1968

TABLE 2–31. SELECTED QUANTITATIVE EFFECTS OF MAN'S ACTIVITIES ON SURFACE EROSION

Initial Status of Land Use	Type of Disturbance	Magnitude of Specific Disturbance[1]	
Forestland	Planting row crops	100 to 1,000	Times
Grassland	Planting row crops	20 to 100	Times
Forestland	Building logging roads	220	Times
Forestland	Woodcutting and skidding	1.6	Times
Forestland	Fire	7 to 1,500	Times
Forestland	Mining	1,000	Times
Row crop	Construction	10	Times
Pastureland	Construction	200	Times
Forestland	Construction	2,000	Times

[1] Relative magnitude of surface erosion from disturbed surface, assuming "I" for the initial status. The first row of the table, for example, indicates that transforming a forestland into row crops may increase surface erosion 100 to 1,000 times.
Source: U.S. Environmental Protection Agency, 1976, Loading Functions for Assessments of Water Pollution from Non-Point Sources, Environmental Protection Technology Series, Washington, D.C.

TABLE 2–32. REPRESENTATIVE RATES OF EROSION FROM VARIOUS LAND USES

Land Use	Amount of Erosion (Tons/Square Mile/Year)	Rate of Erosion Relative to Forest = 1
Forest	24	1
Grassland	240	10
Abandoned Surface Mines	2,400	100
Cropland	4,800	200
Harvested Forest	12,000	500
Active Surface Mines	48,000	2,000
Construction	48,000	2,000

Source: U.S. Environmental Protection Agency, 1973, Methods For Identifying and Evaluating the Nature and Extent of Nonpoint Source of Pollutants, EPA 430/9–73–014, Washington, D.C.

TABLE 2–33. SERIOUS EROSION AREAS IN THE UNITED STATES

The Palouse, covering parts of Washington, Oregon, and Idaho, is dryfarmed to wheat, barley, peas, and lentils. Most of the cropland is hilly, with slopes of from 15–25 percent. Runoff from melting snow and heavy rains causes erosion of 50 to 100 tons per acre.

Southeastern Idaho cropland is planted in hard red wheat one year and left fallow the next to conserve moisture. Erosion occurs during intense summer rainstorms and even more destructive rains in late winter, when a thawed layer of soil moves downhill over frozen subsoil. Annual erosion may reach 16 tons per acre per year on 35 percent slopes.

Texas Blackland Prairie is an important farming area, with two-thirds of the land in crops, mainly cotton and grain sorghum. Rainfall averages 30 to 50 inches a year on the gently rolling land. Many soils in the area are highly erodible, and erosion is appreciably higher than the national average.

Southern Mississippi Valley, including parts of five states, is about one-third cropland, much of it sloping to steep. The soils are deep, fertile, productive, and erodible. Many row crops are grown without adequate conservation practices, and annual soil losses on much of the land reach 20 tons per acre or more.

The Corn Belt States experience some of the highest erosion rates in the country: in 1977, Iowa cropland lost an average of 9.9 tons of soil per acre; Illinois, 6.7 tons per acre, and Missouri, 10.9 tons per acre.

Aroostook County, Maine, is famous for its potatoes. They are grown on slopes ranging from nearly level to 25 percent. The upper 2 feet of soil have been lost since cultivation began, lowering crop yields. Some sloping fields are losing as much as an inch of soil a year.

Source: U.S. Soil Conservation Service, 1981, America's Soil and Water: Condition and Trends

TABLE 2–34. DISSOLVED AND SUSPENDED SEDIMENT LOADS IN SELECTED RIVERS OF THE UNITED STATES

River and Location	Elevation (ft)	Drainage Area (sq mi)	Average Discharge, Q (cfs)	Discharge÷ Drainage Area (cfs/sq mi)	Years of Record in Sample[1]	Average Suspended Load	Average Dissolved Load (millions of tons/yr)	Total Average Suspended and Dissolved Load	Total Average Load ÷Drainage Area (tons sq mi/yr)	Dissolved Load as Percent of Total Load (%)
Little Colorado at Woodruff, Arizona	5,129	8,100	63.3	.0078	6	1.6	.02	1.62	199	1.2
Canadian River near Amarillo, Texas	2,989	19,445	621	.032	1	6.41	.124	6.53	336	1.9
Colorado River near San Saba, Texas	1,096	30,600	1,449	.047	5	3.02	.208	3.23	105	6.4
Bighorn River at Kane, Wyoming	3,609	15,900	2,391	.150	1	1.60	.217	1.82	114	12
Green River at Green River, Utah	4,040	40,600	6,737	.166	26–20	19	2.5	21.5	530	12
Colorado River near Cisco, Utah	4,090	24,100	8,457	.351	25–20	15	4.4	19.4	808	23
Iowa River at Iowa City, Iowa	627	3,271	1,517	.464	3	1.184	.485	1.67	510	29
Mississippi River at Red River Landing, Louisiana		1,144,500[2]	569,500[2]	.497	3	284	101.8	385.8	337	26
Sacramento River at Sacramento, California	0	27,000[3]	25,000[3]	.926	3	2.85	2.29	5.14	190	44
Flint River near Montezuma, Georgia	256	2,900	3,528	1.22	1	.400	.132	.53	183	25
Juniata River near New Port, Pennsylvania	364	3,354	4,329	1.29	7	.322	.566	.89	265	64
Delaware River at Trenton, New Jersey	8	6,780	11,730	1.73	9–4	1.003	.830	1.83	270	45

[1] Computation of load, dissolved or suspended, depends on discharge for same period. Years of record pertain to number of years used for related values of discharge and of suspended and dissolved load. Where two figures are shown, the first is for suspended load and the second is for dissolved load.

[2] From USGS records for Vicksburg, Mississippi station.

[3] Estimated.

Source: Leopold, Wolman, and Miller, Fluvial Processes in Geomorphology, W.H. Freeman and Company, 1964

FIGURE 2–10. CONCENTRATION OF SUSPENDED SEDIMENT IN RIVERS AND DISCHARGE OF SUSPENDED SEDIMENT TO THE COASTAL ZONE IN THE CONTERMINOUS UNITED STATES

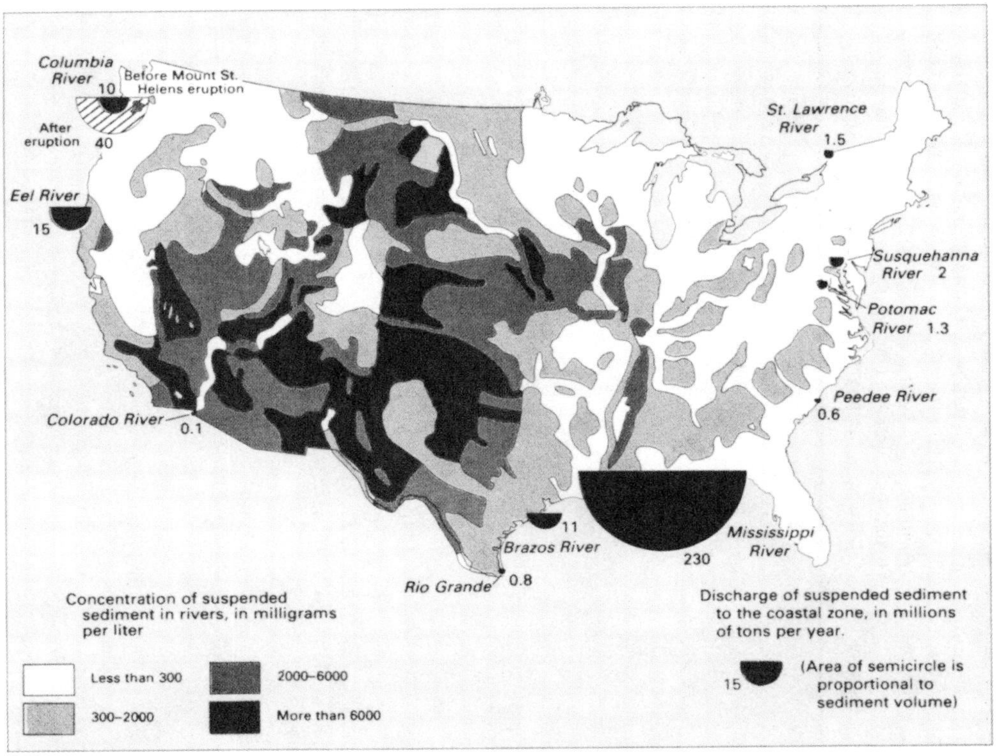

Source: Meade, R.H., and Parker, R.S., 1984, Sediment in the Rivers of the United States, National Water Summary 1984, U.S. Geological Survey Water-Supply Paper 2275

TABLE 2–35. DISCHARGE OF SUSPENDED SEDIMENT TO THE COASTAL ZONE BY 10 MAJOR RIVERS OF THE UNITED STATES, ABOUT 1980

Rivers	Average Annual Sediment Discharge (million ton/yr)
Rivers that discharge the largest sediment loads:	
Mississippi	230[1]
Copper	80
Yukon	65
Susitna	25
Eel	15
Brazos	11
Columbia:	
Before Mount St. Helens eruption	10
Since Mount St. Helens eruption-approximate	40
Rivers with large drainage areas:	
St. Lawrence	1.5
Rio Grande	.8
Colorado	.1

[1] Includes Atchafalaya River.

Source: Meade, R.H., and Parker, R.S., 1985, Sediment in Rivers of the United States, U.S. Geological Survey Water-Supply Paper 2275, National Water Summary 1984

TABLE 2–36. DIMENSION AND RATE OF FORMATION OF MODERN DELTAS

River	Dimension of Subaerial Delta, Statute Mi		Amount of Sediment Discharged		Annual Extension of Subaerial Delta	
	Length	Breadth	River Water by Weight (avg), ppm	Annual Volume of Sediment, mi³	Measurement Period, Years	Approximate Distance, Ft
Mississippi's present bird-foot delta	12	30	550	0.068	1838–1947	250
Hwang-Ho	300	470[1]	50,000[2]		1870–1937	950
Ganges-Brahmaputra	220	200	870	0.043 (Ganges only)		
Rhone into Mediterranean Sea	30	47	400–590	0.005	1737–1870	190
Danube	46	46	310	0.008		40
Nile (prior to barrages)	96	145	1600	0.001	1100–1870	45
Colorado above Hoover Dam	43	0.05–0.6	8300	0.032	1936–1948	3.6 mi (gorge)
Euphrates Tigris	350	90			1793–1853	180

[1] Includes 100 mi of nondeltaic Shantung Peninsula.
[2] Maximum is 400,000 ppm.
Source: McGraw-Hill Encyclopedia of the Geological Sciences, Copyright 1978. Reprinted with permission

TABLE 2–37. WATER STORAGE CAPACITY AND THE CAPACITY LOST ANNUALLY DUE TO SEDIMENTATION IN THE CONTERMINOUS UNITED STATES

[Reservoirs with 5000 acre-feet or more total capacity]

Farm Production Region	Total Water Storage Capacity (million ac-ft)	Usable Water Storage Capacity (million ac-ft)	Estimated Water Storage Capacity Lost (%)	Estimated Water Storage Capacity Lost (thousand ac-ft)	Stream Sediment Originating on Cropland (%)	Reservoir Sedimentation from Cropland (thousand ac-ft)
Northeast	36.5	25.2	0.08	28.1	29	8.2
Appalachian	59.5	30.6	0.13	75.5	29	21.9
Southeast	73.6	47.6	0.17	127.3	33	42.0
Lake States	29.3	19.5	0.27	79.1	64	50.6
Corn Belt	39.7	15.2	0.26	104.8	63	66.0
Delta States	42.7	20.1	0.21	87.5	41	35.9
Northern Plains	78.9	54.4	0.23	184.6	36	66.5
Southern Plains	110.3	46.6	0.19	207.4	19	39.4
Mountain	167.1	138.1	0.18	302.5	8	24.2
Pacific	90.7	74.7	0.49	441.6	9	39.7
United States (Lower 48)	728.3	472.1	0.22	1,638.5		394.4

Source: Crowder, B. M., 1987, Economic Costs of Reservoir Sedimentation: A Regional Approach to Estimating Cropland Erosion Damages, Journal of Soil and Water Conservation, Soil Conservation Society of America

TABLE 2–38. RIVERBANK TREATMENT IN THE UNITED STATES

[Values in miles as of 1966; Corps of Engineers projects only; for regions see Fig. 2–6]

Region	Mattress	Jetties	Training Walls	Riprap	Revetment	Dikes	Other	Total Treated
North Atlantic	–	18	18	–	–	–	–	36
South Atlantic-Gulf	35	67	–	–	–	–	–	102
Great Lakes	–	–	–	–	–	–	–	–
Ohio	–	–	–	50	10	–	–	60
Tennessee	–	–	–	–	–	–	–	–
Upper Mississippi	–	–	–	–	–	–	–	–
Lower Mississippi	573	–	–	–	–	75	5,836	6,484
Souris-Red-Rainy	–	–	–	–	–	–	–	–
Missouri	–	735	–	–	735	–	–	1,470
Arkansas-White-Red	250	1	80	–	–	140	–	471
Texas-Gulf	0	0	0	0	0	0	0	0
Rio Grande	–	250	–	–	–	–	–	250
Upper Colorado	–	–	–	–	–	–	–	–
Lower Colorado	–	–	–	–	–	–	–	–
Great Basin	0	0	0	0	0	0	0	0
Columbia-North Pacific	–	–	–	261	–	–	–	261
California	–	–	–	–	–	–	19	19
Alaska	–	–	6	–	–	–	–	6
Hawaii	0	0	0	0	0	0	0	0
Puerto Rico	–	–	–	–	–	–	–	–

Source: U.S. Water Resources Council, 1968

TABLE 2–39. SEDIMENT YIELD FROM DRAINAGE AREAS OF 100 SQUARE MILES OR LESS OF THE UNITED STATES

[For regions see Fig. 2–6]

Region	Estimated Sediment Yield			Region	Estimated Sediment Yield		
	High	Low	Average		High	Low	Average
		tons/sq mi/yr				*tons/sq mi/yr*	
North Atlantic	1,210	30	250	Arkansas-White-Red	8,210	260	2,200
South Atlantic-Gulf	1,850	100	800	Texas-Gulf	3,180	90	1,800
Great Lakes	800	10	100	Rio Grande	3,340	150	1,300
Ohio	2,110	160	850	Upper Colorado	3,340	150	1,800
Tennessee	1,560	460	700	Lower Colorado	1,620	150	600
Upper Mississippi	3,900	10	800	Great Basin	1,780	100	400
Lower Mississippi	8,210	1,560	5,200	Columbia-North Pacific	1,100	30	400
Souris-Red-Rainy	470	10	50	California	5,570	80	1,300
Missouri	6,700	10	1,500				

Source: U.S. Water Resources Council, 1968

TABLE 2-40. PERMISSIBLE VELOCITIES FOR CHANNELS LINED WITH VEGETATION[1]

[Values apply to average, uniform stands of each type of cover]

Cover	Slope Range[2]	Permissible Velocity: Erosion Resistant Soils	Permissible Velocity: Easily Eroded Soils
	Percent	*Ft. per sec.*	*Ft. per sec.*
Bermudagrass	0–5	8	6
	5–10	7	5
	over 10	6	4
Buffalograss Kentucky bluegrass Smooth brome Blue grama	0–5	7	5
	5–10	6	4
	over 10	5	3
Grass mixture	0–5[2]	5	4
	5–10	4	3
Lespedeza sericea Weeping lovegrass Yellow bluestem Kudzu Alfalfa Crabgrass	0–5[3]	3.5	2.5
Common lespedeza[4] Sudangrass[4]	0–5[5]	3.5	2.5

[1] Use velocities exceeding 5 feet per second only where good covers and proper maintenance can be obtained.
[2] Do not use on slopes steeper than 10 percent except for side slopes in a combination channel.
[3] Do not use on slopes steeper than 5 percent except for side slopes in a combination channel.
[4] Annuals — used on mild slopes or as temporary protection until permanent covers are established.
[5] Use on slopes steeper than 5 percent is not recommended.
Source: U.S. Soil Conservation Service

TABLE 2-41. PERMISSIBLE VELOCITIES FOR CHANNELS WITH LININGS OTHER THAN VEGETATION

[Values apply to aged straight channels with mild bed slopes]

Original Material Excavated	Clear, Water, No Detritus	Water Transporting Colloidal Silts	Water Transporting Noncolloidal Silts, Sands, Gravels, or Rock Fragments
	Ft. per sec.	*Ft. per sec.*	*Ft. per sec.*
Fine sand, noncolloidal	1.50	2.50	1.50
Sandy loam, noncolloidal	1.75	2.50	2.00
Silt loam, noncolloidal	2.00	3.00	2.00
Alluvial silts, noncolloidal	2.00	3.50	2.00
Ordinary firm loam	2.50	3.50	2.25
Volcanic ash	2.50	3.50	2.00
Fine gravel	2.50	5.00	3.75
Stiff clay, very colloidal	3.75	5.00	3.00
Graded, loam to cobbles, noncolloidal	3.75	5.00	5.00
Alluvial silts, colloidal	3.75	5.00	3.00
Graded, silt to cobbles, colloidal	4.00	5.50	5.00
Coarse gravel, noncolloidal	4.00	6.00	6.50
Cobbles and shingles	5.00	5.50	6.50
Shales and hardpans	6.00	6.00	5.00

Source: Fortier and Scobey, Trans. Amer. Soc. Civil Engrs., 1926

TABLE 2–42. SETTLING VELOCITIES OF SAND AND SILT IN STILL WATER

[Temperature 50°F; all particles assumed to have a specific gravity of 2.65]

Diameter of Particle	Order of Size	Settling Velocity	Time Required to Settle 1 Foot
mm.		*mm./sec.*	
10.0	Gravel	1,000	0.3 seconds
1.0	Coarse Sand	100	3.0 seconds
0.8	Coarse Sand	83	
0.6	Coarse Sand	63	
0.5	Coarse Sand	53	
0.4	Coarse Sand	42	
0.3	Coarse Sand	32	
0.2	Coarse Sand	21	
0.15	Coarse Sand	15	
0.10	Fine Sand	8	38.0 seconds
0.08	Fine Sand	6	
0.06	Fine Sand	3.8	
0.05	Fine Sand	2.9	
0.04	Fine Sand	2.1	
0.03	Fine Sand	1.3	
0.02	Fine Sand	0.62	
0.015	Fine Sand	0.35	
0.010	Silt	0.154	33.0 minutes
0.008	Silt	0.098	
0.006	Silt	0.065	
0.005	Silt	0.0385	
0.004	Silt	0.0247	
0.003	Silt	0.0138	
0.002	Silt	0.0062	
0.0015	Silt	0.0035	
0.001	Bacteria	0.00154	55.0 hours
0.0001	Clay Particles	0.0000154	230.0 days
0.00001	Colloidal Particles	0.000000154	63.0 years

Source: Amer. Water Works Assoc.

TABLE 2–43. CLASSIFICATION OF ALLUVIAL CHANNELS BASED ON CHANNEL STABILITY AND ON MODE OF SEDIMENT TRANSPORT

Mode of Sediment Transport	Silt-Clay in Channel Sediment, Percent[1]	Proportion of Total Sediment Load		Channel Stability		
		Suspended Load Percent	Bedload, Percent	Stable (Graded Stream)	Depositing (Excess Load)	Eroding (Deficiency of Load)
Suspended load	30–100	85–100	0–15	Stable suspended-load channel. Width-depth ratio less than 7; sinuosity greater than 2.1; gradient relatively gentle.	Depositing suspended load channel. Major deposition on banks cause narrowing of channel; streambed deposition minor.	Eroding suspended-load channel. Streambed erosion predominant; channel widening minor.
Mixed load	8–30	65–85	15–35	Stable mixed-load channel. Width-depth ratio greater than 7 less than 25; sinuosity, less than 2.1 greater than 1.5; gradient moderate.	Depositing mixed-load channel. Initial major deposition on banks followed by streambed deposition.	Eroding mixed-load channel. Initial streambed erosion followed by channel widening.
Bedload	0–8	30–65	35–70	Stable bedload channel. Width-depth ratio greater than 25; sinuosity, less than 1.5; gradient relatively steep.	Depositing bedload channel. Streambed deposition and island formation.	Eroding bedload channel. Little streambed erosion; channel widening predominant.

[1] The percentage of sediment finer than 0.074 mm in the perimeter of the channel.
Source: U.S. Geological Survey

TABLE 2–44. DRAINAGE AREA, WATER AND SUSPENDED SEDIMENT DISCHARGES FOR MAJOR RIVERS OF THE WORLD

			Sediment Discharge Millions of Tons per Year		
River	Drainage Area (x $10^6 km^2$)	Water Discharge (km^3 yr)	Strakhov (1961) and Lisitzin (1972)	Holeman (1968)	Milliman and Meade (1983)
North America					
St. Lawrence (Canada)	1.03	447	4	4	4
Hudson (USA)	.02	12	36	...	1
Mississippi (USA) (including Atchafalaya)	3.27	580	500	349	210
Brazos (USA)	.11	7	32	32	16
Colorado (Mexico)	.64	20	135	135	.1
Eel (USA)	.008	...	...	16	14
Columbia (USA)	.67	251	36	9	8
Fraser (Canada)	.22	112	...	...	20
Yukon (USA)	.84	195	88	...	60
Copper (USA)	.06	39	...	...	70
Susitna (USA)	.05	40	...	...	25
MacKenzie (Canada)	1.81	306	15	...	100
Total North America	9.57				528
South America					
Chira (Peru)	.02	5	...	...	4–75
Magdalena (Colombia)	.24	237	...	...	220
Orinoco (Venezuela)	.99	1100	86	86	210
Amazon (Brazil)	6.15	6300	498	364	900
Sao Francisco (Brazil)	.64	97	...	...	6
La Plata (Argentina)	2.83	470	129	82	92
Negro (Argentina)	.10	30	...	...	13
Total South America	10.85				1420
Europe					
Rhone (France)	.09	49	31	...	10
Po (Italy)	.07	46	18	15	15
Danube (Romania)	.81	206	67	19	67
Semani (Albania)	...	...	...	22	?
Drini (Albania)	.01	...	...	15	?
Total Europe	.97				92
Eurasian Arctic					
Yana (USSR)	.22	29	3	...	3
Ob (USSR)	2.50	385	16	15	16
Yenisei (USSR)	2.58	560	13	...	13
Severnay Dvina (USSR)	.35	106	4.5	...	4.5
Lena (USSR)	2.50	514	15	...	12
Kolyma (USSR)	.64	71	6	...	6
Indigirka (USSR)	.36	55	14	...	14
Total Euras. Arctic	9.15				68
Asia					
Amur (USSR)	1.85	325	25	...	52
Liaohe (China)	.17	6	...	...	41
Daling (China)	.02	1	...	...	36
Haiho (China)	.05	2	...	...	81
Yellow (Huangho) (China)	.77	49	1890	1890	1080

TABLE 2–44. DRAINAGE AREA, WATER AND SUSPENDED SEDIMENT DISCHARGES FOR MAJOR RIVERS OF THE WORLD (Continued)

			Sediment Discharge Millions of Tons per Year		
River	Drainage Area (x 10^6 km^2)	Water Discharge (km^3 yr)	Strakhov (1961) and Lisitzin (1972)	Holeman (1968)	Milliman and Meade (1983)
Yangtze (China)	1.94	900	500	502	478
Huaihe (China)	.26	...	...	...	14
Pearl (Zhu Jiang) (China)	.44	302	...	27	69
Hungho (Vietnam)	.12	123	130	130	130
Mekong (Vietnam)	.79	470	170	170	160
Irrawaddy (Burma)	.43	428	299	300	265
Ganges/Brahmaputra (Bangladesh)	1.48	971	2180	2180	1670
Mehandi (India)	.13	67	...	62	2
Damodar (India)	.02	10	...	28	?
Godavari (India)	.31	84	...	...	96
Indus (Pakistan)	.97	238	435	440	100
Tigris-Euphrates (Iraq)	1.05	46	105	53	?
Total Asia	9.74				4334
Africa					
Nile (Egypt)	2.96	30	110	111	0
Niger (Nigeria)	1.21	192	67	4	40
Zaire (Zaire)	3.82	1250	65	64	43
Orange (S. Africa)	1.02	11	153	...	17
Zambesi (Mozambique)	1.20	223	100	...	20
Limpopo (Mozambique)	.41	5	...	...	33
Rufiji (Tanzania)	.18	9	...	...	17
Tana (Kenya)	.032	...	...	...	32
Total Africa (minus Nile)	7.48				175
Oceania					
Murray (Aust.)	1.06	22	32	32	30
Waiapu (N.Z.)	...	...	...	...	28
Haast (N.Z.)	.001	6	...	...	13
Fly (New Guinea)	.061	77			30
Purari (New Guinea)	.031	77			80
Choshui (Taiwan)	.003	6	...	...	66
Kaoping (Taiwan)	.003	9	...	...	39
Tsengwen (Taiwan)	.001	2	...	...	28
Hualien (Taiwan)	.002	4	...	...	19
Peinan (Taiwan)	.002	4	...	...	17
Hsiukuluan (Taiwan)	.002	4	...	...	16
Total Oceania (excluding Murray)	1.074	39			336

Source: Milliman, J.D., and Meade, R.H., 1983, World-Wide Delivery of River Sediment to the Oceans, Copyright J. of Geology, Vol. 91, No.1, pp. 1–21. Reproduced with permission

SECTION I. TRANSPIRATION

TABLE 2–45. TRANSPIRATION RATIOS FOR CROPS

[Transpiration ratios are weighted values in pounds of water per pound of crop product. Data measured at Akron, Colorado.]

Crop	Varieties Tested, No.	Transpiration Ratio	
		Range	Mean
Grains:			
Proso	2	531–603	567
Millet	3	863–1,117	959
Buckwheat	1	......	969
Sorgo	3	863–1,804	1,237
Grain sorghum	4	750–1,050	868
Barley	4	1,128–1,464	1,241
Corn	2	821–1,998	1,405
Oats	4	1,379–1,915	1,627
Wheat, emmer	1	......	1,167
Wheat, durum	6	1,365–1,622	1,475
Wheat, common	11	1,244–3,398	1,872
Wheat, hybrids	2	1,995–2,163	2,079
Rye	1	......	2,142
Flax	6	2,010–5,162	3,252
Legumes:			
Clover	2	636–759	698[1]
Clover, sweet	1	......	731[1]
Vetch	5	562–899	708[1]
Alfalfa	10	626–920	844[1]
Cowpeas	1	......	1,632
Beans	2	1,583–1,815	1,699
Beans, soy	1	......	1,974
Chickpeas	1	......	1,685
Peas, Canadian field	1	......	2,153
Lupinus albus	1	......	4,734
Grasses:			
Sudan grass	1	......	380[1]
Wheat grass	1	......	678[1]
Brome grass	1	......	977[1]
Miscellaneous:			
Cotton	1	......	568[1]
Sugar beets	1	......	629
Potatoes	2	1,325–2,877	2,101
Cabbage	1	......	518[1]
Rape	1	......	714[1]
Watermelons	1	......	1,102
Cantaloupes	1	......	1,754
Turnips	1	......	1,471
Cucumbers	1	......	1,549

[1] Based on total dry matter

Source: Shantz and Piemeisel, Jour. Agric. Research, 1927

TABLE 2–46. TRANSPIRATION RATIOS FOR WEEDS AND NATIVE PLANTS

[Transpiration ratios are weighted values in pounds of water per pound of dry matter. Data measured at Akron, Colorado.]

Weeds or Native Plants	Transpiration Ratio	Weeds or Native Plants	Transpiration Ratio
Weeds:		Native weeds (continued):	
Tumbleweed	260	Sunflower	623
Pigweed	305	Mountain sage	654
Russian thistle	314	Verbena	702
Lamb's quarters	658	Fetid marigold	847
Polygonum	678		
Native weeds:		Native plants:	
Purslane	281	Buffalo grass	296
Cocklebur	415	Buffalo and grama grass	338
Nightshade	487	Clammyweed	483
Buffalo bur	536	Iva	534
Gumweed	585	Western ragweed	912
		Western wheat grass	1,035
		Franseria	1,131

Source: Shantz and Piemeisel, Jour. Agric. Research, 1927

TABLE 2–47. TRANSPIRATION RATIOS FOR TREES

[Transpiration ratios are expressed as pounds of water per pound of dry-leaf matter. Data measured by Hohnel, 1879-1880.]

Tree	Scientific Name	Transpiration Ratio
Ash	*Fraxinus excelsior*	981
White birch	*Betula alba*	849
Beech	*Fagus sylvatica*	1,043
Hornbeam	*Carpinus betulus*	787
Field elm	*Ulmus campestris*	738
Stiel oak	*Quercus pedunculus*	454
Trauben oak	*Quercus sessilifolia*	790
Zerr oak	*Quercus cerris*	669
Black alder	*Alnus glutinosa*	840
Gray alder	*Alnus incana*	678
Sycamore maple	*Acer platanoides*	520
Mountain maple	*Acer pseudoplat*	635
Field maple	*Acer campestria*	1,281
Linden	*Tilia grandifolia*	1,038
Aspen	*Populus tremula*	873
Service berry	*Sorbus tormin*	1,748
Larch	*Larix europea*	1,165
Spruce	*Abies excelsa*	242
Fir	*Abies pectinata*	96
Scotch white pine	*Pinus silvestris*	110
Black Austrian pine	*Pinus larico*	123

Source: U.S. Weather Bureau

SECTION J. EVAPORATION

TABLE 2–48. RESERVOIR EVAPORATION AT SELECTED STATIONS IN THE UNITED STATES

[Mean monthly computed values in inches]

Station	Month												
	Jan.	Feb.	Mar.	Apr.	May	June	July	Aug.	Sept.	Oct.	Nov.	Dec.	Annual
Ariz., Yuma	3.9	4.6	6.5	8.0	9.8	11.5	13.4	12.9	10.7	8.0	6.1	4.5	100
Calif., Sacramento	0.8	1.4	2.5	3.6	5.0	7.1	8.9	8.6	7.1	4.8	2.6	1.2	54
Colo., Denver	1.6	1.8	2.5	3.7	5.0	7.4	8.8	8.4	6.7	4.6	3.0	1.9	55
Fla., Miami	3.0	3.4	4.1	4.9	5.0	4.8	5.3	5.1	4.3	4.1	4.3	2.7	51
Ga., Macon	1.7	2.2	3.1	4.3	5.1	6.2	6.3	5.8	5.2	4.2	2.8	1.8	49
Me., Eastport	0.8	0.7	0.9	1.1	1.4	1.7	2.0	2.1	2.0	1.6	1.1	0.7	16
Minn., Minneapolis	0.3	0.4	0.9	1.7	3.2	4.4	6.0	5.8	4.6	3.0	1.3	0.4	32
Miss., Vicksburg	1.3	1.9	2.9	4.2	5.0	5.7	5.8	5.5	5.2	4.4	2.9	1.6	46
Mo., Kansas City	0.9	1.1	1.7	3.1	4.4	6.1	8.0	7.8	6.0	4.5	2.5	1.0	47
Mont., Havre	0.5	0.5	1.1	2.5	4.5	6.1	8.2	8.3	5.6	3.3	1.5	0.7	43
Nebr., North Platte	0.8	1.1	2.2	3.7	5.0	6.5	8.6	8.4	6.9	4.6	2.6	1.1	51
N. Mex., Roswell	2.1	3.2	4.9	6.8	8.3	9.8	9.4	8.3	6.9	5.5	3.5	2.5	71
N. Y., Albany	0.6	0.7	1.1	2.0	3.2	4.3	5.2	4.7	3.4	2.4	1.4	0.8	30
N. Dak., Bismarck	0.4	0.5	1.0	2.3	4.0	5.3	7.3	7.7	5.8	3.3	1.3	0.5	39
Ohio, Columbus	0.6	0.8	1.1	2.3	3.5	4.6	5.6	5.1	4.1	3.0	1.6	0.6	33
Okla., Oklahoma City	1.5	1.9	3.1	4.7	5.5	7.8	10.2	10.7	8.8	6.3	3.5	2.0	66
Ore., Baker	0.5	0.7	1.4	2.5	3.4	4.4	6.9	7.3	4.9	2.9	1.5	0.6	37
S.C., Columbia	1.6	2.4	3.2	4.5	5.4	6.3	6.6	6.0	5.5	4.4	3.0	1.9	51
Tenn., Nashville	0.9	1.3	1.9	3.3	4.1	5.1	5.8	5.4	4.9	3.7	2.1	1.1	39
Tex., Galveston	0.9	1.3	1.6	2.6	4.1	5.6	6.2	6.1	5.7	4.6	2.7	1.3	43
Tex., San Antonio	2.2	3.1	4.5	5.6	6.5	8.4	9.4	9.4	7.6	5.8	3.7	2.4	69
Utah, Salt Lake City	0.8	1.0	2.0	3.5	5.1	7.9	10.6	10.4	7.3	3.9	2.0	1.0	55
Va., Richmond	1.3	1.7	2.2	3.5	4.1	5.0	5.6	4.9	4.1	3.2	2.4	1.5	39
Wash., Seattle	0.8	0.8	1.4	2.1	2.7	3.4	3.9	3.4	2.6	1.6	1.1	0.7	24
Wis., Milwaukee	0.6	0.7	0.9	1.3	2.1	3.2	5.0	5.4	4.7	3.2	1.6	0.6	29
Gulf off Texas Coast	4.0	4.0	3.5	3.5	4.0	4.5	5.0	5.5	6.5	6.5	6.0	5.0	58
Gulf Stream off Cape Hatteras, N.C.	9.0	9.5	8.5	7.0	5.5	3.5	3.5	3.5	5.5	9.0	9.5	10.0	84
Ocean off Massachusetts	3.0	2.5	2.0	1.5	1.0	1.5	1.5	2.0	2.5	3.0	3.5	4.0	28

Source: Minnesota Resources Commission

TABLE 2–49. EVAPORATION EQUATIONS

[The following equations enable pan and lake evaporation to be computed from climatic data at first-order weather stations. Daily values of evaporation are obtained using mean daily temperature and vapor pressure data together with data on solar radiation and wind movement as specified.]

For pan evaporation, the expression is:

$$\begin{aligned} E_p = \; & \{\exp[(T_a-212)(0.1024-0.01066 \ln R)] - 0.0001 \\ & + 0.025(e_s-e_a)^{0.88}(0.37+0.0041\,U_p)\} \\ & \times \{0.025 + (T_a+398.36)^{-2}\, 4.7988\times10^{10} \\ & \exp[-7482.6/(T_a+398.36)]\}^{-1} \end{aligned}$$

For lake evaporation, the expression is:

$$\begin{aligned} E_L \; & \{\exp[(T_a-212)(0.1024-0.01066 \ln R)] -0.0001 \\ & + 0.0105(e_s-e_a)^{0.88}(0.37-0.0041\,U_p)\} \\ & \times \{0.015 + (T_a+398.36)^{-2}\, 6.8554\times10^{10} \\ & \exp\{-7482.6/(T_a+398.36)]\}^{-1} \end{aligned}$$

The terms in these expressions are:

E_p = pan evaporation, inches

E_L = lake evaporation, inches

T_a = air temperature, degrees Fahrenheit

e_a = vapor pressure, inches of mercury at temperature T_a

e_s = vapor pressure, inches of mercury at temperature T_d

T_d = dew point temperature, degrees Fahrenheit

R = solar radiation, langleys per day

U_p = wind movement, miles per day

Source: U.S. Weather Bureau, 1962

FIGURE 2-11. MEAN ANNUAL LAKE EVAPORATION IN THE UNITED STATES

[Values in inches for period 1946–55.]

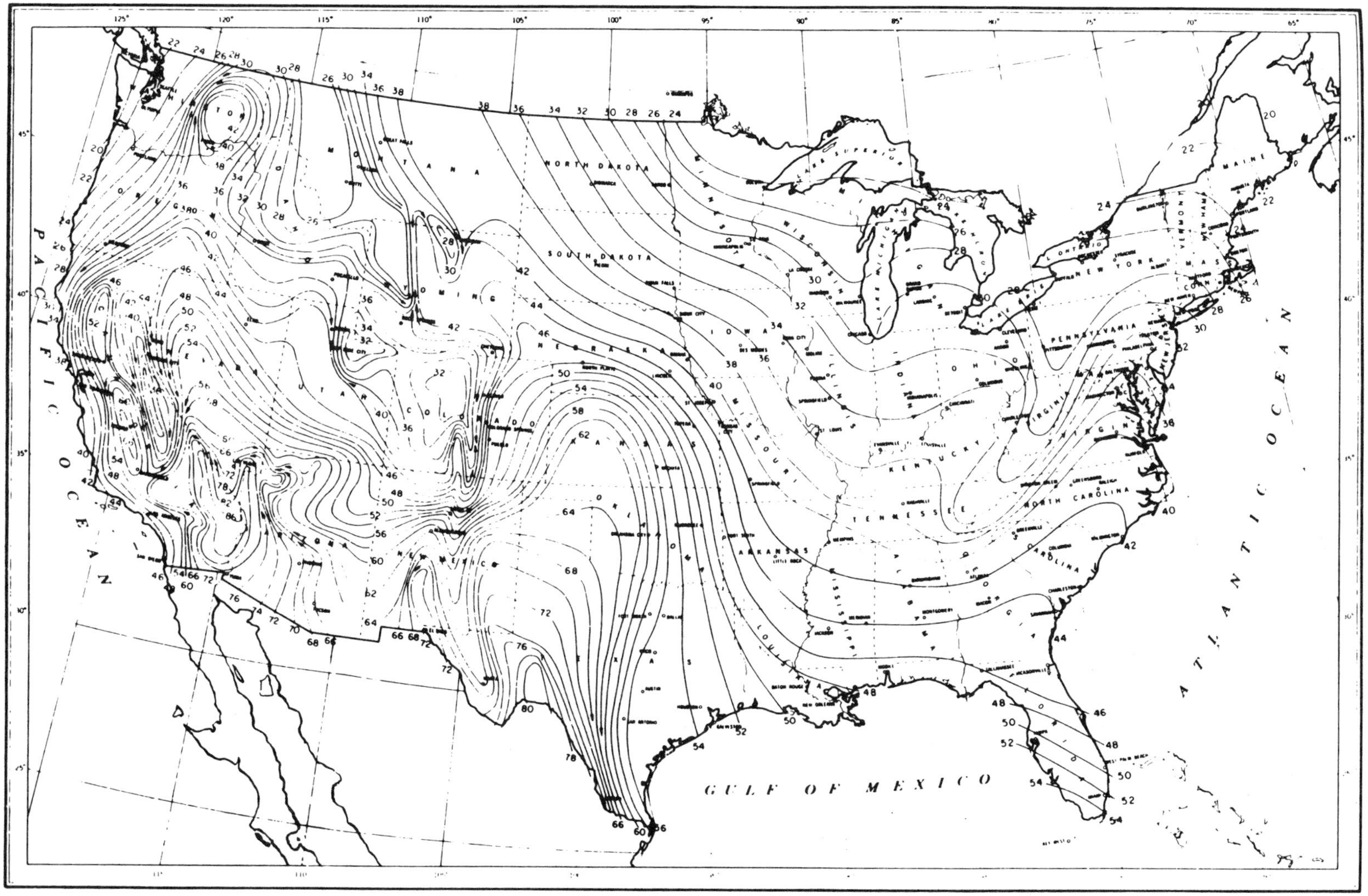

Source: U.S. Weather Bureau

FIGURE 2–12. ANNUAL EVAPORATION IN THE WORLD (in mm)

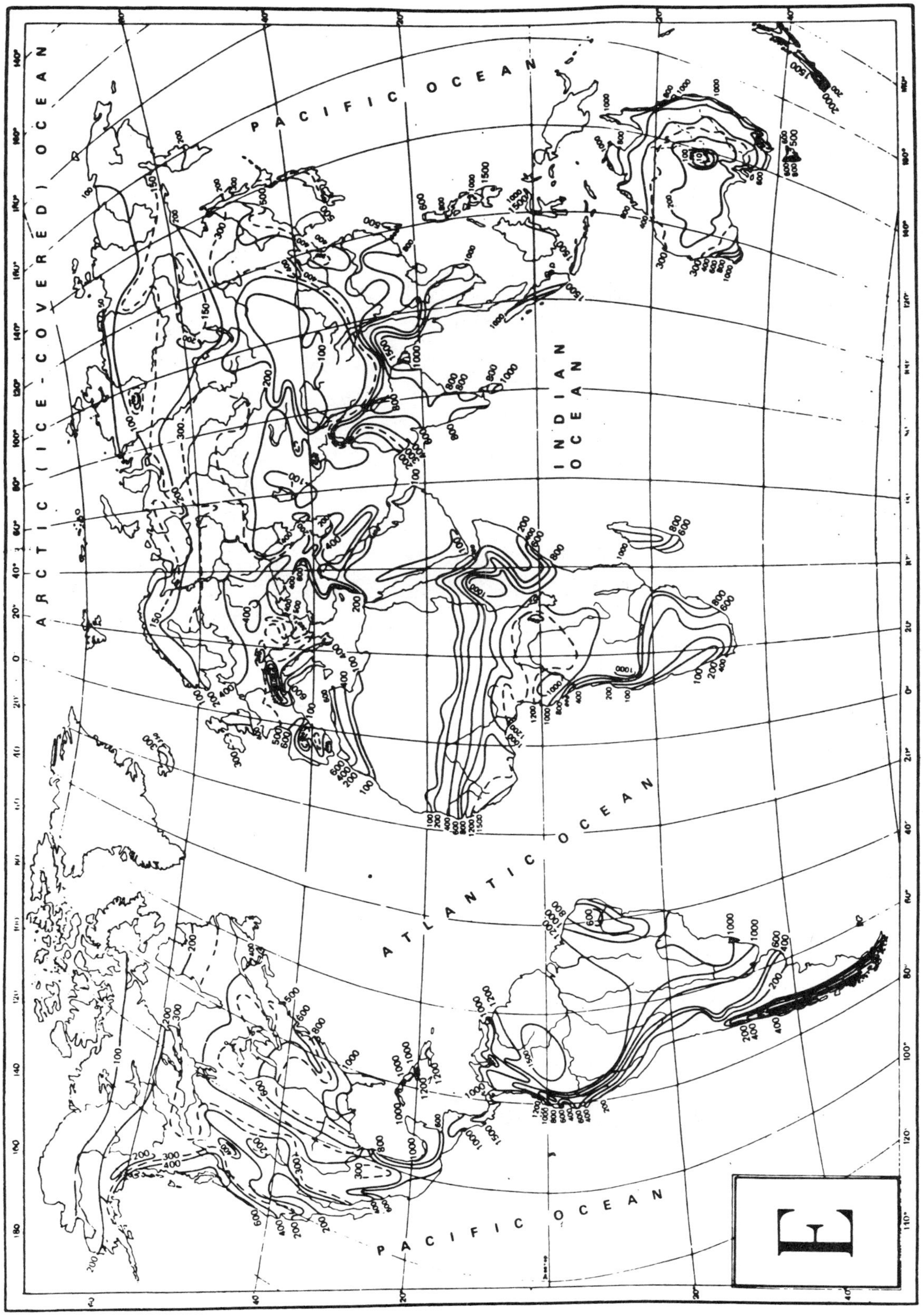

Source: Lvovitch, M.I., EOS, Vol. 54, No. 1., 1973, Copyright by American Geophysical Union

FIGURE 2–13. AREAS OF NATURAL WATER SURPLUS AND NATURAL WATER DEFICIENCY

[Computed by subtracting values of potential evapotranspiration from average precipitation]

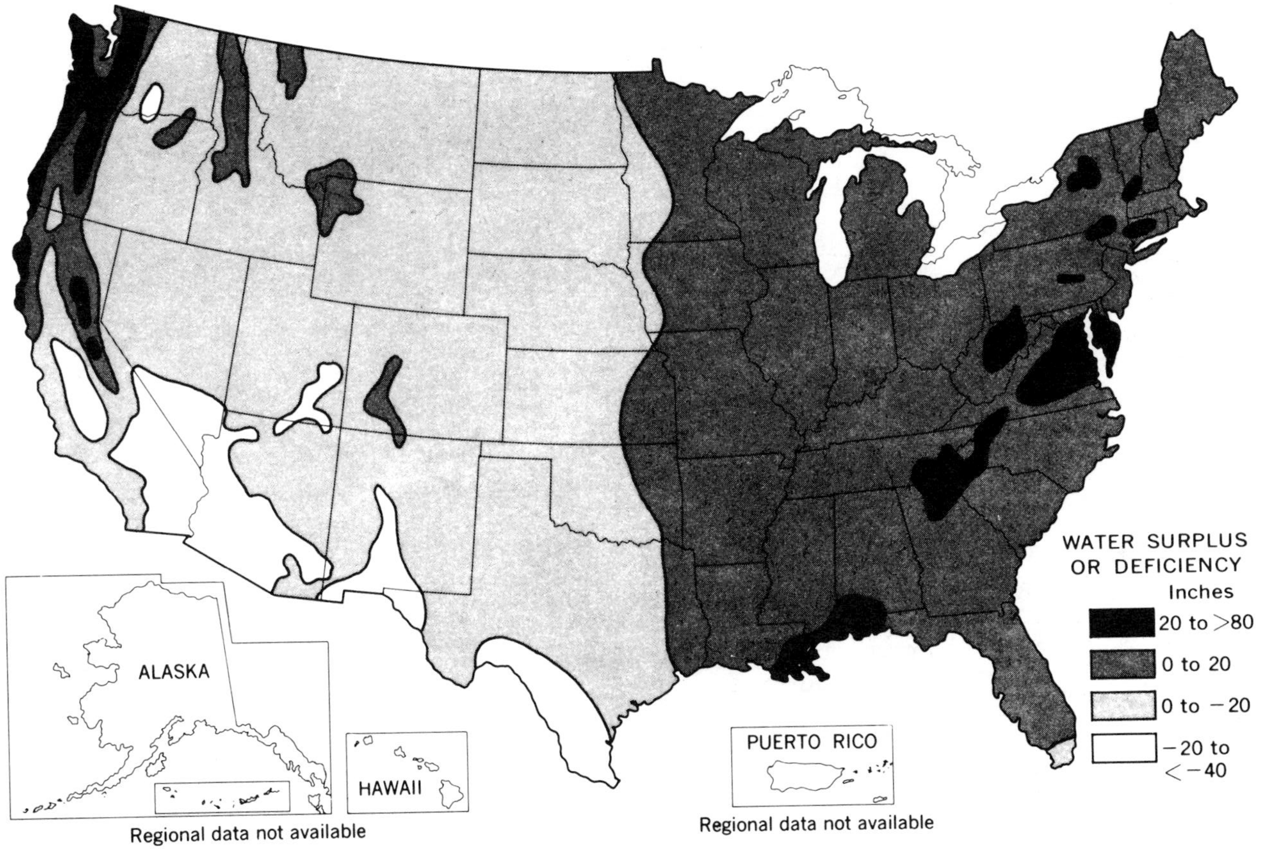

Source: U.S. Water Resources Council, 1968, The Nation's Water Resources

SECTION K. CONSUMPTIVE USE

TABLE 2–50. CONSUMPTIVE USE BY IRRIGATED CROPS IN THE WESTERN UNITED STATES

[Data for irrigation season only]

Location	Crop	Consumptive Use (Evapotranspiration), Inches							
		Apr.	May	June	July	Aug.	Sept.	Oct.	Total
Arizona, Mesa	Alfalfa	5.0	6.5	9.0	12.0	10.0	6.0	4.0	52.5
	Dates	6.2	7.6	8.3	9.2	8.4	7.2	5.7	52.6
California:									
Los Angeles[1]	Lemons	2.1	2.6	3.3	3.9	3.7	3.4	2.8	21.8
	Oranges	2.2	2.2	3.1	3.4	3.7	3.1	2.9	20.6
	Walnuts	3.8	5.0	5.9	6.1	5.0	2.8	2.0	30.6
	Alfalfa	3.3	6.7	5.4	7.8	4.2	5.6	4.4	37.4
Coastal	Alfalfa	4.9	4.9	4.3	5.2	5.9	5.5	4.7	35.4
Ontario	Peaches	1.0	3.5	6.7	8.0	6.5	2.7	1.4	29.8
Shafter	Cotton	0.5	1.0	4.0	8.5	9.7	5.8	3.2	32.7
Firebaugh	Cotton	–	0.8	1.1	7.3	7.8	3.6	2.0	22.6
	Cotton	–	0.4	0.7	8.4	9.5	3.0	2.5	24.5
Delta[2]	Alfalfa	3.6	4.8	6.0	7.8	6.6	6.0	1.2	36.0
	Potatoes	–	1.8	4.6	6.2	3.6	1.8	–	18.0
	Truck	1.2	3.0	6.0	5.4	5.4	3.6	1.8	26.4
	Sugar beets	1.6	3.8	6.1	7.3	6.4	2.4	–	27.6
	Beans	1.9	2.4	1.7	2.9	6.9	4.4	–	20.2
	Fruit	2.2	3.8	6.0	6.8	4.8	2.8	0.8	27.2
	Onions	1.6	3.2	5.9	5.2	2.4	1.9	–	19.8
Davis	Sugar beets	–	5.2	5.7	7.1	5.8	–	–	23.8
	Tomatoes	–	–	3.2	6.2	4.9	4.7	–	22.3
	Alfalfa	–	6.8	7.9	8.3	7.1	4.3	–	–
	Prunes	–	5.8	6.0	7.6	6.5	5.0	–	–
	Peaches	–	5.4	6.4	7.9	7.2	5.0	–	–
	Walnuts	–	6.6	6.7	8.4	7.2	4.8	–	–
	Grapes	–	4.6	4.9	6.2	5.3	4.3	–	–
Winters	Apricots	–	–	5.6	6.8	6.5	4.9	–	–
Nebraska,									
Scottsbluff	Alfalfa	1.4	4.0	7.0	7.1	6.4	3.0	–	28.9
	Beets	1.9	3.3	5.2	6.9	5.8	1.1	–	24.2
	Potatoes	–	–	–	3.4	5.8	4.4	–	–
	Oats	–	3.0	6.1	5.1	–	–	–	14.2

[1] In San Fernando Valley, City of Los Angeles, Calif.

[2] In Sacramento-San Joaquin Delta, Calif.

Source: U.S. Dept. of Agriculture

TABLE 2–51. CONSUMPTIVE USE BY PRINCIPAL CROPS IN THE CENTRAL VALLEY, CALIFORNIA

[Values in inches; S.V. is Sacramento Valley, and S.J.V. is San Joaquin Valley]

Month	Improved Pasture	Alfalfa	Sugar Beets S.V.	Sugar Beets S.J.V.	Cotton [1]	Cotton [2]	Deciduous Orchard	Rice S.V.
January	1.0	1.0	0.9	–	–	–	–	(0.8)[3]
February	1.8	1.8	1.2	–	–	–	–	(1.5)
March	3.0	2.8	–	1.6	–	–	1.8	(1.4)
April	4.7	4.2	–	3.6	–	–	3.3	4.8
May	6.1	5.4	1.8	5.6	1.0	0.6	4.9	7.6
June	7.8	7.0	6.0	7.7	6.0	3.6	6.7	9.6
July	8.2	7.6	8.5	8.5	9.7	8.8	7.5	10.0
August	7.1	6.8	7.3	5.1	8.5	7.8	6.4	8.5
September	5.2	5.1	5.6	1.9	5.5	5.3	4.5	6.4
October	3.5	3.5	3.5	–	2.1	1.8	2.7	3.2
November	1.6	1.6	1.6	–	–	–	–	(1.4)
December	0.8	0.8	0.8	–	–	–	–	(0.8)
Total	50.8	47.6	37.2	34.0	32.8	27.9	37.8	56.0

[1] Solid planting or one row skipped in three.
[2] Planting of two rows skipped in four.
[3] Values in parentheses are for nongrowing season. Values may change with differences in rainfall.
Source: Calif. Dept. of Water Resources, 1967

TABLE 2–52. CONSUMPTIVE USE BY CROPS IN THE SACRAMENTO-SAN JOAQUIN DELTA, CALIFORNIA

[Depths in feet. Figures in parentheses show estimated losses by soil evaporation and weed transpiration.]

Month	Alfalfa	Asparagus	Beans	Beets	Celery	Corn	Fruit	Grain and Hay	Onions	Potatoes
January	(0.06)	0.05	(0.06)	(0.06)	(0.04)	(0.04)	(0.04)	(0.04)	(0.04)	(0.06)
February	(0.08)	0.05	(0.08)	(0.08)	(0.04)	(0.04)	(0.04)	(0.04)	(0.04)	(0.08)
March	0.10	0.05	(0.08)	(0.08)	(0.04)	(0.04)	(0.04)	0.07	0.08	(0.08)
April	0.30	0.05	(0.16)	0.13	(0.08)	(0.08)	0.18	0.60	0.13	(0.16)
May	0.40	0.08	(0.20)	0.32	(0.10)	(0.10)	0.32	0.83	0.27	0.15
June	0.50	0.14	0.14	0.51	0.10	0.24	0.50	0.20	0.49	0.38
July	0.65	0.40	0.24	0.61[1]	0.10	0.85	0.57	(0.14)	0.43	0.52
August	0.55	0.68	0.58	0.53[1]	0.20	0.84[1]	0.40	(0.23)	0.20	0.30
September	0.50	0.55	0.37	0.20[1]	0.25	0.40[1]	0.23	(0.21)	(0.16)	0.15
October	0.20	0.42	(0.09)	(0.13)	0.30	0.10	0.07	(0.14)	(0.13)	(0.09)
November	(0.10)	0.12	(0.07)	(0.10)	0.20	(0.10)	(0.07)	(0.07)	(0.10)	(0.07)
December	(0.07)	0.10	(0.05)	(0.07)	0.05	(0.07)	(0.05)	(0.05)	(0.07)	(0.05)
Growing season	3.20	2.69	1.33	2.30	1.20	2.43	2.27	1.70	1.60	1.50
Year	3.51	2.69	2.12	2.82	1.50	2.90	2.51	2.62	2.14	2.09

[1] Including additional use of water by weeds.
Source: Calif. Dept. of Public Works

TABLE 2–53. TOTAL CONSUMPTIVE USE AND PEAK DAILY USE, WESTERN UNITED STATES

Crops	SOUTHERN COASTAL				SOUTH PACIFIC COASTAL INTERIOR AND NORTH COASTAL							
	300 Days Plus		250-300 Days		250-300 Days		210-250 Days		180-210 Days		150-180 Days	
	Season Use in.	Daily Use in./d	Season Use in.	Daily Use in./d	Season Use in.	Daily Use in./d	Season Use in.	Daily Use in./d	Season Use in.	Daily Use in./d	Season Use in.	Daily Use in./d
Alfalfa	36.0	0.20	30.0	0.17	37.0	0.27	32.0	0.22	26.0	0.20	22.0	0.18
Pasture	33.5	0.20	28.0	0.17	33.0	0.27	30.0	0.22	24.0	0.20	20.0	0.18
Grain – small	16.0	0.18	14.0	0.16	17.0	0.22	14.5	0.20	12.0	0.20	10.0	0.18
Beets – sugar	29.0	0.20	25.0	0.18	30.0	0.27	26.0	0.22	–	–	–	–
Beans – field	12.0	0.18	10.0	0.16	13.0	0.22	11.0	0.18	–	–	–	–
Corn – field	–	–	–	–	–	–	19.0	0.25	16.0	0.22	14.0	0.20
Potatoes	–	–	–	–	24.0	0.20	20.0	0.18	–	–	–	–
Peas – green	10.0	0.18	8.0	0.16	11.0	0.20	9.0	0.18	8.0	0.16	7.0	0.16
Legume seed	25.0	0.20	22.0	0.18	26.0	0.25	22.0	0.22	20.0	0.20	18.0	0.18
Tomatoes	18.0	0.16	15.0	0.16	19.0	0.20	16.0	0.18	13.0	0.16	11.0	0.16
Vegetable seed	16.0	0.16	14.0	0.16	18.0	0.18	16.0	0.18	14.0	0.16	12.0	0.16
Beans – pole	–	–	–	–	18.0	0.20	16.0	0.20	14.0	0.18	12.0	0.16
Corn – sweet	16.0	0.18	14.0	0.16	16.0	0.20	14.0	0.18	12.0	0.18	11.0	0.16
Apples	–	–	–	–	24.0	0.20	22.0	0.20	20.0	0.18	18.0	0.18
Cherries	–	–	–	–	24.0	0.20	22.0	0.20	–	–	–	–
Peaches	–	–	–	–	24.0	0.20	22.0	0.20	–	–	–	–
Prunes	–	–	–	–	22.0	0.20	20.0	0.20	–	–	–	–
Apricots	–	–	–	–	22.0	0.20	20.0	0.20	–	–	–	–
Oranges	20.0	0.16	18.0	0.16	22.0	0.18	–	–	–	–	–	–
Avocados	18.0	0.16	–	–	–	–	–	–	–	–	–	–
Walnuts	22.0	0.20	18.0	0.18	24.0	0.25	22.0	0.22	20.0	0.20	–	–
Strawberries	22.0	0.20	18.0	0.18	23.0	0.25	18.0	0.22	16.0	0.20	14.0	0.18
Lettuce	4.0	0.16	4.0	0.16	6.0	0.18	5.0	0.18	–	–	–	–
Mint	–	–	–	–	23.0	0.24	21.0	0.22	19.0	0.20	17.0	0.18
Hops	–	–	–	–	20.0	0.22	20.0	0.20	–	–	–	–

Crops	CENTRAL VALLEY – CALIFORNIA AND VALLEYS EAST SIDE OF CASCADE MOUNTAINS											
	250–300 Days		210–250 Days		180–210 Days		150–180 Days		120–150 Days		90–120 Days	
Alfalfa	40.0	0.30	34.0	0.28	30.0	0.25	26.0	0.22	20.0	0.20	14.0	0.18
Pasture	36.0	0.30	30.0	0.28	28.0	0.25	24.0	0.22	18.0	0.20	13.0	0.18
Grain – small	18.0	0.22	16.0	0.22	15.0	0.20	14.0	0.18	13.0	0.18	12.0	0.16
Beets – sugar	33.0	0.30	28.0	0.25	24.0	0.22	20.0	0.20	18.0	0.18	–	–
Beans – field	17.0	0.22	13.0	0.20	13.0	0.20	12.0	0.18	12.0	0.18	–	–
Corn – field	26.0	0.35	22.0	0.32	22.0	0.30	20.0	0.25	18.0	0.22	17.0	0.20
Potatoes – summer	12.0	0.16	–	–	–	–	–	–	–	–	–	–
Potatoes – fall	–	–	19.0	0.25	18.0	0.22	18.0	0.20	17.0	0.18	16.0	0.16
Peas – green	–	–	8.0	0.18	7.0	0.18	7.0	0.18	7.0	0.16	6.0	0.15
Peas – field	–	–	10.0	0.18	9.0	0.18	9.0	0.18	8.0	0.16	8.0	0.15
Tomatoes	20.0	0.20	18.0	0.18	18.0	0.18	17.0	0.17	16.0	0.16	–	–
Cotton	26.0	0.30	22.0	0.28	–	–	–	–	–	–	–	–
Grain – sorghum	15.0	0.20	13.0	0.18	12.0	0.17	10.0	0.16	9.0	0.15	8.0	0.15
Apples	–	–	26.0	0.20	23.0	0.20	21.0	0.18	–	–	–	–

TABLE 2–53. TOTAL CONSUMPTIVE USE AND PEAK DAILY USE, WESTERN UNITED STATES (continued)

Crops	250-300 Days Season Use *in.*	250-300 Days Daily Use *in./d*	210-250 Days Season Use *in.*	210-250 Days Daily Use *in./d*	180-210 Days Season Use *in.*	180-210 Days Daily Use *in./d*	150-180 Days Season Use *in.*	150-180 Days Daily Use *in./d*	120-150 Days Season Use *in.*	120-150 Days Daily Use *in./d*	90-120 Days Season Use *in.*	90-120 Days Daily Use *in./d*
CENTRAL VALLEY – CALIFORNIA AND VALLEYS EAST SIDE OF CASCADE MOUNTAINS (continued)												
Cherries	–	–	24.0	0.20	21.0	0.20	19.0	0.18	–	–	–	–
Peaches	22.0	0.22	22.0	0.20	20.0	0.20	18.0	0.18	–	–	–	–
Apricots	20.0	0.22	17.0	0.20	15.0	0.20	–	–	–	–	–	–
Oranges	28.0	0.18	–	–	–	–	–	–	–	–	–	–
Strawberries	24.0	0.20	20.0	0.20	18.0	0.18	–	–	–	–	–	–
Lettuce – winter	4.0	0.20	–	–	–	–	–	–	–	–	–	–
Mint	–	–	20.0	0.22	18.0	0.20	–	–	–	–	–	–
Hops	–	–	18.0	0.20	16.0	0.18	–	–	–	–	–	–
Grapes	30.0	0.25	25.0	0.22	22.0	0.20	–	–	–	–	–	–
Walnuts	24.0	0.22	20.0	0.20	–	–	–	–	–	–	–	–
Almonds	22.0	0.25	20.0	0.22	–	–	–	–	–	–	–	–
INTERMOUNTAIN, DESERT, AND WESTERN HIGH PLAINS	**250–300 Days**		**210–250 Days**		**180–210 Days**		**150–180 Days**		**120–150 Days**		**90–120 Days**	
Alfalfa	52.0	0.40	44.0	0.32	36.0	0.29	30.0	0.26	24.0	0.22	19.0	0.20
Pasture	48.0	0.40	40.0	0.30	33.0	0.28	28.0	0.25	22.0	0.22	17.0	0.20
Grain – small	21.0	0.25	18.0	0.22	16.0	0.20	16.0	0.20	16.0	0.20	14.0	0.18
Beets – sugar	37.0	0.30	32.0	0.30	30.0	0.28	26.0	0.25	24.0	0.22	18.0	0.20
Beans – field	22.0	0.25	17.0	0.20	14.0	0.20	14.0	0.18	14.0	0.17	12.0	0.15
Corn – field	–	–	30.0	0.35	26.0	0.30	24.0	0.28	22.0	0.24	–	–
Potatoes – fall	–	–	23.0	0.30	21.0	0.28	20.0	0.25	19.0	0.22	17.0	0.20
Peas – field	–	–	–	–	10.0	0.19	10.0	0.18	10.0	0.17	9.0	0.15
Tomatoes	–	–	20.0	0.22	18.0	0.20	17.0	0.18	16.0	0.17	–	–
Cotton	32.0	0.30	30.0	0.28	–	–	–	–	–	–	–	–
Grain – sorghum	19.0	0.25	18.0	0.20	16.0	0.20	14.0	0.18	12.0	0.17	–	–
Apples	–	–	–	–	28.0	0.22	24.0	0.20	20.0	0.18	–	–
Cherries	–	–	–	–	26.0	0.22	–	–	–	–	–	–
Peaches	–	–	29.0	0.25	27.0	0.22	–	–	–	–	–	–
Apricots	26.0	0.25	24.0	0.25	25.0	0.20	–	–	–	–	–	–
Almonds	22.0	0.25	20.0	0.25	–	–	–	–	–	–	–	–
Vineyards	40.0	0.27	32.0	0.25	26.0	0.22	–	–	–	–	–	–
Legume seed	–	–	–	–	–	–	–	–	16.0	0.18	14.0	0.16
Grass seed	–	–	–	–	–	–	–	–	14.0	0.14	12.0	0.14
Potatoes – seed	–	–	–	–	–	–	–	–	16.0	0.16	14.0	0.15
Grapefruit	45.0	0.20	–	–	–	–	–	–	–	–	–	–
Oranges	36.0	0.18	–	–	–	–	–	–	–	–	–	–
Lettuce – winter	6.0	0.18	–	–	–	–	–	–	–	–	–	–
Melons	22.0	0.25	20.0	0.22	18.0	0.20	16.0	0.18	–	–	–	–
Palm dates	60.0	0.30	–	–	–	–	–	–	–	–	–	–
Truck crops	20.0	0.25	18.0	0.22	14.0	0.20	12.0	0.18	12.0	0.16	10.0	0.15

Source: Woodward, Sprinkler Irrigation, Sprinkler Irrigation Assoc., 1959

TABLE 2–54. ACCUMULATED USE OF WATER BY CROPS WITH VARIOUS PLANTING TO MATURITY PERIODS

Percentage of Growing Period[1]	Accumulated Consumptive Use of Water in Percentage of Total Use	Total Period of Growth: 2 Mo.		3 Mo.		4 Mo.		5 Mo.		6 Mo.	
		Days Since Plant-ing	Accum. Water Use, Inches	Days Since Plant-ing	Accum. Water Use, Inches	Days Since Plant-ing	Accum. Water Use, Inches	Days Since Plant-ing	Accum. Water Use, Inches	Days Since Plant-ing	Accum. Water Use, Inches
10	6.0	6	0.3	9	0.4	12	0.6	15	0.8	18	0.9
20	13.8	12	0.7	18	1.0	24	1.4	30	1.7	36	2.0
30	23.5	18	1.1	27	1.7	36	2.3	45	2.9	54	3.5
40	34.5	24	1.7	36	2.6	48	3.4	60	4.3	72	5.1
50	46.5	30	2.3	45	3.4	60	4.6	75	5.7	90	6.8
60	59.4	36	2.9	54	4.4	72	5.8	90	7.3	108	8.7
70	72.3	42	3.5	63	5.3	84	7.1	105	8.9	126	10.6
80	84.3	48	4.1	72	6.2	96	8.3	120	10.3	144	12.4
90	94.5	54	4.7	81	7.0	108	9.3	135	11.6	162	13.9
100	100.0	60	4.9	90	7.4	120	9.8	150	12.3	180	14.7

[1] Growing Period refers to the entire time from planting to the time the plant dies, which is usually longer than the period from planting to harvesting. Flowering will occur at about 50 to 60 percent of the growing period and fruiting after 60 percent.

Source: Israelson and Hansen, Irrigation Principles and Practices, John Wiley & Sons, Copyright 1962

TABLE 2-55. VARIATIONS IN CONSUMPTIVE USE BY CROPS

Crop	No. of Tests	Range in Water Requirements, ft.
A. FARM CROPS IN THE SOUTHWEST		
Alfalfa	369	3.47–5.08
Rhodes grass	12	3.49–4.43
Sudan grass	25	2.88–3.16
Barley	3	1.24–1.83
Oats	2	1.90–2.09
Wheat	46	1.46–2.24
Corn	42	1.44–1.99
Kafir	16	1.32–1.54
Flax	3	1.23–1.59
Broomcorn	9	0.97–1.15
Emmer	6	1.19–1.87
Feterita	8	0.97–1.10
Millet	5	0.91–1.09
Milo	35	0.96–1.67
Sorghum	34	1.69–2.08
Cotton	103	2.35–3.51
Potatoes	12	1.59–2.04
Soybeans	36	1.66–2.81
Sugar beets	5	1.77–2.72
Sugar cane[1]	41	3.48–4.56
B. VEGETABLE CROPS IN THE SOUTHWEST		
Beans, snap	9	0.83–1.44
Beets, table	28	0.87–1.37
Cabbage	21	0.94–1.49
Carrots	6	1.27–1.60
Cauliflower	6	1.43–1.77
Lettuce	49	0.72–1.35
Onions	4	0.73–1.52
Peas	8	1.21–1.56
Melons	3	2.48–3.40
Spinach	12	0.80–1.07
Sweet potatoes	3	1.77–2.25
Tomatoes	17	0.95–1.42
C. CROPS IN THE MISSOURI AND ARKANSAS BASINS		
Forage, including alfalfa	648	1.94–2.62
Barley	335	1.33–1.82
Oats	409	1.35–1.81
Wheat	542	1.36–1.80
Corn	70	1.23–1.83
Kafir corn	15	1.43–1.57
Flax	50	1.47–1.85
Millet	14	0.81–0.94
Milo maize	27	1.09–1.70
Sorghum	26	1.06–1.47
Apples	4	2.10–2.60
Beans	4	1.30–1.60
Buckwheat	3	1.05–1.30
Cantaloupes	10	1.50–2.30
Peas	168	1.36–1.94
Potatoes	350	1.38–1.70
Sugar beets	128	1.60–2.50
Sunflowers	16	1.20–1.40
Tomatoes	6	2.10–2.80
Cucumbers	7	1.73–3.75

[1] Not commonly produced in the Southwest.
Source: U.S. Dept. of Agriculture

TABLE 2–56. CONSUMPTIVE USE OF WATER BY CROPS IN FLORIDA

[Everglades agricultural area; in inches]

Month	Citrus	Pasture[1]	Sugarcane	Rice[2]	Rainfall	Pan Evapo-transpira-tion[3]
January	2.09	2.01	1.42	0	1.97	3.39
February	2.60	2.52	1.10	0	1.97	4.00
March	3.58	3.35	2.52	0	3.21	5.70
April	4.49	4.21	3.39	1.63	2.96	6.54
May	5.31	5.20	4.80	3.07	4.74	7.06
June	4.41	4.25	5.98	5.82	9.08	6.24
July	4.88	4.80	6.50	8.43	8.58	6.36
August	4.80	4.80	6.69	3.05	8.21	6.12
September	4.02	3.86	5.12	(5.00)	8.82	5.31
October	3.59	3.43	5.20	(3.00)	5.65	4.82
November	2.72	2.48	3.19	0	1.74	3.71
December	2.09	1.93	2.59	0	1.80	3.19
Total	44.58	42.84	48.50	22.00	58.76	62.44

[1] Mean monthly values averaged over 5 years and averaged over water table depth of 12, 24, and 36 inches maintained in lysimeters at Ft. Lauderdale. These turfgrass evapotranspiration values are assumed to be valid for pastures adequately supplied with water.
[2] Assuming planting date of April 15 which is approximately the middle of the planting season. Values in parentheses are estimates for a ratoon crop. Rice is not always ratoon cropped.
[3] Pan evapotranspiration is a measure of the capability of the air to evaporate water. A relatively higher reading indicates relatively high consumptive use of water.
Source: Bajwa, R.S., 1985, Analysis of Irrigation Potential in the Southeast: Florida, A Special Report, U.S. Department of Agriculture

FIGURE 2–14. STEPS IN DETERMINING AGRICULTURAL APPLIED WATER DEMAND

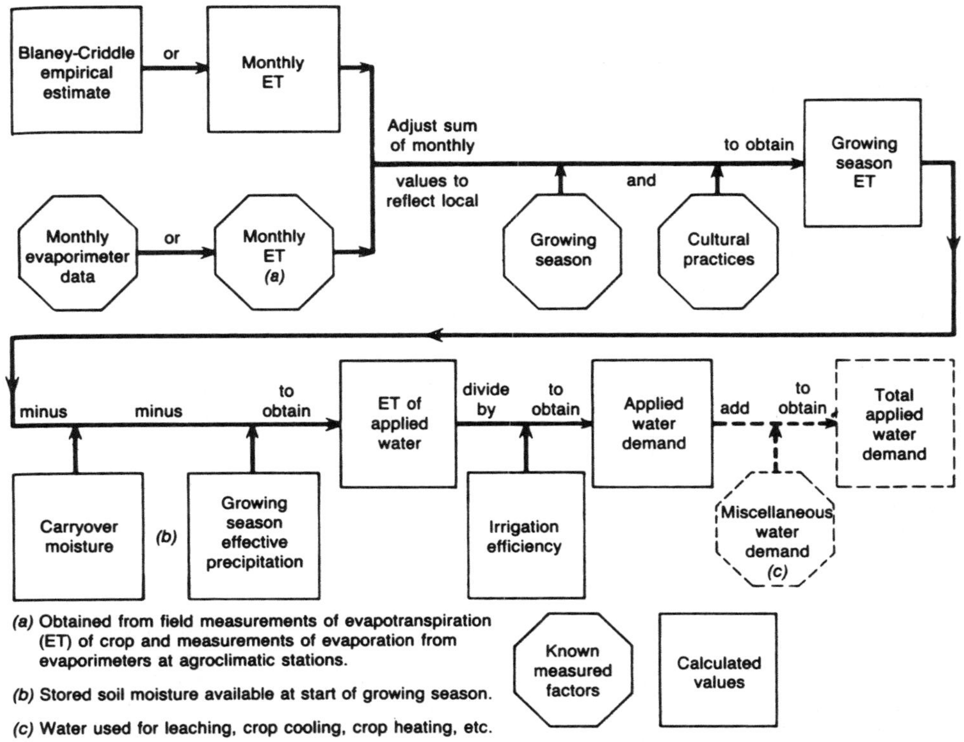

Source: American Society of Civil Engineers, 1972, Copyright, Ground Water Management. Reproduced with permission

TABLE 2–57. ECONOMIC IRRIGATION REQUIREMENTS IN THE WESTERN UNITED STATES

[Estimated minimum quantities of irrigation water required, assuming optimal irrigation practices]

Div. No.	Division Description	State	Irrigation Season, Inclusive	Total Depth, Ft.
A. COLUMBIA RIVER BASIN				
1	Snake River Valley	Idaho	Apr.-Oct.	2.5
2	Upper Snake River Valley	Idaho	Apr.-Sept.	2.3
3	Jackson Lake and Upper Snake basin	Idaho and Wyoming	May-Sept.	1.7
4	Southwest Idaho and north Nevada	Idaho and Nevada	Apr.-Sept.	1.9
5	Salmon River basin	Idaho	May-Aug.	2.0
6	North Idaho	Idaho	May-Sept.	1.5
7	Bitterroot and Missoula river basins	Montana	Apr.-Nov.	2.1
8	Flathead Lake and River basins	Montana	Apr.-Sept.	1.8
9	Owyhee and Malheur River basins	Oregon	Apr.-Sept.	2.4
10	Northeast Oregon	Oregon	Apr.-Sept.	2.0
11	Umatilla, John Day, Deschutes, and Hood basins	Oregon	Apr.-Oct.	2.5
12	Central Oregon	Oregon	May-Aug.	2.4
13	Yakima and Wenatchee river basins	Washington	Apr.-Nov.	2.6
14	Southeast Washington	Washington	Apr.-Oct.	2.1
15	Northeast Washington	Washington	Apr.-Oct.	2.2
16	Okanogan River basin	Washington	Apr.-Nov.	2.3
17	Lower Columbia River basin	Washington	May-Sept.	1.3
18	Willamette River basin	Oregon	May-Sept.	1.2
19	Puget Sound region[1]	Washington	May-Sept.	1.4
B. PACIFIC SLOPE BASINS				
1	Umpqua, Coquill, and Lower Rogue basins	Oregon	Apr.-Sept.	0.85
2	Upper Rogue River basin	Oregon	Mar.-Sept.	1.50
3	Klamath Lake and River basins	Oregon and California	Apr.-Sept.	2.00
4	Northwest California	California	Apr.-Oct.	1.40
5	Pit River basin	California	Apr.-Sept.	1.60
6	Feather, Yuba, and American River basins	California	Mar.-Nov.	1.50
7	Sacramento Valley	California	Mar.-Oct.	2.10
8	Sacramento-San Joaquin Delta	California	May-Sept.	2.00
9	San Francisco Bay basin	California	Mar.-Nov.	1.50
10	Salinas River basin	California	Mar.-Oct.	1.70
11	Santa Maria, Santa Inez, and Santa Clara basins	California	Jan.-Dec.	1.60
12	San Joaquin Valley	California	Feb.-Oct.	2.30
13	West slope of Sierras	California	Feb.-Nov.	1.70
14	East slope of Coast Range	California	Feb.-Oct.	1.80
15	Antelope and Victor Valleys	California	Mar.-Oct.	1.90
16	Los Angeles, San Gabriel, and Santa Ana basins	California	Jan.-Dec.	1.70
17	Upper Santa Ana River Valley	California	Jan.-Dec.	1.80
18	San Diego County	California	Jan.-Dec.	1.40
C. SOUTHWEST				
1	Imperial Valley	California	Jan.-Dec.	3.10
2	South Nevada	Nevada	Jan.-Dec.	2.90
3	Southwest Arizona	Arizona	Jan.-Dec.	3.00
4	Northwest Arizona	Arizona	Mar.-Oct.	2.30
5	Navajo country	Arizona	Mar.-Oct.	2.30
6	Southeast Arizona	Arizona	Feb.-Nov.	2.60
7	San Juan basin	New Mexico	Apr.-Sept.	2.20
8	West New Mexico	New Mexico	Apr.-Oct.	1.70
9	Rio Grande basin	New Mexico	Jan.-Dec.	2.60
10	Pecos River basin	New Mexico	Jan.-Dec.	2.40
11	Northeast New Mexico	New Mexico	Feb.-Nov.	1.60
12	Central Rio Grande basin	Texas	Jan.-Nov.	2.40

TABLE 2–57. ECONOMIC IRRIGATION REQUIREMENTS IN THE WESTERN UNITED STATES (continued)

Div. No.	Division Description	State	Irrigation Season, Inclusive	Total Depth, Ft.
13	Pecos River basin	Texas	Jan.-Nov.	2.25
14	West central Texas	Texas	Jan.-Dec.	1.60
15	Lower Rio Grande basin	Texas	Jan.-Dec.	1.75
16	Upper Nueces and Colorado River basins	Texas	Jan.-Dec.	1.30
17	Upper Brazos and Red River basins	Texas	Jan.-Dec.	1.10
18	Eastern Panhandle	Texas	Mar.-Oct.	1.35
19	Western Panhandle	Texas	Mar.-Oct.	1.65
20	Panhandle	Oklahoma	Apr.-Oct.	1.25
21	West Oklahoma	Oklahoma	Apr.-Oct.	1.00
22	San Luis basin	Colorado	May-Sept.	1.80
23	San Juan basin	Colorado	Apr.-Sept.	1.90
24	Yampa and White River basins	Colorado	May-Aug.	1.35
25	Upper Colorado River basin	Colorado	Apr.-Sept.	1.70
26	Virgin River basin	Utah	Feb.-Nov.	2.25
27	San Juan basin	Utah	Apr.-Sept.	2.10
28	Green River basin	Utah	Apr.-Oct.	2.00
29	Uintah basin	Utah	Apr.-Sept.	1.75
30	Green River basin	Wyoming	May-Aug.	1.60
D. GREAT BASIN				
1	Bear River basin	Idaho and Utah	May-Oct.	2.0
2	Utah Lake and Great Salt Lake Valleys[2]	Utah	May-Oct.	2.2
3	Sevier River basin	Utah	May-Oct.	2.1
4	Irrigable lands, southwest Utah	Utah	Apr.-Oct.	1.8
5	Irrigable lands, southern Nevada	Nevada	Apr.-Oct.	1.7
6	Antelope Valley and Mohave River areas	California and Nevada	Mar.-Oct.	1.8
7	Mono, Owens, and Inyo-Kern valleys	California	Mar.-Oct.	2.1
8	Walker River basin	California and Nevada	Apr.-Oct.	2.0
9	Truckee and Carson River basins	California and Nevada	May-Oct.	2.1
10	Humboldt, Quinn, and White River basins	Nevada	May-Sept.	2.0
11	Honey Lake basin	California and Nevada	Apr.-Sept.	1.7
12	Malheur Lake, Harney Lake, and other basins	Oregon	Apr.-Sept.	1.5
E. MISSOURI AND ARKANSAS BASINS				
1	Northeast Montana	Montana	May-Aug.	1.40
2	North central Montana	Montana	May-Aug.	1.50
3	Central Montana	Montana	May-Aug.	1.70
4	Upper Missouri River basin	Montana	May-Aug.	1.60
5	Upper Yellowstone River basin	Montana	May-Aug.	1.90
6	Southeast Montana	Montana	May-Sept.	1.95
7	Big Horn River basin	Wyoming	May-Sept.	1.65
8	Yellowstone and Missouri River basins	Wyoming	May-Sept.	1.70
9	Upper Platte River basin	Wyoming	May-Sept.	1.60
10	Northeast Colorado	Colorado	Apr.-Sept.	2.05
11	North central Colorado	Colorado	Apr.-Sept.	2.20
12	South central Colorado	Colorado	Apr.-Sept.	2.10
13	Southeast Colorado	Colorado	Apr.-Oct.	2.30
14	West Kansas	Kansas	Apr.-Oct.	1.75
15	Central Nebraska	Nebraska	Apr.-Oct.	1.25
16	West Nebraska	Nebraska	Apr.-Oct.	2.00
17	Western South Dakota	South Dakota	May-Sept.	1.50
18	Western North Dakota	North Dakota	May-Sept.	1.35

[1] Not in the Columbia River Basin.
[2] South of Weber River basin.
Source: U.S. Dept. of Agriculture

TABLE 2–58. ESTIMATED EVAPOTRANSPIRATION FOR TYPES OF VEGETATION IN THE WESTERN UNITED STATES

Vegetation Type	Annual Evapotranspiration, Inches
Forest:	
Lodgepole pine	19
Engelmann spruce-fir	15
White pine-larch-fir	22
Mixed conifer	22
True fir	24
Aspen	23
Pacific Douglas-fir-hemlock-redwood	30
Interior ponderosa pine	17
Interior Douglas-fir	21
Chaparral and woodland:	
Southern California chaparral	20
California woodland-grass	18
Arizona chaparral	17.5
Piñon-juniper	14.5
Semiarid grass and shrub	10.6
Alpine	20

Source: Select Committee on National Water Resources, U.S. Senate, 1960

TABLE 2–59. CONSUMPTIVE USE IN A MUNICIPAL AREA

[Data for Raymond Basin, Los Angeles County, Calif.]

Cultural Classification	Consumptive Use in Feet	Cultural Classification	Consumptive Use in Feet
Estates	2.07	Reservoir sites	1.34
Class A residential	1.92	Park	2.40
Class B residential	1.88	Schools	1.63
Rural residential	1.78	River wash	0.99
Semicommercial	1.32		

Source: California Department of Water Resources

BLANEY-CRIDDLE CONSUMPTIVE USE FORMULA

The consumptive use of an irrigated crop in which ample water supply is available can be estimated by the Blaney-Criddle formula. For a given month the consumptive use is given by

$$u = k\,(tp) \qquad (1)$$

where u is the monthly consumptive use measured in inches, k is a monthly consumptive use coefficient dependent on the crop and location, t is the mean monthly temperature in degrees Fahrenheit, and p is the monthly percentage of daytime hours of the year.

Because t and p can be found from climatic data at a given location, they are often combined into a monthly consumptive use factor f, so that Equation (1) becomes simply

$$u = kf \qquad (2)$$

For an entire growing season the consumptive use is given by

$$U = KF \qquad (3)$$

where U is the seasonal consumptive use measured in inches, K is a seasonal consumptive use coefficient, and F is the sum of the monthly consumptive use factors.

Values for the Western United States of k are given in Table 2–60, K in Table 2–61, p in Table 2–63, and f in Table 2–64.

TABLE 2–60. MONTHLY CONSUMPTIVE USE COEFFICIENTS (k) FOR IRRIGATED CROPS IN THE WESTERN UNITED STATES

[For use with the Blaney-Criddle consumptive use formula]

Crop	Location	Mar.	Apr.	May	June	July	Aug.	Sept.	Oct.	Nov.
Alfalfa	California, coastal	0.60	0.65	0.70	0.80	0.85	0.85	0.80	0.70	0.60
	California, interior	0.65	0.70	0.80	0.90	1.10	1.00	0.85	0.80	0.70
	North Dakota	–	0.84	0.89	1.00	0.86	0.78	0.72		
	Utah, St. George	–	0.88	1.15	1.24	0.97	0.87	0.81		
Corn (maize)	North Dakota	–	–	0.47	0.63	0.78	0.79	0.70		
Cotton	Arizona	–	0.27	0.30	0.49	0.86	1.04	1.03	0.81	
	Texas	0.24	0.22	0.61	0.42	0.50				
Orchard, citrus	Arizona	0.57	0.60	0.60	0.64	0.64	0.68	0.68	0.65	0.62
	California, coastal	–	0.40	0.42	0.52	0.55	0.55	0.55	0.50	0.45
Pasture	California, Murrieta	–	–	0.84	0.84	0.77	0.82	1.09	0.70	
Potatoes	North Dakota	–	–	0.45	0.74	0.87	0.75	0.54		
	South Dakota	–	–	0.69	0.60	0.80	0.89	0.39		
Small grain	North Dakota	–	0.19	0.55	1.13	0.77	0.30			
Wheat	Texas	0.64	1.16	1.26	0.87					
Sorghum	Arizona	–	–	–	–	0.34	0.72	0.97	0.62	0.60
	Kansas	–	–	–	0.80	0.94	1.17	0.86	0.47	
	Texas	–	–	–	0.26	0.73	1.20	0.85	0.49	
Soy beans	Arizona	-	-	-	0.26	0.58	0.92	0.92	0.55	
Sugar beets	California, coastal	–	0.39	0.38	0.36	0.37	0.35	0.38		
	California, interior	–	0.30	0.60	0.86	0.96	0.91	0.41		
	Montana	–	–	–	–	0.83	1.05	1.02		
Truck crops	California, interior	0.19	0.26	0.38	0.55	0.71	0.82	0.69	0.37	0.35

Source: U.S. Dept. of Agriculture

TABLE 2–61. SEASONAL CONSUMPTIVE USE COEFFICIENTS (K) FOR IRRIGATED CROPS IN THE WESTERN UNITED STATES

[For use with the Blaney-Criddle consumptive use formula]

Crop	Length of Normal Growing Season or Period[1]	Consumptive-Use Coefficient (K)[2]
Alfalfa	Between frosts	0.80 to 0.90
Bananas	Full year	.80 to 1.00
Beans	3 months	.60 to .70
Cocoa	Full year	.70 to .80
Coffee	Full year	.70 to .80
Corn (Maize)	4 months	.75 to .85
Cotton	7 months	.60 to .70
Dates	Full year	.65 to .80
Flax	7 to 8 months	.70 to .80
Grains, small	3 months	.75 to .85
Grain, sorghums	4 to 5 months	.70 to .80
Oilseeds	3 to 5 months	.65 to .75
Orchard crops:		
Avocado	Full year	.50 to .55
Grapefruit	Full year	.55 to .65
Orange and lemon	Full year	.45 to .55
Walnuts	Between frosts	.60 to .70
Deciduous	Between frosts	.60 to .70
Pasture crops:		
Grass	Between frosts	.75 to .85
Ladino whiteclover	Between frosts	.80 to .85
Potatoes	3 to 5 months	.65 to .75
Rice	3 to 5 months	1.00 to 1.10
Sisal	Full year	.65 to .70
Sugar beets	6 months	.65 to .75
Sugarcane	Full year	.80 to .90
Tobacco	4 months	.70 to .80
Tomatoes	4 months	.65 to .70
Truck crops, small	2 to 4 months	.60 to .70
Vineyard	5 to 7 months	.50 to .60

[1] Length of season depends largely on variety and time of year when the crop is grown. Annual crops grown during the winter period may take much longer than if grown in the summertime.

[2] The lower values of K for use in the Blaney-Criddle formula, $U = KF$, are for the more humid areas, and the higher values are for the more arid climates.

Source: U.S. Dept. of Agriculture

TABLE 2–62. MONTHLY AND SEASONAL CONSUMPTIVE USE COEFFICIENTS (K) FOR MATURE URBAN LANDSCAPE PLANTINGS

[For use with the Blaney-Criddle consumptive use formula; the higher the monthly or seasonal factor, the greater the planting's water demand in that time period]

Planting Type	Monthly (Maximum K)	Seasonal (Average K)
Warm-Season Grass (bermuda, buffalo), Mesa and Tempe, Arizona (elevation 1,200 feet)	1.17	0.97
Warm-Season Grass (bermuda, buffalo), Laramie and Wheatland, Colorado (elevation 7,200 and 4,700 feet)	1.08	0.79
Cool-Season Grass (Kentucky bluegrass), Mesa and Tempe, Arizona (elevation 1,200 feet)	1.41	1.20
Cool-Season Grass (Kentucky bluegrass), Laramie and Wheatland, Colorado (elevation 7,200 and 4,700 feet)	1.30	0.95
Deciduous Fruit or Nut Tree in Bare Soil, Mulch, or Paved Area	0.90	0.70
Deciduous Fruit or Nut Tree with Cover Crop (turf, ground cover, etc.)	1.25	1.00
Grapefruit in Bare Soil, Mulch, or Paved Area	0.75	0.65
Lemon, Orange	0.65	0.55
Other Trees	0.90	0.80
Grapevine, Mesa and Tempe, Arizona (elevation 1,200 feet)	0.80	0.70
Grapevine, Elsewhere	0.75	0.60
Other Vines	0.80	0.70
Shrubs Over 4 Feet Diameter	0.80	0.70
Shrubs Under 4 Feet Diameter	1.00	0.90
Ground Cover Plants, Other Small Plants	1.00	0.90
Arid Climate Native Plants	0.35	0.25
Pavements, Mulches, Nonliving Soil Covers	0.00	0.00

Source: Ferguson, B.K., 1987, Water Conservation Methods in Urban Landscape Irrigation: An Exploratory Overview, Water Resources Bulletin, Vol. 23, No.1, pp.147–152

TABLE 2–63. MONTHLY PERCENTAGE OF DAYTIME HOURS OF THE YEAR (p)

[Latitudes 60°N to 46°S. For use with the Blaney-Criddle consumptive use formula.]

Latitude North	January	February	March	April	May	June	July	August	September	October	November	December
0	8.50	7.66	8.49	8.21	8.50	8.22	8.50	8.49	8.21	8.50	8.22	8.50
5	8.32	7.57	8.47	8.29	8.65	8.41	8.67	8.60	8.23	8.42	8.07	8.30
10	8.13	7.47	8.45	8.37	8.81	8.60	8.86	8.71	8.25	8.34	7.91	8.10
15	7.94	7.36	8.43	8.44	8.98	8.80	9.05	8.83	8.28	8.26	7.75	7.88
20	7.74	7.25	8.41	8.52	9.15	9.00	9.25	8.96	8.30	8.18	7.58	7.66
25	7.53	7.14	8.39	8.61	9.33	9.23	9.45	9.09	8.32	8.09	7.40	7.42
30	7.30	7.03	8.38	8.72	9.53	9.49	9.67	9.22	8.33	7.99	7.19	7.15
32	7.20	6.97	8.37	8.76	9.62	9.59	9.77	9.27	8.34	7.95	7.11	7.05
34	7.10	6.91	8.36	8.80	9.72	9.70	9.88	9.33	8.36	7.90	7.02	6.92
36	6.99	6.85	8.35	8.85	9.82	9.82	9.99	9.40	8.37	7.85	6.92	6.79
38	6.87	6.79	8.34	8.90	9.92	9.95	10.10	9.47	8.38	7.80	6.82	6.66
40	6.76	6.72	8.33	8.95	10.02	10.08	10.22	9.54	8.39	/7.75	6.72	7.52
42	6.63	6.65	8.31	9.00	10.14	10.22	10.35	9.62	8.40	7.69	6.62	6.37
44	6.49	6.58	8.30	9.06	10.26	10.38	10.49	9.70	8.41	7.63	6.49	6.21
46	6.34	6.50	8.29	9.12	10.39	10.54	10.64	9.79	8.42	7.57	6.36	6.04
48	6.17	6.41	8.27	9.18	10.53	10.71	10.80	9.89	8.44	7.51	6.23	5.86
50	5.98	6.30	8.24	9.24	10.68	10.91	10.99	10.00	8.46	7.45	6.10	5.65
52	5.77	6.19	8.21	9.29	10.85	11.13	11.20	10.12	8.49	7.39	5.93	5.43
54	5.55	6.08	8.18	9.36	11.03	11.38	11.43	10.26	8.51	7.30	5.74	5.18
56	5.30	5.95	8.15	9.45	11.22	11.67	11.69	10.40	8.53	7.21	5.54	4.89
58	5.01	5.81	8.12	9.55	11.46	12.00	11.98	10.55	8.55	7.10	5.31	4.56
60	4.67	5.65	8.08	9.65	11.74	12.39	12.31	10.70	8.57	6.98	5.04	4.22
South												
0	8.50	7.66	8.49	8.21	8.50	8.22	8.50	8.49	8.21	8.50	8.22	8.50
5	8.68	7.76	8.51	8.15	8.34	8.05	8.33	8.38	8.19	8.56	8.37	8.68
10	8.86	7.87	8.53	8.09	8.18	7.86	8.14	8.27	8.17	8.62	8.53	8.88
15	9.05	7.98	8.55	8.02	8.02	7.65	7.95	8.15	8.15	8.68	8.70	9.10
20	9.24	8.09	8.57	7.94	7.85	7.43	7.76	8.03	8.13	8.76	8.87	9.33
25	9.46	8.21	8.60	7.84	7.66	7.20	7.54	7.90	8.11	8.86	9.04	9.58
30	9.70	8.33	8.62	7.73	7.45	6.96	7.31	7.76	8.07	8.97	9.24	9.85
32	9.81	8.39	8.63	7.69	7.36	6.85	7.21	7.70	8.06	9.01	9.33	9.96
34	9.92	8.45	8.64	7.64	7.27	6.74	7.10	7.63	8.05	9.06	9.42	10.08
36	10.03	8.51	8.65	7.59	7.18	6.62	6.99	7.56	8.04	9.11	9.51	10.21
38	10.15	8.57	8.66	7.54	7.08	6.50	6.87	7.49	8.03	9.16	9.61	10.34
40	10.27	8.63	8.67	7.49	6.97	6.37	6.76	7.41	8.02	9.21	9.71	10.49
42	10.40	8.70	8.68	7.44	6.85	6.23	6.64	7.33	8.01	9/9.26	9.82	10.64
44	10.54	8.78	8.69	7.38	6.73	6.08	6.51	7.25	7.99	9.31	9.94	10.80
46	10.69	8.86	8.70	7.32	6.61	5.92	6.37	7.16	7.96	9.37	10.07	10.97

Source: U.S. Dept. of Agriculture

TABLE 2–64. MONTHLY CONSUMPTIVE USE FACTORS (f) FOR LOCATIONS IN THE WESTERN UNITED STATES AND HAWAII

[For use with the Blaney-Criddle consumptive use formula]

Month	Arizona			California
	Phoenix	Safford	Yuma	Bakersfield
	f	*f*	*f*	*f*
January	3.64	3.14	3.90	3.33
February	3.82	3.40	4.07	3.58
March	5.07	4.51	5.36	4.74
April	5.89	5.35	6.10	5.55
May	7.28	6.73	7.36	6.87
June	8.17	7.61	8.16	7.61
July	8.85	8.20	8.91	8.38
August	8.25	7.52	8.41	7.69
September	6.91	6.26	6.98	6.20
October	5.58	5.04	5.81	5.13
November	4.20	3.67	4.42	3.68
December	3.62	3.15	3.87	3.38
Total	71.28	64.58	73.35	65.84
Frost-free period	2/5–12/6	4/5–11/4	1/12–12/26	2/21–11/25

Month	California			
	El Centro	Escondido	Merced	Red Bluff
	f	*f*	*f*	*f*
January	3.88	3.70	3.16	3.06
February	4.00	3.70	3.38	3.35
March	5.37	4.72	4.45	4.53
April	6.17	5.26	5.27	5.33
May	7.58	6.20	6.57	6.70
June	8.20	6.49	7.35	7.58
July	9.07	7.22	8.08	8.34
August	8.56	6.93	7.39	7.61
September	7.19	5.95	6.06	6.14
October	5.95	5.14	4.95	4.98
November	4.44	4.06	3.67	3.60
December	3.93	3.73	3.14	3.20
Total	74.34	63.10	53.47	64.24
Frost-free period	1/29–12/9	3/9–11/25	3/9–11/20	3/5–12/5

Month	California		Colorado	
	Sacramento	Santa Ana	Fort Collins	Grand Junction
	f	*f*	*f*	*f*
January	3.13	3.77	1.76	1.72
February	3.39	3.78	1.89	2.29
March	4.52	4.77	3.02	3.57
April	5.18	5.27	4.10	4.67
May	6.30	6.17	5.49	6.20
June	6.93	6.50	6.45	7.22
July	7.42	7.05	7.11	7.98
August	6.92	6.71	6.50	7.19
September	5.81	5.81	4.98	5.60
October	4.89	5.11	3.73	4.22
November	3.64	4.15	2.42	2.71
December	3.06	3.80	1.81	1.92
Total	61.19	62.89	49.26	55.29
Frost-free period	2/6–12/10	2/7–12/7	5/7–9/29	4/13–10/25

TABLE 2–64. MONTHLY CONSUMPTIVE USE FACTORS (f) FOR LOCATIONS IN THE WESTERN UNITED STATES AND HAWAII (continued)

Month	Colorado	Idaho		
	Montrose	Boise	Idaho Falls	Lewiston
	f	*f*	*f*	*f*
January	1.68	1.82	1.26	2.09
February	2.15	2.22	1.55	2.42
March	3.32	3.44	2.79	3.77
April	4.32	4.44	4.05	4.84
May	5.70	5.74	5.45	6.25
June	6.64	6.68	6.26	7.12
July	7.31	7.58	7.19	8.13
August	6.62	6.87	6.45	8.01
September	5.20	5.15	4.79	5.40
October	3.89	3.83	3.60	4.02
November	2.56	2.58	2.18	2.64
December	1.77	1.90	1.45	2.12
Total	51.16	52.25	47.02	56.81
Frost-free period	5/6–10/6	4/23–10/17	5/15–9/19	4/5–10/26

Month	Idaho	Kansas		Montana
	Twin Falls	Garden City	Wichita	Agricultural College
	f	*f*	*f*	*f*
January	1.77	2.12	2.21	1.30
February	2.16	2.32	2.39	1.48
March	3.35	3.65	3.80	2.50
April	4.39	4.81	4.98	3.76
May	5.75	6.32	6.45	5.15
June	6.53	7.30	7.45	6.03
July	7.57	7.97	8.10	6.87
August	6.67	7.37	7.53	6.21
September	5.00	5.83	5.99	4.53
October	3.93	4.41	4.65	3.35
November	2.48	2.93	3.13	2.03
December	9.84	2.17	2.32	1.41
Total	51.44	57.20	59.00	44.63
Frost-free period	5/18–9/26	4/25–10/16	4/10–10/27	5/24–9/16

Month	Montana	Nebraska		New Mexico
	Missoula	McCook	Scottsbluff	Albuquerque
	f	*f*	*f*	*f*
January	1.16	1.85	1.72	2.40
February	1.54	2.11	1.88	2.69
March	2.77	3.34	3.02	3.85
April	4.01	4.59	4.20	4.87
May	5.46	6.14	5.71	6.23
June	6.34	7.17	6.79	7.09
July	7.17	7.97	7.58	7.65
August	6.32	7.25	6.87	6.98
September	4.51	5.56	5.19	5.65
October	3.26	4.20	3.83	4.46
November	1.98	2.64	2.45	3.01
December	1.27	1.91	1.81	2.54
Total	45.79	54.73	51.05	57.39
Frost-free period	5/18–9/23	5/3–10/6	5/11–9/26	4/13–10/28

TABLE 2–64. MONTHLY CONSUMPTIVE USE FACTORS (f) FOR LOCATIONS IN THE WESTERN UNITED STATES AND HAWAII (continued)

Month	New Mexico		Nevada	
	Carlsbad	State College	Carson City	Yerington
	f	*f*	*f*	*f*
January	3.18	2.96	2.20	2.06
February	3.37	3.12	2.40	2.46
March	4.64	4.29	3.45	3.56
April	5.56	5.15	4.25	4.40
May	6.89	6.29	5.49	5.65
June	7.61	7.29	6.23	6.37
July	7.93	7.72	7.04	7.17
August	7.55	7.17	6.42	6.63
September	6.15	5.94	5.00	5.09
October	5.04	4.77	3.87	3.95
November	3.70	3.45	2.69	2.67
December	3.06	2.85	2.24	2.11
Total	64.68	61.00	51.28	52.12
Frost-free period	3/29–11/4	4/6–10/31	5/25–9/19	5/23–9/18

Month	Oklahoma	Oregon		
	Altus	Baker	Hood River	Medford
	f	*f*	*f*	*f*
January	2.74	1.60	2.09	2.50
February	3.09	1.90	2.40	2.81
March	4.43	3.12	3.61	3.90
April	5.49	4.10	4.57	4.69
May	6.86	5.33	5.83	5.91
June	7.78	6.11	6.47	6.74
July	8.33	6.92	7.15	7.54
August	7.82	6.26	6.50	6.85
September	6.34	4.72	5.04	5.33
October	5.07	3.55	3.92	4.12
November	3.65	2.32	2.65	2.90
December	2.88	1.68	2.15	2.42
Total	64.48	47.64	52.38	55.71
Frost-free period	3/28–11/9	5/12–10/3	4/20–10/20	5/6–10/4

Month	Texas			Utah
	Amarillo	Fort Stockton	Lubbock	Logan
	f	*f*	*f*	*f*
January	2.33	3.45	2.85	1.59
February	2.48	3.61	3.09	1.86
March	3.78	4.87	4.28	3.05
April	4.75	5.74	5.25	4.28
May	6.07	7.08	6.57	5.63
June	6.98	7.69	7.37	6.53
July	7.54	7.94	7.80	7.53
August	6.98	7.47	7.28	6.86
September	5.67	6.25	5.95	5.19
October	4.39	5.25	4.83	3.87
November	2.87	3.93	3.47	2.46
December	2.43	3.39	2.84	1.69
Total	56.27	66.67	61.58	50.54
Frost-free period	4/11–11/2	4/1–11/3	4/12–11/3	5/7–10/11

TABLE 2–64. MONTHLY CONSUMPTIVE USE FACTORS (f) FOR LOCATIONS IN THE WESTERN UNITED STATES AND HAWAII (continued)

Month	Utah	Washington		Wyoming
	Salt Lake City	Prosser	Yakima	Cheyenne
	f	*f*	*f*	*f*
January	1.96	1.95	1.72	1.70
February	2.26	2.36	2.22	1.82
March	3.47	3.80	3.71	2.75
April	4.45	4.82	4.64	3.67
May	5.77	6.21	6.06	5.08
June	6.83	7.03	6.88	6.13
July	7.77	7.74	7.63	6.87
August	7.13	6.92	6.79	6.29
September	5.40	5.24	5.21	4.78
October	4.06	3.97	3.91	3.46
November	2.75	2.55	2.41	2.32
December	2.06	2.01	1.92	1.84
Total	53.91	54.55	53.11	46.71
Frost-free period	4/13–10/22	4/28–10/4	4/15–10/22	5/14–10/2

Month	Wyoming	Hawaii	
	Worland	Honolulu W.B. Airport	Waianai
	f	*f*	*f*
January	0.97	5.62	5.57
February	1.40	5.15	5.16
March	2.79	6.04	6.08
April	4.08	6.26	6.45
May	5.57	6.87	7.05
June	6.58	6.98	7.23
July	7.41	7.20	7.49
August	6.60	7.02	7.35
September	4.87	6.49	6.65
October	3.57	6.27	6.43
November	2.07	5.67	5.74
December	1.19	5.54	5.61
Total	47.10	75.07	76.81
Frost-free period	5/10–9/27	–	–

Source: U.S. Dept. of Agriculture

SECTION L. PHREATOPHYTES

TABLE 2–65. PHREATOPHYTE AREAS AND THEIR CONSUMPTIVE USE IN SELECTED WESTERN STATES

[Estimates as of 1953]

State	Area (Acres)	Annual Use (Acre-Feet)
Arizona	405,000	1,280,000
California[1]	317,000	1,150,000
Colorado[1]	737,000	1,056,000
Idaho	500,000	1,000,000
Montana	1,600,000	3,200,000
Nebraska[1]	515,000	709,000
Nevada	2,801,000	1,500,000
New Mexico	300,000	900,000
North Dakota	1,035,000	1,660,000
Oregon[1]	40,800	21,200
South Dakota	850,000	1,240,000
Texas[1]	262,000	436,500
Utah	1,200,000	1,500,000
Wyoming	527,000	1,100,000
Total (approximate)[2]	11,090,000	16,750,000

[1] Partial data, from published reports on areas within the State.
[2] Partial data.
Source: U.S. Geological Survey

TABLE 2–66. CONSUMPTIVE USE BY SOME COMMON PHREATOPHYTES IN THE WESTERN UNITED STATES

Plant	Annual Rate Including Precipitation	Volume Density	Depth to Water	Locality
	Acre-feet per acre	*Percent*	*Feet*	
Alder	5.3	–	–	Santa Ana River, Calif.
Batamote or seepwillow	4.7	100	6	Safford Valley, Ariz.
Cottonwood	7.6–5.2	100	3–4	San Luis Rey River, Calif.
Do	6.0	100	6	Safford Valley, Ariz.
Greasewood	2.5–0.08	–	–	Escalante Valley, Utah
Mesquite	3.3	100	10	Safford Valley, Ariz.
Sacaton	4.0–3.5	–	–	Pecos River Valley, N. Mex.
Saltcedar	5.5–4.7	–	–	Do.
Do	9.2–7.3	100	4–7	Safford Valley, Ariz.
Saltgrass	4.1–1.1	–	1.5–5	Owens Valley, Calif.
Do	2.9–1.1	–	2–4	Santa Ana, Calif.
Do	2.3–1.1	–	.3–2.1	San Luis Valley, Colo.
Do	4.5	–	2.0	Carlsbad, N. Mex.
Do	2.6	–	.65	Isleta, N. Mex.
Do	4.0–0.8	–	.4-3.1	Los Griegos, N. Mex.
Do	1.9	–	2.2	Mesilla Dam, N. Mex.
Do	2.3–1.6	–	1.9–2.6	Escalante Valley, Utah
Do	2.0	–	2.0	Vernal, Utah
Willow	4.4	–	2.0	Santa Ana, Calif.
Do	2.5	–	1.1	Isleta, N. Mex.

Source: Select Committee on National Water Resources, U.S. Senate, 1960

TABLE 2–67. PHREATOPHYTE AREAS AND THEIR CONSUMPTIVE USE IN RIVER BASINS OF THE WESTERN UNITED STATES

State and River or Valley	Area in Acres	Estimated Annual Water Use in Acre-Feet	Year of Survey or Estimate	Principal Species of Phreatophytes
Arizona:				
Little Colorado River:				
Main stem only St. Johns to mouth	49,560	64,720	1951	Saltcedar, cottonwood, willow, brush.
Total Little Colorado including Zuni and Puerco Rivers	65,310	74,360	1951	Do.
Gila River:				
Virden, N. Mex. to Clifton, Ariz.	5,600	6,310	1951	Cottonwood, willow, seepwillow
Head of Safford Valley to Coolidge Dam	25,520	61,500	1951	Saltcedar, mesquite, willow, seepwillow, cottonwood, arrowweed.
Coolidge Dam to Kelvin	3,970	7,460	1951	Do.
Kelvin to Gillespie Dam downstream from Granite Reef Dam on Salt River, Rillito on Santa Cruz River and Lake Pleasant on Agua Fria River	142,880	282,770	1951	Do.
Gillespie Dam to Dome	51,270	69,220	1951	Saltcedar, mesquite, seepwillow.
Total Gila River including San Pedro River, Palominas to mouth, Verde River, Bartlett Dam to mouth, Salt River, Stewart Mountain Dam to Granite Reef Dam	300,710	522,510	1951	
Salt River: Stewart Mountain Dam to Granite Reef Dam	2,710	2,070	1951	Saltcedar, mesquite.
San Pedro River: Palominas to Gila at mouth	42,510	69,420	1951	Do.
Verde River: Bartlett to Gila at mouth	3,790	5,720	1951	Saltcedar, mesquite, seepwillow.
California:				
San Luis Rey River: In San Diego County	6,390	17,800	1945	Cottonwood, willow.
Santa Ana River: Below Prado Dam	1,071	3,000	1946	Do.
Kings River: Kings River Soil Conservation District	1,080	6,500	1958	Do.
Idaho: Malad Valley	10,900	32,400	1953	Grasses and rushes.
New Mexico:				
Pecos River: Alamogordo Dam to Texas State line	42,500	117,000	1956	Saltcedar.
Rio Grande: Bernardo to San Marcial	52,000	77,800	1955	Do.
Nevada:				
Little Humboldt River: Paradise Valley	36,500	23,000	1947	Greasewood, saltgrass, willows.
Muddy River: To mouth	4,900	10,330	1951	Saltcedar.
Nevada-Utah-Arizona-California:				
Virgin River: Littlefield to mouth	7,360	27,160	1951	Do.
Lower Colorado River:				
Main stem:				
Lee Ferry to Hoover Dam	1,480	5,920	1951	Saltcedar, willow.
Hoover Dam to Davis Dam	4,500	11,450	1951	Do.
Davis Dam to Parker Dam	57,490	209,290	1951	Do.
Parker Dam to Laguna Dam	191,890	314,360	1951	Do.
Laguna Dam to International Boundary	33,510	54,210	1951	Do.
Total main stem	288,920	595,230	1951	
Total Lower Colorado River Basin, including Little Colorado River, Virgin River, and Gila River Basins	667,200	1,230,090	1951	

Source: Select Committee on National Water Resources, U.S. Senate, 1960

TABLE 2–68. PHREATOPHYTES IN THE WESTERN UNITED STATES

[The quality of the ground water with respect to its suitability for crop growth is indicated by numerals as follows: 1, excellent to good; 2, good to poor; 3, poor to unsatisfactory. The use of ground water, including precipitation, unless otherwise stated is presumed to be for a plant growth of 100-percent volume density]

Scientific Name	Common Name	Occurrence as a Phreatophyte	Relation to Ground Water: Depth to Water Below Land Surface (feet)	Relation to Ground Water: Quality	Relation to Ground Water: Use (acre feet per acre)	Remarks
Acacia greggii A. Gray.	Catclaw, devilsclaw, una de gato.	Southern California to western Texas.	–	1	–	Uses more water than mesquite. Forms thickets along streams and washes.
Acer negundo Linnacus.	Boxelder	Canada to Oklahoma and Arizona.	–	–	–	Occurs in moist places and along streams, chiefly in mountains. Observed in the flood plain of the North Canadian River near Oklahoma City, Okla. Widely used as a shade tree.
Alhagi camelorum Fischer.	Camelthorn	Arizona	–	–	–	Introduced into Southwestern United States from Asia Minor. Poor browse plant. Observed growing as a phreatophyte along Little Colorado River between Holbrook and Winslow, Ariz., in localities where the depth to water ranged from 4 to 6 ft. Aggressive and thicket forming, root system deep and extensive.
Allenrolfea occidentalis (S. Watson) Kuntze.	Pickleweed, iodinebush	California to western Texas	1–20	3	–	
Alnus	Alder	–	–	–	5.3	Occurs along streams, river bottom land, and other wet sites. The use of 5.3 ft. was for the period May to October 1932 in Coldwater Canyon, altitude 2,400 ft., San Bernardino Mountains, Calif., where alder constituted 82 percent of the vegetation.
Anemopsis californica (Nuttall) Hooker and Arnott	Yerba mansa	So. Calif., So. Nevada to Utah and Texas	Shallow	3	–	Used by Pima Indians as a herbal remedy. Common in saline and wet lowlands.
Aster spinosus Bentham.	Spiny aster	Arizona	–	–	–	Identified as phreatophyte in bottom land of lower Safford Valley, Ariz.
Atriplex canescens (Pursh) Nuttall.	Fourwing saltbush, chamiso, chamiza.	South Dakota to Oregon, south to Mexico	8–62	1–2	–	Tolerates alkali. Valuable browse plant. Useful in erosion control. Taproots 30–40 ft. deep. May not always occur as a phreatophyte.
hastata Linnaeus.	–	Oregon and California to Kansas and New Mexico.	–	3	–	Occurs in saline soils, especially around alkaline lakes, in salt marshes, and in other water-soaked soils.
lentiformis (Torrey) Watson	Quailbrush, lenscale Nevada saltbush	Southern Utah and Nevada to California and Sonora, Mexico.	6–15	3	–	High tolerance for alkali and saline soil. Fair browse plant. Reaches height of 10 ft where water table is shallow.
Baccharis emoryi A. Gray.	Emory baccharis	Texas to southern California and southern Utah.	–	2		
glutinosa Persoon.	Batamote, seepwillow, water motie, waterwillow.	Colorado and Texas to California and Mexico.	2–15	2	4.7	Evapotranspiration for plants grown in tanks ranged from 10.3 ft with water level at 2 ft. to 4.6 ft with water level of 6 ft. Safford Valley, Ariz.

TABLE 2–68. PHREATOPHYTES IN THE WESTERN UNITED STATES (continued)

Scientific Name	Common Name	Occurrence as a Phreatophyte	Relation to Ground Water: Depth to Water Below Land Surface (feet)	Relation to Ground Water: Quality	Relation to Ground Water: Use (acre feet per acre)	Remarks
sarothroides A. Gray.	Broom baccharis, desertbroom, rosinbrush.	Southern California, Arizona, southwestern New Mexico.	–	–	–	Occurs along streams in draws, in canyon bottoms and wet alkaline sites.
sergiloides A. Gray.	Squaw baccharis, waterweed.	Arizona, southern California, southern Nevada, southwestern Utah.	–	2	–	Occurs as a phreatophyte in lower Safford Valley, Ariz.
viminea Crandolle.	Mulefat	Southwestern Utah, southern California, Nevada, Arizona.	–	–	–	Useful in erosion control.
Bigelovia hartwegii, probably *Aplopappus heterophyllus* A. Gray.	Rayless goldenrod		–	–	–	Will grow in dry places but thrives where ground water is within reach.
Celtis reticulata Torrey.	Hackberry, cumaru, kom.	Arizona	–	–	–	A large tree that may reach 3 ft in diameter and 50 ft in height. Usually occurs along streams.
Cercidium floridum Bentham.	Blue palo verde	Southwestern Arizona, southeastern California.	–	–	–	Common along washes, canyons, valleys, alluvial plains, grassland at sites where ground water is plentiful.
Chilopsis linearis Sweet.	Desert-willow	Western Texas to southern Nevada, Arizona, southern California.	To 50	–	–	May not always occur as a phreatophyte.
Chrysothamnus pumilus (Nuttall).	Rabbitbrush	Mud Lake, Idaho	–	–	–	
nauseosus consimilis (Greene).	Rubber rabbitbrush	Nevada, Utah, Idaho, Wyoming.	–	2–3	–	
nauseosus graveolens (Nuttall).	Do	Montana, Idaho, Utah, Nevada, New Mexico.	2.5–15	2–3	–	
nauseosus mohavensis (Greene).	Do	Northern California, Nevada.	–	2–3	–	
nauseosus oreophilus (Al Nelson).	Do	Wyoming, Colorado, Utah	–	–	–	
nauseosus viridulus	Do	Colorado to Oregon; Nevada, New Mexico.	–	–	–	
Cowania Stansburiana (Torrey)	Vanadium bush	Arizona, Idaho, Utah	–	–	–	Used as an indicator of vanadium-uranium deposits by prospectors in the Colorado Plateau. Able to grow in highly mineralized ground and to absorb large amounts of uranium

TABLE 2–68. PHREATOPHYTES IN THE WESTERN UNITED STATES (continued)

Scientific Name	Common Name	Occurrence as a Phreatophyte	Relation to Ground Water: Depth to Water Below Land Surface (feet)	Quality	Use (acre feet per acre)	Remarks
Dalea spinosa Gray.	Smoketree, smoke-thorn.	Southeastern California, southwestern Arizona.	–	–	–	Its persistent occurrence in gravelly and sandy washes suggests it depends upon ground-water underflow and occurs as a phreatophyte.
Dasiphora fruticosa Linnaeus.	Bush or shrubby cinquefoil.	Locally in Idaho but widespread in Oregon, Washington, Utah, Nevada and Arizona.	Shallow	–	–	Occurs as a phreatophyte in Pahsimeroi Valley, Idaho. Grows on subalpine meadows, along streams, about cold springs in peaty, sandy, or clayey loams.
Distichlis spicata (Linnaeus) Greene.	Seashore saltgrass	Western United States	–	–	–	
stricta (Torrey) Rydberg.	Saltgrass, or desert saltgrass.	All Western States	2–14	1–3	–	
Elymus condensatus Presl.	Giant wildrye	All Western States except New Mexico	1–12	1–2	–	Fair forage. Killed by overgrazing. Extensive root system.
triticoides Buckley.	Creeping wildrye	Western United States	–	1–2	–	Good forage. Frequently cut for hay. Associated with giant wildrye along Humboldt River, Nev.
Eragrostis obtusiflora (Fournier) Scribner.	Mexican saltgrass, alkali lovegrass.	Southeastern Arizona, southwestern New Mexico.	4–15	2–3	–	Common locally in saline soil near Wilcox, Ariz. Observed growing in Sulphur Springs Valley, Ariz., where depth to water table was from 4–15 ft.
Fraxinus velutina Torrey.	Velvet ash, Arizona ash.	Southwestern Utah, southern Nevada, California, Arizona, New Mexico, and western Texas.	–	1	–	Prominent stream-bank and canyon tree; restricted to areas with a permanent ground-water supply. Popular as a shade tree in Arizona and California.
Hedysarum boreale Nuttall.	Sweet vetch	Colorado, Utah	–	–	–	Deep tap root. Identified as a phreatophyte in Colorado.
Heliotropium curassavicum Linnaeus.	Heliotrope, Chinese pusley.	Southwestern Utah to southern California.	Shallow	2–3	–	High tolerance for alkali. Occurs on moist saline soil.
Hymenoclea monogyra Torrey and Gray.	Burrobush	Western Texas to southern California.	Shallow	1–2	–	Occurs largely along streams, washes, and in bottom lands; aggressive. often forming thickets. Unpalatable to livestock.
salsola Torrey and Gray.	White burrobush	Utah to Arizona and California.	–	–	–	Occurs in sandy desert.
Juglans microcarpa Berlandier.	Walnut, nogal, butternut.	Arizona, New Mexico	2–20	1	–	Occurs along watercourses and washes; intolerant of shade. Deep tap root.
Juncus balticus Willdenow	Wirerush, wiregrass	Western United States	–	–	7.8	Grows in wet sites where ground water is shallow, also in shallow ponds. Appears to occur both as phreatophyte and hydrophyte. Deep root system. Fair to good forage.

TABLE 2–68. PHREATOPHYTES IN THE WESTERN UNITED STATES (continued)

Scientific Name	Common Name	Occurrence as a Phreatophyte	Relation to Ground Water: Depth to Water Below Land Surface (feet)	Quality	Use (acre feet per acre)	Remarks
Juncus cooperi Engelmann	Desertrush	Southern Utah to California	–	2–3	–	Occurs on the margins of salt marshes and alkaline meadows, common in Death Valley, Calif., along the edge of the playa often associated with saltgrass.
Juniperus scopulorum?	Rocky Mountain juniper; locally "swampcedar."	Nevada	10	1–2	–	Occurs locally as a phreatophyte in White River and Spring Valleys, Nev. May be a hybrid between *J. scopulorum* and *J. utahensis.*
Leptochola fascicularis (Lamarck) A. Gray.	Sprangletop	Western United States	–	1–3	–	Occurs along ditches and in moist waste places, often in brackish marshes; most places in alkali plains. Often invades rice fields.
Medicago sativa Linnaeus.	Alfalfa	Western United States	4+	1–2	–	
Phragmites communis Trinius.	Reed, giant reed-grass, carrizo.	Western United States	0–8±	1–2	-	Occurs also as a hydrophyte in the shallow water of streams, lakes, ponds and marshes.
Picea engelmanni Parry.	Engelmann spruce	Mountain areas of western United States.	–	1	–	Requires a good water supply and depends upon ground water in many localities. Shallow root system.
Platanus wrightii Watson.	Arizona sycamore	Southern Arizona, southeastern and southwestern California, New Mexico.	–	1	–	Common along stream and rocky canyons, in foothills and mountains, upper desert, desert grassland, and oak woodland zones. Valuable in erosion control.
Pluchea sericea Coville.	Arrowweed	Texas to southern Utah and southern California.	0–10±	1–2	–	Occurs along streams and flood plains. Abundant along lower reaches of Colorado River and tributaries. Arrowweed may grow where depth to water is 25 ft.
Populus	Cottonwood	Western United States	–	1–2	–	
tremuloides aurea Tidestrom.	Quaking aspen	Mountainous areas of Western United States.	–	1	–	Considered a phreatophyte when it grows along streams, around springs, and in other wet areas. Shallow root system.
Prosopis juliflora (Swartz.)	Mesquite, honey mesquite.	Southern Kansas to southeastern California and Mexico.	–	1–2	–	Extensive root development. Reported to penetrate 60 feet below surface.
velutina Wooton.	Velvet mesquite	Southern Arizona	10+	1–2	3.3	Occurs in bottom lands. Extensive root development.
pubescens Bentham.	Screwbean mesquite, tornillo.	Western Texas to southern Nevada and southern California.	–	–	–	Characteristic of bottom lands along desert streams and water holes of Mojave and Colorado Deserts.
Quercus agrifolia Nee.	California live oak	California	35±	1	-	Occurrence related to depth to water table.
lobata Nee.	Roble oak	California	10–20	–	–	
Salicornia europaea Linnaeus	Glasswort	–	–	3	–	Frequently occurs in salt flats with salts approximately 1.0 percent of weight of soil.

TABLE 2–68. PHREATOPHYTES IN THE WESTERN UNITED STATES (continued)

Scientific Name	Common Name	Occurrence as a Phreatophyte	Relation to Ground Water: Depth to Water Below Land Surface (feet)	Relation to Ground Water: Quality	Relation to Ground Water: Use (acre feet per acre)	Remarks
rubra Linnaeus.	Glasswort	Colorado, New Mexico, Nevada, Utah.	–	3	–	Some value as waterfowl feed. In Nevada occurs along edges of channels draining into playas.
utahensis Tidestrom.	Do	Utah	–	3	–	Occurs on borders of salt lakes and alkaline places.
Salix	Willow	Western United States	–	–	–	
Sambucus	Elder, elderberry	Western United States	–	1	–	Eleven species reported to grow in Western United States. Grows along streams, in canyons and in moist sites.
Sarcobatus vermiculatus (Hook). Torrey.	Big greasewood	Western United States	60±	-	-	
Sequoia gigantea (Lindley).	Giant or bigtree sequoia.	California	–	1	–	Appears to prefer localities where a water table is within reach of its roots but will grow elsewhere.
Sesuvium portulocostrum	Lowland purslane	Southern Nevada and California.	–	3	–	Reported as a plant that grows on moist alkaline soils. Indicative of ground water but usually of poor quality. Alkali resistant.
verrucosum Rafinesque	Warty sesuvium, sea purslane.	Southern Arizona, California, and New Mexico.	Shallow	3	–	
Shepherdia	Buffalo berry	Arizona, New Mexico, Nevada, Oregon, Black Hills.	–	1	–	Fruit edible. Grows in moist sites and along streams and river bottoms. One species reported growing as a phreatophyte in Big Smoky Valley, Nev. Occurs also as a phreatophyte in Mason Valley, Nev.
Sporobolus airoides Torrey.	Alkali sacaton	Western United States	5–25±	1–3	3.7	Most common in the Southwest where it is important as forage; deep, coarse root system. Prefers moist alkali flats. Grows in very saline or saline-alkali soils. Soil salinity may range from 0.3 to 3.0 percent. Grows best in range 0.3 to 0.5 percent.
wrightii Munro.	Sacaton	Arizona to western Texas	–	1–2	–	Occurs in alluvial flats and bottom lands. Will not grow on highly alkaline soils.
Suaeda depressa Watson.	Seepweed, saltwort	Southwest	–	3	–	Browsed when other forage is scarce. Occurs on saline or saline-alkali soils with salt content in first foot as much as 3.2 percent.
Suaeda suffrutescens Watson.	Desert seepweed	Western Texas, New Mexico, Arizona	–	3	–	Browsed when other forage is scarce. Occurs on saline or saline-alkali soils with salt content in first foot as much as 3.2 percent.
torreyana Watson.	Torrey seepweed, iodineweed, inkweed.	Eastern Oregon to New Mexico and California	4–15	3	–	Do.
Tamarix aphylla Linnaeus.	Athel tree	Southwest	–	1–3	–	
gallica Linnaeus.	Saltcedar, French tamarisk	Do	–	1–3	–	
Washingtonia filifera Wendland.	Fan palm, California palm	Southern Arizona, California, SE New Mexico.	–	1–3	–	Highly tolerant to alkali. Generally grows where ground water is at shallow depth.

Source: U.S. Geological Survey

SECTION M. REFERENCE BOOKS

TABLE 2–69. HYDROLOGY AND WATER RESOURCES REFERENCE BOOKS

Bear, Jacob, 1972, *Dynamics of Fluids in Porous Media,* Am. Elsevier Publ. Co., New York, NY, 764 pp.

Bear, Jacob, 1980, *Hydraulics of Ground Water,* McGraw-Hill Book Co., New York, NY, 567 pp.

Bear, Jacob, Zaslavsky, D., and Irmay, S., 1968, *Physical Principles of Water Percolation and Seepage.* Arid Zone Research XXIX, UNESCO, Paris, 465 pp.

Bear, Jacob, and Corapcioglu, M. Y., 1985, *Fundamentals of Transport Phenomena in Porous Media,* Kluwer Acad. Publ., Hingham, MA, 1,003 pp.

Bouwer, Herman, 1978, *Groundwater Hydrology,* McGraw-Hill Book Co., New York, NY, 480 pp.

Bowen, Robert, 1980, *Ground Water,* John Wiley & Sons, Inc., Somerset, NJ, 227 pp.

Bowen, Robert, 1982, *Surface Water,* John Wiley & Sons, Inc., New York, NY, 289 pp.

Butler, S. S., 1957, *Engineering Hydrology,* Prentice-Hall, Inc., Englewood Cliffs, NJ.

Cedergren, H. R., 1977, *Seepage, Drainage and Flow Nets,* John Wiley & Sons, Inc., Somerset, NJ, 534 pp.

Chow, V. T. (ed.), 1964, *Handbook of Applied Hydrology,* McGraw-Hill Book Co., New York, NY, 1453 pp.

Chow, V. T. (ed.), Mays, L.W., and Maidment, D.R., 1988, *Applied Hydrology,* McGraw-Hill Book Co., New York, NY, 570 pp.

Custodio, Emilio, and Llamas, M. R., 1975, *Hidrología Subterránea,* 2 vols., Ediciones Omega, Barcelona, 2,359 pp.

Davis, S. N., and DeWiest, R. J. M., 1966, *Hydrogeology,* John Wiley & Sons, Inc., New York, NY, 463 pp.

DeWiest, R. J. M., 1965, *Geohydrology,* John Wiley & Sons, Inc., New York, NY, 366 pp.

DeWiest, R. J. M., (ed.), 1969, *Flow Through Porous Media,* Academic Press, New York, NY, 530 pp.

Driscoll, Fletcher, 1986, *Ground Water and Wells,* Johnson Division, Signal Environmental Systems, Inc., St. Paul, MN, 1100 pp.

Fetter, C.W., 1980, *Applied Hydrogeology,* Charles E. Merrill Publ. Co., Columbus, OH, 488 pp.

Freeze, R. A., and Cherry, J. A. 1979, *Ground Water,* Prentice-Hall, Inc., Englewood Cliffs, NJ.

Glover, R. E., 1974, *Transient Ground Water Hydraulics,* Water Resources Publications, Fort Collins, CO, 413 pp.

Gray, D. M., 1973, *Handbook on the Principles of Hydrology,* Water Information Center, Inc., Plainview, NY, 720 pp.

Greenkorn, R. A., 1985, *Flow Phenomena in Porous Media,* Marcel Dekker, Inc., New York, NY, 560 pp.

Grigg, N.S., 1985, *Water Resources Planning,* McGraw-Hill Book Co., New York, NY, 328 pp.

Grigg, N.S., 1986, *Urban Water Infrastructure: Planning, Management, and Operations,* John Wiley & Sons, Inc., New York, NY, 328 pp.

Harr, M. E., 1977, *Ground Water and Seepage,* McGraw-Hill Book Co. Inc., New York, NY, 315 pp.

Heath, R. C., and Trainer, F. W., 1968, *Introduction to Ground-Water Hydrology,* John Wiley & Sons, Inc., New York, NY, 284 pp; Revised ed. 1981, Water Well Journal Publ. Co., Dublin, OH.

Heath, R.C., 1983, *Basic Ground-Water Hydrology,* U.S. Geol. Survey, Water-Supply Paper 2220.

Hilliel, D., 1971, *Soil and Water, Physical Principles and Process,* Academic Press, New York, NY, 288 pp.

Hofkes, E.H., 1981, *Small Community Water Supplies: Technology of Small Water Supply Systems in Developing Countries,* John Wiley & Sons, Inc., New York, NY, 488 pp.

Imhoff, K., and others, 1988, *Handbook of Urban Drainage and Sewage Disposal,* John Wiley & Sons, Inc., New York, NY, 350 pp.

Kashef, A. I, 1986, *Groundwater Engineering,* McGraw-Hill Book Co., New York, NY, 512 pp.

Kazmann, R. G., 1972, *Modern Hydrology,* Harper & Row, Inc., New York, NY, 635 pp.

Lencastre, A., 1987, *Handbook of Hydraulic Engineering,* John Wiley & Sons, Inc., New York, NY, 540 pp.

Linsley, R. K., and Kohler, M.A., 1982, *Hydrology for Engineers,* 3rd ed., McGraw-Hill Book Co., New York, NY, 512 pp.

Luthin, J. N. (ed.), 1957, *Drainage of Agricultural Lands,* American Society of Agronomy, Madison, WI. Water Information Center, Plainview, NY.

Marino, M. A., and Luthin, J. N., 1981, *Seepage and Groundwater,* Elsevier Science Publ. Co., Inc., New York, NY, 492 pp.

Montgomery, J.M., 1985, *Water Treatment Principles and Design,* John Wiley & Sons, Inc., New York, NY, 696 pp.

Powers, J. P., 1984, *Construction Dewatering: A Guide to Theory and Practice,* John Wiley & Sons, Somerset, NJ, 494 pp.

Rethati, L., 1984, *Groundwater in Civil Engineering,* Elsevier Science Publ. Co., Inc., New York, NY, 474 pp.

Rodda, J.C. (ed.), 1985, *Facets of Hydrology,* 2 vols., John Wiley & Sons, Inc., New York, NY.

Rydzewski, J.R., 1987, *Irrigation Development Planning: An Introduction for Engineers,* John Wiley & Sons, Inc., New York, NY, 265 pp.

Speidel, D.H. (ed.), 1988, *Perspectives on Water Uses and Abuses,* Oxford Univ. Press, New York, NY, 388 pp.

Schwille, F., 1988, *Dense Chlorinated Solvents in Porous and Fractured Media,* translated from German by J.F. Pankow, Lewis Publishers, Inc., Chelsea MI, 146 pp.

Todd, D. K., 1980, *Groundwater Hydrology,* John Wiley & Sons, Inc., New York, NY, 535 pp.

van der Leeden, Frits, 1975, *Water Resources of the World,* Water Information Center, Inc., Plainview, NY, 568 pp.

Verruijt, A., 1970, *Theory of Ground Water Flow,* Gordon & Breach, Inc., New York, NY.

Walton, W. C., 1970, *Groundwater Resource Evaluation,* McGraw-Hill Book Co., Inc., New York, NY, 664 pp.

Wisler, C. O., and Brater, E. F., 1959, *Hydrology,* John Wiley & Sons, Inc., New York, NY.

CHAPTER 3

Surface Water

SECTION A. RIVERS

TABLE 3–1. LARGEST RIVERS IN THE UNITED STATES, IN DISCHARGE, DRAINAGE AREA, OR LENGTH

[Of the 32 rivers listed here, the 20 largest in three categories—discharge, drainage basin, and length—are ranked from 1 to 20; these ranks are shown in parentheses. Abbreviations: ft^3/s=cubic feet per second; mi^2=square miles. All data have been rounded to no more than three significant figures. Sources of data: Stream discharge and drainage area—mainly U.S. Geological Survey reports and files; length—publications and files of U.S. Geological Survey, U.S. Army Corps of Engineers, U.S. Environmental Protection Agency, and the Tennessee Valley Authority; data for the St. Lawrence River from "Facts from Canadian Maps," Canada Department of Energy, Mines and Resources, 1972. Period of record for most rivers is 1951–80. Some data are provisional and subject to revision.

River	Location of Mouth	Average Discharge at Mouth (1,000 ft^3/s)		Drainage Area (1,000 mi^2)		Length from Source to Mouth (miles)		Source Stream (Name and Location)	Water-Resources Region Number at–	
									Source	Mouth
Arkansas	Arkansas	41.0	(16)	161	(9)	1,460	(6)	East Fork Arkansas River, Colorado (Lake County).	11	8
Atchafalaya (excluding about 167,000 ft^3/s diverted from Mississippi River).[1]	Louisiana	58.0	(11)	95.1	(11)	1,420	(8)	Tierra Blanca Creek, New Mexico (Curry County).	11	8
Brazos	Texas	(*)	---	45.6	(19)	1,280	(11)	Blackwater Draw, New Mexico (Curry County).	12	12
Canadian	Oklahoma	(*)	---	46.9	(18)	906	(16)	Canadian River, Colorado (Las Animas County).	11	11
Colorado	Mexico	(*)	---	246 (U.S.-Mexico)	(7)	1,450	(7)	Colorado River, Colorado (Grand County).	14	---
Colorado (of Texas)	Texas	(*)	---	42.3	---	862	(18)	Colorado River (of Texas), Texas (Dawson County).	12	12
Columbia	Oregon-Washington	265	(4)	258 (U.S.-Canada)	(6)	1,240	(12)	Columbia River, British Columbia Canada.	---	17
Copper	Alaska	59	(10)	24.4	---	286	---	Copper River at terminus of Copper Glacier, Alaska.	19	19
Gila	Arizona	(*)	---	58.2 (U.S.-Mexico)	(16)	649	---	Middle Fork Gila River, New Mexico (Catron County).	15	15
Kansas	Kansas	(*)	---	59.5	(15)	743	---	Arikaree River, Colorado (Elbert County).	10	10
Kuskokwim	Alaska	67	(9)	48	(17)	724	---	South Fork Kuskokwim River at terminus of unnamed glacier, Alaska.	19	19
Mississippi (excluding Atchafalaya-Red River basin).[1, 2]	Louisiana	593	(1)	1,150 (U.S.-Canada)	(1)	2,350	(2)	Mississippi River, Minnesota (Clearwater County).	7	8
Missouri[2]	Missouri	76.2	(6)	529 (U.S.-Canada)	(2)	2,540	(1)	Red Rock Creek, Montana (Beaverhead County).	10	10
Mobile	Alabama	67.2	(8)	44.6	---	774	(20)	Tickanetley Creek, Georgia (Gilmer County).	3	3
North Canadian	Oklahoma	(*)	---	17.6	---	800	(19)	Corrumpa Creek, New Mexico (Union County).	11	11
Nushagak	Alaska	36	(20)	13.4	---	285	---	Nushagak River, Alaska.	19	19
Ohio	Illinois-Kentucky	281	(3)	203	(8)	1,310	(9)	Allegheny River, Pennsylvania (Potter County).	5	5
Pecos	Texas	(*)	---	44.3	---	926	(15)	Pecos River, New Mexico (Mora County).	13	13
Platte	Nebraska	(*)	---	84.9	(13)	990	(14)	Grizzly Creek, Colorado (Jackson County).	10	10
Porcupine	Alaska	23	---	45.1 (U.S.-Canada)	(20)	569	---	Porcupine River, Yukon Territory, Canada.	---	19
Red[1]	Louisiana	56.0	(13)	93.2	(12)	1,290	(10)	Tierra Blanca Creek, New Mexico (Curry County).	11	8
Rio Grande	Mexico-Texas	(*)	---	336 (U.S.-Mexico)	(4)	1,760	(5)	Rio Grande, Colorado (San Juan County).	13	13
St. Lawrence (-Great Lakes)	Canada	348	(2)	396 (U.S.-Canada)	(3)	1,900	(4)	North River, Minnesota (Lake County).	4	---
Snake	Washington	56.9	(12)	108	(10)	1,110	(13)	Snake River, Wyoming (Teton County).	17	17
Stikine	Alaska	56	(13)	20 (U.S.-Canada)	---	379	---	Stikine River, British Columbia Canada.	---	19
Susitna	Alaska	51	(15)	20	---	313	---	Susitna River at terminus of Susitna Glacier, Alaska.	19	19
Susquehanna	Maryland	38.2	(18)	27.2	---	447	---	Hayden Creek, New York (Otsego County).	2	2

TABLE 3–1. LARGEST RIVERS IN THE UNITED STATES, IN DISCHARGE, DRAINAGE AREA, OR LENGTH (continued)

River	Location of Mouth	Average Discharge at Mouth (1,000 ft^3/s)		Drainage Area (1,000 mi^2)		Length from Source to Mouth (miles)		Source Stream (Name and Location)	Water-Resources Region Number at–	
									Source	Mouth
Tanana	Alaska	41	(16)	44.5	---	659	---	Nabesna River at terminus of Nabesna Glacier, Alaska.	19	19
Tennessee	Kentucky	68.0	(7)	40.9	---	883	(17)	North Fork French Broad River, North Carolina (Transylvania County).	6	6
Willamette	Oregon	37.4	(19)	11.4	---	309	---	Middle Fork Willamette River, Oregon (Douglas County).	17	17
Yellowstone	North Dakota	(*)	---	70.0	(14)	692	---	Yellowstone River, Wyoming (Park County).	10	10
Yukon	Alaska	225	(5)	328	(5)	1,980	(3)	Nisutlin River, Yukon Territory, Canada.	---	19

*Less than 15,000 ft^3/s, and therefore not among the largest rivers in terms of discharge.

[1]In east-central Louisiana 50 miles northwest of Baton Rouge, the Red River flows into the Atchafalaya River, a distributary of the Mississippi River. The discharge of the Atchafalaya River, as shown in the table above, includes the entire discharge of the Red River, but excludes all water diverted into the Atchafalaya River from the Mississippi River. Thus, the respective discharges represent drainage from corresponding drainage areas.

[2]The total discharge from the entire 1,250,000-mi^2 Mississippi River system, including the Atchafalaya, Red, and Missouri River basins, averages 651,000 cubic feet per second. For the Mississippi River system as a whole, the longest continuous river channel is from the Missouri River headwater source in Montana to the mouth of the Missouri to the Gulf of Mexico, a combined length of about 3,710 miles.

Source: U.S. Geological Survey, National Water Summary 1985—Hydrologic Events and Surface-Water Resources, Water-Supply Paper 2300

FIGURE 3–1. LARGE RIVERS IN THE UNITED STATES

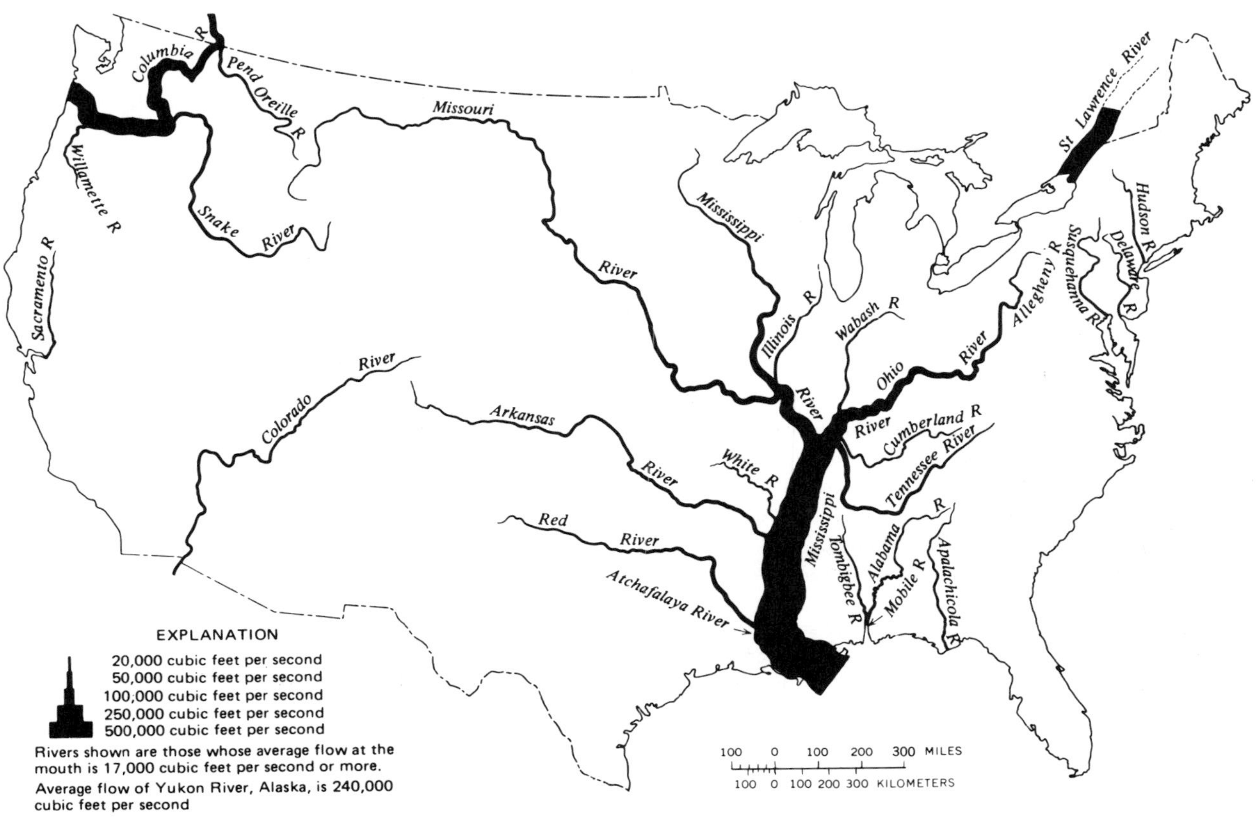

Source: Iseri, K.T., and W.B. Langbein, 1974, Large Rivers of the United States, U.S. Geol. Survey Circular 686

TABLE 3–2. AVERAGE DISCHARGE AT DOWNSTREAM GAGING STATIONS ON LARGE RIVERS OF THE UNITED STATES, 1931–1960, AND 1941–1970

River	Gaging-Station Location	Drainage Area (square miles)	Average Discharge (1931–60) (cubic feet per second)	Average Discharge (1941–70) (cubic feet per second)
Alabama	At Claiborne, Ala	22,000	31,140	31,510
Allegheny	At Natrona, Pa	11,410	19,200	18,810
Apalachicola	At Chattahoochee, Fla	17,200	20,700	21,700
Arkansas	At Little Rock, Ark	158,000	41,300	42,130
Atchafalaya[1]	At Krotz Springs, La	93,320	[2]160,800	[2]180,800
Colorado	Below Hoover Dam, Ariz.–Nev[3]	167,800	[4]14,580	14,530
Columbia	At The Dalles, Oreg	237,000	183,000	189,000
Cumberland	Near Grand Rivers, Ky	17,598	26,900	[5]28,030
Delaware	At Trenton, N.J[6]	9,397	16,100	[7]14,500
Hudson	At Green Island, N.Y	8,090	....	[8]12,520
Illinois	At Merdosia, Ill	25,300	20,500	20,670
Mississippi	At Alton, Ill	171,500	91,300	98,300
Mississippi	At Vicksburg, Miss	1,144,500	554,000	565,300
Missouri	At Hermann, Mo	528,200	69,200	76,200
Ohio	At Metropolis, Ill	203,000	257,000	257,200
Pend Oreille	At international boundary	25,200	26,900	28,420
Red	At Alexandria, La	67,412	32,470	32,100
Sacramento	At Verona, Calif[9]	....	25,700	27,200
St. Lawrence	At Cornwall, Ontario–near Massena, N.Y[10]	299,000	[11]233,000	[11]239,000
Snake	Near Clarkston, Wash	103,200	48,600	48,960
Susquehanna	At Marietta, Pa	25,990	36,100	35,060
Tennessee	Near Paducah, Ky	40,200	63,400	[5]64,050
Tombigbee	At Jackson Lock and Dam near Coffeeville, Ala	18,500	25,200	25,130
Wabash	At Mount Carmel, Ill	28,600	26,400	26,600
White	At Clarendon, Ark	25,497	29,490	29,360
Willamette	At Salem, Oreg	7,280	23,870	24,780
Yukon	At Ruby, Alaska	259,000	....	[12]170,000

[1]Continuation of Red River.
[2]Includes diversion from Mississippi River through Old River or Old River diversion channel.
[3]Very little tributary flow downstream. Downstream station located at Yuma, Ariz., drainage area 242,900 square miles. The greater part of the natural flow is diverted for irrigation and other uses in the basin above Yuma. Average flow at Yuma, 1963–70, is less than 1,000 cubic feet per second.
[4]For the period 1934–60.
[5]Interbasin diversion beginning June 1966 between Lake Barkley on Cumberland River and Lake Kentucky on Tennessee River through Barkley–Kentucky Canal.
[6]Five tributaries below Trenton have been added.
[7]Unadjusted for diversion by New York City reservoirs since 1954.
[8]October 1946 to September 1970 (24 years).
[9]American River and Yolo bypass have been added.
[10]Formerly at Ogdensburg, N.Y.
[11]Furnished by the U.S. Army Corps of Engineers through International St. Lawrence River Board of Control.
[12]Average is for 1957–70; station operated only since 1956.
Source: Iseri, K.T., and W.B. Langbein, 1974, Large Rivers of the United States, U.S. Geol. Survey Circular 686

TABLE 3–3. LENGTH OF PRINCIPAL RIVERS IN THE UNITED STATES AND CANADA

[Comprises rivers 600 miles or more in length. Length represents distance to designated outflow from (a) original headwater of named river where name applies to entire length of channel, or (b) upper limit of channel so named, usually the junction of two tributaries or headwater streams.]

RIVER/OUTFLOW	LENGTH Miles	LENGTH Kilometers
Alabama/Mobile River	735	1,183
Albany/James Bay	610	981
Arkansas/Mississippi River	1,459	2,348
Black/Chantrey Inlet	600	965
Brazos/Gulf of Mexico	870	1,400
Canadian/Arkansas River	906	1,458
Churchill/Hudson Bay	1,000	1,609
Cimarron/Arkansas River	600	965
Colorado (U.S.-Mex.)/Gulf of California	1,450	2,333
Colorado (Texas)/Matagorda Bay	840	1,352
Columbia/Pacific Ocean	1,243	2,000
Columbia, Upper/to mouth of Snake River	890	1,432
Cumberland/Ohio River	720	1,158
Fraser/Strait of Georgia	850	1,368
Gila/Colorado River	630	1,014
Green (UT-WY)/Colorado River	730	1,175
Hamilton/Atlantic Ocean	600	965
James (ND-SD)/Missouri River	710	1,142
Kuskokwim/Kuskokwim Bay	680	1,094
Liard/Mackenzie River	693	1,115
Mackenzie/Arctic Ocean	900	1,448
Milk/Missouri River	625	1,006
Mississippi/Mouth of SW Pass	2,348	3,778
Mississippi, Upper/to mouth of Missouri River	1,171	1,884
Mississippi-Missouri-Red Rock/Mouth of SW Pass	3,710	5,969
Missouri/Mississippi River	2,315	3,725
Missouri-Red Rock/Mississippi River	2,533	4,076
Mobile-Alabama-Coosa/Mobile Bay	780	1,255
North Canadian/Canadian River	760	1,223
North Platte/Platte River	618	994
Ohio/Mississippi River	981	1,578
Ohio-Allegheny/Mississippi River	1,306	2,101
Ottawa/St. Lawrence	790	1,271
Ouachita/Red River	605	973
Peace/Slave River	1,195	1,923
Pecos/Rio Grande	735	1,183
Red (OK-TX-LA)/Mississippi River	1,270	2,043
Rio Grande/Gulf of Mexico	1,885	3,033
St. Lawrence/Lake Ontario	800	1,287
Saskatchewan N./Lake Winnipeg	1,100	1,770
Saskatchewan S./Lake Winnipeg	1,205	1,939
Severn (Ontario)/Hudson Bay	610	981
Snake/Columbia River	1,038	1,670
Tanana/Yukon River	620	998
Tennessee/Ohio River	652	1,049
Tennessee-French Broad/Ohio River	900	1,448
White (AR-MO)/Mississippi River	720	1,158
Yellowstone/Missouri River	671	1,080
Yukon/Bering Sea	1,979	3,185

Source: Statistical Abstract of the United States 1986

TABLE 3–4. FLOW OF SELECTED STREAMS IN THE UNITED STATES

[Gaging station: Period of analysis is for the water years used to compute average discharge and may differ from that used to compute other streamflow characteristics. Streamflow characteristics: The 7-day, 10-year low flow is a discharge statistic; the lowest average discharge during 7 consecutive days of a year will be equal to or less than this value, on the average, once every 10 years. The average discharge is the arithmetic average of average annual discharges during the period of analysis. The 100-year flood is the peak flow that has a 1-percent chance of being equaled or exceeded in a given year. The degree of regulation is the effect of dams on the natural flow of the river. Abbreviations: Do. =ditto; mi^2 =square miles; ft^3/s=cubic feet per second; . . .=insufficient data or not applicable. Sources: Reports of the U.S. Geological Survey and State agencies]

ALABAMA

	Gaging Station			Streamflow Characteristics		
	Name	Drainage Area (mi^2)	Period of Analysis	7-day, 10-year Low Flow (ft^3/s)	Average Discharge (ft^3/s)	100-year Flood (ft^3/s)
	SOUTH ATLANTIC-GULF REGION CHOCTAWHATCHEE-ESCAMBIA SUBREGION					
1.	Choctawhatchee River, Newton	686	1923–26 1937–83	88	983	40,900
2.	Conecuh River, Brantley	500	1937–83	31	680	27,300
	ALABAMA SUBREGION					
3.	Coosa River, Childersburg	8,392	1915–68	2,000	13,860	157,600
			1969–78	1,330	13,860	144,900
4.	Tallapoosa River, Wadley	1,675	1923–83	140	2,594	73,800
5.	Alabama River, Montgomery	15,087	1927–68	5,240	24,260	317,000
			1969–83	3,860	24,260	219,500
	MOBILE-TOMBIGBEE SUBREGION					
6.	Cahaba River, Centreville	1,027	1902–07, 1931, 1937–83	143	1,633	117,000
7.	Mulberry Fork, Garden City	365	1928–83	4.9	681	51,300
8.	Black Warrior River, Northport	4,820	1895–1902, 1929–60	90	8,041	221,000
			1961–83	504	8,041	305,400
9.	Tombigbee River, Coatopa	15,385	1928–83	685	23,500	
	TENNESSEE REGION MIDDLE TENNESSEE-ELK SUBREGION					
10.	Flint River, Chase	342	1930–83	66	554	75,200
11.	Tennessee River, Florence	30,810	1894–1983	7,490	51,900	. . . .

ALASKA

	Name	Drainage Area (mi^2)	Period of Analysis	7-day, 10-year Low Flow (ft^3/s)	Average Discharge (ft^3/s)	100-year Flood (ft^3/s)
	ALASKA REGION SOUTHEAST ALASKA SUBREGION					
1.	Stikine River, Wrangell	19,920	1976–83	[1]4,500	56,674	[1]299,600
2.	Fish Creek, Ketchikan	32.1	[2]1915–36 1938–83	31	421	5,420
	SOUTH-CENTRAL ALASKA SUBREGION					
3.	Copper River, Chitina	20,600	1955–83	3,040	37,670	321,000

[1]Less than 10 years of record. Minimum discharge and maximum instantaneous discharge for period of record are shown.
[2]Record interrupted.

TABLE 3–4. FLOW OF SELECTED STREAMS IN THE UNITED STATES (continued)

ALASKA (continued)

	Gaging Station			Streamflow Characteristics		
	Name	Drainage Area (mi²)	Period of Analysis	7-day, 10-year Low Flow (ft³/s)	Average Discharge (ft³/s)	100-year Flood (ft³/s)
	SOUTH–CENTRAL ALASKA SUBREGION (continued)					
4.	Susitna River, Gold Creek	6,160	1949–83	723	9,724	115,000
5.	Susitna River, Susitna Station	19,400	1974–83	[1]5,000	49,940	[1]230,000
	SOUTHWEST ALASKA SUBREGION					
6.	Kvichak River, Igiugig	6,500	1967–83	7,380	18,060	65,500
7.	Nuyakuk River, Dillingham	1,490	1953–83	1,100	6,156	36,200
8.	Nushagak River, Ekwok	9,850	1977–83	[1]6,000	23,840	[1]89,200
9.	Kuskokwim River, Crooked Creek	31,100	1951–83	7,850	41,220	445,000
	YUKON SUBREGION					
10.	Yukon River, Eagle	113,500	[2]1911–13, 1950–83	10,500	82,660	605,000
11.	Porcupine River, Fort Yukon	29,500	1964–79	[2]6	14,230	476,000
12.	Chena River, Fairbanks	1,980	1948–83	150	1,384	[3]38,800
13.	Tanana River, Nenana	25,600	1962–83	4,740	23,550	[4]153,000
14.	Koyukuk River, Hughes	18,700	1960–82,	267	14,540	332,000
15.	Yukon River, Pilot Station	321,000	1975–83	[1]37,000	219,600	[1]751,000
	NORTHWEST ALASKA SUBREGION					
16.	Kobuk River, Kiana	9,520	1976–83	[1]1,300	15,270	[1]152,000
	ARCTIC SUBREGION					
17.	Kuparuk River, Deadhorse	3,130	1971–83	No flow	1,367	218,000

ARIZONA

LOWER COLORADO REGION
LOWER COLORADO RIVER BASIN

	Name	Drainage Area (mi²)	Period of Analysis	7-day, 10-year Low Flow (ft³/s)	Average Discharge (ft³/s)	100-year Flood (ft³/s)
1.	Colorado River, Lees Ferry	111,800	1912–62	1,670	17,850	189,500
			1965–84	....	....	....
2.	Colorado River, below Hoover Dam	171,700	1935–84	2,550	13,590	....
3.	Bill Williams River, below Alamo Dam	4,730	1940–68	0.72	92.3	325,000
4.	Colorado River, above Morelos Dam	246,700	1950–84	541	....	....

[1]Less than 10 years of record. Minimum discharge and maximum instantaneous discharge for period of record are shown.
[2]Record Interrupted.
[3]Adjusted for no-flow periods.
[4]Adjusted for high-outlier in period of record. Did not use 1981 peak because it was regulated.
[5]Adjusted for high-outlier in period of record.

TABLE 3–4. FLOW OF SELECTED STREAMS IN THE UNITED STATES (continued)

ARIZONA (continued)

	Gaging Station			Streamflow Characteristics		
	Name	Drainage Area (mi²)	Period of Analysis	7-day, 10-year Low Flow (ft³/s)	Average Discharge (ft³/s)	100-year Flood (ft³/s)
	LITTLE COLORADO SUBREGION					
5.	Little Colorado River, Cameron	26,500	1947–84	....	244	32,800
	UPPER GILA SUBREGION					
6.	Gila River, Clifton	4,010	1928–84	8.15	192	30,600
7.	Gila River, Solomon	7,896	1914–84	22.0	468	86,800
	MIDDLE GILA SUBREGION					
8.	San Pedro River, Palominas	741	1950–81	0.03	32.1	21,800
9.	San Pedro River, Winkelman	4,471	1966–79	....	57.1	....
10.	Gila River, Kelvin	18,011	1912–84	0.82	494	244,000
11.	Santa Cruz River, Tucson	2,222	1915–81	....	22.7	20,300
	SALT SUBREGION					
12.	Black River, Fort Apache	1,232	1958–84	16.7	412	56,100
13.	White River, Fort Apache	632	1958–84	4.80	201	11,900
14.	Salt River, Roosevelt	4,306	1925–84	81.9	888	164,000
15.	Verde River, above Horseshoe Dam	5,872	1945–84	72.5	564	158,000

ARKANSAS

	Name	Drainage Area (mi²)	Period of Analysis	7-day, 10-year Low Flow (ft³/s)	Average Discharge (ft³/s)	100-year Flood (ft³/s)
	LOWER MISSISSIPPI REGION MISSISSIPPI RIVER MAIN STEM					
1.	Mississippi River, Memphis, Tenn.	932,800	1933–81	99,000	474,200	1,860,000
	LOWER MISSISSIPPI-ST. FRANCIS SUBREGION St. Francis River basin					
2.	St. Francis Bay, Riverfront	....	1936–75, 1978–81,	57	5,274	....
			1944–75, 1978–81	83	....	....
	LOWER RED–OUACHITA SUBREGION					
3.	Ouachita River, Malvern	1,585	1928–84	105	2,380	194,000
			1954–84	244		
4.	Ouachita River, Camden	5,357	1928–84	236	7,490	299,000
			1954–84	548	....	....
5.	Smackover Creek, Smackover	385	1961–83	0.35	374	39,700
6.	Saline River, Rye	2,102	1937–83	12.6	2,590	102,000
7.	Bayou Bartholomew, McGehee	576	1939–42, 1946–84	6.5	676	6,930

TABLE 3–4. FLOW OF SELECTED STREAMS IN THE UNITED STATES (continued)

ARKANSAS (continued)

	Gaging Station			Streamflow Characteristics		
	Name	Drainage Area (mi²)	Period of Analysis	7-day, 10-year Low Flow (ft³/s)	Average Discharge (ft³/s)	100-year Flood (ft³/s)
	ARKANSAS-WHITE-RED REGION UPPER WHITE SUBREGION White River basin					
8.	Buffalo River, St. Joe	829	1939–84	16.5	1,027	176,000
9.	White River, Calico Rock	9,978	1939–83	894	9,830	352,000
			1945–83	973	....	....
			1958–83	1,120	....	....
10.	Spring River, Imboden	1,183	1936–83	279	1,360	163,000
11.	Black River, Black Rock	7,369	1929–31, 1939–83	1,980	8,410	176,000
			1950–83	1,990	....	....
12.	Middle Fork Little Red River, Shirley	302	1939–83	<0.19	467	140,000
13.	White River, Clarendon	25,555	1928–81	4,090	29,510	291,000
			1945–81	5,050	....	....
			1958–81	6,020	....	....
	LOWER ARKANSAS SUBREGION Arkansas River basin					
14.	Poteau River, Cauthron	203	1939–83	<0.1	215	47,100
			1950–83	<0.1	....	....
15.	Mulberry River, Mulberry	373	1938–83	<0.16	534	82,400
16.	Big Piney Creek, Dover	274	1950–83	0.15	398	112,000
17.	Petit Jean River, Danville	764	1916–84	0.74	809	91,900
			1949–84	1.9	....	....
18.	Arkansas River, Murray Dam	158,030	1927–84	1,230	40,290	588,000
	RED-SULPHUR SUBREGION Red River basin					
19.	Red River, Index	48,030	1936–84	812	11,710	190,000
			1945–84	934	....	....
			1969–84	1,110	....	....
	CALIFORNIA					
	CALIFORNIA REGION SACRAMENTO SUBREGION					
1.	Feather River, Nicolaus	5,921	1944–69	169	7,957	521,000
			1970–83	1,061	9,424	332,000
2.	Sacramento River, Verona	21,251	1930–69	1,618	18,240	[1]77,700
			1970–83	5,732	22,680	[1]94,700
3.	American River Fair Oaks	1,888	1906–55	64	3,735	257,000
			1956–83	426	3,942	150,000
	TULARE—BUENA VISTA LAKES AND SAN JOAQUIN SUBREGIONS					
4.	Kern River, Kernville	846	1912–84	104	762	45,800
5.	Kings River, Trimmer	1,342	1953–83	111	2,177	135,000
6.	Merced River, Stevinson	1,273	1941–83	52	733	14,400
7.	San Joaquin River, Vernalis	13,536	1930–83	241	4,783	99,900

[1]Sutter and Yolo Bypasses carry much of floodflow past Verona gage.

TABLE 3–4. FLOW OF SELECTED STREAMS IN THE UNITED STATES (continued)

CALIFORNIA (continued)

	Gaging Station			Streamflow Characteristics		
	Name	Drainage Area (mi²)	Period of Analysis	7-day, 10-year Low Flow (ft³/s)	Average Discharge (ft³/s)	100-year Flood (ft³/s)
	SOUTHERN CALIFORNIA COASTAL SUBREGION					
8.	San Diego River, Santee	377	1914–43	0.1	42.3	54,900
			1944–82	1.0	13.7	5,400
9.	Santa Margarita River, Ysidora	740	1924–48	0	43.3	46,000
			1949–83	0	31.0	32,000
10.	Santa Ana River, Santa Ana	1,700	1942–84	0	52.8	33,800
11.	Los Angeles River, at Long Beach	827	1930–40	.1	110	192,000
			1941–82	3.8	222	118,000
12.	Santa Clara River, Los Angeles—Ventura County Line	625	1953–71	.1	36.2	161,000
			1972–84	2.9	67.8	58,500
	CENTRAL CALIFORNIA COASTAL SUBREGION					
13.	Salinas River, Spreckels	4,156	1930–41	0.1	659	[1]145,000
			1942–65	.5	262	[1]145,000
			1966–84	.6	590	[1]145,000
14.	San Lorenzo River, Big Trees	106	1937–84	9.2	140	39,600
	KLAMATH-NORTHERN CALIFORNIA COASTAL SUBREGION					
15.	Russian River, Guerneville	1,338	1940–58	77	2,230	108,000
			1959–83	40	2,435	93,400
16.	Eel River, Scotia	3,113	1911–84	43	7,412	608,000
17.	Klamath River, Klamath	12,100	1911–84	1,859	18,110	556,000
18.	Smith River, Crescent City	609	1932–84	191	3,891	231,000
	GREAT BASIN REGION CENTRAL LAHONTAN SUBREGION					
19.	Truckee River, Tahoe City	507	1910–84	2.4	240	2,830

COLORADO

	Name	Drainage Area (mi²)	Period of Analysis	7-day, 10-year Low Flow (ft³/s)	Average Discharge (ft³/s)	100-year Flood (ft³/s)
	MISSOURI REGION NORTH AND SOUTH PLATTE SUBREGIONS					
1.	North Platte River, Northgate	1,431	1915–84	35	440	7,870
2.	South Platte River, Hartsel	880	1933–84	3.3	79.1	2,410
3.	South Platte River, Kersey	9,598	1901–84	51	834	40,400
4.	South Platte River, Julesburg	23,138	1902–84	7.6	524	62,300
	ARKANSAS—WHITE—RED REGION UPPER ARKANSAS SUBREGION					
5.	Arkansas River, Canon City	3,117	1888–1981	129	715	14,300
6.	Arkansas River, La Junta	12,210	1912–73	4.8	244	96,300
			1974–84	3.8	233	19,300
7.	Purgatoire River, Trinidad	795	1895–1976	2.7	83.3	34,400
			1977–81	....	64.3	....
8.	Purgatoire River, Las Animas	3,503	1922–31, 1948–76	0.34	116	94,000
			1977–84	....	81.0	....
9.	Arkansas River Lamar	19,780	1913–42	1.1	301	131,000
			1948–84	0.63	93.6	35,500

[1]Regulation has little effect on high floodflows.

TABLE 3–4. FLOW OF SELECTED STREAMS IN THE UNITED STATES (continued)

COLORADO (continued)

	Gaging Station			Streamflow Characteristics		
	Name	Drainage Area (mi²)	Period of Analysis	7-day, 10-year Low Flow (ft³/s)	Average Discharge (ft³/s)	100-year Flood (ft³/s)
	RIO GRANDE REGION RIO GRANDE HEADWATERS SUBREGION					
10.	Rio Grande, Del Norte	1,320	1889–1984	107	901	13,400
11.	Rio Grande, Lobatos	7,700	1899–1984	7.1	575	19,900
	UPPER COLORADO REGION COLORADO HEADWATERS SUBREGION					
12.	Colorado River, near Dotsero	4,394	1940–84	536	2,136	23,800
13.	Colorado River, Cameo	8,050	1933–84	997	3,900	41,900
	GUNNISON SUBREGION					
14.	Gunnison River, Gunnison	1,012	1910–28	148	888	11,500
			1944–84	115	709	9,000
15.	Gunnison River Grand Junction	7,928	1896–1965	265	2,611	38,100
			1968–84	495	2,659	30,500
	WHITE—YAMPA SUBREGION					
16.	Yampa River, Maybell	3,410	1916–84	39	1,573	19,900
17.	White River, Meeker	755	1901–84	179	626	6,570
	SAN JUAN SUBREGION					
18.	Animas River, Durango	692	1912–84	128	819	15,500

CONNECTICUT

	Name	Drainage Area (mi²)	Period of Analysis	7-day, 10-year Low Flow (ft³/s)	Average Discharge (ft³/s)	100-year Flood (ft³/s)
	NEW ENGLAND REGION CONNECTICUT SUBREGION Connecticut River basin					
1.	Connecticut River, Thompsonville	9,661	1928–83	2,200	16,400	209,000
2.	Burlington Brook, Burlington	4.10	1931–83	0.7	8.3	1,250
3.	Farmington River, Rainbow	590	1928–60	144	1,030	44,000
			1961–83	101	1,040	24,000
4.	Salmon River, East Hampton	100	1928–83	5.2	184	16,600
	CONNECTICUT COASTAL SUBREGION Thames River basin					
5.	Mount Hope River, Warrenville	28.6	1940–83	0.9	51.2	5,620
6.	Shetucket River, Willimantic	404	1928–52	46.5	667	25,000
			1953–83	44.2	734	22,500
7.	Quinebaug River, Jewett City	713	1918–58	119	1,250	29,500
			1959–83	90.0	1,330	26,500
8.	Yantic River, Yantic	89.3	1930–83	5.2	165	10,800
	Quinnipiac River basin					
9.	Quinnipiac River, Wallingford	115	1930–83	32.6	211	6,340

TABLE 3–4. FLOW OF SELECTED STREAMS IN THE UNITED STATES (continued)

CONNECTICUT (continued)

	Gaging Station			Streamflow Characteristics		
	Name	Drainage Area (mi²)	Period of Analysis	7-day, 10-year Low Flow (ft³/s)	Average Discharge (ft³/s)	100-year Flood (ft³/s)
	Housatonic River basin					
10.	Housatonic River, Falls Village	634	1912–83	119	1,090	24,000
11.	Shepaug River, Roxbury	132	1930–71	6.2	236	24,000
12.	Pomperaug River, Southbury	75.1	1932–83	6.0	128	19,900
13.	Housatonic River, Stevenson	1,544	1928–83	160	2,600	95,100
14.	Naugatuck River, Beacon Falls	260	1928–59	61.2	484	46,000
			1960–83	59.4	557	23,000
	Saugatuck River basin					
15.	Saugatuck River, Westport	79.8	1932–67	2.25	119	13,400
	DELAWARE					
	MID-ATLANTIC REGION DELAWARE SUBREGION Christina River basin					
1.	Christina River, Coochs Bridge	20.5	1943–84	1.5	28.8	4,840
2.	Brandywine Creek, Wilmington	314	1946–84	75	488	34,300
	UPPER CHESAPEAKE SUBREGION Indian River basin					
3.	Stockley Branch, Stockley	5.24	1943–84	0.66	7.04	200
	Nanticoke River basin					
4.	Nanticoke River, Bridgeville	75.4	1943–84	15	92.8	3,570
	FLORIDA					
	SOUTH ATLANTIC-GULF REGION ALTAMAHA-ST. MARYS SUBREGION					
1.	St. Marys River, Macclenny	700	1927–83	18	672	40,500
	ST. JOHNS SUBREGION					
2.	St. Johns River, Christmas	1,539	1934–83	24	1,310	18,500
3.	St. Johns River, DeLand	3,066	1934–83	0	3,120	21,900
4.	Oklawaha River, Rodman Dam	2,747	1944–68	788	2,020	12,900
			1969–83	....	1,550	....
	SOUTHERN FLORIDA SUBREGION					
5.	Fisheating Creek, Palmdale	311	1932–83	0	257	21,400
6.	Kissimmee River, S-65E near Okeechobee	2,886	1929–62	809	2,190	29,800
			1964–83	36	1,390	

TABLE 3–4. FLOW OF SELECTED STREAMS IN THE UNITED STATES (continued)

FLORIDA (continued)

	Gaging Station			Streamflow Characteristics		
	Name	Drainage Area (mi²)	Period of Analysis	7-day, 10-year Low Flow (ft³/s)	Average Discharge (ft³/s)	100-year Flood (ft/s)
	PEACE-TAMPA BAY SUBREGION					
7.	Peace River, Arcadia	1,367	1932–83	57	1,150	34,400
8.	Hillsborough River, Zephyrhills	220	1940–83	53	259	10,300
9.	Withlacoochee River, Holder	1,825	1932–83	158	1,090	9,750
	SUWANNEE SUBREGION					
10.	Suwannee River, Branford	7,880	1932–83	1,790	6,940	68,000
11.	Santa Fe River, Fort White	1,017	1928–29, 1933–83	730	1,610	16,400
12.	Suwannee River, Wilcox	9,640	1931, 1942–83	4,020	10,400	66,400
	OCHLOCKONEE SUBREGION					
13.	Ochlockonee River, Havana	1,140	1927–83	30	1,030	41,200
	APALACHICOLA SUBREGION					
14.	Apalachicola River, Chattahoochee	17,200	1929–83	7,000	22,400	264,000
	CHOCTAWHATCHEE-ESCAMBIA SUBREGION					
15.	Choctawhatchee River, Bruce	4,384	1931–83	1,630	7,140	128,000
16.	Yellow River, Milligan	624	1939–83	184	1,170	45,900
17.	Shoal River, Crestview	474	1939–83	291	1,100	33,600
18.	Escambia River, Century	3,817	1935–83	777	6,360	179,000
19.	Perdido River, Barrineau Park	394	1942–83	221	766	34,200
	GEORGIA					
	SOUTH ATLANTIC—GULF REGION OGEECHEE—SAVANNAH SUBREGION					
1.	Broad River, Bell	1,430	1927–32 1937	200	1,809	60,400
2.	Savannah River, Augusta	7,508	1960–81	5,500	10,200	....
	ALTAMAHA—ST MARYS SUBREGION					
3.	Oconee River, Greensboro	1,090	1903–32, 1936–78	150	1,446	50,700
4.	Altamaha River, Doctortown	13,600	1931–83	2,250	13,770	225,000
5.	Penholoway Creek, Jesup	210	1958–83	0	201	7,180
	SUWANNEE SUBREGION					
6.	Alapaha River, Statenville	1,400	1931–83	25	1,044	24,200

TABLE 3–4. FLOW OF SELECTED STREAMS IN THE UNITED STATES (continued)

GEORGIA (continued)

	Gaging Station			Streamflow Characteristics		
	Name	Drainage Area (mi²)	Period of Analysis	7-day, 10-year Low Flow (ft³/s)	Average Discharge (ft³/s)	100-year Flood (ft³/s)
	APALACHICOLA SUBREGION					
7.	Chattahoochee River, Atlanta	1,450	1959–81	860	2,840	....
8.	Flint River, Culloden	1,850	1911–23, 1928–31, 1937–83	180	2,402	99,100
9.	Flint River, Albany	5,310	1901–21, 1929–83	1,000	6,303	94,600
	ALABAMA SUBREGION					
10.	Etowah River, Allatoona Dam	1,120	1950–81	240	1,944	....
	TENNESSEE REGION MIDDLE TENNESSEE—HIAWASSEE SUBREGION					
11.	Toccoa River, Dial	177	1912–83	125	498	16,600

HAWAII

	Name	Drainage Area (mi²)	Period of Analysis	7-day, 10-year Low Flow (ft³/s)	Average Discharge (ft³/s)	100-year Flood (ft³/s)
	HAWAII REGION KAUAI SUBREGION					
1.	East Branch of North Fork Wailua River near Lihue	6.27	1916–83	10.4	48.6	10,400
	OAHU SUBREGION					
2.	Kalihi Stream, near Honolulu	2.61	1917–83	0.29	6.74	10,400
	MAUI SUBREGION					
3.	Honopou Stream near Huolo	0.64	1911–83	0.26	4.69	4,410
	HAWAII SUBREGION					
4.	Waiakea Stream near Mountain View	17.4	1931–83	0.10	11.8	1,140

IDAHO

	Name	Drainage Area (mi²)	Period of Analysis	7-day, 10-year Low Flow (ft³/s)	Average Discharge (ft³/s)	100-year Flood (ft³/s)
	GREAT BASIN REGION BEAR SUBREGION Bear River basin					
1.	Bear River, Preston	4,545	1944–84	80	937	8,190
	PACIFIC NORTHWEST REGION KOOTENAI—POND OREILLE—SPOKANE SUBREGION Pend Oreille River basin					
2.	Priest River, Priest River	902	1904, 1930–84	200	1,686	11,500
	Spokane River basin					
3.	Spokane River, Post Falls	3,340	1913–84	180	6,297	46,000

TABLE 3–4. FLOW OF SELECTED STREAMS IN THE UNITED STATES (continued)

IDAHO (continued)

	Gaging Station			Streamflow Characteristics		
	Name	Drainage Area (mi²)	Period of Analysis	7-day, 10-year Low Flow (ft³/s)	Average Discharge (ft³/s)	100-year Flood (ft³/s)
	UPPER SNAKE SUBREGION					
4.	Snake River, Irwin	5,225	1950–84	560	6,691	31,700
5.	Henrys Fork, Rexburg	2,920	1910–84	400	2,088	12,100
6.	Portneuf River, Pocatello	1,250	1913–16, 1918–84	14	280	2,650
7.	Snake River, Milner	17,180	1910–26, 1927–84	5	2,711	42,400 28,300
8.	Big Lost River, below Mackay Reservoir, Mackay	813	1905, 1913–14, 1920–84	36	314	3,280
9.	Big Wood River below Magic Dam, Richfield	1,600	1913–84	2	480	10,400
	MIDDLE SNAKE SUBREGION					
10.	Snake River, King Hill	35,800	1910–26 1927–84	6,000	10,910	54,600 39,100
11.	Bruneau River, Hot Spring	2,630	1909–14, 1943–84	47	409	7,500
12.	Boise River, Boise	2,680	1953–84	1	2,951	10,000
13.	Payette River, Payette	3,240	1936–84	400	3,183	10,000
14.	Weiser River, Weiser	1,460	1953–84	54	1,132	26,000
15.	Snake River, Weiser	69,200	1911–84	6,600	18,490	10,000
	LOWER SNAKE SUBREGION					
16.	Salmon River, White Bird	13,550	1911–17, 1920–84	2,400	11,420	126,000
17.	Clearwater River, Spalding	9,570	1910–13, 1925–84	1,500	15,550	188,000

ILLINOIS

UPPER MISSISSIPPI REGION
UPPER AND LOWER ILLINOIS SUBREGIONS

	Gaging Station			Streamflow Characteristics		
	Illinois River main stem					
1.	Illinois River, Marseilles	8,259	1919–83 1940–83	 3,180	 9,791	91,100 ....
2.	Illinois River, Meredosia	26,028	1921–83 1940–83	 3,630	 21,976	132,300 ....
	Illinois River basin—tributaries					
3.	Kankakee River, Wilming- ton	5,150	1915–83	463	4,233	68,100
4.	Des Plaines, River Riverside	630	1914–83 1943–83 1974–83	.... 6.0 48	.... 471	7,830 ....
5.	Fox River, Dayton	2,642	1915–83 1974–83	176 366	1,703	37,400 ....
6.	Vermilion River, Leonore	1,251	1931–83 1973–83	.... 9.6	822	40,700 ....
7.	Mackinaw River, Congerville	767	1945–83	1.3	511	43,900
8.	Spoon River, Seville	1,636	1914–83	20	1,054	37,600
9.	Sangamon River, Oakford	5,093	1910–83 1974–83	147 263	3,335	82,800 ....
10.	La Moine River, Ripley	1,293	1921–83	10	802	27,500

TABLE 3–4. FLOW OF SELECTED STREAMS IN THE UNITED STATES (continued)

ILLINOIS (continued)

	Gaging Station			Streamflow Characteristics		
	Name	Drainage Area (mi^2)	Period of Analysis	7-day, 10-year Low Flow (ft^3/s)	Average Discharge (ft^3/s)	100-year Flood (ft^3/s)
	ROCK SUBREGION Rock River basin					
11.	Pecatonica River, Freeport	1,326	1914–83	191	900	21,300
12.	Kishwaukee River, Perryville	1,099	1940–83	68	713	25,000
13.	Rock River, Joslin	9,549	1940–83	1,270	6,020	58,800
14.	Green River, Geneseo	1,003	1936–83	40	610	13,000
	UPPER MISSISSIPPI-KASKASKIA-MERAMEC SUBREGION Kaskaskia and Big Muddy River basins					
15.	Kaskaskia River, Vandalia	1,940	1908–69	14	1,412	33,000
			1970–83	34	1,769	30,400
16.	Big Muddy River, Murphysboro	2,169	1916–70	....	....	39,300
			1931–70	2.3	1,788	....
			1971–83	47	1,888	41,000
	OHIO REGION WABASH AND LOWER OHIO SUBREGIONS Embarras and Little Wabash River basins					
17.	Embarras River, Ste. Marie	1,516	1910–83	14	1,224	53,700
18.	Little Wabash, River, Carmi	3,102	1940–83	6.2	2,529	45,300

INDIANA

	Name	Drainage Area (mi^2)	Period of Analysis	7-day, 10-year Low Flow (ft^3/s)	Average Discharge (ft^3/s)	100-year Flood (ft^3/s)
	OHIO REGION GREAT MIAMI SUBREGION Whitewater River basin					
1.	Whitewater River, Alpine	522	1928–83	48	551	49,000
	WABASH SUBREGION Wabash River main stem-White River basin-Patoka River basin					
2.	Muscatatuck River, Deputy	293	1947–83	0.0	348	41,200
3.	South Fork Patoka River, Spurgeon	42.8	1964–83	2.2	51.9	5,990
4.	Eagle Creek, Indianapolis	174	1938–68	0.5	148	18,400
			1969–83	6.0	168	11,800
5.	Driftwood River, Edinburgh	1,060	1940–83	91	1,144	49,500
6.	Wabash River, Peru	2,686	1943–67	92	2,290	74,300
			1970–83	155	2,500	31,000
7.	Wabash River, Mount Carmel, Ill.	28,635	1927–83	2,280	27,440	315,000
	UPPER MISSISSIPPI REGION UPPER ILLINOIS SUBREGION Kankakee River basin					
8.	Kankakee River, Shelby	1,779	1922–83	417	1,619	6,950
9.	Iroquois River, Foresman	449	1948–83	11	383	5,660

TABLE 3–4. FLOW OF SELECTED STREAMS IN THE UNITED STATES (continued)

INDIANA (continued)

	Gaging Station			Streamflow Characteristics		
	Name	Drainage Area (mi^2)	Period of Analysis	7-day, 10-year Low Flow (ft^3/s)	Average Discharge (ft^3/s)	100-year Flood (ft^3/s)
	GREAT LAKES REGION SOUTHESTERN LAKE MICHIGAN SUBREGION St. Joseph River basin					
10.	St. Joseph River, Elkhart	3,370	1947–83	818	3,177	21,500
11.	Pigeon Creek, Angola	106	1947–83	5.8	78.5	843
	WESTERN LAKE ERIE SUBREGION Maumee River basin					
12.	Maumee River, New Haven	1,967	1956–83	72	1,645	25,600

IOWA

	Name	Drainage Area (mi^2)	Period of Analysis	7-day, 10-year Low Flow (ft^3/s)	Average Discharge (ft^3/s)	100-year Flood (ft^3/s)
	UPPER MISSISSIPPI REGION MISSISSIPPI RIVER MAIN STEM					
1.	Mississippi River, Clinton	85,600	1873–1983	10,050	47,390	[1]295,000
	Northeast Iowa River basin[2]					
2.	Upper Iowa River, Decorah	511	1951–83	32	327	22,400
3.	Turkey River, Garber	1,545	1913–16, 1919–27, 1929–30, 1932–83	81	949	33,100
4.	Maquoketa River, Maquoketa	1,553	1913–83	160	1,027	47,700
5.	Wapsipinicon River, De Witt	2,330	1934–83	98	1,537	31,600
	Iowa—Cedar River basin[3]					
6.	Iowa River, Iowa City	3,271	1903–58 1959–83	60 93	1,470 2,180	43,700 17,400
7.	English River, Kalona	573	1939–83	2.3	370	25,300
8.	Shell Rock River, Shell Rock	1,746	1953–83	64	974	42,800
9.	Cedar River, Waterloo	5,146	1940–83	284	2,984	98,900
10.	Cedar River, Cedar Rapids	6,510	1902–83	347	3,414	83,500
11.	Iowa River, Wapello	12,499	1914–58 1959–83	555 893	5,950 8,650	102,000 116,000
	Skunk River basin[3]					
12.	South Skunk River, Oskaloosa	1,635	1945–83	10	916	25,800
13.	North Skunk River, Sigourney	730	1945–83	2.3	436	27,400
14.	Skunk River, Augusta	4,303	1914–83	31	2,407	55,200

[1]From Upper Mississippi River Basin Commission, 1978.
[2]Within the Upper Mississippi—Black—Root, Upper Mississippi—Maquoketa—Plum, and Upper Mississippi—Iowa—Skunk—Wapsipinicon Subregion (Seaber and others, 1984).
[3]Within the Upper Mississippi—Iowa—Skunk—Wapsipinicon Subregions (Seaber and others, 1984).

TABLE 3–4. FLOW OF SELECTED STREAMS IN THE UNITED STATES (continued)

IOWA (continued)

	Gaging Station			Streamflow Characteristics		
	Name	Drainage Area (mi²)	Period of Analysis	7-day, 10-year Low Flow (ft³/s)	Average Discharge (ft³/s)	100-year Flood (ft³/s)
			Des Moines River basin[1]			
15.	Des Moines River, Stratford	5,452	1920–83	40	1,882	54,600
16.	North Raccoon River, Jefferson	1,619	1940–83	8.3	708	27,200
17.	South Raccoon River, Redfield	988	1940–83	26	449	32,200
18.	Raccoon River, Van Meter	3,441	1915–83	34	1,346	46,500
19.	Des Moines River, Keosauqua	14,038	1903–06, 1911–68,	143	5,160	123,000
			1969–83	245	7,860	90,600
			MISSOURI REGION MISSOURI RIVER MAIN STEM[2]			
20.	Missouri River, Sioux City	314,600	[3]1897–1956,	3,810	30,000	437,000
			1957–83	6,570	28,700	[4]144,500
			Western Iowa River basin[5]			
21.	Big Sioux River, Akron	9,030	1928–83	19	901	71,000
22.	Floyd River, James	886	1934–83	2.7	197	34,300
23.	Little Sioux River, Correctionville	2,500	1918–25, 1928–32, 1936–83	14	766	32,600
24.	Boyer River, Logan	871	1918–25, 1937–83	6.5	315	31,800
			Southern Iowa River basin[6]			
25.	Nishnabotna River, Hamburg	2,806	1922–23, 1928–83	28	1,057	40,700
26.	Nodaway River, Clarinda	762	1918–25, 1936–83	5.8	338	37,900
27.	Thompson River, Davis City	701	1918–25, 1941–83	1.6	370	25,500
28.	Chariton River, Rathbun	549	1956–69,	.25	303	40,327
			1970–83	4.0	382	2,130

KANSAS

	Name	Drainage Area (mi²)	Period of Analysis	7-day, 10-year Low Flow (ft³/s)	Average Discharge (ft³/s)	100-year Flood (ft³/s)
			MISSOURI REGION REPUBLICAN AND SMOKY HILL SUBREGIONS Republican and Smoky Hill River basins			
1.	Republican River, Clay Center	24,542	1917–83	[7]75	990	[7]76,000
2.	Smoky Hill River, Elkader	3,555	1940–83	0.0	30	70,000
3.	Solomon River, Niles	6,770	1897–1903, 1917–83	[7]33	550	[8]51,000
4.	Smoky Hill River, Enterprise	19,260	1935–83	[7]120	1,600	[8]85,000

[1]Within the Minnesota and Des Moines Subregions (Seaber and others, 1984).
[2]Within the Missouri—Big Sioux, Missouri—Little Sioux, and Missouri—Nishnabotna Subregions (Seaber and others, 1984).
[3]Flow parameters based only on 1929–31 and 1939–56 water years.
[4]From U.S. Army Corps of Engineers, February 1978.
[5]Within the Missouri—Big Sioux, Missouri—Little Sioux, and Missouri—Nishnabotna Subregions (Seaber and others, 1984).
[6]Within the Missouri-Nishnabotna, Chariton-Grand, and Upper Mississippi-Salt Subregions (Seaber and others, 1984).
[7]Based on period of analysis since regulation began. These values are not based on detailed analyses, are approximate estimates, and are for information purposes only.
[8]From flood-insurance hydrology study. Based on detailed analyses of regulated-flow conditions.

TABLE 3–4. FLOW OF SELECTED STREAMS IN THE UNITED STATES (continued)

KANSAS (continued)

	Gaging Station			Streamflow Characteristics		
	Name	Drainage Area (mi²)	Period of Analysis	7-day, 10-year Low Flow (ft³/s)	Average Discharge (ft³/s)	100-year Flood (ft³/s)
	KANSAS, GASCONADE-OSAGE, AND MISSOURI-NISHNABOTNA SUBREGIONS **Kansas, Osage, and Missouri River basins**					
5.	Kansas River, Fort Riley	44,870	1964–83	[1]240	2,600	[2]140,000
6.	Big Blue River, Manhattan	9,640	1955–83	[1]18	2,000	[2]50,000
7.	Kansas River, De Soto	59,756	1917–83	[1]800	7,000	[2]230,000
8.	Marais des Cygnes River, Kansas-Missouri State line	3,230	1959–83	[1]2.5	2,000	[1]67,000
9.	Missouri River, St. Joseph, Mo.	420,300	1929–83	[1]6,100	42,000	....
	ARKANSAS–WHITE–RED REGIONS **MIDDLE ARKANSAS, UPPER CIMARRON, AND ARKANSAS-KEYSTONE SUBREGIONS** **Arkansas River basin**					
10.	Arkansas River, Syracuse	25,763	1902–06, 1921–83	[1]0.3	310	[1]130,000
11.	Little Arkansas River, Valley Center	1,327	1922–83	10	280	43,000
12.	Arkansas River, Arkansas City	43,713	1902–06, 1922–83	[1]170	1,800	[2]99,000
	MIDDLE ARKANSAS AND NEOSHO–VERDIGRIS SUBREGIONS **Walnut, Verdigris, and Neosho River basins**					
13.	Verdigris River, Independence	2,892	1895–1904, 1921–83	[1]9.0	1,700	[2]72,000
14.	Neosho River, Parsons	4,905	1922–83	[1]7.5	2,500	[1]56,000

KENTUCKY

	Name	Drainage Area (mi²)	Period of Analysis	7-day, 10-year Low Flow (ft³/s)	Average Discharge (ft³/s)	100-year Flood (ft³/s)
	OHIO REGION **MIDDLE AND LOWER OHIO SUBREGIONS** **Ohio River main stem**					
1.	Ohio River, Greenup Dam	62,000	1968–83	7,400	92,530	699,000
2.	Ohio River, Louisville	91,170	1928–83	8,200	115,700	862,000
3.	Ohio River, Metropolis, Ill.	203,000	1928–83	46,000	271,000	1,580,000
	Salt River basin					
4.	Salt River, Shepherdsville	1,197	1938–83	0.22	1,572	61,900
5.	Rolling Fork, Boston	1,299	1938–83	2.3	1,801	65,600
	BIG SANDY-GUYANDOTTE SUBREGION					
6.	Levisa Fork, Pikeville	1,232	1937–83	5.8	1,474	76,400
	KENTUCKY-LICKING SUBREGION **Licking River basin**					
7.	Licking River, Catawba	3,300	1914–83	13	4,143	84,900

[1]Based on period of analysis since regulation began. These values are not based on detailed analyses, are approximate estimates, and are for information purposes only.
[2]From flood-insurance hydrology study. Based on detailed analyses of regulated-flow conditions.

TABLE 3–4. FLOW OF SELECTED STREAMS IN THE UNITED STATES (continued)

KENTUCKY (continued)

	Gaging Station			Streamflow Characteristics		
	Name	Drainage Area (mi^2)	Period of Analysis	7-day, 10-year Low Flow (ft^3/s)	Average Discharge (ft^3/s)	100-year Flood (ft^3/s)
			Kentucky River basin			
8.	Middle Fork Kentucky River, Tallega	537	1930–83	0.64	730	51,400
9.	Kentucky River, Salvisa	5,102	1925–83	136	6,737	125,000
			GREEN SUBREGION **Green River basin**			
10.	Green River, Munfordville	1,673	1915–83	73	2,722	70,300
11.	Pond River, Apex	194	1940–83	0	267	25,800
			CUMBERLAND SUBREGION **Cumberland River basin**			
12.	Cumberland River, Williamsburg	1,607	1959–83	22	2,736	54,000
13.	Little River, Cadiz	244	1940–83	11	349	18,200
			TENNESSEE REGION **LOWER TENNESSEE SUBREGION**			
14.	Tennessee River, Paducah	40,200	1889–1983	8,190	[1]64,060 [2]65,450	. . . .

LOUISIANA

	Name	Drainage Area (mi^2)	Period of Analysis	7-day, 10-year Low Flow (ft^3/s)	Average Discharge (ft^3/s)	100-year Flood (ft^3/s)
			SOUTH ATLANTIC—GULF REGION **PEARL SUBREGION** **Pearl River basin**			
1.	Pearl River, Bogalusa	6,573	1939–83	1,320	9,887	129,000
2.	Bogue Chitto, Bush	1,213	1938–83	460	1,915	93,200
			LOWER MISSISSIPPI REGION **Mississippi River main stem[3]**			
3.	Mississippi River, Vicksburg, Miss.	1,118,160	1929–83	127,000	578,800	2,203,000
4.	Mississippi River, Tarbert Landing, Miss.	1,124,900	1939–83	142,000	514,200	. . . .
			LOWER RED—OUACHITA SUBREGION **Ouachita River basin**			
5.	Big Creek, Pollock	51	1943–83	7.4	61.4	37,200
			LOWER MISSISSIPPI—LAKE MAUREPAS SUBREGION			
6.	Amite River, Denham Springs	1,280	1939–83	304	2,021	136,000
7.	Tangipahoa River, Robert	646	1939–83	284	1,154	81,900

[1]Prior to opening of Barkley-Kentucky Canal (1889–1965).
[2]Since the opening of Barkley-Kentucky Canal (1965–83).
[3]Includes all or parts of the Lower Mississippi—Yazoo, Lower Mississippi—Big Black, Lower Mississippi—Lake Maurepas, and the Lower Mississippi Subregions (Seaber, Kapinos, and Knapp, 1984).

TABLE 3–4. FLOW OF SELECTED STREAMS IN THE UNITED STATES (continued)

LOUISIANA (continued)

	Gaging Station			Streamflow Characteristics		
	Name	Drainage Area (mi^2)	Period of Analysis	7-day, 10-year Low Flow (ft^3/s)	Average Discharge (ft^3/s)	100-year Flood (ft^3/s)
	LOUISIANA COASTAL SUBREGION **Atchafalaya—Teche—Vermillion and Calcasieu—Mermentau River basin**					
8.	Atchafalaya River, Simmesport	87,570	1939–83	26,000	196,700	....
9.	Calcasieu River, Oberlin	753	1923–24, 1939–83	37	1,147	58,900
10.	Calcasieu River, Kinder	1,700	1923–24, 1939–57, 1962–83	202	2,568	121,000
	ARKANSAS-WHITE-RED REGION **RED—SULPHUR SUBREGION** **Red River basin**					
11.	Red River, Shreveport	60,613	1929–83	1,150	24,030	297,000
12.	Red River, Alexandria	67,500	1929–83	1,650	30,870	251,000
13.	Saline Bayou, Lucky	154	1941–83	4.5	162	17,200
	TEXAS—GULF REGION **SABINE SUBREGION** **Sabine River basin**					
14.	Sabine River, Ruliff, Tex.	9,329	1925–83	432	7,491	90,700
	MAINE					
	NEW ENGLAND REGION **ST. JOHN SUBREGION**					
1.	St. John River, Ninemile Bridge	1,341	1950–85	96	2,330	47,900
2.	St. John River, Fort Kent	5,665	1926–85	747	9,730	167,000
3.	Aroostook River, Washburn	1,654	1930–85	143	2,670	51,500
	MAINE COASTAL SUBREGION					
4.	St Croix River, Baring	1,374	1958–85	484	2,760	31,000
5.	Narraguagus River, Cherryfield	227	1948–85	29	503	11,300
6.	Sheepscot River, North Whitefield	148	1938–85	8.8	249	7,080
	PENOBSCOT SUBREGION					
7.	Penobscot River, Dover-Foxcroft	298	1902–85	19	603	25,400
8.	Penobscot River, West Enfield	6,671	1901–85	2,970	11,960	150,000
	KENNEBEC SUBREGION					
9.	Kennebec River, Bingham	2,715	1907–10, 1930–85	1,310	4,450	59,200
10.	Carrabassett River, North Anson	353	1902–07, 1925–85	45	717	39,500

TABLE 3–4. FLOW OF SELECTED STREAMS IN THE UNITED STATES (continued)

MAINE (continued)

	Gaging Station			Streamflow Characteristics		
	Name	Drainage Area (mi²)	Period of Analysis	7-day, 10-year Low Flow (ft³/s)	Average Discharge (ft³/s)	100-year Flood (ft³/s)
	ANDROSCOGGIN SUBREGION					
11.	Swift River, Roxbury	96.9	1929–85	6.9	199	21,100
12.	Little Androscoggin River, South Paris	75.8	1913–24, 1931–85	2.6	139	6,700
13.	Androscoggin River, Auburn	3,263	1928–85	1,690	6,140	99,700
	SACO SUBREGION					
14.	Royal River, Yarmouth	141	1949–85	24	275	11,000
15.	Saco River, Cornish	1,293	1916–85	386	2,710	36,800
	MARYLAND (AND THE DISTRICT OF COLUMBIA)					
	MID-ATLANTIC REGION POTOMAC SUBREGION					
1.	Conoccocheague Creek, Fairview	494	1928–83	53	590	26,800
2.	Antietam Creek, Sharpsburg	281	1899–1983	66	275	14,400
3.	Monocacy River, Frederick	817	1929–83	50	926	65,900
	UPPER CHESAPEAKE SUBREGION					
4.	Pocomoke River, Willards	60.5	1949–83	3.4	71	1,830
5.	Choptank River, Greensboro	113	1948–83	5.4	132	9,360
6.	Patuxent River, Unity	34.8	1944–83	2.8	39	26,900
	SUSQUEHANNA SUBREGION					
7.	Susquehanna River, Conowingo	27,100	1968–83	....	42,180	....
	OHIO REGION MONONGAHELA SUBREGION					
8.	Youghiogheny River, Oakland	134	1941–83	5.9	297	12,800
	MASSACHUSETTS					
	NEW ENGLAND REGION CONNECTICUT SUBREGION					
1.	Millers River, Erving	372	1915–83	47	630	....
2.	North River, Shattuckville	89.0	1940–83	8.1	183	17,000
3.	Deerfield River, West Deerfield	557	1941–83	97	1,285	56,000
4.	Connecticut River, Montague City	7,860	1905–83	1,700	13,760	....
5.	Ware River, Barre	96.3	1929–83	6.4	167	....
6.	East Branch Swift River, Hardwick	43.7	1938–83	0.2	69.5	3,100
7.	Chicopee River, Indian Orchard	689	1929–83	130	903	....

TABLE 3–4. FLOW OF SELECTED STREAMS IN THE UNITED STATES (continued)

MASSACHUSETTS (continued)

	Gaging Station			Streamflow Characteristics		
	Name	Drainage Area (mi²)	Period of Analysis	7-day, 10-year Low Flow (ft³/s)	Average Discharge (ft³/s)	100-year Flood (ft³/s)
	NEW ENGLAND REGION CONNECTICUT SUBREGION (continued)					
8.	West Branch Westfield River, Huntington	94.0	1936–83	5.7	190	29,000
9.	Westfield River, Westfield	497	1915–83	84	921	....
	MERRIMACK SUBREGION					
10.	Nashua River, East Pepperell	316	1936–83	46	568	16,000
11.	Concord River, Lowell	307	1937–83	33	630	6,000
12.	Merrimack River, Lowell	4,423	1924–83	937	7,530	....
	MASSACHUSETTS-RHODE ISLAND COASTAL SUBREGION					
13.	Parker River, Byfield	21.3	1946–83	0.2	36.7	610
14.	Ipswich River, Ipswich	125	1931–83	2.0	187	3,120
15.	Charles River, Dover	183	1938–83	13	302	3,800
16.	Indian Head River, Hanover	30.2	1967–83	1.4	62.4	1,800
17.	Wading River, Norton	43.3	1926–83	2.3	73.3	1,500
	CONNECTICUT COASTAL SUBREGION					
18.	Housatonic River, Great Barrington	280	1914–83	69	526	11,000
	MICHIGAN					
	GREAT LAKES REGION NORTHWESTERN LAKE MICHIGAN AND SOUTHEASTERN LAKE MICHIGAN SUBREGIONS					
1.	St. Joseph River, Niles	3,666	1931–84	945	3,260	20,400
2.	Kalamazoo River, Fennville	1,600	1930–36, 1938–84	335	1,420	12,300
3.	Red Cedar River, East Lansing	355	1903, 1932–84	9.79	205	6,890
4.	Grand River, Lansing	1,230	1902–06, 1935–84	80.2	833	8,800
5.	Grand River, Grand Rapids	4,900	1902–05, 1931–84	721	3,570	53,000
6.	Escanaba River, Cornell	870	1904–12, 1951–84	168	892	13,000
	NORTHEASTERN LAKE MICHIGAN-LAKE MICHIGAN SUBREGION					
7.	Muskegon River, Evart	1,450	1931, 1934–84	314	998	9,060
8.	Muskegon River, Newaygo	2,350	1910–14, 1917–19, 1931–84	672	1,970	14,100
9.	Manistee River, Manistee	1,780	1952–84	1,210	2,000	8,240

TABLE 3-4. FLOW OF SELECTED STREAMS IN THE UNITED STATES (continued)

MICHIGAN (continued)

	Gaging Station			Streamflow Characteristics		
	Name	Drainage Area (mi²)	Period of Analysis	7-day, 10-year Low Flow (ft³/s)	Average Discharge (ft³/s)	100-year Flood (ft³/s)
	SOUTHWESTERN LAKE HURON-LAKE HURON SUBREGION					
10.	Shiawassee River, Fergus	637	1940–84	42.1	420	9,330
11.	Flint River, Fosters	1,188	1940–84	66.4	743	16,700
12.	Cass River, Frankenmuth	841	1936, 1940–84	20.4	490	20,800
13.	Tittabawassee River, Midland	2,400	1937–84	187	1,680	44,600
	SOUTHERN LAKE SUPERIOR—LAKE SUPERIOR AND ST. CLAIR—DETROIT SUBREGIONS					
14.	Ontonagon River, Rockland	1,340	1943–84	308	1,430	32,400
15.	Sturgeon River, Sidnaw	171	1913–15, 1944–84	8.19	216	4,830
16.	Tahquamenon River, Paradise	790	1954–84	196	936	7,660
17.	Clinton River, Mt. Clemens	734	1935–84	61.4	531	23,200
18.	Huron River, Ann Arbor	729	1905–84	43.6	456	5,940

MINNESOTA

	Name	Drainage Area (mi²)	Period of Analysis	7-day, 10-year Low Flow (ft³/s)	Average Discharge (ft³/s)	100-year Flood (ft³/s)
	UPPER MISSISSIPPI REGION MISSISSIPPI RIVER BASIN[1]					
1.	Mississippi River, Anoka	19,000	1932–83	1,194	7,655	98,000
2.	Crow Wing River, Pillager	3,300	1968–83	173	1,264	15,300
3.	Sauk River, St. Cloud	925	1910–12 1931 1935–81	13.1	276	10,000
4.	Crow River, Rockford	2,520	1910–17 1931 1935–83	14.7	664	19,000
5.	Rum River, St. Francis	1,360	1931 1934–83	64.4	602	14,000
6.	Cannon River, Welch	1,320	1911–13 1931–71	61.6	501	34,000
7.	Zumbro River, Zumbro Falls	1,130	1910–17 1931–80	77.7	517	40,200
8.	Root River, Houston	1,270	1910–17 1931–83	178	696	51,500
	MINNESOTA SUBREGION					
9.	Minnesota River, Jordan	16,200	1935–83	171	3,520	115,000
10.	Lac qui Parle River, Lac qui Parle	983	1913,1932 1934–83	0.20	120	19,300
11.	Chippewa River, Milan	1,870	1938–83	2.90	269	12,400
12.	Cottonwood River, New Ulm	1,280	1912–13 1936–37 1939–83	2.77	289	33,000
13.	Blue Earth River, Rapidan	2,430	1940–45 1950–83	14.9	895	34,600
	ST. CROIX SUBREGION					
14.	St. Croix River, St. Croix Falls	6,240	1903–83	1,099	4,235	61,000

[1]Includes the Mississippi Headwaters and the Upper Mississippi Black-Root Subregions.

TABLE 3–4. FLOW OF SELECTED STREAMS IN THE UNITED STATES (continued)

MINNESOTA (continued)

	Gaging Station			Streamflow Characteristics		
	Name	Drainage Area (mi²)	Period of Analysis	7-day, 10-year Low Flow (ft³/s)	Average Discharge (ft³/s)	100-year Flood (ft³/s)
	SOURIS-RED-RAINY REGION RED SUBREGION Red Lake River basin					
15.	Otter Tail River, Orwell Dam Fergus Falls	1,830	1931–83	12.3	304	4,800
16.	Red River of the North, Grand Forks	30,100	1883–1983	71.4	2,558	89,000
17.	Red Lake River, Crookston	5,280	1902–83	31.6	1,130	31,000
	RAINY SUBREGION Little Fork and Big Fork River basins					
18.	Rainy River, Manitou Rapids	19,400	1929–83	3,597	12,830	80,000
19.	Little Fork River, Littlefork	1,730	1912–16 1929–83	40.3	1,053	27,400
20.	Big Fork River, Big Falls	1,460	1929–79 1983	33.7	715	21,800
	GREAT LAKES REGION WESTERN LAKE SUPERIOR SUBREGION					
21.	Pigeon River, Grand Portage	600	1924–83	44.5	506	13,600
22.	Baptism River, Beaver Bay	140	1928–83	3.45	169	8,820
23.	St. Louis River, Scanlon	3,430	1909–83	316	2,313	38,000

MISSISSIPPI

	Name	Drainage Area (mi²)	Period of Analysis	7-day, 10-year Low Flow (ft³/s)	Average Discharge (ft³/s)	100-year Flood (ft³/s)
	LOWER MISSISSIPPI REGION LOWER MISSISSIPPI—YAZOO SUBREGION Yazoo River basin					
1.	Yazoo River, Greenwood[1]	7,450	1907–12, 1927–39	831	9,330	. . . .
			1940–84	741	10,900	45,000
2.	Big Sunflower River, Sunflower[1]	767	1935–84	87	1,070	14,100
	LOWER MISSISSIPPI—BIG BLACK SUBREGION Big Black River basin					
3.	Big Black River, Bovina	2,810	1936–84	84	3,800	73,400
	SOUTH ATLANTIC—GULF REGION PEARL SUBREGION Pearl River basin					
4.	Pearl River, Monticello	4,993	1938–60	324	6,110	. . . .
			1961–84	365	7,530	97,100
	MOBILE—TOMBIGBEE SUBREGION Tombigbee River basin					
5.	Tombigbee River, Columbus	4,463	1899–1912, 1928–82	233	6,520	223,000

[1]Data furnished by U.S. Army Corps of Engineers.

TABLE 3–4. FLOW OF SELECTED STREAMS IN THE UNITED STATES (continued)

MISSISSIPPI (continued)

	Gaging Station			Streamflow Characteristics		
	Name	Drainage Area (mi²)	Period of Analysis	7-day, 10-year Low Flow (ft³/s)	Average Discharge (ft³/s)	100-year Flood (ft³/s)
	PASCAGOULA SUBREGION **Pascagoula River basin**					
6.	Pascagoula River, Merrill	6,590	1930–68 1969–84	865 1,080	9,350 11,800	 221,000

MISSOURI

	Name	Drainage Area (mi²)	Period of Analysis	7-day, 10-year Low Flow (ft³/s)	Average Discharge (ft³/s)	100-year Flood (ft³/s)
	UPPER MISSISSIPPI REGION **UPPER MISSISSIPPI-KASKASKIA-MERAMEC SUBREGION**					
1.	Salt River, New London	2,480	1922–83	1.7	1,700	87,000
2.	Mississippi River, St. Louis	697,000	1951–83	43,000	183,000	1,000,000
3.	Meramec River, Eureka	3,788	1921–83	280	3,100	144,000
4.	Mississippi River, Thebes, Ill.	713,200	1951–83	47,100	198,000	1,100,000
	LOWER MISSISSIPPI REGION **LOWER MISSISSIPPI-ST. FRANCIS SUBREGION** **St. Francis River basin**					
5.	St. Francis River, Patterson	956	1920–83	15	1,100	89,000
6.	Little River, Morehouse	450	1945–83	33	530	11,000
	MISSOURI REGION **GASCONADE-OSAGE AND CHARITON-GRAND SUBREGIONS** **Osage and Grand River basins**					
7.	Missouri River, Kansas City	485,200	1955–83	6,400	51,000	. . . .
8.	Grand River, Gallatin	2,250	1921–83	4.0	1,200	72,000
9.	Osage River, St. Thomas	14,500	1931–83	480	9,900	. . . .
10.	Gasconade River, Jerome	2,840	1923–83	320	2,500	106,000
11.	Missouri River, Hermann	524,200	1955–83	11,000	72,000	. . . .
	ARKANSAS-WHITE-RED REGION **UPPER WHITE SUBREGION** **White River basin**					
12.	James River, Galena	987	1921–83	38	940	69,000
13.	White River, Branson	4,020	1956–83	78	3,500	. . . .
14.	Current River, Doniphan	2,038	1918–83	940	2,700	104,000
15.	Spring River, Waco	1,164	1924–83	18	840	80,000

MONTANA

	Name	Drainage Area (mi²)	Period of Analysis	7-day, 10-year Low Flow (ft³/s)	Average Discharge (ft³/s)	100-year Flood (ft³/s)
	MISSOURI REGION **Missouri River basin**[1]					
1.	Beaverhead River, Barretts	2,737	1907–83	122	430	3,040

[1]Includes the Saskatchewan, the Missouri Headwaters, the Missouri-Marias, the Missouri-Musselshell, the Milk, and the Missouri-Poplar Subregions.

TABLE 3–4. FLOW OF SELECTED STREAMS IN THE UNITED STATES (continued)

MONTANA (continued)

	Gaging Station			Streamflow Characteristics		
	Name	Drainage Area (mi²)	Period of Analysis	7-day, 10-year Low Flow (ft³/s)	Average Discharge (ft³/s)	100-year Flood (ft³/s)
	MISSOURI REGION **Missouri River basin[1] (continued)**					
2.	Missouri River, Fort Benton	24,749	1890–1983	2,230	7,827	96,000
3.	Marias River, Shelby	3,242	1902–04, 1905–06, 1907–08, 1911–83	65	940	....
4.	Musselshell River, Mosby	7,846	1929, 1930–32, 1934–83	0.00	301	34,600
5.	Milk River, Nashua	22,332	1939–83	13	710	36,600
6.	Missouri River, Culbertson	91,557	1941–51, 1958–83	1,520	11,000	54,900
	Yellowstone River basin[2]					
7.	Yellowstone River, Billings	11,795	1928–83	1,090	7,074	80,000
8.	Bighorn River, Bighorn	22,885	1945–83	767	3,939	41,700
9.	Tongue River, Miles City	5,379	1938–42, 1946–83	3.3	440	17,800
10.	Powder River, Locate	13,194	1938–83	1.6	612	48,100
11.	Yellowstone River, Sidney	69,103	1910–31, 1933–83	1,410	13,080	156,000
	PACIFIC NORTHWEST REGION **Clark Fork basin[3]**					
12.	Clark Fork, St. Regis	10,709	1910–83	1,440	7,583	79,800
13.	Bitterroot River, Darby	1,049	1937–83	123	931	13,200
14.	Flathead River, Columbia Falls	4,464	1928–83	1,090	9,737	84,000
15.	Clark Fork, Plains	19,958	1910–83	4,440	20,010	145,000
	Kootenai River basin[3]					
16.	Kootenai River, Libby	10,240	1911–70, 1973–83	1,610 2,560	12,100 11,740	116,000 76,300

NEBRASKA

	Name	Drainage Area (mi²)	Period of Analysis	7-day, 10-year Low Flow (ft³/s)	Average Discharge (ft³/s)	100-year Flood (ft³/s)
	MISSOURI REGION **Missouri River main stem[4]**					
1.	Missouri River, Fort Randall Dam, S.D.	263,500	1947–83	[5]1,450	[5]25,230	[5]99,000
2.	Missouri River, Rulo	414,900	1950–83	[5]6,210	[5]40,190	[5]241,000

[1]Includes the Saskatchewan, the Missouri Headwaters, the Missouri-Marias, the Missouri-Musselshell, the Milk, and the Missouri-Poplar Subregions.
[2]Includes the upper Yellowstone, the Big Horn, the Powder-Tongue, the lower Yellowstone, and the Missouri-Little Missouri Subregions.
[3]Contained within the Kootenai-Pend Oreille-Spokane Subregion.
[4]Within the Missouri-Big Sioux, Missouri-Little Sioux, and Missouri-Nishnabotna Subregions.
[5]Analyses based on period of record since regulation began.

TABLE 3–4. FLOW OF SELECTED STREAMS IN THE UNITED STATES (continued)

NEBRASKA (continued)

	Gaging Station			Streamflow Characteristics		
	Name	Drainage Area (mi^2)	Period of Analysis	7-day, 10-year Low Flow (ft^3/s)	Average Discharge (ft^3/s)	100-year Flood (ft^3/s)
			NIOBRARA SUBREGION			
3.	Niobrara River, Norden	8,390	1953–63	516	952	....
			1964–83	398	810	10,900
			NORTH PLATTE SUBREGION			
4.	Pumpkin Creek, Bridgeport	1,020	1932–83	0.35	28.3	3,320
5.	North Platte River, North Platte	30,900	1896–1940	....	2,720	36,700
			1941–83	135	713	10,700
			SOUTH PLATTE SUBREGION			
6.	South Platte River, North Platte	24,300	1918–46	....	435	77,300
			1947–83	78	402	57,300
			PLATTE SUBREGION			
7.	Platte River, Overton	57,700	1915–40	....	2,860	60,700
			1941–83	46	1,470	32,800
8.	Platte River, Louisville	85,800	1954–83	430	5,980	169,000
			LOUP SUBREGION			
9.	Middle Loup River, Dunning	1,850	1946–83	260	401	1,100
10.	Loup River, Genoa	14,400	1943–83	0.96	574	130,000
			ELKHORN SUBREGION			
11.	Elkhorn River, Waterloo	6,900	1929–83	119	1,120	83,500
			MISSOURI-NISHNABOTNA SUBREGION			
12.	Big Nemaha River, Falls City	1,340	1945–83	11.4	587	80,700
			REPUBLICAN SUBREGION			
13.	Medicine Creek, Harry Strunk Lake	770	1951–83	16.0	65.9	23,700
14.	Republican River, Cambridge	14,520	1950–83	[1]18.0	[1]279	[1]16,800
			KANSAS SUBREGION Blue River basin			
15.	Big Blue River, Barneston	4,447	1933–83	35.1	787	50,100
16.	Little Blue River, Fairbury	2,350	1911–15, 1930–83	48.3	369	48,500

NEVADA

	Name	Drainage Area (mi^2)	Period of Analysis	7-day, 10-year Low Flow (ft^3/s)	Average Discharge (ft^3/s)	100-year Flood (ft^3/s)
			LOWER COLORADO REGION LOWER COLORADO—LAKE MEAD SUBREGION			
1.	Virgin River, Littlefield, Ariz.	5,090	1929–83	48	243	35,300

[1]Analyses based on period of record since regulation began.

TABLE 3-4. FLOW OF SELECTED STREAMS IN THE UNITED STATES (continued)

NEVADA (continued)

	Gaging Station			Streamflow Characteristics		
	Name	Drainage Area (mi^2)	Period of Analysis	7-day, 10-year Low Flow (ft^3/s)	Average Discharge (ft^3/s)	100-year Flood (ft^3/s)
	LOWER COLORADO REGION LOWER COLORADO—LAKE MEAD SUBREGION (continued)					
2.	Muddy River, Moapa	3,820	1913–15, 1916–18, 1928–31, 1944–83	31	41.5	5,000
3.	Lee Canyon, Charleston Park	9.20	1963–83	0	0.025	4,300
4.	Las Vegas Wash, Henderson	2,125	1957–83	....	46.6	6,500
	GREAT BASIN REGION BLACK ROCK DESERT-HUMBOLDT SUBREGION Humboldt River basin					
5.	Humboldt River, Palisade	5,010	1902–06, 1911–83	8.9	385	7,700
6.	Humboldt River, Imlay	15,700	1911–83	0.3	235	5,700
	CENTRAL LAHONTAN SUBREGION Walker Lake basin					
7.	Walker River, Wabuska	2,600	1902–04, 1920–24, 1925–35, 1939–41, 1942–43, 1944–83	3.8	170	6,700
	Carson River basin					
8.	Carson River, Carson City	886	1939–83	4.8	418	28,300
	Truckee River basin					
9.	Truckee River, Nixon	1,827	1957–83	14	538	28,300
	CENTRAL NEVADA DESERT BASINS SUBREGION					
10.	Newark Valley tributary, Hamilton	157	1962–83	0	0.325	1,100
11.	South Twin River, Round Mountain	20	1965–83	0.78	7.06	260

NEW HAMPSHIRE

	Name	Drainage Area (mi^2)	Period of Analysis	7-day, 10-year Low Flow (ft^3/s)	Average Discharge (ft^3/s)	100-year Flood (ft^3/s)
	NEW ENGLAND REGION ANDROSCOGGIN SUBREGION					
1.	Androscoggin River, Gorham	1,361	1913–83	1,280	2,465	20,900
	SACO SUBREGION					
2.	Saco River, Conway	385	1903–09, 1929–83	93	933	53,800
3.	Lamprey River, Newmarket	183	1934–83	4.9	282	6,310

[1]Based on record to 1981.

TABLE 3-4. FLOW OF SELECTED STREAMS IN THE UNITED STATES (continued)

NEW HAMPSHIRE (continued)

	Gaging Station			Streamflow Characteristics		
	Name	Drainage Area (mi^2)	Period of Analysis	7-day, 10-year Low Flow (ft^3/s)	Average Discharge (ft^3/s)	100-year Flood (ft^3/s)
	MERRIMACK SUBREGION					
4.	Pemigewasset River, Plymouth	622	1903–83	115	1,358	60,800
5.	Blackwater River, Webster	129	1918–20, 1927–83	13	213	....
6.	Soucook River, Concord	76.8	1951–83	3.7	112	4,080
7.	Merrimack River, Goffs Falls Manchester	3,092	1936–83	663	5,280	....
	CONNECTICUT SUBREGION					
8.	Connecticut River, Pittsburg	254	1956–83	35	571	....
9.	Upper Ammonoosuc River, Groveton	232	1940–80, 1982–83	49	473	10,800
10.	Ammonoosuc River, Bethlehem Junction	87.6	1939–83	27	208	13,600
11.	Connecticut River, Wells River, Vt.	2,644	1949–83	632	4,731	....
12.	Sugar River, West Claremont	269	1928–83	40	404	13,800
13.	Connecticut River, North Walpole	5,493	1942–83	993	9,380	....
14.	Ashuelot River, Hinsdale	420	1907–11 1914–83	46	671	....

NEW JERSEY

	MID-ATLANTIC REGION LOWER HUDSON-LONG ISLAND SUBREGION Hackensack and Passaic River basins					
1.	Hackensack River, New Milford	113	1922–84	0	103	5,570
2.	Passaic River, Chatham	100	[2]1904–84	3.7	172	3,730
3.	Passaic River, Little Falls	762	1898–1984	32	1,168	22,500
4.	Saddle River, Lodi	54.6	1924–84	13	102	5,750
	Raritan River basin					
5.	South Branch Raritan River, High Bridge	65.3	1919–84	22	123	6,600
6.	Stony Brook, Princeton	44.5	1954–84	0.1	65.1	8,390
7.	Raritan River below Calco Dam Bound Brook	785	[2]1904–84	72	1,293	40,800
8.	Green Brook, Plainfield	9.75	1939–84	0	12.9	3,280
	DELAWARE AND LOWER HUDSON-LONG ISLAND SUBREGIONS Atlantic coastal basins					
9.	Swimming River, Red Bank	49.2	1923–84	0	[3]80.8	11,000

[1]Based on record to 1981.
[2]Period of record not continuous.
[3]Adjusted for diversion and change in reservoir contents.

TABLE 3–4. FLOW OF SELECTED STREAMS IN THE UNITED STATES (continued)

NEW JERSEY (continued)

	Gaging Station			Streamflow Characteristics		
	Name	Drainage Area (mi²)	Period of Analysis	7-day, 10-year Low Flow (ft³/s)	Average Discharge (ft³/s)	100-year Flood (ft³/s)
	DELAWARE AND LOWER HUDSON-LONG ISLAND SUBREGIONS Atlantic coastal basins (continued)					
10.	Manasquan River, Squankum	44.0	1932–84	18	75.9	2,870
11.	Oyster Creek, Brookville	7.43	1966–84	13	28.7	514
12.	Great Egg Harbor River, Folsom	57.1	1926–84	22	86.8	1,230
	DELAWARE SUBREGION Delaware River basin and streams tributary to Delaware Bay					
13.	Maurice River, Norma	112	1933–84	37	168	2,880
14.	Flat Brook, Flatbrookville	64.0	1924–84	7.8	110	7,070
15.	Delaware River, Trenton	6,780	1914–84	[2]1,800	11,740	[2]217,000
16.	Crosswicks Creek, Extonville	81.5	[1]1941–84	24	136	5,800
17.	McDonalds Branch, Lebanon State Forest	2.35	1954–84	0.9	2.32	49
18.	Cooper River, Haddonfield	17.0	1964–84	8.6	36.3	3,840
	NEW MEXICO					
	ARKANSAS-WHITE-RED REGION[3] UPPER CANADIAN SUBREGION Canadian River basin					
1.	Canadian River, Logan	11,141	1904–83	0.0	[4]392	333,000
				....	[5]257	....
				....	[6]39.7	....
	RIO GRANDE REGION UPPER AND LOWER PECOS SUBREGIONS Pecos River basin					
2.	Pecos River, Pecos	189	1919–83	12.0	98.1	3,070
3.	Delaware River, Red Bluff, Texas	689	1912–83	0.0	13.0	82,500
	Rio Grande River basin (main stem)[7]					
4.	Rio Grande, Albuquerque	117,440	1941–83	0.3	[8]1,068	22,000
				[9]1,232	....	....
	UPPER COLORADO REGION SAN JUAN SUBREGION					
5.	San Juan River, Shiprock	12,900	1927–83	53.6	2,181	67,200

[1]Period of record not continuous.
[2]Analysis based on regulated period 1955–84.
[3]Also includes parts of the Upper Arkansas, Upper Cimarron, Lower Canadian, North Canadian, and Red Headwaters Subregions.
[4]Fifteen years, prior to completion of Conchas Dam.
[5]Twenty-four years, prior to completion of Ute Dam.
[6]Twenty-one years (1963–83), subsequent to completion of Ute Dam.
[7]Includes all or parts of Rio Grande Headwaters, Rio Grande-Elephant-Butte, Rio Grande-Mimbres, and Rio Grande Closed Basins Subregions.
[8]Thirty-two years, prior to closure of Cochiti Dam.
[9]Ten years (1974–83), subsequent to closure of Cochiti Dam.

TABLE 3–4. FLOW OF SELECTED STREAMS IN THE UNITED STATES (continued)

NEW MEXICO (continued)

	Gaging Station			Streamflow Characteristics		
	Name	Drainage Area (mi²)	Period of Analysis	7-day, 10-year Low Flow (ft³/s)	Average Discharge (ft³/s)	100-year Flood (ft³/s)
	LOWER COLORADO REGION[1]					
6.	Gila River, Gila	1,864	1927–83	19.7	141	24,900

NEW YORK

	Name	Drainage Area (mi²)	Period of Analysis	7-day, 10-year Low Flow (ft³/s)	Average Discharge (ft³/s)	100-year Flood (ft³/s)
	MID—ATLANTIC REGION UPPER HUDSON SUBREGION					
1.	Sacandaga River, Stewarts Bridge	1,055	1907–29	106	2,230	30,400
			1931–84	757	2,090	14,800
2.	Mohawk River, Cohoes	3,456	1917–84	772	5,750	128,000
3.	Hudson River, Green Island	8,090	1946–84	2,810	13,700	191,000
4.	Wappinger Creek, Wappingers Falls	181	1928–84	6.5	253	18,500
	DELAWARE SUBREGION					
5.	East Branch Delaware River, Fishs Eddy	784	1912–54	89	1,670	73,700
			1955–84	111	1,100	40,100
6.	Delaware River, Port Jervis	3,070	1904–54	416	5,570	184,000
			1963–84	832	4,750	170,000
	SUSQUEHANNA SUBREGION					
7.	Susquehanna River, Waverly	4,773	1937–84	385	7,580	139,000
8.	Chemung River, Chemung	2,506	1903–84	104	2,530	143,000
	GREAT LAKES REGION SOUTHWESTERN AND SOUTHEASTERN LAKE ONTARIO SUBREGIONS					
9.	Genesee River, Rochester	2,467	1919–51	511	2,780	44,100
			1952–84	311	2,880	30,600
10.	Oswego River, Oswego	5,100	1933–84	980	6,690	38,600
	NORTHEASTERN LAKE ONTARIO—LAKE ONTARIO—ST. LAWRENCE SUBREGION					
11.	Black River, Watertown	1,874	1920–84	825	4,020	41,000
12.	West Branch Oswegatchie River, Harrisville	244	1916–84	43	515	7,290
13.	St. Lawrence River, Massena	298,000	1860–1984	179,000	243,000	358,000

NORTH CAROLINA

	Name	Drainage Area (mi²)	Period of Analysis	7-day, 10-year Low Flow (ft³/s)	Average Discharge (ft³/s)	100-year Flood (ft³/s)
	SOUTH ATLANTIC-GULF REGION CHOWAN-ROANOKE SUBREGION					
1.	Roanoke River, Roanoke Rapids	8,386	1911–49	1,010	8,085	215,000
			1950–84	1,310	7,700	66,800
	NEUSE-PAMLICO SUBREGION					
2.	Tar River, Tarboro	2,183	1896–1900 1931–84	90	2,234	45,500
3.	Neuse River, Kinston	2,692	1930–81	210	2,892	43,100
			1981–84	....	....	33,000

[1]Includes parts of the Little Colorado, Upper Gila, and Sonora Subregions.

TABLE 3–4. FLOW OF SELECTED STREAMS IN THE UNITED STATES (continued)

NORTH CAROLINA (continued)

	Gaging Station			Streamflow Characteristics		
	Name	Drainage Area (mi²)	Period of Analysis	7-day, 10-year Low Flow (ft³/s)	Average Discharge (ft³/s)	100-year Flood (ft³/s)
			CAPE FEAR SUBREGION			
4.	Cape Fear River, Lillington	3,464	1923–75 1975–81	75 600	3,300 3,300	117,000 80,000
			PEE DEE SUBREGION			
5.	South Yadkin River, Mocksville	306	1939–84	61	340	15,700
			TENNESSEE REGION **TENNESSEE SUBREGION**			
6.	French Broad River, Asheville	945	1896–1984	455	2,093	49,100

NORTH DAKOTA

	Name	Drainage Area (mi²)	Period of Analysis	7-day, 10-year Low Flow (ft³/s)	Average Discharge (ft³/s)	100-year Flood (ft³/s)
			SOURIS-RED-RAINY REGION **SOURIS AND RED SUBREGIONS** **Souris River and Red River of the North basins**			
1.	Souris River, Minot	[1]10,600 [2]6,700	1904–83	<0.1	171	11,500
2.	Red River of the North, Wahpeton	[1]4,010	1943–83	12.3	519	11,000
3.	Big Coulee, Churchs Ferry	[1]2,510 [2]690	1951–79	0	44.1	3,370
4.	Sheyenne River, West Fargo	[1]8,870 [2]5,780	1904–05 1930–83	12.9	176	5,280
5.	Red River of the North Grand	[1]30,100 [2]3,800	1883–1983	69.1	2,558	89,000
			MISSOURI REGION **MISSOURI-LITTLE MISSOURI AND MISSOURI-OAHE SUBREGIONS** **Missouri River main stem and tributary river basins**			
6.	Little Missouri River, Watford City	[1]8,310	1935–83	<0.1	593	78,700
7.	Knife River, Hazen	[1]2,240	1930–33, 1938–83	2.7	181	36,200
8.	Missouri River, Bismarck	[1]186,400	1921–83	6,570	22,740	63,700
9.	Heart River, Mandan	[1]3,310	1929–32 1938–83	<0.3	268	49,500
10.	Cannonball River, Breien	[1]4,100	1935–83	<0.1	256	60,000
11.	James River, Jamestown	[1]2,820 [2]1,650	1929–34	1.1	62.2	4,800

OHIO

	Name	Drainage Area (mi²)	Period of Analysis	7-day, 10-year Low Flow (ft³/s)	Average Discharge (ft³/s)	100-year Flood (ft³/s)
			OHIO REGION **MUSKINGUM SUBREGION**			
1.	Tuscarawas River, Massillon	518	1937–84	71	441	8,670
2.	Tuscarawas River, Newcomerstown	2,443	1921–37, 1938–84	253	2,541	66,900 23,000
3.	Muskingum River, McConnelsville	7,422	1921–37 1938–84	641	7,596	183,000 97,100

[1]Approximate.
[2]Noncontributing.

TABLE 3–4. FLOW OF SELECTED STREAMS IN THE UNITED STATES (continued)

OHIO (continued)

	Gaging Station			Streamflow Characteristics		
	Name	Drainage Area (mi²)	Period of Analysis	7-day, 10-year Low Flow (ft³/s)	Average Discharge (ft³/s)	100-year Flood (ft³/s)
	SCIOTO SUBREGION					
4.	Scioto River, Prospect	567	1925–32, 1939–84	9.3	454	13,900
5.	Olentangy River, Delaware	393	1923–34, 1938–51, 1951–84	5.2	351	22,000 6,280
6.	Scioto River, Higby	5,131	1930–84	296	4,579	184,000
	GREAT MIAMI SUBREGION					
7.	Great Miami River, Sidney	541	1914–84	21	477	27,300
8.	Stillwater River, Englewood	650	1925–84	15	579	10,500
9.	Mad River, Dayton	635	1914–84	131	629	21,100
10.	Great Miami River, Hamilton	3,630	1907–18, 1927–84	284	3,279	140,000 95,900
	GREAT LAKES REGION WESTERN LAKE ERIE SUBREGION Maumee River basin					
11.	Blanchard River, Findlay	346	1923–35, 1940–84	2.3	251	14,300
12.	Auglaize River, Defiance	2,318	1915–84	11	1,718	62,700
13.	Maumee River, Waterville	6,330	1898–1901, 1921–35, 1939–84	95	4,926	97,800
	SOUTHERN LAKE ERIE SUBREGION Cuyahoga River basin					
14.	Cuyahoga River, Hiram Rapids	151	1927–35, 1944–84	16	207	4,410 3,690
15.	Cuyahoga River, Independence	707	1921–23, 1927–35, 1940–84	63	817	18,000

OKLAHOMA

	Name	Drainage Area (mi²)	Period of Analysis	7-day, 10-year Low Flow (ft³/s)	Average Discharge (ft³/s)	100-year Flood (ft³/s)
	ARKANSAS-WHITE-RED REGION Arkansas River basin, Salt Fork Arkansas River and Cimarron River basin, Verdigris River and Grand (Neosho) River basins, and Canadian River basins[1]					
1.	Salt Fork Arkansas River, Tonkawa	4,528	1942–82	6.68	730	69,400
2.	Cimarron River Perkins	17,852	1940–82	8.73	1,180	174,000
3.	Arkansas River, Tulsa	74,615	1926–64 1965–82	155 346	6,550 6,940	324,000 165,000
4.	Verdigris River, Claremore	6,534	1936–62 1965–82	3.06 15.4	3,720 3,720	178,000 46,400
5.	Illinois River, Tahlequah	959	1936–82	16.8	867	141,000
6.	Little River, Sasakwa	865	1943–65 1966–82	.69 .08	398 242	66,800 23,300
7.	Beaver River, Beaver	7,955	1938–82	.03	95.9	68,100
8.	Fourche Maline, Red Oak	122	1939–63 1966–82	.10 .12	126 133	51,400 17,900

[1]Includes parts or all of the Upper Cimarron, Arkansas-Keystone, Lower Cimarron, Lower Arkansas, Neosho-Verdigris, Lower Canadian and North Canadian Subregions.

TABLE 3–4. FLOW OF SELECTED STREAMS IN THE UNITED STATES (continued)

OKLAHOMA (continued)

	Gaging Station			Streamflow Characteristics		
	Name	Drainage Area (mi²)	Period of Analysis	7-day, 10-year Low Flow (ft³/s)	Average Discharge (ft³/s)	100-year Flood (ft³/s)
	Red River basin[1], Washita River basin					
9.	North Fork Red River, Headrick	4,244	1946–82	0.39	266	54,200
10.	Red River, Gainesville	30,782	1947–82	97.6	2,750	180,000
11.	Washita River, Dickson	7,202	1929–60 1962–82	33.7 4.87	1,540 1,140	117,000 64,200
12.	Muddy Boggy Creek, Farris	1,087	1938–82	.14	880	61,200
13.	Red River, Arthur City, Tex.	44,531	1945–82	375	7,890	174,000
	OREGON					
	PACIFIC NORTHWEST REGION OREGON CLOSED BASINS SUBREGION					
1.	Silvies River, Burns	934	1923–83	1.5	175	4,900
2.	Donner und Blitzen River, Burns	200	1912–13, 1915–16, 1918–21, 1939–83	20	125	4,200
	MIDDLE SNAKE SUBREGION					
3.	Owyhee River, Owyhee Reservoir	11,160	1930–83	1.7	380	....
	MIDDLE COLUMBIA SUBREGION					
4.	Umatilla River, Umatilla	2,290	1928–83	1.3	456	19,700
5.	John Day River, McDonald Ferry	7,580	1906–83	28	2,036	37,800
6.	Deschutes River, Moody	10,500	1897–99, 1907–83	3,610	5,846	....
	WILLAMETTE SUBREGION					
7.	Santiam River, Jefferson	1,790	1909–53 1967–82	323 1,150	7,821 7,821	....
8.	Willamette River, Salem	7,280	1911–41, 1969–82	2,720 5,160	23,650	....
	OREGON-WASHINGTON COASTAL SUBREGION Rogue River basin					
9.	Wilson River, Tillamook	161	1932–83	51	1,205	36,700
10.	Umpqua River, Elkton	3,683	1906–83	797	7,517	276,000
11.	Rogue River, Raygold	2,053	1905–83	870	2,978	139,000
	CALIFORNIA REGION KLAMATH-NORTHERN CALIFORNIA COASTAL SUBREGION Klamath River basin					
12.	Sprague River, Chiloquin	1,580	1922–83	127	584	13,300
13.	Williamson River, Chiloquin	3,000	1918–82	414	1,049	14,100
14.	Klamath River, Keno	3,920	1905–12, 1930–83	165	1,684	13,000

[1]Includes parts or all of the Red Headwaters, Red-Washita, and Red-Sulphur Subregions.

TABLE 3–4. FLOW OF SELECTED STREAMS IN THE UNITED STATES (continued)

PENNSYLVANIA

	Gaging Station			Streamflow Characteristics		
	Name	Drainage Area (mi²)	Period of Analysis	7-day, 10-year Low Flow (ft³/s)	Average Discharge (ft³/s)	100-year Flood (ft³/s)
	MID-ATLANTIC REGION DELAWARE SUBREGION Delaware River main stem					
1.	Bush Kill Shoemakers	117	1908–83	7.6	235	10,800
2.	Delaware River, Belvidere, N.J.	4,535	1922–83	920	7,890	220,000
3.	Delaware River Trenton, N.J.	6,780	1913–83	....	11,685	270,000
	Schuylkill River basin					
4.	Schuylkill River, Landingville	133	1947–83	28	292	14,600
5.	Schuylkill River, Pottstown	1,147	1926–83	260	1,891	74,000
	SUSQUEHANNA SUBREGION Susquehanna River main stem					
6.	Susquehanna River, Towanda	7,797	1913–83	550	10,600	105,000
7.	Susquehanna River, Danville	1,220	1899–1983	980	15,320	260,000
8.	Susquehanna River, Sunbury	18,300	1937–83	1,600	26,520	530,000
9.	Susquehanna River, Harrisburg	24,100	1890–1983	2,556	34,350	750,000
	West Branch Susquehanna River basin					
10.	West Branch Susquehanna River, Lewisburg	6,847	1939–83	655	10,810	280,000
	Juniata River basin					
11.	Juniata River, Newport	3,354	1899–1983	380	4,295	145,000
	POTOMAC SUBREGION					
12.	Tonoloway Creek, Needmore	10.7	1965–83	0.27	12.4	1,590
	OHIO REGION ALLEGHENY SUBREGION					
13.	Allegheny River, Port Allegany	248	1974–83	24	476	9,300
14.	Oil Creek, Rouseville	300	1932–83	29	535	19,800
15.	Allegheny River, Franklin	5,982	1914–83	511	10,470	125,000
	MONONGAHELA SUBREGION					
16.	Monongahela River, Elizabeth	5,340	1933–83	698	9,109	170,000
17.	Monongahela River, Braddock	7,337	1938–83	1,150	12,460	230,000
	UPPER OHIO SUBREGION Ohio River main stem					
18.	Connoquenessing Creek, Zelienople	356	1919–83	11	464	19,450

TABLE 3–4. FLOW OF SELECTED STREAMS IN THE UNITED STATES (continued)

PUERTO RICO

	Gaging Station			Streamflow Characteristics		
	Name	Drainage Area (mi^2)	Period of Analysis	7-day, 10-year Low Flow (ft^3/s)	Average Discharge (ft^3/s)	100-year Flood (ft^3/s)
			CARIBBEAN REGION **PUERTO RICO SUBREGION** **North Coast area**			
1.	Rio Culebrinas, Moca	71.2	1967–85	20.0	299	111,000
2.	Rio Grande de Arecibo, Central Cambalache	[1]200	1969–84	90.0	510	....
3.	Rio Grande de Manati, Manati	[2]197	1970–85	60.0	375	255,000
4.	Rio Cibuco, Vega Baja	[3]99.1	1973–85	[1]8.2	125	45,800
5.	Rio de la Plata, Toa Alta	[4]200	1960–85	7.8	276	202,000
6.	Rio Grande de Loiza, Caguas	89.8	1960–85	14.0	219	131,000
7.	Rio Herrera, Colonia Dolores	2.75	1966–73	[1]1.4	9.47	[1]5,430
8.	Rio Espirutu Santo, Rio Grande	8.62	1966–85	5.0	57.0	22,400
			East Coast area			
9.	Rio Fajardo, Fajardo	14.9	1961–85	3.5	68.9	45,400
			South Coast area			
10.	Rio Grande de Patillas, Patillas	18.3	1966–85	5.4	60.9	40,500
11.	Rio Inabon, Real Abajo	9.70	1964–70 1971–85	1.3	18.6	15,100
12.	Rio Cerrillos, Ponce	17.8	1964–85	3.1	35.8	22,200
13.	Rio Portugues, Ponce	8.82	1964–85	1.5	18.2	21,800
			West Coast area			
14.	Rio Guanajibo, Hormigueros	120	1973–85	[1]5.9	220	[1]160,000
15.	Rio Grande de Anasco, San Sebastian	[5]134	1963–85	38.0	304	83,600

RHODE ISLAND

	Name	Drainage Area (mi^2)	Period of Analysis	7-day, 10-year Low Flow (ft^3/s)	Average Discharge (ft^3/s)	100-year Flood (ft^3/s)
			NEW ENGLAND REGION **MASSACHUSETTS-RHODE ISLAND COASTAL SUBREGION** **Blackstone River basin**			
1.	Branch River, Forestdale	91.2	1941–83	13	171	7,110
2.	Blackstone River, Woonsocket	416	1930–83	100	763	19,200

[1]Estimated.
[2]Drainage area includes 38 mi^2 which are partly or entirely noncontributing and excludes 6.0 mi^2 upstream from Lago El Guineo and Lago de Matrullas.
[3]Drainage area includes 25.4 mi^2 which do not contribute directly to surface runoff.
[4]Drainage area excludes 8.2 mi^2 upstream form Lago Carite, flow from which is diverted to the Rio Guamani.
[5]Drainage area includes 39.7 mi^2 from headwaters of Lago Yahuecas (17.05 mi^2), Lago Guayo (9.67 mi^2), Lago Prieto (9.50 mi^2), and Lago Toro (3.5 mi^2) which does not contribute to surface runoff except at high stages.

TABLE 3–4. FLOW OF SELECTED STREAMS IN THE UNITED STATES (continued)

RHODE ISLAND (continued)

	Gaging Station			Streamflow Characteristics		
	Name	Drainage Area (mi²)	Period of Analysis	7-day, 10-year Low Flow (ft³/s)	Average Discharge (ft³/s)	100-year Flood (ft³/s)
	Pawtuxet River basin					
3.	South Branch Pawtuxet River, Washington	63.8	1942–83	16	130	2,890
4.	Pawtuxet River, Cranston	200	1941–83	73	345	5,220
	Pawcatuck River basin					
5.	Pawcatuck River, Wood River Jct.	100	1942–83	28	194	2,090
6.	Wood River, Hope Valley	72.4	1942–83	20	156	2,630
7.	Pawcatuck River, Westerly	295	1942–83	67	576	6,850

SOUTH CAROLINA

	Name	Drainage Area (mi²)	Period of Analysis	7-day, 10-year Low Flow (ft³/s)	Average Discharge (ft³/s)	100-year Flood (ft³/s)
	SOUTH ATLANTIC-GULF REGION PEE DEE SUBREGION Lower Pee Dee River basin					
1.	Pee Dee River, Peedee	8,830	1938–83	[1]1,500	9,850	[1]160,000
2.	Lynches River, Effingham	1,030	1925–83	132	1,035	22,100
3.	Little Pee Dee River, Galivants Ferry	2,790	1943–83	315	3,243	31,300
4.	Black River, Kingstree	1,252	1920–83	5.7	942	39,100
5.	Waccamaw River, Longs	1,110	1950–83	0.99	1,223	17,300
	EDISTO-SANTEE SUBREGION Santee River basin					
6.	North Pacolet River, Fingerville	116	1931–83	43	215	13,100
7.	Broad River, Richtex	4,850	1925–83	[1]970	6,250	[1]210,000
8.	Saluda River, Columbia	2,520	1925–83	[1]260	2,929	[1]70,000
9.	Wateree River, Camden	5,070	1904–10, 1925–83	[1]490	6,444	[1]225,000
10.	Congaree River, Columbia	7,850	1939–83	[1]1,800	9,425	[1]220,000
11.	Lake Marion-Moultrie Diversion Canal	....	1943–83	2,320	15,125	....
	Edisto-South Carolina Coastal basin					
12.	Edisto River, Givhans	2,730	1939–83	442	2,711	29,200
13.	Salkehatchie River, Miley	341	1951–83	33	356	4,390
	OGEECHEE-SAVANNAH SUBREGION Savannah River basin					
14.	Savannah River, Augusta, Ga.	7,508	1883–1891, 1896–1906, 1925–1983	[1]4,700	10,300	....

[1]Analysis based on records collected since regulation began.

TABLE 3–4. FLOW OF SELECTED STREAMS IN THE UNITED STATES (continued)

SOUTH DAKOTA

	Gaging Station			Streamflow Characteristics			
	Name	Drainage Area (mi²)	Period of Analysis	7-day, 10-year Low Flow (ft³/s)	Average Discharge (ft³/s)	100-year Flood (ft³/s)	
	MISSOURI REGION **Missouri River main stem[1]**						
1.	Missouri River, Mobridge	208,700	[2]1928–62	3,500	21,560	471,000	
2.	Missouri River, Pierre	243,500	1929–65	[3]2,100	21,860	[3]97,500	
3.	Missouri River, Fort Randall Dam	263,500	1947–83	[3]1,450	25,230	[3]99,000	
4.	Missouri River, Yankton	279,500	1930–83	[3]5,980	26,430	[3]92,400	
5.	Missouri River, Sioux City, Iowa	314,600	1929–83	[3]6,380	29,360	[3]115,000	
	Western tributaries[4]						
6.	Little Missouri River, Camp Crook	1,970	[5]1903–83	0.2	136	13,300	
7.	Grand River, Little Eagle	5,370	1958–83	[3]0.3	238	[3]24,400	
8.	Moreau River, Whitehorse	4,880	1954–83	0.0	202	44,900	
9.	Cheyenne River, Cherry Creek	23,900	1960–83	[3]26.1	827	[3]84,600	
10.	Bad River, Fort Pierre	3,107	1928–83	0.0	147	47,000	
11.	White River, Oacoma	10,200	1928–83	0.5	531	49,200	
12.	Keya Paha River, Wewela	1,070	[5]1937–83	3.6	68.9	8,680	
	Eastern tributaries[6]						
13.	James River, Scotland	20,300	1928–83	1.5	372	23,600	
14.	Vermillion River, Wakonda	1,680	1945–83	0.9	125	6,050	
15.	Big Sioux River, Akron, Iowa	8,360	1928–83	18.8	901	73,200	

TENNESSEE

	Name	Drainage Area (mi²)	Period of Analysis	7-day, 10-year Low Flow (ft³/s)	Average Discharge (ft³/s)	100-year Flood (ft³/s)	
	OHIO REGION **CUMBERLAND SUBREGION** **Cumberland basin**						
1.	New River, New River	382	1934–85	0.47	741	63,500	None
2.	Wolf River, Byrdstown	106	1942–85	5.19	192	31,400	....do....
3.	Cumberland River, Celina	7,307	1924–85	850	11,830	78,000	Appreciable
4.	West Fork Stones River, Smyrna	237	1965–85	9.0	440	57,400	None
5.	Harpeth River, Kingston Springs	681	1924–85	25.4	986	69,700	....do....
6.	Red River, Port Royal	935	1961–85	66.6	1,351	18,000	....do....

[1]Within the Missouri-Oahe, Missouri-White, and Missouri-Big Sioux Subregions.
[2]Station discontinued subsequent to construction of Oahe Dam in 1962.
[3]Analysis based on period of record after regulation began.
[4]Within the Missouri-Oahe, Missouri-Little Missouri, Cheyenne, Missouri-White, and Niobrara Subregions.
[5]Period of record not continuous.
[6]Within the James and Missouri-Big Sioux Subregions.

TABLE 3–4. FLOW OF SELECTED STREAMS IN THE UNITED STATES (continued)

TENNESSEE (continued)

	Gaging Station			Streamflow Characteristics		
	Name	Drainage Area (mi^2)	Period of Analysis	7-day, 10-year Low Flow (ft^3/s)	Average Discharge (ft^3/s)	100-year Flood (ft^3/s)
	TENNESSEE REGION UPPER TENNESSEE, MIDDLE TENNESSEE-HIWASSEE, MIDDLE TENNESSEE-ELK, AND LOWER TENNESSEE SUBREGIONS Tennessee basin					
7.	Nolichucky River, Embreeville	805	1919–85	224	1,370	72,600
8.	Little River, Maryville	269	1951–85	54.8	535	37,200
9.	Obed River, Lancing	518	1958–68, 1974–85	1.3	1,062	84,400
10.	South Chickamauga Creek, Chickamauga	428	1928–78, 1980–85	88.3	698	35,100
11.	Tennessee River, Chattanooga	21,400	1874–1985	10,000	37,100	257,000
12.	Elk River, Prospect	1,784	1905–07, 1920–85	330	3,076	128,000
13.	Duck River, Hurricane Mills	2,557	1925–85	303	4,121	114,000
14.	Buffalo River, Lobelville	707	1927–85	174	1,196	88,900
15.	Big Sandy River, Bruceton	205	1929–85	35.5	294	18,900
	LOWER MISSISSIPPI REGION LOWER MISSISSIPPI-HATCHIE SUBREGION Lower Mississippi basin					
16.	Obion River, Obion	1,852	1929–58, 1966–85	266	2,702	92,800
17.	Hatchie River, Bolivar	1,480	1929–85	126	2,428	68,000
18.	Loosahatchie River, Arlington	262	1969–85	71	364	24,000
19.	Wolf River, Germantown	699	1969–85	200	1,040	42,100
	TEXAS					
	ARKANSAS-WHITE-RED REGION Canadian-Red River basin[1]					
1.	Canadian River, Amarillo	15,376	1939–83	0.3	331	135,000
2.	Red River Terral, Okla.	22,787	1939–83	76.4	2,117	....
	TEXAS-GULF REGION Sabine-Neches-Trinity-San Jacinto River basin[2]					
3.	Trinity River, Dallas	6,106	1903–83	20.5	1,530	....
4.	Trinity River, Romayor	17,186	1969–83	64	7,417	....
5.	Neches River, Rockland	3,636	1962–83	27.6	1,974	68,400

[1]Within the Upper Canadian, Lower Canadian, North Canadian, Red Headwaters, Red-Washita, and Red-Sulphur Subregions.
[2]Within the Sabine, Neches, Trinity, and Galveston Bay-San Jacinto Subregions.

TABLE 3–4. FLOW OF SELECTED STREAMS IN THE UNITED STATES (continued)

TEXAS (continued)

	Gaging Station			Streamflow Characteristics		
	Name	Drainage Area (mi²)	Period of Analysis	7-day, 10-year Low Flow (ft³/s)	Average Discharge (ft³/s)	100-year Flood (ft³/s)
			Brazos-Colorado River basin[1]			
6.	Salt Fork Brazos River, Aspermont	2,496	1940–83	0.0	108	52,900
7.	Brazos River, South Bend	13,107	1939–83	0.0	836	....
8.	North Bosque River, Clifton	968	1968–83	0.0	167	73,200
9.	Colorado River, Colorado City	1,585	1953–83	0.0	38.9	....
10.	Llano River, Junction	1,849	1916–83	17.8	194	363,000
11.	Colorado River, Wharton	30,600	1939–83	224	2,685	....
			Lavaca-Guadalupe-Nueces River basin[2]			
12.	Guadalupe River, Spring Branch	1,315	1923–83	0.1	311	158,000
13.	Nueces River, Laguna	737	1924–83	9.6	148	408,000
14.	Nueces River, Three Rivers	15,427	1916–83	0.0	848	116,000
			RIO GRANDE REGION **Rio Grande basin[3]**			
15.	Pecos River, Girvin	29,560	1940–83	3.3	84.2	23,300

U.S. VIRGIN ISLANDS

	Name	Drainage Area (mi²)	Period of Analysis	7-day, 10-year Low Flow (ft³/s)	Average Discharge (ft³/s)	100-year Flood (ft³/s)
			CARIBBEAN REGION **U.S. VIRGIN ISLANDS SUBREGION** **St. Thomas**			
1.	Bonne Resolution Gut, Bonne Resolution	0.49	1963–68, 1979–81, 1982	0	0.24	[4]1,650
2.	Turpentine Run, Mariendal	2.97	1963–69, 1979–80, 1982	0	1.07	[4]9,710
			St. John			
3.	Guinea Gut, Bethany	0.37	1963–67, 1983	0	0.08	[4]946
			St. Croix			
4.	Jolly Hill Gut, Jolly Hill	2.10	1963–69, 1983	0	0.02	[5]223

[1]Within the Brazos Headwaters, Middle Brazos, Lower Brazos, Upper Colorado, and Lower Colorado-San Bernard Coastal Subregions.
[2]Within the Central Texas Coastal and Nueces-Southwestern Texas Coastal Subregions.
[3]Within the Rio Grande-Mimbres, Rio Grande Amistad, Rio Grande Closed Basins, Upper Pecos, Lower Pecos, Rio Grande-Falcon, and Lower Rio Grande Subregions.
[4]Discharge represents highest recorded. Data available are not adequate to determine a discharge-frequency relation, but it is estimated to have exceeded the 100-year flood.
[5]Discharge represents highest recorded.

TABLE 3–4. FLOW OF SELECTED STREAMS IN THE UNITED STATES (continued)

GUAM, AMERICAN SAMOA, AND THE TRUST TERRITORY OF THE PACIFIC ISLANDS

	Gaging Station			Streamflow Characteristics		
	Name	Drainage Area (mi²)	Period of Analysis	7-day, 10-year Low Flow (ft³/s)	Average Discharge (ft³/s)	100-year Flood (ft³/s)
			AIPAN			
1.	S. F. Talofofo Stream	0.64	1968–84	0.0	1.35	....
			GUAM			
2.	Ugum River	5.76	1977–84	3.6	23.3	....
3.	Ylig River	6.48	1952–84	0.2	28.0	5,980
4.	Pago River	5.67	1951–82	0.2	26.3	12,300
			PALAU			
5.	Diongradid River	4.45	1969–84	3.2	32.4	2,870
6.	Tabecheding River	6.07	1970–84	1.7	48.4	4,910
			YAP			
7.	Oaringeel Stream	0.24	1968–84	0.1	1.07	696
			TRUK			
8.	Wichen River	0.57	1968–83	0.02	3.05	1,060
			POHNPEI			
9.	Nanpil River	3.00	1970–84	1.8	44.6	10,000
			KOSRAE			
10.	Malem River	0.76	1971–81, 1982–84	0.3	6.71	2,760
			AMERICAN SAMOA			
11.	Aasu Stream	1.03	1958–84	0.4	6.05	586
12.	Afuelo Stream	0.25	1958–84	0.03	1.45	683
			UTAH			
			UPPER COLORADO REGION Colorado River main stem[1]			
1.	Colorado River, Cisco	[2]24,100	1895–1984	1,100	7,563	87,600
			UPPER COLORADO—DOLORES SUBREGION Dolores River basin			
2.	Dolores River, Cisco	[2]4,580	1951–84	19	785	23,600
			GREAT DIVIDE—UPPER GREEN AND LOWER GREEN SUBREGIONS Green River basin			
3.	Green River, Jensen	[2]29,660	[3]1904–84	480 [4]743	4,396 [4]4,456	38,200

[1]Within the Upper Colorado-Dolores and Upper Colorado-Dirty Devil Subregions.
[2]Approximate.
[3]Period of analysis not continuous.
[4]Since completion of Flaming Gorge Reservoir in 1963.

TABLE 3–4. FLOW OF SELECTED STREAMS IN THE UNITED STATES (continued)

UTAH (continued)

	Gaging Station			Streamflow Characteristics		
	Name	Drainage Area (mi²)	Period of Analysis	7-day, 10-year Low Flow (ft³/s)	Average Discharge (ft³/s)	100-year Flood (ft³/s)
	GREAT DIVIDE—UPPER GREEN AND LOWER GREEN SUBREGIONS Green River basin (continued)					
4.	Green River, Green River	[2]44,850	[1]1895–1984	730 [4]1,214	6,316 [4]5,977	66,600
5.	Duchesne River, Randlette	4,247	1943–84	14	582	12,100
6.	White River, Watson	[2]4,020	[3]1904–79	130	695	9,100
7.	Price River, Woodside	1,540	1946–84	.2	115	10,200
8.	San Rafael River, Green River	1,628	[3]1909–84	0	152	11,400
	UPPER COLORADO—DIRTY DEVIL AND SAN JUAN SUBREGIONS					
9.	Dirty Devil River Hanksville	4,159	1948–84	0	99.1	35,200
10.	San Juan River, Bluff	[2]23,000	1915–84	60	2,542	62,300
	LOWER COLORADO REGION LOWER COLORADO—LAKE MEAD SUBREGION					
11.	Virgin River, Virgin	[3]934	1909–84	38	208	20,400
	GREAT BASIN REGION BEAR AND GREAT SALT LAKE SUBREGIONS					
12.	Bear River, Randolph	1,616	1944–84	5.8	204	3,040
13.	Bear River, Collinston	6,267	1889–1984	11	1,510	11,000
14.	Little Bear River, Paradise	198	1937–84	12	92.8	1,990
15.	Logan River, State Dam, Logan	214	1896–1984	15	273	2,160
16.	Weber River, Oakley	162	1905–84	37	221	3,450
17.	Weber River, Gateway	1,627	[3]1890–1984	58	581	7,370
18.	Ogden River Pineview Dam, Ogden	321	1938–59	.2	86.2	2,600
19.	Jordan River, Lehi	3,010	1913–84	1.0	381	1,240
20.	Spanish Fork, Lakeshore	675	[3]1904–84	0	91.4	2,110
21.	Provo River, Provo	673	[3]1903–84	.5	198	2,320
	ESCALANTE DESERT—SEVIER LAKE SUBREGION Sevier River basin					
22.	Sevier River, Hatch	340	[3]1911–84	36	127	1,760
23.	Sevier River, Sigurd	3,375	1914–84	.4	102	1,660
24.	Sevier River, Lynnyl	5,966	[3]1914–84	7.2	210	2,470
25.	Beaver River, Beaver	91.0	1914–84	12	52.2	1,270

[1]Within the Upper Colorado-Dolores and Upper Colorado-Dirty Devil Subregions.
[2]Approximate.
[3]Period of analysis not continuous.
[4]Since completion of Flaming Gorge Reservoir in 1963.

TABLE 3–4. FLOW OF SELECTED STREAMS IN THE UNITED STATES (continued)

VERMONT

	Gaging Station			Streamflow Characteristics		
	Name	Drainage Area (mi²)	Period of Analysis	7-day, 10-year Low Flow (ft³/s)	Average Discharge (ft³/s)	100-year Flood (ft³/s)
			NEW ENGLAND REGION CONNECTICUT SUBREGION			
1.	Connecticut River, North Stratford, N.H.	799	1930–83	165	1,583	....
2.	Moose River, Victory	75.2	1947–83	5.7	143	4,460
3.	Passumpsic River, Passumpsic	436	1928–83	88	739	17,200
4.	White River, West Hartford	690	1915–83	87	1,184	56,700
5.	Saxtons River, Saxtons River	72.2	1940–82	3.8	120	9,510
6.	Connecticut River, North Walpole, N.H.	5,493	1942–83	993	9,380	....
			ST. FRANCOIS SUBREGION			
7.	Black River, Coventry	122	1951–83	19	201	4,220
			MID-ATLANTIC REGION RICHELIEU SUBREGION			
8.	Otter Creek, Center Rutland	307	1928–83	79	551	13,700
9.	Otter Creek, Middlebury	628	1904–06, 1911–19, 1929–83	157	987	10,600
10.	Dog River, Northfield Falls	76.1	1934–83	3.2	122	11,400
11.	Winooski River, Essex Junction	1,044	1928–83	149	1,706	....
12.	Lamoille River, East Georgia	686	1929–83	158	1,239	23,900
13.	Missisquoi River, East Berkshire	479	1911–83	57	925	21,000
			UPPER HUDSON SUBREGION			
14.	Batten Kill, Arlington	152	1928–80	52	339	9,420

VIRGINIA

	Name	Drainage Area (mi²)	Period of Analysis	7-day, 10-year Low Flow (ft³/s)	Average Discharge (ft³/s)	100-year Flood (ft³/s)
			MID-ATLANTIC REGION POTOMAC SUBREGION			
1.	South Fork Shenandoah River, Front Royal	1,642	1899–1906, 1930–84	243	1,590	143,000
2.	Accotink Creek, Annandale	23.5	1947–84	0.84	27.7	14,500
			LOWER CHESAPEAKE SUBREGION			
3.	Pamunkey River, Hanover	1,081	1941–71	33	915	38,500
			1972–84	70	1,144	....
4.	Cowpasture River, Clifton Forge	461	1925–84	54	525	28,000
5.	James River, Cartersville	6,257	1898–1979	584	7,060	264,000
			1980–84	670	....	....

[1]Based on record to 1981.

TABLE 3–4. FLOW OF SELECTED STREAMS IN THE UNITED STATES (continued)

VIRGINIA (continued)

	Gaging Station			Streamflow Characteristics		
	Name	Drainage Area (mi²)	Period of Analysis	7-day, 10-year Low Flow (ft³/s)	Average Discharge (ft³/s)	100-year Flood (ft³/s)
	SOUTH ATLANTIC-GULF REGION CHOWAN-ROANOKE SUBREGION					
6.	Nottoway River, Stony Creek	579	1929–84	12.8	564	25,700
7.	Blackwater River, Franklin	617	1944–84	1.4	643	11,700
8.	Smith River, Philpott	216	1946–50 1951–84	 58	354 268	....
	OHIO REGION KANAWHA SUBREGION					
9.	New River, Allisonia	2,202	1929–84	725	3,220	131,600
	BIG SANDY-GUYANDOTTE SUBREGION					
10.	Russell Fork, Haysi	286	1926–50 1951–84	0.81 2.3	326 337	41,700 74,300
	TENNESSEE REGION UPPER TENNESSEE SUBREGION					
11.	North Fork Holston River, Saltville	222	1907–08, 1920–84	24	302	20,500
	WASHINGTON					
	PACIFIC NORTHWEST REGION OREGON-WASHINGTON COASTAL SUBREGION Chehalis River basin					
1.	Chehalis River, Grand Mound	895	1930–83	116	2,853	55,800
	Quinault River basin					
2.	Quinault River, Quinault Lake	264	1912–83	350	2,861	54,100
	PUGET SOUND SUBREGION Duwamish River basin					
3.	Green River, Auburn	399	1937–61 1962–83	113 199	1,337 1,377	33,700 12,200
	Skagit River basin					
4.	Skagit River, Mt. Vernon	3,093	1941–83	4,770	16,700	157,000
	UPPER COLUMBIA SUBREGION Colville River basin and Columbia River main stem					
5.	Columbia River, international boundary	59,700	1939–72 1973–83	26,800 36,700	101,300 98,880	571,000 342,000
6.	Colville River, Kettle Falls	1,007	1924–83	19	306	3,720
7.	Columbia River, Grand Coulee Dam	74,700	1939–83	35,100	111,500	616,000
	YAKIMA SUBREGION					
8.	Yakima River, Kiona	5,615	1934–83	704	3,640	52,000

TABLE 3–4. FLOW OF SELECTED STREAMS IN THE UNITED STATES (continued)

WEST VIRGINIA

	Gaging Station			Streamflow Characteristics		
	Name	Drainage Area (mi²)	Period of Analysis	7-day, 10-year Low Flow (ft³/s)	Average Discharge (ft³/s)	100-year Flood (ft³/s)
	MID-ATLANTIC REGION POTOMAC SUBREGION					
1.	Patterson Creek, Headsville	219	1939–83	2.94	167	18,500
2.	South Branch Potomac River, Springfield	1,471	1929–83	70.9	1,296	132,000
3.	Opequon Creek, Martinsburg	272	1948–83	34.1	227	22,900
	OHIO REGION MONONGAHELA SUBREGION					
4.	Tygart Valley River, Colfax	1,366	1940–83	197	2,653	22,900
5.	Cheat River, Rowlesburg	972	1924–83	38.7	2,274	72,200
6.	Big Sandy Creek, Rockville	200	1922–83	2.42	417	22,000
	UPPER OHIO SUBREGION					
7.	Wheeling Creek, Elm Grove	282	1941–83	0.64	336	27,000
8.	Middle Island Creek, Little	458	1929–83	0.51	659	28,000
9.	Little Kanawha River, Palestine	1,515	1940–78 1979–83	4.10 38.0	2,089 2,287	56,200 51,200
	KANAWHA SUBREGION					
10.	Greenbrier River, Alderson	1,364	1896–1983	53.9	1,994	74,500
11.	Kanawha River, Kanawha Falls	8,371	1878–1938 1939–83	1,333 1,818	12,700 12,840	318,000 167,000
12.	Elk River, Queen Shoals	1,145	1929–59 1960–83	6.0 67.2	1,951 2,179	71,800 65,000
13.	Coal River, Tornado	862	1962–83	13.3	1,233	50,700
	BIG SANDY—GUYANDOTTE AND MIDDLE OHIO SUBREGIONS					
14.	Guyandotte River, Baileysville	306	1969–83	33.3	447	50,900
15.	East Fork Twelvepole Creek, Dunlow	38.5	1965–83	0.07	54.4	5,170
16.	Tug Fork, Kermit	1,188	1935–83	40.2	1,414	101,000

WISCONSIN

	Name	Drainage Area (mi²)	Period of Analysis	7-day, 10-year Low Flow (ft³/s)	Average Discharge (ft³/s)	100-year Flood (ft³/s)
	UPPER MISSISSIPPI REGION ST. CROIX SUBREGION					
1.	St. Croix River, St. Croix Falls	6,240	1902–85	1,080	4,235	58,300
	CHIPPEWA SUBREGION					
2.	Chippewa River, Chippewa Falls	5,650	1888–1983	798	5,134	92,000
	UPPER MISSISSIPPI—BLACK—ROOT SUBREGION					
3.	Black River, Neillsville	749	1905–09, 1913–85	6.6	593	39,300

TABLE 3–4. FLOW OF SELECTED STREAMS IN THE UNITED STATES (continued)

WISCONSIN (continued)

	Gaging Station			Streamflow Characteristics		
	Name	Drainage Area (mi^2)	Period of Analysis	7-day, 10-year Low Flow (ft^3/s)	Average Discharge (ft^3/s)	100-year Flood (ft^3/s)
	WISCONSIN SUBREGION					
4.	Wisconsin River, Merrill	2,760	1902–85	775	2,681	31,700
5.	Wisconsin River, Muscoda	10,400	1913–85	2,790	8,662	80,400
	ROCK SUBREGION					
6.	Rock River, Afton	3,340	1914–85	200	1,800	14,700
7.	Pecatonica River, Martintown	1,034	1939–85	170	714	18,600
	GREAT LAKES REGION WESTERN LAKE SUPERIOR SUBREGION					
8.	Bad River, Odanah	597	1914–22, 1948–85	65	620	22,100
	SOUTHWESTERN LAKE MICHIGAN SUBREGION					
9.	Milwaukee River, Milwaukee	696	1914–85	24	411	14,000
	NORTHWESTERN LAKE MICHIGAN SUBREGION					
10.	Peshtigo River, Peshtigo	1,080	1953–83	211	1,240	9,790
11.	Fox River, Rapide Croche Dam, Wrights-town	6,010	1898–1985	950	4,200	27,200
12.	Wolf River, New London	2,260	1896–1985	466	1,740	17,100

WYOMING

	Name	Drainage Area (mi^2)	Period of Analysis	7-day, 10-year Low Flow (ft^3/s)	Average Discharge (ft^3/s)	100-year Flood (ft^3/s)
	MISSOURI REGION UPPER YELLOWSTONE SUBREGION					
1.	Clarks Fork Yellowstone River, Belfry, Mont.	1,154	1921–84	87	953	12,600
	BIGHORN SUBREGION					
2.	Wind River, Riverton	2,309	1906–84	46	876	13,700
3.	Fivemile Creek, Shoshoni	418	1941–42, 1948–83	20	157	4,530
4.	Bighorn River, Kane	15,765	1928–84	329	2,285	31,900
	POWDER-TONGUE SUBREGION					
5.	Piney Creek, Ucross	267	1917–19, 1950–83	2.6	86.9	3,640
	CHEYENNE SUBREGION					
6.	Beaver Creek, Newcastle	1,320	1945–84	0.05	31.3	8,530

TABLE 3–4. FLOW OF SELECTED STREAMS IN THE UNITED STATES (continued)

WYOMING (continued)

	Gaging Station			Streamflow Characteristics		
	Name	Drainage Area (mi²)	Period of Analysis	7-day, 10-year Low Flow (ft³/s)	Average Discharge (ft³/s)	100-year Flood (ft³/s)
			NORTH PLATTE SUBREGION			
7.	Encampment River, Encampment	265	1940–84	17	247	4,600
8.	North Platte River, Wyoming-Nebraska state line	22,218	1929–84	....	2,766	17,700
			UPPER COLORADO REGION GREAT DIVIDE-UPPER GREEN SUBREGION			
9.	Green River, Daniel	468	1932–84	66	511	5,100
10.	Blacks Fork, Millburne	152	1940–84	9.5	163	2,760
			GREAT BASIN REGION BEAR SUBREGION			
11.	Smiths Fork, Border	165	1942–84	50	200	1,680
			PACIFIC NORTHWEST REGION UPPER SNAKE SUBREGION			
12.	Snake River, Reservoir, Alpine	3,465	1937–39, 1954–84	1,031	4,638	32,200

Source: U.S. Geological Survey, National Water Summary 1985—Hydrologic Events and Surface-Water Resources, Water-Supply Paper 2300

FIGURE 3-2. MONTHLY FLOWS OF SELECTED MAJOR RIVERS IN THE UNITED STATES

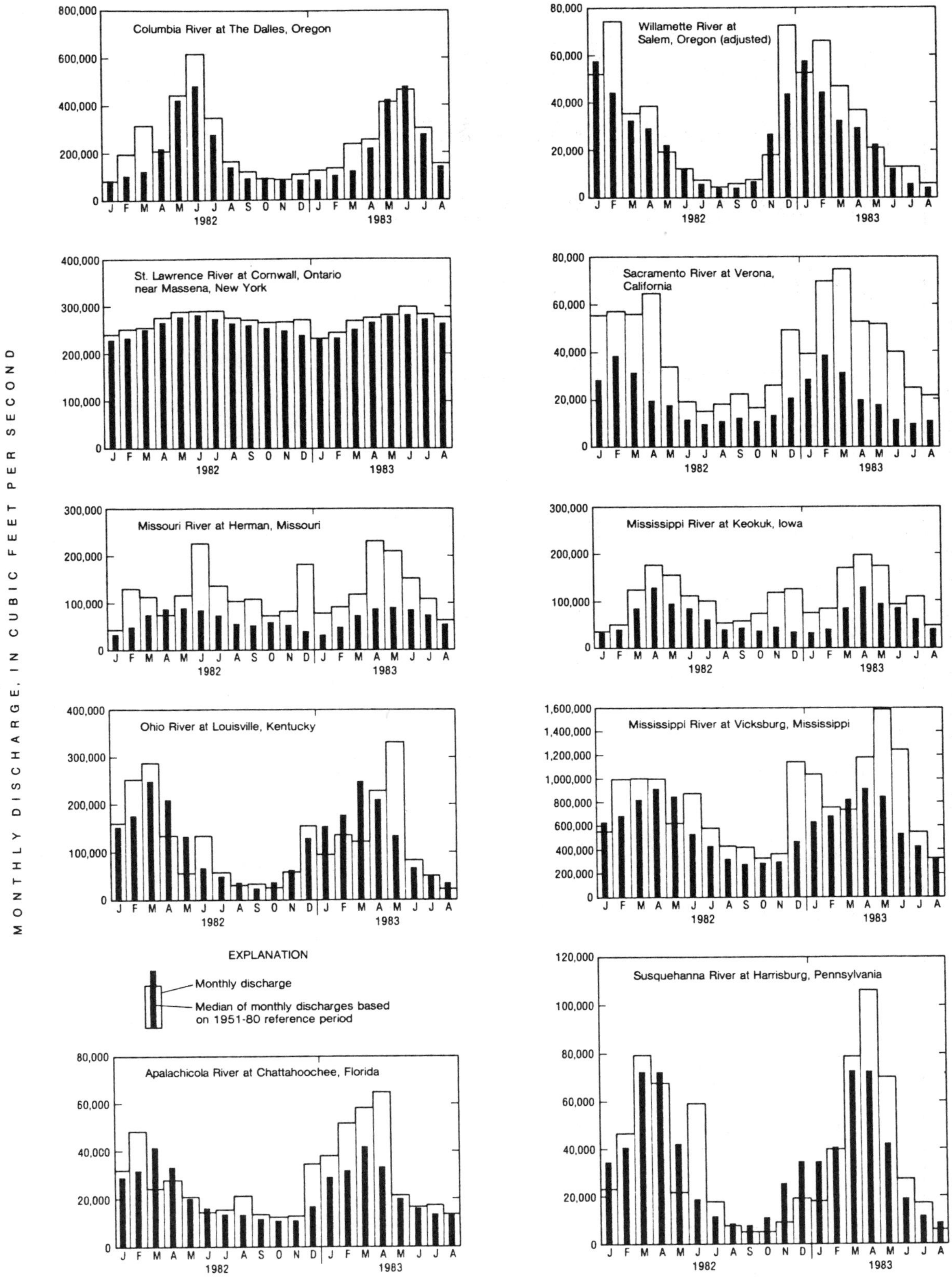

[Diagrams show monthly median flows for the period 1951–80 and monthly flow in the period January 1982-August 1983]

Source: U.S. Geological Survey, 1984, National Water Summary 1983 - Hydrologic Events and Issues, Water-Supply Paper 2250

TABLE 3–5. MONTHLY DISCHARGE OF PRINCIPAL RIVERS IN THE UNITED STATES

River and Station	Basin Area km²	Mean Monthly Discharge, m³/s													Period of Record
		Jan.	Feb.	Mar.	Apr.	May	Jun.	Jul.	Aug.	Sep.	Oct.	Nov.	Dec.	Year	
Penobscot, West Enfield(Me.)	17,000	214	186	290	809	685	345	221	181	176	224	328	282	327	1902–65
Kennebec, Bingham(Me.)	7,040	94.4	99.5	116	202	259	136	94.8	89.6	89.0	88.2	93.3	92.9	121	1907–10
Androscoggin, Auburn(Me.)	8,436	123	111	194	428	336	161	93.5	81.7	93.4	107	149	137	168	1929–65
Merrimack, Lowell(Mass.)	12,000	187	188	348	542	320	165	86.6	71.3	90.8	98.4	171	193	205	1923–65
Connecticut, Thompsonville(Conn.)	25,020	385	341	677	1,300	763	365	194	172	202	237	390	393	451 [1]	1928–65
Delaware, Trenton(N.J.)	17,600	351	353	611	653	387	235	197	170	157	179	292	334	327	1912–65
Susquehanna, Harrisburg(Pa.)	62,400	1,070	1,100	2,210	2,090	1,290	703	422	332	306	457	696	877	963 [1]	1890–1965
Potomac, Washington(D.C.)	29,940	369	461	642	552	379	221	132	143	110	150	161	245	296	1930–65
James, Richmond(Va.)	17,500	260	300	358	316	192	126	83.5	106	85.1	95.3	105	185	184	1934–65
Roanoke, Roanoke Rapids(N.C.)	21,800	304	329	362	315	233	182	175	183	158	155	153	221	230	1911–65
Cape Fear, Lillington(N.C.)	8,910	143	181	169	140	66.6	49.7	64.9	61.3	72.7	59.0	57.6	87.7	95.9	1923–65
Pee Dee, Pee Dee(S.C.)	22,900	316	431	471	396	241	179	190	183	210	182	171	224	265	1938–65
Santee, Pineville(S.C.)	38,100	58.6	123	161	139	47.8	24.8	22.2	18.8	63.6	72.9	31.6	43.0	67.0	1942–65
Savannah, Clyo(Ga.)	25,500	418	438	519	520	338	240	244	253	222	277	249	313	336	1929–33;37–65
Altamaha, Doctortown(Ga.)	35,200	477	636	795	782	388	239	231	214	177	177	162	295	385	1931–65
St. Johns, De Land(Fla.)	8,080	78.8	69.0	73.4	70.8	46.2	46.8	80.2	100	124	168	144	103	92.2	1933–65
Suwannee, Branford(Fla.)	20,000	169	203	287	316	207	145	148	180	185	175	141	142	191	1931–65
Apalachicola,Chattahoochee(Fla.)	44,300	752	876	1,150	1,040	652	465	477	429	341	361	359	550	620 [1]	1928–65
Escambia, Century(Fla.)	9,886	223	246	339	337	166	108	120	123	95.6	74.6	90.5	160	173	1934–65
Alabama, Claiborne(Ala.)	57,000	1,280	1,550	1,830	1,770	899	531	519	484	369	367	433	834	902 [1]	1930–65
Tombigbee, Leroy(Ala.)	49,500	1,190	1,580	1,750	1,530	714	318	330	208	175	162	347	661	742 [1]	1928–60
Pascagoula, Merrill(Miss.)	17,000	381	492	561	515	279	140	170	124	100	78.5	133	269	269	1930–65
Pearl, Bogalusa(La.)	17,200	332	483	554	502	298	138	144	103	74.8	70.0	101	222	250	1938–65
Ohio, Louisville(Ky.)	236,100	4,880	5,470	7,210	6,020	3,730	2,280	1,610	1,190	743	738	1,500	2,780	3,120 [1]	1928–65
Wabash, Mount Carmel(Ill.)	74,100	1,130	1,110	1,370	1,380	1,130	742	528	270	194	217	332	537	744 [1]	1928–65
Cumberland, Smithland(Ky.)	46,395	1,210	1,580	1,740	1,390	718	446	348	272	217	188	383	809	779	1939–65
Tennessee, Paducah(Ky.)	104,000	2,890	3,590	3,160	2,100	1,520	1,170	1,140	1,080	1,050	987	1,370	2,130	1,820 [1]	1939–65
Ohio, Metropolis(Ill.)	526,000	10,900	13,400	14,900	13,400	8,650	5,480	4,070	3,040	2,300	2,180	3,640	6,200	7,300 [1]	1928–65
Fox, Wrightstown(Wis.)	15,900	109	112	132	199	171	143	95.4	76.2	73.1	86.4	99.8	105	117	1896–1965
Grand, Grand Rapids(Mich.)	12,700	90.7	111	213	185	128	88.9	54.4	41.0	46.0	54.0	64.8	70.9	95.2	1904–65
Maumee, Waterville(Ohio)	16,350	197	203	317	268	157	92.4	47.4	22.7	22.4	31.5	58.7	108	127	1899–1901; 1921–35;39–65
St. Lawrence, Ogdensburg(N.Y.)	764,600	6,230	6,150	6,430	6,970	7,230	7,340	7,290	7,100	6,860	6,630	6,510	6,450	6,760	1860–1965
Red of the North,Grand Forks(N.D.)	78,000	19.5	18.0	48.1	238	138	108	80.3	44.1	36.0	35.2	31.4	24.8	68.5	1882–1965
Mississippi,Clinton(Iowa)	222,000	677	730	1,330	2,420	2,280	1,990	1,560	1,030	1,040	1,100	1,030	708	1,330 [1]	1874–1965
Mississippi, Alton(Ill.)	444,200	1,790	2,120	3,540	4,870	4,290	3,680	2,840	1,740	1,650	1,660	1,790	1,530	2,620 [1]	1927–65
Missouri,Culbertson(Mont.)	237,130	187	171	212	259	205	184	236	363	389	377	249	197	252	1941–51;58–65
Yellowstone, Sidney(Mont.)	178,220	143	173	307	310	527	1,200	690	257	197	228	197	153	365 [1]	1910–31;33–65
Missouri, Yankton(S.D.)	723,900	267	299	600	1,120	829	1,250	1,060	737	696	663	485	263	689 [1]	1930–65

Platte,South Bend(Nebr.)	221,000	84.7	148	249	233	218	267	134	88.6	93.7	87.6	109	93.4	150	1953–65
Missouri,Nebraska City(Nebr.)	1,073,000	374	532	1,060	1,540	1,200	1,690	1,360	918	848	807	674	384	948[1]	1929–65
Kansas,Bonner Springs(Kans.)	155,100	67.0	106	161	221	261	426	332	169	180	136	99.1	67.5	185	1917–65
Missouri, Herman(Missouri)	1,368,000	1,080	1,420	2,430	3,360	3,020	4,160	3,390	1,930	1,820	1,600	1,440	1,050	2,220[1]	1897–1965
White, DeValls Bluff(Ark.)	60,686	850	943	1,140	1,270	1,230	829	505	339	291	305	402	542	720[1]	1928–45;50–65
Arkansas,Tulsa(Okla.)	193,250	73.6	90.9	118	261	375	368	277	151	164	185	116	82.6	188	1925–65
Canadian, Whitefield(Okla.)	123,220	61.5	102	147	255	430	291	164	69.0	88.6	117	87.0	76.9	157	1938–65
Arkansas, Little Rock(Ark.)	409,741	987	1,160	1,330	1,810	2,320	1,780	1,070	545	555	762	695	721	1,140[1]	1927–65
Mississippi, Vicksburg(Miss.)	2,964,300	16,200	21,100	24,800	27,900	23,600	18,500	14,600	9,340	7,320	7,090	8,220	11,100	15,800[1]	1928–65
Red, Alexandria(La.)	175,000	1,100	1,330	1,380	1,350	1,600	1,150	555	305	260	342	432	673	880[1]	1928–65
Ouachita, Monroe(La.)	39,622	627	879	1,010	1,039	940	524	272	103	90.1	109	174	316	507[1]	1932–65
Mississippi, Tarbert Landing	3,923,800	17,900	23,000	27,800	30,800	27,900	21,400	16,500	10,500	8,100	7,930	8,750	12,100	17,700[1]	1928–65
Sabine, Ruliff(Tex.)	24,160	399	416	406	366	381	252	120	82.9	57.5	55.5	116	221	239	1924–65
Neches, Evadale(Tex.)	20,590	273	297	308	308	325	188	73.4	37.3	28.2	36.6	81.1	164	176	1904–06;21–65
Trinity, Romayor(Tex.)	44,512	247	262	291	278	439	295	118	41.4	53.6	86.4	122	185	202	1924–65
Brazos, Richmond(Tex.)	114,000	200	234	231	257	459	303	145	68.9	99.2	153	140	188	207	1903–06;22–65
Colorado, Wharton(Tex.)	107,200	67.6	79.5	67.5	101	133	131	82.8	47.0	72.2	77.2	80.6	69.8	80.7	1919–25.38–65
Nueces, Mathis(Tex.)	43,150	8.6	12.0	11.4	15.0	43.5	38.9	29.5	6.9	42.7	38.2	10.8	4.2	21.9	1939–65
Pecos, Shumla(Tex.)	91,069	5.42	5.28	4.82	6.57	13.4	9.84	7.23	5.28	16.6	12.7	6.31	5.44	8.24	1954–65
Rio Grande, Laredo(Tex.)	352,178	73.7	72.2	61.9	69.7	121	158	136	143	269	188	96.8	82.5	123	1900–14;22–65
Green, Green River(Utah)	105,000	49.3	64.8	124	222	485	600	250	108	76.8	77.4	68.5	50.6	181	1895–99;1905–65
Colorado, Lees Ferry(Ariz.)	279,500	150	192	274	561	1,280	1,600	657	313	236	245	210	163	489[1]	1911–65
Colorado, Yuma(Ariz.)	629,100	222	262	293	367	665	1,040	586	306	230	214	200	213	383	1902–65
Sevier, Juab(Utah)	13,300	0.67	0.35	1.28	6.86	20.4	13.6	14.2	9.13	5.98	2.36	1.05	0.38	6.36	1911–65
Humboldt, Imlay(Nev.)	40,700	2.09	3.53	6.60	9.83	14.2	13.9	10.8	2.54	1.05	0.71	0.92	1.71	5.66	1935–41;45–65
San Joaquin, Vernalis(Calif.)	35,070	126	181	187	199	239	212	64.9	28.1	33.8	46.8	58.6	99.2	124	1922–65
Sacramento,Sacramento(Calif.)	60,940	940	1,180	1,020	960	850	501	299	289	323	320	419	681	647[1]	1948–65
Eel, Scotia(Calif.)	8,063	508	566	349	262	108	35.3	9.42	4.08	3.57	22.1	121	370	200	1910–65
Klamath, Klamath(Calif.)	31,300	806	1,050	749	793	648	381	162	94.2	93.6	148	348	629	486	1910–26;50–65
Chehalis, Porter(Wash.)	3,351	279	251	182	131	64.5	31.0	15.8	10.9	12.4	38.9	176	239	119	1952–65
Pend Oreille,Newport(Wash.)	62,700	377	382	439	723	1,540	2,060	1,210	505	339	340	402	391	726[1]	1903–41;52–65
Columbia,International Bdry.	155,000	1,080	1,110	1,200	1,930	5,040	8,140	5,920	3,130	1,940	1,600	1,430	1,190	2,810	1937–65
Snake, Clarkston(Wash.)	267,300	866	1,010	1,320	2,300	3,540	3,140	1,130	555	549	666	777	887	1,390[1]	1915–65
Columbia, Dalles(Ore.)	614,000	2,750	3,020	3,590	5,700	10,300	14,000	9,630	5,270	3,470	2,810	2,800	2,820	5,520[1]	1878–1965
Willamette, Salem(Ore.)	18,900	1,300	1,250	971	824	624	412	205	134	147	253	772	1,130	665[1]	1909–16;23–65
Cowlitz, Castle Rock(Wash.)	5,796	371	354	297	323	345	291	150	76.3	65.2	122	302	415	259	1927–65
Umpqua,Elkton(Ore.)	9,539	443	450	350	278	191	113	50.9	33.6	33.4	55.8	194	361	212	1905–65
Rogue, Agness(Ore.)	10,200	312	352	260	229	181	94.2	46.7	37.1	35.9	64.1	154	427	182	1960–65
Copper, Chitina(Alaska)	53,300	160	130	124	147	848	2,320	3,250	2,920	1,460	615	303	200	1,040[1]	1955–65
Kuskokwim,Crooked Creek(Alaska)	80,500	385	328	284	315	2,080	2,740	2,000	2,420	2,470	1,300	599	457	1,280	1951–65
Yukon, Eagle(Alaska)	294,000	475	433	394	408	3,220	6,160	5,130	3,980	2,980	1,930	980	598	2,220[1]	1911–14;50–65
Yukon, Rampart(Alaska)	516,400	674	591	502	517	5,420	11,500	7,950	6,480	5,090	2,780	1,220	799	3,630[1]	1955–65
Yukon, Kaltag(Alaska)	767,000	1,320	1,100	915	930	7,980	18,300	13,100	11,800	9,820	5,400	2,350	1,490	6,210[1]	1956–65

[1] Monthly and yearly averages rounded to three significant figures.
Source: UNESCO, 1971

TABLE 3–6. FLOWING WATER RESOURCES OF THE UNITED STATES

[Based on stream order and channel morphology]

Stream Order[1]	Number Streams	Average Length Miles	Total Length Miles (L)	Drainage Area Sq. Miles (A_d)	Mean Flow For Area Drained (CFS)	Mean Width Feet (W)	Mean Depth Feet (D)	Mean Velocity Feet/Sec. (V)	Calculated Discharge CFS = WDV	Total Surface Area, A_s Sq. Miles (thousands)	Total Channel Storage Acre Feet (millions)
1	1,570,000	1	1,570,000	1	0.65	4	0.15	1.0	0.60	1.2	0.11
2	350,000	2.1	810,000	4.7	3.1	10	0.29	1.3	3.7	1.5	0.29
3	80,000	5.3	420,000	23	15.0	18	0.58	1.5	15.6	1.4	0.53
4	18,000	12	220,000	109	71.0	37	1.10	1.8	73	1.5	1.1
5	4,200	28	116,000	518	340	75	2.20	2.3	380	1.6	2.4
6	950	64	61,000	2,500	1,600	160	4.1	2.7	1,800	1.8	4.9
7	200	147	30,000	12,000	7,600	320	8.0	3.3	8,500	1.8	9.3
8	41	338	14,000	56,000	36,000	650	15.0	3.9	38,000	1.7	16.5
9	8	777	6,200	260,000	171,000	1,300	29.0	5.6	211,000	1.5	28.3
10	1	1,800	1,800	1,250,000	810,000	2,800	55.0	5.9	900,000	1.0	34.3
Total	2,023,000		3,249,000							15.0	97.0

[1] Stream order classification based on river characteristics. A first order stream has no tributary channels; a second order stream is formed when two first order streams merge. When two second order streams merge, a third order stream is formed, and so on, downstream in the drainage basin until the water is discharged to the sea.

Source: Keup, L.E., 1985, Flowing Water Resources, Water Resources Bulletin, v.21, no.2. Reprinted with permission

FIGURE 3–3. INADEQUATE SURFACE WATER SUPPLY FOR INSTREAM USE IN THE UNITED STATES

(Data as of 1975)

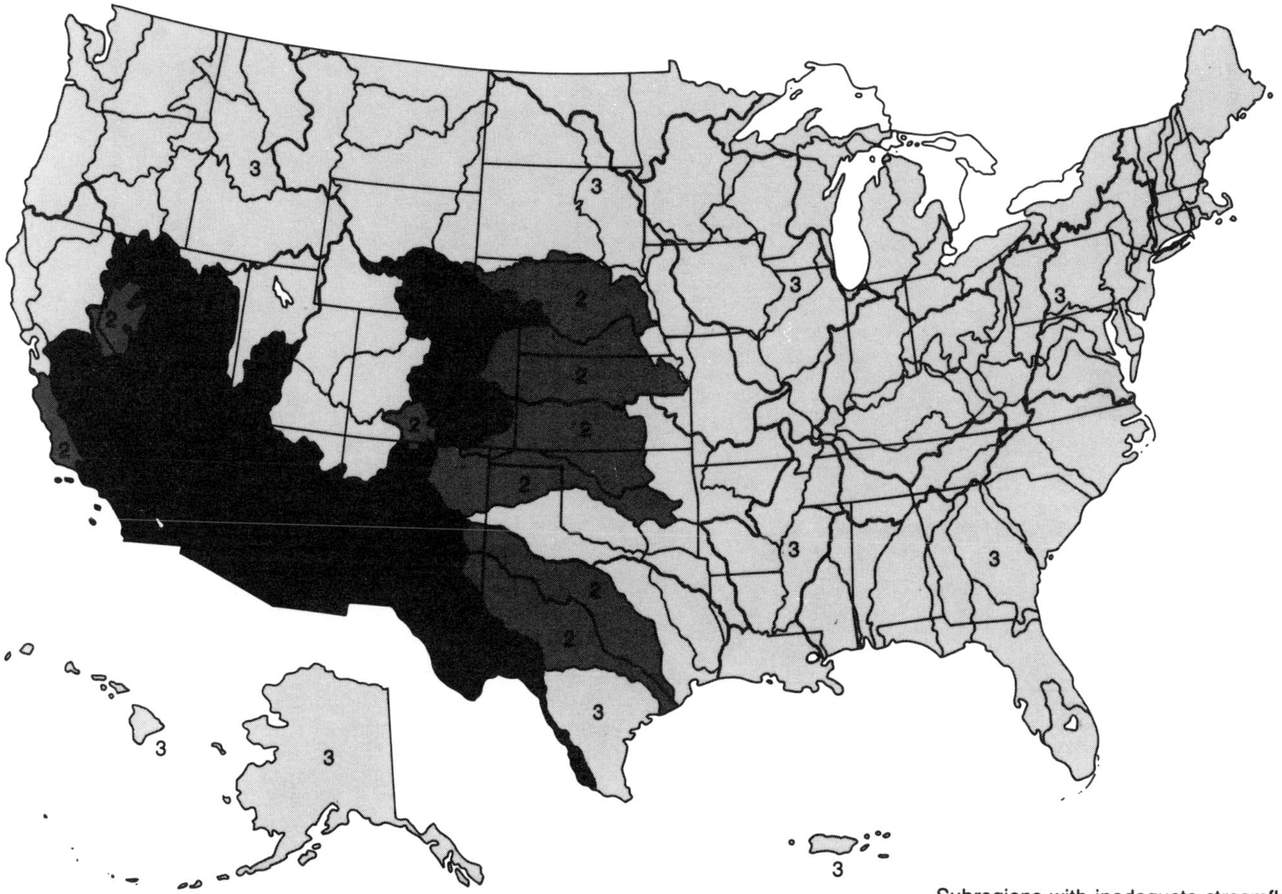

Subregions with inadequate streamflow

1	70% or more depleted in average year
2	70% or more depleted in dry year
3	Less than 70 % depleted

Increased demand is causing significant competition among major users of water. Streamflow in 14 of the 106 subregions is inadequate to support navigation, hydropower, recreation, fish, wildlife, and other instream uses in an average year. Inadequate means that 70% or more of the water is consumed offstream during a given year. In a dry year, nine more subregions are in the 70% or more depletion category.

Source: Council on Environmental Quality, 1981, Environmental Trends

TABLE 3–7. VELOCITY OF LOW FLOWS AND AVERAGE LENGTH OF STREAMS IN THE UNITED STATES

[Velocities are estimated for discharges which are exceeded 95 percent of the time. Stream lengths are estimated for representative streams in each region. For location of river basins see Figure 2–6]

Water Resource Region	Mean Velocity (Miles per Hour)	Average Stream Length (Miles)	Mean Depth at Velocity Given in Col. 1 (Feet)	Mean Flow at Velocity Given in Col. 1 (Cubic Feet per Second)
New England	1 1/2	100	4.0	1,500
Delaware-Hudson	1 1/2	75	3.0	500
Eastern Great Lakes	1 1/2	50	3.0	800
Western Great Lakes	1 1/2	50	3.0	800
Chesapeake Bay	1 1/2	75	5.0	2,500
Ohio	1	100	4.0	1,500
Cumberland	1 1/2	100	3.0	500
Tennessee	1 1/2	75	4.0	1,200
Southeast	1 1/2	150	5.0	2,500
Upper Mississippi	1	150	3.0	500
Lower Mississippi	1 1/2	150	4.0	1,200
Upper Missouri	1 1/2	250	3.0	500
Lower Missouri	1	125	3.5	900
Upper Arkansas-Red	1 1/2	200	2.0	200
Lower Arkansas, Red, and White	1 1/2	175	3.5	1,000
Western Gulf	1	300	4.0	1,500
Rio Grande and Pecos	1 1/2	150	1.5	100
Colorado	1	150	2.0	300
Great Basin	1	100	1.0	50
Pacific Northwest	1	150	4.0	1,500
Central Pacific	1 1/2	100	4.0	1,500
South Pacific	1	50	.5	20

Source: U.S. Geological Survey

TABLE 3–8. ANNUAL RIVER FLOW RATES IN CANADA

[For map of river basin regions see Figure 2–8]

Ocean Basin Region		River Basin Region	Annual Flow Rates[a] (Cubic Meters per Second) Reliable[b] (Low)	Mean	High[c]
Pacific	1.	Pacific Coastal	12,570	16,390	20,200
	2.	Fraser-Lower Mainland[d]	3,044	3,972	4,900
	3.	Okanagan-Similkameen[e]	31	74	116
	4.	Columbia[e]	1,644	2,009	2,373
	5.	Yukon[e]	1,806	2,506	3,206
Arctic	6.	Peace-Athabasca	1,862	2,903	3,946
	7.	Lower Mackenzie[f]	6,114	7,337	8,561
	8.	Arctic Coast-Islands	5,920	10,251	14,582
Gulf of Mexico	9.	Missouri[e]	3	12	41
Hudson Bay	10.	North Saskatchewan	160	234	373
	11.	South Saskatchewan	147	239	418
	12.	Assiniboine-Red[e]	16	50	188
	13.	Winnipeg[de]	382	758	1,137
	14.	Lower Saskatchewan-Nelson[ef]	1,108	1,911	2,714
	15.	Churchill[d]	323	701	1,070
	16.	Keewatin	2,945	3,876	4,806
	17.	Northern Ontario[d]	3,733	5,995	8,258
	18.	Northern Quebec[d]	12,820	16,830	20,830
Atlantic	19.	Great Lakes	2,403	3,067	3,733
	20.	Ottawa	1,390	1,990	2,590
	21.	St. Lawrence[ef]	1,504	2,140	2,777
	22.	North Shore-Gaspé	6,437	8,706	10,980
	23.	St. John-St. Croix[e]	507	779	1,050
	24.	Maritime Coastal	2,079	3,081	4,085
	25.	Newfoundland-Labrador	6,908	9,324	11,739
CANADA			75,856	105,135	134,674

Note:

[a] From recorded flows except in Prairie basins where natural flows have been estimated.
[b] Flow equalled or exceeded in 19 years out of 20.
[c] Flow equalled or exceeded in 1 year out of 20.
[d] Excludes flow transferred into neighboring basin region; because this flow is recorded in importing basin, transfers have little effect on national total.
[e] Excludes inflow from United States portion of basin region.
[f] Excludes inflow from upper basin region.

Source: Pearse, P.H., 1985, Currents of Change, Final Report Inquiry on Federal Water Policy, Ottawa, Canada

TABLE 3–9. LONGEST RIVERS OF THE WORLD

Name	Outflow	Length* miles	kilometers	Rank in World (first 100)
World				
Nile	Mediterranean Sea	4,132	6,650	1
Amazon	South Atlantic Ocean	4,000	6,400	2
Yangtze	East China Sea	3,915	6,300	3
Mississippi	Gulf of Mexico	3,710	5,971	4
Yenisey	Kara Sea	3,442	5,540	5
Huang Ho (Yellow)	Gulf of Chihli	3,395	5,464	6
Ob	Gulf of Ob	3,362	5,410	7
Paraná	Río de la Plata	3,032	4,880	8
Congo	South Atlantic Ocean	2,900	4,700	9
Amur	Sea of Okhotsk	2,761	4,444	10
Africa				
Nile	Mediterranean Sea	4,132	6,650	1
Congo	South Atlantic Ocean	2,900	4,700	9
Niger	Bight of Biafra	2,600	4,200	16
Zambezi	Mozambique Channel	2,200	3,500	24
Kasai	Congo River	1,338	2,153	63
Orange	South Atlantic Ocean	1,300	2,100	66
White Nile (al-Bahr al-Abyad)	Nile River	1,295	2,084	67
Lualaba	Congo River	1,100	1,800	87
Limpopo	Mozambique Channel	1,100	1,800	87
Jubba (Juba)	Indian Ocean	1,030	1,658	94
Sénégal	South Atlantic Ocean	1,020	1,641	95
Okavango (Kubango)	Okavango Swamp	1,000	1,600	100
Lomami	Congo River	930	1,500	
Blue Nile (al-Bahr al-Azraq)	White Nile River	907	1,460	
Chari (Shari)	Lake Chad	870	1,400	
Ubangi-Uele	Congo River	870	1,400	
Awash	Lake Abe	750	1,200	

*Conversions of rounded figures are rounded to nearest hundred miles or kilometres.

TABLE 3–9. LONGEST RIVERS OF THE WORLD (continued)

Name	Outflow	Length* miles	Length* kilo-meters	Rank in World (first 100)
America, North				
Mississippi-Missouri-Red Rock	Gulf of Mexico	3,710	5,971	4
Mackenzie-Slave-Peace	Beaufort Sea	2,635	4,241	15
Missouri-Red Rock	Mississippi River	2,533	4,076	17
St. Lawrence-Great Lakes	Gulf of Saint Lawrence	2,500	4,000	19
Mississippi	Gulf of Mexico	2,348	3,779	20
Missouri	Mississippi River	2,315	3,726	21
Yukon-Nisutlin	Bering Sea	1,979	3,185	28
Rio Grande	Gulf of Mexico	1,885	3,034	29
Yukon	Bering Sea	1,875	3,018	31
Nelson-Saskatchewan	Hudson Bay	1,600	2,575	46
Arkansas	Mississippi River	1,459	2,348	55
Colorado	Gulf of California	1,450	2,333	56
Ohio-Allegheny	Mississippi River	1,306	2,102	65
Red	Mississippi River	1,270	2,044	68
Columbia	North Pacific Ocean	1,243	2,000	71
Saskatchewan	Lake Winnipeg	1,205	1,939	76
Peace	Slave River	1,195	1,923	79
Snake	Columbia River	1,038	1,670	93
Churchill	Hudson Bay	1,000	1,609	99
Ohio	Mississippi River	981	1,579	
Canadian	Arkansas River	906	1,456	
Tennessee-French Broad	Ohio River	900	1,448	
Upper Columbia	Columbia River	890	1,432	
Brazos	Gulf of Mexico	870	1,400	
South Saskatchewan	Saskatchewan River	865	1,392	
Fraser	Strait of Georgia	850	1,368	
Colorado (of Texas)	Matagorda Bay	840	1,352	
St. Lawrence	Lake Ontario	800	1,287	
North Saskatchewan	Saskatchewan River	800	1,287	
Ottawa	St. Lawrence	790	1,271	
North Canadian	Canadian River	760	1,223	
Pecos	Rio Grande River	735	1,183	
Kuskokwim	Bering Sea	680	1,094	
America, South				
Amazon-Ucayali-Apurimac	South Atlantic Ocean	4,000	6,400	2
Paraná	Río de la Plata	3,032	4,880	8
Madeira-Mamoré-Guaporé	Amazon River	2,082	3,350	25
Jurua	Amazon River	2,040	3,283	26
Purus	Amazon River	1,995	3,211	27
São Francisco	South Atlantic Ocean	1,811	2,914	33
Japurá (Caquetá)	Amazon River	1,750	2,816	37
Ucayali-Apurimac	Amazon River	1,701	2,738	40
Orinoco	South Atlantic Ocean	1,700	2,736	41
Tocantins	Pará River	1,677	2,699	42
Araguaia	Tocantins River	1,632	2,627	44
Paraguay	Paraná River	1,584	2,550	47
Pilcomayo	Paraguay River	1,550	2,500	51
Negro (Guainia)	Amazon River	1,400	2,253	59
Xingu	Amazon River	1,300	2,100	66
Tapajos-Teles Pires	Amazon River	1,238	1,992	72
Mamoré	Guaporé River	1,200	1,931	77
Marañón-Huallaga	Amazon River	1,184	1,905	80
Guaporé (Iténez)	Mamore River	1,087	1,749	90
Parnaiba	South Atlantic Ocean	1,056	1,700	92
Madre de Dios	Beni River	1,056	1,700	92
Putumayo (Iça)	Amazon River	1,000	1,609	99
Solimões	Amazon River	1,000	1,609	99
Uruguay	Río de la Plata	990	1,593	
Magdalena	Caribbean Sea	930	1,497	
Guaviare	Orinoco River	930	1,497	
Ucayali	Marañón River	910	1,465	
Teles Pires	Tapajós River	870	1,400	
Grande	Mamoré River	845	1,360	
Cauca	Magdalena River	838	1,349	
Iguaçu	Paraná River	808	1,300	
Asia				
Yangtze	East China Sea	3,915	6,300	3
Yenisey-Baikal-Selenga	Kara Sea	3,442	5,540	5
Huang Ho (Yellow)	Gulf of Chihli	3,395	5,464	6
Ob-Irtysh	Gulf of Ob	3,362	5,410	7
Amur-Argun	Sea of Okhotsk	2,761	4,444	10
Lena	Laptev Sea	2,734	4,400	11
Mekong	South China Sea	2,700	4,350	12
Ob-Katun	Gulf of Ob	2,696	4,338	13
Irtysh-Chorny Irtysh	Ob River	2,640	4,248	14
Yenisey	Kara Sea	2,549	4,102	18
Ob	Gulf of Ob	2,268	3,650	22
Syrdarya-Arabelsu	Aral Sea	1,876	3,019	30
Nizhnyaya Tunguska	Yenisey River	1,857	2,989	32
Brahmaputra	Jamuna River	1,800	2,900	34
Indus	Arabian Sea	1,800	2,900	34
Amur	Sea of Okhotsk	1,755	2,824	36
Euphrates	Shatt-al-Arab	1,740	2,800	38
Vilyuy	Lena River	1,647	2,650	43
Amu Darya-Pyandzh	Aral Sea	1,578	2,540	48
Kolyma-Kulu	East Siberian Sea	1,562	2,513	49
Ganges	Padma River	1,560	2,510	50
Ishim	Irtysh River	1,522	2,450	52
Salween	Gulf of Martaban	1,500	2,400	54
Olenyok	Laptev Sea	1,424	2,292	57
Aldan	Lena River	1,412	2,273	58
Syrdarya	Aral Sea	1,374	2,212	60
Chu Chiang (Pearl)-Hsi	South China Sea	1,365	2,197	62
Kolyma (Kolima)	East Siberian Sea	1,323	2,129	64
Tarim	Lop Nor Basin	1,261	2,030	69
Chulym-Bely lyus	Ob River	1,257	2,023	70
Irrawaddy	Andaman Sea	1,238	1,992	72
Vitim-Vitimkan	Lena River	1,229	1,978	73
Indigirka-Khastakh	East Siberian Sea	1,228	1,977	74
Hsi	South China Sea	1,216	1,957	75
Sungari	Amur River	1,197	1,927	78
Tigris	Shatt-al-Arab	1,180	1,900	81
Podkamennaya Tunguska	Yenisey River	1,159	1,865	83
Vitim	Lena River	1,141	1,837	84
Chulym	Ob River	1,118	1,799	88
Angara	Yenisey River	1,105	1,779	89
Indigirka	East Siberian Sea	1,072	1,726	91
Khatanga-Kotuy	Laptev Sea	1,017	1,636	96
Ket	Ob River	1,007	1,621	97
Argun	Amur River	1,007	1,620	98
Shilka-Onon	Amur River	989	1,592	
Tobol-Kokpektysay	Irtysh River	989	1,591	
Alazeya-Kadylchan	East Siberian Sea	988	1,590	
Han Shui	Yangtze River	952	1,532	
Yana-Sartang	Laptev Sea	927	1,492	
Godavari	Bay of Bengal	910	1,465	
Amga	Aldan River	908	1,462	
Sutlej	Indus River	900	1,450	
Ili-Tekes	Lake Balkhash	894	1,439	
Olyokma	Lena River	892	1,436	
Amu Darya	Aral Sea	879	1,415	
Taz	Gulf of Taz	871	1,401	
Yamuna	Ganges River	855	1,376	
Kura	Caspian Sea	848	1,364	
Tavda-Lozva	Tobol River	843	1,356	
Liao	Gulf of Liaotung	836	1,345	
Taseyeva-Chuna	Angara River	820	1,319	
Vyatka	Kama River	817	1,314	
Krishna	Bay of Bengal	800	1,290	
Narmada	Gulf of Cambay	800	1,290	
Zeya	Amur River	772	1,242	
Chu	Betpak Dala Plateau	663	1,067	
Europe				
Volga	Caspian Sea	2,193	3,530	23
Danube	Black Sea	1,770	2,850	35
Ural	Caspian Sea	1,509	2,428	53
Dnepr	Black Sea	1,367	2,200	61
Don	Sea of Azov	1,162	1,870	82
Pechora	Barents Sea	1,124	1,809	85
Kama	Volga River	1,122	1,805	86
Oka	Volga River	932	1,500	
Belaya	Kama River	889	1,430	
Rhine	North Sea	865	1,392	
Dnestr	Black Sea	840	1,352	
Northern Dvina-Sukhona	White Sea	809	1,302	
Oceania				
Darling	Murray River	1,702	2,739	39
Murray	Great Australian Bight	1,609	2,589	45
Murrumbidgee	Murray River	981	1,579	
Lachlan	Murrumbidgee River	992	1,484	
Cooper Creek	Lake Eyre	882	1,420	

*Conversions of rounded figures are rounded to nearest hundred miles or kilometres.
Source: Reprinted with permission from Encyclopaedia Britannica, 15th edition, Copyright 1988 by Encyclopaedia Britannica, Inc.

TABLE 3–10. LARGE RIVERS OF THE WORLD

River	Country	Drainage Area (Thousands of Sq Mi)	Average Discharge at Mouth (Thousands of cfs)	Rank
		North America		
Mississippi [1]	U.S.A. and Canada	1,244	611	7
St. Lawrence	U.S.A. and Canada	498	500	11
Mackenzie	Canada	697	280	17
Columbia	U.S.A. and Canada	258	256	19
Yukon	Canada	360	180	24
Frazer	Canada	92	113	32
Nelson	Canada	414	80	37
Mobile	U.S.A.	42	58	43
Susquehanna	U.S.A.	28	38	48
		South America		
Amazon	Brazil	2,231	7,500[2]	1
Orinoco	Venezuela	340	600	8
Parana	Argentina	890	526	10
Tocantins	Brazil	350	360	16
Magdalena	Colombia	93	265	18
Uruguay	[3]	90	136	26
Sao Francisco	Brazil	260	100	34
		Africa		
Congo	Congo	1,550	1,400	2
Zambezi	Mozambique	500	250	20
Niger	Nigeria	430	215	22
Nile	Egypt	1,150	100	33
		Asia		
Yangtze	China	750	770	3
Brahmaputra	Bangladesh	361	700	4
Ganges	India	409	660	5
Yenisei	U.S.S.R.	1,000	614	6
Lena	U.S.S.R.	936	547	9
Irrawaddy	Burma	166	479	12
Ob	U.S.S.R.	959	441	13
Mekong	Thailand	310	390	14
Amur	U.S.S.R.	712	388	15
Indus	Pakistan	358	196	23
Kolyma	U.S.S.R.	249	134	27
Sankai (Si)	China	46	127	28
Godavari	India	115	127	29
Hwang Ho (Yellow)	China	260	116	31
Pyasina	U.S.S.R.	74	90	36
Krishna	India	119	69	39
Indigirka	U.S.S.R.	139	64	40
Salween	Burma	108	53	44
Shatt-al Arab [4]	Iraq	209	51	45
Yana	U.S.S.R.	95	35	49
		Europe		
Danube	Romania	315	218	21
Pechora	U.S.S.R.	126	144	25
Dvina (Northern)	U.S.S.R.	139	124	30
Neva	U.S.S.R.	109	92	35
Rhine	Netherlands and Germany	56	78	38
Dnepr	U.S.S.R.	194	59	41
Rhone	France	37	59	42
Po	Italy	27	51	46
Vistula	Poland	76	38	47

[1] Includes Atchafalaya River.
[2] Department of Interior News Release, Feb. 24, 1964.
Source: Young, L.L., U.S. Geological Survey, 1964
[3] Argentina and Uruguay.
[4] Tigris, Euphrates and Karun.

SECTION B. LAKES

TABLE 3–11. NATURAL FRESH-WATER LAKES OF THE UNITED STATES OF 10 SQUARE MILES OR MORE

[Lakes are arranged by states; the Great Lakes are excluded.]

Name	Latitude	Longitude	Area (sq mi)
Alaska:			
Iliamna	59°35	155°00	1,000
Becharof	57°50	156°25	458
Teshekpuk	70°35	153°30	315
Naknek	58°35	156°00	242
Tustumena	60°25	150°20	117
Clark	60°10	154°00	110
Dall	60°15	163°45	100
Inland[1]	66°30	159°50	95
Imuruk Basin[1]	65°05	165°40	80
Upper Ugashik	57°50	156°25	75
Kukaklek	59°35	155°00	72
Lower Ugashik	57°30	156°55	72
Nerka	59°20	158°45	69
Nuyakuk	59°50	158°50	64
Aropuk	61°10	163°45	57
Tazlina	61°50	146°30	57
Nanwhyenuk or Nonvianuk	59°00	155°25	56
Nunavakpak	60°45	162°40	53
Kaghasuk[1]	60°55	163°40	52
Skilak	60°25	150°20	38
Chauekuktuli	60°05	158°50	34
Chikuminuk	60°15	158°55	34
Beverly	59°40	158°45	33
Whitefish	61°20	160°00	33
Aleknagik	59°20	158°45	31
Brooks	58°30	155°55	31
Kgun	61°35	163°50	31
Nonvianuk	59°00	155°30	31
Takslesluk	61°05	162°55	31
George	61°15	148°35	29
Nunavak Anukslak	61°05	162°30	29
Unnamed	60°55	164°00	28
Grosvenor	58°40	155°15	27
Tetlin	63°05	142°45	27
Chakachamna	61°10	152°30	26
Imuruk	65°35	163°10	26
Nunavakanuk	62°05	164°40	25
Louise	62°20	146°30	23
Minchumina	63°55	152°15	23
Klutina	61°40	145°30	22
Unnamed	61°30	164°30	22
do	71°05	156°30	21
Beluga	61°25	151°30	20
Unnamed	60°05	164°00	20
do	61°40	160°25	20
Kenai	60°25	149°35	19
Kyigayalik	61°00	162°30	19
Alaska–Continued:			
Tikchik	59°55	158°20	19
Bering	60°20	144°20	17
Kulik	59°50	158°50	17
Upnuk	60°05	158°55	17
Unnamed slough	62°40	163°30	17
Teloquana	60°55	153°55	16
Unnamed	60°30	161°40	16
do	61°00	163°45	16
do	61°30	164°55	16
Five Day Slough	62°05	162°00	15
Togiak	59°35	159°35	15
Unnamed	59°55	163°15	15
Black	56°25	159°00	14
Ualik	59°05	159°30	14
Walker	67°05	154°25	14
Unnamed	60°20	164°25	14
do	60°50	163°30	14
do	59°50	163°30	14
Amanka	59°05	159°10	13
Whitefish	60°55	154°55	13
Unnamed	71°00	156°00	13
Crosswind	62°20	146°00	12
Kakhomak	59°30	154°10	12
Karluk	57°20	154°05	12
Mother Goose	57°10	157°20	12
Unnamed	60°25	164°10	12
do	59°50	163°25	12
do	62°15	162°20	12
do	70°50	153°30	12
Coleville	58°45	155°40	11
Harlequin	59°25	138°55	11
Unnamed	60°25	164°10	11
do	60°55	162°20	11
Bear	56°00	160°15	10
Chignik	56°15	158°50	10
Ewan	62°25	145°50	10
Kontrashibuna	60°10	154°00	10
Kukaklik	61°40	160°30	10
Kulik	58°55	155°00	10
do	61°45	160°40	10
Miles	60°40	144°45	10
Susitna	62°25	146°40	10
Unnamed	60°20	162°00	10
do	60°25	162°00	10
do	60°55	162°10	10
do	59°55	163°15	10
do	62°00	162°00	10

TABLE 3–11. NATURAL FRESH-WATER LAKES OF THE UNITED STATES OF 10 SQUARE MILES OR MORE (continued)

Name	County	Area (sq mi)
California:		
Tahoe[2]	Placer, Eldorado	193
Clear	Lake	65
Eagle[3]	Lassen	41
Florida:		
Okeechobee	Hendry, Glades, Okeechobee, Martin, Palm Beach.	700
George	Putnam, Marion, Volusia, Lake.	70
Kissimmee	Osceola, Polk	55
Apopka	Orange	48
Istokpoga	Highlands	43
Tsala Apopka	Citrus	30
Tohopekaliga	Osceola	29
Harris	Lake	27
Orange	Alachua, Marion	26
East Tohopekaliga	Osceola	19
Griffin	Lake	14
Monroe	Seminole, Volusia	14
Jessup	Seminole	13
Weohyakapka	Polk	12
Talquin	Gadsden, Leon	11
Eustis	Lake	11
Blue Cypress	Osceola, Indian River.	10
Hatchineha	Polk, Osceola	10
Lochloosa	Alachua	10
Idaho:		
Pend Oreille	Bonner	148
Bear[4]	Bear Lake	[5]110
Coeur d'Alene	Kootenai	50
Priest	Bonner	37
Grays[6]	Bonneville, Caribou	34
Henrys	Fremont	10
Iowa:		
Spirit	Dickinson	12
Louisiana:		
White[7]	Vermilion	83
Grand	Iberia, St. Mary, St. Martin.	64
Caddo[8]	DeSoto	60
Catahoula[9]	LaSalle	32
Grand	Cameron	32
Six Mile	St. Martin, St. Mary	30
Fausse Pointe	St. Mary, Iberia	24
Lac des Allemands.	St. John the Baptist.	23
Verret	Assumption	22
Polourde	St. Martin, St. Mary, Assumption.	18
Maine:		
Moosehead	Piscataquis, Somerset.	117
Sebago	Cumberland	45
Maine—Continued		
Chesuncook[10]	Piscataquis	43
West Grand	Washington	37
Flagstaff	Somerset, Franklin	28
Spedni[11]	Washington	28
Grand Falls[11]	do	27
East Grand[11]	Washington, Aroostook.	26
Mooselookme-guntic.	Oxford, Franklin	26
Twin	Penobscot, Piscataquis.	25
Chamberlain and Telos.	Piscataquis	22
Graham	Hancock	19
Churchill and Eagle.	Piscataquis	17
Baskahegan	Washington	16
Umbagog[12]	Oxford	16
Brassua	Somerset	15
Square	Aroostook	14
Millinocket	Penobscot, Piscataquis.	14
Great	Kennebec	13
Richardson	Oxford	13
Schoodic	Piscataquis	11
Sebec	do	11
Aziscohos	Oxford	10
Canada Falls	Somerset	10
Rangeley	Oxford	10
Michigan:		
St Clair[13]		460
Houghton	Roscommon	31
Torch	Antrim, Kalkaska	29
Charlevoix[14]	Charlevoix	27
Burt	Cheboygan	27
Mullet	do	26
Gogebic	Ontonagon, Gogebic.	21
Manistique	Mackinac, Luce	16
Black	Cheboygan, Presque Isle.	16
Crystal	Benzie	15
Portage	Houghton	15
Higgins	Crawford, Roscommon.	15
Hubbard	Alcona	14
Leelanau	Leelanau	13
Indian	Schoolcraft	12
Elk	Antrim, Grand Traverse.	12
Glen	Leelanau	10
Minnesota:		
Lake of the Woods.[15]	Lake of the Woods.	1,485
Upper and Lower. Red.	Beltrami	451

TABLE 3–11. NATURAL FRESH-WATER LAKES OF THE UNITED STATES OF 10 SQUARE MILES OR MORE (continued)

Name	County	Area (sq mi)
Minnesota—Continued:		
Rainy[15]	Koochiching, St. Louis.	345
Mille Lacs	Aitken, Crow Wing, Mille Lacs.	207
Leech	Cass	176
Winnibigoshish	Itasca, Cass	109
Vermilion	St. Louis	77
Lac La Croix[15]	do	53
Cass	Cass, Beltrami	46
Basswood[15]	Lake	46
Namakan[15]	St. Louis	44
Kabetogama	Itasca	40
Pepin[16]	Goodhue, Wabasha	39
Mud	Marshall	37
Saganaga[15]	Cook	32
Pokegama	Itasca	24
Minnetonka	Hennepin, Carver	22
Otter Tail	Otter Tail	22
Gull	Cass, Crow Wing	20
Pelican	St. Louis	19
Traverse[17]	Traverse	18
Big Stone[17]	Big Stone	17
Crooked[15]	St. Louis, Lake	17
Sandy	Aitkin	15
Swan	Nicollet	15
Island	St. Louis	14
Bowstring	Itaska	14
Burntside	St. Louis	14
Sand Point[15]	do	14
Trout	do	14
St. Croix[16]	Washington	13
Lac qui Parle	Chippewa, Lac qui Parle.	13
Pelican	Crow Wing	13
Dead	Otter Tail	12
Minnewaska	Pope	12
Thief	Marshall	12
Nett	St. Louis, Koochiching.	12
Osakis	Douglas, Todd	10
Bemidji	Beltrami	10
Lida	Otter Tail	10
Montana:		
Flathead	Lake, Flathead	[18]197
Medicine	Sheridan	[19]15
McDonald	Flathead	10
Nevada:		
Tahoe[2]	Ormsby, Douglas	193
New Hampshire:		
Winnipesaukee	Belknap, Carroll	72
Umbagog[12]	Coos	16
Squam	Gafton, Carroll	11
New York:		
Champlain[20]	Clinton, Essex	[21]490
Oneida	Oswego, Oneida	80
Seneca	Seneca, Schuyler	67
Cayuga	Cayuga, Seneca, Tompkins.	66
George	Warren	44
Chautauqua	Chautauqua	21
Black	St. Lawrence	17
Canandaigua	Ontario, Yates	17
Skaneateles	Onondaga, Cayuga	14
Owasco	Cayuga	10
North Carolina:		
Mattamuskeet[22]	Hyde	67
Phelps	Washington	25
Waccamaw	Columbus	14
Oregon:		
Upper Klamath	Klamath	[23]142
Crater	do	21
South Dakota:		
Traverse[17]	Roberts	18
Big Stone[17]	do	17
Tennessee:		
Reelfoot	Lake, Obion	22
Texas:		
Caddo[8]	Marion	60
Utah:		
Utah	Utah	140
Bear[4]	Rich	110
Vermont:		
Champlain[20]	Chittenden, Franklin.	[21]490
Washington:		
Chelan	Chelan	55
Washington	King	35
Ozette	Clallam	12
Wisconsin:		
Winnebago	Winnebago, Calumet, Fond du Lac.	215
Pepin[16]	Pierce, Pepin	39
Poygan	Winnebago	17
Koshkonong	Jefferson	16
Mendota	Dane	15
St. Croix[16]	St. Croix	12
Green	Green Lake	11
Wyoming:		
Yellowstone	Yellowstone National Park.	[24]137
Jackson	Teton	[25]39
Shoshone	Yellowstone National Park.	11

(See footnotes on next page.)

TABLE 3–11. (Footnotes)

[1] May be salt water.
[2] California and Nevada.
[3] Mildly saline, less than 1,000 ppm.
[4] Idaho and Utah.
[5] 136 sq mi including Mud Lake.
[6] Submerged marsh.
[7] Originally brackish; now kept fresh by controls on salt water intrusion.
[8] Louisiana and Texas.
[9] Shrinks to small area at extremely low stages.
[10] Includes Ripogenus and Caribou.
[11] Maine and Quebec.
[12] Maine and New Hampshire.
[13] Michigan and Ontario.
[14] Formerly called Pine.
[15] Minnesota and Ontario.
[16] Minnesota and Wisconsin.
[17] Minnesota and South Dakota.
[18] At normal high water; 188 sq mi at medium low water; lake regulated for power between these limits.
[19] Includes 4 islands having area of about 1 sq mi.
[20] New York, Vermont, and Quebec.
[21] Includes islands totaling about 55 sq mi.
[22] The lake originally landlocked, was drained and provided with outlet and is fresh water; level regulated to some extent by control works on canals draining the area.
[23] At upper level; dam at outlet allows regulation so that area varies between 93 and 142 sq mi.
[24] Includes islands totaling 3 sq mi.
[25] Enlarged by dam; original area, 30 sq mi.

Source: U.S. Geological Survey

TABLE 3–12. NATURAL FRESH-WATER LAKES OF THE UNITED STATES OF 100 SQUARE MILES OR MORE

[The Great Lakes are excluded]

Name	Location	Area (sq mi)	Name	Location	Area (sq mi)
Lake of the Woods	Minnesota and Ontario.	1,485	Mille Lacs	Minnesota	207
Iliamna	Alaska	1,000	Flathead	Montana	197
Okeechobee	Florida	700	Tahoe	California and Nevada.	193
Champlain	New York, Vermont, and Quebec.	490	Leech	Minnesota	176
St. Clair	Michigan and Ontario.	460	Pend Oreille	Idaho	148
Becharof	Alaska	458	Upper Klamath	Oregon	142
Upper and Lower Red.	Minnesota	451	Utah	Utah	140
Rainy	Minnesota and Ontario.	345	Yellowstone	Wyoming	137
Teshekpuk	Alaska	315	Tustumena	Alaska	117
Naknek	do	242	Moosehead	Maine	117
Winnebago	Wisconsin	215	Clark	Alaska	110
			Bear	Idaho and Utah	110
			Winnibigoshish	Minnesota	109
			Dall	Alaska	100

Source: U.S. Geological Survey

TABLE 3–13. NATURAL FRESH-WATER LAKES OF THE UNITED STATES, 250 FEET DEEP OR MORE

[The Great Lakes are excluded]

Name	Location	Depth (feet)	Name	Location	Depth (feet)
Crater	Oregon	1,932	Cooper	Alaska	>400
Tahoe	California and Nevada.	1,645	Champlain	New York, Vermont, and Quebec.	400
Chelan	Washington	1,605	Kasnyku	Alaska	393
Pend Oreille	Idaho	1,200	Chakachamna	do	380
Nuyakuk	Alaska	930	Ozette	Washington	331
Deer	do	877	Aleknagik	Alaska	330
Chauekuktuli	do	700	Sebago	Maine	316
Crescent	Washington	624	Swan	Alaska	>314
Seneca	New York	618	Baranoff	do	303
Clark	Alaska	606	Payette	Idaho	>300
Beverley	do	500	Quinault	Washington	About 300
Nerka	do	475	Crescent	Alaska	291
Tokatz	do	474	Wallowa	Oregon	283
Long	do	470	Chilkoot	Alaska	282
Lower Sweetheart.	do	459	Odell	Oregon	279
Cayuga	New York	435	Silver	Alaska	278
Crater	Alaska	414	Grant	do	>250

Source: U.S. Geological Survey

TABLE 3–14. LARGEST LAKE IN EACH STATE OF THE UNITED STATES

State	Largest Entirely Within State	Largest Partly in Another State	Shared With	Origin	Area in Square miles	Feet Above Sea Level	Maximum Depth feet	Shoreline Length miles
Ala.	Wheeler			Man-made	104.84	556	58	1,063
		Guntersville	Tenn.	Man-made	107.97	595	60	962
Alaska	Illamna			Natural	1,033	50	.....	188
Ariz.	San Carlos*			Man-made	30.6	2,523	249	.....
		Powell	Utah	Man-made	252	3,700	580	.....
Ark.	Ouachita			Man-made	62.65	578	207	690
		Bull Shoals	Mo.	Man-made	111.31	695	243	1,050
Calif.	Salton Sea			Natural	360	-231	46	.....
		Tahoe	Nev.	Natural	192	6,229	1,685	71
Colo.	John Martin*			Man-made	28.72	3,765	118	86
Conn.	Candlewood			Man-made	8.46	429	85	65
Del.	Lum's Pond			Man-made	.31	44	12	3.5
Fla.	Okeechobee			Natural	700	18.7	.....	110
Ga.	Sidney Lanier			Man-made	57.96	1,035	180	540
		Clark Hill*	S.C.	Man-made	111.09	330	190	1,057
Hawaii	Koloa*			Man-made	.66	233	22.5	3.3
Idaho	Pend Oreille			Natural	133	2,063	1,150	111.3
		Bear	Utah	Natural	136	5,930	30	51.5
Ill.	Crab Orchard			Man-made	10.96	405	33	103
		Michigan	Wis., Ind., Mich.	Natural	22,400	578.8	923	1,660
Ind.	Wawasee			Natural	4.09	859	68	18
		Michigan	Wis., Ill., Mich.	Natural	22,400	578.8	923	1,660
Iowa	Spirit			Natural	8.84	1,402	.....	.....
Kan.	Tuttle Creek*			Man-made	24.68	1,075	56	112
Ky.	Cumberland			Man-made	78.51	723	.....	1,255
		Kentucky	Tenn.	Man-made	247.34	375	145	2,380
La.	Pontchartrain			Natural	630	S.L.	15	113
Me.	Moosehead			Natural	117	1,028	246	.....
Md.	Deep Creek			Man-made	7.03	2,462	.....	.....
Mass.	Quabbin*			Man-made	38.6	530	150	118
Mich.	Houghton			Natural	31.3	1,139	20	30
		Superior	Wis., Ont., Minn.	Natural	31,800	600	1,333	2,980
Minn.	Red			Natural	451	1,175	31	123
		Superior	Wis., Ont., Mich.	Natural	31,800	600	1,333	2,980
Miss.	Sardis			Man-made	15.31	234	.....	60
Mo.	Lake of the Ozarks			Man-made	93.75	.....	125	1,375
		Bull Shoals	Ark.	Man-made	111.31	695	243	1,050
Mont.	Fort Peck*			Man-made	382.81	2,250	220	1,600
Nebr.	McConaughy			Man-made	55	3,276	150	50
Nev.	Pyramid			Natural	187.5	3,800	330	70
		Mead	Ariz.	Man-made	247	1,221	589	550
N.H.	Winnipesaukee			Natural	71.55	504	120	128
N.J.	Hopatcong			Man-made	4.19	924	58	35
N.M.	Elephant Butte*			Man-made	58.85	4,450	193	250
N.Y.	Oneida			Natural	80	370	50	52
		Erie	Mich., Pa., Ohio, Ont.	Natural	9,910	570	210	856
N.C.	Norman			Man-made	50.78	760	115	520
		John H. Kerr*	Va.	Man-made	76.4	300	100	800

TABLE 3–14. LARGEST LAKE IN EACH STATE OF THE UNITED STATES (continued)

State	Largest Entirely Within State	Largest Partly in Another State	Shared With	Origin	Area in Square miles	Feet Above Sea Level	Maximum Depth feet	Shoreline Length miles
N.D.	Garrison*			Man-made	609.38	1,850	200	1,600
Ohio	Grand			Man-made	20	869	12	60
		Erie	Mich., Pa., N.Y., Ont.	Natural	9,910	570	210	856
Okla.	Eufaula*			Man-made	160.16	585	87	600
		Texoma	Texas	Man-made	149.06	617	94	540
Ore.	Upper Klamath (incl. Agency Lake)			Natural	140.63	4,139	40	105
Pa.	Wallenpaupack			Man-made	9	1,182	50	45
		Erie	Mich., N.Y., Ohio, Ont.	Natural	9,910	570	210	856
R.I.	Scituate*			Man-made	5.68	284	80	38
S.C.	Marion			Man-made	157.03	75	35	299
		Clark Hill*	Ga.	Man-made	111.09	330	190	1,057
S.D.	Francis Case			Man-made	160.31	1,375	140	540
Tenn.	Watts Bar			Man-made	60.31	745	80	783
		Kentucky	Ky.	Man-made	247.34	375	145	2,380
Texas	Texarkana			Man-made	46.56	225	39	141
		Texoma	Okla.	Man-made	149.06	617	94	540
Utah	Great Salt			Natural	1,500	4,200	48	350
		Powell	Ariz.	Man-made	252	3,700	580	.....
Vt.	Bomoseen			Natural	3.69	411	.....	.....
		Champlain	N.Y. Que.	Natural	430	100	399	.....
Va.	Smith Mountain			Man-made	31.25	795	200	500
		John H. Kerr*	N.C.	Man-made	76.4	300	100	800
Wash.	F.D. Roosevelt			Man-made	123.44	1,288	375	302
W. Va.	Tygart			Man-made	5.37	1,010	.....	106
		Bluestone*	Va.	Man-made	3.07	1,409	42	33
Wis.	Winnebago			Natural	215.26	.....	21.6	91.96
		Superior	Minn., Mich., Ont.	Natural	31,800	600	1,333	2,980
Wyo.	Yellowstone			Natural	137	7,735	.....	.....

*Reservoir

Source: National Geographic Society

FIGURE 3–4. THE GREAT LAKES

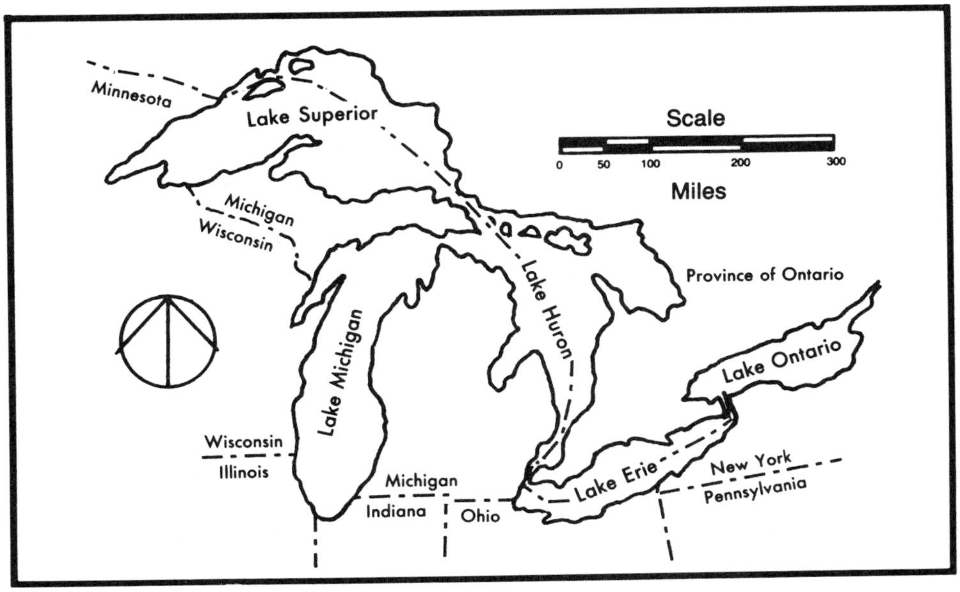

TABLE 3–15. SELECTED FACTS ABOUT THE GREAT LAKES SYSTEM

Lake	Volume (mi^3)	Water Surface Area (mi^2)	Drainage Area (mi^2)	Shoreline Length (Includes Islands) (mi)	Outlet	Remarks
Superior	2,900	31,700	49,300	2,726	St. Marys River to Lake Huron.	Largest surface area of all the freshwater lakes in the world. Outflow controlled by St. Marys River Compensating works.
Michigan	1,180	22,300	45,600	1,638	Straits of Mackinac to Lake Huron.	Sixth largest surface area of world's freshwater lakes.
Huron.................	850	23,000	51,700	3,827	St. Clair River to Lake St. Clair.	Fifth largest surface area of world's freshwater lakes.
St. Clair..............	1	430	4,800	257	Detroit River to Lake Erie.	Shallowest lake in the Great Lakes system.
Erie......................	116	9,910	22,720	871	Niagara River and Falls to Lake Ontario.	Eleventh largest surface area of world's freshwater lakes.
Ontario	393	7,340	23,400	712	St. Lawrence River to Atlantic Ocean.	Outflow controlled by St. Lawrence Seaway and Power Project.
Total	5,440	94,680	197,520	10,031		

Source: U.S. Geological Survey, National Water Summary 1985—Hydrologic Events and Surface-Water Resources, Water-Supply Paper 2300

FIGURE 3–5. WATER LEVEL FLUCTUATIONS IN THE GREAT LAKES, 1950–1985

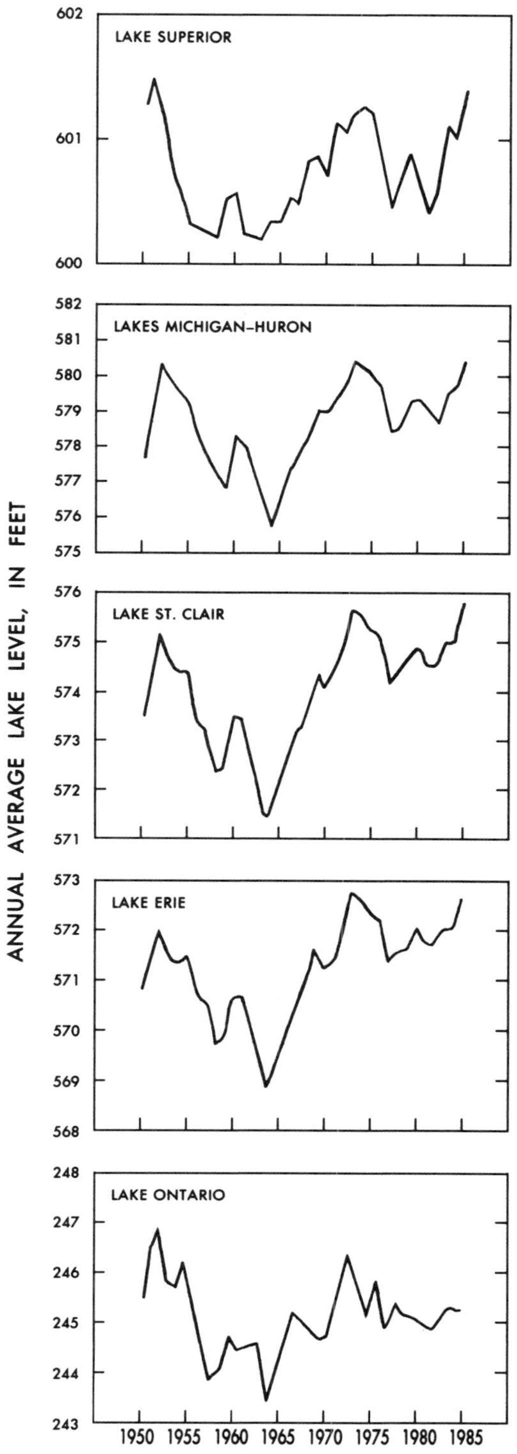

LEVELS REFERENCED TO INTERNATIONAL GREAT LAKES DATUM 1955

Source: U.S. Geological Survey, National Water Summary 1985—Hydrologic Events and Surface-Water Resources, Water-Supply Paper 2300

FIGURE 3–6. PROFILE OF THE GREAT LAKES–ST. LAWRENCE RIVER DRAINAGE SYSTEM

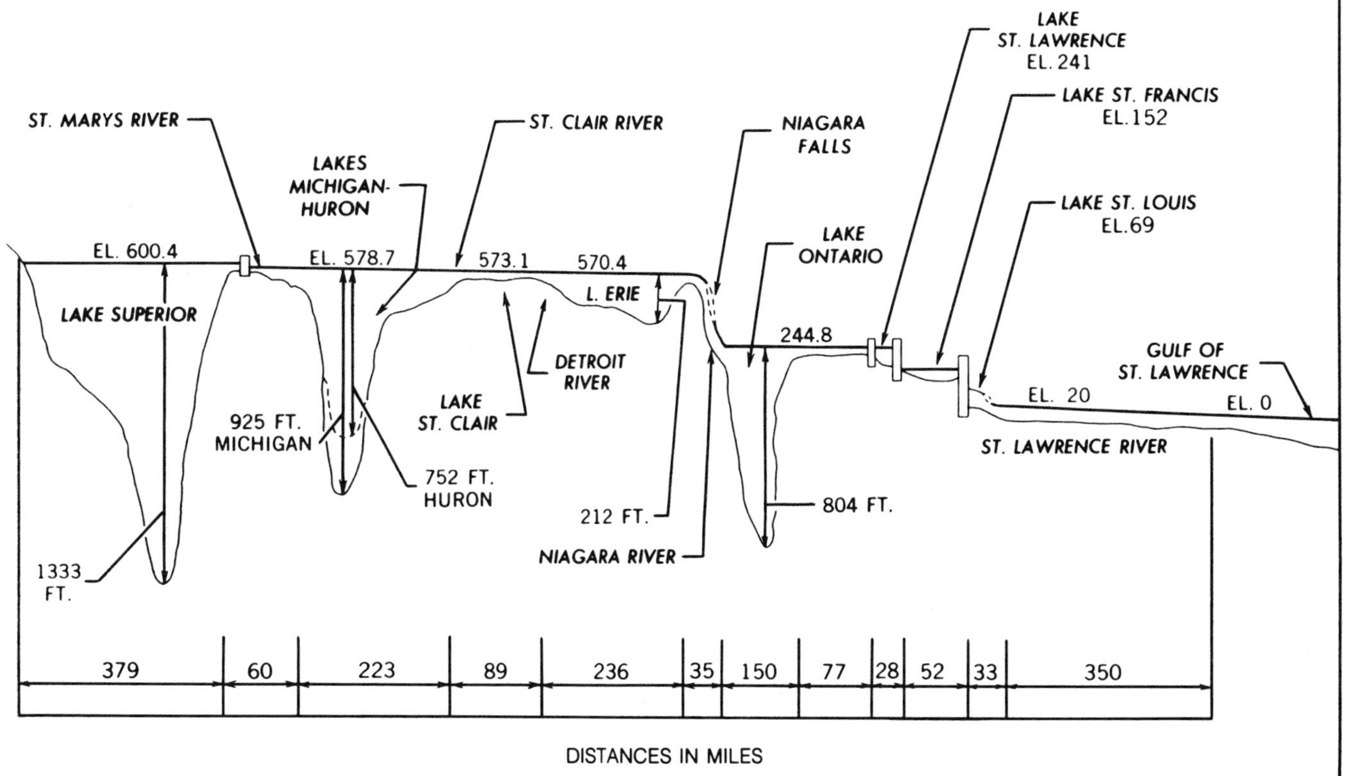

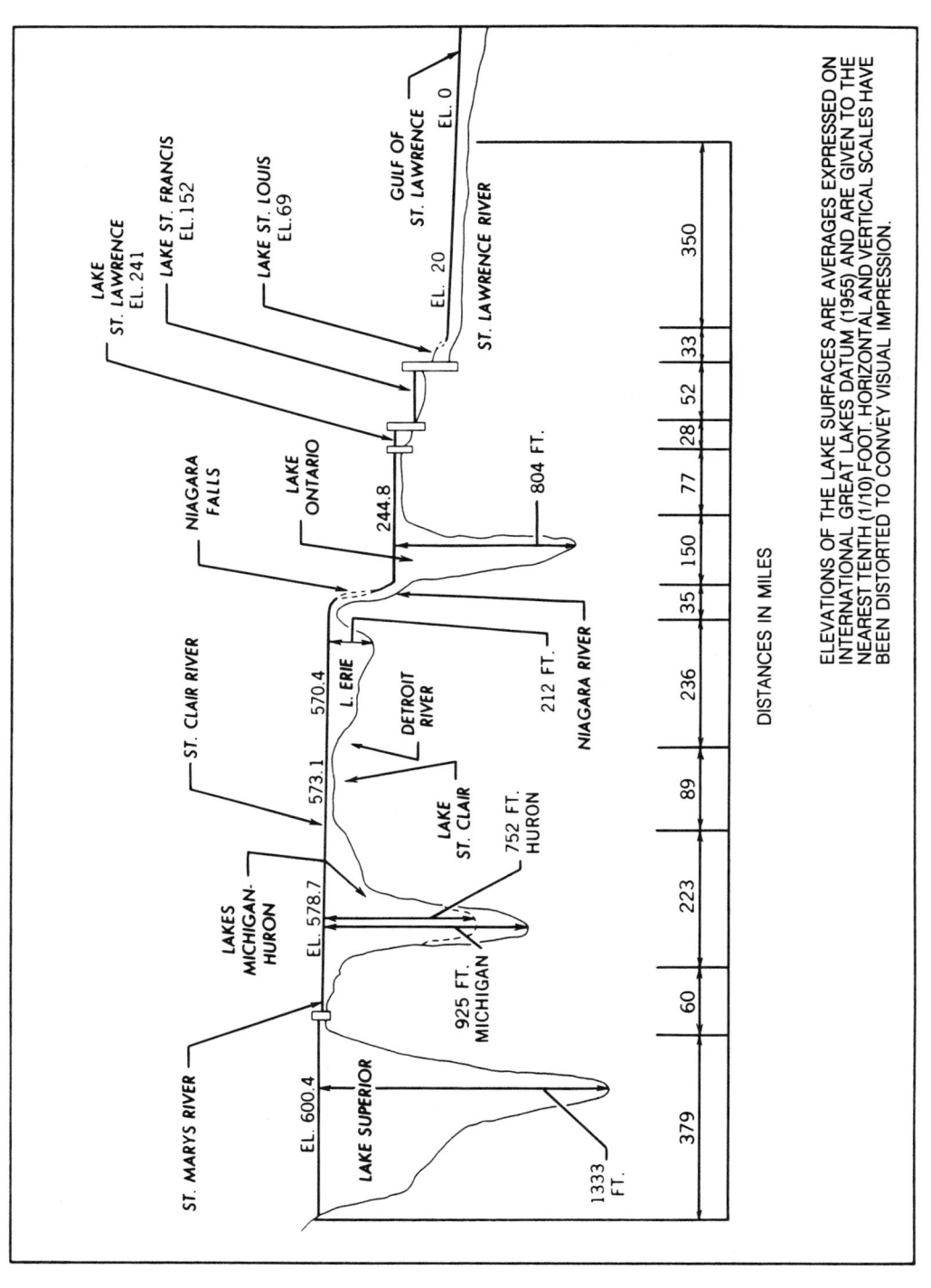

ELEVATIONS OF THE LAKE SURFACES ARE AVERAGES EXPRESSED ON INTERNATIONAL GREAT LAKES DATUM (1955) AND ARE GIVEN TO THE NEAREST TENTH (1/10) FOOT. HORIZONTAL AND VERTICAL SCALES HAVE BEEN DISTORTED TO CONVEY VISUAL IMPRESSION.

Source: International Great Lakes Levels Board, 1973. Regulation of Great Lakes Water Levels, Report to the International Joint Commission, Washington, D.C.

TABLE 3–16. HYDROLOGIC CHARACTERISTICS OF THE GREAT LAKES

Lake	Lake Surface Elevation, in Feet, 1900–1984: Monthly Mean			Monthly Range (From Winter Low to Summer High)			Total Dissolved Solids, 1968 ppm	Mean Discharge m³/s
	Average	Maximum	Minimum	Average	Maximum	Minimum		
Superior	600.59	602.02	598.23	1.2	2.1	0.4	52	2,076
Michigan-Huron .	578.27	581.04	575.35	1.2	2.1	.4	—	—
Michigan	—	—	—	—	—	—	150	1,558
Huron	—	—	—	—	—	—	118	5,038
St. Clair	573.34	576.23	569.86	1.7	3.3	.6	—	—
Erie	570.44	573.51	567.49	1.6	2.8	.9	198	5,545
Ontario	244.71	248.06	241.45	2.0	3.6	.7	194	6,624

LEVELS REFERENCED TO INTERNATIONAL GREAT LAKES DATUM 1955

Source: U.S. Geological Survey, National Water Summary 1985—Hydrologic Events and Surface-Water Resources, Water-Supply Paper 2300 and U.S. Army Corps of Engineers data

TABLE 3–17. PRINCIPAL SALINE LAKES OF THE UNITED STATES

Lake	Present Area (square miles)	Remarks
California:		
Salton Sea	350	About 650 sq mi at highest stage in 1905–07.
Owens.........................	Dry at times each year since 1943.	110 sq mi in 1872, prior to diversions from Owens River; 35 sq mi in 1943.
Mono...........................	76	Maximum, 89 sq mi in 1919.
Goose (in California and Oregon)	About 100	Maximum, 186 sq mi, 125 in California and 61 in Oregon; overflowed into Pitt River in 1869 and 1881; dry in 1930; 150 sq mi in 1958.
Eagle	41	Some question as to whether Eagle Lake should be considered saline or fresh water. It has no surface outlet (Martin, 1962), but since 1924 it has been tapped by tunnel to Willow Creek. Salinity is considerably less than 1,000 ppm, according to California Dept. of Water Resources. During period 1895–1925 lake rose to highest level since at least 1650 (Harding, 1935); rise believed due to closing of subterranean outlet by earthquake in 1890 (Antevs, 1938).
Honey..........................	Dry	90 sq mi in 1867, possibly higher in 1890; dry in 1903; high in 1904; dry in 1924. Contained some water April 1958 to September 1960, and early in 1962.
Louisiana:		
Pontchartrain..............	625	These lakes are connected with the Gulf of Mexico, and are subject to tidal fluctuation.
Sabine (Louisiana and Texas)	95	Do.
Calcasieu	90	Do.
Maurepas....................	90	Do.
Salvador	70	Do.
Nevada:		
Pyramid	180	Maximum size, 220 sq mi. Low until 1860; reached extreme high level in 1862 and 1868 or 1869; nearly as high in 1890; began to drop in 1917 (Hardman and Venstrom, 1941).
Walker.........................	107	Maximum size, 125 sq mi.
Winnemucca	Dry	Maximum size, 180 sq mi. Dry in 1840, but began to fill shortly thereafter (Zones, 1961). According to Russell (1885) the lake rose more than 50 ft and approximately doubled its area between 1867 and 1882. Was 87 ft deep in 1882. Dry since 1945.
Carson.........................	Nearly dry	Maximum size, 41 sq mi. A few water-filled pot holes remain. Once called South Carson Lake; received flow of Carson River before Lahontan Reservoir was built.

TABLE 3–17. PRINCIPAL SALINE LAKES OF THE UNITED STATES (continued)

Lake	Present Area (square miles)	Remarks
Carson Sink	Dry	A shallow playa some 250 square miles in area shown on some maps as a body of water. Russell (1885) called it North Carson Lake. Dry in 1882, but probably has had some water at times since. Once received water from both Carson and Humboldt Rivers.
Ruby	....	Maximum size, 37 sq mi. Shown as swamp on recent maps of Army Map Service and Nevada Dept. of Highways.
Franklin	....	Maximum size, 32 sq mi. Shown as swamp on recent maps of the Army Map Service and Nevada Dept. of Highways.
North Dakota:		
Devils	24	140 sq mi in 1867; 70 sq mi in 1883; 45 sq mi in 1900; 10 sq mi in 1940. Since 1940 lake has been rising.
Oregon:		
Malheur and Harney	Probably dry	Malheur, the larger of the two lakes, overflows into Harney, which has no outlet. Maximum combined size, 125 sq mi. Reported dry in 1931; high in late 1950's; about 1 sq mi in 1961, and expected to go dry in 1962.
Goose (see California)	....	
Abert	52	Maximum size, 60 sq mi. Dry in 1930 or thereabouts, but fairly high in 1958.
Summer	Probably dry	Maximum size, 70 sq mi. Nearly dry in 1961.
Silver	Dry	Maximum size, 15 sq mi. Dry in 1961. Because of the transient nature of the lake, the water—whenever there is any—is relatively fresh; hay is raised on the dry lake bed.
Warner	Probably less than 10	A series of shallow lakes; combined area about 30 sq mi in 1953, a wet year, estimated from Army Map Service map based on aerial photographs taken in 1953. Present lakes are all that is left of Pleistocene Warner Lake, which covered about 300 sq mi and was about 270 ft deep.
Utah:		
Great Salt	About 1,000	Maximum size since 1851, 2,400 sq mi in 1870's; minimum, 950 sq mi in October 1961; seasonal high in 1962 was 1,050 sq mi in June.
Sevier	Dry	Maximum size, 125 sq mi; has been dry for several years.

Source: U.S. Geological Survey, 1963

FIGURE 3–7. WATER LEVELS, EVAPORATION AND SURFACE INFLOW OF THE SALTON SEA, CALIFORNIA

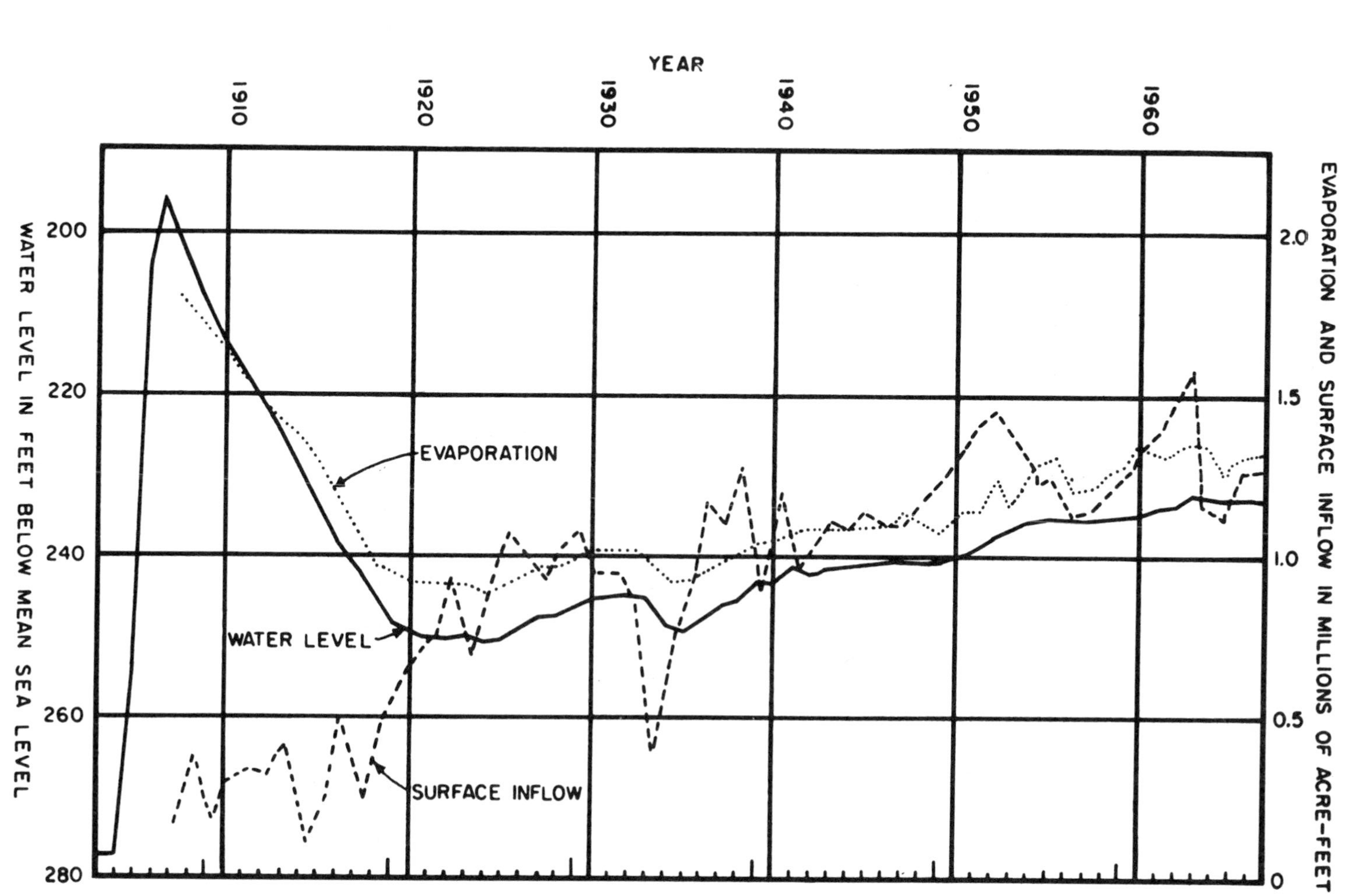

Source: U. S. Geological Survey, 1966, Professional Paper 486-C and California Department of Water Resources Bulletin 143–7, Geothermal Wastes and the Water Resources of the Salton Sea Area

FIGURE 3–8. WATER-LEVEL FLUCTUATION OF GREAT SALT LAKE, UTAH 1847–1986

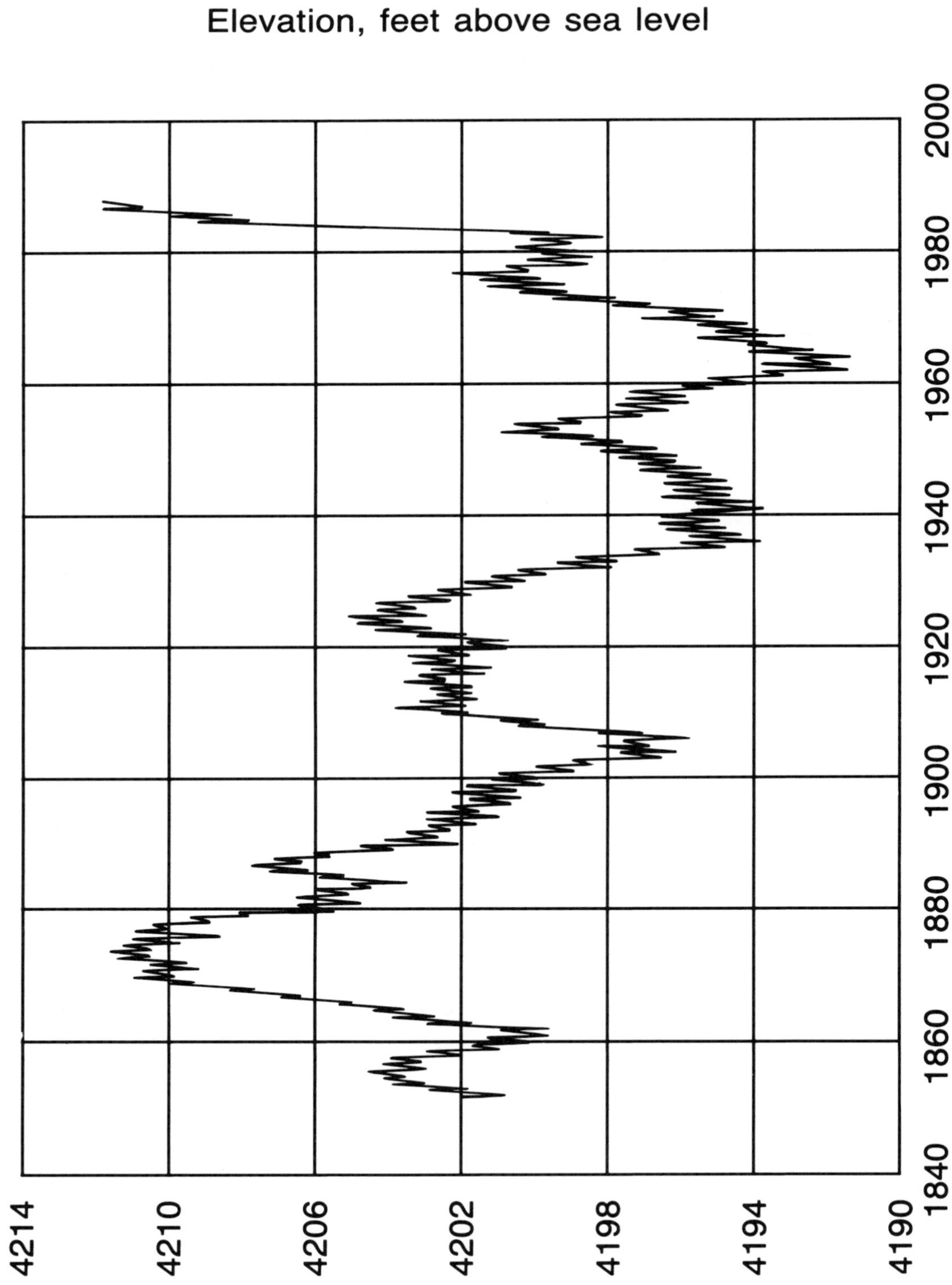

Source: U.S. Geological Survey Circular 913 and Utah Division of Water Resources

TABLE 3–18. HYDROLOGIC DATA FOR GREAT SALT LAKE AND WEST DESERT PUMPING PROJECT

Great Salt Lake

Dimensions	80 x 35 mi.
Average depth	22 ft.
Maximum depth	42 ft.
Contents in dissolved minerals	4–5 billion tons.

(mainly chloride, sodium, sulfate, magnesium, potassium with lesser amounts of calcium, lithium, bromium and boron[a])

Rock and earthfill Southern Pacific Transportation Co. railroad causeway separates the lake into two parts.

Southern part of lake (60% of total area) receives 90% of lake's freshwater inflow.

Total annual inflow 1931–76 averaged 2.9 million acre-ft.

Northern part of lake receives most of its water as brine flowing through culverts and causeway from southern part of lake.

Lake salinity varies with lake level.

Northern lake is 16% salt and about 3 times saltier than southern lake (June 1987).

West Desert Pumping Project

Flood control project by State of Utah to lower water level of Great Salt Lake.

Start of construction July 1986.

Cost of construction and 1st year of operation	$60 million.
Projected volume of diversion	2 million acre-ft/yr.

Water is lifted by 3 large pumps (capacity 1,000 cfs each) through 4.1 mi. long outlet canal to evaporation pond (west pond).

Surface area of west pond	500 sq. mi.
Rate of evaporation from west pond	825,000 acre-ft/yr.
Salinity of water in west pond	350 g/L.

[a] For chemical analysis of brine see Table 6-8.

Source: Compiled from information provided by Utah Division of Water Resources, 1987 and U.S. Geological Survey Circular 913

TABLE 3–19. HYDROLOGIC DATA FOR CLOSED LAKES

[These lakes occupy topographic sinks with no discharges by surface streams or seepage and with a ground-water gradient toward the lake.]

Lake	Drainage Area (sq mi)	Evaporation (ft per year)		Coefficient of Variation of Lake Area**	Response Time† (years)	Overflow Expressed as Depth Over Tributary Area (feet)	Salinity		Mean Depth (ft)	Lake Area (sq mi)
		Gross	Net*				Date	ppm		
Devils Lake, N. Dak.	3,000	2.5	1.2	0.40	14	2.5	1899	8,470	13	45
							1923	15,210	10	26
							1948	25,000	4.5	14
							1952	8,680	10	20
Basin Lake, Saskatchewan	105	2.25	1.0	.07	25	......	1938–41	11,900	20	16
Quill Lakes, Saskatchewan	2,700	2.0	.75	.15	20	......	1938–41	25,000	10	230
Redberry Lake, Saskatchewan.	120	2.25	1.0	.038	50	......	1938–41	14,000	43	27
Great Salt Lake, Utah	21,000	3.3	2.7	.125	9	1.0	1877	138,000	18	2,200
							1932	276,000	13	1,300
Sevier Lake, Utah	16,000	3.7	3.2	.35	3	4	1872	86,400	8	188
Pyramid Lake, Nev.	2,650	4.2	3.7	.04	65	9	1882	3,486	167	200
Walker Lake, Nev.	3,500	4.2	3.8	.075	45	13	1882	2,500	120	110
Mono Lake, Calif.	600	4.1	3.3	.043	35	200	1882	51,170	61	85
Elsinore Lake, Calif.	717	4.5	3.2	.68	3.0	.3	1949	8,880	5	5
Owens Lake, Calif.	2,900	5.5	5.0	.10	10	12	1876	60,000	24	105
							1905	213,700	11	76
Omak Lake, Wash.	100	3.2	2.2	.067	30	19	([1])	5,704	50	5.5
Lake Abert, Oregon	900	3.5	2.5	.5	6	11	1902	76,000	5	50
							1912	30,000		......
							1956–59	20,000	10	60
Summer Lake, Oregon	330	3.5	2.5	1.0	2	61	1901	36,000	3	30
							1912	18,000		......
Harney Lake, Oregon	5,300	3.3	2.5	.8	2	1.0	1912	22,380	4.8	47
Lake Eyre, Australia	550,000	7.5	7.0	2.5	1.5	10	1950	[2]40,000	8.5	3,100
							1951	[2]240,000	2.8	740
Lake Corangamite, Australia	1,300	4.0	2.0	.30	10	3.8	1933	105,000	3.5	74
							1950	50,000	6.0	88
							1956	12,000	12	140
Aral Sea, U.S.S.R.	625,000	3.0	2.6	.10	35	10	......	10,700	52	25,000
Caspian Sea, Asia	1,400,000	3.3	2.8	.015	300	22	......	11,000	600	170,000
Dead Sea, Palestine	12,000	5.1	4.8	.03	40	175	......	220,000	460	390
Lake of Urmia, Iran	20,000	3.0	2.5	.11	9		......	148,000	16	1,800
Lake Van, Turkey	6,000	3.3	2.0	.02	150	53	1944	[2]22,400	175	1,450
Tuz Golu, Turkey	4,400	3.4	2.4	.5	1	86	1959	250,000	2	650
Elton Lake, U.S.S.R.	......	......	3.0	1.0	1.0	......	......	300,000	2.3	110
Baskuntschak Lake, U.S.S.R.	......	......	3.0	2.0	.5	......	......	260,000	1.15	50

[1]Before 1924.
[2]Milligrams per liter.
*Net evaporation is gross evaporation minus precipitation.
**Coefficient of variation of lake area is equal to the standard deviation of lake volume divided by the area of the lake.
†Response time is the ratio of a change in lake volume to the corresponding change in rate of discharge.
Source: U.S. Geological Survey

TABLE 3–20. WATER BALANCE OF THE MAJOR LAKES OF THE WORLD

Lake	Observation Period	Volume km^3	Inflow km^3/yr	Precipitation mm/yr	Outflow km^3/yr	Evaporation mm/yr	Inflow Factor[1]	Outflow Factor[2]	Retention Time years
Caspian Sea	1940–1966	78200	289	246	10.8	994	0.76	0.03	204
Michigan and Huron	1959–1966	8200	142	780	161	750	0.61	0.65	33
Superior	1959–1966	11600	47.6	760	69.7	470	0.43	0.64	107
Victoria	1925–1959	2700	17.9	1630	21.9	1570	0.14	0.17	21
Aral Sea	1959–1969	1020	49.5	173	0	1050	0.82	0	15
Tanganyika	long-term	18900	25.8	1000	3.0	1690	0.44	0.05	322
Baikal	1901–1970	23000	60.3	405	59.5	416	0.82	0.82	317
Nyasa	long-term	7720	34.2	1220	6.3	2130	0.48	0.09	107
Great Slave	long-term	1070	136	350	141	166	0.93	0.97	7.4
Erie	1959–1966	545	190	860	182	920	0.90	0.88	2.6
Ontario	1959–1966	1710	210	900	210	800	0.93	0.94	7.6
Balkhash	1911–1966	112	15.7	154	0	1020	0.85	0	6.0
Ladoga	1932–1958	908	69.1	606	73.7	344	0.87	0.92	11
Chad	1954–1962	44	45.8	378	0	2260	0.85	0	0.9
Eyre	long-term	—	4.2	150	0	dries up	0.77	0	—
Maracaibo	long-term	—	19.6	977	4.9	2080	0.60	0.15	—
Onega	long-term	295	15.9	575	18.0	350	0.74	0.84	14
Rudolf	long-term	—	16.0	750	0.0	2610	0.71	0	—
Titicaca	long-term	710	7.7	625	0.6	1500	0.60	0.05	55

[1] Percentage of inflow of the sum of inflow and lake precipitation.

[2] Percentage of outflow of the sum of outflow and lake evaporation.

Source: Kuusisto, E.E., 1985, Lakes Their Physical Aspects, in Facets of Hydrology II, John C. Rodda, Editor, John Wiley and Sons. Reproduced with permission.

TABLE 3–21. MAJOR LAKES IN THE WORLD

Lake	Country	Surface Area, km²	Maximum depth, m	Volume, km³
		Europe		
Caspian Sea*	USSR, Iran	374,000	1,025	78,200
Ladozskoje	USSR	17,700	230	908
Onezskoje	"	9,630	127	295
Vänern	Sweden	5,550	100	180
Cudskoje with Pskovskoje	USSR	3,550	15	25
Vättern	Sweden	1,900	119	72
Saimaa	Finland	1,800	58	36
Beloje	USSR	1,290	20	5.2
Vygozero	"	1,140	18	7.1
Mälaren	Sweden	1,140	64	10
Il'men'	USSR	1,100	10	12
Päyänne	Finland	1,065	93	—
Inari	"	1,000	80	28
Imandra	USSR	900	67	11
Balaton	Hungary	596	12	1.9
Lac de Geneve	Switzerland, France	581	310	90
Bodensee	German Federal Republic, Switzerland, Austria	538	252	48
Hälmaren	Sweden	484	22	—
Stor Sjön	Sweden	464	74	8.0
Kubenskoje	USSR	407	13	1.7
Loch Ness	Great Britain	396	31	—
Garda	Italy	370	346	50
Mjøsa	Norway	363	434	56
Skadarsko	Albania, Yugoslavia	362	10	2.2
Ohridsko	Albania, Yugoslavia	350	256	61
Sniardwy	Poland	331	47	2.8
Torne Träsk	Sweden	330	168	17
Neusiedler See	Austria, Hungary	323	2	—
Prespansko	Greece, Albania, Yugoslavia	288	54	4.0
Neuchâtel	Switzerland	216	152	—
Lago Maggiore	Italy, Switzerland	214	372	—
Femund	Norway	202	131	6.0
Como	Italy	146	410	—
		Asia		
Aral'skoje More*	USSR	64,100	68	1,020
Bajkal	"	31,500	1,741	23,000
Balchas	"	18,200	26	112
Tonle Sap	Cambodia	10,000[1]	12	40
Issyk-Kul'	USSR	6,200	702	1,730
Dongtinghu	China	6,000[2]	10	—
Rizaiyeh (Urumiyeh)*	Iran	5,800	16	45
Zajsan	USSR	5,510	8.5	53
Tajmyr	"	4,560	26	13
Kukunor*	China	4,220	38	—
Chanka	USSR, China	4,190	10.6	18.5
Van*	Turkey	3,760	145	—
Lob Nor*	China	3,500	5	(5)
Ubsa Nor*	Mongolia	3,350	—	—
Poyanghu	China	2,700	20	—
Alakol'	USSR	2,650	54	58.6
Chövsgöl Nuur	Mongolia	2,620	270	480
Cany	USSR	2,500	10	4.5
Tuz*	Turkey	2,500	—	—
Namru Tso*	China	2,460	—	—
Taihu	"	2,210	—	—
Char Us Nuur	Mongolia	1,760	—	—
Tengiz*	USSR	1,590	8	—

TABLE 3–21. MAJOR LAKES IN THE WORLD (continued)

Lake	Country	Surface Area, km²	Maximum depth, m	Volume, km³
Ebi Nor*	China	1,420	—	—
Chirgis Nuur	Mongolia	1,480	—	—
Sevan	USSR	1,230	86	38
Dalai Nur	China	1,100	—	—
Ulyungur Nor	"	1,000	—	—
Dead Sea*	Israel, Jordan	940	400	188
Seletyteniz	USSR	777	3.2	1.5
Sasykkol'	"	736	—	—
P'asino	"	735	10	—
Kulundinskoje*	"	728	4.9	—
Biwa ko	Japan	688	103	27.5
Gandhi	India	663	64	39.2
Karnaphuli	Bangladesh, India	656	33	13.8
Buir Nuur	Mongolia	610	11	—
Markakol'	USSR	449	30	—
Ubinskoje	"	440	3	—
Karakul'*	"	380	238	—
Tungabharda	India	378	47	12.4
Fumibhal	Thailand	300	123	29.7
Kronockoje	USSR	245	128	—
Teleckoje	"	223	325	40
	Africa			
Victoria	Tanzania, Kenya, Uganda	69,000	92	2,700
Tanganyika	Tanzania, Zaire, Zambia, Burundi, Rwanda	32,900	1,435	18,900
Nyasa	Malawi, Mozambique, Tanzania	30,900	706	7,725
Chad	Chad, Niger, Nigeria	16,600[3]	12	44.4
Rudolf	Kenya	8,660	73	—
Mobutu Sese Seko	Uganda, Zaire	5,300	57	64.0
Mweru	Zambia, Zaire	5,100	15	32.0
Bangweulu	Zambia	4,920[4]	5	5.00
Rukwa	Tanzania	4,500	—	—
Tana	Ethiopia	3,150	14	28.0
Idi Amin Dada	Zaire, Uganda	2,500	131	78.2
Kivu	Zaire, Rwanda	2,370	496	569
Mai Ndombe	Zaire	2,325	6	—
Kamnit	Nigeria	1,270	60	14.0
Abaya	Ethiopia	1,160	13	8.20
Shirwa	Malawi	1,040	2.6	45.0
Tumba	Zaire	765	—	—
Faguibine	Mali	620	14	3.72
Gab el Aulia	Sudan	600	12	—
Chamo	Ethiopia	551	12.7	—
Upemba	Zaire	530	3.5	0.90
Zwai	Ethiopia	434	7	1.10
Shala	"	409	266	37.0
Langana	"	230	46.2	3.82
L. de Guiers	Senegal	213	7	0.64
Hora Abyata	Ethiopia	205	14.2	1.56
Naivasha	Kenya	140	—	—
Awusa	Ethiopia	130	21	1.34
	North America			
Superior	Canada, USA	82,680	406	11,600
Huron	Canada, USA	59,800	229	3,580
Michigan	USA	58,100	281	4,680
Great Bear	Canada	30,200	137	1,010
Great Slave	"	27,200	156	1,070

TABLE 3-21. MAJOR LAKES IN THE WORLD (continued)

Lake	Country	Surface Area, km²	Maximum depth, m	Volume, km³
Erie	Canada, USA	25,700	64	545
Winnipeg	Canada	24,600	19	127
Ontario	Canada, USA	19,000	236	1,710
Nicaragua	Nicaragua	8,430	70	108
Athabaska	Canada	7,900	60	110
Dear Lake	"	6,300	—	—
Winnipegosis	"	5,470	12	16
Nipigon	"	4,800	162	
Manitoba	"	4,720	28	17
Great Salt*	USA	4,660	14	19
Forest	Canada, USA	4,410	21	—
Dubawnt	Canada	4,160	—	—
Mistassini	"	2,190	120	—
Managua	Nicaragua	1,490	80	—
Saint Clair	Canada	1,200	7.2	5.3
Lesser Slave	"	1,190	3	—
Chapala	Mexico	1,080	10	10.2
Winnibago	USA	818	6	4.1
Marion	"	465	—	2.8
Winnipesaukee	"	181	55	3.8
	South America			
Maracaibo	Venezuela	13,300	35	—
Titicaca	Peru, Bolivia	8,110	230	710
Poopó*	Bolivia	2,530	3	2
Buenos-Aires	Chile, Argentina	2,400	—	—
Lago Argentina	Argentina	1,400	300	—
Valencia	Venezuela	350	—	—
	Australia			
Eyre*		up to 15,000	20	—
Amadeus*		8,000	—	—
Torrens*		5,800	—	—
Gairdner*		4,780	—	—
George		145	3	0.3
	New Zealand			
Taupo		611	159	—
Te Anau		352	276	—
Wakatipu		293	378	—
Wanaka		194	—	—
Manapouri		130	—	—
Hawea		119	—	—

[1] At low levels 3000 km², at high levels 30,000 km².
[2] At low levels 4000 km², at high levels 12,000 km².
[3] At low levels 7000–10,000 km², at high levels 18,000–22,000 km².
[4] At low levels 4000 km², at high levels 15,000 km².
* Salt lakes.
Source: USSR National Committee for the International Hydrological Decade, Atlas of World Water Balance UNESCO, 1977

SECTION C. WATERFALLS

TABLE 3–22. MAJOR WATERFALLS OF THE WORLD

[Height—total drop in one or more leaps; †—falls of more than one leap; *—falls that diminish greatly seasonally; **—falls that reduce to a trickle or are dry for part of each year. If river names not shown, they are same as the falls. R.—river; L.—lake; (C)—Cascade-type.]

Name and Location	Ft.
AFRICA	
Angola	
Duque de Braganca, Lucala R.	344
Ruacana, Cunene R.	406
Ethiopia	
Baratieri, Ganale Dorya R.	459
Dal Verme, Ganale Dorya R.	98
Fincha	508
*Tesissat, Blue Nile R.	140
Lesotho	
Maletsunyane	630
Rhodesia-Zambia	
*Victoria, Zambezi R.	355
South Africa	
*Aughrabies, Orange R.	400
Howick, Umgeni R.	311
†Tugela (5 falls)	3,110
Highest fall	1,350
Tanzania-Zambia	
*Kalambo	726
Uganda	
Murchison, Victoria Nile R.	130
Zambia	
Chirombo, leisa R.	880
ASIA	
India—**Cauvery	330
†**Gersoppa (Jog), Sharavati R.	830
Japan	
**Kegon, L. Chuzenji	330
Yudaki, L. Yuno	335
AUSTRALASIA	
Australia	
New South Wales	
†Wentworth	518
Highest fall	360
Wollomombi	1,100
Queensland	
Coomera	210
Tully	450
New Zealand	
*Bowen (from Glaciers)	540
Helena	890
Stirling	505
†Sutherland, Arthur R.	1,904
EUROPE	
Austria—Upper Gastein	207
Lower Gastein	280
(Both on Ache R.)	
†Golling, Schwarzbach R.	200
Krimml (Krimmler)	1,250
France—†Gavarnie (C).	1,385
Great Britain—Wales	
Pistyll Cain, Afon Gain R.	150
Pistyll Rhaiadr	240
Scotland	
Glomach	370
Iceland—Detti, Jokul R.	144
Gull, Hvita R.	101
Italy—Toce (C)	470
Norway—	
†Eastern Mardalsfoss	1,696
Highest fall	974
Western Mardalsfoss	1,535
(Both on L. Eikesdal)	
Skjeggedal	525
Skykkje, Skykkjua R.	820
Vettis, Morkedöla R.	1,214
Highest fall	889
Vöring, Bjoreia R.	597
Sweden	
†Handöl, Handöl Cr.	345
†*Stora Sjöfallet, Lule R.	130
Tannforsen, Are R.	120
Switzerland	
†Giétroz (Glacier) (C)	1,640
†Diesbach	394
†Giessbach	1,312
Handegg, Aare R.	151
Iffigen	394
Pissevache, La Salanfe R.	213
†Reichenbach	656
Rhine	65
†Simmen, Simme R.	459
Stäuber	590
Staubbach	984
†Trümmelbach	1,312
NORTH AMERICA	
Canada	
British Columbia	
†Takakkaw (Daly Glacier)	1,650
Highest fall	1,200
Panther, Nigel Cr.	600
Labrador	
Churchill Falls, Churchill R.	245
Mackenzie District	
Virginia, S. Nahanni R.	315
Quebec	
Montmorency	251
Canada-United States	
Ontario-New York	
Niagara: American	193
Horseshoe	186

TABLE 3–22. MAJOR WATERFALLS OF THE WORLD (continued)

[Height—total drop in one or more leaps; †—falls of more than one leap; *—falls that diminish greatly seasonally; **—falls that reduce to a trickle or are dry for part of each year. If river names not shown, they are same as the falls. R.—river; L.—lake; (C)—Cascade-type.]

Name and Location	Ft.
United States	
Arizona	
Mooney, Havasu Cr.	220
California	
Feather, Fall R.	640
Illilouette	370
Nevada, Merced R.	594
**Ribbon	1,612
Silver Strand	1,170
Vernal, Merced R.	317
†Yosemite	2,425
Bridalveil	620
*Yosemite (upper)	1,430
*Yosemite (lower)	320
Colorado	
Seven	266
Georgia	
†Tallulah	251
Idaho	
Henry's Fork (upper)	96
Henry's Fork (lower)	70
**Shoshone, Snake R.	195
**Twin, Snake R.	125
Kentucky	
Cumberland	68
Maryland	
Great Potomac R. (C)	90
Minnesota	
**Minnehaha	54
Montana	
Missouri	75
New Jersey	
**Passaic	70
New York	
Taughannock	215
Oregon	
†Multnomah	620
Highest fall	542
Tennessee	
Fall Creek	256
Rock House Creek	125
Washington	
Fairy Falls	700
Mt. Rainer Nat. Pk	
Narada, Paradise R.	168
Sluiskin, Paradise R.	300
Palouse	198
Snoqualmie	270
Wisconsin	
Manitou, Black R.	165
Wyoming	
Yellowstone National Pk.	
Tower	132
Yellowstone (upper)	109
Yellowstone (lower)	308
Mexico—El Salto	
**Juanacatlán, Rio Grande de Santiago	66
SOUTH AMERICA	
Argentina-Brazil	
†Iguazú	230
Brazil—Glass	1,325
Herval	400
Paulo Afonso, São Francisco R.	275
Patos-Maribondo, Rio Grande	115
Urubupunga, Alto Paraná R.	40
Brazil-Paraguay	
Sete Quedas, or Guaira Alto Paraná R.	130
Colombia—Tequendama,	
Bogotá R.	427
Catarata de Candelas, Cusiana R.	984
Ecuador	
Agoyan, Pastaza R.	200
Guyana	
Kaieteur, Potaro R.	741
King Edward VIII, Semang R.	840
King George VI, Utshi R.	1,600
†Marina, Ipobe R.	500
Highest fall	300
Venezuela—†Angel	3,212
Highest fall	2,648
Cuquenán	2,000

Source: National Geographic Society

SECTION D. GLACIERS AND ICE

TABLE 3–23. GLACIAL ICE COVERAGE OF THE WORLD

Land Area	Square Miles
Continental Europe	3,880
Continental Asia	43,270
Continental North America	30,900
Continental South America	9,600
South polar regions	5,020,450
North polar regions	721,150
Africa	8
New Zealand	386
New Guinea	6
Total	5,829,650

Source: Huberty and Flock, Natural Resources, McGraw-Hill, Copyright 1959

TABLE 3–24. GLACIERS IN THE UNITED STATES

State	Approximate Number of Glaciers	Total Glaciated Area (square miles)	Glacier Contribution to July–August Streamflow (Estimated)	
			Thousand acre-feet	Million gallons
Alaska	(unknown)	29,000	150,000	49,000,000
Washington	950	160	870	280,000
California	290	19	65	21,000
Wyoming	100	19	80	26,000
Montana	200	16	65	21,000
Oregon	60	8	40	13,000
Colorado	25	.6	2	650
Idaho	20	.6	2	650
Nevada	5	.1	.4	130
Utah	1	.04	.1	33

Source: U.S. Geological Survey National Water Summary 1985-Hydrologic Perspectives. Water-Supply Paper 2300

TABLE 3–25. SEASONAL CHANGE IN GLACIER-RUNOFF CHARACTERISTICS

Season	Snowpack Thickness	Albedo	Diurnal Fluctuation in Streamflow	Amount of Runoff	Characteristics of Direct Precipitation-Runoff
Winter	Moderate to high	Very high	Nil	Slight	All precipitation stored
Spring	Highest	High	Slight	Moderate	Subdued, delayed
Summer	Moderate	Moderate to low	High	High	Slight delay
Fall (before a snowpack accumulates)	Low	Low	Moderate	Moderate	No delay, very "flashy"

Source: Meier, in Chow, Handbook of Applied Hydrology, McGraw-Hill, Copyright 1964

SECTION E. FLOODS

FIGURE 3–9. MEAN ANNUAL FLOOD POTENTIAL IN THE UNITED STATES

[In thousands of cfs; for a typical 300 mi² (780 km²) drainage basin; a mean annual flood is one that will be exceeded in about half the years; the probability of a mean annual flood in any given year is about 50%]

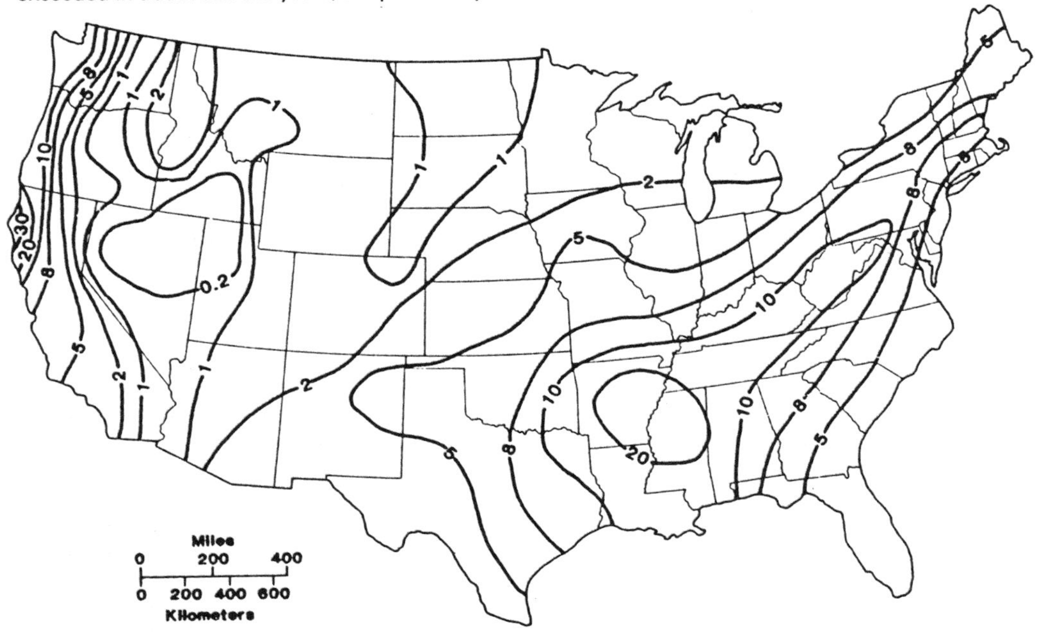

Source: U.S. Geological Survey, 1965. National Atlas Map 121

FIGURE 3–10. TEN-YEAR FLOOD POTENTIAL IN THE UNITED STATES

[In thousands of cfs; for a typical 300 mi² (780 km²) drainage basin; a ten-year flood will be exceeded at irregular intervals that average ten years; the probability of a 10-year flood in any given year is 10%]

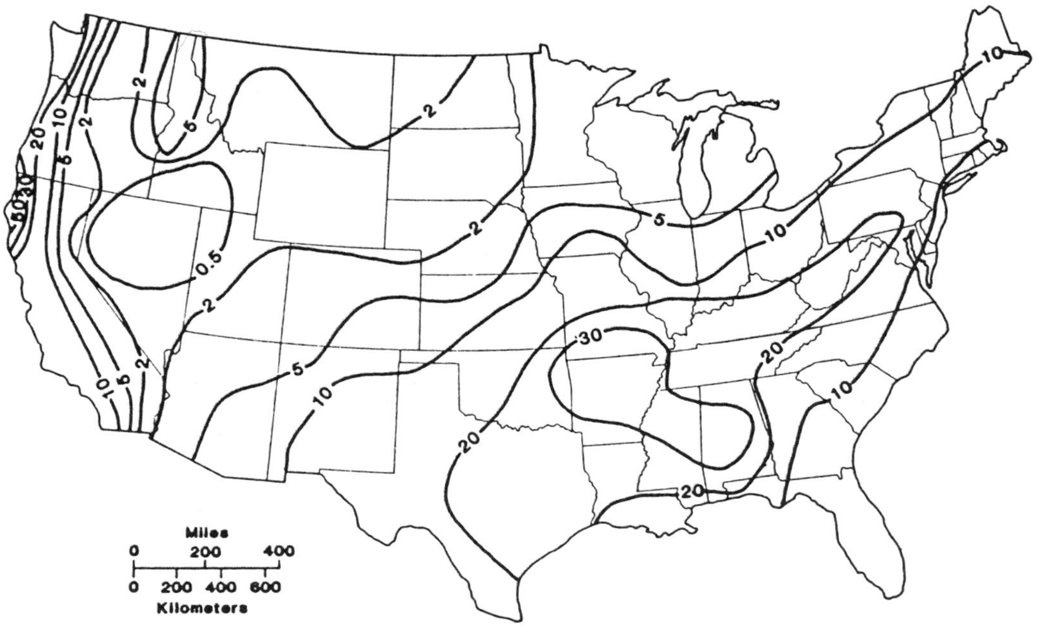

Source: U.S. Geological Survey, 1965. National Atlas Map 121

FIGURE 3–11. FLASH FLOOD WARNING SYSTEM IN THE UNITED STATES

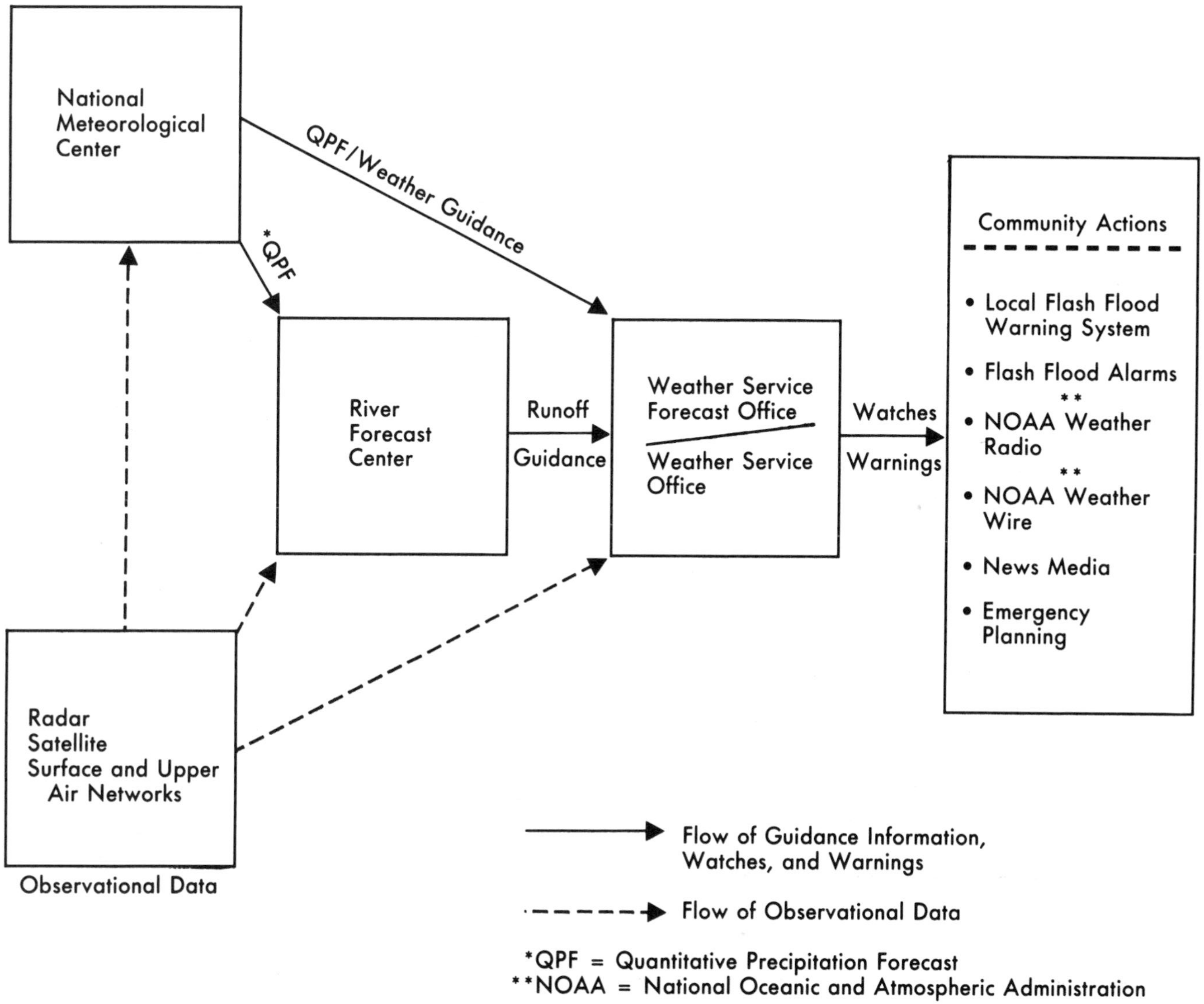

Source: National Weather Service, 1985, Operations of the National Weather Service. U.S. Department of Commerce

TABLE 3–26. GREAT FLOODS IN THE UNITED STATES SINCE 1889

Number[a]	Type of Flood	Date	Location	Lives Lost	Estimated Damages (Millions of Dollars)
1	b	May 1889	Johnstown, Pennsylvania, dam failure	3,000	—
2	c	September 8, 1900	Hurricane—Galveston, Texas	6,000	30
3	d	May–June 1903	Kansas, Lower Missouri, and Upper Mississippi River	100	40
4	d	March 1913	Ohio River and Tributaries	467	147
5	c	September 14, 1919	Hurricane—south of Corpus Christi, Texas	600–900	22
6	b,e	June 1921	Arkansas River, Colorado	120	25
7	d	September 1921	Texas rivers	215	19
8	d	Spring of 1927	Mississippi River valley	313	284
9	d	November 1927	New England rivers	88	46
10	b	March 12–13, 1928	St. Francis Dam failure, southern California	450	14
11	f	September 13, 1928	Lake Okeechobee, Florida	1,836	26
12	d	May–June 1935	Republican and Kansas Rivers	110	18
13	d	March–April 1936	Rivers in Eastern United States	107	270
14	d	January–February 1937	Ohio and Lower Mississippi River basins	137	418
15	d	March 1938	Streams in southern California	79	25
16	d	September 21, 1938	New England	600	306
17	e	July 1939	Licking and Kentucky Rivers	78	2
18	d	May–July 1947	Lower Missouri and Middle Mississippi River basins	29	235
19	d	June–July 1951	Kansas and Missouri	28	923
20	d	August 1955	Hurricane Diane floods—Northeastern United States	187	714
21	d	December 1955	West coast rivers	61	155
22	d	June 27–30, 1957	Hurricane Audrey—Texas and Louisiana	390	150
23	d	December 1964	California and Oregon	40	416
24	d	June 1965	South Platte River basin, Colorado	16	415
25	c	September 10, 1965	Hurricane Betsy—Florida and Louisiana	75	1,420
26	d	January–February 1969	Floods in California	60	399
27	c,d	August 17–18, 1969	Hurricane Camille—Mississippi, Louisiana, and Alabama	256	1,421
28	c	July 30–August 5, 1970	Hurricane Celia—Texas	11	453
29	b	February 1972	Buffalo Creek, West Virginia	125	10
30	e	June 1972	Black Hills, South Dakota	237	165
31	c,d	June 1972	Hurricane Agnes floods—Eastern United States	105	4,020
32	d	Spring 1973	Mississippi River basin	33	1,155
33	d	June–July 1975	Red River of the North basin	<10	273
34	c,d	September 1975	Hurricane Eloise floods—Puerto Rico and Northeastern United States	50	470
35	b	June 1976	Teton Dam failure, southeast Idaho	11	1,000
36	e	July 1976	Big Thompson River, Colorado	139	30
37	e	April 1977	Southern Appalachian Mountains area	22	424
38	b,e	July 1977	Johnstown-western Pennsylvania	78	330
39	d	April 1979	Mississippi and Alabama	<10	500
40	c	September 12–13, 1979	Hurricane Frederic floods—Mississippi, Alabama, and Florida	13	2,000

[a] Number corresponds to those shown on Figure 3–12.
[b] Dam break flood.
[c] Tidal flood.
[d] Riverine flood.
[e] Flash flood.
[f] Flood wave generated in Lake Okeechobee by hurricane.

Source: Hays, W.W., 1981, Facing Geologic and Hydrologic Hazards, Earth-Science Considerations, U.S. Geological Survey Professional Paper 1240-B

FIGURE 3-12. DISTRIBUTION OF GREAT FLOODS IN THE CONTERMINOUS UNITED STATES SINCE 1889

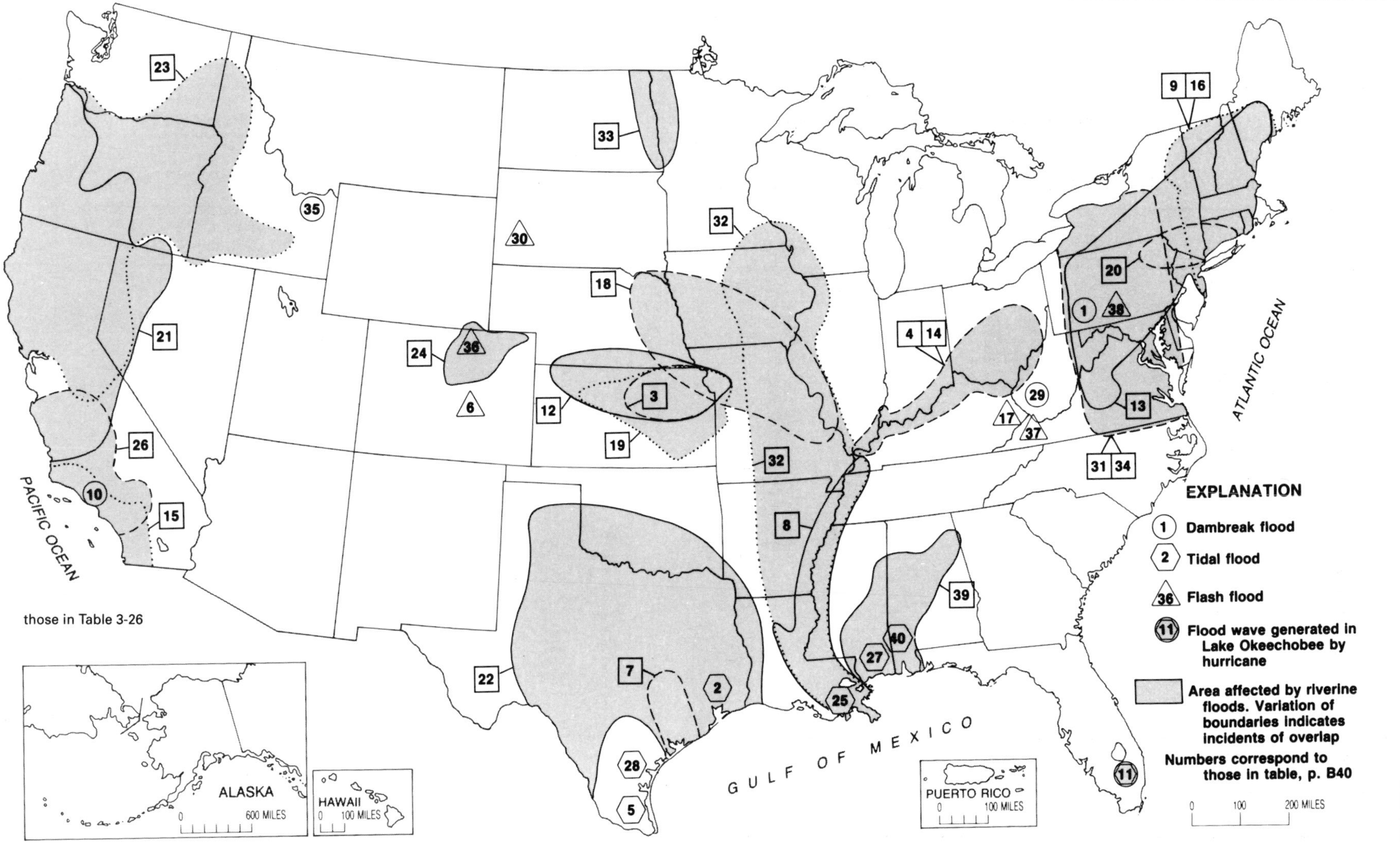

Source: Hays, W.W., 1981, Facing Geologic and Hydrologic Hazards, Earth-Science Considerations, U.S. Geological Survey Professional Paper 1240-B

TABLE 3–27. PROJECTED ANNUAL FLOOD DAMAGES IN THE UNITED STATES

[Values in millions of dollars; for regions see Fig. 2–6. Upstream refers to those streams above a point where the total area drained is 250,000 acres or less; downstream refers to the stream pattern below that point.]

Region	1957		1966		1980[1]		2000[1]		2020[1]	
	Down-stream	Up-stream	Down-stream	Up-stream	Down-stream	Up-stream	Down-stream	Up-stream	Down-stream	Up-stream
North Atlantic	64.3	62.6	63.1	70.7	75.6	91.2	89.8	120.9	116.2	163.3
South Atlantic-Gulf	46.7	109.6	44.1	123.8	55.8	183.2	74.8	267.3	90.4	383.7
Great Lakes	12.3	29.8	13.0	33.7	15.8	43.8	21.0	57.2	27.7	76.1
Ohio	78.7	49.2	73.9	55.6	99.5	68.3	151.0	90.9	237.0	116.6
Tennessee	3.5	27.3	4.9	30.9	7.6	42.6	8.3	58.3	[2]	80.2
Upper Mississippi	60.0	52.0	64.5	68.5	96.0	101.9	151.0	143.2	218.0	197.2
Lower Mississippi	66.0	38.5	86.8	43.5	117.2	55.3	164.2	73.5	224.5	100.1
Souris-Red-Rainy	5.8	13.5	5.6	15.3	6.4	18.4	7.5	24.1	8.6	32.2
Missouri	101.4	148.1	44.0	167.3	69.0	222.3	118.0	302.7	221.0	430.1
Arkansas-White-Red	48.6	129.2	50.0	146.0	61.6	184.0	90.6	245.3	127.0	330.0
Texas-Gulf	32.5	49.5	28.2	55.9	39.5	86.1	59.3	125.3	86.4	178.4
Rio Grande	12.2	10.4	14.7	11.8	14.8	19.5	15.8	30.9	18.8	44.9
Upper Colorado	0.9	16.9	13.0	19.1	19.0	27.4	30.3	42.1	57.0	62.1
Lower Colorado	5.4	25.8	10.0	29.1	20.2	59.3	42.2	93.3	96.7	141.3
Great Basin	3.0	8.4	4.1	9.5	6.7	17.5	10.2	27.5	14.1	42.0
Columbia-North Pacific	52.2	106.7	52.1	120.6	73.6	170.1	120.6	235.3	197.7	325.8
California	36.9	67.2	61.6	75.9	102.1	134.3	185.9	211.0	262.6	311.2
Alaska	3.2	[3]	4.3	[3]	5.6	[3]	8.4	[3]	12.4	[3]
Hawaii	1.2	10.6	1.8	12.2	2.2	16.8	2.8	23.7	3.6	34.0
Puerto Rico-Virgin Islands	2.6	3.8	2.9	4.3	3.2	6.0	3.5	8.5	4.0	12.0
Total[4]	637	959	643	1,094	891	1,548	1,355	2,181	2,024	3,061

[1] Projected damages based on existing flood control works. [2] Not reported. [3] Not available. [4] Rounded.

Source: U.S. Water Resources Council, 1968

TABLE 3–28. ESTIMATED INCREASE IN MEAN ANNUAL FLOOD AS A RESULT OF URBANIZATION IN A ONE-SQUARE MILE AREA

[Data are expressed as the ratio of discharge after urbanization to discharge under previous conditions.]

Percentage of Area Served by Storm Sewerage	Percentage of Area Made Impervious					
	0	20	40	60	80	100
0	1.0	1.2	1.5	1.8	2.0	2.4
20	1.3	1.5	2.1	2.5	2.9	3.7
40	1.4	2.0	2.5	2.9	3.8	4.8
60	1.5	2.2	2.8	3.6	4.6	5.6
80	1.6	2.4	3.0	4.2	5.1	6.2
100	1.7	2.5	3.2	4.4	5.6	6.8

Source: U.S. Geological Survey

TABLE 3–29. MAJOR FLOOD DISASTERS OF THE WORLD

Date	Location	Deaths
1228	Holland	100,000
1642	China	300,000
1887	Huang He River, China	900,000
1889 May 31	Johnstown, Pa.	2,200
1900 Sept. 8	Galveston, Tex.	5,000
1903 June 15	Heppner, Ore.	325
1911	Chang Jiang River, China	100,000
1913 Mar. 25–27	Ohio, Indiana	732
1915 Aug. 17	Galveston, Tex.	275
1928 Mar. 13	Collapse of St. Francis Dam, Santa Paula, Cal.	450
1928 Sept. 13	Lake Okeechobee, Fla	2,000
1931 Aug.	Huang He River, China	3,700,000
1937 Jan. 22	Ohio, Miss. Valleys	250
1939	Northern China	200,000
1946 Apr. 1	Hawaii, Alaska	159
1947	Honshu Island, Japan	1,900
1951 Aug.	Manchuria	1,800
1953 Jan. 31	Western Europe	2,000
1954 Aug. 17	Farahzad, Iran	2,000
1955 Oct. 7–12	India, Pakistan	1,700
1959 Nov. 1	Western Mexico	2,000
1959 Dec. 2	Frejus, France	412
1960 Oct. 10	Bangladesh	6,000
1960 Oct. 31	Bangladesh	4,000
1962 Feb. 17	German North Sea coast	343
1962 Sept. 27	Barcelona, Spain	445
1963 Oct. 9	Dam collapse, Vaiont, Italy	1,800
1966 Nov. 3–4	Florence, Venice, Italy	113
1967 Jan. 18–24	Eastern Brazil	894
1967 Mar. 19	Rio de Janeiro, Brazil	436
1967 Nov. 26	Lisbon, Portugal	464
1968 Aug. 7–14	Gujarat State, India	1,000
1968 Oct. 7	Northeastern India	780
1969 Jan. 18–26	So. Cal.	100
1969 Mar. 17	Mundau Valley, Alagoas, Brazil	218
1969 Aug. 20–22	Western Virginia	189
1969 Sept. 15	South Korea	250
1969 Oct. 1–8	Tunisia	500
1970 May 20	Central Romania	160
1970 July 22	Himalayas, India	500
1971 Feb. 26	Rio de Janeiro, Brazil	130
1972 Feb. 26	Buffalo Creek, W. Va.	118
1972 June 9	Rapid City, S.D.	236
1972 Aug. 7	Luzon Is., Philippines	454
1973 Aug. 19–31	Pakistan	1,500
1974 Mar. 29	Tubaro, Brazil	1,000
1974 Aug. 12	Monty-Long, Bangladesh	2,500
1976 June 5	Teton Dam collapse, Ida.	11
1976 July 31	Big Thompson Canyon, Col.	139
1976 Nov. 17	East Java, Indonesia	136
1977 July 19–20	Johnstown, Pa.	68
1978 June-Sept.	Northern India	1,200
1979 Jan.-Feb.	Brazil	204
1979 July 17	Lomblem Is., Indonesia	539
1979 Aug. 11	Morvi, India	5,000–15,000
1980 Feb. 13–22	So. Cal., Ariz.	26
1981 Apr.	Northern China	550
1981 July	Sichuan, Hubei Prov., China	1,300
1982 Jan. 23	Nr. Lima, Peru	600
1982 May 12	Guangdong, China	430
1982 June 6	So. Conn.	12
1982 Sept. 17–21	El Salvador, Guatemala	1,300+
1982 Dec. 2–9	Ill., Mo., Ark.	22
1983 Feb.-Mar.	Cal. coast	13
1983 Apr. 6–12	Ala., La., Miss., Tenn.	15
1984 May 27	Tulsa, Okla.	13
1984 Aug.-Sept.	S. Korea	200+
1985 July 19	Northern Italy, dam burst	361

Source: The World Almanac and Book of Facts 1988. Copyright Pharos Books, a Scripps Howard Co., New York. Reproduced with permission

TABLE 3–30. MAXIMUM FLOOD FLOWS IN THE WORLD

[Arranged by size of drainage basin]

Country	Station	Basin Area km²	Maximum Discharge m³/s	K[1] Value	Year
U.S.A (CALIFORNIA)	San Rafael San Rafael	3.2	250	5.194	1973
U.S.A. (CALIFORNIA)	L. San Gorgonio Beaumont	4.5	311	5.226	1969
U.S.A. (HAWAII)	Halawa	12	762	5.494	1965
U.S.A. (HAWAII)	Waãilua Lihue	58	2,470	5.819	1963
CUBA	Buey San Miguel	73	2,060	5.623	1963
TAHITI	Papenoo	78	2,200	5.650	1983
MEXICO	San Bartolo	81	3,000	5.859	1976
NEW CALEDONIA	Ouinne Embouchure	143	4,000	5.845	1975
TAIWAN	Cho Shui	259	7,780	6.225	1979
NEW CALEDONIA	Ouaãième derniers rapides	330	10,400	6.389	1981
NEW CALEDONIA	Yaté	435	5,700	5.810	1981
U.S.A. (NEW YORK)	Little Nemaha Syracuse	549	6,370	5.826	1950
NEW ZEALAND	Haast Roaring Billy	1,020	7,690	5.765	1979
U.S.A. (CALIFORNIA)	M.F. American	1,360	8,780	5.770	1964
MEXICO	Cithuatlan Paso del Mojo	1,370	13,500	6.156	1959
AUSTRALIA	Pioneer Pleystowe	1,490	9,840	5.840	1918
TAIWAN	Hualien Hualien Bridge	1,500	11,900	6.011	1973
JAPAN	Nyodo Ino	1,560	13,510	6.111	1963
JAPAN	Kiso Imujama	1,680	11,150	5.910	1961
U.S.A. (TEXAS)	W. Nueces Bracketville	1,800	15,600	6.156	1959
INDIA	Macchu	1,900	14,000	6.060	1979
TAIWAN	Tam Shui Taipei Bridge	2,110	16,700	6.199	1963
JAPAN	Shingu Oga	2,350	19,025	6.290	1959
U.S.A. (TEXAS)	Pedernales Johnson City	2,450	12,500	5.873	1952
NORTH KOREA	Daeryong Gang	3,020	13,500	5.830	1975
JAPAN	Yoshino Iwazu	3,750	14,470	5.844	1974

TABLE 3–30. MAXIMUM FLOOD FLOWS IN THE WORLD (continued)

[Arranged by size of drainage basin]

Country	Station	Basin Area km²	Maximum Discharge m³/s	K[1] Value	Year
PHILIPPINES	Cagayan Echague Isabella	4,244	17,550	5.980	1959
JAPAN	Tone Yattajima	5,110	16,900	5.871	1947
U.S.A. (TEXAS)	Nueces Uvalde	5,504	17,400	5.870	1935
U.S.A. (CALIFORNIA)	Eel Scotia	8,060	21,300	5.917	1964
U.S.A. (TEXAS)	Pecos Comstock	(9,100)	26,800	6.110	1954
MADAGASCAR	Betsiboka Ambodiroka	11,800	22,000	5.780	1927
NORTH KOREA	Toedong Gang Mirim	12,175	29,000	6.060	1967
SOUTH KOREA	Han Koan	23,880	37,000	6.047	1925
PAKISTAN	Jhelum Mangla	29,000	31,100	5.739	1929
CHINA	Hanjiang Hankang	41,400	40,000	5.868	1583
MADAGASCAR	Mangoky Banyan	50,000	38,000	5.698	1933
INDIA	Narmada Garudeshwar	88,000	69,400	6.210	1970
CHINA	Chang Jiang Yitchang	1,010,000	110,000	5.197	1870
U.S.S.R.	Lena Kusur	2,430,000	189,000	5.520	1967
BRAZIL	Amazonas Obidos	4,640,000	370,000	6.760	1953

[1] Flood coefficient k = 10 [(1-(log(Q)-6)/(log(A)-8)] where Q is the largest flood in m³/sec; A is the basin area in km².
Source: Rodier, J.A., and Roche, M., 1984, World Catalogue of Maximum Observed Floods, International Assoc. Hydrological Sciences Publ. No. 143

FIGURE 3–13. RELATIONSHIP BETWEEN MAXIMUM FLOOD FLOWS IN THE WORLD AND SIZE OF DRAINAGE BASIN

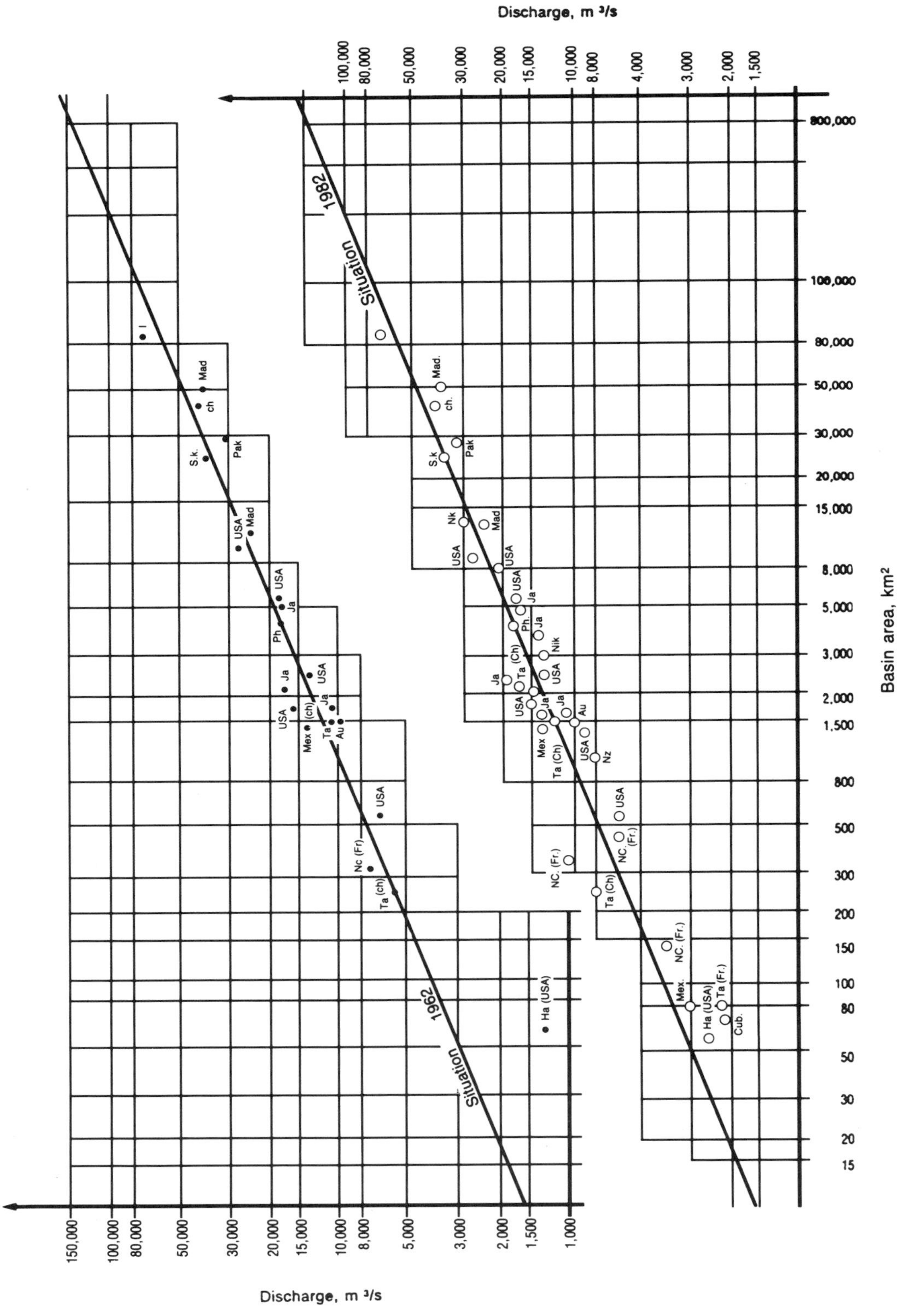

Source: Rodier, J.A., and Roche, M. 1984, World Catalogue of Maximum Observed Floods, International Assoc. Hydrological Sciences Publ. No. 143

TABLE 3–31. TORNADOES, FLOODS, AND TROPICAL CYCLONES IN THE UNITED STATES, 1931–1984

ITEM	1931–1935	1936–1945	1946–1955	1956–1965	1966–1975	1976	1977	1978	1979	1980	1981	1982	1983	1984
Tornadoes, number[1]	830	1,514	2,969	6,572	8,030	835	852	788	852	866	783	1,046	931	907
Lives lost, total	909	1,896	1,751	924	1,172	44	43	53	84	28	24	64	34	122
Most in a single tornado	37	216	169	44	58	5	22	16	42	5	5	10	3	16
Property loss of $500,000 and over	15	56	130	191	428	46	46	59	73	92	55	92	95	125
Floods: Lives lost	368	953	808	557	1,528	187	212	125	103	97	90	155	200	126
Property loss (mil. dol.)	187	1,484	3,350	2,721	10,225	1,000	1,393	1,000	4,000	1,500	1,000	3,500	4,100	4,000
North Atlantic tropical cyclones and hurricanes:[2]														
Number reaching U.S. coast	21	41	40	33	25	2	1	2	5	2	2	1	2	2
Hurricanes only	12	19	21	14	13	1	1	-	3	1	-	-	1	1
Lives lost in U.S.	494	768	495	692	504	9	-	35	11	2	-	-	22	4

- Represents zero. [1] A violent, rotating column of air descending from a cumulonimbus cloud in the form of a tubular- or funnel-shaped cloud, usually characterized by movements along a narrow path and wind speeds from 100 to over 300 miles per hour. Also known as a "twister" or "waterspout." [2] Tropical cyclones have maximum winds of 39 to 73 miles per hour; hurricanes have maximum winds of 74 miles per hour or higher.

Source: Bureau of the Census, Statistical Abstract of the United States 1987 and data from the U.S. National Oceanic and Atmospheric Administration

TABLE 3–32. DEATHS, INJURIES, AND DAMAGE CAUSED BY FLOODS IN THE UNITED STATES, 1965–1985

Fiscal Year	No. of Events	Persons Killed	Persons Injured	Dwellings Destroyed	Dwellings Damaged	Dwellings Destroyed & Damaged
1965–66	67	22	102	91	9,131	9,222
1966–67	NA	16	161	108	22,353	22,461
1967–68	NA	38	824	84	14,224	14,308
1968–69	NA	24	284	71	17,674	17,745
1969–70	NA	51	783	83	33,769	33,852
1970–71	49	22	58	105	6,993	7,098
1971–72	77	519	16,587	7,346	133,805	141,151
1972–73	78	105	1,559	3,229	81,467	84,696
1973–74	83	71	366	1,417	31,309	32,726
1974–75	90	48	500	803	25,008	25,811
1975–76	70	55	2,071	1,377	26,179	27,556
1976–77	58	165	1,469	3,581	35,942	39,523
1977–78	106	196	3,712	1,489	48,508	49,997
1978–79	148	143	3,842	2,659	56,646	59,305
1979–80	122	79	1,121	887	37,439	38,326
1980–81	115	NA	NA	NA	NA	19,578
1981–82	133	70	2,561	NA	NA	46,256
1982–83	149	69	1,988	NA	NA	48,874
1983–84	121	65	1,478	NA	NA	41,578
1984–85	48	9	29	NA	NA	2,308
TOTALS	—	—	—	—	—	762,371

Note: Based on American National Red Cross data which are by fiscal year (July 1–June 30).

Source: Rubin, C.B., Yezer, A.M., Hussain, Q, and Webb, A., 1986, Summary of Major Natural Disaster Incidents in the U.S. 1965–85, Natural Hazards Research and Applications Information Center, George Washington University Spec. Publ. 17

TABLE 3–33. DEATHS CAUSED BY FLOODS IN THE UNITED STATES IN 1987

[By location; flash floods and floods]

State	Boat	Open	Other	Outside	Perm. Home	Play-ing	Auto	All
AL	0	0	0	0	0	0	2	2
AR	0	0	0	0	0	0	4	4
GA	0	0	0	0	0	0	4	4
HI	0	1	0	2	0	0	0	3
IL	0	0	0	0	0	0	1	1
IN	0	0	0	1	0	0	1	2
KY	0	0	0	1	0	1	0	2
MA	0	0	0	1	0	0	0	1
MI	0	0	0	0	0	0	1	1
MN	0	0	0	0	1	0	1	2
NY	0	0	0	1	0	0	10	11
OK	0	0	0	1	0	0	2	3
OR	1	0	0	0	0	0	0	1
PA	0	1	0	0	0	0	0	1
PR	0	0	0	0	0	0	7	7
SC	0	0	0	0	0	1	0	1
TN	0	0	0	0	0	0	3	3
TX	0	1	1	1	0	0	14	17
VA	0	0	0	2	0	0	1	3
VT	0	0	0	0	0	0	1	1
Total	1	3	1	10	1	2	52	70
Percent	1%	4%	1%	14%	1%	3%	74%	99%*

* Rounding to the nearest percent causes the column to sum to less than 100 percent.
Source: Peters, B.E., 1988, National Weather Service, Fort Worth, TX

TABLE 3–34. MAJOR NATURAL DISASTERS IN THE UNITED STATES, 1965–1985

[Federally-declared disasters, by type]

Type of Disaster	Number	Federal Outlay (thousands of current dollars)	Federal Outlay (thousands of 1982 dollars)
Ice and snow events	19	151,427	205,511
Hurricanes/tropical storms	39	1,173,141	1,947,939
Earthquakes	7	203,881	405,706
Dam and levee failures	7	55,764	80,806
Rains, storms & flooding*	337	1,684,702	2,439,852
High winds & waves	2	125,313	120,536
Coastal storms & flooding	7	158,261	205,357
Tornadoes	109	441,685	648,352
Drought/water shortage	4	1,134	5,344
TOTALS	531	3,995,308	6,059,403

*Includes land, mud, and debris flows and slides.
Source: Rubin, C.B., Yezer, A.M., Hussain, Q, and Webb, A., 1986, Summary of Major Natural Disaster Incidents in the U.S. 1965–85, Natural Hazards Research and Applications Information Center, George Washington University Spec. Publ. 17

TABLE 3–35. DEATHS, INJURIES, AND DAMAGE CAUSED BY HURRICANES IN THE UNITED STATES, 1965–1985

Fiscal Year	No. of Events	Persons Killed	Persons Injured	Dwellings Destroyed	Dwellings Damaged	Dwellings Destroyed & Damaged
1965–66	5	72	25,202	2,059	148,607	150,666
1966–67	NA	0	13	6	316	322
1967–68	NA	19	11,396	388	29,405	29,793
1968–69	NA	2	45	1	705	706
1969–70	NA	272	9,062	6,046	48,734	54,780
1970–71	5	9	4,498	1,887	34,442	36,329
1971–72	4	2	235	36	24,258	24,294
1972–73	0	0	0	0	0	0
1973–74	0	0	0	0	0	0
1974–75	2	3	8	45	2,514	2,559
1975–76	3	32	4,409	4,642	31,670	36,312
1976–77	1	2	23	15	498	513
1977–78	3	0	8	6	142	148
1978–79	1	0	0	1	3	4
1979–80	6	20	6,765	6,897	65,033	71,930
1980–81	2	NA	NA	NA	NA	14,865
1981–82	1	0	0	NA	NA	3
1982–83	2	2	961	NA	NA	7,454
1983–84	4	16	3,094	NA	NA	18,663
1984–85	0	0	0	0	0	0
TOTALS	—	—	—	—	—	449,341

Note: Based on American National Red Cross data which are by fiscal year (July 1–June 30)
Source: Rubin, C.B., Yezer, A.M., Hussain, Q, and Webb, A., 1986, Summary of Major Natural Disaster Incidents in the U.S. 1965–85, Natural Hazards Research and Applications Information Center, George Washington University Spec. Publ. 17

TABLE 3–36. PUBLIC AND PRIVATE OUTLAYS FOR HURRICANE DAMAGE IN THE UNITED STATES, 1965–1985

State	Year	Federal Outlay (in thousands of dollars)	Insurance Payment (in thousands of dollars)	States Affected
LA	1965	38,543	500,000	LA,FL,MS
FL	1965	1,706		
MS	1965	1,783		
		42,032		
TX	1967	9,925	34,800	TX
FL	1968	640	2,580	FL
MS	1969	74,524	165,300	MS,LA,AL,FL
LA	1969	15,167		
AL	1969	918		
		90,609		
TX	1970	35,808	309,950	TX
LA	1971	1,160	4,730	LA,MS
FL	1972	3,361	97,853	FL,NY,VA,PA MD,WV,OH,GA SC,NC,MI,DE DC,NJ,CT,RI MA,VT,ME
NY	1972	98,098		
VA	1972	16,815		
PA	1972	351,531		
MD	1972	23,309		
WV	1972	1,294		
OH	1972	1,453		
		495,861		
LA	1974	4,565	14,721	LA
NY	1976	6,773	22,697	NY,NJ,CT,MA
CA	1976	8,507	NA	—

TABLE 3–36. PUBLIC AND PRIVATE OUTLAYS FOR HURRICANE DAMAGE IN THE UNITED STATES, 1965–1985 (continued)

State	Year	Federal Outlay (in thousands of dollars)	Insurance Payment (in thousands of dollars)	States Affected
AL	1979	189,893	752,510	AL,MS,FL,GA
MS	1979	33,684		SC,NC,VA,MD
FL	1979	3,691		DC,DE,PA,NJ
		227,268		NY,CT,MA
TX	1980	31,817	57,911	TX
TX	1980	386	NA	—
HI	1982	11,920	137,000	HI
TX	1983	40,038	675,520	TX
NC	1984	3,460	36,000	NC,SC
MS	1985	18,929	543,304	MS,AL,FL,LA
AL	1985	4,647		
FL	1985	13,933		
		37,509		
PA	1985	9,233	418,750	PA,CT,RI,NJ
CT	1985	21,359		NY,MA,NC,VA
RI	1985	5,846		MD,DE,NH,VT
NJ	1985	4,613		ME
NY	1985	38,750		
MA	1985	13,862		
		93,663		
LA	1985	23,962	44,000	LA,MS,AL,FL
FL	1985	7,238	77,600	FL,GA
TOTAL NO = 39		1,173,141	3,895,226	

Source: Rubin, C.B., Yezer, A.M., Hussain, Q, and Webb, A., 1986, Summary of Major Natural Disaster Incidents in the U.S. 1965–85, Natural Hazards Research and Applications Information Center, George Washington University Spec. Publ. 17

FIGURE 3–14. TRENDS IN DISTRIBUTION OF ANNUAL FLOOD LOSSES IN THE UNITED STATES, 1975–2000

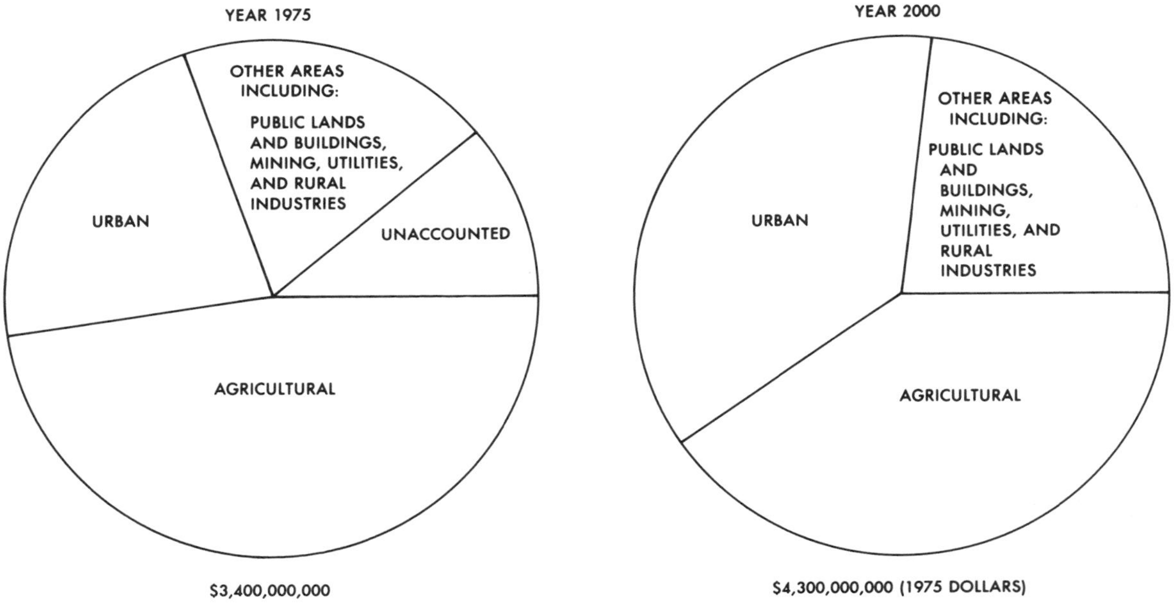

Source: Hays, W.W., 1981, Facing Geologic and Hydrologic Hazards, Earth-Science Considerations, U.S. Geological Survey Professional Paper 1240-B

TABLE 3–37. FLOOD TOLERANCE OF PLANTS

[Cultivated species; effect of flooding with 4 to 10 in. of water for 10 days during the growing season in June 1972]

Common Name	Scientific Name
TOLERANT - NO APPARENT DAMAGE OR MORTALITY	
Shade Trees	
Red maple	*Acer rubrum*
Cornelian cherry	*Cornus mas*
White ash	*Fraxinus americana*
Thornless honey locust	*Gleditsia inermis* (= *G. triacanthos* var. *inermis*)
Black walnut	*Juglans nigra*
Dolgo crabapple	*Malus pumila* 'Dolgo'
White mulberry	*Morus alba*
American sycamore	*Platanus occidentalis*
Cottonwood	*Populus deltoides*
White willow	*Salix alba*
Pussy willow	*S. discolor*
European littleleaf linden	*Tilia cordata*
Evergreens	
Red cedar	*Juniperus virginiana*
Pfitzer juniper	*J. chinensis* var. *pfitzeriana*
Shrubs	
Japanese barberry	*Berberis thunbergii*
Gray-stem dogwood	*Cornus paniculata*
Regel privet	*Ligustrum obtusifolium* var. *regelianum*
Arrowwood	*Viburnum dentatum*
Sweet viburnum	*V. lentago*
American cranberry bush	*V. trilobum*
INTOLERANT - DEFOLIATION OR DEATH	
Shade and Ornamental Trees	
Sugar maple	*Acer saccharum*
Norway maple	*A. platanoides*
Paper birch	*Betula papyrifera*
Gray birch	*B. populifolia*
Redbud	*Cercis canadensis*
Yellowwood	*Cladrastis lutea*
White flowering dogwood	*Cornus florida*
	C. florida 'Cloud 9'
	C. florida 'Cherokee Chief'
Red flowering dogwood	*C. florida* var. *rubra*
Washington hawthorn	*Crataegus phaenopyrum*
Lavalle hawthorn	*C. lavallei*
Saucer magnolia	*Magnolia soulangeana*
Apple	*Malus* sp. 'Lodi,' McIntosh,' 'Radiant,' 'Hope,' 'Bechtel,'
Flowering peach	*Prunus persica*
Black cherry	*P. serotina*
Weeping cherry	*P. subhirtella* var. *pendula*
Red oak	*Quercus rubra*
Black locust	*Robinia pseudoacacia*
European mountain ash	*Sorbus aucuparia*
Evergreens	
Norway spruce	*Picea abies*
Colorado spruce	*P. pungens*
Colorado blue spruce	*P. pungens* var. *glauca*
Upright yew	*Taxus cuspidata*
Spreading yew	*T. cuspidata* var. *expansa*
Hicks yew	*T. media* 'Hicksii'
American arborvitae	*Thuja occidentalis*
Hemlock	*Tsuga canadensis*

Source: White, R.M., 1973. Plant Tolerance in Standing Water: An Assessment, Cornell Plantations 28:50–52. Reproduced with permission

SECTION F. FLOOD PREVENTION

TABLE 3–38. STRATEGIES AND TOOLS FOR ACHIEVING FLOOD HAZARD REDUCTION

Nonstructural

A. Modify susceptibility to flood damage and disruption
  1. Floodplain regulations
    a. State regulations for flood hazard areas
    b. Local regulations for flood hazard areas
      (1) Zoning
      (2) Subdivision regulations
      (3) Building codes
      (4) Housing codes
      (5) Sanitary and well codes
      (6) Other regulatory tools
  2. Development and redevelopment policies
    a. Design and location of services and utilities
    b. Land-right acquisition and open-space use
    c. Redevelopment and renewal
    d. Permanent evacuation
  3. Disaster preparedness and response planning
  4. Floodproofing
  5. Flood forecasting and warning systems and emergency plans

B. Modify the impact of flooding on individuals and the community
  1. Information and education
  2. Flood insurance
  3. Tax adjustments
  4. Flood emergency measures
  5. Postflood recovery

Structural

C. Modify flooding
  1. Dams and reservoirs
  2. Dikes, levees, and floodwalls
  3. Channel alterations
  4. High-flow diversions and spillways
  5. Land treatment measures
  6. On-site detention measures

Source: U.S. Water Resources Council, 1981

TABLE 3–39. STRUCTURAL ADJUSTMENTS AS FLOODPROOFING MEASURES

Measure	Material Protected	Class of Measure	Prerequisites: Structural	Prerequisites: Hydrologic
Seepage control	St-Co	P-C	Well constructed	None
Sewer adjustment	St-Co	P-C	None	H-W
Permanent closure	St-Co	P	Impervious walls	H-S
Openings protected	St-Co	C-E	Impervious walls	H-S-W
Interiors protected	St	P-C	None	S-W
Protective coverings	St-Co	P-C-E	None	H-W-F
Fire protection	St-Co	P	None	None
Appliance protection	Co	E	None	W
Utilities service	Co	P-C-E	None	S-W-V
Roadbed protection	St	P-E	Sound structure	H-W-V-D
Elevation	St-Co	P-C-E	Sound structure	S-W-V-F
Temporary removal	Co	E	None	W-F
Rescheduling	Co	E	Alternatives	W
Proper salvage	Co	. . . . .	None	None
Watertight caps	Co	P-C	None	W
Proper anchorage	St-Co	P-C	Sound structure	S-W-V-D
Underpinning	St	P	Sound structure	V
Timber treatment	St	P	None	None
Deliberate flooding	St-Co	E	None	None
Structural design	St-Co	P	Design	H-S
Reorganized use	Co	P	Alternatives	None

St = structure; Co= content
P = permanent; C = contingent; E = emergency
H = hydrostatic pressure; S = stage of flood; W = warning
F = flood-to-peak interval; V = velocity of flow; D = duration of flood

Source: Schaeffer, Univ. Chicago, Dept. Geography Research Paper, 1960

TABLE 3–40. FLOOD SAFETY RULES

BEFORE THE FLOOD:

1. Keep on hand materials like sandbags, plywood, plastic sheeting, and lumber.
2. Install check valves in building sewer traps, to prevent flood water from backing up in sewer drains.
3. Arrange for auxiliary electrical supplies for hospitals and other operations which are critically affected by power failure.
4. Keep first aid supplies at hand.
5. Keep your automobile fueled; if electric power is cut off, filling stations may not be able to operate pumps for several days.
6. Keep a stock of food which requires little cooking and no refrigeration; electric power may be interrupted.
7. Keep a portable radio, emergency cooking equipment, lights and flashlights in working order.

WHEN YOU RECEIVE A FLOOD WARNING:

8. Store drinking water in clean bathtubs, and in various containers. Water service may be interrupted.
9. If forced to leave your home and time permits, move essential items to safe ground; fill tanks to keep them from floating away; grease immovable machinery.
10. Move to a safe area before access is cut off by flood water.

DURING THE FLOOD:

11. Avoid areas subject to sudden flooding.
12. Do not attempt to cross a flowing stream where water is above your knees.
13. Do not attempt to drive over a flooded road—you can be stranded, and trapped.

AFTER THE FLOOD:

14. Do not use fresh food that has come in contact with flood waters.
15. Test drinking water for potability; wells should be pumped out and the water tested before drinking.
16. Seek necessary medical care at nearest hospital. Food, clothing, shelter, and first aid are available at Red Cross shelters.
17. Do not visit disaster area; your presence might hamper rescue and other emergency operations.
18. Do not handle live electrical equipment in wet areas; electrical equipment should be checked and dried before returning to service.
19. Use flashlights, not lanterns or torches, to examine buildings; flammables may be inside.
20. Report broken utility lines to appropriate authorities.

During any flood emergency, stay tuned to your radio or television station. Information from NOAA and civil emergency forces may save your life.

Source: Environmental Science Services Administration, 1966

TABLE 3–41. METHODS OF FLOOD CONTROL AND ORGANIZATION

Solutions for the flood problem fall into two distinct classes. The first includes those aimed at preventing the overflow of valley lands. The second embraces measures for human adjustment to the flood hazard.

The overflow of the valley lands may be prevented, or reduced in frequency and extent, by:

1. Providing an additional or an alternative channel to carry flood flows;
2. Increasing the capacity of the existing channel, so that the same flood may be passed downstream at lesser heights, thus reducing flood damages—a solution commonly known as channel improvement;
3. Reducing flood heights and damages by holding back a part of the floodwaters by means of reservoirs;
4. Constructing levees and flood walls to prevent the spread of floodwaters, or
5. Any combination of the above.

Measures of the second class, aimed at adjustment to the hazard include:

1. Zoning of the flood plain to inhibit the development of high damageable values in hazardous areas;
2. Abandonment of efforts to use parts of the flood plain;
3. Use of flood forecasting so that damage may be minimized by removal of people and movable property;
4. Use of flood insurance, not to reduce flood damages, but to spread out the cost of floods over a period of years and thus minimize economic shock;
5. Flood relief in the event of disasters.

Source: Task Force on Water Resources and Power, 1955

SECTION G. FLOOD CONTROL WORKS

TABLE 3–42. UPSTREAM FLOOD CONTROL WORKS IN THE UNITED STATES

[Data covers the existing and approved program of the Soil Conservation Service only. In addition, there have been many projects in upstream areas constructed by the Corps of Engineers, the Bureau of Reclamation, and the Bureau of Land Management. Neither the totals nor regional data are available, but this construction amounted to over 1,000 projects, with an estimated cost of about $1 billion. For regions see Fig. 2–6.]

Region	Projects	Watershed Area	Flood Prevention Cost			
			Land Treatment	Reservoirs	Channel Improvement	Total
	No.	*1,000 acres*	*Mil. dol.*	*Mil. dol.*	*Mil. dol.*	*Mil. dol.*
North Atlantic	120	4,906	27.0	108.8	52.5	181.3
South Atlantic-Gulf	163	8,897	93.9	88.6	75.5	258.0
Great Lakes	21	879	11.3	6.4	7.8	25.5
Ohio	89	4,442	45.5	78.8	30.7	155.0
Tennessee	21	1,440	11.8	18.8	10.1	40.7
Upper Mississippi	49	2,052	10.9	23.8	8.4	43.1
Lower Mississippi	104	6,895	80.6	65.9	65.9	212.4
Souris-Red-Rainy	21	2,882	3.8	4.0	15.8	23.6
Missouri	184	4,938	118.9	176.4	31.1	326.4
Arkansas-White-Red	153	12,438	94.4	274.1	27.5	369.0
Texas-Gulf	83	14,616	65.2	113.6	24.9	203.7
Rio Grande	29	1,512	4.7	14.3	2.0	21.0
Upper Colorado	6	517	3.2	9.7	4.3	17.2
Lower Colorado	10	884	5.6	17.0	7.7	30.3
Great Basin	13	871	5.6	8.8	1.2	15.6
Columbia-North Pacific	24	768	5.7	15.0	6.8	27.5
California	31	1,402	11.2	25.7	77.2	114.1
Alaska	0	0	0	0	0	0
Hawaii	5	278	0.5	0.9	10.3	11.7
Puerto Rico-Virgin Islands	3	252	2.7	3.2	7.6	13.5
Total	1,129	70,869	602.5	1,046.8	467.3	2,089.6

Source: U.S. Water Resources Council, 1968

TABLE 3–43. DOWNSTREAM FLOOD CONTROL WORKS IN THE UNITED STATES

[For regions see Fig. 2–6]

Region	Reservoirs			Levees and Floodwalls			Channel Improvement		
	Projects	Storage	Cost[1]	Projects	Structures	Cost	Projects	Improvement	Cost
	No.	1,000 af	Mil. dol.	No.	Miles	Mil. dol.	No.	Miles	Mil. dol.
North Atlantic	5	492	36.0	36	132	144.4	25	54	27.3
South Atlantic-Gulf	7	3,090	49.8	[2]	876	154.4	23	185	5.1
Great Lakes	1	377	23.4	6	7	1.3	9	28	7.1
Ohio	36	12,500	600.0	65	252	202.0	[2]	138	[2]
Tennessee	17	11,590	180.2	0	0	0	0	0	0
Upper Mississippi	14	3,020	54.0	65	861	242.0	19	70	8.7
Lower Mississippi	5	4,400	76.7	[2]	1,621	841.0	[2]	3,348	980.0
Souris-Red-Rainy	5	1,030	1.8	2	2	2.7	7	224	2.9
Missouri	56	20,700	656.0	50	1,130	193.6	7	75	22.2
Arkansas-White-Red	43	24,800	635.5	56	1,023	52.4	39	563	54.0
Texas-Gulf	20	8,600	234.2	5	128	121.4	8	106	94.4
Rio Grande[3]	4	795	31.1	5	205	7.6	7	114	10.5
Upper Colorado	3	1,500	5.5	8	5	0.2	1	1	20.0
Lower Colorado	6	12,100	59.6	2	7	3.2	1	4	0.5
Great Basin	6	386	7.4	4	33	1.6	3	23	1.4
Columbia-North Pacific	24	15,210	320.7	103	546	28.1	27	55	27.7
California	11	3,720	186.5	13	1,515	260.6	12	84	91.4
Alaska	0	0	0	2	3	0.6	1	1	0.1
Hawaii	0	0	0	4	6	0.2	4	3	0.2
Puerto Rico-Virgin Is.	0	0	0	0	0	0	0	0	0
Total[4]	263	124,310	3,158	426+	8,352	2,257	193+	5,076	1,354+

[1] Does not include cost of flood control storage in Bureau of Reclamation projects.
[2] Not reported
[3] Does not include some facilities constructed by the International Boundary and Water Commission.
[4] Rounded.
Source: U.S. Water Resources Council, 1968

SECTION H. WATER AREAS-UNITED STATES

TABLE 3–44. TOTAL, LAND, AND WATER AREAS OF THE UNITED STATES, 1980

[By region, state, or other areas]

REGION AND STATE OR OTHER AREA	TOTAL AREA			LAND AREA [1]		WATER AREA [2]	
	Rank	Sq. mi.	Sq. km.	Sq. mi.	Sq. km.	Sq. mi.	Sq. km.
United States	(X)	3,618,770	9,372,614	3,539,289	9,166,759	79,481	205,856
New England	(X)	66,672	172,681	63,012	163,201	3,660	9,480
Maine	39	33,265	86,156	30,995	80,277	2,270	5,879
New Hampshire	44	9,279	24,032	8,993	23,292	286	739
Vermont	43	9,614	24,900	9,273	24,017	341	883
Massachusetts	45	8,284	21,456	7,824	20,265	460	1,191
Rhode Island	50	1,212	3,140	1,055	2,732	158	408
Connecticut	48	5,018	12,997	4,872	12,618	147	380
Middle Atlantic	(X)	102,203	264,707	99,733	258,308	2,470	6,398
New York	30	49,108	127,190	47,377	122,707	1,731	4,483
New Jersey	46	7,787	20,169	7,468	19,342	319	827
Pennsylvania	33	45,308	117,348	44,888	116,260	420	1,088
East North Central	(X)	248,540	643,719	243,961	631,859	4,579	11,860
Ohio	35	41,330	107,044	41,004	106,201	325	843
Indiana	38	36,185	93,720	35,932	93,064	253	656
Illinois	24	56,345	145,934	55,645	144,120	700	1,814
Michigan	23	58,527	151,586	56,954	147,511	1,573	4,075
Wisconsin	26	56,153	145,436	54,426	140,964	1,727	4,472
West North Central	(X)	517,825	1,341,166	508,132	1,316,063	9,693	25,104
Minnesota	12	84,402	218,601	79,548	206,030	4,854	12,571
Iowa	25	56,275	145,753	55,965	144,950	310	803
Missouri	19	69,697	180,516	68,945	178,568	752	1,948
North Dakota	17	70,702	183,119	69,300	179,486	1,403	3,633
South Dakota	16	77,116	199,730	75,952	196,715	1,164	3,014
Nebraska	15	77,355	200,350	76,644	198,508	711	1,842
Kansas	14	82,277	213,098	81,778	211,805	499	1,293
South Atlantic	(X)	278,926	722,420	266,910	691,296	12,017	31,123
Delaware	49	2,045	5,295	1,932	5,005	112	290
Maryland	42	10,460	27,092	9,837	25,477	623	1,615
District of Columbia	(X)	69	178	63	162	6	16
Virginia	36	40,767	105,586	39,704	102,832	1,063	2,754
West Virginia	41	24,232	62,760	24,119	62,468	112	291
North Carolina	28	52,669	136,413	48,843	126,504	3,826	9,909
South Carolina	40	31,113	80,582	30,203	78,227	909	2,355
Georgia	21	58,910	152,576	58,056	150,365	854	2,211
Florida	22	58,664	151,939	54,153	140,256	4,511	11,683
East South Central	(X)	181,947	471,243	178,824	463,154	3,123	8,090
Kentucky	37	40,410	104,660	39,669	102,743	740	1,917
Tennessee	34	42,144	109,152	41,155	106,591	989	2,561
Alabama	29	51,705	133,915	50,767	131,487	938	2,428
Mississippi	32	47,689	123,515	47,233	122,333	457	1,183
West South Central	(X)	437,701	1,133,646	427,271	1,106,633	10,430	27,013
Arkansas	27	53,187	137,754	52,078	134,883	1,109	2,872
Louisiana	31	47,752	123,677	44,521	115,310	3,230	8,366
Oklahoma	18	69,956	181,186	68,655	177,817	1,301	3,369
Texas	2	266,807	691,030	262,017	678,623	4,790	12,407
Mountain	(X)	863,563	2,236,628	855,193	2,214,951	8,369	21,677
Montana	4	147,046	380,848	145,388	376,555	1,658	4,293
Idaho	13	83,564	216,432	82,412	213,447	1,153	2,985
Wyoming	9	97,809	253,326	96,989	251,202	820	2,125
Colorado	8	104,091	269,596	103,595	268,311	496	1,285
New Mexico	5	121,593	314,925	121,335	314,258	258	667
Arizona	6	114,000	295,260	113,508	293,986	492	1,274
Utah	11	84,899	219,889	82,073	212,569	2,826	7,320
Nevada	7	110,561	286,352	109,894	284,624	667	1,728

TABLE 3–44. TOTAL, LAND, AND WATER AREAS OF THE UNITED STATES, 1980 (continued)

[By region, state, or other areas]

REGION AND STATE OR OTHER AREA	TOTAL AREA			LAND AREA [1]		WATER AREA [2]	
	Rank	Sq. mi.	Sq. km.	Sq. mi.	Sq. km.	Sq. mi.	Sq. km.
Pacific	(X)	921,392	2,386,406	896,253	2,321,295	25,140	65,112
Washington	20	68,139	176,479	66,511	172,264	1,627	4,215
Oregon	10	97,073	251,419	96,184	249,117	889	2,302
California	3	158,706	411,049	156,299	404,814	2,407	6,235
Alaska	1	591,004	1,530,700	570,833	1,478,458	20,171	52,243
Hawaii	47	6,471	16,759	6,425	16,641	46	118
Other areas:							
Puerto Rico	(X)	3,515	9,104	3,459	8,959	56	145
American Samoa	(X)	77	199	77	199	-	-
Guam	(X)	209	541	209	541	-	-
Virgin Islands of the U.S.	(X)	132	342	132	342	1	3
Pacific Islands, Trust Territory of the[3]	(X)	533	1,381	533	1,381	-	-
No. Mariana Islands[3]	(X)	184	477	184	477	-	-

- Represents zero. X Not applicable. [1] Dry land and land temporarily or partially covered by water, such as marshland, swamps, etc.; streams and canals under one-eighth statute mile wide; and lakes, reservoirs, and ponds under 40 acres in area. [2] Permanent inland water surface, such as lakes, reservoirs, and ponds having an area of 40 acres or more; streams, sloughs, estuaries, and canals one-eighth statute mile or more in width; deeply indented embayments and sounds, and other coastal waters behind or sheltered by headlands or islands separated by less than 1 nautical mile of water, and islands under 40 acres in area. Excludes areas of oceans, bays, sounds, etc., lying within U.S. jurisdiction but not defined as inland water.
[3] Under trusteeship.
Source: Statistical Abstract of the United States, 1987

TABLE 3–45. COASTLINE OF THE UNITED STATES

[By State]

STATE	GENERAL COASTLINE[1]		TIDAL SHORELINE[2]		STATE	GENERAL COASTLINE[1]		TIDAL SHORELINE[2]	
	Statute miles	Kilo-meters	Statute miles	Kilo-meters		Statute miles	Kilo-meters	Statute miles	Kilo-meters
U.S.	12,383	19,924	88,633	142,610	**Gulf coast**	1,631	2,624	17,141	27,580
Atlantic coast	2,069	3,329	28,673	46,135	Alabama	53	85	607	977
Connecticut	-	-	618	994	Florida	770	1,239	5,095	8,198
Delaware	28	45	381	613					
Florida	580	933	3,331	5,360	Louisiana	397	639	7,721	12,423
Georgia	100	161	2,344	3,771	Mississippi	44	71	359	578
Maine	228	367	3,478	5,596	Texas	367	591	3,359	5,405
Maryland	31	50	3,190	5,133					
Massachusetts	192	309	1,519	2,444	**Pacific coast**	7,623	12,265	40,298	64,839
					Alaska	5,580	8,978	31,383	50,495
New Hampshire	13	21	131	211	California	840	1,352	3,427	5,514
New Jersey	130	209	1,792	2,883					
New York	127	204	1,850	2,977	Hawaii	750	1,207	1,052	1,693
North Carolina	301	484	3,375	5,430	Oregon	296	476	1,410	2,269
Pennsylvania	-	-	89	143	Washington	157	253	3,026	4,869
Rhode Island	40	64	384	618					
South Carolina	187	301	2,876	4,627	**Arctic coast, Alaska**	1,060	1,706	2,521	4,056
Virginia	112	180	3,315	5,334					

- Represents zero. [1] Figures are lengths of general outline of seacoast. Measurements were made with a unit measure of 30 minutes of latitude on charts as near the scale of 1:1,200,000 as possible. Coastline of sounds and bays is included to a point where they narrow to width of unit measure, and includes the distance across at such point. [2] Figures obtained in 1939–1940 with a recording instrument on the largest-scale charts and maps then available. Shoreline of outer coast, offshore islands, sounds, bays, rivers, and creeks is included to the head of tidewater or to a point where tidal waters narrow to a width of 100 feet.
Source: Statistical Abstract of the United States, 1987

TABLE 3–46. WATER AREAS OTHER THAN INLAND WATER OF THE UNITED STATES

[Includes only that portion of body of water under the jurisdiction of the United States, excluding Alaska and Hawaii. Excludes inland waters]

BODY OF WATER	AREA Sq. mi.	AREA Sq. km.
Total	74,364	192,603
Atlantic coastal water	2,298	5,952
Florida	37	96
Georgia	48	124
Maine	1,102	2,854
Massachusetts	959	2,484
Rhode Island	14	36
South Carolina	138	357
Chesapeake Bay	3,237	8,384
Maryland	1,726	4,470
Virginia	1,511	3,913
Delaware Bay	665	1,722
Delaware	350	907
New Jersey	315	816
Lake Erie	5,002	12,955
Michigan	216	559
New York	594	1,538
Ohio	3,457	8,954
Pennsylvania	735	1,904
Straits of Georgia and Juan de Fuca: Washington	1,610	4,170
Lake Huron: Michigan	8,975	23,245
Long Island Sound	1,298	3,364
Connecticut	573	1,484
New York	726	1,880
Gulf of Mexico coastal water	3,837	9,938
Alabama	560	1,450
Florida	1,698	4,398
Louisiana	1,016	2,631
Mississippi	556	1,440
Texas	7	18
Lake Michigan	22,178	57,441
Illinois	1,526	3,952
Indiana	228	591
Michigan	13,037	33,766
Wisconsin	7,387	19,132
New York Harbor	92	238
New Jersey	69	179
New York	23	60
Lake Ontario: New York	3,033	7,855
Pacific coastal water	343	888
California	69	179
Oregon	48	124
Washington	226	585
Puget Sound: Washington	561	1,453
Lake St. Clair: Michigan	116	300
Lake Superior	21,118	54,696
Michigan	16,231	42,038
Minnesota	2,212	5,729
Wisconsin	2,675	6,928

Source: Statistical Abstract of the United States, 1987

FIGURE 3–15. EXTENT OF WETLANDS IN THE CONTERMINOUS UNITED STATES (mid-1970's)

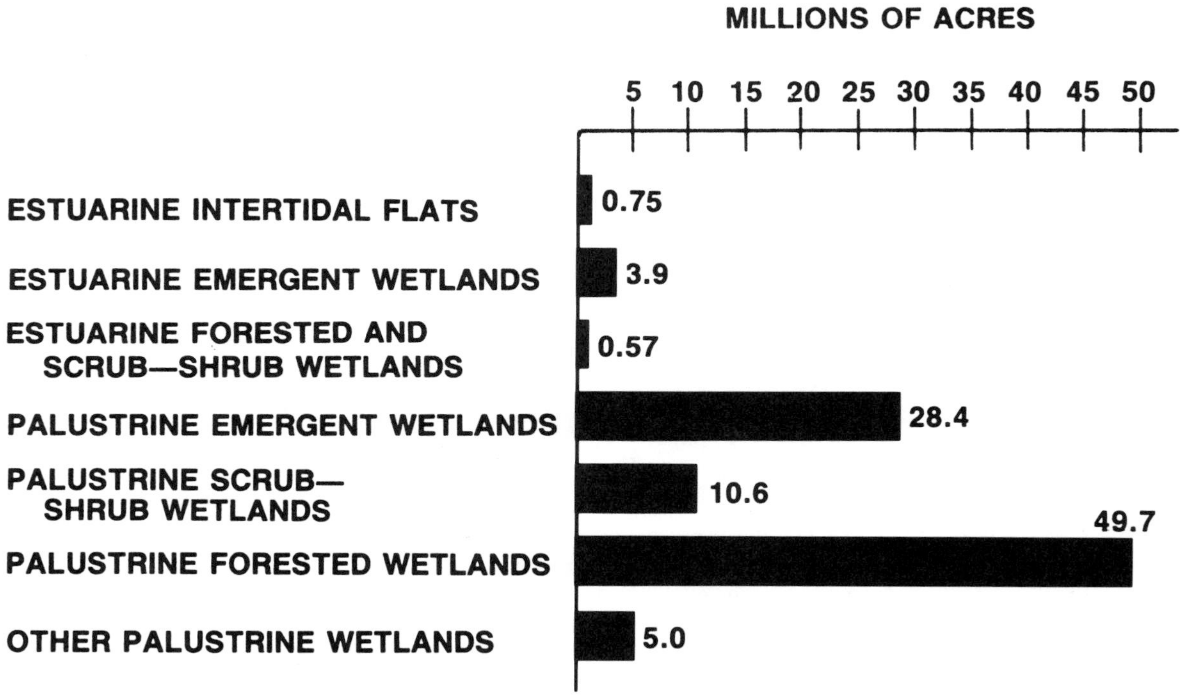

Source: Tiner, R.W., Jr., 1984, Wetlands of the United States: Current Status and Trends, U.S. Department of the Interior, Fish and Wildlife Service

FIGURE 3–16. MAJOR WETLAND TYPES IN THE UNITED STATES

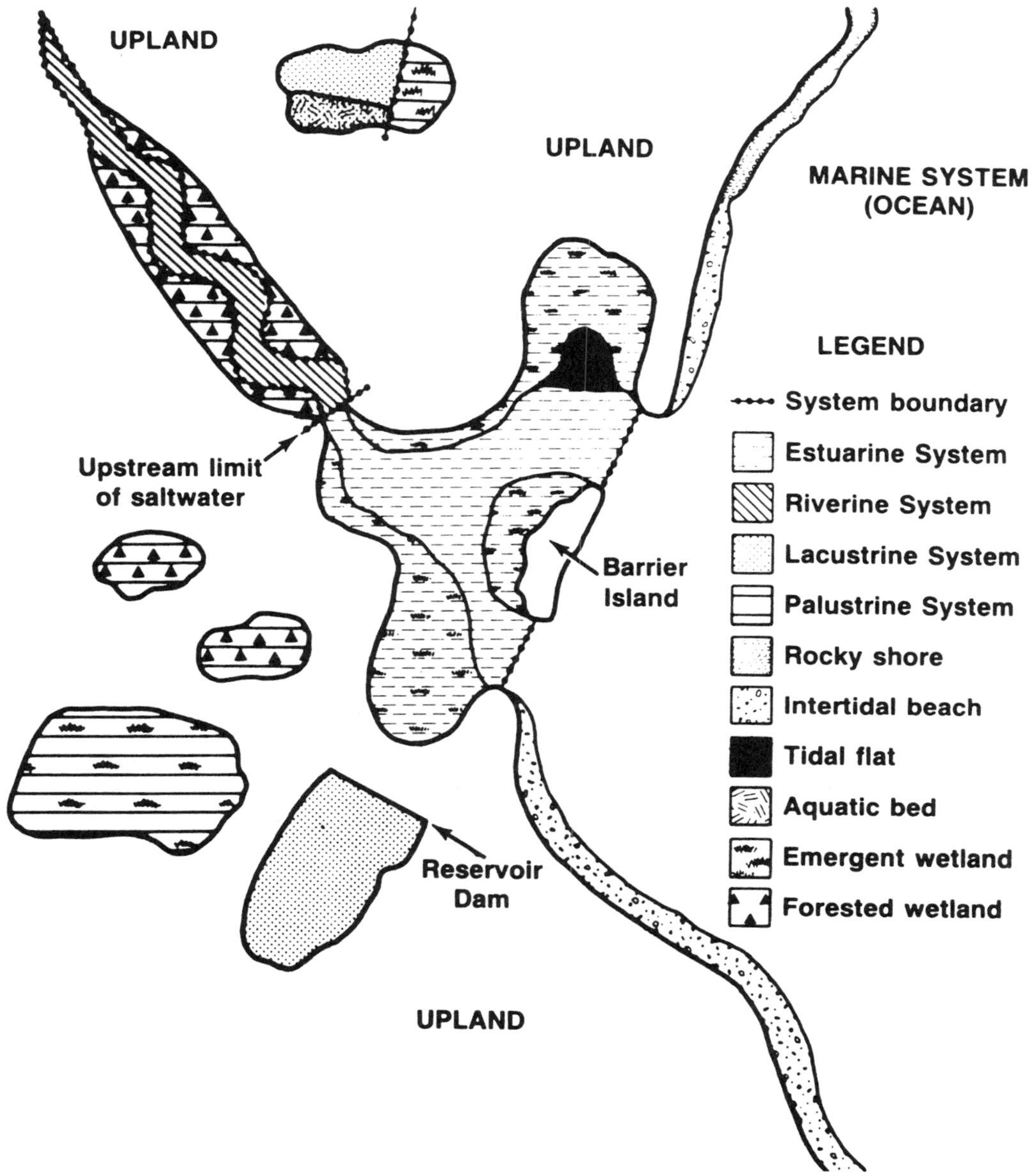

Wetlands occur in every state of the country and due to regional differences in climate, vegetation, soil and hydrologic conditions, they exist in a variety of sizes, shapes and types. Although more abundant in other areas, wetlands even exist in deserts.

Wetlands and deepwater habitats are divided into five ecological systems: (1) Marine, (2) Estuarine, (3) Riverine, (4) Lacustrine, and (5) Palustrine. The Marine System generally consists of the open ocean and its associated coastline. It is mostly a deepwater habitat system, with marine wetlands limited to intertidal areas like beaches, rocky shores and some coral reefs. The Estuarine System includes coastal wetlands like salt and brackish tidal marshes, mangrove swamps, and intertidal flats, as well as deepwater bays, sounds and coastal rivers. The Riverine System is limited to freshwater river and stream channels and is mainly a deepwater habitat system. The Lacustrine System is also a deep water dominated system, but includes standing waterbodies like lakes, reservoirs and deep ponds. The Palustrine System encompasses the vast majority of the country's inland marshes, bogs and swamps and does not include any deepwater habitat.

Source: Tiner, R.W., Jr., 1984, Wetlands of the United States: Current Status and Recent Trends, U.S.Department of the Interior, Fish and Wildlife Service

TABLE 3–47. WETLANDS LOST IN THE UNITED STATES

State or Region	Original Wetlands (acres)	Wetlands in 1984 (acres)	% of Wetlands Lost
Iowa's Natural Marshes	2,333,000	26,470	99
California	5,000,000	450,000	91
Nebraska's Rainwater Basin	94,000	8,460	91
Mississippi Alluvial Plain	24,000,000	5,200,000	78
Michigan	11,200,000	3,200,000	71
North Dakota	5,000,000	2,000,000	60
Minnesota	18,400,000	8,700,000	53
Louisiana's Forested Wetlands	11,300,000	5,635,000	50
Connecticut's Coastal Marshes	30,000	15,000	50
North Carolina's Pocosins	2,500,000	1,503,000*	40
South Dakota	2,000,000	1,300,000	35
Wisconsin	10,000,000	6,750,000	32

* Only 695,000 acres of pocosins remain undisturbed; the rest are partially drained, developed or planned for development.

Source: Tiner, R.W., Jr., 1984, Wetlands of the United States: Current Status and Recent Trends, U.S. Department of the Interior, Fish and Wildlife Service

FIGURE 3–17. CAUSES OF RECENT WETLAND LOSSES IN THE CONTERMINOUS UNITED STATES

(mid-1950's to mid-1970's)

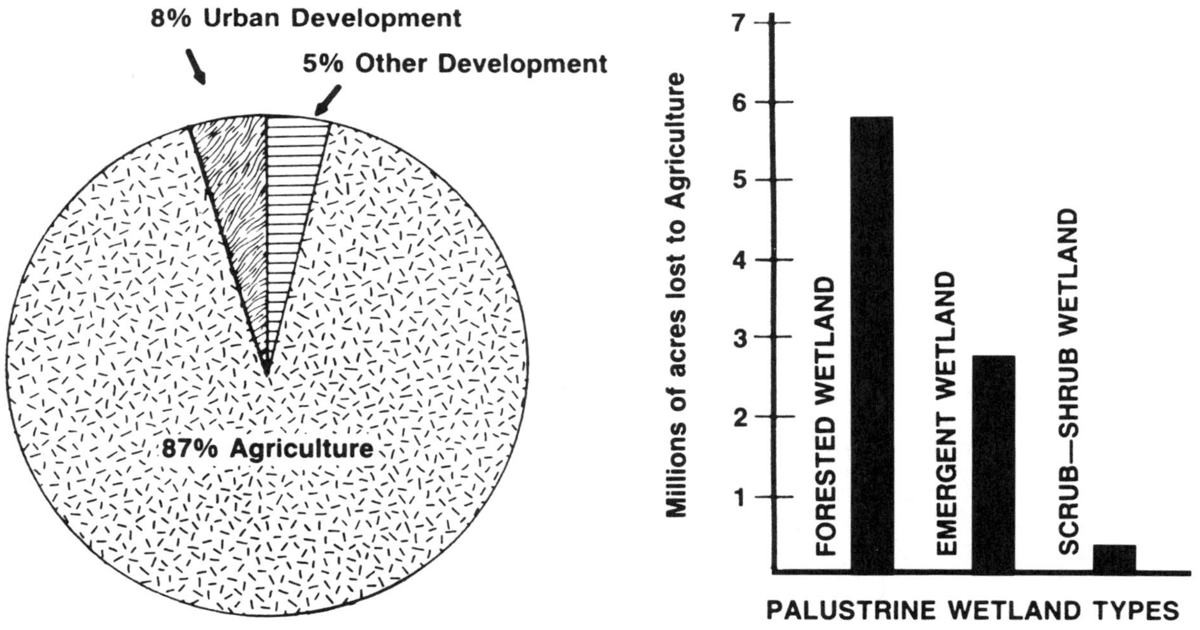

Source: Tiner, R.W., Jr., 1984, Wetlands of the United States: Current Status and Recent Trends, U.S. Department of the Interior, Fish and Wildlife Service

FIGURE 3–18. PRINCIPAL WATERFOWL HABITAT AREAS IN THE UNITED STATES

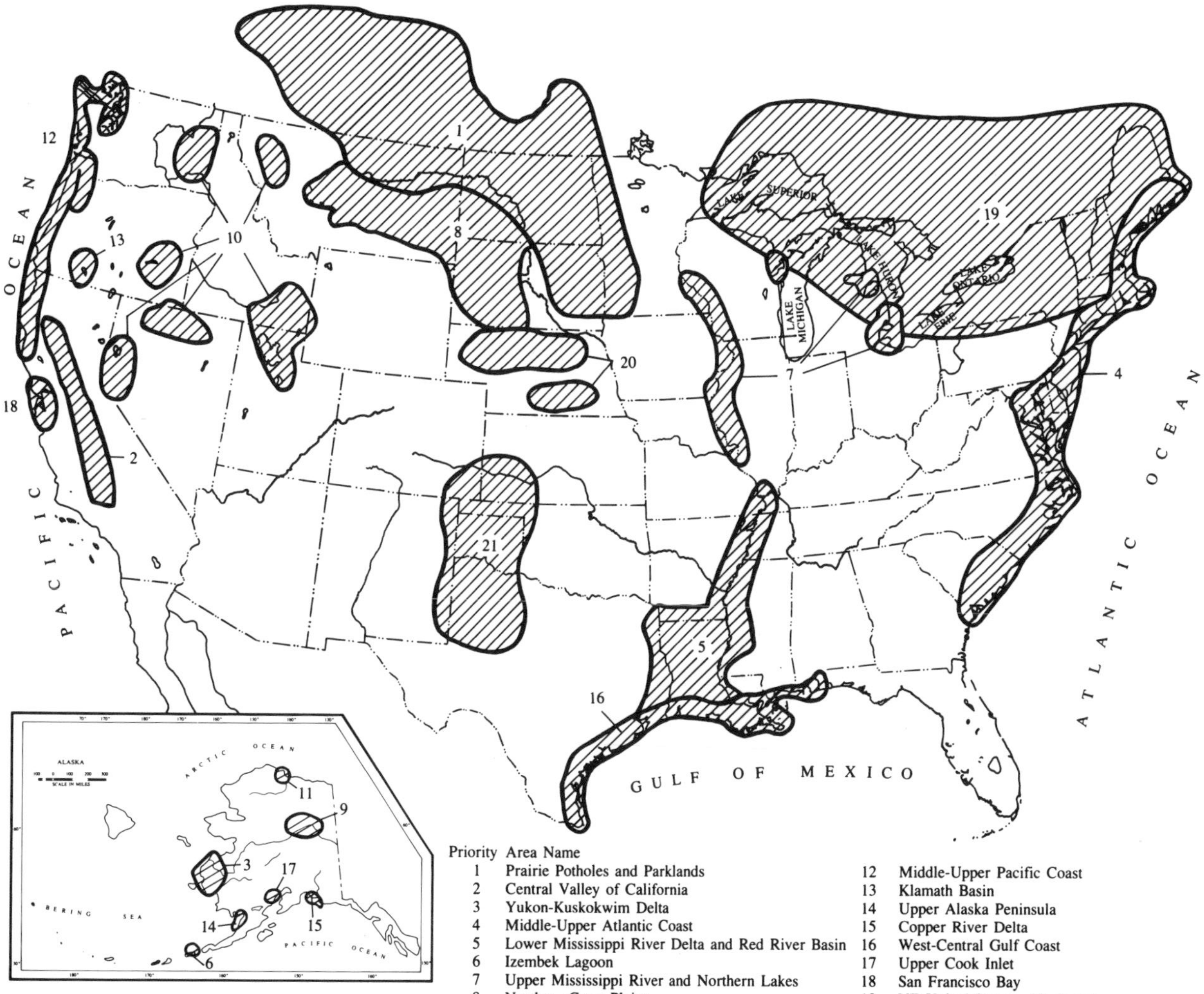

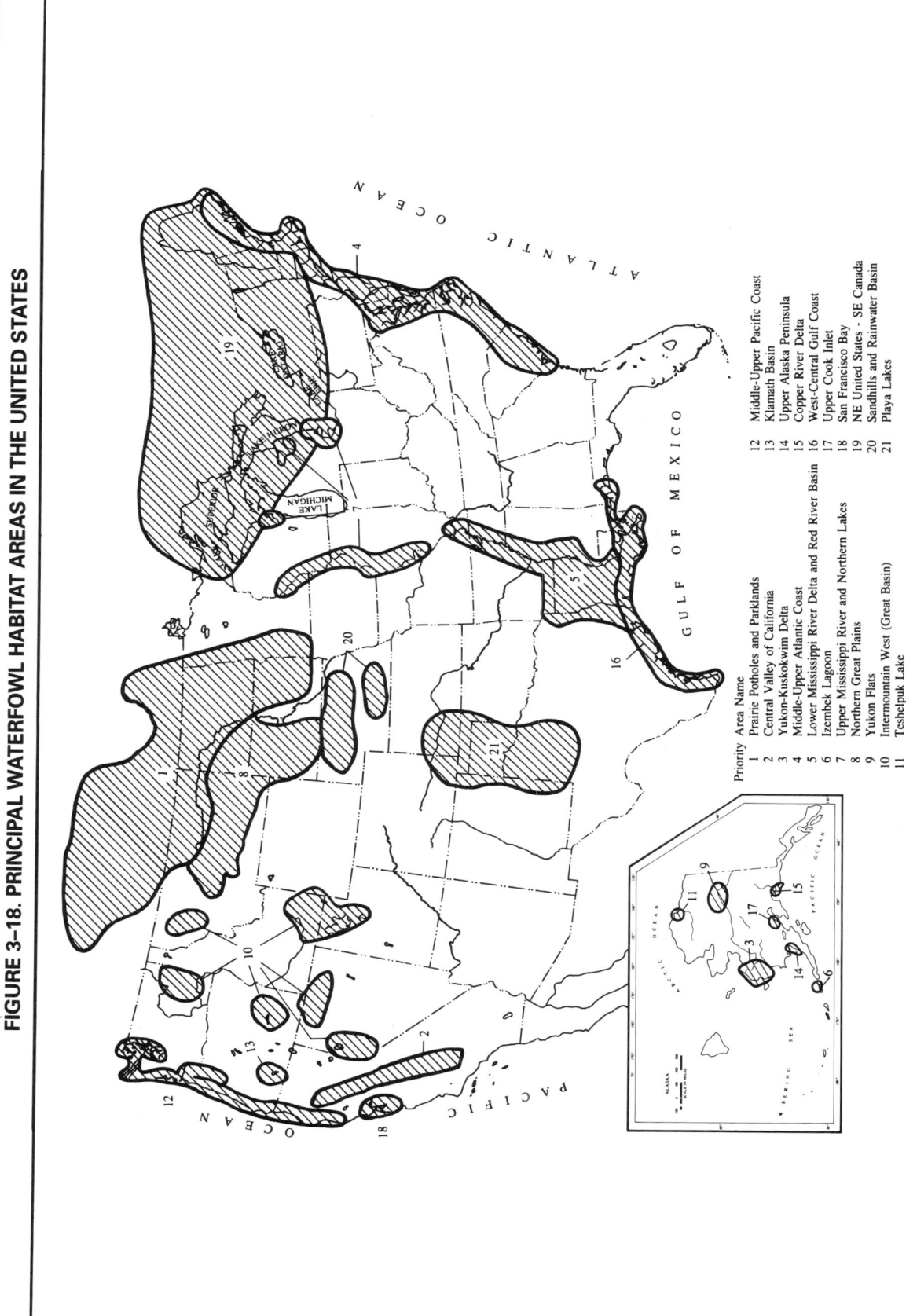

Priority	Area Name	Priority	Area Name
1	Prairie Potholes and Parklands	12	Middle-Upper Pacific Coast
2	Central Valley of California	13	Klamath Basin
3	Yukon-Kuskokwim Delta	14	Upper Alaska Peninsula
4	Middle-Upper Atlantic Coast	15	Copper River Delta
5	Lower Mississippi River Delta and Red River Basin	16	West-Central Gulf Coast
6	Izembek Lagoon	17	Upper Cook Inlet
7	Upper Mississippi River and Northern Lakes	18	San Francisco Bay
8	Northern Great Plains	19	NE United States - SE Canada
9	Yukon Flats	20	Sandhills and Rainwater Basin
10	Intermountain West (Great Basin)	21	Playa Lakes
11	Teshelpuk Lake		

Source: Tiner, R.W., Jr., 1984, Wetlands of the United States: Current Status and Recent Trends, U.S. Department of the Interior, Fish and Wildlife Service

TABLE 3–48. MAJOR CAUSES OF WETLAND LOSS AND DEGRADATION

Human Threats

Direct:

1. Drainage for crop production, timber production and mosquito control.
2. Dredging and stream channelization for navigation channels, flood protection, coastal housing developments, and reservoir maintenance.
3. Filling for dredged spoil and other solid waste disposal, roads and highways, and commercial, residential and industrial development.
4. Construction of dikes, dams, levees and seawalls for flood control, water supply, irrigation and storm protection.
5. Discharges of materials (e.g., pesticides, herbicides, other pollutants, nutrient loading from domestic sewage and agricultural runoff, and sediments from dredging and filling, agricultural and other land development) into waters and wetlands.
6. Mining of wetland soils for peat, coal, sand, gravel, phosphate and other materials.

Indirect:

1. Sediment diversion by dams, deep channels and other structures.
2. Hydrologic alterations by canals, spoil banks, roads and other structures.
3. Subsidence due to extraction of groundwater, oil, gas, sulphur, and other minerals.

Natural Threats:

1. Subsidence (including natural rise of sea level)
2. Droughts
3. Hurricanes and other storms
4. Erosion
5. Biotic effects, e.g., muskrat, nutria and goose "eat-outs."

Source: Tiner, R.W., Jr., 1984, Wetlands of the United States: Current Status and Trends, U.S. Fish and Wildlife Service

TABLE 3–49. MAJOR WETLAND VALUES

Fish and Wildlife Values

- Fish and shellfish habitat
- Waterfowl and other bird habitat
- Furbearer and other wildlife habitat

Environmental Quality Values

- Water quality maintenance
- Pollution filter
- Sediment removal
- Oxygen production
- Nutrient recycling
- Chemical and nutrient absorption
- Aquatic productivity
- Microclimate regulator
- World climate (ozone layer)

Socio-economic Values

- Flood control
- Wave damage protection
- Erosion control
- Groundwater recharge and water supply
- Timber and other natural products
- Energy source (peat)
- Livestock grazing
- Fishing and shellfishing
- Hunting and trapping
- Recreation
- Aesthetics
- Education and scientific research

Source: Tiner, R.W., Jr., 1984, Wetlands of the United States: Current Status and Trends, U.S.Fish and Wildlife Service

TABLE 3–50. EFFECT OF WETLANDS ON FLOOD PEAK REDUCTION IN WISCONSIN

Wetland Present in a Basin (percent)	Storm Recurrence Interval (Years) 2	25	50	100
	Percent of Flood Peak Reduction			
1	19	22	26	27
2	28	33	38	39
3	34	39	45	46
5	42	48	54	55
10	51	58	63	64
15	56	63	70	71
30	64	71	77	79

Source: Conger, D.H., 1971, Estimating Magnitude and Frequency of Floods in Wisconsin, U.S. Geological Survey Open -File Rept.

TABLE 3–51. WATER FEATURES AND PATTERNS

[F - usually associated with fluvial expressions]
[L - usually associated with lacustrine expressions]

Antidune (F)	Mouth
Arm (L)	Narrows
Backwater	Neck cutoff (F)
Bay (L)	Neckfluve (L)
Bay head (L)	Outlet (L)
Bayou	Pass
Bend (F)	Passage
Boil (F)	Plunge pool (F)
Braids (F)	Pool (F)
Branch (F)	Pothole (F)
Cascade (F)	Rapid (F)
Cataract (F)	Reach
Channel	Riffle (F)
Chute	Rift (F)
Cove (L)	Ripple
Cutoff (F)	Rollers (F)
Deadwater (F)	Scour pools (F)
Debouchure (F)	Shallows
Disciplined waters (F)	Shooting water (F)
Eddy (F)	Sinuous channel (F)
Entrenched meander (F)	Slack water
Fork (F)	Slough
Free meander (F)	Sound (L)
Geysers	Standing water
Goosenecks (F)	Straight channel (F)
Gut (L)	Strait
Harbor	Streaming water (F)
Headwaters	Swirls
Inclosed meanders (F)	Torrent (F)
Inlet (L)	Waterfalls - Falls (F)
Lagoon	Waves (L)
Meander (F)	White Water (F)
	Wind streaks

Source: Litton, R.B., Tetlow, R.J., Sorensen, J.,and Beatty, R.A., 1974, Water and Landscape, Water Information Center. Reprinted with permission

TABLE 3–52. TERMS DESCRIPTIVE OF WATER LANDSCAPES

Fluvial Types

Bayou
Braided stream
Brook
Canal
Connecting stream
Creek
Disappearing stream
Feeder stream
Fluvial lakes
Freshet
Inlet
Influent
Intermittent stream
Interrupted stream
Misfit river
Outlet
Raft
Rill
River
Slough
Spring
Stream
Torrent
Vigorously meandering stream
Watercourse
Waterway

Lacustrine Types

Aestival ponds
Alkali lakes
Alluvial dam lakes
Alpine lakes
Bar lake
Barrier lake
Bayou
Blind lake
Blowout pond
Bog lake
Borrow pit pond
Caldera lake
Chain of lakes
Charco
Cirque lake
Clear lake
Closed lake
Crater lake
Dead lake
Deflation lake
Delta lake
Doline lake
Drainage lake
Dry lake
Dugout pond
Dune lake
Dystrophic lake
Effluent lake
Evanescent lake
Extinct lake
Farm pond
Finger lake
Fission lake
Fluvial lake
Fluviatile lake
Fosse lake
Glacial lake
Grass lake
Headwaters lake
Holding pond
Holm lake
Hot springs
Impoundment
Intermittent lake
Kettle lake
Lagoon
Lake
Lakelet
Landslide lake
Laguna
Lateral lake
Marl lake
Marsh lake
Meadow lake
Mesotrophic lake
Mill pond
Mirror lake
Moat lake
Morainal lake
Nova lake
Oligotrophic lake
Open lake
Oriented lake
Oxbow
Palodolac
Perched lake
Perennial lake
Pit lake
Playa
Pond
Pool
Pothole
Puddle
Quarry pond
Raft lake
Reflection basin
Rejuvenated lake
Reservoir
Ria lakes
Rift lakes
Riverine lakes
Rock lakes
Sag pond
Salt lakes
Satellite lakes
Scour lakes
Seepage lakes
Senescent lake
Sink lakes
Slough
Snag lake
Strath lake
Swarm of lakes
Tailing pond
Tanks
Tarn
Thaw lakes
Tundra lakes
Vernal & autumnal ponds
Walled lakes

Source: Litton, R.B., Tetlow, R.J., Sorensen, J., and Beatty, R.A., 1974, Water and Landscape, Water Information Center, Inc. Reprinted with permission

TABLE 3–53. AESTHETIC EVALUATION OF RIVERS

Aesthetic factors:

- Landscape
  - views and vistas
  - diversity of flora and geologic features
  - color
  - form and contrasts
- Sensual stimuli
  - temperature regime
  - winds and other aerial features
  - sounds
  - odors
  - visual patterns
- Intellectual interests
  - opportunities for interpretive programs
  - ecology
  - geology
  - wildlife
  - range and diversity of subjects available for study
- Emotional interest
  - physical stimuli
  - intellectual potentials
  - possibility for adventure
  - interaction of flora, fauna, and people
  - access
  - climatic factors
- Obstacles or discomforts
  - troublesome flora and fauna
  - access
  - climatic factors
- Culture
  - quality of land use management construction
  - scenic pollution
  - historic artifacts

Subjective analysis of most important factors in viewing river - most significant in deriving pleasurable feelings:

- vista
- color
- vegetation (amount and variety)
- spaciousness
- serenity
- naturalness
- riffles in water
- turbidity
- lack of pollution

Source: Morisawa, M., and Murie, M., 1969, Evaluation of Natural Rivers. Antioch College, Water Resources Research, Yellow Springs, OH, in Litton, R.B. and others, 1974, Water and Landscape, Water Information Center, Inc. Reprinted with permission

SECTION I. OCEANS AND SEAS

TABLE 3–54. DIMENSIONS OF THE OCEANS

Ocean	Area ($10^9 m^2$)	Mean Depth (meters)	Volume ($10^{15} m^3$)
Arctic	14,090	1205	17.0
North Pacific	83,462	3858	322.0
South Pacific	65,521	3891	254.9
North Atlantic	46,772	3285	153.6
South Atlantic	37,364	4091	152.8
Indian	81,602	4284	349.6
Antarctic	32,249	3730	120.3

Source: U.S. Naval Oceanographic Office, 1966

TABLE 3–55. MAXIMUM DEPTHS OF THE OCEANS

Name of Area	Location		Depth		
			Meters	Fathoms	Feet
	Pacific Ocean				
Mariana Trench	11°20'N	142°12'E	10,924	5,973	35,840
Tonga Trench	23°16'S	174°44'W	10,800	5,906	35,433
Philippine Trench	10°38'N	126°36'E	10,057	5,499	32,995
Kermadec Trench	31°53'S	177°21'W	10,047	5,494	32,963
Bonin Trench	24°30'N	143°24'E	9,994	5,464	32,788
Kuril Trench	44°15'N	150°34'E	9,750	5,331	31,988
Izu Trench	31°05'N	142°10'E	9,695	5,301	31,808
New Britain Trench	06°19'S	153°45'E	8,940	4,888	29,331
Yap Trench	08°33'N	138°02'E	8,527	4,663	27,976
Japan Trench	36°08'N	142°43'E	8,412	4,600	27,599
Peru-Chile Trench	23°18'S	71°14'W	8,064	4,409	26,457
Palau Trench	07°52'N	134°56'E	8,054	4,404	26,424
Aleutian Trench	50°51'N	177°11'E	7,679	4,199	25,194
New Hebrides Trench	20°36'S	168°37'E	7,570	4,139	24,836
North Ryukyu Trench	24°00'N	126°48'E	7,181	3,927	23,560
Mid America Trench	14°02'N	93°39'W	6,662	3,643	21,857
	Atlantic Ocean				
Puerto Rico Trench	19°55'N	65°27'W	8,605	4,705	28,232
So. Sandwich Trench	55°42'S	25°56'E	8,325	4,552	27,313
Romanche Gap	0°13'S	18°26'W	7,728	4,226	25,354
Cayman Trench	19°12'N	80°00'W	7,535	4,120	24,721
Brazil Basin	09°10'S	23°02'W	6,119	3,346	20,076
	Indian Ocean				
Java Trench	10°19'S	109°58'E	7,125	3,896	23,376
Ob Trench	09°45'S	67°18'E	6,874	3,759	22,553
Diamantina Trench	35°50'S	105°14'E	6,602	3,610	21,660
Vema Trench	09°08'S	67°15'E	6,402	3,501	21,004
Agulhas Basin	45°20'S	26°50'E	6,195	3,387	20,325
	Arctic Ocean				
Eurasia Basin	82°23'N	19°31'E	5,450	2,980	17,881
	Mediterranean Sea				
Ionian Basin	36°32'N	21°06'E	5,150	2,816	16,896

Source: The World Almanac and Book of Facts 1987. Copyright Pharos Books, a Scripps Howard Co., New York. Reprinted with permission

TABLE 3–56. WATER RESIDENCE TIMES OF THE OCEANS

	North Polar	Atlantic	Pacific	Indian	Total
A. Ocean volume (km^3 x 10^4)	8.85	350	695	295	1,349
B. Surface ocean volume (200 m) (km^3 x 10^6)	1.7	19.6	35.4	15.5	72.2
C. Ocean flow compensation (km^3/year)	3,000	-17,100	+28,000	-13,900	0
D. Total compensation time: A/C (year)	2,950	20,500	25,000	21,200	
E. Surface-only compensation time: B/C (year)	570	1,150	1,250	1,100	
F. Stream runoff to oceans (km^3/year)	2,600	19,400	21,100	5,600	39,700
G. Runoff residence time: A/F (year)	3,400	18,000	57,500	52,700	34,000
H. Surface runoff residence time: B/F (year)	650	1,000	2,900	2,800	1,800
I. Atmospheric cycling (precipitation minus evaporation) (km^3/year)	400	-36,500	15,900	-19,500	-39,700
J. Whole-ocean atmospheric-cycling residence time: A/I (year)	22,125	9,600	43,700	15,100	34,000
K. Atmospheric-cycling surface ocean residence time: B/I (year)	4,250	500	2,200	1,000	1,800

Source: Speidel, D.H., and Agnew, A.F., 1979, The Natural Geochemistry of our Environment, in An Overview of Research in Biogeochemistry and Environmental Health, Committee Print 825, Committee on Science and Technology. U.S. House of Representatives, 77–239

TABLE 3–57. WATER BALANCE OF THE OCEANS

Ocean	Area, thou. km^3	Precipitation		Evaporation		Inflow of Water from Continents		Water Balance	
		mm	thou. km^3	mm	thou. km^3	mm	thou. km^3	mm	thou. km^3
Pacific	178,700	1,460	260	1,510	269.7	83	14.8	30	5.1
Southern sector	25,300	1,140	28.9	684	17.3	72	1.8	530	13.4
Without southern sector	153,400	1,510	231.1	1,640	252.4	85	13.0	-50	-8.4
Atlantic	91,700	1,010	92.7	1,360	124.4	226	20.8	-120	-10.9
Southern sector	15,500	1,190	18.4	466	7.2	37	0.6	760	11.8
Without southern sector	76,200	975	74.3	1,540	117.2	265	20.6	-300	-22.7
Indian	76,200	1,320	100.4	1,420	108.0	81	6.1	-20	-1.5
Southern sector	28,500	1,240	35.4	688	19.6	30	0.8	580	16.6
Without southern sector	47,700	1,360	65.0	1,850	88.4	111	5.3	-380	-18.1
Arctic	14,700	361	5.3	220	8.2	355	5.2	500	7.3
World	361,300	1,270	458.0	1,400	505.0	130	47.0	0	0

Source: UNESCO, 1977, Atlas of World Water Balance. Reproduced with permission

FIGURE 3–19. DISTRIBUTION OF PRECIPITATION, SURFACE WATER RUNOFF, AND GROUND WATER DISCHARGE TO WORLD'S OCEANS
[BY LATITUDINAL LAND ZONES]

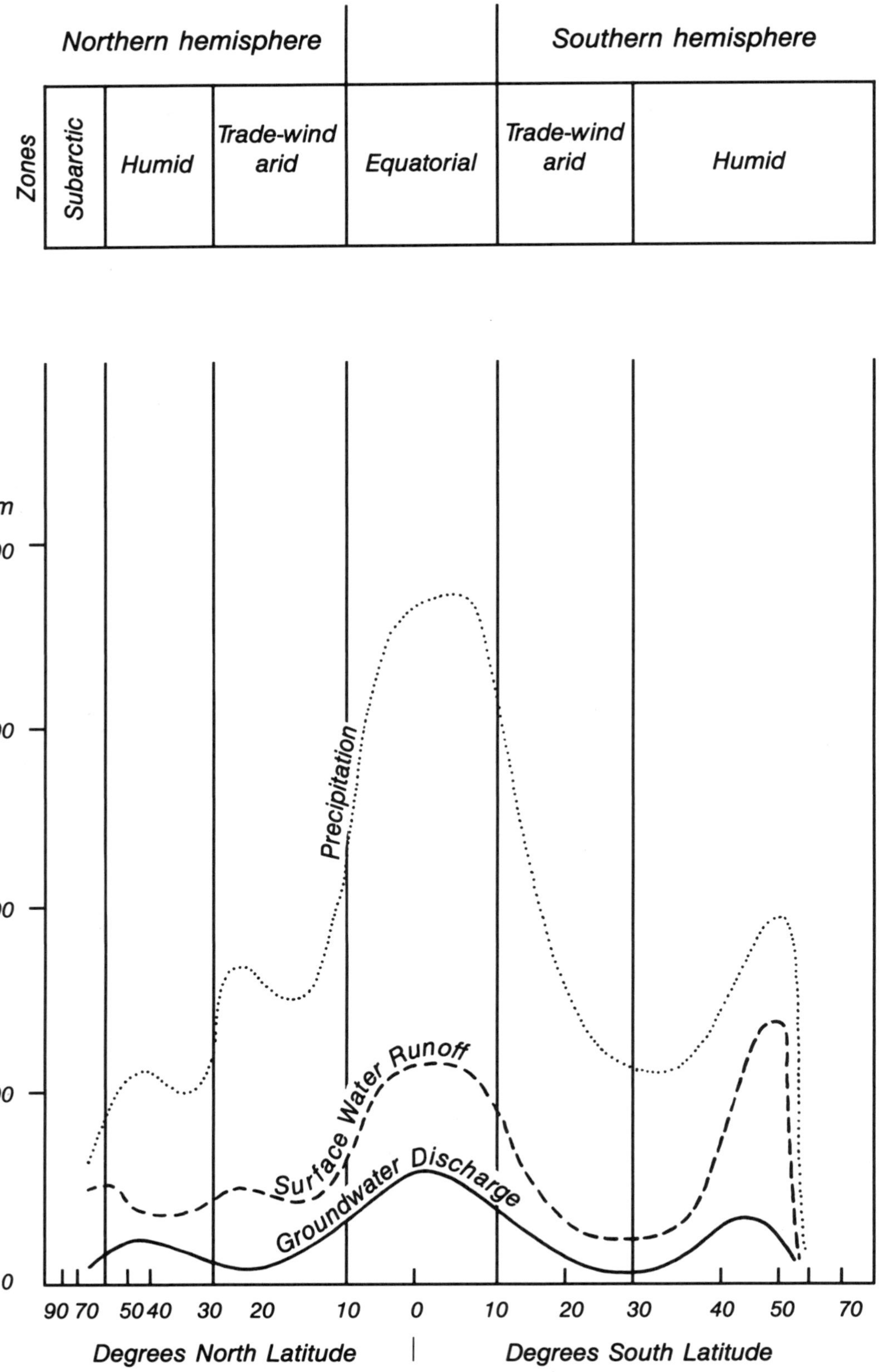

Source: Zektser, I.S. and others, 1984, The Effect of Groundwater on the Salt

TABLE 3–58. DIMENSIONS OF INDIVIDUAL SEAS

Sea	Area ($10^9 m^2$)	Mean Depth (meters)	Volume ($10^{12} m^3$)
Tributary to Arctic Ocean			
Norwegian Sea	1383	1742	2408
Greenland Sea	1205	1444	1740
Barents Sea	1405	229	322
White Sea	90	89	8
Kara Sea	883	118	104
Laptev Sea	650	519	338
East Siberian Sea	901	58	53
Chukchi Sea	582	88	51
Beaufort Sea	476	1004	478
Baffin Bay	689	861	593
Tributary to North Atlantic			
North Sea	600	91	55
Baltic Sea	386	86	33
Mediterranean Sea	2516	1494	3758
Black Sea	461	1166	537
Marmara	11	357	4
Azov	40	9	0.4
Caribbean Sea	2754	2491	6860
Gulf of Mexico	1543	1512	2332
Gulf of St. Lawrence	238	127	30
Hudson Bay	1232	128	158
Tributary to South Atlantic			
Gulf of Guinea	1533	2996	4592
Tributary to Indian Ocean			
Red Sea	450	558	251
Persian Gulf	241	40	10
Arabian Sea	3863	2734	10561
Bay of Bengal	2172	2586	5616
Andaman Sea	602	1096	660
Great Australian Bight	484	950	459
Tributary to North Pacific			
Gulf of California	177	818	145
Gulf of Alaska	1327	2431	3226
Bering Sea	2304	1598	3683
Okhotsk Sea	1590	859	1365
Japan Sea	978	1752	1713
Yellow Sea	417	40	17
East China Sea	752	349	263
Sulu Sea	420	1139	478
Celebes Sea	472	3291	1553
In both North and South Pacific			
South China Sea	3685	1060	3907
Makassar Strait	194	967	188
Molukka Sea	307	1880	578
Ceram Sea	187	1209	227
Tributary to South Pacific			
Java Sea	433	46	20
Bali Sea	119	411	49
Flores Sea	121	1829	222
Savu Sea	105	1701	178
Banda Sea	695	3064	2129
Ceram Sea	187	1209	227
Timor Sea	615	406	250
Arafura Sea	1037	197	204
Coral Sea	4791	2394	11470

Source: U.S. Naval Oceanographic Office, 1966; amended

TABLE 3–59. AVERAGE RISE AND FALL OF TIDES IN THE UNITED STATES AND CANADA

Location	Feet
East Coast	
Quebec	13.7
Halifax, N.S.	4.4
St. John, N.B.	20.8
Eastport, ME	18.2
Portland, ME	9.0
Boston, MA	9.5
Newport, RI	3.5
New London, CT	2.6
Bridgeport, CT	6.7
New York (The Battery)	4.5
Port Jefferson, NY	6.6
Albany, NY	4.6
Newark, NJ	5.1
Sandy Hook, NJ	4.6
Philadelphia, PA	5.9
Cape May, NJ	4.3
Washington, DC	2.9
Cape Hatteras, NC	3.6
Wilmington, NC	4.2
Charlotte, SC	5.2
Savannah, GA	7.4
Miami, FL	2.5
Key West, FL	1.3
Mobile, AL	1.5[a]
Galveston, TX	1.4
San Juan, PR	1.1
West Coast	
Vancouver, B.C.	10.6[a]
Seattle, WA	14.4
San Francisco, CA	9.0
Los Angeles, CA	6.6
San Diego, CA	6.5

[a] Diurnal range
Source: National Oceanic and Atmospheric Administration, National Ocean Survey Tide Tables

TABLE 3–60. TEMPERATURES AND SALINITIES OF THE OCEANS

[Temperatures in degrees Centigrade; salinities in parts per thousand]

NORTH ATLANTIC		Temperature	Salinity
1.	North Polar water	-1 to +2	34.9
2.	Subarctic water	+3 to +5	34.7 to 34.9
3.	North Atlantic central water	+4 to +17	35.1 to 36.2
4.	North Atlantic deep water	+3 to +4	34.9 to 35.0
5.	North Atlantic bottom water	+1 to +3	34.8 to 34.9
6.	Mediterranean water	+6 to +10	35.3 to 36.4
SOUTH ATLANTIC			
1.	South Atlantic central water	+5 to +16	34.3 to 35.6
2.	Antarctic intermediate water	+3 to +5	34.1 to 34.6
3.	Subantarctic water	+3 to +9	33.8 to 34.5
4.	Antarctic circumpolar water	+0.5 to +2.5	34.7 to 34.8
5.	South Atlantic deep and bottom water	0 to +2	34.5 to 34.9
6.	Antarctic bottom water	-0.4	34 to 36
INDIAN OCEAN			
1.	Equatorial water	4 to 16	34.8 to 35.2
2.	Indian central water	6 to 15	34.5 to 35.4
3.	Antarctic intermediate water	2 to 6	34.4 to 34.7
4.	Subantarctic water	2 to 8	34.1 to 34.6
5.	Indian Ocean deep and antarctic circumpolar water	0.5 to 2	34.7 to 34.75
6.	Red Sea water	9	35.5
SOUTH PACIFIC			
1.	Eastern South Pacific water	9 to 16	34.3 to 35.1
2.	Western South Pacific water	7 to 16	34.5 to 35.5
3.	Antarctic intermediate water	4 to 7	34.3 to 34.5
4.	Subantarctic water	3 to 7	34.1 to 34.6
5.	Pacific deep water and Antarctic circumpolar water	(-1) to 3	34.6 to 34.7
NORTH PACIFIC			
1.	Subarctic water	2 to 10	33.5 to 34.4
2.	Pacific equatorial water	6 to 16	34.5 to 35.2
3.	Eastern North Pacific water	10 to 16	34.0 to 34.6
4.	Western North Pacific water	7 to 16	34.1 to 34.6
5.	Arctic intermediate water	6 to 10	34.0 to 34.1
6.	Pacific deep water and Arctic circumpolar water	(-1) to 3	34.6 to 34.7

Source: U.S. Naval Oceanographic Office, 1966

TABLE 3–61. COMPOSITION OF SEA WATER

Constituent:	Concentration (ppm)	Constituent	Concentration (ppm)
Chloride	18,980	Lead	.004–.005
Sodium	10,560	Selenium	.004
Sulfate	2,560	Arsenic	.003–.024
Magnesium	1,272	Copper	.001–.09
Calcium	400	Tin	.003
Potassium	380	Iron	.002–.02
Bicarbonate	142	Cesium	~.002
Bromide	65	Manganese	.001–.01
Strontium	13	Phosphorous	.001–.10
Boron	4.6	Thorium	=<.0005
Fluoride	1.4	Mercury	.0003
Rubidium	.2	Uranium	.00015–.0016
Aluminum	.16–1.9	Cobalt	.0001
Lithium	.1	Nickel	.0001–.0005
Barium	.05	Radium	$8.x10^{-11}$
Iodide	.05	Beryllium	—
Silicate	.04–8.6	Cadmium	—
Nitrogen	.03–.9	Chromium	—
Zinc	.005–.014	Titanium	Trace

Source: U.S. Geological Survey

TABLE 3–62. APPROXIMATE MINERAL CONTENT OF ONE CUBIC MILE OF SEA WATER

Mineral :	Weight, in tons	Mineral :	Weight, in tons
Sodium Chloride	120,000,000	Fluorine	6,400
Magnesium Chloride	18,000,000	Barium	900
Magnesium Sulfate	8,000,000	Iodine	100 to 12,000
Calcium Sulfate	6,000,000	Arsenic	50 to 350
Potassium Sulfate	4,000,000	Rubidium	200
Calcium Carbonate	550,000	Silver	up to 45
Magnesium Bromide	350,000	Copper, Manganese,	
Bromine	300,000	Zinc, Lead	10 to 30
Strontium	60,000	Gold	up to 25
Boron	21,000	Radium	about 1/6 (ounce)
		Uranium	7

Source: Smith, The Sun, the Sea, and Tomorrow; Potential Sources of Food, Energy and Minerals from the Sea, Charles Scribners, 1954

CHAPTER 4

Ground Water

PAGE

SECTION A. GROUND WATER—UNITED STATES

FIGURE 4–1. PRINCIPAL AQUIFERS OF THE UNITED STATES

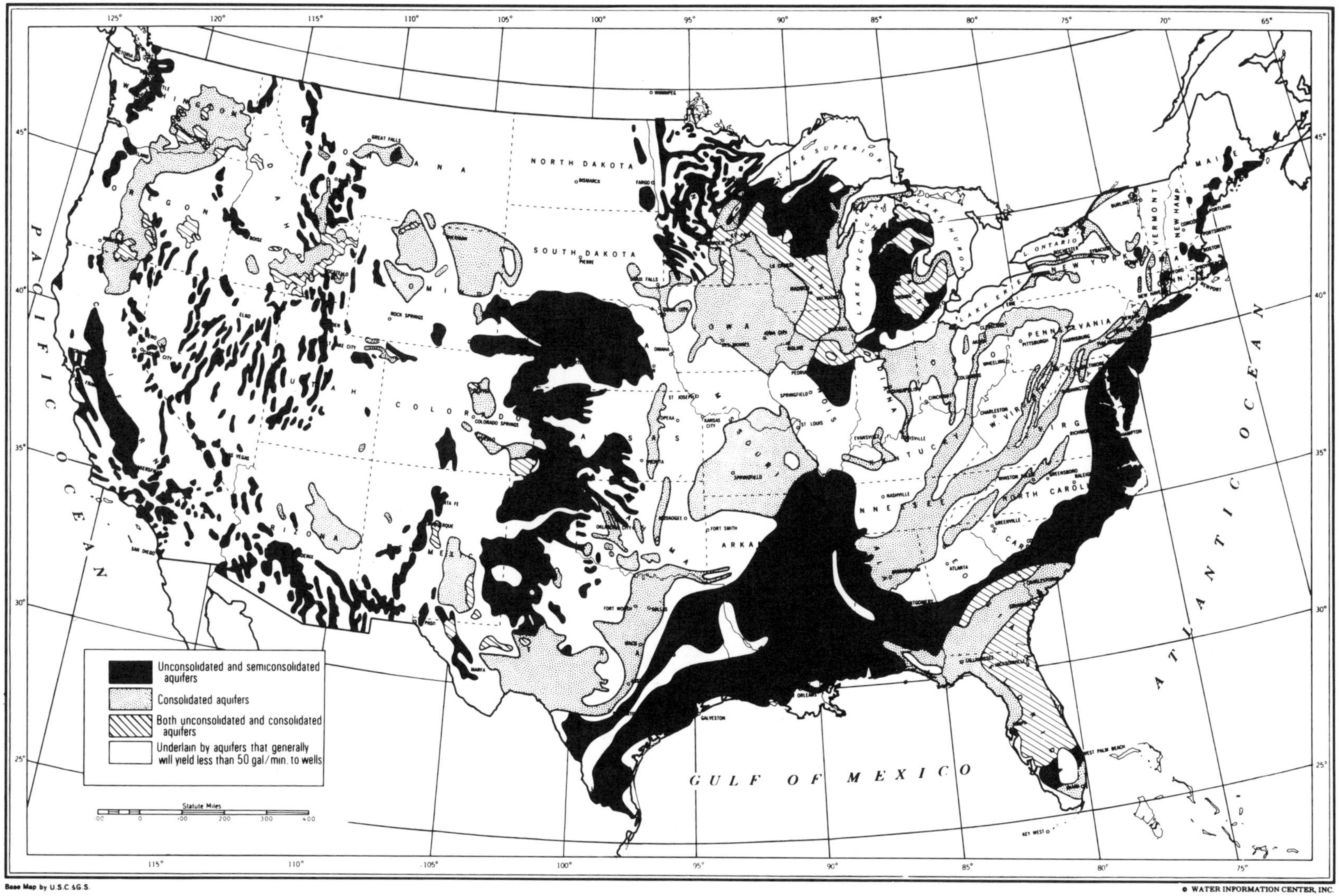

Source: Water Information Center, 1973, Water Atlas of the United States. Original Source: H.E. Thomas, The Conservation of Ground Water, McGraw-Hill, 1951

FIGURE 4–2. RIVER VALLEY AQUIFERS IN THE UNITED STATES

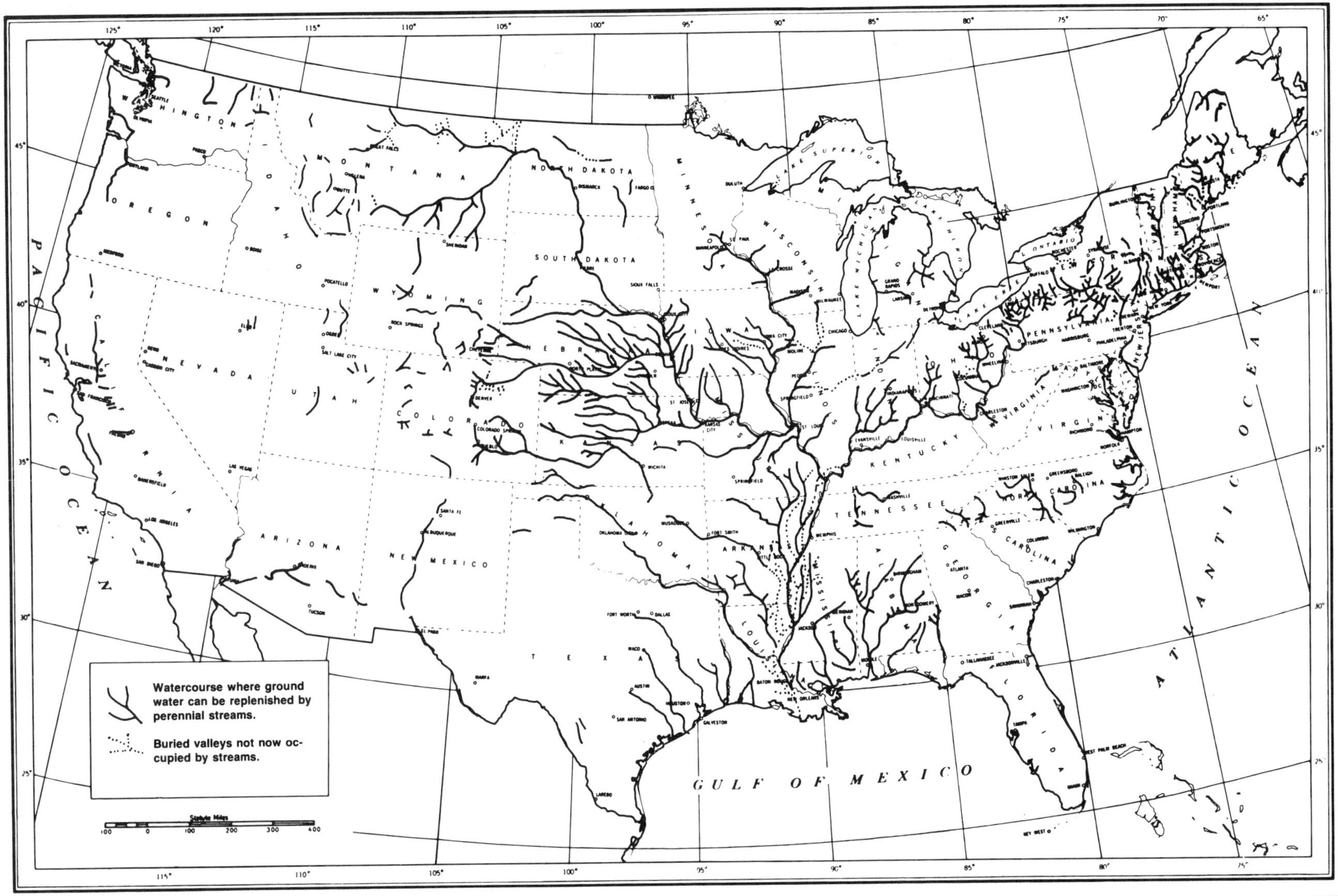

Source: Water Information Center, 1973, Water Atlas of the United States. Original source: H.E.Thomas, The Conservation of Ground Water, McGraw-Hill, 1951

TABLE 4–1. OCCURRENCE OF AQUIFERS IN THE UNITED STATES

[Abbreviations: (1) Aquifers: P, principal aquifer in region; I, important aquifer in region; M, minor aquifer in region; U, unimportant as an aquifer in region. (2) Rock terms; S, sand, Ss, sandstone; G, gravel; C, conglomerate; Sh, shale; Ls, limestone; Fm, formation; Gp, group.]

(1) Geologic Age and Rock Type	(2) Western Mountain Ranges	(3) Arid Basins	(4) Columbia Lava Plateau	(5) Colorado Plateau	(6) High Plains	(7) Unglaciated Central Region	(8) Glaciated Central Region	(9) Unglaciated Appalachian Region	(10) Glaciated Appalachian Region	(11) Atlantic and Gulf Coastal Plain	(12) Special´ Comments
Cenozoic Quaternary Alluvium and related deposits (primarily. Recent and Pleistocene sediments and may include some of Pliocene age)	S and G deposits in valleys and along stream courses. Highly productive but not greatly developed — P to M.	S and G deposits in valleys and along stream courses. Highly developed with local depletion. Storage large but perennial recharge limited-P.	S and G deposits along streams, interbedded with basalt—I to M.	U	S and G along water courses. Sand dune deposits—P (in part).	S and G along water courses and in terrace deposits—I (limited).	S and G along water courses —M.	S and G along water courses and in terrace deposits. Not developed.		S and G along water courses and in terrace and littoral deposits, especially in the Mississippi and tributary valleys. Not highly developed in East and South. Some depletion in Gulf Coast—I.	The most widespread and important aquifers in the United States. Well over one-half of all ground-water pumped in United States is withdrawn from these aquifers. Many are easily available for artificial recharge and induced infiltration. Subject to salt-water contamination in coastal areas.
Glacial drift, especially outwash (Pleistocene)	S and G deposits in northern part of region—I.	S and G deposits especially in northern part of region and in some valleys—I.	S and G outwash, especially in Spokane area—I.	U	S and G outwash, much of it reworked (see above)—I.	S and G outwash especially along northern boundary of region—I.	S and G outwash, terrace deposits and lenses in till throughout region—P (in part).	S and G outwash in northern part. Not highly developed —M.	S and G outwash, terrace deposits and lenses in till. Locally highly developed—I.	S and G outwash in Mississippi Valley (see above)—I.	
Other Pleistocene sediments	Alluvial Fm and other basin deposits in the southern part—M to P (see Alluvium above).		U	U	Alluviated plains and valley fills—M to I.		U	U	U	Coquina, limestone, sand, and marl Fms in Florida—M.	
Tertiary Sediments, Pliocene	S and G in valley fill and terrace deposits. Not highly developed—M.	Some S and G in valley fill—M.	U	U	Ogalalla Fm in High Plains. Extensive S and G with huge storage but little recharge locally. Much depletion—P (in part).	U	U	Absent	Absent	Dewitt Ss in Texas. Citronelle and LaFayette Fms in Gulf States—I.	

Miocene	Ellensburg Fm in Washington—I; elsewhere—U.	U	Ellensburg FM in Washington—I; elsewhere—U.	U	Arikaree Fm—M.	Arikaree Fm—M.	Flaxville and other terrace deposits. S and G in northwestern part—M.	Absent	Absent	New Jersey, Maryland, Delaware, Virginia—Cohansey and Calvert Fms—I. Delaware to North Carolina—St. Marys and Calvert Fms—I. Georgia and Florida—Tampa Ls, Alluvium Bluff Gp, and Tamiami Fm—I. Eastern Texas—Oakville and Catahoula Ss—I.	Aquifers in coastal areas subject to salt-water encroachment and contamination.
Oligocene	U	U	U	U	Brule clay, locally—I; elsewhere—U.	U	U	Absent	Absent	Suwannee Fm, Byram Ls, and Vicksburg Gp—I.	
Eocene	Knight and Almy Fm in southwest Wyoming—M.	U	U	Knight and Almy Fm in southwest Wyoming, Chuska Ss, and Tohatchi Sh in northwest Arizona and northeast New Mexico—M.	U	Claibourne and Wilcox Gp in southern Illinois (?), Kentucky, and Missouri—M; elsewhere—U.	Absent	Absent	Absent	New Jersey, Maryland, Delaware, Virginia—Pamunkey Gp—I. North Carolina to Florida—Ocala Ls and Castle Hayne Marl—P (in part). Florida—Avon Park Ls., South Carolina to Mexican border, Claibourne Gp, Wilcox Gp—I.	Includes the principal formations (Ocala Ls, especially) of the great Floridan aquifer. Subject to saltwater contamination in coastal areas but source of largest groundwater supply in southeastern United States.
Paleocene	U	U	U	U	Ft. Union Gp—M.	Ft. Union Gp—M.	Ft. Union Gp—M.	Absent	Absent	Clayton Fm in Georgia—I.	
Volcanic rocks, primarily basalt	U	Local flows—M.	Many interbedded basalt flows from Eocene to Pliocene—P.	Local flows—M.	Absent	Absent	Absent	Absent	Absent	Absent	

TABLE 4–1. OCCURRENCE OF AQUIFERS IN THE UNITED STATES (continued)

(1) Geologic Age and Rock Type	(2) Western Mountain Ranges	(3) Arid Basins	(4) Columbia Lava Plateau	(5) Colorado Plateau	(6) High Plains	(7) Unglaciated Central Region	(8) Glaciated Central Region	(9) Unglaciated Appalachian Region	(10) Glaciated Appalachian Region	(11) Atlantic and Gulf Coastal Plain	(12) Special Comments
Mesozoic Upper Cretaceous	U	Ss lenses in southern California—M; elsewhere—U.	U	Dakota Ss and other not clearly distinguishable Ss a notable source of water from Minnesota and Iowa to the Rocky Mountains and south into New Mexico; also in Utah and Arizona—I. In northwestern part of region Fox Hills and related Ss (Lennep, Colgate, etc.) locally valuable as water sources—M. Ss of Montana Gp—M. Ss members of Mesaverde Gp in Wyoming, Colorado, Utah, New Mexico, and Arizona—M. In Texas aquifers listed under col. 11—I.				U	U	New Jersey, Maryland, Delaware—Magothy and Raritan Fm—I. North and South Carolina—Peedee and Black Creek Fms—I. Alabama and Georgia—Ripley and Eutah Fms—I. Tennessee, Kentucky, Illinois—McNairy Ss—I. Arkansas to Texas—Navarro Gp and Taylor Fm—I.	In coastal areas subject to saltwater encroachment and contamination. Ss aquifers of the Central Regions and the West primarily valuable when water from other sources is unavailable.
Lower Cretaceous	U	U	U	In northern part of these regions Lakota, Cloverly, and Kootenai Ss—M. In southern part Purgatoire and Dakota(?) Ss—M. Texas aquifers listed in col. 11—I.			U	U	U	Texas—Woodbine Ss—I. New Jersey, Maryland, Delaware—Patapsco and Patuxent Fms—I. West of Mississippi River, especially in Texas—Edwards Ls and Ss in Trinity Gp—I.	
Jurassic	Locally—Ss Fm—M.	Locally—Ss Fm—M.	U	Ss Fms. Some may not be developed—I.	U	U	Absent	Absent	Absent	U	
Triassic	Locally—Ss and C Fms—M.		U	Ss and C Fms used locally. Shinarump C and correlatives give rise to springs—I.	U	U	Absent	Ss, C, jointed shale, and basalt beds of Newark Gp in Massachusetts, Connecticut, New Jersey, Pennsylvania, Maryland, Virginia, and North Carolina—M.			Water from Ss, C, and Ls Fms west of Mississippi River, especially valuable when water from other sources is unavailable.

Paleozoic Permian	U	U	U	DeChelly Ss—I. Kaibab Ls—M.	U	San Andres Ls in Roswell Basin—P. Quartermaster Gp gives rise to many springs—M. Other Ss and Ls in Kansas, Oklahoma, and Texas—M.	U	U	Absent	U	
Pennsylvanian	Tensleep Ss in Wyoming and other Ss elsewhere—M.	U	U	U	U	Ss and C beds from the Appalachians to Iowa and eastern Kansas—M to I.			Jointed and weathered Sh, Ss, and C in Rhode Island and Massachusetts —M.	U	
Mississippian	Ls locally but little developed; springs arise from Ls in Rocky Mountains—M.	A few springs arise from Ls locally—U.	U	Some springs arise from Ls locally—U.	U	In Illinois, Iowa, Missouri, and Kentucky the Burlington, Keokuk, and St. Louis Ls—I. Some Ss (primarily Chester)—M. In Alabama and Tennessee—the Ft. Payne chert, Gaspar Fm, and St. Genevieve and Tuscumbia Ls—I. In Kentucky many springs arise in Ls.			U	U	Do.
Devonian	U	U	U	U	U	U, except locally in Michigan (Traverse Fm), Illinois, Missouri, Ohio (Columbia Ls), and Kentucky—M.		Jointed Ls, Ss, and Sh, some highly metamorphosed. M locally and little used.		U	

TABLE 4–1. OCCURRENCE OF AQUIFERS IN THE UNITED STATES (continued)

(1) Geologic Age and Rock Type	(2) Western Mountain Ranges	(3) Arid Basins	(4) Columbia Lava Plateau	(5) Colorado Plateau	(6) High Plains	(7) Unglaciated Central Region	(8) Glaciated Central Region	(9) Unglaciated Appalachian Region	(10) Glaciated Appalachian Region	(11) Atlantic and Gulf Coastal Plain	(12) Special Comments
Silurian	U	U	U	U	U	Ls and dolomite Fms in New York, Kentucky, Tennessee, Ohio, Illinois, and Iowa. Better-known aquifers include Monroe dolomite and related carbonate Fms in Ohio—I; "Niagaran" dolomite in Illinois—P (in part).		U	U	U	
Ordovician	U	U	U	U	U	In Arkansas, Missouri, Iowa, Illinois, eastern Indiana, southern Wisconsin, southeastern Minnesota, the St. Peter Ss—I. Overlying and subjacent Ls and Ss where present in above states and in Kansas, Oklahoma, and New York—M to I. In Kentucky and Tennessee—Ls Fm—M to I.		Locally Ls and Ss Fms; not highly developed—M.		U	
Cambrian	U	U	U	U	U	Ss beds in Wisconsin, Minnesota, Iowa, and Illinois include Jordan Ss, "Dresbach Fm" (Galesville Ss, Eau Claire Fm, Mt. Simon Ss)—P (in part). Ls and Ss Fms in Missouri and Arkansas give rise to many large springs and yield water to many wells—P.		Ls Fms give rise to large springs in southern Appalachians. Otherwise—U.	Eastern New York and New England Ss Fms—M; otherwise—U.	U	
Precambrian (including crystalline rocks which may be younger)	Weathered and jointed rocks locally—M.		U	U	U	U	Weathered and jointed rocks locally in Minnesota, Wisconsin, northern Michigan, Piedmont Plateau, New England—M to I. Some Ss in North Central States.			U	Do.

Source: Maxey, in Chow, Handbook of Applied Hydrology, McGraw-Hill, Copyright 1964

FIGURE 4–3. GROUND-WATER REGIONS OF THE UNITED STATES

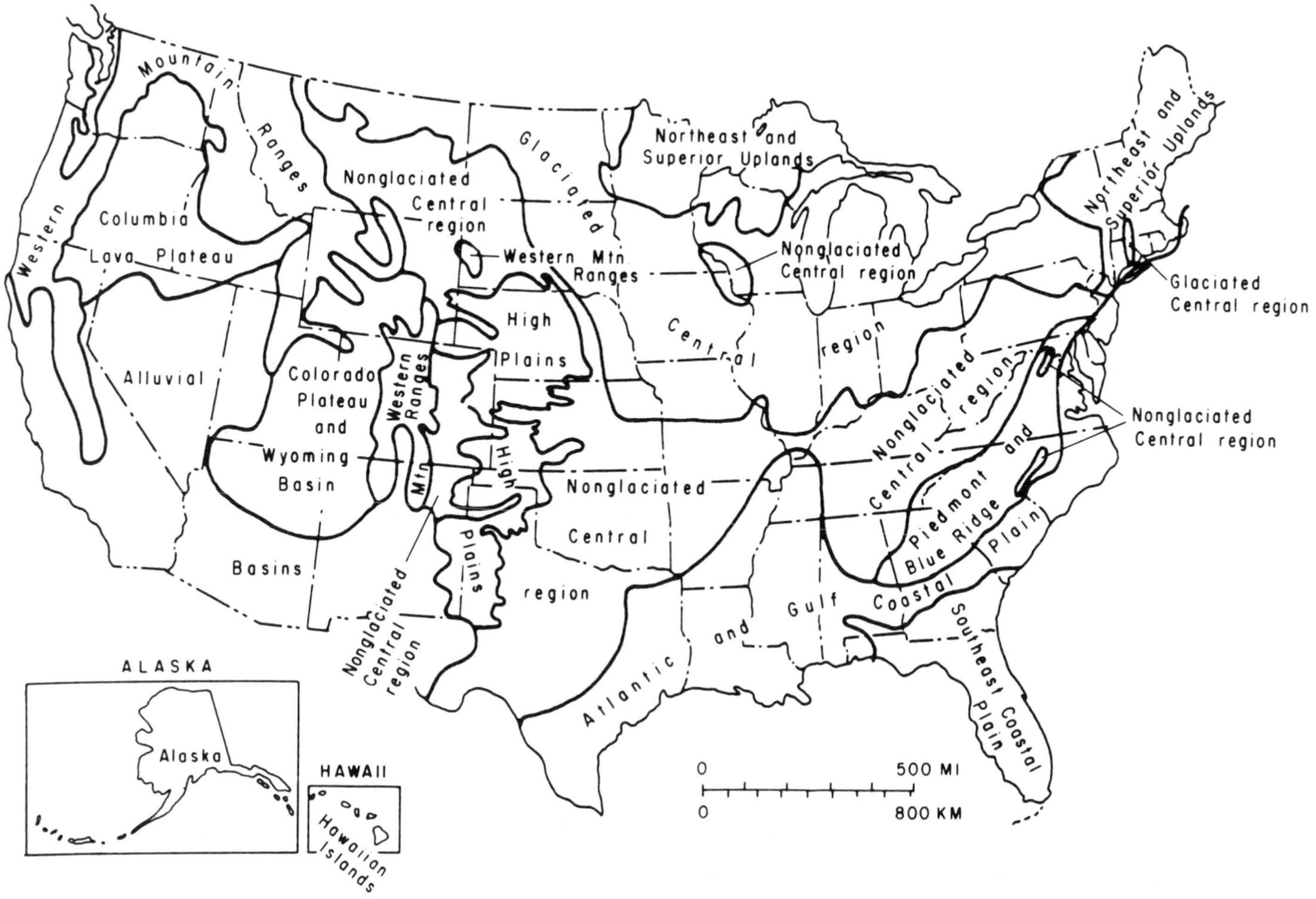

Source: Heath, R.C., 1982, Classification of Ground-Water Regions of the United States, Groundwater, v.20, no. 4. Reprinted with permission

TABLE 4–2. PRINCIPAL PHYSICAL AND HYDROLOGIC CHARACTERISTICS OF GROUND-WATER REGIONS IN THE UNITED STATES

[For map of regions, see Figure 4–3]

Name	Components of the system															Characteristics of the dominant aquifers																				
																Water-bearing openings					Compo-sition			Storage and transmission properties							Recharge and discharge conditions					
	Unconfined aquifer			Confining beds				Confined aquifers			Presence and arrangement					Primary			Secon-dary		Degree of solubility			Porosity			Transmissivity				Recharge			Discharge		
	Hydrologically insignificant	Minor aquifer	Dominant aquifer	Hydrologically insignificant	Thin, discontinuous, or v. leaky	Thick, impermeable	Interlayered with aquifers	Hydrologically insignificant	Not highly productive	Multiple aquifers	A single, dominant aquifer	Single dominant unconfined aquifer	Two interconnected aquifers	Unconfined aquifer, confining bed, confined aquifer	Complex interbedded sequence	Pores in unconsolidated deposits	Pores in semiconsolidated rocks	Tubes and cooling cracks in lava	Fractures and faults	Solution enlarged openings	Insoluble	Mixed soluble and insoluble	Soluble	Large (>0.2)	Moderate (0.01-0.2)	Small (<0.01)	Large (>2500 m^2 day^{-1})	Moderate (250-2500 m^2 day^{-1})	Small (25-250 m^2 day^{-1})	Very small (<25 m^2 day^{-1})	Uplands between streams	Losing streams	Leakage thru confining beds	Springs and surface seepage	Evaporation and basin sinks	Into other aquifers
Western Mountain Ranges		X		X					X				X			X			X		X				X				X		X	X		X		
Alluvial Basins		X					X			X					X	X					X			X			X					X			X	X
Columbia Lava Plateau		X					X			X					X	X		X			X					X	X					X		X		
Colorado Plateau and Wyoming Basin	X						X			X					X		X		X		X					X			X		X		X	X		
High Plains			X	X				X				X				X					X			X			X					X		X		
Nonglaciated Central Region		X					X			X					X		X		X	X		X				X		X			X		X	X		X
Glaciated Central Region		X			X					X					X	X	X		X	X		X			X			X					X			X
Piedmont and Blue Ridge		X		X					X				X			X			X		X					X				X	X			X		
Northeast and Superior Uplands		X			X				X				X			X			X		X					X			X		X			X		
Atlantic and Gulf Coastal Plain		X					X			X					X	X				X		X			X			X			X		X	X		X
Southeast Coastal Plain		X				X					X			X		X				X			X	X			X				X		X	X		
Alluvial Valleys			X		X			X				X				X					X			X			X				X			X		
Hawaii			X		X			X				X						X			X			X			X				X	X		X		
Alaska			X		X				X						X	X			X			X			X			X				X		X		

Source: Heath, R.C., 1982, Classification of Ground-Water Regions of the United States, Groundwater, v. 20, no.4. Reprinted with permission

TABLE 4–3. BASIC DATA REQUIRED FOR GROUND-WATER STUDIES

A. Maps, Cross Sections, and Fence Diagrams

1. Planimetric
2. Topographic
3. Geologic
    a. structure
    b. stratigraphy
    c. lithology
4. Hydrologic
    a. location of wells, observation wells, and springs
    b. ground-water table and potentiometric contours
    c. depth to water
    d. quality of water
    e. recharge, discharge, and contributing areas
5. Vegetative cover, location of wetlands
6. Soils
7. Aerial photographs

B. Data on Wells and Springs

1. Location, depth, diameter, types of well, and logs
2. Static and pumping water level, hydrographs, yield, specific capacity, quality of water
3. Present and projected ground-water development and use
4. Corrosion, incrustation, well interference, and similar operation and maintenance problems
5. Location, type, geologic setting, and hydrographs of springs
6. Observation well networks
7. Water sampling sites

C. Aquifer Data

1. Type, such as unconfined, artesian, or perched
2. Thickness, depths, and formational designation
3. Boundaries
4. Transmissivity, storativity, and permeability
5. Specific retention
6. Discharge and recharge
7. Ground and surface water relationships
8. Aquifer models

D. Climatic Data

1. Precipitation
2. Temperature
3. Evapotranspiration

E. Surface Water

1. Use
2. Quality
3. Runoff distribution, reservoir capacities, inflow and outflow data
4. Return flows, section gain or loss
5. Recording stations
6. Low flow data

F. Environment

1. Location of hazardous waste sites or other potential sources of pollution
2. Use of herbicides, pesticides, fertilizers, and road salt
3. Site history

G. Local Drilling Facilities and Practices

1. Size and types of drilling rigs locally available
2. Logging services locally available
3. Locally used materials, well designs, and drilling practices
4. State or local rules and regulations

Source: U.S. Bureau of Reclamation, 1977, Ground Water Manual; amended

FIGURE 4–4. REGIONAL AQUIFER STUDIES IN THE UNITED STATES

The Regional Aquifer-System Analysis (RASA) Program is a systematic effort by the U.S. Geological Survey to study a number of regional ground-water systems that represent a significant part of the Nation's water supply. A regional aquifer system, as the term is used here, may be of two general types: (1) aquifers that are of regional extent, such as those underlying the Great Plains and the Atlantic Coastal Plain; and (2) groups of aquifers that share so many characteristics that investigation of a few of these aquifers can establish common principles and hydrologic factors controlling the occurrence, movement, and quality of ground water throughout similar aquifer systems.

Purpose:

- Provide the basic information and knowledge needed to manage ground-water development from a regional perspective.

Activities:

- Determine availability and chemical quality of water stored in and being transmitted through each aquifer system.
- Evaluate discharge-recharge characteristics of each aquifer system.
- Evaluate geologic, hydrologic, and chemical controls that govern the responses of the aquifer systems to stresses.
- Develop computer-based flow-simulation models to assist in understanding the aquifer systems and their response to such human activities as pumping or irrigation.

Source: Cardin, C.W. and others, 1986, Water Resources Division in the 1980's, U.S. Geological Survey Circular 1005

FIGURE 4–5. DEPTH TO SALINE GROUND WATER IN THE UNITED STATES

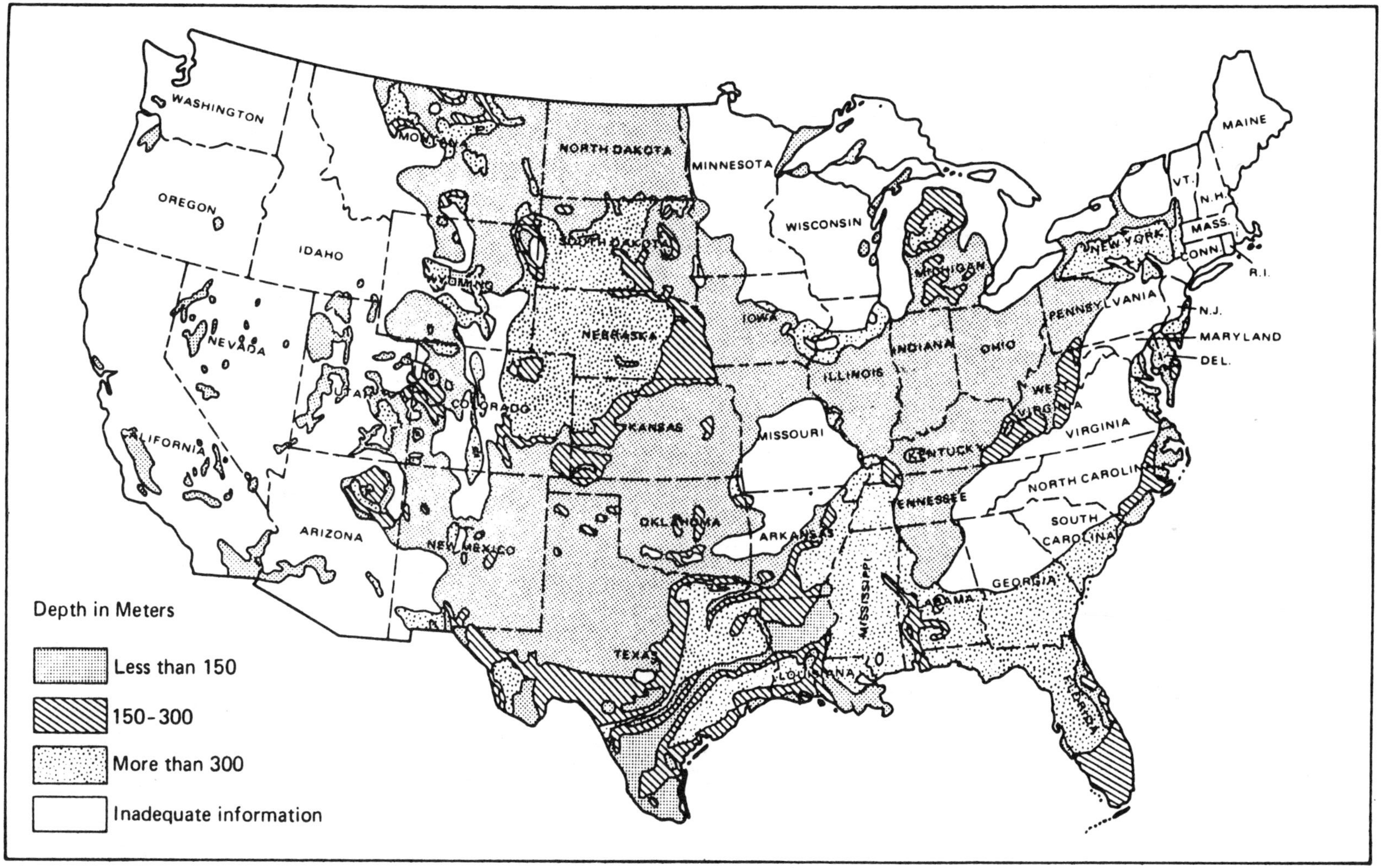

Source: Todd, D.K.,1980, Groundwater Hydrology. Copyright John Wiley & Sons. Reprinted with permission. Based on map prepared by J.H. Feth, U.S. Geological Survey, Hydrologic Inv. Atlas HA-199, 1965

FIGURE 4–6. AVERAGE TEMPERATURE OF SHALLOW GROUND WATER IN THE UNITED STATES

[Ground water temperatures in °F in wells ranging from 50' to 150' in depth]

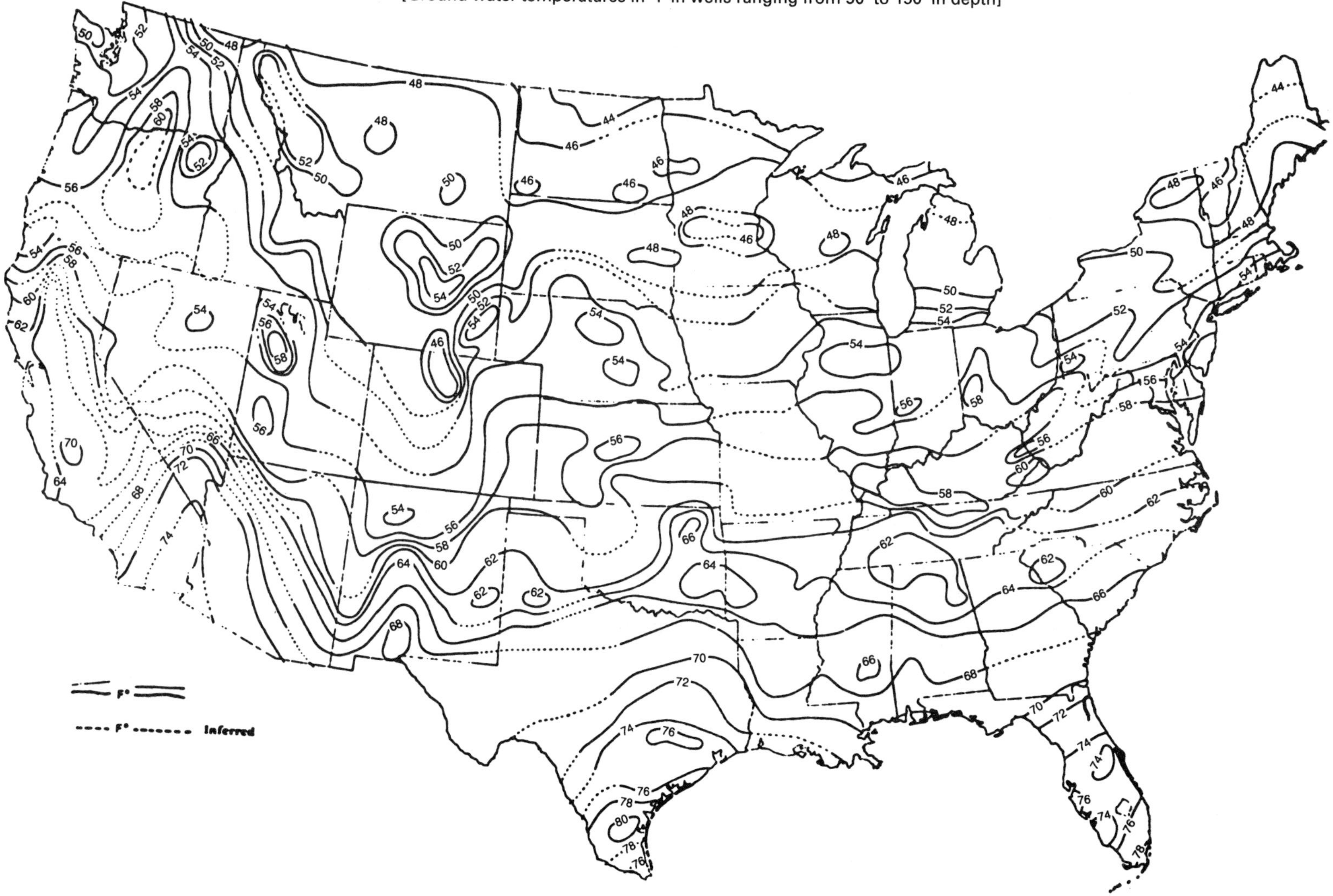

Source: Water Well Journal, September 1979. Copyright National Water Well Association. Reprinted with permission

FIGURE 4–7. GROUND-WATER POTENTIAL IN CANADA

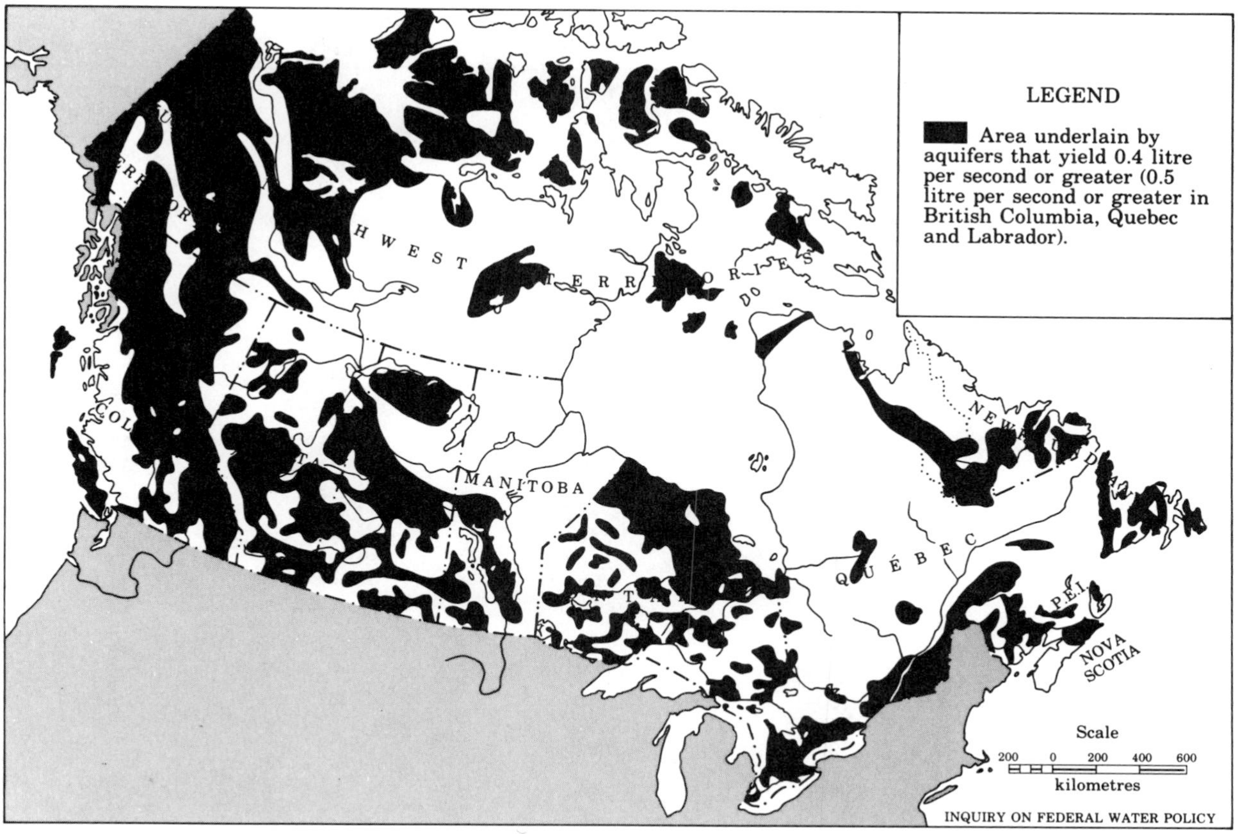

Source: Pearce, P.H., 1985, Currents of Change, Final Report, Inquiry on Federal Water Policy, Ottawa

TABLE 4–4. ESTIMATED GROUND WATER IN STORAGE, BY CONTINENT

[In millions of km³; based on publications by Soviet hydrologists]

Continent	0–100 m	100–200 m	200–2000 m	Total
Europe	0.2	0.3	1.1	1.6
Asia	1.3	2.1	4.4	7.8
Africa	1	1.5	3.0	5.5
North America	0.7	1.2	2.4	4.3
South America	0.3	0.9	1.8	3
Australia	0.1	0.2	0.9	1.2
Total	3.6	6.2	13.6	23.4

Source: Castany, G., 1981, Hydrogeology of Deep Aquifers, Episodes, v.1981 no.3

SECTION B. WATER WELLS—UNITED STATES

TABLE 4–5. NUMBER AND TYPE OF WATER WELLS IN THE UNITED STATES, 1988

State	Irrigation	Public Supply[1]	Community Supply	Household
Alabama	N/A	1,706	1,013	267,202
Alaska	10	1,283	454	32,391
Arizona	6,125	2,533	1,650	43,226
Arkansas	21,078	3,220	1,073	213,672
California	67,770	8,143	3,320	359,584
Colorado	17,809	3,116	1,465	84,459
Connecticut	N/A	5,373	1,073	241,130
Delaware	559	1,141	497	52,701
Florida	29,017	11,337	4,650	573,059
Georgia	4,492	4,139	2,460	405,078
Hawaii	N/A	219	218	536
Idaho	7,371	3,114	1,312	88,853
Illinois	1,107	10,018	1,492	443,681
Indiana	N/A	16,100	1,805	546,381
Iowa	2,210	5,052	3,674	236,709
Kansas	18,658	2,456	2,405	116,567
Kentucky	N/A	1,144	386	247,506
Louisiana	4,558	3,162	1,996	200,446
Maine	N/A	3,433	595	149,331
Maryland	N/A	4,955	953	252,142
Massachussets	N/A	2,894	1,384	132,119
Michigan	N/A	12,188	1,138	934,184
Minnesota	4,250	13,163	2,467	382,572
Mississippi	N/A	3,109	2,257	150,816
Missouri	3,700	3,996	1,633	305,853
Montana	997	2,504	999	80,817
Nebraska	61,361	1,558	1,501	112,740
Nevada	2,332	1,156	697	24,142
New Hampshire	N/A	2,009	733	110,712
New Jersey	N/A	7,765	2,256	227,326
New Mexico	8,031	2,478	1,545	70,157
New York	651	18,068	5,381	659,973
North Carolina	530	15,972	5,094	821,995
North Dakota	808	1,261	718	54,008
Ohio	N/A	13,306	3,508	692,062
Oklahoma	4,351	3,265	2,181	164,506
Oregon	9,241	3,330	1,267	178,407
Pennsylvania	N/A	17,477	5,578	800,292
Rhode Island	N/A	916	169	33,987
South Carolina	185	3,376	2,174	297,435
South Dakota	1,266	1,244	872	56,512
Tennessee	320	3,125	1,006	258,997
Texas	59,636	13,297	9,207	490,453
Utah	2,295	1,961	1,164	14,511
Vermont	N/A	1,739	653	58,380
Virginia	N/A	6,771	2,869	455,556
Washington	5,853	6,130	3,810	195,132
West Virginia	N/A	2,771	834	181,069
Wisconsin	N/A	22,982	2,239	521,579
Wyoming	1,409	1,373	647	30,900
TOTALS	348,116	282,827	98,472	13,101,846
				GRAND TOTAL 13,732,680

N/A-Not available

[1] Includes community supply (systems with at least 15 service connections used by year-round residents or regularly serving at least 25 year-round residents.

Source: National Water Well Association, 1988

TABLE 4–6. NUMBER OF WATER WELLS DRILLED IN THE UNITED STATES, 1960 TO 1984

State	Estimated Number of Wells Drilled[1]							Percent of Total Drilled in 1984	Percentage change Between Annual Totals[2]		
	1960	1964	1980	1981	1982	1983	1984		1960 and 1964	1964 and 1984	1980 and 1984
Alabama	4,000	4,500	5,960	6,420	5,920	5,570	6,260	1.6	+13	+39	+5
Alaska	726	1,000	2,440	2,400	2,400	2,800	2,700	.68	+38	+170	+11
Arizona	1,400	1,520	2,190	2,220	2,380	2,710	2,760	.7	+8.6	+82	+26
Arkansas	5,000	5,000	4,010	5,910	2,750	3,320	4,200	1.1	NC	-16	+5
California	9,100	10,000	17,100	15,900	11,300	11,000	14,300	3.6	+10	+43	-16
Colorado	3,100	5,910	4,910	4,570	4,390	4,360	4,060	1.0	+91	-31	-17
Connecticut	6,500	6,500	5,470	5,410	4,500	5,140	5,780	1.5	NC	-11	+6
Delaware	3,800	3,440	2,000	2,680	2,290	2,700	3,100	.78	-9.5	-10	+55
District of Columbia	12	12	ND	0	0	0	ND	ND	NC	ND	ND
Florida	33,900	55,000	40,200	40,500	38,900	43,200	45,600	11	+62	-17	+13
Georgia	10,500	10,000	11,000	13,400	10,100	10,800	12,200	3.1	-4.8	+22	+11
Hawaii	17	21	11	11	7	7	2	<0.1	+24	-90	-82
Idaho	1,400	1,400	2,880	1,470	2,400	1,590	1,630	.35	NC	+16	-43
Illinois	21,000	19,500	14,000	12,200	13,400	13,600	15,300	3.9	-7.1	-22	+9.3
Indiana	17,700	15,000	9,670	8,180	9,700	9,180	10,300	2.6	-15	-31	+6.5
Iowa	9,000	15,000	5,890	6,850	4,120	3,780	3,140	.79	+67	-79	-47
Kansas	4,700	5,500	4,530	5,050	3,380	3,420	3,910	1.0	+17	-29	-14
Kentucky	9,880	9,620	5,060	5,100	4,800	5,440	5,740	1.4	-2.6	-40	+13
Louisiana	974	2,620	6,050	6,830	6,580	5,180	5,560	1.4	+170	+110	-8.1
Maine	1,500	1,700	2,860	2,570	2,440	3,470	3,900	1.0	+13	+130	+36
Maryland	4,020	6,900	7,200	8,000	6,700	8,800	8,300	2.1	+72	+20	+15
Massachusetts	8,000	9,000	6,330	6,270	5,370	6,820	7,670	1.9	+12	-15	+21
Michigan	25,000	25,000	24,000	20,000	16,000	17,000	18,500	4.7	NC	-26	-23
Minnesota	13,000	9,000	14,400	10,500	10,800	11,100	12,500	3.1	-31	+39	-13
Mississippi	5,300	5,900	2,670	3,550	2,540	2,400	2,640	.66	+11	-55	-1
Missouri	6,380	9,990	10,900	8,530	7,830	10,200	11,500	2.9	+57	+15	+5.5
Montana	1,900	2,000	3,580	6,410	6,260	2,360	2,560	.64	+5.3	+28	-28
Nebraska	5,510	6,000	4,500	5,940	3,470	3,260	3,660	.92	+8.9	-39	-19
Nevada	824	825	775	765	503	639	718	.18	NC	-13	-7.4
New Hampshire	3,600	4,400	3,050	4,190	2,630	5,210	5,860	1.5	+22	+33	+92
New Jersey	3,800	3,440	8,620	8,540	8,580	10,900	13,100	3.3	-9.5	+280	+52
New Mexico	2,290	3,150	2,750	2,880	3,370	3,430	3,110	.78	+38	-1.3	+13
New York	25,000	25,000	16,800	17,000	15,600	17,800	20,000	5.0	NC	-20	+19
North Carolina	20,000	25,000	10,500	12,000	13,500	15,900	17,100	4.3	+25	-32	+63
North Dakota	4,200	3,760	1,710	2,190	1,450	1,480	1,450	.36	-10	-61	-15
Ohio	17,100	18,600	16,700	14,300	14,200	14,000	15,700	4.0	+8.8	-160	-6.0
Oklahoma	4,400	5,000	7,980	7,630	6,500	5,870	6,590	1.8	+14	+32	-17
Oregon	3,500	4,500	7,500	6,620	3,800	3,550	3,530	.89	+29	-22	-53
Pennsylvania	13,500	16,200	15,600	12,400	9,620	8,140	10,800	2.7	+20	-33	-31
Rhode Island	200	250	319	240	206	387	548	.14	+25	+120	+72
South Carolina	5,300	5,400	11,400	5,340	4,640	7,780	8,740	2.2	+1.9	+62	-23
South Dakota	6,080	5,430	2,210	1,820	1,590	1,330	1,500	.38	-11	-72	-32
Tennessee	10,000	8,000	7,080	7,130	6,710	7,600	8,020	2.0	-20	NC	+13
Texas	19,000	25,000	16,200	17,700	21,700	17,700	21,200	5.3	+32	-15	+31
Utah	630	650	630	547	507	488	548	.14	+3.2	-16	-13
Vermont	1,240	1,460	3,100	2,280	1,900	2,330	3,050	.77	+18	+110	-1.6
Virginia	8,500	10,000	10,900	8,830	9,060	15,300	16,900	4.3	+18	+69	+55
Washington	1,400	1,700	5,040	4,290	3,550	4,320	4,030	1.0	+21	+137	-20
West Virginia	5,500	5,900	3,280	3,510	2,730	2,580	2,900	.73	+7.3	-51	-12
Wisconsin	11,000	12,000	11,600	9,900	9,590	10,400	11,700	2.9	+9.1	-2.5	+.86
Wyoming	1,000	1,000	3,010	3,680	2,970	2,500	2,520	.6	NC	+152	-16
Totals	381,000	434,000	387,000	371,000	336,000	359,000	397,000	100	+14	-8.5	+2.6

NC-No change
ND-No Data available

[1] Numbers rounded to three significant figures.
[2] Numbers rounded to two significant figures.
Source: Hindall, S.M., Eberle, Michael, 1987, National and Regional Trends in Water-Well Drilling in the United States 1964–1984, U.S. Geological Survey, Open File Report 87–247

TABLE 4–7. REGIONAL TRENDS IN WATER-WELL CONSTRUCTION IN THE UNITED STATES, 1960 TO 1984

Region	Number of Wells Drilled							Average Annual Total 1980 through 1984
	1960	1964	1980	1981	1982	1983	1984	
Northeast (includes D.C.)	63,300	68,000	62,100	60,600	50,800	60,200	70,800	60,900
Southeast	95,300	126,000	102,000	106,000	93,600	112,000	121,000	107,000
Great Lakes and Central Appalachians	130,000	123,000	106,000	95,300	87,900	90,900	101,000	96,200
South-Central	54,800	77,200	63,200	62,100	60,600	57,300	63,300	61,300
Northern Rockies and Northern Great Plains	20,100	19,600	17,900	16,700	18,100	12,500	13,300	15,700
Southwest (includes Hawaii)	12,000	13,000	20,700	17,700	15,100	14,800	18,300	17,300
Pacific Northwest (includes Alaska)	5,620	7,200	15,000	11,400	9,750	8,670	10,300	11,000
Totals	381,000	434,000	387,000	370,000	336,000	359,000	397,000	370,000

Region	Percentage of Total Wells Drilled,		Percentage Change Between Annual Totals		
	1964	1984	1960 and 1964	1964 and 1984	1980 and 1984
Northeast (includes D.C.)	16	18	+7.4	+4.1	+14
Southeast	29	30	+32	-3.9	+19
Great Lakes and Central Appalachians	28	25	-5.4	-18	-4.7
South-Central	18	16	+41	-18	NC
Northern Rockies and Northern Great Plains	4.5	3.3	-2.5	-32	-26
Southwest (includes Hawaii)	3.0	4.6	+8.3	+41	-12
Pacific Northwest (includes Alaska)	1.7	2.6	+28	+43	-31
Totals	100	100	+14	-8.5	+2.6

Source: Hindall, S.M., Eberle, Michael, 1987, National and Regional Trends in Water-Well Drilling in the United States 1964–1984, U.S. Geological Survey, Open File Report 87–247

FIGURE 4–8. NUMBER OF WATER WELLS DRILLED IN THE UNITED STATES AND RELATION TO MAJOR EVENTS IN UNITED STATES HISTORY

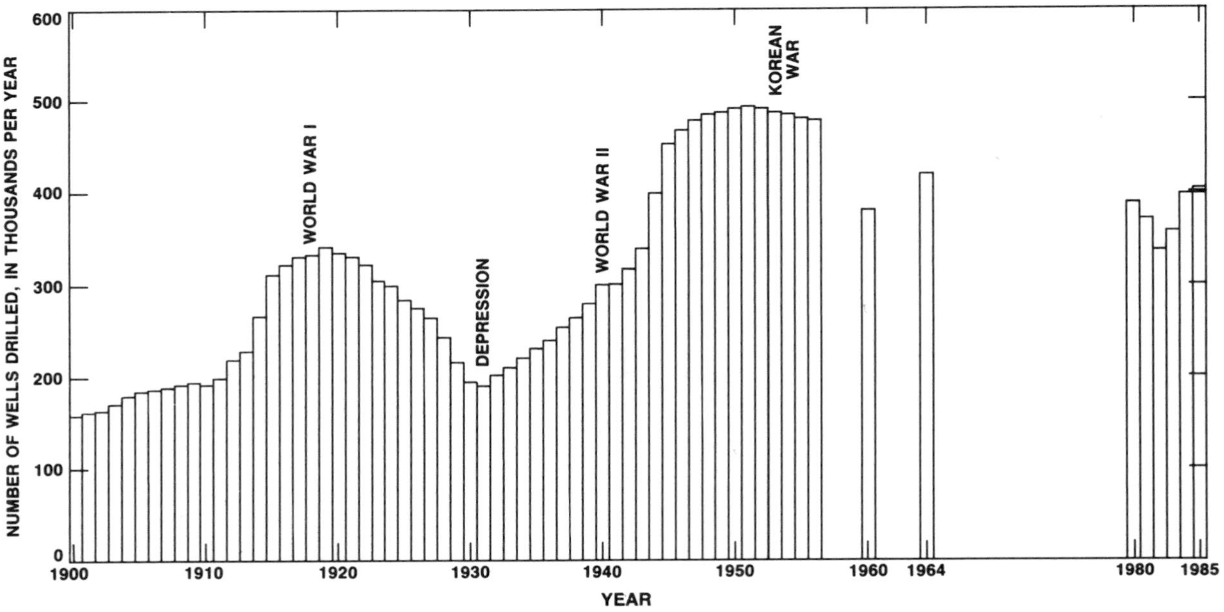

Source: Hindall, S.M., Eberle, Michael, 1987, National and Regional Trends in Water-Well Drilling in the United States 1964–1984, U.S. Geological Survey, Open File Report 87–247; 1985 data from National Water Well Association

TABLE 4–8 AND FIGURE 4–9. NUMBER AND TYPE OF WATER WELLS AND BOREHOLES CONSTRUCTED IN THE UNITED STATES IN 1985

[Based on Water Well Journal survey of 8,043 firms]

Application/type	1985
Commercial/industrial	49,379
Heat pump supply/return	18,029
Agricultural irrigation	21,583
Private household	488,918
Public supply	20,010
Monitoring	121,294
Livestock watering*	29,343
Lawn/turf irrigation*	27,036
Other	34,482
Total	810,074

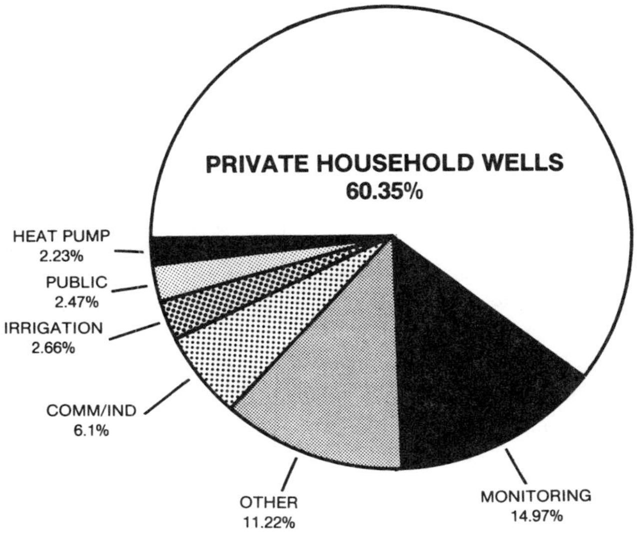

* Included in "Other" category on pie chart.
Source: McCray, Kevin. Copyright Water Well Journal September 1986. Reprinted with permission

FIGURE 4–10. DENSITY OF HOUSING UNITS USING ON SITE DOMESTIC WATER SUPPLY SYSTEMS IN THE UNITED STATES

[By county]

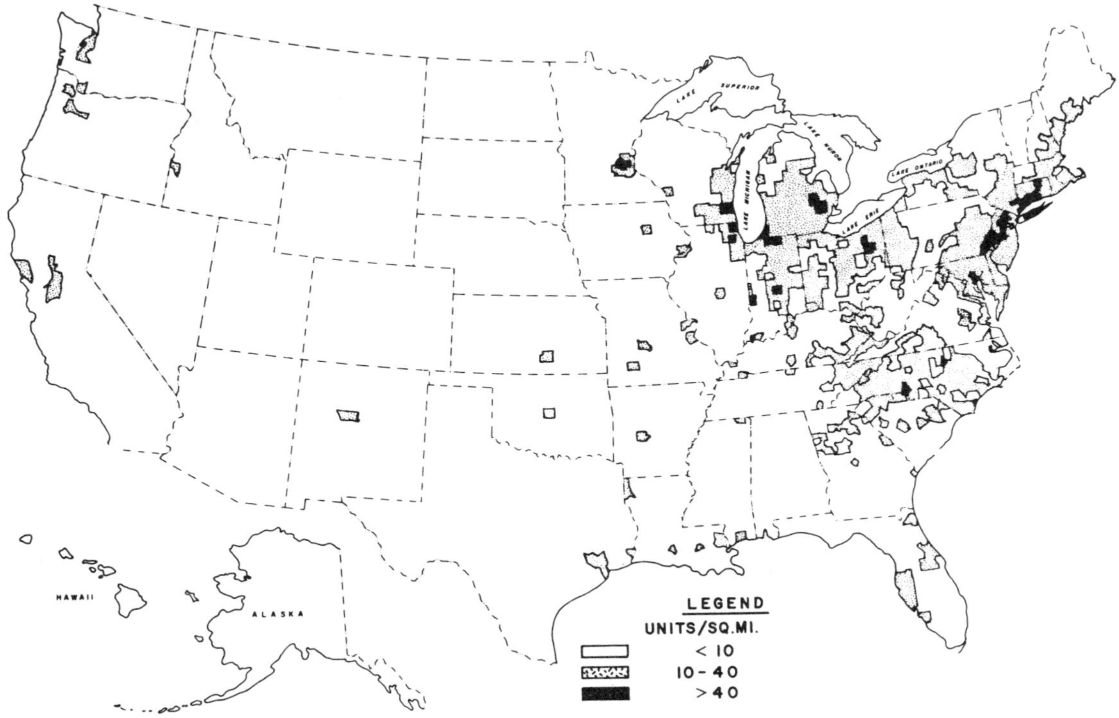

Source: U.S. Environmental Protection Agency, Office of Water Supply, Office of Solid Waste Management Programs, 1977, The Report to Congress: Waste Disposal Practices and Their Effects on Ground Water

SECTION C. WATER WELLS

TABLE 4–9. WATER WELL CONSTRUCTION METHODS AND APPLICATIONS

Method	Materials for Which Best Suited	Water Table Depth for Which Best Suited, m	Usual Maximum Depth, m	Usual Diameter Range, cm	Usual Casing Material	Customary Use	Yield, m³/day[a]	Remarks
Augering Hand auger	Clay, silt, sand, gravel less than 2 cm	2–9	10	5–20	Sheet metal	Domestic, drainage	15–250	Most effective for penetrating and removing clay. Limited by gravel over 2 cm. Casing required if material is loose.
Power auger	Clay, silt, sand, gravel less than 5 cm	2–15	25	15–90	Concrete, steel or wrought-iron pipe	Domestic, irrigation, drainage	15–500	Limited by gravel over 5 cm, otherwise same as for hand auger.
Driven Wells Hand, air hammer	Silt, sand, gravel less than 5 cm	2–5	15	3–10	Standard weight pipe	Domestic, drainage	15–200	Limited to shallow water table, no large gravel.
Jetted Wells Light, portable rig	Silt, sand, gravel less than 2 cm	2–5	15	4–8	Standard weight pipe	Domestic, drainage	15–150	Limited to shallow water table, no large gravel.
Drilled Wells Cable tool	Unconsolidated and consolidated medium hard and hard rock	Any depth	450[b]	8–60	Steel or wrought-iron pipe	All uses	15–15,000	Effective for water exploration. Requires casing in loose materials. Mud-scow and hollow rod bits developed for drilling unconsolidated fine to medium sediments.
Rotary	Silt, sand, gravel less than 2 cm; soft to hard consolidated rock	Any depth	450[b]	8–45	Steel or wrought-iron pipe	All uses	15–15,000	Fastest method for all except hardest rock. Casing usually not required during drilling. Effective for gravel envelope wells.
Reverse-circulation rotary	Silt, sand, gravel, cobble	2–30	60	40–120	Steel or wrought-iron pipe	Irrigation, industrial, municipal	2500–20,000	Effective for large-diameter holes in unconsolidated and partially consolidated deposits. Requires large volume of water for drilling. Effective for gravel envelope wells.
Rotary-percussion	Silt, sand, gravel less than 5 cm; soft to hard consolidated rock	Any depth	600[b]	30–50	Steel or wrought-iron pipe	Irrigation, industrial, municipal	2500–15,000	Now used in oil exploration. Very fast drilling. Combines rotary and percussion methods (air drilling) cuttings removed by air. Would be economical for deep water wells.

[a]Yield influenced primarily by geology and availability of ground water.
[b]Greater depths reached with heavier equipment.
Source: U.S. Soil Conservation Service, 1969, Engineering Field Manual for Conservation Practices

TABLE 4–10. RELATIVE PERFORMANCE OF DIFFERENT DRILLING METHODS IN VARIOUS TYPES OF GEOLOGIC FORMATIONS

Type of Formation	Cable Tool	Direct Rotary (with fluids)	Direct Rotary (with air)	Direct Rotary (Down-the-hole air hammer)	Direct Rotary (Drill-through casing hammer)	Reverse Rotary (with fluids)	Reverse Rotary (Dual Wall)	Hydraulic Percussion	Jetting	Driven	Auger
Dune sand	2	5	Not recommended	Not recommended	6	5*	6	5	5	3	1
Loose sand and gravel	2	5	Not recommended	Not recommended	6	5*	6	5	5	3	1
Quicksand	2	5	Not recommended	Not recommended	6	5*	6	5	5	Not recommended	1
Loose boulders in alluvial fans or glacial drift	3–2	2–1	Not recommended	Not recommended	5	2–1	4	1	1	Not recommended	1
Clay and silt	3	5	Not recommended	Not recommended	5	5	5	3	3	Not recommended	3
Firm shale	5	5	Not recommended	Not recommended	5	5	5	3	Not recommended	Not recommended	2
Sticky shale	3	5	Not recommended	Not recommended	5	3	5	3	Not recommended	Not recommended	2
Brittle shale	5	5	Not recommended	Not recommended	5	5	5	3	Not recommended	Not recommended	Not applicable
Sandstone—poorly cemented	3	4	Not recommended	Not recommended	Not applicable	4	5	4	Not recommended	Not recommended	Not applicable
Sandstone—well cemented	3	3	5	Not recommended	Not applicable	3	5	3	Not recommended	Not recommended	Not applicable
Chert nodules	5	3	3		Not applicable	3	3	5	Not recommended	Not recommended	Not applicable
Limestone	5	5	5	6	Not applicable	5	5	5	Not recommended	Not recommended	Not applicable
Limestone with chert nodules	5	3	5	6	Not applicable	3	3	5	Not recommended	Not recommended	Not applicable
Limestone with small cracks or fractures	5	3	5	6	Not applicable	2	5	5	Not recommended	Not recommended	Not applicable
Limestone, cavernous	5	3–1	2	5	Not applicable	1	5	1	Not recommended	Not recommended	Not applicable
Dolomite	5	5	5	6	Not applicable	5	5	5	Not recommended	Not recommended	Not applicable
Basalts, thin layers in sedimentary rocks	5	3	5	6	Not applicable	3	5	5	Not recommended	Not recommended	Not applicable
Basalts—thick layers	3	3	4	5	Not applicable	3	4	3	Not recommended	Not recommended	Not applicable
Basalts—highly fractured (lost circuiation zones)	3	1	3	3	Not applicable	1	4	1	Not recommended	Not recommended	Not applicable
Metamorphic rocks	3	3	4	5		3	4	3			
Granite	3	3	5	5		3	4	3			

Rate of Penetration:
1 Impossible
2 Difficult
3 Slow
4 Medium
5 Rapid
6 Very rapid

* Assuming sufficient hydrostatic pressure is available to contain active sand (under high confining pressures)

Source: Driscoll, F.G., 1986, Groundwater and Wells. Copyright Johnson Division. Reprinted with permission

TABLE 4–11. DATA ON STANDARD AND LINE PIPE COMMONLY USED FOR WATER WELL CASING

Nominal Size, Inches	Outside Diameter, Inches	Outside Diameter Couplings, Inches	Schedule or Class[1]	Wall Thickness, Inches	Weight per Foot-Plain End, Pounds	Inside Diameter, Inches	[2]Suggested Maximum Setting, Feet
4	4.500	5.200	--------	0.219	10.10	4.062	1,190
			40	.237	10.79	4.026	1,060
6	6.625	7.390	--------	.250	17.02	6.125	705
			40(S)	.280	18.97	6.065	850
8	8.625	9.625	20	.250	22.36	8.125	420
			30	.277	24.70	8.071	525
			40(S)	.322	28.55	7.981	695
10	10.750	11.750	20	.250	28.04	10.250	235
			30	.307	34.24	10.136	410
			40(S)	.365	40.48	10.020	580
12	12.750	14.000	20	.250	33.38	12.250	140
			30	.330	43.77	12.090	320
			S	.375	49.56	12.000	435
			40*	.406	53.56	11.938	515
14	14.000	15.000	10	.250	36.71	13.500	105
			20	.312	45.68	13.376	195
			30(S)	.375	54.57	13.250	350
			40	.438	63.37	13.124	495
16	16.000	17.000	10	.250	42.05	15.500	70
			20	.312	52.36	15.376	140
			30(S)	.375	62.58	15.250	240
			40	.500	82.77	15.000	495
18	18.000	19.000	10	.250	47.39	17.500	50
			20	.312	59.03	17.376	100
			S	.375	70.59	17.250	170
			30	.438	82.06	17.124	270
			40	.562	104.76	16.876	495
20	20.000	21.000	10	.250	52.73	19.500	35
			20(S)	.375	78.60	19.250	125
			30	.500	104.13	19.000	295
			40*	.594	123.06	18.802	445
22	22.000	--------	10	.250	58.07	21.500	30
			20(S)	.375	86.61	21.250	95
			30	.500	114.81	21.000	220
24	24.000	--------	10	.250	63.41	23.500	20
			20(S)	.375	94.62	23.250	70
			30	.562	140.80	22.876	240
			40	.688	171.17	22.624	410
26	26.000	--------	10	.312	85.73	25.376	30
			S	.375	102.63	25.250	55
			20	.500	136.17	25.000	135
28	28.000	--------	10*	.312	92.41	27.376	25
			(S)	.375	110.41	27.250	45
			20	.500	146.85	27.000	105
			30	.625	182.73	26.750	210
30	30.000	--------	10*	.312	99.08	29.376	20
			(S)	.375	118.65	29.250	35
			20	.500	157.53	29.000	85
			30	.625	196.08	28.750	170

TABLE 4–11. DATA ON STANDARD AND LINE PIPE COMMONLY USED FOR WATER WELL CASING (continued)

Nominal Size, Inches	Outside Diameter, Inches	Outside Diameter Couplings, Inches	Schedule or Class[1]	Wall Thickness, Inches	Weight per Foot-Plain End, Pounds	Inside Diameter, Inches	[2]Suggested Maximum Setting, Feet
32	32.000	--------	10*	.312	105.76	31.376	20
			(S)	.375	126.66	31.250	30
			20	.500	168.21	33.000	70
			30	.625	209.43	32.750	140
34	34.000	--------	10*	.312	112.43	33.376	15
			S	.375	134.67	33.250	25
			20	.500	178.89	33.000	60
			30	.625	222.78	32.750	115
36	36.000	--------	10*	.312	119.11	35.376	10
			(S)	.375	142.68	35.250	20
			20	.500	189.57	35.000	50
			30	.625	236.13	34.750	100

[1] ASA Standard B36.10 schedule numbers (S) indicates standard weight pipe; * indicates a non-API standard.
[2] Maximum settings were estimated for the worst possible conditions in unconsolidated formation. A design factor of approximately 1.5 was used for steel with yield strength less than 40,000 lb/in^2. A 50-percent increase in depth of setting beyond those given is considered safe under favorable conditions.
Source: Bureau of Reclamation, 1977, Ground Water Manual

TABLE 4–12. RECOMMENDED CASING DIAMETERS FOR WATER WELLS

[For line shaft vertical turbine pumps 1800 rpm]

Yield, Gallons per Minute	Recommended Casing Size, Inches
Less than 100	6 I.D.[a]
75–175	8 I.D.
150–400	10 I.D.
350–600	12 I.D.
600–1,300	16 O.D.[b]
1,300–1,800	20 O.D.
1,800–3,000	24 O.D.
3,000–4,500	30 O.D.
Over 4,500	30 O.D.

[a] I.D.=Insided Diameter
[b] O.D.=Outside Diameter
Source: U.S. Environmental Protection Agency, Manual of Water Well Construction Practices, EPA-570/9–75–001

TABLE 4–13. RECOMMENDED CASING SIZES FOR DOMESTIC WATER WELLS

Yield at 50' Drawdown	Recommended Casing Diameter Inches	Pump Type: Jet	Pump Type: Double Jet	Pump Type: Submersible
Less than 8 gpm	2	X	X	
	3	X	X	X
	4	X	X	X
	5		X	X
	6			X
8 to 16.5 gpm	2	X	X	
	3	X	X	X
	4	X	X	X
	5		X	X
	6			X
Greater than 16.5 gpm	3	X		
	4	X	X	X
	5		X	X
	6			X

Source: U.S. Environmental Protection Agency, Manual of Water Well Construction Practices, EPA-570/9–75–001

TABLE 4–14. RECOMMENDED MAXIMUM DEPTH OF SETTING FOR CALIFORNIA STOVEPIPE CASING

[Includes similar sheet steel and steel-plate fabricated casing; in feet]

Diameter, Inches	Gauge[1]						Thickness, Inches			
	12		10		8	6	3/16	1/4	5/16	3/8
	D	S	D	S	D	D				
8	340	125	750	260	X	X	X	X	X	X
10	150	60	390	135	X	X	320	750	X	X
12	100	35	225	75	390	X	180	435	875	X
14	60	20	140	45	250	X	115	270	530	X
16	40	15	90	30	165	275	75	180	360	630
18	30		65	20	115	190	55	125	260	445
20	20		45		85	140	35	90	180	320
22			35		60	105	X	X	X	X
24			25		45	80	20	50	100	185
26			20		35	60	X	X	X	X
30			10		25	40	10	25	50	95

D=Telescoping.
S=Single thickness.
X=Not commonly made in these sizes.

[1] U.S. Standard Gauge.
Source: Bureau of Reclamation, 1977, Ground Water Manual

TABLE 4–15. RECOMMENDED DIAMETER AND THICKNESS OF PVC CASING FOR WATER WELLS

Well diameters 1.5 inches through 4 inches-ASTMD 2241–73* SDR 21 (Type 1120–1220)

Nominal Size	Outside Diameter	Inside Diameter	Minimum Wall Thickness
	Inches		
1.5	1.900	1.720	0.090
2	2.375	2.149	0.113
2.5	2.875	2.601	0.137
3	3.500	3.166	0.177
4	4.500	4.072	0.214

Well diameters 5 inches through 12 inches-ASTMD 1785–73* Schedule 40 (Type 1120–1220)

Nominal Size	Outside Diameter,	Inside Diameter	Minimum Wall Thickness
	Inches		
5	5.563	5.047	0.258
6	6.625	6.065	0.280
8	8.625	7.981	0.322
10	10.750	10.020	0.365
12	12.750	11.938	0.406

*New ASTM Standards are currently under review.
Source: U.S. Environmental Protection Agency, Manual of Water Well Construction Practices, EPA-570/9–75–001

TABLE 4–16. WELL SCREEN SELECTION CHART FOR SMALL-CAPACITY WELLS

GRADATION OF SAND	AVERAGE SLOT SIZE	MINIMUM SUGGESTED LENGTH for corresponding screen diameter* and desired well yield					
	Thousandths of an inch	1¼" SCREEN	2" SCREEN	3" SCREEN	4" SCREEN	5" SCREEN	6" SCREEN
VERY FINE SAND 6 – 7 – 8 Slot About the finest material that can be utilized for a water supply. A line composed of 12 grains would measure about 1/16".	7	300 gph — 5 ft. 450 gph — 8 ft. 600 gph — 12 ft.	450 gph — 6 ft. 600 gph — 9 ft. 900 gph — 13 ft.	600 gph — 8 ft. 900 gph — 11 ft. 1200 gph — 14 ft.	600 gph — 6 ft. 1200 gph — 10 ft. 1800 gph — 14 ft.	900 gph — 6 ft. 1200 gph — 9 ft. 1800 gph — 11 ft.	1200 gph — 8 ft. 2000 gph — 12 ft. 2400 gph — 15 ft.
FINE SAND 9 – 10 – 12 Slot Often called "sugar sand". Line of 6 or 7 average grains measures 1/16".	10	300 gph — 4 ft. 450 gph — 6 ft. 600 gph — 9 ft.	450 gph — 4 ft. 600 gph — 6 ft. 900 gph — 9 ft.	600 gph — 6 ft. 900 gph — 8 ft. 1200 gph — 10 ft.	600 gph — 4 ft. 1200 gph — 7 ft. 1800 gph — 10 ft.	900 gph — 5 ft. 1200 gph — 7 ft. 1800 gph — 8 ft.	1200 gph — 6 ft. 2000 gph — 9 ft. 2400 gph — 11 ft.
MEDIUM SAND 16 – 18 – 20 Slot Average grain size is about 4 grains to 1/16".	18	300 gph — 4 ft. 450 gph — 5 ft. 600 gph — 7 ft.	600 gph — 5 ft. 900 gph — 7 ft. 1200 gph — 9 ft.	600 gph — 4 ft. 1200 gph — 9 ft. 1800 gph — 13 ft.	600 gph — 3 ft. 1200 gph — 6 ft. 1800 gph — 9 ft.	900 gph — 4 ft. 1200 gph — 6 ft. 1800 gph — 7 ft.	1200 gph — 5 ft. 2000 gph — 8 ft. 2400 gph — 10 ft.
MEDIUM AND COARSE SAND MIXED Average grain size a little less than 1/32", or between 2 and 3 grains to 1/16".	25	300 gph — 3 ft. 450 gph — 5 ft. 600 gph — 6 ft.	600 gph — 5 ft. 900 gph — 6 ft. 1200 gph — 8 ft.	600 gph — 4 ft. 1200 gph — 7 ft. 1800 gph — 11 ft.	600 gph — 3 ft. 1200 gph — 5 ft. 1800 gph — 8 ft.	900 gph — 4 ft. 1200 gph — 5 ft. 1800 gph — 6 ft.	1200 gph — 5 ft. 2000 gph — 7 ft. 2400 gph — 9 ft.
COARSE SAND Average grain size a little over 1/32" (2 grains to 1/16").	35	450 gph — 4 ft. 600 gph — 5 ft. 900 gph — 7 ft.	600 gph — 4 ft. 900 gph — 5 ft. 1200 gph — 7 ft.	900 gph — 4 ft. 1200 gph — 6 ft. 1800 gph — 10 ft.	900 gph — 3 ft. 1200 gph — 4 ft. 1800 gph — 7 ft.	1200 gph — 4 ft. 1800 gph — 6 ft. 2000 gph — 8 ft.	1200 gph — 4 ft. 2000 gph — 7 ft. 2400 gph — 8 ft.
COARSE SAND AND FINE GRAVEL MIXED Average grain size about 1/16". In coarser gravels, No. 80 and No. 100 slot are often used.	50	450 gph — 4 ft. 600 gph — 5 ft. 900 gph — 7 ft.	600 gph — 4 ft. 900 gph — 5 ft. 1200 gph — 6 ft.	900 gph — 4 ft. 1200 gph — 6 ft. 1800 gph — 10 ft.	900 gph — 3 ft. 1200 gph — 4 ft. 1800 gph — 7 ft.	1200 gph — 4 ft. 1800 gph — 6 ft. 2000 gph — 7 ft.	1200 gph — 4 ft. 2000 gph — 6 ft. 2400 gph — 8 ft.

* Nominal size of screen.
Source: Edward E. Johnson, Inc.

TABLE 4–17. RECOMMENDED MINIMUM SCREEN ASSEMBLY DIAMETERS

	Minimum Nominal Screen Assembly Diameter, Inches
Discharge, gal/min:	
Up to 50	2
50 to 125	4
125 to 350	6
350 to 800	8
800 to 1,400	10
1,400 to 2,500	12
2,500 to 3,500	14
3,500 to 5,000	16
5,000 to 7,000	18
7,000 to 9,000	20

Source: U.S. Bureau of Reclamation, 1977, Ground Water Manual

TABLE 4–18. COST OF WATER WELL SCREENS

[U.S. Dollars; prices as of May, 1986; per foot]

	Nominal Screen Size Diameter (Inches)							
Type of Screen	**2**	**4**	**6**	**8**	**10**	**12**	**14**	**16**
Telescope-size screens								
Stainless steel	–	39.10	64.00	80.00	103.00	133.80	162.70	191.50
Galvanized low-carbon steel	–	27.40	38.90	45.30	52.30	59.90	75.50	87.10
Pipe-size screens								
Stainless steel	34.70	54.80	71.40	88.50	121.00	164.50	191.40	213.70
Galvanized low-carbon steel	19.20	34.90	42.50	46.50	60.40	77.30	88.80	98.10
PVC plastic screen	14.50	23.80	33.90	52.60	–	–	–	–

Source: Johnson Division Signal Environmental Systems, Inc., St. Paul, MN

TABLE 4–19. INTAKE AREAS OF WELL SCREENS

WIRE-WOUND TELESCOPIC SCREENS

INTAKE AREAS (SQ. IN. PER FT OF SCREEN)

NOM. DIAM. (Inches)	SLOT OPENING SIZE							
	10-SLOT	20-SLOT	40-SLOT	60-SLOT	80-SLOT	100-SLOT	150-SLOT	250-SLOT
3	15	26	41	52	59	65	73	82
4	20	35	57	71	81	88	101	115
5	26	45	72	90	102	112	112	132
6	30	53	85	106	100	112	132	156
8	28	51	87	113	133	149	160	194
10	36	65	108	141	166	186	200	243
12	42	77	130	143	171	195	237	265
14	37	68	97	132	161	185	232	292
16	42	60	108	148	180	208	261	327
18	36	69	124	169	206	237	298	375
20	41	77	139	189	229	264	280	366
24	61	113	131	182	226	265	343	449
26	63	118	138	191	237	278	360	471
30	75	138	161	224	278	325	422	552
36	84	157	184	255	317	371	481	629

PVC PLASTIC SCREENS

	INTAKE AREAS (SQ. IN. PER FT OF SCREEN)									
SIZE (IN.)	SLOT OPENING SIZE									
	6-SLOT	8-SLOT	10-SLOT	12-SLOT	15-SLOT	20-SLOT	25-SLOT	30-SLOT	35-SLOT	40-SLOT
$1^1/_4$	3.0	3.4	4.8	6.0	7.0	8.9	10.8	12.5	14.1	15.6
$1^1/_2$	3.4	4.5	5.5	6.5	8.1	10.2	12.3	14.2	16.2	17.9
2	4.3	5.5	6.8	8.1	10.0	12.8	15.4	17.9	20.3	22.4
3	5.4	7.1	8.8	10.4	12.8	16.5	20.0	23.2	26.5	29.3
4	7.0	9.0	11.3	13.5	16.5	21.2	25.8	30.0	33.9	37.7
5	8.1	10.6	13.1	15.5	19.1	24.7	30.0	34.9	39.7	44.2
6	8.1	10.6	13.2	15.6	19.2	25.0	30.5	35.8	40.7	45.4
8	13.4	17.6	21.7	25.7	31.5	40.6	49.3	57.4	65.0	72.3

Note: The maximum transmitting capacity of the screen can be derived from these figures. To determine GPM per ft. of screen, multiply the intake area in square inches by 0.31. It must be remembered that this is the maximum capacity of the screen under ideal conditions with an entrance velocity of 0.1 ft. per sec.
Source: Johnson Division of Signal Environmental Systems, Inc., St. Paul, MN

TABLE 4–20. OPTIMUM WELL SCREEN ENTRANCE VELOCITIES

Coefficient of permeability (gallons per day per square foot)	Optimum screen entrance velocities (feet per minute)
>6000	12
6000	11
5000	10
4000	9
3000	8
2500	7
2000	6
1500	5
1000	4
500	3
<500	2

Source: Illinois State Water Survey, 1962

TABLE 4–21. CHLORINATED LIME REQUIRED TO DISINFECT A WELL OR SPRING

[Values provide a dosage of approximately 50 parts per million of available chlorine.]

Capacity of well or spring in gallons	Chlorinated lime required (25% available chlorine)		Approximate volume of water, in gallons to be used in preparing chlorine solution
	Pounds	*and Ounces*	
50	–	1.5	5
100	–	3.0	5
200	–	6.0	5
300	–	9.0	5
400	–	12.0	5
500	–	15.0	5
1,000	1	14.0	10
2,000	3	12.0	15
3,000	5	10.0	20

Source: U.S. Public Health Service

TABLE 4–22. VOLUME OF WATER IN WELL PER FOOT OF DEPTH

Nominal Casing Size, Inches	Schedule No.	Volume, Gallons per Foot of Depth
4	40	0.66
5	40	1.04
6	40	1.50
8	30	2.66
10	30	4.19
12	30	5.80
14	30	7.16
16	30	9.49
18	30	11.96
20	30	14.73
22	30	17.99
24	30	21.58

Source: U.S. Bureau of Reclamation, 1977, Ground Water Manual

SECTION D. INJECTION WELLS

TABLE 4–23. STATISTICAL ANALYSIS OF INJECTION WELL DATA

Distribution of Injection Wells By Industry Type

Type of Industry	Percentage of Existing Wells
Refineries and Natural Gas Plants	20%
Chemical, Petrochemical, and Pharmaceutical companies	55%
Metal products companies	7%
Other	18%

Total Depth of Injection Wells

Total Well Depth	Percentage of Wells
0 – 1,000 ft.	5%
1,001 – 2,000 ft.	32%
2,001 – 4,000 ft.	27%
4,001 – 6,000 ft.	28%
6,001 –12,000 ft.	6%
over 12,000 ft.	2%

Type of Rock Used for Injection

Rock Type	Percentage of Wells
Sand	33%
Sandstone	41%
Limestone and dolomite	22%
Other	4%

Rate of Injection

Injection Rate	Percentage of Wells
0– 50 gpm	23%
51–100 gpm	11%
101–200 gpm	25%
201–400 gpm	19%
401–800 gpm	3%
over 800 gpm	1%
unknown	18%

Pressure at Which Waste Is Injected

Injection Pressure	Percentage of Wells
Gravity flow	11%
Gravity – 150 psi	19%
151 – 300 psi	15%
301 – 600 psi	6%
601 –1,500 psi	13%
over 1,500 psi	2%
unknown	34%

Source: Water Well Journal, 1968

TABLE 4–24. CLASSIFICATION OF INJECTION WELLS IN THE UNITED STATES

Class I	1. Wells used by generators of hazardous wastes or owners or operators of hazardous waste management facilities to inject hazardous waste, other than class IV wells. 2. Other industrial and municipal disposal wells which inject fluids beneath the lowermost formation containing, within one-quarter mile of the well bore, an underground source of drinking water.
Class II	Wells which inject fluids: 1. Which are brought to the surface in connection with conventional oil or natural gas production; 2. For enhanced recovery of oil or natural gas; and 3. For storage of hydrocarbons which are liquid at standard temperature and pressure.
Class III	Wells which inject for extraction of minerals or energy, including: 1. Mining of sulfur by the Frasch process; 2. Solution mining of minerals; *Note.*—Solution mining of minerals includes sodium chloride, potash, phosphate, copper, uranium, and any other mineral which can be mined by this process. 3. In situ combustion of fossil fuel; and *Note.*—Fossil fuels include coal, tar sands, oil shale, and any other fossil fuel which can be mined by this process. 4. Recovery of geothermal energy to produce electric power. *Note.*—Class III wells include the recovery of geothermal energy to produce electric power, but do not include wells used in heating or aquaculture, which fall under class V.
Class IV	Wells used by generators of hazardous wastes or of radioactive wastes, by owners or operators of hazardous waste management facilities, or by owners or operators of radioactive waste disposal sites to dispose of hazardous wastes or radioactive wastes into or above a formation which within one-quarter mile of the well contains an underground source of drinking water.
Class V	Injection wells not included in classes I, II, III, or IV. *Note.*—Class V wells include: 1. Air conditioning return flow wells used to return to the supply aquifer the water used for heating or cooling in a heat pump; 2. Cesspools or other devices that receive wastes, which have an open bottom and sometimes have perforated sides. The Underground Injection Control (UIC) requirements do not apply to single family residential cesspools; 3. Cooling water return flow wells used to inject water previously used for cooling; 4. Drainage wells used to drain surface fluid, primarily storm runoff, into a subsurface formation; 5. Dry wells used for the injection of wastes into a subsurface formation; 6. Recharge wells used to replenish the water in an aquifer; 7. Saltwater intrusion barrier wells used to inject water into a freshwater aquifer to prevent the intrusion of salt-water into the freshwater; 8. Sand backfill wells used to inject a mixture of water and sand, mill tailings, or other solids into mined-out portions of subsurface mines; 9. Septic system wells used: To inject the waste or effluent from a multiple dwelling, business establishment, community, or regional business establishment septic tank; or For a multiple dwelling, community, or regional cesspool. The UIC requirements do not apply to single family residential waste disposal systems;

(continued)

TABLE 4–24. CLASSIFICATION OF INJECTION WELLS IN THE UNITED STATES (continued)

10. Subsidence control wells (not used for the purpose of oil or natural gas production) used to inject fluids into a non-oil- or non-gas-producing zone to reduce or eliminate subsidence associated with the overdraft of freshwater;
11. Wells used for the storage of hydrocarbons which are gases at standard temperature and pressure;
12. Geothermal wells used in heating and aquaculture; and
13. Nuclear disposal wells.

Source: U.S. Environmental Protection Agency, 1980

TABLE 4–25. DISTRIBUTION OF INJECTION WELLS IN THE UNITED STATES

USEPA Region	State	All Wells	Class I Wells	Class II Wells	Class III Wells	Class IV Wells	Class V Wells
1	Connecticut	173	9	—	—	7	157
	Massachusetts	18,252	—	—	—	—	18,252
	Maine	18	—	—	—	—	18
	New Hampshire	27	—	—	—	—	27
	Rhode Island	42	—	—	—	—	42
	Vermont	1	—	—	—	—	1
2	New Jersey	1,327	—	—	—	—	1,327
	New York	6,348	11	3,853	149	184	2,151
3	Delaware	3	—	—	—	—	3
	Maryland	968	—	—	—	3	965
	Pennsylvania	8,760	5	4,607	—	31	4,117
	Virginia	1,676	—	1	—	3	1,672
	West Virginia	2,034	7	319	17	—	1,691
4	Alabama	169	8	152	9	—	—
	Florida[1]	7,075	52	80	3	3	6,937
	Georgia	4	—	—	—	—	4
	Kentucky	4,642	—	4,357	—	—	285
	Mississippi	1,348	7	1,223	—	—	118
	North Carolina	33	3	—	—	3	27
	South Carolina	63	—	—	—	30	33
	Tennessee	57	—	13	—	11	33
5	Illinois	18,503	10	18,493	—	—	—
	Indiana	3,669	76	3,565	—	—	28
	Michigan	4,207	97	1,275	110	—	2,725
	Minnesota	19	—	—	—	—	19
	Ohio	6,417	—	3,601	2	—	2,814
6	Arkansas	871	23	808	—	—	40
	Louisiana	4,544	80	4,249	215	—	—
	Oklahoma	11,291	13	11,278	—	—	—
	Texas	65,470	129	41,859	23,124	—	358
	Indian lands within the region	3,300	—	3,300	—	—	—
7	Iowa	14	—	—	—	—	14
	Kansas	16,298	57	15,175	394	—	672
	Missouri	223	—	223	—	—	—
	Nebraska	1,983	—	1,983	—	—	—
8	Colorado	1,069	1	1,001	59	2	6
	Montana	1,448	—	1,447	—	—	1
	North Dakota	434	1	429	4	—	—
	South Dakota	8	—	8	—	—	—
	Utah	541	—	504	30	—	7
	Wyoming	4,924	—	4,016	898	—	10
9	Arizona	509	—	3	484	5	17
	California	13,844	—	13,844	—	—	—
	Guam	136	—	—	—	—	136
	Indian lands within the region	519	—	518	—	—	1
10	Alaska	164	—	160	—	1	3
	Idaho	581	—	—	—	1	580
	Oregon	712	—	—	—	—	712
	Washington	5,640	1	—	—	10	5,629
	Total	220,358	590	142,344	25,498	294	51,632

[1] Number of wells in Florida adjusted to reflect a more recent Florida Department of Environmental Regulation inventory of injection wells.
Source: U.S. Environmental Protection Agency Federal Underground Injection Control Reporting System, June 21, 1983

TABLE 4–26. SUMMARY OF DEEP-WELL INJECTION SYSTEMS IN FLORIDA

Location	Injected Effluent		First Year of Operation	Design Capacity		Injection Wells			Well Casings		
	Type	Pretreatment		ML/d	mgd	Number of Wells	Depth m	Depth ft*	Number of Casings	Depth m	Depth ft*
Belle Glade, Palm Beach County	industrial	cooling	1966	5.6	1.5	2	975	3200	4	883	2900
Sunset Park, South Miami	municipal	secondary	1971	22.7	6.0	1	914	3000	3	563	1850
Mulberry, Polk County	industrial	none	1971	1.3	0.35	1	1371	4500	3	1219	4000
Kendale Lakes, South Miami	municipal	secondary	1973	22.7	6.0	1	975	3200	3	670	2200
Margate, Broward County	municipal	secondary	1975	56.7	15.0	1	975	3200	3	731	2400
St. Petersburg, Southwest plant	municipal	tertiary	1976	75.7	20.0	3	304	1000	3	274	900
Gainesville, Kanapaha plant	municipal	advanced wastewater treatment	1976	28.3	7.5	3	304	1000	3	152	500
West Palm Beach	municipal	secondary	1978	302.8	80.0	5	1097	3600	4	914	3000
Vero Beach, Indian River County	industrial	neutralization	1979	1.1	0.3	1	914	3000	4	731	2400
Miami-Dade Water & Sewer Authority	municipal	secondary	1983	423.9	112.0	9	944	3100	4	792	2600

*Rounded to nearest 100 ft

Source: Garcia-Bengochea, J.I., 1983, Protecting Water Supply Aquifers in Areas Using Deep-Well Wastewater Disposal, J.Am. Water Works Assoc., v.75, no.6. Copyright AWWA. Reprinted with permission

SECTION E. PUMPING OF WATER

TABLE 4–27. USEFUL FACTORS IN PRELIMINARY PLANNING OF SMALL PUMPING PLANTS

Pump or Pipe Size, In.	Gallons per Minute	Acre-Inches per 24 Hours	Pipe Velocity, Feet per Second	Velocity Head, $\frac{V^2}{2g}$ Feet	Friction in Feet per 100 Feet of Pipe	Horsepower Required for 10 Feet Total Head. Pump and Transmission Efficiency = 70 Percent
6	400	21.2	4.54	0.32	2.21	1.4
6	600	31.8	6.72	0.70	4.7	2.2
6	800	42.4	9.08	1.28	8.0	2.9
6	1,000	53.0	11.32	1.99	12.0	3.6
8	900	47.7	5.75	0.52	2.46	3.2
8	1,100	58.3	7.03	0.77	3.51	4.0
8	1,300	68.9	8.32	1.07	4.72	4.7
8	1,500	79.5	9.60	1.43	6.27	5.4
10	1,200	63.6	4.91	0.38	1.46	4.3
10	1,600	84.8	6.56	0.67	2.35	5.8
10	2,000	106.1	8.10	1.02	3.65	7.2
10	2,400	127.3	9.73	1.47	5.04	8.7
12	2,000	106.1	5.60	0.48	1.43	7.2
12	2,500	132.6	7.00	0.77	2.28	9.0
12	3,000	159.1	8.40	1.10	3.15	10.8
12	3,500	185.6	9.80	1.49	4.10	12.6
14	2,000	106.1	4.20	0.27	0.66	7.2
14	3,000	159.1	6.30	0.61	1.47	10.8
14	4,000	212.1	8.40	1.09	2.47	14.4
14	5,000	265.2	10.50	1.71	3.92	18.0
16	3,600	190.9	5.74	0.51	1.10	13.0
16	4,400	233.3	7.01	0.76	1.58	15.9
16	5,200	275.8	8.29	1.06	2.16	18.8
16	6,000	318.2	9.56	1.42	2.60	21.6
18	4,500	238.6	5.70	0.50	0.93	16.2
18	5,500	291.7	6.96	0.75	1.32	19.8
18	6,500	344.7	8.22	1.05	1.82	23.4
18	8,000	424.2	10.02	1.56	2.65	28.9
20	5,000	265.2	5.13	0.41	0.68	18.0
20	6,500	344.7	6.66	0.69	1.06	23.4
20	8,000	424.2	8.17	1.03	1.63	28.9
20	10,000	530.3	10.40	1.68	2.53	36.1
24	8,000	424.2	5.68	0.50	0.66	28.9
24	10,000	530.3	7.07	0.78	0.98	36.1
24	12,000	636.4	8.50	1.12	1.40	43.3
24	14,000	742.4	9.95	1.54	1.87	50.5
30	12,000	636.4	5.44	0.46	0.47	43.3
30	16,000	848.5	7.36	0.84	0.83	57.7
30	20,000	1061.0	9.09	1.29	1.22	72.2
30	24,000	1273.0	10.90	1.86	1.71	86.6

Source: U.S. Dept. of Agriculture

TABLE 4–28. CHARACTERISTICS OF PUMPS FREQUENTLY EMPLOYED IN WELLS

Type of Pump	Practical Suction Lift[a]	Usual Well-Pumping Depth	Usual Pressure Heads	Advantages	Disadvantages
Reciprocating:					
Shallow well	6–7 m	6–7 m	30–60 m	Positive action; discharge against variable heads; pumps water containing sand and silt; especially adapted to low capacity and high lifts.	Pulsating discharge; subject to vibration and noise; maintenance cost may be high; may cause destructive pressure if operated against closed valve
Deep well	6–7 m	Up to 180 m	Up to 180 m above cylinder		
Centrifugal:					
Shallow well straight centrifugal (single stage)	6 m max	3–6 m	30–45 m	Smooth, even flow; pumps water containing sand and silt; pressure on system is even and free from shock; low-starting torque; usually reliable and good service life	Loses prime easily; efficiency depends on operating under design heads and speed
Regenerative vane turbine type (single impeller)	8 m max	8 m	30–60 m	Same as straight centrifugal except not suitable for pumping water containing sand or silt; self-priming	Same as straight centrifugal except maintains priming easily
Deep well					
Vertical line shaft turbine (multistage)	Impellers submerged	15–90 m	30–250 m	Same as shallow well turbine	Efficiency depends on operating under design head and speed; requires straight well large enough for turbine bowls and housing; lubrication and alignment of shaft critical; abrasion from sand

(continued)

TABLE 4–28. CHARACTERISTICS OF PUMPS FREQUENTLY EMPLOYED IN WELLS (continued)

Type of Pump	Practical Suction Lift[a]	Usual Well-Pumping Depth	Usual Pressure Heads	Advantages	Disadvantages
Submersible turbine (multistage)	Pump and motor submerged	15–120 m	15–120 m	Same as shallow well turbine; easy to frost-proof installation; short pump shaft to motor	Repair to motor or pump requires pulling from well; sealing of electrical equipment from water vapor critical; abrasion from sand
Jet:					
Shallow well	4–6 m below ejector	Up to 4–6 m below ejector	25–45 m	High capacity at low heads; simple in operation; does not have to be installed over well; no moving parts in well	Capacity reduces as lift increases; air in suction or return line will stop pumping
Deep well	4–6 m below ejector	7–35 m 60 m max.	25–45 m	Same as shallow well jet	Same as shallow well jet
Rotary:					
Shallow well .... (gear type)	7 m	7 m	15–75 m	Positive action; discharge constant under variable heads; efficient operation	Subject to rapid wear if water contains sand or silt; wear of gears reduces efficiency
Deep well .... (helical rotary type).	Usually submerged	15–150 m	30–150 m	Same as shallow well/rotary; only one moving pump device in well	Same as shallow well rotary except no gear wear

[a] Practical suction lift at sea level. Reduce lift 0.3 m for each 300 m above sea level.
Source: U.S. Public Health Service, 1962, Manual of Individual Water-Supply Systems, Publ. No. 24

TABLE 4–29. SELECTION OF PUMP SIZE AND DIAMETER OF WELLS

[ID, inside diameter; OD, outside diameter]

Anticipated Well Yield			Nominal Size of Pump Bowls (in.)	Optimum Well Diameter (in.)
In gal min^{-1}	In ft^3 min^{-1}	In m^3 min^{-1}		
Less than 100	Less than 13	Less than 0.38	4	6 ID
75–175	10–23	.28–.66	5	8 ID
150–400	20–53	.57–1.52	6	10 ID
350–650	47–87	1.33–2.46	8	12 ID
600–900	80–120	2.27–3.41	10	14 OD
850–1,300	113–173	3.22–4.93	12	16 OD
1,200–1,800	160–240	4.55–6.82	14	20 OD
1,600–3,000	213–400	6.06–11.37	16	24 OD

Source: Health, R.C., 1983, Basic Ground-Water Hydrology, U.S. Geological Survey Water-Supply Paper 2220

TABLE 4–30. PUMPING PLANT PERFORMANCE STANDARDS

Type of Power Unit	Standard Consumption of Fuel or Energy per Water Horsepower*
Diesel engine	0.091 gal per hr
Gasoline engine	0.116 gal per hr
Propane engine	0.145 gal per hr
Natural gas	160 cu ft per hr
Electric motor	0.885 kw-hr per hr

* Based on pump efficiency of 75 percent.
Source: College of Agriculture, University of Nebraska

TABLE 4–31. STANDARD FUEL REQUIREMENTS FOR GOOD PUMPING PLANTS

[Based on performance standards in Table 4–30]

Pumping Rate, in gpm	Head, in Feet	Water Horse-power	Fuel or Energy Required				
			Diesel, gal per hr	Gasoline, gal per hr	Propane, gal per hr	Natural Gas, cu ft per hr	Electricity, kwh per hr
500	100	13	1 1/4	1 1/2	2	190	14
	150	19	1 3/4	2 1/4	2 3/4	280	21
	200	25	2 1/4	3	3 3/4	380	29
700	100	18	1 3/4	2	2 3/4	270	20
	150	27	2 1/2	3 1/4	4	400	30
	200	35	3 1/4	4 1/4	5 1/4	530	40
800	100	20	1 3/4	2 1/2	3	300	23
	150	30	2 3/4	3 1/2	4 1/2	450	34
	200	40	3 3/4	4 3/4	6	610	46
1000	100	25	2 1/4	3	3 3/4	380	29
	150	38	3 1/2	4 1/2	5 3/4	570	43
	200	50	4 1/2	6	7 1/2	760	57

Source: College of Agriculture, University of Nebraska

TABLE 4–32. WINDMILL PUMPING CAPACITY

[Based on wind velocity of 20 mph]

Representative Cylinder Size, inches	6 1/2 foot mill		8 foot mill		10 foot mill	
	Depth to Water, ft.	Gals. per Hour	Depth to Water, ft.	Gals. per Hour	Depth to Water, ft.	Gals. per Hour
1–11/16	144	100	224	130	384	165
2	100	160	156	200	243	240
2–1/2	70	230	108	260	169	300
3	48	330	75	400	117	475
3–1/2	30	475	52	550	81	625
4	–	–	40	750	63	850
4–1/2	–	–	–	–	42	1000

Source: Water Systems Council, Manual of Water Supply Equipment

FIGURE 4–11. GROUND WATER OVERDRAFT IN THE UNITED STATES

[Data as of 1975; by water resource subregion]

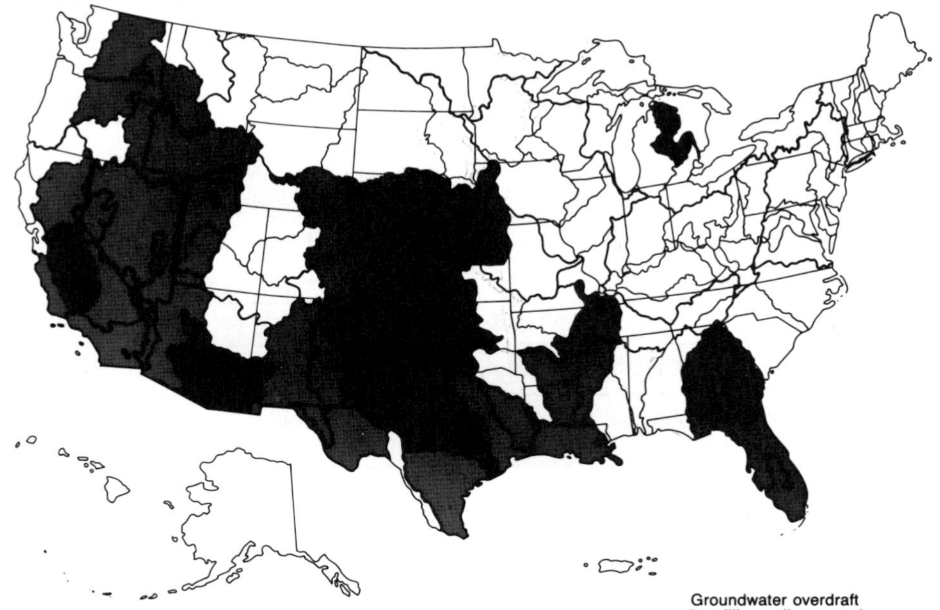

Groundwater overdraft in million gallons per day

- Critical (more than 500)
- Moderate (21-500)
- No overdraft

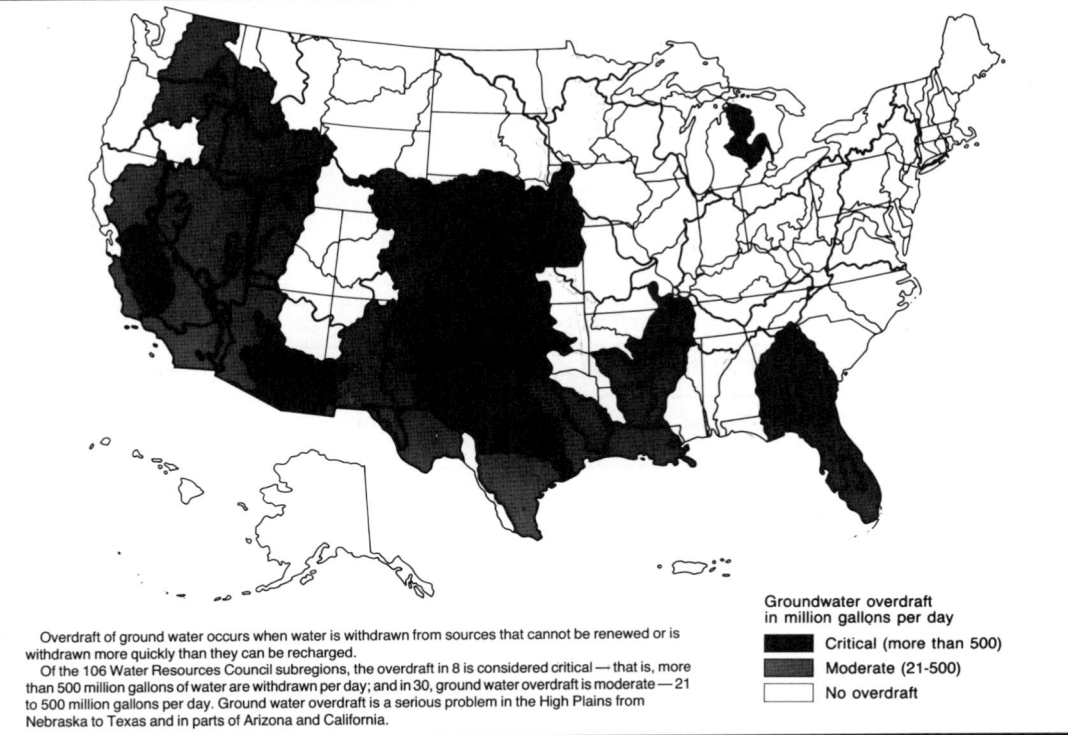

Overdraft of ground water occurs when water is withdrawn from sources that cannot be renewed or is withdrawn more quickly than they can be recharged.

Of the 106 Water Resources Council subregions, the overdraft in 8 is considered critical — that is, more than 500 million gallons of water are withdrawn per day; and in 30, ground water overdraft is moderate — 21 to 500 million gallons per day. Ground water overdraft is a serious problem in the High Plains from Nebraska to Texas and in parts of Arizona and California.

Source: Council on Environmental Quality, 1981, Environmental Trends

FIGURE 4–12. AREAS OF WATER-TABLE OR ARTESIAN WATER-LEVEL DECLINE IN EXCESS OF 40 FEET IN THE UNITED STATES

[Decline in at least one aquifer since predevelopment]

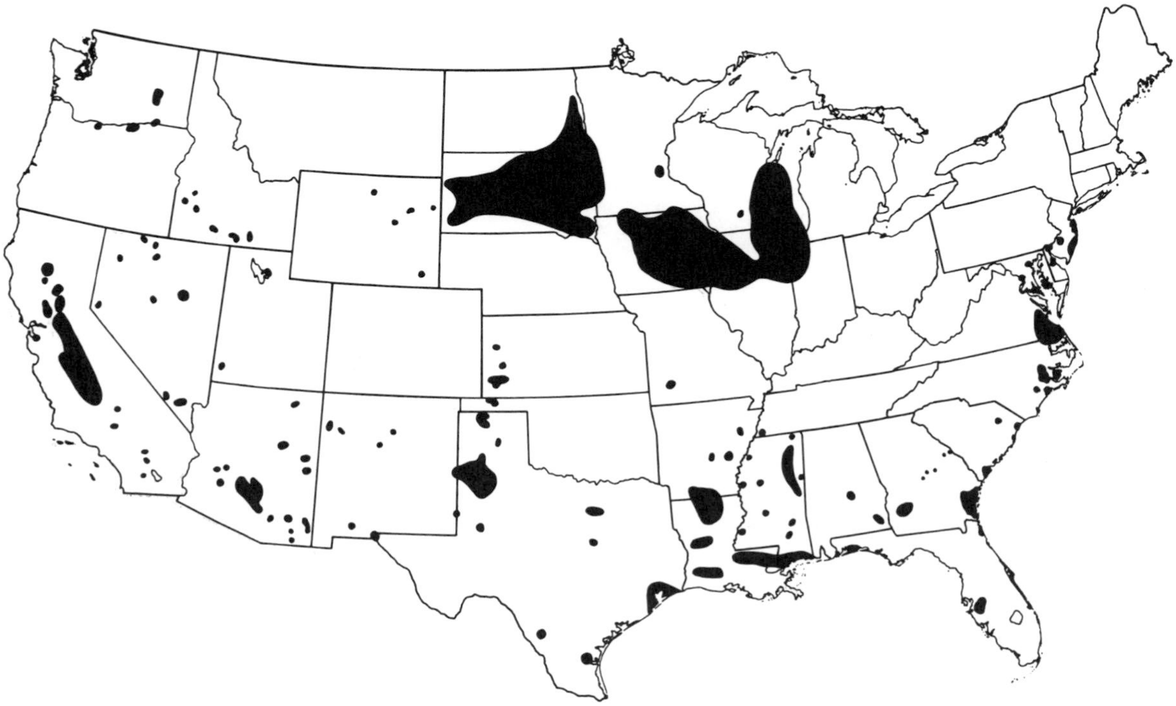

Source: U.S. Geological Survey, 1984, National Water Summary 1983 - Hydrologic Events and Issues, Water-Supply Paper 2250

SECTION F. SUBSIDENCE

FIGURE 4–13. AREAS OF LAND SUBSIDENCE FROM GROUND-WATER WITHDRAWAL IN THE UNITED STATES

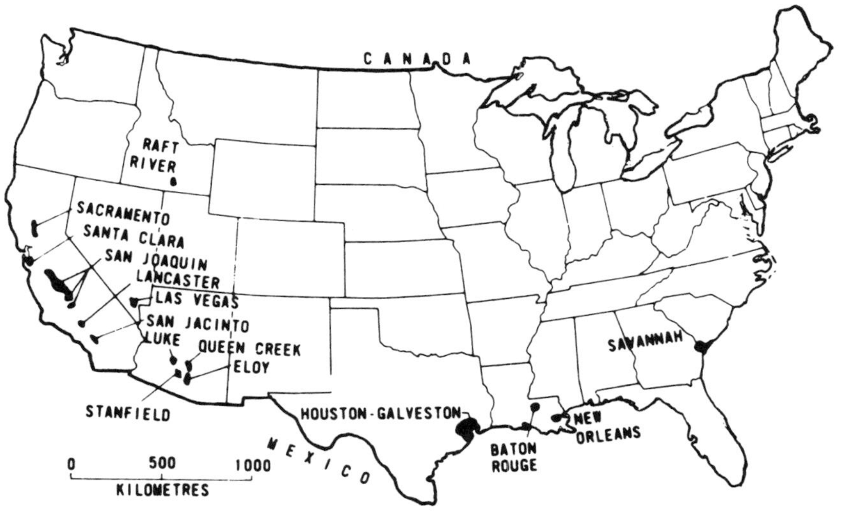

Source: Poland, J.F., 1981, J. Irrigation and Drainage Division ASCE v. 107, IR2. Copyright American Society of Civil Engineers. Reprinted with permission

FIGURE 4–14. MAGNITUDE OF LAND SUBSIDENCE FROM GROUND-WATER WITHDRAWAL IN THE UNITED STATES

[numbers in columns represent area in square kilometers]

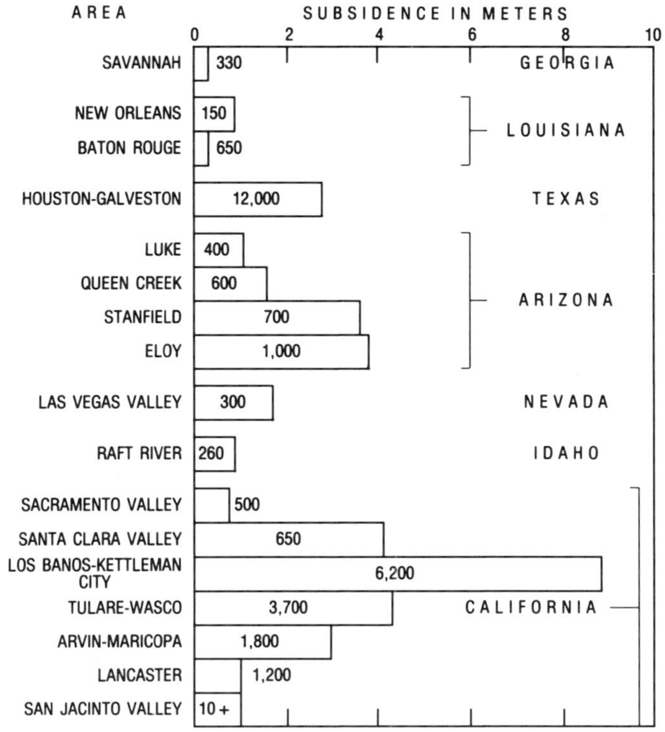

Source: Poland, J.F., 1981, J. Irrigation and Drainage division ASCE, v. 107, IR2. Copyright American Society of Civil Engineers. Reprinted with permission

TABLE 4–33. AREAS OF MAJOR LAND SUBSIDENCE DUE TO GROUNDWATER OVERDRAFT

Location	Depositional Environment and Age	Depth Range of Compacting Beds, m	Maximum Subsidence, m	Area of Subsidence, sq km	Time of Principal Occurrence
Japan					
Osaka	Alluvial and shallow marine; Quaternary	10–400	3	190	1928–1968
Tokyo	As above	10–400	4	190	1920–1970+
Mexico					
Mexico City	Alluvial and lacustrine; late Cenozoic	10–50	9	130	1938–1970+
Taiwan					
Taipei basin	Alluvial and lacustrine; Quaternary	10–240	1.3	130	1961–1969+
United States					
Arizona, central	Alluvial and lacustrine; late Cenozoic	100–550	2.3	650	1948–1967
California					
Santa Clara Valley	Alluvial and shallow marine; late Cenozoic	55–300	4	650	1920–1970
San Joaquin Valley (three subareas)	Alluvial and lacustrine; late Cenozoic	60–1000	2.9–9	11,000 (>0.3 m)	1935–1970+
Lancaster area	Alluvial and lacustrine; late Cenozoic	60–300(?)	1	400	1955–1967+
Nevada					
Las Vegas	Alluvial; late Cenozoic	60–300	1	500	1935–1963
Texas					
Houston-Galveston area	Fluvial and shallow marine; late Cenozoic	60–600(?)	1–1.5	6,860 (>0.15 m)	1943–1964+
Louisiana					
Baton Rouge	Fluvial and shallow marine; Miocene to Holocene	50–600(?)	0.3	650	1934–1965+

Source: Poland, J.F., 1972, Subsidence and its Control, in Underground Waste Management and Environmental Implications, Amer. Assoc. Petr. Geologists, Memoir 18

SECTION G. AQUIFER CHARACTERISTICS

FIGURE 4–15. TYPES OF WATER-BEARING OPENINGS IN DOMINANT AQUIFERS OF THE UNITED STATES

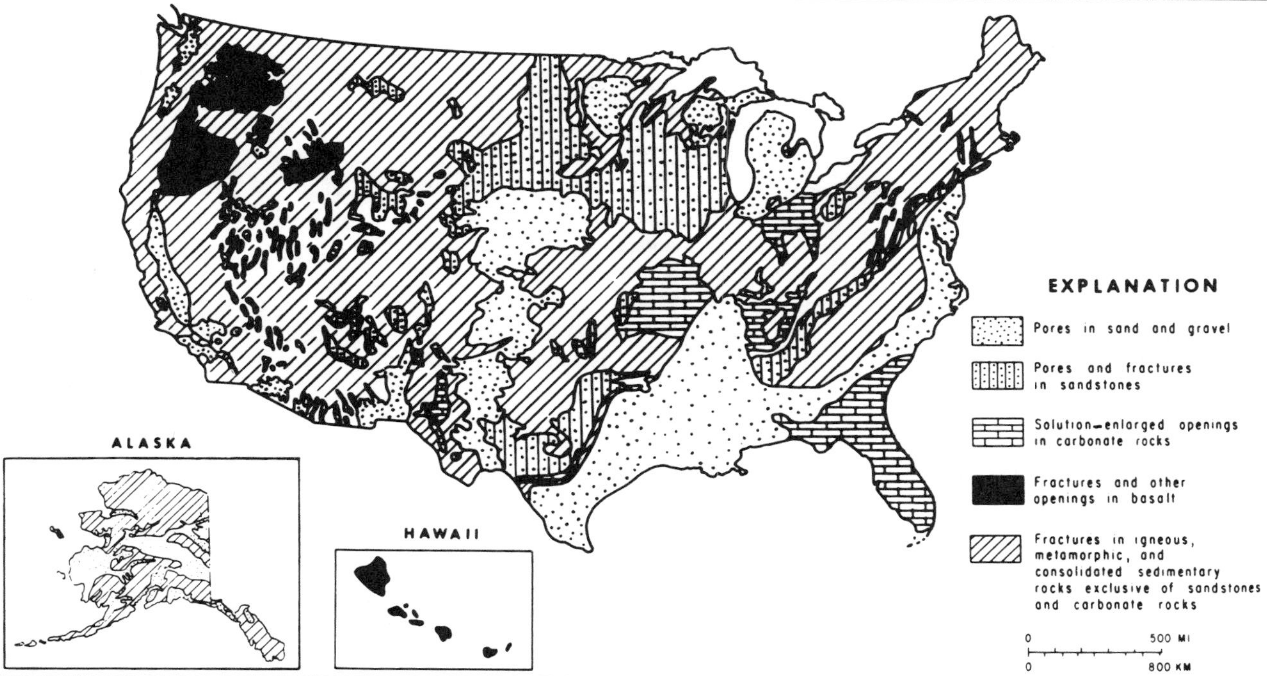

Source: Heath, R.C., 1982, Classification of Ground-Water Regions of the United States, Groundwater, v.20, no.4. Reprinted with permission

TABLE 4–34. FEATURES OF GROUND-WATER SYSTEMS USEFUL IN THE CLASSIFICATION OF GROUND-WATER REGIONS

Feature	Aspect	Range in Conditions	Significance of Feature
Component of the system	Unconfined aquifer	Thin, discontinuous, hydrologically insignificant Minor aquifer, serves primarily as a storage reservoir and recharge conduit for underlying aquifer The dominant aquifer	Affects response of the system to pumpage and other stresses. Affects recharge and discharge conditions for the system. Determines susceptibility of the system to pollution.
	Confining beds	Not present, or hydrologically insignificant Thin, markedly discontinuous, or very leaky Thick, extensive, and impermeable Complexly interbedded with aquifers or productive zones	
	Confined aquifers	Not present, or hydrologically insignificant Thin or not highly productive Multiple thin aquifers interbedded with non productive zones The dominant aquifer—thick and productive	
	Presence and arrangement of components	A single, hydrologically-dominant, unconfined aquifer Two interconnected aquifers of essentially equal hydrologic importance A three-unit system consisting of an unconfined aquifer, a confining bed, and a confined aquifer A complexly interbedded sequence of aquifers and confining beds	

(continued)

TABLE 4–34. FEATURES OF GROUND-WATER SYSTEMS USEFUL IN THE CLASSIFICATION OF GROUND-WATER REGIONS (continued)

Feature	Aspect	Range in Conditions	Significance of Feature
Water-bearing openings of dominant aquifer	Primary openings	▲ Pores in unconsolidated deposits Pores in semiconsolidated rocks Pores, tubes, and cooling fractures in volcanic (extrusive-igneous) rocks ▼	Controls water-storage and transmission characteristics. Affects dispersion and dilution of wastes.
	Secondary openings	▲ Fractures and faults in crystalline and consolidated sedimentary rocks Solution-enlarged openings in limestones and other soluble rocks ▼	
Composition of rock matrix of dominant aquifer	Insoluble Soluble	▲ Essentially insoluble Both relatively insoluble and soluble constituents Relatively soluble ▼	Affects water-storage and transmission characteristics. Has major influence on water quality.
Storage and transmission characteristics of dominant aquifer	Porosity	▲ Large, as in well-sorted, unconsolidated deposits Moderate, as in poorly-sorted unconsolidated deposits and semiconsolidated rocks Small, as in fractured crystalline and consolidated sedimentary rocks. ▼	Controls response to pumpage and other stresses. Determines yield of wells. Affects long-term yield of system. Affects rate at which pollutants move.
	Transmissivity	▲ Large, as in cavernous limestones, lava flows with flow tubes, and clean gravels Moderate, as in well-sorted, coarse-grained sands, and semiconsolidated limestones Small, as in poorly-sorted, fine-grained deposits and fractured rocks Very small, as in confining beds, which are commonly clay-rich. ▼	
Recharge and discharge conditions of dominant aquifer	Recharge	▲ In upland areas between streams Through channels of losing streams Largely or entirely by leakage across confining beds from adjacent aquifers ▼	Affects (a) response to stress and (b) long-term yields. Determines susceptibility to pollution. Affects water quality.
	Discharge	▲ Through springs or by seepage to stream channels, lakes, estuaries, or the ocean By evaporation on flood plains and in basin "sinks" By seepage across confining beds into adjacent aquifers ▼	

Source: Heath, R.C., 1982, Classification of Ground-Water Systems of the United States, Groundwater, v.20, no.4. Reprinted with permission

TABLE 4–35. GEOLOGIC ORIGIN OF AQUIFERS BASED ON TYPE OF POROSITY AND ROCK TYPE

Type of Porosity	Sedimentary		Igneous and Metamorphic	Volcanic	
	Consolidated	Unconsolidated		Consolidated	Unconsolidated
Intergranular		Gravelly sand Clayey sand Sandy clay	Weathered zone of granite-gneiss	Weathered zone of basalt	Volcanic ejecta, blocks, and fragments Ash
Intergranular and fracture	Breccia Conglomerate Sandstone Slate		Zoogenic limestone Oolitic limestone Calcareous grit	Volcanic tuff Cinder Volcanic breccia Pumice	
Fracture			Limestone Dolomite Dolomitic limestone Granite Gneiss Gabbro Quartzite Diorite Schist Mica schist	Basalt Andesite Rhyolite	

Source: United Nations Department of Economic and Social Affairs, 1975, Ground-Water Storage and Artificial Recharge, Natural Resources, Water Series No.2

TABLE 4–36. ROCKS OF GREATEST IMPORTANCE IN GROUND-WATER HYDROLOGY

Sedimentary Rocks			Igneous Rocks	
Unconsolidated (Pores)	**Consolidated (Pores, Fractures, and Solution Openings)**	**Metamorphic Rocks (Fractures)**	**Intrusive (Fractures)**	**Extrusive (Pores, Tubes, Rubble Zones, and Fractures)**
GRAVEL[a]	Conglomerate[b]	Gneiss	Granite and other coarse-grained igneous rocks	BASALT and other fine-grained igneous rocks
SAND	SANDSTONE	Quartzite-schist		
Silt	Siltstone	Schist		
Clay[c]	*Shale*	Slate-schist		
Till	Tillite (rare)			
Marl	LIMESTONE-DOLOMITE	Marble		
Coquina				

[a] Capitalized names indicate rocks that are major sources of large ground water supplies.

[b] Lowercase names indicate rocks of relatively wide extent that are sources of small to moderate ground water supplies.

[c] Italic names indicate rocks that function primarily as confining beds.

Source: Heath, R.C., 1988, Ground Water, in Perspectives on Water, Uses and Abuses, D. Speidel, editor, Copyright Oxford University Press. Reprinted with permission

TABLE 4–37. COMMON RANGES IN HYDRAULIC CHARACTERISTICS OF GROUND-WATER REGIONS IN THE UNITED STATES

[All values rounded to one significant figure; for map of regions, see Figure 4–6]

Region No.	Region	Geologic Situation	Common ranges in hydraulic characteristics of the dominant aquifers							
			Transmissivity		Hydraulic Conductivity		Recharge Rate		Well Yield	
			$m^2\ day^{-1}$	$ft^2\ day^{-1}$	$m\ day^{-1}$	$ft\ day^{-1}$	$mm\ yr^{-1}$	$in\ yr^{-1}$	$m^3\ min^{-1}$	$gal\ min^{-1}$
1	Western Mountain Ranges	Mountains with thin soils over fractured rocks, alternating with narrow alluvial and, in part, glaciated valleys	0.5–100	5–1,000	0.0003–15	0.001–50	3–50	0.1–2	0.04–0.4	10–100
2	Alluvial Basins	Thick[1] alluvial (locally glacial) deposits in basins and valleys bordered by mountains	20–20,000	2,000–200,000	30–600	100–2,000	0.03–30	0.001–1	.4–20	100–5,000
3	Columbia Lava Plateau	Thick lava sequence interbedded with unconsolidated deposits and overlain by thin soils	2,000–500,000	20,000–5,000,000	200–3,000	500–10,000	5–300	0.2–10	0.4–80	100–20,000
4	Colorado Plateau and Wyoming Basin	Thin[1] soils over fractured sedimentary rocks	0.5–100	5–1,000	0.003–2	0.01–5	0.3–50	0.01–2	0.04–2	10–1,000
5	High Plains	Thick alluvial deposits over fractured sedimentary rocks	1,000–10,000	10,000–100,000	30–300	100–1,000	5–80	0.2–3	0.4–10	100–3,000
6	Nonglaciated Central region	Thin regolith over fractured sedimentary rocks	300–10,000	3,000–100,000	3–300	10–1,000	5–500	0.2–20	0.4–20	100–5,000
7	Glaciated Central region	Thick glacial deposits over fractured sedimentary rocks	100–2,000	1,000–20,000	2–300	5–1,000	5–300	0.2–10	0.2–2	50–500
8	Piedmont and Blue Ridge	Thick regolith over fractured crystalline and metamorphosed sedimentary rocks	9–200	100–2,000	0.001–1	0.003–3	30–300	1–10	0.2–2	50–500
9	Northeast and Superior Uplands	Thick glacial deposits over fractured crystalline rocks	50–500	500–5,000	2–30	5–100	30–300	1–10	0.1–1	20–200
10	Atlantic and Gulf Coastal Plain	Complexly interbedded sands, silts and clays	500–10,000	5,000–100,000	3–100	10–400	50–500	2–20	0.4–20	100–5,000
11	Southeast Coastal Plain	Thick layers of sand and clay over semi-consolidated carbonate rocks	1,000–100,000	10,000–1,000,000	30–3,000	100–10,000	30–500	1–20	4–80	1,000–20,000
12	Alluvial Valleys	Thick sand and gravel deposits beneath flood-plains and terraces of streams	200–50,000	2,000–500,000	30–2,000	100–5,000	50–500	2–20	0.4–20	100–5,000
13	Hawaiian Islands	Lava flows segmented by dikes, interbedded with ash deposits, and partly overlain by alluvium	10,000–100,000	100,000–1,000,000	200–3,000	500–10,000	30–1,000	1–40	0.4–20	100–5,000
14	Alaska	Glacial and alluvial deposits in part perennially frozen and overlying crystalline, metamorphic, and sedimentary rocks	100–10,000	1,000–100,000	30–600	100–2,000	3–300	0.1–10	0.04–4	10–1,000

[1] An average thickness of about five meters was used as the break point between thick and thin.

Source: Heath, R.C., 1982, Classification of Ground-Water Regions of the United States, Groundwater, v. 20, no. 4. Reprinted with permission

TABLE 4–38. REPRESENTATIVE VALUES OF POROSITY

Material	Porosity, Percent	Material	Porosity, Percent
Gravel, coarse	28[a]	Loess	49
Gravel, medium	32[a]	Peat	92
Gravel, fine	34[a]	Schist	38
Sand, coarse	39	Siltstone	35
Sand, medium	39	Claystone	43
Sand, fine	43	Shale	6
Silt	46	Till, predominantly silt	34
Clay	42	Till, predominantly sand	31
Sandstone, fine-grained	33	Tuff	41
Sandstone, medium-grained	37	Basalt	17
Limestone	30	Gabbro, weathered	43
Dolomite	26	Granite, weathered	45
Dune sand	45		

[a] These values are for repacked samples; all others are undisturbed.
Source: Johnson, A.I., 1967, Specific Yield-Compilation of Specific Yields for Various Materials, U.S. Geological Survey Water-Supply Paper 1662-D

TABLE 4–39. REPRESENTATIVE VALUES OF SPECIFIC YIELD

Material	Specific Yield, Percent
Gravel, coarse	23
Gravel, medium	24
Gravel, fine	25
Sand, coarse	27
Sand, medium	28
Sand, fine	23
Silt	8
Clay	3
Sandstone, fine-grained	21
Sandstone, medium-grained	27
Limestone	14
Dune sand	38
Loess	18
Peat	44
Schist	26
Siltstone	12
Till, predominantly silt	6
Till, predominantly sand	16
Till, predominantly gravel	16
Tuff	21

Source: Johnson, A.I., 1967, Specific Yield-Compilation of Specific Yields for Various Materials, U.S. Geological Survey Water-Supply Paper 1662-D

TABLE 4–40. DRILLERS' TERMS USED IN ESTIMATING SPECIFIC YIELD

Crystalline Bedrock (fresh)
Specific yield zero

Granite
Hard boulders
Hard granite
Hard rock
Graphite and rocks
Rock (if in area of known crystalline rocks)

Clay and Related Materials
Specific yield 3 percent

Adobe
Brittle clay
Caving clay
Cement
Cement ledge
Choppy clay
Clay
Clay, occasional rock
Crumbly clay
Cube clay
Decomposed granite
Dirt
Good clay
Gumbo clay
Hard clay
Hardpan (H.P.)
Hardpan shale
Hard shale
Hard shell
Joint clay
Lava
Loose shale
Muck
Mud
Packed clay
Poor clay
Shale
Shell
Slush
Soapstone
Soapstone float
Soft clay
Squeeze clay
Sticky
Sticky clay
Tiger clay
Tight clay
Tule mud
Variable clay
Volcanic rock

Clay and Gravel, Sandy Clay, and Similar Materials
Specific yield 5 percent

Cemented gravel (cobbles)
Cemented gravel and clay
Cemented gravel, hard
Cement and rocks (cobbles)
Clay and gravel (rock)
Clay and boulders (cobbles)
Clay, pack sand, and gravel
Cobbles in clay
Conglomerate
Dry gravel (below water table)
Gravel and clay
Gravel (cement)
Gravel and sandy clay
Gravel and tough shale
Gravelly clay
Rocks in clay
Rotten cement
Rotten concrete mixture
Sandstone and float rock
Silt and gravel
Soil and boulders
Clay and sandy clay
Clay and silt
Clay, cemented sand
Clay, compact loam and sand
Clay to coarse sand
Clay, streaks of hard packed sand
Clay, streaks of sandy clay
Clay, water
Clay with sandy pocket
Clay with small streaks of sand
Clay with some sand
Clay with streaks of fine sand
Clay with thin streaks of sand
Porphyry clay
Quicksandy clay
Sand—clay
Sand shell
Shale and sand
Solid clay with strata of cemented sand
Sticky sand and clay
Tight muddy sand
Very fine tight muddy sand

Cemented sand
Cemented sand and clay
Clay sand
Dry hard packed sand
Dry sand (below water table)
Dry sand and dirt
Fine muddy sand
Dry sandy silt
Fine sandy loam
Fine sandy silt
Ground surface
Loam
Loam and clay
Sandy clay loam

Clay and Gravel, Sandy Clay, and Similar Materials
(continued)

Fine sand, streaks of clay
Fine tight muddy sand
Hard packed sand, streaks of clay
Hard sand and clay
Hard set sand and clay
Muddy sand and clay
Packed sand and clay
Packed sand and shale
Sand and clay mix
Sand and tough shale
Sand rock
Sandstone
Sandstone and lava
Set sand and clay
Set sand, streaks of clay
Cemented sandy clay
Hard sandy clay (tight)
Sandy clay
Sandy clay with small sand streaks, very fine
Sandy shale
Set sandy clay
Silty clay
Soft sandy clay
Clay and fine sand
Clay and pumice streaks
Ash
Caliche
Chalk
Hard lava formation
Sediment
Silt
Silt and clay
Silty clay loam
Silty loam
Soft loam
Soil
Soil and clay
Soil and mud
Soil and sandy shale
Surface formation
Top hardpan soil
Topsoil
Topsoil and sandy silt
Topsoil—silt

Decomposed hardpan
Hardpan and sandstone
Hardpan and sandy clay
Hardpan and sandy shale
Hardpan and sandy stratas
Hard rock (alluvial)
Sandy hardpan
Semi-hardpan
Washboard

Hard pumice
Porphyry
Seepage soft clay
Volcanic ash

Fine Sand, Tight Sand, Tight Gravel, and Similar Materials
Specific yield 10 percent

Sand and clay
Sand and clay strata (traces)
Sand and dirt
Sand and hardpan
Sand and hard sand
Sand and lava
Sand and pack sand
Sand and sandy clay
Sand and soapstone
Sand and soil
Sand and some clay
Sand, clay, and water
Sand crust
Sand-little water
Sand, mud, and water
Sand (some water)
Sand streaks, balance clay
Sand, streaks of clay
Sand with cemented streaks
Sand with thin streaks of clay

Coarse, and sandy
Loose sandy clay
Medium sandy
Sandy
Sandy and sandy clay

Sandy loam
Sandy loam, sand, and clay
Sandy silt
Sandy soil
Surface and fine sand

Cloggy sand
Coarse pack sand
Compacted sand and silt
Dead sand
Dirty sand
Fine pack sand
Fine quicksand with alkali streak
Fine sand
Fine sand, loose
Hard pack sand
Hard sand
Hard sand and streaks of sandy clay

Hard sand rock and some water sand
Hard sand, soft streaks
Loamy fine sand
Medium muddy sand
Milk sand
More or less sand
Muddy sand

(continued)

TABLE 4–40. DRILLERS' TERMS USED IN ESTIMATING SPECIFIC YIELD (continued)

Fine Sand, Tight Sand, Tight Gravel, and Similar Materials
Specific yield 10 percent (continued)

Sandy clay, sand, and clay
Sandy clay—water bearing
Sandy clay with streaks of sand
Sandy formation
Sandy muck
Sandy sediment
Very sandy clay

Boulders, cemented sand
Cement, gravel, sand, and rocks
Clay and gravel, water bearing
Clay & rock, some loose rock
Clay, sand and gravel
Clay, silt, sand, and gravel
Conglomerate, gravel, and boulders
Conglomerate, sticky clay, sand and gravel
Dirty gravel
Fine gravel, hard
Gravel and hardpan strata
Gravel, cemented sand
Gravel with streaks of clay
Hard gravel
Hard sand and gravel

Pack sand
Poor water sand
Powder sand
Pumice sand
Quicksand
Sand, mucky or dirty
Set sand
Silty sand
Sloppy sand
Sticky sand
Streaks fine and coarse sand
Surface sand and clay
Tight sand

Brittle clay and sand
Clay and sand
Clay, sand, and water
Clay with sand
Clay with sand streaks
More or less clay, hard sand and boulders
Mud and sand
Mud, sand, and water
Sand and mud with chunks of clay
Silt and fine sand
Silt and sand
Soil, sand, and clay

Fine Sand, Tight Sand, Tight Gravel, and Similar Materials
Specific yield 10 percent (continued)

Packed gravel
Packed sand and gravel
Quicksand and cobbles
Rock sand and clay
Sand and gravel, cemented streaks
Sand and silt, many gravel
Sand, clay, streaks of gravel
Sandy clay and gravel
Set gravel
Silty sand and gravel (cobbles)
Tight gravel

Topsoil and light sand
Water sand sprinkled with clay

Float rock (stone)
Laminated
Pumice
Seep water
Soft sandstone
Strong seepage

Gravel, Sand, Sand and Gravel, and Similar Materials
Specific yield 25 per cent

Boulders
Coarse gravel
Coarse sand
Cobbles
Cobble stones
Dry gravel (if above water table
Float rocks
Free sand
Gravel
Loose gravel
Loose sand
Rocks

Gravel and sand
Gravel and sandrock
Medium sand
Rock and gravel
Running sand
Sand
Sand, water
Sand and boulders
Sand and cobbles
Sand and fine gravel
Sand and gravel
Sandy gravel
Water gravel

Source: U.S. Geological Survey

TABLE 4–41. REPRESENTATIVE VALUES OF HYDRAULIC CONDUCTIVITY

Material	Hydraulic Conductivity ft/day	m/day	Type of Measurement[a]
Gravel, coarse	490	150	R
Gravel, medium	890	270	R
Gravel, fine	1,500	450	R
Sand, coarse	150	45	R
Sand, medium	40	12	R
Sand, fine	8.2	2.5	R
Silt	0.62	0.08	H
Clay	0.00066	0.0002	H
Sandstone, fine-grained	0.66	0.2	V
Sandstone, medium-grained	10	3.1	V
Limestone	3	0.94	V
Dolomite	0.0033	0.001	V
Dune sand	66	20	V
Loess	0.26	0.08	V
Peat	19	5.7	V
Schist	0.66	0.2	V
Slate	0.00026	0.00008	V
Till, predominantly sand	1.6	0.49	R
Till, predominantly gravel	100	30	R
Tuff	0.66	0.2	V
Basalt	0.033	0.01	V
Gabbro, weathered	0.66	0.2	V
Granite, weathered	4.6	1.4	V

[a] H is horizontal hydraulic conductivity, R is a repacked sample, and V is vertical hydraulic conductivity.
Source: Morris, D.A., and Johnson, A.I., 1967, U.S. Geological Survey Water-Supply Paper 1839-D

FIGURE 4–16. HYDRAULIC CONDUCTIVITY OF SELECTED CONSOLIDATED AND UNCONSOLIDATED AQUIFERS

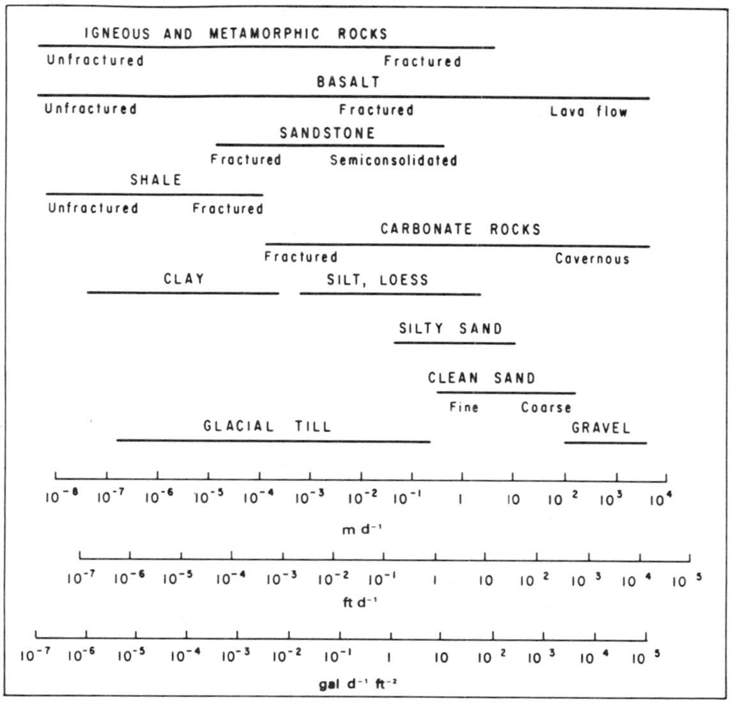

Source: Heath, R.C., 1983, Basic Ground-Water Hydrology, U.S. Geological Survey Water-Supply Paper 2220

TABLE 4–42. REPRESENTATIVE PERMEABILITY RANGES FOR SEDIMENTARY MATERIALS

Material	Permeability, gal. per day/ft^2	Material	Permeability, gal. per day/ft^2
Clay	$10^{-5} - 10^{-3}$	Very fine sand	$1 - 10^{2}$
Silty clay	$10^{-5} - 10^{-3}$	Fine sand	$10^{1} - 10^{3}$
Sandy clay	$10^{-4} - 10^{-2}$	Medium sand	$10^{2} - 10^{3}$
Silty clay loam	$10^{-3} - 10^{-1}$	Coarse sand	$10^{2} - 10^{4}$
Sandy clay loam	$10^{-2} - 1$	Gravel and sand	$10^{2} - 10^{4}$
Silt	$10^{-2} - 1$	Gravel	$10^{2} - 10^{4}$
Silt loam	$10^{-2} - 1$	Sandstone	$10^{1} - 10^{3}$
Loam	$10^{-2} - 1$	Limestone[a]	$1 - 10^{2}$
Sandy loam	$10^{-1} - 10^{1}$	Shale	$1 - 10^{2}$

[a] Excluding cavernous limestone

TABLE 4–43. TEMPERATURE CORRECTION FOR PERMEABILITY

[To convert coefficient of permeability computed at water temperature shown in table to coefficient of permeability at 60°F, multiply by appropriate factor Tc]

°F	T_c	°F	T_c	°F	T_c	°F	T_c
40	1.37	53	1.11	66	0.92	79	0.78
41	1.35	54	1.09	67	0.91	80	0.77
42	1.33	55	1.08	68	0.89	81	0.76
43	1.31	56	1.06	69	0.88	82	0.75
44	1.28	57	1.04	70	0.87	83	0.74
45	1.26	58	1.03	71	0.86	84	0.73
46	1.24	59	1.01	72	0.85	85	0.72
47	1.22	60	1.00	73	0.84	86	0.71
48	1.20	61	0.99	74	0.83	87	0.70
49	1.18	62	0.97	75	0.82	88	0.69
50	1.16	63	0.96	76	0.81	89	0.68
51	1.15	64	0.95	77	0.80	90	0.67
52	1.13	65	0.93	78	0.79		

FIGURE 4–17. GROUND WATER FLOW VELOCITY RANGES

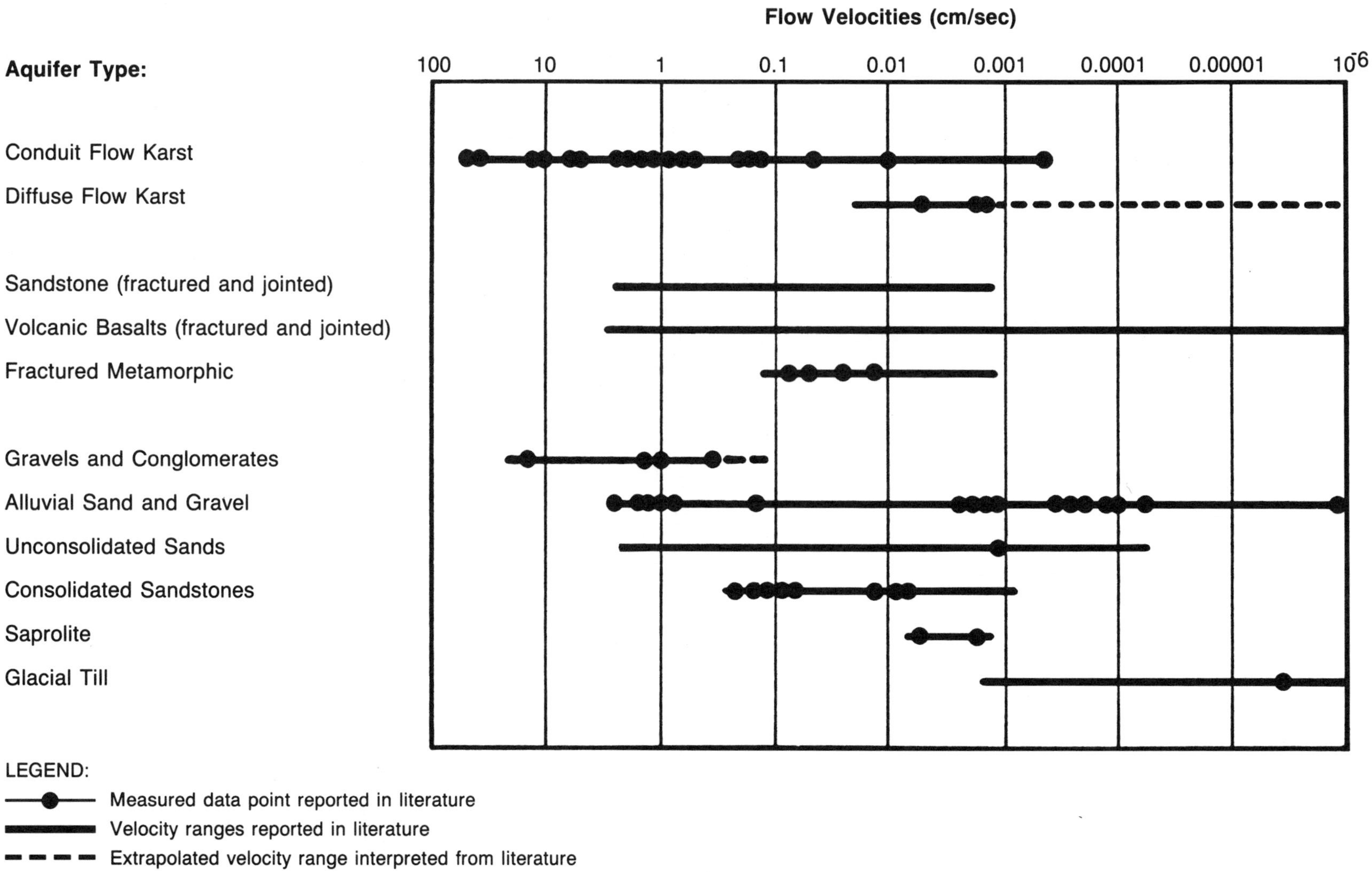

Source: U.S. Environmental Protection Agency, 1987, Guidelines for Delineation of Wellhead Protection Areas, PB88–111430.
Original source: Everett, A.G., 1987

SECTION H. SOIL MOISTURE

FIGURE 4–18. SOIL-MOISTURE FORMS AND PROPERTIES FOR AN ASSUMED FINE SANDY LOAM SOIL

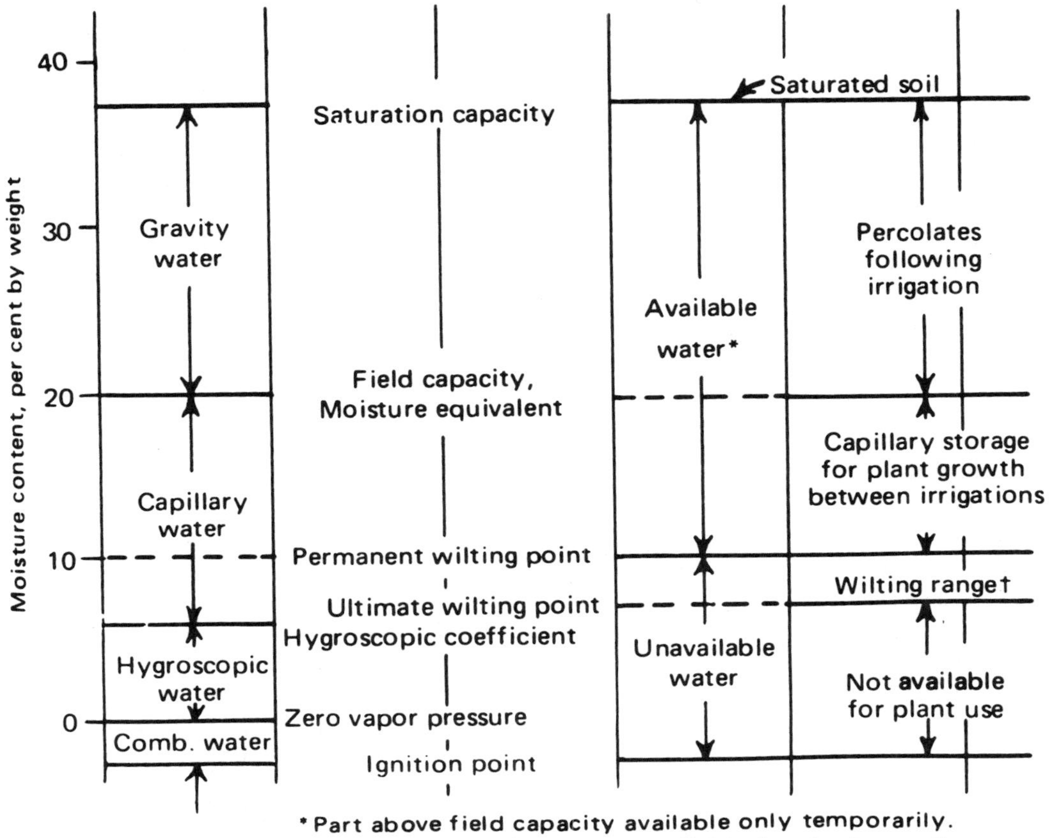

* Part above field capacity available only temporarily.
† Sustains plant life but not available for plant growth.

Source: Houk, Irrigation Engineering, v.1, John Wiley & Sons, Copyright 1951

TABLE 4–44. GUIDE FOR JUDGING HOW MUCH OF THE AVAILABLE MOISTURE HAS BEEN REMOVED FROM SOIL

Soil Moisture Deficiency	Feel or Appearance of Soil and Moisture Deficiency in Inches of Water Per Foot of Soil: Coarse Texture	Moderately Coarse Texture	Medium Texture	Fine and Very Fine Texture
0% (Field capacity)	Upon squeezing, no free water appears on soil but wet outline of ball is left on hand 0.0	Upon squeezing, no free water appears on soil but wet outline of ball is left on hand. 0.0	Upon squeezing, no free water appears on soil but wet outline of ball is left on hand. 0.0	Upon squeezing, no free water appears on soil but wet outline of ball is left on hand. 0.0
0–25%	Tends to stick together slightly, sometimes forms a very weak ball under pressure. 0.0 to 0.2	Forms weak ball, breaks easily, will not slick. 0.0 to 0.4	Forms a ball, is very pliable, slicks readily if relatively high in clay. 0.0 to 0.5	Easily ribbons out between fingers, has slick feeling. 0.0 to 0.6
25–50%	Appears to be dry, will not form a ball with pressure. 0.2 to 0.5	Tends to ball under pressure but seldom holds together. 0.4 to 0.8	Forms a ball somewhat plastic, will sometimes slick slightly with pressure. 0.5 to 1.0	Forms a ball, ribbons out between thumb and forefinger. 0.6 to 1.2
50–75%	Appears to be dry, will not form a ball with pressure.[1] 0.5 to 0.8	Appears to be dry, will not form a ball.[1] 0.8 to 1.2	Somewhat crumbly but holds together from pressure. 1.0 to 1.5	Somewhat pliable, will ball under pressure.[1] 1.2 to 1.9
75–100% (100% is permanent wilting)	Dry, loose, single-grained, flows through fingers. 0.8 to 1.0	Dry, loose, flows through fingers. 1.2 to 1.5	Powdery, dry, sometimes slightly crusted but easily broken down into powdery condition. 1.5 to 2.0	Hard, baked, cracked, sometimes has loose crumbs on surface. 1.9 to 2.5

[1] Ball is formed by squeezing a handful of soil very firmly.
Source: Israelson and Hansen, Irrigation Principles and Practices, John Wiley & Sons, Copyright 1962

TABLE 4–45. APPROXIMATE LIMITS OF MOISTURE CONDITIONS IN MOST IRRIGATION SOILS

Item	Soil-Moisture Condition	Approximate Limits, per cent by weight: Lower	Upper	Item	Soil-Moisture Condition	Approximate Limits, per cent by weight: Lower	Upper
1	Hygroscopic moisture content	1–	15	7	Ultimate wilting point	1	25
2	Hygroscopic coefficient	1–	15	8	Moisture in wilting range	1–	5
3	Saturation capacity	15	60	9	Available moisture capacity	5	20
4	Field capacity	7	40	10	Maximum available storage	1[a]	3
5	Moisture equivalent	5	50	11	Gravity water in saturated soils	8	40
6	Permanent wilting point	2	30				

[a] Inches per foot of soil depth.
Source: Houk, Irrigation Engineering, v.1, John Wiley & Sons, Copyright 1951

TABLE 4–46. WATER-HOLDING CHARACTERISTICS OF VARIOUS SOILS

Soil Type	Approximate Depth of Water per Foot Depth of Soil in Plant Root Zone, Inches: Field Capacity	Irrigation Desirable	Wilting Point
Sand	1.2	0.6	0.3
Fine Sand	1.5	0.7	0.4
Sandy Loam	1.9	1.0	0.6
Fine Sandy Loam	2.5	1.3	0.8
Loam	3.2	1.7	1.2
Silt Loam	3.5	2.0	1.4
Light Clay Loam	3.7	2.2	1.6
Clay Loam	3.8	2.4	1.8
Heavy Clay Loam	3.8	2.6	2.1
Clay	3.9	2.8	2.4

Source: U.S. Dept. of Agriculture

TABLE 4–47. REPRESENTATIVE PHYSICAL PROPERTIES OF SOILS

[Normal ranges are shown in parentheses]

Soil Texture	Infiltration[1] and Permeability Inches/hour	Total Pore Space %	Apparent Specific Gravity	Field Capacity %	Permanent Wilting %	Total Available Moisture[2]: Dry Weight Basis %	Volume Basis %	Inches per Foot
Sandy	2 (1–10)	38 (32–42)	1.65 (1.55–1.80)	9 (6–12)	4 (2–6)	5 (4–6)	8 (6–10)	1.0 (0.8–1.2)
Sandy Loam	1 (0.5–3)	43 (40–47)	1.50 (1.40–1.60)	14 (10–18)	6 (4–8)	8 (6–10)	12 (9–15)	1.4 (1.1–1.8)
Loam	0.5 (0.3–0.8)	47 (43–49)	1.40 (1.35–1.50)	22 (18–26)	10 (8–12)	12 (10–14)	17 (14–20)	2.0 (1.7–2.3)
Clay Loam	0.3 (0.1–0.6)	49 (47–51)	1.35 (1.30–1.40)	27 (23–31)	13 (11–15)	14 (12–16)	19 (16–22)	2.3 (2.0–2.6)
Silty Clay	0.1 (0.01–0.2)	51 (49–53)	1.30 (1.25–1.35)	31 (27–35)	15 (13–17)	16 (14–18)	21 (18–23)	2.5 (2.2–2.8)
Clay	0.2 (0.05–0.4)	53 (51–55)	1.25 (1.20–1.30)	35 (31–39)	17 (15–19)	18 (16–20)	23 (20–25)	2.7 (2.4–3.0)

[1] Intake rates vary greatly with soil structure and structural stability, even beyond the normal ranges shown above.
[2] Readily available moisture is approximately 75% of the total available moisture.
Source: Israelson and Hansen, Irrigation Principles and Practices, John Wiley & Sons, Copyright 1962

SECTION I. SPRINGS

TABLE 4–48. CLASSIFICATION OF SPRINGS ACCORDING TO MAGNITUDE OF DISCHARGE

Magnitude	Old System English Units	New System Metric Units
First	Greater than 100 ft^3 sec	>10 m^3/s
Second	10 to 100 ft^3 sec	1–10 m^3/s
Third	1 to 10 ft^3 sec	0.1–1 m^3/s
Fourth	100 gal/min to 1 ft^3/sec	10–100 l/s
Fifth	10 to 100 gal/min	1–10 l/s
Sixth	1 to 10 gal/min	0.1–1 l/s
Seventh	1 pt/min to 1 gal/min	10–100 ml/s
Eighth	Less than 1 pt/min	<10 ml/s

Source: U.S. Geological Survey

TABLE 4–49. SPRINGS OF THE FIRST MAGNITUDE IN THE UNITED STATES

[Springs having a probable average discharge of 100 cfs or more.]

Region	Ground-Water Province	Number of First Magnitude Springs (Having Probable Average Discharge of 100 second-feet or more)	Kind of Rock	Age of Rock
Florida and adjacent parts of Georgia and Alabama	Atlantic Coastal Plain	11	Limestone	Tertiary
Northern Alabama and adjacent areas	South-central Paleozoic and Blue Ridge-Appalachian Valley	1	do	Paleozoic
Ozark region of Missouri and Arkansas	South-central Paleozoic	8	do	Early Paleozoic
Balcones fault belt in Texas	Atlantic Coastal Plain	4	do	Cretaceous
Snake River Basin in Idaho	Columbia Lava Plateau	15	Volcanic rock (also gravel)	Tertiary and Quaternary
Sacramento River Basin, Calif.	Lava-covered areas chiefly in the Southwestern Bolson province	7	do	do
Deschutes River Basin, Oreg.	Columbia Lava Plateau	8	do	do
Klamath River Basin, Oreg.	Lava-covered areas chiefly in the Southwestern Bolson province	2	do	do
Willamette and Umpqua River Basins, Oreg.	Lava-covered areas in or adjacent to Columbia Lava Plateau	5	do	do
Interior basins of Oregon	do	1	Lake beds overlying volcanic rock	Tertiary
Montana	Montana-Arizona Plateau	3	Sandstone	Cretaceous and Jurassic
Northeastern Utah	Northern Rocky Mountain	0	Limestone	Cambrian(?)
		65		

Source: U.S. Geological Survey, 1927

TABLE 4–50. THERMAL SPRINGS IN THE UNITED STATES

State	Developed		Not Used	Total
	Resorts	Used for Bathing, Irrigation, or Water Supply		
Arizona	5	9	7	21
Arkansas	2	1	3	6
California	53	89	42	184
Colorado	17	16	12	45
Georgia	2	5	1	8
Idaho	16	67	120	203
Massachusetts	–	1	–	1
Montana	18	20	2	40
Nevada	15	93	66	174
New Mexico	5	15	18	38
New York	1	–	–	1
North Carolina	1	–	–	1
Oregon	14	73	18	105
Pennsylvania	–	1	–	1
South Dakota	1	2	–	3
Texas	–	3	–	3
Utah	7	43	13	63
Virginia	7	13	–	20
Washington	8	5	3	16
West Virginia	4	6	–	10
Wyoming:				
Outside of Yellowstone National Park	5	10	5	20
In Yellowstone National Park	3	–	93	96
Total	184	472	403	1,059

Source: U.S. Geological Survey, 1937

TABLE 4–51. NATURAL HEAT FLOWS OF SOME HOT SPRING AREAS OF THE WORLD

Area	Approximate Size[1] (km^2)	Maximum Recorded Temperature[2] (°C)	Total Heat Flow[3] (10^6 cal/sec)
BRITISH WEST INDIES			
Qualibou, St. Lucia	~ .1	(S)185	8.6
St. Vincent	~ 1	(S)>27	18
Dominica	~ 1	(S) 90	17
Montserrat	~ .1	(S) 97	1.6
EL SALVADOR			
Total of country	–	–	200
Northern belt, total	–	–	50
Southern belt, total	–	–	>150
Ahuachapán group	80	(D)174	80
El Playón de Ahuachapán	~ .25	(S) boiling	.46
Agua Shuca	~ .25	(S) boiling	.32
FIJI ISLANDS			
Savusavu	~ 1	(S)100	2
ICELAND			
Steam fields, total heat flow	–	–	630
			1,000
Hengill, total	50	(D)230	55–80
Do	–	–	25–125
Hengill, southern part only	–	(D)230	28
Torfajökull	100	(S) boiling	500
Reykjanes	1	?	5–25
Trölladyngja	5	?	5–25
Krysuvik	10	(D)230	5–25
Kerlingafjöll	5	(S) boiling	25–125
Vonarskard	?	?	5–125
Grimsvötn	12	?	125–750
Kverkfjöll	10	?	25–125
Askja	25	?	5–25
Námafjall	2.5	?	25–125
Krafla	.5	(S) boiling	5–25
Theistareykir	2.5	(S) boiling	5–25
Low temperature areas; about 250 areas.	–	(D)146	100
Six lines of thermal springs, each.	–	(S)100	5–25
Reykjavík	~ 5	(D)146	1.7
Reykir	~ 5	(D) 98	11
Deilartunga line, total	–	(S)100	25–125
Deilartunga spring	–	(S)100	24
ITALY			
(Larderello)	(~50)	((D)240)	(5)
Ischia and Flegreian Fields	~10	(D)296	?
(Monta Amiata)	(~ 3)	((D)165)	(?)
Vulcano	~ 1	(D)194	?
JAPAN			
Otaki, Kyushu	–	(D)185	?
Atami, Shizuoka-ken	5	(D)180	16
Do	–	–	22
Ito, Shizuoka-ken	–	–	44
Obama, Nagasaki-ken	1.5	(D)180	57
Beppu, Oita-ken	~10	(D)150	19
Kawayu, Hokkaido	.7	(S) 65	8
Yunokawa, Hokkaido	~ 1	(S) 66	4.0
Yachigashira, Hokkaido	?	(S) 69	.5
Shikabe, Hokkaido	~ .5	(D)113	1.2
Toyako, Hokkaido	~ 3	(S) 55	2.2
Noboribetsu, Hokkaido:			
Hot Lake area, total	~ .2	(D)112	14
Jigokudani Valley (variable)	~ .3	(D)160	~6–11.2
Matsukawa, N. Honshu	–	(D)189	?
Onikobe, N. Honshu	~80	(D)185	?
Narugo, N. Honshu	–	(D)175	?

TABLE 4–51. NATURAL HEAT FLOWS OF SOME HOT SPRING AREAS OF THE WORLD (continued)

Area	Approximate Size[1] (km^2)	Maximum Recorded Temperature[2] (°C)	Total Heat Flow[3] (10^6 cal/sec)
MEXICO			
Pathé, Hidalgo	~ 2	(D)155	?
Ixtlan, Michoacan	–	(D)150	?
NEW ZEALAND			
Wairakei, 1951, 1952	7	(D)266	133
1954	7	(D)266	82
1956?	7	(D)266	143
1958, 1959	7	(D)266	163
1958	7	(D)266	101
Waiotapu	~15	(D)295	272
Orakei Korako	~ 5	(S) boiling	130
Tikitere	5	(S) boiling	40
Tokopia –	–	–	30
Waikiti	–	(S) 91	20
Ngatamariki	~ 1	(S) hot	12.6
Rotokaua	~ 5	(S) boiling	52
Ohaki	~ 1	(S) boiling	12.8
Taupo Spa	~ 3	(S) boiling?	36
Kawerau (Onepu) 1959?	?	(D)277	25
1962	?	(D)285	18
Rotorua	?	(D)>160	–
UNION OF SOUTH AFRICA			
Seven scalding springs	?	(S) 64	1.7
UNITED STATES			
California			
The Geysers	~ 1	(D)208	0.4
Sulphur Bank	~ 2	(D)136	.2
Wilbur Springs area	~ 5	(S) 69	.4
Casa Diablo—Hot Creek	>25(?)	(D)180	70
Alkali Lakes area	–	–	–
(Salton Sea)	(~50)	(D)>270	(4)
Nevada			
Steamboat Springs	5	(D)187	7
Bradys Springs	~ 2	(D)168	?
Beowawe	~ 3	(D)207	?
Wyoming			
Yellowstone Park, Wyoming	9,000	–	–
Total, discharging water	~70	(S)138	207
Total, calculated	~70	(D)205	500
Norris Geyser Basin	~ 3	(D)205	8
Upper Geyser Basin	~10	(D)180	90
Mammoth-Hot River	~ 8	(S) 73	34
U.S.S.R.			
Pauzhetsk, Kamchatka	~ 1	(D)195	18
Total			~2,700

[1] The limits of a hydrothermal area are very difficult to define and meaningful criteria are difficult to apply. Depending upon the definition, the "limits" of an area can vary by at least an order of magnitude. The definition used here is: "The rather broad boundaries containing specific areas with some surface evidence for abnormally high temperatures at depth. The evidence can consist of one or more of the following: hot springs, fumaroles, active hydrothermal alteration, and abnormally high near-surface geothermal gradient. Closely spaced 'hot spots' not separated by areas of approximately 'normal' gradient for the region are included in a single thermal area."

[2] (S) indicates temperatures measured at the surface; (D) temperatures from drill holes.

[3] Most heat flows are relative to mean annual surface temperatures but a few are relative to 0° or 4°C; such differences are small compared to the uncertainties and have not been modified. 1×10^6 cal/cm^2/sec approximates the "normal" heat flow from 60 to 70 km.[2]

Source: U.S. Geological Survey, 1965

SECTION J. ARTIFICIAL RECHARGE

TABLE 4–52. ARTIFICIAL RECHARGE PROJECTS IN THE UNITED STATES AND OTHER COUNTRIES

[Number of projects reported based on return of American Society of Civil Engineers' questionnaires, 1988]

United States

Alabama	2	Mississippi	1
Arizona	3	Missouri	1
Arkansas	2	Montana	2
California	42	Nebraska	4
Colorado	3	Nevada	1
Connecticut	1	New Jersey	3
Florida	1	New Mexico	1
Georgia	2	New York	1
Hawaii	1	North Carolina	3
Illinois	1	Ohio	3
Indiana	2	Oklahoma	4
Iowa	1	Pennsylvania	3
Kansas	2	South Carolina	2
Kentucky	1	Tennessee	1
Louisiana	3	Texas	9
Maine	1	Utah	1
Maryland	5	Virginia	4
Massachusetts	3	Washington	4
Michigan	3	Wisconsin	1
Minnesota	2		

Other Countries

Denmark	1	Morocco	1
Fed. Rep. of Germany	10	Namibia	1
Finland	1	Netherlands	20
Greece	4	New Zealand	2
India	3	Oman	1
Israel	2	Qatar	1
Italy	5	Switzerland	1
Jamaica	1	Thailand	2
Japan	3		

Source: A.I. Johnson, personal communication, October 30, 1988

TABLE 4–53. OPERATION AND MAINTENANCE PROBLEMS OF ARTIFICIAL RECHARGE PROJECTS

Problem	Manifestation	Corrective Actions To Be Considered
Silt	Lodging of particles within interstices of soil near the surface area, reducing the infiltration rate.	(1) Desilt in retention reservoir and/or in uppermost series of basins. Flocculating agent such as "Separan" has been used with success. (2) Bypass water until concentration of silt will not be detrimental, with concentration depending upon soil condition. Ditches and furrows generally can accept waters containing higher concentrations of silt if sufficient velocity is maintained through the project to carry silt back to the main canal. (3) Scrape, harrow, and/or disc after proper drying. Period of drying usually ranges from one to seven days depending upon soil and weather conditions. (4) Remove silt after drying. Silt may be used to build up levees of basins or bridges of ditches or furrows. (5) Sustain vegetative growth. (6) Sluice the silt out of ditches and furrows, and from channels, with due regard to erosion problems. (7) Pump injection well to loosen silt from interstices and remove silt from the well.
Weeds	Increases percolation rate and shortens drying period required for working an area or removing silt from the basin. There is a disadvantage in that vegetative growth may be a fire hazard.	(1) Control by chemical means and/or remove when weeds become a fire hazard, especially around structures. Consider use of hand labor instead of mechanical means in order to maintain infiltration rates. However, if possible, leave vegetation undisturbed in wetted area. (2) Prolonged deep submergence will kill vegetation. (3) The control of weeds is generally not considered a problem in the operation of pits and shafts, or injection wells.
Rodents	Leaks and failures of dikes and levees. Public nuisance near urban area.	(1) Set out poison about twice a year. (2) Use of traps.
Public health and safety	Rodents and mosquito problem and possible injury to individuals. Potential problems of injury is greatest when depth of water is large in basins and pits.	(1) Enclose area with fence and gates with locks. (2) Patrol area with particular attention to children and structural failures, before and during operation, especially near inhabited area. (3) Vector control by use of mosquito fish, chemicals, and/or drying. (4) Rodent control by poisoning or traps. (5) Proper posting of signs when using chemical which is poisonous.
Maintaining of percolation rates	Reduction of percolation rates will decrease efficiency of system, increasing the unit cost of the amount of water actually recharged.	(1) Proper treatment of water. Desilt water to concentration desired. Use chlorine or copper sulfate for control of bacterial slime and algae. Use of chemicals to reduce the possibility of chemical incrustation, which is usually deposition of calcium carbonate. (2) Schedule intermittent drying periods to prevent problems due to swelling of soil particles. Permit growth of vegetation to decrease the drying period by removing the water in the root zone and loosening the soil. Studies have shown that bermuda grass has been successfully used to maintain rates, even under prolonged periods of deep submergence. (3) Prevent aeration of water, especially when operating recharge wells, pits, and shafts. (4) Increase head of water generally by increasing depth of water. (5) Use hand labor whenever possible to decrease the possibility of using heavy equipment which will cause surface compaction especially when soil is wet. (6) Scrape, harrow, and/or disc after proper drying. (7) Remove silt, chemical incrustation, and/or any material decreasing infiltration rates after proper drying period. (8) Maintain the design velocity to reduce silting to a minimum in use of ditches and furrows. (9) Recondition injection wells by use of dry ice, hydrochloric acid, and/or sulfuric acid. (10) Prevent freezing of water during winter by continuous spreading. (11) Check the possibility of base exchange reactions. (12) Soil can be reconditioned by using organic material such as cotton gin trash, or chemical agents, such as krilium.
Maintenance to diversion structures	Breakdown of spreading operations.	(1) Systematic and routine maintenance check as well as patrolling when in operation. Attention should be given to wooden structures since they deteriorate faster due to frequent wetting and drying cycles. Also attention should be given to settlement of structures thus changing flow condition. (2) Attention should be given to undercutting of structure particularly on the downstream end with preventive maintenance primarily in the form of riprapping. (3) Sluicing of channel to remove silt and debris which have accumulated near and at diversion structure.

Source: Richter and Chun, Proc. Am. Soc. Civil Engrs., 1959

SECTION K. GEOPHYSICAL LOGGING

TABLE 4–54. BOREHOLE GEOPHYSICAL LOGGING METHODS AND THEIR USES IN HYDROLOGIC STUDIES

Method	Uses	Recommended Conditions
Electric logging: Single-electrode resistance	Determining depth and thickness of thin beds. Identification of rocks, provided general lithologic information is available, and correlation of formations. Determining casing depths.	Fluid-filled hole. Fresh mud required. Hole diameter less than 8 to 10 inches. Log only in uncased holes.
Short normal (electrode spacing of 16 inches)	Picking tops of resistive beds. Determining resistivity of the invaded zone. Estimating porosity of formations (deeply invaded and thick interval). Correlation and identification, provided general lithologic information is available.	Fluid-filled hole. Fresh mud. Ratio of mud resistivity to formation-water resistivity should be 0.2 to 4. Log only in uncased part of hole.
Long normal (electrode spacing of 64 inches)	Determining true resistivity in thick beds where mud invasion is not too deep. Obtaining data for calculation of formation-water resistivity.	Fluid-filled hole. Ratio of mud resistivity to formation-water resistivity should be 0.2 to 4. Log only in uncased part of hole.
Deep lateral (electrode spacing approximately 19 feet)	Determining true resistivity where mud invasion is relatively deep. Locating thin beds.	Fluid-filled uncased hole. Fresh mud. Formations should be of thickness different from electrode spacing and should be free of thin limestone beds.
Limestone sonde (electrode spacing of 32 inches)	Detecting permeable zones and determining porosity in hard rock. Determining formation factor in situ.	Fluid-filled uncased hole. May be salty mud. Uniform hole size. Beds thicker than 5 feet.
Laterolog	Investigating true resistivity of thin beds. Used in hard formations drilled with very salty muds. Correlation of formations, especially in hard-rock regions.	Fluid-filled uncased hole. Salty mud satisfactory. Mud invasion not too deep.
Microlog	Determining permeable beds in hard or well-consolidated formations. Detailing beds in moderately consolidated formations. Correlation in hard-rock country. Determining formation factor in situ in soft or moderately consolidated formations. Detailing very thin beds.	Fluid required in hole. Log only in uncased part of hole. Bit-size hole (caved sections may be logged, provided hole enlargements are not too great).
Microlaterolog	Determining detailed resistivity of flushed formation at wall of hole when mudcake thickness is less than three-eighths inch in all formations. Determining formation factor and porosity. Correlation of very thin beds.	Fluid-filled uncased hole. Thin mud cake. Salty mud permitted.
Spontaneous potential	Helps delineate boundaries of many formations and the nature of these formations. Indicating approximate chemical quality of water. Indicate zones of water entry in borehole. Locating cased interval. Detecting and correlating permeable beds.	Fluid-filled uncased hole. Fresh mud.
Radiation logging: Gamma ray	Differentiating shale, clay, and marl from other formations. Correlations of formations. Measurement of inherent radioactivity in formations. Checking formation depths and thicknesses with reference to casing collars before perforating casing. For shale differentiation when holes contain very salty mud. Radioactive tracer studies. Logging dry or cased holes. Locating cemented and cased intervals. Logging in oil-base muds. Locating radioactive ores. In combination with electric logs for locating coal or lignite beds.	Fluid-filled or dry cased or uncased hole. Should have appreciable contrast in radioactivity between adjacent formations.

TABLE 4–54. BOREHOLE GEOPHYSICAL LOGGING METHODS AND THEIR USES IN HYDROLOGIC STUDIES (continued)

Method	Uses	Recommended Conditions
Neutron	Delineating formations and correlation in dry or cased holes. Qualitative determination of shales, tight formations, and porous sections in cased wells. Determining porosity and water content of formations, especially those of low porosity. Distinguishing between water- or oil-filled and gas-filled reservoirs. Combining with gamma-ray log for better identification of lithology and correlation of formations. Indicating cased intervals. Logging in oil-base muds.	Fluid-filled or dry cased or uncased hole. Formations relatively free from shaly material. Diameter less than 6 inches for dry holes. Hole diameter similar throughout.
Induction logging	Determining true resistivity, particularly for thin beds (down to about 2 feet thick) in wells drilled with comparatively fresh mud. Determining resistivity of formations in dry holes. Logging in oil-base muds. Defining lithology and bed boundaries in hard formations. Detection of water-bearing beds.	Fluid-filled or dry uncased hole. Fluid should not be too salty.
Sonic logging	Logging acoustic velocity for seismic interpretation. Correlation and identification of lithology. Reliable indication of porosity in moderate to hard formations; in soft formations of high porosity it is more responsive to the nature rather than quantity of fluids contained in pores.	Not affected materially by type of fluid, hole size, or mud invasion.
Temperature logging	Locating approximate position of cement behind casing. Determining thermal gradient. Locating depth of lost circulation. Locating active gas flow. Used in checking depths and thickness of aquifers. Locating fissures and solution openings in open holes and leaks or perforated sections in cased holes. Reciprocal-gradient temperature log may be more useful in correlation work.	Cased or uncased hole. Can be used in empty hole if logged at very slow speed, but fluid preferred in hole. Fluid should be undisturbed (no circulation) for 6 to 12 hours minimum before logging; possibly several days may be required to reach thermal equilibrium.
Fluid-conductivity logging	Locating point of entry of different quality water through leaks or perforations in casing or opening in rock hole. (Usually fluid resistivity is determined and must be converted to conductivity.) Determining quality of fluid in hole for improved interpretation of electric logs. Determining fresh-water-salt-water interface.	Fluid required in cased or uncased hole. Temperature log required for quantitative information.
Fluid-velocity logging	Locating zones of water entry into hole. Determining relative quantities of water flow into or out of these zones. Determine direction of flow up or down in sections of hole. Locating leaks in casing. Determine approximate permeability of lithologic sections penetrated by hole, or perforated section of casing.	Fluid-filled cased or uncased hole. Injection, pumping, flowing, or static (at surface) conditions. Flange or packer units required in large diameter holes. Caliper (section gage) logs required for quantitative interpretation.
Casing-collar locator	Locating position of casing collars and shoes for depth control during perforating. Determining accurate depth references for use with other types of logs.	Cased hole.

(continued)

TABLE 4–54. BOREHOLE GEOPHYSICAL LOGGING METHODS AND THEIR USES IN HYDROLOGIC STUDIES (continued)

Method	Uses	Recommended Conditions
Caliper (section gage) survey	Determining hole or casing diameter. Indicates lithologic character of formations and coherency of rocks penetrated. Locating fractures, solution openings, and other cavities. Correlation of formations. Selection of zone to set a packer. Useful in quantitative interpretation of electric, temperature, and radiation logs. Used with fluid-velocity logs to determine quantities of flow. Determining diameter of underreamed section before placement of gravel pack. Determining diameter of hole for use in computing volume of cement to seal annular space. Evaluating the efficiency of explosive development of rock wells. Determining construction information on abandoned wells.	Fluid-filled or dry cased or uncased hole. (In cased holes does not give information on beds behind casing.)
Dipmeter survey	Determining dip angle and dip direction (from magnetic north) in relation to well axis in the study of geologic structure. Correlation of formations.	Fluid-filled uncased hole. Carefully picked zones needing survey, because of expense and time required. Directional survey required for determination of true dip and strike (generally obtained simultaneously with dipmeter curves).
Directional (inclinometer) survey	Locating points in a hole to determine deviation from the vertical. Determining true depth. Determining possible mechanical difficulty for casing installation or pump operation. Determining true dip and strike from dipmeter survey.	Fluid-filled or dry uncased hole.
Magnetic logging	Determining magnetic field intensity in bore-hole and magnetic susceptibility of rocks surrounding hole. Studying lithology and correlation, especially in igneous rocks.	Fluid-filled or dry uncased hole.

Source: U.S. Geological Survey, 1968

CHAPTER 5

Water Use

SECTION A. WATER USE—UNITED STATES

FIGURE 5-1. THE MANY USES OF WATER

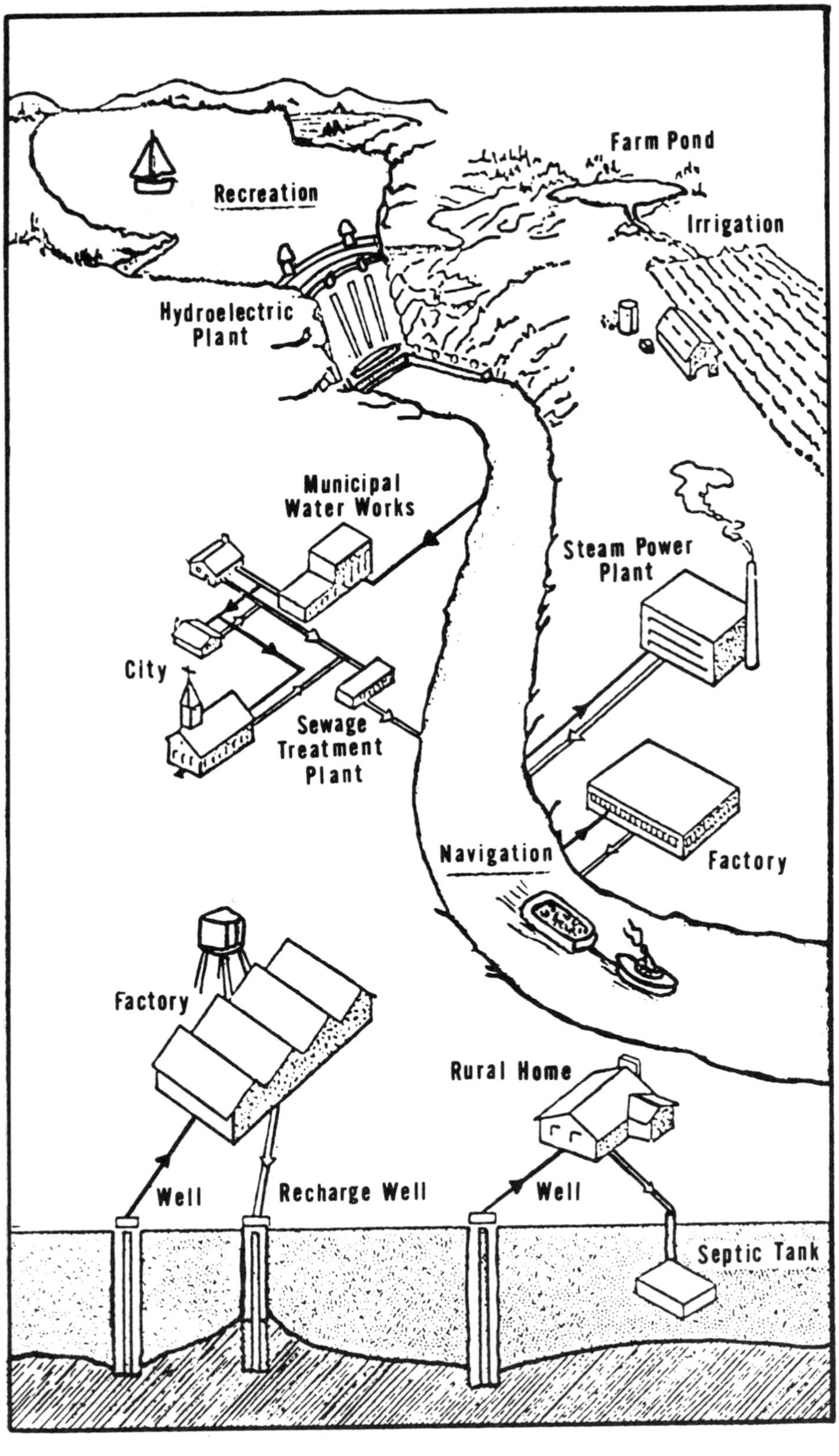

Source: Laas and Beicos, The Water in Your Life, Popular Library, Copyright 1967

FIGURE 5–2. TOTAL WATER WITHDRAWALS FOR ALL OFFSTREAM WATER-USE CATEGORIES IN THE UNITED STATES, 1985

[By water-resources region]

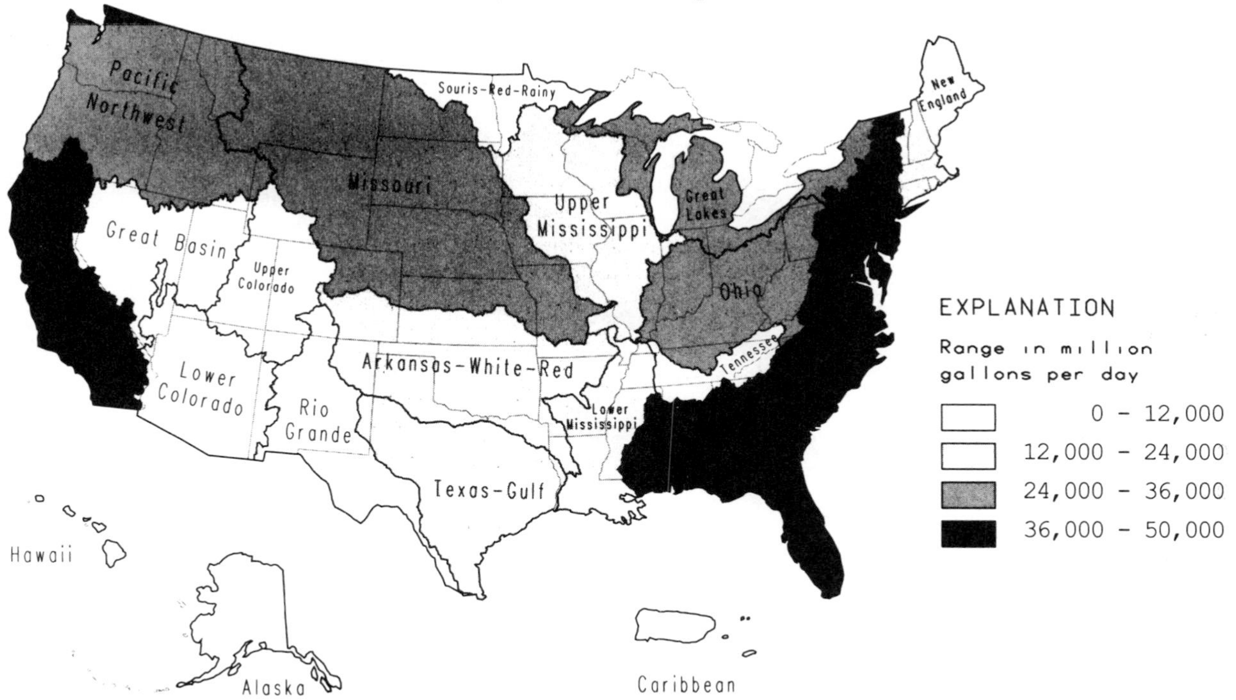

Source: Solley, W.B., Merk, C.F., and Pierce, R.R., 1988, Estimated Water Use in the United States in 1985, U.S. Geological Survey Circular 1004

FIGURE 5–3. TOTAL WATER WITHDRAWAL FOR PUBLIC SUPPLY, RURAL, IRRIGATION, THERMOELECTRIC POWER, AND OTHER INDUSTRIES IN THE UNITED STATES, 1950–1985

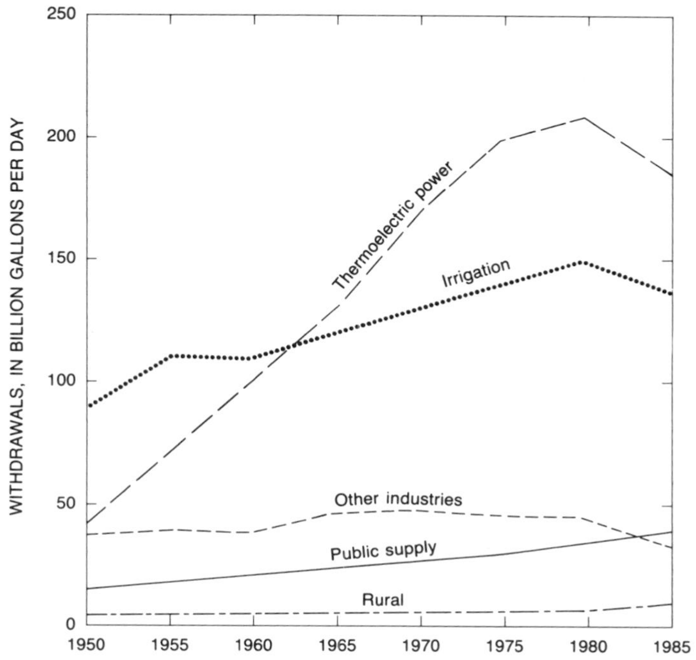

Source: Solley, W.B., and others, 1986, Water Use in the United States, U.S. Geological Survey Water Inv. Rept. 86–4182 and Circular 1004 (1985 data)

TABLE 5–1. TRENDS IN WATER USE IN THE UNITED STATES, 1950–1985

[In thousands of million gallons per day; data are rounded to two significant figures; percentages are calculated from unrounded numbers]

	Year								Percentage Change
	1950[1]	1955[1]	1960[2]	1965[2]	1970[3]	1975[4]	1980[4]	1985[4]	1980–85
Population, in millions	150.7	164.0	179.3	193.8	205.9	216.4	229.6	242.4	+6
Offstream use:									
Total withdrawals	180	240	270	310	370	420	440[5]	400	-10
Public supply	14	17	21	24	27	29	34	37	+7
Rural domestic and livestock	3.6	3.6	3.6	4.0	4.5	4.9	5.6	7.8	+39
Irrigation	89	110	110	120	130	140	150	140	-6
Industrial:									
Thermoelectric power use	40	72	100	130	170	200	210	190	-13
Other industrial use	37	39	38	46	47	45	45	31	-33
Source of water:									
Ground:									
Fresh	34	47	50	60	68	82	83[5]	73	-12
Saline	([6])	.6	.4	.5	1	1	.9	.7	-29
Surface:									
Fresh	140	180	190	210	250	260	290	260	-8
Saline	10	18	31	43	53	69	71	60	-16
Reclaimed sewage	([6])	.2	.6	.7	.5	.5	.5	.6	+22
Consumptive use	([6])	([6])	61	77	87[7]	96[7]	100[7]	92[7]	-9
Instream use:									
Hydroelectric power	1,100	1,500	2,000	2,300	2,800	3,300	3,300	3,100	-7

[1] 48 States and District of Columbia.
[2] 50 States and District of Columbia.
[3] 50 States, District of Columbia, and Puerto Rico.
[4] 50 States, District of Columbia, Puerto Rico, and Virgin Islands.
[5] Revised.
[6] Data not available.
[7] Freshwater only.

Source: Solley, W.B., Merk, C.F., and Pierce, R.R., 1988, Estimated Water Use in the United States in 1985, U.S. Geological Survey Circular 1004

FIGURE 5–4. TRENDS IN GROUND-WATER USE IN THE UNITED STATES, 1950–1985

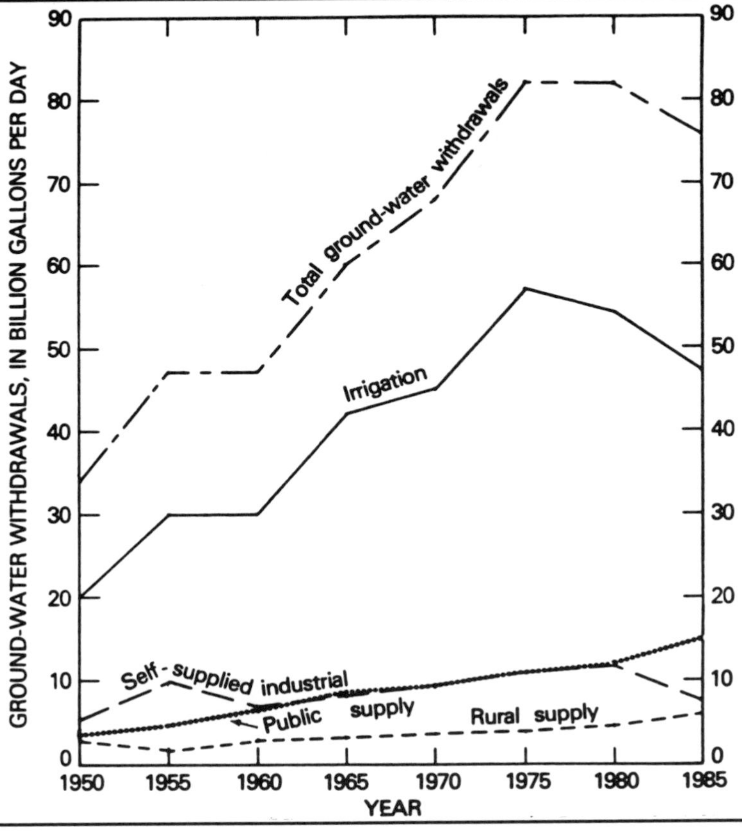

Source: U.S. Environmental Protection Agency, 1987, Prevention of Ground-Water Contamination in the United States. Based on data provided by Wayne B. Solley, U.S. Geological Survey

TABLE 5–2. PERCENT OF UNITED STATES POPULATION RELYING ON GROUND WATER AS A SOURCE OF DRINKING WATER

ALABAMA	44
ALASKA	79
ARIZONA	65
ARKANSAS	63
CALIFORNIA	70
COLORADO	21
CONNECTICUT	32
DELAWARE	61
FLORIDA	91
GEORGIA	47
HAWAII	93
IDAHO	91
ILLINOIS	48
INDIANA	66
IOWA	81
KANSAS	59
KENTUCKY	34
LOUISIANA	57
MAINE	47
MARYLAND	33
MASSACHUSETTS	36
MICHIGAN	34
MINNESOTA	80
MISSISSIPPI	93
MISSOURI	47
MONTANA	53
NEBRASKA	90
NEVADA	40
NEW HAMPSHIRE	57
NEW JERSEY	49
NEW MEXICO	90
NEW YORK	34
NORTH CAROLINA	58
NORTH DAKOTA	63
OHIO	44
OKLAHOMA	35
OREGON	43
PENNSYLVANIA	42
PUERTO RICO	30
RHODE ISLAND	24
SOUTH CAROLINA	43
SOUTH DAKOTA	85
TENNESSEE	51
TEXAS	48
UTAH	66
VERMONT	56
VIRGIN ISLANDS	58
VIRGINIA	37
WASHINGTON	54
WEST VIRGINIA	51
WISCONSIN	70
WYOMING	64
TOTAL UNITED STATES, including P.R. and V.I.	56

Source: Abstracted from Solley, W.B., Merk, C.F., and Pierce, R.R., 1988, Estimated Water Use in the United States in 1985, U.S. Geological Survey Circular 1004

FIGURE 5–5. SOURCE, USE, AND DISPOSITION OF FRESHWATER IN THE UNITED STATES, 1985

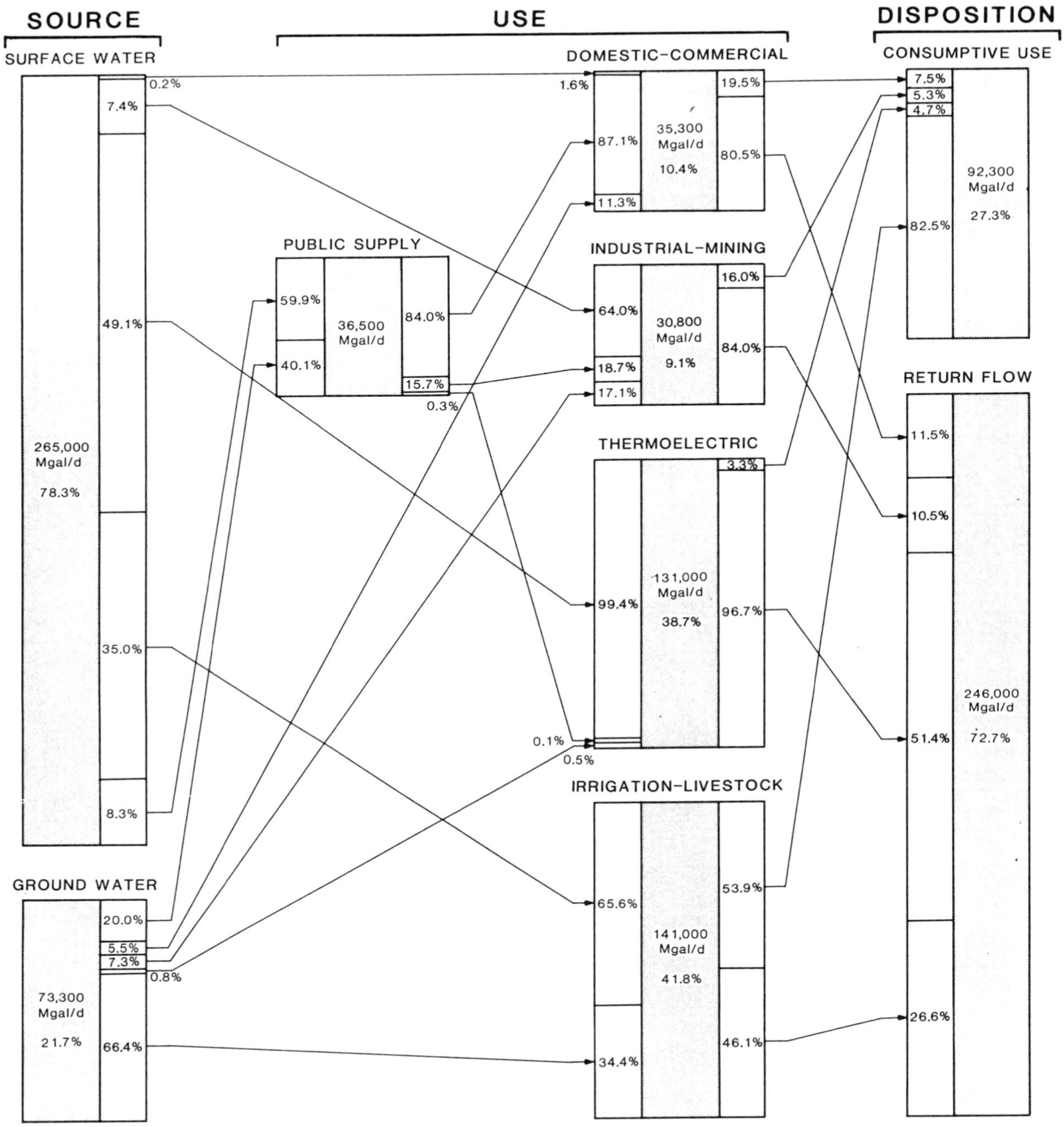

For each water-use category, this diagram shows the relative proportion of water source and disposition and the general distribution of water from source to disposition. The lines and arrows indicate the distribution of water from source to disposition for each category; for example, surface water was 78.3 percent of total freshwater withdrawn, and, going from the "Source" to "Use" columns, the line from the surface-water block to the domestic and commercial block indicates that 0.2 percent of all surface water withdrawn was the source for 1.6 percent of total water (self-supplied withdrawals and public-supply deliveries) for domestic and commercial purposes. In addition, going from the "Use" to "Disposition" columns, the line from the domestic and commercial block to the consumptive use block indicates that 19.5 percent of the water for domestic and commercial purposes was consumptive use; this represented 7.5 percent of total consumptive use by all water-use categories.

Source: Solley, W.B., Merk, C.F., and Pierce, R.R., 1988, Estimated Water Use in the United States in 1985, U.S. Geological Survey Circular 1004

TABLE 5–3. SURFACE-WATER WITHDRAWALS FOR OFFSTREAM WATER-USE CATEGORIES IN THE UNITED STATES, 1985

[By state; figures may not add to totals because of independent rounding. All values in million gallons per day]

State	Public Supply	Domestic	Commercial	Irrigation	Livestock	Industrial		Mining		Thermoelectric		Total	
	Fresh	Fresh	Fresh	Fresh	Fresh	Fresh	Saline	Fresh	Saline	Fresh	Saline	Fresh	Saline
Alabama	442	.0	.0	51	29	804	.0	.0	.0	6920	.0	8250	.0
Alaska	35	.7	.0	.0	146	106	.0	19	.0	26	.0	334	.0
Arizona	233	1.4	.1	3020	36	.7	.0	13	.0	21	4.6	3330	4.6
Arkansas	156	.0	1.9	541	198	108	.0	2.3	.0	1090	.0	2100	.0
California	1570	15	5.7	20300	159	105	254	58	25	412	11700	22600	12000
Colorado	651	.0	.3	10300	45	113	.0	30	.0	96	.0	11200	.0
Connecticut	296	.0	.0	2.6	2.0	59	68	1.4	.0	694	2510	1060	2580
Delaware	49	.0	.0	8.0	.0	3.1	391	.0	.0	.0	1120	60	1520
D.C.	218	.0	.0	.0	.0	.0	.0	.0	.0	130	.0	348	.0
Florida	185	.0	1.0	1320	8.1	54	27	22	.0	633	10700	2230	10700
Georgia	631	.0	4.6	145	22	283	31	2.9	.0	3280	46	4370	77
Hawaii	31	.3	.0	570	3.1	4.8	.0	.0	.0	3.8	880	613	880
Idaho	27	13	.0	17300	.0	26	.0	135	.0	.0	.0	17500	.0
Illinois	1320	.0	74	.0	.0	385	.0	53	.0	11700	.0	13500	.0
Indiana	304	.0	.1	11	.0	2550	.0	83	.0	4450	.0	7400	.0
Iowa	92	.0	9.0	15	37	126	.0	12	.0	1800	.0	2090	.0
Kansas	158	.0	.0	260	26	11	.0	8.3	.0	403	.0	866	.0
Kentucky	356	4.8	11	7.4	48	175	.0	22	.0	3370	.0	3990	.0
Louisiana	352	.0	.0	775	127	1790	.0	1.2	5.0	5440	494	8480	499
Maine	84	.0	28	1.7	26	207	30	3.2	.0	432	642	783	673
Maryland	702	.0	6.3	15	9.8	54	275	8.2	.5	397	5030	1190	5300
Massachusetts	586	.0	171	12	.9	103	22	.0	.0	5070	3380	5940	3400
Michigan	1030	.1	26	119	5.4	1200	.0	52	.0	8390	.0	10800	.0
Minnesota	208	.0	1.8	78	9.5	108	.0	271	.0	1470	.0	2150	.0
Mississippi	37	.0	.0	161	12	69	5.7	.6	.0	430	191	736	197
Missouri	474	.0	.0	23	30	52	.0	3.3	.0	4890	.0	5470	.0
Montana	96	.8	.0	8220	34	27	.0	3.6	.0	67	.0	8450	.0
Nebraska	39	.0	.0	2090	19	7.2	.0	111	.0	2190	.0	4450	.0
Nevada	193	.6	.4	2600	20	7.8	.0	2.7	.0	7.5	.0	2830	.0
New Hampshire	61	.0	.0	.6	.7	204	.2	1.1	.0	336	207	603	207
New Jersey	641	.0	1.1	95	.0	127	801	68	.0	722	3820	1650	4620
New Mexico	28	.0	.0	1650	39	.4	.0	13	.0	48	.0	1780	.0
New York	2330	.0	66	18	7.3	760	.0	50	.0	4720	6150	7950	6150
North Carolina	507	.0	7.9	123	5.1	371	6.0	38	.0	6400	866	7450	872
North Dakota	39	.0	.0	90	8.9	6.8	.0	.9	.0	891	.0	1040	.0
Ohio	1020	.0	.4	9.4	16	451	.0	.4	.0	10500	.0	12000	.0
Oklahoma	414	1.2	6.9	65	2.6	84	.0	.4	.0	133	.0	707	.0
Oregon	332	9.6	.1	5240	21	263	.0	7.2	.0	12	.0	5880	.0
Pennsylvania	1340	.0	.0	9.2	8.4	1910	.0	30	.0	10200	.0	13500	.0
Rhode Island	101	.0	.0	3.1	.8	13	.2	2.3	.0	.0	261	120	261
South Carolina	283	.0	32	13	5.0	1090	.0	2.4	.0	5180	6.3	6600	6.3
South Dakota	16	.9	5.2	347	28	.9	.0	25	.0	2.6	.0	425	.0
Tennessee	384	.0	.0	6.6	34	1510	.0	11	.0	6060	.0	8010	.0
Texas	1760	.0	7.7	2700	150	834	1430	18	.0	7400	3550	12900	4980
Utah	148	1.6	.0	3200	7.2	7.5	.0	5.5	108	24	.0	3390	108
Vermont	36	.0	.0	.5	1.7	50	.0	1.1	.0	.4	.0	89	.0
Virginia	504	.0	3.8	46	25	477	81	14	.0	3460	2300	4530	2380
Washington	616	.0	.3	4310	8.8	411	37	.6	.0	427	.0	5780	37
West Virginia	114	.0	.5	3.4	9.6	853	.0	24	.0	4210	.0	5210	.0
Wisconsin	301	.0	.0	2.0	3.1	424	.0	.0	.0	5440	.0	6170	.0
Wyoming	50	1.3	7.1	5360	13	3.6	.0	39	.0	224	.0	5700	.0
Puerto Rico	307	9.1	.0	107	.0	.0	.0	.0	.0	.0	2000	423	2000
Virgin Islands	4.3	1.0	.4	.0	.0	.0	14	.0	.0	.0	103	5.7	117
Total	21,900	62	481	91,300	1,450	18,400	3,480	1,270	138	130,000	56,000	265,000	59,600

Source: Solley, W.B., Merk, C.F., and Pierce, R.R., 1988, Estimated Water Use in the United States in 1985, U.S. Geological Survey Circular 1004

TABLE 5–4. GROUND-WATER WITHDRAWALS FOR OFFSTREAM WATER-USE CATEGORIES IN THE UNITED STATES, 1985

[By state; figures may not add to totals because of independent rounding. All values in million gallons per day]

State	Public Supply	Domestic	Commercial	Irrigation	Livestock	Industrial		Mining		Thermoelectric	Total	
	Fresh	Fresh	Fresh	Fresh	Fresh	Fresh	Saline	Fresh	Saline	Fresh	Fresh	Saline
Alabama	173	39	3.2	18	67	34	.0	10	3.4	.0	343	3.4
Alaska	41	9.0	.4	.0	10	7.7	.0	.1	.0	4.3	72	.0
Arizona	385	26	17	2500	25	44	8.3	67	.0	32	3090	8.4
Arkansas	101	60	6.0	3330	242	64	.0	1.0	.0	1.1	3810	.0
California	3730	125	47	10400	41	326	7.9	108	276	68	14800	284
Colorado	86	17	8.1	2130	16	7.4	.0	29	32	14	2310	32
Connecticut	66	39	8.6	.2	6.4	18	.0	.3	.0	5.8	144	.0
Delaware	29	10	2.0	19	1.9	16	.0	.0	.0	.9	79	.0
D.C.	.0	.0	.0	.0	.0	.0	.0	.0	.0	.0	.0	.0
Florida	1490	259	55	1590	58	340	.0	236	.0	19	4050	.0
Georgia	205	99	24	308	25	323	.0	16	.0	5.0	1000	.0
Hawaii	172	11	33	336	.7	16	.0	.0	.0	86	655	.0
Idaho	185	76	16	3310	1040	172	.0	.0	.0	.0	4800	.0
Illinois	467	130	33	71	57	150	.0	14	38	6.7	930	38
Indiana	271	139	1.0	36	48	109	.0	7.4	.0	24	635	.0
Iowa	259	65	29	53	135	71	.0	50	.0	9.2	671	.0
Kansas	158	42	.0	4470	42	70	.0	5.3	.0	12	4800	.0
Kentucky	49	43	5.1	.3	2.5	66	.0	3.4	.0	36	205	.0
Louisiana	276	46	.2	709	75	297	5.3	.7	.3	30	1430	5.6
Maine	24	19	10	.2	3.0	8.3	.0	.8	.0	.0	66	.0
Maryland	70	63	19	20	13	20	.0	12	.0	1.8	219	.0
Massachusetts	181	35	67	3.9	.4	26	.0	2.0	.0	.0	315	.0
Michigan	222	123	8.0	91	19	121	3.7	8.5	.8	2.3	596	4.5
Minnesota	265	131	24	131	53	76	.0	1.7	.0	1.2	685	.0
Mississippi	275	16	4.1	725	373	131	.0	3.1	.0	50	1580	.0
Missouri	171	54	17	283	10	37	.0	24	.3	44	640	.3
Montana	62	15	.0	80	16	29	.0	.6	.0	.0	203	.0
Nebraska	208	24	.3	5180	101	41	.0	8.5	.0	25	5590	.0
Nevada	94	12	7.0	750	5.9	2.3	.0	19	2.8	16	905	2.8
New Hampshire	28	22	.0	.0	.6	33	.0	.1	.0	.0	84	.0
New Jersey	406	64	14	37	3.1	129	.1	12	.0	3.4	667	.1
New Mexico	198	38	7.3	1170	11	.4	.0	69	.0	11	1510	.0
New York	535	191	65	20	12	272	.0	.0	.0	.0	1100	.0
North Carolina	88	169	15	9.8	29	43	.0	81	.0	.0	435	.0
North Dakota	30	15	.1	64	13	2.2	.0	2.8	.0	.5	127	.0
Ohio	395	139	51	7.3	25	11	.0	78	.1	25	730	.1
Oklahoma	106	25	25	380	2.0	22	.0	6.7	.0	1.0	568	.0
Oregon	83	70	1.5	471	3.8	29	.0	.3	.0	.0	660	.0
Pennsylvania	258	184	27	1.5	62	149	.0	118	.0	.0	799	.0
Rhode Island	15	5.6	.0	.3	1.5	3.9	.0	.4	.0	.0	27	.0
South Carolina	76	62	8.5	21	5.2	38	.0	2.9	.0	1.1	214	.0
South Dakota	65	15	12	113	19	7.8	.0	16	.0	1.7	249	.0
Tennessee	243	70	4.9	2.4	32	89	.0	2.4	.0	.0	444	.0
Texas	1230	105	17	5420	111	149	.4	103	229	52	7180	229
Utah	299	4.5	.4	384	31	20	.0	50	21	.0	790	25
Vermont	17	12	.0	.0	3.8	4.0	.0	.0	.0	.4	37	.0
Virginia	75	112	18	6.5	29	99	.2	2.1	.0	.1	341	.2
Washington	339	98	20	629	21	108	.0	2.4	.0	.7	1220	.0
West Virginia	37	21	.3	.3	16	33	.0	119	.0	.0	227	.0
Wisconsin	275	84	3.8	82	87	38	.0	.0	.0	1.6	570	.0
Wyoming	48	12	9.9	300	3.2	7.8	.0	111	23	12	504	23
Puerto Rico	84	9.1	.0	50	8.6	18	.0	.0	.0	5.1	175	.0
Virgin Islands	.2	.6	.5	.0	.0	.0	.2	.0	.0	.0	1.4	.2
Total	14,600	3,250	746	45,700	3,020	3,930	26	1,410	626	608	73,300	656

Source: Solley, W.B., Merk, C.F., and Pierce, R.R., 1988, Estimated Water Use in the United States in 1985, U.S. Geological Survey Circular 1004

TABLE 5–5. TOTAL WATER USE FOR ALL OFFSTREAM WATER-USE CATEGORIES IN THE UNITED STATES, 1985

[By state; figures may not add to totals because of independent rounding. Mgal/d = million gallons per day; gal/d = gallons per day]

State	Population, in thousands	Per Capita Use, Freshwater, in gal/d	Withdrawals, in Mgal/d (includes irrigation conveyance losses)									Reclaimed Sewage, in Mgal/d	Conveyance Losses, in Mgal/d	Consumptive Use Freshwater, in Mgal/d
			By Source and Type						Total					
			Ground Water			Surface Water								
			Fresh	Saline	Total	Fresh	Saline	Total	Fresh	Saline	Total			
Alabama	4021	2140	343	3.4	347	8250	.0	8250	8590	3.4	8600	.0	.0	541
Alaska	558	727	72	.0	72	334	.0	334	406	.0	406	.0	.0	27
Arizona	3279	1960	3090	8.4	3100	3330	4.6	3330	6420	13	6430	34	1180	3700
Arkansas	2359	2500	3810	.0	3810	2100	.0	2100	5910	.0	5910	.0	275	3210
California	26354	1420	14800	284	15100	22600	12000	34600	37400	12300	49700	238	929	21100
Colorado	3231	4190	2310	32	2340	11200	.0	11200	13500	32	13600	4.9	2880	4850
Connecticut	3198	375	144	.0	144	1060	2580	3640	1200	2580	3780	.0	.0	106
Delaware	622	222	79	.0	79	60	1520	1580	138	1520	1650	.0	.0	39
D.C.	626	556	.0	.0	.0	348	.0	348	348	.0	348	.0	.0	24
Florida	11328	554	4050	.0	4050	2230	10700	13000	6280	10700	17000	51	57	2730
Georgia	5976	899	1000	.0	1000	4370	77	4440	5370	77	5450	28	.0	838
Hawaii	1152	1100	655	.0	655	613	880	1490	1270	880	2150	1.4	91	132
Idaho	1005	22200	4800	.0	4800	17500	.0	17500	22300	.0	22300	.0	6430	5290
Illinois	11585	1250	930	38	968	13500	.0	13500	14400	38	14500	.0	.0	686
Indiana	5467	1470	635	.0	635	7400	.0	7400	8030	.0	8030	.0	.0	454
Iowa	2881	960	671	.0	671	2090	.0	2090	2770	.0	2770	.0	.0	473
Kansas	2451	2310	4800	.0	4800	866	.0	866	5670	.0	5670	.0	287	4710
Kentucky	3726	1130	205	.0	205	3990	.0	3990	4200	.0	4200	.0	.3	260
Louisiana	4480	2210	1430	5.6	1440	8480	499	8980	9920	505	10400	.0	146	2090
Maine	1157	733	66	.0	66	783	673	1460	848	673	1520	.0	.0	203
Maryland	4392	321	219	.0	219	1190	5300	6490	1410	5300	6710	81	.0	423
Massachusetts	5822	1070	315	.0	315	5940	3400	9340	6260	3400	9660	.0	.0	316
Michigan	9007	1270	596	4.5	600	10800	.0	10800	11400	4.5	11400	21	.0	611
Minnesota	4193	676	685	.0	685	2150	.0	2150	2830	.0	2830	4.8	19	768
Mississippi	2613	885	1580	.0	1580	736	197	933	2310	197	2510	.0	92	661
Missouri	5029	1210	640	.3	640	5470	.0	5470	6110	.3	6110	.0	.0	504
Montana	823	10500	203	.0	203	8450	.0	8450	8650	.0	8650	.0	4270	1900
Nebraska	1605	6250	5590	.0	5590	4450	.0	4450	10000	.0	10000	.0	2610	4910
Nevada	968	3860	905	2.8	908	2830	.0	2830	3740	2.8	3740	11	732	1890
New Hampshire	998	688	84	.0	84	603	207	810	687	207	894	.0	.0	76
New Jersey	7559	307	667	.1	668	1650	4620	6270	2320	4620	6940	.0	.0	279
New Mexico	1418	2320	1510	.0	1510	1780	.0	1780	3280	.0	3280	.0	.0	1530
New York	17783	508	1100	.0	1100	7950	6150	14100	9040	6150	15200	.0	.0	1400
North Carolina	6256	1260	435	.0	435	7450	872	8320	7890	872	8760	.0	.0	439
North Dakota	688	1690	127	.0	127	1040	.0	1040	1160	.0	1160	.0	8.4	201
Ohio	10752	1180	730	.1	730	12000	.0	12000	12700	.1	12700	.0	.0	396
Oklahoma	3302	386	568	.0	568	707	.0	707	1270	.0	1270	.0	3.3	576
Oregon	2676	2450	660	.0	660	5880	.0	5880	6540	.0	6540	5.2	772	2600
Pennsylvania	11853	1210	799	.0	799	13500	.0	13500	14300	.0	14300	.0	.0	589
Rhode Island	968	152	27	.0	27	120	261	381	148	261	409	.0	.0	23
South Carolina	3348	2040	214	.0	214	6600	6.3	6610	6810	6.3	6820	.0	.0	340
South Dakota	706	956	249	.0	249	425	.0	425	675	.0	675	.0	120	361
Tennessee	4762	1770	444	.0	444	8010	.0	8010	8450	.0	8450	.1	.0	275
Texas	16361	1230	7180	229	7410	12900	4980	17900	20100	5210	25300	93	747	8650
Utah	1645	2540	790	25	815	3390	108	3500	4180	133	4320	5.8	305	2130
Vermont	535	235	37	.0	37	89	.0	89	126	.0	126	.0	.0	26
Virginia	5706	853	341	.2	341	4530	2380	6910	4870	2380	7250	1.5	5.2	269
Washington	4384	1600	1220	.0	1220	5780	37	5810	7000	37	7030	.0	39	4700
West Virginia	1936	2810	227	.0	227	5210	.0	5210	5440	.0	5440	.0	.0	877
Wisconsin	4804	1400	570	.0	570	6170	.0	6170	6740	.0	6740	.0	.0	321
Wyoming	509	12200	504	23	526	5700	.0	5700	6200	23	6220	.0	1610	2670
Puerto Rico	3390	176	175	.0	175	423	2000	2430	598	2000	2600	.0	16	163
Virgin Islands	104	68	1.4	.2	1.6	5.7	117	123	7.1	117	124	.0	.0	1.2
Total	242,351	1,400	73,300	656	74,000	265,000	59,600	325,000	338,000	60,300	399,000	579	23,600	92,300

Source: Solley, W.B., Merk, C.F., and Pierce, R.R., 1988, Estimated Water Use in the United States in 1985, U.S. Geological Survey Circular 1004

TABLE 5–6. COMMON USES OF WATER IN RELATION TO CONSUMPTIVE AND NONCONSUMPTIVE USES

Common Uses of Water	Consumptive	Nonconsumptive
Steam generation (locomotive or stationary).	Steam released to atmosphere.	([1])
Air conditioning:		
Evaporative	Cooling achieved by evaporation.	([1])
Recirculating	Some water evaporated with each use.	([1])
Other cooling (recirculating)	do.	([1])
Storage in surface reservoirs	Evaporation from water surface.	Seepage underground.
Irrigation by sprinkling	Evaporation and transpiration.	Little seepage.
Irrigation by flooding	Evaporation from ponds, transpiration.	Seepage varies.
Cooking	Steam to atmosphere	Contributes to sewage.
Processing foods, beverages, plastics.	Some water goes into manufactured products.	Carries organic compounds.
Processing:		
Petroleum products	Proportion of consumptive use is increased by reuse of the nonconsumptive water.	Carries chemicals.
Paper and pulp	do.	Carries pulp and chemicals.
Chemicals	do.	Carries toxic or other chemicals.
Metal products	do.	Carries sludge and soluble chemicals.
Atomic fission	Evaporation from tanks	Carries radioactive materials.
Stock watering	Evaporation from tanks and ponds.	Organic wastes into ground.
Drinking	Perspiration.	Organic wastes into sewage.
Irrigation by furrow	Evapotranspiration.	Dissolves chemicals from soil.
Washing	Evaporation in drying.	Carries sediment and soluble matter.
Mining (metals, coal, oil)		Carries natural brines and acids, sediments.
Cooling (once-through)	([2])	Water temperature increased by use.
Air conditioning (once-through).	([2])	do.
Fish culture	([3])	
Steam heating	([4])	Steam condenses and is reused.
Year-round heat exchange	([4])	Requires storage of water and heat from one season to another.
Sanitation (bath, toilet, dishwasher).	([4])	Sewage carries chiefly organic wastes.
Hydroelectric power	([5])	Takes water toward oceans.
Navigation	([5])	Inland waterways require maintenance of flow.

[1] With efficient operation, nonconsumptive use is limited to that required for cooling and/or cleaning equipment.
[2] Increase in water temperature may cause increased evaporation.
[3] Consumptive use by evaporation from water surfaces; may be increased by aerators.
[4] Consumptive use is limited to losses through leaking pipes, valves, etc.
[5] Consumptive use is limited to evaporation from lakes, reservoirs, etc., that are required for continuous operation.
Source: House of Representatives, U.S. Congress

SECTION B. WATER USE—WORLD

TABLE 5–7. WATER SUPPLIES AND DEMANDS IN CANADA, 1981 AND 2011

[By river basin (see Figure 2–8); millions of cubic meters per year]

River Basin Region	Current Reliable Annual Flows[1]	1981 Withdrawals[2]	1981 Consumption	2011 Withdrawals Low Estimate	2011 Withdrawals High Estimate
Pacific Coastal	396,400	2,134	69	2,025	4,398
Fraser-Lower Mainland	96,000[3]	975	219	974	2,047
Okanagan-Similkameen	971[4]	312	146	300	681
Columbia	51,850[4]	305	33	292	654
Yukon	56,950[4]	16	1	33	50
Peace-Athabasca	58,720	251	155	304	585
Lower Mackenzie	192,800[5]	26	2	46	68
Arctic Coast-Islands	186,670	—	—	—	—
Missouri	105[4]	156	38	215	365
North Saskatchewan	5,046	1,405	154	2,041	3,267
South Saskatchewan	4,636[3,4]	2,578	1,680	2,601	6,058
Assiniboine-Red	497[4]	1,012	207	1,354	2,284
Winnipeg	12,040[3,4]	143	3	152	312
Lower Sask.-Nelson	34,937[4,5]	122	25	132	310
Churchill	10,190[3]	1	—	—	—
Keewatin	92,870	1	1	—	—
Northern Ontario	117,700[3]	128	2	120	269
Northern Quebec	404,300[3]	121	9	143	268
Great Lakes	75,780[4]	20,850	567	28,471	46,655
Ottawa	43,840	638	77	626	1,304
St. Lawrence	47,430[4,5]	3,320	343	3,317	6,774
North Shore-Gaspé	203,000	443	39	394	876
St. John-St. Croix	15,990[4]	892	54	1,164	1,786
Maritime Coastal	65,564	1,736	68	2,238	3,565
Newfoundland-Labrador	217,900	299	14	344	542
CANADA	2,392,186	37,864	3,906	47,738	84,039

Note:

[1] Flow equaled or exceeded, on average, in 19 out of 20 years.

[2] For municipal, rural residential, industrial, mining, agricultural, thermoelectric power uses. Rural residential uses are excluded from 2011 forecasts.

[3] Excludes flow transferred into neighboring basin region.

[4] Excludes inflow from United States portion of basin region.

[5] Excludes inflow from upper basin region.

Source: Pearse, P.H., 1985, Currents of Change, Final Report, Inquiry on Federal Water Policy, Ottawa

TABLE 5–8. WATER WITHDRAWAL AND CONSUMPTION IN CANADA, 1981

[In millions of cubic meters per year; for location of river basin regions see Figure 2–8]

River Basin Region	Withdrawals							Total Consumption
	Municipal[1]	Rural Residential	Agriculture	Mining	Manufacturing	Thermo Electric	Total Withdrawal	
Pacific Coastal	102	8	—	32	1,632	360	2,134	69
Fraser-Lower Mainland	380	23	258	51	263	—	975	219
Okanagan-Similkameen	29	3	254	19	7	—	312	146
Columbia	36	2	33	15	219	—	305	33
Yukon	5	1	—	10	—	—	16	1
Peace-Athabasca	5	10	17	84	132	4	251	155
Lower Mackenzie	—	1	—	24	1	—	26	2
Arctic Coast-Islands	—	—	—	—	—	—	—	—
Missouri	—	—	46	—	—	110	156	38
North Saskatchewan	132	17	94	55	90	1,018	1,405	154
South Saskatchewan	264	20	1,963	8	121	202	2,578	1,680
Assiniboine-Red	155	21	188	14	24	610	1,012	207
Winnipeg	17	1	1	29	96	—	143	3
Lower Sask.-Nelson	13	4	28	5	72	—	122	25
Churchill	—	1	—	—	—	—	1	—
Keewatin	—	—	—	1	—	—	1	1
Northern Ontario	15	2	—	10	101	—	128	2
Northern Quebec	—	1	—	46	74	—	121	9
Great Lakes	1,313	92	131	75	4,407	14,832	20,850	567
Ottawa	182	15	28	61	352	—	638	77
St. Lawrence	1,207	63	65	3	1,674	308	3,320	343
North Shore-Gaspé	104	8	8	22	301	—	443	39
St. John-St. Croix	111	10	2	—	133	635	892	54
Maritime Coastal	120	34	8	34	421	1,118	1,736	68
Newfoundland-Labrador	68	15	1	50	81	84	299	14
CANADA WITHDRAWALS	4,263	347	3,125	648	10,201	19,281	37,864	—
CANADA CONSUMPTION	640	—	2,412	178	507	168	—	3,906

Note:

[1] Municipalities withdrew an additional 811 m^3 for mining, manufacturing and thermoelectric water uses. These are recorded in the appropriate columns.

Source: Pearse, P.H., 1985, Currents of Change, Final Report Inquiry on Water Policy, Ottawa, Canada

TABLE 5-9. AVERAGE ANNUAL WATER USE IN SELECTED COUNTRIES

[Total and per-capita]

	Water Withdrawal		Share Withdrawn by Sector (percent)			
Country	Total (km³)	Per Capita (m³)	Public	Industry	Electric Cooling	Agriculture/Irrigation
United States	472.000	1,986	10	11	38	41
Canada	30.000	1,172	13	39	39	10
Egypt	45.000	962	1	0	0	98
Finland	4.610	946	7	85	0	8
Belgium	8.260	836	6	37	47	10
USSR	226.000	812	8	15	14	63
Panama	1.300	596	12	11	0	77
India	380.000	499	3	1	3	93
China	460.000	460	6	7	0	87
Poland	15.900	423	14	21	40	25
Libya[p]	1.470	408	17	0	0	83
Oman	0.043	350	2	0	0	98
South Africa[p]	9.200	284	17	0	0	83
Nicaragua[i]	0.890	272	18	45	0	37
Barbados	0.027	102	45	35	0	20
Malta	0.023	60	100	0	0	0

[p] = Public and Industry
[i] = Industry and Electric Cooling
Source: World Resources Institute, 1986. World Resources 1986, Copyright Basic Books. Reprinted with permission

TABLE 5-10. TOTAL WATER WITHDRAWAL BY MAJOR USES IN SELECTED COUNTRIES, 1980

	Total		Public Water		Industry	Electrical
	Million m³	Per Capita (m³/ cap.)	Supply (%)	Irrigation (%)	No Cooling (%)	Cooling (%)
Canada[3]	36153	1509	12.6	7.8	11.6	36.8
USA	525053	2306	8.9	39.5	10.3	39.8
Japan[11]	84831	726	16.1	66.8	15.7	1.0
Australia[1,13]	17800	1211	12.1	57.1	4.6	. .
New Zealand	1200	383	41.7	. .	. .	. .
Austria[9]	2240	298	26.8	1.8	24.6	46.9
Belgium	9030	917	7.3	. .	10.0	50.9
Denmark[1]	1210	236	. .	. .	. .	. .
Finland[7]	3702	774	10.5	1.4	38.3	2.6
France[5,6,11]	32648	608	13.2	15.9	16.8	36.8
Germany[13]	42206	686	12.0	0.4	6.8	60.4
Greece[2]	6945	720	10.8	82.7	1.3	0.8
Italy[9]	56200	985	14.2	57.3	14.2	12.5
Netherlands	14956	1057	7.0	. .	2.5	67.5
Norway	1999	489	19.5	2.0	70.0	. .
Portugal[11]	10500	1062	15.0	46.9	37.5	. .
Spain[1]	39920	1068	. .	. .	. .	. .
Sweden[14]	2853	343	32.7	2.2	64.7	0.3
Switzerland[1]	690	108	. .	. .	. .	. .
Turkey[10,11]	29928	669	12.5	77.8	9.7	. .
UK[8,11]	12433	251	47.3	0.3	13.6	38.4
Yugoslavia[4]	8767	392	17.2	7.9	2.9	31.9

[1] Total and per capita data refer to 1975 for withdrawal and to 1980 for population.
[2] Irrigation: data refer to total agriculture water withdrawal.
[3] Public water supply data are based on residential and commercial withdrawal only; industrial (municipal) withdrawal is included in industry water use.
[4] Industry: water consumption only.
[5] Public water supply: non-irrigation agriculture uses included.
[6] 1975 data; per capita figure: 1980 data for population and 1975 data for withdrawal.
[7] Irrigation: data representing mean withdrawal during dry years which occur 2 to 3 times per decade.
[8] England and Wales only.
[9] Public water supply: data are estimates.
[10] Irrigation: small scale removal excluded.
[11] Industry: cooling water included.
[12] Industry: industrial and electrical cooling included.
[13] Withdrawal by sector: 1979 data.
[14] 1983.
NOTE: Withdrawals from the 4 sectors do not necessarily add up to 100%, since 'other agricultural uses than irrigation', 'industrial cooling' and 'other uses' are not covered in this table.
Source: Organisation for Economic Cooperation and Development, Environmental Data Compendium 1987. Copyright OECD. Reprinted with permission

TABLE 5-11. TOTAL AND PER-CAPITA WATER WITHDRAWAL IN SELECTED COUNTRIES, 1960–1985

	Total Water Withdrawal (million m³)/					Water Withdrawal per Capita (m³/inh.)				
	1960	1970	1975	1980	1985	1960	1970	1975	1980	1985
Canada	16152	23308	28128	36153	..	902	1093	1238	1509	..
USA[15]	293935	439606	472500	525053	..	1627	2144	2188	2306	..
Japan[2]	..	82611	84964	84831	..	..	799	762	726	..
Australia	..	..	17800	..	..	..	..	1281	..	..
New Zealand[6,7]	..	990	1045	1200	1900	..	351	339	382	584
Austria[5,14]	..	..	2620	2240	..	..	..	348	298	..
Belgium[23]	8981	9481	..	9030	..	981	984	..	917	..
Denmark	..	..	1210	..	..	..	..	239	..	..
Finland[17]	4016	4190	3990	3702	2390	907	910	847	774	488
France[10]	..	..	27000	32648	..	..	..	512	608	..
Germany[3]	18712	29488	33544	42206	41216	338	486	543	686	671
Greece[4]	..	4254	5847	6945	..	..	484	646	720	..
Italy[9]	..	41900	..	56200	..	..	781	..	985	..
Netherlands[1,18]	11721	13270	13734	14956	..	994	1018	1005	1057	..
Norway	..	..	2380	1999	2235	..	..	594	489	539
Portugal	..	7900	9200	10500	..	..	874	976	1062	..
Spain[11]	15620	24600	36080	39920	45250	513	726	1016	1068	1172
Sweden[12,24]	..	..	3979	..	2853	..	..	486	..	342
Switzerland	..	..	690	..	..	..	..	108	..	..
Turkey[16]	..	11760	16041	29928	..	..	330	398	669	..
UK[13]	..	15583	13085	12433	11511	..	318	264	251	231
Yugoslavia[8]	..	..	7370	8767	..	..	..	345	392	..

[1] Agriculture withdrawal and irrigation excluded.
[2] 1970 and 1980 data are composite totals including estimated data for nonspecified uses.
[3] Do not include agriculture except irrigation. 1980 and 1985 data refer to 1979 and 1983.
[4] Withdrawal for cooling power stations excluded.
[5] Water used by agriculture excluded but water used by irrigation included. 1980 data are estimates.
[6] Withdrawal for irrigation, industry and power plant cooling excluded.
[7] 1980 data is a composite total including 1975 agriculture without irrigation, 1975 industry no cooling, 1980 public water supply.
[8] Secretariat estimates.
[9] Does not include data for agriculture except irrigation. 1970 and 1980 data for industry cooling were estimated in 1973.
[10] 1975 data are estimates based on 4 basins out of 6.
[11] Withdrawals for agriculture and industry cooling excluded.
[12] 1975 data are composite totals.
[13] England and Wales only. 1970 and 1985 data refer to 1971 and 1984.
[14] Based on rough estimates.
[15] 1975 and 1980 data include Puerto Rico and Virgin Islands. 1960, 1970 and 1980 data exclude industry cooling.
[16] Data exclude agriculture except irrigation and cooling of electrical power plants.
[17] 1960, 1970, and 1975 data are Secretariat estimates. 1985 data refer to 1982.
[18] 1960, 1970, 1975, and 1980 data refer to 1962, 1972, 1976, and 1981.
[19] Amounts of surface water and ground water do not add up to the total because the origin of some water in the total was not indicated.
[20] Data are rough estimates. Surface water withdrawals exclude agriculture withdrawal, irrigation, and industry process except cooling. Ground water withdrawals exclude agriculture withdrawals except cooling, industry cooling, and cooling of electrical power plants.
[21] Surface water withdrawals: data do not include agriculture except irrigation. Secretariat estimates for the years 1960, 1970, 1975.
[22] England and Wales only.

Source: Organisation for Economic Cooperation and Development, Environmental Data Compendium 1987.

TABLE 5-12. SURFACE AND GROUND-WATER WITHDRAWAL IN SELECTED COUNTRIES, 1960–1985

	Surface Water (million m³)					% of Total Latest Year Avail.	Ground Water (million m³)					% of Total Latest Year Avail.
	1960	1970	1975	1980	1985		1960	1970	1975	1980	1985	
Canada[19]	14663	21643	26128	30508	..	97.7	1489	1660	2000	1487	..	2.3
USA[15]	230644	346604	359200	403617	..	76.9	63291	93002	110714	121436	..	23.1
Japan[2]	..	70519	72835	73066	..	86.1	..	12092	12130	11766	..	13.9
Australia[19]	..	..	15000	..	..	84.3	..	..	2460	..	..	13.8
Austria[20]	..	..	..	1055	..	47.1	..	..	..	1185	..	52.9
Belgium[23]	8205	8710	..	8251	..	91.4	776	771	..	778	..	8.6
Finland[17,19,22]	3932	4045	3790	3412	2120	88.7	65	115	165	200	268	11.2
France[10]	..	..	21300	27407	..	80.6	..	..	5700	5601	..	16.5
Germany[3]	12747	21906	25971	35344	34225	83.0	5965	7582	7573	6862	6991	17.0
Greece	..	3167	4088	4968	..	71.5	..	1087	1759	1977	..	28.5
Netherlands[18]	5506	10930	12163	13296	..	93.6	799	1119	1153	1020	..	7.2
Norway	..	..	..	1599	2127	95.2	..	..	..	..	108	4.8
Portugal	..	6300	7400	8500	..	81.0	..	1600	1800	2000	..	19.0
Spain[11]	13320	21000	31140	34800	39840	88.0	2300	3600	4940	5120	5410	12.0
Sweden[12,24]	..	..	3471	..	2374	83.2	..	..	508	..	479	16.8
Turkey[8,16]	..	..	9500	19928	..	66.6	..	..	6541	10000	..	33.4
UK[22]	..	..	..	9946	9159	80.0	..	..	..	2487	2352	20.0

Notes: See preceding table.
Source: Organisation for Economic Cooperation and Development, Environmental Data Compendium 1987.

TABLE 5–13. WORLD-WIDE FRESHWATER AVAILABILITY AND USE, 1970s

		Availability		Withdrawal		Sectoral Use (percent)			
	Year of Information	Total (cubic kilometers per year)	Per Capita (thousand cubic meters per year)	Total (cubic kilometers per year)	Per Capita (thousand cubic meters per year)	Public	Industry (self-supplied)	Electricity Generation Facilities (cooling)	Agriculture (irrigation)
WORLD									
AFRICA									
Algeria	1970	25.00	1.82	2.00	0.15	13	6	0	81
Angola	X	X	X	X	X	X	X	X	X
Benin	1970	26.00	9.60	X	X	X	X	X	X
Botswana	X	X	X	X	X	X	X	X	X
Burkina Faso	X	X	X	X	X	X	X	X	X
Burundi	X	X	X	X	X	X	X	X	X
Cameroon	X	208.00	27.43	X	X	X	X	X	X
Cape Verde	X	X	X	X	X	X	X	X	X
Central African Rep	X	X	X	X	X	X	X	X	X
Chad	X	X	X	X	X	X	X	X	X
Comoros	X	X	X	X	X	X	X	X	X
Congo	X	X	X	X	X	X	X	X	X
Djibouti	1973	X	X	0.01	0.03	X	X	X	X
Egypt	1976	56.00	1.47	45.00	1.18	1	0	0	98
Equatorial Guinea	X	X	X	X	X	X	X	X	X
Ethiopia	X	110.00	3.79	X	X	X	X	X	X
Gabon	X	X	X	X	X	X	X	X	X
Gambia	X	X	X	X	X	X	X	X	X
Ghana	1970	53.00	6.15	0.30	0.03	44	3	0	54
Guinea	X	X	X	X	X	X	X	X	X
Guinea-Bissau	X	X	X	X	X	X	X	X	X
Ivory Coast	X	74.00	10.94	X	X	X	X	X	X
Kenya	X	14.80	1.08	X	X	X	X	X	X
Lesotho	X	X	X	X	X	X	X	X	X
Liberia	X	X	X	X	X	X	X	X	X
Libya	1977–78	0.70	0.26	1.47	0.55	17	0	0	83
Madagascar	X	40.00	5.26	X	X	X	X	X	X
Malawi	X	X	X	X	X	X	X	X	X
Mali	X	X	X	X	X	X	X	X	X
Mauritania	1978	X	X	0.73	0.47	2	0	0	98
Mauritius	1974	2.20	2.53	0.36	0.41	X	X	X	X
Morocco	1972	25.00	1.63	8.00	0.52	4	3	0	94
Mozambique	X	X	X	X	X	X	X	X	X
Niger	X	X	X	X	X	X	X	X	X
Nigeria	X	X	X	X	X	X	X	X	X
Rwanda	X	X	X	X	X	X	X	X	X
Senegal	X	X	X	X	X	X	X	X	X
Sierra Leone	X	X	X	X	X	X	X	X	X
Somalia	X	X	X	X	X	X	X	X	X
South Africa	1970	50.00	2.20	9.20	0.40	17	0	0	83
Sudan	1970	18.50	1.33	18.15	1.31	2	0	0	98
Swaziland	X	X	X	X	X	X	X	X	X
Tanzania, United Rep	1970	X	X	0.48	0.04	38	0	0	63
Togo	X	11.50	5.11	0.05	0.02	90	0	0	10
Tunisia	1977	3.35	0.57	1.07	0.18	19	5	0	77
Uganda	1970	X	X	0.20	0.02	43	0	0	57
Zaire	X	X	X	X	X	X	X	X	X
Zambia	1970	X	X	0.36	0.09	72	0	0	28
Zimbabwe	X	X	X	X	X	X	X	X	X

TABLE 5–13. WORLD-WIDE FRESHWATER AVAILABILITY AND USE, 1970s (continued)

	Year of Information	Availability: Total (cubic kilometers per year)	Availability: Per Capita (thousand cubic meters per year)	Withdrawal: Total (cubic kilometers per year)	Withdrawal: Per Capita (thousand cubic meters per year)	Sectoral Use (percent): Public	Sectoral Use (percent): Industry (self-supplied)	Sectoral Use (percent): Electricity Generation Facilities (cooling)	Sectoral Use (percent): Agriculture (irrigation)
NORTH AMERICA									
Barbados	1962	0.05	0.23	0.03	0.12	45	35	0	20
Canada	1977	3122.00	134.24	30.00	1.29	13	39	39	10
Costa Rica	1970	95.00	54.85	1.35	0.78	0	8	0	92
Cuba	1975	34.50	3.70	8.10	0.87	14	4	0	83
Dominican Rep	1975	20.00	4.04	X	X	X	X	X	X
El Salvador	1975	18.95	4.57	1.00	0.24	17	0	0	83
Guatemala	1970	116.00	21.67	0.73	0.14	0	18	0	82
Haiti	X	11.00	2.13	X	X	X	X	X	X
Honduras	1970	102.00	38.65	1.34	0.51	0	4	0	96
Jamaica	1975	8.30	4.06	0.32	0.16	3	6	0	91
Mexico	1975	357.40	5.94	54.20	0.90	5	7	0	88
Nicaragua	1975	175.00	72.67	0.89	0.37	18	45	0	37
Panama	1975	144.00	82.38	1.30	0.74	12	11	0	77
Trinidad and Tobago	1975	X	X	0.15	0.14	0	50	0	50
United States	1975	2478.00	11.47	472.00	2.19	10	11	38	41
SOUTH AMERICA									
Argentina	1976	694.00	26.64	27.60	1.06	9	8	10	73
Bolivia	1959	X	X	X	X	1	1	0	97
Brazil	X	5190.00	48.04	X	X	X	X	X	X
Chile	1975	X	X	16.80	1.65	5	4	0	92
Colombia	1960	1070.00	68.86	X	X	14	0	0	86
Ecuador	1973	314.00	45.57	X	X	X	X	X	X
Guyana	1971	X	X	5.40	7.62	1	0	0	99
Paraguay	X	X	X	X	X	X	X	X	X
Peru	1975	40.00	2.64	X	X	7	0	0	93
Suriname	X	X	X	X	X	X	X	X	X
Uruguay	1965	X	X	0.65	0.24	15	8	0	77
Venezuela	1970	856.00	78.09	4.10	0.37	37	4	0	59
ASIA									
Afghanistan	1970	50.00	4.01	X	X	X	X	X	X
Bahrain	1975	0.00	0.00	0.20	0.74	10	6	0	84
Bangladesh	X	1357.00	17.72	X	X	X	X	X	X
Bhutan	X	X	X	X	X	X	X	X	X
Burma	1970	1082.00	39.57	X	X	X	X	X	X
China	1970	2680.00	3.23	X	X	X	X	X	X
Cyprus	1972	1.00	1.63	0.55	0.89	5	2	0	93
India	1975	1850.00	2.99	380.00	0.61	3	1	3	93
Indonesia	1978	2530.00	17.45	X	X	95	5	0	0
Iran	1975	117.50	3.52	45.40	1.36	3	0	0	97
Iraq	1969	84.50	9.03	42.30	4.52	1	5	0	93
Israel	1975	1.65	0.48	1.72	0.50	17	6	0	77
Japan	1975	547.00	4.90	117.90	1.06	17	33	0	50
Jordan	1975	0.85	0.33	0.38	0.15	1	2	0	97
Kampuchea, Dem	1970	88.10	12.70	X	X	X	X	X	X
Korea, Dem People's Rep	X	X	X	X	X	X	X	X	X
Korea, Rep	1976	63.00	1.23	10.70	0.21	11	13	0	75
Kuwait	1974	0.00	0.00	0.13	0.13	35	4	0	61
Lao People's Dem Rep	X	270.00	78.76	X	X	X	X	X	X
Lebanon	1975	4.30	1.55	0.75	0.27	13	0	0	87

TABLE 5–13. WORLD-WIDE FRESHWATER AVAILABILITY AND USE, 1970s (continued)

		Availability		Withdrawal		Sectoral Use (percent)			
	Year of Information	Total (cubic kilometers per year)	Per Capita (thousand cubic meters per year)	Total (cubic kilometers per year)	Per Capita (thousand cubic meters per year)	Public	Industry (self-supplied)	Electricity Generation Facilities (cooling)	Agriculture (irrigation)
ASIA (continued)									
Malaysia	1975	456.00	37.05	9.42	0.77	X	X	X	X
Mongolia	X	24.60	17.04	X	X	X	X	X	X
Nepal	1970	170.00	14.80	X	X	X	X	X	X
Oman	1975	0.66	0.86	0.43	0.56	2	0	0	98
Pakistan	1975	298.00	3.96	153.40	2.04	X	X	X	X
Philippines	1975	323.00	7.59	29.50	0.69	X	X	X	X
Qatar	1975	0.02	0.13	0.04	0.23	33	0	0	67
Saudi Arabia	1975	2.20	0.30	2.33	0.32	36	6	0	58
Singapore	1975	0.60	0.27	0.19	0.08	X	X	X	X
Sri Lanka	1970	43.20	3.45	6.30	0.50	0	2	0	98
Syrian Arab Rep	1976	35.30	4.75	7.00	0.94	6	0	0	94
Thailand	1975	110.00	2.66	X	X	1	0	0	99
Turkey	1970	167.00	4.73	11.80	0.33	7	2	7	85
United Arab Emirates	1975	0.22	0.43	0.36	0.71	9	0	0	91
Viet Nam	X	X	X	X	X	X	X	X	X
Yemen	X	X	X	X	X	X	X	X	X
Yemen, Dem	1975	1.50	0.91	1.93	1.17	1	0	0	99
EUROPE									
Albania	1967	27.50	14.71	0.20	0.11	1	1	98	0
Austria	1975	90.00	11.97	2.75	0.37	31	69	0	0
Belgium	1971	12.50	1.30	8.26	0.86	6	37	47	10
Bulgaria	1971	197.00	23.14	X	X	33	67	0	0
Czechoslovakia	1975	90.00	6.08	5.02	0.34	24	72	0	5
Denmark	1970	12.90	2.62	0.77	0.16	43	38	1	18
Finland	1972	104.00	22.58	4.61	1.00	7	85	0	8
France	1975	180.00	3.42	27.00	0.51	16	20	44	19
German Dem Rep	1975	26.20	1.55	8.30	0.49	10	77	0	13
Germany, Fed Rep	1975	160.00	2.59	33.20	0.54	10	35	55	0
Greece	1975	55.00	6.08	4.28	0.47	14	2	1	84
Hungary	1972	120.00	11.59	5.60	0.54	9	34	9	48
Ireland	1972	43.70	14.79	1.68	0.57	11	14	69	6
Italy	1970	167.00	3.12	36.00	0.67	19	4	5	71
Luxembourg	1973	3.36	9.28	0.06	0.17	47	50	0	3
Malta	1977	0.03	0.07	0.02	0.06	X	X	X	X
Netherlands	1972	90.50	6.94	14.41	1.11	5	24	40	32
Norway	1972	383.00	98.79	1.40	0.36	14	84	0	2
Poland	1976	58.80	1.73	15.90	0.47	14	21	40	25
Portugal	1977	87.50	9.11	6.46	0.67	7	38	0	55
Romania	1971	192.00	9.43	X	X	15	85	0	0
Spain	1975	110.00	3.09	25.00	0.70	7	5	17	72
Sweden	1975	183.00	22.34	5.00	0.61	50	50	0	0
Switzerland	1973	50.00	7.81	2.50	0.39	44	48	0	8
United Kingdom	1972	162.70	2.93	17.66	0.32	23	41	35	1
Yugoslavia	1976	244.00	11.43	8.88	0.42	16	38	39	7
USSR	1975	4714.00	18.60	226.00	0.89	8	15	14	63
OCEANIA									
Australia	1975	343.00	25.17	16.90	1.24	X	X	X	X
Fiji	X	X	X	X	X	X	X	X	X
New Zealand	1967–68	397.00	140.78	1.01	0.36	52	11	23	14
Papua New Guinea	X	X	X	X	X	X	X	X	X
Solomon Islands	X	X	X	X	X	X	X	X	X

0 = zero or less than one-half the unit of measure; X = not available.
Source: World Resources Institute and the International Institute for Environment and Development, World Resources 1986.
 Original source: Bureau of Geological and Mining Research, National Geological Survey, France

SECTION C. PUBLIC WATER SUPPLY—UNITED STATES

TABLE 5–14. PUBLIC WATER-SUPPLY SYSTEMS IN THE UNITED STATES, 1985

	Very Small		Small			Medium		Large				Very Large	TOTAL
Population Served	**25–100**	**101–500**	**501–1,000**	**1,001–2,500**	**2,501–3,300**	**3,301–5,000**	**5,001–10,000**	**10,001–25,000**	**25,001–50,000**	**50,001–75,000**	**75,001–100,000**	**Over 100,000**	
Number of systems	19,717	18,321	6,254	6,432	1,432	1,980	2,128	1,782	697	230	94	279	59,346
Percent of systems	33.2	30.9	10.5	10.8	2.4	3.3	3.6	3.0	1.2	0.4	0.2	0.5	100.0
Principal source													
Ground water	18,182	15,858	4,881	4,380	894	1,114	1,164	823	273	70	17	44	47,700
Surface water	1,535	2,463	1,373	2,052	538	794	964	959	424	160	78	235	11,151
Ownership													
Public	1,440	7,199	4,944	5,325	1,216	1,644	1,771	1,491	507	187	79	209	25,803
Private	17,986	10,800	1,236	1,064	0	330	356	290	190	43	15	47	32,357
Indian	291	322	74	43	6	6	1	1	—	—	—	—	744
Approximate population served (Millions)	1.089	4.550	4.646	10.496	4.158	7.831	15.351	28.496	25.034	14.063	8.099	95.409	219.22
(Percent)	0.5	2.1	2.1	4.8	1.9	3.6	7.0	13.0	11.4	6.4	3.7	43.5	100.0

Source: U.S. Environmental Protection Agency

FIGURE 5–6. NUMBER AND TYPE OF PUBLIC WATER SYSTEMS IN THE UNITED STATES

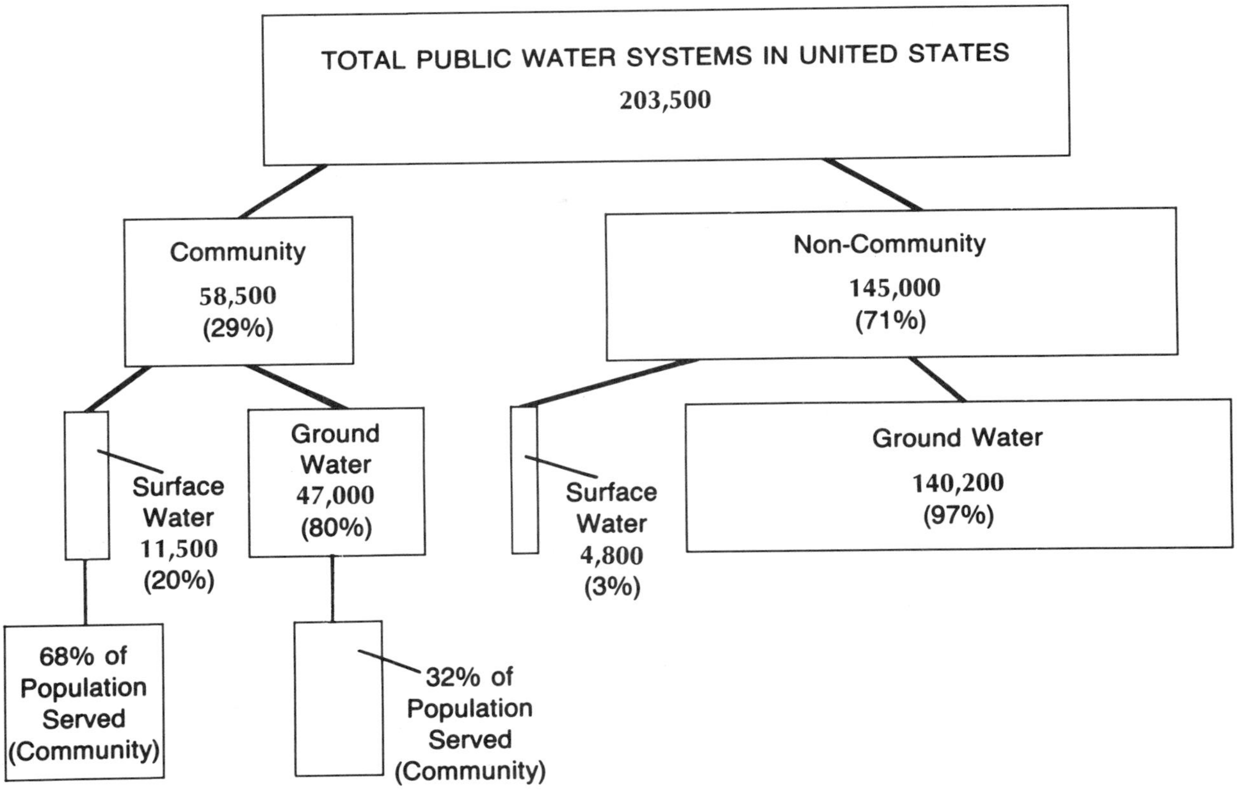

Source: American Water Works Association, 1988, New Dimensions in Safe Drinking Water.

FIGURE 5–7. SIZE DISTRIBUTION OF COMMUNITY WATER SYSTEMS IN THE UNITED STATES

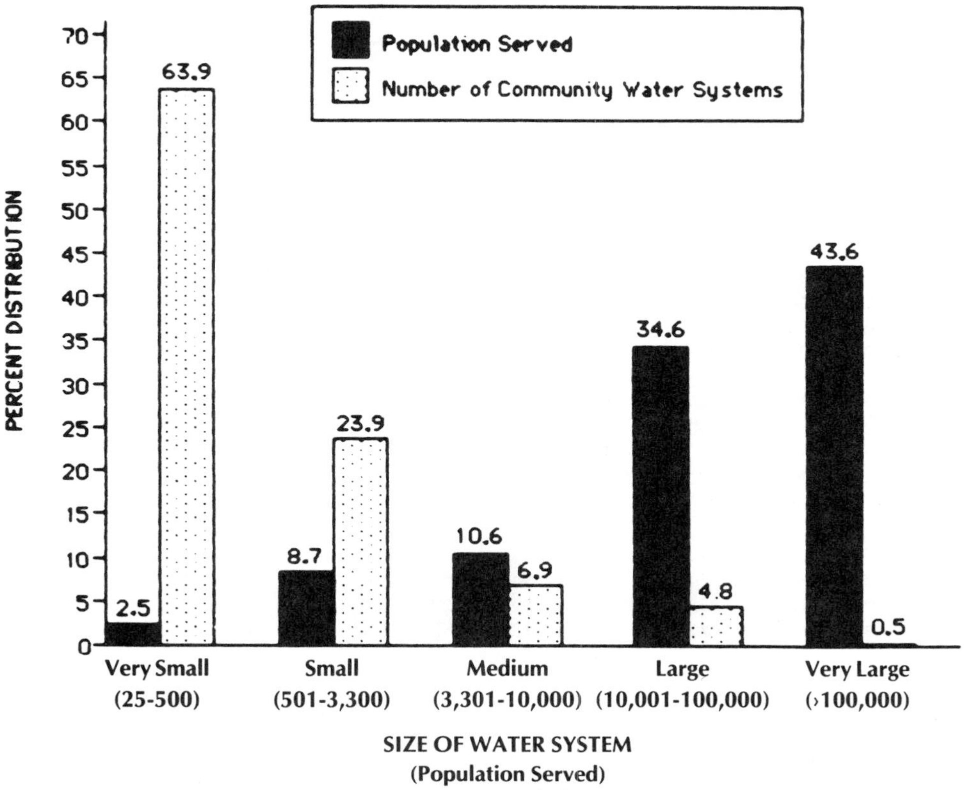

Source: American Water Works Association, 1988, New Dimensions in Safe Drinking Water. Copyright AWWA. Reprinted with permission

TABLE 5–15. PUBLIC-SUPPLY FRESHWATER USE IN THE UNITED STATES, 1985

[Water withdrawn by public and private water suppliers and delivered to multiple users for domestic, commercial, industrial, and thermoelectric power uses; water systems furnish water to at least 25 people or have a minimum of 15 hookups; by state; figures may not add to totals because of independent rounding. Mgal/d = million gallons per day; gal/d = gallons per day]

State	Population Served, in thousands			Water Withdrawals, in Mgal/d			Water Deliveries, by Type of Use, in Mgal/d					Per Capita Use, in gal/d
	Source			Source								
	Ground Water	Surface Water	Total	Ground Water	Surface Water	Total	Domestic	Commercial	Industrial	Thermoelectric Power	Public Use and Losses[1]	
Alabama	1240	2270	3510	173	442	615	332	61	221	.0	.0	175
Alaska	231	117	348	41	35	76	30	31	7.8	.3	7.6	217
Arizona	1930	1160	3090	385	233	618	449	90	79	.0	.0	200
Arkansas	801	880	1680	101	156	257	170	87	.1	.3	.0	153
California	16300	8000	24300	3730	1570	5310	3240	1220	494	31	325	218
Colorado	447	2560	3010	86	651	737	456	112	18	13	138	245
Connecticut	518	2170	2680	66	296	362	178	49	62	1.3	72	135
Delaware	274	240	514	29	49	77	36	12	18	.6	11	151
D.C.	.0	626	626	.0	218	218	174	44	.0	.0	.0	348
Florida	8680	1060	9740	1490	185	1680	1200	248	142	4.8	84	172
Georgia	1460	3200	4660	205	631	836	545	142	135	.0	14	179
Hawaii	1050	80	1130	172	31	204	132	51	6.4	.0	15	181
Idaho	611	92	704	185	27	212	200	6.1	6.7	.0	.0	302
Illinois	3850	5980	9830	467	1320	1780	850	471	255	.9	206	181
Indiana	1790	1880	3670	271	304	575	423	78	73	.0	.8	157

TABLE 5–15. PUBLIC-SUPPLY FRESHWATER USE IN THE UNITED STATES, 1985 (continued)

[Water withdrawn by public and private water suppliers and delivered to multiple users for domestic, commercial, industrial, and thermoelectric power uses; water systems furnish water to at least 25 people or have a minimum of 15 hookups; by state; figures may not add to totals because of independent rounding. Mgal/d = million gallons per day; gal/d = gallons per day]

State	Population Served, in thousands			Water Withdrawals, in Mgal/d			Water Deliveries, by Type of Use, in Mgal/d					Per Capita Use, in gal/d
	Source			Source								
	Ground Water	Surface Water	Total	Ground Water	Surface Water	Total	Domestic	Commercial	Industrial	Thermoelectric Power	Public Use and Losses[1]	
Iowa	1570	557	2130	259	92	350	289	4.2	41	1.6	14	164
Kansas	994	1000	2000	158	158	316	150	83	41	1.0	41	158
Kentucky	309	2460	2770	49	356	404	179	19	167	.0	40	146
Louisiana	1960	1940	3900	276	352	629	564	7.6	1.5	.0	55	161
Maine	216	613	829	24	84	108	96	1.7	11	.0	.2	130
Maryland	619	2950	3560	70	702	771	365	57	55	.0	294	216
Massachusetts	1610	3720	5330	181	586	767	415	276	69	4.2	3.1	144
Michigan	1400	5970	7370	222	1030	1250	630	339	247	.0	36	170
Minnesota	1850	848	2700	265	208	473	401	23	46	1.5	1.5	175
Mississippi	2070	189	2260	275	37	312	165	47	28	1.5	71	138
Missouri	1470	2670	4140	171	474	645	355	60	133	.3	97	156
Montana	228	386	614	62	96	158	90	29	1.2	.0	38	257
Nebraska	1170	159	1320	208	39	248	149	50	49	.0	.0	187
Nevada	303	579	882	94	193	288	189	54	6.3	2.4	36	326
New Hampshire	208	429	637	28	61	89	63	9.1	16	.0	.0	139
New Jersey	2850	3860	6710	406	641	1050	503	136	240	.6	167	156
New Mexico	857	147	1000	198	28	226	179	42	2.8	.0	2.8	225
New York	4170	11700	15900	535	2330	2860	1470	282	1010	.0	95	180
North Carolina	803	2650	3450	88	507	595	315	137	128	.0	17	172
North Dakota	256	256	512	30	39	69	40	14	2.3	.0	13	135
Ohio	2840	6060	8900	395	1020	1420	467	326	340	.3	283	159
Oklahoma	690	2140	2830	106	414	521	158	58	204	2.2	98	184
Oregon	418	1520	1940	83	332	416	246	45	53	.0	72	214
Pennsylvania	1320	6850	8170	258	1340	1600	539	186	246	.0	629	196
Rhode Island	152	732	884	15	101	116	59	15	20	.0	23	132
South Carolina	614	1910	2520	76	283	359	189	5.7	86	.0	79	142
South Dakota	431	117	548	65	16	80	61	14	5.4	.0	.0	147
Tennessee	1340	2320	3660	243	384	627	303	163	98	.6	63	172
Texas	6890	8510	15400	1230	1760	2990	2200	105	284	23	373	194
Utah	1020	553	1570	299	148	447	340	50	15	.0	43	285
Vermont	106	237	343	17	36	53	34	5.2	13	.0	.0	154
Virginia	598	3620	4210	75	504	579	337	70	57	.0	115	137
Washington	1480	2040	3530	339	616	955	516	133	306	.0	.0	271
West Virginia	363	947	1310	37	114	151	81	21	22	.0	27	115
Wisconsin	1700	1430	3130	275	301	575	169	98	153	.2	155	184
Wyoming	146	183	329	48	50	98	61	14	3.2	1.9	18	298
Puerto Rico	544	2390	2930	84	307	391	178	30	12	1.4	170	133
Virgin Islands	3.5	44	47	.2	4.3	4.5	2.1	.2	.0	.7	1.5	95
Total	84,800	115,000	200,000	14,600	21,900	36,500	21,000	5,710	5,730	96	4,040	183

[1] Includes transfers from adjacent areas.

Source: Solley, W.B., Merk, C.F., and Pierce, R.R., 1988, Estimated Water Use in the United States in 1985, U.S. Geological Survey Circular 1004

TABLE 5–16. WATER PRODUCED BY MAJOR PUBLIC SUPPLY SYSTEMS IN THE UNITED STATES

[Selected water systems serving a population of 100,000 or more]

	ANNUAL WATER PRODUCED—PURCHASED								
	GROUND WATER Mgal	SURFACE WATER Mgal	PURCHASED Mgal	TOTAL Mgal	AVERAGE PRODUCTION Mgal/d	MAXIMUM DAY Mgal/d	GAL/CAP DAY	TOTAL POPULATION	MAX PLANT CAP/Mgal/d
ALABAMA									
BIRMINGHAM		21,316	10,038	31,354	85.90	126.66	115	750,000	155.00
MONTGOMERY	6,570	7,300		13,870	38.00	42.47	190	200,000	52.00
ALASKA									
ANCHORAGE	4,811	3,548		8,359	22.90	32.40	99	230,846	40.50
ARIZONA									
MESA	3,431		12,301	15,732	43.10	69.30	208	207,514	170.00
PHOENIX/ARIZONA WTR CO	5,234			5,234	14.34		106	135,000	50.00
PHOENIX/PHOENIX WTR & SWR DEPT	22,564	61,138		83,702	229.32		244	941,000	508.00
TEMPE	1,073	11,742		12,815	35.11	60.91	251	140,000	98.00
TUCSON	25,908			25,908	70.98	106.80	149	475,000	164.60
ARKANSAS									
FORT SMITH		7,088		7,088	19.42		194	100,000	35.00
LITTLE ROCK		16,987		16,987	46.54	72.80	140	332,984	148.00
N LITTLE ROCK			6,015	6,015	16.48	24.00	122	135,000	
CALIFORNIA									
ANAHEIM	9,866		11,713	21,579	59.12	87.90	247	239,519	93.30
CONCORD/CONTRA COSTA WTR DEPT		39,603		39,603	108.50	60.00	36	3,000,000	74.00
CORTE MADERA/MARIN MUNIC WTR DIST		9,472	1,497	10,968	30.05	47.50	177	170,000	46.00
ESCONDIDO		8,081	6,380	14,461	39.62	63.00	360	110,000	55.00
FREMONT/ALAMEDA CNTY WTR DIST	5,585	3,103	4,307	12,994	35.60	62.00	157	226,157	70.80
FRESNO	28,324			28,324	77.60	146.10	307	252,600	267.10
FULLERTON	5,110		5,475	10,585	29.00	52.00	274	106,000	24.00
GARDEN GROVE	6,607		3,614	10,220	28.00	39.60	217	128,879	34.70
GLENDALE	1,836		8,107	9,943	27.24	35.50	189	144,500	23.00
LA MESA/HELIX WTR DIST	73	2,847	10,439	13,359	36.60	59.64	174	210,346	67.00
LA PUENTE/SOUTHWEST SUBURBAN WTR	11,936		4,271	16,206	44.40	73.20	133	333,000	51.30
LONG BEACH/DOMINGUEZ WTR CORP	3,201		9,297	12,498	34.24	46.00	342	100,000	60.00
LOS ANGELES/LOS ANGELES WTR & POWER	37,595	172,645	9,125	219,365	601.00	800.00	192	3,131,000	600.00
MODESTO	11,388			11,388	31.20	68.60	312	100,000	68.60
OAKLAND/EAST BAY MUNIC UTIL DIST		64,970		64,970	178.00	328.00	165	1,080,000	502.00
OXNARD	51		6,690	6,742	18.47	25.43	154	120,000	164.00
PALM SPRINGS/DESERT WTR AGENCY	10,687	1,347		12,034	32.97	49.54	330	100,000	64.80
PASADENA	5,767	1,387	5,183	12,337	33.80	63.40	360	130,000	33.50
POMONA	5,329	1,205	2,263	8,797	24.10	38.70	236	102,257	34.60
RIVERSIDE	17,703		657	18,360	50.30	71.30	269	187,000	102.00
SAN BERNARDINO	9,957	1,745		11,702	32.06	53.56	245	131,030	98.00
SAN DIEGO		27,397	37,299	64,696	177.25	270.27	184	962,200	275.00
SAN JOSE/CALIFORNIA WTR SERV CO	44,056	621	49,348	94,024	257.60	462.50	199	1,295,200	1100.00
SAN JOSE/SAN JOSE WTR WKS	28,835	3,650	14,965	47,450	130.00	221.00	183	710,000	195.00
SANTA BARBARA	712	7,570		8,282	22.69	29.11	214	106,121	41.70
SUNNYVALE	2,774		7,227	10,001	27.40	55.00	249	110,000	

TABLE 5–16. WATER PRODUCED BY MAJOR PUBLIC SUPPLY SYSTEMS IN THE UNITED STATES (continued)

[Selected water systems serving a population of 100,000 or more]

	ANNUAL WATER PRODUCED—PURCHASED								
	GROUND WATER Mgal	SURFACE WATER Mgal	PURCHASED Mgal	TOTAL Mgal	AVERAGE PRODUCTION Mgal/d	MAXIMUM DAY Mgal/d	GAL/CAP DAY	TOTAL POPULATION	MAX PLANT CAP/Mgal/d
COLORADO									
AURORA	949	15,695	186	16,830	46.11	79.00	220	210,000	79.00
COLORADO SPGS	511	19,710		20,221	55.40	128.10	202	274,700	135.00
DENVER		97,090		97,090	266.00	485.00	303	877,100	625.00
PUEBLO	80	19,345		19,425	53.22	52.00	515	103,392	80.00
CONNECTICUT									
BRIDGEPORT	1,566	20,316		21,882	59.95	79.71	163	367,000	212.00
CLINTON/CONNECTICUT WTR CO	2,394	3,424	29	5,847	16.02	22.80	102	157,000	29.20
GREENWICH/CONNECTICUT-AMERICAN WTR	146	5,913	657	6,716	18.40	23.59	138	133,198	40.60
HARTFORD		21,396		21,396	58.62	96.90	147	400,000	
NEW BRITAIN		4,015		4,015	11.00	14.80	101	109,337	30.00
NEW HAVEN/REGIONAL WTR AUTH	1,832	17,053		18,885	51.74	70.48	140	370,000	153.00
DELAWARE									
NEWARK/ARTESIAN WTR CO	4,015		365	4,380	12.00	15.40	80	150,000	15.00
DISTRICT OF COLUMBIA									
WASHINGTON/WTR & SWR UTIL ADMIN		67,963		67,963	186.20	214.45	292	638,300	
WASHINGTON/WASHINGTON AQUEDUCT		75,920		75,920	208.00	262.33	189	1,100,000	366.00
FLORIDA									
CLEARWATER/PINELLAS CNTY WTR SYS	22,502			22,502	61.65	92.00	114	540,000	
COCOA	5,997			5,997	16.43	30.00	164	100,000	35.00
FORT LAUDERDALE	16,425			16,425	45.00	59.00	188	240,000	90.00
GAINESVILLE	6,570			6,570	18.00	21.20	162	111,000	45.00
HIALEAH			10,512	10,512	28.80		213	135,000	
HOLLYWOOD	6,490			6,490	17.78	17.39	147	121,323	40.00
JACKSONVILLE	20,075		91	20,166	55.25	97.00	137	402,500	206.00
LAKELAND	8,395			8,395	23.00	40.70	207	111,000	51.00
MIAMI	122,640			122,640	336.00	295.90	238	1,409,000	358.00
ORLANDO	20,586			20,586	56.40	92.60	179	315,000	344.00
POMPANO BEACH	9,059			9,059	24.82		138	180,000	81.00
TALLAHASSEE	6,570			6,570	18.00	33.70	133	135,000	48.80
TAMPA	5,475	16,425		21,900	60.00	93.00	150	400,000	140.00
W. PALM BEACH/PALM BEACH CNTY UTIL	5,092		33	5,125	14.04	19.80	109	129,170	39.24
GEORGIA									
ACWORTH/WYCKOFF TREATMENT DIV		10,476		10,476	28.70	41.38	144	200,000	45.00
COLUMBUS		10,950		10,950	30.00	54.40	194	155,000	54.00
MARIETTA/QUARLES TREATMENT DIV		11,315		11,315	31.00	50.10	155	200,000	48.00
MORROW/CLAYTON CNTY WTR AUTH		5,840	438	6,278	17.20	23.80	123	139,532	19.50
STONE MOUNTAIN/DEKALB CNTY WTR		23,360		23,360	64.00	96.00	138	465,000	128.00

HAWAII									
HONOLULU	45,808			45,808	125.50	171.20	170	737,000	309.00
IDAHO									
BOISE		8,990		8,990	24.63	64.87	199	124,000	74.08
ILLINOIS									
ALTON/ALTON DIST/ILL-AMER WTR						12.89			
CHAMPAIGN/NORTHERN ILLINOIS WTR	5,968			5,968	16.35	23.35	154	106,000	32.45
EAST ST. LOUIS/INTERURBAN DIST I		19,706		19,706	53.99	65.66	182	296,783	69.30
EVANSTON		9,512		9,512	26.06		194	134,033	
PEORIA/PEORIA DIST/ILL-AMER WTR	4,081	2,694		6,774	18.56	27.02	109	170,900	45.00
SPRINGFIELD		7,483		7,483	20.50	31.19	145	141,631	38.00
INDIANA									
FORT WAYNE		11,133		11,133	30.50	51.30	135	226,200	72.00
GARY		11,936		11,936	32.70	48.70	121	270,000	76.00
HAMMOND		9,490		9,490	26.00	33.00	112	232,000	50.00
INDIANAPOLIS	365	40,150		40,515	111.00	143.53	157	706,500	182.00
SOUTH BEND	8,395			8,395	23.00	43.00	192	120,000	72.00
IOWA									
CEDAR RAPIDS	8,760			8,760	24.00	33.50	218	110,300	52.50
DAVENPORT/DAVENPORT WTR CO		6,935		6,935	19.00	26.60	146	130,000	30.00
DES MOINES	5,475	7,300		12,775	35.00	54.92	135	259,868	96.00
KANSAS									
KANSAS CITY		10,687		10,687	29.28	46.30	182	161,000	60.00
MISSION/JOHNSON CNTY WTR DIST 1	3,395	7,337		10,731	29.40	60.92	133	221,725	80.00
TOPEKA		8,322		8,322	22.80	44.21	160	142,900	65.00
WICHITA	11,607	7,446		19,053	52.20	95.08	172	303,070	84.00
KENTUCKY									
EDGEWOOD/KENTON CNTY WTR DIST		7,264		7,264	19.90	30.10	133	150,000	45.00
LEXINGTON/KENTUCKY-AMERICAN WTR		12,027		12,027	32.95	49.07	161	204,165	60.00
LOUISVILLE		43,800		43,800	120.00	168.00	179	671,391	300.00
LOUISIANA									
BATON ROUGE/BATON ROUGE WTR WKS	16,633			16,633	45.57	56.04	132	346,400	71.00
JEFFERSON		21,900		21,900	60.00	59.15	218	274,622	70.00
LAFAYETTE	5,475			5,475	15.00	23.00	125	120,000	30.50
MARRERO/JEFFERSON WTR WKS DIST 2		10,950		10,950	30.00	27.11	167	179,970	44.00
NEW ORLEANS		47,585		47,585	130.37	170.00	234	557,515	230.00
MAINE									
PORTLAND	329	7,483		7,811	21.40	30.00	135	158,000	54.00
MARYLAND									
BALTIMORE		96,532		96,532	264.47	340.38	175	1,512,100	465.00
COLUMBIA/HOWARD CNTY DEPT PUB WK			4,490	4,490	12.30	14.20	121	102,000	
GLEN BURNIE/ANNE ARUNDEL CNTY	8,395		730	9,125	25.00		157	159,400	
HYATTSVILLE/WASHINGTON SUBURBAN		52,925		52,925	145.00	199.00	121	1,200,000	225.00

TABLE 5–16. WATER PRODUCED BY MAJOR PUBLIC SUPPLY SYSTEMS IN THE UNITED STATES (continued)

[Selected water systems serving a population of 100,000 or more]

	ANNUAL WATER PRODUCED—PURCHASED								
	GROUND WATER Mgal	SURFACE WATER Mgal	PURCHASED Mgal	TOTAL Mgal	AVERAGE PRODUCTION Mgal/d	MAXIMUM DAY Mgal/d	GAL/CAP DAY	TOTAL POPULATION	MAX PLANT CAP/Mgal/d
MASSACHUSETTS									
SPRINGFIELD		14,418		14,418	39.50	54.60	152	260,370	100.00
MICHIGAN									
ANN ARBOR	1,190	4,595		5,785	15.85	29.05	147	108,000	50.00
DETROIT		237,177		237,177	649.80	1069.60	172	3,769,800	1610.00
GRAND RAPIDS		18,016		18,016	49.36	78.33	188	263,000	100.00
HOLLAND/WYOMING WTR TRTMNT PLANT		7,665		7,665	21.00	61.90	150	140,000	64.00
KALAMAZOO	6,205			6,205	17.00	.53	131	130,000	
LANSING	8,432			8,432	23.10	26.00	157	147,000	50.00
LIVONIA			7,264	7,264	19.90	37.80	197	101,000	
SAGINAW		10,147		10,147	27.80	41.80	146	190,000	54.00
STERLING HTS			5,840	5,840	16.00	30.90	145	110,000	
WARREN			9,132	9,132	25.02		155	161,134	
MINNESOTA									
DULUTH		5,950		5,950	16.30	25.00	148	110,000	30.00
ST. PAUL	1,898	16,425		18,323	50.20	98.40	129	390,000	144.00
MISSISSIPPI									
JACKSON	1,095	10,950		12,045	33.00	37.80	147	224,000	60.00
MISSOURI									
INDEPENDENCE/MISSOURI WTR CO	8,293			8,293	22.72	31.87	91	250,000	36.00
KANSAS CITY/KANSAS CITY, MO WTR	4,417	44,968	37	49,421	135.40	162.80	238	569,827	300.00
SPRINGFIELD	256	6,935		7,191	19.70	29.90	180	109,200	57.60
ST. JOSEPH/ST. JOSEPH DIST/MISS		6,570		6,570	18.00	22.97	174	103,200	38.00
ST. LOUIS/ST. LOUIS WATER DIV		59,495		59,495	163.00	218.90	353	462,000	460.00
ST. LOUIS/ST. LOUIS CNTY WTR CO		54,750		54,750	150.00	255.83	150	1,000,000	325.00
NEBRASKA									
LINCOLN	11,279			11,279	30.90	74.50	172	180,000	77.00
OMAHA	15,038	23,725		38,763	106.20	182.00	206	515,000	218.00
NEVADA									
LAS VEGAS	12,812	36,902		49,713	136.20	218.50	300	453,300	350.00
RENO/SIERRA PACIFIC POWER CO	2,957	15,586		18,542	50.80	102.00	299	170,000	112.00
NEW HAMPSHIRE									
MANCHESTER		5,015		5,015	13.74	20.75	119	115,000	40.00

NEW JERSEY									
CLIFTON/PASSAIC VALLEY WTR COMM		26,944		26,944	73.82	93.60	123	600,000	135.00
ELIZABETH/ELIZABETHTOWN WTR CO	9,114	35,387	88	44,588	122.16	147.20	106	1,150,000	230.00
ELIZABETH/ELIZABETH WTR & SWR			5,274	5,274	14.45	16.38	136	106,000	
HADDON HEIGHTS/NEW JERSEY WTR CO	9,340			9,340	25.59	45.82	90	285,430	49.90
HARRINGTON PARK/HACKENSACK WTR CO	5,329	37,157	73	42,559	116.60	171.30	117	1,000,000	219.00
ISELIN/MIDDLESEX WTR CO	2,081	5,621	1,460	9,162	25.10	35.17	135	186,000	54.00
SHORT HILLS/COMMONWEALTH WTR CO	3,522	2,942	6,143	12,607	34.54	46.76	173	200,080	34.20
SHREWSBURY/MONMOUTH CONSOLIDATED	646	10,293	4	10,943	29.98	46.76	120	249,000	70.42
WANAQUE/NORTH NJ DIST WTR SUPPLY		29,857		29,857	81.80		126	650,000	
NEW MEXICO									
ALBUQUERQUE	32,668			32,668	89.50	157.00	231	386,895	247.00
NEW YORK									
ALBANY		7,665		7,665	21.00	28.00	175	120,000	40.00
BUFFALO/ERIE CNTY WTR AUTH		24,200		24,200	66.30		171	386,948	138.00
EAST MEADOW/HEMPSTEAD WTR DEPT	6,315			6,315	17.30	35.90	115	150,000	72.20
LAKE SUCCESS/JAMAICA WTR CO	24,601		2,774	27,375	75.00	105.60	115	650,000	
LYNBROOK/LONG ISLAND WTR CORP	10,242			10,242	28.06	52.25	120	234,350	70.00
MAMARONECK/WESTCHESTER JOINT WTR			4,325	4,325	11.85	20.02	58	204,733	37.00
NEW YORK/NEW YORK BUR WTR SUPPLY		556,990		556,990	1,526.00	1787.00	216	7,071,300	
OAKDALE/SUFFOLK WTR AUTH	42,340			42,340	116.00	262.00	136	853,000	560.00
ROCHESTER/BUREAU OF WTR		13,593	4,285	17,878	48.98	91.00	126	390,000	48.00
ROCHESTER/MONROE CNTY WTR AUTH		20,075	7,337	27,412	75.10	88.00	162	465,000	140.00
SYRACUSE		18,454	219	18,673	51.16	63.86	218	235,000	67.00
UTICA		7,300		7,300	20.00	26.50	164	122,000	50.00
WEST NYACK/SPRING VALLEY WTR CO	5,329	37,157	73	42,559	116.60	171.30	117	1,000,000	219.00
YONKERS		10,585	4,015	14,600	40.00	43.81	205	195,000	158.00
NORTH CAROLINA									
CHARLOTTE		18,250		18,250	50.00	68.00	130	385,000	
DURHAM		6,975		6,975	19.11	25.04	137	140,000	34.00
GREENSBORO		10,038		10,038	27.50	35.00	183	150,000	50.00
RALEIGH		10,585		10,585	29.00	38.26	112	259,000	75.50
WINSTON-SALEM/CITY/CNTY UTIL COM		13,140		13,140	36.00	40.90	176	205,000	63.00
OHIO									
AKRON		16,878		16,878	46.24	62.35	149	310,000	100.00
CINCINNATI	5,581	44,249		49,830	136.52	185.55	170	803,000	275.00
CLEVELAND		124,830		124,830	342.00	481.70	221	1,548,724	545.00
DAYTON	22,995			22,995	63.00	84.00	210	300,075	144.00
POLAND/OHIO WTR SERVICE CO	2,555	5,913		8,468	23.20	30.00	116	200,000	43.00
TOLEDO		27,813		27,813	76.20	120.00	169	451,000	180.00
YOUNGSTOWN			10,950	10,950	30.00	30.00	200	150,000	50.00
OKLAHOMA									
OKLAHOMA CITY	26	31,536		31,562	86.47	137.50	104	834,088	163.00
TULSA		42,840		42,840	117.37	155.00	276	426,000	195.00

TABLE 5–16. WATER PRODUCED BY MAJOR PUBLIC SUPPLY SYSTEMS IN THE UNITED STATES (continued)

[Selected water systems serving a population of 100,000 or more]

	ANNUAL WATER PRODUCED—PURCHASED								
	GROUND WATER Mgal	SURFACE WATER Mgal	PURCHASED Mgal	TOTAL Mgal	AVERAGE PRODUCTION Mgal	MAXIMUM DAY Mgal/d	GAL/CAP DAY	TOTAL POPULATION	MAX PLANT CAP/Mgal/d
OREGON									
EUGENE		8,979		8,979	24.60	52.00	176	140,000	90.00
PORTLAND		40,880		40,880	112.00	209.00	160	700,700	255.00
PENNSYLVANIA									
ALLENTOWN	3,927	5,512		9,439	25.86	34.42	185	140,000	42.00
BRYN MAWR/PHILADELPHIA SUBURBAN	4,515	24,897	1,960	31,372	85.95	101.52	109	787,000	120.00
CHESTER		9,735		9,735	26.67	30.74	186	143,106	45.00
GREENSBURG/MUN AUTH OF WESTMORELND		22,740		22,740	62.30	43.80	312	200,000	60.00
HERSHEY/KEYSTONE WTR CO	2,091	10,136	281	12,509	34.27	45.58	114	300,000	65.37
LANCASTER		6,033		6,033	16.53	25.40	153	108,000	40.00
PHILADELPHIA		133,006		133,006	364.40	478.20	201	1,813,210	681.00
PITTSBURGH/WESTERN PENNSYLVANIA	1,730	37,861	1,164	40,756	111.66	141.89	91	1,225,000	172.23
PITTSBURGH/WEST VIEW MUNIC AUTH		5,840		5,840	16.00	26.00	80	200,000	40.00
YORK		6,205		6,205	17.00	19.70	133	128,200	29.90
RHODE ISLAND									
PAWTUCKET	1,533	4,198		5,731	15.70	21.00	150	105,000	36.00
SOUTH CAROLINA									
ANDERSON		4,223		4,223	11.57	16.26	116	100,000	24.00
CHARLESTON		16,900		16,900	46.30	65.60	123	375,000	113.50
SPARTANBURG		10,278		10,278	28.16	38.71	171	165,000	64.00
TENNESSEE									
CHATTANOOGA/TENNESSEE-AMERICAN WTR		13,990	4	13,994	38.34	50.59	160	239,339	72.00
KNOXVILLE/KNOXVILLE UTIL BRD		11,315		11,315	31.00	41.00	159	195,000	50.00
MEMPHIS	46,611			46,611	127.70	176.30	177	722,403	200.00
NASHVILLE/METRO DEPT WTR & SWR		26,828		26,828	73.50	98.00	239	307,000	141.00
TEXAS									
ABILENE		8,760		8,760	24.00	42.00	222	108,000	53.00
AMARILLO	7,037	7,574		14,611	40.03	84.00	253	158,500	131.50
ARLINGTON		14,542		14,542	39.84	72.97	166	240,000	136.00
DALLAS		113,011	3,223	116,234	318.45	508.77	228	1,395,800	715.00
EL PASO	26,645	6,570		33,215	91.00	170.00	182	499,000	205.00
FORT WORTH		44,165		44,165	121.00	216.00	156	775,000	250.00
IRVING	1,460		6,570	8,030	22.00	42.00	156	141,000	
LUBBOCK	1,916	11,195		13,111	35.92	60.00	196	183,000	105.00
SAN ANTONIO	62,415			62,415	171.00	255.80	210	812,570	660.60
WACO		3,906		3,906	10.70	44.90	97	110,000	65.00
WICHITA FALLS		9,965		9,965	27.30	44.00	188	145,000	37.70

UTAH									
SALT LAKE CITY/SALT LAKE CITY	2,628	15,330	8,213	26,171	71.70	192.40	172	417,227	140.00
SALT LAKE CITY/METRO WTR DIST		12,906		12,906	35.36	115.00	51	700,000	100.00
SALT LAKE CITY/SALT LAKE CNTY CO	3,219	3,756	7,556	14,531	39.81	116.65	100	400,000	140.00
VIRGINIA									
ALEXANDRIA/VIRGINIA-AMERICAN WTR			5,296	5,296	14.51	18.71	133	109,100	35.10
ARLINGTON			8,724	8,724	23.90	28.00	149	160,000	
CHESTERFIELD/CHESTERFIELD CNTY	15	3,701	1,854	5,570	15.26	21.10	91	168,300	24.00
FALLS CHURCH			5,585	5,585	15.30	23.90	133	115,000	
MERRIFIELD/FAIRFAX CNTY WTR AUTH	303	27,864	617	28,784	78.86	119.40	109	722,000	162.60
NEWPORT NEWS		15,695		15,695	43.00	57.00	130	330,000	66.00
NORFOLK	2,446	20,878		23,324	63.90	85.10	101	630,000	128.00
RICHMOND/CNTY OF HENRICO	1,022		6,680	7,702	21.10	27.50	117	180,200	
RICHMOND/CITY OF RICHMOND		21,170		21,170	58.00	72.00	145	400,000	84.00
VIRGINIA BEACH			8,687	8,687	23.80	32.40	113	210,000	
WOODBRIDGE/PRINCE WILLIAM CNTY	767		2,154	2,920	8.00	11.60	67	120,000	13.50
WASHINGTON									
BELLEVUE			5,475	5,475	15.00	35.00	143	105,000	
EVERETT		18,626		18,626	51.03		232	220,000	
LYNNWOOD/ALDERWOOD WTR DIST			5,256	5,256	14.40	30.60	118	122,000	30.00
SEATTLE/SEATTLE WTR DEPT		62,744		62,744	171.90	338.00	158	1,090,000	
SPOKANE	22,630			22,630	62.00	160.00	350	177,100	280.00
TACOMA	2,942	24,700		27,641	75.73	111.48	379	200,000	132.00
VANCOUVER	6,716			6,716	18.40	33.70	175	105,000	73.50
WEST VIRGINIA									
CHARLESTON/W VA WTR CO - KANAWHA		9,928		9,928	27.20	35.24	150	181,434	
CLARKSBURG					6.80				
FAIRMONT									
HUNTINGTON		5,694		5,694	15.60	18.40	142	110,000	20.00
WISCONSIN									
MADISON	10,778			10,778	29.53	43.97	163	181,000	66.80
MILWAUKEE		51,830		51,830	142.00	210.00	168	844,000	400.00
RACINE		8,191		8,191	22.44	35.74	195	115,000	60.00

Source: American Water Works Association, 1984 Water Utility Operating Data. Copyright AWWA. Reprinted with permission

TABLE 5–17. AVERAGE CHARGE FOR WATER SUPPLIED BY PUBLIC WATER SYSTEMS IN THE UNITED STATES

[Average charge for four use levels by USEPA region; for map of EPA regions, see Fig. 10-2]

USEPA Region	3750 gal	7500 gal	75,000 gal	750,000 gal
I	$8.64	$13.47	$112.69	$ 913.40
II	7.26	9.97	82.80	660.34
III	7.91	10.85	119.41	1098.36
IV	5.45	10.09	77.30	707.06
V	7.16	11.51	96.17	839.11
VI	7.23	11.62	93.37	805.54
VII	6.15	10.54	82.05	712.71
VIII	6.14	9.30	77.48	680.15
IX	5.69	8.72	71.28	663.32
X	6.13	9.40	71.56	663.01
National average	$6.89	$11.14	$ 90.44	$ 806.50

Source: American Water Works Association, 1988. Grigg, N.S., 1984 Water Utility Operating Data, Summary Report. Reprinted with permission. Based on USEPA survey of 58,470 systems serving a population of 59,071,000

TABLE 5–18. COST OF RESIDENTIAL WATER IN THE UNITED STATES

[Average cost per 1000 gallons in 1984; 368 water utilities reporting]

ALABAMA	1.18
ALASKA	1.72
ARIZONA	1.72
ARKANSAS	1.46
CALIFORNIA	1.04
COLORADO	1.27
CONNECTICUT	2.42
DELAWARE	0.28
FLORIDA	1.11
GEORGIA	1.08
HAWAII	1.02
IDAHO	.86
ILLINOIS	1.97
INDIANA	1.68
IOWA	1.32
KANSAS	1.94
KENTUCKY	1.25
LOUISIANA	1.51
MAINE	1.30
MARYLAND	.99
MASSACHUSETTS	1.01
MICHIGAN	.91
MINNESOTA	1.16
MISSISSIPPI	.99
MISSOURI	1.77
MONTANA	NA
NEBRASKA	.60
NEVADA	.92

TABLE 5–18. COST OF RESIDENTIAL WATER IN THE UNITED STATES (continued)

[Average cost per 1000 gallons in 1984; 368 water utilities reporting]

NEW HAMPSHIRE	1.56
NEW JERSEY	1.80
NEW MEXICO	1.69
NEW YORK	1.12
NORTH CAROLINA	.66
NORTH DAKOTA	.96
OHIO	1.62
OKLAHOMA	1.26
OREGON	.87
PENNSYLVANIA	2.29
RHODE ISLAND	1.56
SOUTH CAROLINA	.82
SOUTH DAKOTA	1.17
TENNESSEE	1.66
TEXAS	1.25
UTAH	.58
VERMONT	2.50
VIRGINIA	1.75
WASHINGTON	1.26
WEST VIRGINIA	2.52
WISCONSIN	1.01
WYOMING	1.41
TOTAL UNITED STATES	1.27

Source: American Water Works Association, 1984 Water Utility Operating Data. Copyright. Reprinted with permission

TABLE 3–19. UNACCOUNTED-FOR WATER IN PUBLIC WATER SUPPLY SYSTEMS IN THE UNITED STATES

[By USEPA Region (See Fig. 10-2)]

Region	Unaccounted-for Water percent	Region	Unaccounted-for Water percent
I	15.1	VI	12.7
II	14.3	VII	10.8
III	13.7	VIII	9.6
IV	14.6	IX	6.4
V	13.5	X	3.3

By Population Served:

Population Served	Unaccounted-for Water percent
10,000–25,000	17.1
25,000–50,000	14.5
50,000–100,000	12.3
100,000–500,000	12.2
500,000–1,000,000	12.8
>1,000,000	17.2

Source: Grigg, N.S., 1988, 1984 Water Utility Operating Data, Copyright American Water Works Association. Reproduced with permission

SECTION D. PUBLIC WATER SUPPLY—WORLD

TABLE 5-20. POPULATION SERVED BY PUBLIC WATER-SUPPLY SYSTEMS IN CANADA, 1986

	Alberta	B.C.	Manitoba	N.B.	NFLD.	N.W.T.	N.S.	Ontario	P.E.I.	Québec	Saskatchewan	Yukon	Canada
Number of communities surveyed	454	152	280	88	323	51	113	448	24	1,490	212	15	3,650
Total population surveyed	1,924,564	2,402,949	845,469	409,900	492,588	48,420	553,993	8,274,231	57,828	6,366,662	630,982	24,576	22,032,162
No. of communities with water distribution network	406	142	197	65	249	51	89	430	9	1,034	200	15	2,887
Population served by water distribution network	1,919,408	2,392,649	817,116	366,836	458,385	48,420	524,585	8 248,484	47,303	5,976,823	628,019	24,576	21,452,604
No. of communities with water treatment	348	92	169	21	208	49	77	386	1	535	194	11	2,901
Population served by water treatment	1,905,571	2,041,491	792,957	256,184	375,116	45,732	514,762	7,443,412	21,000	5,568,235	613,533	20,500	19, 598,493

Source: Ministry of Supply and Services, 1987, National Inventory of Municipal Waterworks and Wastewater Systems in Canada, 1986

TABLE 5–21. POPULATION SUPPLIED WITH SAFE DRINKING WATER AND ADEQUATE SANITARY FACILITIES, BY COUNTRY

Region and Country	Percentage of Population with Access to Safe Drinking Water				Percentage of Population with Access to Adequate Sanitary Facilities			
	Year	Total	Urban	Rural	Year	Total	Urban	Rural
	AFRICA							
ALGERIA	1980	77.0	100.0	70.0	1980	. . .	95.0	70.0
ANGOLA	1983	28.0	90.0	12.0	1983	17.8	29.0	15.0
BENIN	1983	14.4	45.0	9.0	1983	10.2	45.0	4.0
BOTSWANA	1984	77.0	98.0	72.0	1983	36.0	79.0	12.5
BURKINA FASO	1984	35.0	50.0	26.0	1984	8.6	38.0	5.0
BURUNDI	1983	23.4	32.8	22.0	1983	52.0	90.0	25.0
CAMEROON	1985	35.9	46.0	30.0	1985	36.0	. . .	. . .
CAPE VERDE	1983	31.2	99.0	21.0	1983	10.4	36.0	9.0
CENTRAL AFRICAN REPUBLIC	1980	16.0	24.0	5.0	1984	19.0	36.0	9.0
CHAD	1984	31.0	27.0	30.0	1984	14.5	. . .	. . .
COMOROS	1982	58.1	99.0	52.0		. . .	. . .	. . .
CONGO	1985	50.0	42.0	7.0	1985	40.0	17.0	. . .
COTE D'IVOIRE	1980	20.0	30.0	10.0	1980	17.1	13.0	20.0
EQUATORIAL GUINEA		. . .	47.0	. . .	1983	. . .	28.0	. . .
ETHIOPIA		. . .	93.0	42.0	1984	. . .	. . .	5.0
GABON	1983	50.0	75.0	34.3	1983	50.0	. . .	. . .
GAMBIA	1983	45.0	100.0	33.0	1983	77.0	. . .	. . .
GHANA	1983	48.9	72.0	39.0	1983	26.0	47.0	17.0
GUINEA	1984	19.8	91.0	2.0	1980	11.6	54.0	1.0
GUINEA-BISSAU	1984	31.0	21.0	37.0	1984	25.0	21.0[1]	13.0[1]
KENYA	1983	27.0	61.0	21.0	1983	44.4	75.0	39.0
LESOTHO		. . .	37.0	14.0	1980	12.4	22.0	11.0
LIBERIA	1983	37.3	50.0	24.0	1983	21.4	24.0	20.0
MADAGASCAR	1983	20.5	73.0	9.0	1983	. . .	8.0	. . .
MALAWI	1984	65.0	82.0	54.0	1984	55.0	. . .	. . .
MALI	1980	6.0	58.0	19.7	1983	21.2	90.0[3]	5.0[3]
MAURITANIA	1984	37.1	80.0	16.0	1983	. . .	7.0	. . .
MAURITIUS	1984	99.0	100.0	98.0	1984	97.0	100.0	95.0
MOZAMBIQUE	1980	9.2	82.0	2.0	1980	10.3	. . .	. . .
NIGER	1983	36.5	48.0	34.2	1983	8.5	36.0	3.0
NIGERIA	1983	36.3	60.0	30.0	1983	. . .	30.0	. . .
RWANDA	1983	60.0	55.0	60.0	1983	60.0	60.0	60.0
SAO TOME AND PRINCIPE	1981	52.0	. . .	. . .	1983	15.0	. . .	. . .
SENEGAL	1983	43.5	63.0	27.0	1984	. . .	87.0	2.0[1]
SEYCHELLES	1984	95.0	. . .	. . .	1984	99.0	. . .	. . .
SIERRA LEONE	1984	24.0	58.0	8.0	1984	21.0	43.0	10.0
SOUTH AFRICA		. . .	. . .	. . .		. . .	. . .	. . .
SWAZILAND	1984	38.0	. . .	. . .	1980	. . .	62.0	10.0[3]
TOGO	1983	34.8	68.0	26.0	1983	13.5	34.0	8.0
UGANDA	1983	16.0	45.0	12.0	1983	13.0	40.0	10.0
UNITED REPUBLIC OF TANZANIA	1984	52.1	85.0	47.0	1984	78.0	91.0	76.0
ZAIRE		. . .	43.0	5.0	1980	. . .	8.0	10.0[2]
ZAMBIA	1984	48.0	70.0	32.0	1984	47.0	56.0	41.0
ZIMBABWE	1984	52.0	100.0	10.0	1984	26.0	100.0[2]	5.0[2]
	AMERICAS							
ANTIGUA AND BARBUDA	1983	95.0	. . .	. . .	1983	100.0	. . .	. . .
ARGENTINA	1983	67.0	72.0	19.0	1983	84.0	93.0	37.0
BAHAMAS	1983	59.0	59.0	. . .	1983	64.0	64.0	. . .
BARBADOS	1983	52.0	100.0	20.0	1983	100.0	40.0	. . .
BOLIVIA	1983	43.0	78.0	12.0	1983	24.0	41.0	9.0
BRAZIL	1983	75.0	. . .	52.0	1983	24.0	33.0	1.0
CANADA	1982	97.0	. . .	. . .	1982	60.0	. . .	. . .
CHILE	1983	85.0	100.0	18.0	1983	83.0	100.0	10.0
COLOMBIA	1983	91.0	100.0	76.0	1983	68.0	96.0	14.0
COSTA RICA	1983	88.0	93.0	86.0	1983	76.0	100.0	40.0
CUBA	1982	61.2	. . .	. . .	1982	31.0	. . .	. . .
DOMINICA	1980	77.0	. . .	. . .	1980	86.0	. . .	. . .
DOMINICAN REPUBLIC	1983	62.0	85.0	33.0	1983	27.0	41.0	10.0
ECUADOR	1983	59.0	98.0	21.0	1983	45.0	64.0	26.0
EL SALVADOR	1983	55.0	71.0	43.0	1983	41.0	52.0	34.0

TABLE 5–21. POPULATION SUPPLIED WITH SAFE DRINKING WATER AND ADEQUATE SANITARY FACILITIES, BY COUNTRY (continued)

Region and Country	Percentage of Population with Access to Safe Drinking Water				Percentage of Population with Access to Adequate Sanitary Facilities			
	Year	Total	Urban	Rural	Year	Total	Urban	Rural
AMERICAS (continued)								
GRENADA	1983	85.0	...	...		...	...	...
GUATEMALA	1983	51.0	90.0	26.0	1983	36.0	53.0	28.0
GUYANA	1983	80.0	100.0	61.0	1983	90.0	54.0	81.0
HAITI	1983	33.0	73.0	25.0	1983	19.0	54.0	12.0
HONDURAS	1983	69.0	91.0	55.0	1983	44.0	44.0	40.0
JAMAICA	1983	73.0	99.0	93.0	1983	90.0	92.0	90.0
MEXICO	1983	74.0	90.0	40.0	1983	56.0	93.0	12.0
NICARAGUA	1983	56.0	98.0	9.0	1983	28.0	73.0	16.0
PANAMA	1983	62.0	97.0	26.0	1983	66.0	61.0	71.0
PARAGUAY	1983	25.0	46.0	10.0	1983	84.0	92.0	95.0
PERU	1983	52.0	73.0	18.0	1983	35.0	57.0	2.0
SAINT CHRISTOPHER AND NEVIS	1983	75.0	...	...	1985	96.0	...	...
SAINT LUCIA	1979	70.0	...	...	1979	62.0	...	...
SAINT VINCENT AND THE GRENADINES	1981	75.0	...	...	1981	88.0	...	...
SURINAME	1983	89.0	93.0	87.0	1983	100.0	100.0	96.0
TRINIDAD AND TOBAGO	1983	87.0	97.0	77.0	1983	99.0	100.0	98.0
UNITED STATES OF AMERICA	1984	100.0	100.0	...	1980	98.2	...	...
URUGUAY	1983	83.0	95.0	27.0	1983	59.0	59.0	59.0
VENEZUELA	1983	83.0	88.0	65.0	1983	45.0	57.0	6.0
SOUTH-EAST ASIA								
BANGLADESH	1983	40.0	29.0	43.0	1983	4.0	21.0	2.0
BHUTAN		...	40.0	14.0		...	...	...
BURMA	1983	25.0	36.0	21.0	1983	20.0	34.0	15.0
DEMOCRATIC PEOPLE'S REPUBLIC OF KOREA	1983	100.0	100.0	100.0	1983	100.0	100.0	100.0
INDIA	1983	54.0	80.0	47.0	1983	8.0	30.0	1.0
INDONESIA	1983–84	33.0	40.0	32.0	1983–84	30.0	30.0	30.0
MALDIVES	1983	17.0	53.0	8.0	1983	15.0	69.0	1.0
MONGOLIA	1983	100.0	100.0	100.0	1983	50.0	...	...
NEPAL	1983	16.0	71.0	11.0	1983	2.0	16.0	1.0
SRI LANKA	1983	37.0	76.0	26.0	1983	66.0	...	...
THAILAND	1983	70.0	70.0	70.0	1983	45.0	50.0	44.0
EASTERN MEDITERRANEAN								
AFGHANISTAN	1984	13.0	30.0	10.0	1984	2.0	...	...
BAHRAIN	1985	100.0	100.0	100.0	1985	100.0	100.0	100.0
CYPRUS	1984	100.0	100.0	100.0	1984	100.0	100.0	100.0
DEMOCRATIC YEMEN	1983	50.0	73.0	39.0	1983	45.0	69.0	33.0
DJIBOUTI	1985	45.0	53.0	20.0	1985	37.0	43.0	19.0
EGYPT	1981	90.0	93.0	61.0	1981	70.0	95.0	49.0
IRAN	1985	71.0	90.0	52.0	1985	65.0	95.0	35.0
IRAQ	1985	80.0	100.0	46.0	1985	69.0	100.0	15.0
JORDAN	1984	97.0	100.0	90.0	1984	98.0	100.0	95.0
KUWAIT	1985	100.0	100.0	100.0	1985	100.0	100.0	100.0
LEBANON	1985	98.0	98.0	98.0	1985	75.0	94.0	18.0
LIBYA	1983	90.0	100.0	77.0	1983	70.0	...	...
OMAN		...	70.0	10.0	1980	...	60.0	...
PAKISTAN	1985	44.0	84.0	28.0	1985	19.0	56.0	5.0
QATAR	1982	95.0	98.0	50.0	1982	35.0	70.0	...
SAUDI ARABIA	1983	93.0	100.0	68.0	1983	86.0	100.0	33.0
SOMALIA	1985	33.0	60.0	20.0	1985	17.0	60.0	5.0
SUDAN	1984	40.0	...	...	1984	5.0	20.0	1.0
SYRIAN ARAB REPUBLIC	1984	71.0	77.0	65.0	1984	...	70.0	...
TUNISIA	1984	89.0	98.0	79.0	1984	46.0	66.0	29.0
UNITED ARAB EMIRATES	1985	100.0	100.0	100.0	1985	86.0	93.0	22.0
YEMEN	1983	31.0	100.0	21.0	1983	12.0	75.0	...
WESTERN PACIFIC								
AUSTRALIA	1982	98.6	...	...	1982	98.6	...	...
BRUNEI	1984	90.0	...	...	1984	80.0	...	...
CHINA		...	...	...		...	...	...

TABLE 5–21. POPULATION SUPPLIED WITH SAFE DRINKING WATER AND ADEQUATE SANITARY FACILITIES, BY COUNTRY (continued)

Region and Country	Percentage of Population with Access to Safe Drinking Water				Percentage of Population with Access to Adequate Sanitary Facilities			
	Year	Total	Urban	Rural	Year	Total	Urban	Rural
WESTERN PACIFIC (continued)								
COOK ISLANDS	1984	80.0[4]	...	...		...	...	...
DEMOCRATIC KAMPUCHEA		...	...	...		...	...	...
FIJI	1980	83.3	...	...		...	...	...
JAPAN		...	...	...		...	...	...
KIRIBATI		...	...	...		...	...	...
LAO PEOPLE'S DEMOCRATIC REPUBLIC		...	...	...		...	...	...
MALAYSIA	1983	71.0	...	...	1980	75.4	...	...
NEW ZEALAND	1984	100.0	...	...	1984	88.0	...	...
PAPUA NEW GUINEA		...	54.0	10.0	1983	...	50.9	3.4
PHILIPPINES	1985	64.5	...	...	1985	56.5	...	...
REPUBLIC OF KOREA	1983	83.0	...	...		...	...	...
SAMOA	1982	80.0	...	...	1983	80.0	...	...
SINGAPORE	1982	100.0	...	...	1984	85.0	...	...
SOLOMON ISLANDS		...	90.9	...	1984	...	86.4	...
TONGA	1984	95.0	...	...	1984	40.0	...	...
VANUATU	1981–85	45.0	...	...	1981–85	30.0	...	...
VIET NAM		...	90.0	30.0	1983	30.0	...	...
EUROPE								
ALBANIA	1980	92.0	...	...	...	...	...	...
AUSTRIA	1980	100.0	...	...	1980	85.0	...	...
BELGIUM	1984	95.0	...	...	1982	99.0	...	...
BULGARIA	1980	96.0	...	...	...	...	...	...
CZECHOSLOVAKIA	1983	74.5	...	...	1983	60.5	...	...
DENMARK	1982	100.0	...	...	1982	100.0	...	...
FINLAND	1984	79.0	...	...	1984	72.0	...	...
FRANCE	1983	98.0	...	...	1980	85.0	...	...
GERMAN DEMOCRATIC REPUBLIC	1985	90.0	...	...	1985	70.0	...	...
GERMANY, FEDERAL REPUBLIC OF	1984	100.0	...	...	1980	88.0	...	...
GREECE		...	...	...	...	...	...	...
HUNGARY	1984	84.0	...	...	1984	60.0	...	...
ICELAND	1984	100.0	...	...	1980	100.0	...	...
IRELAND	1984	96.6	...	...	1984	94.0	...	...
ISRAEL	1984	98.0	...	...	1984	95.0	...	...
ITALY	1981	98.8	...	...	1981	98.7	...	...
LUXEMBOURG	1984	100.0	...	...	1984	100.0	...	...
MALTA	1983	100.0	...	...	1983	100.0	...	...
MONACO		...	...	...	...	...	...	...
MOROCCO	1984	57.0	...	...	1984	46.0	...	...
NETHERLANDS	1982	99.8	...	...	1981	100.0	...	...
NORWAY	1983	99.0	...	...	1983	85.3	...	...
POLAND	1980	67.0	...	...	1980	50.0	...	...
PORTUGAL	1980	57.0	...	...	1980	41.0	...	...
ROMANIA	1980	77.0	...	...	1980	50.0	...	...
SAN MARINO	1983	100.0	...	...	1983	100.0	...	...
SPAIN	1983	95.0	...	...	1984	90.0	...	...
SWEDEN	1983	100.0	...	...	1980	85.0	...	...
SWITZERLAND	1984	99.1	...	...	1980	85.0	...	...
TURKEY	1980	67.0	...	...	1982	10.0	...	...
UNION OF SOVIET SOCIALIST REPUBLICS	1983	100.0	...	...	1980	50.0	...	...
UNITED KINGDOM	1985	100.0	...	...	1985	100.0	...	...
YUGOSLAVIA	1981	67.8	...	...	1981	58.4	...	...

. . . = Not available.

[1] 1980.

[2] 1983.

[3] 1984.

[4] Rarotonga 99%, other islands 80%.

Source: Copyright World Health Organization 1986, World Health Statistics Annual. Reprinted with permission

TABLE 5–22. POPULATION SERVED WITH DRINKING WATER AND SANITATION IN THE WORLD IN 1980

[Population in thousands; percentages shown in brackets]

Region	No. of Reporting Countries/ Territories	Population			Population with Drinking Water				Population with Sanitation			
					Urban			Rural	Urban			Rural
		Total	Urban	Rural	Total	by H.C.[1]	by P.S.[2]		Total	by S.C.[3]	by other	
Africa	21	103 723	20 788	82 935	13 723	5 294	7 705	17 981	11 214	2 273	8 941	14 787
			(20)	(80)	(66)	(29)	(37)	(22)	(54)	(11)	(43)	(20)
Americas	21	333 398	218 109	115 289	169 302	155 375	13 927[4]	48 628[5]	122 822	91 613	31 209	14 675[6]
			(65)	(35)	(78)	(71)	(7)	(42)	(56)	(42)	(14)	(20)
South-East Asia		1 019 961	232 601	787 360	148 834	. . .	. . .	241 664	70 075	. . .	. . .	49 431
			(23)	(77)	(64)			(31)	(30)			(6)
Europe	3	65 185	28 303	36 882	27 258	17 543	9 715	15 580[6]	11 239[6]	2 000[6]	9 000[6]	67[6]
			(43)	(57)	(96)	(62)	(34)	(62)	(57)	(10)	(46)	(84)
Eastern	12	180 607	65 498	115 109	54 117	34 596	19 521	34 532	36 052	16 779	9 155	6 315
Mediterranean			(36)	(64)	(83)	(53)	(30)	(30)	(57)	(42)	(46)	(7)
Western	20	165 075	66 397	98 678	45 446	43 089	. . .	40 075	51 905	9 459	42 528	62 221
Pacific			(40)	(60)	(81)	(77)	. . .	(41)	(93)	(17)	(76)	(63)
Totals	86	1 867 949	631 696	1 236 253	458 680	. . .	. . .	398 460	303 307	. . .	. . .	147 496
			(34)	(66)	(74)	(48)[7]		(33)	(50)	(22)[7]		(13)

[1] H.C. = house connection.
[2] P.S. = public standpost.
[3] S.C. = sewer connection.
[4] Not all countries reported on urban public standposts.
[5] Several countries reported only on rural house connections.
[6] Not all countries for which population is given reported on this item, percentage quoted is based on population of reporting countries.
[7] Approximate percentage based on information from countries reporting on house and sewer connections.

Source: World Health Organization, 1984. The International Drinking Water Supply and Sanitation Decade, WHO Publ. 85. Reprinted with permission

FIGURE 5-8. DECADE TARGETS FOR URBAN AND RURAL WATER SUPPLY AND SANITATION IN SELECTED DEVELOPING COUNTRIES

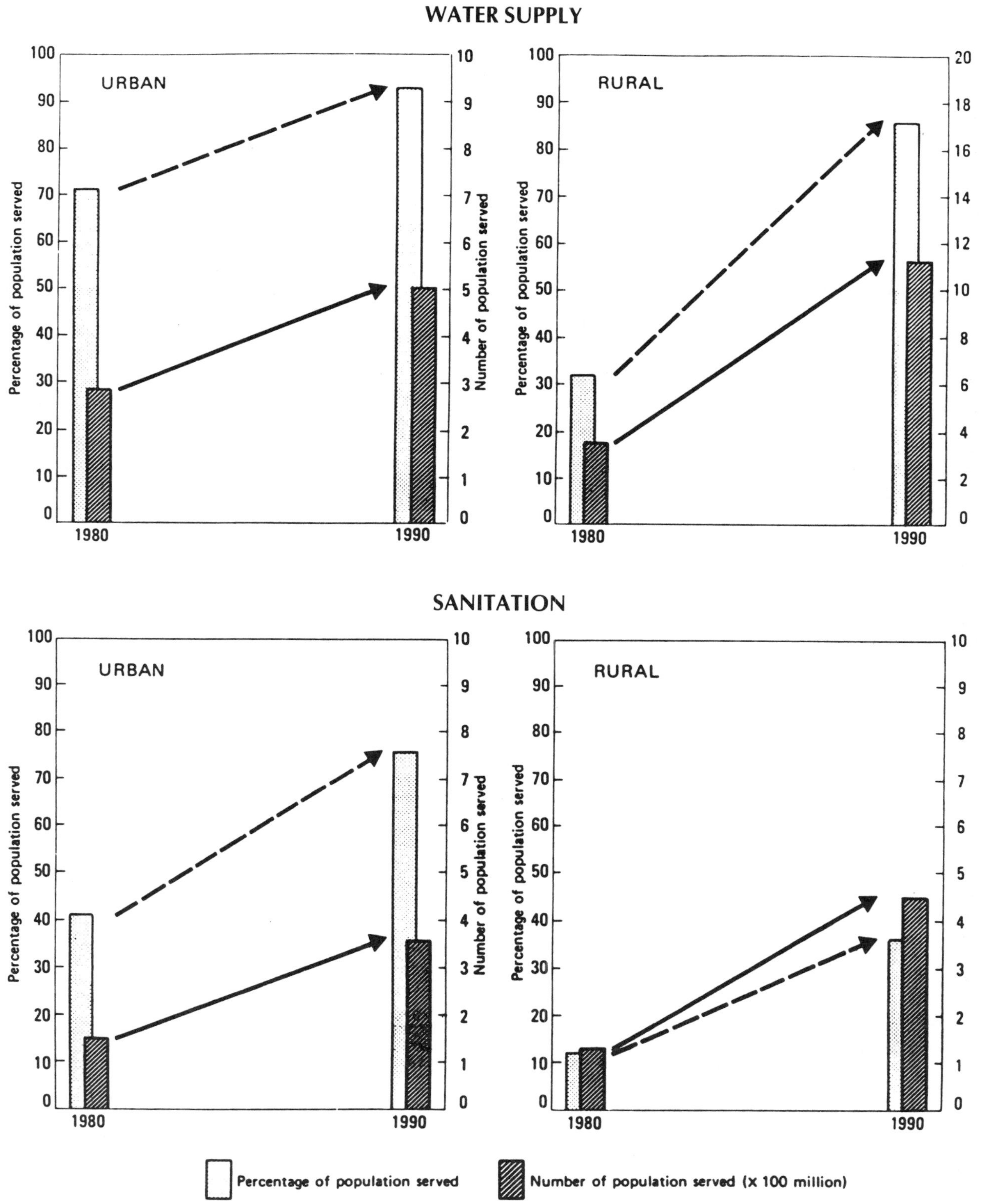

Source: World Health Organization, 1984, The International Drinking Water Supply and Sanitation Decade, Review of National Base-Line Data. Publication No. 85. Reprinted with permission

TABLE 5–23. CONSTRAINTS IN IMPROVING WATER SUPPLY AND SANITATION SERVICES IN DEVELOPING COUNTRIES

[Ranking and frequency of constraints as reported by 87 countries]

Constraints	Number of Countries Indicating Constraint			Ranking Index[1]
	Very Severe	Severe	Moderate	
Insufficiency of trained personnel (professional)	16	40	27	155
Funding limitations	21	31	30	155
Insufficiency of trained personnel (sub-professional)	16	38	29	153
Operation and maintenance[2]	16	36	23	143
Logistics[2]	11	35	23	126
Inadequate cost-recovery framework	11	34	22	123
Inappropriate institutional framework	6	30	35	113
Insufficient health education efforts	7	24	43	112
Intermittent water service	10	19	32	100
Lack of planning and design criteria	6	17	41	93
Non-involvement of communities	6	15	44	92
Inadequate or outmoded legal framework	10	14	34	92
Inappropriate technology	5	18	33	84
Insufficient knowledge of water resources	1	20	39	82
Inadequate water resources	5	11	40	77
Lack of definite government policy for sector	4	10	44	76
Import restrictions	5	12	21	60

[1] Ranking index = (No. very severe x 3) + (No. severe x 2) + (No. moderate x 1).

[2] "Logistics" is ranked ahead of "Operation and maintenance" in the group of Least Developed Countries.

Source: World Health Organization, 1984. The International Drinking Water Supply and Sanitation Decade, WHO Publ. 85. Reprinted with permission

SECTION E. DOMESTIC WATER CONSUMPTION

TABLE 5–24. DOMESTIC FRESHWATER USE IN THE UNITED STATES, 1985

[Includes water for normal household purposes, such as drinking, food preparation, bathing, washing clothes and dishes, flushing toilets and watering lawns and gardens; by state, figures may not add to totals because of independent rounding. Mgal/d = million gallons per day; gal/d = gallons per day]

STATE	Self Supplied					Public Supply			Total	
	Population, in Thousands	Water Withdrawals, in Mgal/d			Per Capita Use, in gal/d	Population Served, in Thousands	Water Deliveries, in Mgal/d	Per Capita Use, in gal/d	Withdrawals and Deliveries, in Mgal/d	Consumptive Use, in Mgal/d
		Source		Total						
		Ground Water	Surface Water							
Alabama	515	39	.0	39	75	3510	332	95	371	62
Alaska	210	9.0	.7	9.7	46	348	30	85	39	7.9
Arizona	189	26	1.4	27	142	3090	449	145	476	275
Arkansas	678	60	.0	60	89	1680	170	101	230	60
California	2010	125	15	140	70	24300	3240	133	3380	879
Colorado	222	17	.0	17	75	3010	456	152	473	145
Connecticut	515	39	.0	39	75	2680	178	66	217	59
Delaware	108	10	.0	10	96	514	36	70	46	4.6
D.C.	.0	.0	.0	.0	.0	626	174	279	174	17
Florida	1590	259	.0	259	163	9740	1200	123	1460	304
Georgia	1310	99	.0	99	75	4660	545	117	643	116
Hawaii	23	11	.3	11	479	1130	132	117	143	71
Idaho	301	76	13	89	296	704	200	284	289	5.3
Illinois	1750	130	.0	130	74	9830	850	86	981	97
Indiana	1800	139	.0	139	77	3670	423	115	562	56
Iowa	751	65	.0	65	87	2130	289	136	354	144
Kansas	453	42	.0	42	93	2000	150	75	192	87
Kentucky	956	43	4.8	47	50	2770	179	65	226	60
Louisiana	576	46	.0	46	80	3900	564	145	610	122
Maine	329	19	.0	19	58	829	96	115	114	80
Maryland	827	63	.0	63	77	3560	365	102	428	43
Massachusetts	496	35	.0	35	71	5330	415	78	450	77
Michigan	1640	123	.1	123	75	7370	630	85	752	98
Minnesota	1490	131	.0	131	88	2700	401	148	532	172
Mississippi	355	16	.0	16	45	2260	165	73	181	36
Missouri	893	54	.0	54	60	4140	355	86	408	114
Montana	209	15	.8	16	78	614	90	147	106	50
Nebraska	281	24	.0	24	87	1320	149	112	173	87
Nevada	86	12	.6	12	141	882	189	215	201	102
New Hampshire	361	22	.0	22	60	637	63	99	85	17
New Jersey	851	64	.0	64	75	6710	503	75	567	103
New Mexico	414	38	.0	38	93	1000	179	178	218	106
New York	1890	191	.0	191	101	15900	1470	93	1660	166
North Carolina	2800	169	.0	169	60	3450	315	91	484	154
North Dakota	176	15	.0	15	83	512	40	77	54	18
Ohio	1850	139	.0	139	75	8900	467	53	606	91
Oklahoma	469	25	1.2	26	56	2830	158	56	185	55
Oregon	735	70	9.6	80	109	1940	246	127	326	89
Pennsylvania	3680	184	.0	184	50	8170	539	66	723	72
Rhode Island	84	5.6	.0	5.6	67	884	59	66	64	14
South Carolina	823	62	.0	62	75	2520	189	75	251	50
South Dakota	158	15	.9	16	103	548	61	110	77	19
Tennessee	1100	70	.0	70	64	3660	303	83	373	37
Texas	958	105	.0	105	110	15400	2200	143	2310	832
Utah	76	4.5	1.6	6.1	81	1570	340	217	346	119
Vermont	192	12	.0	12	60	343	34	100	46	9.1
Virginia	1490	112	.0	112	75	4210	337	80	448	90
Washington	856	98	.0	98	115	3530	516	146	614	82
West Virginia	626	21	.0	21	34	1310	81	62	102	29
Wisconsin	1680	84	.0	84	50	3130	169	54	253	25
Wyoming	180	12	1.3	13	75	329	61	185	74	30
Puerto Rico	458	9.1	9.1	18	40	2930	178	61	196	39
Virgin Islands	57	.6	1.0	1.6	29	47	2.1	44	3.7	.7
Total	42,500	3,250	62	3,320	78	200,000	21,000	105	24,300	5,680

Source: Solley, W.B., Merk, C.F., and Pierce, R.R., 1988, Estimated Water Use in the United States in 1985, U.S. Geological Survey Circular 1004

TABLE 5–25. TYPICAL URBAN WATER USE BY A FAMILY OF FOUR

[Assuming no water delivery losses]

Type of Household Use	Daily Use		
	Per Family		Per Capita
	Gallons per day	%	Gallons per day
Drinking and water used in kitchen	8	2	2.00
Dishwasher (3 loads per day)	15	4	3.75
Toilet (16 flushes per day)	96	28	24.00
Bathing (4 baths or showers per day)	80	23	20.00
Laundering (6 loads per week)	34	10	8.50
Automobile washing (2 car washes per month)	10	3	2.50
Lawn watering and swimming pools (180 hours per year)	100	29	25.00
Garbage disposal unit (1 percent of all other uses)	3	1	0.75
Total	346	100	86.50

Source: U.S. Water Resources Council, Second National Water Assessment, The Nation's Water Resources 1975–2000; percentages added

TABLE 5–26. FLOW RATES FOR CERTAIN PLUMBING, HOUSEHOLD, AND FARM FIXTURES

Location	Flow Pressure[1] (psi)	Flow Rate (gpm)
Ordinary basin faucet	8	2.0
Self-closing basin faucet	8	2.5
Sink faucet, 3/8 in.	8	4.5
Sink faucet, 1/2 in.	8	4.5
Bathtub faucet	8	6.0
Laundry tub faucet, 1/2 in.	8	5.0
Shower	8	5.0
Ball-cock for closet	8	3.0
Flush valve for closet	15	15–40[2]
Flushometer valve for urinal	15	15.0
Garden hose (50 ft, 3/4-in. sill cock)	30	5.0
Garden hose (50 ft, 5/8-in. outlet)	15	3.33
Drinking fountains	15	0.75
Fire hose 1 1/2 in., 1/2-in. nozzle	30	40.0

[1] Flow pressure is the pressure in the supply near the faucet or water outlet while the faucet or water outlet is wide open and flowing.
[2] Wide range due to variation in design and type of closet flush valves.
Source: U.S. Public Health Service, 1962, Manual of Individual Water Supply Systems

TABLE 5–27. WATER REQUIREMENTS FOR VARIOUS TYPES OF ESTABLISHMENTS

Type of Establishment	Gallons per day
Airports (per passenger)	3–5
Apartments, multiple family (per resident)	60
Bath houses (per bather)	10
Camps:	
Construction, semipermanent (per worker)	50
Day with no meals served (per camper)	15
Luxury (per camper)	100–150
Resorts, day and night, with limited plumbing (per camper)	50
Tourist with central bath and toilet facilities (per person)	35
Cottages with seasonal occupancy (per resident)	50
Courts, tourist with individual bath units (per person)	50
Clubs:	
Country (per resident member)	100
Country (per nonresident member present)	25
Dwellings:	
Boarding houses (per boarder)	50
Additional kitchen requirements for nonresident boarders	10
Luxury (per person)	100–150
Multiple family apartments (per resident)	40
Rooming houses (per resident)	60
Single family (per resident)	50–75
Estates (per resident)	100–150
Factories (gal per person per shift)	15–35
Hotels with private baths (two persons per room)	60
Hotels without private baths (per person)	50
Institutions other than hospitals (per person)	75–125
Hospitals (per bed)	250–400
Laundries, self-serviced (gal per washing, i.e., per customer)	50
Motels with bath, toilet, and kitchen facilities (per bed space)	50
With bed and toilet (per bed space)	40
Parks:	
Overnight with flush toilets (per camper)	25
Trailers with individual bath units (per camper)	50
Picnic areas:	
With bath houses, showers, and flush toilets (per picnicker)	20
With toilet facilities only (gal per picnicker)	10
Restaurants with toilet facilities (per patron)	7–10
Without toilet facilities (per patron)	2 1/2–3
With bars and cocktail lounge (additional quantity per patron)	2
Schools:	
Boarding (per pupil)	75–100
Day with cafeteria, gymnasiums, and showers (per pupil)	25
Day with cafeteria but no gymnasiums or showers (per pupil)	20
Day without cafeteria, gymnasiums, or showers (per pupil)	15
Service stations (per vehicle)	10
Stores (per toilet room)	400
Swimming pools (per swimmer)	10
Theaters:	
Drive-in (per car space)	5
Movie (per auditorium seat)	5
Workers:	
Construction (per person per shift)	50
Day (school or offices per person per shift)	15

Source: U.S. Public Health Service, 1962, Manual of Individual Water Supply Systems

SECTION F. BOTTLED WATER

TABLE 5–28. CONSUMPTION OF BOTTLED WATER IN THE UNITED STATES RANKED BY LEADING STATES, 1983–1986

State	1986 Rank	1983 Million Gallons	1983 Share %	1984 Million Gallons	1984 Share %	1985 Million Gallons	1985 Share %	1986 Million Gallons	1986 Share %
California	1	330.4	41.2	373.3	40.0	428.7	40.0	457.5	37.9
Texas	2	65.8	8.2	76.6	8.2	87.8	8.2	87.1	7.2
New York	3	50.6	6.3	59.8	6.4	70.7	6.6	82.7	6.9
Florida	4	61.1	7.6	63.1	6.8	72.2	6.7	78.4	6.5
Illinois	5	26.5	3.3	31.7	3.4	37.5	3.5	51.2	4.2
Arizona	6	18.5	2.3	23.3	2.5	27.9	2.6	43.1	3.6
Louisiana	7	15.3	1.9	21.5	2.3	24.6	2.3	36.1	3.0
Maryland/DC	8	20.9	2.6	24.3	2.6	28.9	2.7	34.1	2.8
Pennsylvania	9	16.1	2.0	20.5	2.2	23.6	2.2	28.5	2.4
Massachusetts	10	9.6	1.2	14.0	1.5	16.1	1.5	24.1	2.0
TOTAL TOP 10 STATES		614.8	76.6	708.1	75.9	818.0	76.3	922.8	76.5
Ohio	11	15.3	1.9	19.6	2.1	21.4	2.0	22.0	1.8
New Jersey	12	12.8	1.6	15.9	1.7	19.3	1.8	20.4	1.7
Connecticut	13	6.4	0.8	11.2	1.2	15.0	1.4	17.6	1.5
Michigan	14	12.0	1.5	14.0	1.5	16.1	1.5	16.3	1.3
Georgia	15	6.4	0.8	7.7	0.8	8.4	0.8	9.3	0.8
TOTAL TOP 15		667.7	83.2	776.5	83.2	898.2	83.8	1,008.4	83.6
All Others		135.1	16.8	157.1	16.8	173.1	16.2	197.8	16.4
SUBTOTAL		802.8	100.0	933.6	100.0	1,071.3	100.0	1,206.2	100.0
Club Soda/Seltzer		143.5		156.5		166.4		176.6	
TOTAL U.S.		946.3		1,090.1		1,237.7		1,382.8	

Source: U.S. Bottled Water Market and Packaging Report 1987. Copyright Beverage Marketing Corp., New York. Reproduced with permission

TABLE 5–29. CONSUMPTION OF BOTTLED WATER IN THE UNITED STATES BY REGION, 1983–1986

[Non-sparkling water]

	1983 Million Gallons	1983 Share %	1984 Million Gallons	1984 Share %	1985 Million Gallons	1985 Share %	1986 Million Gallons	1986 Share %	Share of Population
Pacific	299.2	41.4	338.8	40.2	382.8	40.2	423.8	39.6	14.8
Northeast	135.9	18.8	162.3	19.2	187.8	19.7	220.1	20.5	23.1
Southwest	101.3	14.0	122.9	14.6	139.4	14.6	158.1	14.8	10.3
South	84.7	11.7	102.6	12.2	110.4	11.6	121.3	11.3	21.4
East Central	64.8	9.0	77.9	9.2	89.0	9.3	97.6	9.1	19.7
West Central	19.9	2.8	22.0	2.6	25.1	2.6	27.5	2.6	7.3
West	16.6	2.3	16.8	2.0	18.6	2.0	22.0	2.1	3.4
TOTAL	722.4	100.0	843.3	100.0	953.1	100.0	1,070.4	100.0	100.0

Source: U.S. Bottled Water Market and Packaging Report 1987. Copyright Beverage Marketing Corp., New York. Reproduced with permission

TABLE 5–30. CONSUMPTION OF BOTTLED WATER IN THE UNITED STATES BY TYPE OF WATER, 1976–1986

[Millions of gallons]

Year	Non-Sparkling	Sparkling		Imports	Total	Per Capita Consumption (Gallons)
		Domestic Sparkling	Club/Seltzer			
1976	237.9	16.3	61.8	1.2	317.2	1.5
1977	256.7	22.0	63.6	3.2	345.5	1.6
1978	352.9	49.5	64.0	13.4	479.8	2.2
1979	401.2	58.4	75.4	28.1	563.1	2.5
1980	463.1	61.9	93.3	11.4	629.7	2.8
1981	537.7	62.2	113.7	11.6	725.2	3.2
1982	614.0	64.8	131.7	12.9	823.4	3.6
1983	722.4	66.8	143.5	13.6	946.3	4.0
1984	843.3	73.8	156.5	16.5	1,090.1	4.6
1985	953.1	88.4	166.4	29.8	1,237.7	5.2
1986	1,070.4	105.0	176.6	30.8	1,382.8	5.7

Source: U.S. Bottled Water Market and Packaging Report, 1987. Copyright Beverage Marketing Corp., New York. Reproduced with permission

TABLE 5–31. BOTTLED WATER MARKET IN THE UNITED STATES, 1986

[Estimated wholesale dollars]

Type of Water	Gallons (Millions)	Dollars (Millions)
Non-Sparkling	1,070.4	$ 947.9
Sparkling	105.0	255.0
Imports	30.8	132.4
TOTAL	1,206.2	1,335.3

Source: U.S. Bottled Water Market and Packaging Report, 1987. Copyright Beverage Marketing Corp., New York. Reproduced with permission

TABLE 5–32. PROJECTED CONSUMPTION OF BOTTLED WATER IN THE UNITED STATES, 1986–1996

[Millions of gallons]

	Non-Sparkling	Sparkling		Imports	Total
		Domestic	Club/Seltzer		
1986	1,070.4	105.0	176.6	30.8	1,382.8
1991	1,725.0	173.0	242.0	62.0	2,202.0
1996	2,420.0	260.0	325.0	98.0	3,103.0

Source: U.S. Bottled Water Market and Packaging Report, 1987. Copyright Beverage Marketing Corp., New York. Reproduced with permission

TABLE 5–33. PROJECTED GROWTH OF BOTTLED WATER CONSUMPTION IN THE UNITED STATES, 1986–1996

[By distribution channel]

	Home	Commercial	Retail	Vending
		Non-Sparkling Water		
1986–1991	9.6%	7.7%	12.8%	5.0%
1991–1996	4.8%	4.0	10.1	5.0
		Sparkling Water		
1986–1991	2.1%	4.0%	9.2%	—
1991–1996	1.5	3.5	7.7	—

Projected Per Capita Consumption

Year	Gallons	Year	Gallons
1985	5.2	1991	8.7
1986	5.7	1992	9.3
1987	6.2	1993	9.9
1988	6.8	1994	10.5
1989	7.5	1995	11.0
1990	8.2	1996	11.7

Source: U.S. Bottled Water Market and Packaging Report, 1987. Copyright Beverage Marketing Corp., New York. Reproduced with permission

SECTION G. INDUSTRIAL AND COMMERCIAL WATER USE—UNITED STATES

TABLE 5–34. INDUSTRIAL WATER USE IN THE UNITED STATES, 1985

[By state; figures may not add to totals because of independent rounding. All values in million gallons per day]

STATE	Self-Supplied Withdrawals								Total				
	By source and Type								Public-Supply Deliv-eries	With-drawals and Deliveries	Comsumptive Use		
	Ground Water		Surface Water		Total			Reclaimed Sewage					
	Fresh	Saline	Fresh	Saline	Fresh	Saline	Total		Fresh	Saline	Fresh	Saline	Total
Alabama	34	.0	804	.0	838	.0	838	.0	221	1060	233	.0	233
Alaska	7.7	.0	106	.0	114	.0	114	.0	7.8	121	11	.0	11
Arizona	44	8.3	.7	.0	45	8.3	53	4.9	79	124	78	8.3	87
Arkansas	64	.0	108	.0	172	.0	172	.0	.1	172	21	.0	21
California	326	7.9	105	254	431	262	693	2.9	494	925	301	11	312
Colorado	7.4	.0	113	.0	120	.0	120	.0	18	138	28	.0	28
Connecticut	18	.0	59	68	77	68	145	.0	62	139	13	1.4	14
Delaware	16	.0	3.1	391	19	391	410	.0	18	36	3.6	3.9	7.5
D.C.	.0	.0	.0	.0	.0	.0	.0	.0	.0	.0	.0	.0	.0
Florida	340	.0	54	27	394	27	421	.0	142	537	144	.2	145
Georgia	323	.0	283	31	606	31	637	.0	135	740	74	.5	74
Hawaii	16	.0	4.8	.0	20	.0	20	.0	6.4	27	6.3	.0	6.3
Idaho	172	.0	26	.0	199	.0	199	.0	6.7	205	5.4	.0	5.4
Illinois	150	.0	385	.0	535	.0	535	.0	255	790	273	.0	273
Indiana	109	.0	2550	.0	2660	.0	2660	.0	73	2730	228	.0	228
Iowa	71	.0	126	.0	197	.0	197	.0	41	239	31	.0	31
Kansas	70	.0	11	.0	81	.0	81	.0	41	122	40	.0	40
Kentucky	66	.0	175	.0	241	.0	241	.0	167	408	17	.0	17
Louisiana	297	5.3	1790	.0	2090	5.3	2090	.0	1.5	2090	189	.0	189
Maine	8.3	.0	207	30	215	30	246	.0	11	226	83	4.6	88
Maryland	20	.0	54	275	74	275	350	81	55	130	32	69	101
Massachusetts	26	.0	103	22	129	22	151	.0	69	198	44	.4	44
Michigan	121	3.7	1200	.0	1320	3.7	1330	.0	247	1570	123	.2	124
Minnesota	76	.0	108	.0	184	.0	184	.0	46	231	70	.0	70
Mississippi	131	.0	96	5.7	227	5.7	233	.0	28	255	44	1.0	45
Missouri	37	.0	52	.0	88	.0	88	.0	133	221	30	.0	30
Montana	29	.0	27	.0	56	.0	56	.0	1.2	57	8.8	.0	8.8
Nebraska	41	.0	7.2	.0	48	.0	48	.0	49	97	34	.0	34
Nevada	2.3	.0	7.8	.0	10	.0	10	.0	6.3	16	3.2	.0	3.2
New Hampshire	33	.0	204	.2	238	.2	238	.0	16	254	51	.0	51
New Jersey	129	.1	127	801	256	801	1060	.0	240	496	37	16	53
New Mexico	.4	.0	.4	.0	.8	.0	.8	.0	2.8	3.6	2.1	.0	2.1
New York	272	.0	760	.0	1030	.0	1030	.0	1010	2050	205	.0	205
North Carolina	43	.0	371	6.0	414	6.0	420	.0	128	542	55	.5	56
North Dakota	2.2	.0	6.8	.0	8.9	.0	8.9	.0	2.3	11	9.2	.0	9.2
Ohio	11	.0	451	.0	462	.0	462	.0	340	802	156	.0	156
Oklahoma	22	.0	84	.0	106	.0	106	.0	204	310	21	.0	21
Oregon	29	.0	263	.0	293	.0	293	.6	53	346	22	.0	22
Pennsylvania	149	.0	1910	.0	2060	.0	2060	.0	246	2300	186	.0	186
Rhode Island	3.9	.0	13	.2	17	.2	17	.0	20	37	2.6	.0	2.6
South Carolina	38	.0	1090	.0	1130	.0	1130	.0	86	1210	183	.0	183
South Dakota	7.8	.0	.9	.0	8.7	.0	8.7	.0	5.4	14	1.8	.0	1.8
Tennessee	89	.0	1510	.0	1600	.0	1600	.0	98	1700	187	.0	187
Texas	149	.4	834	1430	983	1430	2420	55	284	1270	507	824	1330
Utah	20	.0	7.5	.0	28	.0	28	.0	15	43	19	.0	19
Vermont	4.0	.0	50	.0	54	.0	54	.0	13	67	13	.0	13
Virginia	99	.2	477	81	576	81	657	.0	57	633	71	9.7	80
Washington	108	.0	411	37	519	37	556	.0	306	825	107	5.0	112
West Virginia	33	.0	853	.0	886	.0	886	.0	22	909	133	.0	133
Wisconsin	38	.0	424	.0	461	.0	461	.0	153	614	58	.0	58
Wyoming	7.8	.0	3.6	.0	11	.0	11	.0	3.2	15	2.0	.0	2.0
Puerto Rico	18	.0	.0	.0	18	.0	18	.0	12	31	6.1	.0	6.1
Virgin Islands	.0	.2	.0	14	.0	14	14	.0	.0	.0	.0	2.0	2.1
Total	3,930	26	18,400	3,480	22,300	3,500	25,800	144	5,730	28,100	4,200	957	5,160

Source: Solley, W.B., Merk, C.F., and Pierce, R.R., 1988, Estimated Water Use in the United States in 1985, U.S. Geological Survey Circular 1004

TABLE 5–35. COMMERCIAL FRESHWATER USE IN THE UNITED STATES, 1985

[Includes water for motels, hotels, restaurants, office buildings, other commercial facilities, and civilian and military institutions; by state, figures may not add to totals because of independent rounding. All values in million gallons per day]

STATE	Self-Supplied Withdrawals: Source: Ground Water	Self-Supplied Withdrawals: Source: Surface Water	Self-Supplied Withdrawals: Total	Public-Supply Deliveries	Total: Withdrawals and Deliveries	Total: Consumptive Use
Alabama	3.2	.0	3.2	61	65	14
Alaska	.4	.0	.4	31	31	4.6
Arizona	17	.1	17	90	107	54
Arkansas	6.0	1.9	8.0	87	95	5.6
California	47	5.7	53	1220	1270	331
Colorado	8.1	.3	8.4	112	120	21
Connecticut	8.6	.0	8.6	49	58	15
Delaware	2.0	.0	2.0	12	14	1.4
D.C.	.0	.0	.0	44	44	4.4
Florida	55	1.0	56	248	305	84
Georgia	24	4.6	28	142	171	31
Hawaii	33	.0	33	51	83	31
Idaho	16	.0	16	6.1	22	.0
Illinois	33	74	107	471	577	64
Indiana	1.0	.1	1.1	78	79	5.5
Iowa	29	9.0	38	4.2	42	5.5
Kansas	.0	.0	.0	83	83	25
Kentucky	5.1	11	16	19	35	1.3
Louisiana	.2	.0	.2	7.6	7.8	1.5
Maine	10	28	38	1.7	40	13
Maryland	19	6.3	25	57	82	8.1
Massachusetts	67	171	238	276	514	112
Michigan	8.0	26	34	339	374	27
Minnesota	24	1.8	26	23	49	11
Mississippi	4.1	.0	4.1	47	51	8.2
Missouri	17	.0	17	60	78	5.4
Montana	.0	.0	.0	29	29	11
Nebraska	.3	.0	.3	50	51	13
Nevada	7.0	.4	7.4	54	61	12
New Hampshire	.0	.0	.0	9.1	9.1	1.8
New Jersey	14	1.1	15	136	151	7.5
New Mexico	7.3	.0	7.3	42	49	25
New York	65	66	130	282	413	40
North Carolina	15	7.9	23	137	160	5.0
North Dakota	.1	.0	.1	14	15	2.4
Ohio	51	.4	51	326	377	19
Oklahoma	25	6.9	32	58	90	6.3
Oregon	1.5	.1	1.6	45	46	9.3
Pennsylvania	27	.0	27	186	214	46
Rhode Island	.0	.0	.0	15	15	.6
South Carolina	8.5	32	41	5.7	47	7.1
South Dakota	12	5.2	17	14	32	3.2
Tennessee	4.9	.0	4.9	163	168	15
Texas	17	7.7	25	105	130	6.5
Utah	.4	.0	.4	50	50	6.6
Vermont	.0	.0	.0	5.2	5.2	1.0
Virginia	18	3.8	22	70	92	12
Washington	20	.3	20	133	153	21
West Virginia	.3	.5	.8	21	22	2.4
Wisconsin	3.8	.0	3.8	98	102	27
Wyoming	9.9	7.1	17	14	31	3.4
Puerto Rico	.0	.0	.0	30	30	5.9
Virgin Islands	.5	.4	.9	.2	1.1	.2
Total	746	481	1,230	5,710	6,940	1,190

Source: Solley, W.B., Merk, C.F., and Pierce, R.R., 1988, Estimated Water Use in the United States in 1985, U.S. Geological Survey Circular 1004

TABLE 5–36. WATER USE IN MINING AND MANUFACTURING IN THE UNITED STATES, 1968 TO 1983, AND BY INDUSTRY GROUP, 1983

[Based on establishments reporting water intake of 20 million gallons. This represented 95 percent and 96 percent of the total water use estimated for mining and manufacturing industries. Water intake refers to that which is used/consumed in the production and processing operations and for sanitary services]

INDUSTRY	Establishments Reporting[1]	Gross Water Used: Total: Quantity (bil. gal.)	Gross Water Used: Total: Average per Establishment (mil. gal.)	Gross Water Used: Water Intake (bil. gal.)	Gross Water Used: Water Recycled[2] (bil. gal.)	Water Discharged: Quantity (bil. gal.)	Water Discharged: Percent Untreated	Water Pollutants Abatement: Capital Expenditures (mil. dol.)	Water Pollutants Abatement: Operating Cost (mil. (dol.)
MINING									
1968	1,801	3,694	2,051	1,408	(NA)	1,365	78.8	(NA)	(NA)
1973	1,687	3,965	2,350	1,665	2,300	1,605	54.3	38	124
1978	1,056	3,554	3,366	1,473	2,430	1,592	67.3	244	201
1983, total	**1,534**	**3,328**	**2,169**	**1,197**	**2,131**	**1,037**	**31.9**	**189**	**499**
Metal mining	135	735	5,444	170	564	133	39.8	22	65
Anthracite mining	16	5	313	2	3	8	12.5	(Z)	1
Bituminous coal, lignite mining	275	119	433	45	73	116	26.7	14	69
Oil and gas extraction	555	1,452	2,616	602	850	476	31.1	131	318
Nonmetallic minerals, exc. fuels	553	1,018	1,841	378	640	304	32.6	22	46
MANUFACTURING									
1968	9,402	35,701	3,797	15,467	(NA)	14,276	69.5	(NA)	(NA)
1973	10,668	43,413	4,069	15,024	([3])	14,144	56.5	511	866
1978	9,605	44,494	4,632	12,992	34,199	11,682	59.7	1,249	2,119
1983, total	**10,262**	**33,835**	**3,297**	**10,039**	**23,796**	**8,914**	**54.9**	**819**	**3,259**
Food and kindred products	2,656	1,406	529	648	759	552	64.5	105	187
Tobacco products	20	34	1,700	5	29	4	(D)	(D)	5
Textile mill products	761	333	438	133	200	116	52.6	(D)	25
Lumber and wood products	223	218	978	86	132	71	63.4	4	23
Furniture and fixtures	66	7	106	3	3	3	100.0	2	4
Paper and allied products	600	7,436	12,393	1,899	5,537	1,768	27.1	66	438
Chemicals and allied products	1,315	9,630	7,323	3,401	6,229	2,980	67.0	187	1,013
Petroleum and coal products	260	6,177	23,758	818	5,359	699	46.2	165	543
Rubber, misc. plastic products	375	328	875	76	252	63	63.5	4	37
Leather and leather products	69	7	101	6	1	6	(D)	(S)	6
Stone, clay, and glass products	602	337	560	155	182	133	75.2	10	38
Primary metal products	776	5,885	7,584	2,363	3,523	2,112	58.1	100	421
Fabricated metal products	724	258	356	65	193	61	49.2	33	100
Machinery, exc. electrical	523	307	587	120	186	105	67.6	19	76
Electric and electronic equipment	678	335	494	74	261	70	61.4	45	108
Transportation equipment	380	1,011	2,661	153	859	139	67.6	55	171
Instruments and related products	154	112	727	30	82	28	50.0	10	45
Miscellaneous manufacturing	80	15	188	4	11	4	(D)	2	7

D=Withheld to avoid disclosing individual company data; NA=Not available; S=Figure does not meet publication standards; Z=Less than $500,000.

[1] Establishments reporting water intake of 20 million gallons or more. These counts do not apply to water pollutants abatement columns for manufacturing in 1983.

[2] Refers to water recirculated and water reused.

[3] Data estimated; not strictly comparable to other years.

Source: U.S. Department of Commerce, Statistical Abstract of the United States, 1987

TABLE 5–37. MINING WATER USE IN THE UNITED STATES, 1985

[By state; figures may not add to totals because of independent rounding. All values in million gallons per day]

State	Withdrawals: By Source and Type: Ground Water: Fresh	Ground Water: Saline	Ground Water: Total	Surface Water: Fresh	Surface Water: Saline	Surface Water: Total	Total: Fresh	Total: Saline	Total: Total	Consumptive Use: Fresh	Consumptive Use: Saline	Consumptive Use: Total
Alabama	10	3.4	14	.0	.0	.0	10	3.4	14	10	.3	10
Alaska	.1	.0	.1	19	.0	19	19	.0	19	.1	.0	.1
Arizona	67	.0	67	13	.0	13	80	.0	80	59	.0	59
Arkansas	1.0	.0	1.0	2.3	.0	2.3	3.3	.0	3.3	3.3	.0	3.3
California	108	276	384	58	25	82	165	301	466	44	79	123
Colorado	29	32	61	30	.0	30	59	32	91	17	4.5	21
Connecticut	.3	.0	.3	1.4	.0	1.4	1.7	.0	1.7	.0	.0	.0
Delaware	.0	.0	.0	.0	.0	.0	.0	.0	.0	.0	.0	.0
D.C.	.0	.0	.0	.0	.0	.0	.0	.0	.0	.0	.0	.0
Florida	236	.0	236	22	.0	22	258	.0	258	100	.0	100
Georgia	16	.0	16	2.9	.0	2.9	19	.0	19	2.3	.0	2.3
Hawaii	.0	.0	.0	.0	.0	.0	.0	.0	.0	.0	.0	.0
Idaho	.0	.0	.0	135	.0	135	135	.0	135	.0	.0	.0
Illinois	14	38	52	53	.0	53	66	38	104	10	38	48
Indiana	7.4	.0	7.4	83	.0	83	91	.0	91	.3	.0	.3
Iowa	50	.0	50	12	.0	12	63	.0	63	.0	.0	.0
Kansas	5.3	.0	5.3	8.3	.0	8.3	14	.0	14	.6	.0	.6
Kentucky	3.4	.0	3.4	22	.0	22	25	.0	25	.7	.0	.7
Louisiana	.7	.3	1.0	1.2	5.0	6.3	2.0	5.3	7.3	.6	4.8	5.4
Maine	.8	.0	.8	3.2	.0	3.2	4.0	.0	4.0	.6	.0	.6
Maryland	12	.0	12	8.2	.5	8.7	21	.5	21	4.1	.1	4.2
Massachusetts	2.0	.0	2.0	.0	.0	.0	2.0	.0	2.0	.5	.0	.5
Michigan	8.5	.8	9.3	52	.0	52	60	.8	61	2.2	.1	2.3
Minnesota	1.7	.0	1.7	271	.0	271	273	.0	273	122	.0	122
Mississippi	3.1	.0	3.1	.6	.0	.6	3.7	.0	3.7	.8	.0	.8
Missouri	24	.3	25	3.3	.0	3.3	28	.3	28	2.8	.0	2.8
Montana	.6	.0	.6	3.6	.0	3.6	4.2	.0	4.2	1.1	.0	1.1
Nebraska	8.5	.0	8.5	111	.0	111	119	.0	119	1.8	.0	1.8
Nevada	19	2.8	22	2.7	.0	2.7	22	2.8	24	18	1.9	20
New Hampshire	.1	.0	.1	1.1	.0	1.1	1.2	.0	1.2	.0	.0	.0
New Jersey	12	.0	12	68	.0	68	80	.0	80	8.0	.0	8.0
New Mexico	69	.0	69	13	.0	13	82	.0	82	35	.0	35
New York	.0	.0	.0	50	.0	50	50	.0	50	5.0	.0	5.0
North Carolina	81	.0	81	38	.0	38	119	.0	119	28	.0	28
North Dakota	2.8	.0	2.8	.9	.0	.9	3.6	.0	3.6	2.5	.0	2.5
Ohio	78	.1	78	.4	.0	.4	78	.1	78	11	.0	11
Oklahoma	6.7	.0	6.7	.4	.0	.4	7.1	.0	7.1	6.8	.0	6.8
Oregon	.3	.0	.3	7.2	.0	7.2	7.5	.0	7.5	1.6	.0	1.6
Pennsylvania	118	.0	118	30	.0	30	148	.0	148	20	.0	20
Rhode Island	.4	.0	.4	2.3	.0	2.3	2.7	.0	2.7	.3	.0	.3
South Carolina	2.9	.0	2.9	2.4	.0	2.4	5.3	.0	5.3	.5	.0	.5
South Dakota	16	.0	16	25	.0	25	40	.0	40	10	.0	10
Tennessee	2.4	.0	2.4	11	.0	11	13	.0	13	1.1	.0	1.1
Texas	103	229	332	18	.0	18	121	229	350	121	.0	121
Utah	50	21	71	5.5	108	113	56	129	185	15	123	138
Vermont	.0	.0	.0	1.1	.0	1.1	1.1	.0	1.1	.0	.0	.0
Virginia	2.1	.0	2.1	14	.0	14	16	.0	16	1.9	.0	1.9
Washington	2.4	.0	2.4	.6	.0	.6	3.0	.0	3.0	.4	.0	.4
West Virginia	119	.0	119	24	.0	24	142	.0	142	29	.0	29
Wisconsin	.0	.0	.0	.0	.0	.0	.0	.0	.0	.0	.0	.0
Wyoming	111	23	134	39	.0	39	150	23	173	24	.0	24
Puerto Rico	.0	.0	.0	.0	.0	.0	.0	.0	.0	.0	.0	.0
Virgin Islands	.0	.0	.0	.0	.0	.0	.0	.0	.0	.0	.0	.0
Total	1,410	626	2,030	1,270	138	1,410	2,670	764	3,440	724	251	975

Source: Solley, W.B., Merk, C.F., and Pierce, R.R., 1988, Estimated Water Use in the United States in 1985, U.S. Geological Survey Circular 1004

TABLE 5–38. WATER USE BY MINERAL INDUSTRIES IN THE UNITED STATES, 1983

Industry Group and Industry	Establishments Reporting Water Intake of 20 million gallons or more during 1982 (number)	Gross Water Used[1]					Water Discharged[2]		
		Total		Water Intake					
		Quantity (billion gallons)	Percent of all Mineral Industries	Quantity (billion gallons)	Percent of all Mineral Industries	Water Recirculated and Reused (billion gallons)	Total (billion gallons)	Untreated (billion gallons)	Treated (billion gallons)
All mineral industries	**1 534**	**3 328.3**	**100**	**1 197.1**	**100**	**2 131.1**	**1 036.7**	**331.2**	**705.5**
Metal mining	**135**	**734.5**	**22**	**170.2**	**14**	**564.3**	**133.0**	**52.8**	**80.2**
Iron ores	15	421.6	13	45.7	4	375.9	51.2	40.3	11.0
Copper ores	29	223.4	7	89.7	7	133.7	17.7	(D)	(D)
Lead and zinc ores	14	(D)	(D)	(D)	(D)	(D)	26.6	.7	25.9
Gold and silver ores	28	12.4	(Z)	7.3	1	5.0	5.5	1.2	4.3
Gold ores	20	(D)	(D)	5.8	(Z)	(D)	4.0	(D)	(D)
Silver ores	8	(D)	(D)	1.6	(Z)	(D)	1.5	(D)	(D)
Miscellaneous metal ores	42	46.0	1	14.4	1	31.6	16.2	6.6	9.7
Uranium-radium-vanadium ores	34	15.5	(Z)	10.5	1	5.0	11.3	6.5	4.8
Metallic ores, n.e.c.	8	30.5	1	3.9	(Z)	26.6	4.9	(Z)	4.9
Anthracite mining	**16**	**5.4**	**(Z)**	**2.2**	**(Z)**	**3.2**	**7.5**	**.9**	**6.5**
Anthracite	16	5.4	(Z)	2.2	(Z)	3.2	7.5	.9	6.5
Bituminous coal and lignite mining	**275**	**118.6**	**4**	**45.3**	**4**	**73.3**	**116.2**	**30.5**	**85.7**
Bituminous coal and lignite mining	275	118.6	4	45.3	4	73.3	116.2	30.5	85.7
Bituminous coal and lignite	275	118.6	4	45.3	4	73.3	116.2	30.5	85.7
Oil and gas extraction	**555**	**1 451.8**	**44**	**601.6**	**50**	**850.1**	**475.6**	**147.5**	**328.1**
Crude petroleum and natural gas	312	672.8	20	436.6	36	236.2	318.8	84.7	234.1
Natural gas liquids	161	750.7	23	146.7	12	604.1	(D)	(D)	(D)
Oil and gas field services	82	28.2	1	18.4	2	9.8	(D)	(D)	(D)
Drilling oil and gas wells	67	27.3	1	17.5	1	9.7	16.7	8.1	8.5
Oil and gas field services, n.e.c.	14	.9	(Z)	.8	(Z)	.1	(D)	(D)	(D)
Nonmetallic minerals, except fuels	**553**	**1 018.1**	**31**	**377.8**	**32**	**640.2**	**304.3**	**99.3**	**205.0**
Dimension stone	3	(D)	(D)	1.3	(Z)	(D)	1.0	(D)	(D)
Crushed and broken stone, including riprap	178	63.1	2	46.5	4	16.6	45.6	19.2	26.4
Crushed and broken limestone	97	52.9	2	39.0	3	13.9	39.7	15.7	24.1
Crushed and broken granite	46	5.1	(Z)	3.3	(Z)	1.8	2.6	(D)	(D)
Crushed and broken stone, n.e.c.	35	5.1	(Z)	4.2	(Z)	.9	3.3	(D)	(D)
Sand and gravel	248	201.4	6	123.1	10	78.3	67.0	25.6	41.4
Construction sand and gravel	215	123.5	4	85.1	7	38.4	57.3	23.3	34.1
Industrial sand	33	77.8	2	38.0	3	39.8	9.7	2.3	7.4
Clay, ceramic, and refractory minerals	35	(D)	(D)	(D)	(D)	(D)	(D)	(D)	(D)
Kaolin and ball clay	11	16.2	(Z)	15.0	1	1.2	13.4	(D)	(D)
Clay, ceramic, and refractory minerals, n.e.c.	16	11.4	(Z)	4.4	(Z)	7.0	2.6	(D)	(D)
Chemical and fertilizer mineral mining	71	684.8	21	149.6	12	535.1	141.3	(D)	(D)
Barite	10	2.7	(Z)	(D)	(D)	(D)	(D)	(D)	(D)
Potash, soda, and borate minerals	20	125.5	4	26.9	2	98.6	12.4	(D)	(D)
Phosphate rock	19	489.5	15	60.2	5	429.3	63.4	.3	63.1
Miscellaneous nonmetallic minerals	17	(D)	(D)	(D)	(D)	1.4	(D)	(D)	.7
Talc, soapstone, and pyrophyllite	5	(D)	(D)	.4	(Z)	(D)	.3	(D)	(D)
Miscellaneous nonmetallic minerals, n.e.c.	12	(D)	(D)	(D)	(D)	(D)	(D)	(D)	(D)

[1] Total gross water used is equal to sum of water intake plus water recirculated and reused without regard to evaporation.
[2] Volume of water discharged may be greater than water intake due to mine water that is drained and discharged.
D=Withheld to avoid disclosing data for individual companies.
Z=Less than half the unit shown.
n.e.c.=Not elsewhere classified.
Source: 1982 Census of Mineral Industries, U.S. Dept. of Commerce Bureau of the Census, 1985

TABLE 5–39. WATER USE BY MANUFACTURING INDUSTRIES IN THE UNITED STATES, 1983

[Billion gallons]

Industry Group and Industry	Establishments Reporting Water Intake of 20 million gallons or more during 1982	Gross Water Used[1]				Water Discharged		
		Total	Water Intake: Total	Water Intake: From Public Water System	Water Recirculated and Reused	Total	Untreated	Treated
All manufacturing industries[2]	**10 262**	**33 835.2**	**10 038.9**	**1 310.7**	**23 796.3**	**8 913.7**	**4 889.8**	**4 023.9**
Food and kindred products	**2 656**	**1 406.2**	**647.7**	**219.0**	**758.6**	**552.0**	**355.8**	**196.1**
Meat products	552	119.5	92.7	49.5	26.8	85.5	28.7	56.8
Meat packing plants	205	56.5	44.7	16.9	11.8	41.9	14.7	27.2
Sausages and other prepared meats	104	13.9	11.3	7.8	2.6	9.8	4.1	5.6
Poultry dressing plants	217	47.3	35.2	24.0	12.1	32.4	9.0	23.4
Poultry and egg processing	26	1.8	1.5	.8	.3	1.4	.9	.6
Dairy products	498	69.8	38.8	17.9	30.9	35.9	29.4	6.4
Creamery butter	17	1.3	1.0	.3	.2	.9	.8	.1
Cheese, natural and processed	111	15.2	10.2	3.9	5.1	9.3	6.6	2.7
Condensed and evaporated milk	67	14.2	9.5	2.4	4.8	8.6	7.1	1.5
Ice cream and frozen desserts	40	2.0	1.4	1.1	.6	1.1	1.1	.1
Fluid milk	263	37.0	16.7	10.3	20.3	16.0	13.9	2.1
Preserved fruits and vegetables	489	201.4	100.1	38.6	101.3	88.6	46.7	41.9
Canned specialties	52	21.5	17.4	10.3	4.1	13.4	7.0	6.5
Canned fruits and vegetables	224	81.3	30.6	13.9	50.7	26.3	13.8	12.5
Dehydrated fruits, vegetables, and soups	26	(D)	5.6	1.8	(D)	4.9	1.8	3.0
Pickles, sauces, and salad dressings	36	7.2	2.4	2.1	4.9	1.9	.8	1.1
Frozen fruits and vegetables	117	70.4	40.0	8.1	30.4	38.8	21.5	17.3
Frozen specialties	34	(D)	4.1	2.5	(D)	3.4	1.8	1.6
Grain mill products	119	147.0	79.3	19.7	67.6	74.3	65.7	8.6
Flour and other grain mill products	19	.9	.8	.5	.1	.5	(D)	(D)
Cereal breakfast foods	19	12.7	5.9	3.6	6.9	3.8	(D)	(D)
Rice milling	10	(D)	.6	.6	(D)	.4	(D)	(D)
Blended and prepared flour	4	(D)	.1	.1	(D)	.1	.1	-
Wet corn milling	21	126.9	68.3	12.9	58.6	66.6	59.7	7.0
Dog, cat, and other pet food	32	(D)	2.9	1.9	(D)	2.1	1.1	1.1
Prepared feeds, n.e.c.	14	1.1	.8	.3	.3	.7	.6	.2
Bakery products	76	7.6	3.0	2.3	4.6	2.0	1.7	.3
Bread, cake, and related products	61	2.8	1.9	(D)	.9	1.1	1.0	.1
Cookies and crackers	15	4.8	1.1	(D)	3.7	.9	.8	.1
Sugar and confectionery products	151	377.2	178.7	12.7	198.5	142.0	93.8	48.3
Raw cane sugar	35	198.9	83.1	(D)	115.8	63.0	40.1	23.0
Cane sugar refining	20	76.0	62.8	(D)	13.3	54.9	39.1	15.9
Beet sugar	37	65.3	14.6	.9	50.7	9.6	(D)	(D)
Confectionery products	44	(D)	(D)	2.1	15.2	(D)	(D)	(D)
Chocolate and cocoa products	10	(D)	(D)	1.0	3.4	(D)	(D)	.3
Fats and oils	154	103.3	34.1	10.4	69.2	28.9	18.4	10.6
Cottonseed oil mills	18	2.8	1.2	(D)	1.5	(D)	(D)	(D)
Soybean oil mills	60	70.4	20.1	(D)	50.4	18.2	13.5	4.7
Vegetable oil mills, n.e.c.	5	.8	.5	(D)	.3	(D)	(D)	(D)
Animal and marine fats and oils	31	6.3	2.8	.5	3.6	1.5	.8	8
Shortening and cooking oils	40	23.0	9.5	3.0	13.5	7.8	3.3	4.5
Beverages	406	308.7	88.5	46.9	220.3	68.2	50.9	17.3
Malt beverages	56	(D)	53.3	32.1	(D)	41.4	30.8	10.7
Malt	20	19.5	7.3	.8	12.1	6.5	4.7	1.8
Wines, brandy, and brandy spirits	27	2.7	2.6	.3	.1	2.3	.8	1.5
Distilled liquor, except brandy	28	17.2	10.3	1.5	6.9	9.0	(D)	(D)
Bottled and canned soft drinks	247	14.3	12.3	10.9	2.0	6.8	5.1	1.7
Flavoring extracts and syrups, n.e.c.	28	(D)	2.7	1.4	(D)	2.3	(D)	(D)

TABLE 5–39. WATER USE BY MANUFACTURING INDUSTRIES IN THE UNITED STATES, 1983 (continued)

[Billion gallons]

Industry Group and Industry	Establishments Reporting Water Intake of 20 million gallons or more during 1982	Gross Water Used[1]				Water Discharged		
		Total	Water Intake		Water Recirculated and Reused	Total	Untreated	Treated
			Total	From Public Water System				
Food and kindred products –Con.								
Miscellaneous foods and kindred products	211	71.9	32.6	20.9	39.3	26.5	20.6	5.9
Canned and cured seafoods	10	(D)	3.0	(D)	(D)	(D)	(D)	(D)
Fresh or frozen packaged fish	43	4.5	4.4	(D)	.1	4.3	3.0	1.3
Roasted coffee	9	(D)	(D)	-	30.4	(D)	(D)	(D)
Manufactured ice	24	(D)	1.2	.4	(D)	.9	.8	.1
Macaroni and spaghetti	7	(D)	.1	.1	(D)	(D)	(D)	-
Food preparations, n.e.c.	118	(D)	(D)	8.7	8.5	11.3	7.4	3.8
Tobacco products	**20**	**33.9**	**5.3**	**(D)**	**28.6**	**4.0**	**(D)**	**(D)**
Cigarettes	11	(D)	(D)	(D)	(D)	3.1	(D)	(D)
Textile mill products	**761**	**332.9**	**132.6**	**70.2**	**200.3**	**115.6**	**61.2**	**54.4**
Weaving mills, cotton	69	70.9	20.7	(D)	50.2	17.9	13.2	4.7
Weaving mills, manmade fiber and silk	135	67.7	18.1	11.0	49.6	14.5	9.4	5.1
Weaving and finishing mills, wool	23	2.9	2.7	.7	.2	2.6	(D)	(D)
Narrow fabric mills	12	.5	.5	(D)	.1	.4	(D)	(D)
Knitting mills	191	48.2	27.2	20.6	21.0	25.3	15.9	9.3
Women's hosiery, except socks	11	(D)	.7	(D)	(D)	.7	.7	(Z)
Hosiery, n.e.c.	20	2.8	.9	.8	1.9	.8	(D)	(D)
Knit outerwear mills	23	2.6	2.2	(D)	.4	2.1	1.5	.7
Knit underwear mills	12	(D)	.6	(D)	(D)	.5	(D)	(D)
Circular knit fabric mills	89	26.2	17.2	13.0	9.0	16.2	9.9	6.3
Warp knit fabric mills	36	(D)	5.6	4.1	(D)	4.9	2.8	2.1
Textile finishing, except wool	125	62.3	39.0	15.4	23.3	33.6	11.5	22.1
Finishing plants, cotton	27	8.9	6.8	1.9	2.1	6.2	1.0	5.2
Finishing plants, manmade	69	45.4	26.9	9.2	18.5	22.5	7.6	14.9
Finishing plants, n.e.c.	29	8.1	5.4	4.2	2.7	4.8	2.8	2.0
Floor covering mills	65	14.1	11.3	8.8	2.8	9.8	4.7	5.1
Tufted carpets and rugs	61	(D)	10.4	(D)	(D)	(D)	(D)	(D)
Carpets and rugs, n.e.c.	4	(D)	1.0	(D)	(D)	(D)	(D)	(D)
Yarn and thread mills	77	40.5	8.1	5.7	32.4	7.4	4.0	3.3
Yarn mills, except wool	49	30.2	4.6	3.2	25.6	4.2	2.8	1.4
Throwing and winding mills	15	(D)	1.8	1.4	(D)	1.6	.9	.7
Thread mills	11	(D)	(D)	1.1	(D)	1.5	.4	1.1
Miscellaneous textile goods	64	25.7	5.0	2.3	20.7	4.3	1.6	2.7
Felt goods, except woven felts and hats	10	(D)	1.4	.2	(D)	1.4	.1	1.2
Processed textile waste	4	.2	.2	.2	-	(D)	(D)	-
Coated fabrics, not rubberized	9	(D)	.6	.3	(D)	.6	(D)	(D)
Tire cord and fabric	8	(D)	.3	(D)	(D)	.2	(D)	(D)
Nonwoven fabrics	22	(D)	1.5	1.1	(D)	1.0	.6	.5
Textile goods, n.e.c.	7	(D)	(D)	.1	(D)	(D)	(Z)	(D)
Lumber and wood products	**223**	**218.2**	**86.0**	**(D)**	**132.2**	**71.0**	**44.9**	**26.1**
Logging camps and logging contractors	9	(D)	.2	.1	(D)	.3	(D)	(D)
Sawmills and planing mills	89	168.5	68.4	(D)	100.1	58.9	39.4	19.5
Sawmills and planing mills, general	89	168.5	68.4	(D)	100.1	58.9	39.4	19.5
Millwork, plywood, and structural members	78	25.9	7.6	1.9	18.3	4.4	(D)	(D)
Millwork	3	(D)	.4	(Z)	(D)	.4	(Z)	.3
Hardwood veneer and plywood	10	.4	.3	.1	.1	.1	(D)	(D)
Softwood veneer and plywood	64	(D)	6.9	1.7	(D)	3.9	(D)	(D)

TABLE 5–39. WATER USE BY MANUFACTURING INDUSTRIES IN THE UNITED STATES, 1983 (continued)

[Billion gallons]

Industry Group and Industry	Establishments Reporting Water Intake of 20 million gallons or more during 1982	Gross Water Used[1] Total	Water Intake Total	Water Intake From Public Water System	Water Recirculated and Reused	Water Discharged Total	Water Discharged Untreated	Water Discharged Treated
Lumber and wood products–Con.								
Miscellaneous wood products	42	(D)	9.6	(D)	(D)	7.3	(D)	(D)
Wood preserving	7	.4	(D)	—	(D)	(D)	(D)	(D)
Particleboard	8	(D)	(D)	(D)	(D)	(D)	(D)	(D)
Wood products, n.e.c.	27	21.2	8.5	1.3	12.7	6.6	(D)	(D)
Furniture and fixtures	**66**	**6.8**	**3.4**	**2.2**	**3.4**	**3.3**	**2.9**	**.4**
Household furniture	26	3.6	1.9	(D)	1.7	1.8	(D)	(D)
Wood household furniture	15	1.7	1.5	(D)	.2	1.5	(D)	(D)
Metal household furniture	6	(D)	.2	.2	(D)	.2	(D)	(D)
Office furniture	11	(D)	.6	(D)	(D)	.6	.4	.2
Metal office furniture	10	(D)	.6	(D)	(D)	.6	.4	.2
Partitions and fixtures	17	(D)	.5	.5	(D)	.5	(D)	(D)
Metal partitions and fixtures	13	(D)	(D)	(D)	(D)	(D)	(D)	(D)
Miscellaneous furniture and fixtures	8	(D)	(D)	(D)	(Z)	(D)	(D)	(D)
Drapery hardware and blinds and shades	7	(D)	(D)	(D)	(Z)	(D)	(D)	(D)
Paper and allied products	**600**	**7 435.8**	**1 899.3**	**257.3**	**5 536.5**	**1 768.1**	**479.0**	**1 289.1**
Pulp mills	36	1 020.0	283.2	49.1	736.8	282.7	45.7	237.0
Paper mills, except building paper	234	3 908.6	1 009.5	92.1	2 899.0	958.2	304.3	653.9
Paperboard mills	151	2 353.9	538.7	107.2	1 815.2	462.3	85.4	377.0
Miscellaneous converted paper products	104	125.2	56.5	(D)	68.7	55.0	36.9	18.1
Paper coating and glazing	41	16.0	6.5	(D)	9.4	6.2	3.7	2.4
Bags, except textile bags	16	26.2	5.6	(D)	20.6	5.5	(D)	(D)
Pressed and molded pulp goods	11	(D)	32.1	(D)	(D)	(D)	(D)	(D)
Sanitary paper products	20	(D)	8.9	(D)	(D)	9.4	(D)	(D)
Converted paper products, n.e.c.	12	(D)	3.3	(D)	(D)	3.2	(D)	(D)
Paperboard containers and boxes	57	16.8	6.6	(D)	10.3	5.9	4.5	1.4
Folding paperboard boxes	17	6.2	4.1	.4	2.1	(D)	(D)	(D)
Corrugated and solid fiber boxes	21	.9	.4	.4	.5	.4	.3	.1
Sanitary food containers	14	(D)	.9	.4	(D)	.9	.8	(Z)
Fiber cans, drums, and similar products	5	(D)	1.1	(D)	(D)	(D)	(D)	(D)
Building paper and board mills	18	11.3	4.8	(D)	6.5	3.9	2.2	1.7
Chemicals and allied products	**1 315**	**9 630.1**	**3 400.7**	**210.5**	**6 229.4**	**2 979.8**	**1 996.3**	**983.6**
Industrial inorganic chemicals	301	2 164.6	885.0	57.0	1 279.6	758.4	556.3	202.1
Alkalies and chlorine	32	287.0	157.4	(D)	129.6	142.9	40.8	102.1
Industrial gases	76	492.4	18.6	8.9	473.8	11.9	7.3	4.6
Inorganic pigments	22	96.0	48.9	(D)	47.1	49.5	26.0	23.5
Industrial inorganic chemicals, n.e.c.	171	1 289.1	660.1	28.6	629.0	554.0	482.2	71.9
Plastics materials and synthetics	206	1 435.7	427.1	23.4	1 008.6	391.7	277.0	114.6
Plastics materials and resins	143	580.5	132.7	16.0	447.9	108.0	62.3	45.8
Synthetic rubber	22	236.6	62.9	(D)	173.7	58.5	(D)	(D)
Cellulosic manmade fibers	7	133.4	71.7	(D)	61.7	67.6	(D)	(D)
Organic fibers, noncellulosic	34	485.2	159.9	4.1	325.3	157.5	123.1	34.3
Drugs	112	240.1	90.5	19.9	149.6	87.1	55.8	31.3
Biological products	17	3.3	.8	.8	2.5	.5	.3	.2
Medicinals and botanicals	23	122.9	55.3	5.3	67.7	54.9	(D)	(D)
Pharmaceutical preparations	72	113.8	34.4	13.9	79.4	31.6	(D)	(D)

TABLE 5-39. WATER USE BY MANUFACTURING INDUSTRIES IN THE UNITED STATES, 1983 (continued)

[Billion gallons]

Industry Group and Industry	Establishments Reporting Water Intake of 20 million gallons or more during 1982	Gross Water Used[1]				Water Discharged		
		Total	Water Intake		Water Recirculated and Reused	Total	Untreated	Treated
			Total	From Public Water System				
Chemicals and allied products–Con.								
Soaps, cleaners, and toilet goods	108	103.6	64.8	8.5	38.8	61.2	53.7	7.5
Soap and other detergents	32	42.0	16.6	4.3	25.4	14.3	(D)	(D)
Polishes and sanitation goods	22	(D)	(D)	(D)	(D)	3.8	(D)	(D)
Surface active agents	25	(D)	(D)	1.3	8.0	(D)	(D)	(D)
Toilet preparations	29	(D)	2.3	(D)	(D)	(D)	(D)	(D)
Paints and allied products	41	3.5	2.1	1.5	1.4	2.1	1.7	.3
Industrial organic chemicals	296	4 122.3	1 515.9	78.6	2 606.4	1 381.0	815.2	565.8
Gum and wood chemicals	10	(D)	(D)	(D)	21.3	6.5	(D)	(D)
Cyclic crudes and intermediates	75	(D)	(D)	7.0	287.2	30.9	(D)	(D)
Industrial organic chemicals, n.e.c.	211	3 765.5	1 467.6	(D)	2 297.9	1 343.5	798.5	545.0
Agricultural chemicals	116	1 381.4	305.0	14.5	1 076.4	202.7	154.6	48.1
Nitrogenous fertilizers	59	757.4	70.5	12.1	686.9	45.2	(D)	(D)
Phosphatic fertilizers	34	559.0	216.1	(D)	342.9	142.2	(D)	(D)
Agricultural chemicals, n.e.c.	18	64.9	18.4	(D)	46.5	15.3	(D)	(D)
Miscellaneous chemical products	135	178.9	110.3	7.1	68.6	95.7	82.0	13.7
Adhesives and sealants	31	(D)	(D)	.6	(D)	(D)	(D)	.9
Explosives	13	53.4	(D)	—	(D)	(D)	(D)	(D)
Carbon black	14	(D)	1.6	(D)	(D)	.3	(D)	(D)
Chemical preparations, n.e.c.	73	93.6	63.3	5.8	30.3	52.4	(D)	(D)
Petroleum and coal products	**260**	**6 177.3**	**818.4**	**137.7**	**5 358.9**	**699.3**	**323.4**	**375.9**
Petroleum refining	202	6 170.3	814.4	135.9	5 355.9	695.1	321.0	374.1
Paving and roofing materials	39	4.9	2.6	1.0	2.3	3.3	(D)	(D)
Paving mixtures and blocks	15	1.8	.8	(D)	1.0	1.9	(D)	(D)
Asphalt felts and coatings	24	3.0	1.8	(D)	1.3	1.4	(D)	(D)
Miscellaneous petroleum and coal products	19	2.2	1.4	.8	.7	.9	(D)	(D)
Lubricating oils and greases	7	(D)	.2	.2	(D)	.1	(D)	(D)
Petroleum and coal products, n.e.c.	12	(D)	1.3	.6	(D)	.7	(D)	(D)
Rubber and miscellaneous plastics products	**375**	**327.8**	**76.0**	**27.4**	**251.8**	**62.6**	**39.8**	**22.8**
Tires and inner tubes	47	121.5	(D)	4.2	(D)	16.5	8.2	8.3
Rubber and plastics footwear	3	.1	.1	(Z)	(Z)	.1	.1	(Z)
Rubber and plastics hose and belting	24	58.0	(D)	3.0	(D)	5.0	(D)	(D)
Fabricated rubber products, n.e.c.	82	26.8	8.3	4.6	18.4	7.4	(D)	(D)
Miscellaneous plastics products	219	121.4	41.9	15.6	79.5	33.6	22.9	10.8
Leather and leather products	**69**	**6.5**	**6.1**	**2.7**	**.4**	**5.7**	**(D)**	**(D)**
Leather tanning and finishing	65	(D)	(D)	(D)	.4	(D)	(D)	(D)
Stone, clay, and glass products	**602**	**336.7**	**154.7**	**24.0**	**181.9**	**132.8**	**99.5**	**33.3**
Flat glass	16	20.9	4.8	(D)	16.0	4.7	(D)	(D)
Glass and glassware, pressed or blown	120	92.2	13.3	8.7	78.9	11.4	7.1	4.3
Glass containers	75	38.3	7.1	4.0	31.2	5.9	3.8	2.1
Pressed and blown glass, n.e.c.	45	53.9	6.2	4.7	47.7	5.5	3.3	2.2
Products of purchased glass	26	(D)	7.2	(D)	(D)	7.1	(D)	(D)
Cement, hydraulic	100	115.2	80.0	2.4	35.2	68.7	58.6	10.1

TABLE 5–39. WATER USE BY MANUFACTURING INDUSTRIES IN THE UNITED STATES, 1983 (continued)

[Billion gallons]

Industry Group and Industry	Establishments Reporting Water Intake of 20 million gallons or more during 1982	Gross Water Used[1]				Water Discharged		
		Total	Water Intake		Water Recirculated and Reused	Total	Untreated	Treated
			Total	From Public Water System				
Stone, clay, and glass products–Con.								
Structural clay products	10	(D)	1.4	.8	(D)	.7	.5	.2
Ceramic wall and floor tile	4	(D)	(D)	—	(D)	(D)	(D)	(D)
Clay refractories	5	(D)	(D)	—	(D)	(D)	(D)	(D)/
Pottery and related products	30	2.1	1.5	1.1	.6	1.1	.5	.6
Vitreous plumbing fixtures	9	(D)	.5	.3	(D)	.4	.1	.3
Vitreous china food utensils	5	.4	.3	.3	.1	.3	.1	.2
Porcelain electrical supplies	11	.9	.6	.4	.3	.4	.3	.2
Concrete, gypsum, and plaster products	160	16.7	10.8	2.7	5.8	6.5	3.9	2.7
Concrete products, n.e.c.	20	(D)	(D)	.1	(D)	.3	(D)	(D)
Ready-mixed concrete	65	4.5	2.1	.7	2.4	1.5	1.0	.4
Lime	23	7.0	4.3	(D)	2.8	3.2	(D)	(D)
Gypsum products	51	4.3	3.7	(D)	.6	1.5	1.0	.5
Cut stone and stone products	15	(D)	1.1	.5	(D)	1.1	(D)	(D)
Miscellaneous nonmetallic mineral products	125	70.4	34.5	(D)	35.9	31.5	22.7	8.7
Abrasive products	17	(D)	3.5	1.0	(D)	3.5	2.6	.9
Asbestos products	13	8.2	1.8	.8	6.3	1.4	1.0	.3
Gaskets, packing, and sealing devices	12	3.6	1.8	.5	1.7	.9	.6	.2
Minerals, ground or treated	22	22.2	18.7	.8	3.5	18.8	(D)	(D)
Mineral wool	42	27.4	6.2	(D)	21.2	4.4	2.7	1.7
Nonclay refractories	12	2.4	2.0	.3	.3	2.0	(D)	(D)
Nonmetallic mineral products, n.e.c.	7	(D)	.5	(D)	(D)	.5	(D)	(D)
Primary metal industries	**776**	**5 885.2**	**2 362.5**	**108.7**	**3 522.8**	**2 112.0**	**1 227.9**	**884.1**
Blast furnace and basic steel products	259	4 990.5	2 077.6	62.2	2 912.9	1 868.1	1 113.7	754.3
Blast furnaces and steel mills	137	4 908.4	2 038.9	53.9	2 869.5	1 829.8	1 090.9	739.0
Electrometallurgical products	13	11.3	1.2	.4	10.1	1.5	(D)	(D)
Steel wire and related products	34	4.5	2.1	1.4	2.5	2.1	(D)	(D)
Cold finishing of steel shapes	32	18.0	11.3	1.4	6.7	11.0	5.8	5.2
Steel pipe and tubes	43	48.3	24.1	5.2	24.1	23.7	(D)	(D)
Iron and steel foundries	154	218.1	69.1	22.6	149.0	51.5	24.2	27.3
Gray iron foundries	93	200.8	63.0	18.8	137.8	45.7	18.7	27.0
Malleable iron foundries	7	(D)	(D)	1.4	(D)	1.4	1.4	-
Steel investment foundries	10	(D)	(D)	.4	(D)	.5	(D)	(D)
Steel foundries, n.e.c.	44	7.4	4.2	2.0	3.2	3.9	(D)	(D)
Primary nonferrous metals	71	417.1	125.6	3.6	291.4	111.6	51.6	60.0
Primary copper	20	77.0	(D)	(D)	(D)	10.6	8.0	2.5
Primary lead	5	5.1	.7	(D)	4.4	.3	(D)	(D)
Primary zinc	5	18.7	7.2	.5	11.5	(D)	(D)	(D)
Primary aluminum	24	(D)	67.8	1.1	(D)	62.9	(D)	(D)
Primary nonferrous metals, n.e.c.	17	(D)	(D)	.5	.8	(D)	(D)	(D)
Secondary nonferrous metals	34	13.5	3.7	1.3	9.8	3.1	2.2	.9
Nonferrous rolling and drawing	170	213.2	79.9	15.7	133.3	71.4	32.9	38.6
Copper rolling and drawing	29	43.8	18.6	5.2	25.2	14.8	8.0	6.8
Aluminum sheet, plate, and foil	23	93.5	37.0	2.1	56.4	34.6	12.4	22.2
Aluminum extruded products	40	9.1	3.9	2.6	5.3	3.7	2.1	1.6
Aluminum rolling and drawing, n.e.c.	6	12.0	7.1	.5	4.9	6.9	3.6	3.3
Nonferrous rolling and drawing, n.e.c.	26	30.6	6.6	3.0	24.0	4.9	1.5	3.4
Nonferrous wire drawing and insulating	46	24.3	6.8	2.5	17.5	6.5	5.3	1.2
Nonferrous foundries	46	15.5	2.4	1.6	13.1	2.2	1.1	1.1
Aluminum foundries	29	14.2	1.7	1.2	12.5	(D)	(D)	(D)
Brass, bronze, and copper foundries	6	(D)	.2	(D)	(D)	(D)	(D)	(D)
Nonferrous foundries, n.e.c.	11	(D)	.5	(D)	(D)	.5	.1	.4

TABLE 5–39. WATER USE BY MANUFACTURING INDUSTRIES IN THE UNITED STATES, 1983 (continued)

[Billion gallons]

Industry Group and Industry	Establishments Reporting Water Intake of 20 million gallons or more during 1982	Gross Water Used[1]				Water Discharged		
			Water Intake					
		Total	Total	From Public Water System	Water Recirculated and Reused	Total	Untreated	Treated
Primary metal industries–Con.								
Miscellaneous primary metal products	42	17.3	4.1	1.8	13.2	4.0	2.1	1.9
Metal heat treating	28	1.9	1.5	1.3	.4	1.5	.5	1.0
Primary metal products, n.e.c.	14	15.4	2.6	.4	12.8	2.6	1.6	1.0
Fabricated metal products	**724**	**257.9**	**65.4**	**38.1**	**192.5**	**61.4**	**29.7**	**31.7**
Metal cans and shipping containers	89	62.1	6.6	5.6	55.5	6.0	1.5	4.4
Metal cans	80	(D)	6.2	5.2	(D)	5.6	(D)	(D)
Metal barrels, drums, and pails	9	(D)	.4	.4	(D)	.4	(D)	(D)
Cutlery, hand tools, and hardware	72	30.8	14.9	4.0	15.8	14.5	(D)	(D)
Cutlery	4	(D)	(D)	.2	(D)	(D)	(D)	(D)
Hand and edge tools, n.e.c.	28	(D)	1.0	.8	(D)	.9	(D)	(D)
Hand saws and saw blades	4	(D)	.4	(D)	(D)	(D)	(D)	(D)
Hardware, n.e.c.	36	(D)	(D)	(D)	(D)	2.9	1.2	1.7
Plumbing and heating, except electric	25	3.7	3.0	2.5	.7	3.0	(D)	(D)
Metal sanitary ware	6	(D)	.6	(D)	(D)	.7	(D)	(D)
Plumbing fittings and brass goods	12	(D)	2.1	(D)	(D)	2.1	(D)	(D)
Heating equipment, except electric	7	(D)	.3	.3	(D)	.2	.2	-
Fabricated structural metal products	76	28.3	7.0	4.4	21.4	6.0	3.8	2.2
Fabricated structural metal	11	(D)	1.2	(D)	(D)	1.1	(D)	(D)
Metal doors, sash, and trim	10	(D)	.6	.6	(D)	.6	.2	.3
Fabricated plate work (boiler shops)	30	22.2	4.0	2.6	18.2	3.3	1.7	1.6
Sheet metal work	11	.7	.3	.2	.4	.3	.2	.1
Prefabricated metal buildings	5	(D)	(D)	(D)	—	(D)	(D)	(D)
Miscellaneous metal work	8	1.6	.6		1.0	.6	(D)	(D)
Screw machine products, bolts, etc.	49	8.4	2.8	1.8	5.6	2.7	1.6	1.1
Screw machine products	5	(D)	.1	.1	(D)	.1	(D)	(D)
Bolts, nuts, rivets, and washers	44	(D)	2.7	1.7	(D)	2.6	(D)	(D)
Metal forgings and stampings	122	81.5	10.1	8.0	71.5	9.7	7.7	2.1
Iron and steel forgings	36	12.0	2.8	2.3	9.2	2.8	1.9	.9
Nonferrous forgings	6	3.7	1.4	1.4	2.3	1.4	1.1	.3
Automotive stampings	42	62.4	3.3	2.8	59.2	3.0	(D)	(D)
Crowns and closures	4	.3	.3	(D)	-	.3	(D)	(D)
Metal stampings, n.e.c.	34	3.1	2.3	(D)	.8	2.2	1.7	.5
Metal services, n.e.c.	138	7.8	6.9	4.8	1.0	6.7	2.9	3.8
Plating and polishing	126	(D)	6.0	(D)	(D)	5.8	2.6	3.2
Metal coating and allied services	12	(D)	.9	(D)	(D)	.9	.3	.5
Ordnance and accessories, n.e.c.	35	9.5	6.7	2.7	2.8	6.0	3.5	2.6
Small arms ammunition	7	(D)	1.4	(D)	(D)	1.2	(D)	(D)
Ammunition, except for small arms, n.e.c.	14	(D)	2.2	(D)	(D)	2.0	.7	1.3
Small arms	6	(D)	.8	(D)	(D)	.8	(D)	(D)
Ordnance and accessories, n.e.c.	8	(D)	2.3	1.7	(D)	2.0	1.6	.4
Miscellaneous fabricated metal products	118	25.8	7.5	4.3	18.3	6.8	(D)	(D)
Steel springs, except wire	4	(D)	.2	.2	(D)	(D)	(D)	(D)
Valves and pipe fittings	54	10.4	2.9	2.0	7.4	2.8	2.1	.7
Wire springs	4	.9	.1	.1	.8	.1	(D)	(D)
Miscellaneous fabricated wire products	8	(D)	.2	.2	(D)	.2	(D)	(D)
Metal foil and leaf	11	(D)	1.7	(D)	(D)	1.4	1.0	.4
Fabricated pipe and fittings	9	.2	.2	.1	(Z)	(D)	(D)	(D)
Fabricated metal products, n.e.c.	28	(D)	2.2	(D)	(D)	2.1	(D)	(D)
Machinery, except electrical	**523**	**306.5**	**120.0**	**52.5**	**186.4**	**104.9**	**71.2**	**33.8**
Engines and turbines	47	66.2	32.0	(D)	34.1	20.7	12.8	7.9
Internal combustion engines, n.e.c.	35	(D)	(D)	3.8	(D)	(D)	(D)	(D)
Farm and garden machinery	41	40.1	32.8	(D)	7.3	32.0	(D)	(D)
Farm machinery and equipment	28	39.2	32.1	(D)	7.1	31.3	(D)	(D)
Lawn and garden equipment	13	.9	.7	(D)	.2	.7	(D)	(D)

TABLE 5–39. WATER USE BY MANUFACTURING INDUSTRIES IN THE UNITED STATES, 1983 (continued)

[Billion gallons]

Industry Group and Industry	Establishments Reporting Water Intake of 20 million gallons or more during 1982	Gross Water Used[1]				Water Discharged		
		Total	Water Intake		Water Recirculated and Reused	Total	Untreated	Treated
			Total	From Public Water System				
Machinery, except electrical–Con.								
Construction and related machinery	69	51.2	11.4	3.5	39.8	11.8	6.8	5.1
Construction machinery	31	42.6	10.3	(D)	32.3	10.8	6.2	4.6
Mining machinery	8	.3	(D)	(D)	(D)	.1	.1	(Z)
Oil field machinery	21	(D)	.8	(D)	(D)	(D)	(D)	(D)
Metalworking machinery	57	(D)	4.0	1.8	(D)	3.7	3.6	.2
Machine tools, metal cutting types	22	3.2	3.0	(D)	.2	2.7	(D)	(D)
Machine tools, metal forming types	3	(D)	(D)	—	—	(D)	(D)	—
Special dies, tools, jigs, and fixtures	6	(D)	.2	.2	(D)	(D)	(D)	—
Machine tool accessories	14	(D)	(D)	—	(D)	.4	(D)	(D)
Power driven hand tools	10	(D)	.3	.2	(D)	.3	.2	.1
Special industry machinery	36	(D)	6.4	(D)	(D)	6.2	(D)	(D)
Food products machinery	9	(D)	(D)	—	(Z)	.3	.2	(Z)
Paper industries machinery	5	(D)	(D)	(D)	.2	.1	.1	(Z)
Printing trades machinery	4	(D)	(D)	—	(D)	.1	(D)	(D)
Special industry machinery, n.e.c.	13	(D)	(D)	.3	(D)	(D)	(D)	(D)
General industrial machinery	101	38.5	7.1	4.1	31.4	6.1	4.7	1.4
Pumps and pumping equipment	26	(D)	1.6	.8	(D)	1.5	1.2	.3
Ball and roller bearings	22	26.5	2.6	1.4	23.9	2.4	1.7	.7
Air and gas compressors	15	(D)	1.4	1.0	(D)	.7	(D)	(D)
Blowers and fans	9	(D)	(D)	—	.1	(D)	(D)	(Z)
Speed changers, drives, and gears	5	(D)	.4	(D)	(D)	.4	(D)	(D)
Power transmission equipment, n.e.c.	15	1.1	.7	.3	.4	.7	(D)	(D)
General industrial machinery, n.e.c.	7	(D)	.2	.2	(D)	.2	(D)	(D)
Office and computing machines	71	39.7	15.0	6.0	24.7	13.3	12.0	1.3
Electronic computing equipment	57	(D)	(D)	—	(D)	12.3	(D)	(D)
Office machines, n.e.c. and typewriters	12	(D)	(D)	—	(D)	(D)	(D)	(D)
Refrigeration and service machinery	70	32.5	9.6	8.7	23.0	9.6	(D)	(D)
Refrigeration and heating equipment	61	(D)	9.2	8.4	(D)	9.3	(D)	(D)
Measuring and dispensing pumps	4	(D)	.2	.2	(D)	.2	.2	—
Service industry machinery, n.e.c.	3	(D)	(D)	—	—	.1	.1	—
Miscellaneous machinery, except electrical	31	18.3	1.7	(D)	16.5	1.5	.8	.7
Carburetors, pistons, rings, valves	18	17.8	1.5	(D)	16.3	1.3	.6	.7
Machinery, except electrical, n.e.c.	13	.5	.2	(D)	.2	.2	.2	(Z)
Electric and electronic equipment	**678**	**334.8**	**74.1**	**55.1**	**260.7**	**70.3**	**42.5**	**27.8**
Electric distributing equipment	43	(D)	3.4	3.2	(D)	4.7	(D)	(D)
Transformers	14	(D)	1.6	(D)	(D)	2.9	(D)	(D)
Switchgear and switchboard apparatus	29	3.3	1.9	(D)	1.4	1.8	(D)	(D)
Electrical industrial apparatus	76	18.5	9.5	6.6	9.1	9.4	7.4	1.9
Motors and generators	30	6.7	4.8	(D)	2.0	5.0	4.5	.5
Industrial controls	18	(D)	(D)	—	(D)	.6	.5	.1
Welding apparatus, electric	6	.4	.2	.2	.2	.2	.1	(Z)
Carbon and graphite products	19	9.8	3.8	3.6	6.1	3.5	2.2	1.3
Electrical industrial apparatus, n.e.c.	3	(D)	(D)	(D)	(D)	.1	.1	—
Household appliances	75	24.1	9.4	8.8	14.7	9.1	5.7	3.4
Household cooking equipment	15	2.8	1.8	1.5	1.0	1.7	(D)	(D)
Household refrigerators and freezers	11	(D)	2.6	(D)	(D)	2.5	(D)	(D)
Household laundry equipment	10	4.5	2.1	2.1	2.3	2.2	1.4	.9
Electric housewares and fans	17	1.6	1.0	(D)	.6	.9	.8	.2
Household vacuum cleaners	4	(D)	(D)	—	(D)	.2	.1	.1
Household appliances, n.e.c.	15	(D)	1.5	1.5	(D)	1.5	1.1	.4

TABLE 5–39. WATER USE BY MANUFACTURING INDUSTRIES IN THE UNITED STATES, 1983 (continued)

[Billion gallons]

Industry Group and Industry	Establishments Reporting Water Intake of 20 million gallons or more during 1982	Gross Water Used[1]				Water Discharged		
			Water Intake					
		Total	Total	From Public Water System	Water Recirculated and Reused	Total	Untreated	Treated
Electric and electronic equipment–Con.								
Electric lighting and wiring equipment	75	17.8	6.9	4.2	10.9	6.0	4.3	1.7
Electric lamps	15	1.2	(D)	—	(D)	.8	.6	.1
Current-carrying wiring devices	15	2.7	2.1	1.4	.5	2.1	1.7	.4
Noncurrent-carrying wiring devices	20	2.8	1.7	1.2	1.1	1.6	1.0	.6
Residential lighting fixtures	10	.4	.3	.3	.1	.3	.2	.1
Commercial lighting fixtures	5	(D)	.1	.1	(D)	.1	(D)	(D)
Lighting equipment, n.e.c.	4	(D)	.2	.2	(D)	(D)	(Z)	(D)
Radio and TV receiving equipment	22	(D)	2.3	(D)	(D)	2.4	(D)	(D)
Radio and TV receiving sets	12	2.6	1.7	(D)	.8	1.8	(D)	(D)
Phonograph records and prerecorded tape	10	(D)	.6	(D)	(D)	.5	(D)	(D)
Communication equipment	122	61.4	13.3	9.1	48.1	11.7	8.3	3.5
Telephone and telegraph apparatus	26	9.2	3.5	2.3	5.7	3.2	2.0	1.3
Radio and TV communication equipment	96	52.2	9.8	6.8	42.4	8.5	6.3	2.2
Electronic components and accessories	193	166.0	23.6	17.6	142.4	21.9	10.3	11.6
Electron tubes, all types	19	(D)	2.8	(D)	(D)	2.6	1.8	.8
Semiconductors and related devices	65	78.0	11.6	9.6	66.4	10.7	3.3	7.3
Electronic capacitors	12	(D)	3.0	.5	(D)	2.9	2.4	.6
Electronic resistors	8	(D)	.4	(D)	(D)	.4	(D)	(D)
Electronic connectors	21	1.5	1.4	.7	.2	1.2	(D)	(D)
Electronic components, n.e.c.	68	63.3	4.6	4.1	58.7	4.2	1.8	2.4
Miscellaneous electrical equipment and supplies	72	26.7	5.6	(D)	21.0	5.1	(D)	(D)
Storage batteries	36	(D)	1.9	(D)	(D)	1.5	.7	.8
X-ray electromedical, and electrotherapeutic apparatus	8	.6	.4	.4	.3	(D)	(D)	(D)
Engine electrical equipment	13	17.0	2.7	(D)	14.2	2.7	(D)	(D)
Electrical equipment and supplies, n.e.c.	7	(D)	(D)	—	(D)	.2	.2	(Z)
Transportation equipment	**380**	**1 011.3**	**152.8**	**82.3**	**858.6**	**139.2**	**94.0**	**45.2**
Motor vehicles and equipment	194	(D)	66.4	42.8	(D)	59.6	30.7	28.9
Motor vehicles and car bodies	55	(D)	22.6	(D)	(D)	21.0	(D)	(D)
Truck and bus bodies	6	(D)	(D)	—	(D)	.2	(D)	(D)
Motor vehicle parts and accessories	130	229.7	43.5	22.0	186.2	38.4	19.2	19.2
Aircraft and parts	93	(D)	58.4	(D)	(D)	54.1	43.6	10.5
Aircraft	26	54.2	18.0	(D)	36.2	(D)	(D)	2.8
Aircraft engines and engine parts	34	(D)	(D)	6.9	(D)	(D)	(D)	(D)
Aircraft equipment, n.e.c.	33	(D)	(D)	4.2	21.5	4.5	(D)	(D)
Ship and boat building and repairing	30	27.6	16.3	(D)	11.3	15.8	12.6	3.2
Ship building and repairing	30	27.6	16.3	(D)	11.3	15.8	12.6	3.2
Railroad equipment	16	8.1	3.1	2.4	5.0	2.8	1.8	1.0
Motorcycles, bicycles, and parts	3	5.0	1.0	.9	4.0	.9	.6	.3
Guided missiles, space vehicles, parts	34	30.2	6.5	6.0	23.8	4.9	4.0	.9
Guided missiles and space vehicles	17	19.7	2.8	(D)	16.9	2.2	1.8	.4
Space propulsion units and parts	12	5.2	3.3	(D)	1.9	2.3	1.9	.5
Space vehicle equipment, n.e.c.	5	5.4	.4	.4	4.9	.4	.3	(Z)
Miscellaneous transportation equipment	10	5.3	1.2	.9	4.1	1.1	.7	.3
Tanks and tank components	5	5.2	1.1	.8	4.1	1.0	.7	.3
Transportation equipment, n.e.c.	4	(D)	(D)	—	—	(Z)	(Z)	(Z)
Instruments and related products	**154**	**112.0**	**29.8**	**9.8**	**82.3**	**27.6**	**13.6**	**13.9**
Engineering and scientific instruments	5	(D)	.2	(D)	(D)	.2	(D)	(D)

TABLE 5–39. WATER USE BY MANUFACTURING INDUSTRIES IN THE UNITED STATES, 1983 (continued)

[Billion gallons]

Industry Group and Industry	Establishments Reporting Water Intake of 20 million gallons or more during 1982	Gross Water Used[1]				Water Discharged		
		Total	Water Intake		Water Recirculated and Reused	Total	Untreated	Treated
			Total	From Public Water System				
Instruments and related products–Con.								
Measuring and controlling devices	57	21.6	4.2	3.2	17.4	3.8	2.6	1.2
Environmental controls	10	(D)	1.0	(D)	(D)	1.0	(D)	(D)
Process control instruments	12	(D)	.8	(D)	(D)	.7	(D)	(D)
Fluid meters and counting devices	11	(D)	.4	.3	(D)	.3	.3	(Z)
Instruments to measure electricity	14	(D)	1.7	1.3	(D)	1.5	.7	.9
Measuring and controlling devices, n.e.c.	10	(D)	.3	(D)	(D)	.3	.2	.1
Optical instruments and lenses	15	2.1	(D)	—	(D)	.4	(D)	(D)
Medical instruments and supplies	40	4.2	2.2	(D)	2.0	2.1	1.5	.5
Surgical and medical instruments	16	(D)	(D)	—	(D)	.8	(D)	(D)
Surgical appliances and supplies	23	(D)	1.4	1.4	(D)	1.3	(D)	(D)
Photographic equipment and supplies	25	77.7	17.3	(D)	60.4	15.7	(D)	(D)
Watches, clocks, and watchcases	5	(D)	(D)	—	(D)	(D)	(D)	(D)
Miscellaneous manufacturing industries	**80**	**15.4**	**4.3**	**3.4**	**11.1**	**4.0**	**(D)**	**(D)**
Toys and sporting goods	28	(D)	1.5	1.3	(D)	1.4	1.2	.2
Games, toys, and children's vehicles	13	(D)	(D)	—	(D)	.4	(D)	(D)
Sporting and athletic goods, n.e.c.	14	(D)	1.0	1.0	(D)	.9	(D)	(D)
Pens, pencils, and office and art supplies	8	(D)	.6	.3	(D)	.6	.6	(Z)
Pens and mechanical pencils	3	(D)	(D)	—	(D)	(D)	(D)	(D)
Costume jewelry and notions	22	(D)	.8	(D)	(D)	.8	(D)	(D)
Costume jewelry	8	(D)	(D)	—	—	(D)	(D)	(D)
Needles, pins, and fasteners	13	(D)	.6	(D)	(D)	(D)	(D)	(D)
Miscellaneous manufactures	15	(D)	1.2	(D)	(D)	1.1	(D)	(D)
Manufacturing industries, n.e.c.	7	(D)	.6	(D)	(D)	(D)	(D)	(D)

[1] Total gross water used is equal to sum of water intake plus water recirculated and reused without regard to evaporation.
[2] Excludes data for establishments classified as Apparel and Other Textile Products; Printing and Publishing; and administrative and auxiliary establishments for all major groups.
D=Withheld to avoid disclosing data for individual companies.
Z=Less than half the unit shown.
n.e.c.=Not elsewhere classified.
Source: 1982 Census of Manufacturers, U.S. Dept. of Commerce Bureau of the Census, 1986

TABLE 5–40. MAKE-UP WATER REQUIRED IN INDUSTRIAL COOLING SYSTEMS

[Estimated amounts in gallons per 1000 gallons per minute recirculation]

Cycles of Concentration	Temperature Drop				
	15°F (gal.)	20°F (gal.)	25°F (gal.)	30°F (gal.)	35°F (gal.)
1.5	45	60	75	90	105
2.0	30	40	50	60	70
2.5	25	33	42	50	58.5
3.0	22.5	30	37.5	45	52.5
3.5	21	28	35	42	49.1
4.0	20	26.7	33.2	40	46.9
4.5	19.2	25.7	32.0	38.7	45.1
5.0	18.7	25.0	31.8	37.5	44
5.5	18.3	24.5	30.7	36.8	43
6.0	18	24	30	36.1	42.1
6.5	17.7	23.7	29.5	35.5	41.5
7.0	17.5	23.3	29.1	35	40.9

Source: National Aluminate Corp.

TABLE 5–41. GEOGRAPHIC DISTRIBUTION OF WATER-INTENSIVE MANUFACTURING INDUSTRIES IN THE UNITED STATES

[1975 Data; for map of Water Resource Regions see Figure 2–6]

Water Resource Region	Paper		Chemicals		Petroleum Refining		Primary Metals	
	ML/day[1]	mgd	ML/day	mgd	ML/day	mgd	ML/day	mgd
New England	9401	2541	1598	432			643	174
Mid-Atlantic	7104	1920	15 736	4253	10 704	2893	10 341	2795
South Atlantic Gulf	31 627	8548	13 590	3673	1417	383	4329	1170
Great Lakes	8156	2196	10 859	2935	6389	1727	38 298	10 351
Ohio	2471	668	21 108	5705	3193	863	33 562	9071
Tennessee	4151	1122	1342	2525	973	263	395	107
Upper Mississippi	2101	568	2882	779	2638	713	2752	744
Lower Mississippi	6471	1749	19 920	5384	10 744	2904	2186	591
Souris-Red-Rainy	510	138						
Missouri	99	27	2527	683	1986	537	381	103
Arkansas-White-Red	1809	489	3455	934	7425	2007	1653	447
Texas Gulf	3814	1031	41 425	11 196	32 996	8918	4495	1215
Rio Grande			1010	273			103	28
Upper Colorado					44	12		
Lower Colorado	432	117	399	108	40	11	136	37
Great Basin			48	13	614	166	1949	527
Columbia-North Pacific	13 819	3735	1798	486	1017	275	1646	445
California	4310	1165	2682	725	8029	2170	795	215
Alaska	1306	353	162	44	51	14		
Hawaii					310	84		
Total	96 658	26 124	148 543	40 147	87 601	23 676	104 251	28 176

[1]Million liters per day.

Source: Kollar, K.L., and MacAuley, P., 1980, Water Requirements for Industrial Development, J. Am. Water Works Assoc., vol. 72, no.1. Copyright AWWA. Reprinted with permission

TABLE 5–42. PERCENTAGE OF GROSS INDUSTRIAL WATER USE BY PURPOSE IN THE UNITED STATES

Industry	Parameters of Water Use	Gross Water Use by Unit of Production	Percentage Noncontact Cooling	Percentage Process and Related	Percentage Sanitary and Miscellaneous
Meatpacking	gal/lb carcass weight	3.6 gal/lb	42	46	12
Poultry dressing	gal/bird poultry slaughter	11.6 gal/bird	12	77	12
Dairy products	gal/lb milk processed	0.85 gal/lb	53	27	19
Canned fruits and vegetables	gal/case 24–303 cans eq	225 gal/case	19	67	13
Frozen fruits and vegetables	gal/lb frozen product	11.2/gal/lb	19	72	8
Wet corn milling	gal/bu corn grind	416 gal/bu	36	63	1
Cane sugar	gal/ton cane sugar	28 100 gal/ton	30	69	1
Beet sugar	gal/ton beet sugar	33 100 gal/ton	31	67	2
Malt beverages	gal/barrel malt beverage	1500 gal/bbl	72	13	15
Textile mills	gal/lb fiber consumption	34 gal/lb	57	37	6
Sawmills	gal/bd ft lumber	5.4 gal/bd ft	58	36	6
Pulp and paper mills	gal/ton pulp and paper	130 000 gal/ton	18	80	1
Paper converting	gal/ton paper converted	6600 gal/ton	18	77	5
Alkalis and chlorine	gal/ton chlorine	29 800 gal/ton	85	14	1
Industrial gases	gal/1000 cu ft industrial gases	636 gal/mcf	86	13	1
Inorganic pigments	gal/ton inorganic pigments	97 800 gal/ton	41	58	1
Industrial inorganic chemicals	gal/ton chemicals 100 percent basic	14 500 gal/ton	83	16	1
Plastic materials and resins	gal/lb plastic	24 gal/lb	93	7	
Synthetic rubber	gal/lb synthetic rubber	55 gal/lb	83	17	Z
Cellulosic man-made fibers	gal/lb fibers	231 gal/lb	69	30	1
Organic fibers, noncellulosic	gal/lb fibers	101 gal/lb	94	6	1
Paints and pigments	gal/gal paint	13 gal/gal	79	17	4
Industrial organic chemicals	gal/ton chemical building blocks	125 000 gal/ton	91	9	1
Nitrogenous fertilizers	gal/ton fertilizer	28 506 gal/ton	92	8	Z
Phosphatic fertilizers	gal/ton fertilizer	35 602 gal/ton	71	28	1
Carbon black	gal/lb carbon black	4.6 gal/lb	57	38	6
Petroleum refining	gal/barrel crude oil input	1851 gal/bbl	95	5	Z
Tires and inner tubes	gal/tire car and truck tires	518 gal/tire	81	16	3
Hydraulic cement	gal/ton cement	1360 gal/ton	82	17	1
Steel	gal/ton steel net production	62 600 gal/ton	56	43	1
Iron and steel foundries	gal/ton ferrous castings	12 400 gal/ton	34	58	8
Primary copper	gal/lb copper	53 gal/lb	52	46	2
Primary aluminum	gal/lb aluminum	49 gal/lb	72	26	2
Automobiles	gal/car domestic automobiles	36 500 gal/car	28	69	3

Z = less than 0.5 percent of gross water use; percentages may not add evenly due to rounding.
Source: Kollar, K.L., and MacAuley, P., 1980, Water Requirements for Industrial Development, J. Am. Water Works Assoc., vol. 72, no.1. Copyright AWWA. Reprinted with permission

TABLE 5–43. INDUSTRIAL WATER USE PER EMPLOYEE IN THE UNITED STATES

Industry Group	Gross Water Use Per Employee		Intake Per Employee	
	L/day	gal/d	L/day	gal/d
Food and kindred products	15 540	4200	10 360	2800
Tobacco manufacturers	22 570	6100	1480	400
Textile mill products	6660	1800	2960	800
Apparel and related products	370	100	370	100
Lumber and wood products	5920	1600	3700	1000
Furniture and fixtures	440	120	370	100
Paper and allied products	43 930	38 900	42 920	11 600
Printing and publishing	370	100	370	100
Chemicals and allied products	149 110	40 300	56 240	15 200
Petroleum and coal products	603 100	163 000	94 350	25 500
Rubber and plastic products	10 730	2900	3700	1000
Leather and leather products	740	200	703	190
Stone, clay, and glass products	11 470	3100	5550	1500
Primary metal industries	78 440	21 200	44 030	11 900
Fabricated metal products	2960	800	1110	300
Machinery, except electrical	3700	1000	1480	400
Electrical machinery	9250	2500	1110	300
Transportation equipment	17 020	4600	2220	600
Instruments and related products	4440	1200	1110	300
Miscellaneous manufacturing	1110	300	2960	200

Source: Kollar, K.L., and MacAuley, P., 1980, Water Requirements for Industrial Development, J. Am. Water Works Assoc., vol. 72, no.1. Copyright AWWA. Reprinted with permission

TABLE 5-44. WATER USE VERSUS INDUSTRIAL UNITS OF PRODUCTION IN THE UNITED STATES

Industry	Parameters of Water Use	Gross Water Used by Unit of Production	Intake by Unit of Production	Consumption by Unit of Production	Discharge by Unit of Production
Meatpacking	gal/lb carcass weight	3.6 gal/lb	2.2 gal/lb	0.1 gal/lb	2.1 gal/lb
Poultry dressing	gal/bird poultry slaughter	11.6 gal/bird	10.3 gal/bird	0.5 gal/bird	9.8 gal/bird
Dairy products	gal/lb milk processed	0.85 gal/lb	0.52 gal/lb	0.03 gal/lb	0.48 gal/lb
Canned fruits and vegetables	gal/case 24–303 cans eq	225 gal/case	107 gal/case	10 gal/case	98 gal/case
Frozen fruits and vegetables	gal/lb frozen product	11.2 gal/lb	7.1 gal/lb	0.2 gal/lb	6.9/ gal/lb
Wet corn milling	gal/bu corn grind	416 gal/bu	223 gal/bu	18 gal/bu	205 gal/bu
Cane sugar	gal/ton cane sugar	28 100 gal/ton	18 250 gal/ton	950 gal/ton	17 300 gal/ton
Beet sugar	gal/ton beet sugar	33 100 gal/ton	11 100 gal/ton	390 gal/ton	10 700 gal/ton
Malt beverages	gal/barrel malt beverage	1500 gal/bbl	420 gal/bbl	90 gal/bbl	330 gal/bbl
Textile mills	gal/lb fiber consumption	34 gal/lb	14 gal/lb	1.4 gal/lb	12.8 gal/lb
Sawmills	gal/bd. ft lumber	5.4 gal/bd ft	3.3 gal/bd ft	0.6 gal/bd ft	2.7 gal/bd ft
Pulp and paper mills	gal/ton pulp and paper	130 000 gal/ton	38 000 gal/ton	1800 gal/ton	36 200 gal/ton
Paper converting	gal/ton paper converted	6600 gal/ton	3900 gal/ton	270 gal/ton	3600 gal/ton
Alkalis and chlorine	gal/ton chlorine	29 800 gal/ton	22 200 gal/ton	700 gal/ton	21 600 gal/ton
Industrial gases	gal/1000 cu ft industrial gases	636 gal/mcf	226 gal/mcf	31 gal/mcf	193 gal/mcf
Inorganic pigments	gal/ton inorganic pigments	97 800 gal/ton	49 400 gal/ton	1600 gal/ton	47 800 gal/ton
Industrial inorganic chemicals	gal/ton chemicals 100 percent basis	14 500 gal/ton	4750 gal/ton	470 gal/ton	4300 gal/ton
Plastic materials and resins	gal/lb plastic	24 gal/lb	6.7 gal/lb	0.6 gal/lb	6.1 gal/lb
Synthetic rubber	gal/lb synthetic rubber	55 gal/lb	6.5 gal/lb	1.4 gal/lb	5.1 gal/lb
Cellulosic man-made fibers	gal/lb fibers	231 gal/lb	68 gal/lb	4.6 gal/lb	63 gal/lb
Organic fibers, noncellulosic	gal/lb fibers	101 gal/lb	38 gal/lb	1.1 gal/lb	37 gal/lb
Paints and pigments	gal/gal paint	13 gal/gal	7.8 gal/gal	0.4 gal/gal	7.4 gal/gal
Industrial organic chemicals	gal/ton chemical building blocks	125 000 gal/ton	54 500 gal/ton	2800 gal/ton	51 700 gal/ton
Nitrogenous fertilizers	gal/ton fertilizer	28 506 gal/ton	4001 gal/ton	701 gal/ton	3299 gal/ton
Phosphatic fertilizers	gal/ton fertilizer	35 602 gal/ton	8461 gal/ton	1277 gal/ton	7184 gal/ton
Carbon black	gal/lb carbon black	4.6 gal/lb	3.9 gal/lb	0.9 gal/lb	3.1 gal/lb
Petroleum refining	gal/barrel crude oil input	1851 gal/bbl	289 gal/bbl	28 gal/bbl	261 gal/bbl
Tires and inner tubes	gal/tire car and truck tires	518 gal/tire	153 gal/tire	14 gal/tire	139 gal/tire
Hydraulic cement	gal/ton cement	1360 gal/ton	830 gal/ton	150 gal/ton	680 gal/ton
Steel	gal/ton steel net production	62 600 gal/ton	38 200 gal/ton	1400 gal/ton	36 800 gal/ton
Iron and steel foundries	gal/ton ferrous castings	12 400 gal/ton	3030 gal/ton	260 gal/ton	2760 gal/ton
Primary copper	gal/lb copper	53 gal/lb	17 gal/lb	4.1 gal/lb	13 gal/lb
Primary aluminum	gal/lb aluminum	49 gal/lb	12 gal/lb	0.2 gal/lb	11.8 gal/lb
Automobiles	gal/car domestic automobiles	36 500 gal/car	11 464 gal/car	649 gal/car	10 814 gal/car

Source: Kollar, K.L., and MacAuley, P., 1980, Water Requirements for Industrial Development, J. Am. Water Works Assoc., vol. 72, no.1. Copyright AWWA. Reprinted with permission

TABLE 5–45. WATER USE VERSUS STANDARDIZED UNITS OF PRODUCTION IN THE UNITED STATES

Industry	Parameters of Water Use	Gross Water Used by Unit of Production	Intake by Unit of Production	Consumption by Unit of Production	Discharge by Unit of Production
Meatpacking	gal/lb carcass weight	7194 gal/ton	4331 gal/ton	78 gal/ton	4253 gal/ton
Poultry dressing	gal/ton ready-to-cook weight	7389 gal/ton	6542 gal/ton	296 gal/ton	6246 gal/ton
Dairy products	gal/ton milk processed	1692 gal/ton	1035 gal/ton	63 gal/ton	964 gal/ton
Canned fruits and vegetables	gal/ton vegetables canned	19 700 gal/ton	9400 gal/ton	850 gal/ton	8550 gal/ton
Frozen fruits and vegetables	gal/ton vegetables frozen	22 500 gal/ton	14 100 gal/ton	300 gal/ton	13 800 gal/ton
Wet corn milling	gal/ton corn ground	14 869 gal/ton	7988 gal/ton	643 gal/ton	7345 gal/ton
Cane sugar	gal/ton cane sugar	28 102 gal/ton	18 256 gal/ton	944 gal/ton	17 312 gal/ton
Beet sugar	gal/ton beet sugar	33 145 gal/ton	11 118 gal/ton	386 gal/ton	10 731 gal/ton
Malt beverages	gal/gal beer and malt liquor	49 gal/gal	14 gal/gal	3 gal/gal	11 gal/gal
Textile mills	gal/ton textile fiber input	69 808 gal/ton	30 016 gal/ton	3008 gal/ton	27 008 gal/ton
Sawmills	gal/bd ft lumber	5.4 gal/bd ft	3.3 gal/bd ft	0.63 gal/bd ft	2.7 gal/bd ft
Pulp and paper mills	gal/ton paper	130 047 gal/ton	37 971 gal/ton	1178 gal/ton	36 193 gal/ton
Paper converting	gal/ton paper converted	6584 gal/ton	3861 gal/ton	273 gal/ton	3588 gal/ton
Alkalis and chlorine	gal/ton chlorine	29 840 gal/ton	22 302 gal/ton	676 gal/ton	21 626 gal/ton
Industrial gases	gal/ton weight of gas	16 080 gal/ton	5700 gal/ton	780 gal/ton	4900 gal/ton
Inorganic pigments	gal/ton pigments	97 800 gal/ton	49 400 gal/ton	1600 gal/ton	47 800 gal/ton
Industrial inorganic chemicals	gal/ton chemical products	14 500 gal/ton	4700 gal/ton	470 gal/ton	4300 gal/ton
Plastic materials and resins	gal/ton plastics	47 061 gal/ton	13 338 gal/ton	1078 gal/ton	12 278 gal/ton
Synthetic rubber	gal/ton synthetic rubber	110 600 gal/ton	13 200 gal/ton	2800 gal/ton	10 373 gal/ton
Cellulosic man-made fibers	gal/ton fibers	462 230 gal/ton	135 100 gal/ton	9200 gal/ton	125 846 gal/ton
Organic fibers, noncellulosic	gal/ton fibers	202 123 gal/ton	76 523 gal/ton	2153 gal/ton	74 369 gal/ton
Paints and pigments	gal/gal paint	13.2 gal/gal	7.8 gal/gal	0.4 gal/gal	7.4 gal/gal
Industrial organic chemicals	gal/ton chemical building blocks	124 700 gal/ton	54 500 gal/ton	2800 gal/ton	51 700 gal/ton
Nitrogenous fertilizers	gal/ton fertilizer	28 506 gal/ton	4001 gal/ton	701 gal/ton	3299 gal/ton
Phosphatic fertilizers	gal/ton fertilizer	35 602 gal/ton	8461 gal/ton	1277 gal/ton	7184 gal/ton
Carbon black	gal/ton carbon black	9200 gal/ton	7885 gal/ton	1771 gal/ton	6114 gal/ton
Petroleum refining	gal/gal crude petroleum input	44 gal/gal	6.9 gal/gal	0.7 gal/gal	6.2 gal/gal
Tires and inner tubes	gal/tire car and truck tires	518 gal/tire	153 gal/tire	14 gal/tire	139 gal/tire
Hydraulic cement	gal/ton cement	1355 gal/ton	831 gal/ton	146 gal/ton	685 gal/ton
Steel	gal/ton steel net tons	62 601 gal/ton	38 200 gal/ton	1400 gal/ton	36 800 gal/ton
Iron and steel foundries	gal/ton ferrous castings	12 407 gal/ton	3024 gal/ton	260 gal/ton	2764 gal/ton
Primary copper	gal/ton copper	106 000 gal/ton	34 000 gal/ton	8200 gal/ton	26 000 gal/ton
Primary aluminum	gal/ton aluminum	96 300 gal/ton	23 900 gal/ton	381 gal/ton	23 500 gal/ton
Automobiles	gal/car automobiles	36 500 gal/car	11 464 gal/car	649 gal/car	10 814 gal/car

Source: Kollar, K.L., and MacAuley, P., 1980, Water Requirements for Industrial Development, J. Am. Water Works Assoc., vol. 72, no.1. Copyright AWWA. Reprinted with permission

TABLE 5–46. TYPICAL WATER USES IN PAPER MILLS

[1000 ton per day integrated bleached kraft paper mill]

Purpose	Gross Water Use		Intake Requirement (Low Reuse)		Intake Requirement (High Reuse)	
	ML/day	mgd	ML/day	mgd	ML/day	mgd
Kraft pulping (process use)	118	32	51	14	22	6
Kraft pulping (cooling system)	44	12	44	12	1.5	0.4
Bleaching	140	38	70	19	18	5
Paper forming (process system)	129	35	44	12	22	6
Paper forming (cooling system)	14	4	14	4	0.7	0.2
Electric power cooling[1]	51	14	51	14	1.8	0.5
Net totals[2]	499	135	225	61	44	12.1

[1] Condenser cooling requirements for a steam electric plant producing half of the total electric power needs.
[2] Intake net totals are less than the sum of the individual components because much of the wastewater from high quality uses is cascaded to lower quality uses.
Source: Kollar, K.L., and MacAuley, P., 1980, Water Requirements for Industrial Development, J. Am. Water Works Assoc., vol 72, no.1. Copyright AWWA. Reprinted with permission

TABLE 5–47. WATER INTAKE REQUIREMENTS IN THE UNITED STATES—AVERAGE PLANTS VERSUS HIGH RECYCLING PLANTS

Industry	Parameters of Water Use	Intake		Recycling Rate[1]	
		1973 Industry Average	BAT[2] With Maximum Feasible Recycling	1973 Industry Average	BAT[2] With Maximum Feasible Recycling
Meatpacking	gal/lb carcass weight	2.2 gal/lb	0.5 gal/lb	1.66	6.67
Poultry dressing	gal/bird poultry slaughter	10.3 gal/bird	1.7 gal/bird	1.13	6.71
Dairy products	gal/lb milk processed	0.52 gal/lb	0.13 gal/lb	1.64	6.67
Canned fruits and vegetables	gal/case 24–303 cans eq	107 gal/case	29 gal/case	2.10	7.75
Frozen fruits and vegetables	gal/lb frozen product	7.1 gal/lb	1.6 gal/lb	1.60	7.25
Wet corn milling	gal/bu corn grind	223 gal/bu	46 gal/bu	1.86	9.09
Cane sugar	gal/ton cane sugar	18 250 gal/ton	5300 gal/ton	1.54	5.26
Beet sugar	gal/ton beet sugar	11 100 gal/ton	6200 gal/ton	2.98	5.38
Malt beverages	gal/barrel malt beverage	420 gal/bbl	105 gal/bbl	3.50	14.3
Textile mills	gal/lb fiber consumption	14 gal/lb	1.8 gal/lb	2.23	18.2
Sawmills	gal/bd ft lumber	3.3 gal/ft	0.8 gal/ft	1.64	6.85
Pulp and paper mills	gal/ton pulp and paper	38 000 gal/ton	10 700 gal/ton	3.42	12.2
Paper converting	gal/ton paper converted	3900 gal/ton	750 gal/ton	1.70	8.93
Alkalis and chlorine	gal/ton chlorine	22 200 gal/ton	860 gal/ton	1.34	34.5
Industrial gases	gal/1000 cu ft industrial gases	226 gal/mcf	18 gal/mcf	2.82	34.5
Inorganic pigments	gal/ton inorganic pigments	49 400 gal/ton	6100 gal/ton	1.98	16.1
Industrial inorganic chemicals	gal/ton chemicals 100 percent basis	4750 gal/ton	470 gal/ton	3.08	31.2
Plastic materials and resins	gal/lb plastic	6.7 gal/lb	0.7 gal/lb	3.53	33.3
Synthetic rubber	gal/lb synthetic rubber	6.5 gal/lb	1.6 gal/lb	8.38	33.3
Cellulosic man-made fibers	gal/lb fibers	68 gal/lb	8.4 gal/lb	3.42	27.8
Organic fibers, noncellulosic	gal/lb fibers	38 gal/lb	5.0 gal/lb	2.64	20.0
Paints and pigments	gal/gal paint	7.8 gal/gal	0.8 gal/gal	1.69	16.1
Industrial organic chemicals	gal/ton chemical building blocks	54 500 gal/ton	4000 gal/ton	2.29	31.2
Nitrogenous fertilizers	gal/ton fertilizer	4000 gal/ton	900 gal/ton	7.12	31.2
Phosphatic fertilizers	gal/ton fertilizer	8500 gal/ton	2400 gal/ton	4.21	14.7
Carbon black	gal/lb carbon black	3.9 gal/lb	0.3 gal/lb	1.17	16.1
Petroleum refining	gal/barrel crude oil input	289 gal/bbl	55 gal/bbl	6.38	33.3
Tires and inner tubes	gal/tire car and truck tires	153 gal/tire	18 gal/tire	3.39	29.4
Hydraulic cement	gal/ton cement	830 gal/ton	180 gal/ton	1.63	7.41
Steel	gal/ton steel net production	38 200 gal/ton	5300 gal/ton	1.64	11.9
Iron and steel foundries	gal/ton ferrous castings	3030 gal/ton	1080 gal/ton	4.10	11.5
Primary copper	gal/lb copper	17 gal/lb	4.5 gal/lb	3.12	11.9
Primary aluminum	gal/lb aluminum	12 gal/lb	2.9 gal/lb	4.11	16.9
Automobiles	gal/car domestic automobiles	11 500 gal/car	2200 gal/car	3.18	16.3

[1] The recycling rate is obtained by dividing gross water use by intake.
[2] Best available technology economically achievable as defined by Water Pollution Control Act amendments of 1972.
Source: Kollar, K.L., and MacAuley, P., 1980, Water Requirements for Industrial Development, J. Am. Water Works Assoc., vol. 72, no.1. Copyright AWWA. Reprinted with permission

TABLE 5–48. WATER RECYCLING IN THE 20 PLANTS WITH THE HIGHEST RATES IN 34 MAJOR WATER-USING INDUSTRIES IN THE UNITED STATES, 1970

Industry	Gross Water Use[1]	Intake[1]	Mean Recycling Rate[2]	Highest Recycling Rate[2]	Tenth Highest Recycling Rate[2]	Twentieth Highest Recycling Rate[2]
Meat packing plants	49.732	20.335	2.45	7.05	2.41	1.85
Poultry dressing	3.473	1.990	1.75	4.28	1.30	1.14
Fluid milk	8.118	0.859	9.45	71.71	7.92	3.96
Canned fruit and vegetables	10.673	3.419	3.12	18.24	2.50	1.76
Frozen fruit and vegetables	17.353	9.259	1.87	7.13	1.97	1.39
Wet corn milling	53.986	32.109	1.68	11.91	2.31	1.11
Beet sugar	58.949	16.829	3.50	22.24	2.97	1.84
Malt liquors	64.350	12.675	5.08	10.00	2.85	1.11
Shortening and cooking oils	48.106	5.425	8.87	113.53	8.23	1.30
Cigarettes	60.765	2.292	26.51	33.39	15.31	1.11
Weaving mills, cotton	74.289	1.186	62.64	285.31	64.25	27.99
Weaving mills, synthetics	88.114	0.717	122.89	558.25	111.27	48.53
Weaving and finishing, wool	19.163	2.637	7.27	93.44	24.19	1.18
Pulp mills	713.440	208.179	3.43	7.57	3.84	1.41
Papermills, except building paper	723.008	71.057	10.18	76.54	8.96	6.06
Paperboard mills	272.670	14.515	18.79	50.00	14.68	8.22
Alkalis and chlorine	198.798	87.167	2.28	25.11	1.79	1.12
Industrial gases	141.450	1.490	94.93	157.80	84.83	46.23
Cyclic intermediate and crudes	327.354	55.446	5.90	160.00	13.45	2.24
Inorganic pigments	120.387	50.222	2.40	15.22	1.53	1.11
Industrial organic chemicals	962.830	35.142	27.40	48.18	23.20	15.80
Industrial inorganic chemicals	505.919	16.670	30.35	70.95	30.10	23.81
Plastic materials and resins	704.229	5.131	137.25	613.60	27.37	13.81
Cellulosic man-made fibers	209.801	48.088	4.36	20.83	4.30	1.37
Organic fibers, noncellulosic	392.335	151.969	2.58	28.06	2.82	1.16
Pharmaceutical preparations	70.621	15.385	4.59	104.73	7.36	1.11
Fertilizers	282.251	23.373	12.08	90.60	9.72	2.45
Petroleum refining	2026.521	30.221	67.06	251.05	44.08	34.36
Cement, hydraulic	20.868	4.320	4.83	97.35	2.58	1.77
Blast furnaces and steel mills	394.549	29.050	13.58	95.13	18.66	6.76
Electrometallurgical products	22.732	1.827	12.44	65.81	25.64	5.07
Gray iron foundaries	35.396	10.254	3.45	15.23	2.86	1.82
Primary copper	78.473	33.218	2.36	9.85	2.23	1.18
Primary aluminum	65.519	15.723	4.17	10.10	3.50	1.66

[1] Billions of gallons per year: 1 bil gal = 3.7 GL.
[2] The recycling rate is obtained by dividing gross water use by intake.
Source: Kollar, K.L., and MacAuley, P., 1980, Water Requirements for Industrial Development, J. Am. Water Works Assoc., vol. 72, no.1. Copyright AWWA. Reprinted with permission

TABLE 5–49. TYPICAL UNIT WATER REQUIREMENTS FOR ENERGY PRODUCTION IN THE UNITED STATES

Fuel and Process	Standard Unit	Gallons per Standard Unit	Gallons per Million Btu	Major Use
Coal:				
Western coal mining	ton	6.0 - 14.7	0.25 - 0.61	Dust control and washing
Eastern coal mining	ton	15.8 - 18.0	0.66 - 0.75	Dust control and washing
Coal gasification	MSCF[1]	72.0 - 158	72 - 158	Process and cooling
Coal liquefaction	barrel	1,134.0 - 1,750	31 - 200	Process and cooling
Petroleum:				
Oil and gas production	barrel	1.7 - 3.0	3.05	Well drilling and recovery
Oil refining	barrel	43.0	7.58	Process and cooling
Oil shale production	barrel	145.4	30.1	Mining, cooling, processing, and waste disposal
Gas processing	MSCF[1]	1.67	1.67	Cooling
Nuclear fuels	—	—	14.3	Mining and processing
Power generation:				
Fossil fuels	kWh	0.41	120.16	Cooling
Nuclear fuels	kWh	0.80	234.46	Cooling
Geothermal	—	—	527	Cooling and extraction

[1] Million standard cubic feet.

Source: United States Federal Energy Administration, 1974, "Project Independence," Project Independence Report, p. 304: U.S. Government Printing Office, Washington, D.C. 20402

TABLE 5–50. THERMOELECTRIC POWER WATER USE, BY ENERGY SOURCE, IN THE UNITED STATES, 1985

[By state; figures may not add to totals because of independent rounding. All values in million gallons per day]

State	Fossil Fuel: Withdrawals, by Source and Type: Ground Water			Fossil Fuel: Consumptive Use		Geothermal: Withdrawals, by Source: Ground Water	Geothermal: Consumptive Use	Nuclear: Withdrawals, by Source and Type: Ground Water			Nuclear: Consumptive Use	
	Ground Water	Surface Water				Ground Water		Ground Water	Surface Water			
	Fresh	Fresh	Saline	Fresh	Saline	Fresh	Fresh	Fresh	Fresh	Saline	Fresh	Saline
Alabama	.0	5640	.0	75	.0	.0	.0	.0	1290	.0	38	.0
Alaska	4.3	26	.0	3.1	.0	.0	.0	.0	.0	.0	.0	.0
Arizona	30	21	.0	51	.0	.0	.0	2.0	.0	4.6	2.0	4.6
Arkansas	1.1	91	.0	26	.0	.0	.0	.0	1000	.0	.0	.0
California	8.7	412	8400	19	5.8	60	42	.0	.2	3340	5.6	.0
Colorado	14	94	.0	37	.0	.0	.0	.0	1.6	.0	.0	.0
Connecticut	4.5	158	1340	3.7	27	.0	.0	1.3	537	1170	11	24
Delaware	.9	.0	1120	.7	.0	.0	.0	.0	.0	.0	.0	.0
D.C.	.0	130	.0	2.0	.0	.0	.0	.0	.0	.0	.0	.0
Florida	18	633	8680	19	11	.0	.0	.7	.0	2010	.5	2.0
Georgia	3.9	3220	46	75	.0	.0	.0	1.1	55	.0	40	.0
Hawaii	86	3.8	880	.9	8.8	.0	.0	.0	.0	.0	.0	.0
Idaho	.0	.0	.0	.0	.0	.0	.0	.0	.0	.0	.0	.0
Illinois	5.9	7950	.0	46	.0	.0	.0	.8	3720	.0	76	.0
Indiana	24	4450	.0	77	.0	.0	.0	.0	.0	.0	.0	.0
Iowa	9.2	1780	.0	54	.0	.0	.0	.0	27	.0	.0	.0
Kansas	12	401	.0	43	.0	.0	.0	.0	2.1	.0	.2	.0
Kentucky	36	3370	.0	124	.0	.0	.0	.0	.0	.0	.0	.0
Louisiana	30	4350	494	188	4.9	.0	.0	.2	1080	.0	17	.0
Maine	.0	103	21	.0	.0	.0	.0	.0	.5	621	.0	.0
Maryland	1.5	397	2600	289	175	.0	.0	.3	.0	2430	.2	.0
Massachusetts	.0	4880	2890	63	116	.0	.0	.0	191	487	2.3	.0
Michigan	2.3	6220	.0	80	.0	.0	.0	.0	2170	.0	28	.0
Minnesota	1.2	712	.0	125	.0	.0	.0	.0	762	.0	15	.0
Mississippi	16	430	191	77	5.7	.0	.0	34	.0	.0	19	.0

TABLE 5–50. THERMOELECTRIC POWER WATER USE, BY ENERGY SOURCE, IN THE UNITED STATES, 1985 (continued)

[By state; figures may not add to totals because of independent rounding. All values in million gallons per day]

State	Fossil Fuel: Withdrawals, by Source and Type: Ground Water			Fossil Fuel: Consumptive Use		Geothermal: Withdrawals, by Source: Ground Water	Geothermal: Consumptive Use	Nuclear: Withdrawals, by Source and Type: Ground Water			Nuclear: Consumptive Use	
	Ground Water	Surface Water						Ground Water	Surface Water			
	Fresh	Fresh	Saline	Fresh	Saline	Fresh	Fresh	Fresh	Fresh	Saline	Fresh	Saline
Missouri	44	4880	.0	88	.0	.0	.0	.0	2.3	.0	1.4	.0
Montana	.0	67	.0	18	.0	.0	.0	.0	.0	.0	.0	.0
Nebraska	25	1390	.0	1.8	.0	.0	.0	.0	794	.0	.0	.0
Nevada	15	7.5	.0	23	.0	1.0	.2	.0	.0	.0	.0	.0
New Hampshire	.0	329	207	.0	.0	.0	.0	.0	6.9	.0	5.3	.0
New Jersey	2.1	722	2100	1.4	12	.0	.0	1.3	.0	1720	.0	.0
New Mexico	11	48	.0	43	.0	.0	.0	.0	.0	.0	.0	.0
New York	.0	3730	4660	37	47	.0	.0	.0	989	1480	890	1340
North Carolina	.0	3710	.0	21	.0	.0	.0	.0	2690	866	16	19
North Dakota	.4	891	.0	23	.0	.2	.0	.0	.0	.0	.0	.0
Ohio	25	10500	.0	61	.0	.0	.0	.0	23	.0	3.2	.0
Oklahoma	1.0	133	.0	45	.0	.0	.0	.0	.0	.0	.0	.0
Oregon	.0	.5	.0	.5	.0	.0	.0	.0	11	.0	2.3	.0
Pennsylvania	.0	7920	.0	142	.0	.0	.0	.0	2280	.0	51	.0
Rhode Island	.0	.0	261	.0	2.6	.0	.0	.0	.0	.0	.0	.0
South Carolina	.2	1400	6.3	25	.0	.0	.0	.9	3780	.0	30	.0
South Dakota	1.7	2.6	.0	.1	.0	.0	.0	.0	.0	.0	.0	.0
Tennessee	.0	5930	.0	.8	.0	.0	.0	.0	128	.0	.0	.0
Texas	52	7400	3550	198	12	.0	.0	.0	.0	.0	.0	.0
Utah	.0	24	.0	22	.0	.0	.0	.0	.0	.0	.0	.0
Vermont	.4	.4	.0	.6	.0	.0	.0	.0	.0	.0	.0	.0
Virginia	.0	1630	779	27	12	.0	.0	.0	1830	1520	27	23
Washington	.6	12	.0	11	.0	.0	.0	.1	415	.0	11	.0
West Virginia	.0	4210	.0	658	.0	.0	.0	.0	.0	.0	.0	.0
Wisconsin	1.6	3970	.0	40	.0	.0	.0	.0	1470	.0	15	.0
Wyoming	12	224	.0	38	.0	.0	.0	.0	.0	.0	.0	.0
Puerto Rico	5.1	.0	2000	1.3	.0	.0	.0	.0	.0	.0	.0	.0
Virgin Islands	.0	.0	103	.1	.0	.0	.0	.0	.0	.0	.0	.0
Total	505	105,000	40,300	3,000	440	61	42	42	25,200	15,700	1,310	1,410

Source: Solley, W.B., Merk, C.F., and Pierce, R.R., 1988, Estimated Water Use in the United States in 1985, U.S. Geological Survey Circular 1004

TABLE 5–51. THERMOELECTRIC POWER (ELECTRIC UTILITY GENERATION) WATER USE, IN THE UNITED STATES, 1985

[By state; figures may not add to totals because of independent rounding. Mgal/d = million gallons per day; GWh = gigawatthour]

State	All Thermoelectric Power Water Use, in Mgal/d: Self-Supplied Withdrawals, by Source and Type: Ground Water				Public Supply Deliveries	Total: Withdrawals and Deliveries	Total: Consumptive Use			Power Generated, in GWh
	Ground Water	Surface Water								
	Fresh	Fresh	Saline	Total	Fresh	Fresh	Fresh	Saline	Total	
Alabama	.0	6920	.0	6920	.0	6920	113	.0	113	65500
Alaska	4.3	26	.0	26	.3	31	3.1	.0	3.1	3430
Arizona	32	21	4.6	25	.0	53	53	4.6	57	26200
Arkansas	1.1	1090	.0	1090	.3	1090	26	.0	26	21300
California	68	412	11700	12200	31	511	67	5.8	72	98900
Colorado	14	96	.0	96	13	123	37	.0	37	26500
Connecticut	5.8	694	2510	3210	1.3	701	15	50	65	26400
Delaware	.9	.0	1120	1120	.6	1.6	.7	.0	.7	8460
D.C.	.0	130	.0	130	.0	130	2.0	.0	2.0	107
Florida	19	633	10700	11300	4.8	656	20	13	33	95200

TABLE 5-51. THERMOELECTRIC POWER (ELECTRIC UTILITY GENERATION) WATER USE, IN THE UNITED STATES, 1985 (continued)

[By state; figures may not add to totals because of independent rounding. Mgal/d = million gallons per day; GWh = gigawatthour]

State	All Thermoelectric Power Water Use, in Mgal/d									Power Generated, in GWh
	Self-Supplied Withdrawals, by Source and Type					Total				
	Ground Water	Surface Water			Public Supply Deliveries	Withdrawals and Deliveries	Consumptive Use			
	Fresh	Fresh	Saline	Total	Fresh	Fresh	Fresh	Saline	Total	
Georgia	5.0	3280	46	3320	.0	3280	114	.0	114	78100
Hawaii	86	3.8	880	884	.0	90	.9	8.8	9.7	6150
Idaho	.0	.0	.0	.0	.0	.0	.0	.0	.0	.0
Illinois	6.7	11700	.0	11700	.9	11700	121	.0	121	107000
Indiana	24	4450	.0	4450	.0	4480	77	.0	77	43300
Iowa	9.2	1800	.0	1800	1.6	1810	54	.0	54	22300
Kansas	12	403	.0	403	1.0	416	43	.0	43	27300
Kentucky	36	3370	.0	3370	.0	3410	124	.0	124	60100
Louisiana	30	5440	494	5930	.0	5470	205	4.9	210	39700
Maine	.0	103	642	746	.0	103	.0	.0	.0	8070
Maryland	1.8	397	5030	5420	.0	399	290	175	465	25500
Massachusetts	.0	5070	3380	8450	4.2	5070	65	116	182	35800
Michigan	2.3	8390	.0	8390	.0	8390	108	.0	108	73800
Minnesota	1.2	1470	.0	1470	1.5	1480	140	.0	140	30100
Mississippi	50	430	191	621	1.5	481	96	5.7	102	30200
Missouri	44	4890	.0	4890	.3	4930	89	.0	89	48500
Montana	.0	67	.0	67	.0	67	18	.0	18	8750
Nebraska	25	2190	.0	2190	.0	2210	1.8	.0	1.8	4140
Nevada	16	7.5	.0	7.5	2.4	26	24	.0	24	12400
New Hampshire	.0	336	207	542	.0	336	5.3	.0	5.3	5710
New Jersey	3.4	722	3820	4540	.6	726	1.4	12	14	34000
New Mexico	11	48	.0	48	.0	59	43	.0	43	27000
New York	.0	4720	6150	10900	.0	4720	927	1380	2310	81600
North Carolina	.0	6400	866	7270	.0	6400	36	19	55	71700
North Dakota	.5	891	.0	891	.0	892	23	.0	23	20200
Ohio	25	10500	.0	10500	.3	10500	64	.0	64	111000
Oklahoma	1.0	133	.0	133	2.2	136	45	.0	45	40100
Oregon	.0	12	.0	12	.0	12	2.8	.0	2.8	7500
Pennsylvania	.0	10200	.0	10200	.0	10200	193	.0	193	134000
Rhode Island	.0	.0	261	261	.0	.0	.0	2.6	2.6	548
South Carolina	1.1	5180	6.3	5180	.0	5180	55	.0	55	51700
South Dakota	1.7	2.6	.0	2.6	.0	4.2	.1	.0	.1	2520
Tennessee	.0	6060	.0	6060	.6	6060	.8	.0	.8	63500
Texas	52	7400	3550	10900	23	7480	198	12	210	217000
Utah	.0	24	.0	24	.0	24	22	.5	23	14400
Vermont	.4	.4	.0	.4	.0	.8	.6	.0	.6	3720
Virginia	.1	3460	2300	5760	.0	3460	55	35	89	41200
Washington	.7	427	.0	427	.0	427	22	.0	22	16200
West Virginia	.0	4210	.0	4210	.0	4210	658	.0	658	80200
Wisconsin	1.6	5440	.0	5440	.2	5440	54	.0	54	35200
Wyoming	12	224	.0	224	1.9	238	38	.0	38	34200
Puerto Rico	5.1	.0	2000	2000	1.4	6.6	1.3	.0	1.3	11500
Virgin Islands	.0	.0	103	103	.7	.7	.1	.0	.1	480
Total	608	130,000	56,000	186,000	96	131,000	4,350	1,850	6,200	2,140,000

Source: Solley, W.B., Merk, C.F., and Pierce, R.R., 1988, Estimated Water Use in the United States in 1985, U.S. Geological Survey Circular 1004

SECTION H. INDUSTRIAL WATER USE—WORLD

TABLE 5–52. WATER USE BY MANUFACTURING INDUSTRIES IN CANADA, 1981

	Water Intake (10^6 m^3)				
Industry Group	**Processing**	**Cooling, Condensing and Steam**	**Sanitary**	**Other**	**Total**
Food and beverage	129	260	36	5	430
Rubber and plastics	19	32	3	—	54
Textiles	46	76	2	—	124
Wood	20	48	5	—	73
Paper and allied	2188	658	43	10	2899
Primary metals	711	1943	54	11	2719
Metal fabricating	16	12	2	—	30
Transportation equipment	48	52	7	2	109
Nonmetallic mineral products	19	58	3	2	82
Petroleum and coal products	38	516	3	6	563
Chemical and chemical products	195	2614	31	13	2853
Total	3429	6269	189	49	9936

Note: Dashes (—) refer to negligible quantities.
The addition of figures may not be possible because of rounding.
Source: Canadian Water Quality Guidelines, 1987; based on Tate, D.M., and Scharf, D.N., 1985. Water Use in Canadian Industry, 1981. Soc. Sci. Ser. No. 19. Environment Canada, Ottawa

TABLE 5–53. WATER REQUIREMENTS FOR SELECTED INDUSTRIES IN THE WORLD

[Water requirements for unit of product produced]

Industry, Product, and Country	**Unit of Product (Ton, Except as Specified)**	**Water Required per Unit (Liters)**		
Food Products				
Bread or pastry, Belgium		1,100		
Bread, United States		2,100	to	4,200
*Bread, Cyprus		600		
Canned food:				
Belgium:				
Fish, canned		400		
Fish, preserved		1,500		
Fruit		15,000		
Vegetables		8,000	to	80,000
Canada:				
*Fruits and vegetables		10,000	to	50,000
Cyprus:				
*Citrus/tomato juice		2,800		
*Grapefruit sections		16,000		
*Peaches/pears		10,000	to	15,000
*Grapes		30,000		
*Tomatoes, whole		2,000		
*Tomato paste		21,000		
*Peas		10,000		
*Carrots		16,000		
*Spinach		30,000		
Israel:				
*Citrus fruits	ton of raw citrus	4,000		
*Vegetables		10,000	to	15,000
United States				
Apricots		21,200		
Asparagus		20,500		
Beans, green		9,300		
Beans, lima		69,800		
Beets, corn and peas		7,000		
Grapefruit juice		2,800		
Grapefruit sections		15,600		
Peaches and pears		18,100		
Pork and beans		9,300		
Pumpkin and squash		7,000		
Sauerkraut		950		
Spinach		49,400		
Succotash		34,800		
Tomato products		20,500		
Tomatoes, whole		2,200		
*Industry average, fruits, vegetables and juices (1965)		24,000		

TABLE 5-53. WATER REQUIREMENTS FOR SELECTED INDUSTRIES IN THE WORLD (continued)

[Water requirements for unit of product produced]

Industry, Product, and Country	Unit of Product (Ton, Except as Specified)	Water Required per Unit (Liters)
Meat:		
*Meat freezing, Cyprus	ton of carcass	500
Meat freezing, New Zealand		3,000 to 8,600
*Meat packing, United States	ton of prepared meat	23,000
*Meat packing, Canada	ton of carcass	8,800 to 34,000
Meat products, Belgium	ton of prepared meat	200
Sausage factory, Finland		20,000 to 35,000
*Sausage factory, Cyprus		25,000
Slaughtering, Finland	ton, live weight	4,000 to 9,000
*Slaughtering, Cyprus	ton of carcass	10,000
*Meat preserving, Israel	ton of prepared meat	10,000
Fish:		
*Fresh and frozen fish, Canada		30,000 to 300,000
*Canned fish, Canada		58,000
*Canning and preserving fish, Israel	ton of raw fish	16,000 to 20,000
Poultry:		
*Poultry, Canada	ton	6,000 to 43,000
*Chickens, Israel	ton of dressed chicken	33,000
*Chickens, United States	per bird	25
*Turkeys, United States	per bird	75
Milk and milk products:		
Butter:		
*New Zealand		20,000
Cheese:		
*Cyprus		10,000
*New Zealand		2,000
*United States		27,500
Milk:		
Belgium	1,000 liters	7,000
Finland		2,000 to 5,000
*Israel		2,700
Sweden		2,000 to 4,000
*United States		3,000
Milk powder:		
*New Zealand		45,000
South Africa		200,000
*Whey, United States		10,000
*Dairy products, general, Canada		12,200
*Ice cream, United States		10,000
*Yogurt, Cyprus		20,000
Sugar:		
*Denmark	ton of sugar beets	4,800 to 15,800
Finland	ton of sugar beets	10,000 to 20,000
*France	ton of sugar beets	10,900
*Germany, Federal	ton of sugar beets	10,400 to 14,000
*Great Britain	ton of sugar beets	14,900
*Israel	ton of sugar beets	1,800
*Italy	ton of sugar beets	10,500 to 12,500
*Republic of China	ton of sugar cane	15,000
*United States	ton of sugar beets (range)	3,200 to 8,300
*United States	ton of sugar beets (average)	6,000
Beverages:		
Beer:		
Belgium	kiloliter	7,000 to 20,000
*Canada	kiloliter	10,000 to 20,000
*Cyprus	kiloliter (incl. cleaning bottles)	22,000 to 30,000
Finland	kiloliter	10,000 to 20,000
*France	kiloliter	14,500
*Israel	kiloliter	13,500
*United Kingdom	kiloliter	6,000 to 10,000
United States	kiloliter	15,200
*Whiskey, United States	kiloliter of proof spirit	2,600 to 76,000
*Distilled spirits, Israel	kiloliter	30,000
*Wine, France	kiloliter	2,900
*Wine, Israel	kiloliter	500

TABLE 5–53. WATER REQUIREMENTS FOR SELECTED INDUSTRIES IN THE WORLD (continued)

[Water requirements for unit of product produced]

Industry, Product, and Country	Unit of Product (Ton, Except as Specified)	Water Required per Unit (Liters)
Miscellaneous Food Products:		
Chocolate, confectionery, Belgium		15,000 to 17,000
Gelatin (edible), United States		55,100 to 83,500
Maize (wet milling), United States	liter of maize	15.0 to 25.5
Maize syrup, United States	liter of maize	3.8 to 4.3
*Wheat milling, Cyprus		2,000
*Wheat milling, Israel		700 to 1,300
Potato flour, Finland	ton of potatoes	10,000 to 20,000
*Potato starch, Canada	ton of starch	80,000 to 150,000
*Macaroni, Cyprus		1,200
Molasses, Belgium	hectoliter of raw material	1,000 to 12,000
Molasses, United States	hectoliter of 100 proof	840
Pulp and Paper		
Groundwood pulp:		
Finland	ton of wood-pulp	30,000 to 40,000
Sulphate pulp:		
*China, Republic of	ton of bleached pulp	340,000
*China, Republic of	ton of unbleached pulp	230,000
Finland	ton of pulp	250,000 to 350,000
*Sweden	ton of unbleached pulp	75,000 to 300,000
*Sweden	ton of bleached pulp	170,000 to 500,000
Sulphite pulp:		
Finland	ton of bleached pulp	450,000 to 500,000
Finland	ton of unbleached pulp	250,000 to 300,000
*Sweden	ton of bleached pulp	300,000 to 700,000
*Sweden	ton of unbleached pulp	140,000 to 500,000
Wood pulp:		
*Sweden	ton of dry pulp	50,000 to 100,000
South Africa		150,000
Blotting paper, Sweden		350,000 to 400,000
Craft, printing and fine paper, Finland		375,000
*Printing paper, Republic of China		340,000
*Newsprint, Republic of China		190,000
*Newsprint, Canada		165,000 to 200,000
*Fine paper, Republic of China		800,000
Fine paper, Sweden		900,000 to 1,000,000
Newsprint paper, Sweden		200,000
Packing and cartridge paper, Sweden		125,000
Press paper, Finland		200,000
Printing paper, Sweden		500,000
Cardboard, Finland		125,000
Paperboard, United States		62,000 to 376,000
Paper and cardboard, Belgium		180,000
Strawboard, United States		109,000
Wallboard, Finland		125,000
*Wallboard, Sweden		50,000
*Industry average, United States	ton of pulp and paper	236,000
*Industry average, United Kingdom	ton of paper and board	90,000[1]
*Industry average, France	ton of pulp and paper	150,000
Petroleum and Synthetic Fuels		
Aviation gasoline, United States	kiloliter	25,000
*Aviation gasoline, Republic of China	kiloliter	25,000
Gasoline, United States	kiloliter	7,000 to 10,000
*Gasoline, Republic of China	kiloliter	8,000
Gasoline, polymerization, United States	kiloliter	34,000
Kerosene, Belgium	ton	40,000
Synthetic gasoline, United States	kiloliter	377,000
Oilfields, United States	kiloliter of crude petroleum	4,000

[1] Does not include cooling water for power generating plants.

TABLE 5–53. WATER REQUIREMENTS FOR SELECTED INDUSTRIES IN THE WORLD (continued)

[Water requirements for unit of product produced]

Industry, Product, and Country	Unit of Product (Ton, Except as Specified)	Water Required per Unit (Liters)	
Petroleum and Synthetic Fuels–Continued			
Oil refineries:			
*Belgium			
*China, Republic of	ton of crude petroleum	30,500	
Sweden	ton of crude petroleum	10,000	
*United States			
Synthetic fuel:			
From coal:			
South Africa		50,100	
United States	kiloliter	265,500	
From natural gas, United States	kiloliter	88,900	
From shale, United States	kiloliter	20,800	
Chemicals			
Acetic acid, United States		417,000	to 1,000,000
Alcohol, 100 proof, United States	liter	138	
Alcohol, 190 proof, United States	liter	52	to 100
Alumina (Bayer process), United States		26,300	
Ammonia, synthetic, United States	ton of liquid NH_3	129,000	
*Ammonia (Naphtha, reforming), Japan		255,000	
Ammonium nitrate, Belgium		52,000	
Ammonium sulphate, United States		835,000	
Calcium carbide, United States		125,000	
Calcium metaphosphate, United States		16,700	
Carbon dioxide, United States		83,500	
*Caustic soda and chlorine, Canada		125,000	
*Caustic soda (Solvay process), United States		60,500	
*Caustic soda (Dual process), Federal Republic of Germany		160,000	
*Caustic soda (Dual process), Republic of China		200,000	
*Caustic soda (Solvay process), Republic of China		150,000	
Cellulose nitrate, United States		41,700	
Charcoal and wood chemicals, United States	ton of crude $CaAc_2$	271,000	
*Chlorine, Federal Republic of Germany		12,600	
*Ethylene, Israel		16,000	
*Gases, compressed and liquified, Canada	cubic meter	60	to 70
Glycerine, United States		4,600	
Gunpowder, United States		401,000	to 835,000
Hydrochloric acid (salt process), United States	ton of 20 Be HCl	12,100	
Hydrochloric acid (synthetic process), United States	ton of 20 Be HCl	2,000	to 4,200
Hydrogen, United States		2,750,000	
Lactose, United States		835,000	to 918,000
Magnesium carbonate, basic,	ton of basic $MgCO_3$	18,000	
United States	ton of $MgCO_3$	163,000	
Oxygen, United States	cubic meter of O_2	243	
*Polyethylene, Federal Republic of Germany		231,000	(incl. 225,000 cooling water)
*Polyethylene, Israel		8,400	
Potassium chloride (sylvinite), United States		167,000	to 209,000
Smokeless powder, United States		209,000	
Soap, Belgium		37,000	
*Soap, Cyprus		4,500	
Soap (laundry), United States		960	to 2,100
Soda ash (ammonia soda process), 58 percent, United States		62,600	to 75,100
Sodium chlorate, United States		250,000	
Sodium silicate, United States	ton of 40 Be water-glass	670	
Stearine, soap and washing agents, Sweden	ton of fat	70,000	to 200,000
Sulfuric acid, Belgium		20,000	to 25,000
Sulfuric acid (chamber process), United States	ton of 100 percent H_2SO_4	10,400	
Sulfuric acid (contact process), United States	ton of 100 percent H_2SO_4	2,700	to 20,300
*Sulfuric acid, Federal Republic of Germany	ton of SO_3	83,500	

TABLE 5–53. WATER REQUIREMENTS FOR SELECTED INDUSTRIES IN THE WORLD (continued)

[Water requirements for unit of product produced]

Industry, Product, and Country	Unit of Product (Ton, Except as Specified)	Water Required per Unit (Liters)
Textiles		
Steeping, dressing, scouring and bleaching:		
Steeping flax, Belgium		30,000 to 40,000
Dressing flax, Sweden		30,000 to 40,000
Scouring wool, Belgium		240,000 to 250,000
Washing wool, Sweden		10,000
Bleaching textiles, Belgium		180,000
Dyeing:		
Textiles, Belgium		200,000
*Textiles, France (range)		52,000 to 560,000
*Textiles, France (average)		180,000
Finishing:		
Wet finishing of textiles, Belgium		100,000 to 150,000
Dyeing and finishing:		
*Cotton yarn, Israel		60,000 to 180,000
*Synthetic yarn, Israel		90,000 to 180,000
*Wool yarn, Israel		70,000 to 140,000
*Fabrics, Israel		60,000 to 100,000
Mills:		
Cotton:		
Finland		50,000 to 150,000
Sweden		10,000 to 250,000
*Canada	square yard	1.0
Wool:		
Finland	ton of cloth or yarn	150,000 to 350,000
Sweden	ton of wool	400,000
Synthetic fibers:		
Artificial silk, Sweden		2,000,000
Rayon:		
Belgium		2,000,000
Finland		1,000,000 to 2,000,000
Rayon staple, Belgium		550,000
*Industrial duck products, Canada		22,000
*Carpets, Canada	square yard	20
Mining and Quarrying		
Gold, South Africa	ton of ore	1,000
Iron ore (brown), United States		4,200
*Bauxite, United States	ton of ore	300
Sulfur, United States		12,500
Copper, Finland		3,750
*Copper, Israel		3,100
*Gravel, Israel		400
Limestone and by-products, Belgium		200 to 6,500
Iron and Steel Products		
Belgium:		
Blast furnace, no recycling		58,000 to 73,000
Blast furnace, with recycling		50,000
*Finished and semi-finished steel, no recycling		61,000
Finished and semi-finished steel, with recycling		27,000
Canada:		
*Pig iron		130,000
*Open hearth steel		22,000
France:		
*Smelting		46,000
*Martin process (Open hearth)		15,000
*Thomas process (Bessemer converter)		10,000
*Electric furnace steel		40,000
*Rolling mills		30,000
Germany, Federal Republic:		
*Steel works		8,000 to 12,000

TABLE 5–53. WATER REQUIREMENTS FOR SELECTED INDUSTRIES IN THE WORLD (continued)

[Water requirements for unit of product produced]

Industry, Product, and Country	Unit of Product (Ton, Except as Specified)	Water Required per Unit (Liters)
Iron and Steel Products–Continued		
South Africa:		
Steel		12,500
Sweden:		
Iron and steel works		10,000 to 30,000
United States (average):		
*Fully integrated mills		86,000
*Rolling and drawing mills		14,700
*Blast furnace smelting		103,000
*Electrometallurgical ferroalloys		72,000
*Industry, consumptive use (est.)		3,800
Miscellaneous Products		
*Automobiles, United States	vehicle	38,000
Boilers, steam, United States	horsepower-hour	15
*Casein, New Zealand		55,000
Cement, Portland:		
*Belgium		1,900
*Cyprus (dry process)		550
Finland		2,500
*United States (wet process)		900
Ceramics and tiles, Belgium		1,800 to 2,000
**Coal:		
*Ruhr (Fed. Rep. of Germany)		1,000 (min.) to 1,750 (avg.)
*Great Britain		less than 3,000
*Netherlands		2,650
**Includes generation of electricity. If this is not included, the quantities above are reduced by about one-half		
Coal, Belgium		5,000 to 6,000
Coal, coke and by-product coke, United States		6,300 to 15,000
Coal washing, United States		840
Condensers, surface, United States	pound of condensed steam	9.1 to 27.3
Distilling, grain:		
Belgium	hectoliter of grain treated	6,000 to 7,000
United States	hectoliter of grain treated	6,450
Distilling, Sweden	kiloliter of 100 percent alcohol	15,000 to 100,000
Electric power (conventional thermal):		
Sweden	ton of coal	200,000 to 400,000
South Africa	kilowatt-hour (consumptive use)	5
*United States	kilowatt-hour	200
*Republic of China	kilowatt-hour	230
Explosives:		
Sweden		800,000
United States		835,000
Fertilizer plant, Finland	ton of saltpeter (25 percent nitrogen)	270,000
Glass, Belgium		68,000
Laundry:		
*Cyprus	ton of washed goods	45,000
Finland	ton of washed goods	20,000
Sweden	ton of washed goods	30,000 to 50,000
Leather, South Africa		50,100
Leather factory, Finland	ton of hides	50,000 to 125,000
*Leather tanning, United States	sq. meter of hide	20 to 2,550 (range)
*Leather tanning, United States	sq. meter of hide	440 (average)
*Leather tanning, Cyprus	sq. meter of small animal skins	110
Non-ferrous metals, raw and semi-finished, Belgium		80,000
Rock wool, United States		16,700 to 20,900
Rubber, synthetic, United States:		
Butadiene		83,500 to 2,750,000
Buna S		125,000 to 2,630,000
Grade GR-S		117,000 to 2,800,000
Starch:		
Belgium	ton of maize	13,000 to 18,000
Sweden	ton of potatoes	10,000

*Figures based on newer data (post-1960).
Other figures based on older data (pre-1950).

Source: Dept. of Economic and Social Affairs, United Nations, 1969

SECTION I. IRRIGATION—UNITED STATES

FIGURE 5-9. IRRIGATION WATER BUDGET OF THE UNITED STATES

[Percent of diversions; billions gallons per day]

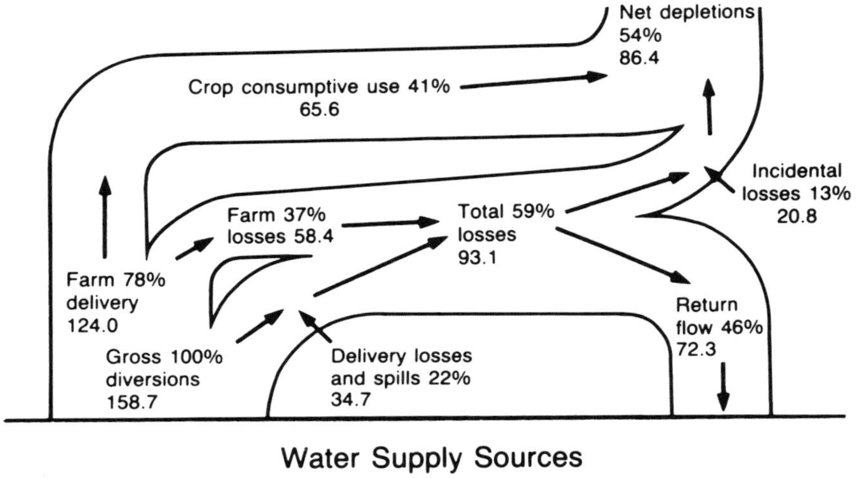

Source: Soil Conservation Service, 1981, America's Soil and Water: Condition and Trends

TABLE 5–54. ACREAGE IRRIGATED WITH ONFARM PUMPED WATER IN THE UNITED STATES

Region and State	Ground Water				Surface Water				Both				Total			
	1974	1977	1980	1983	1974	1977	1980	1983	1974	1977	1980	1983	1974	1977	1980	1983
	1,000 acres															
Northeast:	137	153	166	171	155	151	154	158	0	0	0	0	292	304	320	329
Connecticut	1	2	1	1	8	10	7	7	0	0	0	0	9	12	8	8
Delaware	20	26	35	35	6	4	7	7	0	0	0	0	26	30	42	42
Maine	0	0	0	0	7	7	7	7	0	0	0	0	7	7	7	7
Maryland	8	17	22	27	16	16	18	22	0	0	0	0	24	33	40	49
Massachusetts	1	1	1	1	31	31	31	31	0	0	0	0	32	32	32	32
New Hampshire	0	1	0	0	6	7	7	7	0	0	0	0	6	7	7	7
New Jersey	75	75	75	75	30	30	30	30	0	0	0	0	105	105	105	105
New York	30	30	30	30	29	24	25	25	0	0	0	0	59	54	55	55
Pennsylvania	2	2	2	2	17	17	17	17	0	0	0	0	19	19	19	19
Rhode Island	0	0	0	0	3	3	3	3	0	0	0	0	3	3	3	3
Vermont	0	0	0	0	2	2	2	2	0	0	0	0	2	2	2	2
Lake States:	253	605	906	1,007	158	146	248	262	0	0	0	0	411	751	1,154	1,269
Michigan	56	56	234	253	53	53	137	148	0	0	0	0	109	109	371	401
Minnesota	82	352	430	454	50	45	61	64	0	0	0	0	132	397	491	518
Wisconsin	115	197	242	300	55	48	50	50	0	0	0	0	170	245	292	350
Corn Belt:	274	490	751	787	96	135	172	191	0	7	7	12	370	632	930	990
Illinois	50	40	129	129	0	13	13	13	0	0	0	0	50	53	142	142
Indiana	19	36	55	70	14	20	28	43	0	2	2	3	33	58	85	116
Iowa	50	150	227	235	7	15	15	15	0	0	0	0	57	165	242	250
Missouri	143	248	324	340	55	57	86	103	0	5	5	9	198	310	415	452
Ohio	12	16	16	13	20	30	30	17	0	0	0	0	32	46	46	30
Northern Plains:	6,380	8,977	10,130	10,690	684	676	710	714	186	185	190	190	7,250	9,838	11,030	11,594
Kansas	2,230	3,073	3,489	3,489	65	75	85	85	10	10	15	15	2,305	3,158	3,589	3,589
Nebraska	4,074	5,670	6,316	6,850	505	440	440	440	176	175	175	175	4,755	6,285	6,931	7,465
North Dakota	33	85	127	141	23	11	11	15	0	0	0	0	56	96	138	156
South Dakota	43	149	198	210	91	150	174	174	0	0	0	0	134	299	372	384
Appalachia:	17	23	25	27	175	107	265	315	0	3	10	2	192	223	200	344
Kentucky	1	1	1	1	26	26	26	29	0	0	0	0	27	27	27	30
North Carolina	5	7	3	11	104	110	150	185	0	3	10	2	109	120	163	198
Tennessee	6	7	10	12	11	12	15	18	0	0	0	0	17	19	25	30
Virginia	3	3	11	3	31	47	63	82	0	0	0	0	36	55	74	85
West Virginia	0	0	0	0	3	2	1	1	0	0	0	0	3	2	1	1

TABLE 5–54. ACREAGE IRRIGATED WITH ONFARM PUMPED WATER IN THE UNITED STATES (continued)

Region and State	Ground Water				Surface Water				Both				Total			
	1974	1977	1980	1983	1974	1977	1980	1983	1974	1977	1980	1983	1974	1977	1980	1983
							1,000 acres									
Southeast:	1,058	1,343	2,178	2,437	980	1,359	1,285	1,316	3	8	14	105	2,041	2,710	3,477	3,858
Alabama	8	25	35	43	17	30	95	112	0	5	10	0	25	60	140	155
Florida	960	1,076	1,450	1,610	823	960	817	840	0	0	0	0	1,783	2,036	2,267	2,450
Georgia	80	230	663	725	114	347	323	294	0	0	0	100	194	577	986	1,119
South Carolina	10	12	30	59	26	22	50	70	3	3	4	5	39	37	84	134
Delta States:	1,466	1,486	1,837	2,733	722	676	525	460	500	500	675	567	2,688	2,662	3,037	3,760
Arkansas	900	900	1,075	1,770	296	300	75	63	500	500	675	567	1,696	1,700	1,825	2,400
Louisiana	340	284	395	405	332	276	350	355	0	0	0	0	672	560	745	760
Mississippi	226	302	367	558	94	100	100	42	0	0	0	0	320	402	467	600
Southern Plains:	7,700	7,448	7,091	6,618	1,491	1,569	1,215	1,193	256	256	712	712	9,517	9,273	9,018	8,523
Oklahoma	680	730	746	645	40	118	120	120	0	0	0	0	720	848	866	765
Texas	7,090	6,718	6,345	5,973	1,451	1,451	1,095	1,073	256	256	712	712	8,797	8,425	8,152	7,758
Mountain:	3,537	3,687	3,821	3,821	1,098	1,139	1,208	1,236	1,384	1,475	1,520	1,515	6,020	6,301	6,549	6,574
Arizona	552	550	550	548	0	0	0	0	391	390	390	390	943	940	940	938
Colorado	900	940	940	940	45	50	60	60	700	710	720	720	1,645	1,700	1,720	1,720
Idaho	1,056	1,200	1,250	1,270	478	482	527	537	100	150	180	180	1,634	1,832	1,957	1,987
Montana	40	57	58	60	284	316	389	401	0	0	0	0	324	373	447	461
Nevada	170	170	170	170	34	34	34	34	0	0	0	0	204	204	204	204
New Mexico	634	585	653	633	43	43	43	45	143	175	180	172	820	803	876	850
Utah	60	60	70	70	164	164	80	80	0	0	0	0	224	224	150	150
Wyoming	125	125	130	130	50	50	75	79	50	50	50	53	225	225	255	264
Pacific:	4,561	4,912	4,687	4,771	1,725	1,845	2,078	2,363	0	0	200	200	6,286	6,757	6,965	7,334
California	4,073	4,388	4,065	4,065	380	410	410	410	0	0	200	200	4,453	4,798	4,675	4,675
Oregon	246	264	292	328	644	686	738	811	0	0	0	0	890	950	1,030	1,139
Washington	242	260	330	378	701	749	930	1,142	0	0	0	0	943	1,009	1,260	1,520
Alaska	3	1	1	1	4	1	1	1	0	0	0	0	7	2	2	2
Hawaii	70	80	80	80	6	6	6	6	0	0	0	0	76	86	86	86
Total	25,465	29,205	31,673	33,143	7,294	7,807	7,867	8,215	2,329	2,431	3,328	3,302	35,150	39,443	42,868	44,660

Source: Sloggett, G. 1985, Energy and U.S. Agriculture: Irrigation Pumping, 1974–83, United States Department of Agriculture, Agricultural Report No. 545

TABLE 5-55. GROWTH OF IRRIGATED FARMLAND IN THE UNITED STATES, 1889–1984[1]

Year	Farmland[2]	Irrigated Farmland[3]	Share of Farmland Irrigated	Acreage Irrigated per Farm Irrigated	Change in Irrigated Acreage	Average Annual Growth in Irrigated Acreage
	Million acres		Percent	Acres	Million acres	Percent
1889	623	3.6	0.6	67	NA	NA
1900	839	7.5	.9	70	3.9	10.8
1910	879	14.4	1.6	89	6.9	9.2
1920	956	19.2	2.0	83	4.8	3.3
1930	987	19.5	2.0	74	.3	1.6
1939	1,065	18.0	1.7	60	-1.5	-.8
1949	1,161	25.8	2.2	84	7.8	4.3
1959	1,123	33.0	3.0	108	7.2	2.8
1969	1,063	39.0	3.7	152	6.0	1.8
1978	1,014	50.3	5.0	168	11.7	3.0
1982	987	49.0	4.9	176	-1.3	-.5
1984	na	44.7	na	210	-4.3	-2.1

NA = Not applicable.
na = Not available.
[1] Various estimates of irrigated U.S. land acreage are available. These estimates often differ because of differences in definition, time periods, and statistical bases.
[2] Farmland includes agricultural land used for crops, pasture, or grazing. It also includes woodland and wasteland not actually under cultivation or used for pasture or grazing, provided it is part of the farm operator's total operation.
[3] Does not include double-cropped land.
Source: Day, J.C., and Horner, G.L., 1987, U.S. Irrigation, Extent and Economic Importance, U.S. Department of Agriculture Information Bulletin 523

FIGURE 5–10. GROWTH OF IRRIGATED ACREAGE IN THE UNITED STATES
[By crop group]

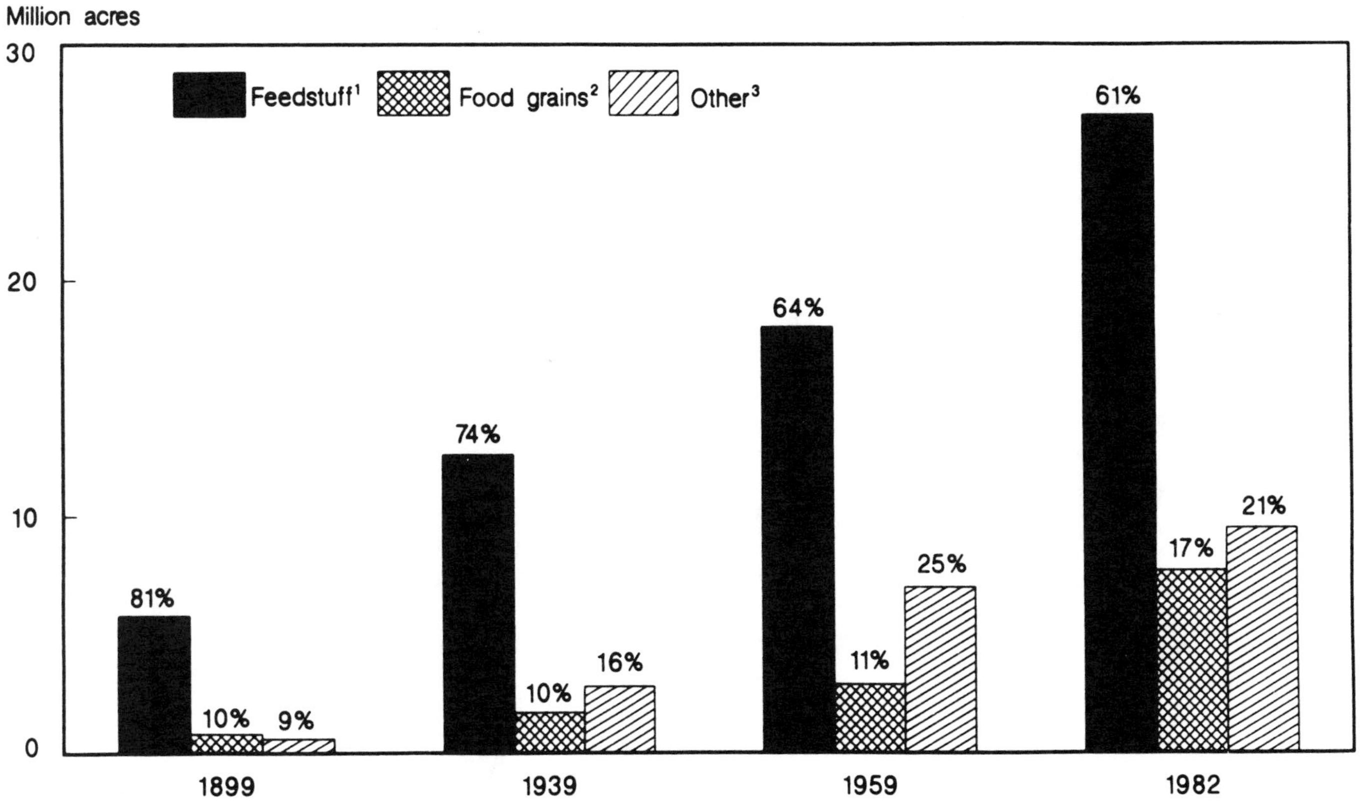

[1] Corn, barley, oats, sorghum, hay, pasture, and silage.
[2] Wheat and rice.
[3] Cotton, sugar, peanuts, tobacco, soybeans, vegetables, and orchards.
Source: Day, J.C., and Horner, G.L., 1987, U.S. Irrigation, Extent and Economic Importance, U.S. Department of Agriculture Information Bulletin 523

TABLE 5–56. ACREAGE AND VALUE OF IRRIGATED CROPLAND IN THE UNITED STATES, 1982

[By crop; contiguous United States]

Crop	Acreage: Irrigated	Acreage: Share of Crop Irrigated	Acreage: Share of Total Cropland[1]	Value: Irrigated	Value: Irrigated Share of Crop Value	Value: Irrigated Share of Total Crop Value[2]	Value: Irrigated Value per Acre	Value: Rainfed Value per Acre
	1,000 acres	Percent	Percent	Million dollars	Percent	Percent	Dollars	Dollars
Corn	9,604	12.3	2.8	3,440	17.2	4.5	358	235
Sorghum	2,295	17.0	.7	901	53.2	1.2	393	66
Wheat	4,650	6.6	1.4	1,144	16.7	1.4	246	84
Barley and oats	2,098	11.8	.6	375	20.0	.5	179	93
Rice	3,233	100.0	1.0	1,226	100.0	1.7	379	NA
Cotton	3,424	35.0	1.0	1,883	58.4	2.5	549	200
Soybeans	2,321	3.6	.7	491	4.4	.6	211	17
Irish potatoes	812	64.0	.2	1,261	81.7	1.7	1,553	546
Hay	8,507	15.0	2.5	2,275	27.7	3.0	267	120
Vegetables and melons	2,029	60.7	.6	3,375	79.7	4.5	1,663	591
Orchard crops	3,343	70.4	1.0	4,732	85.1	6.2	1,415	502
Sugar beets	550	53.2	.2	491	77.7	.6	893	262
Other crops[3]	2,428	17.9	.7	2,424	25.4	3.1	998	638
Total[4]	45,289	NA	13.4	24,047	NA	31.8	531[5]	176[5]

NA = Not applicable.

[1] This is the share each irrigated crop represents of the total acreage of crops produced in the 48 States in 1982.

[2] This is the share each irrigated crop represents of the total value of crops produced in the 48 States in 1982.

[3] Includes peanuts, tobacco, dry edible beans, and the minor acreage crops rye, flax, sugarcane, and dry edible pears.

[4] Includes about 932,000 acres of double-cropped land. Figures might not add to totals due to rounding.

[5] Average weighted by acreage.

Source: Day, J.C., and Horner, G.L., 1987, U.S. Irrigation, Extent and Economic Importance, U.S. Department of Agriculture Information Bulletin 523

TABLE 5–57. IRRIGATION WATER USE IN THE UNITED STATES, 1985

[By state; figures may not add to totals because of independent rounding]

STATE	Irrigated Land By Type, in thousand acres: Spray	Irrigated Land: Flood	Thousand Acre-Feet Per Year: Withdrawals, by Source: Freshwater: Ground	Freshwater: Surface	Total	Reclaimed Sewage	Conveyance Losses	Consumptive Use, Freshwater	Million Gallons Per Day: Withdrawals, by Source: Freshwater: Ground	Freshwater: Surface	Total	Reclaimed Sewage	Conveyance Losses	Consumptive Use, Fresh-Water
Alabama	162	.0	20	57	77	.0	.0	58	18	51	69	.0	.0	51
Alaska	.4	.0	.0	.0	.0	.0	.0	.0	.0	.0	.0	.0	.0	.0
Arizona	465	852	2800	3390	6180	32	1320	3550	2500	3020	5520	29	1180	3170
Arkansas	182	1840	3740	607	4340	.0	308	3200	3330	541	3870	.0	275	2850
California	2000	7580	11600	22700	34400	263	1040	21700	10400	20300	30600	235	929	19300
Colorado	675	2680	2390	11500	13900	5.5	3220	5120	2130	10300	12400	4.9	2880	4570
Connecticut	6.7	.0	.2	2.9	3.0	.0	.0	3.0	.2	2.6	2.7	.0	.0	2.7
Delaware	61	.0	21	8.9	30	.0	.0	30	19	8.0	27	.0	.0	27
D.C.	.0	.0	.0	.0	.0	.0	.0	.0	.0	.0	.0	.0	.0	.0
Florida	804	1110	1780	1490	3270	57	64	2250	1590	1320	2910	51	57	2010
Georgia	1100	.0	346	162	508	31	.0	508	308	145	453	28	.0	453
Hawaii	.6	263	377	639	1020	1.6	102	23	336	570	906	1.4	91	21
Idaho	2770	1330	3710	19400	23100	.0	7210	5810	3310	17300	20600	.0	6430	5180
Illinois	256	.0	80	.0	80	.0	.0	80	71	.0	71	.0	.0	71
Indiana	155	.0	40	12	52	.0	.0	52	36	11	47	.0	.0	47

TABLE 5–57. IRRIGATION WATER USE IN THE UNITED STATES, 1985 (continued)

[By state; figures may not add to totals because of independent rounding]

STATE	Irrigated Land By Type, in thousand acres		Thousand Acre-Feet Per Year: Withdrawals, by Source: Freshwater		Thousand Acre-Feet Per Year				Million Gallons Per Day: Withdrawals, by Source: Freshwater		Million Gallons Per Day			
	Spray	Flood	Ground	Surface	Total	Reclaimed Sewage	Conveyance Losses	Consumptive Use, Freshwater	Ground	Surface	Total	Reclaimed Sewage	Conveyance Losses	Consumptive Use, Fresh-Water
Iowa	139	39	59	17	76	.0	.0	76	53	15	67	.0	.0	67
Kansas	1180	1770	5020	291	5310	.0	321	4990	4470	260	4730	.0	287	4450
Kentucky	21	.4	.3	8.3	8.6	.0	.4	8.2	.3	7.4	7.7	.0	.3	7.3
Louisiana	170	707	795	868	1660	.0	164	1660	709	775	1480	.0	146	1480
Maine	6.3	.0	.2	2.0	2.1	.0	.0	2.1	.2	1.7	1.9	.0	.0	1.9
Maryland	57	.0	22	16	39	.0	.0	39	20	15	34	.0	.0	34
Massachusetts	45	1.8	4.4	14	18	.0	.0	18	3.9	12	16	.0	.0	16
Michigan	325	.0	102	134	236	23	.0	259	91	119	210	21	.0	231
Minnesota	416	31	147	87	234	5.4	21	213	131	78	209	4.8	19	190
Mississippi	258	484	813	180	993	.0	103	439	725	161	886	.0	92	392
Missouri	172	326	318	25	343	.0	.0	248	283	23	306	.0	.0	221
Montana	662	1640	90	9220	9310	.0	4780	1980	80	8220	8300	.0	4270	1760
Nebraska	3330	4150	5800	2340	8140	.0	2930	5220	5180	2090	7270	.0	2610	4650
Nevada	156	687	840	2910	3750	12	821	1940	750	2600	3350	11	732	1730
New Hampshire	1.3	.0	.0	.6	.6	.0	.0	.6	.0	.6	.6	.0	.0	.6
New Jersey	89	3.3	41	107	148	.0	.0	133	37	95	132	.0	.0	119
New Mexico	367	577	1310	1850	3160	.0	.0	1430	1170	1650	2820	.0	.0	1270
New York	59	.0	22	20	42	.0	.0	42	20	18	38	.0	.0	38
North Carolina	222	.0	11	137	148	.0	.0	141	9.8	123	132	.0	.0	126
North Dakota	125	79	72	101	173	.0	9.4	139	64	90	154	.0	8.4	124
Ohio	32	.0	8.1	10	19	.0	.0	17	7.3	9.4	17	.0	.0	15
Oklahoma	280	168	426	73	499	.0	3.7	490	380	65	445	.0	3.3	437
Oregon	1130	911	528	5870	6400	5.2	865	2750	471	5240	5710	4.6	772	2450
Pennsylvania	18	.0	1.7	10	12	.0	.0	12	1.5	9.2	11	.0	.0	11
Rhode Island	8.5	.0	.4	3.5	3.9	.0	.0	3.9	.3	3.1	3.4	.0	.0	3.4
South Carolina	58	.0	24	14	38	.0	.0	38	21	13	34	.0	.0	34
South Dakota	324	73	127	389	516	.0	135	313	113	347	460	.0	120	279
Tennessee	29	.0	2.7	7.4	10	.2	.0	5.9	2.4	6.6	8.9	.1	.0	5.3
Texas	1920	4830	6070	3020	9100	43	837	7530	5420	2700	8120	38	747	6720
Utah	174	937	431	3590	4020	6.5	341	2170	384	3200	3590	5.8	305	1940
Vermont	1.2	.0	.0	.6	.6	.0	.0	.6	.0	.5	.5	.0	.0	.5
Virginia	84	.0	7.3	51	58	1.6	5.8	38	6.5	46	52	1.5	5.2	34
Washington	1270	350	706	4840	5540	.0	44	4980	629	4310	4940	.0	39	4440
West Virginia	4.1	.0	.3	3.9	4.2	.0	.0	4.2	.3	3.4	3.7	.0	.0	3.7
Wisconsin	249	.0	92	2.3	94	.0	.0	94	82	2.0	84	.0	.0	84
Wyoming	211	1600	336	6000	6340	.0	1800	2870	300	5360	5660	.0	1610	2560
Puerto Rico	8.8	29	56	120	176	.0	18	114	50	107	157	.0	16	102
Virgin Islands	.0	.0	.0	.0	.0	.0	.0	.0	.0	.0	.0	.0	.0	.0
Total	22,200	35,000	51,200	102,000	154,000	487	26,500	82,700	45,700	91,300	137,000	434	23,600	73,800

Source: Solley, W.B., Merk, C.F., and Pierce, R.R., 1988, Estimated Water Use in the United States in 1985, U.S. Geological Survey Circular 1004

TABLE 5–58. FEDERAL IRRIGATION PROJECTS, BY TYPE OF SERVICE: 1960 TO 1984

TYPE OF SERVICE	Unit	1960	1970	1975	1979	1980	1981	1982	1983	1984
All projects:										
Irrigable acreage	1,000	8,171	10,198	10,930	12,236	11,944	11,941	12,592	11,901	12,041
Irrigated acreage	1,000	6,900	8,570	9,309	10,234	10,093	10,122	10,635	9,236	9,805
Gross crop value	Mil. dollars	1,158	1,882	4,419	6,909	7,438	6,744	7,323	7,005	7,348
Full service:[1]										
Irrigable acreage	1,000	4,326	4,844	4,925	5,076	5,082	5,089	5,085	5,098	5,107
Irrigated acreage	1,000	3,488	4,037	4,213	4,254	4,275	4,290	4,249	3,978	4,182
Gross crop value	Mil. dollars	581	847	1,858	2,348	2,668	2,553	2,514	2,943	3,077
Supplemental and temporary service:[2]										
Irrigable acreage	1,000	3,845	5,354	6,005	7,160	6,862	6,852	7,507	6,804	6,934
Irrigated acreage	1,000	3,412	4,533	5,096	5,980	5,818	5,832	6,386	5,258	5,623
Gross crop value	Mil. dollars	577	1,035	2,561	4,561	4,770	4,191	4,809	4,062	4,271

[1] Applies to irrigable land receiving its sole irrigation supply through Bureau of Reclamation-constructed facilities and to previously irrigated land in non-Federal projects where a substantial part of the facilities was constructed, rehabilitated, or replaced by the Bureau.
[2] Applies to irrigable land receiving irrigation water through Bureau projects in addition to supply from non-project sources and to land for which water is delivered under temporary arrangements.
Source: U.S. Bureau of Reclamation, Annual Report

TABLE 5–59. IRRIGATION IN AREAS OF THE UNITED STATES WITH DECLINING GROUND-WATER SUPPLIES

[In 11 states with major ground-water irrigation[1]]

State	Total Ground-Water Irrigation 1977	Total Ground-Water Irrigation 1983	Irrigated Area With Declining Water[2] 1977	Irrigated Area With Declining Water[2] 1983	Percent of Irrigated Land Experiencing Declining Water Levels in 1983
	1,000 acres				
Arizona	940	938	—[3]	606	65
Arkansas	1,400	2,337	—[3]	425	18
California	4,388	4,265	—[3]	2,068	48
Colorado	1,650	1,660	570	590	36
Florida	1,076	1,610	—[3]	250	16
Idaho	1,149	1,450	—[3]	223	15
Kansas	3,083	3,504	1,950	2,180	62
Nebraska	5,855	7,025	1,842	2,029[4]	29
New Mexico	760	805	560	560	70
Oklahoma	730	645	507	523	81
Texas	7,846	6,685	6,425	4,565[4]	73
Total	28,877	30,924	—[2]	14,029	45

[1] Total ground-water area irrigated was estimated for 1977 and 1983. Decline area irrigated was estimated from data for the latest year available.
[2] Only areas experiencing at least a six-inch average annual decline are included in these estimates.
[3] Data insufficient to make time comparisons.
[4] Data for 1984.
Source: Ogg, C.W., Hostetter, J.E. and Lee, D.J., 1988, Expanding the Conservation Reserve to Achieve Multiple Environmental Goals, J. of Soil and Water Conservation. Copyright 1988. Soil and Water Conservation Society. Reprinted with permission

TABLE 5–60. TOP 10 STATES IN IRRIGATED AGRICULTURE IN THE UNITED STATES

[In selected categories]

Item	Unit	Ranking by State 1	2	3	4	5	6	7	8	9	10
Number of farms in 1982:											
All farms	1000	TX 185.0	IA 115.4	MO 112.4	KY 101.6	IL 98.5	MN 94.4	TE 90.6	OH 86.9	CA 82.5	WI 82.1
Irrigated farms	do.	CA 58.4	NE 22.2	TX 19.8	ID 17.4	WA 16.3	OR 15.3	CO 15.2	UT 11.2	FL 10.6	MT 9.2
Percentage of farms with irrigated land in 1982	Percent	UT 80.2	NV 79.2	CA 70.8	ID 70.2	AR 60.5	WY 59.6	CO 56.2	NM 51.3	WA 45.0	OR 45.0
Irrigated land in 1982	Mil. acre	CA 8.5	NE 6.0	TX 5.6	ID 3.5	CO 3.2	KS 2.7	MT 2.0	AK 2.0	OR 1.8	WA 1.6
Percentage of harvested cropland irrigated in 1982	Percent	AR 99.6	NV 99.5	UT 73.5	CA 68.8	WY 63.2	ID 62.1	NM 54.1	FL 49.3	OR 41.6	NE 34.8
Market value of agricultural products sold in 1982:											
All farms	$ bil.	CA 12.5	IA 9.8	TX 8.9	IL 7.3	NE 6.6	KS 6.2	MN 5.9	WI 4.9	MO 3.6	FL 3.5
Irrigated farms	do.	CA 10.3	NE 4.1	TX 2.9	FL 2.5	KS 1.8	ID 1.8	WA 1.8	CO 1.7	OR 1.2	AK 1.1
Percentage of crops sold from irrigated farms in 1982	Percent	NV 100	AR 99.9	CA 98.7	UT 91.4	NM 89.7	FL 88.8	ID 86.0	WY 82.7	CO 77.8	OR 76.3
Market value of products sold per farm in 1982:											
Nonirrigated farms	$ thous.	AR 159	CO 103	CA 92	DE 91	IA 84	IL 72	KS 66	NE 66	NM 66	ID 64
Irrigated farms	do.	DE 296	MS 271	KS 254	PA 240	AZ 240	IL 240	FL 236	IA 232	CT 217	WI 208
Major irrigated crops in 1982:											
Corn	1000 acres	NE 4335	KS 755	CO 739	TX 629	CA 292	SD 180	GA 165	MN 160	MI 135	WA 130
Wheat	do.	TX 897	CA 717	KS 715	ID 658	WA 359	CO 202	MT 165	OR 162	OK 160	AZ 134
Hay	do.	CA 1132	CO 1025	MT 995	ID 902	WY 829	OR 694	UT 552	NV 526	NE 398	WA 327

TABLE 5–60. TOP 10 STATES IN IRRIGATED AGRICULTURE IN THE UNITED STATES (continued)

[In selected categories]

Item	Unit	Ranking by State 1	2	3	4	5	6	7	8	9	10
Energy expense for onfarm pumping of irrigation water in 1982											
Total	$ mil.	CA 228.7	TX 189.2	NE 125.9	AZ 62.0	CO 56.2	KS 49.7	ID 46.2	AK 39.4	FL 30.0	WA 25.5
Electricity	do.	CA 211.7	TX 72.0	NE 55.4	AR 49.3	CO 46.2	ID 44.5	WA 25.1	AK 22.1	OR 16.9	FL 13.8
Natural gas	do.	TX 103.8	KS 26.4	NE 19.5	AR 11.7	NM 7.8	CO 7.5	OK 6.3	CA 5.7	LA 3.5	AK 1.7
Acres irrigated by source of water in 1984:											
Wells	Thous. acre	NE 5019	TX 4012	CA 3113	KS 2139	AK 1666	CO 1398	ID 1060	FL 789	AZ 524	NM 442
Onfarm surface	do.	MT 767	CA 685	WY 631	OR 629	CO 427	ID 284	NV 254	TX 223	AK 215	FL 182
Off-farm suppliers	do.	CA 4621	ID 1959	CO 1436	MT 1083	WA 943	WY 843	UT 833	OR 803	TX 699	NE 692
Water applied in irrigation by source in 1984:											
Total	Mil. acre-feet	CA 23.9	TX 6.8	NE 6.3	ID 5.7	CO 5.1	AR 3.9	OR 3.6	MT 3.4	WA 3.2	KS 3.1
Wells	do.	CA 8.2	NE 5.0	TX 4.9	KS 2.9	AK 2.3	AR 2.1	CO 1.9	ID 1.6	FL 1.1	NM .9
Onfarm sources	do.	MT 1.8	CA 1.7	OR 1.3	WY .9	CO .70	TX .49	WA .35	AK .32	LA .315	FL .29
Off-farm suppliers	do.	CA 14.0	ID 3.7	CO 2.5	WA 2.2	UT 1.8	OR 1.72	AR 1.68	MT 1.6	WY 1.5	TX 1.4
Irrigation water applied per acre in 1984	Acre-feet	AR 4.4	CA 3.1	NV 2.6	NM 2.2	WA 2.2	UT 2.1	OR 2.0	FL 1.9	MT 1.8	ID 1.7
Average land value per acre in 1986:											
Dry land	$ per acre	MA 6287	CT 4402	NJ 3688	NH 2800	FL 2197	MD 1724	CA 1541	AZ 1494	LA 1364	PA 1342
Irrigated cropland	do.	NY 10961	FL 7212	CA 2944	AZ 2881	NM 2604	MD 2378	NJ 2230	UT 1704	WA 1600	IN 1507

Source: Bajwa, R.S., Crosswhite, W.M., and Hostetler, J.E., 1987, Agricultural Irrigation and Water Supply, U.S. Department of Agriculture, Economic Research Service, Agriculture Information Bulletin 532

TABLE 5–61. STANDARDS FOR CLASSIFICATION OF LANDS AS IRRIGABLE

Land Characteristics	Minimum Requirements
Soils	
Texture	Loamy sand to permeable clay.
Depth	
To sand, gravel, or cobble	18 inches of good free-working soil of fine sandy loam or heavier, or from 24 to 30 inches of lighter textured soil.
To bedrock	At least 18 inches over shattered bedrock or tilted shale bedrock; or 24 inches over massive bedrock or hardpan.
Topography	
Slopes	Smooth slopes up to no more than 30 percent in general gradient in reasonably large-sized bodies sloping in the same plane; or undulating slopes which are less than 20 percent in general gradient.
Rock cover	No more than enough loose rock and rock outcroppings to moderately reduce productivity and interfere with cultural practices. Varies with soil depth and topographic conditions.
Erosion	No more than moderate erosion, with very few gullies which are not crossable by tillage implements.
Drainage	
Soil and topography	Such that moderate farm drainage may be required, but without excessive cost.
Salinity	Total salts in the soil solution do not exceed 0.5 percent, except in readily drained soils where reclamation appears feasible.
Alkalinity	The pH value is 9.0 or less, unless the soil is calcareous in which case higher values may be allowed. If there is evidence of black alkali a lower pH value may be limiting.

Source: California State Water Resources Board, 1955

TABLE 5–62. SUMMARY OF LOSSES AND WASTE OF IRRIGATION WATER

During Delivery	
Average field evaporation before topsoil dries	0.5 inch per irrigation
Surface waste, allowance for large projects	10 per cent of diversions
Seasonal percolation losses, except on porous soils	0–1.5 acre-feet per acre
Losses of flow in farm ditches	5–50 per cent per mile
Deliveries to farms	1–7 acre-feet per acre
Consumptive use, diversified crops	1–3.5 acre-feet per acre
Irrigation efficiencies, common farm crops	20–50 per cent
Irrigation efficiencies, fruit and special crops	35–70 per cent
Average irrigation efficiencies on large projects	30–50 per cent
During Conveyance	
Evaporation from canal surfaces	Negligible
Evapo-transpiration at canal banks	Negligible
Canal seepage, large projects, mostly unlined canals	15–45 per cent of diversions
Seepage losses, most canals lined	5–15 per cent of diversions
Waste on large projects, ample water supplies	5–30 per cent of diversions
Waste on large projects, limited water supplies	1–10 per cent of diversions
Over-all efficiencies, large projects	20–35 per cent
Diversions for large projects	2–10 acre-feet per acre

Source: Houk, Irrigation Engineering, v.I, John Wiley & Sons, Copyright 1951

TABLE 5–63. TYPICAL WATER-APPLICATION LOSSES AND IRRIGATION EFFICIENCIES FOR DIFFERENT SOIL CONDITIONS

Item	General Soil Type: Open, Porous (%)	Medium Loam (%)	Heavy Clay (%)
Farm-lateral loss[1]	15	10	5
Surface runoff loss	5	10	25
Deep percolation loss	35	15	10
Field-irrigation efficiency[2]	60	75	65
Farm-irrigation efficiency[3]	45	65	60

[1] Unlined ditches (loss in new-lined ditches and pipelines is usually about one percent).
[2] For water measured at the field.
[3] For water measured at the farm headgate.
Source: U.S. Department of Agriculture

TABLE 5–64. ADAPTATIONS AND LIMITATIONS OF COMMON IRRIGATION METHODS

Irrigation Method	Suitabilities and Conditions of Use				Remarks
	Crops	Topography	Water Supply	Soils	
Small rectangular basins	Grain, field crops, orchards, rice	Relatively flat land; area within each basin should be levelled	Can be adapted to streams of various size	Suitable for soils of high or low intake rates; should not be used on soils that tend to puddle	High installation costs. Considerable labour required for irrigating. When used for close-spaced crops, a high percentage of land is used for levees and distribution ditches. High efficiencies of water use possible.
Large rectangular basins	Grain, field crops, rice	Flat land; must be graded to uniform plane	Large flows of water	Soils of fine texture with low intake rates	Lower installation costs and less labour required for irrigation than with small basins. Substantial levees needed.
Contour checks	Orchards, grain, rice, forage crops	Irregular land; slopes less than 2 percent	Flows greater than 30 liters (1 cubic foot) per second	Soils of medium to heavy texture which do not crack on drying	Little land grading required. Checks can be continuously flooded as for rice, water ponded as for orchards, or intermittently flooded as for pastures.
Narrow borders up to 5 metres (16 feet) wide	Pasture, grain, lucerne, vineyards, orchards	Uniform slopes less than 7 percent	Moderately large flows	Soils of medium to heavy texture	Borders should be in direction of maximum slope. Accurate cross-levelling required between guide levees.
Wide borders up to 30 metres (100 feet) wide	Grain, lucerne, orchards	Land graded to uniform plane with maximum slope less than 0.5 percent	Large flows, up to 600 liters (20 cubic feet) per second	Deep soils of medium to fine texture	Very careful land grading necessary. Minimum of labour required for irrigation. Little interference with use of farm machinery.
Wild flooding	Pasture, grain	Irregular surfaces with slopes up to 20 percent	Can utilize small continuous flows on steeper land or large flows on flatter land	Soils of medium to fine texture with stable aggregate which do not crack on drying	Little land grading required. Low initial cost for system. Best adapted to shallow soils since percolation losses may be high on deep permeable soils.
Benched terraces	Grain, field crops, forage crops, orchards, vineyards	Slopes up to 20 percent	Streams of small to medium size	Soils must be sufficiently deep that grading operations will not impair crop growth	Care must be taken in constructing benches and providing adequate drainage channels for excess water. Irrigation water must be properly managed. Misuse of water can result in serious soil erosion.
Straight furrows	Vegetables, row crops, orchards, vineyards	Uniform slopes not exceeding 2 percent for cultivated crops	Flows up to 350 liters (12 cubic feet) per second	Can be used on all soils if length of furrows is adjusted to type of soil	Best suited for crops which cannot be flooded. High irrigation efficiency possible. Well adapted to mechanized farming.
Graded contour furrows	Vegetables, field crops, orchards, vineyards	Undulating land with slopes up to 8 percent	Flows up to 100 liters (3 cubic feet) per second	Soils of medium to fine texture which do not crack on drying	Rodent control is essential. Erosion hazard from heavy rains or water breaking out of furrows. High labour requirement for irrigation.
Corrugations	Close-spaced crops such as grain, pasture, lucerne	Uniform slopes of up to 10 percent	Flows up to 30 liters (1 cubic foot) per second	Best on soils of medium to fine texture	High water losses possible from deep percolation or surface run-off. Care must be used in limiting size of flow in corrugations to reduce soil erosion. Little land grading required.
Basin furrows	Vegetables, cotton, maize and other row crops	Relatively flat land	Flows up to 150 liters (5 cubic feet) per second	Can be used with most soil types	Similar to small rectangular basins, except crops are planted on ridges.
Zigzag furrows	Vineyards, bush berries, orchards	Land graded to uniform slopes of less than 1 percent	Flows required are usually less than for straight furrows	Used on soils with low intake rates	This method is used to slow the flow of water in furrows to increase water penetration into soil.

Source: Bouher, L.J., 1974, Surface Irrigation, FAO Agricultural Development Paper 95. Reprinted with permission

TABLE 5–65. IRRIGATION EFFICIENCIES

A. Field Efficiencies by Method of Irrigation

Method of Irrigation	Range of Efficiency, percent
Graded borders	60 to 75
Basins and level borders	60 to 80
Contour ditch	50 to 55
Furrows	55 to 70
Corrugations	50 to 70
Subsurface	Up to 80

B. Average Efficiencies for Selected Crops in California

Crop	Average Efficiency, percent
Alfalfa and irrigated pasture	85
Citrus	80
Deciduous	85
Truck	70
Vineyard	80
Walnuts	85

C. Sprinkler Efficiencies

Climate	Efficiency, percent
Hot dry	60
Moderate	70
Humid or cool	80

Source: U.S. Soil Conservation Service and California State Water Rights Board

TABLE 5–66. CROP IRRIGATION DEPTHS

[Soil depth in inches]

Crop	Humid Areas	Semiarid to Arid Areas[1]
Alfalfa	36–42	60–120
Beans	....	36–48
Beets (sugar)	....	48–72
Broccoli	....	24
Cabbage	....	24
Clover (ladino)	....	24
Corn (maize)	24–36	48–60
Cotton	24–36	48–72
Grapes	24–30	48–72
Orchards:		
Citrus	....	48–72
Deciduous	36–60	72–96
Pasture	18–36	36–48
Peas	....	36–48
Potatoes (white)	12–24	26–48
Small grain	18–30	48
Sorghum	20–30	
Soybeans	18–36	
Tobacco	15–24	
Tomatoes	....	72–120
Truck crops:		
Shallow-rooted	9–12	
Medium-rooted	12–24	
Deep-rooted	24–30	

[1] Larger figure applies to arid areas.
Source: U.S. Soil Conservation Service

TABLE 5–67. WATER REQUIREMENTS FOR VARIOUS IRRIGATION AND SOIL TYPES

Irrigation Type	Slope Land percent	Coarse Sandy Soils		Light Sandy Loam		Medium Silt Loam		Clay Loam Soils		Very Heavy Clay Soils	
		Q per Unit	Length of Run	Q per Unit	Length of Run	Q per Unit	Length of Run	Q per Unit	Length of Run	Q per Unit	Length or Run
Basins[1]	0–2	20 cfs per acre		7.5 cfs per acre		5 cfs per acre		3 cfs per acre		2 cfs per acre	
Borders[1] or checks	0–2	1.5 cfs per 10' width	220'	.75 cfs per 10' width	440'	.5 cfs per 10' width	550'– 880'	.33 cfs per 10' width	660'– 880'	.3 cfs per 10' width	1,000'
Furrows	0–2	.02 cfs per each	220'	.01 cfs per each	330'	.01 cfs per each	440' 660'	.008 cfs each	660'	.005 cfs per each	880'
	2–5			.005 each	220'	.005 per each	220' 440'	.003 per each	440'	.003 per each	550'
	5–8					.002 per each	110' 220'	.001 per each	330'	.001 per each	330'
Sprinkling	0–2	2" per hour		.75" per hour		.5" per hour		.2" per hour			
	2–5	2" per hour		.75" per hour		.5" per hour		.2" per hour			
	5–8	1.5" per hour		.5" per hour		.4" per hour		.15" per hour			
	8–12	1.0" per hour		.4" per hour		.3" per hour					

[1] The range in slope 0–2 per cent is in itself a very rough picture of field practices where the actual slopes, particularly with borders, tend to be closer to .2 or .3 per cent rather than this higher limit 2 per cent.

Source: Calif. Agric. Exp. Station

TABLE 5–68. BORDER IRRIGATION RELATIONSHIPS FOR VARIOUS SOILS, SLOPES, AND DEPTHS OF APPLICATION

Soil Texture	Slope of Land (percent)	Depth of Application (inches)	Suggested Borderstrip Size		Size of Irrigation Stream (cubic feet) per second)
			Width (feet)	Length (feet)	
Coarse	0.25	2	50	500	8.0
		4	50	800	7.0
		6	50	1,320	6.0
	1.00	2	40	300	2.75
		4	40	500	2.50
		6	40	900	2.50
	2.00	2	30	200	1.25
		4	30	300	1.00
		6	30	600	1.00
Medium	0.25	2	50	800	7.0
		4	50	1,320	6.0
		6	50	1,320	3.5
	1.00	2	40	500	2.5
		4	40	1,000	2.5
		6	40	1,320	2.5
	2.00	2	30	300	1.0
		4	30	600	1.0
		6	30	1,000	1.0
Fine	0.25	2	50	1,320	4.0
		4	50	1,320	2.5
		6	50	1,320	1.5
	1.00	2	40	1,320	2.5
		4	40	1,320	1.25
		6	40	1,320	0.75
	2.00	2	30	660	1.0
		4	30	1,320	1.0
		6	30	1,320	0.67

Source: U.S. Dept. of Agriculture

TABLE 5–69. IRRIGATION FREQUENCY IN RELATION TO SOIL TEXTURE AND DEPTH OF ROOT ZONE WETTED

[Approximate number of days between irrigations assuming water use to be one inch per week]

Soil Depth Irrigated, (inches)	Soil Type		
	Sands	Loams	Clays
6	3	5	8
12	5	10	17
18	8	16	25
24	10	21	35
30	13	26	—
36	15	31	—
42	18	36	—
48	20	—	—

Source: U.S. Dept. of Agriculture

TABLE 5–70. FURROW IRRIGATION RELATIONSHIPS FOR VARIOUS SOILS, SLOPES, AND DEPTHS OF APPLICATION

Soil Texture		Coarse				Medium				Fine			
		Depth or Irrigation Application, inches											
Furrow Slope (percent)	Maximum Allowable Nonerosive Furrow Stream (gallons per minute)	2	4	6	8	2	4	6	8	2	4	6	8
		Maximum Allowable Length of Run, feet											
0.25	40	500	720	875	1,000	820	1,150	1,450	1,650	1,050	1,500	1,750	2,140
0.50	20	345	480	600	680	560	800	975	1,120	730	1,020	1,250	1,460
0.75	13	270	380	480	550	450	630	775	900	580	820	1,000	1,150
1.00	10	235	330	400	470	380	540	650	760	500	750	850	990
1.50	7	190	265	330	375	310	430	530	620	400	570	700	800
2.00	5	160	225	275	320	260	370	450	530	345	480	600	675
3.00	3	125	180	220	250	210	295	360	420	270	385	470	550
5.00	2	95	135	165	190	160	225	270	320	210	290	350	410

Source: U.S. Dept. of Agriculture

TABLE 5–71. PERCENTAGE OF WATER OBTAINED BY PLANTS FROM VARIOUS DEPTHS IN THE SOIL

Item	Crop	Soil	Water Obtained from					
			First Foot (percent)	Second Foot (percent)	Third Foot (percent)	Fourth Foot (percent)	Fifth Foot (percent)	Sixth Foot (percent)
1	Cotton	Heavy clay	66.0	15.9	6.1	5.2	5.7	1.1
2	. . . do	Sandy loam	34.6	27.2	20.7	12.5	4.9	. . .
3	. . . do	Clay loam	33.2	25.6	17.3	12.6	6.7	4.5
4	Alfalfa	Fine sandy loam	47.0	15.0	15.0	12.0	8.0	3.0
5	Orange trees, mature	Sandy loam	37.0	30.0	15.0	11.0	4.0	3.0

Sources: Univ. of California Agric. Extension and U.S. Dept. of Agriculture

TABLE 5–72. IRRIGATION FREQUENCY AND AMOUNT OF WATER TO BE APPLIED BY SPRINKLING IRRIGATION WHEN VARYING AMOUNTS OF AVAILABLE WATER REMAIN ON THE SOIL

[Length of growing period is 3 months; soil texture is coarse.]

Time Since Planting in Days	Consumptive Use of Water, Inches per Day	Average Depth of Rooting, Inches	Maximum Depth of Available Water in Root Zone, Inches	Interval in Days Between Irrigations and Depth in Inches to be Applied When Different Amounts of Available Water Remain in the Soil							
				75%		50%		25%		0%	
				Days	Depth Inches	Days	Depth Inches	Days	Depth Inches	Days	Depth Inches
0–9	0.05	2	0.2	1	0.3	2	0.4	3	0.5	4	0.6
9–18	0.07	6	0.6	2	0.5	5	0.7	8	1.0	10	1.3
18–27	0.08	11	1.1	3	0.6	7	1.1	10	1.5	13	1.9
27–36	0.09	16	1.6	4	0.8	8	1.4	13	2.0	17	2.5
36–45	0.10	21	2.1	5	1.0	10	1.7	15	2.5	20	3.1
45–54	0.11	26	2.6	6	1.1	12	2.1	17	2.9	24	3.7
54–63	0.11	31	3.1	7	1.3	14	2.4	21	3.3	28	4.2
63–72	0.10	35	3.5	8	1.5	17	2.7	25	3.7	34	4.7
72–81	0.09	38	3.8	11	1.6	21	2.9	33	4.0	44	5.1
81–90	0.05	41	4.2	20	1.7	41	3.1	61	4.2	82	5.5

Source: Israelson and Hansen, Irrigation Principles and Practices, John Wiley & Sons, Copyright 1962

TABLE 5–73. METHODS, SOURCES, AND PUMPING COST OF IRRIGATION IN THE UNITED STATES

Item	1984 Farms	1984 Acres	1984 AF/Acre	1979 Farms	1979 Acres	1979 AF/Acre
			Number			
Methods:						
All sprinkler systems	104,641	16,877,412	1.3	121,749	18,380,747	1.4
Center pivot	32,442	9,370,614	1.2	24,078	8,642,454	1.3
Mechanical move	25,475	3,355,164	1.2	29,617	5,052,563	1.4
Hand move	46,885	2,918,241	1.6	67,344	3,705,654	1.4
Solid set and permanent	19,694	1,233,393	2.5	16,334	980,076	2.0
All gravity systems	126,827	27,457,244	2.0	153,946	31,210,020	2.1
Gated pipe	42,826	8,363,825	1.4	38,552	8,400,090	1.7
Ditch with siphon tube	59,255	10,051,246	2.3	47,413	8,654,866	2.2
Flooding	45,045	9,042,173	2.2	85,351	14,155,064	2.5
Drip or trickle	11,651	837,624	1.9	7,134	321,260	2.1
Subirrigation	2,905	623,013	3.8	760	242,522	1.5
Sources:						
Wells	100,703	24,286,826	1.4	120,952	NA	1.5
Onfarm surface sources	35,982	5,886,832	1.8	44,644	NA	1.6
Off-farm water suppliers	98,672	15,647,770	2.3	112,600	NA	2.2
Onfarm energy pumping expenses:	Number		$/acre	Number		$/acre
Total energy expenses	135,319	31,067,689	32	169,658	35,179,033	20
Electricity	96,324	18,106,589	35	113,427	19,071,824	21
Natural gas	15,519	5,800,547	34	21,203	7,427,301	21
LP gas, propane, butane	12,668	1,804,629	22	16,688	2,259,014	15
Diesel fuel	30,339	5,193,599	24	35,268	6,075,127	16
Gasoline and gasohol	6,057	162,325	23	9,975	345,767	18

Source: Bajwa, R.S., Crosswhite, W.M., and Hostetler, J.E., 1987, Agricultural Irrigation and Water Supply, U.S. Department of Agriculture, Economic Research Service, Agriculture Information Bulletin 532

TABLE 5–74. SPRINKLER IRRIGATION IN THE UNITED STATES, 1984

[Contiguous United States]

Region and State	Acreage	Share of Total	Acreage Change, 1979–84
	Acres	Percent	
East:			
Arkansas	172,157	0.9	93.8
Florida	860,293	4.7	30.9
Louisiana	88,352	.5	200.0
All other	2,204,601	12.0	6.5
Subtotal	3,325,403	18.1	16.8
West:			
Arizona	124,701	.6	-35.5
California	2,439,355	13.3	4.8
Colorado	1,127,574	6.5	-13.3
Idaho	1,774,447	9.7	12.2
Kansas	950,794	5.2	-.6
Montana	629,458	3.4	23.1
Nebraska	3,067,434	16.7	-8.3
Nevada	135,726	.7	-1.8
New Mexico	229,289	1.3	5.0
North Dakota	100,279	.5	43.0
Oklahoma	240,377	1.3	5.4
Oregon	896,857	4.9	-6.3
South Dakota	226,184	1.2	-20.6
Texas	1,387,216	7.6	-29.5
Utah	318,905	1.7	17.9
Washington	1,155,582	6.3	-3.6
Wyoming	208,530	1.1	-26.1
Subtotal	15,012,708	81.9	-5.3
Total	18,338,111	100.0	-1.9

Source: Day, J.C., and Horner, G.L., 1987, U.S. Irrigation, Extent and Economic Importance, U.S. Department of Agriculture Information Bulletin 523

TABLE 5–75. WATER APPLICATIONS FOR SELECTED IRRIGATED CROPS IN THE UNITED STATES, 1982

[Contiguous United States]

Crop	Average Application Rate	Irrigated Acreage	Estimated Water Use	Share of Total Water Used
	Acre-feet per acre	1,000 acres	Million acre-feet	Percent
Corn	1.5	9,604	14.4	19
Sorghum	1.2	2,245	2.7	4
Wheat	1.4	4,650	6.5	9
Barley and oats	1.6	2,098	3.4	5
Rice	2.9	3,233	9.4	13
Cotton	2.1	3,424	7.2	10
Soybeans	.8	2,321	1.9	2
Irish potatoes	1.9	812	1.5	2
Hay	2.1	8,507	17.9	24
Vegetables and melons	2.2	2,024	4.5	6
Orchard crops	1.2	3,343	4.0	5
Sugar beets	2.6	550	1.4	2
Total[1]	NA	45,289	74.8	100

NA — Not applicable.

[1] Figures may not add, due to rounding.

Source: Day, J.C., and Horner, G.L., 1987, U.S. Irrigation, Extent and Economic Importance, U.S. Department of Agriculture Information Bulletin 523

TABLE 5–76. ENERGY REQUIREMENTS AND ENERGY COSTS FOR VARIOUS IRRIGATION SYSTEMS IN THE UNITED STATES

[Diesel fuel only]

	Total Lift, Feet	Field Efficiency, Percent	Potential Efficiency	Water Per Acre, Inches	Fuel Per Acre, Gallons	Cost of Energy		
						1973	1979	1980
Traveling Big Guns	389	70	75	17	91.1	$13.67	$74.70	$113.88
Center-Pivot								
(High) 75	273	80	85	15	56.4	8.46	46.25	70.50
(Low) 50	215	80	85	15	44.4	6.66	36.41	55.50
Skid-Tow	215	75	80	16	47.4	7.11	38.87	59.25
Gated Pipe without reuse	120	60	70	20	33.1	4.97	27.14	41.38
Gated Pipe with reuse	120	70	85	17	28.1	4.22	23.04	35.13
Auto-Surface	120	85	92	14	23.1	3.47	18.94	28.88
Drip Trickle	150	85	92	14	28.9	4.34	23.70	36.13

NOTE: Based on 100 foot of lift from groundwater reservoir; 12 inches net of water; pumping plant operating at 75% of Nebraska performance standard of 10.94 water horsepower hours per gallon of diesel fuel or 14.6 brake horsepower hours per gallon of diesel fuel.

Fuel costs per gallon: $0.15–1973, $0.82–1979; $1.25–1980.

Source: Fischbach, P.E., 1980, Energy Requirements of Auto-Surface Irrigation, Copyright, The Irrigation Assoc., 1980 Technical Conference Proceedings. Reprinted with permission

TABLE 5–77. FUEL ENERGY REQUIREMENTS FOR PUMPING ONE ACRE-FOOT OF WATER AT ONE POUND PER SQUARE INCH

Energy	Horsepower Hours[1]	Percentage of Water Pump Efficiency		
		65	60	55
		Unit fuel per acre-foot per psi		
Electricity	1.206 per kWh	4.0503	4.3876	4.7866
Diesel	12.35 per gallon	.4000	.4330	.4659
Gasoline	9.875 per gallon	.5004	.5417	.5830
Natural gas	79 per MCF[2]	.0625	.0677	.0729
LPG	7.9 per gallon	.6254	.6771	.7287

[1] This column refers to the assumed number of horsepower hours produced per unit of fuel.
[2] MCF equals 1,000 cubic feet.
Source: Sloggett, G., 1985, Energy and U.S. Agriculture: Irrigation Pumping, 1974–83, U.S. Dept. of Agriculture Economic Report 545

TABLE 5–78. SELECTED ENERGY PRICES IN THE UNITED STATES, 1973–1983

Item	Unit	1973	1974	1977	1980	1983	Percentage Change, 1973-80
		Dollars per unit					Percent
Electricity	kWh	0.023	0.027	0.035	0.055	0.065	182
Diesel	Gal.	.23	.37	.45	1.00	.99	330
Gasoline	Gal.	.33	.47	.57	1.17	1.26	281
Natural gas[1]	MCF	.50	1.00	1.50	2.50	4.00	700
LPG	Gal.	.20	.30	.39	.62	.77	285

[1] Estimated.
Source: Sloggett, G., 1985, Energy and U.S. Agriculture: Irrigation Pumping, 1974–83, U.S. Dept. of Agriculture Economic Report 545

SECTION J. IRRIGATION—WORLD

TABLE 5-79. LAND AREAS UNDER IRRIGATION IN VARIOUS COUNTRIES OF THE WORLD, 1971 TO 1986

[In thousands of hectares]

Continent and Country	1971	1976	1981	1986
World	172 552	193 933	214 146	227 520
Africa	9 041	9 645	10 218	10 986
Algeria	239F	246F	270F	345F
Benin	3F	4F	5F	6F
Botswana	1F	1F	2F	2F
Burkina Faso	6F	8F	10F	14F
Burundi	35F	48F	58F	68F
Cameroon	7F	10F	16F	22F
Cape Verde	2	2°	2°	2
Chad	5F	6F	6F	10F
Congo	1F	2F	3F	4F
Côte d'Ivoire	22	36F	46F	56F
Egypt	2 852	2 730F	2 470	2 528F
Ethiopia	155F	158F	160F	162F
Gambia	8F	10F	12F	12F
Ghana	8F	8F	8F	8F
Guinea	53F	65F	68F	70F
Kenya	32F	42	40F	40F
Liberia	2°	2	2F	2F
Libya	180F	205F	225F	236F
Madagascar	350°	500F	682F	860F
Malawi	5	15F	18F	18F
Mali	85F	124F	159F	195F
Mauritania	8F	11F	11F	12F
Mauritius	15	15°	16F	17
Morocco	950F	1 090F	1 223F	1 250F
Mozambique	28F	45F	70F	100F
Namibia	4F	4F	4F	4F
Niger	18F	20F	24	25F
Nigeria	804F	814F	830F	850F
Reunion	5F	5F	5	5F
Rwanda	4F	4F	4F	4F
Senegal	119	165F	170F	175F
Sierra Leone	8F	14F	22F	30F
Somalia	115F	120F	150	190F
South Africa	1 017	1 040F	1 128	1 128F
Sudan	1 640F	1 715F	1 790F	1 860F
Swaziland	50F	56F	58F	62F
Tanzania	40F	65F	120F	129F
Togo	5F	6F	6F	7F
Tunisia	95F	130F	163	260F
Uganda	4F	4	6F	9F
Zaire		2F	7F	9F
Zambia	11F	18F	19F	20F
Zimbabwe	50F	80F	130F	180F
North America	21 358	23 291	27 800	25 524
Belize	1°	1	2	2
Canada	437F	517F	625F	775F
Costa Rica	26°	41F	66F	110F
Cuba	480F	650F	793	870F
Dominican Rep	125F	145F	170F	200F
El Salvador	20	30	110F	112F
Guadeloupe	2	1	2F	3F
Guatemala	60	62F	70F	77F
Haiti	65F	70°	70°	70F
Honduras	70F	80F	82F	87F

TABLE 5–79. LAND AREAS UNDER IRRIGATION IN VARIOUS COUNTRIES OF THE WORLD, 1971 TO 1986 (continued)

[In thousands of hectares]

Continent and Country	1971	1976	1981	1986
North America–Con.				
Jamaica	30F	32F	33F	34F
Martinique	1	2	5F	6F
Mexico	3 750F	4 816	5 020	4 900F
Nicaragua	45F	70	80F	83F
Panama	20F	23F	28F	30F
Puerto Rico	39	39	39°	39F
Saint Lucia	1°	1	1	1F
St Vincent	1	1	1°	1F
Trinidad and Tobago	15F	20F	21F	22F
USA	16 170F	16 690	20 582	18 102
South America	5 860	6 784	7 650	8 421
Argentina	1 310F	1 477	1 600F	1 700F
Bolivia	80	120F	150F	160F
Brazil	900F	1 400F	1 900F	2 400F
Chile	1 200F	1 245F	1 257F	1 257F
Colombia	260F	320F	420F	470
Ecuador	470F	510F	525F	542F
Guyana	115F	122F	125F	127F
Paraguay	50°	55F	60F	65F
Peru	1 110	1 140F	1 170F	1 220F
Suriname	29F	34F	43	58
Uruguay	52°	58F	85F	96F
Venezuela	284°	303	315F	326F
Asia	111 768	123 935	133 743	143 975
Afghanistan	2 360F	2 520F	2 660F	2 660F
Bahrain	1	1°	1°	1F
Bangladesh	1 050F	1 406	1 639	2 098
Brunei Darussalem			1	1F
Burma	890	985	1 073	1 059
China	39 036F	43 571F	44 997	44 653
Cyprus	30F	30F	30F	31
Hong Kong	8	6	3	3
India	31 100	34 490	38 805	44 350F
Indonesia	4 490F	4 900	5 418F	7 260
Iran	5 251	5 840	5 315F	5 740F
Iraq	1 500F	1 600F	1 750F	1 750F
Israel	171	187	203	275F
Japan	3 364	3 144	3 031	2 950F
Jordan	34F	36F	37	43F
Kampuchea Dm	89	89F	89F	90F
Korea DPR	500F	1 000°	1 050F	1 150F
Korea Rep	1 020F	1 082	1 160F	1 240F
Kuwait	1	1	1°	1F
Laos	19	50F	116	120F
Lebanon	75F	86F	86	86F
Malaysia	279	312	330	336F
Mongolia	12F	25F	35F	44F
Nepal	117	290F	584	650F
Oman	30F	35F	38	41F
Pakistan	12 986	13 830	15 300	16 040
Philippines	864	1 070F	1 269	1 450F
Saudi Arabia	368F	378F	396	420F
Sri Lanka	439	483F	548	607
Syria	476	547	567	652
Thailand	2 106	2 448	3 171	3 900F
Turkey	1 850F	2 000F	2 080F	2 170F
United Arab Emirates	5°	5°	5°	5F
Viet Nam	980F	1 200F	1 650	1 790F
Yemen Ar	215F	230	245F	247F
Yemen Dem	52F	58F	60F	62F

TABLE 5–79. LAND AREAS UNDER IRRIGATION IN VARIOUS COUNTRIES OF THE WORLD, 1971 TO 1986 (continued)

[In thousands of hectares]

Continent and Country	1971	1976	1981	1986
Europe	11 451	13 338	14 851	16 266
Albania	290F	340F	378	394F
Austria	4°	4°	4°	4°
Belgium-Luxembourg	1°	1°	1°	1F
Bulgaria	1 021	1 147	1 185	1 242
Czechoslovakia	128F	138F	150	257
Denmark	100F	201	388	405F
Finland	20F	45	60F	62F
France	760F	819	1 100F	1 180F
Germany DR	135F	140F	145F	150°
Germany FR	288F	310	315F	322F
Greece	793	939	962	1 110F
Hungary	205	320	184	163
Italy	2 590F	2 750F	2 900F	3 020F
Malta	1	1	1	1F
Netherlands	390F	440F	490F	535F
Norway	35F	44	77	92
Poland	202	198	100F	100F
Portugal	623F	628F	630F	632F
Romania	957	1 729	2 350F	3 000F
Spain	2 625	2 854	3 058	3 220F
Sweden	37F	50F	53	48F
Switzerland	25F	25F	25	25
UK	87	91	145	155F
Yugoslavia	134	124	150	148
Oceania	1 591	1 640	1 855	1 881
Australia	1 470F	1 475	1 654	1 615F
Fiji	1°	1°	1	1F
New Zealand	120	164	200F	265F
USSR	11 483	15 300	18 029	20 467

F — FAO estimate.
° — Data from unofficial source.
Source: Food and Agriculture Organization of the United Nations, 1987 FAO Production Yearbook

TABLE 5–80. AMOUNT OF IRRIGATED LAND PER CAPITA BY WORLD REGION

	Irrigated Area 1984 million ha	Population 1987 millions	Irrigated Area ha/capita
World	219.7	4,997	0.044
Africa	10.4	589	0.018
Americas	35.4	691	0.051
Asia & Oceania	138.8	2,938	0.047
Europe & USSR	35.1	779	0.045

Source: International Commission on Irrigation and Drainage, Editorial, ICID Bulletin v. 36, No. 2, July 1987

TABLE 5–81. GROWTH OF IRRIGATED AREA IN THE WORLD, 1950–1982

[By continent]

Region	Total Irrigated Area, 1982	Growth in Irrigated Area		
		1950–60	1960–70	1970–80[1]
	Million hectares	*Percent*		
Africa	12	25	80	33
Asia[2]	177	52	32	34
Europe[3]	28	50	67	40
North America	34	42	71	17
South America	8	67	20	33
Oceania	2	0	100	0
World	261	49	41	32

[1] Percentage increase between 1970 and 1982 prorated to 1970–80 to maintain comparison by decade.
[2] Includes the Asian portion of the Soviet Union.
[3] Includes the European portion of the Soviet Union.
Source: Postel, S., 1984, Water: Rethinking Management in an Age of Scarcity, Worldwatch Paper 62. Reprinted with permission

TABLE 5–82. IRRIGATION DEVELOPMENT AND REHABILITATION COSTS, 1974 A COMPARISON OF DEVELOPING REGIONS

Region	Cost of New Irrigation Development in $/ha	Costs of Rehabilitation of Existing Projects in $/ha
Africa	2,400	500
Far East	1,466	418
Near East	2,467	560
Latin America	1,500	420

Source: Biswas, A. K., 1987, Irrigation in Africa, Bulletin of the International Commission on Irrigation and Drainage v. 36, no. 2

SECTION K. LIVESTOCK

TABLE 5–83. WATER REQUIREMENTS FOR FARM ANIMALS AND POULTRY

[Gallons per day]

Horse, work	12
Mule	12
Cattle	
Holstein calves (liquid milk or dried milk and water supplied)	
4 weeks of age	1.2–1.4
8 weeks of age	1.6
12 weeks of age	2.2–2.4
16 weeks of age	3.0–3.4
20 weeks of age	3.8–4.3
26 weeks of age	4.0–5.8
Dairy heifers — Pregnant	7.2–8.4
Steers	
Maintenance ration	4.2
Fattening ration	8.4
Range cattle	4.2–8.4
Jersey cows[1]	
Milk production 5–30 lbs./day	7.2–12
Holstein cows[1]	
Milk production 20–50 lbs./day	7.8–22
Milk production 80 lbs./day	23
Dry	11
Pigs	
Body weight — 30 lbs.	0.6–1.2
Body weight — 60–80 lbs.	0.8
Body weight — 75–125 lbs.	1.9
Body weight — 200–380 lbs.	1.4–3.6
Pregnant sows	3.6–4.6
Lactating sows	4.8–6.0
Sheep	
On range or dry pasture	0.6–1.6
On range (salty feeds)	2.0
On rations of hay and grain or hay, roots and grain	0–0.7
On good pasture	Little, if any
Chickens (100 birds)	
1–3 weeks of age	0.4–2.0
3–6 weeks of age	1.4–3.0
6–10 weeks of age	3.0–4.0
9–13 weeks of age	4.0–5.0
Pullets	3.0–4.0
Nonlaying hens	5.0
Laying hens (moderate temperatures)	5.0–7.5
Laying hens (temperature 90°F.)	9.0
Turkeys (100 birds)	
1–3 weeks of age	1.1–2.6
7–4 weeks of age	3.7–8.4
9–13 weeks of age	9–14
15–19 weeks of age	17
21–26 weeks of age	14–15

[1] Allow 15 to 20 additional gallons per day for each cow for flushing stables and washing dairy utensils.
Source: U.S. Dept. of Agriculture

TABLE 5–84. LIVESTOCK FRESHWATER USE IN THE UNITED STATES, 1985

[Includes water for stock watering, feed lots, dairy operations, fish farming and other on farm needs; figures may not add totals because of independent rounding; all values in million gallons per day]

State	Withdrawals			Consumptive Use
	By Source		Total	
	Ground Water	Surface Water		
Alabama	67	29	96	58
Alaska	10	146	156	.2
Arizona	25	36	61	9.8
Arkansas	242	198	440	242
California	41	159	199	155
Colorado	16	45	61	31
Connecticut	6.4	2.0	8.4	1.8
Delaware	1.9	.0	1.9	1.9
D.C.	.0	.0	.0	.0
Florida	58	8.1	66	66
Georgia	25	22	47	47
Hawaii	.7	3.1	3.8	2.2
Idaho	1040	.0	1040	102
Illinois	57	.0	57	49
Indiana	48	.0	48	41
Iowa	135	37	172	172
Kansas	42	26	68	68
Kentucky	2.5	48	50	50
Louisiana	75	127	203	92
Maine	3.0	26	29	25
Maryland	13	9.8	23	11
Massachusetts	.4	.9	1.3	1.3
Michigan	19	5.4	25	22
Minnesota	53	9.5	63	63
Mississippi	373	12	385	85
Missouri	10	30	41	41
Montana	16	34	50	50
Nebraska	101	19	120	116
Nevada	5.9	20	26	6.6
New Hampshire	.6	.7	1.2	.2
New Jersey	3.1	.0	3.1	3.1
New Mexico	11	39	50	49
New York	12	7.3	20	18
North Carolina	29	5.1	34	34
North Dakota	13	8.9	22	22
Ohio	25	16	41	41
Oklahoma	2.0	2.6	4.6	4.6
Oregon	3.8	21	25	25
Pennsylvania	62	8.4	70	61
Rhode Island	1.5	.8	2.3	2.0
South Carolina	5.2	5.0	10	10
South Dakota	19	28	47	47
Tennessee	32	34	65	28
Texas	111	150	261	261
Utah	31	7.2	38	6.9
Vermont	3.8	1.7	5.6	1.1
Virginia	29	25	53	5.4
Washington	21	8.8	30	25
West Virginia	16	9.6	26	22
Wisconsin	87	3.1	90	73
Wyoming	3.2	13	16	16
Puerto Rico	8.6	.0	8.6	8.6
Virgin Islands	.0	.0	.0	.0
Total	3,020	1,450	4,470	2,370

Source: Solley, W.B., Merk, C.F., and Pierce, R.R., 1988, Estimated Water Use in the United States in 1985, U.S. Geological Survey Circular 1004

SECTION L. NAVIGATION AND WATERWAYS

TABLE 5–85. PROJECTED FLOWS REQUIRED FOR EFFICIENT NAVIGATION ON INLAND WATERWAYS OF THE UNITED STATES

Waterway	Critical Flow[1] in cubic feet per second			Comments
	1959	1980	2000	
New England, existing waterways: all	0	0	0	All waterways are in tidal reaches.
Middle Atlantic, existing waterways:				
Great Lakes to Hudson River and Champlain Canals	2,000	2,000	2,000	Available flow 13,000 c.f.s.
Delaware	[2]	[2]	[2]	If depth increased to 45 feet minimum flow will have to be increased by 2,900, c.f.s. to repel salinity.
South Atlantic:				
Existing waterways:				
Cape Fear River above Wilmington, N.C.	50	100	[3]	
Savannah River below Augusta, Ga.	5,000	5,800	[3]	
Altamaha River below junction of Ocmulgee and Oconee Rivers.	1,000	5,000	[3]	
Okeechobee Waterway	50	50	[3]	
Apalachicola below Jim Woodruff lock and dam	9,300	9,300	[3]	
Alabama River below Selma, Ala.	3,000	3,000	[3]	
Warrior system	540	540	[3]	
Possible future waterways:				
Santee-Congaree	......	5,200	[3]	Project might be found feasible if river developed for power.
Ocmulgee River below Macon, Ga.	......	300	[3]	
Cross-Florida Barge Canal	......	700	[3]	Authorized project.
Chattahoochee River below Atlanta, Ga.	......	300	[3]	
Flint River below Albany, Ga.	......	200	[3]	
Coosa River	......	300	[3]	Do.
Tennessee-Tombigbee	......	1,246	[3]	Completed project. (1,246 c.f.s. to be diverted from the Tennessee River.)
Arkansas-White-Red:				
Existing waterways:				
White River (to mile 168.7)	6,500	10,000	10,000	
Ouachita-Black	100	100	100	
Waterways under construction:				
Verdigris River (at Catoosa, Okla.)	......	115	115	Provision of 9-foot waterway as part of multiple-purpose development of the Arkansas River.
Arkansas (at Webber Falls)	......	300	300	Do.
Arkansas (at Short Mountain)	......	530	530	Do.
Arkansas (at Dardanelle)	......	505	505	Do.
Arkansas (Dardenelle to mouth)	......	1,000	1,000	Do.
Possible future waterways: Overton-Red Waterway.	......	300		Authorized project.
Gulf-Southwest:				
Waterways under construction: Guadalupe to Victoria.	......	0–1,800	0–1,800	For periodic flushing.
Possible future waterways:				
Trinity River (below Dallas)	......	400	800	Authorized project.
Trinity River (above Dallas)	......	150	325	Do.
Missouri River Basin, existing waterways:				
Missouri at Sioux City[4]	......	30,000	30,000	Could be reduced to 25,000 by dredging.
Missouri at Omaha[4]	28,000	28,000	28,000	
Missouri at Nebraska City[4]	31,000	31,000	31,000	
Missouri at Kansas City[4]	35,000	32,500	32,500	Could be reduced to 27,500 by dredging.
Missouri at mouth[4]	......	35,000	35,000	

TABLE 5–85. PROJECTED FLOWS REQUIRED FOR EFFICIENT NAVIGATION ON INLAND WATERWAYS OF THE UNITED STATES (continued)

Waterway	Critical Flow[1] in cubic feet per second			Comments
	1959	1980	2000	
Upper Mississippi Basin:				
Existing waterways:				
Main stem, pools 1–10	375	750	940	Not required in winter.
Main stem, pools 11–22	2,000	2,000	2,000	Do.
Main stem, pools 24–26	70,000	25,000	25,000	Required all year.
Main stem, Missouri to Ohio	54,000	75,000	75,000	Do.
Minnesota	0	0	0	Mississippi backwater.
St. Croix	0	0	0	Do.
Illinois Waterway	720[5]	1,826[5,6]	1,826[5,6]	
Fox River	300	300	300	Not required in winter.
Possible future waterways:				
Kaskaskia River	0	200	300	
Big Muddy River	0	200	400	
Tittabawassee River	0	200	200	
Ohio River Basin:				
Existing waterways:				
Ohio River (main stem)	[3]	[3]	[3]	River canalized.
Allegheny River	100	140	150	
Monongahela River	340	440	670	
Kanawha River	300	400	400	
Kentucky River	100	150	200	
Green River	150	250	400	
Cumberland River	100	150	250	
Possible future waterways:				
Big Sandy River (main stem)	0	329	329	
Big Sandy River (Levisa Fork)	0	176	176	
Lake Erie-Ohio River Canal	0	430	430	
Lower Mississippi River:				
Existing waterways:				
Main stem at Cairo	100,000	120,000	120,000	
Main stem below Arkansas	140,000	150,000	150,000	
Ouachita-Black	100	150	150	
Possible future waterways:				
Yazoo below Greenwood	......	100	100	
Columbia River Basin:				
Existing waterways:				
Main stem at Bonneville	40,000	40,000	40,000	
Main stem at The Dallas	700	1,500	2,500	
Willamette, mouth to Salem	6,000	6,000	6,000	
Willamette, Salem to Corvallis	5,000	5,000	5,000	
Snake, mouth to Ice Harbor	300	1,000	2,000	
Waterways under construction:				
Main stem John Day to McNary	110,000	1,500	2,500	
Snake, Ice Harbor to Lewiston	......	1,000	2,000	
Possible future waterways:				
Main stem head McNary Pool to Rock Island	......	36,000	......	Under study.
Snake, Lewiston to mile 174	......	500	1,000	Do.
Snake, mile 174 to mile 188	......	500	1,000	
Snake, mile 188 to mile 232	......	350	700	
North Pacific Coast, Possible future waterways: Skagit, mouth to Concrete.	......	10,000	10,000	
Central Valley, existing waterways:				
Sacramento	5,000	5,000	5,000	Shallow draft.
San Joaquin above Mossdale	100	100	100	Do.

[1] Rate below which streamflow cannot drop without reducing the efficiency of navigation.
[2] Anticipated flows will meet the needs of navigation.
[3] Estimates not available.
[4] Flows required from April through November, assuming continuation of open river navigation. Canalization now being studied. If Missouri canalized these flows would be greatly reduced.
[5] Average annual flow required at Lockport, Ill. Includes 250 c.f.s. to prevent reversal of flow into Lake Michigan when storm runoff occurs. Excludes 120 c.f.s. of present industrial usage which bypasses Lockport lock.
[6] Requirements for recommended duplicate lock system.
Source: Select Committee on National Water Resources, U.S. Senate, 1960; amended

TABLE 5–86. NAVIGATION LOCKS AND DAMS IN THE UNITED STATES

[Operable as of September 30, 1986]

Project	Miles Above Mouth	Community in Vicinity	Locks: Width of Chamber	Locks: Available Length for Full Width	Locks: Lift at Normal Pool Level	Locks: Depth on Sills Upper	Locks: Depth on Sills Lower	Locks: Type[1]	Dams: Length (feet)	Dams: Year Opened	Authorized Channel: Length (miles)	Authorized Channel: Depth (feet)	Authorized Channel: Width (feet)
Alabama-Coosa Rivers, AL													
Claiborne Lock and Dam	81.8[2]	Claiborne, AL	84	600	30	16	13	Movable	3,160[3]	1973	60.5	9	200
Millers Ferry Lock and Dam	142.3[2]	Camden, AL	84	600	45	16	13	Movable	9,900[3]	1969	103.1	9	200
Jones Bluff Lock and Dam	254.4[2]	Benton, AL	84	600	45	16	13	Movable	14,962[3]	1974	88.0	9	200
Allegheny River, PA and NY													
Lock and Dam No. 2	6.7	Aspinwall, PA	56	360	11	11	12	Fixed	1,393	1934[20]	7.8	9	200
Lock and Dam No. 3	14.5	Cheswick, PA	56	360	13	12	11	Fixed	1,436	1934[20]	9.7	9	200
Lock and Dam No. 4	24.2	Natrona, PA	56	360	10	9	10	Fixed	876	1927	6.2	9	200
Lock and Dam No. 5	30.4	Freeport, PA	56	360	12	10	11	Fixed	780	1927	5.9	9	200
Lock and Dam No. 6	36.3	Clinton, PA	56	360	12	11	11	Fixed	1,140	1928	9.4	9	200
Lock and Dam No. 7	45.7	Kittanning, PA	56	360	13	11	10	Fixed	916	1930	6.9	9	200
Lock and Dam No. 8	52.6	Templeton, PA	56	360	18	14	10	Fixed	984	1931	9.6	9	200
Lock and Dam No. 9	62.2	Rimer, PA	56	360	22	11	11	Fixed	950	1938	9.8	9	200
Apalachicola, Chattahoochee, and Flint Rivers, GA, AL and FL													
Jim Woodruff Lock and Dam	107.6[4]	Chattahoochee, FL	82	450	33	14	14	Movable	5,924	1957	46.7	6	100
George W. Andrews Lock and Dam	154.3[5]	Columbia, AL	82	450	25	19	13	Movable	620	1963	28.5	6	100
Walter F. George Lock and Dam	182.8[5]	Fort Gaines, GA	82	450	88	18	13	Movable	13,371	1963	85.0	6	100
Atlantic Intracoastal Waterway													
Albemarle and Chesapeake Canal Route:													
Great Bridge Lock	11.5[6]	Great Bridge, VA	72	530	3	16[7]	16[7]	None		1932		12	90
Dismal Swamp Canal Route:													
Deep Creek Lock	10.6[6]	Deep Creek, VA	52	300	12	12[7]	12[7]	None		1940		6	50
South Mills Lock	33.2[6]	South Mills, NC	52	300	12	12[7]	12[7]	None		1941		6	50
Bayou Teche, LA													
Berwick Lock	1.5[8]	Berwick, LA	45	307	7[9]	9[7]	9[7]	None		1950		8	80
Keystone Lock	72.5[8]	New Iberia, LA	36	162	9	9	8	Fixed	175	1913	34.5	6	80
Black Rock Channel and Tonawanda Harbor, NY													
Black Rock Lock	0.0	Buffalo, NY	68	625	5	21.6	21.6	None		1914	7.1	21	200
Black Warrior, Warrior and Tombigbee Rivers, AL													
Coffeeville (Jackson) Lock and Dam	116.6[10]	Coffeeville, AL	110	600	33	13	13	Movable	1,185	1965	96.6	9	200
Demopolis Lock and Dam	213.2[10]	Demopolis, AL	110	600	40	13	13	Fixed	1,485	1962	47.9	9	200
Armistead I. Selden Lock and Dam	261.1[10]	Eutaw, AL	110	600	22	13	13	Movable	1,832	1962	77.0	9	200
Wm. Bacon Oliver Lock and Dam	338.1[10]	Tuscaloosa, AL	95	460	28	12	12	Fixed	700	1940	8.9	9	200
Holt Lock and Dam	347.0[10]	Holt, AL	110	600	64	19	13	Movable	1,138	1969	18.1	9	200
John Hollis Bankhead Lock and Dam (new lock)	365.1[10]	Adger, AL	110	600	68	13	13	Fixed	1,170	1975	42.7	9	200
Canaveral Harbor, FL													
Canaveral Lock	2.7	Cocoa, FL	90	600	3	14	14	None		1965		12	125
Cape Fear River, NC													
Lock and Dam No. 1	67.0	Kings Bluff, NC	40	200	11	9	9	Fixed	275	1915	32.0	8	100
Lock and Dam No. 2	99.0	Browns Landing, NC	40	200	9	12	12	Fixed	229	1917	24.0	8	100
William O. Huske Lock and Dam	123.0	Tolars Landing, NC	40	300	9	9	9	Fixed	220	1935	20.0	8	100
Central and Southern Florida													
S-61 Lock	0.0	St. Cloud, FL	30	90	2	6	6	None		1963		8	20
S-65 Lock	56.0	Frostproof, FL	30	90	6	6	6	None		1964	28.0	3	30
S-65A Lock	46.0	Avon Park, FL	30	90	6	6	6	None		1967	10.0	3	30
S-65B Lock	32.0	Sebring, FL	30	90	6	6	6	None		1965	14.0	3	30
S-65C Lock	25.0	Sebring, FL	30	90	7	6	6	None		1965	7.0	3	30
S-65D Lock	16.0	Okeechobee, FL	30	90	6	6	6	None		1964	9.0	3	30
S-65E Lock	12.0	Okeechobee, FL	30	90	5	6	6	None		1964	4.0	3	30
S-308B	38.7	Port Mayaca, FL	56	400	14	14	14	Movable		1977	5.0	8	100
S-310	0.0	Clewston, FL	50	60	5	13	13	Movable		1980	.5	6	50

TABLE 5–86. NAVIGATION LOCKS AND DAMS IN THE UNITED STATES (continued)

[Operable as of September 30, 1986]

Project	Miles Above Mouth	Community in Vicinity	Locks: Width of Chamber	Locks: Available Length for Full Width	Locks: Lift at Normal Pool Level	Locks: Depth on Sills Upper	Locks: Depth on Sills Lower	Locks: Type[1]	Dams: Length (feet)	Dams: Year Opened	Authorized Channel: Length (miles)	Authorized Channel: Depth (feet)	Authorized Channel: Width (feet)
Chicago Harbor, IL													
Chicago Lock	0.0	Chicago, IL	80	600	4	23	23	None		1939	.15	21	470
Columbia River, OR and WA													
Bonneville Lock and Dam	146.0	Bonneville, OR	76	500	65	32	24	Movable	2,680	1938	47.5	27	300
The Dalles Lock and Dam	190.0	The Dalles, OR	86	675	88	15	15	Movable	8,735	1957	25.0	14	250
John Day Lock and Dam	215.0	Rufus, OR	86	675	110	15	15	Movable	5,900	1968	76.0	14	250
McNary Lock and Dam	292.0	Umatilla, OR	86	675	75	15	20	Movable	7,365	1953	64.0	14	250
Cross-Florida Barge Canal													
Inglis Lock, Dam and Spillway	168.6[11]	Inglis, FL	84	600	28	18	15	Movable	5,100	1968	11.0	12	150
Eureka Lock and Dam	20.0	Sparr, FL	84	600	20	14		Movable	3,830	1971		12	150
Harry H. Buckman Lock	90.4	Palatka, FL	84	600	20	14	15	None		1972	21.2	12	150
Cumberland River, KY and TN													
Barkley Dam	30.6	Kuttawa, KY	110	800	57	11	13	Movable	9,959	1964	118.1	9	150
Cheatham Lock and Dam	148.7	Ashland City, TN	110	800	26	14	12	Movable	801	1959	67.5	9	150
Old Hickory Lock and Dam	216.2	Old Hickory, TN	84	400	60	14	10	Movable	3,605	1957	97.3	9	150
Cordell Hull Dam and Reservoir	313.5	Carthage, TN	84	400	59	14	13	Movable	1,138	1973	71.9	9	150
Fox River, WI													
DePere Lock	7.1	DePere, WI	36	146	9	10	12			1936	5.9	6	100
DePere Dam	7.2	DePere, WI						Movable	986	1929			
Little Kaukauna Lock	13.0	DePere, WI	36	146	7	8	10			1938	6.2	6	100
Little Kaukauna Dam	13.1	DePere, WI						Movable	588	1926			
Rapide Croche Lock	19.2	Wrightstown, WI	36	146	8	9	10			1934	3.6	6	100
Rapide Croche Dam	19.3	Wrightstown, WI						Movable	461	1930			
Kaukauna Fifth Lock	22.8	Kaukauna, WI	36	144	9	7	9			1898	0.3	6	100
Kaukauna Fourth Lock	23.1	Kaukauna, WI	37	144	10	7	6			1879	0.2	6	100
Kaukauna Third Lock	23.3	Kaukauna, WI	31	144	10	7	6			1879	0.1	6	100
Kaukauna Second Lock	23.4	Kaukauna, WI	35	144	10	6	6			1903	0.2	6	100
Kaukauna First Lock	23.6	Kaukauna, WI	35	144	11	7	6			1883	0.4	6	100
Kaukauna Dam	24.0	Kaukauna, WI						Movable	603	1931			
Kaukauna Guard Lock	24.0	Kaukauna, WI	40		9					1891	1.4	6	100
Little Chute combined Lock													
Lower	25.4	Little Chute, WI	35	147	11	6	9			1879		6	100
Upper	25.4	Little Chute, WI	36	144	11	8	6			1879		6	100
Little Chute Second Lock	26.4	Little Chute, WI	35	144	14	8	6			1881	1.0	6	100
Little Chute First (Guard) Lock	26.5	Little Chute, WI	35		7					1904	0.7	6	100
Little Chute Dam	26.6	Little Chute, WI						Movable	562	1932			
Cedars Lock	27.3	Little Chute, WI	35	144	10	7	7			1888	3.4	6	100
Cedars Lock	27.3	Little Chute, WI	35	144	10	7	7			1888	3.4	6	100
Cedars Dam	27.4	Little Chute, WI						Movable	654	1933			
Appleton Fourth Lock	30.7	Appleton, WI	35	144	8	8	8			1907	0.6	6	100
Appleton Lower Dam	30.9	Appleton, WI						Movable	549	1934			
Appleton Third Lock	31.3	Appleton, WI	35	144	9	6	9			1900	0.3	6	100
Appleton Second Lock	31.6	Appleton, WI	35	145	10	7	6			1901	0.3	6	100
Appleton First Lock	31.9	Appleton, WI	34	145	10	7	6			1884	5.1	6	100
Appleton Upper Dam	32.2	Appleton, WI						Movable	691	1940			
Menasha Lock	37.8	Menasha, WI	34	144	8	7	8			1899	28.3	6	100
Menasha Dam	37.8	Menasha, WI						Movable	401	1937			
Freshwater Bayou, LA													
Freshwater Bayou Lock	19.0[8]	Intracoastal City, LA	84	600		16	16	None	401	1968		12	125
Green and Barren Rivers, KY													
Green River:													
Lock and Dam No. 1	9.1	Spottsville, KY	84	600	12	12	11	Fixed	482	1956	54.0	9	200
Lock and Dam No. 2	63.1	Calhoun, KY	84	600	14	15	12	Fixed	519	1956	45.4	9	200
Gulf Intracoastal Waterway													
Inner Harbor Navigation Channel Lock	92.6[12]	New Orleans, LA	75	640	9[9]	31[7]	31[7]	None		1923		12	150
Harvey Lock	98.3[12]	Harvey, LA	75	425	10[9]	12[7]	12[7]	None		1935		12	125
Algiers Lock	88.0[12]	Algiers, LA	75	797	10[9]	13[7]	13[7]	None		1956		16	150
Bayou Boeuf Lock	93.3	Morgan City, LA	75	1,158	6[9]	13[7]	13[7]	None		1956		16	150
Bayou Sorrel Lock	131.0[13]	Plaquemine, LA	56	800	10[9]	14[7]	14[7]	None		1952		12	125
Port Allen Lock	228.5[12]	Port Allen, LA	84	1,198	45	13[7]	14[7]	None		1961		12	125
Vermilion Lock	162.7[13]	Abbeville, LA	56	1,200	3[9]	11[7]	11[7]	None		1934		16	200
Calcasieu Lock	238.5[13]	Lake Charles, LA	75	1,205	6[9]	13[7]	13[7]	None		1950		16	200
Brazos River Floodgates	404.1[13]	Freeport, TX	75			15[7]	15[7]	None		1943		12	125

TABLE 5–86. NAVIGATION LOCKS AND DAMS IN THE UNITED STATES (continued)

[Operable as of September 30, 1986]

Project	Miles Above Mouth	Community in Vicinity	Locks: Width of Chamber	Locks: Available Length for Full Width	Locks: Lift at Normal Pool Level	Locks: Depth on Sills Upper	Locks: Depth on Sills Lower	Dams: Type[1]	Dams: Length (feet)	Dams: Year Opened	Authorized Channel: Length (miles)	Authorized Channel: Depth (feet)	Authorized Channel: Width (feet)
Colorado River													
East Lock	444.8	Matagorda, TX	75	1,200	5[9]	15[7]	15[7]	None		1954		12	125
West Lock	444.8	Matagorda, TX	75	1,200	5[9]	15[7]	15[7]	None		1954		12	125
Hudson River, NY													
Troy Lock and Dam	153.8	Troy, NY	45	493	17	16	13	Fixed	1,495	1917	2.2	14	200
Illinois Waterway, IL													
LaGrange Lock and Dam	80.2	Beardstown, IL	110	600	10	16	13	Movable	1,066	1939	77.5	9	300
Peoria Lock and Dam	157.7	Peoria, IL	110	600	11	16	12	Movable	536	1939	73.3	9	300
Starved Rock Lock and Dam	231.0	Utica, IL	110	600	19	17	14	Movable	1,280	1933	13.6	9	300
Marseilles Lock	244.6	Marseilles, IL	110	600	24	19	14			1933	26.9	9	300
Marseilles Dam	247.0	Marseilles, IL						Movable	819	1933	24.6		
Dresden Island Lock and Dam	271.5	Morris, IL	110	600	22	17	12	Movable	1,616	1933	14.5	9	300
Brandon Road Lock and Dam	286.0	Joliet, Il	110	600	34	18	14	Movable	2,373	1933	5.1	9	300
Lockport Lock	291.1	Lockport, IL	110	600	40	12	15	None		1933		9	300
Thomas J. O'Brien Lock and Dam	326.5	Chicago, IL	110	1,000	2	14	14	Movable	257	1960	6.9	9	300
Inland Route, MI													
Crooked River Lock and Weir	30.0	Alanson, MI	17.8	66	1	6.9	7.9	SSP Weir	83	1967	4.0	5	30
Kanawha River, WV													
Winfield Lock and Dam	31.1	Winfield, WV	56	360	28	18	12	Movable	834	1937	36.7	9	300
Marmet Lock and Dam	67.8	Marmet, WV	56	360	24	18	12	Movable	707	1934	15.0	9	300
London Lock and Dam	82.8	London, WV	56	360	24	18	12	Movable	707	1934	7.8	9	300
Kaskaskia River, IL													
Kaskaskia Lock and Dam	.8	Ellis Grove, IL	84	600	32	18	11	Movable	130	1973	35.0	9	225
Kentucky River, KY													
Lock and Dam No. 1	4.0	Carrolton, KY	38	145	8	8	15	Fixed	424	1839	27.0	6	100
Lock and Dam No. 2	31.0	Lockport, KY	38	145	14	8	6	Fixed	400	1939	11.0	6	100
Lock and Dam No. 3	42.0	Gest, KY	38	145	13	9	7	Fixed	465	1844	23.0	6	100
Lock and Dam No. 4	65.0	Frankfort, KY	38	145	13	6	6	Fixed	543	1844	17.2	6	100
Lock and Dam No. 5[24]	82.2	Tyrone, KY	38	145	15	10	6	Fixed	556	1844	14.0	6	100
Lock and Dam No. 6[24]	96.2	High Bridge, KY	52	147	14	9	6	Fixed	413	1891	20.8	6	100
Lock and Dam No. 7[24]	117.0	High Bridge, KY	52	147	15	9	7	Fixed	350	1897	22.9	6	100
Lock and Dam No. 8[24]	139.9	Camp Nelson, KY	52	146	19	11	6	Fixed	257	1900	17.6	6	100
Lock and Dam No. 9[24]	157.5	Valley View, KY	52	148	17	11	7	Fixed	362	1907	18.9	6	100
Lock and Dam No. 10[24]	176.4	Ford, KY	52	148	17	9	6	Fixed	472	1907	24.6	6	100
Lock and Dam No. 11[24]	201.0	Irvine, KY	52	148	18	10	6	Fixed	208	1906	19.9	6	100
Lock and Dam No. 12[24]	220.9	Ravenna, KY	52	148	17	10	6	Fixed	240	1910	19.0	6	100
Lock and Dam No. 13[24]	239.9	Willow, KY	52	148	18	10	6	Fixed	248	1915	9.1	6	100
Lock and Dam No. 14[24]	249.0	Heidelberg, KY	52	148	17	9	6	Fixed	248	1917	9.6	6	100
Lake Washington Ship Canal													
Hiram M. Chittenden Lock													
Large Lock	1.3	Seattle, WA	80	760	26	33.5	29	Movable	235	1916	17.0	34	150
Small Lock	1.3	Seattle, WA	28	123	26	16	16						
McClellan-Kerr Arkansas River Navigation System, AR and OK													
Norrell Lock and Dam	10.3	Arkansas Post, AR	110	600	30	16	15	Fixed	4,677	1967	3.0	9	250
Lock No. 2	13.3	Arkansas Post, AR	110	600	20	18	14			1967	36.9	9	250
Dam No. 2	40.5	Arkansas Post, AR						Movable	42,073	1968			
Lock and Dam No. 3	50.2	Grady, AR	110	600	20	18	14	Movable	6,110	1968	15.8	9	250
Lock and Dam No. 4	66.0	Pine Bluff, AR	110	600	14	18	14	Movable	5,745	1968	20.3	9	250
Lock and Dam No. 5	86.3	Redfield, AR	110	600	17	18	14	Movable	7,455	1968	21.8	9	250
David D. Terry Lock and Dam	108.1	Little Rock, AR	110	600	18	18	14	Movable	8,890	1968	17.3	9	250
Murray Lock and Dam	125.4	Little Rock, AR	110	600	18	18	14	Movable	3,930	1969	30.5	9	250
Toad Suck Ferry Lock and Dam	155.9	Conway, AR	110	600	16	18	14	Movable	1,580	1969	21.0	9	250
Lock and Dam No. 9	176.9	Morrilton, AR	110	600	19	18	14	Movable	1,505	1969	28.6	9	250
Dardanelle Lock and Dam	205.5	Russellville, AR	110	600	54	18	14	Movable	1,815	1969	51.3	9	250
Ozark-Jetta Taylor Lock and Dam	256.8	Ozark, AR	110	600	34	18	15	Movable	2,480	1969	36.0	9	250
Lock and Dam No. 13	292.8	Fort Smith, AR	110	600	20	18	14	Movable	4,725	1969	26.8	9	250

TABLE 5–86. NAVIGATION LOCKS AND DAMS IN THE UNITED STATES (continued)

[Operable as of September 30, 1986]

Project	Miles Above Mouth	Community in Vicinity	Locks: Width of Chamber	Locks: Available Length for Full Width	Locks: Lift at Normal Pool Level	Locks: Depth on Sills Upper	Locks: Depth on Sills Lower	Locks: Type[1]	Dams: Length (feet)	Dams: Year Opened	Authorized Channel: Length (miles)	Authorized Channel: Depth (feet)	Authorized Channel: Width (feet)
McClellan-Kerr Arkansas (Continued)													
W. D. Mayo Lock and Dam	319.6	Fort Smith, AR	110	600	20	14	15	Movable	7,400	1970	16.6	9	250
Robert S. Kerr Lock and Dam and Reservoir	336.2	Sallisaw, OK	110	600	48	18	14	Movable	7,230	1970	30.4	9	250
Webbers Falls Lock and Dam	366.6	Webbers Falls, OK	110	600	30	19	16	Movable	4,370	1970	34.8	9	250
Chouteau Lock and Dam	401.2	Muskogee, OK	110	600	21	15	14	Movable	11,490	1970	20.2	9	150
Newt Graham Lock and Dam	421.4	Inola, OK	110	600	21	15	14	Movable	1,629	1970	23.7	9	150
Mississippi River													
Between Ohio and Missouri Rivers													
Lock and Dam No. 27	185.1[14]	Granite City, IL	110	1,200	21	16	15	Fixed	3,240	1953	17.8	9	200
Mississippi River Between Missouri River and Minneapolis, MN													
Lock and Dam No. 26	202.9[15]	Alton, IL	110	600	24	19	10	Movable	1,725	1938	38.5	9	200
				360	24	16	10						
Lock and Dam No. 25	241.4[15]	Cap Au Gris, MO	110	600	15	19	12	Movable	1,296	1939	32.0	9	200
Lock and Dam No. 24	273.4[15]	Clarksville, MO	110	600	15	19	12	Movable	4,280	1940	27.8	9	200
Lock and Dam No. 22	301.2[15]	Saverton, MO	110	600	10	18	14	Movable	1,024	1938	23.7	9	NS
Lock and Dam No. 21	324.9[15]	Quincy, IL	110	600	10	17	12	Movable	1,066	1938	18.3	9	NS
Lock and Dam No. 20	343.2[15]	Canton, MO	110	600	10	15	12	Movable	2,144	1936	21.0	9	NS
Lock and Dam No. 19	364.2[15]	Keokuk, IA	110	1,200	38	15	13	Fixed	4,434	1913	46.3	9	NS
Lock and Dam No. 18	410.5[15]	Burlington, IA	110	600	10	17	14	Movable	1,350	1937	26.6	9	NS
Lock and Dam No. 17	437.1[15]	New Boston, IL	100	600	8	16	13	Movable	921	1939	20.1	9	NS
Lock and Dam No. 16	457.2[15]	Muscatine, IA	110	600	9	17	12	Movable	1,099	1937	25.7	9	NS
Lock and Dam No. 15	482.9[15]	Rock Island, IL	110	600	16	27	11	Movable	1,203	1934	10.4	9	NS
			110	360	16	27	11						
Lock and Dam No. 14	493.1[15]	LeClaire, IA	80	320	11	18	11			1922			
	493.3[15]	LeClaire, IA	110	600	11	21	14	Movable	1,343	1939	29.2	9	NS
Lock and Dam No. 13	522.5[15]	Clinton, IA	110	600	11	19	13	Movable	1,066	1939	34.2	9	NS
Lock and Dam No. 12	556.7[15]	Bellevue, IA	110	600	9	17	13	Movable	849	1938	26.3	9	NS
Lock and Dam No. 11	583.0[15]	Dubuque, IA	110	600	11	19	13	Movable	1,278	1937	32.1	9	NS
Lock and Dam No. 10	615.1[15]	Guttenberg, IA	110	600	8	15	12	Movable	763	1937	32.8	9	NS
Lock and Dam No. 9	647.9[15]	Lynxville, IA	110	600	9	16	13	Movable	811	1937	31.3	9	NS
Lock and Dam No. 8	679.2[15]	Genoa, WI	110	600	11	22	14	Movable	935	1937	23.3	9	NS
Lock and Dam No. 7	702.5[15]	Dresbach, MN	110	600	8	18	12	Movable	940	1937	11.8	9	NS
Between Missouri River and Minneapolis, MN													
Lock and Dam No. 6	714.3[15]	Trempealeau, WI	110	600	7	17	13	Movable	893	1936	14.2	9	NS
Lock and Dam No. 5A	728.5[15]	Winona, MN	110	600	6	18	13	Movable	682	1936	9.6	9	NS
Lock and Dam No. 5	738.1[15]	Minneiska, MN	110	600	9	18	12	Movable	1,619	1935	14.7	9	NS
Lock and Dam No. 4	752.8[15]	Alma, WI	110	600	7	17	13	Movable	1,367	1935	44.1	9	NS
Lock and Dam No. 3	796.9[15]	Red Wing, MN	110	600	8	17	14	Movable	365	1938	18.3	9	NS
Lock and Dam No. 2	815.2[15]	Hastings, MN	110	600	12	22	13	Movable	822	1931	32.4	9	NS
				500	12	16	15			1948			
Lock and Dam No. 1	847.6[15]	Minneapolis-St. Paul, MN	56	400	38	13	8	Fixed	574	1932	5.7	9	NS
St. Anthony Falls													
Lower Lock and Dam	853.3[15]	Minneapolis, MN	56	400	25	14	10	Movable	188	1956	0.6	9	100
Upper Lock and Dam	853.9[15]	Minneapolis, MN	56	400	49	16	14	Fixed	3,584	1963	3.8	9	100
Monongahela River, PA and WV													
Locks and Dam No. 2	11.2	Braddock, PA	56	360	9	16	16	Fixed	748	1951	12.6	9	300
			110	720	9	16	16			1953[21]			
Locks and Dam No. 3	23.8	Elizabeth, PA	56	360	8	12	12	Fixed	670	1907	17.7	9	300
			56	720	8	12	12			1907[21]			
Locks and Dam No. 4	41.5	Monessen, PA	56	360	17	20	11	Movable	535	1932	19.7	9	300
			56	720	17	20	11			1932[21]			
Maxwell Locks and Dam	61.2	Maxwell, PA	84	720	20	21	15	Movable	460	1964	23.8	9	300
			84	720	20	21	15			1964			
Locks and Dam No. 7	85.0	Greensboro, PA	56	360	15	11	10	Fixed	610	1925[21]	5.8	9	300
Locks and Dam No. 8	90.8	Point Marion, PA	56	360	19	15	10	Movable	560	1925[21,22]	11.2	9	300
Morgantown Lock and Dam	102.0	Morgantown, WV	84	600	17	18	15	Movable	410	1950	6.0	9	300
Hildebrand Lock and Dam	108.0	Morgantown, WV	84	600	21	14	15	Movable	530	1959	7.4	9	300
Opekiska Lock and Dam	115.4	Morgantown, WV	84	600	22	18	14	Movable	366	1964	7.0	9	300
Ohio River													
Locks and Dam No. 53	18.4	Mound City, IL	110	600	12	15	10	Movable	3,978	1929	23.7	9	300
			110	1,200	12	15	10	Temporary Lock		1980			

TABLE 5–86. NAVIGATION LOCKS AND DAMS IN THE UNITED STATES (continued)

[Operable as of September 30, 1986]

Project	Miles Above Mouth	Community in Vicinity	Locks: Width of Chamber	Locks: Available Length for Full Width	Locks: Lift at Normal Pool Level	Locks: Depth on Sills Upper	Locks: Depth on Sills Lower	Type[1]	Dams: Length (feet)	Year Opened	Authorized Channel: Length (miles)	Authorized Channel: Depth (feet)	Authorized Channel: Width (feet)
Ohio River (Continued)													
Locks and Dam No. 52	42.1	Brookport, IL	110	600	12	15	11	Movable	3,073	1928	35.8	9	300
			110	1,200	12	15	11	Temporary Lock		1969			
Smithland Locks and Dam	35.3	Smithland, KY	110	1,200	22	34	12	Movable	3,560	1980	99.7	9	300
				1,200	22	34	12						
Uniontown Locks and Dam	135.0	Uniontown, KY	110	1,200	22	34	12	Movable	3,516	1975	69.9	9	300
				600	22	34	12						
Newburgh Locks and Dam	204.9	Newburgh, IN	110	1,200	16	32	16	Movable	2,272	1975	55.4	9	300
			110	600	16	32	16						
Cannelton Locks and Dam	260.3	Cannelton, IN	110	1,200	25	38	13	Movable	1,965	1972	116.3	9	400
			110/	600	25	38	13						
McAlpine Locks and Dam	374.2	Louisville, KY	110	1,200	37	49	12	Movable	8,627	1961	75.3	9	300
			110	600	37	19	11			1921			
			56	360	37	19	11			1930			
Markland Locks and Dam	449.5	Markland, IN	110	1,200	35	50	15	Movable	1,395	1963	95.3	9	300
			110	600	35	50	15			1963		9	300
Captain Anthony Meldahl Locks and Dam	544.8	Chilo, OH	110	1,200	30	45	15	Movable	1,756	1962	95.2	9	300
			110	600	30	45	15			1962			
Greenup Locks and Dam	640.0	Greenup, KY	110	1,200	30	45	15	Movable	1,287	1962	61.8	9	300
			110	600	30	45	15						
Gallipolis Locks and Dam	701.8	Hogsett, WV	110	600	23	18	15	Movable	1,132	1937	41.7	9	300
			110	360	23	18	15						
Racine Locks and Dam	743.5	Letart Falls, OH	110	1,200	22	37	15	Movable	1,202	1970	33.6	9	300
			110	600	22	37	15						
Belleville Locks and Dam	777.1	Reedsville, OH	110	1,200	22	37	15	Movable	1,206	1969	36.4	9	300
			110	600	22	37	15						
Willow Island Locks and Dam	819.3	Waverly, WV	110	1,200	20	35	15	Movable	1,128	1973	35.3	9	1,000
			110	600	20	35	15						
Hannibal Locks and Dam	854.6	New Martinsville, WV	110	1,200	21	38	17	Movable	1,098	1972	36.0	9	300
			110	600	21	38	17						
Pike Island Locks and Dam	896.7	Warwood, WV	110	1,200	21	17	18	Movable	1,306	1965	29.9	9	300
			110	600	21	17	18						
New Cumberland Locks and Dam	926.6	Stratton, OH	110	1,200	21	17	15	Movable	1,315	1961	22.7	9	300
			110	600	21	17	15						
Montgomery Island Locks and Dam	949.3	Industry, PA	110	600	18	16	15	Movable	1,379	1936	18.4	9	300
			56	360	18	16	15						
Dashields Locks and Dam	967.7	Glenwillard, PA	110	600	10	13	18	Fixed	1,585	1929	7.1	9	300
			56	360	10	13	18						
Emsworth Locks and Dam	974.8	Emsworth, PA	110	600	18	17	13	Movable	1,717	1921	6.2	9	300
			56	360	18	16	13						
Okeechobee Waterway, FL													
St. Lucie Lock and Dam	15.1	Stuart, FL	50	250	13	14	12	Movable	170	1941	15.1	8	80
Moore Haven Lock	78.0	Moore Haven, FL	50	250	2	10	11	None		1953	15.6	8	90
Ortona Lock and Dam	93.6	LaBelle, FL	50	250	11	12	11	Movable	104	1937	15.6	8	90
W. P. Franklin Lock and Control Structure	121.4	Fort Myers, FL	56	400	3	13	—	Movable	1,150	1965	7.9	8	90
Old River, LA													
Old River Lock	304.0	Simmesport, LA	75	1,200	35	11	11	None		1963		12	125
Ouachita and Black Rivers, AR & LA													
Jonesville Lock and Dam (Black River)	25.0	Jonesville, LA	84	600	30	18	14	Movable	450	1972	92.2	9	100
Columbia Lock and Dam (Ouachita River)	117.2	Columbia, LA	84	600	18	18	13	Fixed	400	1972	106.2	9	100
Calion Lock (Ouachita River)	281.7		84	600	12	18	13	Movable	350	1984	50.3	9	100
Felsenthal Lock (Ouachita River)	226.8		84	600	13	18	13	Movable	350	1984	55.2	9	100
Pearl River, MS and LA													
Lock 1	29.4[17]	Pearl River, LA	65	310	26.7	10	10	None		1951	11.1	7	80
Lock 2	40.7[17]	Bush, LA	65	310	15	10	10	None		1951	3.2	7	80
Lock 3	43.9[17]	Sun, LA	65	310	11	10	10	None		1951	13.2	7	80

TABLE 5–86. NAVIGATION LOCKS AND DAMS IN THE UNITED STATES (continued)

[Operable as of September 30, 1986]

Project	Miles Above Mouth	Community in Vicinity	Locks: Width of Chamber	Locks: Available Length for Full Width	Locks: Lift at Normal Pool Level	Locks: Depth on Sills Upper	Locks: Depth on Sills Lower	Locks: Type[1]	Dams: Length (feet)	Dams: Year Opened	Authorized Channel: Length (miles)	Authorized Channel: Depth (feet)	Authorized Channel: Width (feet)
Red River Waterway													
Lock 1	43.8		84	685	36	22	13	Fixed	550	1984	44.2	9	200
Sacramento River, CA													
Barge Canal Lock	42.8	West Sacramento, CA	86	600	4	13	13	None		1961	1.5	13	120
Snake River, WA													
Ice Harbor Lock and Dam	9.7	Pasco, WA	86	675	100	15	15	Movable	2,790	1962	31.9	14	250
Lower Monumental Lock and Dam	41.6	Kahlotus, WA	86	675	98	15	15	Movable	3,800	1969	28.7	14	250
Little Goose Lock and Dam	70.3	Starbuck, WA	86	675	98	15	15	Movable	2,670	1970	37.2	14	250
Lower Granite Lock and Dam	107.5	Pomeroy, WA	86	675	100	15	15	Movable	3,200	1975	32.0	14	250
St. Mary's River, MI													
South Canal:													
MacArthur Lock	47.0	Sault Ste. Marie, MI	80	800	22	31	31	None		1943		27	
Poe Lock	47.0	Sault Ste. Marie, MI	110	1,200	22	32	32	None		1968	—		
North Canal:													
Davis Lock	47.0	Sault Ste. Marie, MI	80	1,350	22	24	23	None		1914		—	
Sabin Lock	47.0	Sault Ste. Marie, MI	80	1,350	22	24	23	None		1919		—	
Savannah River, GA													
New Savannah Bluff Lock and Dam	187.2	Augusta, GA	56	360	15	14	12	Movable	360	1936	16.2	9	90
Tennessee River, TN, AL, MS, and KY[18]													
Kentucky Lock and Dam	22.4	Gilbertsville, KY	110	600	56	11	13	Tainter	7,976	1944	184.3	9	300
Pickwick Landing Lock and Dam													
Auxiliary Lock	206.7	Hamburg, TN	110	600	55	10	13	Bulkhead	7,385	1937	52.7	9	300
Main Lock	206.7	Hamburg, TN	110	1,000	55	10	13	Bulkhead	7,385	1984	52.7	9	300
Waterway Connecting Tombigbee and Tennessee Rivers, AL and MS													
Gainesville Lock and Dam	49.1	Gainesville, AL	110	600	36	15	15	Movable	15,460	1978	49.1	9	300
Aliceville Lock and Dam	89.8	Aliceville, AL	110	600	27	15	15	Movable	14,790	1979	40.7	9	300
Columbus Lock and Dam	117.6	Columbus, MS	110	600	27	15	15	Movable	10,040	1981	27.8	9	300
Aberdeen Lock and Dam	140.4	Aberdeen, MS	110	600	27	15	15	Movable	10,640	1984	22.8	9	300
Lock A and Spillway	154.1	Amory, MS	110	600	30	15	15	Movable	—	1985	13.7	9	300
Lock B and Spillway	159.3	Smithville, AL	110	600	25	18	18	Movable	—	1985	14.7	9	300
Lock C and Spillway	174.0	Fulton, MS	110	600	25	18	18	Movable	—	1985	7.4	9	300
Lock D and Spillway	181.4	Fulton, MS	110	600	30	18	18	Movable	—	1985	8.3	9	300
Lock E and Spillway	189.7	Belmont, MS	110	600	30	18	18	Movable	—	1985	5.2	9	300
Bay Springs Lock and Dam	194.9	Tupelo, MS	110	600	84	18	18	None	10,640	1985	64.5	9	300
Wilson Lock and Dam													
Main Lock	259.4	Florence, AL	110	600	94	13	13	Bulkhead	3,728	1959	15.5	9	300
Auxiliary Lock	259.4	Florence, AL	60	292	94	11	11	Bulkhead	3,728	1927	15.5	9	300
General Joe Wheeler Lock and Dam													
Main Lock	274.9	Florence, AL	110	600	48	13	13	Bulkhead	5,738	1963	74.1	9	300
Auxiliary Lock	274.9	Florence, AL	60	400	48	15	13	Bulkhead	5,738	1962	74.1	9	300
Guntersville Lock and Dam													
Main Lock	349.0	Guntersville, AL	110	600	39	13	13	Tainter	3,837	1965	75.7	9	300
Auxiliary Lock	349.0	Guntersville, AL	60	400	39	13	12	Tainter	3,837	1939	75.7	9	300
Nickajack Lock and Dam	424.7	Chattanooga, TN	110	600	39	13	11	Tainter	3,763	1967	46.3	9	300
Chickamauga Lock and Dam	471.0	Chattanooga, TN	60	360	49	10	14	Tainter	5,654	1940	58.9	9	300
Watts Bar Lock and Dam	529.9	Breedenton, TN	60	360	58	12	12	Tainter	2,646	1942	72.4	9	300
Fort Loudon Lock and Dam	602.3	Lenoir City, TN	60	360	72	12	12	Tainter	3,687	1943	49.8	9	300
Melton Hill Lock and Dam (Clinch River)	23.1	Kingston, TN	75	400	54	13	13	Tainter	1,072	1963	38.2	9	300[19]
Willamette River at Willamette Falls, OR													
Lock No. 1	26.0	Oregon City, OR	40	210	20	6	8	None		1872		6	150
Lock No. 2	26.0	Oregon City, OR	40	210	10	6	8	None		1872			
Lock No. 3	26.0	Oregon City, OR	40	210	10	6	8	None		1872			
Lock No. 4	26.0	Oregon City, OR	40	210	10	6	8	None		1872		.4	
Guard Lock	26.4	Oregon City, OR	40	210	10	6	8	None		1872	23.6	6	NS

TABLE 5–86. NAVIGATION LOCKS AND DAMS IN THE UNITED STATES (continued)

[Operable as of September 30, 1986]

[1] Fixed: crest without gates or other facility to control streamflow. Movable: includes any type of crest gates, tainter gates, wickets, or others to control streamflow.
[2] Above mouth of Mobile River.
[3] Includes length of earth dikes.
[4] From mouth of Apalachicola River.
[5] From mouth of Chattahoochee River.
[6] Above Norfolk, VA.
[7] With reference to mean low water, or low mean Gulf as case may be.
[8] From Gulf Intracoastal Waterway.
[9] Average high and low water conditions, lift varying widely, depending on tides and river stages.
[10] From foot of Government Street, Mobile, AL.
[11] From mouth of St. Johns River.
[12] Above Head of Passes.
[13] From New Orleans, LA.
[14] Salt water barrier in down position. Will clear 15-foot vessels in normal position.
[15] Above the Ohio River.
[16] Channel completed to depth of 6.5 feet.
[17] From mouth of West Pearl River.
[18] Tennessee River locks operated by the Corps, dams operated by the Tennessee Valley Authority.
[19] 300-foot width from lock to mouth of Clinch River and 175-foot width upstream from lock to Clinton, TN.
[20] Dates shown represent replacement structures.
[21] Dates shown represent reconstruction for locks and dams Nos. 2–8 inclusive.
[22] Dam was rebuilt in 1959.
[23] Old 110' x 358' lock replaced in 1957.
[24] Inoperable, caretaker status September 7, 1982.
[25] Inoperable, caretaker status October 1, 1981.
Source: Department of the Army, Secretary of the Army on Civil Works Activities, Annual Report FY 1986

TABLE 5–87. MAJOR SHIP CANALS OF THE WORLD

Name and Location	Year Opened to Traffic	Length in Miles	Width, Feet	Controlling Depth, Feet	Number of Locks
Albert, Belgium	1939	80.0	53.0	16.5	6
Amsterdam-North Sea, Netherlands	1876	13.0	164.0	41.0	4
Beaumont-Port Arthur, United States	1916	40.0	200.0	34.0	None
Bruges-Zeebrugge, Belgium	1907	6.3	65.7	18.1	1
Brussels-Rupel (Canal Maritime), Belgium	1922	18.5	52.5	21.0	4
Cape Cod, United States	1914	17.5	450.0	28.9	None
Chesapeake and Delaware, United States	1927	19.0	250.0	27.0	None
Chicago Sanitary and Ship, United States	1900	30.0	110.0	22.0	1
Corinth, Greece	1893	4.0	69.0	26.3	None
Falsterbo, Sweden	1942	1.0	82.0	24.0	2
Ghent-Terneuzen, Belgium	1927	17.0	80.0	28.0	6
Göta, Sweden (reconstructed)	1832	47.0	23.6	9.7	58
Houston, United States (reconstructed)	1914	50.0	200.0	33.0	None
Kiel, Germany	1895	53.3	144.0	37.0	4
Lake Washington, United States	1916	8.0	80.0	30.0	2
Manchester, England	1894	46.5	65.0	28.0	10
Moscow-Volga, U.S.S.R.	1937	80.0	98.4	18.0	11
Panama, Panama Canal Zone	1914	50.0	110.0	41.0	12
Sault Ste. Marie, Canada	1895	1.2	60.0	16.8	1
Sault Ste. Marie, United States	1915	1.6	80.0	25.0	4
Suez, Egypt	1869	87.5	197.0	34.0	None
Trollhätte, Sweden (reconstructed)	1916	54.0	45.0	14.4	6
Welland, Canada	1931	27.6	80.0	25.0	8

Source: World Atlas, Encyclopedia Britannica

SECTION M. WATERBORNE COMMERCE

TABLE 5–88. WATERBORNE COMMERCE OF THE UNITED STATES

[Cargo tonnage in millions of short tons. Includes Puerto Rico and outlying areas]

CLASS	1960	1970	1975	1978	1979	1980	1981	1982	1983	1984
Net total	**1,100**	**1,532**	**1,695**	**2,021**	**2,074**	**1,999**	**1,941**	**1,777**	**1,708**	**1,836**
Domestic commerce[1]	**761**	**951**	**946**	**1,075**	**1,080**	**1,077**	**1,054**	**957**	**957**	**1,033**
Coastwise, between ports	209	238	232	305	305	330	322	311	310	306
Great Lakes, between ports	155	157	129	143	144	115	115	72	83	98
Intra: Seaports, Great Lakes ports,										
and inland waterways ports	104	81	78	90	93	94	93	76	73	81
Ports and river ports[2]	292	474	507	538	539	539	524	498	490	546
Foreign commerce	**339**	**581**	**749**	**946**	**993**	**922**	**887**	**820**	**751**	**803**
Imports, through seaports	198	313	455	616	608	502	457	403	372	409
Imports, Great Lakes ports	13	26	21	27	25	16	20	14	16	18
Exports, through seaports	105	206	237	259	312	359	366	367	331	335
Exports, Great Lakes ports	23	36	35	44	49	45	43	36	32	41

[1] Includes traffic among ports of outlying areas.
[2] Represents traffic mainly utilizing inland waterways.
Source: U.S.Army Corps of Engineers, Waterborne Commerce of the United States, annual

TABLE 5–89. FREIGHT CARRIED ON INLAND WATERWAYS OF THE UNITED STATES

[In billions of ton-miles. Excludes Alaska and Hawaii, except as noted. Includes waterways, canals, and connecting channels]

SYSTEM	1960	1970	1975	1976	1977	1978	1979	1980	1981	1982	1983	1984
Total	**220.3**	**318.6**	**342.2**	**372.9**	**368.3**	**409.3**	**424.6**	**406.9**	**410.2**	**351.2**	**359.0**	**399.0**
Atlantic coast waterways	28.6	28.6	31.8	32.1	30.4	30.5	31.9	30.4	28.3	25.4	22.5	24.7
Gulf coast waterways	16.9	28.6	30.8	34.3	37.6	37.1	38.1	36.6	35.1	31.8	32.4	36.7
Pacific coast waterways[1]	6.0	8.4	9.7	11.3	12.8	13.6	14.1	14.9	14.4	12.8	13.2	20.5
Mississippi River system[2]	69.3	138.5	170.7	189.5	196.9	209.3	218.8	228.9	234.4	218.0	223.0	234.6
Great Lakes System[3]	99.5	114.5	99.2	105.6	90.7	118.9	121.7	96.0	98.0	63.2	67.9	82.5

[1] Includes Alaskan waterways.
[2] Comprises main channels and all tributaries of the Mississippi, Illinois, Missouri, and Ohio Rivers.
[3] Does not include traffic between foreign ports.
Source: U.S. Army Corps of Engineers, Waterborne Commerce of the United States, annual

FIGURE 5–11. PRINCIPAL COMMODITIES CARRIED ON WATERWAYS OF THE UNITED STATES IN 1985

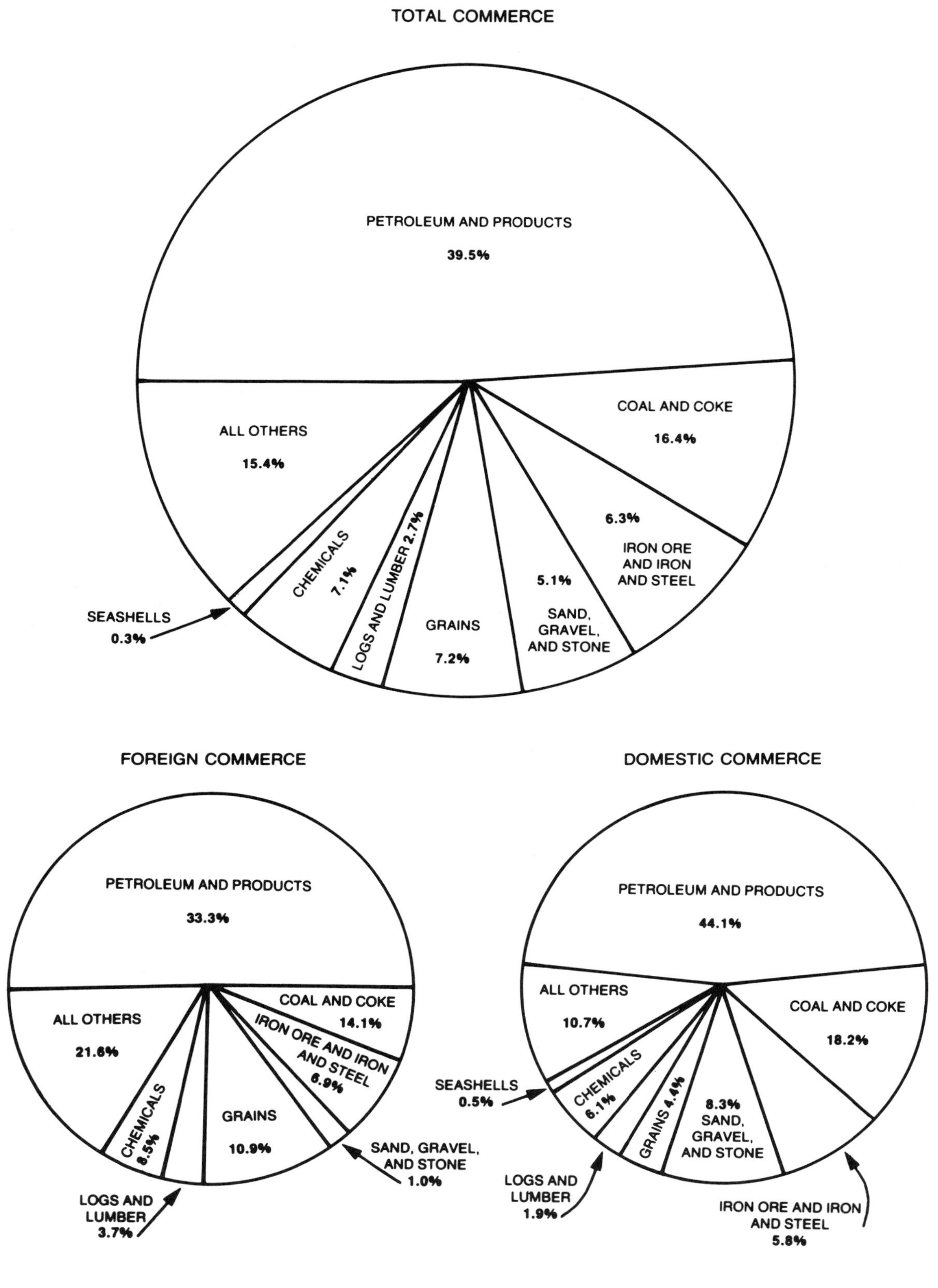

Source: Department of the Army Corps of Engineers, Waterborne Commerce of the United States, 1985

TABLE 5–90. DOMESTIC AND FOREIGN WATERBORNE COMMERCE BY TYPE OF COMMODITY IN THE UNITED STATES

[In millions of short tons. Domestic trade includes all commercial movements between U.S. ports and on inland rivers, Great Lakes, canals, and connecting channels of U.S., Puerto Rico, and Virgin Islands]

COMMODITY	1970		1975		1980		1984			
									Foreign	
	Total	Domestic	Total	Domestic	Total	Domestic	Total	Domestic	Imports	Exports
Net total	**1,531.7**	**950.7**	**1,695.0**	**946.3**	**1,998.9**	**1,077.5**	**1,836.0**	**1,032.7**	**427.1**	**376.2**
Petroleum products[1]	394.7	287.3	407.6	328.5	423.2	339.2	398.8	278.9	85.7	34.2
Gasoline	89.3	88.7	95.7	92.2	87.3	81.0	89.9	76.9	12.6	.4
Distillate fuel oil	78.8	76.9	93.4	87.0	74.6	72.1	78.3	60.5	14.7	3.1
Residual fuel oil	170.5	78.8	172.1	111.5	188.0	141.3	148.3	95.2	40.9	12.2
Crude petroleum	210.5	116.3	333.8	77.9	480.2	174.2	352.4	180.7	171.7	—
Coal and lignite	225.4	154.1	219.0	152.8	256.4	164.1	271.8	189.5	1.0	81.3
Nonmetallic minerals[1, 2]	166.6	132.6	159.2	123.3	157.1	111.6	145.6	103.6	22.1	19.9
Sand and gravel[3]	78.4	73.9	72.4	67.9	65.1	60.8	62.6	58.6	2.2	1.8
Limestone	42.0	34.1	37.8	29.1	34.2	23.9	23.9	21.3	.6	2.0
Phosphate rock	16.6	5.8	20.1	9.4	23.7	9.5	19.1	8.0	—	11.1
Iron and concentrates	129.3	75.2	121.4	69.3	98.4	64.9	76.1	52.3	18.3	5.5
Farm products[1]	90.4	28.5	144.3	43.1	216.8	63.4	211.5	69.4	5.9	136.2
Corn	27.2	11.7	56.8	20.8	98.6	30.8	84.0	31.4	—	52.6
Wheat	25.0	5.8	44.4	10.4	53.4	14.3	63.5	16.9	—	46.6
Soybeans	21.2	8.3	23.2	9.4	39.6	16.1	35.7	15.5	—	20.2
Chemicals and allied products	62.0	39.7	66.5	40.0	91.9	49.4	110.4	54.7	25.7	30.0
Food and kindred products	35.0	11.0	35.5	13.3	54.8	20.4	52.2	18.2	13.8	20.2
Lumber and wood products[4]	49.9	27.3	48.2	24.7	52.0	22.7	45.7	19.8	4.1	21.8
Primary metal products	42.4	10.8	38.1	9.3	28.9	9.0	35.3	8.4	25.5	1.4
Waste and scrap	25.9	16.0	28.0	18.8	31.1	18.8	33.2	21.1	.4	11.7
Other	99.7	51.9	93.5	45.5	108.1	39.8	102.5	35.5	53.0	14.0

- Represents zero.
[1] Includes commodities not shown separately.
[2] Excludes fuels.
[3] Includes crushed rock.
[4] Excludes furniture.
Source: U.S. Army Corps of Engineers, Waterborne Commerce of the United States, annual

TABLE 5–91. FREIGHT CARRIED ON THE MISSISSIPPI RIVER SYSTEM

[Net traffic. Comprises main channels and all tributaries of the Mississippi, Illinois, Missouri, and Ohio rivers]

ITEM	Unit	1960	1970	1975	1976	1977	1978	1979	1980	1981	1982	1983	1984
Total	**Mil. sh. tons**	**234**	**391**	**453**	**498**	**537**	**553**	**582**	**584**	**589**	**540**	**519**	**544**
Inland	Mil. sh. tons	188	297	330	346	352	355	371	366	366	347	352	387
Coastwise	Mil. sh. tons	22	44	30	26	30	32	33	42	45	35	39	37
Foreign:													
Imports	Mil. sh. tons	11	17	46	67	96	99	106	91	80	57	32	34
Exports	Mil. sh. tons	13	34	48	60	60	67	73	86	98	101	96	86
Total	**Bil. ton-miles**	**69.3**	**138.5**	**170.7**	**189.5**	**196.8**	**209.3**[1]	**218.8**	**228.9**	**234.4**	**218.0**	**223.1**	**234.6**
Inland	Bil. ton-miles	62.4	125.9	151.3	164.8	166.9	176.5	184.4	193.5	199.0	187.9	196.7	210.2
Coastwise	Bil. ton-miles	3.2	4.9	4.5	4.1	4.8	5.4	5.7	7.4	7.7	6.2	6.7	6.0
Foreign:													
Imports	Bil. ton-miles	1.8	2.8	8.0	11.7	16.2	16.8	17.6	15.1	13.3	8.8	5.1	5.3
Exports	Bil. ton-miles	1.9	4.9	6.9	8.8	8.8	10.1	11.1	13.0	14.4	15.1	14.6	13.1

[1] Total reflects adjustment not available by type of traffic.
Source: U.S. Army Corps of Engineers, Waterborne Commerce of the United States, annual

TABLE 5–92. U.S. FREIGHT CARRIED ON THE GREAT LAKES SYSTEM

[In billions of ton-miles]

ITEM	1960	1970	1974	1975	1976	1977	1978	1979	1980	1981	1982	1983	1984
Total	**99.5**	**114.5**	**107.5**	**99.2**	**105.6**	**90.7**	**118.9**	**121.7**	**96.0**	**98.0**	**63.2**	**68.0**	**82.4**
Domestic	80.6	80.1	79.7	69.2	71.3	53.0	76.8	78.5	62.3	62.7	35.9	43.5	50.2
Lakewise	79.7	79.4	78.9	68.5	70.7	52.4	76.3	78.0	61.7	62.1	35.4	43.0	49.8
Internal	.7	.6	.6	.6	.5	.4	.4	.4	.5	.5	.4	.4	.3
Local[1]	.2	.1	.1	.1	.1	.1	.1	.1	.1	.1	.1	.1	.1
Foreign	18.9	34.3	27.8	29.9	34.3	37.7	42.1	43.1	33.7	35.3	27.3	24.3	32.2
Canadian	13.8	23.8	19.4	20.7	23.7	23.1	23.1	26.9	23.6	25.8	18.2	15.9	19.0
Overseas	5.1	10.5	8.4	9.3	10.7	14.7	19.0	16.2	10.1	9.5	9.1	8.4	13.2

[1] Includes coastwise traffic.
Source: U.S. Army Corps of Engineers, Waterborne Commerce of the United States, annual

SECTION N. WATER-BASED RECREATION

TABLE 5–93. BOATING ACTIVITIES IN THE UNITED STATES

ACTIVITY	Unit	1970	1975	1979	1980	1981	1982	1983	1984	1985
Recreational boats owned	Million	8.8	9.7	11.6	11.8	12.5	12.9	13.0	13.5	13.9
Outboard boats	Million	5.2	5.7	6.7	6.8	7.0	7.1	7.2	7.3	7.5
Inboard boats	Million	.6	.8	1.3	1.2	1.2	1.3	1.3	1.4	1.5
Sailboats	Million	.6	.8	.9	1.0	1.0	1.1	1.1	1.1	1.2
Canoes	Million			1.2	1.3	1.4	1.5	1.6	1.7	1.8
Rowboats and other	Million	2.4	2.4	1.5	1.5	1.9	1.9	1.8	1.8	1.9
Expenditures, total[1]	Bil. dol	3.4	4.8	7.5	7.4	8.3	8.1	9.4	12.3	13.3
Outboard motors in use	1,000	7,215	7,649	7,958	8,241	8,527	8,776	9,051	9,400	9,733
Motors sold	1,000	430	435	375	315	318	293	337	411	392
Value, retail	Mil. dol	281	411	597	554	698	759	964	1,294	1,319
Outboard boats sold	1,000	276	328	322	290	281	236	273	317	305
Value, retail	Mil. dol	177	263	469	408	431	409	502	708	759
Inboard/outdrive boats sold	1,000	43	70	89	56	51	55	79	108	115
Value, retail	Mil. dol	182	420	827	616	653	686	975	1,442	1,663

[1] Represents estimated expenditures for new and used boats, motors, accessories, safety equipment, fuel, insurance, docking, maintenance, storage, repairs, and other expenses.

Source: Statistical Abstract of the United States and National Marine Manufacturers Association, Chicago, IL

TABLE 5–94. RETAIL SALES OF NEW RECREATIONAL BOATING PRODUCTS IN THE UNITED STATES

	1985	1986	1987
Outboard boats			
Total units sold	305,000	314,000	342,000
Retail value	$759.4 million	$834.6 million	$1.0 billion
Average unit cost	$2,490	$2,658	$2,927
Inboard runabouts			
Total units sold	4,500	5,300	6,600
Retail value	$68.2 million	$85.4 million	$107.3 million
Average unit cost	$15,168	$16,127	$16,270
Inboard cruisers			
Total units sold	12,200	12,700	13,100
Retail value	$1.34 billion	$1.42 billion	$1.66 billion
Average unit cost	$109,900	$111,900	$127,165
Sterndrive cruisers			
Total units sold	115,000	120,000	144,000
Retail value	$1.63 billion	$1.86 billion	$2.45 billion
Average unit cost	$14,464	$15,508	$17,015
Non-powered sailboats			
Total units sold	34,000	33,300	29,500
Retail value	$115.4 million	$88.4 million	$75.8 million
Average unit cost	$3,395	$2,655	$2,570
Powered sailboats 30 feet and shorter			
Total units sold	1,100	1,100	1,100
Retail value	$37.3 million	$40.6 million	$39.3 million
Average unit cost	$33,941	$36,956	$35,754
Powered sailboats longer than 30 feet			
Total units sold	2,700	2,800	2,900
Retail value	$217.6 million	$236.8 million	$267.1 million
Average unit cost	$80,600	$84,600	$92,128
Sailboards			
Total units sold	50,000	60,000	70,000
Inboard and I/O engines			
Total units sold	155,000	161,900	210,800

Source: National Marine Manufacturers Association

TABLE 5-95. FATALITY RATE IN RECREATIONAL BOATING IN THE UNITED STATES

Year	Fatalities	Estimate of Boats (Million)	Fatality Rate (per 100,000 boats)
1961	1218	5.85	20.8
1962	1114	5.95	18.7
1963	1167	6.05	19.3
1964	1192	6.20	19.2
1965	1360	6.35	21.4
1966	1318	6.50	20.3
1967	1312	6.65	19.7
1968	1342	6.85	19.6
1969	1350	7.10	19.0
1970	1418	7.40	19.2
1971	1582	7.85	20.2
1972	1437	8.50	16.9
1973	1754	9.60	18.3
1974	1446	10.75	13.5
1975	1466	11.80	12.4
1976	1264	12.75	9.9
1977	1312	13.15	10.0
1978	1321	13.50	9.8
1979	1400	13.90	10.1
1980	1360	14.30	9.5
1981	1208	14.60	8.3
1982	1178	14.90	7.9
1983	1241	15.30	8.1
1984	1063	15.70	6.8
1985	1116	16.10	6.9
1986	1066	16.50	6.5

Source: United States Coast Guard, 1987, Boating Statistics 1986 COMDTPUB P16754.1

TABLE 5-96. TYPES OF BOATING ACCIDENTS REPORTED TO THE U.S. COAST GUARD IN 1986

[The USCG estimates that it receives reports of only 5% to 10% of all reportable accidents not involving fatalities]

Accident	Total Vessels Involved	Fatalities
Grounding	367	11
Capsizing	629	370
Swamping/flooding	289	89
Sinking	227	28
Fire/explosion (fuel)	379	6
Fire/explosion (other)	83	2
Collision with another vessel	4,096	86
Collision with fixed object	914	79
Collision with floating object	276	8
Falls overboard	451	277
Falls within boat	72	0
Struck by boat or propeller	147	16
Other	418	37
Unknown	51	57
TOTALS	8,399	1,066

NOTE: Type of accident refers only to the first event that occurred. Some accidents involve more than one event, e.g., a grounding followed by a sinking is included here only as a grounding even though the sinking may have directly led to a drowning fatality.
Source: United States Coast Guard, 1987, Boating Statistics 1986 COMTPUB P16754.1

TABLE 5–97. BOATING ACCIDENTS IN THE UNITED STATES

[USCG estimates that it receives reports of only 5% to 10% of reportable accidents not involving fatalities]

	Vessels Involved	Fatalities
TIME OF DAY		
Midnight to 2:30 am	177	33
2:30 am to 4:30 am	100	19
4:30 am to 6:30 am	96	22
6:30 am to 8:30 am	211	29
8:30 am to 10:30 am	468	63
10:30 am to 12:30 pm	888	93
12:30 pm to 2:30 pm	1,370	116
2:30 pm to 4:30 pm	1,698	173
4:30 pm to 6:30 pm	1,304	141
6:30 pm to 8:30 pm	774	84
8:30 pm to 10:30 pm	452	60
10:30 pm to midnight	335	49
Unknown	526	184
DAY OF WEEK		
Friday	1,037	145
Saturday	2,514	302
Sunday	2,439	223
Monday	690	112
Tuesday	560	93
Wednesday	541	101
Thursday	618	90
Unknown	0	0
MONTH OF YEAR		
January	152	34
February	184	38
March	324	78
April	531	95
May	1,068	140
June	1,449	161
July	1,968	164
August	1,551	120
September	565	99
October	308	54
November	219	57
December	80	26
Unknown	0	0

Source: United States Coast Guard, 1987, Boating Statistics 1986 COMDTPUB P16754.1

FIGURE 5–12. BOATING FATALITIES AND ACCIDENTS DURING THE WEEK IN 1986

[Plotted at two-hour increments]

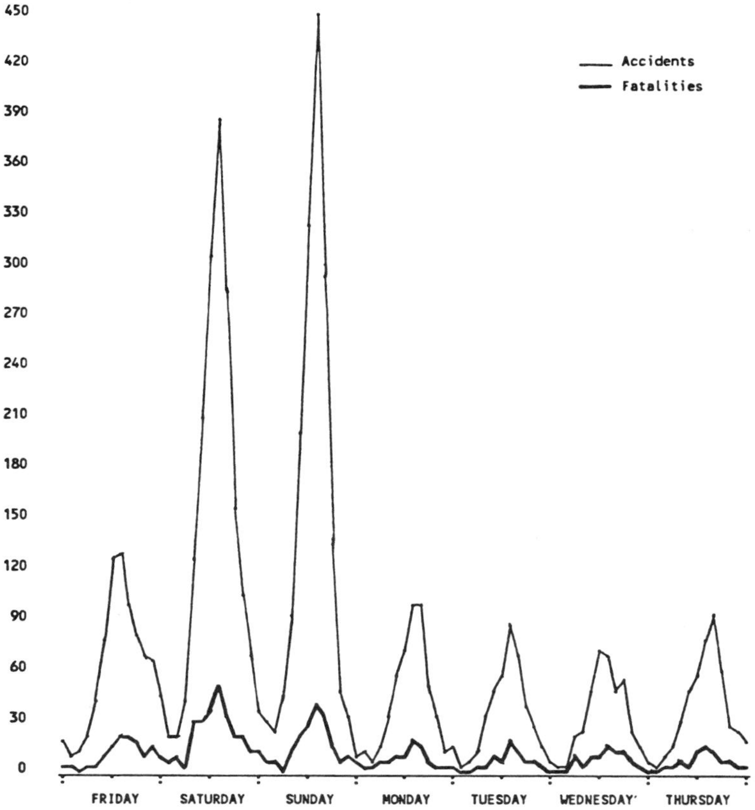

Fatalities (184) and accidents (430) occurring at unknown times are not included. The day of occurrence of all accidents is known.

Source: United States Coast Guard, 1987, Boating Statistics 1986 COMTPUB P16754.1

TABLE 5–98. WATER-BASED RECREATIONAL USE OF NATIONAL FOREST AND PUBLIC LANDS IN 1985

[In thousands of visitor hours for year ending Sept. 30]

National Forests

Activity	
Water travel	101,868
Water skiing and other watersports	6,420
Swimming and scuba diving	56,220
Fishing	190,140

Public Lands administered by Bureau of Land Management

State	Activity		
	Fishing	Boating	Other
AK	281	231	12
AZ	104	131	26
CA	750	2,732	1,136
CO	1,393	922	42
ID	1,748	1,614	394
MT[1]	194	333	35
NV	1509	53	2
NM	309	442	157
OR[2]	6,605	4,674	255
UT	908	549	101
WY	453	29	33
Total	**14,254**	**11,710**	**2,193**

[1] Includes North Dakota and South Dakota.
[2] Includes Washington.
Source: U.S. Bureau of Commerce, Statistical Abstract of the United States 1987

TABLE 5–99. NUMBER AND EXPENDITURES OF SPORT FISHERMEN IN THE UNITED STATES

[Persons 12 years and older]

Activity	Number (1,000)					Expenditures (mil. dol.)				
	1960	1965	1970	1975	1980	1960	1965	1970	1975	1980
Sportsmen[1]	30,435	32,881	36,277	45,773	46,966	3,852	4,046	7,102	16,768	27,214[2]
Fishermen[1]	25,323	28,348	33,158	41,299	41,873	2,691	2,925	4,959	11,798	18,052
Freshwater	21,677	23,962	29,363	36,599	37,081	2,065	2,126	3,734	8,702	14,441
Saltwater	6,292	8,305	9,460	13,738	13,332	626	800	1,225	3,095	3,611

[1] Individuals may participate in more than one form of wildlife-related activity; therefore the sum of the component parts will be greater than the number of participants.
[2] Includes multipurpose licenses issued in 1980.
Source: U.S. Fish and Wildlife Service and U.S. Bureau of the Census, 1980 National Survey of Fishing, Hunting, and Wildlife-Associated Recreation, 1982

TABLE 5–100. NUMBER AND COST OF SPORT FISHING LICENSES IN THE UNITED STATES

Item	Unit	1970	1975	1976	1977	1978	1979	1980	1981	1982	1983	1984
Fishing licenses: Sales.	**Million**	**31.1**	**34.7**	**34.9**	**34.0**	**32.8**	**35.4**	**35.2**	**37.9**	**37.2**	**37.8**	**36.1**
Resident	Million	26.8	30.0	29.8	29.0	27.9	30.3	30.1	32.6	32.1	32.5	31.0
Nonresident	Million	4.3	4.7	5.1	5.0	4.9	5.1	5.1	5.3	5.1	5.3	5.1
Paid license holders[1]	Million	24.4	27.5	27.8	27.3	26.2	27.9	28.0	29.3	29.6	29.1	29.0
Cost to anglers	Mil. dol.	91	142	155	156	159	174	196	213	227	244	260

[1] Resident and nonresident. Includes multiple counting of license holders who bought nonresident licenses as well as a home State license.
Source: U.S. Fish and Wildlife Service, Federal Aid in Fish and Wildlife Restoration, annual

TABLE 5–101. LIFE EXPECTANCY IN WATER

[Exposure by immersion in low temperature water can have serious consequences. Life expectancies for various durations of exposure are indicated by: L—Lethal, 100 percent expectancy of death; M—Marginal, 50 percent expectancy of unconsciousness which will probably result in drowning; S—Safe. It should also be noted that sudden immersion in ice cold water can cause temporary paralysis with resulting helplessness and loss of buoyancy, causing the victim to sink to the bottom.]

Duration, hours	Water Temperature				
	30°F.	40°F.	50°F.	60°F.	70°F.
1	M	M	M	S	S
2	L	L	M	M	S
3	L	L	M	M	S
4	L	L	L	M	S

Source: Pan American Airways and Calif. Dept. of Harbors and Watercraft

SECTION O. FISHERIES

TABLE 5-102. UNITED STATES FISHERIES—QUANTITY AND VALUE OF CATCH, BY STATE, AND CATCH OF PRINCIPAL SPECIES, BY AREA: 1970 TO 1985

[Catch in millions of pounds, live weight, except as indicated; ex-vessel value in millions of dollars. 1980–1985 preliminary]

Area, State, Catch, and Value	1970	1980	1984	1985
New England:				
Catch	531	788	694	590
Value	91	327	433	419
Maine:				
Catch	159	245	179	175
Value	31	93	108	101
New Hampshire:				
Catch	1	19	12	8
Value	1	5	8	5
Massachusetts:				
Catch	287	438	375	296
Value	47	178	233	232
Rhode Island:				
Catch	79	81	120	104
Value	11	46	70	70
Connecticut:				
Catch	5	5	8	7
Value	2	5	14	12
Catch for certain species:				
Cod	53	118	96	83
Flounder	104	118	115	98
Haddock	27	55	26	14
Herring, sea	66	184	74	52
Lobster, American	30	36	42	44
Ocean perch, Atlantic	55	24	12	10
Pollock	9	40	40	43
Whiting	40	18	33	31
Middle Atlantic:				
Catch	140	244	153	151
Value	30	97	109	101
New York:				
Catch	32	39	38	39
Value	16	45	40	38
New Jersey:				
Catch	98	201	112	108
Value	13	50	67	61
Delaware:				
Catch	10	4	3	5
Value	1	2	2	2
Catch for certain species:				
Clams (meats)	63	40	68	73
Menhaden	31	(D)	(D)	(D)
Oysters (meats)	1	3	2	1
Scup or porgy	4	8	7	2
Chesapeake Bay:				
Catch	630	717	663	815
Value	41	130	138	124
Maryland:				
Catch	80	80	89	92
Value	19	45	55	47
Virginia:				
Catch	551	637	574	723
Value	22	85	83	76
Catch for certain species:				
Alewives	21	1	1	1
Clams (meats)	23	36	36	47
Crabs	70	65	98	91
Menhaden	450	(D)	(D)	(D)
Oysters (meats)	25	21	13	8
South Atlantic:				
Catch	280	473	402	311
Value	30	148	150	156
North Carolina:				
Catch	173	356	276	215
Value	9	69	57	65
South Carolina:				
Catch	16	21	15	13
Value	4	20	15	14
Georgia:				
Catch	14	19	16	17
Value	4	20	12	21
Florida (east coast):				
Catch	76	77	95	66
Value	12	39	66	57
Catch for certain species:				
Crabs	43	55	52	48
Menhaden	136	(D)	(D)	(D)
Shrimp	21	33	19	28
Gulf States:				
Catch	1,698	1,979	2,644	2,412
Value	167	463	656	597
Florida (west coast):				
Catch	116	115	112	117
Value	31	86	112	114
Alabama:				
Catch	30	25	27	30
Value	10	25	44	41
Mississippi:				
Catch	298	232	477	471
Value	11	25	47	40
Louisiana:				
Catch	1,107	1,412	1,924	1,693
Value	61	174	263	225
Texas:				
Catch	147	94	104	103
Value	53	153	190	177
Catch for certain species:				
Crabs	36	45	55	53
Menhaden	1,209	(D)	(D)	(D)
Mullet	27	31	19	17
Oysters (meats)	18	17	24	25
Shrimp	230	208	254	263
Great Lakes:[1]				
Catch	72	44	61	54
Value	6	14	14	15
Superior:				
Catch	5	(NA)	(NA)	(NA)
Value	1	(NA)	(NA)	(NA)
Michigan:				
Catch	53	(NA)	(NA)	(NA)
Value	4	(NA)	(NA)	(NA)
Huron:				
Catch	2	(NA)	(NA)	(NA)
Value	(Z)	(NA)	(NA)	(NA)
Erie:				
Catch	10	(NA)	(NA)	(NA)
Value	1	(NA)	(NA)	(NA)
Pacific Coast:				
Catch	1,480	2,140	1,701	1,816
Value	235	1,025	795	863
Washington:				
Catch	134	156	156	167
Value	30	86	76	93
Oregon:				
Catch	98	126	82	101
Value	23	56	34	46
California:				
Catch	703	804	459	363
Value	86	323	176	133
Alaska:				
Catch	545	1,054	1,002	1,185
Value	96	560	509	591
Catch for certain species:				
Anchovies	193	107	18	15
Bonito	9	14	6	5
Crabs	126	347	93	129
Flounder	46	60	65	68
Halibut	34	19	48	61
Herring, sea	22	107	105	142
Jack mackerel	48	44	23	21
Salmon	410	614	670	727
Shrimp	93	98	21	35
Tuna	378	387	177	64
Mississippi River and tributaries:				
Catch	75	85	85	92
Value	10	21	25	29
Hawaii:				
Catch	11	11	35	17
Tuna catch	9	7	29	11
Value	4	12	29	22

D-Withheld to avoid disclosure. NA-Not available. Z-Less than $500,000.

[1] Collected largely by State fishery agencies, and compiled by National Marine Fisheries Service. Includes, in addition to lakes shown, small amounts for Lake Ontario, Lake St. Clair, Lake of the Woods, Namakan Lake, and Rainy Lake.

Source: U.S. Department of Commerce, 1987 Statistical Abstract of the United States

TABLE 5–103. UNITED STATES FISHERIES—EMPLOYMENT, FISHING CRAFT, AND ESTABLISHMENTS, 1970 TO 1984

[In thousands. As of Dec. 31. Data for employment and establishments exclude Alaska]

Item	1970	1974	1975	1976	1977	1978, prel.	1979, prel.	1980, prel.	1981, prel.	1982, prel.	1983, prel.	1984, prel.
Persons employed in U.S.	**227**	**253**	**260**	**267**	**280**	**257**	**267**	**296**	**303**	**314**	**333**	**340**
Fishermen	140	161	168	174	182	172	184	193	198	216	223	230
Shore workers[1]	87	92	92	93	98	85	83	103	105	98	110	110
Craft used	**88**	**101**	**103**	**103**	**107**	**104**	**103**	**113**	**115**	**123**	**127**	**127**
Vessels, 5 net tons and over	14	16	16	17	17	18	18	19	20	20	21	24
Motorboats	72	83	85	84	88	84	84	93	93	102	105	102
Other boats	2	2	2	2	2	2	1	1	2	1	1	1
Fishery shore establishments	**3.7**	**3.5**	**3.6**	**3.6**	**3.6**	**3.3**	**3.4**	**3.6**	**3.6**	**3.6**	**3.9**	**4.0**

[1] Seasonal average.
Source: U.S. Department of Commerce, 1987 Statistical Abstract of the United States

TABLE 5–104. UNITED STATES FISHERIES—DOMESTIC CATCH AND EX-VESSEL VALUE BY AREA, 1980 TO 1985

Area	Catch (mil. lb.)					Value (mil. dol.)				
	1980	1982	1983	1984	1985	1980	1982	1983	1984	1985
United States	6,482	6,367	6,439	6,438	6,258	2,237	2,390	2,355	2,350	2,326
New England States	788	687	711	694	590	327	374	435	434	419
Middle Atlantic States	244	129	128	153	151	97	93	94	109	101
Chesapeake Bay States	717	791	841	663	815	130	120	130	138	124
South Atlantic States	473	427	397	402	311	148	164	173	150	156
Gulf States	1,979	2,300	2,443	2,644	2,412	463	614	615	656	597
Pacific Coast States	2,140	1,872	1,739	1,701	1,816	1,025	964	846	795	863
Great Lakes States	44	36	47	61	54	14	13	14	14	15
Hawaii	11	14	29	35	17	12	14	18	29	22
Other[1]	85	109	103	85	92	21	34	30	25	29

[1] Mississippi River and other areas.
Source: U.S. Department of Commerce, 1987 Statistical Abstract of the United States

SECTION P. WATER IN FOODS

TABLE 5–105. WATER CONTENT OF VARIOUS FOODS

[Percentage by weight]

Tortilla chips	1
Potato chips	2
Peanut butter	2
Popcorn	4
Margarine	14
Butter	16
Croissant	23
Jam	29
Bagel	29
Parmesan cheese	30
Angel food cake	32
Maple syrup	33
Swiss cheese	38
Whole wheat bread	38
English muffin	42
Cheese cake	46
Pizza	48
Brie	49
Apple pie	51
Frankfurter	54
Cream cheese	54
Hamburger	55
Whipping cream	58
Flounder, baked	58
Veal, chuck	59
Chicken, roasted	60
Ice cream	61
Turkey, roasted	62
Tuna, in water	62
Salmon, broiled	63
Ham, smoked, cooked	66
Halibut, broiled	67
Liver, beef, raw	70
Sour cream	71
Lima beans, cooked	71
Avocado	73
Corn, cooked	74
Ricotta cheese	74
Banana	74
Egg, boiled	75
Cottage cheese	79
Potato, raw	80
Clams	81
Grapes	81
Pear	84
Oysters	85
Orange	87
Beets, raw	87
Apple juice	88
Milk, whole	88
Yoghurt, whole milk	88
Carrots, raw	88
Broccoli, raw	89
Mushroom, raw	90
Cantaloupe	90
Milk, skim	91
Beer	92
Asparagus, cooked	94
Tomato, raw	94
Squash, boiled	96
Lettuce, raw	96

Source: Calculated from weight and water content values given in Bowes and Church's Food Values of Portions Commonly Used, 14th ed., Harper & Row

CHAPTER 6

Water Quality

PAGE

SECTION A. WATER QUALITY

TABLE 6–1. SUMMARY OF QUALITY INPUTS TO SURFACE AND GROUND WATERS

[This list includes the types of things that may come from any contributing factor. Not all are present in each specific instance.]

Contributing Factor	Principal Quality Input to Surface Waters

Meteorological water	Dissolved gases native to atmosphere
	Soluble gases from man's industrial activities
	Particulate matter from industrial stacks, dust, and radioactive particles
	Material washed from surface of earth, e.g.:
	  Organic matter such as leaves, grass, and other vegetation in all stages of biodegradation
	  Bacteria associated with surface debris (including intestinal organisms)
	  Clay, silt, and other mineral particles
	  Organic extractives from decaying vegetation
	  Insecticide and herbicide residues

Domestic use (exclusive of industrial)	Undecomposed organic matter, such as garbage ground to sewer, grease, etc. Partially degraded organic matter such as raw wastes from human bodies Combination of above two after biodegradation to various degrees of sewage treatment Bacteria (including pathogens), viruses, worm eggs Grit from soil washings, eggshells, ground bone, etc. Miscellaneous organic solids, e.g., paper, rags, plastics, and synthetic materials Detergents
Industrial use	Biodegradable organic matter having a wide range of oxygen demand Inorganic solids, mineral residues Chemical residues ranging from simple acids and alkalies to those of highly complex molecular structure Metal ions
Agricultural use	Increased concentration of salts and ions Fertilizer residues Insecticide and herbicide residues Silt and soil particles Organic debris, e.g., crop residues
Consumptive use (all sources)	Increased concentration of suspended and dissolved solids by loss of water to atmosphere

Contributing Factor	Principal Quality Input to Groundwaters

Meteorological water	Gases, including O_2 and CO_2, N_2, H_2S, and H
	Dissolved minerals, e.g.:
	  Bicarbonates and sulfates of Ca and Mg dissolved from earth minerals
	  Nitrates and chlorides of Ca, Mg, Na, and K dissolved from soil and organic decay residues
	  Soluble iron, Mn, and F salts

Domestic use (principally via septic tank systems and seepage from polluted surface waters)	Detergents Nitrates, sulfates, and other residues of organic decay Salts and ions dissolved in the public water supply Soluble organic compounds
Industrial use (not much direct disposal to soil)	Soluble salts from seepage of surface waters containing industrial wastes
Agricultural use	Concentrated salts normal to water applied to land Other materials as per meteorological waters
Land disposal of solid wastes (not properly installed)	Hardness-producing leachings from ashes Soluble chemical and gaseous products or organic decay

Source: McGauhey, Engineering Management of Water Quality, McGraw-Hill, Copyright 1968

TABLE 6–2. CONDITIONS THAT MAY CAUSE VARIATIONS IN WATER QUALITY

Climatic conditions	Runoff from snowmelt—muddy, soft, high bacterial count.
	Runoff during drought—high mineral content, hard, groundwater characteristics.
	Runoff during floods—less bacteria than snowmelt, may be muddy (depending upon other factors listed below).
Geographic conditions	Steep headwater runoff differs from lower valley areas in ground cover, gradients, transporting power, etc.
Geologic conditions	Clay soils produce mud.
	Organic soils or swamps produce color.
	Cultivated land yields silt, fertilizers, herbicides, and insecticides.
	Fractured or fissured rocks may permit silt, bacteria, etc., to move with groundwater.
	Mineral content dependent upon geologic formations.
Season of year	Fall runoff carries dead vegetation—color, taste, organic extractives, bacteria.
	Dry season yields dissolved salts.
	Irrigation return water, in growing season only.
	Cannery wastes seasonal.
	Aquatic organisms seasonal.
	Overturn of lakes and reservoirs seasonal.
	Floods generally seasonal.
	Dry period, low flows, seasonal.
Resource management practices	Agricultural soils and other denuded soils are productive of sediments, etc. (See third item under Geologic conditions.)
	Forested land and swamp land yield organic debris.
	Overgrazed or denuded land subject to erosion.
	Continuous or batch discharge of industrial wastes alters shock loads.
	Inplant management of waste streams governs nature of waste.
Diurnal variation	Production of oxygen by planktonic algae varies from day to night.
	Dissolved oxygen in water varies in some fashion.
	Raw sewage flow variable within 24-hour period; treated sewage variation less pronounced.
	Industrial wastes variable—process wastes during productive shift; different material during washdown and cleanup.

Source: McGauhey, Engineering Management of Water Quality, McGraw-Hill, Copyright 1968

TABLE 6–3. OPTIMUM AND MAXIMUM VALUES OF WATER QUALITY CHARACTERISTICS IN RELATION TO TYPE OF BENEFICIAL USE

Characteristics	Domestic Water Supply	Recreation: Bathing and Swimming		Recreation	Wildlife Propagation: Fish		Wildlife Propagation	Shellfish Culture	Irrigation			Industrial: Food Processing		Industrial: Cooling and Other		Aesthetic Enjoy-ment
		Fresh Water	Salt Water	Boating and Fishing	Fresh Water	Salt Water	Fowl Refuge		Truck Garden Vege-tables	Citrus Fruits	Other Crops	Fresh Water	Salt Water	Fresh Water	Salt Water	
1. *Bacterial—per ml.*																
Coliform (opt.)	1.0	none	1.0	10	10	10	100	1.0	1.0	10	100	0.1	1.0	1.0	10	
Coliform (max.)	50	1.0	10	100	100	100	1,000	5	10	100	100	1.0	3.0	10	100	
2. *Organic—ppm.*																
B.O.D. (opt.)	none	5	5	10	10	10	10	5				none	1	5	5	20
B.O.D. (max.)	0.5	10	10	30	30	30	50	20				5	10	10	20	100
D.O. (opt.)	5	5	5	5	5	5	5	5				5	5	3.0	3.0	5.0
D.O. (min.)	2	2	2	2	3	2	2	2				1	1	1.0	1.0	1.0
Oil (opt.)	none	none	none	none	none	none	none	none	none	none	none	none	none	5	5	none
Oil (max.)	2	2	2	5	5	5	5	2	5	5	5	2	5	10	10	10
3. *Reaction*																
pH (opt.)	6.8–7.2	6.8–7.2	6.8–7.2		6.5–8.5	6.5–8.5	6.5–8.5	6.8–7.2	6.5–8.5	6.5–8.5	6.5–8.5	6.5–8.5	6.5–8.5	4.0–10.0	4.0–10.0	
pH (critical)	6.6–8.0	6.5–8.6	6.5–8.6		6.5–8.5	6.5–8.5	6.5–8.5	6.6–8.0	6.0–9.0	6.0–9.0	6.0–9.0	6.0–9.0	6.0–9.0	4.0–10.0	4.0–10.0	
4. *Physical—ppm.*																
Turbid. (opt.)	5	5	5	10	5	5	10	5				5	5			50
Turbid. (max.)	20	20	30	50	10	20	100	50				20	50			
Color (opt.)	10	10	10	10	5	5	10	10				10	10			20
Color (max.)	30	30	30	50	10	20	100	50				30	50			100
Susp. solids (opt.)	10	50	50		10	10	50	10				10	10	50	50	
Susp. solids (max.)	100	100	100		20	50	250	100				50	100	150	150	
Float. solids (opt.)	none	none	none	none	none	none	slight	none				none	none	none	none	slight
Float. solids (max.)		gross	gross	gross	gross	gross	gross	gross				slight	slight	slight	slight	gross

Characteristics	Domestic Water Supply	Recreation: Bathing and Swimming: Fresh Water	Recreation: Bathing and Swimming: Salt Water	Recreation: Boating and Fishing	Wildlife Propagation: Fish: Fresh Water	Wildlife Propagation: Fish: Salt Water	Wildlife Propagation: Fowl Refuge	Shellfish Culture	Irrigation: Truck Garden Vege-tables	Irrigation: Citrus Fruits	Irrigation: Other Crops	Industrial: Food Processing: Fresh Water	Industrial: Food Processing: Salt Water	Industrial: Cooling and Other: Fresh Water	Industrial: Cooling and Other: Salt Water	Aesthetic Enjoy-ment
5. *Chemical—ppm.*																
Total solids (opt.)	500				1,000				500	500	500	500		1,000		
Total solids (max.)	1,500				5,000				1,500	1,500	2,000	1,500		1,500		
Cl (opt.)	250				1,000				200	100	250	500				
Cl (max.)	750				2,500				750	500	750	1,000				
F (opt.)	0.5–1.0								0.5–1.0							
F (max.)	1.5								5							
Toxic metals (opt.)	none	0.1	0.5		0.5	0.5			0.1			none	none			
Toxic metals (max.)	0.05	5	10		10	10		0.1	2.5			0.1	0.5			
Phenol (opt.)	1*	5*	50*	1	0.1	0.5	5	1*	5*			1*	5*			
Phenol (max.)	5*	50*	1	10	1	5	25	10*	20*			10*	50*			
Boron (opt.)										0.5	1.0					
Boron (max.)										1.0	5					
Na ratio† (opt.)									35–50†	35–50†	35–50†			90†		
Na ratio† (max.)									80†	75†	80†			90†		
Hardness (opt.)	100													100		
Hardness (max.)	250													500		
6. *Temp.—°F. (max.)*	60	65	65		60	60		70								
7. *Odor ‡ (max.)*	N	N	N	M	M	M	M	N	O	O	O	M	M	O	O	O
8. *Taste‡ (max.)*	N	M	D		M	M	M	N				M	M			

*Parts per billion.

† Per cent.

‡ Key: D—disagreeable; M—marked; N—noticeable; O—obnoxious.

Source: Calif. State Water Pollution Control Board, 1952

TABLE 6–4. PRINCIPAL CHEMICAL CONSTITUENTS IN WATER—THEIR SOURCES, CONCENTRATIONS, AND EFFECTS UPON USABILITY

Constituent	Major Sources	Concentration in Natural Water	Effect upon Usability of Water
Silica (SiO_2)	Feldspars, ferromagnesium and clay minerals, amorphous silicachert, opal.	Ranges generally from 1.0 to 30 mg/L, although as much as 100 mg/L is fairly common; as much as 4,000 mg/L is found in brines.	In the presence of calcium and magnesium, silica forms a scale in boilers and on steam turbines that retards heat; the scale is difficult to remove. Silica may be added to soft water to inhibit corrosion of iron pipes.
Iron (Fe)	1. Natural sources: Igneous rocks: Amphiboles, ferromagnesian micas, ferrous sulfide (FeS), ferric sulfide or iron pyrite (FeS_2), magnetite (Fe_3O_4). Sandstone rocks: Oxides, carbonates, and sulfides or iron clay minerals. 2. Manmade sources: Well casing, piping, pump parts, storage tanks, and other objects of cast iron and steel which may be in contact with the water. Industrial wastes.	Generally less than 0.50 mg/L in fully aerated water. Ground water having a pH less than 8.0 may contain 10 mg/L; rarely as much as 50 mg/L may occur. Acid water from thermal springs, mine wastes, and industrial wastes may contain more than 6,000 mg/L.	More than 0.1 mg/L precipitates after exposure to air; causes turbidity, stains plumbing fixtures, laundry and cooking utensils, and imparts objectionable tastes and colors to foods and drinks. More than 0.2 mg/L is objectionable for most industrial uses.
Manganese (Mn)	Manganese in natural water probably comes most often from soils and sediments. Metamorphic and sedimentary rocks and mica biotite and amphibole hornblende minerals contain large amounts of manganese.	Generally 0.20 mg/L or less. Ground water and acid mine water may contain more than 10 mg/L. Reservoir water that has "turned over" may contain more than 150 mg/L.	More than 0.2 mg/L precipitates upon oxidation; causes undesirable tastes, deposits on foods during cooking, stains plumbing fixtures and laundry, and fosters growths in reservoirs, filters, and distribution systems. Most industrial users object to water containing more than 0.2 mg/L.
Calcium (Ca)	Amphiboles, feldspars, gypsum, pyroxenes, aragonite, calcite, dolomite, clay minerals.	As much as 600 mg/L in some western streams; brines may contain as much as 75,000 mg/L.	Calcium and magnesium combine with bicarbonate, carbonate, sulfate and silica to form heat-retarding, pipe-clogging scale in boilers and in other heat-exchange equipment. Calcium and magnesium combine with ions of fatty acid in soaps to form soap suds; the more calcium and magnesium, the more soap required to form suds. A high concentration of magnesium has a laxative effect, especially on new users of the supply.
Magnesium (Mg)	Amphiboles, olivine, pyroxenes, dolomite, magnesite, clay minerals.	As much as several hundred mg/L in some western streams; ocean water contains more than 1,000 mg/L, and brines may contain as much as 57,000 mg/L.	
Sodium (Na)	Feldspars (albite); clay minerals; evaporites, such as halite (NaCl) and mirabilite ($Na_2SO_4 \cdot 10H_2O$); industrial wastes.	As much as 1,000 mg/L in some western streams; about 10,000 mg/L in sea water; about 25,000 mg/L in brines.	More than 50 mg/L sodium and potassium in the presence of suspended matter causes foaming, which accelerates scale formation and corrosion in boilers. Sodium and potassium carbonate in recirculating cooling water can cause deterioration of wood in cooling towers. More than 65 mg/L of sodium can cause problems in ice manufacture.
Potassium (K)	Feldspars (orthoclase and microcline), feldspathoids, some micas, clay minerals.	Generally less than about 10 mg/L; as much as 100 mg/L in hot springs; as much as 25,000 mg/L in brines.	

TABLE 6–4. PRINCIPAL CHEMICAL CONSTITUENTS IN WATER—THEIR SOURCES, CONCENTRATIONS, AND EFFECTS UPON USABILITY (continued)

Constituent	Major Sources	Concentration in Natural Water	Effect upon Usability of Water
Carbonate (CO_3)	Limestone, dolomite.	Commonly 0 mg/L in surface water; commonly less than 10 mg/L in ground water. Water high in sodium may contain as much as 50 mg/L of carbonate.	Upon heating, bicarbonate is changed into steam, carbon dioxide, and carbonate. The carbonate combines with alkaline earths—principally calcium and magnesium—to form a crustlike scale of calcium carbonate that retards flow of heat through pipe walls and restricts flow of fluids in pipes. Water containing large amounts of bicarbonate and alkalinity are undesirable in many industries.
Bicarbonate (HCO_3)		Commonly less than 500 mg/L; may exceed 1,000 mg/L in water highly charged with carbon dioxide.	
Sulfate (SO_4)	Oxidation of sulfide ores; gypsum; anhydrite; industrial wastes.	Commonly less than 1,000 mg/L except in streams and wells influenced by acid mine drainage. As much as 200,000 mg/L in some brines.	Sulfate combines with calcium to form an adherent, heat-retarding scale. More than 250 mg/L is objectionable in water in some industries. Water containing about 500 mg/L of sulfate tastes bitter; water containing about 1,000 mg/L may be cathartic.
Chloride (Cl)	Chief source is sedimentary rock (evaporites): minor sources are igneous rocks. Ocean tides force salty water upstream in tidal estuaries.	Commonly less than 10 mg/L in humid regions; tidal streams contain increasing amounts of chloride (as much as 19,000 mg/L) as the bay or ocean is approached. About 19,300 mg/L in sea water; and as much as 200,000 mg/L in brines.	Chloride in excess of 100 mg/L imparts a salty taste. Concentrations greatly in excess of 100 mg/L may cause physiological damage. Food processing industries usually require less than 250 mg/L. Some industries—textile processing, paper manufacturing, and synthetic rubber manufacturing—desire less than 100 mg/L.
Fluoride (F)	Amphiboles (hornblende), apatite, fluorite, mica.	Concentrations generally do not exceed 10 mg/L in ground water or 1.0 mg/L in surface water. Concentrations may be as much as 1,600 mg/L in brines.	Fluoride concentration between 0.6 and 1.7 mg/L in drinking water has a beneficial effect on the structure and resistance to decay of children's teeth. Fluoride in excess of 1.5 mg/L in some areas causes "mottled enamel" in children's teeth. Fluoride in excess of 6.0 mg/L causes pronounced mottling and disfiguration of teeth.
Nitrate (NO_3)	Atmosphere; legumes, plant debris, animal excrement, nitrogenous fertilizer in soil and sewage.	In surface water not subjected to pollution, concentration of nitrate may be as much as 5.0 mg/L but is commonly less than 1.0 mg/L. In ground water the concentration of nitrate may be as much as 1,000 mg/L.	Water containing large amounts of nitrate (more than 100 mg/L) is bitter tasting and may cause physiological distress. Water from shallow wells containing more than 45 mg/L has been reported to cause methemoglobinemia in infants. Small amounts of nitrate help reduce cracking of high-pressure boiler steel.
Dissolved solids	The mineral constituents dissolved in water constitute the dissolved solids.	Surface water commonly contains less than 3,000 mg/L; streams draining salt beds in arid regions may contain in excess of 15,000 mg/L. Ground water commonly contains less than 5,000 mg/L; some brines contain as much as 300,000 mg/L.	More than 500 mg/L is undesirable for drinking and many industrial uses. Less than 300 mg/L is desirable for dyeing of textiles and the manufacture of plastics, pulp paper, rayon. Dissolved solids cause foaming in steam boilers; the maximum permissible content decreases with increases in operating pressure.

Source: U.S. Geological Survey, 1962; amended

TABLE 6–5. RELATIVE ABUNDANCE OF DISSOLVED SOLIDS IN POTABLE WATER

[A classification based upon relative abundance of dissolved solids]

Major Constituents (1.0 to 1000 mg/L)	Secondary Constituents (0.01 to 10.0 mg/L)	Minor Constituents (0.0001 to 0.1 mg/L)	Trace Constituents (generally less than 0.001 mg/L)
Sodium	Iron	Antimony*	Beryllium
Calcium	Strontium	Aluminum	Bismuth
Magnesium	Potassium	Arsenic	Cerium*
Bicarbonate	Carbonate	Barium	Cesium
Sulfate	Nitrate	Bromide	Gallium
Chloride	Fluoride	Cadmium*	Gold
Silica	Boron	Chromium*	Indium
		Cobalt	Lanthanum
		Copper	Niobium*
		Germanium*	Platinum
		Iodide	Radium
		Lead	Ruthenium*
		Lithium	Scandium*
		Manganese	Silver
		Molybdenum	Thallium*
		Nickel	Thorium*
		Phosphate	Tin
		Rubidium*	Tungsten*
		Selenium	Ytterbium
		Titanium*	Yttrium*
		Uranium	Zirconium*
		Vanadium	
		Zinc	

* These elements occupy an uncertain position in the list.
Source: Davis and DeWiest, Hydrogeology, John Wiley & Sons, Copyright 1966

TABLE 6–6. CHARACTERISTICS OF WATER THAT AFFECT WATER QUALITY

Characteristic	Principal Cause	Significance	Remarks
Hardness	Calcium and magnesium dissolved in the water.	Calcium and magnesium combine with soap to form an insoluble precipitate (curd) and thus hamper the formation of a lather. Hardness also affects the suitability of water for use in the textile and paper industries and certain others and in steam boilers and water heaters.	USGS classification of hardness (mg/L as $CaCO_3$): 0–60: Soft 61–120: Moderately hard 121–180: Hard More than 180: Very hard
pH (or hydrogen-ion activity)	Dissociation of water molecules and of acids and bases dissolved in water.	The pH of water is a measure of its reactive characteristics. Low values of pH, particularly below pH 4, indicate a corrosive water that will tend to dissolve metals and other substances that it contacts. High values of pH, particularly above pH 8.5, indicate an alkaline water that, on heating, will tend to form scale. The pH significantly affects the treatment and use of water.	pH values: less than 7, water is acidic; value of 7, water is neutral; more than 7, water is basic.
Specific electrical conductance	Substances that form ions when dissolved in water.	Most substances dissolved in water dissociate into ions that can conduct an electrical current. Consequently, specific electrical conductance is a valuable indicator of the amount of material dissolved in water. The larger the conductance, the more mineralized the water.	Conductance values indicate the electrical conductivity, in micromhos, of 1 cm^3 of water at a temperature of 25°C.
Total dissolved solids	Mineral substances dissolved in water.	Total dissolved solids is a measure of the total amount of minerals dissolved in water and is, therefore, a very useful parameter in the evaluation of water quality. Water containing less than 500 mg/L is preferred for domestic use and for many industrial processes.	USGS classification of water based on dissolved solids (mg/L): Less than 1,000: Fresh 1,000–3,000: Slightly saline 3,000–10,000: Moderately saline 10,000–35,000: Very saline More than 35,000: Briny

Source: Heath, R.C., 1984, Basic Ground-Water Hydrology, U.S. Geological Survey Water-Supply Paper 2220

FIGURE 6–1. DISSOLVED SOLIDS IN SURFACE WATER

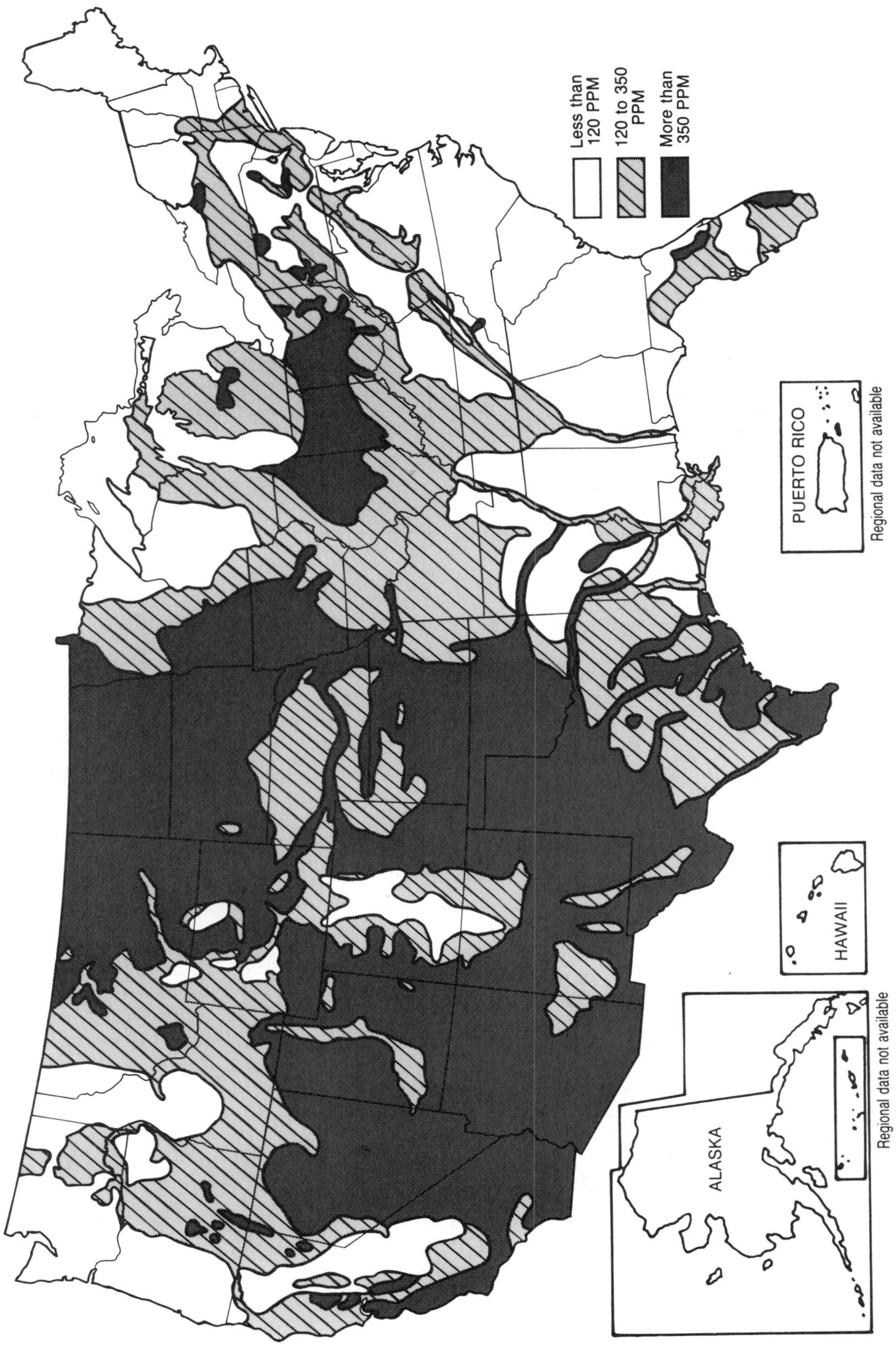

Source: U.S. Water Resources Council, 1968

FIGURE 6–2. WATER-QUALITY TRENDS IN RIVERS OF THE UNITED STATES

(Trends in flow-adjusted concentrations of six common water- quality constituents at NASQAN and NWQSS stations from 1974 to 1981; regional basins are outlined with dashed lines); ▲, increase; ▽, decrease; •, no trend)

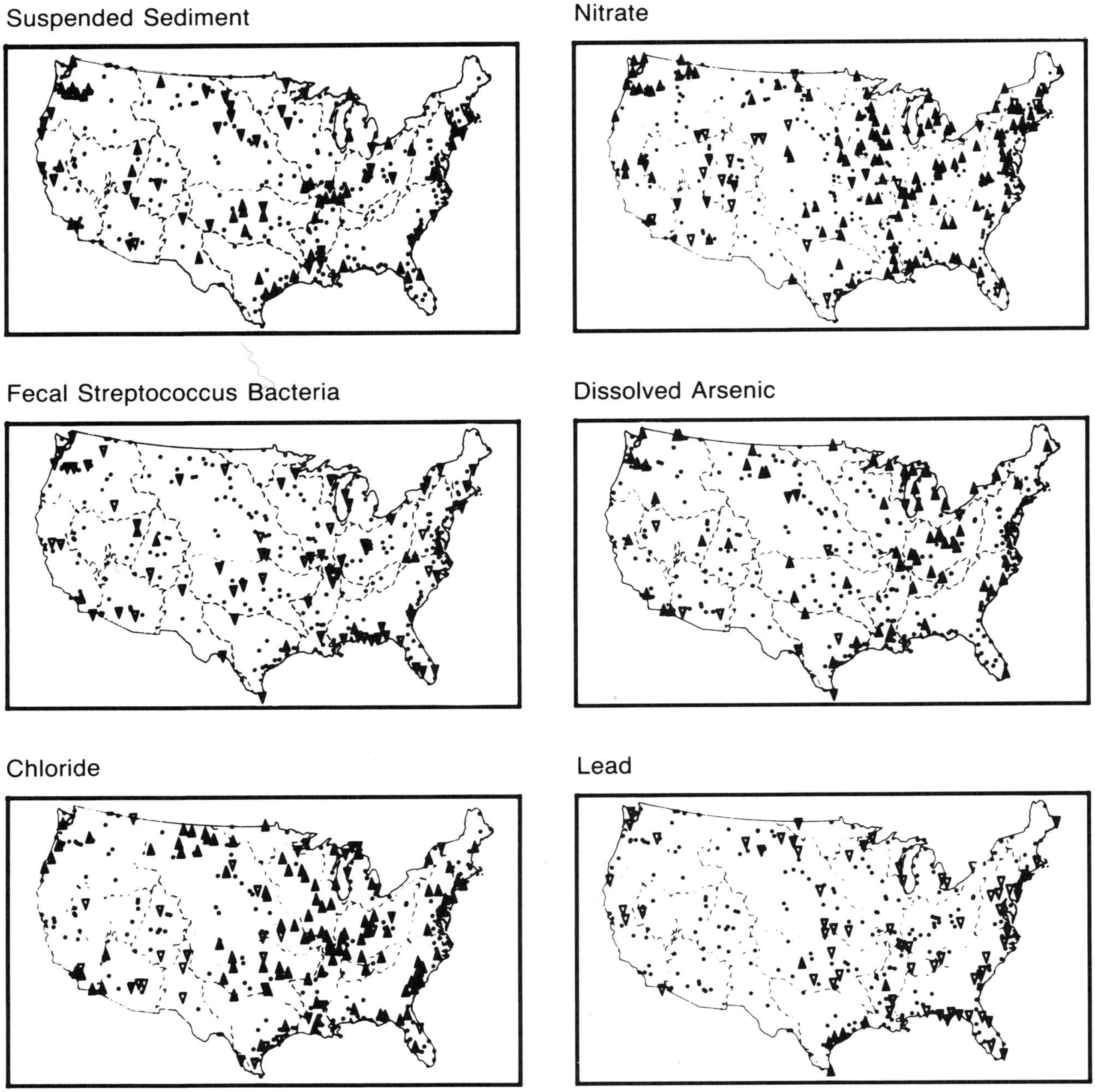

Source: Smith, R.A., 1987, Analysis and Interpretation of Water-Quality Trends in Major U.S. Rivers, U.S. Geol. Survey Water-Supply Paper 2307

TABLE 6–7. TRENDS OF SURFACE-WATER QUALITY IN THE UNITED STATES, 1974–1981

[Selected water-quality constituents and properties at NASQAN stations]

Constituents and Properties	Number of Stations with— Increasing Trends	No Change	Decreasing Trends	Total Stations
Temperature	39	218	46	303
pH	74	174	56	304
Alkalinity	18	207	79	304
Sulfate	82	182	40	304
Nitrate-nitrite	76	203	25	304
Ammonia	31	221	30	282
Total organic carbon	36	230	13	279
Phosphorus	39	232	30	301
Calcium	23	198	83	304
Magnesium	50	208	46	304
Sodium	103	173	28	304
Potassium	69	193	42	304
Chloride	104	164	36	304
Silica	48	213	41	302
Dissolved solids	68	183	51	302
Suspended sediment	44	204	41	289
Conductivity	69	193	43	305
Turbidity	42	199	18	259
Fecal coliform bacteria	19	216	34	269
Fecal streptococcus bacteria	2	190	78	270
Phytoplankton	22	234	44	300
Dissolved trace metals:				
Arsenic	68	228	11	307
Barium	4	81	1	86
Boron	2	15	3	20
Cadmium	32	264	7	303
Chromium	12	152	2	166
Copper	6	83	6	95
Iron	28	258	21	307
Lead	5	232	76	313
Manganese	30	250	19	299
Mercury	8	194	2	204
Selenium	2	201	21	224
Silver	1	32	0	33
Zinc	19	251	32	302

Source: U.S. Geological Survey Water-Supply Paper 2250

TABLE 6–8. WATER QUALITY OF GREAT SALT LAKE, UTAH, 1850–1976

[Composition, in percentage by weight, of dissolved ions in brine]

Date	Silica (SiO_2)	Calcium (Ca)	Magnesium (Mg)	Sodium (Na)	Potassium (K)	Lithium (Li)	Bicarbonate (as CO_3)	Sulfate (SO_4)	Chloride (Cl)	Fluoride (F)	Boron (B)	Bromium (Br)	Total Percent
						Precauseway							
1850	—	—	0.27	38.29	—	—	—	5.57	55.87	—	—	—	100.0
1869	—	0.17	2.52	33.15	1.60	—	—	6.57	55.99	—	—	—	100.0
August 1892	—	1.05	1.23	33.22	1.71	—	—	6.57	56.22	—	—	—	100.0
October 1913	—	.16	2.76	33.17	1.66	—	0.09	6.68	55.48	—	—	—	100.0
March 1930	—	.17	2.75	32.90	1.61	—	.05	5.47	57.05	—	—	—	100.0
						South of causeway							
April 1960	0.002	0.12	2.91	32.71	1.71	—	0.06	6.60	55.88	—	0.01	—	100.0
December 1963	.001	.09	3.29	31.02	1.86	—	.07	9.02	54.64	—	.01	—	100.0
May 1966	.003	.09	3.80	30.56	2.22	0.02	.10	7.99	55.21	0.003	.01	—	100.0
June 1976	—	.17	3.47	31.29	2.66	.02	—	7.22	55.11	—	.01	0.04	100.0
						North of causeway							
December 1963	0.001	0.09	4.66	29.08	2.75	—	0.09	7.28	56.04	—	0.01	—	100.0
May 1966	—	.05	4.38	29.67	2.61	0.02	.09	8.58	54.59	0.002	.01	—	100.0
June 1976	—	.13	3.17	32.04	2.58	.02	—	6.62	55.39	—	.01	0.04	100.0

Source: Arnow, Ted, 1984, Water-Level and Water-Quality Changes in Great Salt Lake, Utah, 1847–1983, U.S. Geological Survey Circ. 913

FIGURE 6–3. GEOCHEMICAL CYCLE OF SURFACE AND GROUND WATER

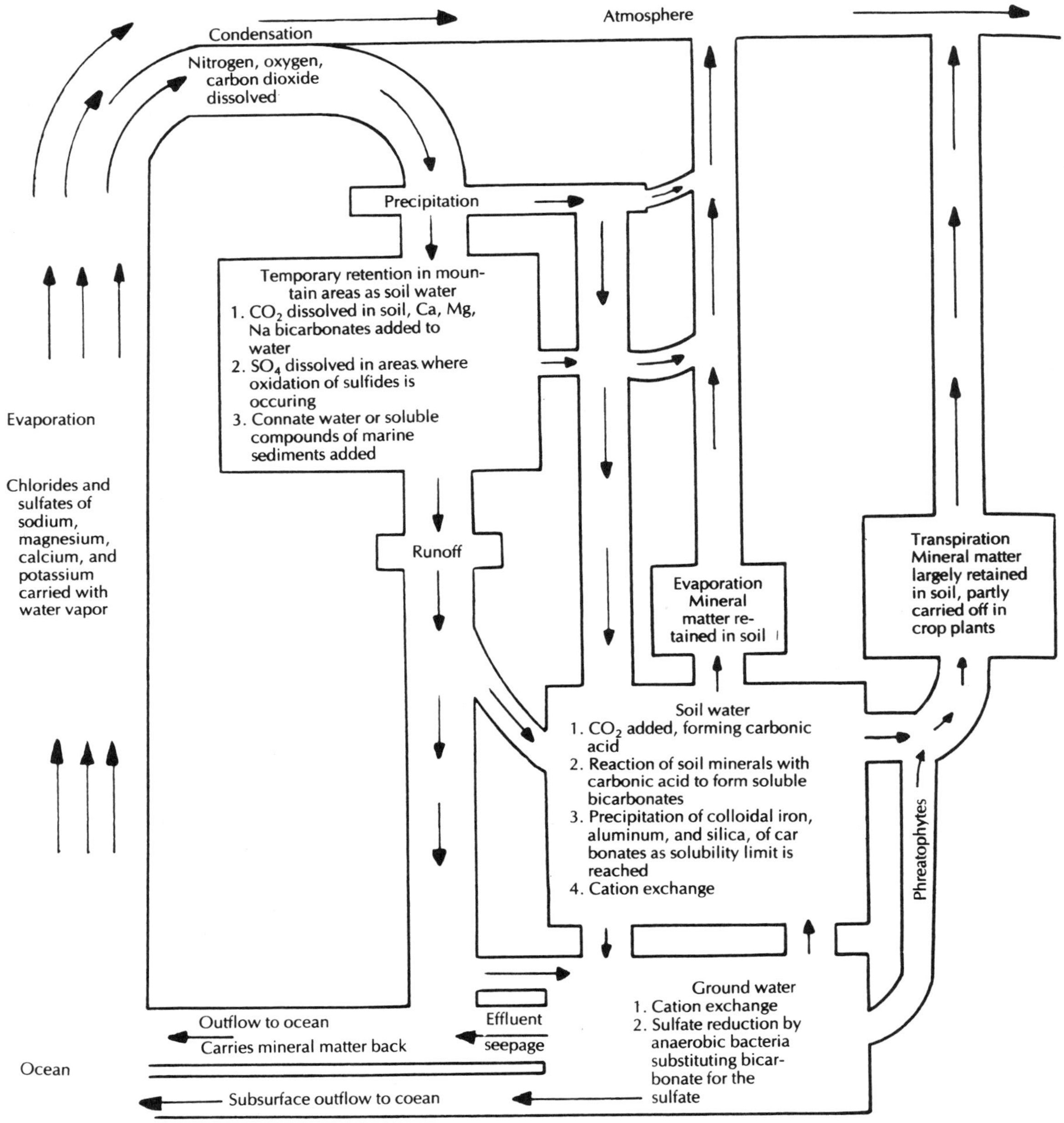

Source: U.S. Geological Survey

TABLE 6–9. NATURAL INORGANIC CONSTITUENTS COMMONLY DISSOLVED IN GROUND WATER THAT ARE MOST LIKELY TO AFFECT USE OF THE WATER

Substance	Major Natural Sources	Effect on Water Use	Concentrations of Significance (mg/L)[1]
Bicarbonate (HCO_3) and carbonate (CO_3)	Products of the solution of carbonate rocks, mainly limestone ($CaCO_3$) and dolomite ($CaMgCO_3$), by water containing carbon dioxide.	Control the capacity of water to neutralize strong acids. Bicarbonates of calcium and magnesium decompose in steam boilers and water heaters to form scale and release corrosive carbon dioxide gas. In combination with calcium and magnesium, cause carbonate hardness.	150–200
Calcium (Ca) and magnesium (Mg)	Soils and rocks containing limestone, dolomite, and gypsum ($CaSO_4$). Small amounts from igneous and metamorphic rocks.	Principal cause of hardness and ofr boiler scale and deposits in hot-water heaters.	25–50
Chloride (Cl)	In inland areas, primarily from seawater trapped in sediments at time of deposition; in coastal areas, from seawater in contact with freshwater in productive aquifers.	In large amounts, increases corrosiveness of water and, in combination with sodium, gives water a salty taste.	250
Fluoride (F)	Both sedimentary and igneous rocks. Not widespread in occurrence.	In certain concentrations, reduces tooth decay; at higher concentrations, causes mottling of tooth enamel.	0.7–1.2[2]
Iron (Fe) and manganese (Mn)	Iron present in most soils and rocks; manganese less widely distributed.	Stain laundry and are objectionable in food processing, dyeing, bleaching, ice manufacturing, brewing, and certain other industrial processes.	Fe>0.3, Mn>0.05
Sodium (Na)	Same as for chloride. In some sedimentary rocks, a few hundred milligrams per liter may occur in freshwater as a result of exchange of dissolved calcium and magnesium for sodium in the aquifer materials.	See chloride. In large concentrations, may affect persons with cardiac difficulties, hypertension, and certain other medical conditions. Depending on the concentrations of calcium and magnesium also present in the water, sodium may be detrimental to certain irrigated crops.	69 (irrigation), 20–170 (health)[3]
Sulfate (SO_4)	Gypsum, pyrite (FeS), and other rocks containing sulfur (S) compounds.	In certain concentrations, gives water a bitter taste and, at higher concentrations, has a laxative effect. In combination with calcium, forms a hard calcium carbonate scale in steam boilers.	300–400 (taste), 600–1,000 (laxative)

[1] A range in concentration is intended to indicate the general level at which the effect on water use might become significant.
[2] Optimum range determined by the U.S. Public Health Service, depending on water intake.
[3] Lower concentration applies to drinking water for persons on a strict diet; higher concentration is for those on a moderate diet.
Source: Heath, R.C., 1982, Basic Ground-Water Hydrology, U.S. Geological Survey Water-Supply Paper 2220

TABLE 6–10. INORGANIC SUBSTANCES FOUND IN GROUND WATER

	Concentration (milligrams per liter)
Aluminum	0.1–1,200
Ammonia	1.0–900
Antimony	—
Arsenic	0.01–2,100
Barium	2.8–3.8
Beryllium	less than 0.01
Boron	—
Cadmium	0.01–180
Calcium	0.5–225
Chlorides	1.0–49,500
Chromium	0.06–2,740
Cobalt	0.01–0.18
Copper	0.01–2.8
Cyanides	1.05–14
Fluorides	0.1–250
Iron	0.04–6,200
Lead	0.01–5.6
Lithium	—
Magnesium	0.2–70
Manganese	0.1–110
Mercury	0.003–0.01
Molybdenum	0.4–40
Nickel	0.05–0.5
Nitrates	1.4–433
Nitrites	—
Palladium	—
Potassium	0.5–2.4
Phosphates	0.4–33
Selenium	0.6–20
Silver	9.0–330
Sodium	3.1–211
Sulfates	0.2–32,318
Sulfites	—
Thallium	—
Titanium	—
Vanadium	243.0
Zinc	0.1–240

Source: Office of Technology Assessment 1984, Protecting the Nation's Ground Water From Contamination, U.S. Congress, Washington D.C.

TABLE 6–11. SUMMARY OF INORGANIC ELEMENTS FOUND IN RURAL WATER SUPPLIES.

[According to survey conducted by United States Environmental Protection Agency]

Element	Level Exceeded (mg/L)	in % of Rural Households				
		Nationwide	West	North-Central	Northeast	South
Mercury	0.002	24.1	10.4	31.8	22.0	25.0
Iron	0.3	18.7	7.0	28.2	16.0	17.0
Cadmium	0.01	16.8	27.1	20.7	1.6	17.3
Lead	0.05	16.6	16.9*	10.8*	9.6*	23.1*
Manganese	0.05	14.2	4.7	19.9	16.9	12.3
Sodium	100	14.2	15.0	19.2	6.0	14.1
Selenium	0.01	13.7	41.3	25.7	0.0	2.1
Silver	0.05	4.7	2.1	3.7	4.8	4.8
Sulfates	250.0	4.0	11.7	7.4	0.5	0.7
Nitrate-N	10.0	2.7	4.0	5.8	0.3	1.3
Fluoride	1.4	2.5	6.2	1.8	0.0	2.7
Arsenic	0.05	0.8	2.1	1.8	0.0	0.0
Barium	1.0	0.3	0.0	0.0	0.0	0.7
Magnesium	125.0	0.1	0.5	0.1	0.0	0.0
Chromium	0.05	**	0.0	0.0	0.0	0.0
Boron	†					

* may be distorted upwards
** not detected
† not tested
Source: U.S. Environmental Protection Agency, 1984, National Statistical Assessment of Rural Conditions, Executive Summary. Office of Drinking Water

TABLE 6–12. WATER QUALITY OF SELECTED RIVERS IN THE WORLD, 1970–83

	Dissolved Oxygen (DO) (milligrams per liter)					Biological Oxygen Demand (BOD) (milligrams per liter)					Nitrates (milligrams per liter)					Phosphorus (milligrams per liter)				
	1970	1975	1980	1983	Latest Three Years of Data (average)	1970	1975	1980	1983	Latest Three Years of Data (average)	1970	1975	1980	1983	Latest Three Years of Data (average)	1970	1975	1980	1983	Latest Three Years of Data (average)
Canada																				
St. Lawrence	8.1	10.0	X	X	X	X	X	X	X	X	0.19	0.23	0.16	0.29	0.19	0.05	0.03	0.07	0.06	0.07
Mackenzie	X	X	X	X	X	X	X	X	X	X	0.08	0.11	0.11	0.08	0.10	X	0.02	0.01	0.01	0.01
Fraser	X	X	X	X	X	X	X	X	X	X	0.05	0.30	0.06	0.12	0.08	0.00	0.11	X	X	X
Nelson	X	X	X	X	X	X	X	X	X	X	0.04	0.40	0.06	0.09	0.06	X	0.02	0.01	0.02	0.01
United States																				
Delaware-Trenton	9.6	10.8	11.9	10.3	11.1	1.9	2.0	2.2	2.6	2.8	a X	0.88	1.08	X	0.97	a X	0.10	0.10	0.09	0.09
Mississippi-St. Franc.	8.4	8.5	8.3	8.9	8.9	2.4	2.2	1.7	1.1	1.8	a X	0.98	1.20	1.58	1.64	a X	0.19	0.23	0.17	0.23
Japan																				
Ishikari	8.7	10.7	10.6	10.6	10.5	1.7	1.4	1.4	1.5	1.4	0.36	X	0.53	X	X	X	0.09	0.09	0.07	0.09
Yodo	8.3	8.9	9.0	8.2	8.0	3.6	3.2	3.8	4.2	4.1	X	0.80	0.76	X	0.66	X	0.19	0.22	0.16	0.18
Australia																				
Brisbane Estuary	X	5.6	6.4	6.0	6.2	X	1.6	1.0	1.2	1.2	X	0.34	0.85	1.05	0.93	X	0.20	0.38	0.48	0.43
New Zealand																				
Waikato	X	9.0	X	X	X	X	1.5	X	X	X	X	0.20	X	X	X	X	0.60	X	X	X
Belgium																				
Meuse-Heer	8.2	10.8	10.6	X	10.7	4.4	6.6	4.2	X	4.7	1.80	7.80	2.18	X	2.09	X	1.23	0.22	X	0.17
Meuse-Lanaye	7.7	8.9	9.5	X	9.2	12.5	4.7	3.9	X	4.0	3.90	9.40	2.52	X	2.50	X	1.41	0.55	X	0.55
Escaut-Bleharies	X	4.0	5.9	X	4.7	X	24.1	10.7	X	9.6	X	0.77	4.17	X	3.58	X	0.70	0.94	X	1.03
Escaut-Doel	6.2	1.3	1.9	X	2.1	4.0	8.2	5.0	X	14.0	3.00	7.35	4.17	X	3.18	X	1.06	0.55	X	0.63
Denmark																				
Gudenaa	X	X	9.7	10.7	10.2	X	X	3.4	4.5	4.3	X	1.25	1.70	2.00	1.84	X	0.30	0.16	0.18	0.13
Skjernaa	X	X	10.5	10.4	10.4	X	X	7.3	8.0	8.3	X	X	3.00	3.18	2.92	X	X	0.22	0.13	0.17
Susaa	X	X	X	8.7	9.1	X	X	X	2.0	2.1	X	5.27	6.73	5.21	6.09	X	0.66	0.26	0.34	0.30
Finland																				
Tornionjoki	11.9	11.9	12.0	X	11.9	1.6	1.6	X	X	1.7	X	X	X	X	X	0.02	0.02	0.02	X	0.03
Kymijoki	9.5	10.8	9.9	X	10.8	3.5	2.4	X	X	2.5	X	X	X	X	X	0.05	0.03	0.03	X	0.03
France																				
Loire-Nantes	10.7	11.1	11.8	12.0	11.6	6.7	4.4	6.6	6.7	6.0	1.58	1.45	1.99	1.85	2.12	X	X	X	X	X
Seine-Tancerville	X	3.3	4.9	5.9	5.2	X	10.2	6.6	4.1	4.9	X	4.18	5.35	5.24	5.18	X	X	X	X	X
Garonne-Bordeaux	9.7	9.9	10.1	9.9	9.9	2.2	1.5	2.3	2.3	2.1	1.15	0.93	1.83	1.67	1.73	X	X	X	X	X
Rhine-Selz	8.2	9.2	10.9	8.6	8.9	X	4.1	4.8	3.0	3.6	X	1.58	2.92	2.94	2.62	X	X	0.47	0.20	0.23
Germany, Fed Rep																				
Rhine-Bimmen	5.6	6.5	8.4	9.1	9.0	6.1	6.9	3.6	3.5	3.6	1.82	3.12	3.80	3.88	3.67	0.52	0.75	0.48	0.59	0.51
Italy																				
Po	X	X	X	X	X	X	7.3	7.3	X	7.2	X	1.35	1.35	X	X	X	0.23	0.23	X	X
Netherlands																				
Meuse-Keizersveer	8.6	9.4	10.0	9.7	9.8	6.2	4.2	2.3	2.0	2.3	3.07	3.69	3.72	3.70	3.62	0.41	0.57	0.50	0.42	0.42
Meuse-Eijsden	9.8	9.5	9.8	8.7	9.1	4.1	3.7	2.8	2.8	2.8	2.45	2.51	2.78	2.71	2.68	0.43	0.73	0.58	0.53	0.50
Scheur-Maasluis	X	7.1	8.1	8.7	8.5	X	3.3	2.2	1.5	1.9	X	3.37	3.84	3.80	3.71	X	0.56	0.65	0.56	0.53
Ijssel-Kampen	6.7	6.7	8.1	8.2	8.4	5.7	6.3	3.9	2.3	2.7	2.76	3.46	4.27	4.33	4.13	0.43	0.62	0.63	0.57	0.57
Norway																				
Skienselva	X	X	X	6.0	X	X	2.5	3.5	X	X	X	0.35	0.35	0.34	0.34	X	0.01	0.01	0.01	0.01
Portugal																				
Tejo	9.0	X	9.0	X	X	1.6	X	2.5	X	X	0.52	X	5.60	X	X	X	X	2.00	X	X
Spain																				
Guadalquivir	X	X	X	X	X	X	12.3	11.8	X	8.2	X	7.54	9.60	X	8.05	X	0.83	0.86	X	0.43
Sweden																				
Dalalven	X	X	X	X	X	X	X	X	X	X	0.12	0.11	0.14	0.12	0.12	0.02	0.02	0.02	0.02	0.02
Switzerland																				
Rhine-Village	11.6	11.2	10.3	10.2	10.4	X	X	X	X	X	X	X	1.39	1.43	1.39	X	0.07	0.17	0.13	0.16
Aare-Brugg	X	10.2	10.2	X	10.4	X	X	X	X	X	X	1.32	1.43	1.38	1.33	X	0.23	0.11	0.09	0.44
Limmat-Baden	X	X	9.1	X	X	X	X	X	X	X	X	X	0.91	0.97	0.92	X	X	0.13	0.10	0.12
Rhone-Port du Sceaux	X	10.7	10.9	X	11.0	X	X	X	X	X	X	0.49	0.52	0.51	0.46	X	0.12	0.10	0.10	0.10
United Kingdom																				
Thames	X	10.8	9.9	9.8	9.9	X	4.2	X	X	4.0	X	6.50	6.89	7.10	6.99	b X	1.07	1.16	X	1.11
Severn	X	10.5	10.3	11.5	10.7	X	3.9	X	X	3.2	X	5.52	5.80	5.45	5.64	b X	0.75	0.54	X	0.58
Clyde	X	7.7	9.4	8.4	8.5	X	7.3	5.6	X	5.3	X	2.66	1.85	2.15	1.94	b X	0.69	0.50	X	0.48
Mersey	X	5.1	6.1	6.1	6.1	X	8.6	X	X	X	X	1.84	2.29	2.19	2.19	X	X	X	X	X
Yugoslavia																				
Dunav	9.6	9.1	9.2	X	9.5	3.4	2.5	3.5	X	3.7	X	1.70	5.80	X	2.50	X	X	X	X	X
Drava	8.8	7.4	10.1	X	10.0	3.4	2.6	3.1	X	3.1	4.60	5.20	7.70	X	6.73	X	X	X	X	X

TABLE 6–12. WATER QUALITY OF SELECTED RIVERS IN THE WORLD, 1970–83 (continued)

	Lead (micrograms per liter)					Cadmium (micrograms per liter)					Chromium (micrograms per liter)					Copper (micrograms per liter)				
	1970	1975	1980	1983	Latest Three Years of Data (average)	1970	1975	1980	1983	Latest Three Years of Data (average)	1970	1975	1980	1983	Latest Three Years of Data (average)	1970	1975	1980	1983	Latest Three Years of Data (average)
Canada																				
St. Lawrence	X	X	1.00	3.00	1.33	X	X	1.00	1.00	1.00	0.00	0.00	X	X	X	0.01	0.01	14.00	17.00	13.67
Mackenzie	10.00	5.00	4.00	3.00	3.67	X	1.00	1.00	1.00	1.00	c X	0.01	X	X	X	10.00	7.00	2.00	14.00	9.33
Fraser	10.00	2.00	X	X	X	X	X	X	X	X	X	X	X	X	X	10.00	4.00	X	X	X
Nelson	X	4.00	4.00	7.00	5.00	X	1.00	1.00	1.00	1.00	X	X	X	X	X	X	4.00	5.00	16.00	7.67
United States																				
Delaware-Trenton	X	6.00	2.00	X	2.00	X	2.00	3.50	X	2.75	d X	27.50	10.00	X	10.00	d X	20.00	3.50	X	3.50
Mississippi-St. Franc.	X	2.00	2.00	X	2.33	X	2.00	2.00	X	2.00	d X	7.80	10.00	X	9.77	d X	4.00	6.30	X	6.10
Japan																				
Ishikari	210.00	1.00	5.00	X	8.00	X	X	X	X	X	X	X	X	X	X	35.00	3.00	18.00	3.00	3.67
Yodo	X	1.00	0.00	X	0.00	X	X	X	X	X	d X	X	X	X	X	d X	0.00	8.00	X	6.67
Australia																				
Brisbane Estuary	c X	5.70	5.30	5.00	5.10	c X	2.30	2.00	2.00	2.00	X	X	X	20.00	X	X	9.70	5.60	5.00	5.20
New Zealand																				
Waikato	X	10.00	X	X	X	X	1.00	X	X	X	X	10.00	X	X	X	X	1.00	X	X	X
Belgium																				
Meuse-Heer	X	1.40	4.00	X	5.33	X	0.80	0.30	X	0.83	X	1.20	1.20	X	2.80	X	4.70	7.00	X	7.50
Meuse-Lanaye	X	5.70	20.00	X	12.00	X	2.60	1.20	X	1.73	X	4.60	2.70	X	3.40	X	4.50	11.30	X	12.23
Escaut-Bleharies	X	11.00	18.00	X	15.67	X	1.60	2.60	X	3.17	X	12.60	9.80	X	12.73	X	6.80	15.60	X	16.67
Escaut-Doel	X	203.50	25.00	X	38.00	X	1.50	5.80	X	5.73	X	15.60	25.10	X	24.37	X	15.50	24.40	X	27.60
Denmark																				
Gudenaa	X	X	X	X	X	X	X	10.00	1.00	4.00	X	X	X	X	X	X	X	X	X	X
Skjernaa	X	X	X	X	X	X	X	X	10.00	X	X	X	X	X	X	X	X	X	X	X
Susaa	X	X	X	X	X	X	X	X	1.00	X	X	X	X	X	X	X	X	X	X	X
Finland																				
Tornionjoki	c X	X	0.06	0.27	0.78	c X	X	0.01	0.02	0.04	c X	1.00	3.00	0.52	2.17	X	2.00	2.80	1.70	1.83
Kymijoki	c X	X	0.28	0.37	0.55	c X	X	0.01	0.02	0.04	X	X	X	0.67	2.22	X	X	1.40	1.50	1.63
France																				
Loire-Nantes	e X	10.00	4.00	5.00	4.67	f X	10.00	1.00	2.00	1.33	e X	10.00	5.00	5.00	5.00	e X	1.00	8.00	8.00	7.33
Seine-Tancerville	e X	26.00	8.00	14.00	9.67	f X	1.00	1.00	1.00	1.33	X	12.00	13.00	34.00	20.67	X	52.00	11.00	17.00	14.33
Garonne-Bordeaux	e X	10.00	5.00	6.00	5.67	f X	10.00	1.00	3.00	1.67	X	10.00	3.00	X	X	X	1.00	6.00	3.00	4.33
Rhine-Selz	X	9.30	12.50	3.20	6.03	X	1.00	0.80	2.00	0.53	X	9.00	16.00	7.60	11.50	X	11.30	15.90	5.10	7.80
Germany, Fed Rep																				
Rhine-Bimmen	X	24.00	7.00	8.00	9.67	X	2.40	1.30	0.40	0.73	X	40.00	19.00	9.00	11.33	X	24.00	16.00	19.00	18.33
Italy																				
Po	d X	0.40	0.55	X	0.27	a X	X	0.05	X	0.06	d X	X	0.60	X	X	d X	0.60	0.85	X	0.91
Netherlands																				
Meuse-Keizersveer	X	12.00	12.00	6.00	7.00	X	0.90	1.50	0.40	0.53	X	7.00	7.00	5.00	5.00	X	9.00	12.00	7.00	6.33
Meuse-Eijsden	X	17.00	23.00	12.00	17.00	X	3.10	3.40	1.40	1.77	X	14.00	10.00	9.00	11.67	X	16.00	11.00	8.00	9.00
Scheur-Maasluis	X	13.00	11.00	2.00	4.33	X	1.00	0.90	0.50	0.67	X	16.00	19.00	6.00	7.00	X	15.00	12.00	6.00	7.33
Ijssel-Kampen	X	17.00	9.00	5.00	5.67	X	1.40	1.30	0.40	0.70	X	25.00	14.00	7.00	8.67	26.00	16.00	9.00	7.00	7.67
Norway																				
Skienselva	c X	1.00	1.00	X	X	c X	1.00	1.00	X	X	X	X	X	X	X	X	4.00	X	X	X
Portugal																				
Tejo	c X	X	50.00	X	X	c X	X	10.00	X	X	c X	X	10.00	X	X	X	X	X	X	X
Spain																				
Guadalquivir	X	20.00	12.70	X	12.00	X	X	X	X	X	X	X	10.00	X	21.20	X	X	2.70	X	7.83
Sweden																				
Dalalven	X	X	X	X	X	X	X	X	X	X	X	X	X	X	X	X	X	17.00	17.00	16.00
Switzerland																				
Rhine-Village	X	1.60	1.40	1.10	1.60	X	0.08	0.14	0.05	0.09	X	1.30	2.00	0.70	1.70	X	1.80	4.20	3.60	4.63
Aare-Brugg	X	X	X	X	X	X	X	X	X	X	X	X	X	X	X	X	X	X	X	X
Limmat-Baden	X	X	X	X	X	X	X	X	X	X	X	X	X	X	X	X	X	X	X	X
Rhone-Port du Sceaux	X	X	3.38	2.13	2.68	X	X	X	X	X	X	X	X	X	X	X	X	3.48	3.82	4.62
United Kingdom																				
Thames	X	X	X	X	X	X	X	X	X	X	X	X	X	X	X	X	X	X	X	X
Severn	X	X	X	X	X	X	X	X	X	X	X	X	X	X	X	X	X	X	X	X
Clyde	X	X	X	X	X	X	X	X	X	X	X	X	X	X	X	X	X	X	X	X
Mersey	X	X	X	X	X	X	X	X	X	X		X	X	X	X	X	X	X	X	X
Yugoslavia																				
Dunav	X	4.41	29.98	X	33.73	X	0.21	6.00	X	6.36	X	X	X	X	X	X	1.39	117.90	X	53.27
Drava	X	X	X	X	X	X	X	X	X	X	X	X	X	X	X	X	X	X	X	X

0 = zero or less than one-half the unit of measure; X = not available.
a = total concentrations; b = orthophosphate concentrations; c = data represent upper limit (actual averages are lower); d = dissolved concentrations; e = 1975 data represent upper limit; f = data for 1975 and 1980 represent upper limit.
Source: World Resources Institute, World Resources 1986.
Original source: Organization for Economic Cooperation and Development

SECTION B. DRINKING WATER QUALITY STANDARDS—UNITED STATES

The U.S. Environmental Protection Agency's National Primary Drinking-Water Regulations and National Secondary Drinking-Water Regulations are summarized in the following tables. The primary regulations specify maximum contaminant levels (MCLs), and health advisories. The MCLs, which are the maximum permissible level of a contaminant in water at the tap, are health related and are legally enforceable. If these concentrations are exceeded or if required monitoring is not performed the public must be notified. The secondary drinking-water regulations specify the secondary maximum contaminant levels (SMCL). The SMCLs are for contaminants in drinking water that primarily affect the esthetic qualities related to public acceptance of drinking water; they are intended to be guidelines for the States and and are not federally enforceable. Health advisories are guidance contaminant levels that would not result in adverse health effects over specified short-time periods for most people.

As provided by the Safe Drinking Water Act of 1974, the U.S. Environmental Protection Agency has the primary responsibility for establishing and enforcing regulations. However, States may assume primacy if they adopt regulations that are at least as stringent as the Federal regulations in levels specified for protection of public health and in provision of surveillance and enforcement. The States may adopt more stringent regulations and may establish regulations for other constituents.

TABLE 6–13. NATIONAL PRIMARY DRINKING WATER STANDARDS

Constituent	MCL mg/L	Constituent	MCL mg/L
INORGANICS		Lindane	0.004
Arsenic (AS)	0.05	Methoxychlor	0.1
Barium (Ba)	1.0	Toxaphene	0.005
Cadmium (Cd)	0.01	Total trihalomethanes	0.10
Chromium (Cr)	0.05	RADIONUCLIDES	
Fluoride (F)	4.0	Beta particle and photon activity, mrem	4 (annual dose equivalent)
Lead (Pb)	0.05	Gross alpha , pCi/L	15
Mercury (Hg)	0.002	Radium-226 and 228, pCi/L	5
Nitrate (as N)	10.0	VOLATILE ORGANIC CHEMICALS	
Selenium (Se)	0.01	Benzene	0.005
Silver (Ag)	0.05	Carbon tetrachloride	0.005
MICROBIOLOGICALS		1,2-Dichloroethane	0.005
Coliforms	1/100 mL	1,1-Dichloroethylene	0.007
PHYSICAL CHARACTERISTICS		1,1,1-Trichloroethane	0.20
Turbidity, NTU	1–5	para-Dichlorobenzene	0.075
ORGANICS		Trichloroethylene	0.005
2, 4-D	0.1	Vinyl chloride	0.002
2,4,5-TP Silvex	0.01		
Endrin	0.0002		

Source: U.S. Environmental Protection Agency

TABLE 6–14. NATIONAL SECONDARY DRINKING WATER STANDARDS

Constituent	SMCL Level (mg/L)	Constituent	SMCL Level (mg/L)
Chloride (Cl)	250	Manganese (Mn)	0.05
Color, color units	15	Odor, threshold odor number	3
Copper (Cu)	1	pH, pH units	6.5–8.5
Corrosivity	Noncorrosive	Sulfate (SO_4)	250
Fluoride	2.0	Total dissolved solids (TDS)	500
Surfactants (MBAS)	0.5	Zinc (Zn)	5.0
Iron (Fe)	0.3		

HEALTH ADVISORY

Constituent	Level (mg/L)
Sodium	20

Source: U.S. Environmental Protection Agency

FIGURE 6–4. TIMETABLE FOR USEPA DRINKING WATER REGULATIONS

(as required by the Safe Drinking Water Act Amendments of 1986)

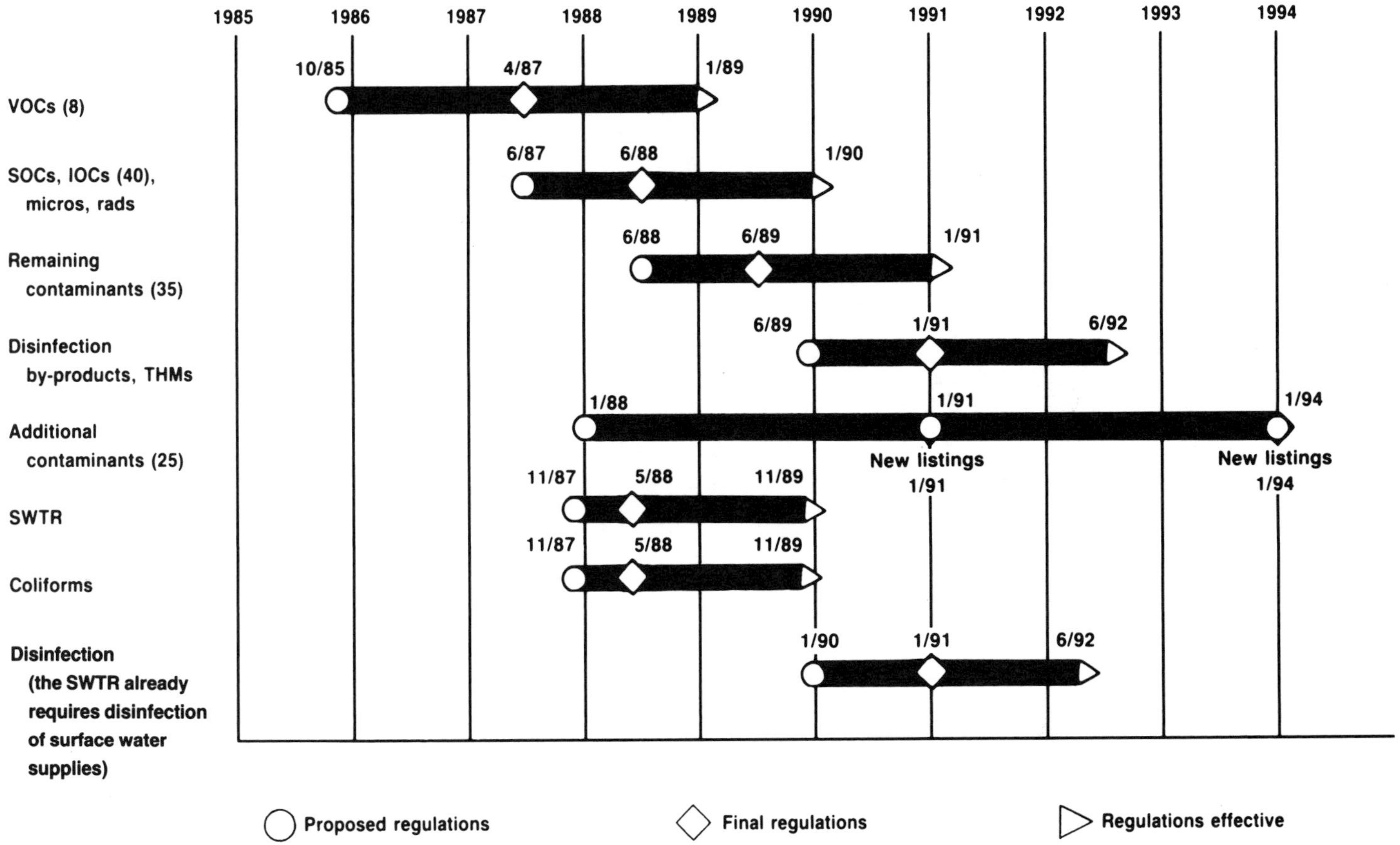

VOC	Volatile Organic Chemicals	micros	microbiological parameters
SOC	Synthetic Organic Chemicals	rads	radionuclides
IOC	Inorganic Chemicals	THM	Trihalomethane
SWTR	Surface Water Treatment Rule		

Source: Dyksen, J.E., Hiltebrand, D.J., and Raczko, R.F., 1988, SDWA Amendments: Effects on the Water Industry. Journal Am. Water Works Assoc., v. 80, no.1. Copyright AWWA. Reprinted with permission

TABLE 6–15. PROPOSED RMCLs FOR MICROBIOLOGICAL AND PARTICULATE CONSTITUENTS IN DRINKING WATER

[RMCL - recommended maximum contaminant level]

Constituent	Proposed RMCL	Constituent	Proposed RMCL
Total coliforms	Zero	Asbestos	7.1 million long fibers/L
Giardia lamblia	Zero	Turbidity	0.1 NTU
Viruses	Zero		

Source: USEPA, November 13, 1985

TABLE 6–16. PROPOSED NATIONAL DRINKING WATER STANDARDS FOR ORGANIC AND INORGANIC CHEMICALS

[MCL - Maximum contaminant level; SMCL - Secondary maximum contaminant Level]

Chemical	Level	Chemical	Level
Proposed MCLs for organic chemicals:		**Proposed MCLs for inorganic chemicals:**	
Acrylamide	treatment technique	Arsenic	0.03 mg/L
Alachlor	0.002 mg/L	Asbestos	7 million fibers/L (longer than 10 μm)
Aldicarb	0.01 mg/L		
Aldicarb sulfoxide	0.01 mg/L	Barium	5 mg/L
Aldicarb sulfone	0.04 mg/L	Cadmium	0.005 mg/L
Atrazine	0.003 mg/L	Chromium	0.1 mg/L
Carbofuran	0.04 mg/L	Mercury	0.002 mg/L
Chlordane	0.002 mg/L	Nitrate**	10.0 mg/L (as N)
Dibromochloropropane	0.0002 mg/L	Nitrite	1.0 mg/L (as N)
o-Dichlorobenzene	0.6 mg/L	Selenium	0.05 mg/L
cis-1,2-Dichloroethylene	0.07 mg/L		
trans-1,2-Dichloroethylene	0.1 mg/L	**Proposed SMCLs:**	
1,2-Dichloropropane	0.005 mg/L		
2,4-D	0.07 mg/L	Aluminum	0.05 mg/L
Epichlorohydrin	treatment technique	o-Dichlorobenzene	0.01 mg/L
Ethylbenzene	0.7 mg/L	p-Dichlorobenzene	0.005 mg/L
Ethylene dibromide	0.00005 mg/L	1,2-Dichloropropane	0.005 mg/L
Heptachlor	0.0004 mg/L	Ethylbenzene	0.03 mg/L
Heptachlor epoxide	0.0002 mg/L	Pentachlorophenol	0.03 mg/L
Lindane	0.0002 mg/L	Silver	0.09 mg/L
Methoxychlor	0.4 mg/L	Styrene	0.01 mg/L
Monochlorobenzene	0.1 mg/L	Toluene	0.04 mg/L
PCBs	0.0005 mg/L	Xylene	0.02 mg/L
Pentachlorophenol	0.2 mg/L		
Styrene*	0.005 mg/L/0.1 mg/L		
Tetrachloroethylene	0.005 mg/L		
Toluene	2 mg/L		
Toxaphene	0.005 mg/L		
2,4,5-TP (Silvex)	0.05 mg/L		
Xylene	10 mg/L		

* EPA proposes MCLs of 0.1 mg/L based on a Group C carcinogen classification and .005 mg/L based on a B_2 classification.
** In addition, MCL for total nitrate and nitrite = 10.0 mg/L.
Source: USEPA Office of Drinking Water, August 1988; amended based on May 22, 1989, Fed. Register Vol. 54, No. 97, pp. 22062-65

TABLE 6–17. CONTAMINANTS TO BE REGULATED UNDER THE SAFE DRINKING WATER ACT AMENDMENTS OF 1986*

Inorganics
- Aluminum
- Antimony
- Arsenic†
- Asbestos
- Barium†
- Beryllium
- Cadmium†
- Chromium†
- Copper
- Cyanide
- Fluoride†
- Lead†
- Mercury†
- Molybdenum
- Nickel
- Nitrate†
- Selenium†
- Silver†
- Sodium
- Sulfate
- Thallium
- Vanadium
- Zinc

Microbiology and turbidity
- *Giardia lamblia*
- *Legionella*
- Standard plate count
- Total coliforms†
- Turbidity†
- Viruses

Organics
- Acrylamide
- Adipates
- Alachlor
- Aldicarb
- Atrazine
- Carbofuran
- Chlordane
- 2,4,-D†
- Dalapon
- Dibromochloropropane
- Dibromomethane
- 1,2-Dichloropropane
- Dinoseb

Organics, continued
- Dioxin
- Diquat
- Endothall
- Endrin†
- Epichlorohydrin
- Ethylene dibromide
- Glyphosate
- Hexachlorocyclopentadiene
- Lindane†
- Methoxychlor†
- Pentachlorophenol
- Phthalates
- Pichloram
- Polychlorinated biphenyls
- Polycyclic aromatic hydrocarbons
- Simazine
- 2,4,5-TP†
- Toluene
- Toxaphene†
- 1,1,2-Trichloroethane
- Vydate
- Xylene

Radionuclides
- Beta particle and photon activity†
- Gross alpha particle activity†
- Radium-226 and radium-228†
- Radon
- Uranium

Volatile organic chemicals
- Benzene†
- Carbon tetrachloride†
- Chlorobenzene
- cis-1,2,-Dichloroethylene
- Dichlorobenzene†
- 1,2-Dichloroethane†
- 1,1-Dichloroethylene†
- Methylene chloride
- Tetrachloroethylene
- trans-1,2,-Dichloroethylene
- Trichlorobenzene
- 1,1,1-Trichloroethane†
- Trichloroethylene†
- Vinyl chloride

* Seven substitutions are permitted.
† Already regulated
Source: Sayre, I.M., 1988, International Standards for Drinking Water, Journal Am. Water Works Assoc., v. 80, no.1. Copyright AWWA. Reprinted with permission; Fed. Reg., 47:43 and 48:194 (Mar. 4, 1982 and Oct. 5, 1983)

TABLE 6–18. MONITORING REQUIREMENTS FOR REGULATED VOLATILE ORGANIC CHEMICALS IN PUBLIC WATER SYSTEMS IN THE UNITED STATES

Initial Monitoring

- All systems monitor at least once in 4 years

- Phase-in by size

>10 000	Begin quarterly sampling January 1988
3300–10 000	Begin quarterly sampling January 1989
<3300	Begin quarterly sampling January 1991

- Ground-water systems

Sample locations	Sample at each entry point to the distribution system representative of each well, after treatment
Number of samples	One sample per well quarterly for 1 year; if first sample detects no VOCs, state can reduce monitoring to that sample; state discretion on confirmation sample

- Surface-water systems

Sample locations	Sample in distribution system at points representative of each source, after treatment
Number of samples	One sample per source quarterly for 1 year; state discretion on confirmation samples

Repeat monitoring

- Repeat monitoring based on vulnerability* and if VOCs were detected† during initial monitoring

	Frequency	
Status§	*Ground Water*	*Surface Water*
VOCs not detected and source not vulnerable	Repeat in 5 years	State discretion
VOCs not detected and source vulnerable		
Systems with >500 service connections	Repeat in 3 years	Repeat in 3 years
Systems with ≤500 service connections	Repeat in 5 years	Repeat in 5 years
VOCs detected	Sample quarterly	Sample quarterly

* State determines.

† Detected is defined as 0.5 μg/L or greater.

§ State will review vulnerability status every 3 years for systems >500 service connections and every 5 years for systems ≤500 service connections.

Source: American Water Works Association, 1988, New Dimensions in Safe Drinking Water. Copyright AWWA. Reprinted with permission. Original source: U.S. Environmental Protection Agency

TABLE 6–19. UNREGULATED CHEMICAL CONTAMINANTS TO BE MONITORED IN PUBLIC SUPPLY SYSTEMS IN THE UNITED STATES

List 1: Monitoring Required for All Systems

Bromobenzene
Bromodichloromethane
Bromoform
Bromomethane
Chlorobenzene
Chlorodibromomethane
Chloroethane
Chloroform
Chloromethane
o-Chlorotoluene
p-Chlorotoluene
Dibromomethane
m-Dichlorobenzene
o-Dichlorobenzene
trans-1,2-Dichloroethylene
cis-1,2-Dichloroethylene
Dichloromethane
1,1-Dichloroethane
1,1-Dichloropropene
1,2-Dichloropropane
1,3-Dichloropropane
1,3-Dichloropropene
2,2-Dichloropropane
Ethylbenzene
Styrene
1,1,2-Trichloroethane
1,1,1,2-Tetrachloroethane
1,1,2,2-Tetrachloroethane
Tetrachloroethylene
1,2,3-Trichloropropane
Toluene
p-Xylene
o-Xylene
m-Xylene

List 2: Monitoring Required for Vulnerable Systems

Ethylene dibromide (EDB)
Dibromochloropropane (DBCP)

List 3: Monitoring Required at the State's Discretion

Bromochloromethane
n-Butylbenzene
Dichlorodifluoromethane
Fluorotrichloromethane
Hexachlorobutadiene
Isopropylbenzene
p-Isopropyltoluene
Naphthalene
n-Propylbenzene
sec-Butylbenzene
tert-Butylbenzene
1,2,3-Trichlorobenzene
1,2,4-Trichlorobenzene
1,2,4-Trimethylbenzene
1,3,5-Trimethylbenzene

Source: American Water Works Association, 1988, New Dimensions in Safe Drinking Water. Copyright AWWA. Reprinted with permission. Original source: Fed. Reg. 52:130 (July 8, 1987)

TABLE 6–20. MONITORING REQUIREMENTS FOR UNREGULATED VOLATILE ORGANIC CHEMICALS IN WATER SYSTEMS IN THE UNITED STATES

Initial Monitoring	
All systems monitor within 4 years	
Phase in by size	
>10 000	Begin quarterly sampling January 1988
3300–10 000	Begin quarterly sampling January 1989
<3300	Begin quarterly sampling January 1991
Ground-water systems	
Sample locations	Sample at each entry point to the distribution systems representative of each well, after treatment
Number of samples	One sample; state discretion on confirmation sample
Surface-water systems	
Sample locations	Sample in distribution system representative of each source, after treatment
Number of samples	One sample each quarter per source for one year; state discretion on confirmation samples
Repeat Monitoring	
SDWA Amendments require monitoring every 5 years	

Source: American Water Works Association, 1988, New Dimensions in Safe Drinking Water. Copyright AWWA. Reprinted with permission. Original source: U.S. Environmental Protection Agency

FIGURE 6–5. FLAVOR WHEEL FOR DRINKING WATER

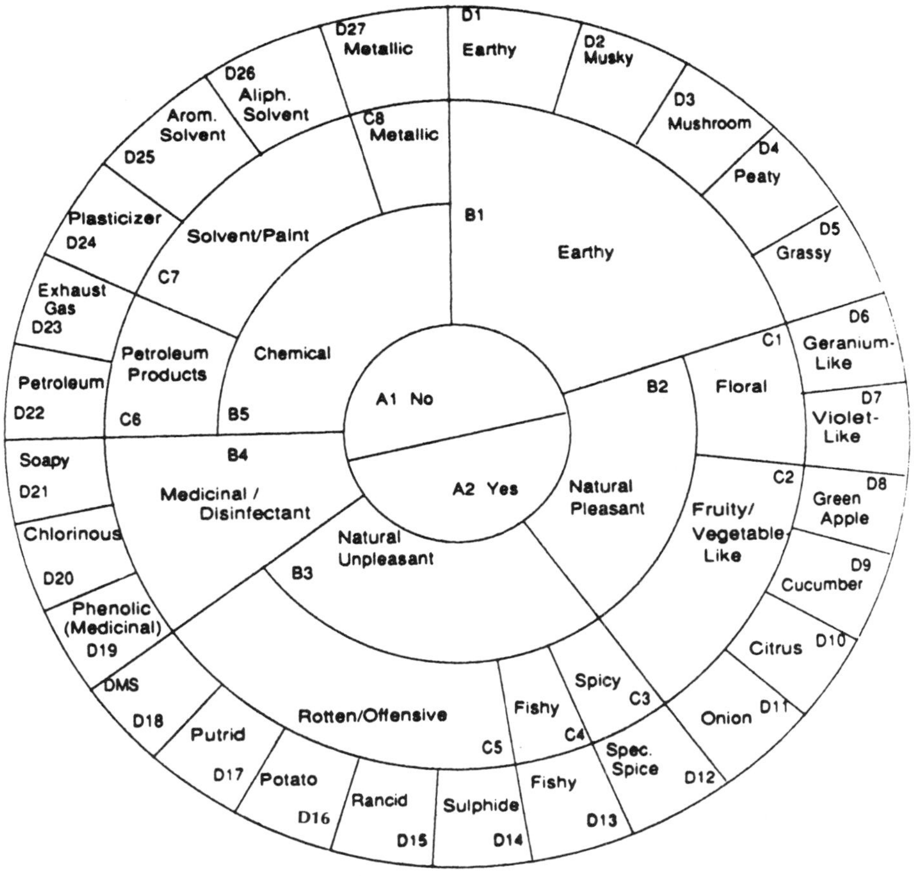

Source: International Association on Water Pollution Research and Control; Water Quality Bulletin, vol. 13 no. 2–3, April-July 1988

SECTION C. DRINKING WATER STANDARDS—WORLD

TABLE 6–21. WORLD HEALTH ORGANIZATION GUIDELINES FOR DRINKING WATER QUALITY

MICROBIOLOGICAL AND BIOLOGICAL QUALITY

Organism	Unit	Guideline Value	Remarks
I. Microbiological quality			
A. Piped water supplies			
A.1 Treated water entering the distribution system			
Faecal coliforms	Number/100 mL	0	Turbidity <1 NTU; for disinfection with chlorine, pH preferably <8.0; free chlorine residual 0.2–0.5 mg/litre following 30 minutes (minimum) contact
Coliform organisms	Number/100 mL	0	
A.2 Untreated water entering the distribution system			
Faecal coliforms	Number/100 mL	0	
Coliform organisms	Number/100 mL	0	In 98% of samples examined throughout the year—in the case of large supplies when sufficient samples are examined
Coliform organisms	Number 100 mL	3	In an occasional sample, but not in consecutive samples
A.3 Water in the distribution system			
Faecal coliforms	Number/100 mL	0	
Coliform organisms	Number/100 mL	0	In 95% of samples examined throughout the year—in the case of large supplies when sufficient samples are examined
Coliform organisms	Number/100 mL	3	In an occasional sample, but not in consecutive samples
B. Unpiped water supplies			
Faecal coliforms	Number/100 mL	0	
Coliform organisms	Number/100 mL	10	Should not occur repeatedly; if occurrence is frequent and if sanitary protection cannot be improved, an alternative source must be found if possible
C. Bottled drinking-water			
Faecal coliforms	Number/100 mL	0	Source should be free from faecal contamination
Coliform organisms	Number/100 mL	0	
D. Emergency water supplies			
Faecal coliforms	Number/100 mL	0	Advise public to boil water in case of failure to meet guideline values
Coliform organisms	Number/100 mL	0	
Enteroviruses		No guideline value set	
II. Biological quality			
Protozoa (pathogenic)	—	No guideline value set	
Helminths (pathogenic)	—	No guideline value set	
Free-living organisms (algae, others)	—	No guideline value set	

TABLE 6–21. WORLD HEALTH ORGANIZATION GUIDELINES FOR DRINKING WATER QUALITY (continued)

INORGANIC CONSTITUENTS OF HEALTH SIGNIFICANCE

Constituent	Unit	Guideline Value	Remarks
Arsenic	mg/L	0.05	
Asbestos	—	No guideline value set	
Barium	—	No guideline value set	
Beryllium	—	No guideline value set	
Cadmium	mg/L	0.005	
Chromium	mg/L	0.05	
Cyanide	mg/L	0.1	
Fluoride	mg/L	1.5	Natural or deliberately added; local or climatic conditions may necessitate adaptation
Hardness	—	No health-related guideline value set	
Lead	mg/L	0.05	
Mercury	mg/L	0.001	
Nickel	—	No guideline value set	
Nitrate	mg/L (N)	10	
Nitrite	—	No guideline value set	
Selenium	mg/L	0.01	
Silver	—	No guideline value set	
Sodium	—	No guideline value set	

ORGANIC CONSTITUENTS OF HEALTH SIGNIFICANCE

Constituent	Unit	Guideline Value	Remarks
Aldrin and dieldrin	μg/L	0.03	
Benzene	μg/L	10[a]	
Benzo[a]pyrene	μg/L	0.01[a]	
Carbon tetrachloride	μg/L	3[a]	Tentative guideline value[b]
Chlordane	μg/L	0.3	
Chlorobenzenes	μg/L	No health-related guideline value set	Odor threshold concentration between 0.1 and 3 μg/L
Chloroform	μg/L	30[a]	Disinfection efficiency must not be compromised when controlling chloroform content
Chlorophenols	μg/L	No health-related guideline value set	Odor threshold concentration 0.1 μg/L
2,4-D	μg/L	100[c]	
DDT	μg/L	1	
1,2-Dichloroethane	μg/L	10[a]	
1,1-Dichloroethene[d]	μg/L	0.3[a]	
Heptachlor and heptachlor epoxide	μg/L	0.1	
Hexachlorobenzene	μg/L	0.01[a]	
Gamma-HCH (lindane)	μg/L	3	
Methoxychlor	μg/L	30	
Pentachlorophenol	μg/L	10	
Tetrachloroethene[a]	μg/L	10[a]	Tentative guideline value[b]
Trichloroethene[d]	μg/L	30[a]	Tentative guideline value[b]
2,4,6-Trichlorophenol	μg/L	10[a,c]	Odor threshold concentration, 0.1 μg/L
Trihalomethanes		No guideline value set	See chloroform

[a] These guideline values were computed from a conservative hypothetical mathematical model which cannot be experimentally verified and values should therefore be interpreted differently. Uncertainties involved may amount to two orders of magnitude (i.e., from 0.1 to 10 times the number).
[b] When the available carcinogenicity data did not support a guideline value, but the compounds were judged to be of importance in drinking-water and guidance was considered essential, a tentative guideline value was set on the basis of the available health-related data.
[c] May be detectable by taste and odor at lower concentrations.
[d] These compounds were previously known as 1,1-dichloroethylene, tetrachloroethylene, and trichloroethylene, respectively.

TABLE 6–21. WORLD HEALTH ORGANIZATION GUIDELINES FOR DRINKING WATER QUALITY (continued)

AESTHETIC QUALITY

Constituent or Characteristic	Unit	Guideline Value	Remarks
Aluminium	mg/L	0.2	
Chloride	mg/L	250	
Chlorobenzenes and chlorophenols	—	No guideline value set	These compounds may affect taste and odor
Color	True color units (TCU)	15	
Copper	mg/L	1.0	
Detergents	—	No guideline value set	There should not be any foaming or taste and odor problems
Hardness	mg/L (as $CaCO_3$)	500	
Hydrogen sulfide	—	Not detectable by consumers	
Iron	mg/L	0.3	
Manganese	mg/L	0.1	
Oxygen—dissolved	—	No guideline value set	
pH	—	6.5–8.5	
Sodium	mg/L	200	
Solids—total dissolved	mg/L	1000	
Sulfate	mg/L	400	
Taste and odor	—	Inoffensive to most consumers	
Temperature	—	No guideline value set	
Turbidity	Nephelometric turbidity units (NTU)	5	Preferably <1 for disinfection efficiency
Zinc	mg/L	5.0	

RADIOACTIVE CONSTITUENTS

Constituent	Unit	Guideline Value	Remarks
Gross alpha activity	Bq/L	0.1	(a) If the levels are exceeded more detailed radionuclide analysis may be necessary. (b) Higher levels do not necessarily imply that the water is unsuitable for human consumption
Gross beta activity	Bq/L	1	

Source: World Health Organization, 1984, Guidelines for Drinking-Water Quality, Vol. 1, Recommendations

TABLE 6–22. CANADIAN GUIDELINES FOR DRINKING WATER QUALITY

Parameter	Maximum Acceptable Concentration in Drinking Water mg/L	Parameter	Maximum Acceptable Concentration in Drinking Water mg/L
INORGANIC PARAMETERS		ORGANIC PARAMETERS	
Antimony	—	Aldrin + dieldrin	0.0007
Arsenic	0.05	Carbaryl	0.07
Asbestos	—	Chlordane (total isomers)	0.007
Barium	1.0	2,4-D	0.1
Boron	5.0	DDT (total isomers)	0.03
Cadmium	0.005	Diazinon	0.014
Chloride	250	Dieldrin + aldrin	0.0007
Chromium	0.05	Endrin	0.0002
Copper	1.0	Heptachlor + heptachlor epoxide	0.003
Cyanide	0.2	Lindane	0.004
Fluoride	1.5	Methoxychlor	0.1
Hardness	—	Methyl parathion	0.007
Iron	0.3	Nitrilotriacetic acid (NTA)	0.05
Lead	0.05	Parathion	0.035
Manganese	0.05	Pesticides (total)	0.1
Mercury	0.001	Phenols	0.002
Nitrate (as N)	10.0	2,4,5-TP	0.01
Nitrite (as N)	1.0	Toxaphene	0.005
pH	6.5–8.5	Trihalomethanes	0.35
Selenium	0.01		
Silver	0.05	PHYSICAL PARAMETERS	
Sulphate	500		
Sulphide (as H_2S)	0.05	Color	15 TCU
Total dissolved solids	500	Odor	—
Uranium	0.02	Taste	—
Zinc	5.0	Temperature	15°C
		Turbidity	5 NTU
RADIOLOGICAL PARAMETERS			
^{137}Cs (Cesium)	50 Bq/L		
^{131}I (Iodine)	10 Bq/L		
^{226}Ra (Radium)	1 Bq/L		
^{90}Sr (Strontium)	10 Bq/L		
^{3}H (Tritium)	40 000 Bq/L		

Source: Canadian Council of Resource and Environment Ministers, March 1987. Canadian Water Quality Guidelines

TABLE 6–23. COMPARISON OF UNITED STATES PRIMARY DRINKING WATER REGULATIONS WITH CANADIAN, EEC, AND WHO GUIDELINES

Substance	US Maximum Contaminant Level* mg/L	Canadian Maximum Acceptable Limit† mg/L	EEC Maximum Admissible Concentration† mg/L	WHO Guideline Value mg/L
Inorganics				
Arsenic	0.05	0.05	0.05	0.05
Barium	1.0	1.0	0.1	NS
Cadmium	0.01	0.005	0.005	0.005
Chromium	0.05	0.05	0.05	0.05
Fluoride	4.0	1.5	NS	1.5
Lead	0.05	0.05	0.05	0.05
Mercury	0.002	0.001	0.001	0.001
Nitrate	10.0	10.0	50	10.0
Selenium	0.01	0.01	0.01	0.01
Silver	0.05	0.05	0.01	NS
Microbials				
Coliforms—*organisms/100 mL*	<1	10	0	0
Turbidity—*ntu*	1–5	5.0	0–4	<1

TABLE 6–23. COMPARISON OF UNITED STATES PRIMARY DRINKING WATER REGULATIONS WITH CANADIAN, EEC, AND WHO GUIDELINES (continued)

Substance	US Maximum Contaminant Level* mg/L	Canadian Maximum Acceptable Limit† mg/L	EEC Maximum Admissible Concentration† mg/L	WHO Guideline Value mg/L
Organics				
2,4-D	0.1	0.1	NS	0.001
Endrin	0.0002	0.0002	NS	NS
Lindane	0.0004	0.004	NS	NS
Methoxychlor	0.1	0.1	NS	0.001
Pesticides (total)	NS	0.1	0.005	NS
Toxaphene	0.005	0.005	NS	NS
2,4,5-TP silvex	0.01	0.01	NS	NS
Trihalomethanes	0.10	0.35	0.001	0.03 ($CHCl_3$ only)
Radionuclides				
Beta particle and photon activity	4 mrem	NS	NS	1.0 Bq/L[§]
Gross alpha particle activity	15 pCi/L	NS	NS	0.1 Bq/L[§]
Radium-226 + radium-228	5 pCi/L	1 Bq/L[§]	NS	NS
Volatile organic chemicals				
Benzene	0.005	NS	NS	0.01
Carbon tetrachloride	0.005	NS	NS	0.003
1,1-Dichloroethylene	0.007	NS	NS	0.003
1,2-Dichloroethane	0.005	NS	NS	0.01
para-Dichlorobenzene	0.075	NS	NS	NS
1,1,1-Trichloroethane	0.2	NS	NS	NS
Trichloroethylene	0.005	NS	NS	0.03
Vinyl chloride	0.002	NS	NS	NS

* Enforceable
† Nonenforceable
NS = No standard
§ Becquerels per liter
EEC = European Economic Community
WHO = World Health Organization
Source: Sayre, I.M., 1988, International Standards for Drinking Water, J. Am. Water Works Assoc., v.80, no.1. Copyright AWWA. Reprinted with permission

TABLE 6–24. COMPARISON OF UNITED STATES SECONDARY DRINKING WATER REGULATIONS WITH CANADIAN, EEC, AND WHO GUIDELINES

Substance	US Secondary Maximum Contaminant Level*	Canadian Maximum Acceptable Limit*	EEC		WHO Guideline Value
			Guide Level*	Maximum Admissible Concentration	
Chloride	250 mg/L	250 mg/L	25 mg/L	NS	250 mg/L
Color	15 cu	15 cu	1 mg Pt-Co/L	20 mg Pt-Co/L	15 cu
Copper	1 mg/L	1.0 mg/L	100 µg at treatment plant; 3,000 µg after 12 hours in piping	NS	1.0 mg/L
Corrosivity	noncorrosive				
Fluoride	2 mg/L	1.5 mg/L		Water should not be aggressive Varies according to average temperature in the area	1.5 mg/L
Foaming agents	0.5 mg/L	NS	NS	NS	
Iron	0.3 mg/L	0.3 mg/L	50 µg/L	300 µg/L	0.3 mg/L
Manganese	0.05 mg/L	0.05 mg/L	20 µg/L	50 µg/L	0.1 mg/L
Odor	3 TON		0 dilution number	2 dilution number at 54°F (12°C)	
pH	6.5–8.5	6.5–8.5	6.5–8.5	NS	6.5–8.5
Sulfate	250 mg/L	500 mg/L	25 mg/L	NS	400 mg/L
Total dissolved solids	500 mg/L	500 mg/L	NS	NS	1,000 mg/L
Zinc	5 mg/L	5 mg/L	100 µg at treatment plant; 5,000 µg after 12 hours in piping	NS	5.0 mg/L

* Nonenforceable
NS = No standard
EEC = European Economic Community
WHO = World Health Organization
Source: Sayre, I.M., 1988, International Standards for Drinking Water, J.Am.Water Works Assoc., v. 80, no.1. Copyright AWWA. Reprinted with permission

SECTION D. MUNICIPAL WATER QUALITY

TABLE 6–25. RANGE IN QUALITY OF FINISHED WATER IN PUBLIC WATER SUPPLIES OF THE 100 LARGEST CITIES IN THE UNITED STATES

[Maximum, median, and minimum values as of 1962; ND means not detected.]

Constituent or Property	Maximum	Median	Minimum
Chemical Analyses [parts per million]			
Silica (SiO_2)	72	7.1	0.0
Iron (Fe)	1.30	.02	.00
Manganese (Mn)	2.50	.00	.00
Calcium (Ca)	145	26	.0
Magnesium (Mg)	120	6.25	.0
Sodium (Na)	198	12	1.1
Potassium (K)	30	1.6	.0
Bicarbonate (HCO_3)	380	46	0
Carbonate (CO_3)	26	0	0
Sulfate (SO_4)	572	26	.0
Chloride (Cl)	540	13	.0
Fluoride (F)	7.0	.4	.0
Nitrate (NO_3)	23	.7	.0
Dissolved solids	1,580	186	22
Hardness as $CaCO_3$	738	90	0
Noncarbonate hardness as $CaCO_3$	446	34	0
Specific conductance ..... micromhos at 25°C	1,660	308	18
pH ..... pH units	10.5	7.5	5.0
Color ..... color units	24	2	0
Turbidity	13	0	0
Spectrographic Analyses [micrograms per liter]			
Silver (Ag)	7.0	0.23	ND
Aluminum (Al)	1,500	54	3.3
Boron (B)	590	31	2.5
Barium (Ba)	380	43	1.7
Chromium (Cr)	35	.43	ND
Copper (Cu)	250	8.3	<.61
Iron (Fe)	1,700	43	1.9
Lithium (Li)	170	2.0	ND
Manganese (Mn)	1,100	5.0	ND
Molybdenum (Mo)	68	1.4	ND
Nickel (Ni)	34	<2.7	ND
Lead (Pb)	62	3.7	ND
Rubidium (Rb)	67	1.05	ND
Strontium (Sr)	1,200	110	2.2
Titanium (Ti)	49	<1.5	ND
Vanadium (V)	70	<4.3	ND
Radiochemical Analyses			
Beta activity ..... picocuries per liter	130	7.2	<1.1
Radium (Ra) ..... do	2.5	<.1	<.1
Uranium (U) ..... micrograms per liter	250	.15	<.1

Source: U.S. Geological Survey

TABLE 6–26. QUALITY LIMITS OF FINISHED WATER IN PUBLIC WATER SUPPLIES OF THE 100 LARGEST CITIES IN THE UNITED STATES

[Data as of 1962; ND means not detected]

Constituent or Property	Water Supplies Having Less Than Stated Concentration	
	Concentration	Percent of Water Supplies
Chemical Analyses [parts per million]		
Silica (SiO_2)	30	94
Iron (Fe)	.25	98
Manganese (Mn)	.10	95
Calcium (Ca)	50	93
Magnesium (Mg)	20	96
Sodium (Na)	50	93
Potassium (K)	5.0	93
Bicarbonate (HCO_3)	150	91
Carbonate (CO_3)	1.0	86
Sulfate (SO_4)	100	93
Chloride (Cl)	50	93
Fluoride (F)	1.0	92
Nitrate (NO_3)	5.0	93
Dissolved solids	500	97
Do	250	86
Hardness as $CaCO_3$	200	94
Noncarbonate hardness as $CaCO_3$	75	94
Specific conductance (micromhos at 25°C)	500	93
pH ... pH units	9.0	90
Color ... color units	10	96
Turbidity	3	94

Constituent or Property	Water Supplies Having Less Than Stated Concentration	
	Concentration	Percent of Water Supplies
Spectrographic Analyses [micrograms per liter]		
Silver (Ag)	0.50	95
Aluminum (Al)	500	87
Boron (B)	100	94
Barium (Ba)	100	94
Chromium (Cr)	5.0	95
Copper (Cu)	100	94
Iron (Fe)	150	94
Lithium (Li)	50	96
Manganese (Mn)	100	97
Molybdenum (Mo)	10	96
Nickel (Ni)	10	95
Phosphorus (P)	ND	92
Lead (Pb)	10	95
Rubidium (Rb)	5.0	91
Strontium (Sr)	500	96
Titanium (Ti)	5.0	96
Vanadium	10	91
Radiochemical Analyses		
Beta activity picocuries per liter	20	92
Radium (Ra) picocuries per liter	.2	91
Uranium (U) micrograms per liter	2.0	93

Source: U.S. Geological Survey

TABLE 6–27. QUALITY OF RAW AND TREATED WATER IN PUBLIC WATER SUPPLIES OF THE 100 LARGEST CITIES IN THE UNITED STATES

[Data as of 1962]

	Raw-Water Supplies[1]		Treated-Water Supplies	
	Population Served (millions)	Number of Cities	Population Served (millions)	Number of Cities
Hardness (ppm):				
Less than 61	21	29	23	30
61–120	15	16	22	41
121–180	16	22	11	16
More than 180	8	27	3.7	13
Dissolved solids (ppm):				
Less than 100	21	27	21	27
101–250	23	38	28	48
251–500	11	29	8	22
More than 500	1.5	6	1	3
pH:				
Less than 7.0	16	18	14	9
7.0–9.0	42	80	38	74
More than 9.0			7	17

[1] A few cities are not included because data are lacking.

Source: U.S. Geological Survey

TABLE 6–28. STANDARDS FOR RAW WATER USED AS SOURCES OF DOMESTIC WATER SUPPLY

Constituents		Excellent Source of Water Supply, Requiring Disinfection Only, as Treatment	Good Source of Water Supply Requiring Usual Treatment Such as Filtration and Disinfection	Poor Source of Water Supply, Requiring Special or Auxiliary Treatment and Disinfection
B.O.D. (5-day) ppm	Monthly Average:	0.75	1.5–2.5	2.0–5.5
	Maximum Day, or sample:	1.0	3.0–3.5	4.0–7.5
Coliform MPN per 100 ml	Monthly Average:	50–100	240–5,000	10,000–20,000
	Maximum Day, or sample:	. . . .	<20%>5,000 <5%>20,000	
Dissolved oxygen	ppm. average	4.0–7.5	2.5–7.0	2.5–6.5
	% saturation	50–75	25–75	. . . .
pH	Average	6.0–8.5	5.0–9.0	3.8–10.5
Chlorides, max.	ppm.	50	250	500
Iron and manganese together	Max. ppm.	0.3	1.0	15
Fluorides	ppm.	1.0	1.0	1.0
Phenolic compounds	Max. ppm.	none	.005	.025
Color	ppm.	0–20	20–70	150
Turbidity	ppm.	0–10	40–250	. . . .

Source: Calif. State Water Pollution Control Board, 1952

TABLE 6–29. SUMMER TEMPERATURES OF SELECTED MUNICIPAL WATER SUPPLIES IN THE UNITED STATES

Surface Water Sources

Location	Temperature at Main Outlet, °F			
	June	July	August	September
Atlanta	78.1	83.5	79.5	77.8
Baltimore	61.0	66.0	70.0	64.0
Birmingham	78.0	82.0	81.0	79.0
Boston	68.3	74.3	73.4	69.4
Buffalo	62.0	71.0	73.0	66.0
Chicago	55.4	68.0	69.4	62.5
Cincinnati	76.0	82.0	81.0	77.0
Cleveland	58.0	68.0	73.5	71.0
Detroit	64.0	75.0	74.0	68.0
Kansas City	84.0	93.0	91.0	85.0
Louisville	77.0	82.0	82.0	77.0
Nashville	84.0	88.0	88.0	84.0
New Orleans	86.0	89.0	90.0	90.0
Oakland	59.0	62.0	64.0	64.0
Philadelphia	71.0	79.0	77.0	72.0
Pittsburgh	75.2	80.6	80.6	75.2
Sacramento	70.7	70.7	80.6	77.0
St. Louis	77.0	85.0	83.0	75.0
Washington	43.0	67.0	73.0	75.0

Ground Water Sources

Location	Temperature at Main Outlet, °F			
	June	July	August	September
Albuquerque, N. M.	72.0	72.0	72.0	72.0
Aurora, Ill.	60.0	60.0	60.0	60.0
Camden, N. J.	58.0	58.0	58.0	58.0
El Paso, Tex.	84.0	85.0	85.0	84.0
Fresno, Calif.	72.0	72.0	72.0	72.0
Houston, Tex.	84.0	84.0	84.0	84.0
Jacksonville, Fla.	84.8	86.3	86.7	82.4
Kalamazoo, Mich.	52.0	52.0	52.0	52.0
Lafayette, Ind.	53.0	53.0	53.0	53.0
Lansing, Mich.	57.5	58.0	59.0	59.0
Lincoln, Neb.	58.0	59.0	59.0	59.0
Lowell, Mass.	50.0	50.0	50.0	50.0
Madison, Wis.	53.0	52.0	52.0	53.0
Marion, Ind.	54.0	54.0	55.0	55.0
Montgomery, Ala.	70.0	70.0	71.0	71.0
Pensacola, Fla.	70.0	70.0	70.0	70.0
Peoria, Ill.	56.0	56.0	56.0	54.0
Pontiac, Mich.	55.0	55.0	55.0	55.0
San Antonio, Tex.	76.0	76.0	76.0	76.0
Sioux Falls, S. D.	55.0	55.0	55.0	55.0

Source: U. S. Dept. of Commerce

TABLE 6–30. QUALITY OF WATER SUPPLIED BY MUNICIPAL WATER SYSTEMS IN THE UNITED STATES–1984

[Average temperature, hardness, alkalinity, and pH; selected systems only]

State and Water Utility	Avg Temp Raw °F	Hardness Raw mg/L	($CaCO_3$) Finished mg/L	Alkalinity Raw mg/L	($CaCO_3$) Finished mg/L	pH Raw	pH Finished
ALABAMA							
BIRMINGHAM	60.8	75	104	73	77	7.9	8.2
MONTGOMERY	54.0	18	30	15	32	6.7	8.6
ALASKA							
ANCHORAGE	37.4	63	61	52	42	7.6	7.2
ARIZONA							
PHOENIX/PHOENIX WTR & SWR DEPT	60.1	190	190	125	105	8.2	7.6
TUCSON	75.2	141	141	150	150	7.5	7.5
ARKANSAS							
FORT SMITH	56.5	39	18	14	28	6.6	9.0
LITTLE ROCK	64.2	9	19	7	11	6.7	7.7
CALIFORNIA							
ANAHEIM	64.0	331	345	123	186	8.3	7.7
BURBANK	60.0	165		159		7.6	
CONCORD/CONTRA COSTA WATER DEPT	63.0	80	68	64	59	8.0	8.5
CORTE MADERA/MARIN MUNIC WTR DIST	58.1	57	67	58	71	7.4	8.7
FREMONT/ALAMEDA CNTY WTR DIST	62.1	97	97	65	65	7.6	8.7
FRESNO	70.0	119	119	123	123	7.8	7.8
GLENDALE	69.8	190	296	156	108	7.9	8.0
LA MESA/HELIX WTR DISTRICT	71.6	300	300	115	115	7.7	8.0
LA PUENTE/SOUTHWEST SUBURBAN WTR	68.0	220	220	180	180	7.4	7.4
LOS ANGELES/LOS ANGELES WTR & POWER	58.1	69	69	89	89	7.9	7.9
OAKLAND/EAST BAY MUNIC UTIL DIST	60.4	26	27	22	22	8.9	8.9
PALM SPRINGS/DESERT WTR AGENCY	64.9	150	150	120	120	7.0	7.0
SAN DIEGO	66.2	224	235	131	135	8.0	8.2
SAN JOSE/CALIFORNIA WTR SERV CO	68.0		181		134		7.9
SANTA BARBARA/SANTA BARBARA PUB	66.2	550	550	210	200	8.1	7.9
SANTA MONICA	68.0	400	180	220	220	7.8	8.1
SUNNYVALE	60.1	300	140	200		8.0	7.3
COLORADO							
BOULDER	54.5	13	23			7.0	7.2
COLORADO SPGS	42.3	34	36	22	29	7.5	7.7
DENVER	50.0	102	99	72	63	8.0	7.5
GREELEY	50.5	34	36	42	41	7.4	7.1
CONNECTICUT							
BRIDGEPORT	55.8	28	45	14	19	6.7	7.1
HARTFORD	55.4		20		8	6.5	7.1
NEW BRITAIN	50.0	50	35	12	14	6.5	8.3
NEW HAVEN/REGIONAL WTR AUTHORITY	53.6	50	69	46	53	7.1	6.9
DELAWARE							
NEWARK/ARTESIAN WTR CO	55.4	77	77	40	45	5.8	7.4
DISTRICT OF COLUMBIA							
WASHINGTON/WASHINGTON AQUEDUCT	62.6	127	137	77	70	8.0	7.7
FLORIDA							
CLEARWATER/PINELLAS CNTY WTR SYS	75.2	215	215	200	210	7.2	7.8
FORT LAUDERDALE	78.8	259	90	231	59	7.6	9.6
JACKSONVILLE	79.0	248	248	139	138	7.5	7.5
MIAMI	77.0	250	66	224	37	7.3	9.0
ORLANDO	75.2	125	125	120	120	7.8	7.4
POMPANO BEACH	68.0	240	63	198	30	7.1	8.8
TALLAHASSEE	68.0		143				7.8
TAMPA	72.5	107	151	90	85	7.5	7.7
W. PALM BEACH/CITY OF WEST PALM	68.2	71	102	58	58	8.1	8.4
WINTER HAVEN	74.5	140	140	125	125	8.1	7.8

TABLE 6–30. QUALITY OF WATER SUPPLIED BY MUNICIPAL WATER SYSTEMS IN THE UNITED STATES–1984 (continued)

[Average temperature, hardness, alkalinity, and pH; selected systems only]

State and Water Utility	Avg Temp Raw °F	Hardness Raw mg/L	($CaCO_3$) Finished mg/L	Alkalinity Raw mg/L	($CaCO_3$) Finished mg/L	pH Raw	pH Finished
GEORGIA							
ACWORTH/WYCKOFF TREATMENT DIV	61.3	17	30	15	22	6.7	7.5
MARIETTA/QUARLES TREATMENT DIV	53.6	15	25	10	18	6.8	8.7
STONE MOUNTAIN/DEKALB CNTY WTR	64.4	14	28	11	17	6.8	8.9
HAWAII							
HONOLULU			60		60		8.0
PEARL HARBOR	72.0	121		69		6.9	
IDAHO							
BOISE			117		223		7.6
ILLINOIS							
CHAMPAIGN/NORTHERN ILLINOIS WTR	53.6	258	80	336	110	7.7	9.0
EAST ST. LOUIS/INTERURBAN DIST	60.8	228	228	159	144	7.7	7.2
PEORIA/PEORIA DIST/ILL-AMER WTR	53.6	250	250	170	161	8.0	7.6
SPRINGFIELD	60.8	195	105	140	25	8.2	9.8
INDIANA							
BLOOMINGTON	59.0	64	80	28	34	7.3	8.0
FORT WAYNE	55.4	272	97	200	25	7.9	9.7
INDIANAPOLIS	57.2		252	170	159	8.1	7.6
SOUTH BEND	51.8	340	340	260	260	8.5	7.8
IOWA							
CEDAR RAPIDS	53.6	276	129	217	67	7.7	9.5
DAVENPORT/DAVENPORT WATER CO	55.9	200	200	155	129	8.2	7.1
DES MOINES	51.1	304	145	239	63	8.3	9.3
SIOUX CITY	54.0	445	445	267	267	7.2	7.5
KANSAS							
KANSAS CITY	62.6	294	272	205	215	8.1	8.0
MISSION/JOHNSON CNTY WTR DIST 1	59.0	245	121	191	54	7.8	9.1
TOPEKA	57.2	250	114	195	79	8.2	9.4
KENTUCKY							
LEXINGTON/KENTUCKY-AMERICAN WTR	59.9	165	148	66	72	7.7	8.1
LOUISVILLE	57.2	138	148	66	65	7.5	8.5
LOUISIANA							
BATON ROUGE/BATON ROUGE WTR WKS		5	5		175	8.7	8.5
JEFFERSON	64.4	154	152	104	99	7.5	7.4
LAKE CHARLES	69.8	110	110	160	175	7.1	8.0
NEW ORLEANS	64.0	161	117	106	61	8.0	10.1
MAINE							
PORTLAND	50.0	9	9	4		6.9	6.7
MARYLAND							
ANNAPOLIS/CITY OF ANNAPOLIS WTR	64.4	25	70	20	50	6.0	8.7
BALTIMORE		47	70	40	46	7.3	8.0
GLEN BURNIE/ANNE ARUNDEL CNTY	53.6	41	56			5.1	8.5
MASSACHUSETTS							
FALL RIVER	46.0	2	3	1	5	5.7	8.5
SPRINGFIELD	45.7	11	11	8	11	6.4	6.9
WEYMOUTH	55.0	21	47	7	11	6.5	8.4
MICHIGAN							
ANN ARBOR	55.8	269	142	210	60	8.0	9.3
GRAND RAPIDS	45.9	140	141	115	106	8.4	7.6
LANSING	51.8	412	88	318	39	7.0	9.4
SAGINAW		96	106	82	84	8.0	8.2

TABLE 6–30. QUALITY OF WATER SUPPLIED BY MUNICIPAL WATER SYSTEMS IN THE UNITED STATES–1984 (continued)

[Average temperature, hardness, alkalinity, and pH; selected systems only]

State and Water Utility	Avg Temp Raw °F	Hardness Raw mg/L	($CaCO_3$) Finished mg/L	Alkalinity Raw mg/L	($CaCO_3$) Finished mg/L	.........pH........... Raw	Finished
MINNESOTA							
BLOOMINGTON	50.0	310	90	315	92	7.6	8.3
DULUTH	39.6	45	45	43	36	7.9	7.1
ST. PAUL	51.8	175	92	166	62	8.1	8.4
MISSISSIPPI							
JACKSON	65.7	19	52	15	17	6.5	8.8
MISSOURI							
INDEPENDENCE/MISSOURI WTR CO	59.5	297	119	227	48	7.3	9.6
KANSAS CITY/KANSAS CITY, MO WTR	55.0	261	173	182	85	8.2	9.5
SPRINGFIELD	61.5	150	152	132	123	7.6	7.3
ST. LOUIS/ST. LOUIS CNTY WTR CO	57.2	221	124	155	48	8.3	9.5
LINCOLN	54.0	240	240	180	180	7.8	7.8
OMAHA	53.2	286	185	177	68	8.2	9.0
NEVADA							
LAS VEGAS	59.4	288	287	128	130	8.0	7.9
RENO/SIERRA PACIFIC POWER CO	53.6	40	40	40	27	7.7	6.9
NEW HAMPSHIRE							
MANCHESTER	59.0	11	11	3	4	6.3	7.2
NASHUA/PENNICHUCK WTR WKS	59.0	30	50	9	16	6.4	7.3
NEW JERSEY							
CLIFTON/PASSAIC VALLEY WTR COMM	54.0	86	85	58	55	7.2	7.1
EAST ORANGE	50.0	280	280	158	158	7.7	7.7
ELIZABETH/ELIZABETHTOWN WTR CO	51.8	73	86	47	40	7.5	7.2
HARRINGTON PARK/HACKENSACK WTR	55.4	120	120	85	80	7.4	7.9
PARSIPPANY	52.0	177	177	139	139	7.1	7.1
SHORT HILLS/COMMONWEALTH WTR CO	55.0	72	135	42	77	7.8	7.4
TOMS RIVER	65.3	12	60	38	70	6.2	7.4
WANAQUE/NORTH NJ DIST WTR SUPPLY	50.0	28	36	16	20	6.7	7.4
NEW MEXICO							
ALBUQUERQUE	69.8	120	120	138	138	7.3	7.3
SANTA FE/SANGRE DE CRISTO WTR CO	59.0	150	150	200	200	7.8	7.8
NEW YORK							
ALBANY	48.4	43	54	38	48	7.3	8.9
BUFFALO/TONAWANDA WTR DEPT	55.4	140	135	95	90	8.2	7.8
EAST MEADOW/HEMPSTEAD WTR DEPT		30	30	5	30	5.6	8.6
LAKE SUCCESS/JAMAICA WATER CO.		50	50	30	50	6.3	7.2
MASSAPEQUA	62.1	8	13	6	30	5.8	7.2
NEW YORK/NEW YORK BUR WTR SUPPLY			65	37		7.3	
ROCHESTER/MONROE CNTY WTR AUTH	55.9	130	130	95	90	7.8	7.4
SYRACUSE	52.0	120	120	100	100	8.2	8.1
YONKERS	59.0	110	100	80	65	7.1	6.5
NORTH CAROLINA							
CHARLOTTE	64.9	13	28	12	19	7.2	9.2
GREENSBORO	64.4	27	44	24	26	7.0	7.7
RALEIGH	68.0	20	40	29	30	6.8	7.4
WINSTON-SALEM/CITY/CNTY UTIL COM	60.8	16	30	16	20	7.0	7.4
NORTH DAKOTA							
FARGO	44.6	289	123	203	84	8.1	9.1
OHIO							
AKRON	54.3	112	112	78	74	7.6	7.3
CINCINNATI	60.1	112	130	44	54	7.6	8.5
CLEVELAND	50.9	125	125	92	84	8.0	7.5
DAYTON		362	149	278	62	7.5	8.6
TOLEDO		127	74	93	38	8.1	9.2

TABLE 6–30. QUALITY OF WATER SUPPLIED BY MUNICIPAL WATER SYSTEMS IN THE UNITED STATES–1984 (continued)

[Average temperature, hardness, alkalinity, and pH; selected systems only]

State and Water Utility	Avg Temp Raw °F	Hardness Raw mg/L	($CaCO_3$) Finished mg/L	Alkalinity Raw mg/L	($CaCO_3$) Finished mg/L	pH Raw	pH Finished
OKLAHOMA							
OKLAHOMA CITY	63.9	154	101	97	40	8.2	10.3
TULSA	66.2	140	136	97	102	8.0	8.2
OREGON							
EUGENE	55.4	24	22	23	22	7.5	7.3
MEDFORD	44.6	33	33	35	35	6.8	6.8
PORTLAND	50.0	12	12	12	10	7.1	6.8
PENNSYLVANIA							
ALLENTOWN	55.9	176	199	128	144	7.7	7.6
BRYN MAWR/PHILADELPHIA SUBURBAN	58.3	109	208	35	38	7.0	7.4
HERSHEY/RIVERTON CONSOL WTR CO	58.6	145	145	121	110	8.0	7.4
LANCASTER	57.2	195	208	135	126	7.8	7.6
PHILADELPHIA	55.4	148	132	64	53	7.6	7.0
PITTSBURGH/WESTERN PENNSYLVANIA	56.8	111	122	21	31	7.3	7.3
RHODE ISLAND							
NEWPORT	60.8	60	70	22	25	6.5	7.5
WEST WARWICK/KENT CNTY WTR AUTH		32	32	13	13	5.8	5.8
SOUTH CAROLINA							
ANDERSON	66.0	6	6	9	10	7.0	7.1
CHARLESTON	68.0	29	29	24	30	6.8	8.2
SPARTANBURG	62.6	11	20	11	11	6.9	7.1
SOUTH DAKOTA							
RAPID CITY	55.0	283	283	180	180	7.7	7.5
TENNESSEE							
CHATTANOOGA/TENNESSEE-AMERICAN	64.9	68	77	53	50	7.1	7.2
KNOXVILLE/KNOXVILLE UTILS BRD	60.8	81	83	78	70	7.6	7.5
MEMPHIS	63.3	47	47	55	55	6.4	7.2
NASHVILLE/METRO DEPT WTR & SWR	60.8	90	90	65	65	7.5	8.0
TEXAS							
ARLINGTON	69.8	109	110	90	90	8.0	8.2
DALLAS	64.4	130	85	110	50	8.0	8.9
EL PASO	68.0	250	150	192	80	8.2	8.7
FORT WORTH	66.2	105	107	87	93	8.0	8.5
LUBBOCK	60.1	223	222	177	166	8.3	7.8
SAN ANTONIO	75.7	250		215		7.2	
WICHITA FALLS	69.8	125	70	125	50	8.1	9.4
UTAH							
OGDEN	53.6	170	125	131	118	7.4	7.1
SALT LAKE CITY/SALT LAKE CNTY CO	46.0	159	156	137	132	7.4	7.3
VERMONT							
BURLINGTON	50.0	59	59	47	47	7.6	8.0
VIRGINIA							
MERRIFIELD/FAIRFAX CNTY WTR AUTH	59.0	60	90	33	44	7.3	7.5
NEWPORT NEWS	60.8	80	80	55	50	7.4	7.1
NORFOLK	64.9	45	69	40	37	6.8	7.0
RICHMOND/CITY OF RICHMOND	57.2	70	70	45	35	7.6	7.5
WASHINGTON							
EVERETT	51.1	12	12	11	20	6.5	7.2
SEATTLE/SEATTLE WTR DEPT	47.3	15	17	11	16	7.2	7.7
TACOMA	50.0	15	15	20	20	6.9	6.9
VANCOUVER	53.0	98	98			6.7	6.7
WEST VIRGINIA							
CHARLESTON/W VA WTR CO - KANAWHA	57.9	47	62	19	26	6.8	8.7
HUNTINGTON	59.0	109	121	39	38	7.4	7.3
WHEELING	56.5	109	129	32	40	7.4	8.9

TABLE 6–30. QUALITY OF WATER SUPPLIED BY MUNICIPAL WATER SYSTEMS IN THE UNITED STATES–1984 (continued)

[Average temperature, hardness, alkalinity, and pH; selected systems only]

State and Water Utility	Avg Temp Raw °F	Hardness Raw mg/L	($CaCO_3$) Finished mg/L	Alkalinity Raw mg/L	($CaCO_3$) Finished mg/L	.........pH........... Raw	Finished
WISCONSIN							
GREEN BAY	46.4		130	132	118	8.2	7.6
MADISON	50.0	350	350	300	300	7.4	7.4
MILWAUKEE		93	93	112	104	8.2	7.5
RACINE	44.3	140	140	112	107	8.3	7.7
WYOMING							
CASPER	48.9	175	208		124	8.1	7.4
CHEYENNE	61.9	110	110	63	65	7.2	7.1

Source: American Water Works Association 1984 Water Utility Operating Data. Copyright AWWA. Reprinted with permission.

TABLE 6–31. RADON IN PUBLIC DRINKING WATER SUPPLIES IN THE UNITED STATES

[Population-weighted average concentrations]

State	^{222}Rn Concentration—*p*Ci/L Systems With <1000 People	System With >1000 People
Alabama	160	160
Alaska	100	100
Arizona	120	320
Arkansas	75	75
California	500	500
Colorado	380	380
Connecticut	1500	770
Delaware	100	126
Florida	1000	148
Georgia	1100	150
Hawaii	50	50
Idaho	256	256
Illinois	100	167
Indiana	105	105
Iowa	250	200
Kansas	250	106
Kentucky	250	110
Louisiana	180	180
Maine	10 000	2000
Maryland	700	450
Massachusetts	1500	770
Michigan	105	105
Minnesota	210	210
Mississippi	150	82
Missouri	300	100
Montana	500	328
Nebraska	300	290
Nevada	550	550
New Hampshire	1400	1183
New Jersey	150	300
New Mexico	200	180
New York	500	132
North Carolina	1100	278
North Dakota	300	150
Ohio	200	169
Oklahoma	250	160
Oregon	300	264
Pennsylvania	1000	720
Rhode Island	3400	1151
South Carolina	1100	276
South Dakota	300	290
Tennessee	100	24
Texas	150	150
Utah	500	360
Vermont	250	656
Virginia	700	450
Washington	300	264
West Virginia	1000	720
Wisconsin	750	234
Wyoming	880	415
American Samoa	50	50
Guam	50	50
Puerto Rico	500	200
United States	780 (29 Bq/L)*	240 (8.9 Bq/L)

*The international unit of activity is the Becquerel (Bq), which is approximately equal to 27 pCi.

Source: Cothern, C.R., 1987, Estimating the Health Risks of Radon in Drinking Water, J. Am. Water Works Assoc., v. 79, no. 4. Copyright AWWA. Reprinted with permission

TABLE 6–32. NUMBER OF PUBLIC DRINKING WATER SUPPLIES IN THE UNITED STATES EXCEEDING VARIOUS LEVELS OF RADON*

Lifetime Risk Level	Radon Concentration pCi/L	Estimated Number That Exceed the Concentration in Column 2	
		Public Drinking Water Supplies	Population thousands
10^{-3}	10 000	500–4000	20–300
10^{-4}	1 000	1000–10 000	200–4000
10^{-5}	100	5000–30 000	10 000–100 000
10^{-6}	10	10 000–40 000	50 000–100 000

*Rounded off to one significant figure.
Source: Cothern, C.R., 1987, Estimating the Health Risk of Radon in Drinking Water, J.Am. Water Works Assoc., v.79, no.4. Copyright AWWA. Reprinted with permission

TABLE 6–33. OCCURRENCE OF RADON IN WELL WATER IN THE UNITED STATES

[Range of detected radon concentrations and corresponding mean radon levels]

State	Number of Wells Sampled	Range of Detected Radon Levels pCi/L	Mean Radon Concentration pCi/L	Associated Error ±pCi/L
Arizona	5	434–681	582	105
California	44	<100–2,003	589	91
Connecticut	3	757–984	841	98
Iowa	6	All <100	12	75
Illinois	16	182–714	449	115
Indiana	28	<100–624	324	106
Massachusetts	28	<100–3,288	1,145	101
New Hampshire	12	880–4,609	1,716	134
New Jersey	113	<100–3,805	394	79
New Mexico	36	<100–678	253	266
Ohio	10	<100–343	148	116
Pennsylvania	64	<100–4,622	1,570	89
Rhode Island	3	640–787	702	61
Virginia	2	465–468	467	53
West Virginia	7	<100–281	93	42
Cumulative	377	<100–4,622	686	104

Source: Dixon, K.L., and Lee, R.G., 1988, Occurrence of Radon in Well Supplies, J.Am. Water Works Assoc., vol 80, no.7. Copyright AWWA. Reprinted with permission

TABLE 6–34. ALUMINUM IN PUBLIC DRINKING WATER SUPPLIES IN THE UNITED STATES

[Finished water; by USEPA region and population category; for map of USEPA regions see Fig. 10-2]

Category	Samples	Samples With Concentrations >0.05 mg/L percent	Samples With Concentrations >0.014 mg/L			Overall Median mg/L
			percent	Median mg/L	Maximum mg/L	
Region						
I	46	13	33	0.043	0.179	<0.014
II	71	25	44	0.066	0.249	<0.014
III	123	37	54	0.070	2.670	0.022
IV	80	54	68	0.161	0.449	0.060
V	100	51	64	0.082	2.160	0.051
VI	35	29	60	0.040	0.889	0.029
VII	53	2	13	0.026	0.051	<0.014
VIII	105	30	47	0.083	2.580	<0.014
IX	89	29	52	0.053	1.167	0.020
X	14	0	7			<0.014
Population served						
25–9999	286	15	28	0.051	1.167	<0.014
10 000–99 999	92	39	61	0.087	2.580	0.023
100 000–999 999	222	48	66	0.094	2.670	0.045
≥1 000 000	116	38	62	0.058	0.402	0.033

Source: Miller, R.G. and others, 1984, The Occurrence of Aluminum in Drinking Water, J. Am. Water Works Assoc., v. 76, no.1. Copyright Am. Water Works Assoc. Reprinted with permission

TABLE 6–35. DETECTIONS OF GIARDIA CYSTS IN SOURCE WATERS OF PUBLIC DRINKING WATER SUPPLIES IN THE UNITED STATES

Classification	Samples	No. Sites	No. Positive Samples	No. Positive Sites	Percent	
					Positive of Samples	Positive of Sites
Creeks	444	75	181	38	41	51
Rivers	449	74	163	38	36	51
Lakes	829	49	138	19	17	39
Springs[a]	84	6	16	2	19	33
Wells[a]	63	40	2	2	3	5

[a] Samples represent finished water. Most water from springs and wells is unfiltered and may or may not be disinfected before consumption.
Source: U.S. Environmental Protection Agency, 1987; Hibler, 1987

TABLE 6–36. DETECTIONS OF GIARDIA CYSTS IN FINISHED DRINKING WATER SUPPLIES OF THE UNITED STATES

Classification	Samples	No. Sites	No. Positive Samples	No. Positive Sites	Percent	
					Positive of Samples	Positive of Sites
Unfiltered, chlorinated	1,214	94	80	16	6.6	17
Direct filtration[a]	615	92	148	17	24.0	18.5
Conventional treatment	357	86	12	5	3.4	5.8
Slow sand and diatomaceous earth filtration	18	3	0	0	0	0
Commercial filters and/or pressure filters	33	12	4	2	12.1	16.7
Cartridge filters	51	13	11	7	21.6	53.8
Infiltration galleries	37	16	7	5	18.9	31.3
Filter type unknown	83	24	15	6	18.0	25.0

[a] May or may not include coagulation or disinfection. Number of systems applying coagulant and/or polymer, or whether disinfection was interrupted, could not be determined.
Source: U.S. Environmental Protection Agency, 1987; Based on data collected from 1979–1986, Hibler, 1987

TABLE 6–37. POPULATION SERVED WITH ADJUSTED AND NATURAL FLUORIDATED WATER IN THE UNITED STATES

[BY USEPA REGION AND STATE; AS OF DECEMBER 31, 1985]

	Total Population*	Population Served By Public Water Supply**	Population Served Fluoridated Water	% Public Water Supply Population Drinking Fluoridated Water	Rank
UNITED STATES	**243,195,000**	**211,730,873**	**130,172,334**	**61**	
REGION I	**12,699,500**	**11,272,500**	**6,998,513**	**62**	
CONNECTICUT	3,182,000	2,558,000	2,365,309	92	10
MAINE	1,169,000	804,000	407,297	51	37
MASSACHUSETTS	5,826,500	5,826,500	3,112,368	53	35
NEW HAMPSHIRE	1,013,000	749,000	125,367	17	47
RHODE ISLAND	971,000	928,000	743,319	80	20
VERMONT	538,000	407,000	244,853	60	30
REGION II	**28,641,500**	**27,540,000**	**15,364,770**	**56**	
NEW JERSEY	7,593,000	7,593,000	1,170,047	15	50
NEW YORK	17,770,500	16,669,000	11,673,411	70	25
PUERTO RICO	3,278,000	3,278,000	2,521,312	77	22
REGION III	**25,233,000**	**21,038,000**	**14,755,219**	**70**	
DELAWARE	627,500	627,500	418,593	67	27
DISTRICT OF COLUMBIA	625,500	625,500	625,500	100	1
MARYLAND	4,427,000	3,361,000	3,226,818	96	5
PENNSYLVANIA	11,879,000	9,883,000	5,290,153	54	34
VIRGINIA	5,748,500	4,774,000	4,043,125	85	15
WEST VIRGINIA	1,927,000	1,767,000	1,151,030	65	28
REGION IV	**42,387,000**	**33,893,373**	**24,015,862**	**71**	
ALABAMA	4,036,000	3,474,000	2,925,570	84	16
FLORIDA	11,529,500	10,255,373	4,860,794	47	38
GEORGIA	6,037,500	4,235,000	4,044,438	96	6
KENTUCKY	3,727,500	2,967,000	2,467,378	83	17
MISSISSIPPI	2,619,000	2,328,000	1,093,800	47	39
NORTH CAROLINA	6,296,000	4,448,000	3,329,736	75	23
SOUTH CAROLINA	3,358,000	2,532,000	2,357,061	93	9
TENNESSEE	4,783,000	3,654,000	2,937,085	80	19
REGION V	**45,889,000**	**38,090,000**	**34,944,342**	**92**	
ILLINOIS	11,545,000	11,407,000	11,105,505	97	2
INDIANA	5,502,000	3,765,000	3,660,005	97	3
MICHIGAN	9,114,000	6,825,000	6,218,797	91	11
MINNESOTA	4,203,000	3,689,000	2,998,883	81	18
OHIO	10,748,000	9,109,000	8,033,399	88	13
WISCONSIN	4,780,000	3,295,000	2,927,753	89	12
REGION VI	**28,166,000**	**26,253,000**	**15,515,286**	**59**	
ARKANSAS	2,366,000	2,003,000	1,173,604	59	33
LOUISIANA	4,493,000	3,896,000	1,978,703	51	36
NEW MEXICO	1,465,000	1,178,000	787,655	67	26
OKLAHOMA	3,306,000	2,639,000	1,576,681	60	32
TEXAS	16,537,000	16,537,000	9,998,643	60	29
REGION VII	**11,971,000**	**10,050,000**	**6,632,825**	**66**	
IOWA	2,866,000	2,290,000	1,990,832	87	14
KANSAS	2,455,000	2,455,000	981,562	40	41
MISSOURI	5,050,000	4,364,000	3,097,162	71	24
NEBRASKA	1,602,000	941,000	563,269	60	31
REGION VIII	**7,624,000**	**6,912,000**	**4,453,381**	**64**	
COLORADO	3,250,000	3,250,000	3,144,949	97	4
MONTANA	821,000	567,000	164,283	29	44
NORTH DAKOTA	682,000	516,000	492,717	95	7
SOUTH DAKOTA	708,000	533,000	502,739	94	8
UTAH	1,655,000	1,655,000	33,522	2	51
WYOMING	509,000	391,000	115,171	29	43
REGION IX	**31,925,000**	**29,816,000**	**4,913,851**	**16**	
ARIZONA	3,235,000	2,980,000	613,479	21	46
CALIFORNIA	26,680,000	24,991,000	4,140,978	17	48
HAWAII	1,058,000	893,000	142,570	16	49
NEVADA	952,000	952,000	16,824	2	52
REGION X	**8,660,000**	**6,866,000**	**2,578,285**	**38**	
ALASKA	527,000	405,000	311,547	77	21
IDAHO	1,003,000	694,000	234,207	34	42
OREGON	2,694,000	1,975,000	483,930	25	45
WASHINGTON	4,436,000	3,792,000	1,548,601	41	40

* Based on 1985 Bureau of Census estimates.

** Federal Reporting Data System PWS service populations exceeded the Bureau of Census estimates for CO, DE, DC, KS, MA, NV, NJ, PR, UT, and TX.

Source: U.S. Public Health Service/Centers for Disease Control, 1988. Fluoridation Census 1985

TABLE 6–38. NUMBER OF PUBLIC WATER SYSTEMS, COMMUNITIES, AND POPULATION USING ADJUSTED OR NATURAL FLUORIDATION IN THE UNITED STATES

[BY USEPA REGION AND STATE; AS OF DECEMBER 31, 1985]

	Using Adjusted			Using Natural		
	Systems*	**Communities**	**Population**	**Systems***	**Communities**	**Population**
UNITED STATES	**8,913**	**7,772**	**121,425,572**	**3,445**	**1,909**	**9,005,262**
REGION I	**267**	**337**	**6,986,286**	**9**	**16**	**12,227**
CONNECTICUT	37	88	2,364,279	4	1	1,030
MAINE	75	90	407,297	0	0	0
MASSACHUSETTS	66	99	3,111,971	3	0	397
NEW HAMPSHIRE	10	8	114,567	2	2	10,800
RHODE ISLAND	14	21	743,319	0	0	0
VERMONT	65	31	244,853	0	0	0
REGION II	**689**	**676**	**15,249,709**	**17**	**22**	**115,061**
NEW JERSEY	28	64	1,056,302	15	20	113,745
NEW YORK	622	570	11,672,095	2	2	1,316
PUERTO RICO	39	42	2,521,312	0	0	0
REGION III	**487**	**856**	**14,750,147**	**271**	**66**	**263,572**
DELAWARE	9	11	413,225	6	2	5,368
DISTRICT OF COLUMBIA	1	1	884,000	0	0	0
MARYLAND	88	121	3,182,845	57	13	49,973
PENNSYLVANIA	140	445	5,290,153	0	0	0
VIRGINIA	121	158	3,828,894	208	50	214,231
WEST VIRGINIA	128	121	1,151,030	0	0	0
REGION IV	**1,748**	**1,301**	**22,646,348**	**276**	**123**	**1,369,514**
ALABAMA	240	174	2,865,939	26	18	59,631
FLORIDA	75	97	4,081,812	35	31	778,982
GEORGIA	288	255	4,028,398	5	4	16,040
KENTUCKY	345	226	2,467,378	0	0	0
MISSISSIPPI	195	105	1,020,704	43	13	73,096
NORTH CAROLINA	237	148	3,248,435	29	23	81,301
SOUTH CAROLINA	173	138	1,996,597	138	34	360,464
TENNESSEE	195	158	2,937,085	0	0	0
REGION V	**3,485**	**2,787**	**33,185,724**	**901**	**538**	**1,758,618**
ILLINOIS	1,338	972	10,155,738	215	153	949,767
INDIANA	450	300	3,397,565	92	78	262,440
MICHIGAN	354	240	6,094,799	96	53	123,998
MINNESOTA	653	641	2,994,771	14	12	4,112
OHIO	370	371	7,791,641	332	118	241,758
WISCONSIN	320	263	2,751,210	152	124	176,543
REGION VI	**683**	**464**	**12,345,239**	**786**	**405**	**3,170,047**
ARKANSAS	182	109	1,156,799	14	7	16,805
LOUISIANA	52	73	1,747,024	94	33	231,679
NEW MEXICO	48	18	542,042	86	41	245,613
OKLAHOMA	158	93	1,467,529	55	36	109,152
TEXAS	243	171	7,431,845	537	288	2,566,798
REGION VII	**595**	**727**	**5,991,585**	**423**	**355**	**641,240**
IOWA	292	297	1,666,149	242	192	324,683
KANSAS	47	170	848,348	83	85	133,214
MISSOURI	203	205	2,932,091	84	65	165,071
NEBRASKA	53	55	544,997	14	13	18,272
REGION VIII	**505**	**319**	**3,487,983**	**567**	**287**	**965,398**
COLORADO	145	51	2,382,836	306	110	762,113
MONTANA	25	9	60,114	65	28	104,169
NORTH DAKOTA	118	109	463,370	83	63	29,347
SOUTH DAKOTA	198	137	460,698	83	67	42,041
UTAH	11	4	27,282	18	8	6,240
WYOMING	8	9	93,683	12	11	21,488
REGION IX	**144**	**94**	**4,489,180**	**120**	**48**	**424,671**
ARIZONA	66	6	203,816	104	44	409,663
CALIFORNIA	55	82	4,140,878	1	0	100
HAWAII	10	3	142,570	0	0	0
NEVADA	13	3	1,916	15	4	14,908
REGION X	**310**	**211**	**2,293,371**	**75**	**49**	**284,914**
ALASKA	190	135	311,547	0	0	0
IDAHO	8	4	48,842	40	18	185,365
OREGON	32	25	444,230	20	19	39,700
WASHINGTON	80	47	1,488,752	15	12	59,849

* Note: System refers to fluoridating systems plus consecutive systems.
Source: U.S. Public Health Service/Centers for Disease Control, 1988, Fluoridation Census 1985

SECTION E. INDUSTRIAL WATER QUALITY

TABLE 6–39. WATER QUALITY TOLERANCE FOR CERTAIN INDUSTRIAL APPLICATIONS

[Milligrams per liter, except as indicated]

Industry	Turbidity	Color	Color + O_2 consumed	D.O.[a] (milliliters per liter)	Odor	Hardness	Alkalinity	pH	Total Solids	Fe	Mn	Fe + Mn	Al_2O_3	SiO_2	Cl	F	CO_3	HCO_3	OH	Na_2SO_4 to Na_2SO_3 (ratio)	General[b]
Air																					
Conditioning[c]	...	...	...	...	...	...	...	...	...	0.5	0.5	0.5	...	...	...	...	...	...	...	...	A,B
Baking	10	10	...	...	Low	[d]	...	...	...	0.2	0.2	0.2	...	...	...	...	...	...	...	...	C
Boiler Feed (pounds per square inch):																					
0–150	20	80	100	2	...	80	...	8.0+	3000–1000	...	...	...	5	40	...	...	200	50	50	1 to 1	...
150–250	10	40	50	0.2	...	40	...	8.5+	2500–500	...	...	...	0.5	20	...	...	100	30	40	2 to 1	...
250–400	5	5	10	0.0	...	10	...	9.0+	1500–100	...	...	...	0.05	5	...	...	40	5	30	3 to 1	...
400 and over	1	2		0.0	...	2	...	9.6+	50	...	...	...	0.01	1	...	...	20	0	15	3 to 1	...
Brewing[e]																					
Light	10	10	...	...	Low	...	75	6.5–7.0	500	0.1	0.1	0.1	...	50	100	1.0	50	...	...	...	C,D,G
Dark	10	10	...	...	Low	...	150	7.0+	1000	0.1	0.1	0.1	...	50	100	1.0	50	...	...	...	C,D,H
Canning:																					
Legumes	10	...	...	...	Low	25–75	...	7.5+	850	0.2	0.2	0.3	...	...	...	1.0	...	...	...	...	C
General	10	...	...	...	Low	50–400	...	7.5+	850	0.2	0.2	0.3	...	...	...	1.0	...	...	...	...	C
Carbonated																					
Beverages[f]	2	10	10	...	Low	250	125	...	850	0.2	0.2	0.3	...	...	250	0.2–1.0	...	...	...	...	C
Confectionery	...	...	...	...	Low	...	...	...[g]	100	0.2	0.2	0.2	...	...	...	...	...	...	...	...	...
Cooling[h]	50	...	...	...	...	50	...	...	...	0.5	0.5	0.5	...	...	...	...	...	...	...	...	A,B
Food,																					
general	10	5–10	...	...	Low	10–250	30–250	...	850	0.2	0.2	0.2	...	...	...	1.0	...	...	...	...	C
Ice (raw																					
water)[i]	1–5	5	...	...	...	...	30–50	...	300	0.2	0.2	0.2	...	10	...	...	...	...	...	...	C
Laundering	...	...	...	...	...	50	60	6.0–6.8	...	0.2	0.2	0.2	...	...	...	...	...	...	...	...	...
Plastics, clear																					
uncolored	2	2	...	...	...	...	...	...	200	0.02	0.02	0.02	...	...	...	...	...	...	...	...	...
Paper and Pulp[j]																					
Groundwood	50	30	...	...	...	200	150	...	500	0.3	0.1	0.3	...	50	75	...	...	...	...	...	E
Kraft, paper,																					
bleached	40	25	...	...	...	100	75	...	300	0.2	0.1	0.2	...	50	200	...	...	...	...	...	E
Soda and sul-																					
fite pulps	25	5	...	...	...	100	75	...	250	.01	0.05	0.1	...	20	75	...	...	...	...	...	E
Fine paper	10	5	...	...	...	100	75	...	200	.01	0.05	0.1	...	20	...	...	...	...	...	...	E

TABLE 6–39. WATER QUALITY TOLERANCE FOR CERTAIN INDUSTRIAL APPLICATIONS (continued)

[Milligrams per liter, except as indicated]

Industry	Turbidity	Color	Color $+O_2$ consumed	D.O.[a] (milliliters per liter)	Odor	Hardness	Alkalinity	pH	Total Solids	Fe	Mn	Fe+ Mn	Al_2O_3	SiO_2	Cl	F	CO_3	HCO_3	OH	Na_2SO_4 to Na_2SO_3 (ratio)	General[b]
Rayon (viscose)																					
Pulp: Producion	5	5	...	...	...	8	50	...	100	0.05	0.03	0.05	8.0	25	5	...	...	...	...	...	F
Manufacture	0.3	...	...	...	...	55	...	7.8–8.3	...	...	...	...	...	...	...	...	...	...	...	...	...
Tanning[k]	20	10–100	...	...	...	50–135	135	6.0–8.0	...	0.2	0.2	0.2	...	...	...	...	...	...	...	...	...
Textiles:																					
General	5	20	...	...	...	20	...	...	...	0.25	0.25	...	...	...	100	...	...	...	...	...	...
Dyeing[l]	5	5–20	...	...	...	20	...	...	...	0.25	0.25	0.25	...	...	...	...	...	...	...	...	...
Wool scouring[m]		70	...	...	...	20	...	...		1.0	1.0	1.0	...	...	...	...	...	...	...	...	...
Cotton bandage[m]	5	5	...	...	Low	20	...	...	...	0.2	0.2	0.2	...	...	...	...	...	...	...	...	...

Sources: American Water Works Association, Water Quality and Treatment, second edition (New York, 1950). Water Quality Criteria, California State Water Quality Control Board, second edition (Sacramento, 1963)

a) Abbreviations as follows:
D.O., dissolved oxygen
ppm, parts per million
pH, hydrogen-ion concentration

b) A - no corrosiveness
B - no slime formation
C - conformity with federal drinking water standards necessary
D - NaCl, 275 ppm
E - free CO_2 less than 10 mg/l
F - copper less than 5 mg/l
G - calcium 100–200 mg/l
H - calcium 200–500 mg/l

c) Water with algae, or hydrogen sulphide odors, is most unsuitable for air conditioning.

d) Some hardness desirable.

e) Water for distilling must meet the same general requirements as for brewing (gin and spirits mashing water of light-beer quality, whiskey mashing water of dark-beer quality).

f) Clear, odorless, sterile water for syrup and carbonization. Water consistent in character. Most high quality filtered municipal water not satisfactory for beverages.

g) Hard candy requires pH of 7.0 or greater, as low value favors inversion of sucrose, causing sticky product.

h) Control of corrosiveness is necessary, as is also control of organisms, such as sulphur and iron bacteria, which tend to form slimes.

i) Ca $(HCO_3)_2$ particularly troublesome. Mg $(HCO_3)_2$ tends to greenish color. CO_2 assists in preventing cracking. Sulphates and chlorides of Ca, Mg, Na should each be less than 300 ppm (white butts).

j) Uniformity of composition and temperature desirable. Iron objectionable since cellulose absorbs iron from dilute solutions. Manganese very objectionable, clogs pipelines and is oxidized to permanganates by chlorine, causing reddish color.

k) Excessive iron, manganese or turbidity creates spots and discoloration in tanning of hides and leather goods.

l) Constant composition; residual alumina <0.5 ppm.

m) Calcium, magnesium, iron, manganese, suspended matter and soluble organic matter may be objectionable.

TABLE 6–40. WATER QUALITY GUIDELINES FOR THE PULP AND PAPER INDUSTRY

	Concentration mg/L					
			Kraft		Chem. Pulp & Paper	
Parameter	Fine Paper	Ground-wood	Bleached	Unbleached	Bleached	Unbleached
pH	—	6–8	—	—	6–8	6–8
Color (HU)	<40	<100	<25	<100	<50	<100
Turbidity (NTU)	<10	<20	<40	<100	<10	<20
Calcium	<20	<20	—	—	<20	<20
Magnesium	<12	<12	—	—	<12	<12
Iron	<0.1	<0.1	<0.2	<1.0	<0.1	<1.0
Manganese	<0.3	<0.1	<0.1	<0.5	<0.5	<0.5
Chloride	—	25–75	<200	<200	<200	<200
Silica	<20	<100	<50	<100	<50	<50
Hardness	<100	<100	<100	<100	<100	<100
Alkalinity	40–75	<150	<75	<150	—	—
Dissolved solids	<200	<250	<300	<500	<200	<250
Suspended solids	<10	—	—	—	<10	<10
Temperature (°C)	—	—	—	—	<36	—
CO_2	<10	<10	<10	<10	—	—
Corrosion tendency	NIL	NIL	NIL	NIL	NIL	NIL
Residual chloride	<2.0	—	—	—	—	—

Source: Canadian Council of Resource and Environment Ministers, Canadian Water Quality Guidelines, March 1987

TABLE 6–41. WATER QUALITY GUIDELINES FOR THE IRON AND STEEL INDUSTRY

	Concentration mg/L				
			Rinse Water		
Parameter	Hot-Rolling, Quenching, Gas Cleaning	Cold-Rolling	Softened	Demineralized	Steel Manufacturing
pH	5.0–9.0	5.0–9.0	6.0–9.0	—	6.8–7.0
Suspended solids	<25	<10	ND[a]	ND	—
Dissolved solids	<1000	<1000	ND	ND	—
Settleable solids	<100	<5.0	ND	ND	—
Dissolved oxygen	(minimum for aerobic conditions)				
Temperature (°C)	<38	<38	<38	<38	<38
Hardness	NS[b,c]	NS[b]	<100	<0.1	<50
Alkalinity	NS[c]	NS[c]	NS[c]	<0.5	—
Sulphate	<200	<200	<200	—	<175
Chloride	<150	<150	<150	ND	<150
Oil	NS	ND	ND	ND	ND
Floating material	NS	ND	ND	ND	ND

[a] ND = not detected.
[b] Controlled by other treatments.
[c] NS = not specified; the parameter has never been a problem at concentrations encountered.
Sources: Canadian Council of Resource and Environment Ministers, Canadian Water Quality Guidelines, March 1987; U.S. Environmental Protection Agency, 1973

TABLE 6–42. WATER QUALITY GUIDELINES FOR THE PETROLEUM INDUSTRY

Parameter	Concentration mg/L[a]
pH units	6.0–9.0
Color	NS[b]
Calcium	<75
Magnesium	<25
Iron	<1
Bicarbonate	NS
Sulphate	NS
Chloride	<200
Nitrate	NS
Fluoride	NS
Silica	NS
Hardness (as $CaCO_3$)	<350
Dissolved solids	<750
Suspended solids	<10

[a] Unless otherwise indicated.
[b] NS = not specified. The parameter has never been a problem at concentrations encountered.
Sources: Canadian Water Quality Guidelines 1987; Federal Water Pollution Control Administration 1968; Ontario Ministry of the Environment 1974

TABLE 6–43. WATER QUALITY GUIDELINES FOR POWER GENERATING STATIONS

Parameter	Concentration mg/L: Cooling Once-Through, Fresh	Cooling Once-Through, Brackish[a]	Boiler Feedwater (10.35–34.48 MPa)	Miscellaneous Uses
Silica	<50	<25	<0.01	—
Aluminum	NS[b]	NS	<0.01	—
Iron	NS	NS	<0.01	<1.0
Manganese	NS	NS	<0.01	—
Calcium	<200	<420	<0.01	—
Magnesium	NS	NS	<0.01	—
Ammonia	NS	NS	<0.07	—
Bicarbonate	<600	<140	<0.5	—
Sulphate	<680	<2700	NS[c]	—
Chloride	<600	<19 000	NS[c]	—
Dissolved solids	<1000	<35 000	<0.5	<1000
Copper	NS	NS	<0.01	—
Hardness	<850	<6250	<0.07	—
Zinc	NS	NS	<0.01	—
Alkalinity (as $CaCO_3$)	<500	<115	<1	—
pH units	5.0–8.3	6.0–8.3	8.8–9.4	5.0–9.0
Organic material:				
Methylene blue active substances	NS	NS	<0.1	<10
Carbon tetrachloride extract	NS[d]	NS[d]	NS	<10
Chemical oxygen demand (COD)	<75	<75	<1.0	—
Dissolved oxygen	—	—	<0.007	—
Suspended solids	<5000	<2500	<0.05	<5

[a] Brackish water—dissolved solids more than 1000 mg/L.
[b] NS = not specified; the parameter has never been a problem at concentrations encountered.
[c] Controlled by treatment for other constituents.
[d] No floating oil.
Source: Canadian Council of Resource and Environment Ministers, Canadian Water Quality Guidelines, March 1987; Krisher, A.S., 1978, Raw Water Treatment in the CPI. Chem. Eng. (N.Y.), v.85, pp.78–98. Chemical Engineering, Aug. 28, 1978. © McGraw-Hill, Inc.

TABLE 6–44. WATER QUALITY GUIDELINES FOR THE FOOD AND BEVERAGE INDUSTRY

Parameter	Concentration mg/L: Baking	Brewing	Carbonate Beverages	Confectionary	Dairy	Food Canning, Freezing, Dried, Frozen Fruits, Vegetables	Food Process (General)	Sugar Manufacturing
pH	—	6.5–7.0	<6.9	>7.0	—	6.5–8.5	—	—
Color (HU)	<10	<5	<10	—	ND	<5	5–10	—
Turbidity (NTU)	<10	<10	1–2	—	—	<5	1–10	—
Taste, odor (units)	low	low	ND[a]	low	ND	ND	low	—
Suspended solids	—	—	—	50–100	<500	<10	—	ND
Dissolved solids	—	<800	<850	50–100	<500	<500	<850	—
Calcium	NS[b,c]	<100	—	—	—	<100	—	<20
Magnesium	—	<30	—	—	—	—	—	<10
Iron	<0.2	0.1–1.0	<0.1	<0.2	0.1–0.3	<0.2	<0.2	<1
Manganese	<0.2[d]	<0.1[d]	<0.05	<0.2[d]	0.03–0.1	<0.2[d]	<0.2	<0.1
Copper	—	—	—	—	ND	—	—	—
Ammonium	—	—	—	—	trace	<0.5	—	—
Bicarbonate	—	ND	—	—	—	—	—	<100
Carbonate	—	<50	<5	—	—	—	—	—
Sulphate	—	<100	<200	—	<60	<250	—	<20
Chloride	—	20–60	<250	<250	<30	<250	—	<20
Nitrate	—	<10	—	—	<20	<10	—	—
Fluoride	—	<1	0.2–1.0	—	—	<1	<1	—
Silica	—	<50	ND	—	—	<50	—	—
Hardness	NS[b]	<70	200–250	—	<180	<250	10–250	<100
Alkalinity	—	<85	50–128	—	—	30–250	30–250	—
Hydrogen sulphide	<0.2	<0.2	<0.2	<0.2	—	—	—	—
Oxygen consumed	—	—	<15	—	—	<1	—	—
Carbon tetrachloride extract	—	—	slight	—	<10	<0.2	—	—
Chloroform extract	—	—	<0.2	—	—	—	—	—
Acidity	—	—	—	—	—	ND	—	—
Phenol	—	ND	ND	—	—	ND	—	ND
Nitrite	—	—	—	—	—	ND	—	—
Organic matter	—	trace	trace	—	—	—	—	trace

[a] ND = not detected.
[b] Some required for yeast action; excess retards fermentation.
[c] NS = not specified.
[d] Total Fe and Mn.
Source: Canadian Council of Resource and Environment Ministers, Canadian Water Quality Guidelines, March, 1987

SECTION F. IRRIGATION WATER QUALITY

TABLE 6–45. RELATIVE TOLERANCE OF CROP PLANTS TO SALT

[The numbers following EC x 10^3 are the electrical conductivity values of the saturation extract in millimhos per centimeter at 25°C associated with a 50-percent decrease in yield. The saturation extract is the solution extracted from a soil at its saturation percentage.]

Field Crops

EC x 10^3 = 16	EC x 10^3 = 10	EC x 10^3 = 4
Barley (grain)	Rye (grain)	Field beans
Sugar beet	Wheat (grain)	
Rape	Oats (grain)	
Cotton	Rice	
	Sorghum (grain)	
	Corn (field)	
	Flax	
	Sunflower	
	Castorbeans	
EC x 10^3 = 10	EC x 10^3 = 6	

Vegetable Crops

EC x 10^3 = 12	EC x 10^3 = 10	EC x 10^3 = 4
Garden beets	Tomato	Radish
Kale	Broccoli	Celery
Asparagus	Cabbage	Green beans
Spinach	Bell pepper	
	Cauliflower	
	Lettuce	
	Sweet corn	
	Potatoes (White Rose)	
	Carrot	
	Onion	
	Peas	
	Squash	
	Cucumber	
EC x 10^3 = 10	EC x 10^3 = 4	EC x 10^3 = 3

Fruit and Nut Crops

High Salt Tolerance	Medium Salt Tolerance	Low Salt Tolerance
Date palm		Almond
		Apple
		Apricot
	Cantaloup	Avocado
	Fig	Blackberry
	Grape	Boysenberry
	Jujube	Cherimoya
	Olive	Cherry, sweet
	Papaya	Cherry, sand
	Pineapple	Currant
	Pomegranate	Gooseberry
		Grapefruit
		Lemon
		Lime
		Loquat
		Mango
		Orange
		Passion fruit
		Peach
		Pear
		Persimmon
		Plum: prune
		Pummelo
		Raspberry
		Rose, apple
		Sapote, white
		Strawberry
		Tangerine

Forage Crops

EC x 10^3 = 18	EC x 10^3 = 12	EC x 10^3 = 4
Alkali sacaton	White sweetclover	White Dutch clover
Saltgrass	Yellow sweetclover	Meadow foxtail
Nuttall alkaligrass	Perennial ryegrass	Alsike clover
Bermuda grass	Mountain brome	Red clover
Rhodes grass	Strawberry clover	Ladino clover
Fescue grass	Dallis grass	Burnet
Canada wildrye	Sudan grass	
Western wheat-grass	Hubam clover	
Barley (hay)	Alfalfa (California common)	
Bridsfoot trefoil	Tall fescue	
	Rye (hay)	
	Wheat (hay)	
	Oats (hay)	
	Orchardgrass	
	Blue grama	
	Meadow fescue	
	Reed canary	
	Big trefoil	
	Smooth brome	
	Tall meadow oat-grass	
	Cicer milkvetch	
	Sourclover	
	Sickle milkvetch	
EC x 10^3 = 12	EC x 10^3 = 4	EC x 10^3 = 2

Source: U.S. Department of Agriculture, 1954; Kandiah, A., FAO, 1987

FIGURE 6–6. QUALITY CRITERIA FOR IRRIGATION WATER

[For explanation of diagram see following page]

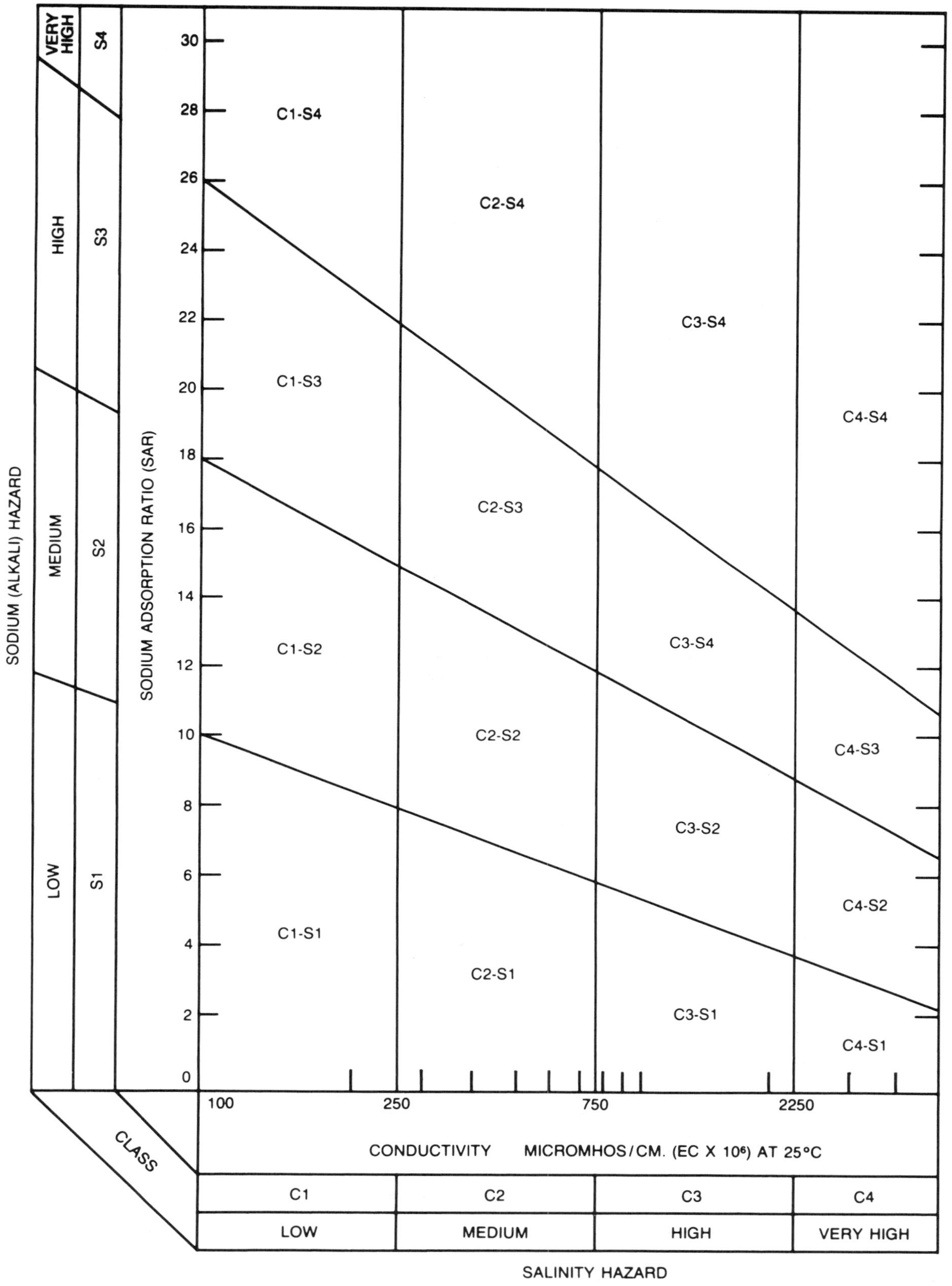

Source: U.S. Dept. of Agriculture

NOTES FOR FIGURE 6–6

Sodium-Adsorption-Ratio SAR = Na/ (Ca + Mg)/2 where concentrations are expressed in millequivalents per liter.

CONDUCTIVITY

LOW-SALINITY WATER (C1) can be used for irrigation with most crops on most soils with little likelihood that soil salinity will develop. Some leaching is required, but this occurs under normal irrigation practices except in soils of extremely low permeability.

MEDIUM-SALINITY WATER (C2) can be used if a moderate amount of leaching occurs. Plants with moderate salt tolerance can be grown in most cases without special practices for salinity control.

HIGH-SALINITY WATER (C3) cannot be used on soils with restricted drainage. Even with adequate drainage, special management for salinity control may be required and plants with good salt tolerance should be selected.

VERY HIGH-SALINITY WATER (C4) is not suitable for irrigation under ordinary conditions, but may be used occasionally under very special circumstances. The soils must be permeable, drainage must be adequate, irrigation water must be applied in excess to provide considerable leaching, and very salt-tolerant crops should be selected.

SODIUM

LOW-SODIUM WATER (S1) can be used for irrigation on almost all soils with little danger of the development of harmful levels of exchangeable sodium. However, sodium-sensitive crops such as stone-fruit trees and avocados may accumulate injurious concentrations of sodium.

MEDIUM-SODIUM WATER (S2) will present an appreciable sodium hazard in fine-textured soils having high cation-exchange-capacity, especially under low-leaching conditions, unless gypsum is present in the soil. This water may be used on coarse-textured or organic soils with good permeability.

HIGH-SODIUM WATER (S3) may produce harmful levels of exchangeable sodium in most soils and will require special soil management—good drainage, high leaching, and organic matter additions. Gypsiferous soils may not develop harmful levels of exchangeable sodium from such waters. Chemical amendments may be required for replacement of exchangeable sodium, except that amendments may not be feasible with waters of very high salinity.

VERY HIGH SODIUM WATER (S4) is generally unsatisfactory for irrigation purposes except at low and perhaps medium salinity, where the solution of calcium from the soil or use of gypsum or other amendments may make the use of these waters feasible.

Another criterion for the evaluation of irrigation water is:

Residual Sodium Carbonate (RSC) = (CO_3 + HCO_3) - (Ca + Mg) where concentrations are expressed in meq/liter.

When RSC>2.5 Probably not suitable for irrigation

1.25–2.5 Marginal

<1.25 Probably safe for irrigation

TABLE 6–46. GUIDES FOR EVALUATING THE QUALITY OF WATER USED FOR IRRIGATION

[MPN is most probable number. Sodium absorption ratio is defined by the formula SAR=Na/ (Ca+Mg)/2 where the concentrations are expressed in milliequivalents per liter. Residual sodium carbonate is the sum of the equivalents of normal carbonate and bicarbonate minus the sum of the equivalents of calcium and magnesium.]

Quality Factor	Threshold Concentration*	Limiting Concentration†
Coliform organisms, MPN per 100 ml	1000‡	§
Total dissolved solids (TDS), mg/liter	500‡	1500‡
Electrical conductivity, µmhos/cm	750‡	2250‡
Range of pH	7.0–8.5	6.0–9.0
Sodium adsorption ratio (SAR)	6.0‡	15
Residual sodium carbonate (RSC), meq	1.25‡	2.5
Arsenic, mg/liter	1.0	5.0
Boron, mg/liter	0.5‡	2.0
Chloride, mg/liter	100‡	350
Sulfate, mg/liter	200‡	1000
Copper, mg/liter	0.1‡	1.0

*Threshold values at which irrigator might become concerned about water quality and might consider using additional water for leaching. Below these values, water should be satisfactory for almost all crops and almost any arable soil.

† Limiting values at which the yield of high-value crops might be reduced drastically, or at which an irrigator might be forced to less valuable crops.

‡ Values not to be exceeded more than 20 percent of any 20 consecutive samples, nor in any 3 consecutive samples. The frequency of sampling should be specified.

§ Aside from fruits and vegetables which are likely to be eaten raw, no limits can be specified. For such crops, the threshold concentration would be limiting.

Source: Calif. State Water Quality Control Board, 1963

TABLE 6–47. FAO GUIDELINES FOR EVALUATING THE QUALITY OF WATER FOR IRRIGATION

Potential Irrigation Problem	Units	Degree of Restriction on Use		
		None	Slight to Moderate	Severe
Salinity (affects crop water availability)[1]				
EC_W	dS/m	<0.7	0.7–3.0	>3.0
or TDS	mg/L	<450	450–2000	>2000
Infiltration (affects infiltration rate of water into the soil. Evaluate using EC_W and SAR together)[2]				
SAR = 0–3 and EC_W =		>0.7	0.7–0.2	<0.2
= 3–6 =		>1.2	1.2–0.3	<0.3
= 6–12 =		>1.9	1.9–0.5	<0.5
= 12–20 =		>2.9	2.9–1.3	<1.3
= 20–40 =		>5.0	5.0–2.9	<2.9
Specific Ion Toxicity (affects sensitive crops)				
Sodium (Na)				
surface irrigation	SAR	<3	3–9	>9
sprinkler irrigation	me/L	<3	>3	
Chloride (Cl)				
surface irrigation	me/L	<4	4–10	>10
sprinkler irrigation	me/L	<3	<3	
Boron (B)	mg/L	<0.7	0.7–3.0	>3.0
Trace elements (see Table 6–48)				
Miscellaneous Effects (affects susceptible crops)				
Nitrogen (NO_3-N)	mg/L	<5	5–30	>30
Bicarbonate (HCO_3) (overhead sprinkling only)	me/L	<1.5	1.5–8.5	>8.5
pH			Normal range 6.5–8.4	

[1] EC_W means electrical conductivity, a measure of the water salinity, reported in deciSiemens per meter at 25°C (dS/m) or in units millimhos per centimeter (mmho/cm). Both are equivalent. TDS means total dissolved solids, reported in milligrams per liter (mg/L).

[2] SAR means sodium absorption ratio.

Source: Food and Agriculture Organization of the United Nations, 1985, Water Quality for Agriculture, Irrigation, and Drainage Paper No. 29. Kandiah, A., Water Quality in Food Production, Water Quality Bulletin, v. 12, no.1, Jan. 1987

TABLE 6–48. FAO RECOMMENDED MAXIMUM CONCENTRATIONS OF TRACE ELEMENTS IN IRRIGATION WATER

Element	Recommended Maximum Concentration[1] (mg/L)	Remarks
Al	5.0	Can cause non-productivity in acid soils (pH <5.5), but more alkaline soils at >pH 7.0 will precipitate the ion and eliminate any toxicity.
As	0.10	Toxicity to plants varies widely, ranging from 12 mg/L for Sudan grass to less than 0.05 mg/L for rice.
Be	0.10	Toxicity to plants varies widely, ranging from 5 mg/L for kale to 0.5 mg/L for bush beans.
Cd	0.01	Toxic to beans, beets and turnips at concentrations as low as 0.1 mg/L in nutrient solutions. Conservative limits recommended due to its potential for accumulation in plants and soils to concentrations that may be harmful to humans.
Co	0.05	Toxic to tomato plants at 0.1 mg/L in nutrient solution. Tends to be inactivated by neutral and alkaline soils.
Cr	0.10	Not generally recognized as an essential growth element. Conservative limits recommended due to lack of knowledge on its toxicity to plants.

TABLE 6–48. FAO RECOMMENDED MAXIMUM CONCENTRATIONS OF TRACE ELEMENTS IN IRRIGATION WATER (continued)

Element	Recommended Maximum Concentration[1] (mg/L)	Remarks
Cu	0.20	Toxic to a number of plants at 0.1 to 1.0 mg/L in nutrient solutions.
F	1.0	Inactivated by neutral and alkaline soils.
Fe	5.0	Not toxic to plants in aerated soils, but can contribute to soil acidification and loss of availability of essential phosphorus and molybdenum. Overhead sprinkling may result in unsightly deposits on plants, equipment and buildings.
Li	2.5	Tolerated by most crops up to mg/L; mobile in soil. Toxic to citrus at low concentrations (0.075 mg/L). Acts similarly to boron.
Mn	0.20	Toxic to a number of crops at a few-tenths to a few mg/L, but usually only in acid soils.
Mo	0.01	Not toxic to plants at normal concentrations in soil and water. Can be toxic to livestock if forage is grown in soils with high concentrations of available molybdenum.
Ni	0.20	Toxic to a number of plants at 0.5 mg/L to 1.0 mg/L; reduced toxicity at neutral or alkaline pH.
Pd	5.0	Can inhibit plant cell growth at very high concentrations.
Se	0.02	Toxic to plants at concentrations as low as 0.025 mg/L and toxic to livestock if forage is grown in soils with relatively high levels of added selenium. An essential element to animals but in very low concentrations.
Sn Ti W		Effectively excluded by plants; specific tolerance unknown.
V	0.10	Toxic to many plants at relatively low concentrations.
Zn	2.0	Toxic to many plants at widely varying concentrations; reduced toxicity at pH >6.0 and in fine textured or organic soils.

[1] The maximum concentration is based on a water application rate which is consistent with good irrigation practices (10 000 m^3/ha/yr). If the water application rate greatly exceeds this, the maximum concentrations should be adjusted downward accordingly. No adjustment should be made for application rates less than 10 000 m^3/ha/yr. The values given are for water used on a continuous basis at one site.

Source: Food and Agriculture Organization of the United Nations, 1985, Water Quality for Agriculture, Irrigation, and Drainage Paper No. 29. Kandiah, A., Water Quality in Food Production, Water Quality Bulletin, v. 12, no. 1, Jan. 1987

TABLE 6–49. LIMITS OF BORON IN IRRIGATION WATER

A. Permissible Limits
[Boron in parts per million]

Class of Water	Crop Group		
	Sensitive	Semitolerant	Tolerant
Excellent	<0.33	<0.67	<1.00
Good	0.33 to 0.67	0.67 to 1.33	1.00 to 2.00
Permissible	0.67 to 1.00	1.33 to 2.00	2.00 to 3.00
Doubtful	1.00 to 1.25	2.00 to 2.50	3.00 to 3.75
Unsuitable	>1.25	>2.50	>3.75

B. Crop Groups of Boron Tolerance
[In each group, the plants first named are considered as being more tolerant; the last named, more sensitive.]

Sensitive	Semitolerant	Tolerant
Pecan	Sunflower (native)	Athel (*Tamarix aphylla*)
Walnut (Black; and Persian, or English)	Potato	Asparagus
Jerusalem-artichoke	Cotton (Acala and Pima)	Palm (*Phoenix canariensis*)
Navy bean	Tomato	Date palm (*P. dactylifera*)
American elm	Sweetpea	Sugar beet
Plum	Radish	Mangel
Pear	Field pea	Garden beet
Apple	Ragged Robin rose	Alfalfa
Grape (Sultanina and Malaga)	Olive	Gladiolus
Kadota fig	Barley	Broadbean
Persimmon	Wheat	Onion
Cherry	Corn	Turnip
Peach	Milo	Cabbage
Apricot	Oat	Lettuce
Thornless blackberry	Zinnia	Carrot
Orange	Pumpkin	
Avocado	Bell pepper	
Grapefruit	Sweet potato	
Lemon	Lima bean	

Source: U.S. Dept. of Agriculture

SECTION G. WATER QUALITY FOR AQUATIC LIFE

TABLE 6–50. WATER QUALITY CRITERIA FOR PROTECTION OF AQUATIC LIFE (EXCLUDING PESTICIDES AND HALOGENATED SPECIES[a])

Pollutant	Criteria	Ref.[b]
Acenaphthene	Acute toxicity occurs as low as 1,700 μg/L in freshwater species and 970 μg/L in saltwater species. Freshwater algae are affected by 520 μg/L, saltwater algae at 500 μg/L. Chronic toxicity occurs in saltwater species as low as 710 μg/L.	1
Acrolein	Acute toxicity occurs as low as 68 μg/L in freshwater species and 55 μg/L in saltwater species. Chronic toxicity occurs in freshwater species as low as 21 μg/L.	1
Acrylonitrile	Acute toxicity occurs as low as 7,550 μg/L in freshwater species. Mortality occurred in freshwater fish exposed for 30 days at 2,600 μg/L.	1
Aluminum	For protection of saltwater species an application factor of 0.01 is recommended to be applied to the 96-hour LC50 for sensitive organisms. Concentrations exceeding 1,500 μg/L constitute a hazard in the marine environment, and levels less than 200 μg/L present minimal risk of deleterious effects.	2
Ammonia (un-ionized)	For marine species, an application factor of 0.1 is recommended. Concentrations equal to or exceeding 400 μg/L constitute a hazard to marine biota. Levels below 10 μg/L present minimal risk of deleterious effects. (Insufficient data for 1984 criterion.)	2
Antimony	Acute toxicity occurs as low as 9,000 μg/L in freshwater species and is toxic to freshwater algae at 610 μg/L. Chronic toxicity occurs in freshwater species as low as 1,600 μg/L.	1
	For protection of saltwater species an application factor of 0.01 is recommended to be applied to the 96-hour LC50 for sensitive organisms. Concentrations exceeding 0.2 μg/L constitute a hazard in the marine environment.	2
Arsenic[c] (trivalent)	For freshwater aquatic life in each 30 consecutive days the average concentration of arsenic shall not exceed 72 μg/L, the maximum concentration shall not exceed 140 μg/L, and the concentration may be between 72 μg/L and 140 μg/L for up to 96 hours. For saltwater aquatic life in each 30 consecutive days the average concentration of arsenic shall not exceed 63 μg/L, the maximum concentration shall not exceed 120 μg/L, and the concentration may be between 63 μg/L and 170 μg/L for up to 96 hours.	3
Barium	For protection of saltwater species an application factor of 0.05 is recommended to be applied to the 96-hour LC50 for sensitive organisms. Concentrations equal to or exceeding 1,000 μg/L constitute a hazard in the marine environment, and levels less than 500 μg/L present minimal risk of deleterious effects.	2
Benzene	Acute toxicity occurs as low as 5,300 μg/L in freshwater species and 5,100 μg/L in saltwater species. Adverse effects occur in saltwater fish exposed for 168 days as low as 700 μg/L.	1
Benzidine	Acute toxicity occurs as low as 2,500 μg/L in freshwater species.	1
Beryllium	Acute toxicity occurs as low as 130 μg/L in freshwater species. Chronic toxicity occurs in freshwater species as low as 5.3 μg/L. Hardness has a substantial effect on acute toxicity.	1
	For protection of saltwater species an application factor of 0.01 is recommended to be applied to the 96-hour LC50 for sensitive organisms. Concentrations equal to or exceeding 1,500 μg/L constitute a hazard in the marine environment, and levels less than 100 μg/L present minimal risk of deleterious effects.	2
Boron	For protection of saltwater species an application factor of 0.1 is recommended to be applied to the 96-hour LC50 for sensitive organisms. Concentrations equal to or exceeding 5,000 μg/L constitute a hazard in the marine environment, and levels less than 5,000 μg/L present minimal risk of deleterious effects.	2
Bromate	It is recommended that ionic bromine in the form of bromate be maintained below 100,000 μg/L in the marine environment.	2

TABLE 6–50. WATER QUALITY CRITERIA FOR PROTECTION OF AQUATIC LIFE (EXCLUDING PESTICIDES AND HALOGENATED SPECIES[a]) (continued)

Pollutant	Criteria	Ref.[b]
Bromine	It is recommended that free (molecular) bromine in the marine environment not exceed 100 μg/L.	(free)
Cadmium	For freshwater aquatic life, the concentration of active cadmium shall not exceed a level equal to 1.16 (ln hardness mg/L) - 3.841 due to acute and chronic toxicities being nearly the same. For saltwater aquatic life in each 30 consecutive days the average concentration of cadmium shall not exceed 12 μg/L, the maximum concentration shall not exceed 38 μg/L, and the concentration may be between 12 μg/L and 38 μg/L for up to 96 hours.	3
Chlorine	For freshwater aquatic life in each 30 consecutive days the average concentration of chlorine shall not exceed 8.3 μg/L, the maximum concentration shall not exceed 14 μg/L, and the concentration may be between 8.3 μg/L and 14 μg/L for up to 96 hours. For saltwater aquatic life in each 30 consecutive days the average concentration of chlorine shall not exceed 7.4 μg/L, the maximum concentration shall not exceed 13 μg/L, and the concentration may be between 7.4 μg/L and 13 μg/L for up to 96 hours.	3
Chromium[c] (hexavalent)	For freshwater aquatic life in each 30 consecutive days the average concentration of chromium shall not exceed 7.2 μg/L, the maximum concentration shall not exceed 11 μg/L, and the contration may be between 7.2 μg/L and 1100 μg/L for up to 96 hours. For saltwater aquatic life in each 30 consecutive days the average concentration of chromium shall not exceed 5.4 μg/L, the maximum concentration shall not exceed 1200 μg/L, and the concentration may be between 5.4 μg/L and 1200 μg/L for up to 96 hours.	3
Chromium[c] (trivalent)	For freshwater aquatic life in each 30 consecutive days the average concentration of chromium shall not exceed 0.819 (ln hardness mg/L) to 537, the maximum concentration shall not exceed 0.819 (ln hardness mg/L) + 3.568. No saltwater criterion were derived, but levels of 10,300 μg/L are lethal to the eastern oyster.	3
Copper[c]	For freshwater aquatic life in each 30 consecutive days the average concentration of copper shall not exceed 0.905 (ln hardness) - 1.705, the maximum concentration shall not exceed 0.905 (ln hardness mg/L) + 3.568, and the concentration may be between the average and the maximum for up to 96 hours. For saltwater aquatic life in each 30 consecutive days the average concentration of copper shall not exceed 2 μg/L, the maximum concentration shall not exceed 3.2 μg/L, and the concentration may be between 2 μg/L and 3.2 g/L for up to 96 hours.	3
Cyanides (sum of HCN and CN^-)	For freshwater aquatic life in each 30 consecutive days the average concentration of cyanides shall not exceed 4.2 μg/L, the maximum concentration shall not exceed 22 μg/L, and the concentration may be between 4.2 μg/L and 22 μg/L for up to 96 hours. For saltwater aquatic life in each 30 consecutive days the average concentration of cyanides shall not exceed 0.57 μg/L, the maximum concentration shall not exceed 1 μg/L, and the concentration may be between 0.57 μg/L and 1 g/L for up to 96 hours.	3
2,4-Dinitrotoluene	Acute toxicity occurs as low as 330 μg/L in freshwater species and 590 μg/L in saltwater species. Chronic toxicity occurs in freshwater species as low as 230 μg/L. A decrease in saltwater algal cell numbers occurs as low as 370 μg/L.	1
1,2-Diphenyl-hydrazine	Acute toxicity occurs as low as 270 μg/L in freshwater species.	1
Ethylbenzene	Acute toxicity occurs as low as 32,000 μg/L in freshwater species and 430 μg/L in saltwater species.	1
Fluoranthene	Acute toxicity occurs as low as 3,980 μg/L in freshwater species and 40 μg/L in saltwater species. Chronic toxicity occurs in saltwater species as low as 16 μg/L.	1
Fluorides	For protection of saltwater species an application factor of 0.1 is recommended to be applied to the marine 96-hour LC50. Concentrations equal to or exceeding 1,500 μg/L constitute a hazard to the marine environment, and levels less than 500 μg/L present minimal risk of deleterious effects.	2

TABLE 6–50. WATER QUALITY CRITERIA FOR PROTECTION OF AQUATIC LIFE (EXCLUDING PESTICIDES AND HALOGENATED SPECIES[a]) (continued)

Pollutant	Criteria	Ref.[b]
Iron	Concentrations equal to or exceeding 300 μg/L constitute a hazard to the marine environment, and levels less than 50 μg/L present minimal risk of deleterious effects.	2
Isophorone	Acute toxicity occurs as low as 117,000 μg/L in freshwater species and 12,900 μg/L in saltwater species.	1
Lead[c]	For freshwater aquatic life in each 30 consecutive days the average concentration of lead shall not exceed 1.34 (ln hardness mg/L) - -5.245, the maximum concentration shall not exceed 1.34 (ln hardness mg/L) - 2.014, and the concentration may be between the average and the maximum for up to 96 hours. For saltwater aquatic life in each 30 consecutive days the average concentration of lead shall not exceed 8.6 μg/L, the maximum concentration shall not exceed 220 μg/L, and the concentration may be between 8.6 μg/L and 220 μg/L for up to 96 hours.	3
Manganese	For protection of saltwater species an application factor of 0.02 is recommended to be applied to the marine 96-hour LC50. Concentrations equal to or exceeding 100 μg/L constitute a hazard to the marine environment, and levels less than 20 μg/L present minimal risk.	2
Mercury[c]	For freshwater aquatic life in each 30 consecutive days the average concentration of mercury shall not exceed 0.2 μg/L, the maximum concentration shall not exceed 1.1 μg/L, and the concentration may be between 0.2 μg/L and 1.1 μg/L for up to 96 hours. For saltwater aquatic life in each 30 consecutive days the average concentration of mercury shall not exceed 0.1 g/L, the maximum concentration shall not exceed 1.9 μg/L, and the concentration may be between 0.1 g/L and 1.9 g/L for up to 96 hours.	3
Molybdenum	It is recommended that the concentration in seawater should not exceed 0.05 of the 96-hour LC50 at any time for the most sensitive species and that the 24-hour average not exceed 0.02 of the 96-hour LC50.	1
Naphthalene	Acute toxicity occurs as low as 2,300 μg/L in freshwater species and 2,350 μg/L in saltwater species. Chronic toxicity in freshwater species occurs as low as 620 μg/L.	1
Nickel	For freshwater aquatic life, total recoverable nickel should not exceed 1,100 μg/L at any time assuming a hardness of 50 mg/L as $CaCO_3$. For saltwater species the concentration should not exceed 140 μg/L at any time. The 24-hour average freshwater criterion is 56 μg/L for a hardness of 50 mg/L. The 24-hour saltwater criterion is 7.1 μg/L.	1
Nitrobenzene	Acute toxicity occurs as low as 27,000 μg/L in freshwater species and 6,680 μg/L in saltwater species.	1
Nitrophenols	Acute toxicity occurs as low as 230 μg/L in freshwater species and 4,850 μg/L in saltwater species. Toxicity to freshwater algae occurs as low as 150 μg/L.	1
Nitrosamines	Acute toxicity occurs as low as 5,850 μg/L in freshwater species and 3,300,000 μg/L in saltwater species.	1
Phenol	Acute toxicity occurs as low as 10,200 μg/L in freshwater species and 5,800 μg/L in saltwater species. Chronic toxicity occurs in freshwater species as low as 2,560 μg/L.	1
2,4-Dimethyl phenol	Acute toxicity occurs as low as 2,120 μg/L in freshwater species.	1
Phenolics (Phenolic compounds)	For freshwater species, an application factor of 0.05 is recommended to be applied to the 96-hour LC50 for important sensitive species. No concentration greater than 100 μg/L is recommended at any time or place.	2
Phthalate Esters	Acute toxicity occurs as low as 940 μg/L freshwater species and 2,944 μg/L in saltwater species. Chronic toxicity occurs in freshwater species as low as 3 μg/L. Toxicity to one species of saltwater algae occurs as low as 3.4 μg/L.	1

TABLE 6–50. WATER QUALITY CRITERIA FOR PROTECTION OF AQUATIC LIFE (EXCLUDING PESTICIDES AND HALOGENATED SPECIES[a]) (continued)

Pollutant	Criteria	Ref.[b]
Phosphorus (elemental)	For protection of saltwater species an application factor of 0.01 is recommended to be applied to the marine 96-hour LC50. Concentrations equal to or exceeding 1 μg/L constitute a hazard to the marine environment.	2
Polychlorinated biphenyls	Acute toxicity probably will only occur at concentrations above 2.0 μg/L for freshwater species and above 10 μg/L for saltwater species. The 24-hour average freshwater criterion is 0.014 μg/L. The 24-hour saltwater criterion is 0.030 μg/L.	1
Polynuclear Aromatic Hydrocarbons	Acute toxicity occurs as low as 300 μg/L for saltwater species.	1
Selenium (as inorganic selenite, Se^{-2})	For freshwater aquatic life, total recoverable inorganic selenite should not exceed 260 μg/L at any time. For saltwater species, the concentration should not exceed 410 μg/L at any time.	1
Selenium (as inorganicselenite, Se^{-2})	The 24-hour average freshwater criterion is 35 μg/L. The 24-hour saltwater criterion is 54 μg/L.	1
Selenium (as inorganic selenate, Se^{+4})	Acute toxicity occurs as low as 760 μg/L in freshwater species.	1
Silver	For freshwater aquatic life, total recoverable silver should not exceed 1.2 μg/L at any time, assuming a hardness of 50 mg/L as $CaCO_3$. For saltwater species the concentration should not exceed 2.3 μg/L at any time. Chronic toxicity in freshwater species occurs as low as 0.12 μg/L.	1
Sulfide	For protection of saltwater species an application factor of 0.1 is recommended to be applied to the marine 96-hour LC50. Concentrations equal to or exceeding 10 μg/L constitute a hazard to the marine environment, and levels less than 5 μg/L present minimal risk of deleterious effects with the pH maintained within a range of 6.5 to 8.5.	2
Hydrogen Sulfide (undissociated)	For freshwater species, a level assumed to be safe for all aquatic organisms including fish is 2 μg/L. It is recommended that the concentration of total sulfides not exceed 2 μg/L at any time or place.	2
Thallium	Acute toxicity occurs as low as 1,400 μg/L in freshwater species and as low as 2,130 μg/L in saltwater species. Chronic toxicity occurs as low as 40 μg/L in freshwater species, and one freshwater fish is affected after 2,600 hours as low as 20 μg/L.	1
	For salt species, because of a chronic effect of long-term exposure, tests should be conducted for at least 20 days to determine harmful, sublethal concentrations. The concentration in seawater should not exceed 0.05 of this concentration. Concentrations equal to or exceeding 100 μg/L constitute a hazard to the marine environment, and levels less than 50 μg/L present minimal risk of deleterious effects.	2
Toluene	Acute toxicity occurs as low as 17,500 μg/L in freshwater species and as low as 6,300 μg/L in saltwater species. Chronic toxicity occurs in saltwater species as low as 5,000 μg/L.	1
Uranium	For protection of saltwater species an application factor of 0.01 is recommended to be applied to the marine 96-hour LC50. Concentrations equal to or exceeding 500 μg/L constitute a hazard to the marine environment, and levels less than 100 μg/L present minimal risk of deleterious effects.	2
Vanadium	It is recommended that the concentration of seawater not exceed 0.05 of the 96-hour LC50 for the most sensitive species.	2
Zinc	For freshwater aquatic life, total recoverable zinc should not exceed 180 μg/L at any time, assuming a hardness of 50 mg/L as $CaCO_3$. For saltwater species the concentration should not exceed 170 μg/L at any time. The 24-hour average criterion for freshwater is 47 μg/L for a hardness of 50 mg/L. The 24-hour saltwater criterion is 58 μg/L.	1

TABLE 6–50. WATER QUALITY CRITERIA FOR PROTECTION OF AQUATIC LIFE (EXCLUDING PESTICIDES AND HALOGENATED SPECIES[a]) (continued)

Footnotes:

[a] In addition to the pollutants listed in the Table, certain pesticides and numerous halogenated organics are addressed by the U.S. EPA 1980 Water Quality Criteria. Criteria for protection of aquatic life and/or levels at which toxicity occurs are specified.

Pesticides	Halogenated Organics
Aldrin/dieldrin	Carbon tetrachloride
Chlordane	Chlorinated benzenes
DDT	Chlorinated ethanes
Endosulfan	Chloroalkyl ethers
Endrin	Chlorinated naphthalene
Heptachlor	Chlorinated phenols
Toxaphene	Chloroform
	2-Chlorophenol
	Dichlorobenzenes
	Dichlorobenzidine
	Dichloroethylenes
	2,4-Dichlorophenol
	Dichloropropanes/propenes
	Haloethanes
	Halomethane
	Hexachlorobutadiene
	Hexachlorocyclohexane
	Hexachlorocyclopentadiene
	Pentachlorophenol
	Tetrachloroethylene
	Trichloroethylene
	Vinyl chloride

[b] References:

1. U.S. EPA, Water Quality Criteria, *Federal Register*, November 28, 1980 (with updates).
2. NAS/NAE. Water Quality Criteria 1972. Prepared for the U.S. Environmental Protection Agency by the National Academy of Sciences, National Academy of Engineering, National Academy of Sciences, Washington, D.C., EPA-R3–73–933.
3. U.S. EPA (1984) Water Quality Criteria; Request for Comments, Federal Register, Volume 49, No. 26, 4551–4554, February 7, 1984.

[c] For arsenic; chromium, copper, lead, and mercury, the chemical is defined as the dissolved fraction that passes through a 0.45 μm membrane filter.

ln=natural logarithm

Source: Kingsbury, G.L., and R.M. Ray, 1986, Reclamation and Redevelopment of Contaminated Land, v. 1 U.S. Case Histories, U.S.EPA/600/2–86/066 Cincinnati, OH 45268

TABLE 6–51. GUIDES FOR EVALUATING THE QUALITY OF WATER FOR AQUATIC LIFE

Determination	Threshold Concentration*	
	Freshwater	Saltwater
Total dissolved solids (TDS), mg/liter	2000†	
Electrical conductivity, μmhos/cm @ 25°C	3000†	
Temperature, maximum °C	34	34
Maximum for salmonoid fish	23	23
Range of pH	6.5–8.5	6.5–9.0
Dissolved oxygen (D.O.), minimum mg/liter	5.0‡	5.0‡
Flotable oil and grease, mg/liter	0	0†
Emulsified oil and grease, mg/liter	10†	10†
Detergent, ABS, mg/liter	2.0	2.0
Ammonia (free), mg/liter	0.5†	
Arsenic, mg/liter	1.0†	1.0†
Barium, mg/liter	5.0†	
Cadmium, mg/liter	0.01†	
Carbon dioxide (free), mg/liter	1.0	
Chlorine (free), mg/liter	0.02	
Chromium, hexavalent, mg/liter	0.05†	0.05†
Copper, mg/liter	0.02†	0.02†
Cyanide, mg/liter	0.02†	0.02†
Fluoride, mg/liter	1.5†	1.5†
Lead, mg/liter	0.1†	0.1†
Mercury, mg/liter	0.01	0.01
Nickel, mg/liter	0.05†	
Phenolic compounds, as phenol, mg/liter	1.0	
Silver, mg/liter	0.01	0.01
Sulfide, dissolved, mg/liter	0.5†	0.5†
Zinc, mg/liter	0.1	

* Threshold concentration is value that normally might not be deleterious to fish life. Waters that do not exceed these values should be suitable habitats for mixed fauna and flora.
† Values not to be exceeded more than 20 percent of any 20 consecutive samples, nor in any 3 consecutive samples. Other values should never be exceeded. Frequency of sampling should be specified.
‡ Dissolved oxygen concentrations should not fall below 5.0 mg/liter more than 20 percent of the time and never below 2.0 mg/liter. (Note: Recent data indicate also that rate of change of oxygen tension is an important factor, and that diurnal changes in D.O. may, in sewage-polluted water, render the value of 5.0 of questionable merit.)
Source: Calif. State Water Quality Control Board, 1963

TABLE 6–52. MAXIMUM CONCENTRATIONS OF COPPER SULFATE SAFE FOR FISH

Fish	Safe Copper Sulfate Concentration	
	ppm.	lb./mil. gal.
Trout	0.14	1.2
Carp	0.30	2.5
Suckers	0.30	2.5
Catfish	0.40	3.5
Pickerel	0.40	3.5
Goldfish	0.50	4.0
Perch	0.75	6.0
Sunfish	1.20	10.0
Black Bass	2.10	17.0

Source: U.S. Public Health Service

TABLE 6–53. OBSERVED LETHAL CONCENTRATION OF SELECTED CHEMICALS IN AQUATIC ENVIRONMENTS

Chemical	Organism Tested	Lethal Concentration, mg/L	Exposure Time, hr
ABS (100 percent)	Fathead minnow	3.5–4.5	96
ABS (100 percent)	Bluegills	4.2–4.4	96
Household syndets	Fathead minnow	39–61	96
Alkyl sulfate	Fathead minnow	5.1–5.9	96
LAS (C12)	Bluegill fingerlings	3	96
LAS (C14)	Bluegill fingerlings	0.6	96
Acetic acid	Goldfish	423	20
Alum	Goldfish	100	12–96
Ammonia	Goldfish	2–2.5 NH_3	24–96
Ammonia	Perch, roach, rainbow trout	3N	2–20
Sodium arsenite	Minnow	17.8 As	36
Sodium arsenate	Minnow	234 As	15
Barium chloride	Goldfish	5000	12–17
Barium chloride	Salmon	158	. . .
Cadmium chloride	Goldfish	0.017	9–18
Cadmium nitrate	Goldfish	0.3 Cd	190
CO_2	Various species	100–200	. . .
CO	Various species	1.5	1–10
Chloramine	Brown trout fry	0.06	. . .
Chlorine	Rainbow trout	0.03–0.08	. . .
Chromic acid	Goldfish	200	60–84
Copper sulfate	Stickleback	0.03 Cu	160
Copper nitrate	Stickleback	0.02 Cu	192
Cyanogen chloride	Goldfish	1	6–48
H_2S	Goldfish	10	96
HCl	Stickleback	pH 4.8	240
HCl	Goldfish	pH 4.0	4–6
Lead nitrate	Minnow, stickleback, brown trout	0.33 Pb	. . .
Mercuric chloride	Stickleback	0.01 Hg	204
Nickel nitrate	Stickleback	1 Ni	156
Nitric acid	Minnow	pH 5.0	. . .
Oxygen	Rainbow trout	3 cc/liter	. . .
Phenol	Rainbow trout	6	3
Phenol	Perch	9	1
Potassium chromate	Rainbow trout	75	60
Potassium cyanide	Rainbow trout	0.13 Cn	2
Sodium cyanide	Stickleback	1.04 Cn	2
Silver nitrate	Stickleback	70 K	154
Sodium fluoride	Goldfish	1000	60–102
Sodium sulfide	Brown trout	15	. . .
Zinc sulfate	Stickleback	0.3 Zn	120
Zinc sulfate	Rainbow trout	0.5	64
Pesticides			
1. Chlorinated hydrocarbons			
A Aldrin	Goldfish	0.028	96
DDT	Goldfish	0.027	96
DDT	Rainbow trout	0.5–0.32	24–36
DDT	Salmon	0.08	36
DDT	Brook trout	0.032	36
DDT	Minnow, guppy	0.75 ppb	29
DDT	Stoneflies (species)	0.32–1.8	96
BHC	Goldfish	2.3	96
BHC	Rainbow trout	3	96

TABLE 6–53. OBSERVED LETHAL CONCENTRATION OF SELECTED CHEMICALS IN AQUATIC ENVIRONMENTS (continued)

Chemical	Organism Tested	Lethal Concentration, mg/L	Exposure Time, hr
Chlordane	Goldfish	0.082	96
Chlordane	Rainbow trout	0.5	24
Dieldrin	Goldfish	0.037	96
Dieldrin	Bluegill	0.008	96
Dieldrin	Rainbow trout	0.05	24
Endrin	Goldfish	0.0019	96
Endrin	Carp	0.14	48
Endrin	Fathead minnow	0.001	96
Endrin	Various species	0.03–0.05 ppb	. . .
Endrin	Stoneflies (species)	0.32–2.4 ppb	96
Heptachlor	Rainbow trout	0.25	24
Heptachlor	Goldfish	0.23	96
Heptachlor	Bluegill	0.019	96
Heptachlor	Redear sunfish	0.017	96
Methoxychlor	Rainbow trout	0.05	24
Methoxychlor	Goldfish	0.056	96
Toxaphene	Rainbow trout	0.05	24
Toxaphene	Goldfish	0.0056	96
Toxaphene	Carp	0.1	. . .
Toxaphene	Goldfish	0.2	24
Toxaphene	Goldfish	0.04	170
Toxaphene	Minnows	0.2	24
2. Organic phosphates			
Chlorothion	Fathead minnow	3.2	96
Dipterex	Fathead minnow	180	96
EPN	Fathead minnow	0.2	96
Guthion	Fathead minnow	0.093	96
Guthion	Bluegill	0.005	96
Malathion	Fathead minnow	12.5	96
Parathion	Fathead minnow	1.4–2.7	96
TEPP	Fathead minnow	1.7	96
3. Herbicides			
Weedex	Young roach }	40–80	1 month
Weeda Zol	and trench }	15–30	1 month
Weeda Zol T.L.		20–40	1 month
Simazine (no plants present)	Minnow	0.5	< 3 days
Atrazine (A361) (plants present)	Minnow	5.0	24
Atrazine in Gesaprime	Minnow	3.75	24
4. Bactericides			
Algibiol	Minnow	20	24
Soricide tetraminol	Minnow	8	48

Source: McGauhey, Engineering Management of Water Quality, McGraw-Hill, copyright 1968

SECTION H. RECREATIONAL WATER QUALITY

TABLE 6–54. GUIDES FOR EVALUATING RECREATIONAL WATERS

Determination	Water Contact		Boating and Aesthetic	
	Noticeable Threshold	Limiting Threshold	Noticeable Threshold	Limiting Threshold
Coliforms, MPN per 100 ml	1000*	†		
Visible solids of sewage origin	None	None	None	None
ABS (detergent), mg/liter	1*	2	1*	5
Suspended solids, mg/liter	20*	100	20*	100
Flotable oil and grease, mg/liter	0	5	0	10
Emulsified oil and grease, mg/liter	10*	20	20*	50
Turbidity, silica scale units	10*	50	20*	‡
Color, standard cobalt scale units	15*	100	15*	100
Threshold odor number	32*	256	32*	256
Range of pH	6.5–9.0	6.0–10.0	6.5–9.0	6.0–10.0
Temperature, maximum °C	30	50	30	50
Transparency, Secchi disk, ft	. . .	. . .	20*	‡

* Value not to be exceeded in more than 20 percent of 20 consecutive samples, nor in any 3 consecutive samples.
† No limiting concentration can be specified in the basis of epidemiological evidence, provided no fecal pollution is evident. (Note: Noticeable threshold represents the level at which people begin to notice and perhaps to complain. Limiting threshold is the level at which recreational use of water is prohibited or seriously impaired.
‡ No concentration likely to be found in surface waters would impede use.
Source: California State Water Quality Control Board, 1963

SECTION I. WATER QUALITY FOR LIVESTOCK

TABLE 6–55. GUIDES FOR EVALUATING THE QUALITY OF WATER USED BY LIVESTOCK

Quality Factor	Threshold Concentration*	Limiting Concentration†
Total dissolved solids (TDS), mg/liter	2500	5000
Cadmium, mg/liter	5	
Calcium, mg/liter	500	1000
Magnesium, mg/liter	250	500‡
Sodium, mg/liter	1000	2000‡
Arsenic, mg/liter	1	
Bicarbonate, mg/liter	500	500
Chloride, mg/liter	1500	3000
Fluoride, mg/liter	1	6
Nitrate, mg/liter	200	400
Nitrite, mg/liter	None	None
Sulfate, mg/liter	500	1000‡
Range of pH	6.0–8.5	5.6–9.0

*Threshold values represent concentrations at which poultry or sensitive animals might show slight effects from prolonged use of such water. Lower concentrations are of little or no concern.
† Limiting concentrations based on interim criteria, South Africa. Animals in lactation or production might show definite adverse reactions.
‡ Total magnesium compounds plus sodium sulfate should not exceed 50 percent of the total dissolved solids.
Source: California State Water Quality Control Board, 1963

TABLE 6–56. QUALITY STANDARDS FOR WATER USED FOR LIVESTOCK

	Threshold Salinity Concentration, ppm.
Poultry	2,860
Pigs	4,290
Horses	6,435
Cattle, dairy	7,150
Cattle, beef	10,000
Adult dry sheep	12,900

Source: Jour. of Agriculture of Western Australia, 1950

TABLE 6–57. WATER QUALITY CRITERIA FOR LIVESTOCK

[Criteria provide a general guide to quality of water acceptable for most livestock; water of different quality may be acceptable because of nature, age, or condition of species being raised or because of special rearing conditions or feed components]

Quality Factor	Limiting Threshold mg/L
Aluminum	5.0
Arsenic	0.2
Boron	5.0
Cadmium	0.05
Chromium	1.0
Cobalt	1.0
Copper	0.5
Fluoride	2.0
Lead	0.1
Mercury	0.01
Nickel	1.0
NO_3-N+NO_2-N	100.0
NO_2-N	10.0
Radionuclides	meeting drinking water objectives
Selenium	0.05
Vanadium	0.1
Zinc	25.0
Salinity (total soluble salts)	3000.0
Toxic algae	no heavy growth
PESTICIDES:	
Aldrin	0.001
Chlordane	0.003
DDT	0.05
Dieldrin	0.001
Endrin	0.0005
Heptachlor	0.0001
Heptachlor epoxide	0.0001
Lindane	0.005
Methoxychlor	1.0
Toxaphene	0.005
Carbamate & organo-phosphorus pesticides	0.1

Source: Ontario Ministry of the Environment, 1984, Water Management

TABLE 6–58. GUIDE TO THE USE OF SALINE WATER FOR LIVESTOCK AND POULTRY

Total Soluble Salts (mg/L)	Comments
Less than 1,000	These waters have a relatively low level of salinity and should present no serious problem to any class of livestock or poultry.
1,000–2,999	These waters should be satisfactory for all classes of livestock and poultry. They may cause temporary and mild diarrhea in livestock not accustomed to them or watery droppings in poultry (especially at the higher levels), but should not affect their health or performance.
3,000–4,999	These waters should be satisfactory for livestock, although they might cause temporary diarrhea or be refused at first by animals not accustomed to them. They are poor waters for poultry, often causing watery feces and (at the higher levels of salinity) increased mortality and decreased growth, especially in turkeys.
5,000–6,999	These waters can be used with reasonable safety for dairy and beef cattle, sheep, swine, and horses. It may be well to avoid the use of those approaching the higher levels for pregnant or lactating animals. They are not acceptable waters for poultry, almost always causing some type of problem, especially near the upper limit, where reduced growth and production or increased mortality will probably occur.
7,000–10,000	These waters are unfit for poultry and probably for swine. Considerable risk may exist in using them for pregnant or lactating cows, horses, sheep, the young of these species, or for any animals subjected to heavy heat stress or water loss. In general, their use should be avoided, although older ruminants, horses, and even poultry and swine may subsist on them for long periods of time under conditions of low stress.
More than 10,000	The risks with these highly saline waters are so great that they cannot be recommended for use under any conditions.

Source: National Academy of Sciences, 1974, Nutrients and Toxic Substances in Water for Livestock and Poultry

SECTION J. WATER TREATMENT PROCESSES

TABLE 6–59. COMMON WATER QUALITY PROBLEMS, EFFECTS, AND TREATMENT

Probable Cause	General Effect	Probable Remedy
Hardness (calcium and magnesium)	Scales in pipes and water heaters; causes "soap curd" on fixtures, tiles, dishes and laundry; low sudsing characteristics.	Removal by ion exchange softener.
Iron, Manganese	Causes discolored water; red, brown, orange or black stains on fixtures, appliances and laundry; dark scale in pipes and water heaters.	Low level (2 ppm) removal by ion exchange softener when hardness is also present; best removed by oxidizing iron filter, aeration and/or chlorination followed by filtration in some cases.
Iron, Manganese, Sulfur Bacteria	Same general effects as above plus slimey deposits that form in pumps, pipes, softeners and toilet tanks.	Low level removal possible by oxidizing iron filter; best removed by chlorination followed by filtration.
Hydrogen Sulfide Gas	Foul rotten-egg odor; corrosion to plumbing; tarnishes silver and stains fixtures and laundry; ruins the taste of foods and beverages.	Best removed by aeration, scrubbing and filtration; also removed by oxidizing filters or chlorination followed by filtration.
Turbidity	Suspended matter in water; examples include mud, clay, silt and sand; can ruin seals and moving parts in appliances.	Removal by backwashing sediment filters; extra fine treatment utilizing sediment cartridge elements.
Acid Water (low pH)	Corrosive water attacks piping and other metals; red and/or green staining of fixtures and laundry.	Best corrected by neutralizing filters or soda ash feeding.
Taste, Odor, Color (organic matter)	Makes water unpalatable; can cause staining.	Depending on the nature of contaminant, aeration followed by filtration; carbon filtration; chlorination followed by filtration.
Tannins, Humic Acid	Can impart an "iced-tea" color to water; causes light staining; can affect the taste of foods and beverages.	Removal by special ion exchange or oxidizing agents and filtration.
Coliform Bacteria	Can cause serious disease and intestinal disorders.	Chlorination and filtration is most widely practiced; iodination, ozonation and ultraviolet treatment are used to a lesser degree.
Organic Halides (e.g., herbicides and pesticides)	Can cause serious disease and/or poisoning.	Most are readily removed by absorption with carbon filters; some can also be removed by hydrolysis and oxidation.
Nitrates, Chlorides and Sulfates	Can cause health-related problems if quantities are high.	Removal by special ion exchange, deionization process or reverse osmosis.
Sodium Salts	Imparts an alkaline or soda taste to water.	Removal by deionization process or reverse osmosis, distillation can be used.

Source: Chandler, J., A Comprehensive Look at Water Treatment, Water Well Journal, May 1988. Copyright Water Well Publ. Co. Reprinted with permission.

TABLE 6–60. SUMMARY OF CONVENTIONAL PROCESSES AND SYSTEMS FOR WATER QUALITY CONTROL

Type or Process	Common Application	Approximate Limit of Quality Input	Principal Change in Quality Factors (Approximate)
		GRAVITY SEPARATION	
Plain sedimentation	Reduction in suspended solids in raw water to be pumped	No theoretical limit: 3000–5000 mg/liter typical maximum in floodwaters	Removes larger and heavier suspended solids
	Primary sewage treatment	Unspecified	50% reduction in suspended solids 35–40% reduction in BOD 50% reduction in turbidity
	Secondary sewage treatment	Unspecified	Unreported
	Concentrating return activated sludge (secondary treatment)	Unspecified	Thickens sludge to 20–25% original volume
	Concentrating or reducing suspended solids in industrial wastes, organic and inorganic	Unspecified	Highly dependent upon nature of waste treated
	Grit removal-raw sewage	Unspecified	Removes heavy suspended solids not transported at velocity of 1 ft/sec
Plain sedimentation plus skimming	Primary sewage treatment	Unspecified	25–40% reduction in BOD 40–70% reduction in suspended solids 25–75% reduction in bacteria 2% reduction in detergents
	Various industrial wastes	Unspecified	Dependent upon nature of waste
Trickling filter plus plain sedimentation	Secondary sewage treatment	0.25–3.0 lb BOD/cu yd/filter	80–95% reduction in BOD 70–92% reduction in suspended solids 90–95% reduction in bacteria 30–35% reduction in ABS 80–90% reduction in LAS
	Organic industrial wastes (e.g., milk process)	Dependent upon waste treated	Dependent upon nature of waste
Activated sludge plus plain sedimentation	Secondary sewage treatment	Unspecified	80–95% reduction in BOD 85–95% reduction in suspended solids 95–98% reduction in bacteria 50% reduction in BAS 90–99% reduction in LAS
Sedimentation after mechanical flocculation	Raw sewage (experimentally)	Unspecified	64% reduction in turbidity 40% reduction in suspended solids 60% reduction in BOD
	Industrial wastes	Unspecified	Variable, depending upon nature of wastes treated
Sedimentation after chemical coagulation	Municipal and industrial water supply Water softening	Unspecified	Seldom evaluated separate from filtration
	Raw sewage (not common)	Unspecified	50–85% reduction in BOD 70–90% reduction in suspended solids 40–80% reduction in bacteria

TABLE 6–60. SUMMARY OF CONVENTIONAL PROCESSES AND SYSTEMS FOR WATER QUALITY CONTROL (continued)

Type or Process	Common Application	Approximate Limit of Quality Input	Principal Change in Quality Factors (Approximate)
		GRAVITY SEPARATION (CONTINUED)	
Sedimentation after chemical coagulation	Industrial wastes	Dependent upon waste	Variable, dependent upon nature of waste
Chemical coagulation plus sedimentation	Municipal water supply	Unspecified	Coalesces and precipitates dispersed clay colloids Reduces turbidity Reduces color
	Phosphate removal from waste waters	Unspecified	Reduces soluble phosphates to trace amounts
	Lime-soda softening of water supplies	Applicable to waters containing Ca and Mg sulfates and bicarbonates; Iron and Mg in natural waters (e.g., maximum from 10 mg/liter; minimum, 3 mg/liter	Reduces hardness to approximately 75 mg/liter; by excess lime to 30–50 mg/liter; by hot process to < 10 mg/liter as $CaCO_3$ Reduces Fe to 0.1 mg/liter (±) Removes CO_2 - requiring restabilization 80–100% reduction in bacteria by excess lime
		FILTRATION	
Slow sand (gravity)	Tertiary treatment of sewage effluent Water reclamation systems	Relatively low turbidity	90–95% reduction in BOD 85–95% reduction in suspended solids 95–98% reduction in bacteria 90–99% reduction in surfactants
	Municipal water supply	Turbidity 40 mg/liter	99% reduction in bacteria 95–100% reduction in turbidity 30% reduction in color Odors and tastes removed 60% reduction in iron
	Industrial wastes	Unspecified	Varies with nature of waste
Rapid sand (gravity)	Municipal and industrial water supply (little used without coagulation)	Low turbidity, e.g., 50 mg/liter, maximum coliform MPN 5000/100 ml	95% reduction in bacteria 90% reduction in turbidity
Rapid sand plus chemical coagulation (gravity)	Municipal and industrial water supply	No limit specified for maximum turbidity Maximum coliform MPN 5,000–20,000/100 ml	90–99% reduction in bacteria 100% (–) reduction in turbidity Color reduction to less than 5 mg/liter Alkali increased 7.7 mg/liter/gr. alum CO_2 increased 6.8 mg/liter/gr. alum Slight reduction in iron Odor and taste partially removed
Rapid sand plus chemical coagulation, chlorination, and activated carbon	Municipal and industrial water supply	No limit specified for maximum turbidity Maximum coliform MPN 5,000–20,000/100 ml	Approximately 100% reduction in bacteria 100% reduction in turbidity Color reduced to near zero Iron and Mn reduced Taste and odor removed
Rapid sand (pressure) (precoat with chemical floc)	Small municipal supply Swimming pools Industrial supply and process Emergency and military use	Generally unspecified Low turbidity desirable	Similar to rapid sand filter but more variable in performance
Diatomaceous earth (pressure and vacuum)	Small municipal supplies Institutional water supply Swimming pools Industrial supply and process Emergency and military use	None specified, but operation depends upon nature of water	Capable of good clarification of water; efficiency, however, not well documented 40–90% reduction in suspended solids 50% reduction in color

TABLE 6–60. SUMMARY OF CONVENTIONAL PROCESSES AND SYSTEMS FOR WATER QUALITY CONTROL (continued)

Type or Process	Common Application	Approximate Limit of Quality Input	Principal Change in Quality Factors (Approximate)
		FILTRATION (CONTINUED)	
Contact filters	Manganese removal Iron removal	None specified	Reduces to USPHS Standards 88% reduction in iron
Bag filters	Swimming pools	Unspecified	Strains out hair and coarser suspended solids, reduces bacteria to level controllable by chlorination practice
Microstraining	Primary clarification of water prior to filtration Clarification of sewage effluents Treatment of industrial wastes	Size of particles to be removed greater than screen size Material suitable for microstraining	87–96% reduction in microscopic organisms 60–90% reduction in microscopic particulates 50–60% reduction in suspended solids trickling filter effluent 30–40% reduction on turbidity
Fine screening	Raw sewage	None specified	5–10% reduction in BOD 2–20% reduction on suspended solids 10–20% reduction in bacteria
	Industrial wastes (e.g., cannery, pulp mill, etc.)	None specified	Varies with nature of waste
Carbon filters	Special municipal and industrial water applications	Very low turbidity, other not specified	Adsorbs exotic organic chemicals, including surfactants Removes tastes and odors Adsorbs miscellaneous gases
		AERATION	
Spray or cascade	Municipal and industrial water supply Industrial waste treatment	Unspecified	Releases gases producing taste and odor Reduces CO_2 in groundwaters to normal surface water levels Partial removal of H_2S Partial removal of gases of decomposition Oxidation and removal of soluble iron in groundwaters; 80–97% reduction observed
Pressure aerators	Treatment of sewage and industrial wastes	Limits variable or unspecified	Grit precipitated Grease concentrated at surface Separates various solids by flotation Maintains aerobic conditions in biological systems, e.g., activated sludge, aerated ponds Reduces ABS or LAS 1–2 mg/liter Reduces septicity of sewage
Oxidation ponds	Treatment of domestic sewage and organic industrial wastes	No toxic substances	75–96% reduction in BOD 90–99% reduction in suspended solids 98–99.9% reduction in bacteria 56–93% reduction in LAS
		DEMINERALIZATION	
Ion exchange (natural or synthetic zeolite)	Softening of groundwater supplies for municipal or industrial use	Hardness (Ca and Mg sulfates and bicarbonates) of natural waters >850–1000 mg/liter $CaCO_3$ Iron < 1.5–2 mg/liter Low in silica CO_2< 15 mg/liter	Increases sodium content by exchange with removed Ca and Mg

TABLE 6–60. SUMMARY OF CONVENTIONAL PROCESSES AND SYSTEMS FOR WATER QUALITY CONTROL (continued)

Type or Process	Common Application	Approximate Limit of Quality Input	Principal Change in Quality Factors (Approximate)
		DEMINERALIZATION (CONTINUED)	
Ion exchange (greensand or styrene base gels)	Iron or Mn removal from groundwater	Iron less than approximately 2.0 mg/liter	90–100% removal of iron Mn partially removed
Ion exchange (organic cation exchangers)	Special water conditioning for industry and commerce	Unspecified	Removes all cations (Na, K, Mg, Fe, Cu, Mn)
Ion exchange (anion exchangers)	Special water conditioning for industry and commerce	Unspecified	Removes SO_4, Cl, NO_3, etc.
Ion exchange (fluoride exchangers)	Defluoridation of public water supply	More than 1.5 mg/liter F in water supply	Approximately 100% removal possible Normally reduced to <1.5 mg/liter
Electrochemical desalting	Reclaiming water from saline sources, public and industrial supplies Demineralizing municipal waste effluents	Applicable to highly saline or brackish waters	Removes anions and cations
Reverse osmosis	Reclamation of water from brackish natural or waste waters (experimental)	Brackish waters, upper limit not specified	Reduces ions depending upon concentration difference across membrane 97–98% reduction in TDS, ABS, and COD
Distillation	Reclamation of water from saline sources Specialty industrial and commercial supplies	No limit	Produces distilled water (may be contamination with NH_3, volatile organics, etc.)
Freezing	Reclamation of water from saline sources Specialty industrial and commercial supplies	No limit	(Experimental)
		CHLORINATION	
Liquid Cl_2 and Cl_2 compounds	Public water supply Industrial water supply	Turbidity low for waters to be sterilized by Cl_2	Reduces bacterial load on filters Oxidizes organic matter Reduces odor Assists in color removal 100% (—) reduction in bacteria Controls plankton growth in reservoirs Reduces Mn concentration in break-point
	Municipal and industrial waste-water treatment and management	Unspecified	Assists in grease removal Controls filter fly nuisance Cleans air stones in aeration systems Removes H_2S Removes NH_3 Controls slime formation in sewers and cooling towers Assists in control of digester foaming Disinfects effluent; 98–99% reduction in bacteria
		DIGESTION	
Anaerobic digestion	Stabilization of sewage solids Stabilization of organic industrial wastes	pH above 6.8 Acids limited No toxic substances in significant amounts Minimum of grit	Reduces organic sludges to humus and relatively stable chemical compounds Produces offensive supernatant

Source: McGauhey, Engineeering Management of Water Quality, McGraw-Hill, Copyright 1968

TABLE 6–61. POTENTIAL WATER TREATMENT EFFICIENCES[a)]

	Conventional Processes					Special Processes						
						Activated Carbon Absorption						
Parameter	Aeration	Chemical Oxidation (Chlorination, etc.)	Coagulation Flocculation	Lime Softening	Filtration	PAC	GAC	Air Stripping	Demineralizing (Reverse Osmosis, etc.)	Ion Exchange	Ozone	Comments
Aldrin	P		P			G	VG				VG	
Antimony			X		A		X					
Arsenic		A	L-G	G-VG	A	P			G-VG	VG		Valencies important
Asbestos			G-VG		G							
Barium			P	G-VG	A	P	P		VG	VG		
Boron			X				G-VG		X	G-VG		
Cadmium			L-G	VG	A		P-L					pH important
Chlordane		P	L	L		VG	VG					
Chloride									VG	VG		
Chromium			G	G	A	P	P		X	X		Valencies important
Color			VG		A						VG	
Copper	A		F-G		A							
Cyanide		VG									VG	
2,4-D		P	P		A	VG	X					
DDT		P	L-VG	F		VG	X				P	
Diazinon							X(L)					
Dieldrin			P-L		A		G-VG	L				
Endrin			L			G-VG	VG					
Fluoride									G	G-VG		
Heptachlor						VG	X(VG)					
Heptachlor Epoxide						VG	X					
Iron	A	A		A	VG					VG		
Lead			G-VG	VG	A		X		G-VG	X		
Lindane		P	P		P	G	G-VG					
Manganese		A	L-G	G	A					VG		
Mercury			G	F-G	A	VG	VG					Form important
Methoxychlor			G	G-VG	A	VG	VG					
Methyl Parathion						X	X				X	
Nitrate									F	F-VG		

TABLE 6–61. POTENTIAL WATER TREATMENT EFFICIENCES[a] (continued)

	Conventional Processes					Special Processes						
Parameter	Aeration	Chemical Oxidation (Chlorination, etc.)	Coagulation Flocculation	Lime Softening	Filtration	Activated Carbon Absorption PAC	Activated Carbon Absorption GAC	Air Stripping	Demineralizing (Reverse Osmosis, etc.)	Ion Exchange	Ozone	Comments
NTA		P									G-VG	
Odor	A	VG				VG	VG				VG	
Parathion		P-VG	P	P	A	VG	L-VG				G-VG	
pH	A		A	A								
Phenol		G	P			G-VG	X				G-VG	
Radionuclides												
^{226}Ra			P	G-VG	A				VG	G-VG		
^{90}Sr			P	G-VG	A					G-VG		
^{137}Cs				P	P-F				VG	VG		
^{131}I			P	L					VG			
Selenium			P-G	P-F	A				F-G	X		Valencies important
Silver			F-G	G-VG	A		P		X	X		
Sulphate									G-VG	G-VG		
Sulphide	F-VG	F-VG									F-VG	pH important
2,4,5-TP		P	X(F)			X(G)	X(G-VG)					
T. Dissolved Solids									G-VG	G-VG		
Toxaphene		P	P			VG	X(VG)	X				
Trihalomethanes							F-G	F-G				Process generated
Turbidity			G-VG		A							
Uranium			L-G	F-G	A		P			VG		
Zinc			P	F-G	A							

VG = 90–100% removal
G = 70–90% removal
F = 50–70% removal
L = 25–50% removal
P = 0.25% removal
A = auxiliary process

X = possible candidate process (data lacking)
PAC = Powdered activated carbon
GAC = Granular activated carbon
[a] = Treatment based on available full-scale, pilot or bench studies and should only be used as potential indicators.
Treatabilty studies and/or site experience should be assessed for specific applications.

Source: Canadian Council of Resource and Environment Ministers, March 1987. Canadian Water Quality Guidelines. Data provided by McDonald & Associates Consulting Engineers, Regina, Saskatchewan

TABLE 6–62. TREATMENT UTILIZED BY MAJOR WATER UTILITIES IN THE UNITED STATES

[Selected utilities serving a population of 100,000 or more]

Water Treatment Process

Water Utility	Lime Softening	Lime/Soda Ash Softening	Ion Exchange Softening	Aeration (Conventional)	Aeration (Removal or VOCs)	Coagulation/Flocculation	Sedimentation	Slow Sand Filtration	Rapid Sand/Anthracite Filtration	Direct Filtration	Pressure Filtration	Filtration Using GAC	PAC Addition	Ozone Disinfection	Chlorine Disinfection	Chlorine Dioxide Disinfection	Chloramine Disinfection	Fluoridation	Organics Removal Using Resins
ALABAMA																			
Birmingham						•	•		•				•		•			•	
Montgomery						•	•		•						•			•	
ALASKA																			
Anchorage						•	•		•						•			•	
ARIZONA																			
Phoenix/Phoenix Wtr & Swr Dept						•	•		•						•				
Tempe						•	•		•						•	•		•	
ARKANSAS																			
Fort Smith						•	•		•						•				
Little Rock						•	•		•						•			•	
CALIFORNIA																			
Concord/Contra Costa Wtr Dept				•		•	•		•				•		•		•	•	
Corte Madera/Marin Cnty Wtr Dist.						•	•		•	•					•			•	
Escondido						•	•		•						•				
Fremont/Alameda Cnty Wtr Dist.						•	•		•			•			•		•	•	
Glendale															•		•		
La Mesa/Helix Wtr Dist.						•	•		•						•		•		•
Los Angeles/Los Angeles Wtr & Power						•									•				
Oakland/East Bay Munic. Util. Dist.				•		•	•		•	•			•		•		•	•	
Palm Springs/Desert Wtr Agency							•								•				
Pasadena						•	•		•						•				
Pomona						•	•		•						•				
San Bernardino									•						•				
San Diego						•	•		•				•				•		
San Jose/ San Jose Wtr Wks						•				•					•				
Santa Barbara/Santa Barbara Pub.				•		•	•		•						•				
COLORADO																			
Aurora						•	•		•	•					•			•	
Colorado Springs						•	•		•	•					•				
Denver						•	•	•	•								•	•	
Pueblo						•	•		•				•				•	•	
CONNECTICUT																			
Bridgeport						•	•		•						•			•	
Clinton/Connecticut Wtr Co.						•				•		•			•			•	
Greenwich/Connecticut-American							•		•						•			•	
Hartford				•		•	•	•	•						•			•	
New Britain						•	•		•				•		•			•	
New Haven/Regional Wtr Authority						•	•	•		•					•			•	
DISTRICT OF COLUMBIA																			
Washington/Washington Aqueduct						•	•	•	•						•			•	
FLORIDA																			
Tampa						•	•		•								•		
West Palm Beach/City of						•	•		•						•				
GEORGIA																			
Acworth/Wyckoff Treatment Div.						•	•		•						•			•	
Columbus						•	•		•						•	•		•	
Marietta/Quarles Treatment Div.						•	•		•						•			•	
Morrow/Clayton Cnty Wtr Auth.				•		•	•		•				•		•	•		•	
Stone Mountain/Dekalb Cnty Wtr						•	•		•						•			•	

TABLE 6–62. TREATMENT UTILIZED BY MAJOR WATER UTILITIES IN THE UNITED STATES (continued)

[Selected utilities serving a population of 100,000 or more]

Water Utility	Lime Softening	Lime/Soda Ash Softening	Ion Exchange Softening	Aeration (Conventional)	Aeration (Removal or VOCs)	Coagulation/Flocculation	Sedimentation	Slow Sand Filtration	Rapid Sand/Anthracite Filtration	Direct Filtration	Pressure Filtration	Filtration Using GAC	PAC Addition	Ozone Disinfection	Chlorine Disinfection	Chlorine Dioxide Disinfection	Chloramine Disinfection	Fluoridation	Organics Removal Using Resins
HAWAII																			
Honolulu								•							•				
ILLINOIS																			
Alton/Alton Dist/Ill.-Amer. Water						•	•		•				•		•		•	•	
East St. Louis/Interurban Dist						•	•		•			•			•	•	•	•	
Evanston						•	•		•						•			•	
Peoria/Peoria Dist/Ill.-Amer. Water						•	•					•			•	•		•	
Springfield	•					•	•		•				•		•		•	•	
INDIANA																			
Fort Wayne		•				•	•		•						•	•	•	•	
Gary						•	•		•				•		•		•	•	
Hammond									•										
Indianapolis						•	•		•						•		•	•	
IOWA																			
Davenport/Davenport Wtr Co.						•	•					•	•		•	•	•	•	
Des Moines		•				•	•		•						•			•	
KANSAS																			
Kansas City						•	•		•				•			•	•	•	
Mission/Johnson Cnty Wtr Dist. 1		•				•			•						•	•	•	•	
Topeka		•				•	•		•				•	•			•	•	
Wichita		•		•		•	•		•								•		
KENTUCKY																			
Edgewood/Kenton Cnty Wtr Dist.						•	•		•						•			•	
Lexington/Kentucky-American Wtr				•		•	•		•						•	•		•	
Louisville		•				•	•		•				•		•		•	•	
LOUISIANA																			
Jefferson						•				•			•				•	•	
Marrero/Jefferson Wtr Wks Dist. 2						•	•		•	•			•				•	•	
New Orleans	•					•	•		•				•				•	•	
MAINE																			
Portland															•		•		
MARYLAND																			
Baltimore						•	•		•						•			•	
Hyattsville/Washington Suburban						•	•		•				•		•			•	
MASSACHUSETTS																			
Springfield						•	•	•	•						•				
MICHIGAN																			
Ann Arbor	•					•	•		•				•				•	•	
Detroit						•	•		•				•		•			•	
Grand Rapids		•				•	•		•						•			•	
Holland/Wyoming Wtr Plant						•	•		•						•			•	
Saginaw						•	•		•				•		•			•	
MINNESOTA																			
Duluth						•	•		•						•		•	•	
St. Paul	•					•	•		•							•	•	•	
MISSISSIPPI																			
Jackson						•	•		•						•		•	•	
MISSOURI																			
Kansas City	•					•	•		•				•		•		•	•	
Springfield						•	•		•						•			•	
St. Joseph						•	•		•				•		•				

TABLE 6–62. TREATMENT UTILIZED BY MAJOR WATER UTILITIES IN THE UNITED STATES (continued)
[Selected utilities serving a population of 100,000 or more]

Water Utility	Lime Softening	Lime/Soda Ash Softening	Ion Exchange Softening	Aeration (Conventional)	Aeration (Removal or VOCs)	Coagulation/Flocculation	Sedimentation	Slow Sand Filtration	Rapid Sand/Anthracite Filtration	Direct Filtration	Pressure Filtration	Filtration Using GAC	PAC Addition	Ozone Disinfection	Chlorine Disinfection	Chlorine Dioxide Disinfection	Chloramine Disinfection	Fluoridation	Organics Removal Using Resins
MISSOURI (continued)																			
St. Louis/St. Louis Water Div.	•					•	•		•				•				•	•	
St. Louis/St. Louis Cnty Wtr Co.		•				•	•		•				•		•		•		
NEBRASKA																			
Omaha		•				•	•		•				•		•			•	
NEVADA																			
Las Vegas						•				•					•	•			
Reno/Sierra Pacific Power Co.						•	•			•					•				
NEW HAMPSHIRE																			
Manchester						•	•		•			•			•				
NEW JERSEY																			
Cliffton/Passaic Valley Wtr Comm.						•	•		•			•	•		•				
Elizabeth/Elizabethtown Wtr Co.						•	•		•				•		•			•	
Harrington Park/Hackensack Wtr Co.				•		•	•		•		•		•		•		•		
Iselin/Middlesex Wtr Co.						•	•		•						•				
Short Hills/Commonwealth Wtr Co.				•		•	•		•				•			•	•		
Shrewsbury/Monmouth Cons.						•	•		•				•		•			•	
Wanaque/North NJ Dist Wtr Co.						•	•		•						•				
NEW YORK																			
Albany		•		•		•	•		•						•				
Buffalo/Erie Cnty Wtr Auth.						•	•		•						•			•	
New York City															•			•	
Rochester/Bureau of Water															•			•	
Rochester/Monroe Cnty Wtr Auth.						•													
Syracuse															•			•	
West Nyack/Spring Valley Wtr Co.				•		•	•		•		•		•		•		•		
Yonkers									•						•				
NORTH CAROLINA																			
Charlotte		•				•	•		•				•		•			•	
Durham						•	•		•						•			•	
Greensboro						•	•		•						•			•	
Raleigh						•	•		•				•		•			•	
Winston-Salem						•	•		•						•			•	
OHIO																			
Akron						•	•		•						•	•		•	
Cincinnati						•	•		•				•		•			•	
Cleveland						•	•		•						•			•	
Poland/Ohio Wtr Svce Co.		•				•	•		•						•			•	
Toledo		•				•	•		•				•		•	•		•	
OKLAHOMA																			
Oklahoma City	•			•		•	•		•				•		•		•	•	
Tulsa		•				•	•		•				•		•			•	
OREGON																			
Eugene		•				•	•		•						•				
Portland																	•		
PENNSYLVANIA																			
Allentown						•	•		•						•				
Bryn Mawr/Phila. Suburban					•	•	•		•				•		•	•	•		
Chester		•				•	•		•			•	•		•		•	•	
Greensburg/Westmoreland M.A.			•	•			•		•						•	•	•		
Hershey/Keystone Wtr Co.						•	•	•					•		•			•	

TABLE 6–62. TREATMENT UTILIZED BY MAJOR WATER UTILITIES IN THE UNITED STATES (continued)

[Selected utilities serving a population of 100,000 or more]

Water Treatment Process

Water Utility	Lime Softening	Lime/Soda Ash Softening	Ion Exchange Softening	Aeration (Conventional)	Aeration (Removal or VOCs)	Coagulation/Flocculation	Sedimentation	Slow Sand Filtration	Rapid Sand/Anthracite Filtration	Direct Filtration	Pressure Filtration	Filtration Using GAC	PAC Addition	Ozone Disinfection	Chlorine Disinfection	Chlorine Dioxide Disinfection	Chloramine Disinfection	Fluoridation	Organics Removal Using Resins
PENNSYLVANIA (continued)																			
Lancaster		•				•	•		•						•			•	
Philadelphia		•		•		•	•		•				•		•		•	•	
Pittsburgh/Western Penn.						•	•	•	•		•	•	•		•	•		•	
Pittsburgh/West View M.A.						•	•					•	•		•	•		•	
York						•	•		•						•		•		
RHODE ISLAND																			
Pawtucket				•		•	•					•			•			•	
SOUTH CAROLINA																			
Anderson						•	•		•						•			•	
Charleston						•	•		•						•	•			
Spartanburg						•	•		•				•		•			•	
TENNESSEE																			
Chattanooga/Tenn.-Am. Wtr Co.						•	•		•			•	•		•			•	
Knoxville/Utilities Board						•	•		•						•			•	
Nashville				•		•	•		•						•	•	•	•	
TEXAS																			
Abilene		•				•	•		•						•		•		
Amarillo						•	•								•				
Arlington				•		•	•		•				•				•	•	
Dallas	•					•	•		•				•		•		•	•	
El Paso		•		•		•	•		•				•		•				
Fort Worth						•	•		•								•	•	
Lubbock						•	•		•										
Waco						•	•		•								•	•	
Wichita Falls						•	•		•								•	•	
UTAH																			
Salt Lake City/City Dept						•	•		•						•				
Salt Lake City/Metro Wtr Dist.				•		•	•		•						•				
Salt Lake City/Salt Lake Cnty						•				•			•		•				
VIRGINIA																			
Chesterfield/County						•	•		•			•	•		•			•	
Merrifield/Fairfax Cnty Wtr A.						•	•		•								•	•	
Newport News						•	•		•						•			•	
Norfolk		•				•	•		•				•		•			•	
Richmond/City of Richmond				•		•	•		•				•		•		•	•	
WASHINGTON																			
Everett		•				•	•			•					•				
Seattle/Seattle Wtr Dept															•			•	
Tacoma															•				
WEST VIRGINIA																			
Beckley				•		•	•		•						•			•	
Charleston/W. Va. Wtr Co.						•	•		•						•	•		•	
Clarksburg						•	•		•			•	•		•			•	
Fairmont									•						•			•	
Huntington		•				•	•					•	•		•			•	
WISCONSIN																			
Milwaukee						•	•		•				•		•		•	•	
Racine						•	•		•						•			•	

Source: Compiled from 1984 Water Utility Operating Data issued by the American Water Works Association. Copyright 1986 AWWA. Printed with permission.

TABLE 6–63. WATER TREATMENT PROCESSES CONSIDERED FOR BEST AVAILABLE TECHNOLOGY

Conventional Processes	Advanced Processes
Coagulation, sedimentation, filtration	Activated alumina
Direct filtration	Adsorption
Diatomaceous earth filtration	GAC
Slow sand filtration	Powdered activated carbon
Lime softening	Resins
Ion exchange	Aeration
Oxidation-disinfection	Packed column
Chlorination	Diffused air
Chlorine dioxide	Spray
Chloramines	Slat tray
Ozone	Mechanical
Bromine	Cartridge filtration
Others	Electrodialysis
	Reverse osmosis
	Ultrafiltration
	Ultraviolet light (UV)
	UV with other oxidants

Source: Dyksen, J.E., Hiltebrand, D.J., and Raczko, R.F., 1988, SDWA Amendments: Effects on the Water Industry. Journal Am. Water Works Assoc., v.80, no.1. Copyright AWWA. Reprinted with permission

TABLE 6–64. CHEMICALS USED FOR TREATMENT BY PUBLIC WATER-SUPPLY SYSTEMS IN THE UNITED STATES AND CANADA

[Based on data from 430 of the largest US utilities and 24 of the 75 largest Canadian utilities]

Chemical	Total Use tons
Quick lime	330,988
Aluminum sulfate	188,986
Chlorine	79,034
Hydrated lime	44,679
Caustic soda	39,030
Carbon dioxide	18,111
Soda ash	13,750
Ferrous sulfate	10,590
Powdered activated carbon	9,016
Ferric sulfate	7,956
Sodium silicofluoride	7,903
Polyelectrolytes	5,915
Ammonia	5,232
Phosphate	3,970
Copper sulfate	2,825
Granular activated carbon	2,587
Potassium permanganate	1,231
Sodium aluminate	1,129
Hypochlorites	1,112
Sodium chloride	828
Clays	133

Source: American Water Works Association, 1988. Grigg, N.S., 1984 Water Utility Operating Data, Summary Report. Reprinted with permission

TABLE 6–65. COSTS OF SOME WATER TREATMENT TECHNOLOGIES

Population Range	Type of Treatment	Cost per Family per Year
501–1,000	conventional coagulation filtration and disinfection to control microbial contaminants	$125
50,001–75,000		$ 50
>1,000,000		$ 25
501–1,000	corrosion control (stabilization with lime) to control lead and other corrosion products	$ 60
50,001–75,000		$ 15
>1,000,000		<$ 10
501–1,000	packed tower aeration to control organic chemicals	$ 55
50,000–75,000		$ 28
>1,000,000		$ 20
501–1,000	granular activated carbon to control synthetic organic chemicals	$190
50,001–75,000		$130
>1,000,000		$ 40

Basic Assumptions:
- 3.2 persons per household
- each person using 180 gallons per day
- total cost per household including operation, maintenance and amortization of capital at 10 percent per year for 20 years

Source: U.S. Environmental Protection Agency; League of Women Voters Education Fund, 1987, Safety on Tap. Reprinted with permission

TABLE 6–66. COST OF TREATING CONTAMINATED GROUND WATER

[Portable treatment systems—1987]

Unit Type	Cost (dollars/unit*)
In situ biological treatment (suspended growth reactor)	$15–40/cu yd, treated
Rotating biodisks	0.20–1.10
Trickling filter	0.08–0.15
Activated sludge	0.10–0.30
Packed towers	0.02–0.10
Aeration basins	0.02–0.08
Carbon adsorption	0.20–0.90
Ultraviolet/hydrogen peroxide	0.04–0.18
Belt press	0.01–0.05
Mixing tanks (including chemicals)	0.03–0.29
Equalization tanks	0.005–0.01
Clarifiers	0.008–0.06
Solidification of solids in situ	0.20–1.00

* All costs per 1,000 gal. except as noted.
Source: Estimated by Geraghty & Miller Inc. Oak Ridge, TN.

TABLE 6–67. TREATMENT COSTS FOR REMOVAL OF TRICHLOROETHYLENE FROM DRINKING WATER

[Raw water concentration 500 ug/L; assuming 99-percent removal]

	Treatment Technique					
	GAC Absorption‡			Packed-Tower Aeration§		
	System Size					
Cost Component†	0.037 mgd	0.95 mgd	36.8 mgd	0.037 mgd	0.95 mgd	36.8 mgd
Capital** (*Thousand $*)	24	240	9000	69	264	4789
Annual O&M (*Thousand $*)	4.5	86	710	1.4	18	617
Total (*¢ per 1000 gal*)	57.0	34.0	14.0	79.0	15.5	9.4

† USEPA estimates in August 1983 dollars.
‡ Based on 10-min empty bed contact time.
§ Does not include air pollution controls.
** Includes site work, engineering, contractor overhead and profit, and contingencies.
Source: American Water Works Association, 1988, New Dimensions in Safe Drinking Water. Copyright AWWA. Reprinted with permission

TABLE 6–68. COST OF REMOVAL OF VOLATILE ORGANIC CHEMICALS IN DRINKING WATER

[US dollars; cost data as of 1984]

Contaminant (1)	Capacity[a] Millions of gallons per day (2)	Capacity[a] Micro-grams per liter (3)	Percent removal (4)	Cost, in Dollars per Thousand Gallons[b] Tower (5)	Cost, in Dollars per Thousand Gallons[b] Aeration basin (6)	Cost, in Dollars per Thousand Gallons[b] Carbon adsorption (7)
Trichloro-ethylene	0.5	100	90	0.273	0.546	0.868
		10	99	0.287	0.793	0.918
		1	99.9	0.296	1.032	1.010
		0.1	99.99	0.303	1.270	1.124
	1	100	90	0.182	0.383	0.637
		10	99	0.191	0.611	0.679
		1	99.9	0.196	0.850	0.765
		0.1	99.99	0.202	1.088	0.867
	10	100	90	0.083	0.207	0.356
		10	99	0.088	0.403	0.390
		1	99.9	0.093	0.587	0.458
		0.1	99.99	0.099	0.755	0.543
Tetrachloro-ethylene	0.5	100	90	0.279	0.637	0.610
		10	99	0.293	0.935	0.660
		1	99.9	0.302	1.228	0.705
		0.1	99.99	0.308	1.486	0.805
	1	100	90	0.186	0.460	0.453
		10	99	0.194	1.752	0.502
		1	99.9	0.201	1.046	0.548
		0.1	99.99	0.206	1.296	0.651
	10	100	90	0.085	0.277	0.197
		10	99	0.091	0.514	0.224
		1	99.9	0.098	0.726	0.251
		0.1	99.99	0.103	0.905	0.313
1.1.1-Tri-chloroethane	0.5	100	90	0.270	0.502	1.445
		10	99	0.289	0.825	1.651
		1	99.9	0.307	1.421	1.945
		0.1	99.99	0.332	2.572	2.605
	1	100	90	0.180	0.348	1.396
		10	99	0.192	0.644	1.500
		1	99.9	0.205	1.234	1.801
		0.1	99.99	0.230	2.313	2.402
	10	100	90	0.082	0.176	0.802
		10	99	0.089	0.430	0.973
		1	99.9	0.102	0.860	1.229
		0.1	99.99	0.122	1.821	1.818
Carbon tetrachloride	0.5	100	90	0.264	0.428	0.942
		10	99	0.287	0.531	1.021
		1	99.9	0.272	0.600	1.132
		0.1	99.99	0.280	0.648	1.340
	1	100	90	0.176	0.292	0.703
		10	99	0.181	0.371	0.775
		1	99.9	0.184	0.427	0.940
		0.1	99.99	0.186	0.470	1.063
	10	100	90	0.081	0.133	0.408
		10	99	0.683	0.196	0.467
		1	99.9	0.084	0.247	0.550
		0.1	99.99	0.085	0.286	0.719

TABLE 6–68. COST OF REMOVAL OF VOLATILE ORGANIC CHEMICALS IN DRINKING WATER (continued)

[US dollars; cost data as of 1984]

Contaminant (1)	Capacity[a]		Percent removal (4)	Cost, in Dollars per Thousand Gallons[b]		
	Millions of gallons per day (2)	Micro-grams per liter (3)		Tower (5)	Aeration basin (6)	Carbon adsorption (7)
Cis-1,2-Dichloro-ethylene	0.5	100	90	0.284	0.727	2.513
		10	99	0.296	1.010	2.791
		1	99.9	0.304	1.281	3.153
		0.1	99.99	0.310	1.572	3.511
	1	100	90	0.189	0.547	2.156
		10	99	0.196	0.828	2.417
		1	99.9	0.202	1.098	2.760
		0.1	99.99	0.206	1.379	3.099
	10	100	90	0.087	0.350	1.735
		10	99	0.093	0.571	1.989
		1	99.9	0.099	0.763	2.327
		0.1	99.99	0.104	0.966	2.660
1,2-Di-chloroethane	0.5	100	90	0.276	0.587	1.286
		10	99	0.285	0.749	1.465
		1	99.9	0.292	0.901	1.748
		0.1	99.99	0.297	1.054	2.322
	1	100	90	0.184	0.415	1.015
		10	99	0.190	0.568	1.177
		1	99.9	0.194	0.720	1.437
		0.1	99.99	0.197	0.871	2.980
	10	100	90	0.084	0.237	0.675
		10	99	0.087	0.368	0.820
		1	99.9	0.090	0.489	1.057
		0.1	99.99	0.094	0.603	1.566
1,1-Di-chloroethylene	0.5	100	90	0.262	0.406	0.880
		10	99	0.265	0.448	0.963
		1	99.9	0.270	0.500	1.066
		0.1	99.99	0.272	0.531	1.243
	1	100	90	0.174	0.274	0.647
		10	99	0.177	0.307	0.721
		1	99.9	0.180	0.348	0.814
		0.1	99.99	0.181	0.371	0.977
	10	100	90	0.080	0.121	0.364
		10	99	0.081	0.144	0.423
		1	99.9	0.082	0.176	0.499
		0.1	99.99	0.083	0.196	0.640

[a] To convert from mgd to m^3/day, multiply by 3,785.
[b] To convert from dollars/1,000 gal to dollars/m^3 multiply by 0.26412.
Source: Clark, R.M., Eilers, R.G., and Goodrich, J.A., 1984, VOC's in Drinking Water: Cost of Removal, U.S. Environmental Protection Agency, Cincinnati, OH 45268; PB85–166429

SECTION K. WATER TREATMENT FACILITIES

TABLE 6–69. NUMBER OF COMMUNITY WATER SYSTEMS (PLANTS) HAVING FILTERED VERSUS UNFILTERED SURFACE WATER SUPPLIES

	Community Water System Size Categories												
	25 to 100	101 to 500	501 to 1,000	1,001 to 3,300	3,301 to 10,000	10,001 to 25,000	25,001 to 50,000	50,001 to 75,000	75,001 to 100,000	1000,001 to 500,000	500,001 to 1,000,000	1,000,000 plus	Total
Number or plants of systems:													
Filtered	523	474	537	814	996	504	303	144	96	166	40	12	4,611
Unfiltered	310	305	217	226	160	65	25	13	10	9	3	3	1,346
Total[1]	833	779	754	1,040	1,156	569	328	157	108	175	43	15	5,957
Estimated population served (millions):													
Filtered	0.08	0.51	1.11	3.87	9.22	11.66	15.02	9.56	9.25	34.78	19.97	18.54	133.56
Unfiltered	0.02	0.08	0.17	0.45	0.97	0.98	0.88	0.76	0.85	1.98	2.41	11.55	21.10
Total[1]	0.09	0.57	1.28	4.33	10.20	12.64	15.91	10.31	10.09	36.77	22.38	30.09	154.66

[1] Totals may not add due to rounding.
Source: U.S. Environmental Protection Agency, 1987

TABLE 6–70. DATA ON SELECTED LARGE RAPID-SAND FILTER PLANTS IN THE UNITED STATES

	Chemical Feed and Mix						Sedimentation			Filtration						
City	Capacity, mgd	Type Feed	Alum Used Grains/gal	Lime used, Grains/gal	Mixing Time, min	Type of Mix	Time, hr	Flow V, ft/min	Basin Depth, ft	Filter Rate, mgd/acre	Unit Size, mgd	Sand Depth, in.	Sand Size, mm	Gravel Depth in.	Wash Rate in./min	Wash Water, per cent
Detroit, MI	320	Dry	0.74	0	3	Baffles	2.0	4.2	16	160	4	30	0.45	17	26–30	2.0
Milwaukee, WI	200	Dry	0.5	0.11	64	Mechanical	4.0	2.32	27	125	6.25	27	0.51	24	24	2.2
St. Louis, MO	160	Solution	1.2	4.7	45	Baffles	36.0	1.5	16–23	125	4	30	0.4	12	24	1.67
Louisville, KY	120	Dry	0.65	0.31	...	None	2.0	5.7	17.5	125	6 and 3	26–30	0.4–0.5	14–24	22–36	2.0
Toledo, OH	80	Dry	1.12	0.5	40	Mechanical	2.8	1.5	15	94	2	22	0.42	18	20	2.43
Denver, CO	64	Dry	0.1–1.0	0.05–0.5	20	Baffles	1.0	12.0	13.5	156	4.5	48*	0.62	15	25	1.4
Atlanta, GA	54	Dry	0.59	0.26	23	Baffles	9.3	0.7	14–24	125	3 and 5	24–27	0.4–0.5	18	30	1.3
Dallas, TX	48	Dry	1.0	5.2	12	Baffles	8.0	1.1	18	120	2	30	0.4–0.45	18	20	1.5
New Orleans, LA	40	Solution	0.75	5.0	60	Baffles	18.0	0.55	13.75	122	4	30	0.33	9	24	0.3
Albany, NY	32	Dry	1.5	0	20	Baffles	2.25	1.5	10–18	125	4	30	0.33	18	24	2.5
Richmond, VA	30	Dry	2.0	0	10	Combined	10.0	6.0	10	120	3	26	0.43	16	24	1.0

* Coal
Source: Cosens, Jour. Amer. Water Works Assoc., 1956

CHAPTER 7

Environmental Problems

PAGE

SECTION A. POLLUTION SOURCES AND PATHWAYS

TABLE 7–1. CAUSES OF DAMAGE TO THE QUALITY OF WATER RESOURCES

Type of Waste	Wastewater Sources	Water Quality Measures	Effects on Water Quality	Effects on Aquatic Life	Effects on Recreation
Disease-carrying agents—human feces, warm-blooded animal feces	Municipal discharges, watercraft discharges, urban runoff, agricultural runoff, feedlot wastes, combined sewer overflows, industrial discharges	Fecal coliform, fecal streptococcus, other microbes	Health hazard for human consumption and contact	Inedibility of shellfish for humans	Reduced contact recreation
Oxygen-demanding wastes—high concentrations of biodegradable organic matter	Municipal discharges, industrial discharges, combined sewer overflows, watercraft discharges, urban runoff, agricultural runoff, feedlot wastes, natural sources	Biochemical oxygen demand, dissolved oxygen, volatile solids, sulfides	Deoxygenation, potential for septic conditions	Fish kills	If severe, eliminated recreation
Suspended organic and inorganic material	Mining discharges, municipal discharges, industrial discharges, construction runoff, agricultural runoff, urban runoff, silvicultural runoff, natural sources, combined sewer overflows	Suspended solids, turbidity, biochemical oxygen demand, sulfides	Reduced light penetration, deposition on bottom, benthic deoxygenation	Reduced photosynthesis, changed bottom organism population, reduced fish production, reduced sport fish population, increased non-sport fish population	Reduced game fishing, aesthetic appreciation
Inorganic materials, mineral substances—metal, salts, acids, solid matter, other chemicals, oil	Mining discharges, acid mine drainage, industrial discharges, municipal discharges, combined sewer overflows, urban runoff, oil fields, agricultural runoff, irrigation return flow, natural sources, cooling tower blowdown, transportation spills, coal gasification	pH, acidity, alkalinity, dissolved solids, chlorides, sulfates, sodium, specific metals, toxicity bioassay, visual (oil spills)	Acidity, salination, toxicity of heavy metals, floating oils	Reduced biological productivity, reduced flow, fish kills, reduced production, tainted fish	Reduced recreational use, fishing, aesthetic appreciation
Synthetic organic chemicals—dissolved organic material, e.g., detergents, household aids, pesticides	Industrial discharges, urban runoff, municipal discharges, combined sewer overflow, agricultural runoff, silvicultural runoff, transportation spills, mining discharges	Cyanides, phenols, toxicity bioassay	Toxicity of natural organics, biodegradable or persistent synthetic organics	Fish kills, tainted fish, reduced reproduction, skeletal development	Reduced fishing, inedible fish for humans
Nutrients—nitrogen, phosphorus	Municipal discharges, agricultural runoff, combined sewer overflows, industrial discharges, urban runoff, natural sources	Nitrogen, phosphorus	Increased algal growth, dissolved oxygen reduction	Increased production, reduced sport fish population, increased non-sport fish population	Tainted drinking water, reduced fishing and aesthetic appreciation
Radioactive materials	Industrial discharges, mining	Radioactivity	Increased radioactivity	Altered natural rate of genetic mutation	Reduced opportunities
Heat	Cooling water discharges, industrial discharges, municipal discharges, cooling tower blowdown	Temperature	Increased temperature, reduced capacity to absorb oxygen	Fish kills, altered species composition	Possible increased sport fishing by extended season for fish which might otherwise migrate

Source: Council of Environmental Quality, 1981, Environmental Trends

TABLE 7–2. POINT- AND NONPOINT SOURCES OF WATER POLLUTION

Sources	Common Pollutant Categories
POINT SOURCES	
Municipal sewage treatment plants	BOD; bacteria; nutrients; ammonia; toxics
Industrial facilities	Toxics; BOD
Combined sewer overflows	BOD; bacteria; nutrients; turbidity; total dissolved solids; ammonia; toxics; bacteria
NONPOINT SOURCES	
Agricultural runoff	Nutrients; turbidity; total dissolved solids; toxics; bacteria
Urban runoff	Turbidity; bacteria; nutrients; total dissolved solids; toxics
Construction runoff	Turbidity; nutrients; toxics
Mining runoff	Turbidity; acids; toxics; total dissolved solids
Septic systems	Bacteria; nutrients
Landfills/spills	Toxics; miscellaneous substances
Silvicultural runoff	Nutrients; turbidity; toxics

Source: U.S. Environmental Protection Agency, National Water Quality Inventory, 1986 Report to Congress

TABLE 7–3. CONTAMINATION SOURCES REPORTED BY PUBLIC WATER-SUPPLY SYSTEMS IN THE UNITED STATES

[Number of utilities reporting in each category]

Type of Contamination	Water-Supply Source		
	Groundwater	River/Stream	Lake/Reservoir
Industrial/Commercial Discharges	62	97	38
Leaking Underground Tanks	81	33	23
Urban Runoff	35	91	24
Landfills	67	49	22
Synthetic or Volatile Organics	83	56	18
Hazardous Waste Site(s)	37	31	8
Land Development	36	76	32
Underground Waste Injection	27	5	3
Agricultural Runoff (pesticides, fertilizers, etc.)	49	126	86
Algae/Bacteria	15	117	124
Overdraft	40	7	4
Water Rights Disputes	16	22	12
Natural Contamination (radionuclides, salinity, etc.)	52	56	35

Source: American Water Works Association, 1984 Water Utility Operating Data; Copyright AWWA. Reprinted with permission

FIGURE 7–1. POLLUTANT PATHWAYS FROM SOIL TO MAN

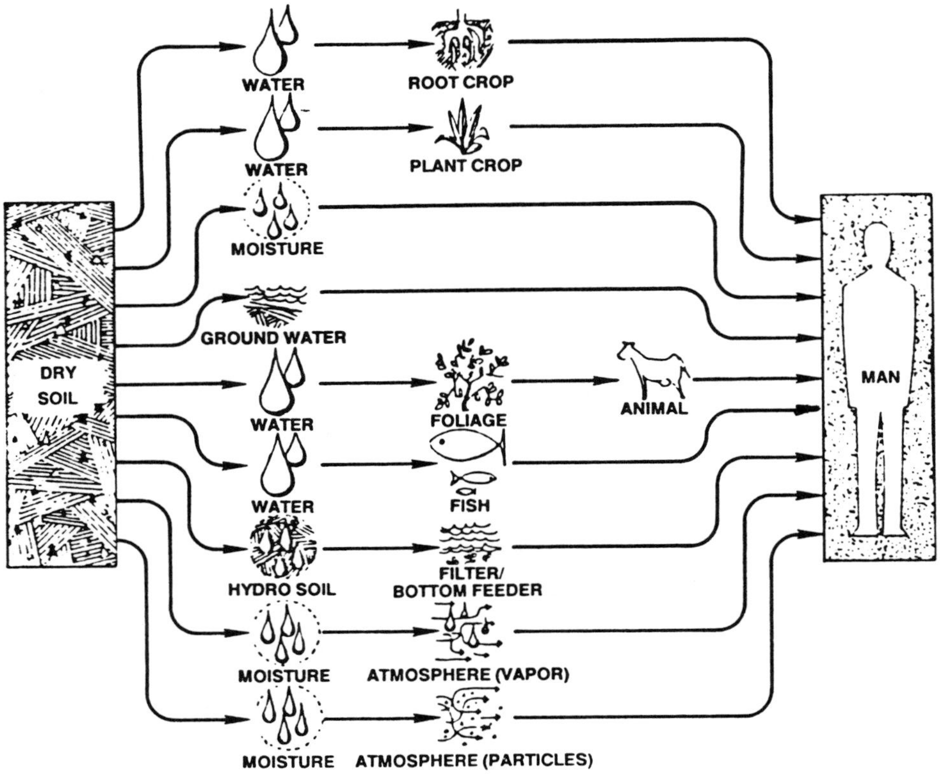

Source: Dacre, I. C., Rosenblatt, D. H. and Cogley D. R., 1980, Preliminary Pollutant Limit Values for Human Health Effects, Environmental Science and Technology 14: 778–783, Copyright American Chemical Society, Washington D.C. Reprinted with permission

SECTION B. SURFACE WATER POLLUTION

TABLE 7-4. TOTAL ASSIMILATIVE CAPACITY OF STREAMS OF DIFFERENT ORDERS

Stream Order	Average Dis-charge (cfs)	Average Depth (ft)	Average Velocity (ft per sec)	Coefficient of Reaera-tion (day $^{-1}$)	Total Length of Streams (miles)	Total Assimilative Capacity (tons per day per unit deficiency in dissolved oxygen)	U.S. Rivers Repre-sentative of Each Order
1	0.6	....	....	....	1,570,000	....	....
2	2.8	....	....	....	810,000	....	....
3	14	0.55	1.2	9.3	420,000	16,300	....
4	65	.95	1.6	5.5	220,000	19,000	....
5	310	1.8	1.8	2.6	116,000	20,000	Pecos
6	1,500	2.7	2.0	1.8	61,000	30,000	Shenandoah, Raritan
7	7,000	5	2.5	1.0	30,000	31,000	Allegheny, Kansas, Rio Grande
8	33,000	12	3.0	.37	14,000	21,000	Tennessee, Wabash
9	160,000	25	4.0	.19	6,200	18,000	Columbia, Ohio
10	700,000	45	5.0	.10	1,800	9,400	Mississippi

Source: U.S. Geological Survey, 1967

TABLE 7-5. SURFACE WATERS IMPACTED BY NONPOINT SOURCES IN THE UNITED STATES, 1985

[Assessed waters only]

Status	Rivers (thousands of miles)	Lakes (millions of acres)	Estuaries (thousands of square miles)	Ocean (thousands of shore-line miles)
No use impairment	230	7.20	11.1	6.30
Threatened	48	3.70	3.0	0.10
Moderately impaired	87	3.50	1.6	0.34
Severely impaired	30	0.88	0.8	0.01
Undistinguished	9	0.09	2.5	0.07
Total assessed	404	15.37	19.0	6.82

Note: The total assessed waters does not include data for Alaska. Excluding Alaska, states report there are 1.48 million river miles, 26.6 million lake acres, 32,000 estuary square miles and 23,000 ocean coastline miles.

Source: The Association of State and Interstate Water Pollution Control Administrators, in cooperation with the U.S. Environmental Protection Agency. 1985. America's Clean Water: The States' Nonpoint Source Assessment, 1985. Washington, D.C.; U.S. Geological Survey, National Water Summary 1986

TABLE 7-6. CHANGES IN SURFACE WATER QUALITY IN THE UNITED STATES, 1972 TO 1982

[Assessed waters only]

Water Quality Status	Rivers & Streams (thousands of miles)	Lakes (millions of acres)	Estuaries (thousands of square miles)
Maintained	296.00	10.13	12.80
Improved	47.00	0.39	3.80
Degraded	11.00	1.65	0.56
Unknown	90.00	4.15	0.17
Total assessed	444.00	16.32	17.33

Note: The total assessed waters does not include data for Alaska.

Source: The Association of State and Interstate Water Pollution Control Administrators, in cooperation with the U.S. Environmental Protection Agency. 1984. America's Clean Water: The States' Evaluation of Progress, 1972–1982. Washington, D.C.; U.S. Geological Survey, National Water Summary 1986

TABLE 7–7. NATIONAL AMBIENT WATER QUALITY IN RIVERS AND STREAMS—VIOLATION RATE: 1975 TO 1984

[In percent. Violation level based on U.S. Environmental Protection Agency water quality criteria. Violation rate represents the proportion of all measurements of a specific water quality pollutant which exceeds the "violation level" for that pollutant. "Violation" does not necessarily imply a legal violation. Data based on U.S. Geological Survey's National Stream Quality Accounting Network (NASQAN) data system; Years refer to water years. A water year begins in Oct. and ends in Sept. μg=micrograms. mg=milligrams]

Pollutant	Violation Level	1975	1976	1977	1978	1979	1980	1981	1982	1983	1984
Fecal coliform bacteria	Above 200 cells per 100 mL.	36	32	34	35	34	31	30	33	34	32
Dissolved oxygen	Below 5 mg per liter	5	6	11	5	4	5	4	5	4	2
Phosphorus, total, as phosporous	Above 1.0 mg per liter	5	5	5	5	3	4	4	3	3	4
Lead, dissolved	Above 50 μg per liter	(B)	(B)	(B)	(B)	13	5	3	2	5	(Z)
Cadmium, dissolved	Above 10 μg per liter	(B)	(B)	(B)	(B)	4	1	1	1	1	(Z)

B=Base figure too small to meet statistical standards for reliability of derived figures; Z=Less than 1.
Source: U.S. Department of Commerce, Statistical Abstracts of the United States 1987

TABLE 7–8. POINT- AND NONPOINT POLLUTION REPORTED BY 20 STATES AS REASONS FOR NOT ATTAINING WATER-QUALITY GOALS IN 1982

	Point Source				Nonpoint Source		
State	General	Municipal	Industrial	Combined Sewer	General	Natural	Other
Arkansas					X		
California[a]					X		
Connecticut			X	X	X		
Georgia		X		X	X		
Kansas						X	X
Louisiana							X
Maine		X		X	X		
Maryland							X
Massachusetts		X	X	X	X		
New Hampshire		X	X	X			X
New Mexico	X				X		
North Dakota						X	
Ohio	X						X
Pennsylvania							X
Rhode Island		X	X	X			
Tennessee	X				X		X
Texas		X	X				
Vermont		X		X	X		
Virginia[a]			X		X		
Washington	X				X	X	

[a] 1980 estimates.
Source: U.S. Environmental Protection Agency, National Water Quality Inventory: Report to Congress, 1984

TABLE 7–9. PROBABLE SOURCES OF WATER QUALITY PROBLEMS IN THE NATION'S STREAMS

[Expressed in total stream miles and as percentages of total miles]

Probable Source	Stream Miles	Percentage
Total nonpoint source contribution	367,244	38.4
Agricultural sources	281,241	29.5
Natural sources	212,389	22.2
Total point source contribution	117,684	12.3
Silviculture/logging	71,736	7.5
Municipal point sources	63,816	6.7
Feed lots	59,947	6.3
Individual sewage disposal	47,823	5.0
Industrial point sources	47,097	4.9
Urban runoff	40,376	4.2
Mining (nonpoint)	31,847	3.3
Combined sewers	29,246	3.1
Construction activity	29,110	3.1
Mining (point)	28,686	3.0
Grazing	21,970	2.3
Other	19,445	2.0
Dam releases	19,314	2.0
Landfill leachate	5,504	0.6
Bedload movement	5,299	0.6
Roads	3,569	0.4

Source: Judy, R.D., and others, 1984, 1982 National Fisheries Survey, U.S. Fish and Wildlife Service., FWS/OBS-84/06

TABLE 7–10. SOURCES OF WATER QUANTITY PROBLEMS ADVERSELY AFFECTING THE NATION'S STREAMS

[Expressed in total stream miles and as percentages of total miles]

Source	Stream Miles	Percentage
Natural conditions	477,791	50.1
Diversions (agricultural)	130,223	13.6
Dam(s) (water storage)	32,901	3.5
Dam(s) (flood control)	28,002	2.9
Dam(s) (power)	24,821	2.6
Other	18,851	2.0
Diversions (municipal)	10,694	1.1
Channelization	10,629	1.1
Floods/low flows	10,527	1.1
Irrigation	8,897	0.9
Logging	6,271	0.7
Ditches	5,335	0.6
Diversions (industrial)	3,292	0.3

Source: Judy, R.D., and others, 1984. 1982 National Fisheries Survey, U.S. Fish and Wildlife Service, FWS/OBS-84/06

TABLE 7–11. HEAT GENERATED AND DISCHARGED TO THE NATION'S FRESH AND SALINE SURFACE WATERS, 1975–2000

	1975		1985		2000	
	Btus x 10^{15}	Percent of Total Discharged	Btus x 10^{15}	Percent of Total Discharged	Btus x 10^{15}	Percent of Total Discharged
Electric power generation:						
Heat generated	11.0	—	24.3	—	57.1	—
Heat discharged to water	5.7	66	7.8	98	7.4	98
(Fresh)	(3.9)	(45)	(3.9)	(49)	(2.8)	(37)
(Saline)	(1.8)	(21)	(3.9)	(49)	(4.6)	(61)
Manufacturing:						
Heat discharged to water	3.0	34	0.2	2	0.2	2
(Fresh)	(2.2)	(73)	(0.2)	(2)	(0.2)	(2)
(Saline)	(0.8)	(27)	(0)	(0)	(0)	(0)
Total heat discharged	8.7	100	8.0	100	7.6	100
(Fresh)	(6.1)	(70)	(4.1)	(51)	(3.0)	(39)
(Saline)	(2.6)	(30)	(3.9)	(49)	(4.6)	(61)

Source: U.S. Water Resources Council, 1978, The Nation's Water Resources 1975–2000, Second National Water Assessment

TABLE 7–12. POINT SOURCE LOADINGS TO RECEIVING WATERS IN THE UNITED STATES

[Millions of tons per year for TSS, BOD, and phosphorus and millions of pounds per year for metals; mid-1980s]

Industry	TSS	BOD	Phosphorus	Metals
Minerals & metals	0.355	0.006	0.255	4.931
Chemical & manufacturing	0.086	0.125	0.267	2.919
Agriculture & fisheries	0.277	0.404	92.800	0.000
POTWs	1.594	1.830	49.555	2.838
Total	2.312	2.365	142.877	10.688

[Note: Mineral & metals includes aluminum forming, coal mining, copper forming, foundries, iron and steel, metal finishing, nonferrous metals mining, nonferrous metals forming, ore mining and petroleum refining industries. Chemical & manufacturing includes battery manufacturing, coil coating, electrical and electronic components, organic and inorganic chemicals, plastics, synthetic fibers, pesticide manufacturing, pharmaceuticals manufacturing, plastics molding and forming, porcelain enameling, leather tanning, pulp and paper, and textile industries. Agriculture & fisheries includes animal feedlots, fish hatcheries, food and beverages, fruits and vegetables, meat packing, and seafood industries. POTWs=publicly owned treatment works. TSS=total suspended solids. BOD=biochemical oxygen demand. Metals=cadmium, copper, lead, mercury and zinc. Industrial loadings are direct discharges based on long-term average concentrations and total industry flows at Best Available Technology (BAT) presented in U.S. EPA (1983), except as follows: loadings for electrical and electronic components reflect current level of treatment (U.S. EPA, 1983); conventional loadings for agriculture & fisheries industries represent post-BAT treatment levels; and conventional and toxic pollutant loadings for POTWs include indirect industrial and residential/commercial loadings not removed by the POTWs. Phosphorus loadings for POTWs represent effluent levels of 5 mg/L.]

Source: U.S. Environmental Protection Agency, Effluent Technology Division, unpublished data. 1979. Washington, D.C. U.S. Geological Survey, National Water Summary 1986

TABLE 7–13. POINT SOURCE DISCHARGES TO WATER IN THE UNITED STATES, 1977

[By Sector]

Total Suspended Solids		Total Dissolved Solids		Biochemical Oxygen Demand	
Sector	**Million pounds per year**	**Sector**	**Million pounds per year**	**Sector**	**Million pounds per year**
Municipal sewage plants	3,850.0	Organic chemicals	36,540.4	Municipal sewage plants	3,800.0
Powerplants	1,165.7	Municipal sewage plants	30,255.2	Pulp & paper mills	530.2
Pulp & paper mills	781.8	Powerplants	18,418.1	Organic chemicals	107.6
Feedlots	422.0	Pulp & paper mills	16,825.8	Feedlots	95.9
Iron & steel mills	254.3	Misc. chemicals	8,176.4	Seafoods	86.9
Organic chemicals	144.0	Misc. food & beverages	7,420.2	Misc. food & beverages	54.8
Misc. food & beverages	91.9	Oil & gas extraction	6,077.0	Cane sugar mills	50.4
Textiles	61.7	Petroleum refining	2,389.8	Iron & steel mills	37.8
Mineral mining	52.7	Coal mining	1,328.7	Misc. chemicals	35.2
Seafoods	50.0	Iron & steel mills	1,324.0	Textiles	24.8
Total, top 10 sectors	6,874.1	Total, top 10 sectors	128,755.6	Total, top 10 sectors	4,823.6
Total, all sectors	13,746.0	Total, all sectors	170,759.0	Total, all sectors	6,944.0
Top 10 sectors as percent of all sectors	50%	Top 10 sectors as percent of all sectors	75%	Top 10 sectors as percent of all sectors	69%

Nitrogen		Phosphorus		Dissolved Heavy Metals	
Sector	**Million pounds per year**	**Sector**	**Million pounds per year**	**Sector**	**Million pounds per year**
Municipal sewage plants	813.5	Municipal sewage plants	73.9	Powerplants	24.4
Pharmaceuticals	87.6	Feedlots	21.8	Municipal sewage plants	9.3
Organic chemicals	41.1	Misc. food & beverages	4.7	Iron & steel mills	7.6
Feedlots	39.9	Meat packing	3.4	Petroleum refining	6.0
Meat packing	36.0	Laundries	3.3	Organic chemicals	3.6
Petroleum refining	15.5	Fertilizers	2.6	Ore mining	2.5
Misc. food & beverages	12.3	Petroleum refining	1.5	Electroplating	0.5
Seafoods	9.5	Seafoods	1.4	Machinery	0.5
Pesticides	8.9	Organic chemicals	1.4	Oil & gas extraction	0.4
Leather tanning	7.1	Poultry	1.2	Foundries	0.1
Total, top 10 sectors	1,071.4	Total, top 10 sectors	115.2	Total, top 10 sectors	54.9
Total, all sectors	1,237.0	Total, all sectors	191.0	Total, all sectors	59.0
Top 10 sectors as percent of all sectors	87%	Top 10 sectors as percent of all sectors	60%	Top 10 sectors as percent of all sectors	93%

Source: Council on Environmental Quality, 1981, Environmental Trends

TABLE 7–14. NONPOINT SOURCE CONTRIBUTIONS TO RECEIVING WATERS IN THE UNITED STATES

[By source; millions of tons per year; 1980]

Source	TSS	BOD	Nitrogen	Phosphorus
Cropland	1870.00	9.00	4.30	1.56
Pasture and rangeland	1220.00	5.00	2.50	1.08
Forest land	256.00	0.80	0.39	0.09
Construction sites	197	NA	NA	NA
Mining sites	59	NA	NA	NA
Urban runoff	20	0.5	0.15	0.019
Rural roadways	2	0.004	0.0005	0.001
Small feedlots	2	0.05	0.17	0.032
Landfills	NA	0.3	0.026	NA
Background	1260.00	5.00	2.50	1.10
Total	4886.00	20.65	10.04	3.88

Note: TSS=total suspended solids. BOD=biochemical oxygen demand. Excluded from the survey area are 207 million acres in public land (14% of the contiguous United States), mostly in the Rocky Mountains, because of inadequacy of information. Urban runoff includes storm water sewers only.

Source: U.S. Environmental Protection Agency, Office of Water Regulations and Standards, Nonpoint Source Control Section. 1986. Estimated pollutant contributions to surface waters from selected nonpoint sources in the contiguous 48 states (1980). Washington, D.C. U.S. Geological Survey, National Water Summary 1986

TABLE 7–15. STATE ESTIMATES OF SURFACE WATER POLLUTION SOURCES IN 1982

[Expressed in stream miles]

State by EPA Region		Miles Assessed	Miles Not Fully Supporting Designated Use	Miles Allotted to Each Source[a]			
				Nonpoint	Municipal	Industrial	Other
I	Connecticut	839	301	9	211	0	81
	Maine	2,011	736	199	405	132	0
	Massachusetts	1,611	831	291	416	83	42
	New Hampshire	14,544	507	5	456	46	0
	Rhode Island	724	57	5	29	23	0
	Vermont	4,863	165	7	116	9	33
	Total	24,592	2,597	516	1,631	294	156
II	New Jersey	1,100	735	257	257	184	37
	New York	3,400	1,224	0	789	435	0
	Total	4,500	1,959	257	1,047	618	37
III	Delaware	419	182	49	4	46	84
	District of Columbia	35	28	23	5	0	0
	Maryland	7,440	602	301	181	30	90
	Pennsylvania	12,962	2,743	2,030	466	247	0
	Virginia	4,500	3,117	1,029	1,091	623	374
	West Virginia	5,262	2,146	687	558	86	815
	Total	30,618	8,818	4,118	2,304	1,032	1,364
IV	Alabama	12,101	729	98	487	144	0
	Florida	11,585	5,720	2,860	1,144	229	1,487
	Georgia	17,000	853	9	836	9	0
	Kentucky	3,338	2,838	710	710	710	710
	Mississippi	10,274	1,014	732	228	54	0
	North Carolina	39,150	7,067	3,887	2,120	1,060	0
	South Carolina	2,765	1,348	337	431	162	418
	Tennessee	4,065	1,120	616	336	168	0
	Total	100,278	20,689	9,248	6,292	2,535	2,615

TABLE 7–15. STATE ESTIMATES OF SURFACE WATER POLLUTION SOURCES IN 1982 (continued)

[Expressed in stream miles]

State by EPA Region		Miles Assessed	Miles Not Fully Supporting Designated Use	Miles Allotted to Each Source[a]			
				Nonpoint	Municipal	Industrial	Other
V	Illinois	7,270	4,478	448	3,582	448	0
	Indiana	45,000	1,000	150	750	100	0
	Michigan	811	463	[b]	[b]	[b]	[b]
	Minnesota	2,708	932	699	233	0	0
	Ohio	6,387	3,700	[b]	[b]	[b]	[b]
	Wisconsin	18,500	485	223	184	78	0
	Total	80,676	11,058	1,520	4,750	625	0
VI	Arkansas	6,000	3,000	2,700	150	150	0
	Louisiana[c]						
	New Mexico	3,500	50	0	0	0	50
	Oklahoma	11,985	2,066	289	1,074	702	0
	Texas	14,003	4,324	0	2,979	353	991
	Total	35,488	9,440	2,989	4,204	1,206	1,041
VII	Iowa	943	149	64	14	14	57
	Kansas	3,404	1,566	532	431	431	172
	Missouri	18,448	393	169	162	42	21
	Nebraska[d]						
	Total	22,795	2,108	765	606	487	250
VIII	Colorado	10,000	606	97	158	30	321
	Montana	17,251	152	152	0	0	0
	North Dakota	5,109	591	236	89	30	236
	South Dakota	3,987	1,487	937	268	283	0
	Utah	3,531	2,582	775	129	129	1,549
	Wyoming	19,655	590	112	18	18	443
	Total	59,533	6,008	2,309	661	489	2,549
IX	Arizona	497	180	20	59	101	0
	California	9,184	451	203	135	23	90
	Hawaii	39	39	28	0	0	11
	Nevada	1,325	518	104	409	5	0
	Total	11,045	1,188	355	604	129	101
X	Alaska	3,279	3,271	425	33	2,813	0
	Idaho	7,070	3,095	2,816	186	93	0
	Oregon	4,479	1,170	667	117	35	351
	Washington	5,952	1,039	831	104	104	0
	Total	20,780	8,575	4,740	439	3,045	351
	National Total[e]	390,305	72,440	26,817	22,538	10,459	12,626
	% of Total		100%	37%	31%	14%	17%

[a] Rows and columns may not total because of rounding.
[b] Did not specify causes for failure to meet support.
[c] Anomalous data; deleted to reduce distortion in results.
[d] Not available.
[e] The 4,163 miles unspecified in Michigan and Ohio are in "other."

Source: U.S. General Accounting Office, 1986, The Nation's Water. Original source: ASIWPCA, America's Clean Water: The States' Evaluation of Progress 1972–82, 1984

TABLE 7–16. RELATIVE IMPACT OF POLLUTION SOURCES IN RIVERS AND STREAMS WITH IMPAIRED USES IN THE UNITED STATES

[In percent]

State	Industrial	Municipal	Combined Sewers	Nonpoint Sources	Natural	Other/ Unknown
Alaska	85	1	0	12	0	2
Arizona	26	10	0	20	0	44
California	0	16	0	64	0	20
Connecticut	0	40	20	9	0	31
Delaware	8	6	8	59	19	0
Florida	25	29	0	40	2	4
Georgia	1	95	0	4	0	0
Idaho	2	3	0	78	17	0
Indiana	2	56	30	10	0	2
Iowa	0	3	0	97	0	0
Kansas	7	36	0	25	28	4
Kentucky	26	20	0	54	0	0
Louisiana	7	26	0	46	17	4
Maine	0	100	0	0	0	0
Maryland	5	30	0	50	15	0
Massachusetts	6	26	16	26	14	12
Minnesota	0	42	0	51	0	7
Mississippi	5	23	0	72	0	0
Missouri	0	1	0	99	0	0
Montana	2	3	0	95	0	0
Nebraska	1	7	0	92	0	0
New Hampshire	12	64	6	18	0	0
New Jersey	25	35	0	35	0	5
New Mexico	1	5	0	81	2	11
New York	20	40	13	11	0	16
North Carolina	12	17	0	71	0	0
Ohio	16	36	11	30	0	7
Oregon	3	10	0	57	30	0
Pennsylvania	7	13	1	71	3	5
Rhode Island	42	24	0	19	0	15
South Carolina	12	60	0	26	0	2
South Dakota	4	9	0	34	49	4
Tennessee	5	8	0	76	0	11
Texas	4	71	0	14	11	0
Vermont	11	22	0	50	11	6
Virginia	4	34	1	51	10	0
West Virginia	4	26	0	64	6	0
Wisconsin	1	1	0	98	0	0
Wyoming	10	4	0	43	43	0
Puerto Rico	11	21	0	63	0	5
Guam	5	10	15	50	20	0
Average (weighted)	9	17	1	65	6	2

Source: U.S. Environmental Protection Agency, National Water Quality Inventory, 1986 Report to Congress

TABLE 7–17. RIVER MILES MEETING THE FISHABLE/SWIMMABLE GOAL OF THE CLEAN WATER ACT, BY STATE

State	No. of Miles Assessed	No. of Miles Fishable	No. of Miles Swimmable	No. of Miles Fishable and Swimmable	No. of Miles Expected Fishable Swimmable in 5–10 Years	No. of Miles Never Fishable Swimmable	No. of Miles Designated More Stringent than F/S	No. of Miles Designated Less Stringent than F/S
Alabama	12,101	6,873	—	2,984	2,144	144	929	629
Arizona	1,412	654	930	—	—	—	81	
Connecticut	880	—	—	596	127	2	289	0
Delaware	516	419	392	—	—	—	—	—
Florida	6,575	—	—	6,118	426	31	265	31
Idaho	7,310	6,422	6,650	6,105	1,032	346	4,075	345
Iowa	4,365	72	—	—	0	4,293	110	3,157[a]
Kansas	4,495	4,170	—	—	559	7	—	—
Kentucky	5,683	—	—	3,130	—	—	—	—
Louisiana	2,500	2,140	2,040	—	—	—	—	—
Maine	31,672	—	—	30,695	977	0	—	—
Maryland	7,440	—	—	6,852	588	0	—	—
Massachusetts	1,676	1,374	802	802	—	—	—	—
Michigan	36,350	35,960	—	—	—	—	—	—
Minnesota	1,896	1,583	688	—	—	—	—	283
Mississippi	10,274	—	—	9,260	1,009	5	48	5
Missouri	20,536	—	—	10,390	—	—	—	—
Montana	19,505	—	—	19,054	230	221	20,422	64
Nebraska	4,794	3,493	1,067[b]	2,717	—	—	—	—
New Hampshire	1,320	—	—	934	107	47	65	47
New Jersey	780	690	225	225	—	—	—	—
New Mexico	3,500	3,164	3,490	—	—	—	—	—
New York	3,400	—	—	2,667	733	—	567	225
North Carolina	37,359	—	—	25,160	5,750	6,476	6,380	0
Ohio	6,628	—	—	4,048	—	—	—	—
Oregon	11,855	—	—	8,773	—	—	—	—
Pennsylvania	6,225	—	—	3,332	—	—	—	—
Rhode Island	724	—	—	588	55	81	276	81
South Carolina	2,442	2,229	1,934	1,748	11	16	—	—
South Dakota	3,987	2,716	1,054	—	—	—	0	6,290
Tennessee	5,748	—	—	3,785	—	—	3,201	0
Texas	15,942	—	—	14,832	—	—	—	—
Vermont	1,167	1,151	855	839	—	217	72	310
Wyoming	19,655	18,549	887[c]	—	49	1,057	7,459	—
Puerto Rico	2,243	1,311	331	283	1,552	1,096	0	0

[a] Iowa—3,157 miles are not designated for swimmable uses.

[b] Nebraska—only 2,347 stream miles have been assigned primary contact as a use; 3,748 miles were assessed.

[c] Wyoming—only 917 stream miles are designated for swimming in the State.

Source: U.S. Environmental Protection Agency, National Water Quality Inventory - 1986 Report to Congress. EPA-440/4–87–008

TABLE 7–18. DESIGNATED USE SUPPORT IN RIVERS AND STREAMS OF THE UNITED STATES

State	Total River Miles	River Miles Assessed	Miles Fully Supporting	Miles Partly Supporting	Miles Not Supporting	Assessed Miles Unknown
Alaska	365,000	5,025	2,662	2,363	0	0
Alabama	40,600	12,101	10,835	804	462	0
Arkansas	11,438	11,438	5,914	5,524	0	0
Arizona	17,537	1,412	615	391	406	0
California	26,959	9,627	6,163	1,518	335	1,611
Connecticut	8,400	880	597	232	51	0
Delaware	579	516	309	184	23	0
District of Columbia	40	40	7	18	10	5
Florida	9,320	6,575	4,448	1,670	457	0
Georgia	20,000	17,000	16,185	458	357	0
Idaho	7,310	7,310	6,046	572	692	0
Illinois	14,080	3,395	1,861	1,457	77	0
Iowa	18,000	4,365	72	3,077	1,216	0
Kansas	20,600	4,495	3,512	359	435	188
Kentucky	40,000	5,683	3,130	1,877	675	1
Louisiana	14,180	2,500	1,240	800	460	0
Maine	31,672	31,672	30,695	513	464	0
Maryland	9,300	7,440	6,852	449	139	0
Massachusetts	10,704	1,676	802	572	302	0
Michigan	36,350	36,350	35,696	0	497	157
Mississippi	10,274	10,274	9,260	1,014	0	0
Missouri	20,536	20,536	10,390	10,075	71	0
Montana	20,532	19,505	12,184	6,934	387	0
Nebraska	24,000	4,794	2,717	1,135	942	0
New Hampshire	14,544	1,320	981	259	80	0
New Jersey	6,450	780	225	465	90	0
New Mexico	3,500	3,500	3,140	360	0	0
New York	70,000	3,400	2,667	246	487	0
North Carolina	37,359	37,359	25,156	10,171	1,867	165
Ohio	43,900	6,628	4,048	1,977	603	0
Oregon	90,000	11,855	9,665	1,915	275	0
Pennsylvania	50,000	6,225	3,332	1,242	1,651	0
Rhode Island	724	724	655	34	35	0
South Carolina	9,679	2,442	2,127	194	121	0
South Dakota	9,937	3,987	1,865	1,130	992	0
Tennessee	19,124	5,748	3,794	1,118	847	0
Texas	80,000	15,942	14,966	0	976	0
Vermont	4,863	1,167	882	269	16	0
Virginia	27,240	4,716	948	1,536	2,232	0
West Virginia	22,819	18,244	10,225	6,631	1,388	0
Wyoming	19,655	19,655	17,386	297	1,972	0
Puerto Rico	3,469	2,243	283	1,076	884	0
Total	1,290,674	370,544	274,537	70,916	22,974	2,127

Note: Reporting from 14 states and territories (Colorado, Hawaii, Indiana, Minnesota, Nevada, North Dakota, Oklahoma, Utah, Washington, Wisconsin, Virgin Islands, Guam, Northern Mariana Islands, and American Samoa) did not allow determination of overall use support.

Source: U.S. Environmental Protection Agency, National Water Quality Inventory, 1986 Report to Congress

TABLE 7–19. POLLUTANT DISCHARGES INTO COASTAL WATERS OF THE UNITED STATES, 1980 TO 1985

[Million tons per year]

Coastal region	BOD	TSS	TN	TP
Northeast	0.57	5.33	0.14	0.05
Mid-Atlantic	0.54	5.42	0.20	0.05
Southeast	0.25	3.18	0.14	0.04
Gulf of Mexico	2.03	149.00	0.88	0.22
West Coast	1.52	101.12	0.64	0.76

Note: BOD=biochemical oxygen demand; TSS=total suspended solids; TN=total nitrogen, TP=total phosphorus.

Source: U.S. Department of Commerce, National Oceanic and Atmospheric Administration, National Ocean Survey, Ocean Assessments Division, Strategic Assessment Branch. 1986. Pollutant discharges from East Coast and Gulf of Mexico coastal counties, circa 1980–1985 and unpublished data compiled from the National Coastal Pollutant Discharge Inventory database. Rockville, MD.; U.S. Geological Survey, National Water Summary 1986

TABLE 7–20. POINT SOURCES OF POLLUTANTS TO COASTAL WATERS OF THE UNITED STATES, 1980 TO 1985

[Number of wastewater discharge facilities by region]

Coastal Region	Municipal Wastewater Treatment Facilities	Industrial Wastewater Treatment Facilities
Northeast	540	1025
Mid-Atlantic	666	1518
Southeast	1053	458
Gulf of Mexico	1651	830
West Coast	521	920

Source: U.S. Department of Commerce, National Oceanic and Atmospheric Administration, National Ocean Survey, Ocean Assessments Division, Strategic Assessment Branch. 1986. Pollutant discharges from East Coast and Gulf of Mexico coastal counties, circa 1980–1985 and unpublished data compiled from the National Coastal Pollutant Discharge Inventory database. Rockville, MD; U.S. Geological Survey, National Water Summary 1986

TABLE 7–21. CLASSIFICATION OF SHELLFISH-GROWING WATERS IN THE UNITED STATES

[In thousands of acres as of 1985]

Region and State	Approved for Harvest	Harvest Limited Areas			% of Total Productive Waters Approved	Nonshellfish/ Nonproductive	Total
		Prohibited	Conditionally Approved	Restricted			
Northeast							
Maine	936	87	13	10	89	0	1,046
New Hampshire	4	6	0	0	40	0	10
Massachusetts	255	41	1	5	84	500	802
Rhode Island	96	20	12	0	75	0	128
Connecticut	309	78	6	0	79	0	393
New York	828	192	1	0	81	0	1,021
New Jersey	236	118	20	21	60	0	395
Pennsylvania	0	0	0	0	—	6	6
Delaware	209	19	3	0	90	44	275
Maryland	1,369	64	0	0	96	97	1,530
Virginia	1,295	174	33	0	86	2	1,504
Subtotal	5,537	799	89	36	86	649	7,110
Southeast							
North Carolina	1,755	370	0	0	83	0	2,125
South Carolina	200	72	9	0	71	0	281
Georgia	61	144	0	0	30	0	205
Florida	40	36	37	0	35	748	861
Subtotal	2,056	622	46	0	75	748	3,472

TABLE 7–21. CLASSIFICATION OF SHELLFISH-GROWING WATERS IN THE UNITED STATES (continued)

[In thousands of acres as of 1985]

Region and State	Approved for Harvest	Harvest Limited Areas			% of Total Productive Waters Approved	Nonshellfish/ Nonproductive	Total
		Prohibited	Conditionally Approved	Restricted			
Gulf of Mexico							
Florida	266	260	306	0	32	578	1,410
Alabama	74	103	195	0	20	2	374
Mississippi	123	96	171	0	32	0	390
Louisiana	0	31	3,462	0	—	0	3,493
Texas	1,310	358	0	0	79	2	1,670
Subtotal	1,773	848	4,134	0	26	582	7,337
West Coast							
California	2	263	12	1	1	248	526
Oregon	14	14	0	12	35	44	84
Washington	147	49	45	0	61	1,795	2,036
Subtotal	163	326	57	13	29	2,087	2,646
U.S. Total	9,529	2,595	4,326	49	58	4,066	20,565

Source: U.S. Environmental Protection Agency, National Water Quality Inventory-1986 Report to Congress; NOAA, 1985 National Shellfish Register of Classified Estuarine Waters, 1985

TABLE 7–22. RIVER AND STREAM MILES SUPPORTING USES IN THE UNITED STATES, 1972 TO 1982

[Thousands of miles and percentage of waters assessed]

Status	1972 Miles	Percent	1982 Miles	Percent
Supporting uses	272	36	488	64
Partially supporting uses	46	6	167	22
Not supporting uses	30	4	35	5
Unknown or not reported	410	54	68	9

Note: Forty-nine (49) states reported on water quality conditions between 1972 and 1982 for 758,000 river and stream miles. Some proportion of the 1972 data unknown or not reported fell into each of the levels of use support.

Source: The Association of State and Interstate Water Pollution Control Administrators, in cooperation with the U.S. Environmental Protection Agency. 1984. America's clean water: The states' evaluation of progress, 1972–1982. Washington, D.C.; U.S. Geological Survey, National Water Summary 1986

TABLE 7–23. CONDITION OF PERENNIAL STREAMS RELATED TO THEIR ABILITY TO SUPPORT FISH IN THE UNITED STATES, 1977 TO 1982

[Thousands of miles]

Condition (Worst to Best)	1977	1982
0	29.87	29.87
1	48.79	49.31
2	170.07	166.31
3	222.02	228.66
4	155.57	156.24
5	38.20	36.13

Source: U.S. Department of the Interior, Fish and Wildlife Service. 1984. 1982 National Fisheries Survey, vol. 1. FWS/OBS-84/06. Washington, D.C.; U.S. Geological Survey, National Water Summary 1986

TABLE 7–24. WATER QUALITY FACTORS AFFECTING THE NATION'S FISHERIES

[Expressed in total stream miles and as percentages of total miles]

Factor	Stream Miles	Percentage
Turbidity	328,261	34.4
High water temperature	250,187	26.2
Nutrient surplus	119,519	12.5
Toxic substances	93,602	9.8
Dissolved oxygen problem	91,022	9.5
Nutrient deficiency	40,603	4.3
Low water temperature	29,877	3.1
Other	26,685	2.8
pH too acidic	24,793	2.6
Low flow	24,364	2.6
Salinity	17,217	1.8
Sedimentation	14,378	1.5
Siltation	9,644	1.0
Gas supersaturation	5,500	0.6
Intermittent water	4,839	0.5
Herbicides and pesticides	4,356	0.5
pH too basic	3,998	0.4
Channelization	2,937	0.3

Source: Judy, R.D., and others, 1984. 1982 National Fisheries Survey, U.S. Fish and Wildlife Service, FWS/OBS-84/06

TABLE 7–25. LIMITING FACTORS ADVERSELY AFFECTING THE NATION'S FISH COMMUNITIES

[Expressed in total stream miles and as percentages of total miles]

Factor	Stream Miles	Percentage
Fish kills	115,435	12.1
Contamination	81,927	8.6
Overharvest	35,566	3.7
Poaching	28,145	2.9
Diseases/parasites	21,873	2.3
Fish stocking	19,350	2.0
Other	18,063	1.9
Habitat	14,213	1.5
Underharvest	12,714	1.3
Competition	10,836	1.1
Water quality	5,879	0.6
Tumors/lesions	5,101	0.5
Low flow	3,194	0.3
Small channel capacity	1,657	0.2

Source: Judy, R.D., and others, 1984. 1982 National Fisheries Survey, U.S. Fish and Wildlife Service, FWS/OBS-84/06

TABLE 7–26. TROPHIC STATUS OF LAKES IN THE UNITED STATES

[Lakes classified according to degree of eutrophication; oligothropic lakes have the lowest nutrient levels and the least amount of plant and algae production; eutrophic lakes are those with the highest level of organic enrichment; mesotrophic lakes are those in an intermediate stage of productivity]

State	Number of Lakes Assessed	Number Eutrophic	Number Mesotrophic	Number Oligotrophic	Number of Other/ Unknown
Connecticut	70	18	44	8	0
Florida	135	39	22	36	38
Illinois	36	32	4	0	0
Indiana	554	499	55	0	0
Iowa	107	107	0	0	0
Kansas	154	125	29	0	0
Kentucky	92	50	28	14	0
Massachusetts	462	62	276	124	0
Michigan	160	28	113	19	0
Minnesota	12,034	7,822	3,009	1,203	0
Mississippi	34	29	5	0	0
Montana	1,880	371	428	452	629[a]
Nebraska	24	22	2	0	0
New Hampshire	418	76	158	141	43
New York	3,340	84	132	85	3,039
North Carolina	25	10	13	2	0
Pennsylvania	26	23	3	0	0
Rhode Island	113	17	52	7	37
Tennessee[b]	64	36	18	10	0
Vermont	223	25	80	118	0
Virginia	220	61	64	10	85[c]
Washington	140	45	24	58	13
Wisconsin	2,925	802	1,518	605	0
Total	23,236	10,383	6,077	2,892	3,884

[a] Includes 127 dystrophic lakes.

[b] Non-federally managed lakes only.

[c] Includes one dystrophic lake.

Source: U.S. Environmental Protection Agency, National Water Quality Inventory- 1986 Report to Congress. EPA-440/4–87–008

TABLE 7–27. LAKE ACRES MEETING THE FISHABLE/SWIMMABLE GOAL OF THE CLEAN WATER ACT, BY STATE

[Excluding the Great Lakes]

State	No. Acres Assessed	No. Acres Fishable	No. Acres Swimmable	No. Acres Fishable and Swimmable	No. Acres Expected Fishable Swimmable in 5–10 years	No. Acres Never Fishable Swimmable	No. Acres Designated More Stringent than F/S	No. Acres Designated Less Stringent than F/S
Alabama	505,336	448,746	—	56,590	—	—	56,590	0
Arizona	34,811	32,431	31,508	—	—	—	0	—
Connecticut	38,884	—	—	36,858	951	0	20,000	0
Florida	796,800	—	—	732,800	64,000	0	339,840	0
Idaho	362,718	362,718	362,718	362,718	—	—	279,250	0
Iowa	73,771	53,899	—	51,827	3,321	17,476	24,658	7,380
Kansas	152,810	139,214	137,430	137,430	—	21,452	0	15,139
Kentucky	362,403	—	—	326,483	—	—	—	—
Louisiana	467,738	—	—	407,215	60,523	0	—	—
Maine	994,560	—	—	958,395	36,165	0	—	—
Maryland	32,583	—	—	32,574	9	—	—	—
Michigan	840,960	820,480	—	—	—	—	—	—
Mississippi	495,191	—	—	476,374	18,817	0	38,000	—

TABLE 7–27. LAKE ACRES MEETING THE FISHABLE/SWIMMABLE GOAL OF THE CLEAN WATER ACT, BY STATE (continued)

[Excluding the Great Lakes]

State	No. Acres Assessed	No. Acres Fishable	No. Acres Swimmable	No. Acres Fishable and Swimmable	No. Acres Expected Fishable Swimmable in 5–10 years	No. Acres Never Fishable Swimmable	No. Acres Designated More Stringent than F/S	No. Acres Designated Less Stringent than F/S
Montana	663,363	650,113	663,363	—	—	13,250	134,000	—
Nebraska	105,840	103,100	105,840	—	—	—	—	—
New Hampshire	144,918	—	—	137,672	—	—	23,000	9
New Jersey	18,923	—	—	13,625	—	—	—	—
New Mexico	5,725	5,524	5,725	5,524	—	—	—	—
New York	729,000	—	—	579,600	149,400	0	545,000	—
North Carolina	320,506	—	—	311,296	7,450	1,760	177,515	—
Oregon	192,000	—	—	112,700	—	—	—	—
Rhode Island	16,520	—	—	15,950	35	535	7,203	436
South Carolina	405,555	405,555	404,465	—	—	0	—	—
South Dakota	655,227	—	—	573,121	—	—	0	0
Vermont	224,066	224,024	224,063	224,021	—	45	—	—
Wyoming	411,284	409,223	230,335	—	14,398	—	184,821	0

Source: U.S. Environmental Protection Agency, National Water Quality Inventory - 1986 Report to Congress. EPA-440/4–87–008

TABLE 7–28. ESTIMATED PHOSPHORUS LOADINGS TO THE GREAT LAKES, 1976–1985

[Metric tons per year]

Year	Lake Superior	Lake Michigan	Lake Huron	Lake Erie	Lake Ontario
1976	3550	6656	4802	18480	12695
1977	3661	4666	3763	14576	8935
1978	5990	6245	5255	19431	9547
1979	6619	7659	4881	11941	8988
1980	6412	6574	5307	14855	8579
1981	3412	4091	3481	10452	7437
1982	3160	4084	4689	12349	8891
1983	3407	4515	3978	9880	6779
1984	3642	3611	3452	12874	7948
1985	2864	3958	5758	11195	7075

Note: The 1978 Great Lakes Water Quality Agreement set target loadings for each lake (in metric tons per year): Lake Superior, 3400; Lake Michigan, 5600; Lake Huron, 4360; Lake Erie, 11000; and Lake Ontario, 7000.

Source: Great Lakes Water Quality Board. Great Lakes Water Quality, 1976; 1977. Appendix B. Surveillance Subcommittee Report to the International Joint Commission, Canada and United States. Windsor, ON and Detroit, MI.; Great Lakes Water Quality Board. Report on Great Lakes Water Quality, 1981; 1983; 1985; 1987. Report to the International Joint Commission, Canada and United States. Windsor, ON and Detroit, MI.; U.S. Geological Survey, National Water Summary 1986

TABLE 7–29. ATMOSPHERIC INPUT OF SOME ORGANIC CONTAMINANTS TO THE GREAT LAKES

[Metric tons per year]

Compound	Lake Superior	Lake Michigan	Lake Huron	Lake Erie	Lake Ontario
Total PCB	9.8	6.9	7.2	3.1	2.3
Dieldrin	0.5	0.4	0.6	0.2	0.1
Total PAH	163	114	118	51	38
Total DDT	0.6	0.4	0.4	0.2	0.1
p,p^1-Methoxychlor	8.3	5.9	6.1	2.6	1.9

Source: Great Lakes Water Quality Board, 1985 Report on Great Lakes Water Quality. Original Source: Eisenreich et al., 1981

TABLE 7–30. SOURCES OF PCBs TO LAKE SUPERIOR

Source	kg/yr	Percent of Total
Atmosphere	6600–8300	82–86
Tributary	1300	13–16
Municipal Discharges	66	1
Industrial Discharges	2	1
Total	8000–9000	

Source: Great Lakes Water Quality Board, 1985 Report on Great Lakes Water Quality. Original source: Eisenreich et al., 1981

TABLE 7–31. CONCENTRATION OF PHOSPHORUS AND NITROGEN IN SELECTED LAKES IN THE WORLD

[Annual mean concentration in mg/L; period 1970 to 1985]

			Total Phosphorous				Total Nitrogen			
			1970	1975	1980	1985	1970	1975	1980	1985
Canada	Ontario		0.022	0.022	0.017	0.013	0.226	0.283	0.308	0.572
	Erie	a	0.017	0.023	0.018	..	..	0.278	0.163	..
USA	Cayuga (NY)	d,e,f	0.020	0.020	..	..	0.370	0.510	..	..
	W. Twin (Ohio)	d,g,h	0.150	0.100	..	..	1.930	..	..	..
Japan	Biwa (North)	k,l	0.012	0.008	0.010	0.008	0.190	0.290	0.290	0.250
	Biwa (South)	k,l	0.027	0.027	0.027	0.022	0.450	0.530	0.410	0.370
	Kasumigaura	l	..	0.040	0.080	0.070	..	1.200	1.000	1.100
Australia	Gordon (Tas.)		..	..	..	..	..	0.270	0.200	..
Austria	Zeller-See		..	0.017	0.018	..	..	..	..	..
Denmark	Knud Soe	b,i	..	0.060	0.050	0.028	..	2.000	3.000	2.900
Finland	Paeijaenne	j	0.010	0.011	0.009	0.010	0.500	0.460	0.460	0.490
France	Aydat		..	0.053	..	..	..	0.694	..	..
	Pavin		..	0.282	..	..	..	..	..	..
Germany	Bodensee		0.061	0.099	0.099	0.071	0.755	0.763	0.856	0.875
Ireland	Ennel	m	..	0.089	0.029	0.032	..	0.270	0.470	0.388
	Owel	m	..	..	0.020	0.015	..	..	0.079	0.047
	Derg	m,n	..	0.025	0.020	0.058	..	0.840	1.200	1.040
	Sheelin	m,l	..	0.022	0.049	0.020	..	..	..	..
Italy	Maggiore		..	0.026	0.036	0.019	..	..	..	..
	Como		..	0.068	0.078	0.052	0.640	0.710	0.800	0.800
	Garda		..	0.009	0.020	0.011	0.310	0.300	0.390	0.350
	Orta		..	..	0.011	0.006	13.000	9.620	9.500	7.110
Netherlands	Ijssel		..	0.350	0.350	0.290	..	4.025	4.385	4.140
Norway	Mjoesa		0.014	0.008	0.008	0.008	0.400	0.400	0.420	0.470
Portugal	Lagoa Obidos		..	..	..	..	..	0.100	0.200	..
Spain	Alcantara	c	..	0.387	2.570	..	..	1.341	2.864	..
Sweden	Malaren		0.029	0.024	0.034	0.031	0.918	0.735	0.708	0.859
	Vaettern		0.008	0.009	0.009	0.006	0.594	0.562	0.625	0.681
Switzerland	Leman		0.104	0.082	0.090	0.073	..	..	0.520	0.570
	Constance	l	0.055	0.078	0.077	0.065	..	0.861	0.900	0.920

[a] 1975 and 1980 data for nitrogen refer to 1978 and 1979.
[b] 1975 data refer to 1976.
[c] 1975 and 1980 data refer to 1974 and 1977.
[d] Mean inorganic nitrogen (NH4+NO3+NO2 as N).
[e] 1970 and 1975 data refer to 1972 and 1973.
[f] Samples collected at 3–5 sampling stations, at surface, 2, 5, and 10 meters, at weekly intervals during June-August, biweekly intervals during mid-April-May and September-October, and monthly intervals the rest of the year, down the long axis of the lake.
[g] 1970 and 1975 data refer to 1971 and 1974.
[h] Samples obtained from the deepest point in each lake, generally weekly from late spring-early fall, and less frequently the rest of the year at 0.1, 2, 4, 7 and 10 meters.
[i] 1985 data refer to 1983.
[j] 1985 data refer to 1982.
[k] 1970 data for phosphorous refer to 1971.
[l] 1985 data refer to 1984.
[m] Oxidized nitrogen.
[n] 1985 figure refers to upper limit.

Source: Organisation of Economic Cooperation and Development, Environmental Data Compendium 1987. Copyright OECD. Reprinted with permission

TABLE 7–32. POLLUTING INCIDENTS REPORTED IN AND AROUND U.S. WATERS: 1970 TO 1984

Item	Incidents	Gallons (1,000)
1970	3,711	15,253
1971	8,736	8,840
1972	9,931	18,806
1973	13,328	24,315
1974	14,432	19,422
1975	12,781	22,243
1976	13,930	36,608
1977	15,330	11,248
1978	14,495	17,557
1979	13,134	13,661
1980	11,155	15,093
1981	10,564	19,773
1982	10,414	23,154
1983	11,346	30,076
1984	10,745	19,749
BREAKDOWN OF INCIDENTS IN 1984		
Vessel	2,466	4,915
Tankship	254	1,830
Tank barge	545	2,646
Other	1,667	439
Nontransport facilities	2,577	6,060
Onshore	1,108	5,501
Offshore	198	499
Pipeline	554	1,363
Marine facilities:		
Onshore/offshore	521	342
Land vehicles	707	623
Land facilities	176	178
Other or unknown	3,744	6,266
Type of pollutant:		
Crude oil	1,983	1,422
Diesel oil	2,069	7,106
Other oil	4,859	5,477
Other or unknown	1,834	5,744
Location:		
Atlantic Coast	2,212	6,411
Gulf Coast	2,306	436
Pacific Coast	1,694	1,877
Great Lakes	78	1,251
Inland	4,380	9,717

Source: U.S. Department of Commerce, Statistical Abstracts of the United States 1987

TABLE 7–33. POLLUTION DISCHARGES IN NAVIGABLE WATERS OF THE UNITED STATES IN 1984

[In gallons; by type of source]

Source	Oil				Hazardous				Other				Total			
	#	%	Quantity	%	#	%	Quantity	%	#	%	Quantity	%	#	%	Quantity	%
VESSELS																
Tank Ships	242	2.6	1,906,666	11.9	9	2.7	2,837	.4	9	1.1	20,126	2.7	260	2.5	1,929,629	11.1
Tank Barges	524	5.7	2,469,002	15.5	18	5.5	52,902	7.9	13	1.6	22,005	2.9	555	5.4	2,543,909	14.6
Dry Cargo Barges	11	.1	2,900	.0	0	.0	0	.0	0	.0	0	.0	11	.1	2,900	.0
Dry Cargo Ships	168	1.8	1,469,828	9.2	0	.0	0	.0	6	.7	22	.0	174	1.7	1,469,850	8.5
Combatants	281	3.1	173,850	1.1	2	.6	2	.0	7	.8	190	.0	290	2.8	174,042	1.0
Other	1,170	12.7	212,987	1.3	4	1.2	74	.0	122	14.7	413	.1	1,296	12.5	213,474	1.2
TOTAL	2,396	26.0	6,235,233	39.0	33	10.1	55,815	8.3	157	18.9	42,756	5.7	2,586	25.0	6,333,804	36.4
LAND VEHICLES																
Rail	38	.4	26,704	.2	4	1.2	328	.0	4	.5	106	.0	46	.4	27,138	.2
Highway	409	4.4	145,874	.9	24	7.3	31,355	4.7	40	4.8	28,839	3.9	473	4.6	206,068	1.2
Other	139	1.5	149,118	.9	7	2.1	268	.0	10	1.2	5,741	.8	156	1.5	155,127	.9
TOTAL	586	6.4	321,696	2.0	35	10.7	31,951	4.7	54	6.5	34,686	4.6	675	6.5	388,333	2.2
NON-TRANSPORTATION																
Refinery	74	.8	86,220	.5	6	1.8	4,737	.7	5	.6	8,555	1.1	85	.8	99,512	.6
Bulk Storage	295	3.2	1,456,765	9.1	12	3.7	74,978	11.1	14	1.7	35,695	4.8	321	3.1	1,567,438	9.0
Onshore Production	99	1.1	24,178	.2	2	.6	128	.0	3	.4	1,198	.2	104	1.0	25,504	.1
Offshore Production	1,170	12.7	56,555	.4	8	2.4	99	.0	13	1.6	1,097	.1	1,191	11.5	57,751	.3
Other	227	2.5	126,465	.8	35	10.7	74,631	11.1	52	6.3	198,324	26.5	314	3.0	399,420	2.3
TOTAL	1,865	20.3	1,750,183	11.0	63	19.2	154,573	22.9	87	10.5	244,869	32.7	2,015	19.5	2,149,625	12.4
TRANS/RELATED PIPELINES	54	.6	2,613	.0	1	.3	2	.0	1	.1	42	.0	56	.5	2,657	.0
MARINE FACILITIES																
Fuel Transfer	27	.3	14,193	.1	0	.0	0	.0	0	.0	0	.0	27	.3	14,193	.1
Bulk Transfer	150	1.6	74,754	.5	10	3.0	14,222	2.1	8	1.0	12,043	1.6	168	1.6	101,019	.6
Non-Bulk Transfer	3	.0	26	.0	0	.0	0	.0	0	.0	0	.0	3	.0	26	.0
Other	96	1.0	27,024	.2	7	2.1	5,563	.8	8	1.0	18,057	2.4	111	1.1	50,644	.3
TOTAL	276	3.0	115,997	.7	17	5.2	19,785	2.9	16	1.9	30,100	4.0	309	3.0	165,882	1.0
LAND FACILITIES	69	.8	15,413	.1	8	2.4	4,450	.7	14	1.7	34,058	4.5	91	.9	53,921	.3
MISC/UNKNOWN	3,952	43.0	7,529,425	47.1	171	52.1	406,989	60.4	500	60.3	362,334	48.4	4,623	44.6	8,298,748	47.7
GRAND TOTAL	9,198	100.0	15,970,560	100.0	328	100.0	673,565	100.0	829	100.0	748,845	100.0	10,355	100.0	17,392,970	100.0

Source: U.S. Coast Guard, Polluting Incidents In and Around U.S. Waters, Calendar Year 1983 and 1984. COMDTINST M16450.2G; NTIS 87–186821

TABLE 7–34. POLLUTION DISCHARGES IN NAVIGABLE WATERS OF THE UNITED STATES IN 1984

[In gallons: by type of location]

Location	Oil				Hazardous				Other				Total			
	#	%	Quantity	%	#	%	Quantity	%	#	%	Quantity	%	#	%	Quantity	%
INLAND																
Open Sheltered Waters	513	5.6	122,776	.8	8	2.4	1,518	.2	22	2.7	364	.0	543	5.2	124,658	.7
River Channels	1,475	16.0	2,637,095	16.5	42	12.8	86,228	12.8	144	17.4	134,281	17.9	1,661	16.0	2,857,604	16.4
Ports and Harbors	312	3.4	932,854	5.8	4	1.2	2,431	.4	21	2.5	204	.0	337	3.3	935,489	5.4
Other	0	.0	0	.0	0	.0	0	.0	0	.0	0	.0	0	.0	0	.0
TOTAL	2,300	25.0	3,692,725	23.1	54	16.5	90,177	13.4	187	22.6	134,849	18.0	2,541	24.5	3,917,751	22.5
ATLANTIC																
Open Sheltered Waters	134	1.5	59,270	.4	1	.3	1,000	.1	16	1.9	0	.0	151	1.5	60,270	.3
River Channels	976	10.6	832,019	5.2	13	4.0	10,186	1.5	137	16.5	4,104	.5	1,126	10.9	846,309	4.9
Ports and Harbors	560	6.1	6,069,657	38.0	4	1.2	13	.0	32	3.9	157	.0	596	5.8	6,069,827	34.9
Territorial Sea (Shore-3Mi)	169	1.8	48,620	.3	5	1.5	7,092	1.1	22	2.7	2,969	.4	196	1.9	58,681	.3
Contiguous Zone (3–12Mi)	15	.2	2,283	.0	0	.0	0	.0	5	.6	55	.0	20	.2	2,338	.0
High Seas (12Mi or more)	8	.1	7,111	.0	0	.0	0	.0	1	.1	0	.0	9	.1	7,111	.0
Other	11	.1	291	.0	0	.0	0	.0	0	.0	0	.0	11	.1	291	.0
TOTAL	1,873	20.4	7,019,251	44.0	23	7.0	18,291	2.7	213	25.7	7,285	1.0	2,109	20.4	7,044,827	40.5
PACIFIC																
Open Sheltered Waters	394	4.3	84,377	.5	4	1.2	224	.0	40	4.8	2,286	.3	438	4.2	86,887	.5
River Channels	73	.8	60,994	.4	2	.6	124	.0	4	.5	10,050	1.3	79	.8	71,168	.4
Ports and Harbors	892	9.7	325,863	2.0	4	1.2	297	.0	58	7.0	34,789	4.6	954	9.2	360,949	2.1
Territorial Sea (Shore-3Mi)	128	1.4	77,098	.5	2	.6	3,000	.4	23	2.8	13,514	1.8	153	1.5	93,612	.5
Contiguous Zone (3–12Mi)	45	.5	1,272,142	8.0	0	.0	0	.0	1	.1	2	.0	46	.4	1,272,144	7.3
High Seas (12Mi or more)	15	.2	122,050	.8	0	.0	0	.0	0	.0	0	.0	15	.1	122,050	.7
Other	0	.0	0	.0	0	.0	0	.0	0	.0	0	.0	0	.0	0	.0
TOTAL	1,547	16.8	1,942,524	12.2	12	3.7	3,645	.5	126	15.2	60,641	8.1	1,685	16.3	2,006,810	11.5
GULF																
Open Sheltered Waters	56	.6	9,563	.1	2	.6	350	.1	10	1.2	20,263	2.7	68	.7	30,176	.2
River Channels	211	2.3	132,840	.8	12	3.7	44,837	6.7	17	2.1	2,317	.3	240	2.3	179,994	1.0
Ports and Harbors	229	2.5	277,026	1.7	13	4.0	3,785	.6	15	1.8	723	.1	257	2.5	281,534	1.6
Territorial Sea (Shore-3Mi)	224	2.4	33,921	.2	5	1.5	113	.0	7	.8	3,453	.5	236	2.3	37,487	.2

TABLE 7–34. POLLUTION DISCHARGES IN NAVIGABLE WATERS OF THE UNITED STATES IN 1984 (continued)

[In gallons: by type of location]

Location	Oil				Hazardous				Other				Total			
	#	%	Quantity	%	#	%	Quantity	%	#	%	Quantity	%	#	%	Quantity	%
GULF (continued)																
Contiguous Zone (3–12Mi)	473	5.1	10,381	.1	2	.6	10	.0	3	.4	20	.0	478	4.6	10,411	.1
High Seas (12Mi or more)	963	10.5	95,170	.6	3	.9	88	.0	20	2.4	1,694	.2	986	9.5	96,952	.6
Other	29	.3	3,773	.0	0	.0	0	.0	1	.1	15	.0	30	.3	3,788	.0
TOTAL	2,185	23.8	562,674	3.5	37	11.3	49,183	7.3	73	8.8	28,485	3.8	2,295	22.2	640,342	3.7
GREAT LAKES																
Open Sheltered Waters	8	.1	295	.0	0	.0	0	.0	16	1.9	0	.0	24	.2	295	.0
River Channels	28	.3	2,551	.0	3	.9	200,040	29.7	25	3.0	20	.0	56	.5	202,611	1.2
Ports and Harbors	5	.1	50	.0	0	.0	0	.0	0	.0	0	.0	5	.0	50	.0
Other	37	.4	1,008,612	6.3	2	.6	1,380	.2	16	1.9	200	.0	55	.5	1,010,192	5.8
TOTAL	78	.8	1,011,508	6.3	5	1.5	201,420	29.9	57	6.9	220	.0	140	1.4	1,213,148	7.0
OTHER																
Open Sheltered Waters	0	.0	0	.0	0	.0	0	.0	0	.0	0	.0	0	.0	0	.0
Ports and Harbors	0	.0	0	.0	0	.0	0	.0	0	.0	0	.0	0	.0	0	.0
Territorial Sea (Shore-3Mi)	1,213	13.2	1,741,858	10.9	197	60.1	310,849	46.1	173	20.9	517,365	69.1	1,583	15.3	2,570,072	14.8
Contiguous Zone (3–12Mi)	0	.0	0	.0	0	.0	0	.0	0	.0	0	.0	0	.0	0	.0
High Seas (12Mi or more)	0	.0	0	.0	0	.0	0	.0	0	.0	0	.0	0	.0	0	.0
Other	2	.0	20	.0	0	.0	0	.0	0	.0	0	.0	2	.0	20	.0
TOTAL	1,215	13.2	1,741,878	10.9	197	60.1	310,849	46.1	173	20.9	517,365	69.1	1,585	15.3	2,570,092	14.8
GRAND TOTALS																
Open Sheltered Waters	1,105	12.0	276,281	1.7	15	4.6	3,092	.5	104	12.5	22,913	3.1	1,224	11.8	302,286	1.7
River Channels	2,763	30.0	3,665,499	23.0	72	22.0	341,415	50.7	327	39.4	150,772	20.1	3,162	30.5	4,157,686	23.9
Ports and Harbors	1,998	21.7	7,605,450	47.6	25	7.6	6,526	1.0	126	15.2	35,873	4.8	2,149	20.8	7,647,849	44.0
Territorial Sea (Shore-3Mi)	1,734	18.9	1,901,497	11.9	209	63.7	321,054	47.7	225	27.1	537,301	71.8	2,168	20.9	2,759,852	15.9
Contiguous Zone (3–12Mi)	533	5.8	1,284,806	8.0	2	.6	10	.0	9	1.1	77	.0	544	5.3	1,284,893	7.4
High Seas (12Mi or more)	986	10.7	224,331	1.4	3	.9	88	.0	21	2.5	1,694	.2	1,010	9.8	226,113	1.3
Other	79	.9	1,012,696	6.3	2	.6	1,380	.2	17	2.1	215	.0	98	.9	1,014,291	5.8
GRAND TOTAL	9,198	100.0	15,970,560	100.0	328	100.0	673,565	100.0	829	100.0	748,845	100.0	10,355	100.0	17,392,970	100.0

Source: U.S. Coast Guard, Polluting Incidents In and Around U.S. Waters, Calendar Year 1983 and 1984. COMDTINST M16450.2G; NTIS 87–186821

TABLE 7–35. POLLUTION DISCHARGES IN NAVIGABLE WATERS OF THE UNITED STATES IN 1984

[In gallons; by state]

State	Oil				Hazardous				Other				Total			
	#	%	Quantity	%	#	%	Quantity	%	#	%	Quantity	%	#	%	Quantity	%
Alabama	130	1.4	11,418	.1	11	3.4	8,430	1.3	16	1.9	371	.0	157	1.5	20,219	.1
Alaska	299	3.3	342,879	2.1	1	.3	9	.0	15	1.8	396	.1	315	3.0	343,284	2.0
Arkansas	30	.3	1,160,780	7.3	4	1.2	9,762	1.4	2	.2	21	.0	36	.3	1,170,563	6.7
California	988	10.7	1,744,049	10.9	16	4.9	52,842	7.8	122	14.7	45,794	6.1	1,126	10.9	1,842,685	10.6
Colorado	35	.4	49,956	.3	2	.6	53,826	8.0	4	.5	3,213	.4	41	.4	106,995	.6
Connecticut	63	.7	9,988	.1	0	.0	0	.0	6	.7	132	.0	69	.7	10,120	.1
Delaware	8	.1	270	.0	0	.0	0	.0	2	.2	375	.1	10	.1	645	.0
District of Columbia	14	.2	3,163	.0	0	.0	0	.0	3	.4	0	.0	17	.2	3,163	.0
Florida	596	6.5	158,971	1.0	2	.6	26	.0	66	8.0	2,421	.3	664	6.4	161,418	.9
Georgia	29	.3	4,194	.0	1	.3	8	.0	3	.4	1,013	.1	33	.3	5,215	.0
Hawaii	176	1.9	11,911	.1	0	.0	0	.0	15	1.8	25,531	3.4	191	1.8	37,442	.2
Illinois	99	1.1	75,031	.5	19	5.8	28,511	4.2	16	1.9	7,332	1.0	134	1.3	110,874	.6
Indiana	35	.4	77,681	.5	1	.3	5	.0	7	.8	1,770	.2	43	.4	79,456	.5
Iowa	32	.3	127,421	.8	6	1.8	940	.1	22	2.7	28,908	3.9	60	.6	157,269	.9
Kansas	45	.5	161,916	1.0	8	2.4	63,726	9.5	4	.5	79,983	10.7	57	.6	305,625	1.8
Kentucky	110	1.2	55,365	.3	7	2.1	1,653	.2	9	1.1	11,055	1.5	126	1.2	68,073	.4
Louisiana	948	10.3	780,256	4.9	14	4.3	4,325	.6	33	4.0	70,394	9.4	995	9.6	854,975	4.9
Maine	126	1.4	3,704	.0	2	.6	2	.0	2	.2	0	.0	130	1.3	3,706	.0
Maryland	210	2.3	42,463	.3	7	2.1	1,494	.2	17	2.1	4,397	.6	234	2.3	48,354	.3
Massachusetts	260	2.8	30,989	.2	1	.3	5	.0	29	3.5	14,422	1.9	290	2.8	45,416	.3
Michigan	23	.3	5,812	.0	2	.6	200,030	29.7	0	.0	0	.0	25	.2	205,842	1.2
Minnesota	37	.4	45,771	.3	9	2.7	1,098	.2	12	1.4	25,440	3.4	58	.6	72,309	.4
Mississippi	78	.8	49,815	.3	1	.3	1	.0	8	1.0	5	.0	87	.8	49,821	.3
Missouri	106	1.2	137,548	.9	24	7.3	6,121	.9	21	2.5	6,294	.8	151	1.5	149,963	.9
Montana	2	.0	5,130	.0	0	.0	0	.0	0	.0	0	.0	2	.0	5,130	.0
Nebraska	40	.4	29,168	.2	12	3.7	1,576	.2	5	.6	22,661	3.0	57	.6	53,405	.3
New Hampshire	11	.1	322	.0	0	.0	0	.0	0	.0	0	.0	11	.1	322	.0
New Jersey	157	1.7	67,315	.4	2	.6	1,100	.2	34	4.1	9,000	1.2	193	1.9	77,415	.4
New York	153	1.7	7,021,347	44.0	3	.9	57	.0	68	8.2	3,812	.5	224	2.2	7,025,216	40.4
North Carolina	79	.9	718,821	4.5	0	.0	0	.0	7	.8	19	.0	86	.8	718,840	4.1
North Dakota	4	.0	8,442	.1	1	.3	1	.0	2	.2	16,823	2.2	7	.1	25,266	.1
Ohio	186	2.0	113,021	.7	39	11.9	42,503	6.3	24	2.9	305,942	40.9	249	2.4	461,466	2.7
Oklahoma	89	1.0	578,291	3.6	3	.9	6	.0	3	.4	12,102	1.6	95	.9	590,399	3.4
Oregon	77	.8	912,076	5.7	1	.3	0	.0	3	.4	1,000	.1	81	.8	913,076	5.2
Pennsylvania	152	1.7	72,594	.5	7	2.1	757	.1	8	1.0	2,418	.3	167	1.6	75,769	.4
Puerto Rico	28	.3	11,543	.1	1	.3	10	.0	0	.0	0	.0	29	.3	11,553	.1
Rhode Island	93	1.0	3,166	.0	0	.0	0	.0	13	1.6	0	.0	106	1.0	3,166	.0
South Carolina	85	.9	1,517	.0	3	.9	8,031	1.2	5	.6	0	.0	93	.9	9,548	.1

TABLE 7–35. POLLUTION DISCHARGES IN NAVIGABLE WATERS OF THE UNITED STATES IN 1984 (continued)

[In gallons; by state]

State	Oil				Hazardous				Other				Total			
	#	%	Quantity	%	#	%	Quantity	%	#	%	Quantity	%	#	%	Quantity	%
South Dakota	7	.1	5,270	.0	2	.6	10	.0	1	.1	30	.0	10	.1	5,310	.0
Tennessee	73	.8	110,911	.7	30	9.1	48,377	7.2	13	1.6	4,384	.6	116	1.1	163,672	.9
Texas	725	7.9	788,873	4.9	42	12.8	101,722	15.1	70	8.4	36,448	4.9	837	8.1	927,043	5.3
Utah	5	.1	19,208	.1	0	.0	0	.0	1	.1	126	.0	6	.1	19,334	.1
Virginia	415	4.5	58,399	.4	9	2.7	10,067	1.5	7	.8	515	.1	431	4.2	68,981	.4
Washington	372	4.0	60,301	.4	3	.9	190	.0	15	1.8	32	.0	390	3.8	60,523	.3
West Virginia	298	3.2	76,727	.5	15	4.6	23,742	3.5	14	1.7	3,685	.5	327	3.2	104,154	.6
Wisconsin	2	.0	35	.0	6	1.8	519	.1	0	.0	0	.0	8	.1	554	.0
Wyoming	51	.6	125,811	.8	6	1.8	1,985	.3	2	.2	3	.0	59	.6	127,799	.7
Guam	10	.1	1,733	.0	0	.0	0	.0	1	.1	40	.0	11	.1	1,773	.0
Virgin Islands	20	.2	11,340	.1	0	.0	0	.0	0	.0	0	.0	20	.2	11,340	.1
Other	1,587	17.3	77,848	.5	5	1.5	98	.0	99	11.9	538	.1	1,691	16.3	78,484	.5
TOTAL	9,198	100.0	15,970,560	100.0	328	100.0	673,565	100.0	829	100.0	748,845	100.0	10,355	100.0	17,392,970	100.0

Source: U.S. Coast Guard, Polluting Incidents In and Around U.S. Waters, Calendar Year 1983 and 1984. COMDTINST M16450.2G; NTIS PB 87–186821

TABLE 7–36. POLLUTION DISCHARGES IN NAVIGABLE WATERS OF THE UNITED STATES IN 1984

[In gallons; by type of material discharged]

Materials	Oil				Hazardous				Other				Total			
	#	%	Quantity	%	#	%	Quantity	%	#	%	Quantity	%	#	%	Quantity	%
Crude Oil	3,069	33.4	1,516,776	9.5	0	.0	0	.0	0	.0	0	.0	3,069	29.6	1,516,776	8.7
Gasoline	642	7.0	593,768	3.7	0	.0	0	.0	0	.0	0	.0	642	6.2	593,768	3.4
Kerosene/Fuel Oil	157	1.7	579,169	3.6	0	.0	0	.0	0	.0	0	.0	157	1.5	579,169	3.3
Diesel Oil	2,142	23.3	7,220,236	45.2	0	.0	0	.0	0	.0	0	.0	2,142	20.7	7,220,236	41.5
Fuel Oil	428	4.7	3,131,237	19.6	0	.0	0	.0	0	.0	0	.0	428	4.1	3,131,237	18.0
Asphalt/Tar/Pitch	35	.4	57,151	.4	0	.0	0	.0	0	.0	0	.0	35	.3	57,151	.3
Other Distillate	73	.8	21,793	.1	4	1.2	1,031	.2	0	.0	0	.0	77	.7	22,824	.1
Solvents	6	.1	154	.0	9	2.7	343	.1	0	.0	0	.0	15	.1	497	.0
Animal/Veg Oil	196	2.1	2,277,334	14.3	0	.0	0	.0	0	.0	0	.0	196	1.9	2,277,334	13.1
Other Oil	2,436	26.5	566,884	3.5	0	.0	0	.0	0	.0	0	.0	2,436	23.5	566,884	3.3
Chemical	0	.0	0	.0	315	96.0	672,191	99.8	0	.0	0	.0	315	3.0	672,191	3.9
Other Substances	14	.2	6,058	.0	0	.0	0	.0	829	100.0	748,845	100.0	843	8.1	754,903	4.3
TOTAL	9,198	100.0	15,970,560	100.0	328	100.0	673,565	100.0	829	100.0	748,845	100.0	10,355	100.0	17,392,970	100.0

Source: U.S. Coast Guard, Polluting Incidents In and Around U.S. Waters, Calendar Year 1983 and 1984. COMDTINST M16450.2G; NTIS 87–186821

TABLE 7–37. POLLUTING INCIDENTS FROM VESSELS IN CANADIAN WATERS, 1974 TO 1983

[Tankers, bulk carriers and other vessels; metric tons]

	Transfer Accident		Collision, Ground, Sinking		Other		Total	
Year	Events	Tons	Events	Tons	Events	Tons	Events	Tons
1974	60	371	21	4 277	60	248	141	4 896
1975	52	116	13	613	28	886	93	1 615
1976	53	206	13	1 613	19	160	85	1 979
1977	47	249	11	931	38	294	96	1 474
1978	51	154	15	1 343	33	73	99	1 570
1979	49	108	6	948	33	8 186	88	9 242
1980	68	145	12	121	68	213	148	479
1981	75	97	13	2 296	33	931	121	3 324
1982	58	199	16	2 106	27	989	101	3 294
1983	28	73	7	504	22	404	57	981
Total	541	1 718	127	14 752	361	12 384	1 029	28 854
Percent	53	6	12	51	35	43	100	100

Source: Environment Canada, Summary of Spill Events in Canada, 1974–1983, EPS 5/SP/1

TABLE 7–38. PCBs AND PAHs IN SEDIMENTS FROM SELECTED ESTUARIES IN THE UNITED STATES, 1984

[Parts per billion]

Estuary	Total PCBs	Total PAHs
Casco Bay, ME	95.28	7,320.00
Merrimack River, MA	52.97	1,730.00
Salem Harbor, MA	533.58	10,220.00
Boston Harbor, MA	17,104.86	26,440.00
Buzzards' Bay, MA	308.46	1,710.00
Narragansett Bay, RI	159.96	2,350.00
East Long Island Sound, NY	10.00	48,560.00
West Long Island Sound, NY	234.43	8,430.00
Raritan Bay, NJ	443.89	5,010.00
Delaware Bay, DE	2.50	330.00
Lower Chesapeake Bay, VA	51.00	410.00
Pamlico Sound, NC	ND	219.25
Charleston Harbor, SC	9.10	802.98
Sapelo Sound, GA	ND	22.28
St. Johns River, FL	140.00	1,926.91
Charlotte Harbor, FL	ND	26.51
Tampa Bay, FL	ND	27.10
Apalachicola Bay, FL	12.00	200.25
Mobile Bay, AL	ND	96.79
Round Island, MS	ND	52.36
Mississippi River Delta, LA	34.00	603.41
Barataria Bay, LA	ND	106.08
Galveston Bay, TX	ND	68.05
San Antonio Bay, TX	ND	8.98
Corpus Christi Bay, TX	ND	28.17
Lower Laguna Madre, TX	ND	0.00
San Diego Harbor, CA	422.10	5,000.00
San Diego Bay, CA	6.74	0.00
Dana Point, CA	7.06	22.87
Seal Beach, CA	46.71	257.96
San Pedro Canyon, CA	159.56	527.00
Santa Monica Bay, CA	14.00	68.25
San Francisco Bay, CA	123.46	5,976.03
Bodega Bay, CA	4.18	11.00
Coos Bay, OR	3.19	234.67
Columbia River Mouth, OR/WA	8.77	145.03

TABLE 7–38. PCBs AND PAHs IN SEDIMENTS FROM SELECTED ESTUARIES IN THE UNITED STATES 1984 (continued)

[parts per billion]

Estuary	Total PCBs	Total PAHs
Nisqually Reach, WA	4.23	0.00
Commencement Bay, WA	20.60	1,200.00
Elliott Bay, WA	329.87	4,700.00
Lutak Inlet, AK	5.50	0.00
Nahku Bay, AK	6.60	100.00

Note: PCBs=Polychlorinated biphenyls; PAHs=Polycyclic aromatic hydrocarbons. ND=Not detected

Source: U.S. Department of Commerce, National Oceanic and Atmospheric Administration, National Ocean Survey, Ocean Assessments Division. 1987. National status and trends program for marine environmental quality. Progress report on preliminary assessment of findings of the benthic surveillance project, 1984. Rockville, MD.; U.S. Geological Survey, National Water Summary 1986

TABLE 7–39. SEWAGE INDICATORS IN SEDIMENTS FROM SELECTED ESTUARIES IN THE UNITED STATES, 1984

[Cells per gram and nanograms per gram]

Estuary	Clostridium perfringens (cells/gram)	Coprostanol (nanogram/gram)
Casco Bay, ME	710.00	221.27
Merrimack River, MA	670.06	206.47
Salem Harbor, MA	57,000.00	7,040.23
Boston Harbor, MA	79,000.00	9,000.00
Buzzards' Bay, MA	413.00	1,376.60
Narragansett Bay, RI	220.00	647.80
East Long Island Sound, NY	290.00	17.00
West Long Island Sound, NY	2,090.00	957.11
Raritan Bay, NJ	24,373.52	5,402.00
Delaware Bay, DE	91.00	148.00
Lower Chesapeake Bay, VA	4.00	781.56
Pamlico Sound, NC	120.00	1,100.00
Charleston Harbor, SC	1,600.00	1,253.23
Sapelo Sound, GA	270.00	510.00
St. Johns River, FL	1,400.00	790.07
Charlotte Harbor, FL	33.00	692.47
Tampa Bay, FL	3.00	350.92
Apalachicola Bay, FL	74.00	687.56
Mobile Bay, AL	100.43	304.90
Round Island, MS	75.00	256.87
Mississippi River Delta, LA	1,200.00	523.98
Barataria Bay, LA	45.00	395.68
Galveston Bay, TX	34.00	278.57
San Antonio Bay, TX	6.00	109.78
Corpus Christi Bay, TX	1.00	248.84
Lower Laguna Madre, TX	27.00	240.00
San Diego Harbor, CA	2,600.00	600.00
San Diego Bay, CA	121.79	33.00
Dana Point, CA	103.27	98.00
Seal Beach, CA	832.72	180.00
San Pedro Canyon, CA	6,596.78	780.00
Santa Monica Bay, CA	471.53	230.00
San Francisco Bay, CA	5,093.47	1,860.00
Bodega Bay, CA	29.00	120.00
Coos Bay, OR	30.96	230.00
Columbia River Mouth, OR/WA	261.05	580.00
Nisqually Reach, WA	35.49	5.33
Commencement Bay, WA	5,300.00	1,900.00
Elliott Bay, WA	7,725.95	370.00
Lutak Inlet, AK	91.78	83.33
Nahku Bay, AK	212.13	310.00

Source: U.S. Department of Commerce, National Oceanic and Atmospheric Administration, National Ocean Survey, Ocean Assessments Division. 1987. National Status and Trends Program for Marine Environmental Quality. Progress Report on Preliminary Assessment of Findings of the Benthic Surveillance Project, 1984. Rockville, MD.; U.S. Geological Survey, National Water Summary 1986

TABLE 7–40. TRACE METALS IN SEDIMENTS FROM SELECTED ESTUARIES IN THE UNITED STATES, 1984

[Parts per million]

Estuary	Chromium	Copper	Lead	Zinc	Cadmium	Silver	Mercury
Casco Bay, ME	92.10	16.97	29.13	76.27	0.15	0.09	0.12
Merrimack River, MA	41.15	6.47	23.25	35.75	0.07	0.05	0.08
Salem Harbor, MA	2296.67	95.07	186.33	238.00	5.87	0.88	1.19
Boston Harbor, MA	223.67	148.00	123.97	291.67	1.61	2.64	1.05
Buzzards' Bay, MA	73.66	25.02	30.72	97.72	0.23	0.37	0.12
Narragansett Bay, RI	93.60	78.95	60.25	144.43	0.35	0.56	0.00
East Long Island Sound, NY	37.63	11.26	22.13	58.83	0.11	0.15	0.09
West Long Island Sound, NY	131.50	111.00	69.75	243.00	0.73	0.68	0.48
Raritan Bay, NJ	181.00	181.00	181.00	433.75	2.74	2.06	2.34
Delaware Bay, DE	27.76	8.34	15.04	49.66	0.24	0.11	0.09
Lower Chesapeake Bay, VA	58.50	11.32	15.70	66.23	0.38	0.08	0.10
Pamlico Sound, NC	79.67	14.13	30.67	102.67	0.33	0.09	0.11
Sapelo Sound, GA	51.80	5.93	16.00	38.33	0.09	0.02	0.03
St. Johns River, FL	37.67	9.77	26.00	67.67	0.18	0.11	0.07
Charlotte Harbor, FL	26.47	1.17	4.33	7.20	0.08	0.01	0.02
Tampa Bay, FL	23.70	4.97	4.67	9.10	0.15	0.08	0.03
Apalachicola Bay, FL	69.17	16.93	30.67	111.67	0.05	0.06	0.06
Mobile Bay, AL	93.00	17.40	29.67	161.00	0.11	0.11	0.12
Mississippi River Delta, LA	72.27	19.40	22.67	90.00	0.47	0.17	0.06
Barataria Bay, LA	52.07	10.50	18.33	59.33	0.19	0.09	0.05
Galveston Bay, TX	41.13	8.03	18.33	33.97	0.05	0.09	0.03
San Antonio Bay, TX	39.43	5.57	11.33	32.00	0.07	0.09	0.02
Corpus Christi Bay, TX	31.43	6.63	13.00	56.00	0.19	0.07	0.04
Lower Laguna Madre, TX	24.53	5.83	11.33	36.00	0.09	0.07	0.03
San Diego Harbor, CA	178.00	218.67	50.97	327.67	0.99	0.76	1.04
San Diego Bay, CA	49.70	7.67	11.61	58.67	0.04	0.76	0.04
Dana Point, CA	39.80	10.03	18.80	53.67	0.22	0.80	0.13
Seal Beach, CA	108.33	26.00	27.37	125.00	0.17	1.27	0.59
San Pedro Canyon, CA	106.50	31.33	17.33	118.33	1.17	1.20	0.32
Santa Monica Bay, CA	53.53	10.53	33.37	46.67	0.18	0.51	0.01
San Francisco Bay, CA	1466.67	160.71	67.39	501.66	0.51	0.37	0.25
Bodega Bay, CA	246.33	0.06	2.17	38.33	0.18	1.74	0.14
Coos Bay, OR	110.30	1.47	4.65	32.00	0.62	0.31	0.11
Columbia River Mouth, OR/WA	29.53	17.00	15.90	107.67	0.86	2.14	0.25
Nisqually Reach, WA	118.07	13.33	24.57	105.33	0.68	2.62	0.32
Commencement Bay, WA	69.50	51.33	34.63	101.00	0.77	5.90	0.01
Elliott Bay, WA	114.37	96.00	20.23	166.00	0.84	1.18	0.11
Lutak Inlet, AK	58.27	26.67	15.90	180.33	0.96	0.09	0.24
Nahku Bay, AK	23.27	9.80	43.30	191.33	1.09	4.37	0.23
Charleston, SC	86.33	16.03	27.33	72.67	—	—	—

Source: Young, D. & J. Means. 1987. Trace Metals in Sediments. IN: U.S. Department of Commerce, National Oceanic and Atmospheric Administration, National Ocean Survey, Ocean Assessments Division. National Status and Trends Program for Marine Environmental Quality. Progress Report on Preliminary Assessment of Findings of the Benthic Surveillance Project, 1984. Rockville, MD.; U.S. Geological Survey, National Water Summary 1986

SECTION C. GROUND-WATER CONTAMINATION

FIGURE 7–2. WASTE DISPOSAL PRACTICES AND CONTAMINATION OF GROUND WATER

[Movement of contaminants in unsaturated zone, alluvial aquifer, and bedrock shown by dark shading]

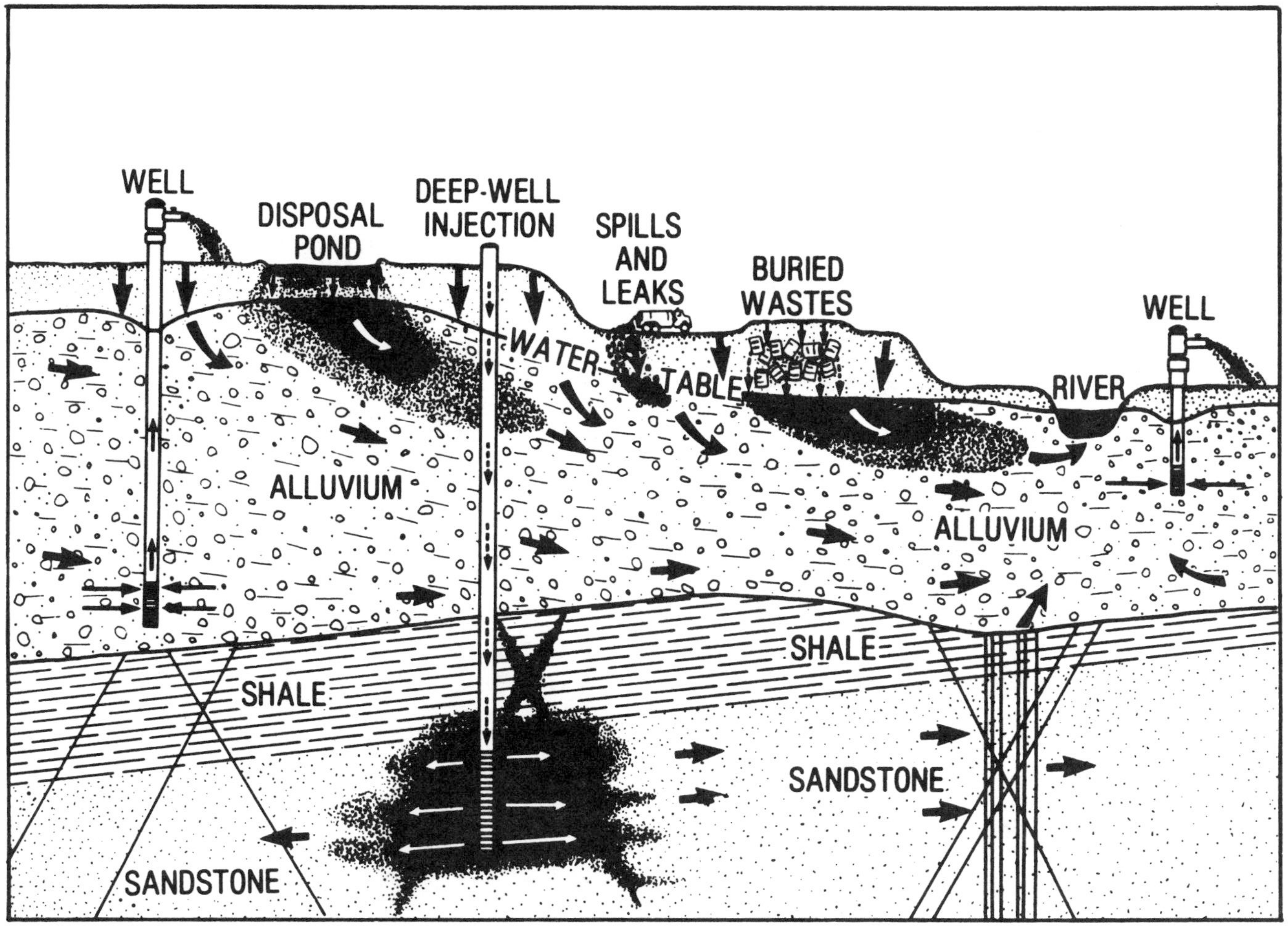

Source: U.S. Geological Survey Water Fact Sheet. Toxic Waste, Ground-Water Contamination, 1983

TABLE 7–41. MAJOR SOURCES OF GROUND-WATER CONTAMINATION REPORTED BY STATES

Source	No. of States Reporting Source[a]	% of States Reporting Source	No. of States Reporting as Primary Source[b]
Septic Tanks	46	89%	9
Underground Storage Tanks	43	83%	13
Agricultural Activities	41	79%	6
On-site Landfills	34	65%	5
Surface Impoundments	33	64%	2
Municipal Landfills	32	62%	1
Abandoned Waste Sites	29	56%	3
Oil and Gas Brine Pits	22	42%	2
Saltwater Intrusion	19	37%	4
Other Landfills	18	35%	0
Road Salting	16	31%	1
Land Application of Sludge	12	23%	0
Regulated Waste Sites	12	23%	1
Mining Activities	11	21%	1
Underground Injection Wells	9	17%	0
Construction Activities	2	4%	0

[a] Based on a total of 52 States and territories which reported ground-water contamination sources in their 305(b) submittals.
[b] Some States did not indicate a primary source.
Source: U.S. Environmental Protection Agency, National Water Quality Inventory- 1986 Report to Congress. EPA-440/4–87–008

TABLE 7–42. MAJOR GROUND-WATER CONTAMINANTS REPORTED BY STATES

Contaminant	Reported as a Major Contaminant: No. of States[a]	Reported as a Major Contaminant: % of States
Sewage	46	89%
Inorganic Chemicals:		
Nitrates	42	75%
Brine/Salinity	36	69%
Arsenic	19	37%
Fluorides	18	35%
Sulfur Compounds	7	14%
Organic Chemicals:		
Synthetic	37	71%
Volatile	36	69%
Metals	34	65%
Pesticides	31	60%
Petroleum	21	40%
Radioactive Materials	12	23%

[a] Based on a total of 52 States and territories which cited ground-water contaminants in their 305(b) submittals.
Source: U.S. Environmental Protection Agency, National Water Quality Inventory-1986 Report to Congress. EPA-440/4–87–008

TABLE 7–43. ACTIVITIES CONTRIBUTING TO GROUND-WATER CONTAMINATION IN THE UNITED STATES

Activity	States Citing	Estimated Sites[1]	Contaminants Frequently Cited as Result of Activity	Remarks
Waste disposal:				
Septic systems	41	22 million.	Bacteria, viruses, nitrate, phosphate, chloride, and organic compounds such as trichloroethylene.	Between 820 and 1,460 billion gallons per year discharged to shallowest aquifers (Office of Technology Assessment, 1984).
Landfills (active)	51	16,400	Dissolved solids, iron, manganese, trace metals, acids, organic compounds, and pesticides.	Traditional disposal method for municipal and industrial solid waste. Unknown number of abandoned landfills.
Surface impoundments	32	191,800	Brines, acidic mine wastes, feedlot wastes, trace metals, and organic compounds.	Used to store oil/gas brines (125,100 sites), mine wastes (19,800), agricultural wastes (17,200), industrial liquid wastes (16,200), municipal sewage sludges (2,400), other wastes (11,100) (U.S. Environmental Protection Agency, 1987).
Injection wells	10	280,800	Dissolved solids, bacteria, sodium, chloride, nitrate, phosphate, organic compounds, pesticides, and acids.	Wells used for injecting waste below drinking-water sources (550), oil/gas brine disposal (161,400), solution mining (22,700), injecting waste into or above drinking-water sources (14), and storm-water disposal, agricultural drainage, heat pumps (96,100) (U.S. Environmental Protection Agency, 1987).
Land application of wastes	12	19,000 land application units.	Bacteria, nitrate, phosphate, trace metals, and organic compounds.	Waste disposal from municipal waste-treatment plants (11,900), industry (5,600), oil/gas production (730), petroleum and wood-preserving wastes (250), others (620) (U.S. Environmental Protection Agency, 1987).
Storage and handling of materials:				
Underground storage tanks	39	2.4–4.8 million.	Benzene, toluene, xylene, and petroleum products.	Useful life of steel tanks, 15–20 years. About 25–30 percent of petroleum tanks may leak (Conservation Foundation, 1987).
Above-ground storage tanks	16	Unknown.	Organic compounds, acids, metals, and petroleum products.	Spills/overflows may contaminate ground water.
Material handling and transfers.	29	10,000–16,000 spills per year.	Petroleum products, aluminum, iron, sulfate, and trace metals.	Includes coal storage piles, bulk chemical storage, containers, and accidental spills.
Mining activities:				
Mining and spoil disposal—coal mines.	23	15,000 active; 67,000 inactive.	Acids, iron, manganese, sulfate, uranium, thorium, radium, molybdenum, selenium, and trace metals.	Leachates from spoil piles of coal, metal, and nonmetallic mineral mining contain a variety of contaminants. Coal mines are sources of acid drainage.
Oil and gas activities:				
Wells	20	550,000 production; 1.2 million abandoned.	Brines.	Contamination from improperly plugged wells and oil brine stored in ponds or injected underground.
Agricultural activities:				
Fertilizer and pesticide applications.	44	363 million acres.[2]	Nitrate, phosphate, and pesticides.	Fertilizer applied 1982–83, 42.3 million tons per year (U.S. Bureau of the Census, 1984); active ingredients of pesticides applied 1982, 660 million pounds (Gianessi, 1987).
Irrigation practices	22	376,000 wells; 49 million acres irrigated.[3]	Dissolved solids, nitrate, phosphate, and pesticides.	Salts, fertilizers, pesticides can concentrate in ground water. Improperly plugged abandoned wells contamination source.
Animal feedlots	17	1,900	Nitrate, phosphate, and bacteria.	Primarily in the Corn Belt and High Plains States (Office of Technology Assessment, 1984).
Urban activities:				
Runoff	15	47.3 million acres urban land.[4]	Bacteria, hydrocarbons, dissolved solids, lead, cadmium, and trace metals.	Infiltration from detention basins, drainage wells, pits, shafts can reach ground water. Karst areas particularly vulnerable.
Deicing chemical storage and use.	14	Not reported.	Sodium chloride, ferric ferrocyanide, sodium ferrocyanide, phosphate, and chromate.	Winter 1983, 9.35 million tons dry salts/abrasives, 7.78 million gallons of liquid salts applied (Office of Technology Assessment, 1984).
Other:				
Saline intrusion or upconing	29	Not reported.	Dissolved solids and brines.	Present in coastal areas and in many inland areas.

[1] Estimated number of sites from U.S. Environmental Protection Agency (1987) unless otherwise indicated.
[2] U.S. Bureau of the Census, 1984, p. 658, 1982 data.
[3] U.S. Bureau of the Census, 1984, p. 639, 1982 data.
[4] U.S. Bureau of the Census, 1984, p. 195, 1980 data.
Source: U.S. Geological Survey, National Water Survey 1986, Water-Supply Paper 2325

TABLE 7–44. PROFILE OF SOURCES OF GROUND-WATER CONTAMINATION IN THE UNITED STATES

Source	Potential Contaminants	Number/Volume	Geographic Distribution
Subsurface percolation systems	Organics, metals, nitrates, phosphates, microorganisms	22 million domestic systems 25,000 industrial systems	Highest concentration in eastern third of country and portions of west coast
Injection wells	Organics, metals, inorganic acids, microorganisms, radionuclides	280,752 active wells	Varies by well type — Class I (hazardous waste) - Gulf Coast and Great Lakes — Class II (oil/gas) - throughout the U.S. — Class III (mining) - Southwest — Class V - agricultural wells in IA, ID, TX, CA; industrial wells in NY and NJ
Land application	Nitrogen, phosphorous, metals organics, microorganisms	2,463 POTWs—sludge application 1,000 POTWs—land treatment 250 hazardous waste land treatment units 18,889 non-hazardous units	Unknown
Landfills	Organics, inorganics, microorganisms, radionuclides	16,416 landfills 9,284 municipal 3,155 industrial	Urban locations nationwide
Open dumps	Organics, inorganics, microorganisms	1,856 – 2,396 dumps	55 states and territories
Residential disposal	Organics, metals, other inorganics, microorganisms	Unknown	Nationwide
Surface impoundments	Organics, metals and other inorganics, microorganisms, radionuclides	191,822 surface impoundments — 16,232 industrial — 2,426 municipal — 17,159 agricultural — 19,813 mining — 125,074 oil and gas — 11,118 other	70% in hydrogeologically vulnerable areas 37% over current ground-water sources of drinking water Highest number of non-hazardous are in AR, KS, LA, MN, OH, OK, PA, TX, WV
Waste tailings and piles	Arsenic, sulfuric acid, copper, selenium, molybdenum, uranium, thorium, radium, lead, manganese, vanadium	Total mining - 2.3 billion tons/yr. — Metal - 250 million tons/yr. — Uranium - 215 million tons/yr. Hazardous waste - 0.39 billion tons	Unknown
Material stockpiles	Metals, inorganics, radionuclides	Annual materials production - 3.4 billion tons/yr. Stockpiles - 700 million tons/yr.	Nationwide
Graveyards	Metals, nonmetals, microorganisms	Unknown	Nationwide
Animal burial		Unknown	Unknown
Aboveground storage	Organics, inorganics, microorganisms, radionuclides	Unknown	Nationwide
Underground storage	Organics, inorganics, microorganisms, radionuclides	Steel - 2.4 - 4.8 million tanks Fiberglass - 0.1 million tanks Total capacity - 25 billion gallons Hazardous storage - 2,032 tanks	Nationwide
Pipelines	Microorganisms, organics, inorganics	175,000 miles of petroleum product pipelines (1976) carrying 9.63 billion bbls 700,000 miles of sewer pipeline (1976) carrying 5.6 trillion gallons	Nationwide

TABLE 7–44. PROFILE OF SOURCES OF GROUND-WATER CONTAMINATION IN THE UNITED STATES (continued)

Source	Potential Contaminants	Number/Volume	Geographic Distribution
Materials transport	Organics, inorganics, microorganisms	10,000–16,000 spills per year; spills account for approximately 0.35 percent of 4 billion tons shipped annually (1984)	Nationwide
Mining/mine drainage	Acids, metals, radionuclides	15,000 active coal mines (1986) 67,000 inactive coal mines phospate mines; metalic ore mines	Varies by mining type
Production wells	Organics, inorganics, microorganisms	548,000 oil wells produced approximately 3.1 billion bbls crude oil (1980) Up to 1.2 million abandoned wells 376,000 irrigation wells for 126,000 farms	Oil Wells - nationwide Geothermal wells - primarily CA, NV, ID Water wells - mostly in the Southwest, Central Plains, Idaho, and Florida
Other wells (monitoring and exploration)	Organics, inorganics, microorganisms radionuclides	Unknown	Unknown
Pesticide application	Organics - 1,200–1,400 active ingredients	552 million pounds of active ingredients applied to crops in 1982	17 pesticides confirmed in 23 states (1986) due to normal agricultural application
	Approximately 280 million acre-treatments annually		
Fertilizer applications	Nitrates, phosphates	Fertilizer use has declined from 54 million tons to 42.3 million tons (1980–1983); fertilizers in 1981–1982 contained 11 million tons of nitrogen, 4.8 million tons of phosphates, 5.6 million tons of potash	Highest fertilizer use in 1981–1982: CA, IL, IN, IO, TX
Deicing	Salts	9.35 million tons dry salts, and abrasives; 1.78 million gallons liquid salts applied to U.S. highways (1982–1983)	Northeast, Mid-Atlantic, Midwest
Irrigation practices	Fertilizers, pesticides, naturally occurring contaminants (e.g., selenium), sediment	14 percent of cropland is irrigated	West, Central, and South Plains, Arkansas, Florida
Percolation of atmospheric pollutants	Sulphur and nitrogen compounds, asbestos, heavy metals	Unknown	Acid rain around Great Lakes, Northeast Distribution of other pollutants varies
Ground water/surface water interaction	Organics, inorganics, microorganisms, radionuclides	Unknown	Unknown
Natural leaching	Inorganics, radionuclides	Unknown	Unknown, very localized
Salt water intrusion	Inorganics, radionuclides	Unknown	Predominantly coastal areas—CA, TX, LA, FL, NY, Southwest, Central Plains

Source: U.S. Environmental Protection Agency, Office of Ground-Water Protection, 1987, EPA Activities Related to Ground-Water Contamination

TABLE 7–45. CONCENTRATIONS OF TOXIC ORGANIC COMPOUNDS FOUND IN DRINKING WATER WELLS AND SURFACE WATER IN THE UNITED STATES

Chemical	Highest Ground Water Concentration Reported (ppb)	State	Highest Surface Water Concentration Reported (ppb)
Trichloroethylene (TCE)	27,300	Pennsylvania	
	14,000	Pennsylvania	
	3,800	New York	160
	3,200	Pennsylvania	
	1,530	New Jersey	
	900	Massachusetts	
Toluene	6,400	New Jersey	
	260	New Jersey	6.1
	55	New Jersey	
1,1,1-Trichloroethane	5,440	Maine	
	5,100	New York	5.1
	1,600	Connecticut	
	965	New Jersey	
Acetone	3,000	New Jersey	NI
Methylene chloride	3,000	New Jersey	
	47	New York	13
Dioxane	2,100	Massachusetts	NI
Ethyl benzene	2,000	New Jersey	NI
Tetrachloroethylene	1,500	New Jersey	
	740	Connecticut	21
	717	New York	
Cyclohexane	540	New York	NI
Chloroform	490	New York	
	420	New Jersey	700
	67	New York	
Di-n-butyl-phthalate	470	New York	NI
Carbon tetrachloride	400	New Jersey	30
	135	New York	
Benzene	330	New Jersey	
	230	New Jersey	4.4
	70	Connecticut	
	30	New York	
1,2-Dichloroethylene	323	Massachusetts	
	294	Massachusetts	9.8
	91	New York	
Ethylene dibromide (EDB)	300	Hawaii	
	100	Hawaii	NI
	35	California	

TABLE 7–45. CONCENTRATIONS OF TOXIC ORGANIC COMPOUNDS FOUND IN DRINKING WATER WELLS AND SURFACE WATER IN THE UNITED STATES (continued)

Chemical	Highest Ground Water Concentration Reported (ppb)	State	Highest Surface Water Concentration Reported (ppb)
Xylene	300	New Jersey	
	69	New York	24
Isopropyl benzene	290	New York	NI
1,1-Dichloroethylene	280	New Jersey	
	118	Massachusetts	
	70	Maine	0.5
1,2-Dichloroethane	250	New Jersey	4.8
Bis (2-ethylhexyl) phthalate	170	New York	NI
DBCP (Dibromochloropropane)	137	Arizona	
	95	California	NI
	68	California	
Trifluorotrichloroethane	135	New York	NI
	35	New York	
Dibromochloromethane	55	New York	317
	20	Delaware	
Vinyl chloride	50	New York	9.8
Chloromethane	44	Massachusetts	12
Butyl benzyl-phthalate	38	New York	NI
gamma-BHC (Lindane)	22	California	NI
1,1,2-Trichloroethane	20	New York	NI
Bromoform	20	Delaware	280
1,1-Dichloroethane	7	Maine	0.2
alpha-BHC	6	California	NI
Parathion	4.6	California	0.4
delta-BHC	3.8	California	NI

NI = Not investigated.

Source: U.S. Council on Environmental Quality, 1981, Contamination of Ground Water by Toxic Organic Chemicals

TABLE 7–46. PUBLIC WATER-SUPPLY WELLS IN THE UNITED STATES CLOSED BECAUSE OF CONTAMINATION AS OF 1984

State and Utility	Water-Supply Wells					
	In Service or Standby		Closed by Man-Made or Chemical Contamination		Closed by Natural Contamination	
	Shallow (<100')	Deep (>100')	Shallow (<100')	Deep (>100')	Shallow (<100')	Deep (>100')
ARIZONA						
Mesa		24		2		0
Phoenix Wtr & Swr Dept.		115		8		5
Tempe		6		2		0
Tucson		272		7		0
CALIFORNIA						
Alhambra		14		3		0
Anaheim		31		4		0
Burbank		4		6		0
Fresno		100		1		0
Fullerton		11		0		1
Garden Grove		14		0		17
Glendale		8		4		0
La Puente/SW Suburban		40		10		2
Long Beach/Dominguez Wtr Corp.		15		0		2
Los Angeles Wtr & Power		180		25		0
Modesto	1	43	0	2	0	0
Pomona		31		2		0
Riverside	4	70	4	28	0	0
Sacremento/Arcade Cnty Wtr		60		1		0
San Bernardino	2	34	0	8	0	0
San Jose/Great Oaks Wtr Co.		10		3		0
Santa Barbara/Goleta Wtr Dist.		9		0		1
Santa Barbara/Santa Barbara Pub.		7		0		2
COLORADO						
Colorado Springs	8		0		1	
CONNECTICUT						
Clinton/Conn. Wtr Co.	24	17	6	2	0	0
DELAWARE						
Newark/Artesian Wtr Co.	7	35	0	3	0	0
FLORIDA						
Boca Raton		50		0		1
Daytona Beach		21		0		5
Englewood	47	9	0	0	7	0
Hollywood	17	6	0	1	0	0
Miami	73		6		0	
Naples	52	7	0	0	2	0
Ocala/Gen. Dev. Util.		4		1		0
Palm Bay/Gen. Dev. Util.		21		1		0
Palm Beach Gdns/Seacoast Util.	39		2		0	
Tallahassee		24		0		2
HAWAII						
Honolulu	60	73	0	9	0	0
ILLINOIS						
Elgin		11		0		2
INDIANA						
Anderson	4	9	2	0	0	2
Richmond/Ind. Am. Wtr. Co.	5		2		0	
South Bend		23		7		0

TABLE 7–46. PUBLIC WATER-SUPPLY WELLS IN THE UNITED STATES CLOSED BECAUSE OF CONTAMINATION AS OF 1984 (continued)

State and Utility	Water-Supply Wells					
	In Service or Standby		Closed by Man-Made or Chemical Contamination		Closed by Natural Contamination	
	Shallow (<100')	Deep (>100')	Shallow (<100')	Deep (>100')	Shallow (<100')	Deep (>100')
IOWA						
Sioux City	5	15	0	0	0	1
LOUISIANA						
Baton Rouge/B.R. Wtr Co.		50		0		2
MICHIGAN						
Kalamazoo	10	108	0	11	0	0
Lansing	3	124	0	1	0	0
MINNESOTA						
Rochester		19		3		0
NEBRASKA						
Grand Island	9	30	0	2	0	0
NEVADA						
Reno/Sierra Pac. Power Co.	0	16	0	0	3	0
NEW JERSEY						
Brick	7	4	1	0	0	0
Harrington Park/Hackensack Wtr Co.	19	68	0	1	1	2
Merchantville		15		1		1
NEW MEXICO						
Albuquerque		86		1		0
Santa Fe/Sangre de Cristo Wtr Co.		12		1		1
NEW YORK						
East Meadow/Hempstead Wtr Dept.		35		3		3
Elmira	4	0	0	1	0	0
Farmingdale		11		0		1
Lake Success/Jamaica Wtr Co.	30	55	14	3	0	0
Oakdale/Suffolk Wtr Auth.	31	357	14	20	0	2
West Nyack/Spring Valley Wtr Co.	19	68	0	1	1	2
OHIO						
Mansfield	3	6	0	0	0	1
OKLAHOMA						
Oklahoma City		4		0		2
PENNSYLVANIA						
Pittsburgh/West View M.A.	24		12		0	
RHODE ISLAND						
Bristol/Bristol Cnty Wtr Co.	2		0		1	
West Warwick/Kent Cnty W.A.	1	4	0	0	1	0
TEXAS						
El Paso		140		1		0
Lubbock		330		3		3
UTAH						
Ogden		8		0		1
Provo		10		0		2
Salt Lake City/Salt Lake Wtr Dept.	7	14	0	0	0	1
VIRGINIA						
Woodbridge/Prince William Cnty		26		1		0
WYOMING						
Cheyenne		37		1		0

Source: Compiled from 1984 Water Utility Operating Data issued by the American Water Works Association.

TABLE 7–47. POLLUTING INCIDENTS REPORTED IN CANADA, 1974 TO 1983

[Number of spills by sector]

Year	Petroleum	Transport	Chemical	Mining & Metallurgy	Government	Pulp & Paper, Forestry	Other	Total
1974	502	169	43	26	63	26	312	1 141
1975	1 135	194	65	45	50	51	229	1 769
1976	1 212	266	73	54	38	60	241	1 944
1977	1 428	382	84	56	68	72	283	2 373
1978	1 498	352	91	42	75	46	324	2 428
1979	1 628	515	81	61	66	41	399	2 791
1980	1 648	506	87	84	52	49	435	2 861
1981	1 178	432	133	109	43	62	419	2 376
1982	1 107	317	78	105	49	29	304	1 989
1983	1 093	275	60	107	104	38	238	1 915
Total	12 429	3 408	795	689	608	474	3 184	21 587
%	58	16	4	3	3	2	15	100

Source: Environment Canada, 1987, Summary of Spill Events, 1974–1983

TABLE 7–48. HYDROCARBON SPILLS REPORTED IN CANADA, 1974 TO 1983

[By type; volume in metric tons]

Year	Condensates and Gases	Crude Oil	No. 2 Fuel	No. 6 Fuel	No. 4 & 5 Fuel	Gasoline	Other Oils	Waste Oil	Asphalt	Total
1974	3 623	14 823	1 046	1 106	5 594	810	222	631	82	27 935
1975	600	11 530	836	2 499	2 844	3 095	256	321	266	22 250
1976	7 429	10 901	1 650	2 084	2 231	2 156	220	38	372	27 085
1977	20 865	10 699	3 753	804	1 587	1 638	551	85	103	40 085
1978	845	12 067	2 801	3 288	1 932	1 237	454	72	476	23 170
1979	3 306	12 540	4 724	9 054	1 801	1 567	339	319	392	34 045
1980	705	15 274	3 517	585	649	918	278	108	479	22 510
1981	1 160	8 041	12 589	763	1 184	1 767	1 013	1 886	287	28 690
1982	281 181	10 658	4 602	915	1 067	847	609	46	147	300 070
1983	768	8 553	1 467	803	422	460	300	69	337	13 180
Total	320 482	115 086	36 985	21 901	19 311	14 495	4 242	3 575	2 941	539 020
%	59	21	7	4	4	3	1	1	1	100

Source: Environment Canada, 1987, Summary of Spill Events, 1974–1983

TABLE 7–49. NONHYDROCARBON SPILLS REPORTED IN CANADA, 1974 TO 1983

[By type; volume in metric tons]

Year	Industrial Wastes	Industrial Chemicals	Saline Water	Minerals	Acids	Bases	Fertilizers	Paints & Dyes	Metals	Radioactive Material	Pesticides	Others	Total
1974	73 041	8 656	15 132	1 437	95	732	-	-	14	-	1	1 508	100 615
1975	16 053	25 847	13 420	5 545	2 953	734	15	2	19	-	42	398 365	462 990
1976	69 694	2 777	21 019	53 060	1 848	126	150	3	1 545	14	28	232 212	382 475
1977	36 612	9 159	17 545	651	2 977	777	60	22	168	2	30	2 835	70 835
1978	70 423	12 865	34 000	1 835	4 371	672	3 362	4 627	1 000	-	97	10 551	143 800
1979	2 175	42 814	37 431	2 394	276	794	400	102	77	2	44	4 579	91 085
1980	298 535	19 980	17 423	6 293	536	316	708	34	3	-	49	883	344 760
1981	452 705	463 616	5 386	2 964	496	1 649	844	19	-	57	88	3 070	930 890
1982	23 418	7 455	47 857	5 900	3 396	48	160	21	10	729	3	14 688	103 690
1983	4 450	2 645	10 654	3 939	955	550	345	2	16	422	102	475 223	499 300
Total	1 047 106	595 814	219 867	84 018	17 903	6 398	6 044	4 838	2 852	1 226	484	1 143 914	3 130 440
%	33	19	7	3	1	0	0	0	0	0	0	37	100

Source: Environment Canada, 1987, Summary of Spill Events, 1974–1983

SECTION D. DOMESTIC SEWAGE

TABLE 7–50. PHYSICAL CHARACTERISTICS OF DOMESTIC SEWAGE

Characteristic	Cause	Significance	Measurement
Temperature	Ambient air temperature. Hot water discharged into sewer from home or industry.	Influences rate of biological activity. Governs solubility of oxygen and other gases. Affects magnitude of density, viscosity, surface tension, etc.	Standard centigrade or Fahrenheit scale.
Turbidity	Suspended matter such as sewage solids, silt, clay, finely divided organic matter of vegetable origin, algae, microscopic organisms.	Excludes light, thus reducing growth of oxygen-producing plants. Impairs aesthetic acceptability of water. May be detrimental to aquatic life.	Light scatter and absorption on an arbitrary standard scale.
Color	Dissolved matter such as organic extractives from leaves and other vegetation (tannins, glucosides, iron, etc.), industrial wastes.	Harmless generally, but impairs aesthetic quality of water.	Light absorption on a standard arbitrary scale.
Odor	Volatile substances, dissolved gases, often produced by decomposition of organic matter. In water it may result from the essential oils in microorganisms.	May indicate presence of decomposing sewage. Affects aesthetic quality of water. As a test of sewage it may serve, for example, as a guide to condition of sewage when it reaches the treatment plant.	Human sense of smell, qualitative scale, and concentration at threshold of odor.
Taste	Materials producing odors. Dissolved matter and various ions.	Impairs aesthetic quality of water.	Not measured in unpotable water.
Solid matter	Dissolved and suspended organic and inorganic solids.	Measures amount of organic solids, silts, etc., hence is a measure of the extent of sewage pollution or the concentration of a sewage.	By gravimetric analysis techniques for the following: Total solids, total volatile solids, total fixed solids, suspended solids, and dissolved solids.[a]

[a] Total solids: As a measure of the amount of total dissolved and suspended matter. Total volatile solids: As a measure of the decomposable organic matter. Total fixed solids: As a measure of inorganic grit and dissolved inorganic matter plus ash of organic matter. Suspended solids: As a measure of material which might be removed by settling. Volatile and fixed suspended solids: As a measure of the decomposable organic matter and of inorganic matter. Dissolved solids: As a measure of material to be removed by secondary sewage treatment processes. Volatile solids: As a measure of organic matter which may be decomposed (may exert BOD). Fixed solids: As a measure of residues which may add to the burden of the effluent (lower its quality).

Source: McGauhey, Engineering Management of Water Quality, McGraw-Hill, Copyright 1968

TABLE 7–51. COMPOSITION OF DOMESTIC SEWAGE

[All values in milligrams per liter]

Constituent	Strong	Medium	Weak
Solids, total	1000	500	200
Volatile	700	350	120
Fixed	300	150	80
Suspended, total	500	300	100
Volatile	400	250	70
Fixed	100	50	30
Dissolved, total	500	200	100
Volatile	300	100	50
Fixed	200	100	50
BOD (5-day, 20°C)	300	200	100
Oxygen consumed	150	75	30
Dissolved oxygen	0	0	0
Nitrogen, total	86	50	25
Organic	35	20	10
Free ammonia	50	30	15
Nitrites (NO_2)	0.10	0.05	0
Nitrates (NO_3)	0.40	0.20	0.10
Chlorides	175	100	15
Alkalinity	200	100	50
Fats	40	20	0

Source: Babbitt and Baumann, Sewerage and Sewage Treatment, John Wiley & Sons, Copyright 1958

TABLE 7–52. RANGE OF MINERAL PICKUP IN DOMESTIC SEWAGE

[Summary of incremental increases in mineral constituents between source and disposal for 15 California cities.]

Mineral Constituent or Property	Normal Range, parts per million
Dissolved solids	100–300
Boron (B)	0.1–0.4
Percent Sodium	5–15[a]
Sodium (Na)	40–70
Potassium (K)	7–15
Magnesium ($CaCO_3$)	15–40
Calcium ($CaCO_3$)	15–40
Total Nitrogen (N)	20–40
Phosphate (PO_4)	20–40
Sulfate (SO_4)	15–30
Chloride (Cl)	20–50
Total Alkalinity ($CaCO_3$)	100–150

[a] In percent

Source: Calif. State Water Pollution Control Board, 1954

TABLE 7–53. RECOMMENDED SIZES OF SEPTIC TANKS

Number of Bedrooms	Maximum Number of Persons Served	Liquid Capacity of Tank in Gallons	Recommended Dimensions: Width		Length		Liquid Depth		Total Depth	
			Ft.	In.	Ft.	In.	Ft.	In.	Ft.	In.
2 or fewer	4	500	3	0	6	0	4	0	5	0
3	6	600	3	0	7	0	4	0	5	0
4	8	750	3	6	7	6	4	0	5	0
5	10	900	3	6	8	6	4	6	5	6
6	12	1,100	4	0	8	6	4	6	5	6
7	14	1,300	4	0	10	0	4	6	5	6
8	16	1,500	4	6	10	0	4	6	5	6

Source: U. S. Dept. of Agriculture

SECTION E. MUNICIPAL SEWERAGE

TABLE 7–54. NUMBER OF SEWAGE TREATMENT PLANTS BY FLOW RANGE IN THE UNITED STATES

Flow Ranges (mgd)	Treatment Plants in Operation in 1986: Number of Plants	Flow Capacity (mgd)
0.01 to 0.10	4,960	251
0.11 to 1.00	7,003	2,671
1.01 to 10.00	2,898	9,372
10.01 and greater	577	24,383
Total	15,438	36,677

Source: U.S. Environmental Protection Agency, 1987. 1986 Needs Survey, Report to Congress. PB87–157251

TABLE 7–55. MUNICIPAL SEWAGE TREATMENT FACILITY DATA BY LEVEL OF TREATMENT, 1984 AND 1986

	Number of Facilities			Design Capacity (mgd)		
Level of Treatment	1984	1986	Net Change	1984	1986	Net Change
Raw discharge	202	149	–53	N/A	N/A	N/A
Less than secondary	2,617	2,112	–505	6,510	5,529	–981
Secondary	8,070	8,403	+333	14,603	15,714	+1,111
Greater than secondary	2,965	3,115	+150	13,874	14,373	+499
No discharge	1,726	1,762	+36	938	973	+35
Other	0	46	+46	0	88	+88
Total	15,580	15,587	+7	35,925	36,677	+752

Source: U.S. Environmental Protection Agency, 1984 and 1986 Needs Survey Report to Congress

TABLE 7–56. PRESENT AND PROJECTED OPERATIONAL DATA OF SEWAGE TREATMENT PLANTS IN THE UNITED STATES

	1986	When All Identified Year 2005 Needs Are Met
Number of treatment plants	15,438	16,980
Number of people served	172,205,000	243,723,000
Percent of the nation	73.1	86.6
Removal Efficiencies[a]		
Flow (mgd)	27,692 mgd	43,023 mgd
Influent BOD	23,220 tpd	38,558 tpd
Effluent BOD	3,705 tpd	3,889 tpd
Percent removal	84.0	89.9
Influent solids	26,270 tpd	40,995 tpd
Effluent solids	3,593 tpd	4,544 tpd
Percent removal	86.3	88.9

[a] Facilities with no discharge or raw discharge are not included.
Source: U.S. Environmental Protection Agency, 1987. 1986 Needs Survey, Report to Congress, PB87–157251

TABLE 7–57. MUNICIPAL SEWAGE TREATMENT FACILITY DATA BY LEVEL OF TREATMENT, 1986/ALL NEEDS MET IN 2005

Level of Treatment	No. of Facilities		Population Served (millions)		Design Capacity (mgd)	
	1986	When Needs Are Met	1986	When Needs Are Met	1986	When Needs Are Met
Raw discharge	149	0	1.6	0	N/A	N/A
Less than secondary[a]	2,112	45[b]	28.8	2.9	5,529	387
Secondary	8,403	9,675	72.3	107.4	15,714	18,844
Greater than secondary	3,115	4,906	54.9	85.8	14,373	21,996
No discharge	1,762	2,273	5.7	10.9	973	1,686
Other	46	81	10.5	36.7	88	110
Total	15,587	16,980	173.8	243.7	36,677	43,023

[a] The Needs Survey definition of less than secondary includes trickling filter and lagoon systems.
[b] These treatment plants have applied for a waiver from the secondary treatment requirements in accordance with section 301(h) of the Clean Water Act. All have received at least tentative approval.
Source: U.S. Environmental Protection Agency, 1986 Needs Survey Report to Congress

TABLE 7–58. NEEDS FOR PUBLICLY-OWNED SEWAGE TREATMENT WORKS IN THE UNITED STATES

[In billions of 1986 dollars]

Category of Need		Needs for 1986 Population	Needs for Population Growth	Design Year (2005) Needs
I.	Secondary Treatment	$17.8	$6.1	$23.9
II.	Advanced Treatment	3.3	1.0	4.3
IIIA.	Infiltration/Inflow Correction	2.6	0.0	2.6
IIIB.	Major Sewer Rehabilitation	3.0	0.0	3.0
IVA.	New Collector Sewers	9.0	3.8	12.8
IVB.	New Interceptor Sewers	9.4	5.0	14.4
V.	Combined Sewer Overflows	15.2	0.0	15.2
Total		60.3	15.9	76.2

Note: All needs are based on a documented water quality or public health problem.
Source: U.S. Environmental Protection Agency, 1986 Needs Survey Report to Congress

TABLE 7–59. UNITED STATES POPULATION SERVED BY SEWER SYSTEMS, BY DEGREE OF TREATMENT: 1960 TO 1984

[In millions, except percent. 1960 and 1970 not strictly comparable with later years]

ITEM	1960	1970	1980	1982	1984
U.S. population	180	204	227	232	236
Not served by sewers	70	59	68	68	66
Percent of population	39	29	30	29	28
Served by sewers	110	145	159	164	170
Not treated	70	59	2	2	1
Treated	40	86	157	162	169
Degree of treatment:					
Less than secondary	36	NA	37	34	34
No discharge			4	4	5
Secondary			63	68	71
Advanced secondary	4	NA	48	51	53
Tertiary			5	5	5

NA-Not available.
Source: U.S. Department of Commerce, Statistical Abstracts of the United States 1987

TABLE 7–60. POPULATION SERVED BY MUNICIPAL SEWER SYSTEMS, BY DEGREE OF TREATMENT, IN THE UNITED STATES AND OTHER AREAS

[Population served represents persons receiving treatment for their generated wastewater at a facility operated by an established publicly-owned sewerage authority. Data based on an inventory of municipal wastewater facilities. Facilities with no discharge are lagoon systems and systems that dispose of their effluent by recycling, reuse, spray irrigation or groundwater discharge; those with less than secondary provide preliminary treatment plus primary sedimentation; secondary treatment provides for preliminary treatment followed by biological processes; advance secondary and tertiary treatment involves the removal of such pollutants as phosphorous and ammonia]

State	Resident Population 1980 (1,000)	Percent Served By Treatment Designed For–				
		No Discharge	Less than Secondary	Secondary	Advance Secondary	Tertiary
Total	230,210	2.4	14.9	30.2	23.0	2.9
Alabama	3,890	—	1.0	42.0	12.4	—
Alaska	400	—	45.3	20.2	1.1	—
Arizona	2,718	8.5	2.4	74.8	.7	.2
Arkansas	2,286	.2	23.7	23.2	10.1	.3
California	23,669	8.1	40.1	20.5	18.0	7.1
Colorado	3,024	1.7	4.0	43.1	48.8	—
Connecticut	3,108	—	7.5	49.7	3.7	1.0
Delaware	595	—	.8	.4	71.3	.4
District of Columbia	638	—	—	—	205.6	—
Florida	9,740	11.8	6.7	25.1	12.6	9.1
Georgia	5,464	1.7	2.2	40.6	15.7	.5
Hawaii	965	.2	54.3	6.5	2.5	—
Idaho	944	4.2	14.0	27.7	8.5	.7
Illinois	11,418	—	2.8	13.6	71.2	.1
Indiana	5,490	—	.9	15.8	47.5	.4
Iowa	2,913	.1	25.2	41.7	5.5	—
Kansas	2,363	4.2	24.9	46.8	3.5	—
Kentucky	3,661	—	1.8	33.2	13.1	—
Louisiana	4,204	.1	12.9	50.1	4.0	2.3
Maine	1,125	.1	7.6	36.6	1.4	—
Maryland	4,216	—	1.5	10.9	42.0	5.4
Massachusetts	5,737	—	41.1	20.9	6.9	1.1
Michigan	9,258	1.0	5.0	3.6	59.2	4.6
Minnesota	4,077	.1	3.9	62.1	6.8	2.0
Mississippi	2,521	1.6	6.3	43.5	6.7	—
Missouri	4,917	—	28.1	43.8	2.9	—
Montana	787	2.6	6.3	51.6	1.3	—
Nebraska	1,570	4.5	32.6	43.0	—	—
Nevada	799	9.3	2.5	9.3	78.7	—
New Hampshire	921	.1	15.4	30.6	.9	—
New Jersey	7,364	.3	18.9	52.9	9.8	3.0
New Mexico	1,300	12.9	.5	45.0	1.1	—
New York	17,577	—	19.8	39.1	11.6	1.0
North Carolina	5,874	—	1.9	19.0	13.9	9.2
North Dakota	653	1.7	1.3	66.2	—	—
Ohio	10,797	—	7.7	22.2	37.0	5.3
Oklahoma	3,025	4.6	11.9	53.2	9.3	—
Oregon	2,633	1.6	1.2	16.5	39.7	2.8
Pennsylvania	11,867	—	16.6	40.5	19.1	1.3
Rhode Island	947	—	28.3	38.5	1.1	—
South Carolina	3,119	.3	5.3	34.7	10.2	—
South Dakota	690	3.4	19.8	37.5	5.5	—
Tennessee	4,591	—	6.2	27.2	16.0	.9
Texas	14,228	6.5	10.3	38.3	36.0	1.9
Utah	1,461	7.1	—	81.1	6.7	—
Vermont	511	.2	16.1	26.4	5.1	—
Virginia	5,346	—	12.7	14.0	22.3	17.0
Washington	4,130	.7	31.0	17.1	10.7	—
West Virginia	1,950	—	6.7	26.5	8.8	4.6
Wisconsin	4,705	1.4	.7	15.3	54.5	.2
Wyoming	450	2.0	17.6	63.0	3.4	—
Am. Samoa	33	—	16.2	—	—	—
Guam	110	4.7	53.2	6.7	—	—
N. Marianas	17	—	6.2	—	—	—
Puerto Rico	3,197	—	47.4	2.0	—	—
Trust Terr. of Pacific Is	118	—	2.0	11.6	.8	—
Virgin Is	99	—	53.5	13.3	—	—

— Represents zero.

Source: U.S. Environmental Protection Agency, unpublished data from 1984 Needs Survey; Statistical Abstract of the United States 1986

TABLE 7–61. NUMBER OF OPERATIONAL SEWAGE TREATMENT PLANTS AND COLLECTION SYSTEMS IN THE UNITED STATES IN 1986

State	Treatment Plants	Collection Systems	State	Treatment Plants	Collection Systems
Alabama	231	275	New Jersey	222	500
Alaska	45	52	New Mexico	102	113
Arizona	117	133	New York	466	858
Arkansas	288	311	North Carolina	501	586
California	576	774	North Dakota	291	293
Colorado	275	336	Ohio	681	856
Connecticut	103	139	Oklahoma	498	513
Delaware	20	35	Oregon	202	247
Dist. of Columbia	1	1	Pennsylvania	663	1,243
Florida	255	325	Rhode Island	20	31
Georgia	388	490	South Carolina	225	288
Hawaii	30	35	South Dakota	270	272
Idaho	149	168	Tennessee	235	261
Illinois	724	965	Texas	1,291	1,554
Indiana	358	402	Utah	97	153
Iowa	684	714	Vermont	91	100
Kansas	567	577	Virginia	242	323
Kentucky	223	274	Washington	245	308
Louisiana	318	354	West Virginia	171	243
Maine	104	164	Wisconsin	572	720
			Wyoming	108	115
Maryland	158	213	American Samoa	2	2
Massachusetts	112	204	Guam	7	7
Michigan	371	603	Mariana Islands	2	2
Minnesota	509	635	Puerto Rico	33	33
Mississippi	303	349	Trust Territories	6	6
Missouri	543	608	Virgin Islands	4	4
Montana	160	164			
Nebraska	450	516			
Nevada	51	56	Total	15,438	19,604
New Hampshire	78	101			

Source: U.S. Environmental Protection Agency, 1987. 1986 Needs Survey Report to Congress, PB87–157251

TABLE 7–62. SEWAGE TREATMENT WATER RELEASES TO SURFACE WATER IN THE UNITED STATES

[By State, 1985; figures may not add to totals because of independent rounding]

State	Number of Facilities		Total Public Releases, million gallons per day
	Public	Other	
Alabama	259	0	320
Alaska	20	38	51
Arizona	197	29	125
Arkansas	277	1708	287
California	892	745	2770
Colorado	255	399	336
Connecticut	88	50	371
Delaware	16	39	86
D.C.	2	6	280
Florida	774	125	1120
Georgia	403	223	485
Hawaii	29	123	139
Idaho	122	75	140
Illinois	764	0	2160
Indiana	330	593	717

TABLE 7–62. SEWAGE TREATMENT WATER RELEASES TO SURFACE WATER IN THE UNITED STATES (continued)

[By State, 1985; figures may not add to totals because of independent rounding]

State	Number of Facilities		Total Public Releases, million gallons per day
	Public	Other	
Iowa	626	392	313
Kansas	382	302	247
Kentucky	120	2746	220
Louisiana	257	108	488
Maine	96	7	103
Maryland	142	228	482
Massachusetts	203	511	834
Michigan	515	916	1610
Minnesota	667	19	454
Mississippi	243	209	196
Missouri	1660	572	885
Montana	229	118	202
Nebraska	477	35	170
Nevada	83	82	129
New Hampshire	65	80	252
New Jersey	264	938	1330
New Mexico	63	71	82
New York	485	0	3070
North Carolina	306	886	481
North Dakota	301	98	29
Ohio	1013	37	146
Oklahoma	536	152	287
Oregon	252	86	336
Pennsylvania	1087	3115	1690
Rhode Island	26	8	113
South Carolina	276	399	270
South Dakota	235	6	48
Tennessee	245	854	536
Texas	1690	2363	1680
Utah	152	38	341
Vermont	92	30	80
Virginia	270	416	551
Washington	232	118	436
West Virginia	2194	0	233
Wisconsin	511	82	2860
Wyoming	85	281	43
Puerto Rico	114	125	152
Virgin Islands	9	0	1.4
Total	20,631	20,581	30,800

Source: Solley, W.B., Merk, C.F., and Pierce, R.R., 1988, Estimated Water Use in the United States in 1985, U.S. Geological Survey Circular 1004

TABLE 7–63. ESTIMATED PERCENT OF WASTEWATER IN MUNICIPAL DRINKING WATER SOURCES

State	City	Source Water	Population (Thousands)	Conservative Estimate of Percent of Wastewater[a]
Pennsylvania	Philadelphia	Schuylkill River	1,950	3.5
South Carolina	Columbia	Saluda River	228	16
Pennsylvania	Bryn Mawr	Neshaminy Creek	820	3.8
Indiana	Indianapolis	White River	680	4.3
Pennsylvania	Middletown	Neshaminy Creek	759	3.8
Texas	Dallas	Lake Ray Hubbard	878	2.8
Alabama	Birmingham	Cabaha River	650	3.6
New Jersey	Elizabeth	Delaware-Raritan Canal	500	4.3
Maryland	Baltimore	Susquehanna River	1,755	1.2
New Jersey	Little Falls	Passaic River	333	4.9
Missouri	St. Louis	Mississippi River	600	2.2
Louisiana	New Orleans	Mississippi River	550	1.8
New Jersey	Milburn	Passaic River	234	4.8
Ohio	Cincinnati	Ohio River	850	1.3
Louisiana	Marrero	Mississippi River	600	1.8
Maryland	Hyattsville/Wash., DC	Potomac River	1,300	0.8
Kentucky	Louisville	Ohio River	700	1.4
Missouri	Kansas City	Missouri River	750	1.2
Texas	Fort Worth	Ben Brook Lake	490	1.6
Missouri	St. Louis	Missouri River	600	1.3
Colorado	Pueblo	Arkansas River	95	7.6
South Carolina	Rock Hill	Catawba River	426	1.4
New Jersey	Elizabeth	Raritan and Millstone	500	1.1
Illinois	East St. Louis	Mississippi River	215	2.2
Georgia	Columbus	Lake Oliver	134	3.2

[a] Cumulative percent wastewater estimates under average source water flow conditions.
Source: Arber, Richard, 1986, State of the Art of Potable Water Reuse, AWWA Seminar Proceedings Implementation of Water Reuse, Denver, CO-June 22, 1986. Copyright Am. Water Works Assoc. Reprinted with permission. Original source NSF/RA-790224

TABLE 7–64. SEWERED AND UNSEWERED POPULATION IN THE UNITED STATES

[Numbers in hundreds]

State or Group of States	Population			Unsewered Population Served	
	Total	Sewered	Unsewered	by Septic Tank/ Cesspool	by Other Means
Alabama	38,272	19,292	18,980	17,152	1,828
Arkansas	22,410	12,593	9,817	8,925	892
California	231,606	206,436	25,170	23,918	1,252
Colorado	28,185	24,297	3,888	3,704	184
Connecticut	29,965	19,333	10,632	10,479	153
District of Columbia	6,042	5,994	48	17	31
Florida	95,915	65,744	30,171	29,555	616
Georgia	53,253	30,643	22,610	21,069	1,541
Illinois	111,745	93,588	18,157	17,536	621
Indiana	53,335	33,518	19,817	19,258	559
Kentucky	35,578	17,905	17,673	14,715	2,958
Louisiana	41,271	29,118	12,153	11,416	737
Michigan	90,804	63,762	27,042	26,584	458

TABLE 7–64. SEWERED AND UNSEWERED POPULATION IN THE UNITED STATES (continued)

[Numbers in hundreds]

State or Group of States	Population			Unsewered Population Served	
	Total	Sewered	Unsewered	by Septic Tank/ Cesspool	by Other Means
Mississippi	24,640	13,516	11,124	9,203	1,921
New Jersey	72,427	61,498	10,929	10,641	288
New York	171,088	129,203	41,885	40,980	905
North Carolina	56,692	25,331	31,361	29,309	2,052
Ohio	105,759	77,221	28,538	27,170	1,368
Oklahoma	29,353	20,750	8,603	8,220	383
Oregon	25,809	16,887	8,922	8,753	169
Pennsylvania	115,819	83,494	32,325	30,940	1,385
South Carolina	30,218	15,077	15,141	13,955	1,186
Tennessee	44,830	24,287	20,543	18,906	1,637
Texas	139,458	113,181	26,277	24,321	1,956
Virginia	51,901	33,131	18,770	16,417	2,353
Washington	40,274	25,768	14,506	14,292	214
West Virginia	19,164	9,110	10,054	8,628	1,426
Wisconsin	45,828	31,713	14,115	13,715	400
Maine, New Hampshire, Vermont	24,606	11,562	13,044	12,573	471
Massachusetts, Rhode Island	64,503	44,527	19,976	19,727	249
Minnesota, Iowa, Missouri, Kansas, Nebraska, South Dakota, North Dakota	166,893	120,782	46,111	43,463	2,648
Maryland, Delaware	47,225	36,412	10,813	10,097	716
Montana, Idaho, Wyoming	21,573	13,779	7,794	7,515	279
Utah, Nevada	22,427	19,340	3,087	3,008	79
Arizona, New Mexico	39,496	30,640	8,856	7,800	1,056
Alaska, Hawaii	13,186	9,921	3,265	2,797	468

Source: U.S. Environmental Protection Agency, 1987, Statistical Abstract of the Unsewered U.S. Population (based on 1980 Census). PB88–113352

TABLE 7–65. SEWERED AND UNSEWERED HOUSING UNITS IN THE UNITED STATES

[Numbers in hundreds]

State or Group of States	All Structures[a]	Housing Units			Unsewered Housing Units	
		Total	Sewered	Unsewered	Septic Tank/ Cesspool	Other Means
Alabama	15,458	14,681	7,687	6,994	6,252	742
Arkansas	9,507	8,991	5,166	3,825	3,415	410
California	98,599	92,935	82,827	10,108	9,541	567
Colorado	12,699	11,956	10,266	1,690	1,505	185
Connecticut	12,493	11,603	7,829	3,774	3,708	66
District of Columbia	3,104	2,785	2,762	23	6	17

TABLE 7–65. SEWERED AND UNSEWERED HOUSING UNITS IN THE UNITED STATES (continued)

[Numbers in hundreds]

State or Group of States	All Structures[a]	Housing Units			Unsewered Housing Units	
		Total	Sewered	Unsewered	Septic Tank/ Cesspool	Other Means
Florida	45,788	43,831	31,457	12,374	12,009	365
Georgia	21,722	20,299	12,109	8,190	7,578	612
Illinois	45,920	43,210	36,575	6,635	6,278	357
Indiana	22,388	20,933	13,655	7,278	6,967	311
Kentucky	14,720	13,688	7,335	6,353	5,172	1,181
Louisiana	16,457	15,494	11,067	4,427	4,091	336
Michigan	37,926	35,949	24,865	11,084	10,531	553
Mississippi	9,770	9,131	5,120	4,011	3,297	714
New Jersey	29,092	27,739	23,987	3,752	3,614	138
New York	73,222	68,734	53,062	15,672	14,996	676
North Carolina	24,724	22,777	10,501	12,276	11,292	984
Ohio	43,376	41,090	31,176	9,914	9,232	682
Oklahoma	13,265	12,365	9,058	3,307	3,112	195
Oregon	11,422	10,844	7,386	3,458	3,316	142
Pennsylvania	48,947	45,975	33,667	12,308	11,312	996
South Carolina	12,605	11,545	6,098	5,447	4,973	474
Tennessee	18,591	17,479	9,882	7,597	6,863	734
Texas	58,950	55,544	44,813	10,731	9,750	981
Virginia	21,996	20,230	13,231	6,999	6,052	947
Washington	18,003	16,910	11,271	5,639	5,442	197
West Virginia	7,832	7,484	3,808	3,676	3,037	639
Wisconsin	19,959	18,671	12,818	5,853	5,348	505
Maine, New Hampshire, Vermont	12,005	11,112	5,016	6,096	5,532	564
Massachusetts, Rhode Island	28,127	25,807	18,364	7,443	7,288	155
Minnesota, Iowa, Missouri, Kansas, Nebraska, South Dakota, North Dakota	73,825	68,589	50,336	18,253	16,430	1,823
Maryland, Delaware	19,210	18,108	14,219	3,889	3,612	277
Montana, Idaho, Wyoming	9,416	8,937	5,815	3,122	2,857	265
Utah, Nevada	8,682	8,305	7,169	1,136	1,064	72
Arizona, New Mexico	16,961	16,215	12,528	3,687	3,181	506
Alaska, Hawaii	5,535	4,975	3,769	1,206	915	291

[a] The numbers indicated under all structures include group quarters. All group quarters are sewered.

Source: U.S. Environmental Protection Agency, 1987, Statistical Abstract of the Unsewered U.S. Population (based on 1980 Census). PB88–113352

TABLE 7–66. SEWER PIPE LENGTH AND ESTIMATED FLOWS IN THE CONTERMINOUS UNITED STATES

Year	Northeast		North Central		South		West		Total	
	Pipe Length[a] (miles)	Estimated Flow[b] (mgd)	Pipe Length[a] (miles)	Estimated Flow[b] (mgd)	Pipe Length[a] (miles)	Estimated Flow[b] (mgd)	Pipe Length[a] (miles)	Estimated Flow[b] (mgd)	Pipe Length[a] (miles)	Estimated Flow[b] (mgd)
1940	40,115	2,310	30,474	2,020	40,740	1,930	23,599	810	134,918	7,070
1950	53,619	2,580	41,687	2,390	54,812	2,350	41,062	1,240	191,180	8,560
1960	81,025	3,060	67,540	3,050	89,510	3,040	69,472	1,920	307,547	11,070
1970 (est)	116,121	3,670	107,636	3,910	135,478	3,820	112,826	2,600	472,061	14,000
1980 (est)	173,718	4,460	161,577	4,880	202,923	4,870	168,774	3,580	706,992	17,790

[a] Total residential sewer pipe in service regionally.
[b] Based on 100 gpd per capita.
Source: Kollar, K.L., 1966, Regional Requirements for Sewer Pipe in Sewerage Utilities. U.S. Department of Commerce

TABLE 7–67. POPULATION SERVED BY MUNICIPAL SEWER SYSTEMS IN CANADA

Province or Territory	Total Population Surveyed	Percentage Served by Sewers	Percentage Served by Sewage Treatment
Newfoundland	497,018	60.0	12.8
Prince Edward Island	57,587	100.0	94.4
Nova Scotia	536,604	83.3	21.4
New Brunswick	409,900	91.8	60.7
Quebec	6,685,434	81.9	6.2
Ontario	7,641,607	86.1	83.5
Manitoba	839,158	94.9	94.7
Saskatchewan	611,072	99.0	99.0
Alberta	1,852,714	99.3	99.3
British Columbia	2,175,754	77.5	77.5
Yukon Territory	21,888	88.0	86.2
Northwest Territories	43,953	91.6	52.5
TOTAL	21,372,690	85.3	57.3

Source: Pearse, P.H., and others, 1985, Currents of Change, Final Report on Federal Water Policy, Ottawa, Canada

TABLE 7–68. POPULATION SERVED BY WASTEWATER TREATMENT PLANTS IN SELECTED COUNTRIES, 1970–1985

		Primary Only				Primary + Secondary and/or Tertiary				Total Served			
		1970	1975	1980	1985	1970	1975	1980	1985	1970	1975	1980	1985
Canada	k,c	..	..	13.0	10.0	..	..	43.0	47.0	..	49.0	56.0	57.0
USA	c,d,l	..	23.0	17.0	15.0	..	44.0	53.0	59.0	42.0	67.0	70.0	74.0
Japan	d,m	..	..	..	..	16.0	23.0	30.0	36.0	16.0	23.0	30.0	36.0
New Zealand	n	..	9.0	10.0	8.0	..	47.0	49.0	80.0	52.0	56.0	59.0	88.0
Austria	a	11.0	..	10.0	..	5.0	..	30.0	..	16.0	..	40.0	..
Belgium	b	0.0	0.0	0.0	..	3.8	5.5	22.9	..	3.8	5.5	22.9	..
Denmark	e,o	31.9	29.0	..	20.0	22.4	41.6	..	70.0	54.3	70.6	..	90.0
Finland	f,o	5.0	3.0	0.0	0.0	22.0	47.0	65.0	69.0	27.0	50.0	65.0	69.0
France	o,p	..	..	2.5	..	..	..	59.0	..	40.0	..	61.5	63.7
Germany	g	20.5	18.4	10.2	7.5	41.3	56.4	71.6	79.0	61.8	74.8	81.8	86.5
Greece		..	..	0.0	..	..	..	0.5	..	..	..	0.5	..
Ireland		..	..	0.2	..	..	..	11.0	..	..	..	11.2	..
Italy	h	8.0	..	..	..	6.0	..	..	..	14.0	..	30.0	..
Luxembourg		23.0	..	16.0	14.0	5.0	..	65.0	69.0	28.0	..	81.0	83.0
Netherlands	d	..	8.0	7.0	4.0	..	37.0	61.0	77.0	..	45.0	68.0	81.0
Norway	o	1.0	2.0	2.0	3.0	21.0	25.0	37.0	48.0	22.0	27.0	39.0	51.0
Portugal	o	1.0	2.0	3.0	3.5	2.1	4.0	7.0	8.5	3.1	6.0	10.0	12.0
Spain		..	7.0	8.8	13.2	..	7.3	9.1	15.8	..	14.3	17.9	29.0
Sweden	i	12.0	4.0	1.0	1.0	66.0	94.0	98.0	98.0	78.0	98.0	99.0	99.0
Switzerland	o	..	..	..	..	35.0	55.0	70.0	81.0	35.0	55.0	70.0	81.0
Turkey		..	..	1.6	1.6	..	..	0.4	1.8	..	..	2.0	3.3
UK	j	..	..	6.0	6.0	..	..	76.0	77.0	..	..	82.0	83.0

[a] Estimated data.
[b] 1980 data refer to 1979.
[c] 1975 data refer to 1976.
[d] 1985 data refer to 1984.
[e] 1975 data refer to 1977.
[f] Networks serving fewer than 200 inhab. are excluded. Primary: mechanical treatment; Secondary: biological tr.; Tertiary: combined chemical-biological treatment.
[g] 1970, 1980 and 1985 data refer to 1969, 1979 and 1983.
[h] 1970 data refer to 1971.
[i] Urban population only (85% of total). Primary: removal of sediments; Secondary: chemical or biological treatm.; Tertiary: chemical, biological and complementary treatment.
[j] England and Wales only. Primary: removal of gross solids; Secondary: aerobic removal of organic material or bacteria; Tertiary: removal of suspended solids following secondary treatment.
[k] Secondary: usually includes private treatment and includes waste stabilization ponds. Tertiary: refers to secondary treatment with phosphorous removal.
[l] 1980 and 85 data for the second category include 1% and 2% of non-discharge treatment. 1980 and 1984 data were determined using different methods than previous data and therefore may not be comparable. Primary treatment: may provide some biological treatment. Secondary: preliminary treatment + biological process, with no additional treatment except disinfection.
[m] Data for the second category may include data for primary treatment only.
[n] 1970 data are Secretariat estimates.
[o] 1985 data refer to 1983.
[p] 1970 total: less than 40%.

Source: Organisation for Economic Cooperation and Development, Environmental Data Compendium 1987. Copyright OECD. Reprinted with permission

SECTION F. INDUSTRIAL WASTEWATER

TABLE 7–69. SUMMARY OF INDUSTRIAL WASTE: ITS ORIGIN, CHARACTER, AND TREATMENT

Industries Producing Wastes	Origin of Major Wastes	Major Characteristics	Major Treatment and Disposal Methods
	FOOD AND DRUGS		
Canned goods	Trimming, culling, juicing, and blanching of fruits and vegetables	High in suspended solids, colloidal and dissolved organic matter	Screening, lagooning, soil absorption or spray irrigation
Dairy products	Dilutions of whole milk, separated milk, buttermilk, and whey	High in dissolved organic matter, mainly protein, fat, and lactose	Biological treatment, aeration, trickling filtration, activated sludge
Brewed and distilled beverages	Steeping and pressing of grain, residue from distillation of alcohol, condensate from stillage evaporation	High in dissolved organic solids, containing nitrogen and fermented starches or their products	Recovery, concentration by centrifugation and evaporation, trickling filtration; use in feeds
Meat and poultry products	Stockyards, slaughtering of animals, rendering of bones and fats, residues in condensates, grease and wash water, picking of chickens	High in dissolved and suspended organic matter, blood, other proteins, and fats	Screening, settling and/or flotation, trickling filtration
Beet sugar	Transfer, screening and juicing waters, draining from lime sludge, condensates after evaporator, juice, extracted sugar	High in dissolved and suspended organic matter, containing sugar and protein	Reuse of wastes, coagulation, and lagooning
Pharmaceutical products	Mycelium, spent filtrate, and wash waters	High in suspended and dissolved organic matter, including vitamins	Evaporation and drying, feeds
Yeast	Residue from yeast filtration	High in solids (mainly organic) and BOD	Anaerobic digestion, trickling filtration
Pickles	Lime water; brine, alum and turmeric, syrup, seeds and pieces of cucumber	Variable pH, high suspended solids, color, and organic matter	Good housekeeping, screening, equalization
Coffee	Pulping and fermenting of coffee bean	High BOD and suspended solids	Screening, settling, and trickling filtration
Fish	Rejects from centrifuge, pressed fish, evaporator and other wash water wastes	Very high BOD, total organic solids, and odor	Evaporation of total waste, barge remainder to sea
Rice	Soaking, cooking, and washing of rice	High in BOD, total and suspended solids (mainly starch)	Lime coagulation, digestion
Soft drinks	Bottle washing, floor and equipment cleaning, syrup-storage-tank drains	High pH, suspended solids and BOD	Screening, plus discharge to municipal sewer
	APPAREL		
Textiles	Cooking of fibers, desizing of fabric	Highly alkaline, colored, high BOD and temperature, high suspended solids	Neutralization, chemical precipitation, biological treatment, aeration and/or trickling filtration
Leather goods	Unhairing, soaking, deliming and bating of hides	High total solids, hardness, salt, sulfides, chromium, pH precipitated lime and BOD	Equalization, sedimentation, and biological treatment
Laundry trades	Washing of fabrics	High turbidity, alkalinity, and organic solids	Screening, chemical precipitation, flotation, and adsorption
	CHEMICALS		
Acids	Dilute wash waters; many varied dilute acids	Low pH, low organic content	Upflow or straight neutralization, burning when some organic matter is present

TABLE 7–69. SUMMARY OF INDUSTRIAL WASTE: ITS ORIGIN, CHARACTER, AND TREATMENT (continued)

Industries Producing Wastes	Origin of Major Wastes	Major Characteristics	Major Treatment and Disposal Methods
	CHEMICALS (Continued)		
Detergents	Washing and purifying soaps and detergents	High in BOD and saponified soaps	Flotation and skimming, precipitation with $CaCl_2$
Cornstarch	Evaporator condensate, syrup from final washes, wastes from "bottling up" process	High BOD and dissolved organic matter; mainly starch and related material	Equalization, biological filtration
Explosives	Washing TNT and guncotton for purification, washing and pickling of cartridges	TNT, colored, acid, odorous, and contains organic acids and alcohol from powder and cotton, metal, acid, oils, and soaps	Flotation, chemical precipitation, biological treatment, aeration, chlorination of TNT, neutralization
Insecticides	Washing and purification products such as 2,4-D and DDT	High organic matter, benzene ring structure, toxic to bacteria and fish, acid	Dilution, storage, activated carbon adsorption, alkaline chlorination
Phosphate and phosphorus	Washing, screening, floating rock, condenser bleed-off from phosphate reduction plant	Clays, slimes and tall oils, low pH, high suspended solids, phosphorus, silica, and fluoride	Lagooning, mechanical clarification, coagulation and settling of refined waste
Formaldehyde	Residues from manufacturing synthetic resins, and from dyeing synthetic fibers	Normally has high BOD and HCHO, toxic to bacteria in high concentrations	Trickling filtration, adsorption on activated charcoal
	MATERIALS		
Pulp and paper	Cooking, refining, washing of fibers, screening of paper pulp	High or low pH; colored; high suspended, colloidal, and dissolved solids; inorganic fillers	Settling, lagooning, biological treatment, aeration, recovery of byproducts
Photographic products	Spent solutions of developer and fixer	Alkaline, contains various organic and inorganic reducing agents	Recovery of silver, plus discharge of wastes into municipal sewer
Steel	Coking of coal, washing of blast-furnace flue gases, and pickling of steel	Low pH, acids, cyanogen, phenol, ore, coke, limestone, alkali, oils, mill scale, and fine suspended solids	Neutralization, recovery and reuse, chemical coagulation
Metal-plated products	Stripping of oxides, cleaning and plating of metals	Acid, metals, toxic, low volume, mainly mineral matter	Alkaline chlorination of cyanide, reduction and precipitation of chromium, and lime precipitation of other metals
Iron-foundry products	Wasting of used sand by hydraulic discharge	High suspended solids, mainly sand; some clay and coal	Selective screening, drying of reclaimed sand
Oil	Drilling muds, salt, oil, and some natural gas, acid sludges and miscellaneous oils from refining	High dissolved salts from field, high BOD, odor, phenol, and sulfur compounds from refinery	Diversion, recovery, injection of salts; acidification and burning of alkaline sludges
Rubber	Washing of latex, coagulated rubber, exuded impurities from crude rubber	High BOD and odor, high suspended solids, variable pH, high chlorides	Aeration, chlorination, sulfonation, biological treatment
Glass	Polishing and cleaning of glass	Red color, alkaline non-settleable suspended solids	Calcium chloride precipitation
Naval stores	Washing of stumps, drop solution, solvent recovery, and oil recovery water	Acid, high BOD	Byproduct recovery, equalization, recirculation and reuse, trickling filtration
	ENERGY		
Steam power	Cooling water, boiler blowdown, coal drainage	Hot, high volume, high inorganic and dissolved solids	Cooling by aeration, storage of ashes, neutralization of excess acid wastes
Coal processing	Cleaning and classification of coal, leaching of sulfur strata with water	High suspended solids, mainly coal; low pH, high H_2SO_4 and $FeSO_4$	Settling, froth flotation, drainage control, and scaling of mines
Nuclear power and radioactive materials	Processing ores, laundering of contaminated clothes, research-lab wastes, processing of fuel, power-plant cooling waters	Radioactive elements; can be very acid and "hot"	Concentration and containing, or dilution and dispersion

Source: Nemerow, Theory and Practice of Industrial Waste Treatment, Addison-Wesley, Copyright 1963

TABLE 7–70. WASTEWATER DISCHARGED BY INDUSTRIES IN THE UNITED STATES

Industry	Water Discharged	
	Quantity billion gallons	Percent Untreated
MINING		
1968	1,365	78.8
1973	1,605	54.3
1978	1,592	67.3
1983, total	**1,037**	**31.9**
Metal mining	133	39.8
Anthracite mining	8	12.5
Bituminous coal, lignite mining	116	26.7
Oil and gas extraction	476	31.1
Nonmetallic minerals, exc. fuels	304	32.6
MANUFACTURING		
1968	14,276	69.5
1973	14,144	56.5
1978	11,682	59.7
1983, total	**8,914**	**54.9**
Food and kindred products	552	64.5
Tobacco products	4	(D)
Textile mill products	116	52.6
Lumber and wood products	71	63.4
Furniture and fixtures	3	100.0
Paper and allied products	1,768	27.1
Chemicals and allied products	2,980	67.0
Petroleum and coal products	699	46.2
Rubber, misc. plastic products	63	63.5
Leather and leather products	6	(D)
Stone, clay, and glass products	133	75.2
Primary metal products	2,112	58.1
Fabricated metal products	61	49.2
Machinery, exc. electrical	105	67.6
Electric and electronic equipment	70	61.4
Transportation equipment	139	67.6
Instruments and related products	28	50.0
Miscellaneous manufacturing	4	(D)

D = Withheld to avoid disclosing individual company data.

Source: U.S. Department of Commerce, Statistical Abstract of the United States 1987

TABLE 7–71. SIGNIFICANT CHEMICALS IN INDUSTRIAL WASTE WATERS

Chemical	Industry
Acetic acid	Acetate rayon, pickle and beetroot manufacture
Alkalies	Cotton and straw kiering, cotton manufacture, mercerizing, wool scouring, laundries
Ammonia	Gas and coke manufacture, chemical manufacture
Arsenic	Sheep-dipping, fell mongering
Chlorine	Laundries, paper mills, textile bleaching
Chromium	Plating, chrome tanning, aluminum anodizing
Cadmium	Plating
Citric acid	Soft drinks and citrus fruit processing
Copper	Plating, pickling, rayon manufacture
Cyanides	Plating, metal cleaning, case-hardening, gas manufacture
Fats, oils, grease	Wool scouring, laundries, textiles, oil refineries
Fluorides	Gas and coke manufacture, chemical manufacture, fertilizer plants, transistor manufacture, metal refining, ceramic plants, glass etching
Formalin	Manufacture of synthetic resins and penicillin
Hydrocarbons	Petrochemical and rubber factories
Hydrogen peroxide	Textile bleaching, rocket motor testing
Lead	Battery manufacture, lead mining, paint manufacture, gasoline manufacture
Mercaptans	Oil refining, pulp mills
Mineral acids	Chemical manufacture, mines, Fe and Cu pickling, DDT manufacture, brewing, textiles, photo-engraving, battery manufacture
Nickel	Plating
Nitro compounds	Explosives and chemical works
Organic acids	Distilleries and fermentation plants
Phenols	Gas and coke manufacture; synthetic resin manufacture; textiles; tanneries; tar, chemical, and dye manufacture; sheep-dipping
Silver	Plating, photography
Starch	Food, textile, wallpaper manufacture
Sugars	Dairies, foods, sugar refining, preserves, wood process
Sulfides	Textiles, tanneries, gas manufacture, rayon manufacture
Sulfites	Wood process, viscose manufacture, bleaching
Tannic acid	Tanning, sawmills
Tartaric acid	Dyeing; wine, leather, and chemical manufacture
Zinc	Galvanizing, plating, viscose manufacture, rubber process

Source: Klein, River Pollution. 2: Causes and Effects, Butterworth & Co., Copyright 1962

TABLE 7–72. BIOCHEMICAL OXYGEN DEMAND (BOD) OF WASTES FROM SELECTED INDUSTRIES

Source of Waste	5-day, 20°C BOD of Waste, mg/liter
Beet sugar refining	450–2,000
Brewery	500–1,200
Beer slop	11,500
Cannery	300–4,000
Grain distilling	15,000–20,000
Molasses distilling	20,000–30,000
Laundry	300–1,000
Milk processing	300–2,000
Meat packing	600–2,000
Pulp and paper	
Sulfite	20
Sulfite-cooker	16,000–25,000
Tannery	500–5,000
Textiles	
Cotton processing	50–1,750
Wool scouring	200–10,000

Source: McGauhey, Engineering Management of Water Quality, McGraw-Hill, Copyright 1968

TABLE 7–73. GENERATION OF HAZARDOUS WASTE BY SELECTED INDUSTRIES IN THE UNITED STATES, 1977

Industry Group	Millions of Tons, Wet
Organic chemicals	12.9
Primary metals	9.9
Electroplating	4.5
Inorganic chemicals	4.4
Textiles	2.1
Petroleum refining	2.0
Rubber and plastics	1.1
Miscellaneous (7 industries)	1.1
Total	38.0

Source: U.S. Environmental Protection Agency, 1979, Hazardous Waste Fact Sheet, EPA Journal: Waste Alert, Vol. 5

TABLE 7–74. WATER POLLUTION POTENTIAL OF SURFACE IMPOUNDMENTS IN THE UNITED STATES

Impoundment Category	Number	High Potential to Contaminate Ground Water[a] Percent	Potential to Contaminate Water Wells Percent	Potential to Contaminate Surface Water Percent
Municipal	37,185	41	27	58
Industrial	27,912	39	29	56
Agricultural	19,437	26	28	61
Mining	25,038	25	17	64
Oil and gas	65,488	8	17	68

[a] Data for "high potential to contaminate groundwater" are independent of data for other two columns.
Source: U.S. Environmental Protection Agency, 1982. Surface Impoundment Assessment, Office of Drinking Water

TABLE 7–75. TYPES OF SURFACE IMPOUNDMENTS IN THE UNITED STATES

	Active Sites[a]	Active Impoundments	Abandoned Sites	Abandoned Impoundments
Agricultural	14,677	19,167	173	270
Municipal	19,116	36,179	630	1,006
Industrial	10,819	25,749	941	2,163
Mining	7,100	24,451	264	587
Oil and gas	24,527	64,951	463	537
Other	1,500	5,745	53	168
Total	77,739	176,242	2,524	4,731

Total active and abandoned sites: 80,263
Total active and abandoned impoundments: 180,973

[a] A site may have more than one impoundment.
Source: U.S. Environmental Protection Agency, Surface Impoundment Assessment National Report, EPA 570/9–84–002, December 1983

TABLE 7–76. PURPOSE OF SURFACE IMPOUNDMENTS IN THE UNITED STATES

[By percent[a] and number]

Category	Storage		Disposal		Treatment	
	Percent	Number	Percent	Number	Percent	Number
Agricultural	55	10,542	26	4,983	19	3,642
Municipal	5	1,809	31	11,215	64	23,155
Industrial	17	4,377	31	7,982	52	13,390
Mining	18	4,401	27	6,602	56	13,693
Oil and gas	29	18,836	67	43,517	4	2,598
Total		39,965		74,299		56,478

[a] Percent storage, disposal, and treatment per category.
Source: U.S. Environmental Protection Agency, Surface Impoundment Assessment National Report, EPA 570/9–84–002, December 1983

SECTION G. WASTEWATER TREATMENT AND RECLAMATION

FIGURE 7–3. STAGES IN WASTEWATER TREATMENT

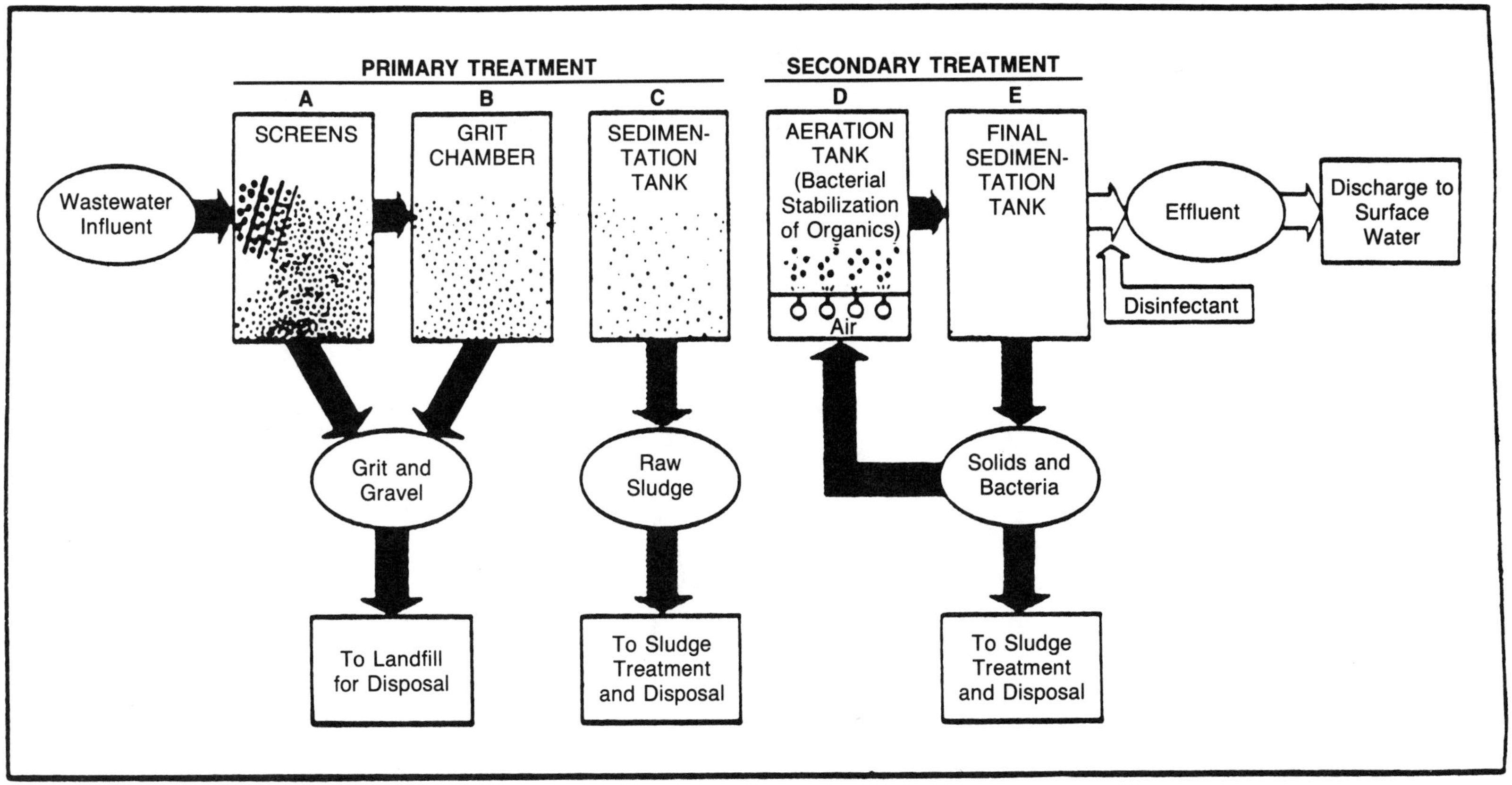

Source: U.S. Environmental Protection Agency. For explanation, see adjacent page

Municipal sewage treatment plants accept wastewater from households, commercial establishments and industries. These wastes are conveyed through a series of pipes and channels (sewage collection system) to the sewage treatment plant. At the sewage treatment plant, wastewaters are treated and discharged to surface waters.

Sewage treatment technology can be divided into several broad categories. First, there is **primary treatment**. In this stage wastewater flows through a screen to remove large floating objects which may interfere with the later treatment processes. After screening, the wastewater passes to a grit chamber where sand, grit, and small stones are removed. From the grit chamber the wastewater flows into a primary clarifier where particulate matter is allowed to settle. Primary treatment removes about 30%-40% of the pollutants in the influent wastewater.

The next stage of wastewater treatment is called **secondary treatment** system. A secondary treatment system removes about 90% of the pollutants in the influent. Federal law mandates secondary treatment as the minimum treatment which must be given to all municipal sewage. In secondary treatment, the wastewater from primary treatment enters a tank where organic matter is broken down by use of bacteria. The most common system is called activated sludge. In this process, wastewater enters an aeration tank where it is mixed with air and sludge loaded with bacteria and allowed to remain for several hours. During this time the bacteria break down the pollutants. The sludge, now activated with millions of bacteria and other tiny organisms, can be used again by returning it to an aeration tank for mixing with new sewage and ample amounts of air from the aeration tank. The wastewater flows into another clarifier/sedimentation tank where the activated sludge settles out. In most cases, the wastewater is then disinfected (usually with chlorine) to remove or reduce pathogens (disease causing organisms).

Where secondary treatment will not provide enough pollutant reduction to protect the water quality of the stream receiving the discharge, Federal law requires that additional treatment necessary to meet the State's water quality standards be installed. The process is called **advanced wastewater treatment**. Advanced treatment can take a variety of forms. Most commonly, another treatment process very similar to a secondary process is added at the end of a secondary treatment plant. Other technologies such as sand filter or carbon filters can also be used. Advanced treatment usually removes more than 95% of the pollutants in the influent.

Source: U.S. Environmental Protection Agency, Office of Water, Municipal Compliance with the Clean Water Act. Press Briefing July 27, 1988

TABLE 7–77. POLLUTANT REMOVAL BY WASTEWATER TREATMENT PROCESSES

Constituent	Secondary Treatment (Biological)	Chemical Precipitation: Lime	Chemical Precipitation: Ferric Chloride	Chemical Precipitation: Alum	Activated Carbon Adsorption	Comments on Activated Carbon	Residual Concentration Level (μg/liter)	General Comments on Removal
Total dissolved solids	P	P	—	—	P	—	—	Generally increased total dissolved solids with treatment; reverse osmosis effective in removal
Ammonia nitrogen	VG	P	—	—	P	—	—	Biological nitrification most effective; breakpoint chlorination and stripping towers F to VG
Nitrate nitrogen	VG	P	—	—	P to G	Depends on anaerobic bioactivity	—	Biological denitrification most feasible
Phenol	P	P	—	—	P	Limited by driving force to about 1 mg/liter	>1	Treatment methods effective in reducing phenol to 1 μg/liter limit
Trace organics	G	G	—	—	G	Removal depends on specific organics	5,000 (total organic carbon and chemical oxygen demand)	Chlorinated organics may be increased with breakpoint chlorination; ammonia stripping effective in removing volatile refractory organics

TABLE 7–77. POLLUTANT REMOVAL BY WASTEWATER TREATMENT PROCESSES (continued)

Constituent	Secondary Treatment (Biological)	Chemical Precipitation: Lime	Chemical Precipitation: Ferric Chloride	Chemical Precipitation: Alum	Activated Carbon Adsorption	Comments on Activated Carbon	Residual Concentration Level (μg/liter)	General Comments on Removal
Arsenic	P to F	P to G	G	—	P	Reacts with sulfide	3	Depends on influent level, pH, and redox potential
Barium	F	P to G	—	G	P to G	Due to highly soluble nature	>30	Enhanced precipitation as sulfate concentration increases
Boron	P	P	—	—	P	—	>290	Generally negligible
Cadmium	P to G	F to VG	—	F	P to VG	Old carbon better	2	High removals due to precipitation of sulfide and hydroxide forms
Chromium	F to G	G	VG	G(Cr^{+6}) VG(Cr^{+3})	P to G (Cr^{+3}) VG(Cr^{+6})	Reduction with bioactivity Cr^{+3} less soluble than Cr^{+6}	20	Depends on influent level and oxidation state
Copper	F to G	P to G	—	G	G to VG	Enhanced sorption better with new carbon	70	Influenced by influent concentration
Iron	P to F	F to G	—	VG	P to G	Sulfide complexes ppt, but anaerobic bioactivity causes reduction to soluble Fe^{+2}	>40	Depends on influent level, pH, and redox potential
Lead	F to G	F to G	—	VG	P to G	—	>5	Enhanced precipitation with higher sulfate levels
Manganese	P	G to VG	—	P	P	Bioactivity on the carbon reduces Mn^{+4} to Mn^{+2}	5	Depends on pH and redox potential
Mercury	P to G	P to G	VG	G	P to G	Variability due to biological activity	5	Removal is a function of pH, initial concentration, and degree of complexation
Selenium	F	P to G	G	F	P to G	Variability due to highly soluble characteristics	2	Depends on influent concentration
Silver	P to G	P to VG	VG	VG	P to G	High affinity for sulfhydryl groups	2	Depends on influent level
Zinc	F to G	P to F	—	F	P to G	Zinc sulfide precipitate	>60	Depends on influent and sulfate levels

NOTE: P = Poor (<30%); F = Fair (30%-60%); G = Good (60%-90%); VG = Very Good (>90%).
Source: National Research Council, 1982, Quality Criteria for Water Reuse, National Academy Press, Washington, D.C. Original source: Englande, A.J., and Reimers, R.S., 1979, Wastewater Reuse-Persistence of Chemical Pollutants, Am. Water Works Research Foundation, Denver, CO

FIGURE 7–4. WATER RECLAMATION AS A PLANNED ENTERPRISE SEPARATE FROM SEWAGE DISPOSAL

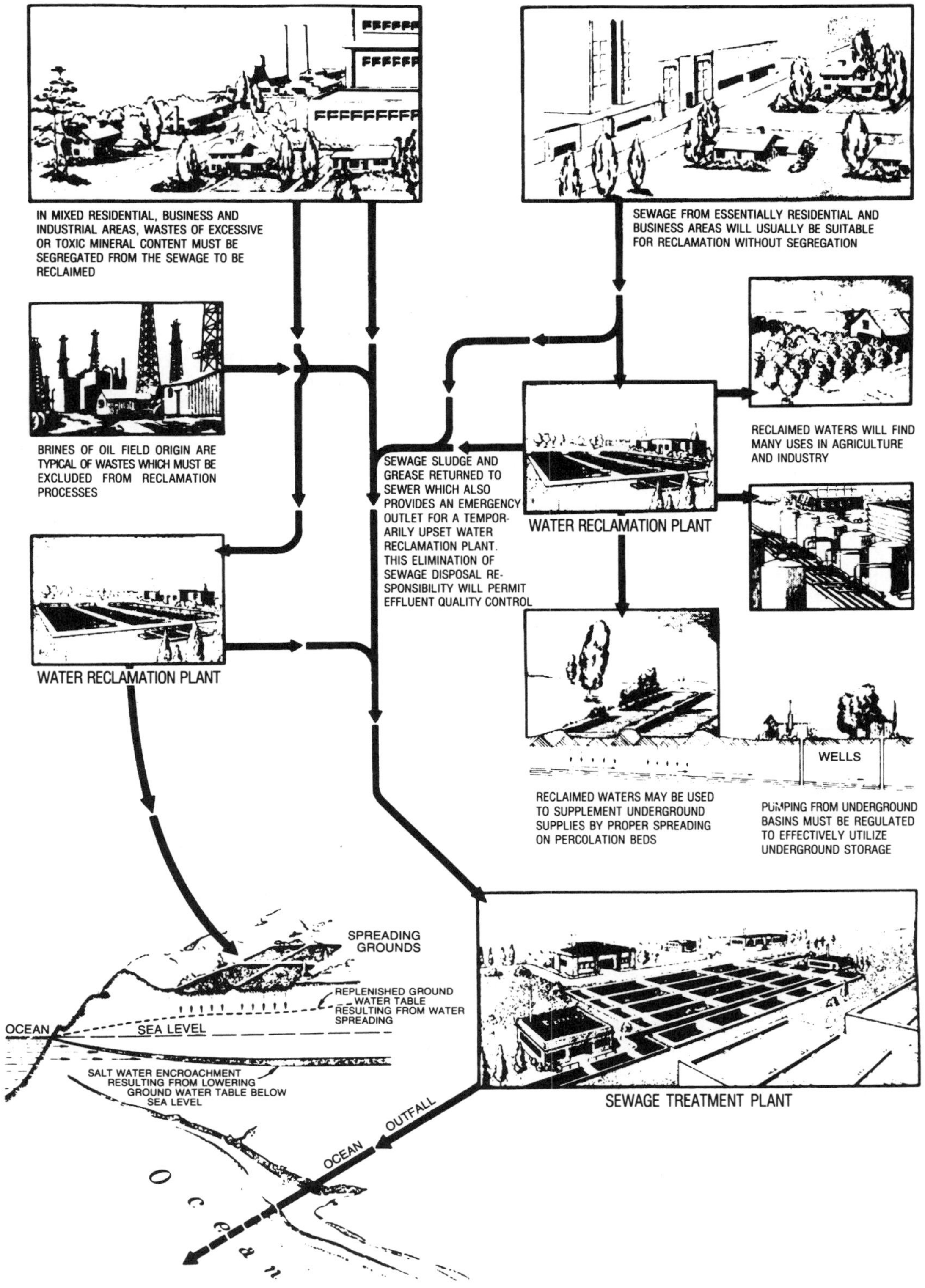
IN MIXED RESIDENTIAL, BUSINESS AND INDUSTRIAL AREAS, WASTES OF EXCESSIVE OR TOXIC MINERAL CONTENT MUST BE SEGREGATED FROM THE SEWAGE TO BE RECLAIMED
SEWAGE FROM ESSENTIALLY RESIDENTIAL AND BUSINESS AREAS WILL USUALLY BE SUITABLE FOR RECLAMATION WITHOUT SEGREGATION
BRINES OF OIL FIELD ORIGIN ARE TYPICAL OF WASTES WHICH MUST BE EXCLUDED FROM RECLAMATION PROCESSES
SEWAGE SLUDGE AND GREASE RETURNED TO SEWER WHICH ALSO PROVIDES AN EMERGENCY OUTLET FOR A TEMPORARILY UPSET WATER RECLAMATION PLANT. THIS ELIMINATION OF SEWAGE DISPOSAL RESPONSIBILITY WILL PERMIT EFFLUENT QUALITY CONTROL
RECLAIMED WATERS WILL FIND MANY USES IN AGRICULTURE AND INDUSTRY
WATER RECLAMATION PLANT
WATER RECLAMATION PLANT
WELLS
RECLAIMED WATERS MAY BE USED TO SUPPLEMENT UNDERGROUND SUPPLIES BY PROPER SPREADING ON PERCOLATION BEDS
PUMPING FROM UNDERGROUND BASINS MUST BE REGULATED TO EFFECTIVELY UTILIZE UNDERGROUND STORAGE
SPREADING GROUNDS
REPLENISHED GROUND WATER TABLE RESULTING FROM WATER SPREADING
OCEAN
SEA LEVEL
SALT WATER ENCROACHMENT RESULTING FROM LOWERING GROUND WATER TABLE BELOW SEA LEVEL
SEWAGE TREATMENT PLANT
OCEAN OUTFALL
Ocean

Source: House of Representatives, U.S. Congress

TABLE 7–78. WASTEWATER REUSE IN THE UNITED STATES IN 1978

[By type of use]

		Thousand Acre-Feet/Year
Irrigation		471
Agricultural	223	
Landscape	37	
Not Specified	211	
Industrial		240
Process	74	
Cooling	159	
Boiler Feed	7	
Ground Water Recharge		38
Other (recreation, fish and wildlife, etc.)		11
TOTAL		760

Source: Office of Water Research and Technology, 1979, Water Reuse and Recycling, Volume I: Evaluation of Needs and Potential, PB 297 283

TABLE 7–79. WASTEWATER REUSE IN THE UNITED STATES IN 1978

[By Water Resources Region; for map of regions see Figure 2–6]

Region	Number of Projects	Thousand Acre-Feet/Year
Lower Colorado	36	204.4
California	283	192.3
Mid-Atlantic	7	130.5
Arkansas-White-Red	31	84.1
Texas Gulf	102	57.4
Pacific Northwest-Columbia	14	29.0
Rio Grande	24	22.0
Great Lakes	13	10.8
South Atlantic-Gulf	6	7.4
Missouri Basin	9	6.8
Great Basin	3	5.6
Upper Mississippi	2	3.0
New England	1	3.0
Upper Colorado	1	1.2
Ohio	2	1.1
Hawaii	1	1.0
Caribbean	1	0.4
TOTAL	536	760

Source: Office of Water Research and Technology, 1979, Water Reuse and Recycling, Volume I: Evaluation of Needs and Potential, PB 297 283

TABLE 7–80. PROJECTED WASTEWATER REUSE IN THE UNITED STATES IN 2000

[By type of use]

		Quantity, bgd
Water withdrawals capable of wastewater reuse		290.3
Agricultural irrigation	177.8	
Landscape irrigation	1.4	
Steam electric	80.1	
Industrial cooling	16.9	
Industrial—other	14.1	
Treated wastewater discharges available for reuse		109.6
Steam electric	69.5	
Industry	12.7	
Municipal	26.7	
Fish hatcheries	0.7	
Untreated agricultural discharges		11.5
Wastewater reused		4.8
Wastewater recycled		865.5

Source: Office of Water Research and Technology, 1979, Water Reuse and Recycling, Volume I: Evaluation of Needs and Potential, PB 297 283

TABLE 7–81. AGRICULTURAL AND RECREATIONAL USE OF SEWAGE EFFLUENT IN THE UNITED STATES

Location	Use
Arizona:	
Tucson	Cotton, grain, and pasture
Grand Canyon	Lawn
California:	
Bakersfield	Animal feed crops and cotton
Barstow	Golf course
Calaveras (State park)	Trees
Camp Pendleton	Golf course
China Lake (Naval Ordnance Test Station)	Golf course
Delano	Animal feed crops and cotton
El Cajon	Golf course
El Toro (Marine base)	Golf course
Fontana	Steel industry–process and cooling water
Fresno	Animal feed crops and cotton
Hanford	Animal feed crops and cotton
Mount Vernon (Sn. Distr.)	Animal feed crops and cotton
Nebo (Marine base)	Golf course
Oceanside	(Under planning)
Ontario	Animal feed crops and grain
Orange County	Hay, sugar beets, dry beans, dry peppers
Palo Alto	Golf course
Pomona	Animal feed crops and citrus
San Bernardino	Animal feed crops and orchards, berries
San Francisco	Park and lawn, lake
Twentynine Palms (Marine base)	Recreational fields
Taft	Cotton–maize seed and cotton
Tulare	Animal feed crops and cotton
Visalia	Grain–walnuts, fruit, grain
Wasco	Cotton–maize seed and cotton
New Mexico:	
Carlsbad	Golf course
Jal	Golf course
Los Alamos	Golf course and recreation fields
Santa Fe	Golf course
Texas, over 200 towns, including:	
Abilene	Animal feed crops
Andrews	Golf course (planned)
Brownsville	Animal feed crops
Kingsville	Animal feed crops and cotton
Lubbock	Grains, cotton
Midland	Cotton, golf course
San Angelo	Animal feed crops
San Antonio	Animal feed crops, lake for wildfowl
Nevada: Las Vegas	Golf course, lawns
Washington: Ephrata	Corn, hay

Source: Select Committee on National Water Resources, U. S. Senate, 1960

TABLE 7–82. ECONOMIC COMPARISON OF POTENTIAL WATER RECLAMATION PROJECTS IN THE LOS ANGELES METROPOLITAN AREA

Sewerage System and Project	Annual Plant Yield, in acre-feet	Capital costs $			Average annual costs $			Unit Costs, in Dollars per Acre-Foot		
		Treatment	Conveyance	Total	Treatment	Conveyance	Total	Treatment	Conveyance	Total
Los Angeles City										
Hyperion-Vernon	45,000	—	10,000,000	—	—	810,000	—	10.60	18.00	28.60
Hyperion-Torrance	17,000	—	1,820,000	—	—	156,000	—	10.60	9.20	19.80
Recharge of West Coast Basin	54,000	—		—	—		—	15.60	3.00	18.60[a]
Griffith Park	3,000	1,120,000	170,000	1,290,000	123,000	15,000	138,000	41.00	5.00	46.00
Los Angeles County										
Whittier Narrows	72,000	15,000,000	5,810,000[b]	20,810,000	1,350,000	312,000[b]	1,662,000	18.80	4.30[b]	23.10
South Whittier	4,400	—	1,330,000	—	—	99,000	—	18.80[c]	22.50	41.30[d]
Orange County										
Talbert Water	2,800	—	—	—	—	—	—	0.50[e]	5.35	5.85
District	2,800	—	—	—	—	—	—	1.00[f]	5.35	6.35

[a] Does not include cost of distribution or injection.
[b] Cost of conveyance of waste water to Whittier Narrows Reclamation Plant. Does not include cost of spreading.
[c] Unit cost of treatment assumed the same as at Whittier Narrows Reclamation Plant.
[d] If this project is combined with the Whittier Narrows Reclamation Plant, the total unit cost would remain at $23.10 for the expanded project.
[e] Initial unit cost of effluent delivered by County Sanitation Districts of Orange County.
[f] Maximum unit cost of effluent delivered by County Sanitation Districts of Orange County.
Source: Calif. Dept. of Water Resources, 1961

TABLE 7–83. PRINCIPAL WATER QUALITY REQUIREMENTS FOR POTENTIAL REUSES

[All units in mg/L unless otherwise noted]

Parameter	Agricultural Irrigation	Landscaped Areas		Recirculating Cooling Water
		Restricted	Open	
pH	4.5–9.0	4.5–9.0	4.5–9.0	6.9–9.0
Fecal coliform—*cfu/100 mL*				
Geometric mean	1000	200	25	
Single sample maximum	4000	1000	75	
Turbidity—*ntu*			5	50
Enteric virus—*pfu*[a]			125/40L	
Total suspended solids				25–100
Total dissolved solids				500–1650
Chemical oxygen demand				75
Biochemical oxygen demand				25
Organics (methylene blue activated substances)				1–2
Ammonium				4
Phosphate				1
Aluminum				0.1
Arsenic	2.0			
Barium				
Boron	1.0			
Cadmium	0.05			
Calcium				50
Chromium (total)	1.0			
Copper	5.0			
Iron				0.5
Lead	10.0			
Magnesium				0.5
Manganese	10.0			0.5
Selenium	0.02			
Silver				
Zinc	10.0			
Chlorides				100–500
Hardness				50–130 650
Alkalinity				20 350
Silica				50
Bicarbonate				24
Sulfate				200

[a] pfu—plaque-forming units
Source: Goff, J.D., and Busch, P.L., 1985, Reclaiming Desert Lands Through Water Reuse, J. Am. Water Works Assoc., vol. 77, no.7. Copyright AWWA. Reprinted with permission

TABLE 7–84. LAND AREA REQUIRED TO REUSE ALL EFFLUENT FOR IRRIGATION

Average Daily Flow		Land Area Required			
		With Storage[a]		Without Storage[b]	
mgd	ML/d	acres	ha	acres	ha
5	19	950	384	5 700	2307
10	38	1900	789	11 400	4614
15	57	2850	1153	17 100	6920
20	76	3800	1538	22 800	9227

[a] Based on average irrigation demand of 6 ft (1.8 m) per year and use of reclaimed water storage to equalize seasonal peaks in irrigation demand.
[b] Based on minimum irrigation demand of 1 in. (2.5 cm) per month during winter months and use of other supplies blended with reclaimed water during nonwinter months.
Source: Goff, J.D. and Busch, P.L., 1985, Reclaiming Desert Lands Through Water Reuse, J.Am. Water Works Assoc., vol. 77, no.7.

TABLE 7–85. EFFLUENT WATER QUALITY FOR ALTERNATIVE LEVELS OF TREATMENT

Parameter	Maximum Concentrations Allowed for Each Level of Treatment[e]				
	1	2	3	4	5
Biochemical oxygen demand—*mg/L*	30	30	10	10	5
Total suspended solids—*mg/L*	30	30	10	5	1
Nitrogen (as nitrate)—*mg/L*[a]	25	5	25	25	25
Total phosphorus—*mg/L*	10	10	10	1	1
Total dissolved solids—*mg/L*	1500	1500	1500	1500	500
Turbidity—*ntu*	[b]	[b]	5	1	1
Fecal coliform—*cfu/100 mL* (geometric mean)	1000[c]	1000[c]	25	2.2[d]	2.2
pH	6.5–9	6.5–9	6.5–9	6.5–9	6.5–9

[a] Assumes full nitrification will occur in secondary treatment; thus, NO_3 will be the predominant nitrogen form.
[b] Not applicable
[c] Can be reduced to 200, if necessary, by increasing the chlorine dosage
[d] This constraint may be relaxed if effluent is used for recreational bodies of water.
[e] Level 1 is appropriate for most agricultural irrigation applications and restricted-access landscape irrigation;
Level 2 would be applicable to agricultural irrigation of crops with low nitrogen tolerances;
Level 3 provides water acceptable for use in open-access landscape irrigation;
Level 4 effluent is suitable for use in recreational lakes or recirculating industrial cooling-water systems.;
Level 5 treatment is required for use of the effluent in most industrial process water applications.
Source: Goff, J.D. and Busch, P.L., 1985, Reclaiming Desert Lands Through Water Reuse, J.Am. Water Works Assoc., vol. 77, no.7.

TABLE 7–86. ISSUES SURROUNDING WATER REUSE FOR IRRIGATION

Resource Issues

1. Effluent quality
   - Nutrient content
   - Heavy metal content
   - Pathogen content
2. Soil productivity
   - Salt buildup
   - Toxicity buildup
   - Viral contamination
   - Physical degradation
3. Crop production
   - Fertilizer and water requirements
   - Crop growth and yields
   - Crop uptake of nutrients
   - Crop uptake of toxics and pathogens
4. Animal health
   - Animal uptake of nutrients
   - Animal transmission of pathogens to human consumers
5. Ground water quality
   - Path of water to water table
   - Quality of water reaching ground water
6. Air quality (with sprinkler irrigation)
   - Health effects for workers and nearby residents
   - Odor considerations

Social and Economic Issues

1. Human health effects
   - Contact with effluent by farmworkers
   - Contact with plant and animal products by consumers
2. Social factors
   - Public attitudes toward application
   - Public attitudes by consumers of products
   - Attitudes of nearby residents
3. Economic considerations
   - Water pricing
   - Transportation costs
   - Subsidies for those who use water
   - Facilities for water storage
   - Value in alternate uses
   - Type of material contained in water

Institutional Issues

1. Water treatment facilities
   - Adequacy and reliability of treatment prior to application
   - Adequacy of storage facilities during periods of nonapplication
2. Monitoring
   - Need for monitoring air, effluent, ground water, crop, and soil quality
3. Legal issues
   - Ownership and sale of water
   - Water rights
   - Liability for damages
   - Responsibility for monitoring
   - Guidelines for water reuse (e.g., crops to be grown, amount of water to be applied)
   - Effect on downstream users (third parties), if water previously was part of return flows

Source: Speidel, D.H., 1988, Perspectives on Water Uses and Abuses, Copyright Oxford University Press; Bruvold, 1982, Agricultural Use of Reclaimed Water, National Science Foundation, Office of Technology Assessment

TABLE 7–87. MAJOR PROJECTS UTILIZING DOMESTIC WASTEWATER AS POTABLE WATER SOURCE

Location	Date Started	Purpose	Pre-treat-ment	Advanced Treatment
Whittier Nar-rows, Calif.	1962	Groundwater recharge basins	AS	Filt, Cl_2
Windhoek, Namibia (formerly South-West Africa)	1969 (mod. 1978)	Direct potable reuse	TF, MP	Lime, Air, CO_2, Cl_2, Filt, Cl_2, GAC, Cl_2
Daspoort, South Africa	1970	Experimental	AS	Lime, Air, CO_2, Filt, Cl_2, GAC, Cl_2
Orange County, Calif.	1976	Groundwater injection	AS	Lime, Air, CO_2, Cl_2, Filt, GAC, RO
Palo Alto, Calif.	1977	Experimental; groundwater injection	AS	Lime, Air, CO_2, O_3, Filt, GAC, Filt, Cl_2
Tahoe-Truckee, Calif.	1978	Water supply; stream discharge	AS	Lime, $C0_2$, Filt, GAC, Ion, Land
Fairfax County, Va.	1978	Water supply; reservoir discharge	AS	Lime, $C0_2$, Filt, GAC, Ion, Cl_2
Washington, D.C.	1981	Experimental; direct	AS	Air, Alum, Cl_2, Filt, GAC, Cl_2
Denver, Colorado	1982	Experimental; direct potable reuse	AS	Lime, $C0_2$, Filt, Ion, Cl_2, GAC, 0_3, RO, $C10_2$
East Meadow, Long Island, NY	1982–84	Experimental; groundwater recharge basins and injection wells	AS	Filt, GAC, Cl_2, N
El Paso, Texas	1985	Groundwater	PAC AS	Lime, $C0_2$, Filt, O_3, GAC

PAC = powdered activated carbon
AS = activated sludge
TF = trickling filter
MP = maturation ponds
Lime = lime treatment
Air = air striping
$C0_2$ = recarbonation
Cl_2 = chlorination
0_3 = ozonation
$C10_2$ = chlorine dioxide disinfection
Filt = filtration
GAC = granular activated carbon
RO = reverse osmosis
Alum = alum coagulation and settling
Ion = ion exchange
Land = land application
N = biological nitrate removal
Source: Arber, Richard, 1986, State of the Art Potable Water Reuse, in AWWA Proceedings; Implementation of Water Reuse. Copyright AWWA. Reprinted with permission; amended by F. van der Leeden

FIGURE 7–5. ADVANCED WASTEWATER TREATMENT PROCESSES AT WATER FACTORY 21, FOUNTAIN VALLEY, CALIFORNIA

[Water Factory 21 is a 15 million gallons per day treatment plant operated by the Orange County Water District]

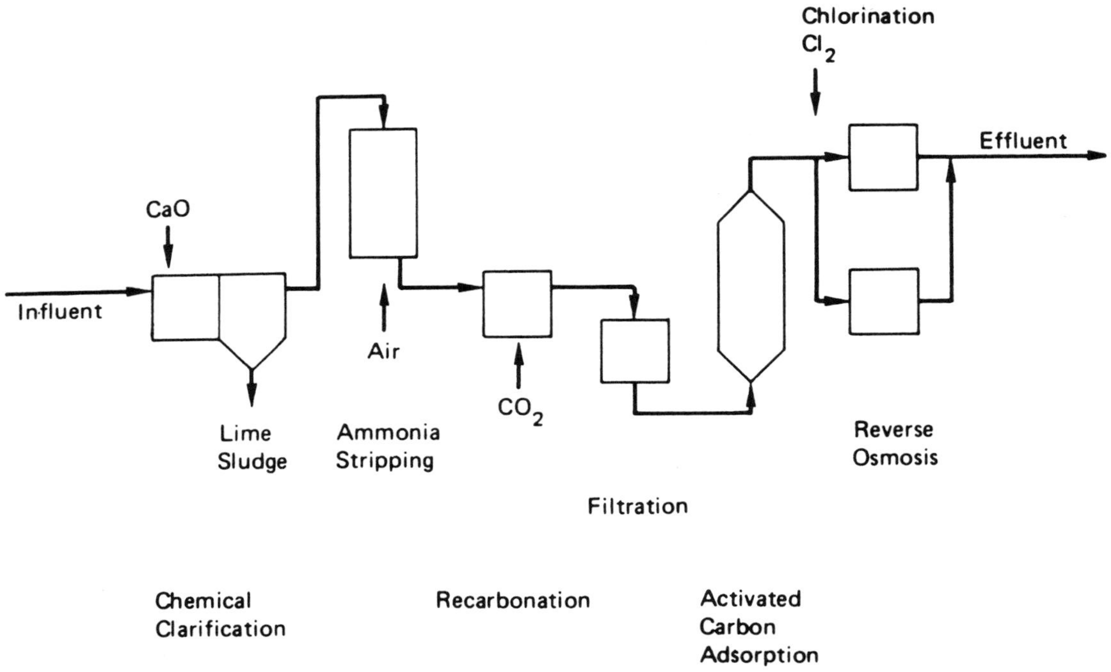

Source: National Research Council, 1982, Quality Criteria for Water Reuse, National Academy Press, Washington, D.C.

TABLE 7–88. CONCENTRATIONS OF HEAVY METALS AND EFFECTIVENESS OF TREATMENT AT WATER FACTORY 21 IN REMOVING THEM

Contaminants	Concentration at AWT Plant (μg/liter) Influent	Effluent	Percent Removal AWT[a]	Dominant Removal Processes[b]
Arsenic	<5	<5	—	
Barium	30	7.4	75	ppt, RO
Cadmium	33	9.5	71	ppt, RO
Chromium	48	3.1	94	ppt, GAC, RO
Copper	72	16	78	ppt, GAC, RO
Lead	7.1	1.0	86	ppt, GAC
Iron	98	42	57	ppt, RO
Mercury	<1	<1	—	
Manganese	29	1.7	94	ppt
Silver	1.2	0.7	42	ppt
Selenium	<5	<5	—	
Zinc	127	100	21	ppt, GAC

[a] AWT refers to advanced wastewater treatment processes except reverse osmosis (RO).
[b] ppt = chemical precipitation with lime.
GAC = granular activated carbon.
Source: National Research Council, 1982, Quality Criteria for Water Reuse, National Academy Press, Washington, D.C.. Original source: McCarty et al., 1980, U.S. Environmental Protection Agency

TABLE 7–89. WASTEWATER PRETREATMENT FOR IRRIGATION OF CERTAIN CROPS AS PRACTICED IN SELECTED COUNTRIES

Crop	California: Surface Irrigation	California: Spray Irrigation	Israel	South Africa	Federal Republic of Germany	Engelberg Report[a]
Unrestricted use for crops eaten raw	Less than 2.2 F.C./100 ml	Coagulation, filtration, disinfection, turbidity, less than 10 units	Not allowed except for peeled-eaten fruits	—	Irrigation only in floration phase	Less than 100 F.C./100 ml; less than 1 nematode egg/liter
Crops eaten after cooking or processing	Primary effluents	Secondary effluent and disinfection, less than 23 F.C./100 ml	Less than 1000 F.C./100 ml in 80% of samples	Tertiary effluent	Tertiary effluent irrigation allowed up to 4 weeks before harvesting	Less than 1 nematode egg/liter
Fodder, fiber & seed crops	Primary effluent	Primary effluent	Secondary effluent	Tertiary effluent	If spray irrigation: tertiary effluent and chlorination	Less than 1 nematode egg/liter
Fruit trees & vineyards	Primary effluent	Not allowed	Secondary effluent	Tertiary effluent, chlorine disinfection; no spray irrigation	No spray irrigation	Less than 1 nematode egg/liter

[a] IBRF, WHO, IRCWD, UNDP, 1985, Health Aspects of Wastewater and Excreta Use in Agriculture: The Engelberg Report, Engelberg, Switzerland. F.C. = Fecal Coliforms.
Source: Saenz, R., 1987, Use of Wastewater Treated in Stabilization Ponds for Irrigation- Evaluation of Mirobiological Aspects, Water Quality Bull., vol. 12, no. 2

TABLE 7–90. SELECTED CONTAMINANT CONCENTRATIONS AND EFFECTIVENESS OF TREATMENT AT WATER FACTORY 21 IN REMOVING THEM

Contaminant or Other Factor Measured	Concentration as Noted: AWT Plant[a] Influent	AWT Plant[a] Effluent	RO[b] Effluent	Percent Removal: AWT[a]	AWT[a] and RO	Dominant Removal Processes[c]
Chemical oxygen demand (COD), mg/liter	47	12	1.3	74	97	ppt, GAC, RO
Total organic carbon (TOC), mg/liter	12	6	2.6	50	78	ppt, GAC, RO
Total dissolved solids (TDS), mg/liter	900	850	77	5	91	RO
Electroconductivity (EC), μS/cm	1,500	1,320	156	12	90	RO
Nitrogen, mg N/liter:						
Organic	2.0	1.1		45		ppt
Ammonia	4.0	0.8		80		strip, Cl_2
Nitrate	2.8	7.7	3.3	−175	−18	
Boron, mg/liter	0.74	0.53		28		
Fluoride, mg/liter	1.3	0.81		38		
Coliforms, MPN[d]/100 ml						
Total	1.6×10^6	0.05		100		ppt, Cl_2
Fecal	0.55×10^6	<1		100		ppt, Cl_2
Phenol, μg/liter	4.9					
Cyanide, μg/liter	25	6.9		72		
Color, units	37	0.8		98		
Methylene blue active substance (MBAS), mg/liter	0.25	0.08		68		

Note: See Figure 7–5 for schematic of processes.
[a] AWT refers to all advanced wastewater treatment processes except reverse osmosis (RO).
[b] RO = Reverse osmosis.
[c] ppt = Chemical precipitation with lime.
GAC = Granular activated carbon.
Strip = Air stripping.
[d] MPN = Most probable number.
Source: National Research Council, 1982, Quality Criteria for Water Reuse, National Academy Press, Washington, D.C. Original source: McCarty et al., 1980, U.S. Environmental Protection Agency

TABLE 7–91. CONCENTRATIONS OF ORGANIC CONTAMINANTS AND EFFECTIVENESS OF TREATMENT AT WATER FACTORY 21 IN THEIR REMOVAL

	Concentration (μg/liter)			Percent Removal		
Contaminant	**AWT[a] Plant Influent**	**AWT[a] Plant Effluent**	**RO Effluent**	**AWT[a]**	**AWT[b] & RO**	**Dominant Removal Processes[b]**
Aromatic hydrocarbons						
Ethylbenzene	0.043	0.014	0.019	67	56	NR[c]
m-Xylene	0.035	0.020	0.024	43	31	NR
p-Xylene	0.015	0.012	0.014	20	7	NR
Naphthalene	0.033	0.010	0.028	70	15	NR
1-Methylnaphthalene	0.008	0.009	0.00	–12	88	NR
2-Methylnaphthalene	0.010	0.02	0.008		20	NR
Styrene	0.048	0.003		94		NR
Synthetic chlorinated compounds						
Carbon tetrachloride	0.033	0.16	0.008	–380	76	Strip
1,1,1-Trichloroethane	3.25	0.20	0.083	94	97	Strip
Trichloroethylene	0.74	0.10	0.1	86	86	Strip
Tetrachloroethylene	1.67	0.83	0.20	50	88	Strip
Chlorobenzene	0.14	0.049	0.034	65	76	Strip, GAC
1,2-Dichlorobenzene	0.64	0.02	0.001	97	99.9	Strip, GAC
1,3-Dichlorobenzene	0.16	0.02	0.004	88	97.5	Strip, GAC
1,4-Dichlorobenzene	1.85	0.012	0.015	99	99	Strip, GAC
1,2,4-Trichlorbenzene	0.11	0.02	0.02	82	82	Strip, GAC
Lindane	0.14	0.05	0.05	64	64	ppt, GAC
PCB (Aroclor 1242)	0.47	0.3	0.3	36	36	ppt, GAC
Chlorination products						
Chloroform	3.1	8.6	0.97	–177	69	Strip, GAC
Dichlorobromomethane	0.53	2.7	0.24	–410	55	Strip, GAC
Chlorodibromomethane	0.69	1.3	0.13	–88	81	Strip, GAC
Bromoform	0.40	0.38	0.007	5	98	Strip, GAC
Phthalate esters						
Dimethylphthalate	4.8	0.47	1.0	90	79	ppt, GAC
Diethylphthalate	0.097	0.3	0.3			
Di-n-butylphthalate	0.79	0.33	1.1	58	–39	ppt, GAC
Diisobutylphthalate	4.7	0.27	0.23	94	95	ppt, GAC
Bis-(2-ethylhexyl) phthalate	11.0	3.1	2.9	72	74	ppt

[a] AWT refers to all advanced wastewater treatment processes except RO.
[b] Strip = chemical precipitation with time
GAC = granular activated carbon
ppt = chemical precipitation with lime
[c] NR = not reported.
Source: National Research Council, 1982, Quality Criteria for Water Reuse, National Academy Press, Washington, D.C. Original source: McCarty et al., 1980, Advanced Treatment for Wastewater Reclamation at Water Factory 21, Stanford University

TABLE 7–92. COST OF ADVANCED WASTEWATER TREATMENT AT WATER FACTORY 21 FOR ONE YEAR

Type of Treatment	Cost (Dollars/1,000 m³) Capital	Operations and Maintenance	Total
Lime treatment and recarbonation	$17.5	$35.7	$53.2
Air stripping[a]	15.6	15.2	30.8
Mixed-media filtration	4.8	2.9	7.6
GAC treatment and regeneration	16.2	19.7	35.9
Chlorination	1.2	11.1	12.2
SUBTOTAL	$55.3	$84.6	$139.7
Reverse osmosis	45.3	118.5	163.8
Injection	4.5	9.2	13.7
TOTAL[b]	$105.1	$212.3	$317.2

[a] With fans operating.
[b] Assuming all water treated by reverse osmosis.
Source: National Research Council, 1982, Quality Criteria for Water Reuse, National Academy Press, Washington, D.C. Original source: Argo, D.G., 1980, Cost of Water Reclamation by Advanced Wastewater Treatment, J. Water Pollution Control Fed, v.52

TABLE 7–93. REMOVAL OF REPRESENTATIVE INFECTIOUS AGENTS BY CONVENTIONAL WASTEWATER TREATMENT

Infectious Agents	Primary Treatment (% Removed)	Secondary Treatment (% Removed) Activated Sludge	Trickling Filter
Salmonella	15	90–99	90–99.9
Mycobacterium	40–60	5–90	70–99
Shigella	15	80–90	85–99
Amoebic cysts	Limited	Limited	10–99.9
Helminth ova	70–95	Limited	60–75
Viruses	Limited	75–99	0–86

Source: National Research Council, 1982, Quality Criteria for Water Reuse, National Academy Press, Washington, D.C.. Original source: Engelbrecht, R.S., and Lund, E., 1975, Biological Properties of Wastewater Sludge and its Potential Health Risk, International Water Conservancy Exhibition, Jonkoping, Sweden

TABLE 7–94. AVERAGE CONCENTRATIONS OF SELECTED CONTAMINANTS IN MUNICIPAL WASTEWATER FOLLOWING SECONDARY BIOLOGICAL TREATMENT

Contaminant (Concentrations in μg/L unless otherwise indicated)	Washington D.C.[a]	Orange County, CA Water District		Phoenix, AZ[b]	Palo Alto, CA
		2nd Period	3rd Period		
Total Organic Carbon (TOC), mg/L	4.5	30	16	9	11
Total Organic Halides (TOX)	85		131	87	192
Trihalomethanes:					
$CHCl_3$	1.5	1.6	3.5	3.5	13
$CHBrCl_2$	<0.3	0.1	0.5	0.3	0.2
$CHBr_2Cl$	<0.2	0.2	0.7	0.2	0.1
$CHBr_3$		0.1	0.5	0.1	0.0
Total	2.0	2.0	5.2	4.1	13.3
Other Chlorinated Organics:					
1,1,1-Trichloroethane	<0.2	4.7	4.8	1.4	65
Trichloroethylene	<0.1	0.9	1.1	0.4	25
Tetrachloroethylene	<0.8	0.6	3.6	1.7	44
Chlorobenzene		2.5	0.1		0.3
o-Dichlorobenzene	<0.05	2.4	0.7	2.4	2.7
m-Dichlorobenzene	0.08	0.7	0.2	0.4	3.6
p-Dichlorobenzene	<0.11	2.1	1.9	1.8	5.4
1,2,4-Trichlorobenzene	<0.02	0.5	0.3	0.4	11.3
Nonchlorinated Organics:					
Toluene	<0.12				
Ethylbenzene	<0.02	1.4	0.04	0.2	0.03
o-Xylene	<0.04			0.4	
m-Xylene	<0.05		0.01	0.8	0.2
p-Xylene	<0.05		0.05	0.2	0.06
Naphthalene	<0.04	0.6	0.1	0.2	3.3

[a] After mixing with a 1:1 Potomac River Water and Blue Plains treated effluent.
[b] Samples taken from spreading basins after secondary treatment.
Source: California Water Resources Control Board, Department of Water Resources and Department of Health, 1987, Report of the Scientific Advisory Panel on Groundwater Recharge with Reclaimed Wastewater

TABLE 7–95. TYPES OF USES FOR RECLAIMED WASTEWATER IN CALIFORNIA

Uses	Number of Installations
Irrigation of fodder, fiber, and seed crops	190
Landscape irrigation—golf courses and similar installations	85
Landscape irrigation—parks and similar installations	29
Orchard and vineyard irrigation	26
Food crop irrigation	10
Industrial uses	9
Restricted recreational impoundments	6
Landscape impoundments	6
Groundwater recharge	5
Nonrestricted recreational impoundments	1
Other	16
Total	383

Source: Crook, J., 1985, Water Reuse in California, J.Am.Water Works Assoc., vol. 77, no.7. Copyright AWWA. Reprinted with permission

TABLE 7–96. POTENTIAL MARKET FOR RECLAIMED WATER IN CALIFORNIA BY YEAR 2000

[In cubic decameters and (acre-feet)]

Area	Potential Market	Reclaimed Water Source[a]
San Francisco Bay Area	49 000 – 123 000 (40,000 – 100,000)	San Francisco Bay Region
Sacramento-San Joaquin Delta	0 – 271,000 (0 – 220,000)	San Francisco Bay Region
South Coast: Ventura County and Santa Barbara	62 000 (50,000)	Ventura Regional County Sanitation District and Cities of Goleta and Santa Barbara
Los Angeles, City of	46 000 (37,000)	Los Angeles City Department of Public Works
Los Angeles County	107 000 (87,000)	Los Angeles County Sanitation District Treatment Facilities
Orange County	30 000 (24,000)	Orange County Sanitation District/Orange County Water District Facilities
San Bernardino County	25 000 (20,000)	Chino Basin Municipal Water District
Riverside County	NEA[b]	—
San Diego Area	37 000 (30,000)	San Diego City/County
San Joaquin Valley	30 000[c] (24,000)	Agricultural waste water
Coachella Valley	30 000[d] (24,000)	Agricultural waste water
TOTALS - Low	416 000 (337,000)	
High	761 000 (616,000)	

[a] Municipal wastewater, unless otherwise stated.
[b] No estimate available.
[c] Desalting required. If feasible, potential for several plants.
[d] Desalting required.
Source: California State Water Resources Control Board, Office of Water Recycling, and the Department of Water Resources, 1981, Bulletin 4, 1982

TABLE 7–97. REPORTED INTENTIONAL USE OF RECLAIMED WATER IN CALIFORNIA

[Data as of 1979; in acre-feet]

Hydrologic Study Area	Industrial		Irrigation			Other Uses					TOTAL
	Power Plant Cooling	Other	Crops	Landscape	Golf Course	Ornamental Lakes	Ground Water Recharge	Recreation	Wildlife Habitat	Unclassified	
North Coast	—	800	8,000	—	—	—	—	—	—	600	9,400
San Francisco Bay	—	300	6,500	1,200	200	—	—	—	100	2,100	10,400
Central Coast	—	—	8,800	200	—	—	—	—	—	100	9,100
Los Angeles	—	—	1,700	11,400	13,600	—	12,800	—	—	6,300	45,800
Santa Ana	—	400	3,700	—	—	—	10,400	—	—	14,500	29,000
San Diego	—	—	100	—	2,000	900	700	—	3,700	1,700	9,100
Sacramento	—	1,600	14,200	200	—	1,100	—	—	—	—	17,100
San Joaquin	200	—	20,300	100	—	—	—	—	—	—	20,600
Tulare Lake	—	—	34,500	—	—	—	—	—	—	—	34,500
North Lahontan	—	—	5,400	—	—	—	—	—	200	—	5,600
South Lahontan	—	400	2,000	300	800	—	—	200	—	—	3,700
Colorado River	—	900	1,700	—	700	—	—	—	—	—	3,300
TOTAL	200	4,400	106,900	13,400	17,300	2,000	23,900	200	4,000	25,300	197,600

Data in this table are based on responses to a 1980 survey of California waste water treatment plants by the Department of Water Resources. The table is not a complete accounting of intentional use.
Source: California Department of Water Resources, 1983, The California Water Plan–Projected Use and Available Water Supplies to 2010, Bulletin 160–83

TABLE 7–98. PROJECTED INCREMENTAL INCREASE IN USE OF RECLAIMED WASTEWATER IN CALIFORNIA[a]

[By major urban areas; in acre-feet]

Region	1990	2000	2010	Increase 1980–2010
San Luis Obispo County	4,500	2,000	0	6,500
Santa Barbara County	10,000	0	0	10,000
Ventura County	15,700	3,900	0	19,600
Orange-Los Angeles Counties	48,200	118,100	76,400	242,700
San Bernardino-Riverside Counties	11,000	0	0	11,000
San Diego County	20,000	10,000	0	30,000
TOTAL	109,400	134,000	76,400	319,800

[a] Assumes some relaxation of Department of Health Services' restrictions on recharge of ground water basins.
Source: California Department of Water Resources, 1983, The California Water Plan–Projected Use and Available Water Supplies to 2010, Bulletin 160–83

TABLE 7–99. CALIFORNIA STATE DEPARTMENT OF HEALTH STANDARDS FOR THE SAFE AND DIRECT USE OF RECLAIMED WASTEWATER FOR IRRIGATION AND RECREATIONAL IMPOUNDMENTS

	Description of Minimum Required Wastewater Characteristics			
Use of Reclaimed Wastewater	**Primary[a]**	**Secondary and Disinfected**	**Secondary Coagulated Filtered[b] and Disinfected**	**Coliform MPN/100 mL Median (daily sampling)**
Irrigation				
Fodder crops	x			No requirement
Fiber crops	x			No requirement
Seed crops	x			No requirement
Produce eaten raw, surface irrigated		x		2.2
Produce eaten raw, spray irrigated			x	2.2
Processed produce, surface irrigated	x			No requirement
Processed produce, spray irrigated		x		23
Landscapes, parks, etc.		x		23
Creation of impoundments				
Lakes (aesthetic enjoyment only)		x		23
Restricted recreational lakes		x		2.2
Non-restricted recreational lakes			x	2.2

[a] Effluent not containing more than 1.0 mL/L/h settlable solids.
[b] Effluent not containing more than 10 turbidity units.
Source: Ongerth, H.J., and Jopling, W.F., 1977 Water Reuse in California. In Shuval, H.I., 1987, Wastewater Reuse for Irrigation: Evolution of Health Standards, Water Quality Bulletin, v. 12, no. 2

TABLE 7–100. COMPARISON OF UNIT COSTS OF TREATED MUNICIPAL SEWAGE EFFLUENT AND DESALINATED WATER IN WESTERN ASIA AND EGYPT

Location	Year	Treated Effluent Unit Cost Range US$/cu. m.	Desalinated Water Unit Cost Range US$/cu. m.	Remarks
Bahrain	1984	0.28		Without a reverse osmosis unit.
	1984	0.84		With a reverse osmosis unit.
Egypt	1980	0.05–2.90		Primary treatment cost of industrial wastewaters.
Kuwait	1981	0.18–0.31	2.50 (1984)	
Qatar	1981	0.24	1.14–1.16	Energy supplied at zero cost.
	1981		1.45–1.64	Energy supplied at market cost.
Saudi Arabia	1980		0.04	Desalination of brackish water by reverse osmosis.
	1979–82		0.71–1.23	Desalination of seawater by multistage flash system.
United Arab Emirates	1982	0.30–0.41	1.00–1.45	

Source: Arar, A., 1987, Irrigation with Sewage Effluent: its Application in the Near East Region (Western Asia), Water Quality Bull., v.12, no.2. Original source: Treatment and Use of Sewage Effluent for Irrigation in Western Asia, Natural Resources, Science and Technology Division, E/ECWA/NR/84/2, ECWA, Baghdad, Iraq

SECTION H. SOLID WASTE

TABLE 7–101. MUNICIPAL SOLID WASTE GENERATION, RECOVERY, AND DISPOSAL IN THE UNITED STATES, 1960 TO 1984

[In millions of tons, except as indicated. Covers post-consumer residential and commercial solid wastes which comprise the major portion of typical municipal collections. Excludes mining, agricultural and industrial processing, demolition and construction wastes, sewage sludge, and junked autos and obsolete equipment wastes. Based on material-flows estimating procedure and wet weight as generated]

Item and Material	1960	1965	1970	1975	1976	1977	1978	1979	1980	1981	1982	1983	1984
Gross waste generated	82.3	98.3	118.3	122.7	130.4	133.3	138.0	140.7	139.1	140.9	137.8	144.1	148.1
Per person per day (lb.)	2.50	2.77	3.16	3.11	3.28	3.32	3.40	3.43	3.35	3.36	3.25	3.37	3.43
Materials recovered	5.9	6.2	8.0	9.1	10.8	11.6	11.8	13.0	13.4	13.2	12.9	13.9	15.1
Per person per day (lb.)	.18	.17	.21	.23	.27	.29	.29	.32	.32	.31	.30	.32	.35
Percent of gross discards recovered:													
Paper and paperboard	18.0	15.0	16.9	19.2	19.8	20.7	19.9	20.9	21.9	20.6	20.6	20.7	20.8
Glass	1.5	1.2	1.2	2.7	3.4	3.5	3.3	4.0	5.0	5.0	5.2	5.9	7.2
Ferrous metals	.5	1.0	1.2	1.9	2.1	2.4	2.7	3.1	3.2	3.1	2.8	2.7	2.8
Aluminum	(NA)	(NA)	1.6	12.4	11.8	13.1	13.7	13.0	19.6	28.4	30.9	29.3	29.4
Processed for energy recovery	(NA)	.2	.4	.7	.9	1.4	1.5	2.3	2.7	2.3	3.5	5.0	6.5
Per person per day (lb.)	(NA)	.01	.01	.02	.02	.03	.04	.06	.06	.05	.08	.12	.15
Net waste disposed of	76.4	91.9	109.9	112.8	118.7	120.3	124.7	125.4	123.0	125.4	121.4	125.2	126.5
Per person per day (lb.)	2.32	2.59	2.94	2.86	2.98	2.99	3.07	3.05	2.96	2.99	2.86	2.92	2.93
Percent distribution of net discards:[1]													
Paper and paperboard	32.1	35.0	33.1	30.4	32.9	33.1	33.9	34.4	33.6	34.5	33.2	35.3	37.1
Glass	8.4	9.2	11.3	11.6	11.3	11.4	11.6	11.3	11.3	11.3	11.0	10.4	9.7
Metals	13.7	11.6	12.2	11.8	11.1	10.9	10.5	10.6	10.3	10.0	10.1	9.9	9.6
Plastics	.5	1.5	2.7	3.9	4.7	5.3	5.8	6.4	6.0	6.1	6.7	7.0	7.2
Rubber and leather	2.2	2.4	2.7	3.3	3.1	2.8	2.8	3.3	3.3	3.2	3.0	2.6	2.5
Textiles	2.6	2.4	2.0	2.2	2.1	2.0	2.2	2.3	2.3	2.4	2.4	2.3	2.1
Wood	3.9	3.8	3.6	3.8	3.8	3.9	3.7	2.6	3.9	3.5	4.0	4.0	3.8
Food wastes	14.6	13.1	11.5	11.8	11.0	10.6	10.0	9.6	9.2	8.9	8.8	8.5	8.1
Yard wastes	20.3	19.2	19.0	19.5	18.3	18.2	17.7	17.7	18.2	18.2	18.7	18.1	17.9
Other wastes	1.7	1.7	1.7	1.9	1.8	1.8	1.8	1.8	1.9	1.9	2.0	1.9	1.9

NA-Not available.

[1] Net discards after materials recovery and before energy recovery.

Source: U.S. Department of Commerce, 1987 Statistical Abstract of the United States. Original source Franklin Associates, Ltd., Prairie Village, KS, Characterization of Municipal Solid Waste in the United States, 1960 to 2000, 1986. Prepared for the U.S. Environmental Protection Agency

FIGURE 7–6. COMPOSITION OF MUNICIPAL SOLID WASTE IN THE UNITED STATES IN 1986

[In percent of total]

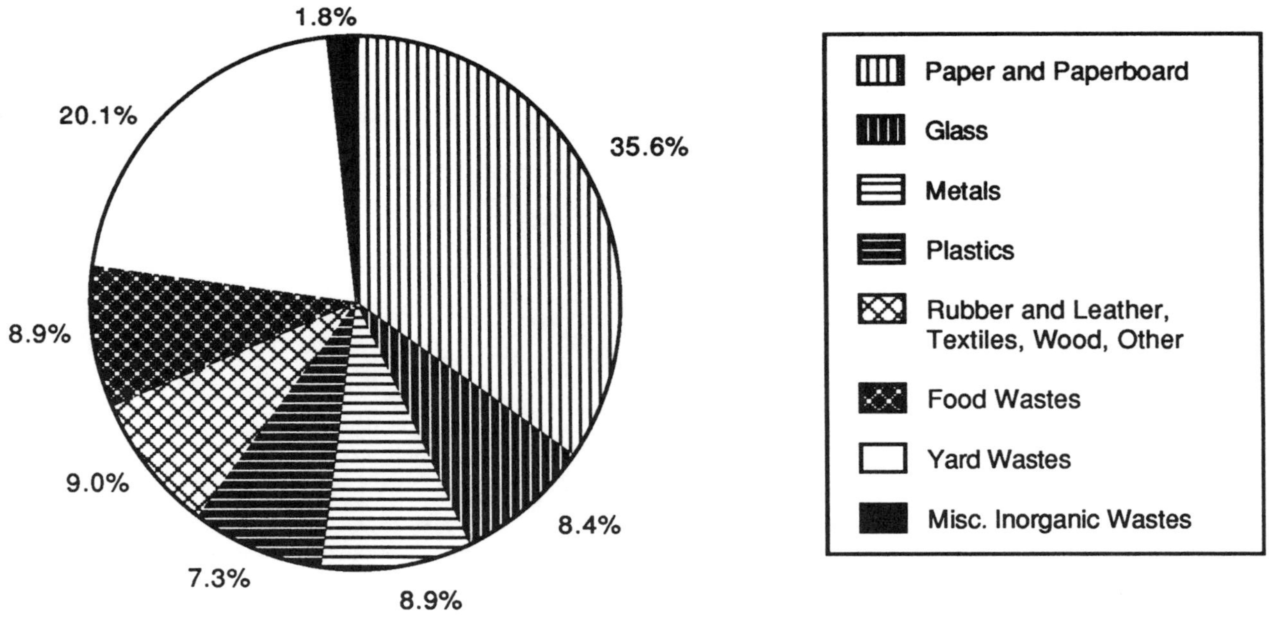

Source: U.S. Environmental Protection Agency, 1988, Characterization of Municipal Solid Waste in the United States 1960 to 2000 (Update 1988), PB88–232780. Based on study by Franklin Associates, Ltd

TABLE 7–102. GROSS DISCARDS, RECOVERY, AND NET DISCARDS OF PRODUCTS IN MUNICIPAL SOLID WASTE IN THE UNITED STATES, 1986

[In millions of tons and percent]

Products	Gross Discards		Materials Recovery		Net Discards[2]	
	Quantity	% of Gross Discards	Quantity	% of Gross Discards	Quantity	% of Net Discards
DURABLE GOODS						
Major Appliances	2.8	1.8	0.2	7.1	2.6	1.8
Rubber Tires	1.8	1.1	0.1	5.6	1.7	1.2
Other Durables	14.9	9.4	0.1	0.7	14.9	10.6
TOTAL DURABLES	19.5	12.4	0.4	2.1	19.1	13.6
NONDURABLE GOODS						
Newspapers	12.6	8.0	3.8	30.2	8.8	6.3
Books and Magazines	4.8	3.0	0.4	8.3	4.4	3.1
Office Papers	6.1	3.9	1.1	18.0	5.0	3.6
Commercial Printing	3.7	2.3	0.5	13.5	3.2	2.3
Other Nonpackaging Paper	8.5	5.4	0.2	2.4	8.3	5.9
Other Miscellaneous Nondurables	3.8	3.7	0.0	0.0	5.8	4.1
TOTAL NONDURABLE GOODS	41.5	26.3	6.0	14.5	35.5	25.2
CONTAINERS AND PACKAGING						
Glass Containers						
Beer & Soft Drink	5.5	3.5	1.1	20.0	4.4	3.1
Other Glass Containers	6.3	4.0	Neg.	0.0	6.3	4.5
Subtotal - Glass	11.8	7.5	1.1	9.3	10.7	7.6
Steel Containers						
Beer & Soft Drink Cans	0.1	0.1	0.0	0.0	0.1	0.1
Food Cans	1.8	1.1	0.1	5.6	1.7	1.2
Other Steel Packaging	0.9	0.6	Neg.	0.0	0.9	0.6
Subtotal - Steel	2.8	1.8	0.1	3.6	2.7	1.9
Aluminum						
Beer and Soft Drink Cans	1.3	0.8	0.6	46.2	0.7	0.5
Other Aluminum Packaging	0.4	0.3	Neg.	0.0	0.4	0.3
Subtotal - Aluminum	1.7	1.1	0.6	35.3	1.0	0.7
Paper and Paperboard						
Corrugated Containers	19.4	12.3	8.0	41.2	11.4	8.1
Other Paperboard	5.4	3.4	0.3	5.6	5.1	3.6
Paper Packaging	4.2	2.7	0.3	7.1	3.9	2.8
Subtotal - Paper	29.0	18.4	8.6	29.7	20.4	14.5
Plastics						
Plastic Containers	2.9	1.8	0.1	3.4	2.8	2.0
Other Plastic Packaging	2.8	1.8	Neg.	0.0	2.8	2.0
Subtotal - Plastics	5.7	3.6	0.1	1.8	5.6	4.0
Wood Packaging	2.1	1.3	Neg.	0.0	2.1	1.5
Other Miscellaneous Packaging	0.2	0.1	Neg.	0.0	0.2	0.1
TOTAL CONTAINERS AND PACKAGING	53.3	33.8	10.6	19.9	42.7	30.3
TOTAL NONFOOD PRODUCT WASTE	114.3	72.5	17.0	14.9	97.4	69.2
OTHER WASTES[1]						
Food Wastes	12.5	7.9	Neg.	0.0	12.5	8.9
Yard Wastes	28.3	17.9	Neg.	0.0	28.3	20.1
Miscellaneous Inorganic Wastes	2.6	1.6	Neg.	0.0	2.6	1.8
GRAND TOTAL	157.7	100.0	17.0	10.8	140.8	100.0

[1] Some of these wastes are composted or otherwise recovered, but this is not estimated here.

[2] Discards after materials recovery and before energy recovery.

Neg. - Less than 100,000 tons.

Details may not add to totals due to rounding.

Source: U.S. Environmental Protection Agency, 1988, Characterization of Municipal Solid Waste in the United States 1960 to 2000 (Update 1988), PB88–232780. Based on study by Franklin Associates, Ltd

TABLE 7–103. COMPOSITION OF MUNICIPAL SOLID WASTE DISCARDS BY ORGANIC AND INORGANIC FRACTIONS IN THE UNITED STATES, 1960 TO 2000

[In percent of total net discards; Discards after materials recovery has taken place, and before energy recovery]

Year	Organics	Inorganics
1960	77.8	22.3
1965	78.3	21.7
1970	75.2	24.8
1975	75.5	24.5
1980	77.1	22.9
1981	77.5	22.5
1982	77.8	22.2
1983	78.7	21.3
1984	79.6	20.4
1985	80.4	19.6
1986	80.8	19.2
1990[a]	80.8	19.2
1995[a]	81.7	18.3
2000[a]	82.5	17.5

[a] Estimate

Source: U.S. Environmental Protection Agency, 1988, Characterization of Municipal Solid Waste in the United States 1960 to 2000 (Update 1988), PB88–232780. Based on study by Franklin Associates, Ltd

TABLE 7–104. CHARACTERISTICS OF LANDFILL LEACHATE AND DOMESTIC WASTEWATER

Constituent	Range[a] (mg/L)	Range[b] (mg/L)	Range[c] (mg/L)	Leachate[d] Fresh	Leachate[d] Old	Typical[d] Domestic Wastewater[d]	Ratio Between Leachate and Domestic Wastewater[d]
Chloride (Cl)	34–2,800	100–2,400	600–800	742	197	50	15
Iron (Fe)	0.2–5,500	200–1,700	210–325	500	1.5	0.1	5,000
Manganese (Mn)	.06–1,400	—	75–125	49	—	0.1	490
Zinc (Zn)	0–1,000	1–135	10–30	45	0.16	—	—
Magnesium (Mg)	16.5–15,600	—	160–250	277	81	30	9
Calcium (Ca)	5–4,080	—	900–1,700	2,136	254	50	43
Potassium (K)	2.8–3,770	—	295–310	—	—	—	—
Sodium (Na)	0–7,700	100–3,800	450–500	—	—	—	—
Phosphate (P)	0–154	5–130	—	7.35	4.96	10	0.7
Copper (Cu)	0–9.9	—	0.5	0.5	0.1	—	—
Lead (Pb)	0–5.0	—	1.6	—	—	—	—
Cadmium (Cd)	—	—	0.4	—	—	—	—
Sulfate (SO_4)	1–1,826	25–500	400–650	—	—	—	—
Total N	0–1,416	20–500	—	989	7.51	40	25
Conductivity (μmhos)	—	—	6,000–9,000	9,200	1,400	700	13
TDS	0–42,276	—	10,000–14,000	12,620	1,144	—	—
TSS	6–2,685	—	100–700	327	266	200	1.6
pH	3.7–8.5	4.0–8.5	5.2–6.4	5.2	7.3	8.0	—
Alkalinity as $CaCO_3$	0–20,850	—	800–4,000	—	—	—	—
Hardness, total	0–22,800	200–5,250	3,500–5,000	—	—	—	—
BOD_5	9–54,610	—	7,500–10,000	14,950	—	200	75
COD	0–89,520	100–51,000	16,000–22,000	22,650	81	500	45

[a] Office of Solid Waste Management Programs, Hazardous Waste Management Division. An Environmental Assessment of Potential Gas and Leachate Problems at Land Disposal Sites. Environmental Protection Publication SW-110 of. [Cincinnati], U.S. Environmental Protection Agency, 1973.

[b] Steiner, R. C., A. A. Fungaroli, R. J. Schoenberger, and P. W. Purdom. Criteria for Sanitary Landfill Development. Public Works, 102(2): 77–79, Mar. 1971.

[c] Gas and Leachate from Land Disposal of Municipal Solid Waste; Summary Report. Cincinnati, U.S. Environmental Protection Agency, Municipal Environmental Research Laboratory, 1975.

[d] Brunner, D. R., and R. A. Carnes. Characteristics of Percolate of Solid and Hazardous Waste Deposits. Presented at AWWA American Water Works Association 94th Annual Conference, June 17, 1974. Boston, Massachuetes. 23 p.

Source: U.S. Environmental Protection Agency, 1977, Procedures Manual for Ground Water Monitoring at Solid Waste Disposal Sites, EPA/530/SW-611

TABLE 7–105. COMPOSITION OF MUNICIPAL SOLID WASTE IN THE UNITED STATES, 1960 TO 2000

[In millions of tons]

Materials	1960	1965	1970	1975	1980	1981	1982	1983	1984	1985	1986	1990	1995	2000
Paper and Paperboard	24.5	32.2	36.5	34.4	42.0	43.6	41.4	45.8	49.4	48.7	50.1	54.9	60.2	66.0
Glass	6.4	8.5	12.5	13.2	14.2	14.3	11.8	13.3	12.8	12.2	11.8	12.3	12.2	12.0
Metals														
Ferrous	9.9	10.0	12.4	12.0	11.2	11.1	11.0	11.1	11.0	10.4	10.6	11.1	11.3	11.3
Aluminum	0.4	0.5	0.8	1.0	1.4	1.4	1.3	1.5	1.5	1.6	1.7	2.0	2.4	2.7
Other Nonferrous	0.2	0.2	0.3	0.3	0.4	0.4	0.3	0.3	0.3	0.3	0.3	0.3	0.3	0.4
Plastics	0.4	1.4	3.0	4.4	7.6	7.8	8.4	9.1	9.6	9.7	10.3	11.8	13.7	15.6
Rubber and Leather	1.7	2.2	3.0	3.7	4.1	4.1	3.8	3.4	3.3	3.4	3.9	3.5	3.6	3.8
Textiles	1.7	1.9	2.0	2.2	2.6	3.4	2.8	2.8	2.8	2.8	2.8	3.0	3.1	3.3
Wood	3.0	3.5	4.0	4.4	4.9	4.4	5.0	5.2	5.1	5.4	5.8	5.3	5.7	6.1
Other	0.0	0.0	0.1	0.1	0.1	0.1	0.1	0.1	0.1	0.1	0.1	0.1	0.1	0.1
TOTAL NONFOOD PRODUCT WASTES	48.2	60.5	74.7	75.6	88.6	90.5	87.8	92.6	95.9	94.5	97.4	104.2	112.5	121.3
Food Wastes	12.2	12.4	12.8	13.4	11.9	12.1	12.0	12.0	12.2	12.3	12.5	12.5	12.4	12.3
Yard Wastes	20.0	21.6	23.2	25.2	26.5	26.7	27.0	27.5	27.8	28.0	28.3	29.5	31.0	32.0
Miscellaneous Inorganic Wastes	1.3	1.6	1.8	2.0	2.2	2.3	2.4	2.4	2.4	2.5	2.6	2.8	3.0	3.2
TOTAL WASTES DISCARDED[1]	81.7	96.1	112.5	116.2	129.2	131.6	129.1	134.5	138.3	137.3	140.8	149.0	158.9	168.8
ENERGY RECOVERY[2]	0.0	0.2	0.4	0.7	2.7	2.3	3.5	5.0	6.5	7.6	9.6	13.3	22.5	32.0
NET WASTES DISCARDED	81.7	95.9	112.1	115.5	126.5	129.3	125.6	129.5	131.8	129.7	131.2	135.7	136.4	136.8

[1] Wastes discarded after materials recovery has taken place.
[2] Municipal solid waste consumed for energy recovery. Does not include residues.
Details may not add to totals due to rounding.
Source: U.S. Environmental Protection Agency, 1988, Characterization of Municipal Solid Waste in the United States 1960 to 2000 (Update 1988), PB88–232780. Based on a study prepared by Franklin Associates, Ltd.

TABLE 7–106. SUMMARY DATA ON SOLID WASTE FACILITIES IN THE UNITED STATES

Percent of uncontrolled sites that are solid waste facilities:	
Of 1,389 sites with actual or presumed problems of releases of hazardous substances	18%
Of 550 sites on National Priority List	20%
Two most prevalent effects at problem solid waste sites:	
Leachate migration, groundwater pollution: at 89% of sites	
Drinking water contamination: at 49% of sites	
Mean size of problem solid wastes sites	67.4 acres
Median hazard ranking score:[a]	
Solid waste sites on the NPL	40.8
All NPL sites	42.2
Estimates for national number of solid waste sites:	
Operating sanitary, municipal landfills	14,000
Closed sanitary, municipal landfills	42,000
Operating industrial landfills	75,000
Closed industrial landfills	150,000
Operating surface impoundments	170,000
Closed surface impoundments	170,000
Total	621,000
Estimate of need for future cleanup:	
Low: 5% landfills, 1% impoundments likely to release toxic substances	17,400
High: 10% landfills, 2% impoundments likely to release toxic substances	34,800
Conservative figure used for cleanup by Superfund	5,000

[a] 28.5 required for placement on National Priorities List; current highest site score is 75.6.
Source: Office of Technology Assessment, 1985

TABLE 7–107. NUMBER OF SOLID WASTE FACILITIES IN THE UNITED STATES

[By state and USEPA Region; non-hazardous waste facilities governed by Subtitle D of the Resource Conservation and Recovery Act; some landfills, however, may contain hazardous materials]

State	Number of All Landfills				
	1981	1982	1983[b]	Number of Open Dumps[a,b]	Open Dumps to Upgrade[a]
Region 1:					
Connecticut	170	155	151	36	24
Maine	336	328	308	45	NA
Massachusetts	286	273	283	81	NA
New Hampshire	450	250	101	26	0
Rhode Island	35	22	18	4	1
Vermont	73	85	92	4	0
Region 2:					
New Jersey	240	185	185	5	1
New York	641	525	525	56	38
Puerto Rico	68	NA	NA	NA	NA
Region 3:					
Delaware	NA	4	35	4	4
District of Columbia	2	3	NA	NA	NA
Maryland	NA	64	47	0	0
Pennsylvania	1,400	847	925	94	75
Virginia	NA	250	209	50	34
West Virginia	228	127	127	41	36
Region 4:					
Alabama	146	135	135	12	11
Florida	209	214	248	55	17
Georgia	517	299	284	6	4
Kentucky	210	112	128	34	NA
Mississippi	286	120	253	133	10
North Carolina	170	225	167	1	0
South Carolina	128	78	225	0	0
Tennessee	134	160	161	6	2
Region 5:					
Illinois	260	450	329	42	0
Indiana	149	129	348	191	2
Michigan	470	NA	362	150	0
Minnesota	371	105	185	60	0
Ohio	680	235	318	54	NA
Wisconsin	1,050	1,100	1,085	66	10
Region 6:					
Arkansas	490	141	311	78	NA
Louisiana	NA	532	532	532	95
New Mexico	228	231	228	0	0
Oklahoma	215	225	225	66	60
Texas	250	1,043	1,075	11	8
Region 7:					
Iowa	NA	95	94	0	0
Kansas	243	220	224	1	0
Missouri	107	108	128	2	1
Nebraska	277	500	400	2	0
Region 8:					
Colorado	205	206	206	32	26
Montana	221	250	222	16	13
North Dakota	84	97	130	0	0
South Dakota	30	NA	200	140	5
Utah	300	290	296	26	8
Wyoming	NA	86	210	0	0
Region 9:					
Arizona	130	122	116	28	27
California	450	443	542	40	31
Hawaii	24	NA	25	9	4
Nevada	114	120	99	52	10
Region 10:					
Alaska	80	NA	NA	NA	NA
Idaho	220	130	132	42	20
Oregon	226	249	226	28	3
Washington	NA	136	136	36	18
Guam	3	NA	NA	NA	NA
Total	12,606	11,704	12,991	2,396	598

[a] Data for 1983.
[b] There may be some overlap between these column entries.
Source: Congress of the United States, Office of Technology Assessment, 1985, Superfund Strategy

TABLE 7–108. SOLID WASTE DISPOSAL BY SELECTED INDUSTRIES IN THE UNITED STATES, 1975–1983

[In millions of short tons. Excludes recovered materials. Data include both wet and dry weight figures. Excludes apparel and other textile, and, beginning 1978, establishments with less than 20 employees]

Industry Group	1975	1976	1977	1978	1979	1980	1981	1982	1983		
									Total	Hazardous Waste[1])	Nonhazardous Waste
All industries[2])	**139.1**	**156.8**	**160.0**	**160.8**	**163.7**	**149.9**	**145.8**	**99.4**	**89.0**	**8.0**	**81.0**
Food	12.6	15.0	13.1	13.4	14.0	14.4	13.2	9.8	9.6	.2	9.4
Lumber and wood	8.1	9.3	6.3	6.7	6.5	5.9	6.4	3.7	4.0	(Z)	4.0
Paper	9.1	10.1	10.6	10.9	13.3	12.3	11.3	11.5	13.7	.1	13.5
Chemicals	38.7	50.3	55.7	48.8	45.4	43.4	43.7	36.1	18.8	3.6	15.2
Petroleum	2.0	2.6	2.9	3.6	3.1	4.9	4.7	4.4	3.6	1.5	2.1
Stone, clay, glass	11.3	11.1	12.6	12.7	14.1	13.3	12.1	5.8	6.2	.2	5.9
Primary metal	42.7	42.4	41.7	46.1	47.8	37.5	36.0	16.7	17.7	1.0	16.7
Fabricated metals	1.9	2.1	2.0	2.0	2.0	1.9	1.8	1.4	2.0	.2	1.7
Machinery exc. electrical	2.7	3.1	3.6	3.4	3.5	3.0	2.8	1.6	1.8	.2	1.7
Electric equipment	1.5	1.5	1.5	1.8	2.3	2.1	1.7	1.3	1.7	.3	1.4
Transportation equipment	3.8	4.3	4.7	5.2	4.3	4.2	4.0	2.9	3.0	.3	2.7

Z: Less than 50,000 short tons.

[1] Covers waste, which because of its quantity, concentration, or physical, chemical, or infectious characteristics, may cause, or significantly contribute to an increase in serious irreversible, or incapacitating reversible illness; or pose a substantial present or potential hazard to human health or the environment when improperly treated, stored, transported, or disposed of or managed. See Resource Conservation and Recovery Act 1976, Public Law 94–580, for listing of hazardous wastes.

[2]) Includes industries not shown separately.

Source: U.S. Department of Commerce, Statistical Abstract of the United States 1987

FIGURE 7–7. HAZARDOUS WASTE SITES IN THE UNITED STATES IN 1986

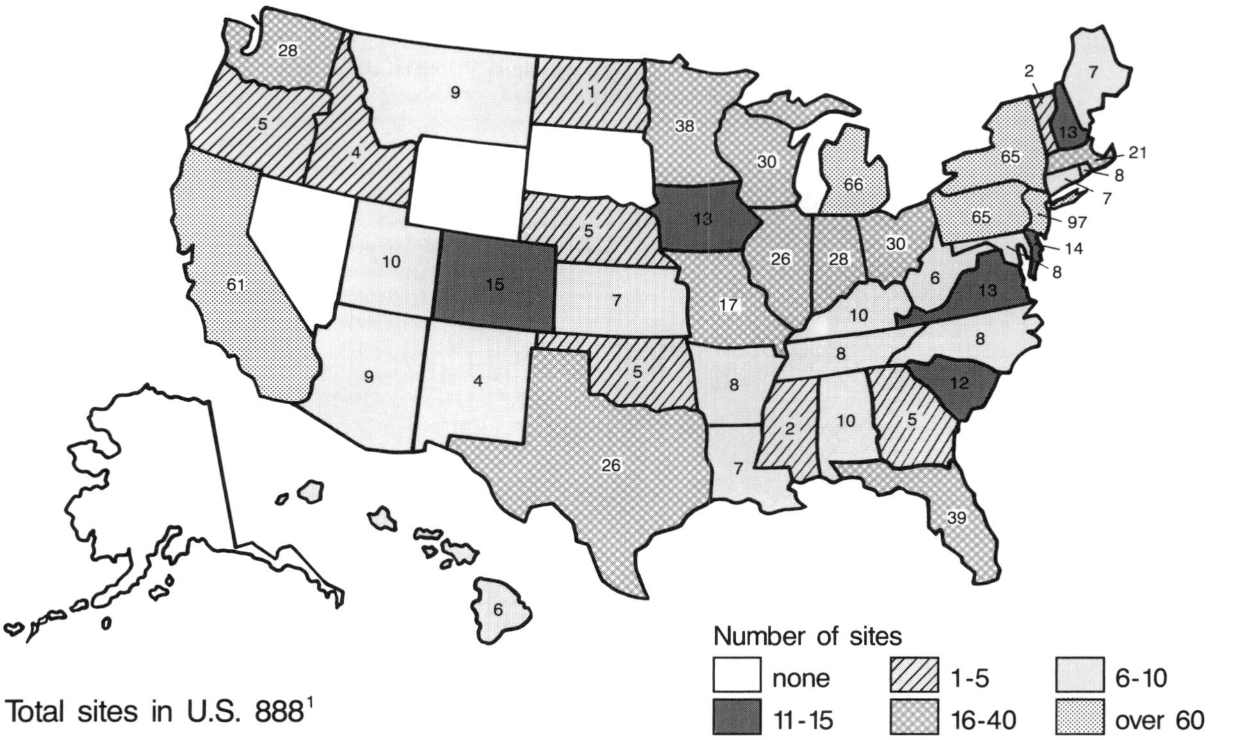

Total sites in U.S. 888[1]

Represents final and proposed sites on National Priority List.

[1]Includes eight in Puerto Rico; and one in Guam.

Source: U.S. Environmental Protection Agency, National Priorities List Fact Book, June 1986

TABLE 7–109. TYPES OF ACTIVITIES AT HAZARDOUS WASTE SITES IN THE UNITED STATES

[Includes 1177 final and proposed sites placed on the National Priorities List as of June 1988]

Activity	Final	Proposed	Total
Surface impoundments	295	137	432
Landfills, commercial/industrial	299	113	412
Containers/drums	229	64	293
Other manufacturing/industrial	102	137	239
Landfills, municipal	157	56	213
Spills	111	73	184
Chemical processing/manufacturing	82	78	160
Waste piles	73	46	119
Leaking containers	80	36	116
Tanks, above-ground	80	28	108
Tanks, below-ground	46	42	88
Ground-water plumes	63	12	75
Electroplating	36	27	63
Wood-preserving	39	16	55
Waste-oil processing	34	16	50
Ore processing/refining/smelting	27	9	36
Open burning	24	12	36
Solvent recovery	24	11	35
Outfall, surface water	20	15	35
Military ordnance production/storage/disposal	19	14	33
Military testing & maintenance	16	10	26
Landfarm, land treatment/spreading	18	7	25
Battery recycling	17	6	23
Incinerators	17	1	18
Mining sites, surface	11	4	15
Underground injection	11	2	13
Drum recycling	8	4	12
Sand and gravel pits	7	3	10
Mining sites, subsurface	6	3	9
Road oiling	7	1	8
Laundries/dry cleaners	2	5	7
Sinkholes	6	1	7
Explosive disposal/detonation	2	1	3
Tire storage/recycling	2	0	2
Total Sites:[1]	799	378	1177

[1] Since each site may have more than one activity, the number of activities is greater than the number of sites.
Source: U.S. Environmental Protection Agency, Office of Emergency Response, Washington, DC 20460

FIGURE 7–8. TYPES OF ACTIVITIES AT HAZARDOUS WASTE SITES IN THE UNITED STATES

[Percent of 1177 final and proposed sites on the National Priorities List as of June 1988; a site may have more than one type of activity]

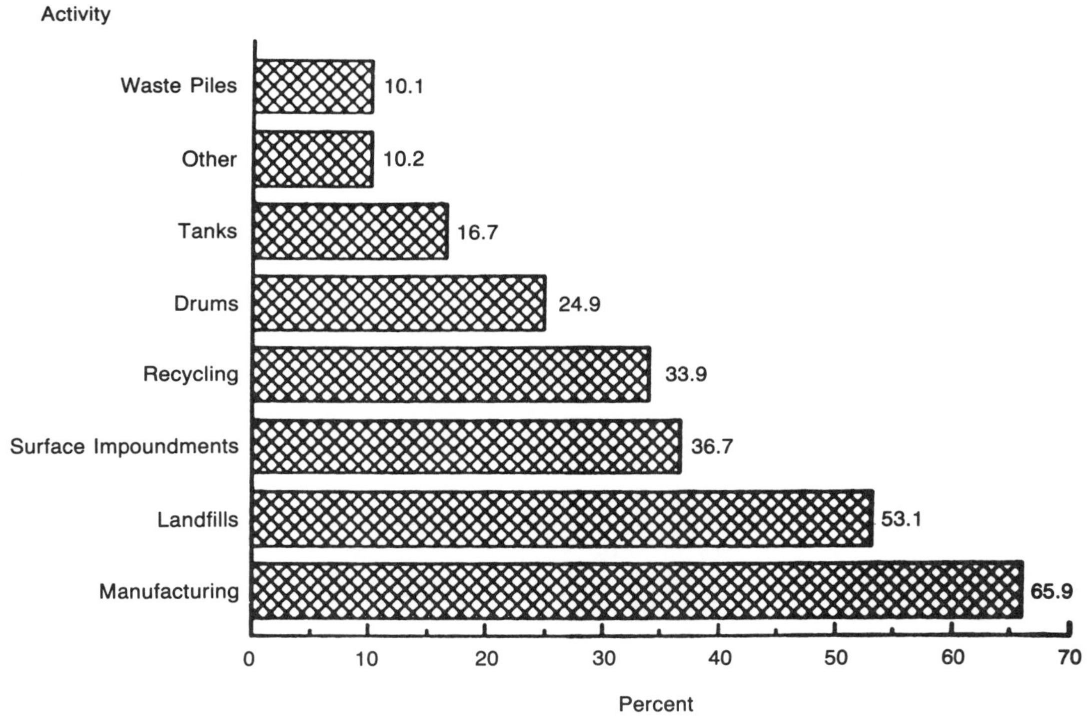

Source: U.S. Environmental Protection Agency, Office of Emergency Response, Washington DC 20460

FIGURE 7–9. OBSERVED CONTAMINATION AT HAZARDOUS WASTE SITES IN THE UNITED STATES

[Percent of 1177 final and proposed sites on the National Priorities List as of June 1988; a site may have more than one type of contamination]

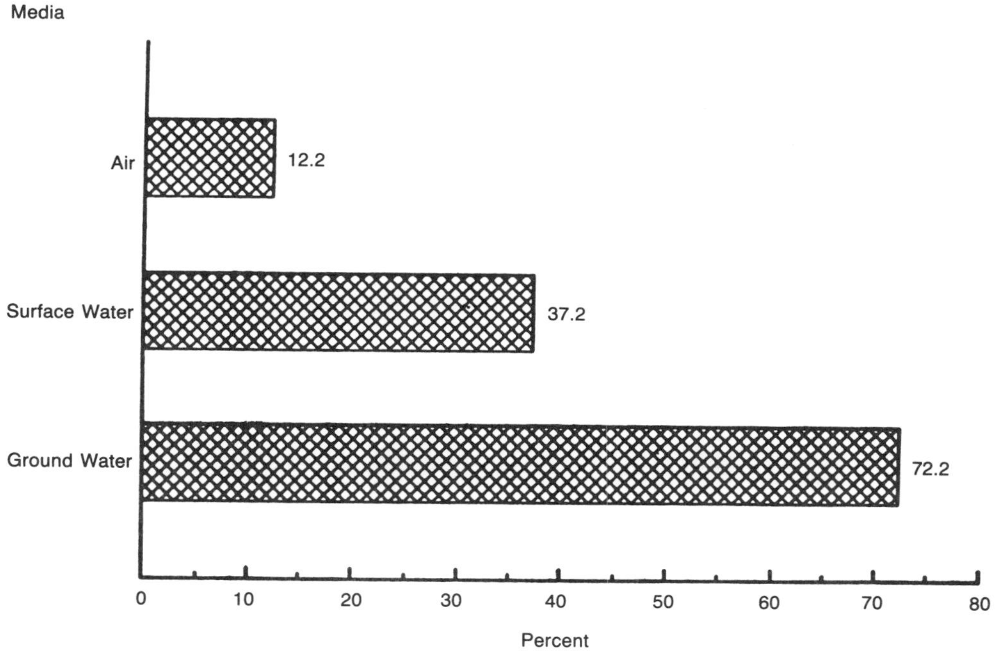

Source: U.S. Environmental Protection Agency, Office of Emergency Response, Washington DC 20460

TABLE 7–110. HAZARDOUS WASTE SITES IN THE UNITED STATES— NATIONAL PRIORITIES LIST

[Final and proposed sites per state/territory—by total sites—June 1988]

State/Territory	Final		Proposed		Total
	Non-Federal	Federal	Non-Federal	Federal	
New Jersey	94	2	13	1	110
Pennsylvania	60	1	34	2	97
California	40	8	37	3	88
Michigan	58	0	23	0	81
New York	62	1	13	0	76
Florida	34	0	17	0	51
Washington	20	3	12	8	43
Minnesota	39	1	0	0	40
Illinois	15	2	20	2	39
Wisconsin	32	0	7	0	39
Indiana	24	0	13	0	37
Ohio	28	0	3	1	32
Texas	20	1	6	1	28
Iowa	7	0	16	0	23
Massachusetts	21	0	1	0	22
Virginia	10	1	11	0	22
Delaware	12	0	8	1	21
Missouri	12	2	7	0	21
North Carolina	9	0	11	1	21
South Carolina	12	0	9	0	21
Kentucky	10	0	7	0	17
Colorado	12	1	2	1	16
New Hampshire	13	0	2	0	15
Connecticut	7	0	7	0	14
Georgia	3	1	9	0	13
Tennessee	8	1	4	0	13
Alabama	8	1	2	1	12
Kansas	7	0	4	0	11
Louisiana	6	0	4	1	11
Oklahoma	5	1	5	0	11
Utah	3	2	5	1	11
Arkansas	9	0	1	0	10
Maryland	6	0	2	2	10
Montana	8	0	2	0	10
New Mexico	4	0	4	2	10

TABLE 7–110. HAZARDOUS WASTE SITES IN THE UNITED STATES—NATIONAL PRIORITIES LIST (continued)

[Final and proposed sites per state/territory—by total sites—June 1988]

State/Territory	Final Non-Federal	Final Federal	Proposed Non-Federal	Proposed Federal	Total
Arizona	5	0	3	1	9
Puerto Rico	8	0	0	1	9
Rhode Island	8	0	1	0	9
Maine	5	1	2	0	8
Vermont	2	0	6	0	8
Oregon	4	1	2	0	7
Hawaii	0	0	6	0	6
West Virginia	5	0	1	0	6
Nebraska	2	1	2	0	5
Idaho	4	0	0	0	4
Mississippi	2	0	1	0	3
North Dakota	1	0	1	0	2
Wyoming	1	0	1	0	2
Alaska	0	0	1	0	1
Guam	1	0	0	0	1
South Dakota	1	0	0	0	1
American Samoa	0	0	0	0	0
Commonwealth of Marianas	0	0	0	0	0
District of Columbia	0	0	0	0	0
Nevada	0	0	0	0	0
Trust Territories	0	0	0	0	0
Virgin Islands	0	0	0	0	0
Total	767	32	348	30	1177

Source: U.S. Environmental Protection Agency, Office of Emergency Response, Washington, DC 20460

TABLE 7–111. FUTURE USE OF CONTAINMENT TECHNOLOGIES FOR CLEANUP OF HAZARDOUS WASTE SITES IN THE UNITED STATES

Technique	Applicability	Effectiveness	Confidence	Capital Cost	Cap/O&M	Projected Level of Use
Barriers:						
Slurry wall	2	1	2	2	1	Extensive
Grout curtain	2–3	1	2	2–3	1	Limited
Vibrating beam	2	1	2–3	2–3	1	Moderate
Sheet pile	3	1–2	2	2–3	1	Nil-Limited
Block displacement	3	1	4	3	1	Nil
Hydraulic controls (wells)	2	1,3	1	1	3	Extensive
Subsurface drains	2	1	2	1	2	Moderate
Runon/runoff controls	1	3	1	1	2	Extensive
Surface seals and caps	1	2,3	2	1	1	Extensive
Solidification, etc.	2	1,3	3–4	2	1	Moderate-Limited

KEY:
Applicability:
1 = Very broadly applicable; little or no site dependency
2 = Broadly applicable; some sites unfavorable
3 = Limited to sites of specific characteristics
Effectiveness:
1 = Can produce "leak-tight" containment
2 = Can reduce migration—some leakage likely
3 = Used as supporting technique in conjunction with other elements
Confidence:
1 = Well proven—long-term effectiveness—high
2 = Well proven—long-term effectiveness—unknown
3 = Limited experience; used in other applications
4 = Developmental; little data
Capital cost for function provided:
1 = Low
2 = Normal
3 = High
Capital to operation and maintenance (O&M) cost ratio:
1 = Capital higher than O&M
2 = Capital about same as O&M
3 = Capital lower than O&M
Source: U.S. Congress, Office of Technology Assessment, 1985, Superfund Strategy. Original source: A. D. Little, "Evaluation of Available Cleanup Technologies for Uncontrolled Sites," contractor report prepared for the Office of Technology Assessment, Nov. 15, 1984

TABLE 7–112. FUTURE USE OF TREATMENT TECHNOLOGIES FOR CLEANUP OF HAZARDOUS WASTE SITES IN THE UNITED STATES

Technique	Applicability	Effectiveness	Confidence	Capital Cost	Cap/O&M	Secondary Disposal	Projected Level Use of
Biological treatment	Or, 1–2	2	1	1	1–2	3	Moderate
Chemical treatment:							
Neutralization/precipitation	In, 1	1	1	1	2	4	Moderate-Extensive
Wet air oxidation	Or, 2	2	2	3	1–2	1	Limited
Chlorination	In, 3	1	2	2	2	1	Limited
Ozonation	Or, 3	2	3	3	2–3	2	Nil
Reduction (Cr)	In, 3	1	2	2	2	3	Limited
Physical treatment:							
Carbon adsorption	Or, In, 1	1	1	2	2–3	2–3	Moderate-Extensive
Sedimentation/filtration	Or, In, 1	1	1	1	2–3	4	Moderate-Extensive
Stripping	Or, 2	1	1	1	2	4	Moderate
Flotation	Or, 2	2	1	1	1	4	Limited
Ion exchange	In, 3	1–3	3	3	3	4	Nil
Reverse osmosis	Or, In, 3	1–2	3	3	3	4	Nil
Gas stream controls:							
Thermal oxidation	Or, 1	1	1	3	3	1	Limited-Moderate
Carbon adsorption	Or, 1	1	1	3	2–3	2–3	Limited-Moderate
Incineration							
Onsite	Or, 1	1	2	3	1	3[a]	Limited
Offsite	Or, 1	1	1	3	NA	3[a]	Moderate
In situ biodegradation	Or, 3	2	3	2	3	1	Limited

NOTES:
[a] Must dispose solid residues.
[b] Depends on reactive material used.

KEY:
Applicability:
Class:
Or = Organic compounds
In = Inorganic compounds
Range:
1 = Broadly applicable to compounds in indicated class
2 = Moderated applicable: depends on waste composition concentration
3 = Limited to special situations
Effectiveness:
1 = Highest levels available
2 = Output may need further treatment; may have pockets untreated (in-situ)
Confidence:
1 = Well proven—easily transferable to site cleanup
2 = Well proven—but not in clean-up settings
3 = Limited experience
4 = Developmental; little data
Capital cost for function provided:
1 = Low
2 = Normal
3 = High
Capital to operations and maintenance (O&M) cost basis:
1 = Capital higher than O&M
2 = Capital about the same
3 = Capital lower than O&M
Secondary treatment or disposal:
1 = None
2 = Minor
3 = Major, but does not require hazardous waste techniques.
4 = Basically a separation process; must be used with subsequent hazardous waste treatment or secure disposal step.

Source: U.S. Congress, Office of Technology Assessment, 1985, Superfund Strategy. Original source: A. D. Little, "Evaluation of Available Cleanup Technologies for Uncontrolled Sites," contractor report prepared for the Office of Technology Assessment, Nov. 15, 1984

TABLE 7–113. COST ESTIMATES FOR CLEANING UP UNCONTROLLED HAZARDOUS WASTE SITES IN THE UNITED STATES

[According to U.S. Congress, Office of Technology Assessment "Superfund Strategy", April 1985]

EPA (1984)[1]	EPA (1983)[2]	GAO[3]	Department of Commerce[4]	ASTSWMO[5]	CMA[6]	National Audubon Society[7]
Number of sites requiring cleanup:						
1,500–2,500	1,400–2,200 23–56% require groundwater response	1,270–2,546 23–56% require groundwater response	546 NPL 1,250 non-NPL 41 municipal	7,113 (43 States surveyed) 1,500 most serious	1,000 (27 States surveyed) 3,681 (potential)	2,200–7,000 38–56% require groundwater response
Total average cleanup costs per site (million):						
$6.7–$13.3	$6–$12 including groundwater response	$2.25–$6.75 constr. $5.25–$15.75 constr. including groundwater response $1.5-other costs	$9.7 NPL $3.2 non-NPL $30 municipal[d]	$1–$6 $6 serious sites	$4–$7 studies, removal, and containment $17–$30 studies, removal, containment, and groundwater response	$8 including O&M $17 including ground-water response
Total costs (unadjusted) (billion):						
$10.0–$33.3	$10.3–$20.6	$5.6–$33.8	$10.5	$14.6–$42.7	NA	$8–$92
Total costs to Fund (billion):						
$7.6–$22.7[a]	$8.4–$16[b]	$5.3–$26[c]	($1.5 surplus)-$1.5[f]	NA	$4.5[e]	NA
Projected years to clean sites:						
NA	14 for 1800 sites	NA	10–15	16–23 if constrained by personnel 28–90 if constrained financially	NA	17–26 for 2,200 sites 53–84 for 7,000 sites

NOTES:

[a] Assumes 40 to 60 percent of sites cleaned by Principal Responsible Parties (PRPs); Federal cost share is 90 percent; cost recovery is 47 percent for removals and 30 percent for remedial actions; 8.5 percent interest earned quarterly on previous year's balance; and 6.5 percent inflation on removal actions, assumed 190 per year at $75 million per year.

[b] PRP lead actions deducted.

[c] Assumes RPRs clean 29 to 44 percent of sites.

[d] Annual O&M costs are: $31,500, $20,900, and $117,600 for NPL, non-NPL, and municipal sites, respectively.

[e] Statement of E. C. Holmer, on behalf of the Chemical Manufacturers Association, June 13, 1984. (This estimate assumes no groundwater cleanup. Also might include estimate that only 10 percent are orphan sites.)

[f] Assumes 45 to 55 percent cost recovery; 1 to 15 percent of sites cleaned by PRPs; 6.5 percent annual construction inflation; 5 percent annual general inflation; and 8.5 percent annual interest on cash balances.

[g] Low estimates reflect $1.2 billion per year budget; high estimates are for $1.5 billion per year budget.

SOURCES:

(1) U.S. Environmental Protection Agency, "Extent of the Hazardous Release Problem and Future Funding Needs, CERCLA Section 301(a)(1)(c) Study," December 1984.
(2) U.S. Environmental Protection Agency, Superfund Task Force Preliminary Assessment, December 1983.
(3) U.S. General Accounting Office, EPA's Preliminary Estimates of Future Hazardous Waste Cleanup Costs Are Uncertain, GAO/RCED-84–152, May 7, 1984.
(4) U.S. Department of Commerce, "Estimated Costs and Expenditures for Cleanup of the Nation's Uncontrolled Hazardous Waste Sites" (draft), Feb. 22, 1984.
(5) Association of State and Territorial Solid Waste Management Officials, "State Cleanup Programs for Hazardous Substance Sites and Spills," Dec. 21, 1984.
(6) Arthur D. Little, Inc., Report to the Chemical Manufacturers' Association, "An Analysis of the Number of Inactive Hazardous Waste Sites That Will Use Superfund," July 1983.
(7) National Audubon Society, Testimony of Leslie Dach before the House Subcommittee on Commerce, Transportation, and Tourism, Mar. 1, 1984.

SECTION I. AGRICULTURAL ACTIVITIES/FERTILIZERS AND PESTICIDES

TABLE 7–114. TRENDS IN REGIONAL AGRICULTURAL ACTIVITY AND SOIL LOSS, 1975 AND 2000

[High growth scenario]

	Acres in Crop Production				Soil Loss				
	1975		2000		1975		2000		
Region	Quantity (10^6 Acres)	Percent of National Total	Percent of 1975 Value	Percent of National Total	Quantity (10^6 Tons)	Percent of National Total	Percent of 1975 Value	Percent of National Total	Annual Average Soil Loss Per Acre (Tons)
I. New England	[a]	[b]	129	[b]	2	[b]	126	[b]	13
II. New York-New Jersey	1	1	119	1	17	[b]	120	[b]	14
III. Middle Atlantic	5	2	127	2	180	5	127	5	36
IV. Southeast	27	12	121	12	900	24	119	24	33
V. Great Lakes	65	29	132	33	880	24	136	27	14
VI. South Central	31	14	108	13	590	16	114	15	20
VII. Central	58	26	107	24	960	26	113	24	18
VIII. Mountain	26	12	105	10	72	1	100	2	3
IX. West	4	2	140	2	10	[b]	144	[b]	3
X. Northwest	7	3	129	3	86	2	129	2	13
Total[c]	223	100	118	100	3,700	100	121	100	17

[a] Less than 0.5 million acres.
[b] Less than 0.5 percent.
[c] Rounding may create inconsistencies in addition.
Source: U.S. Environmental Protection Agency, 1980, Environmental Outlook 1980

FIGURE 7–10. SOURCES OF SEDIMENT DISCHARGE TO SURFACE WATERS IN THE UNITED STATES, 1977

[Estimated percent of sediment yield]

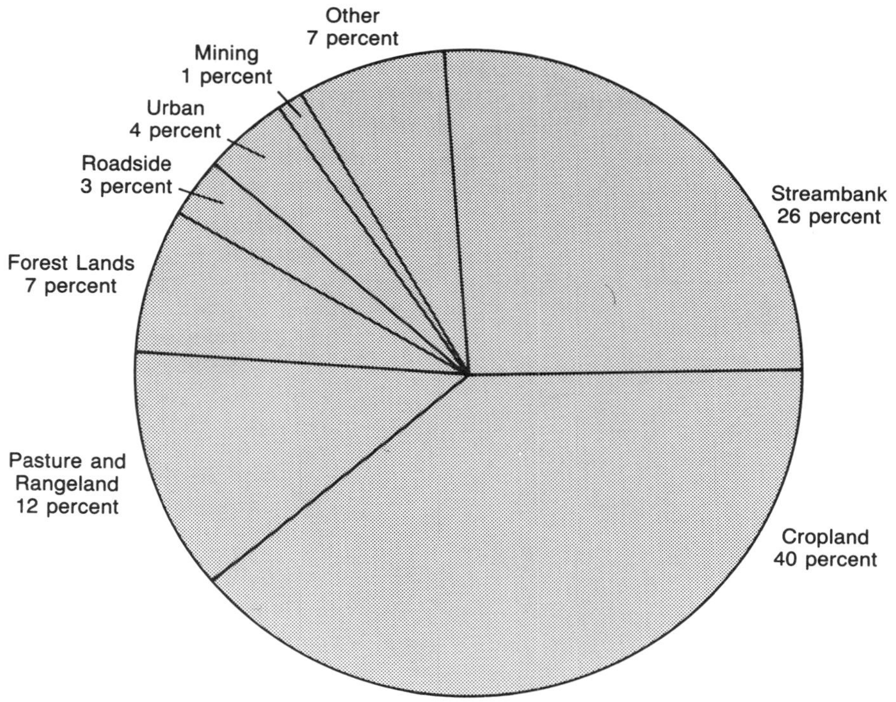

Source: U.S. Department of Agriculture, Soil Conservation Service, 1978, Environmental Impact Statement: Rural Clean Water Program

TABLE 7–115. REPRESENTATIVE VALUES FOR NUTRIENT EXPORT AND INPUT RATES FOR VARIOUS LAND USES IN THE UNITED STATES

[All values are medians and are only approximations owing to the highly variable nature of data on these rates]

Land Use	Total Phosphorus	Total Nitrogen
A. Export rates (kg/ha/yr)[1,2]		
Forest	0.2	2.5
Nonrow crops	0.7	6.0
Pasture	0.8	14.5
Mixed agriculture	1.1	5.0
Row crops	2.2	9.0
Feedlot, manure storage	255.0	2920.0
B. Total atmospheric input rates (kg/ha/yr)[1,2]		
Forest	0.26	6.5
Agricultural/rural	0.28	13.1
Urban industrial	1.01	21.4
C. Wastewater input rates (kg/capita/yr)[2]		
Septic tank input[3]	1.45	4.65

[1] Values in this table are all in kg/ha/yr, which is the standard for such measurements. To convert to pounds per acre per year, multiply by 0.892.
[2] Source: Reckhow et al. 1980.
[3] This is prior to absorption to soil during infiltration; generally, soils will absorb 80 percent or more of this phosphorus.
Source: U.S. Environmental Protection Agency, 1988, The Lake and Reservoir Restoration Guidance Manual, EPA 440/5–88–002

TABLE 7–116. CONTRIBUTION OF TOTAL NITROGEN AND PHOSPHOROUS FROM VARIABLE NONPOINT SOURCES

[In pounds per acre per year]

Source	Areal Loading Rates	
	Nitrogen (N)	Phosphorus (P)
Precipitation	5.0 – 9.8	0.04 – 0.05
Forest land	2.7 – 12	.03 – .8
Range land	——	.07 – .08
Crop land	.10 – 12	.05 – 2.7
Land receiving manure	3.6 – 12	0.7 – 2.7
Irrigation return flows:		
Surface	3.0 – 27	1.0 – 4.1
Subsurface	4.0 – 18	2.9 – 10
Urban land drainage	6.4 – 8.9	1.1 – 5.4
Animal feedlot runoff	90 – 1,400	8.9 – 630

Source: U.S. Geological Survey, 1984, National Water Summary 1983-Hydrologic Events and Issues, Water Supply Paper 2250; based on data from R.C. Loehr, 1974

TABLE 7–117. TRENDS IN GROSS NUTRIENT DISCHARGES IN AGRICULTURAL RUNOFF IN THE UNITED STATES, 1975 AND 2000

Pollutant	1975 (Tons)	2000 (Tons)	
		High Growth	Low Growth
Nitrogen	5,700	8,800	7,600
Phosphorus	3,000	5,700	4,900
Potassium	1,400	2,600	2,200
Total[a]	10,000	17,000	15,000

[a] Rounding may create inconsistencies in addition.
Source: U.S. Environmental Protection Agency, Environmental Outlook 1980

TABLE 7–118. TRENDS IN GROSS DISCHARGES OF SEDIMENT IN AGRICULTURAL RUNOFF IN THE UNITED STATES, 1975 AND 2000

Pollutant	1975 (10^3 Tons)	2000 (10^3 Tons)	
		High Growth	Low Growth
Total Suspended Solids	94,000	110,000	97,000
Total Dissolved Solids	40,000	49,000	42,000
Biochemical Oxygen Demand	630	760	650

Note: Conservative estimates; discharges may uniformly be too low by a factor of four or five.
Source: U.S. Environmental Protection Agency, Environmental Outlook 1980

TABLE 7–119. FARM FERTILIZER USE IN THE UNITED STATES, 1939 TO 1986

[Millions of tons of primary nutrients]

Year	Fertilizer Applied	Year	Fertilizer Applied	Year	Fertilizer Applied
1939	1.6	1955	6.1	1971	17.2
1940	1.8	1956	6.1	1972	17.2
1941	1.9	1957	6.4	1973	18.0
1942	2.1	1958	6.5	1974	19.3
1943	2.4	1959	7.4	1975	17.6
1944	2.6	1960	7.5	1976	20.8
1945	2.7	1961	7.8	1977	22.1
1946	3.1	1962	8.4	1978	20.6
1947	3.3	1963	9.5	1979	22.6
1948	3.6	1964	10.5	1980	23.1
1949	3.9	1965	11.0	1981	23.7
1950	4.1	1966	12.4	1982	21.4
1951	4.7	1967	14.0	1983	18.1
1952	5.2	1968	15.0	1984	21.8
1953	5.6	1969	15.5	1985	21.7
1954	5.9	1970	16.1	1986	19.6

Source: U.S. Department of Commerce, Bureau of the Census, 1976. Historical Statistics of the United States: Colonial Times to 1970, Series K 193. Washington, D.C.; U.S. Department of Commerce, Bureau of the Census. 1985. Statistical Abstracts of the United States: 1986, no. 1161, p. 654. Washington, D.C.; U.S. Department of Agriculture, Economic Research Service. 1987. Inputs Situation and Outlook Report. AR-5. Washington, D.C.; U.S. Geological Survey National Water Summary 1986

TABLE 7–120. NITROGEN FERTILIZER USE IN MAJOR AGRICULTURAL STATES,[a] 1982

State	Nitrogen Consumption (Thousand Metric Tons of Nutrients)[b]	Intensity of Fertilizer Use (Kilograms/Hectare)[c]
Alabama	111	74
Arizona	72	149
Arkansas	179	55
California	432	112
Colorado	126	34
Florida	209	174
Georgia	194	89
Idaho	170	71
Illinois	707	72
Indiana	376	73
Iowa	671	66
Kansas	477	41
Kentucky	144	67
Louisiana	106	52
Maryland	47	73
Michigan	186	59
Minnesota	368	43
Mississippi	137	54
Missouri	276	50
Montana	108	17
Nebraska	523	61
New Mexico	25	41
New York	74	40
North Carolina	188	90
North Dakota	197	17
Ohio	340	77
Oklahoma	224	52
Oregon	131	75
Pennsylvania	71	39
South Carolina	71	62
South Dakota	72	10
Tennessee	106	53
Texas	594	54
Utah	26	45
Virginia	73	58
Washington	210	67
Wisconsin	174	39
Wyoming	37	40

[a] Defined as states containing at least .4 million hectares (1 million acres) of cropland used for crops in 1982.
[b] To convert to short tons, multiply amount by 1.102.
[c] To convert to pounds per acre, multiply amount by .89236.
Source: U.S. Department of Agriculture, 1983, Commercial Fertilizer Consumption for Year Ending June 30, 1983

TABLE 7–121. PRODUCTION AND SALES OF SYNTHETIC ORGANIC PESTICIDES IN THE UNITED STATES, 1960 TO 1984

[Includes a small quantity of soil conditioners]

ITEM	Unit	1960	1965	1970	1975	1976	1977	1978	1979	1980	1981	1982	1983	1984
Production, total	Mil. lb	648	877	1,034	1,603	1,364	1,388	1,416	1,429	1,468	1,430	1,113	1,017	1,189
Herbicides	Mil. lb	102	263	404	788	656	674	664	657	806	839	623	570	716
Insecticides	Mil. lb	366	490	490	660	566	570	605	617	506	448	379	324	350
Fungicides	Mil. lb	179	124	140	155	142	143	147	155	156	143	111	123	123
Production value[1]	Mil. dol	307	577	1,058	2,900	2,880	3,116	3,342	3,685	4,269	5,136	4,331	3,993	5,056
Sales, total	Mil. lb	570	764	881	1,317	1,193	1,263	1,300	1,369	1,406	1,291	1,147	1,017	1,108
Sales value	Mil. dol	262	497	870	2,359	2,410	2,808	3,041	3,631	4,078	4,652	4,432	4,054	4,730

[1] Manufacturers unit value multiplied by production.
Source: U.S. Department of Commerce, Statistical Abstract of the United States 1987

FIGURE 7–11. USE OF PESTICIDES IN THE UNITED STATES, 1964–1984

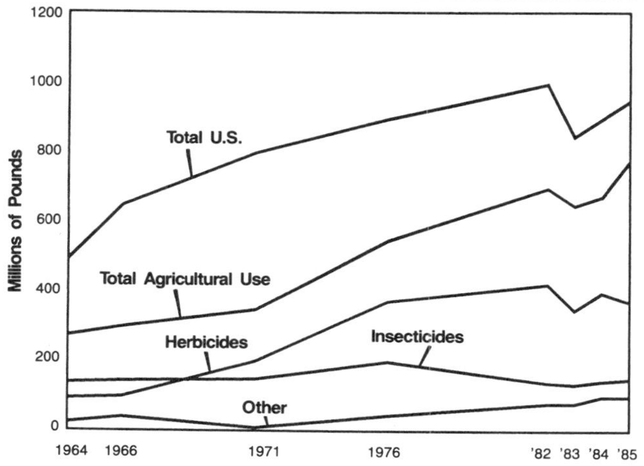

Note: Excludes wood preservatives, disinfectants, and sulfur.

Source: U.S. Environmental Protection Agency, Office of Pesticides and Toxic Substances, 1987, Agricultural Chemicals in Ground Water: Proposed Pesticide Strategy

TABLE 7–122. VOLUME OF PESTICIDES USED IN THE UNITED STATES IN 1985

[Millions of pounds of active ingredients]

	Herbicides[1]	Insecticides[2]	Fungicides[3]	Other[4]	Total
Agriculture	525	225	51	60	861.0
Ind./Comm./Govt.	115	40	21	.1	176.1
Home & Garden	30	35	12	.1	75.1
Total	670	300	84	60.2	1112.2

[1] Includes plant growth regulators.
[2] Includes miticides and contract nematicides.
[3] Does not include wood preservatives.
[4] Includes rodenticides, fumigants, and molluscides.
Source: U.S. Environmental Protection Agency, Office of Pesticides and Toxic Substances, 1987, Agricultural Chemicals in Ground Water: Proposed Pesticide Strategy

TABLE 7–123. PESTICIDE USE IN THE CONTERMINOUS UNITED STATES

[Active ingredients, by state]

State	Consumption (thousand lb)
Alabama	8,688
Arizona	4,481
Arkansas	20,357
California	31,159
Colorado	7,246
Connecticut	664
Delaware	1,887
District of Columbia	40
Florida	15,536
Georgia	18,035
Idaho	6,377
Illinois	64,818
Indiana	32,967
Iowa	74,214
Kansas	14,019
Kentucky	12,035
Louisiana	14,758
Maine	1,398
Maryland	5,030
Massachusetts	990
Michigan	17,827

TABLE 7–123. PESTICIDE USE IN THE CONTERMINOUS UNITED STATES (continued)

[Active ingredients, by state]

State	Consumption (thousand lb)
Minnesota	45,928
Mississippi	18,817
Missouri	17,874
Montana	5,243
Nebraska	31,340
Nevada	271
New Hampshire	249
New Jersey	2,258
New Mexico	1,541
New York	9,052
North Carolina	17,863
North Dakota	13,395
Ohio	31,926
Oklahoma	4,500
Oregon	5,270
Pennsylvania	10,538
Rhode Island	146
South Carolina	9,340
South Dakota	11,016
Tennessee	8,917
Texas	26,656
Utah	851
Vermont	238
Virginia	6,309
Washington	8,454
West Virginia	933
Wisconsin	18,692
Wyoming	1,019

Source: Gianessi, L.P., 1986, A National Pesticide Usage Data Base, U.S. Environmental Protection Agency, Office of Standards and Regulations. Resources for the Future, Washington, D.C.

TABLE 7–124. SELECTED CHARACTERISTICS AND USES OF PESTICIDES MONITORED BY THE U.S. GEOLOGICAL SURVEY-U.S. ENVIRONMENTAL PROTECTION AGENCY PESTICIDE MONITORING NETWORK, 1975 TO 1980

[µg/L, microgram per liter; lb/yr = pounds per year; nd, no available data; nr, none reported]

Chemical	Characteristics					Uses					
	Detection Limit[1] (µg/L)	Water-Quality Criteria[2] (µg/L) Human Health	Water-Quality Criteria[2] (µg/L) Aquatic Life	Solubility[3] (µg/L)	Relative Persistence within Pesticide Group[4]	Principal Uses and Sources	National Use on Farms[5] (million lb/yr) 1966	1971	1976	1982	Total use, 1981[6] (million lb/yr)
Organochlorine insecticides											
Aldrin	0.01	0.0007	0.002	13	Low	Corn	15	7.9	0.9	nr	0.8
							(Most farm uses cancelled 1974)				
Dieldrin	.03	.0007	.002	22	Medium	Termite control, degradation product of aldrin.	.7	.3	nr	nr	0
							(Most farm uses cancelled 1974)				
Chlordane	.15	.005	.004	56	High	Corn, termites, general purpose.	.5	1.9	nr	nr	9.6
							(Most farm uses cancelled 1974)				
DDD	.05	.0002	.001	5	High	Fruits and vegetables, degradation product of DDT.	2.9	.2	nr	nr	0
							(Cancelled 1972)				
DDE	.03	.0002	.001	10	High	Degradation product of DDT and DDD.	nr	nr	nr	nr	0
DDT	.05	.0002	.001	17	High	Cotton, fruits, vegetables, general purpose.	27	.1	nr	nr	0
							(Cancelled 1972)				

TABLE 7–124. SELECTED CHARACTERISTICS AND USES OF PESTICIDES MONITORED BY THE U.S. GEOLOGICAL SURVEY-U.S. ENVIRONMENTAL PROTECTION AGENCY PESTICIDE MONITORING NETWORK, 1975 TO 1980 (continued)

[µg/L, microgram per liter; lb/yr = pounds per year; nd, no available data; nr, none reported]

Chemical	Characteristics					Uses					
	Detection Limit[1] (µg/L)	Water-Quality criteria[2] (µg/L) Human Health	Water-Quality criteria[2] (µg/L) Aquatic Life	Solubility[3] (µg/L)	Relative Persistence within Pesticide Group[4]	Principal Uses and Sources	National Use on Farms[5] (million lb/yr) 1966	1971	1976	1982	Total Use, 1981[6] (million lb/yr)
Organochlorine insecticides (continued)											
Endrin	.05	*1	.002	14	nd	Cotton, wheat	.6	1.4	.8	nr	.3
Heptachlor epoxide	.01	.003	.004	30	Low	Degradation product of heptachlor which is used on corn, and termite control.	1.5	1.2	.6	nr	2.0
Lindane	.01	*4	.08	150	Medium	Livestock, seed treatment, general purpose.	.7	.7	.2	nr	.8
Methoxychlor	.10	*100	*.03	3	nd	Livestock, alfalfa, general purpose.	2.6	3.0	3.8	.6	5.0
Toxaphene	.25	.007	.013	400	nd	Cotton, livestock	35	37	33	5.9	16
Organophosphate insecticides											
Diazinon	.10	nd	nd	40,000	High	Corn, general purpose.	5.6	3.2	1.6	.3	9.0
Ethion	.25	nd	nd	2,000	nd	Citrus fruits	2.0	2.3	nr	nr	2.0
Malathion	.25	nd	.1	145,000	Low	General purpose	5.2	3.6	2.8	1.6	28
Methyl parathion	.25	nd	nd	57,000	Low	Cotton and wheat	8.0	28	23	11	20
Methyl trithion	.50	nd	nd	nd	nd	Not identified	nr	nr	nr	nr	.1
Parathion	.25	nd	.04	24,000	Low	Wheat, corn, sorghum.	8.5	9.5	6.6	4.0	5.0
Trithion	.50	nd	nd	340	nd	General purpose	nr	nr	nr	nr	.1
Chlorophenoxy and triazine herbicides											
Atrazine	.5	nd	nd	33,000	High	Corn	24	54	90	76	92
2,4-D	.5	*100	nd	900,000	Low	Wheat, rangeland, general purpose.	4	31	38	23	60
2,4,5-T	.5	*10	nd	240,000	Medium	Rice, rangeland, general purpose.	.8	nr	nr	.2	2.2
Silvex	.5	nd	nd	140,000	nd	Sugarcane, rice, rangeland.	nr	nr	nr	nr	.4

[1] Detection limits shown are for water samples. Bed-sediment reporting limits are 10 times greater and are expressed in units micrograms per kilogram (Lucas and others, 1980).
[2] All criteria are from U.S. Environmental Protection Agency (1980), except for values marked by asterisks, which are from U.S. Environmental Protection Agency (1976). The human-health criteria for all pesticides except endrin, lindane, methoxychlor, 2,4-D, and 2,4,5-T represent the estimated average concentrations associated with an incremental increase in cancer risk of 10^{-5} (one additional cancer per 100,000 people over a lifetime of exposure). The aquatic-life criteria are for freshwater and are 24-hour average concentrations.
[3] Data from Kenaga and Goring (1980).
[4] Relative persistence within each pesticide group as estimated from Hiltbold (1974) and Wauchope (1978).
[5] Data for 1966, from Eichers and others (1970); for 1971, Andrilenas (1974); for 1976, Eichers and others (1978); for 1982, U.S. Department of Agriculture (1983). Data for 1982 do not include use on livestock or use in California, Colorado, Connecticut, Maine, Massachusetts, Nevada, New Hampshire, New Jersey, New Mexico, Oregon, Rhode Island, Utah, Vermont, West Virginia, and Wyoming.
[6] Data from Mark H. Glaze (U.S. Environmental Protection Agency, written commun., 1983).
* See footnote 2.
Source: U.S. Geological Survey, 1985. National Water Summary 1984. Water-Supply Paper 2275

TABLE 7–125. SUMMARY OF DETECTIONS OF PESTICIDES IN WATER AND BED SEDIMENTS AT THE U.S. GEOLOGICAL SURVEY-U.S. ENVIRONMENTAL PROTECTION AGENCY PESTICIDE MONITORING NETWORK STATIONS, 1975 TO 1980

Chemical	Water				Bed Material			
	Stations		Samples		Stations		Samples	
	Number Monitored	Percentage with Detections	Number Collected	Percentage with Detections	Number Monitored	Percentage with Detections	Number Collected	Percentage with Detections
	Organochlorine insecticides							
Aldrin	177	2.3	2,946	0.2	171	2.9	1,015	0.6
Dieldrin	177	2.3	2,945	.2	172	29	1,017	12
Chlordane	177	.6	2,943	.0	171	30	1,014	9.9
DDD	177	4.0	2,720	.3	171	31	990	12
DDE	177	.6	2,715	.0	172	42	989	17
DDT	177	2.8	2,721	.4	171	26	992	8.5
Endrin	180	1.1	2,950	.1	171	2.3	1,015	.6
Heptachlor epoxide	177	4.5	2,946	.3	171	5.3	1,017	1.0
Lindane	177	8.5	2,945	1.1	171	.6	1,018	.1
Methoxychlor	172	.0	2,761	.0	160	.6	941	.1
Toxaphene	177	2.8	2,946	.4	171	3.5	1,014	.6
	Organophosphate insecticides							
Diazinon	174	9.8	2,859	1.2	164	1.2	929	.2
Ethion	174	.6	2,823	.1	163	.6	928	.4
Malathion	174	.6	2,859	.1	163	.0	929	.0
Methyl parathion	174	2.7	2,861	.1	163	.0	929	.0
Methyl trithion	174	.0	2,822	.0	163	.0	928	.0
Parathion	174	.6	2,856	.0	163	.0	928	.0
Trithion	174	1.1	2,819	.1	163	.0	925	.0
	Chlorophenoxy and triazine herbicides							
Atrazine	144	24	1,363	4.8	126	.0	347	.0
2,4-D	186	2.4	1,764	.2	142	1.4	487	.4
2,4,5-T	186	.6	1,765	.1	142	.7	486	.2
Silvex	167	.6	1,768	.1	142	1.4	488	.4

Source: U.S. Geological Survey, 1985. National Water Summary 1984. Water-Supply Paper 2275

TABLE 7–126. DETECTION OF PESTICIDES IN STREAM AND RIVER WATER AND SEDIMENT IN THE UNITED STATES, 1975–1980

[Average number of detections per 100 samples]

Year	Organochlorine Insecticides		Organophosphate Insecticides	
	In Sediment	In Water	In Sediment	In Water
1975	65.3	2.4	BD	2
1976	82.3	2.9	BD	5
1977	74.2	3.2	BD	7
1978	41.1	3.1	BD	5
1979	52.1	1.2	BD	4
1980	BD	0.9	BD	6

Note: BD=below detection.

Source: Gilliom, R.J., R.B. Alexander & R.A. Smith, 1985. Pesticides in the Nation's Rivers, 1975–1980, and Implications for Future Monitoring. U.S. Department of the Interior, Geological Survey Water-Supply Paper 2271. Reston, VA.; U.S. Geological Survey National Water Summary 1986

TABLE 7–127. POTENTIAL SOURCES OF PESTICIDE CONTAMINATION OF GROUND WATER

	Manufacturer/ Formulator	Dealer	Industrial User	Land Application
SPILLS AND LEAKS				
Storage Areas	x	x	x	x
Storage Tanks/ Pipelines	x	x	x	x
Loading/Unloading	x	x	x	x
Transport Accidents	x	x	x	x
DISPOSAL				
Process Waste	x		x	
Off-specification material	x			
Cancelled Products	x	x	x	x
Containers	x	x	x	x
Rinsate			x	x
LAND APPLICATION				x
Leaching				x
Backflow to Irrigation Well				x
Run-in to Wells, Sinkholes				x
Mixing/loading areas				x
Feed Lots				x

Source: U.S. Environmental Protection Agency, Office of Pesticides and Toxic Substances, 1987, Agricultural Chemicals in Ground Water: Proposed Pesticide Strategy

TABLE 7–128. PESTICIDES FOUND IN GROUND WATER IN THE UNITED STATES

Pesticide	Use[a]	State	Typical Positive, ppb[b]
Alachlor	H	MD, IA, NE, PA	0.1–10
Aldicarb (sulfoxide & sulfone)	I,N	AR, AZ, CA, FL, MA, ME, NC, NJ, NY, OR, RI, TX, VA, WA, WI	1–50
Atrazine	H	PA, IA, NE, WI, MD	0.3–3
Bromacil	H	FL	300
Carbofuran	I,N	NY, WI, MD	1–50
Cyanazine	H	IA, PA	0.1–1.0
DBCP	N	AZ, CA, HI, MD SC	0.02–20
DCPA (and acid products)	H	NY	50–700
1,2-Dichloropropane	N	CA, MD, NY, WA	1–50
Dinoseb	H	NY	1–5
Dyfonate	I	IA	0.1
EDB	N	CA, FL, GA, SC, WA, AZ, MA, CT	0.05–20
Metolachlor	H	IA, PA	0.1–0.4
Metribuzin	H	IA	1.0–4.3
Oxamyl	I,N	NY, RI	5–65
Symazine	H	CA, PA, MD	0.2–3.0
1,2,3-Trichloropropane	N (impurity)	CA, HI	0.1–5.0

[a] H = herbicide; I = insecticide; N = nematicide

[b] ppb = parts per billion; 1 ppb = 1/1000 ppm; 1 ppm = 1 mg/l

Source: U.S. Environmental Protection Agency, Office of Ground-Water Protection, 1986, Pesticides In Ground Water: Background Document

TABLE 7–129. CONSTITUENTS OF LIVESTOCK WASTE

[Some constituents of waste of a 1,000-pound bovine on a daily and feeding period basis, and of 360 head per acre on an annual basis]

	Per day, lb.	140 days, lb.	360 head/acre/year tons
Wet manure and urine	64	8,960	4,200
Dry mineral matter	2.1	294	144
Dry organic matter	8.2	1,148	540
Water	53.7	7518	30.7 in.
Total nitrogen	0.380	55.0	24.9
Total phosphorus	0.048	6.7	3.2
Total potassium	0.260	36.4	16.8

Source: Hansen, R.W., 1971, Livestock Waste Disposal and Water Pollution Control, Colorado State University Cooperative Extension Service Bulletin 480a

TABLE 7–130. FERTILIZER ELEMENTS OF ANIMAL EXCREMENTS

[Fertilizer elements of various complete animal excrements per 1,000 pounds of liveweight]

	Dairy Cattle lb./day	Beef Cattle lb./day	Hens lb./day	Hogs lb./day	Sheep lb./day
Wet manure	88	64	59	50	37
Total mineral matter	1.80	2.1	4.5	1.3	1.5
Organic matter	7.20	8.2	12.9	5.9	6.9
Nitrogen (N)	0.36	0.38	2	0.4	
Phosphorus ($P_2 0_5$)	0.10	0.048	0.69	0.18	
Potassium (K_2O)	0.15	0.26	0.34	0.1	

Source: Hansen, R.W., 1971, Livestock Waste Disposal and Water Pollution Control, Colorado State University Cooperative Extension Service Bulletin 480a

SECTION J. URBAN RUNOFF/DEICING MATERIALS

TABLE 7–131. TRENDS IN GROSS DISCHARGES OF MAJOR POLLUTANTS IN URBAN RUNOFF IN THE UNITED STATES

[Thousands of tons]

	High Growth				Low Growth		
Pollutant	**1975**	**1985**	**1990**	**2000**	**1985**	**1990**	**2000**
Biochemical Oxygen Demand	370	400	420	450	390	400	420
Chemical Oxygen Demand	3,400	3,400	3,900	4,200	3,600	3,800	3,900
Suspended Solids	6,300	6,800	7,100	7,600	6,500	6,900	7,100
Dissolved Solids	4,000	4,400	4,600	4,900	4,300	4,400	4,600
Phosphorus	5	5	6	6	5	5	6
Nitrogen	58	63	65	70	61	63	65
Oil and Grease	86	94	98	110	92	95	98

Note: Conservative estimates; discharges may uniformly be too low by a factor of two to five.
Source: U.S. Environmental Protection Agency, Environmental Outlook 1980

TABLE 7–132. AVERAGE CONCENTRATIONS OF HEAVY METALS IN STREET SWEEPINGS AND URBAN RUNOFF

Heavy Metal	**Street Sweepings mg/kg**	**Urban Runoff μg/L**
Cadmium (Cd)	3.4	18
Chromium (Cr)	211	33
Copper (Cu)	104	45
Iron (Fe)	22,000	—
Lead (Pb)	1,810	235
Manganese (Mn)	418	—
Nickel (Ni)	35	24
Zinc (Zn)	370	236

Source: U.S. Geological Survey, 1984, National Water Summary 1983-Hydrologic Events and Issues, Water-Supply Paper 2250. Based on data from W.L. Bradford, 1977, and the U.S. Environmental Protection Agency

TABLE 7–133. SNOW AND ICE CONTROL MATERIALS USE IN THE UNITED STATES, WINTER 1982–1983

[As reported by State Departments of Transportation/Highways]

State	Population 1983 (thousands)	Total Lane miles[1]	Bare Pavement miles[1]	Control Materials Used: Salt (metric tons)	Calcium Chloride: Dry (metric tons)[2]	Calcium Chloride: Liquid (liters)[3]	Abrasives (metric tons)[2]
Alaska	100	1,020	900	322	227	757,000	16,534
Arizona	2,718	16,650	0	375	23	0	50,803
Arkansas	2,178	34,852	0	777	223	0	8,305
California	22,000	53,000	12,900	—	—	—	—
Colorado	2,500	32,000	32,000	9,885	0	0	394,484
Connecticut	3,100	10,160	0	47,115	544	0	153,859
Delaware	611	9,971	2,543	6,400	0	0	6,998
Florida	9,740	—	—	—	—	—	—
Georgia	5,463	41,809	13,900	7,439	54	0	7,076
Idaho	944	11,512	0	9,979	0	0	174,182
Illinois	11,114	38,515	—	186,883	472	435,843	—
Indiana	5,300	31,036	14,559	105,825	40	341	98,539
Iowa	2,900	24,300	0	54,795	2,019	35,958	105,235
Kansas	2,364	22,371	5,688	28,695	0	0	58,968
Kentucky	3,661	53,846[5]	4,966[5]	29,905	97	0	0
Louisiana	5,000	38,191	0	21	0	0	0
Maine	1,125	7,877	1,178	44,636	485	—	428,652
Maryland	4,500	14,600	0	74,843	887	—	27,995
Massachusetts	5,737	12,000	—	161,935	2,578	0	86,184
Michigan	9,258	13,667	4,695	207,749	242	0	9,072
Minnesota	4,077	28,724	4,162	116,083	254	0	262,084
Mississippi	2,500	23,391	0	259	—	—	—
Missouri	4,917	69,664	—	68,141	3,060	0	—
Montana	700	18,790[1]	18,790	2,944	25	0	181,440
Nebraska	1,600	22,000	—	22,588	504	98,312	78,685
Nevada	799	12,608	11,340	8,919	0	0	46,267
New Hampshire	921	8,630	8,406	85,107	207	0	137,279
New Jersey	7,364	10,366	10,366	32,387	526	862,980	3,810
New Mexico	1,400	27,450	20,000	20,866	0	0	72,576
New York	17,557	29,780	29,780	272,160	272	0	417,312
North Carolina	5,847	112,573	23,342	33,179	145	0	—
North Dakota	600	15,800	—	7,910	197,588[4]	0	34,474
Ohio	10,797	42,192	0	167,234	177	428,806	117,007
Oklahoma	3,025	25,935	0	17,028	0	—	65,318
Oregon	2,656	17,895	—	414	—	1,608,580	—
Pennsylvania	11,867	77,000	—	209,563	4,536	0	594,216
Rhode Island	947	3,015	3,015	26,578	120	0	40,824
South Carolina	3,122	84,450	3,450	814	259	0	4,309
South Dakota	688	18,216	—	3,354	5	0	40,826
Tennessee	4,591	25,087	25,087	—	—	0	—
Utah	2,000	22,000	—	72,322	0	0	113,400
Vermont	511	6,079	6,079	59,555	0	0	96,757
Virginia	5,347	112,814	17,350	86,184	907	0	231,336
Washington	4,130	16,778	0	6,804	0	0	184,643
West Virginia	1,950	70,000	21,000	47,818	285	0	162,064
Wisconsin	4,705	25,774	25,774	208,477	588	151,400	23,508
Wyoming	430	15,743	0	5,752	0	0	115,214
Total	205,361	1,410,131	321,270	2,560,019	217,349	4,379,220	4,650,235
				(2,821,141 tons)	(239,519 tons)	(1,156,990) gallons)	(5,124,559) tons)

[1] To convert to kilometers, multiply amount by 1.609.
[2] To convert to short tons, multiply amount by 1.102.
[3] To convert to gallons, multiply amount by 0.2642.
[4] Sodium chloride liquid used.
[5] Includes parkways
Source: Salt Institute, 1983, Survey of Salt, Calcium Chloride and Abrasives in the United States and Canada for 81/82 and 82/83

SECTION K. AIR EMISSIONS/ACID RAIN/SEA LEVEL RISE

TABLE 7–134. NATIONAL AMBIENT CONCENTRATIONS OF SELECTED AIR QUALITY INDICATORS IN THE UNITED STATES, 1975–1984

[Parts per million and milligrams per cubic meter]

Year	Sulfur Dioxide (ppm)	Carbon Monoxide (ppm)	Ozone (ppm)	Nitrogen Oxide (ppm)	Total Suspended Solids (mg/g)	Lead (mg/g)
1975	0.0154	11.96	0.1532	0.0289	61.90	1.04
1976	0.0156	11.32	0.1533	0.0286	62.78	1.05
1977	0.0150	10.68	0.1526	0.0292	62.04	1.12
1978	0.0134	10.10	0.1534	0.0301	61.42	0.92
1979	0.0127	9.70	0.1374	0.0307	62.20	0.63
1980	0.0114	9.00	0.1417	0.0296	63.71	0.41
1981	0.0111	8.93	0.1272	0.0278	59.26	0.34
1982	0.0100	8.04	0.1253	0.0265	49.76	0.29
1983	0.0098	8.02	0.1408	0.0258	48.92	0.19
1984	0.0098	7.92	0.1271	0.0261	49.73	0.183

Source: U.S. Geological Survey National Water Summary 1986; U.S. Environmental Protection Agency, Office of Air Quality Planning and Standards. 1986. National Air Quality and Emissions Trends Report, 1984. EPA-450/4–86–001. Research Triangle Park, NC

TABLE 7–135. HISTORICAL SULFUR DIOXIDE EMISSIONS IN THE UNITED STATES, 1900–1980

[By source; million metric tons]

Year	Electric Utility	Industrial Process	Commercial/ Residential	Transportation	Other
1900	1.315	4.322	1.267	2.828	1.254
1905	1.917	6.197	1.793	4.068	1.378
1910	2.286	7.830	2.319	5.051	1.516
1915	2.690	9.426	3.027	5.590	1.583
1920	3.039	9.266	3.164	6.157	1.630
1925	3.322	10.997	3.660	5.762	1.847
1930	2.812	9.215	4.006	5.147	2.033
1935	2.072	7.266	3.351	4.047	1.938
1940	3.276	9.524	3.577	4.352	1.841
1945	4.453	9.963	5.548	6.367	2.278
1950	5.609	8.735	3.160	3.067	2.752
1955	7.623	9.308	2.182	1.165	2.694
1960	10.446	8.509	2.306	0.564	2.671
1965	14.084	9.920	2.009	0.506	2.906
1970	19.603	9.747	2.126	0.626	3.038
1975	21.526	6.211	1.477	0.704	2.553
1978	19.080	5.972	1.363	0.914	2.643
1980	19.852	5.554	1.019	1.016	2.476

Source: U.S. Geological Survey National Water Summary 1986, Water-Supply Paper 2325; Gschwandtner, and others. 1986. Historic Emissions of Sulfur and Nitrogen Oxides in the United States from 1900 to 1980. Journal of the Air Pollution Control Association, vol. 36, no. 2

TABLE 7–136. HISTORICAL NITROGEN OXIDE EMISSIONS IN THE UNITED STATES, 1900–1980

[By source; million metric tons]

Year	Electric Utility	Industrial Process	Commercial/ Residential	Transpor- tation	Other
1900	0.564	0.759	0.202	0.347	0.996
1905	0.799	1.063	0.231	0.507	1.047
1910	0.913	1.575	0.271	0.649	1.103
1915	1.029	1.896	0.315	0.788	1.113
1920	1.125	2.045	0.323	1.059	1.120
1925	1.124	2.536	0.356	2.786	1.227
1930	0.887	2.596	0.382	3.197	1.755
1935	0.667	2.066	0.339	2.867	1.361
1940	1.056	2.507	0.370	3.188	1.189
1945	1.397	3.105	0.531	4.059	1.409
1950	1.777	3.229	0.469	4.076	1.789
1955	2.346	3.980	0.399	4.506	1.488
1960	2.981	4.822	0.756	5.117	1.295
1965	3.920	5.793	0.782	6.195	1.546
1970	5.694	4.795	0.914	8.193	1.746
1975	6.929	4.191	0.822	9.494	1.114
1978	7.466	4.681	0.845	10.600	1.063
1980	8.752	3.947	0.784	10.071	0.931

Source: U.S. Geological Survey National Water Summary 1986, Water-Supply Paper 2325; Gschwandtner, and others. 1986. Historic Emissions of Sulfur and Nitrogen Oxides in the United States from 1900 to 1980. Journal of the Air Pollution Control Association, vol. 36, no. 2

TABLE 7–137. TOTAL SULFUR DIOXIDE EMISSIONS AND CHANGES IN EMISSIONS IN THE UNITED STATES, 1900–1984

[By state; metric tons]

State	1900	1950	1977	1980	1984	Change 00–77	Change 77–84
Alabama	112.5	281.6	933.0	866.8	690.0	820.5	-243.0
Arizona	148.3	1200.1	1175.0	872.3	579.2	1026.7	-595.8
Arkansas	87.8	89.1	98.8	96.8	106.9	11.0	8.1
California	154.0	652.3	535.7	631.4	301.6	381.7	-234.1
Colorado	80.7	67.1	131.9	165.0	131.2	51.2	-0.7
Connecticut	110.2	180.4	42.3	68.2	70.2	-67.9	27.9
Delaware	5.2	34.1	141.7	111.1	117.8	136.5	-23.9
Florida	63.9	306.9	853.8	1096.7	703.2	789.9	-150.6
Georgia	95.9	196.9	635.4	899.8	917.3	539.5	281.9
Idaho	46.6	115.5	46.2	30.8	48.9	-0.4	2.7
Illinois	1604.9	2932.6	1632.8	1736.9	1292.0	27.9	-340.8
Indiana	495.9	1310.1	1701.1	2283.6	1748.4	1205.2	47.3
Iowa	503.4	575.3	294.5	371.8	258.0	-208.9	-36.5
Kansas	287.0	210.1	197.0	309.1	214.5	-90.0	17.5

TABLE 7–137. TOTAL SULFUR DIOXIDE EMISSIONS AND CHANGES IN EMISSIONS IN THE UNITED STATES, 1900–1984 (continued)

[By state; metric tons]

State	1900	1950	1977	1980	1984	Change 00–77	Change 77–84
Kentucky	228.6	436.7	1383.4	1302.4	753.3	1154.8	-630.1
Louisiana	154.6	198.0	500.3	456.5	331.5	345.8	-168.8
Maine	60.7	61.6	109.9	72.6	65.8	49.2	-44.1
Maryland	122.7	319.0	265.0	401.5	287.4	142.4	22.4
Massachusetts	248.6	306.9	222.8	380.6	267.7	-25.8	44.9
Michigan	418.8	1076.9	1112.6	995.5	769.1	693.8	-343.5
Minnesota	222.5	238.7	267.2	205.7	154.6	44.7	-112.6
Mississippi	89.1	138.6	335.6	290.4	191.6	246.5	-144.0
Missouri	493.9	557.7	1294.3	1411.3	1141.2	800.4	-153.1
Montana	341.2	330.0	197.6	179.3	123.4	-143.6	-74.2
Nebraska	211.4	110.0	51.1	79.2	63.9	-160.3	12.8
Nevada	7.9	172.7	193.6	242.0	133.6	185.7	-60.0
New Hampshire	49.1	31.9	75.6	95.7	86.4	26.5	10.8
New Jersey	301.2	662.2	284.9	361.9	201.6	-16.3	-83.3
New Mexico	44.7	239.8	342.8	312.4	286.0	298.1	-56.8
New York	764.0	1354.1	882.7	1027.4	716.5	118.8	-166.2
North Carolina	56.9	338.8	558.1	641.3	433.4	501.2	-124.7
North Dakota	16.9	71.5	86.2	133.1	120.7	69.3	34.5
Ohio	1080.0	2431.0	2942.5	3095.4	2388.8	1862.5	-553.7
Oklahoma	73.7	256.3	63.6	161.7	147.8	-10.1	84.2
Oregon	14.3	83.6	44.4	67.1	36.3	30.1	-8.1
Pennsylvania	819.2	1930.5	1698.6	2212.1	1464.3	879.4	-234.3
Rhode Island	54.3	80.3	9.7	14.3	7.3	-44.6	-2.4
South Carolina	35.0	156.2	285.6	334.4	228.1	250.6	-57.5
South Dakota	16.7	29.7	37.1	51.7	39.3	20.4	2.2
Tennessee	164.5	412.5	1344.2	1294.7	809.0	1179.8	-535.2
Texas	427.8	869.0	899.8	1443.2	986.5	472.0	86.7
Utah	67.3	657.8	103.8	125.4	81.9	36.5	-21.9
Vermont	38.1	13.2	7.7	7.7	6.7	-30.4	-1.0
Virginia	101.9	426.8	386.8	344.3	265.1	284.9	-121.7
Washington	39.8	220.0	352.4	308.0	232.2	312.6	-120.2
West Virginia	92.3	408.1	1080.1	1325.5	966.7	987.8	-113.4
Wisconsin	281.3	501.6	627.7	733.7	531.9	346.4	-95.8
Wyoming	31.1	51.7	177.6	267.3	197.1	146.5	19.5

Source: U.S. Geological Survey National Water Summary 1986, Water-Supply Paper 2325; Gschwandtner and others. 1986. Historic Emissions of Sulfur and Nitrogen Oxides in the United States from 1900 to 1980. Journal of the Air Pollution Control Association, vol. 36, no. 2.; Knudson, D.A. 1986. Estimated Monthly Emissions of Sulfur Dioxide and Oxides of Nitrogen for the 49 Contiguous States, 1975–1984. ANL/EES-TM-318, vol. 1. Argonne National Laboratory, Argonne, IL

TABLE 7–138. TOTAL NITROGEN OXIDE EMISSIONS AND CHANGES IN EMISSIONS IN THE UNITED STATES, 1900–1984

[By state; metric tons]

State	1900	1950	1977	1980	1984	Change 00–77	Change 77–84
Alabama	101.4	169.4	440.3	559.9	393.0	338.9	-47.3
Arizona	3.0	67.1	218.6	358.6	233.7	215.6	15.1
Arkansas	173.8	140.8	178.5	244.2	212.2	4.7	33.7
California	53.8	863.5	1309.3	1544.4	1098.5	1255.5	-210.8
Colorado	28.8	90.2	266.7	341.0	279.6	237.9	12.9
Connecticut	36.0	99.0	135.5	156.2	117.9	99.5	-17.6
Delaware	3.0	20.9	60.3	66.0	64.9	57.3	4.6
Florida	58.0	237.6	658.3	772.2	628.1	600.3	-30.2
Georgia	79.2	217.8	500.8	616.0	565.7	421.6	64.9
Idaho	7.4	35.2	84.9	122.1	80.0	77.5	-4.9
Illinois	226.5	685.3	907.1	1118.7	811.9	680.6	-95.2
Indiana	80.7	336.6	688.3	942.7	710.2	607.6	21.9
Iowa	62.9	184.8	257.8	343.2	272.3	194.9	14.5
Kansas	36.0	227.7	434.4	493.9	409.9	398.4	-24.5
Kentucky	41.7	162.8	565.1	622.6	551.9	523.4	-13.2
Louisiana	46.9	367.4	685.6	968.0	576.9	638.7	-108.7
Maine	26.3	42.9	60.0	68.2	50.7	33.7	-9.3
Maryland	39.4	161.7	263.2	336.6	254.2	223.8	-9.0
Massachusetts	85.1	184.8	266.0	293.7	255.0	180.9	-11.0
Michigan	80.9	387.2	660.1	799.7	613.1	579.3	-47.0
Minnesota	54.3	148.5	348.7	339.9	301.2	294.4	-47.5
Mississippi	52.8	150.7	256.7	303.6	228.5	203.9	-28.2
Missouri	89.0	251.9	488.7	595.1	493.2	399.7	4.5
Montana	23.1	81.4	123.7	143.0	125.5	100.6	1.8
Nebraska	31.5	84.7	151.0	189.2	105.9	119.5	-45.1
Nevada	2.5	41.8	95.8	121.0	102.8	93.3	7.0
New Hampshire	21.3	23.1	52.3	60.5	50.9	31.0	-1.4
New Jersey	102.5	353.1	521.1	452.1	419.7	418.6	-101.4
New Mexico	5.9	101.2	218.3	359.7	216.1	212.4	-2.2
New York	276.0	735.9	756.3	779.9	645.3	480.3	-111.0
North Carolina	35.8	213.4	468.6	616.0	428.5	432.9	-40.1
North Dakota	7.8	44.0	105.8	155.1	143.9	98.0	38.1
Ohio	190.5	678.7	1125.6	1299.1	1009.0	935.1	-116.6
Oklahoma	14.2	261.8	371.3	534.6	375.9	357.1	4.6
Oregon	22.3	126.5	215.5	222.2	211.4	193.2	-4.1
Pennsylvania	304.0	811.8	891.8	1289.2	853.1	587.8	-38.7
Rhode Island	23.3	35.2	35.0	39.6	29.9	11.7	-5.1
South Carolina	24.1	99.0	2414.0	298.1	231.0	2389.9	-2183.0
South Dakota	9.2	35.2	68.8	85.8	65.7	59.6	-3.1
Tennessee	46.1	193.6	496.4	639.1	435.0	450.3	-61.4
Texas	65.9	1304.6	2691.4	2803.9	2439.7	2625.5	-251.7
Utah	8.0	81.4	120.3	192.5	127.7	112.3	7.4
Vermont	18.2	15.4	32.9	27.5	33.3	14.8	0.4
Virginia	44.2	226.6	364.1	452.1	325.8	319.9	-38.3
Washington	20.6	130.9	281.0	327.8	247.3	260.4	-33.7
West Virginia	26.7	190.3	408.7	550.0	414.7	382.0	6.0
Wisconsin	64.7	184.8	366.8	481.8	370.2	302.1	3.4
Wyoming	7.9	53.9	204.6	359.7	234.1	196.7	29.5

Source: U.S. Geological Survey National Water Summary 1986, Water-Supply Paper 2325; Gschwandtner and others. 1986. Historic Emissions of Sulfur and Nitrogen Oxides in the United States from 1900 to 1980. Journal of the Air Pollution Control Association, vol. 36, no. 2. Knudson, D.A. 1986. Estimated Monthly Emissions of Sulfur Dioxide and Oxides of Nitrogen for the 49 Contiguous States, 1975–1984. ANL/EES-TM-318, vol. 1. Argonne National Laboratory, Argonne, IL

FIGURE 7–12. AVERAGE pH OF PRECIPITATION IN THE UNITED STATES AND CANADA, 1981

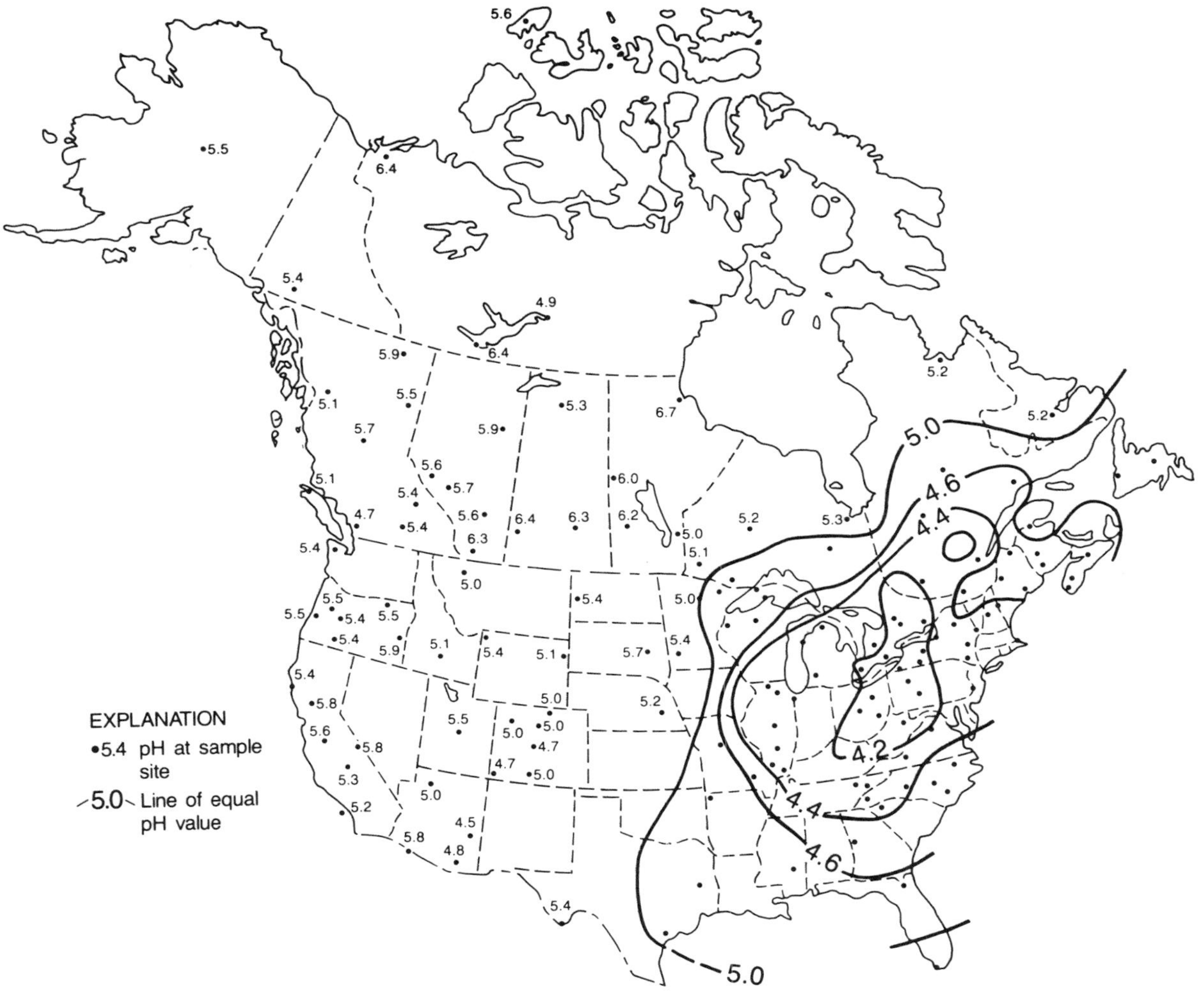

Source: U.S. Geological Survey, 1984, National Water Summary 1983-Hydrologic Events and Issues, Water-Supply Paper 2250

TABLE 7–139. ACID NEUTRALIZING CAPACITY AND pH OF LAKES IN THE WESTERN AND EASTERN UNITED STATES

Western Lake Survey Results

State	Estimated Number of Lakes	Number of Lakes Sampled	Acid Neutralizing Capacity (μeqL^{-1})[a,b] ≤50	≤200	pH[b] ≤6.0
California	2,390	147	880	2,078	32
Colorado	1,476	132	70	591	0
Idaho	972	72	189	599	0
Montana	1,597	80	160	824	0
Oregon	551	55	113	461	10
Utah	548	30	20	484	0
Washington	1,338	117	219	822	31
Wyoming	1,480	83	94	1,068	30

Eastern Lake Survey Results[c]

State	Estimated Number of Lakes	Number of Lakes Sampled	Acid Neutralizing Capacity (μeqL^{-1})[a,b]			pH[b]	
			≤0	≤50	≤200	≤5.0	≤6.0
Connecticut	346	24	47	47	145	19	47
Florida	2,088	138	453	732	1,146	249	677
Georgia	155	54	10	10	49	10	10
Massachusetts	926	97	52	239	578	54	180
Maine	1,966	225	8	200	1,337	8	90
Michigan	2,073	160	107	368	704	103	330
Minnesota	3,026	174	0	143	1,124	0	103
North Carolina	55	30	0	4	35	0	1
New Hampshire	639	69	17	171	537	17	126
New York	2,041	191	168	577	1,200	128	384
Pennsylvania	616	106	20	79	284	13	58
Rhode Island	113	15	13	33	86	0	20
South Carolina	40	12	0	0	10	0	0
Vermont	258	29	0	19	90	0	11
Wisconsin	3,402	253	41	801	1,690	27	386

[a] μeqL^{-1} = microequivalents per liter
[b] Estimate of number of lakes in sampled portion of States
[c] Includes only States in which more than 10 lakes were sampled.
Source: U.S. Environmental Protection Agency, National Water Quality Inventory - 1986 Report to Congress. EPA-440/4–87–008

FIGURE 7–13. PROBABLE EFFECTS OF ACID DEPOSITION ON WATER SUPPLIES

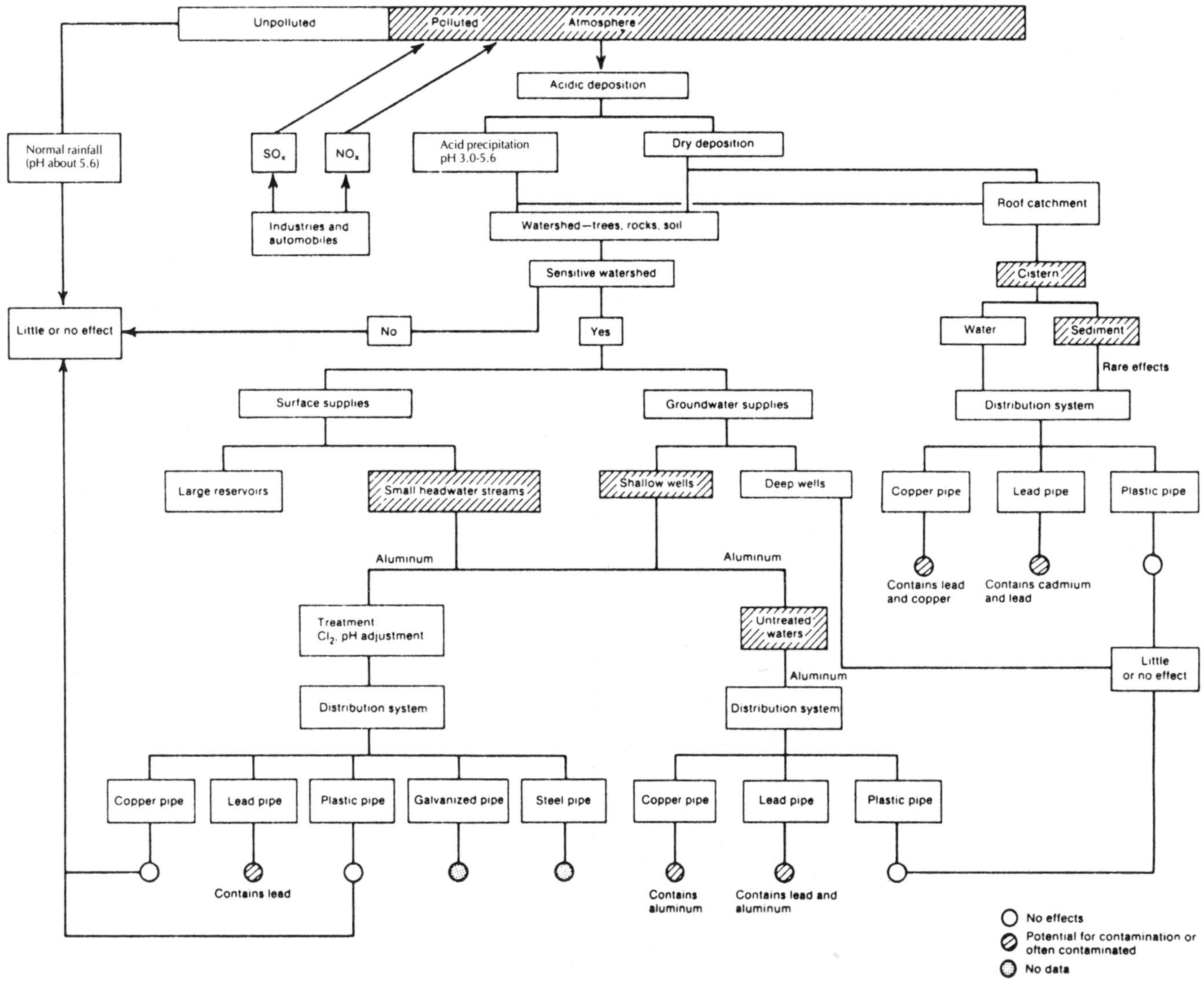

Source: Perry, J.A., 1984, Current Research on the Effects of Acid Deposition, J. Am. Water Works Assoc., v. 76, no. 3. Copyright AWWA. Reprinted with permission

FIGURE 7–14. MAJOR PROCESSES RELATING GREENHOUSE WARMING TO AVERAGE WORLDWIDE SEA LEVEL

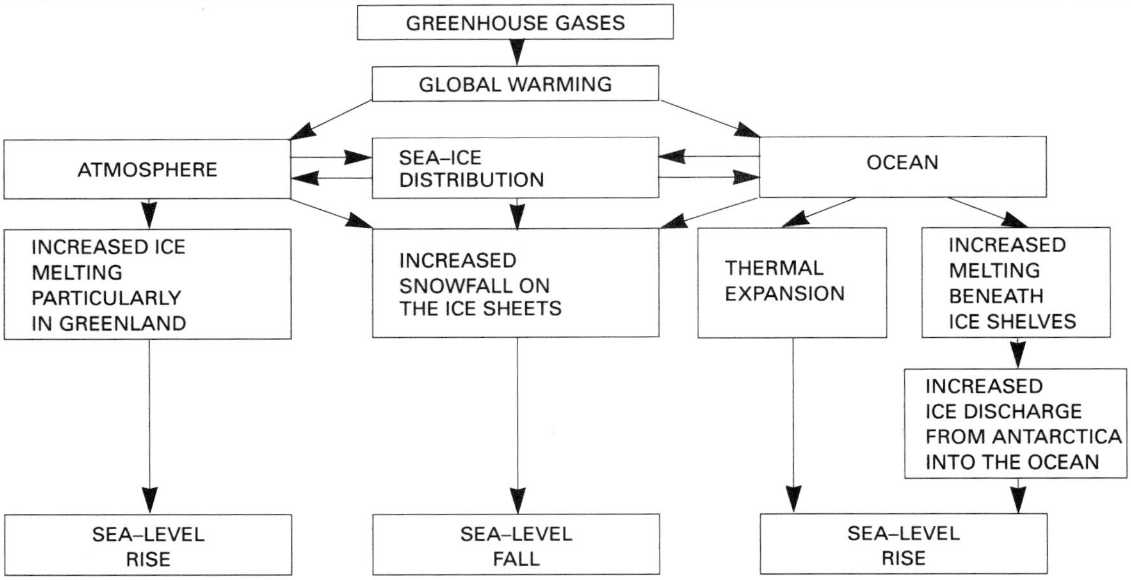

Heat trapped by greenhouse gases raises the temperature of the atmosphere and the ocean. The response of sea level to this warming is strongly determined by the partition of available heat between these two processes. If most of the heat remains in the atmosphere, air temperatures rise rapidly and sea level is affected most by increased melting of ice. Alternatively, rapid transfer of heat into the sea would increase ocean temperatures, and sea level would rise because of thermal expansion and by accelerated Antarctic ice discharge associated with increased melting from beneath the floating ice shelves. Moreover, sea-ice distribution both influences, and is affected by, thermal interactions between atmosphere and ocean.

Source: Thomas, Robert, 1986, Future Sea-Level Rise and its Early Detection by Satellite Remote Sensing, in Effects of Changes in Stratospheric Ozone and Global Climate, Vol. 4: Sea-Level Rise, U.S. Environmental Protection Agency

FIGURE 7–15. TOTAL ESTIMATED SEA-LEVEL RISE, 1980–2100

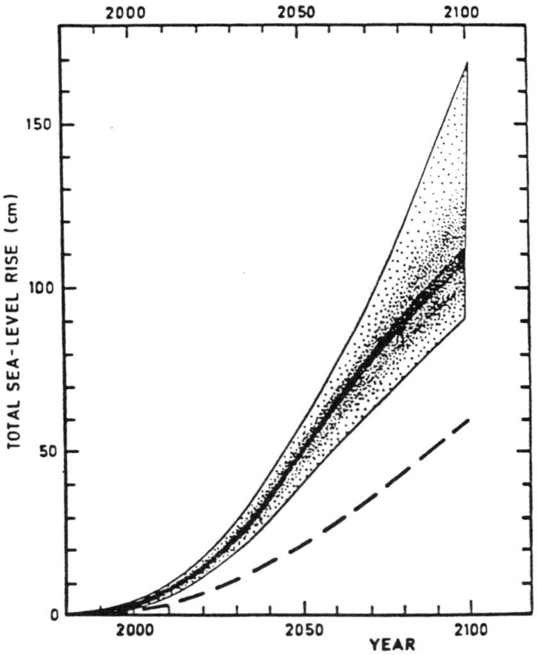

The dark shading indicates the most probable response to the climate scenario shown in Figure 7–14. The broken line depicts the response to a warming trend delayed 100 years by thermal inertia of the ocean. A global warming of 6°C by 2100, which represents an extreme upper limit, would result in a sea level rise of about 2.3 m, but errors on this estimate are very large.

Source: Thomas, Robert, 1986, Future Sea-Level Rise and its Early Detection by Satellite Remote Sensing, in Effects of Changes in Stratospheric Ozone and Global Climate, Vol. 4: Sea-Level Rise, U.S. Environmental Protection Agency

TABLE 7–140. ESTIMATED SEA LEVEL CHANGE BY YEAR 2100, AS A RESULT OF ICE WASTAGE IN A CARBON DIOXIDE-ENHANCED ENVIRONMENT

Ice Mass Contributing to Sea-Level Change	Estimated Sea Level Change (range, in feet)
Glaciers and small ice caps	+0.3 to 1.0
Greenland ice sheet	+0.3 to 1.0
Antarctic ice sheet	-0.3 to +3[1]

[1] Most likely the change will range from 0 to 0.7 foot.
Source: National Academy of Sciences, Committee on Glaciology, 1985

TABLE 7–141. LOSS OF COASTAL WETLANDS IN THE UNITED STATES FOR ONE-METER RISE IN SEA LEVEL

	Current Wetlands Area (mi^2)	All Dryland Protected (%)	Current Development Protected (%)	No Protection and (%) Presumed Lost
Northeast	600	16	10	2
Mid-Atlantic	746	70	46	38
South Atlantic	3,813	64	44	39
S/W Florida	1,869	44	8	7
Louisiana[a]	4,835	77	77	77
Other Gulf	1,218	85	76	75
West	64	56	gain[b]	gain[b]
USA	13,145	50–82	29–69	26–66

[a] Louisiana projections do not consider potential benefits of restoring flow of sediment and freshwater.
[b] Potential gain in wetland acreage not shown because principal author suggested that no confidence could be attributed to those estimates. West coast sites constituted less than 0.5% of wetlands in study sample.
Source: U.S. Environmental Protection Agency, 1988, The Potential Effects of Global Climate Change on the United States. Draft report to Congress. Adapted from Park and others

FIGURE 7–16. LOSS OF DRYLAND IN THE UNITED STATES BY 2100 (A) IF NO SHORES ARE PROTECTED AND (B) IF DEVELOPED AREAS ARE PROTECTED FOR SEA LEVEL RISE

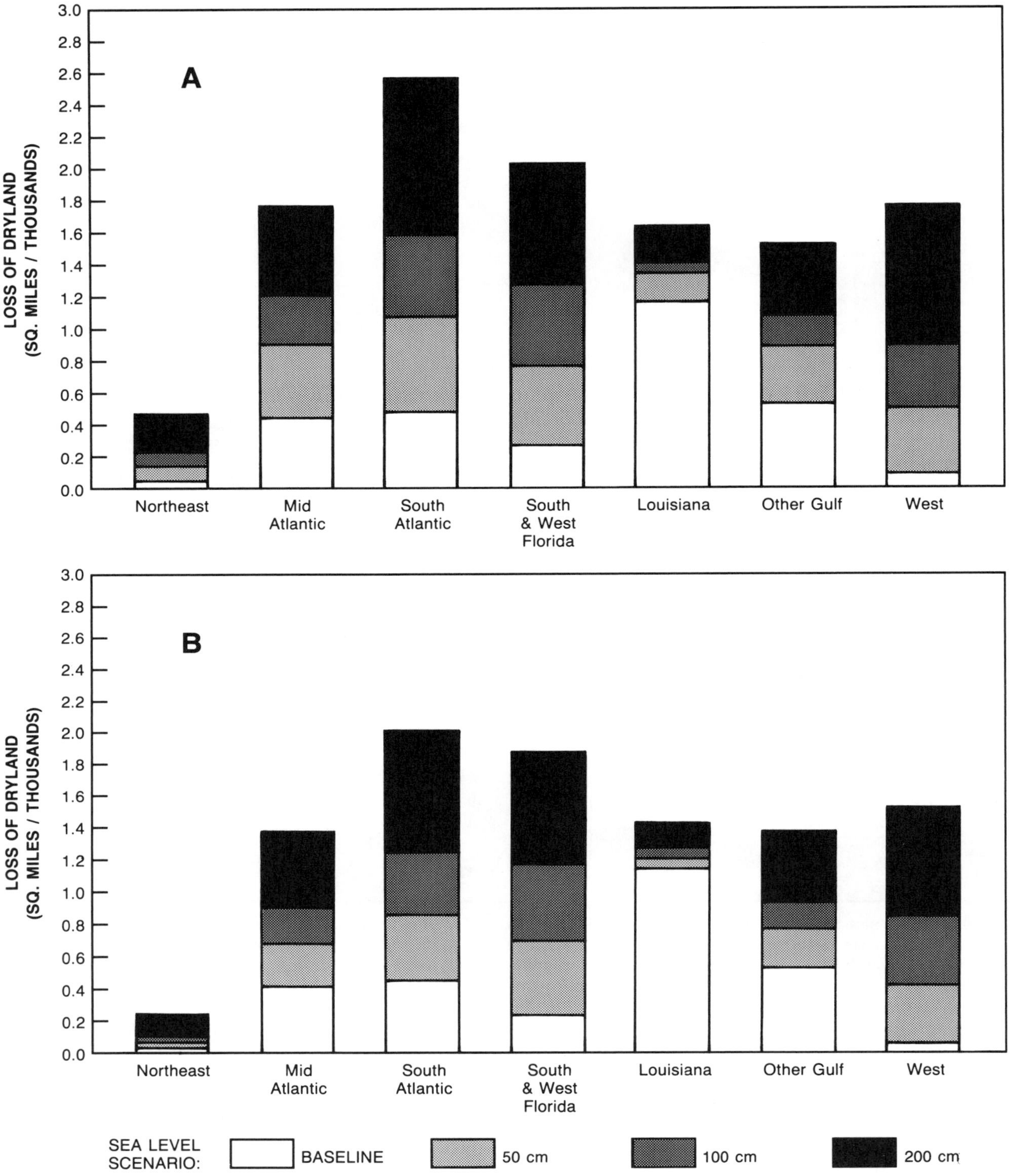

Source: U.S.Environmental Protection Agency, 1988, The Potential Effects of Global Climate Change on the United States. Draft report to Congress. Prepared by Titus and Greene, adapted from Park and others

TABLE 7–142. NATIONWIDE IMPACTS OF SEA-LEVEL RISE IN THE UNITED STATES

	Sea-Level Rise		
	50 cm	**100 cm**	**200 cm**
If Densely Developed Areas Are Protected			
Shore protection costs ($ billions)	32–43	73–111	169–309
Dryland lost (mi^2)	2,200–6,100	4,100–9,200	6,400–13,500
Wetlands lost (%)	20–45	29–69	33–80
If No Shores Are Protected			
Dryland lost (mi^2)	3,300–7,300	5,100–10,300	8,200–15,400
Wetlands lost (%)	17–43	26–66	29–76
If All Shores Are Protected			
Wetlands lost (%)	38–61	50–82	66–90

Source: U.S.Environmental Protection Agency, 1988, The Potential Effects of Global Climate Change on the United States, Draft Report to Congress. Data assembled by Titus and Greene

SECTION L. OFFSHORE WASTE DISPOSAL

TABLE 7–143. OFFSHORE WASTE DISPOSAL IN THE UNITED STATES, 1973–1983

[Thousands of tons]

	1973	1974	1975	1976	1977	1978	1979	1980	1981	1982	1983
ATLANTIC (A)											
Industrial Waste	3,643	3,642	3,322	2,633	1,784	2,548	2,577	2,928	2,271	1,063	283
Sewage Sludge	4,898	5,010	5,040	5,271	5,134	5,535	6,442	7,309	6,703	7,670	8,312
Construction Debris	974	770	396	315	379	241	107	89	0	0	0
Solid Waste/Chemicals Incinerated	0	0	0	0	0	0	0	0	0	0	0
Explosives	0	0	0	0	0	0	0	0	.0003	0	0
Wood Incinerated	11	16	6	9	15	18	45	11	15	13	31
GULF OF MEXICO (B)											
Industrial Waste	1,408	938	120	100	60	0.17	0	0	0	0	0
Sewage Sludge/Construction Debris Solid Waste/Explosives/Wood Incinerated	0	0	0	0	0	0	0	0	0	0	0
Chemicals Incinerated	0	12.3	4.1	0	17.6	0	0	0	700[a]	800[a]	0
PACIFIC (C)											
Industrial Waste	0	0	0	0	0	0	0	.26	23.3	18.8	21.5
Sewage Sludge/Construction Debris Explosives/Chemicals Incinerated	0	0	0	0	0	0	0	0	0	0	0
Solid Waste	240	200	0	0	0	0	0	0	0	0	0
Wood Incinerated	0	0	0	0	12.1	0	0	0	0	0	0
TOTALS OF (A), (B), (C)											
Industrial Waste	5,051	4,580	3,452	2,733	1,844	2,548.17	2,577	2,928.26	2,294.3	1,081.8	304.5
Sewage Sludge	4,890	5,010	5,040	5,271	5,134	5,535	6,442	7,309	6,703	7,670	8,312
Const. Debris	974	770	396	315	379	241	107	89	0	0	0
Solid Waste	240	200	0	0	0	0	0	0	0	0	0
Explosives	0	0	0	0	0	0	0	0	.0003	0	0
Wood Incinerated	11	16	6	9	15	18	45	11	15	13	31
Chemical Incinerated	0	12.3	4.1	0	17.6	0	0	0	700[a]	800[a]	0

[a] Thousand gallons (prior to incineration).

Source: U.S. Environmental Protection Agency, Report to Congress January 1981- December 1983

FIGURE 7–17. SEWAGE SLUDGE AND INDUSTRIAL WASTE DUMPED IN UNITED STATES OCEAN WATERS FROM 1973 TO 1986

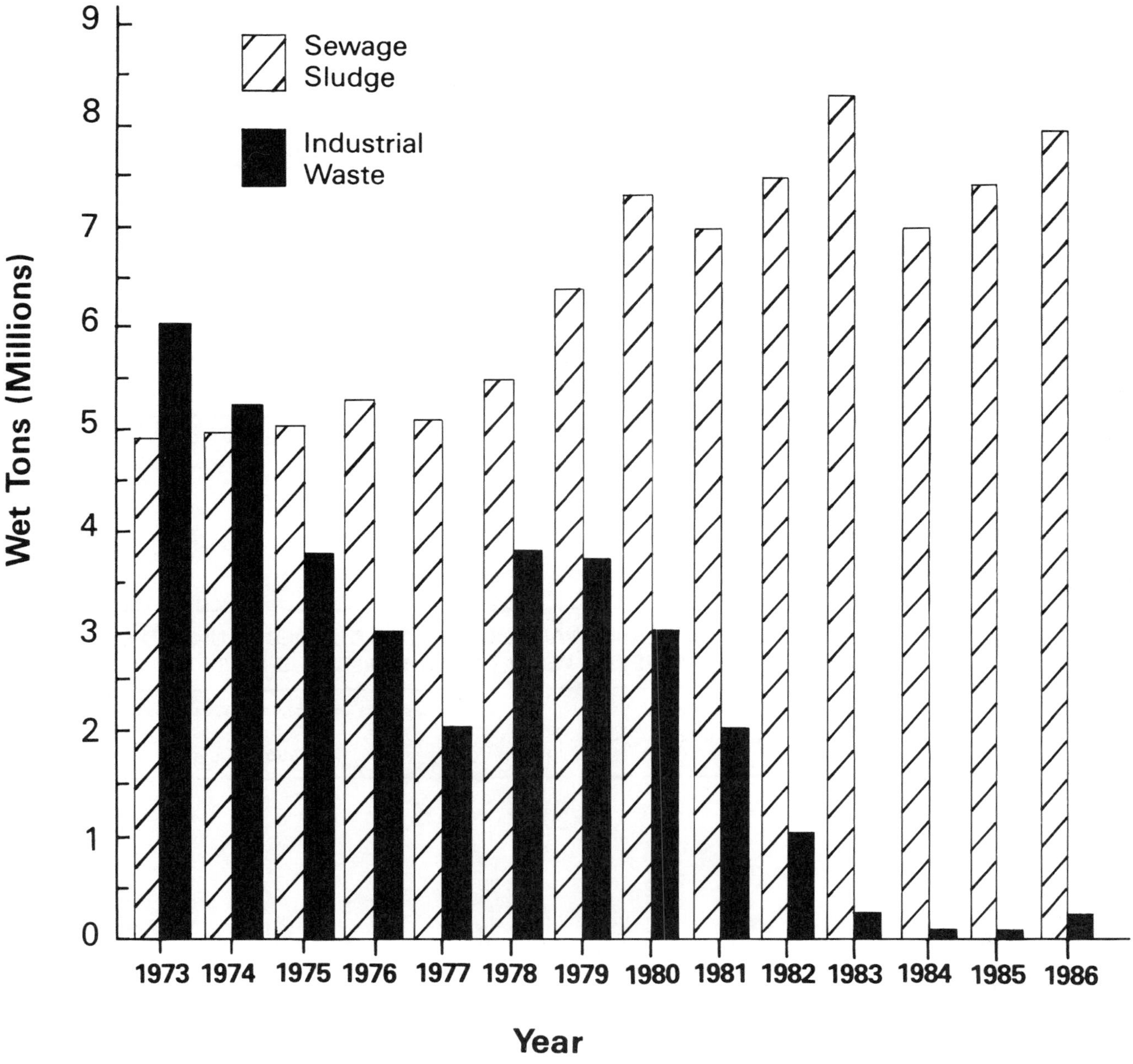

Note: For the purpose of this graph, Industrial Waste Category also includes Fish Waste and Construction Debris

Source: U.S.Environmental Protection Agency, 1988, Report to Congress on Administration of the Marine Protection, Research, and Sanctuaries Act of 1972, as Amended, 1984–1986, EPA–503/8–88/002

TABLE 7–144. QUANTITIES OF SLUDGE DUMPED BY SEWAGE AUTHORITIES IN UNITED STATES OCEAN WATERS IN 1984, 1985, AND 1986

[Nine municipal sewage authorities which had previously held interim permits are dumping sewage sludge pursuant to court orders issued by United States district courts in New York and New Jersey. These authorities have been required to submit permit applications to the USEPA, and currently are shifting their dumping from the 12-Mile Site to the Deepwater Municipal Sludge Dump Site, also known as the 106-Mile Site.]

	Quantities in Thousand Wet Tons		
	1984	**1985**	**1986**
Sewage Authorities			
Bergen County Utilities Authority NJ	255	309	353
Joint Meeting of Essex and Union Counties NJ	385	341	238
Linden Roselle Sewerage Authority NJ	235	95	93
Middlesex County Utilities Authority NJ	966	1,039	1,018
Nassau County Dept. of Public Works NY	520	576	709
New York City Dept. of Environmental Protection NY	3,085	3,345	3,591
Passaic Valley Sewerage Commission NJ	854	884	1,317
Rahway Valley Sewerage Authority NJ	160	187	98
Westchester County Dept. of Environmental Facilities NY	539	470	506
	6,999	7,246	7,923

Source: U.S.Environmental Protection Agency, 1988, Report to Congress on Administration of the Marine Protection, Research, and Sanctuaries Act of 1972, as Amended, 1984–1986, EPA–503/8–88/002

TABLE 7–145. QUANTITIES OF INDUSTRIAL WASTE MATERIALS DUMPED IN UNITED STATES OCEAN WATERS BY SPECIAL PERMIT IN 1984, 1985, AND 1986

	Quantities in Thousand Wet Tons		
	1984	**1985**	**1986**
USEPA Region II			
Acid Waste Site (NY Bight Apex):			
Allied Chemical Corp.[1] NY	40	40	34
Deepwater Industrial Waste Site:			
DuPont — Edge Moor[2] DE	19	0	140
DuPont — Grasselli[3] NJ	146	100	73
USEPA Region IX			
Fish Wastes Site			
Samoa Packing, American Samoa	8	4.6	21.4
Star Kist, American Samoa	7.9	20.3	24.1
Oil Drilling Muds and Cuttings			
THUMS Long Beach, CA	—[4]	2.7	13.6
	220.9	167.6	306.1

[1] Hydrochloric acid waste.
[2] Aqueous iron and miscellaneous chlorides and hydrochloric acid wastes.
[3] Solution of alkaline sodium wastes.
[4] No permit issued.
Source: U.S.Environmental Protection Agency, 1988, Report to Congress on Administration of the Marine Protection, Research, and Sanctuaries Act of 1972, as Amended, 1984–1986, EPA–503/8–88/002

SECTION M. ENERGY DEVELOPMENT

TABLE 7–146. WATER-QUALITY IMPACTS OF VARIOUS ENERGY PROCESSES

[Frequency: H = high, M = medium, L = low. Time frame: L = longer than 10 years, M = 1 to 10 years, S = less than 1 year. Severity: 5 = direct threat to human life, 4 = hazardous to human health, 3 = severe economic damage, 2 = damage to biota, 1 = aesthetic or other intangible harm.]

Process	Water-Quality Impacts	Frequency and Areal Scale		Time Frame	Severity	Effectiveness of Controls
		Regional	Local			
Extraction and on-site processing						
Coal mining						
Underground mining	Most damaging problems are acid mine drainage and disruption of aquifers, which affect pH, dissolved solids and specific ion content, and thus impair utility of streams and ground waters for other uses.					
	a. Surface waters	H		L	3	Poor
	b. Ground waters		M	L	3	Ineffective
Surface mining	Surface disturbance results in high sediment transport potential. Discharge from mines may impair water quality through increase in dissolved solids and specific ions.					
	a. Surface waters	H		M	3	Fair
	b. Ground waters		L	S	1	Good
Beneficiation	Release of chemical and physical treatment materials to streams can impair water quality. Leaching of solid wastes results in pollution similar to acid drainage.					
	a. Surface waters	H		M	2	Fair
	b. Ground waters		L	M	1	Fair
Oil and gas extraction						
Primary recovery	Principal problems are handling of saline waste waters. Leaks in casings, pipes, and storage ponds can release brines to ground waters and streams.					
	a. Surface waters	M		S	2	Good
	b. Ground waters		M	M	3	Fair
Secondary and tertiary recovery	Principal concerns are escape of oil and formation waters through casing, pipe, and storage tank leaks releasing organic and inorganic contaminants to the environment.					
	a. Surface waters		L	S	2	Good
	b. Ground waters		L	L	3	Good
Offshore operation	Blowouts with resulting massive oil contamination are a rare but catastrophic problem.	L		S	3	Good
Oil-shale extraction and processing	Most significant concerns relate to potential for escape of noxious organic and inorganic contaminants to streams. Disruption of aquifers likely, low hazard due to limited occurrence of oil shale.					
Underground mining	Concerns center on disruption of aquifers and disposal of sometimes saline dewatering by injection.					
	a. Surface waters		L	L	1	Good
	b. Ground waters		L	L	1	Good
Surface mining	Surface disturbance results in high sediment transport potential.					
	a. Surface waters		L	S	2	Excellent
	b. Ground waters		L	M	1	Excellent
Surface retorting	Concerns center on potential for escape of organic and inorganic contaminants from plant site due to accidental leaks and spills. A more significant concern is escape of contaminants from waste piles through leaching.					
	a. Surface waters		M	L	2	Good to Fair
	b. Ground waters		L	L	1	Good

TABLE 7–146. WATER-QUALITY IMPACTS OF VARIOUS ENERGY PROCESSES (continued)

[Frequency: H = high, M = medium, L = low. Time frame: L = longer than 10 years, M = 1 to 10 years, S = less than 1 year. Severity: 5 = direct threat to human life, 4 = hazardous to human health, 3 = severe economic damage, 2 = damage to biota, 1 = aesthetic or other intangible harm.]

Process	Water-Quality Impacts	Frequency and Areal Scale: Regional	Frequency and Areal Scale: Local	Time Frame	Severity	Effectiveness of Controls
Oil-shale extraction and processing (cont'd)						
In-situ recovery	Underground effects mainly involve contamination of ground waters by organic and inorganic compounds produced in combustion; where applicable surface effects are similar to those of surface retorting.					
	a. Surface waters		L	S	2	Excellent
	b. Ground waters		M	L	3	Untested
Tar-sands extraction and processing						
Surface mining and processing	Main concerns are accidental release of organic contaminants to streams and potential for failure of waste impoundment structures leading to massive downstream damage from fine waste.					
	a. Surface waters		M	S	2	Good to Fair
	b. Ground waters		L	L	1	Untested
In-situ recovery	Concerns center on potential for escape of noxious organic and inorganic chemicals to ground waters.					
	a. Surface waters		L	S	2	Untested
	b. Ground waters		M	L	3	Untested
Geothermal extraction						
Vapor-dominated systems	Main concerns are escape of noxious inorganic contaminants to ground water from waste disposal, blowouts, and leaks in casings and pipes.					
	a. Surface waters		L	S	2	Good
	b. Ground waters		M	L	1	Fair
Water-dominated systems	Main problems involve escape of noxious and toxic constituents of thermal waters to surface and ground waters from production operations, waste disposal, blowouts, and leaks in casings and pipes.					
	a. Surface waters	H		M	4	Fair
	b. Ground waters		H	L	3	Fair
Uranium mining and milling						
Underground mining	Escape of radioactive and other inorganic contaminants to the environment through disposal of dewatering waste and escape from tailings ponds can seriously impair downstream water uses.					
	a. Surface waters	M		S	4	Good
	b. Ground waters		M	L	4	Poor
Surface Mining	Some concern about high sediment transport potential, but main source of concern is potential for radioactive contamination of streams and ground waters through leakage from tailings disposal ponds.					
	a. Surface waters	M		S	4	Good
	b. Ground waters		M	L	4	Poor
Solution mining	Main concern centers on escape of radioactive and inorganic process chemicals to off-site ground waters.					
	a. Surface waters		L	S	1	Excellent
	b. Ground waters		L	L	4	Excellent
Transportation						
Coal slurry lines	Main concern centers on pipeline breaks and the potential for contamination of streams.					
	a. Surface waters		L	S	2	Excellent
	b. Ground waters		L	L	1	Excellent
Oil pipelines	Most significant problems are pipeline breaks and resulting oil pollution of streams.					
	a. Surface waters		M	S	2	Fair
	b. Ground waters		L	L	1	Good

TABLE 7–146. WATER-QUALITY IMPACTS OF VARIOUS ENERGY PROCESSES (continued)

[Frequency: H = high, M = medium, L = low. Time frame: L = longer than 10 years, M = 1 to 10 years, S = less than 1 year. Severity: 5 = direct threat to human life, 4 = hazardous to human health, 3 = severe economic damage, 2 = damage to biota, 1 = aesthetic or other intangible harm.]

Process	Water-Quality Impacts	Frequency and Areal Scale: Regional	Frequency and Areal Scale: Local	Time Frame	Severity	Effectiveness of Controls
Transportation (cont'd)						
Oil tankers	Escape of oil to marine environment as result of shipwrecks can be catastropic to marine life over wide areas.	L		M	3	Poor
Refining						
Oil refining	Controlled release of waste water and accidental releases of organic and inorganic contaminants are most significant issues; concerns center on impairment of water supplies of other water users.					
	a. Surface waters		M	M	3	Good
	b. Ground waters		L	L	3	Good
Nuclear fuel cycle	Accidental releases of radioactive materials to surface and ground waters from processing and reprocessing plants are main concern; both high- and low-level waste disposal also have potential for escape of radioactivity to the water environment. Controlled release of nonradioactive inorganic chemicals adds to chemical load of receiving waters.					
	a. Surface waters		M	S	4	Good
	b. Ground waters		M	L	4	Fair to poor
Conversion						
Fossil-fueled steam electric generation	Controlled release of cooling-system blowdown to streams and/or leakage from cooling ponds add dissolved solids and treatment chemicals to stream loads. Once-through cooling contributes to thermal pollution.					
	a. Surface waters	M		M	2	Good
	b. Ground waters		L	L	1	Good
Nuclear steam-electric generation	Small controlled releases of radioactive materials and discharge of cooling-system blowdown add radioactivity, dissolved solids, and treatment chemicals to stream loads. Accidental release of radioactivity through reactor containment failure could endanger human life over wide area.					
	a. Surface waters	M		M	5	Good
	b. Ground waters		L	L	4	Good
Geothermal electric generation	Disposal of waste and condensate containing noxious inorganic compounds and thermal load to streams impair downstream uses and damages aquatic life.	H		M	4	Fair
	a. Surface waters		H	L	3	Fair
	b. Ground waters					
Hydroelectric generation	Changes in stream temperature and dissolved gases due to storage and reservoir releases seriously alter the aquatic environment.					
	a. Surface waters	H		L	2	Fair
	b. Ground waters		L	S	1	Good
Coal Conversion Processes	Controlled release of cooling system blowdown and accidental releases of organic and inorganic contaminants, as in oil refining and with similar concern about impairment of other water uses.					
	a. Surface waters		M	M	3	Good
	b. Ground waters		L	L	3	Good

Source: Davis, G.H., 1985, Water and Energy: Demand and Effects, Unesco Studies and Reports in Hydrology 42.

SECTION N. WATERBORNE DISEASES/HEALTH HAZARDS

TABLE 7–147. MAGNITUDE OF WATERBORNE DISEASE OUTBREAKS IN THE UNITED STATES, 1920–1980

Size of Outbreak (Cases of Illness)	Frequency of Occurrence (Number of Outbreaks)			
	Community Systems	Noncommunity Systems	Individual Systems	All Systems
<2	3	0	3	6
2–5	26	35	95	156
6–10	71	50	81	202
11–25	145	119	63	327
26–50	94	124	34	252
51–100	68	82	16	166
101–200	63	50	6	119
201–300	28	14	2	44
301–500	29	14	1	44
501–1000	29	9	1	39
1001–3000	28	3	0	31
3001–5000	9	0	0	9
5001–10,000	5	0	0	5
>10,000	5	0	0	5
Total	603	500	302	1405

Source: Craun, G.F., 1986, Waterborne Diseases in the United States. Copyright CRC Press, Inc., Boca Raton, FL. Reprinted with permission

TABLE 7–148. ETIOLOGY OF WATERBORNE DISEASE OUTBREAKS IN THE UNITED STATES, 1920–1984

Time Period	Disease	Outbreaks	Cases	Deaths	Time Period	Disease	Outbreaks	Cases	Deaths
1920–1925	Typhoid fever	127	7,294	435		Chemical poisoning	5	30	6
	Gastroenteritis	11	27,756	0		Salmonellosis	3	16,425	3
1926–1930	Typhoid fever	100	3,072	234		Giardiasis	1	123	0
	Gastroenteritis	17	63,902	0		Paratyphoid fever	1	5	0
1931–1935	Typhoid fever	85	2,114	140	1966–1970	Gastroenteritis	21	5,922	0
	Gastroenteritis	25	7,664	0		Hepatitis A	19	562	1
	Amebiasis	1	1,412	98		Shigellosis	14	1,215	0
	Hepatitis A	1	28	0		Typhoid fever	4	45	0
1936–1940	Gastroenteritis	91	77,403	2		Salmonellosis	4	226	0
	Typhoid fever	60	1,281	80		Toxigenic E. coli AGI	4	188	4
	Shigellosis	10	3,308	0		Chemical poisoning	4	15	0
	Chemical poisoning	1	92	0		Amebiasis	3	39	2
	Amebiasis	1	4	0		Giardiasis	2	53	0
1941–1945	Gastroenteritis	126	36,118	3	1971–1975	Gastroenteritis	63	17,752	0
	Typhoid fever	56	1,450	46		Shigellosis	14	2,803	0
	Shigellosis	10	2,817	6		Hepatitis A	14	368	0
	Salmonellosis	1	12	0		Giardiasis	13	5,136	0
	Paratyphoid fever	2	14	0		Chemical poisoning	13	513	0
	Chemical poisoning	1	30	0		Typhoid fever	4	222	0
1946–1950	Gastroenteritis	87	10,718	0		Salmonellosis	2	37	0
	Typhoid fever	18	264	5		Toxigenic E, coli AGI	1	1,000	0
	Hepatitis A	5	173	0	1976–1980	Gastroenteritis	114	22,093	0
	Shigellosis	4	2,321	1		Giardiasis	26	14,416	0
	Paratyphoid fever	1	5	0		Chemical poisoning	25	3,081	1
	Leptospirosis	1	9	0		Shigellosis	10	2,392	0
	Tularemia	1	4	0		Viral gastroenteritis	10	3,147	0
1951–1955	Gastroenteritis	31	5,297	0		Salmonellosis	6	1,113	0
	Typhoid fever	7	103	0		Campylobacterosis	3	3,821	0
	Hepatitis A	7	340	0		Hepatitis A	2	95	0
	Shigellosis	4	732	1	1981–1984	Gastroenteritis,			
	Amebiasis	1	31	2		undetermined etiology	59	20,772	0
	Salmonellosis	1	2	0		Giardiasis	48	4,048	0
	Poliomyelitis	1	16	0		Chemical poisoning	11	179	0
1956–1960	Gastroenteritis	21	2,306	0		Shigellosis	7	532	0
	Typhoid fever	13	128	3		Hepatitis A	7	274	0
	Hepatitis A	11	417	0		Viral gastroenteritis,			
	Shigellosis	7	3,081	0		Norwalk agent	7	1,077	0
	Chemical poisoning	3	14	4		Salmonellosis	2	1,150	0
	Salmonellosis	2	17	0		Campylobacteriosis	6	993	0
	Amebiasis	1	5	0		Viral gastroenteritis,			
	Tularemia	1	2	0		rotavirus	1	1,761	0
1961–1965	Gastroenteritis	18	20,627	0		Cholera	1	17	0
	Typhoid fever	11	63	0		Yersiniosis	1	16	0
	Hepatitis A	10	334	0		Cryptosporidium	1	117	0
	Shigellosis	7	520	4		Entamoeba	1	4	0

Source: Craun, G.F., 1986, Waterborne Diseases in the United States. Copyright CRC Press, Inc., Boca Raton, FL. Reprinted with permission; amended with statistics from Center for Disease Control Annual Summaries, 1981–84

TABLE 7–149. WATERBORNE DISEASE OUTBREAKS CAUSED BY USE OF CONTAMINATED, UNTREATED SURFACE WATER IN THE UNITED STATES, 1920–1980

	Type of Water System							
	Community		Non-community		Individual		All	
Deficiency	OB[a]	cases	OB[a]	cases	OB[a]	cases	OB[a]	cases
Contamination on watershed	26	3,498	3	57	12	257	41	3,812
Use of surface water for supplemental source	7	3,613	7	245	2	115	16	3,973
Overflow of sewage or outfall near water intake	3	103	3	39	5	87	11	229
Flooding, heavy rains	2	125	1	93	1	77	4	295
Dead animals in reservoir	—	—	1	100	—	—	1	100
Insufficient data	27	1,228	24	726	28	436	79	2,390
Total	65	8,567	39	1,260	48	972	152	10,799

[a] Number of outbreaks.
Source: Craun, G.F., 1986, Waterborne Diseases in the United States. Copyright CRC Press, Inc., Boca Raton, FL. Reprinted with permission

TABLE 7–150. WATERBORNE DISEASE OUTBREAKS CAUSED BY USE OF CONTAMINATED, UNTREATED GROUND WATER (SPRINGS) IN THE UNITED STATES, 1920–1980

	Type of Water System							
	Community		Non-community		Individual		All	
Deficiency	OB[a]	Cases	OB[a]	Cases	OB[a]	Cases	OB[a]	Cases
Overflow or seepage of sewage	8	238	3	35	5	39	16	312
Surface runoff	11	265	5	162	7	75	23	502
Flooding	2	76	2	123	—	—	4	199
Creviced limestone	1	200	3	213	—	—	4	413
Contamination of raw water transmission line	2	284	1	7	—	—	3	291
Improper construction	—	—	1	26	1	9	2	35
Insufficient data	12	508	18	1961	20	415	50	2884
Total	36	1571	33	2527	33	538	102	4636

[a] Number of outbreaks.
Source: Craun, G.F., 1986, Waterborne Diseases in the United States. Copyright CRC Press, Inc., Boca Raton, FL. Reprinted with permission

TABLE 7–151. WATERBORNE DISEASE OUTBREAKS CAUSED BY USE OF CONTAMINATED, UNTREATED GROUND WATER (WELLS) IN THE UNITED STATES, 1920–1980

	Type of Water System							
	Community		Non-community		Individual		All	
Deficiency	OB[a]	Cases	OB[a]	Cases	OB[a]	Cases	OB[a]	Cases
Overflow or seepage of sewage	28	14,915	104	10,236	52	675	184	25,826
Surface runoff, heavy rains	25	2,492	26	947	34	824	85	4,263
Creviced limestone, fissured rock	9	1,404	19	2,044	12	660	40	4,108
Improper construction, faulty well casing	8	342	10	414	9	141	27	897
Flooding	9	5,883	3	107	5	211	17	6,201
Chemical contamination	3	77	2	16	10	68	15	161
Contamination by stream or river	3	445	6	392	3	48	12	885
Contamination of raw water transmission line	8	10,481	—	—	—	—	8	10,481
Seepage from abandoned well	3	144	1	50	—	—	4	194
Animal in well	1	34	1	238	2	19	4	291
Insufficient data	19	18,480	67	3,309	40	413	126	22,202
Total	116	54,697	239	17,753	167	3,059	522	75,509

[a] Number of outbreaks.
Source: Craun, G.F., 1986, Waterborne Diseases in the United States. Copyright CRC Press, Inc. Boca Raton, FL. Reprinted with permission

TABLE 7–152. WATERBORNE DISEASE OUTBREAK AND DISEASE RATES ATTRIBUTED TO SOURCE CONTAMINATION AND TREATMENT INADEQUACIES IN COMMUNITY SYSTEMS IN THE UNITED STATES USING SURFACE WATER SOURCES, 1971–1985

Type of Community Water System	Waterborne Disease Outbreaks per 1,000 Water Systems	Waterborne Illnesses per Million-Person Years
Untreated	32.5	370.9
Disinfected only	40.5	66.3
Filtered and disinfected water	5.0	4.7

Source: U.S. Environmental Protection Agency, 1987; Craun, G.F., 1987

TABLE 7–153. WATER SUPPLY DEFICIENCIES RESPONSIBLE FOR WATERBORNE OUTBREAKS IN THE UNITED STATES, 1971–1985

Source of Deficiency	Out-breaks	Reported Illnesses
Surface water source:		
No treatment	31	1,647
Disinfection only, or inadequate disinfection	67	23,028
Disinfection with other treatment (but no filtration)	5	969
Filtration and disinfection	20	9,852
Totals	123	35,496
Groundwater source:		
No treatment	154	11,266
Inadequate disinfection	90	40,893
Disinfection with other treatment	1	22
Totals	245	52,181
Distribution system:		
Cross-connection	44	8,124
Contamination of mains/plumbing	14	3,413
Contamination of storage	11	6,244
Corrosive water	10	147
Totals	79	17,928
Grand Total (Reported):		
Outbreaks		447
Illnesses		105,605

Source: U.S. Environmental Protection Agency, 1987; Craun, G.F., 1987

FIGURE 7–18. SEASONAL DISTRIBUTION OF WATERBORNE DISEASE OUTBREAKS IN SURFACE WATER SYSTEMS IN THE UNITED STATES, 1971–1985

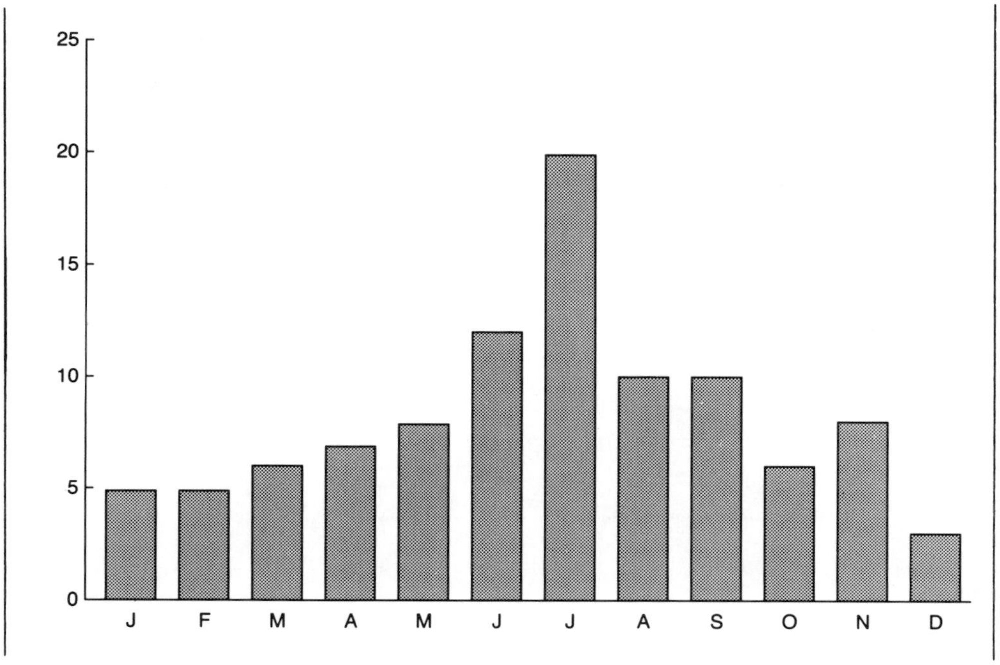

Source: Craun, G.F., 1988, Surface Water Supplies and Health, J. American Water Works Assoc., v. 80, no.2. Copyright AWWA. Reproduced with permission

FIGURE 7–19. ETIOLOGY OF WATERBORNE DISEASE OUTBREAKS IN UNTREATED, DISINFECTED-ONLY, AND FILTERED SURFACE WATER SYSTEMS IN THE UNITED STATES, 1971–1985

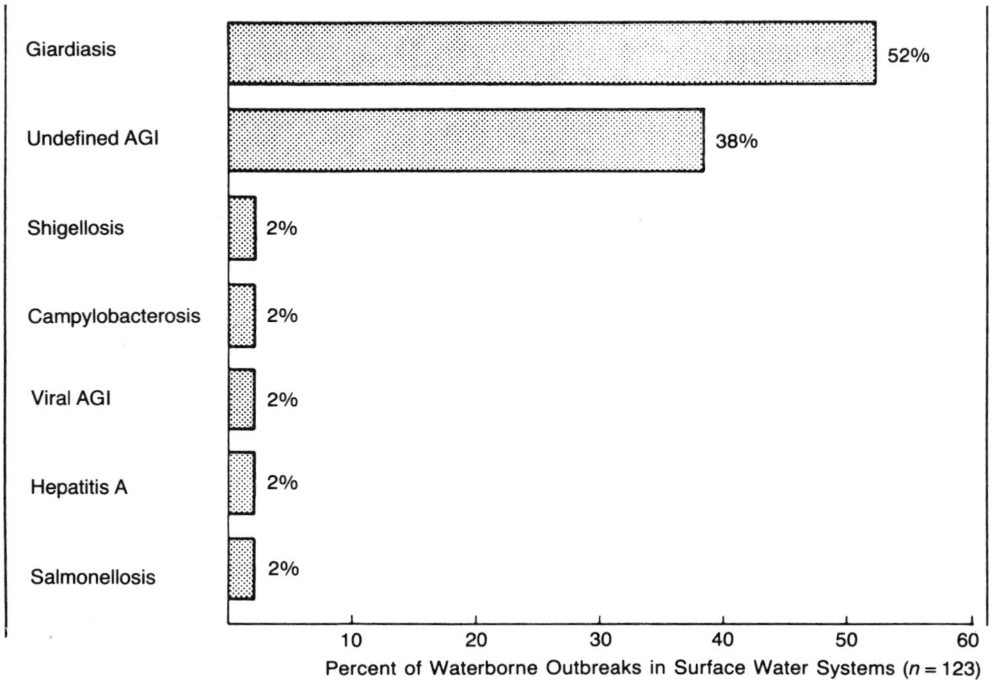

AGI: Acute gastrointestinal illness.
Source: Craun, G.F., 1988, Surface Water Supplies and Health, J. American Water Works Assoc., v. 80, no.2. Copyright AWWA. Reproduced with permission

FIGURE 7–20. MAJOR WATER SUPPLY DEFICIENCIES RESPONSIBLE FOR WATERBORNE DISEASE OUTBREAKS IN THE UNITED STATES, 1971–1985

[Percent of all waterborne outbreaks of Giardiasis and all other illnesses, respectively]

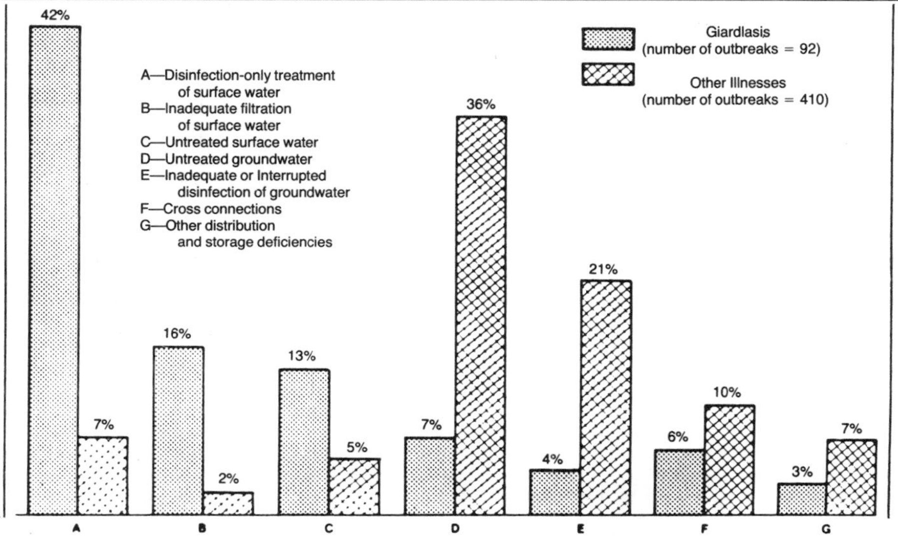

Source: Craun, G.F., 1988, Surface Water Supplies and Health, J. American Water Works Assoc., vol. 80, no.2. Copyright AWWA. Reproduced with permission

TABLE 7–154. EPIDEMIOLOGICAL CHARACTERISTICS OF THE PRINCIPAL PATHOGENIC AGENTS IN WASTEWATER

Agents	Quantity Excreted per g/feces	Latency[a]	Survival[b]	Multiplication in the Environment	Infecting Dose ID 50[c]
Virus					
Enterovirus (including polio, echo and coxsackie)	10^7	0	3 months	no	100
Hepatitis A	10^6?	0	?	no	?
Rotavirus	10^6?	0	?	no	?
Bacteria					
Colibacilli	10^8	0	3 months	yes	$\pm 10^9$
Salmonella thyphi	10^8	0	2 months	yes	10^7
Other salmonellas	10^8	0	2–3 months	yes	10^6
Shigella	10^7	0	1 month	yes	10^4
Campylobacter	10^7	0	7 days	yes	10^6
Cholera	10^7	0	1 month	yes	10^8
Yersinia enterocolitica	10^5	0	3 months	yes	10^9
Leptospira	urine	0	7 days	no	low
Parasites					
Dysentery amoeba	10^7	0	25 days	no	10 to 100
Giardia	10^5	0	25 days	no	25 to 100
Balantidium coli	?	0	20 days?	no	25 to 100
Ascaris	10^4	10 days	1 year	no	several units
Ancyclostoma	10^2	7 days	3 months	no	1
Anguillula	10	3 days	3 weeks	yes	1
Trichocephalus	10^3	20 days	9 months	no	several units
Hymenolepis	?	0	10 days	no	1
Taenia	10^4	2 months	9 months	no	1
Fasciola hepatica	?	2 months	4 months	yes	several units
Other flukes	10^2	6 to 8 weeks	life of host	yes	several units

[a] Period necessary for excreted pathogenic agent to become infectious to receiving or susceptible individual; (0 = immediate).

[b] In environment, outside final host (man or animal).

[c] Dose sufficient to provoke the appearance of clinical symptoms in 50% of individuals tested.

Source: Prost, A., 1987, Health Risks Stemming from Wastewater Reutilization, Water Quality Bulletin, v. 12, no. 2

TABLE 7–155. WATERBORNE DISEASES DUE TO MICROBES

Waterborne Disease	Organism	Health Effect
Gastroenteritis	Various pathogens	Acute diarrhea and vomiting
Typhoid	Salmonella typhosa (bacteria)	Inflamed intestine, enlarged spleen, high temperatures—fatal
Bacillary dysentery	Shigella (bacteria)	Diarrhea, rarely fatal
Cholera	Vibrio Comma (bacteria)	Vomiting, severe diarrhea, rapid dehydration, mineral loss—high mortality
Infectious hepatitis	Virus	Yellow skin, enlarged liver, abdominal pain—low mortality—lasts up to four months
Amebic dysentery	Entamoeba histolytica (protozoa)	Mild diarrhea, chronic dysentery
Giardiasis	Giardia lamblia (protozoa)	Diarrhea, cramps, nausea and general weaknesses—lasts one week to 30 weeks—not fatal

Source: American Water Works Association, 1979, Principles and Practices of Water-Supply Operations, Vol. 1: Introduction to Water Sources and Transmission

TABLE 7–156. OCCURRENCE OF CRYPTOSPORIDIUM OOCYSTS IN VARIOUS WATERS THROUGHOUT THE WESTERN UNITED STATES

Water Sampled	Number of Samples	Number of Samples Positive	Percent Positive	Oocysts/L[a]
Raw sewage	11	10	91	28.4
Treated sewage[b]	22	20	91	17
Reservoir, lake	32	24	75	0.91
Stream, river	58	45	77	0.94
Filtered drinking water	10	2	20	0.001
Nonfiltered drinking water	4	2	50	0.006

[a] Geometric means.
[b] Activated sludge.
Source: Craun, G.F., 1988, Surface Water Supplies and Health, J.Am. Water Works Assoc., v. 80, no.2.

FIGURE 7–21. WATERBORNE OUTBREAKS OF GIARDIASIS IN THE UNITED STATES, 1965–1985

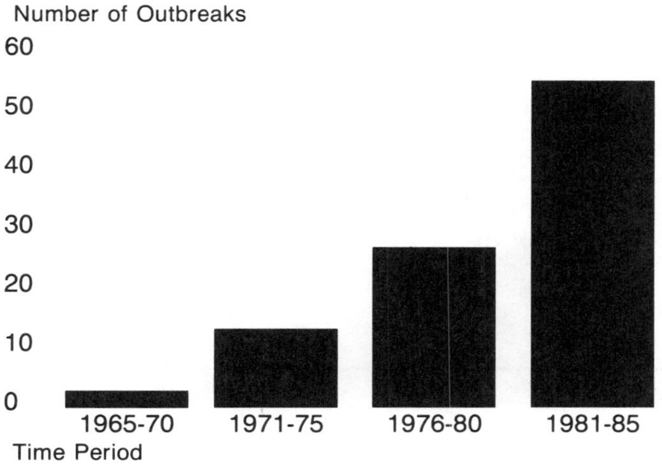

Source: Craun, Gunther F., 1988, Waterborne Outbreaks of Giardiasis: Why They Happen, How to Prevent Them, Health and Environment Digest, v. 2, no. 1

FIGURE 7–22. WATERBORNE OUTBREAKS OF GIARDIASIS BY TYPE OF WATER SYSTEM IN THE UNITED STATES, 1965–1985

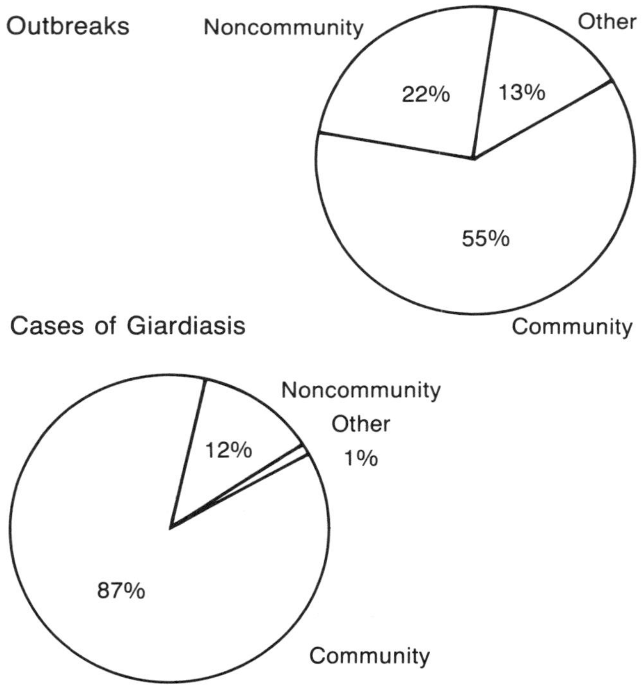

Source: Craun, Gunther F., 1988, Waterborne Outbreaks of Giardiasis: Why They Happen, How to Prevent Them, Health and Environment Digest, v. 2, no. 1

TABLE 7–157. WATER- AND SANITATION-RELATED INFECTIONS IN DEVELOPING COUNTRIES AND THEIR CONTROL

Infections	Importance of Alternate Control Methods[a]							Public Health Importance[a]
	Water Quality	Water Availability	Excreta Disposal	Excreta Treatment	Personal and Domestic Cleanliness	Drainage and Sullage Disposal	Food Hygiene	
Diarrhoeal diseases and enteric fevers								
Viral agents	2	3	2	1	3	0	2	3
Bacterial agents	3	3	2	1	3	0	3	3
Protozoal agents	1	3	2	1	3	0	2	2
Poliomyelitis and hepatitis A	1	3	2	1	3	0	1	3
Worms with no intermediate host								
Ascaris and *Trichuris*	0	1	3	2	1	1	2	2
Hookworms	0	1	3	2	1	0	1	3
Beef and pork tapeworms	0	0	3	3	0	0	3	2
Worms with intermediate aquatic stages								
Schistosomiasis	1	1	3	2	1	0	0	3
Guinea worm	3	0	0	0	0	0	0	2
Worms with two aquatic intermediate stages	0	0	2	2	0	0	3	1
Skin, eye and louse-borne infections	0	3	0	0	3	0	0	2
Infections spread by water-related insects								
Malaria	0	0	0	0	0	1	0	3
Yellow fever and dengue	0	0	0	0	0	1	0	3
Bancroftian filariasis	0	0	3	0	0	3	0	3

[a] 0 = no importance; 1 = little importance; 2 = moderate importance; 3 = great importance

Source: Feachem, R.G., 1984, The Health Dimension of the Decade, in World Water '83: the World Problem, Proc. July 1983 Conference of Institution of Civil Engineers, London

TABLE 7–158. INFECTIONS AND DEATHS DUE TO ABSENCE OF SAFE WATER AND SANITATION IN DEVELOPING COUNTRIES

	Infection	Infections Thousands per year	Deaths Thousands per year	Average No. of Days Lost per Case	Relative Disability[a]
WATER-BORNE DISEASES spread by drinking or washing hands, food or utensils in contaminated water, which acts as a passive vehicle for the infecting agent.	Amebiasis	400,000	30	7–10	3
	Diarrhoeas	3–5,000,000	5–10,000	3–5	2
	Polio	80,000	10–20	3,000+	2
	Typhoid	1,000	25	14–28	2
WATER-WASHED DISEASES spread by poor personal hygiene and insufficient water for washing. Lack of proper facilities for human waste disposal is another contributing factor.	Ascariasis (roundworm)	800,000–1,000,000	20	7–10	3
	Leprosy	12,000	Very low	500–3,000	2–3
	Trichuriasis (whipworm)	500,000	Low	7–10	3
WATER-BASED DISEASES transmitted by a vector which spends a part of its life cycle in water. Contact with water thus infected conveys the disease-causing parasite through the skin or mouth.	Schistosomiasis (bilharzia)	200,000	500–1000	600–1000	3–4
DISEASES WITH WATER-RELATED VECTORS contracted through infection-carrying insects which breed in water and bite near it, especially when it is stagnant.	African trypanosomiasis (sleeping sickness)	1,000	5	150	1
	Malaria	800,000	1,200	3–5	2
	Onchocerciasis (river blindness)	30,000	20–50	3,000	1–2
FECAL DISPOSAL DISEASES caused by organisms that breed in excreta when sanitation is defective.	Hookworm	7–9,000,000	50–60	100	4

[a] 1 means the sufferer is bedridden; 2 able to function to some extent; 3 able to work; 4 experiences minor effects.
Source: United Nations Development Programme, 1987, Decade Dossier; after Julia A. Walsh and Kenneth S. Warren, Selective Primary Health Care: An Interim Strategy for Disease Control in Developing Countries, The New England Journal of Medicine, vol. 301, no. 18, November 1, 1979, p. 967

TABLE 7–159. REDUCTION IN DISEASE MORBIDITY OR INFECTION RATE DUE TO IMPROVEMENT IN WATER SUPPLY AND SANITATION IN DEVELOPING COUNTRIES

Diseases	Interventions, improvements			Percentage Reduction Median	Range	Number of Cases
Diarrhoeal diseases						
	Water quality—			16	0–90	9
	Water availability—			25	0–100	17
	Water quality and availability—			37	0–82	8
	Excreta disposal—			22	0–48	10
	Water supply or excreta disposal			22	0–100	53
	Small service improvement	Adult literacy rate (%)	<40	18		11
			40–75	20		4
			>75	16		10
	Large service improvement	Adult literacy rate (%)	<40	46		7
			40–75	39		8
			>75	32		13
Cholera	Water supply or excreta disposal			41	0–91	11
Shigellosis				48	0–81	27
Entamoeba histolytica				2	0–80	17
Giardia lamblia				0	0–20	10

Source: World Health Organization, 1986, Statistics Quarterly, vol. 39, no.1; after Esrey, S.A., and others, 1985, Interventions for the Control of Diarrhoeal Diseases among Young Children: Improving Water Supplies and Excreta Disposal Facilities, Bull. World Health Organization, vol. 63, no.4, pp. 757–772

TABLE 7–160. CLINICAL SYNDROMES AND INCUBATION PERIODS OF INFECTIOUS AND CHEMICAL AGENTS CAUSING ACUTE WATERBORNE DISEASE IN THE UNITED STATES

Agent	Incubation Period	Clinical Syndrome
Bacteria		
Campylobacter jejuni	2–5 days	Gastroenteritis, often with fever
Enterotoxigenic *Escherichia coli*	6–36 hr	Gastroenteritis
Salmonella	6–48 hr	Either gastroenteritis (often with fever), enteric fever, or extraintestinal infection
Salmonella typhi	10–14 days	Enteric fever—fever, anorexia, malaise, transient rash, splenomegaly, and leukopenia
Shigella	12–48 hr	Gastroenteritis, often with fever and bloody diarrhea
Vibrio cholerae 01	1–5 days	Gastroenteritis, often with significant dehydration
Yersinia enterocolitica	3–7 days	Either gastroenteritis, mesenteric lymphadenitis, or acute terminal ileitis; may mimic appendicitis
Viruses		
Hepatitis A	2–6 weeks	Hepatitis—nausea, anorexia, jaundice, dark urine
Norwalk virus	24–48 hr	Gastroenteritis, of short duration
Rotavirus	24–72 hr	Gastroenteritis, often with significant dehydration
Parasites		
Entamoeba histolytica	2–4 weeks	Varies from mild gastroenteritis to acute fulminating dysentery with fever and bloody diarrhea
Giardia lamblia	1–4 weeks	Chronic diarrhea, epigastric pain, bloating, malabsorption, and weight loss
Chemicals		
Fluoride	<1 hr	Nausea, vomiting, and abdominal cramps
Heavy metals		
Antimony		
Cadmium		
Copper		
Lead		
Tin		
Zinc, etc.	<1 hr	Nausea, vomiting, and abdominal cramps, often accompanied by a metallic taste
Others		
Pesticides		
Petroleum products, etc.	Variable	Variable

Source: Craun, G.F., 1986, Waterborne Diseases in the United States, Copyright CRC Press, Inc. Boca Raton, FL. Reprinted with permission

TABLE 7–161. SURVIVAL OF BACTERIA IN VARIOUS MEDIA

Organism	Survival Time	Media
E. coli	63 days	Recharge well
E. coli	3–3.5 months	Groundwater in the field
E. coli	4–4.5 months	Groundwater held in lab
Coliforms	17 hours—50 percent reduction	Well water
Salmonella	44 days	Water infiltrating sand column
S. typhi	2–85 days	Soil
S. typhi	24–27 days	Septic tank
S. typhi	25–41 days	Soil
Shigella	24 days	Water infiltrating sand column
S. flexneri	26.8 hours—50 percent reduction	Well water
Vibrio cholerae	7.2 hours—50 percent reduction	Well water

Source: McGinnis, J.A. and DeWalle, F.B., 1983, The Movement of Typhoid Organisms in Saturated Permeable Soil, J.Am. Water Works Assoc., vol. 75, no.6. Copyright AWWA. Reprinted with permission

TABLE 7–162. SURVIVAL TIMES OF ORGANISMS

Organism	Media	Survival Time days
Ascaris ova	Vegetables	27–35
	Soil	730–2010
Entamoeba histolytica	Vegetables	3
	Soil	6–8
	Water	60+
Mycobacterium tuberculosis	Soil	180+
	Grass	10–40
	Water	30–90
Salmonella (spp.)	Vegetables	3–40+
	Soil	15–280+
	Pasture	200+
	Grass	100+
Salmonella typhi	Vegetables	10–53
	Lettuce	18–21
	Soil	2–120
	Water	87–104
Shigella (spp.)	Vegetables	7
	Grass	42
Shigella sonnei	Tomatoes	2–10
Streptococcus faecalis	Soil	26–77
Vibrio cholerae	Vegetables	5–14
Vibrio comma	Water	32
Poliovirus	Water	20

Source: Crook, James, 1985, Water Reuse in California, Journal American Water Works Association, v. 77, no.7. Copyright AWWA. Reprinted with permission

TABLE 7–163. MOVEMENT OF BACTERIA IN SOIL IN RELATION TO GROUNDWATER VELOCITY

Pollution	Organism	Length of Travel—m	Medium	Velocity m/d
River water	*E. coli*	0.30	Dune sand	0.11
Sewage	Bacteria	0.31	Dense soil	0.01[a]
Sewage	Bacteria	0.61	Porous soil	0.10[a]
Sewage	Coliforms	0.91	Coarse sand	0.08
Sewage	Bacteria	1.50	Fine sand	0.03[a]
Sewage	*E. coli*	2.00	Fine sand	0.128
River water	*E. coli*	3.00	Sand	0.004
Sewage	*B. coli*	3.10	Fine sand	0.50
Sewage	Bacteria	4.50	Fine sand	0.03[a]
Wastewater	Coliforms	6.10	Fine sand	0.03[a]
River water	Bacteria	7.50	Sand	0.08
Sewage	*B. coli*	10.70	Sand gravel	1.50
Sewage	*C. welchii*	15.20	Fine sand	0.50
Polluted water	Bacteria	18.30	Sand gravel	0.53[a]
Sewage	*B. coli*	19.80	Fine sand	0.11
Sewage	*B. coli*	24.40	Sand gravel	4.10
Pure culture	*B. stearothermophilis*	28.70	Crystalline rock	25.50
Sewage	Coliforms	30.50	Sand	21.80
Sewage	Bacteria	61.00	Sand gravel	30.50
Sewage[b]	*S. typhi*	64.00	Gravel loam	42.70
Sewage	Bacteria	70.70	Fine sand	0.11
Sewage	Coliforms	91.00	Sand gravel	0.24
Sewage	Coliforms	457.20	Coarse gravel	30.50
Sewage	Coliforms	830.00	Sand gravel	1.60
Polluted water	Bacteria	850.00	Coarse gravel	5.50[a]
Polluted water	Bacteria	1000.00	Fractured limestone	5.50[a]

[a] Estimated from soil description, using a hydraulic gradient of 0.01. Other groundwater velocity values were measured by other authors.
[b] Yakima outbreak.
Source: McGinnis, J.A. and DeWalle, F.B., 1983, The Movement of Typhoid Organisms in Saturated Permeable Soil, J.Am. Water Works Assoc., vol. 75, no.6. Copyright AWWA. Reprinted with permission

TABLE 7–164. MIGRATION OF VIRUSES BENEATH LAND TREATMENT SITES

Site Location	Site Type	Maximum Distance of Virus Migration (m)	
		Depth	Horizontal
St. Petersburg, FL	S	6.0	—
Gainesville, FL	S	3.0	7.
Lubbock, TX	S	30.5	—
Kerrville, TX	S	1.4	—
Muskegon, MI	S	10.0	—
San Angelo, TX	S	27.5	—
East Meadow, NY	R	11.4	3.0
Holbrook, NY	R	6.1	45.7
Sayville, NY	R	2.4	3.
12 Pines, NY	R	6.4	—
North Massapequa, NY	R	9.1	—
Babylon, NY	R	22.8	408.
Ft. Devons, MA	R	28.9	183.
Vineland, NJ	R	16.8	250.
Lake George, NY	R	45.7	400.
Phoenix, AZ	R	18.3	3.
Dan Region, Israel	R	31–67	60–270

R = Rapid infiltration; S = Slow-rate infiltration
Adapted from: Gerba, C. P. and Goyal, S. M., 1985, Pathogen Removal From Wastewater During Groundwater Recharge, in Asano, T., Artificial Recharge of Groundwater, Copyright Butterworth Publ, Boston. Reprinted with permission

TABLE 7–165. SELECTED HUMAN HEALTH AND ENVIRONMENTAL EFFECTS FROM TOXIC CHEMICALS

Chemical	Human Health Effects[1]			Environmental Effects
	Carcinogen[2]	Teratogen[2]	Others	
Aldrin/dieldrin	o		tremors, convulsions, kidney damage	toxic to aquatic organisms, reproductive failure in birds and fish, bioaccumulation in aquatic organisms
Arsenic	o	o	vomiting, poisoning, liver and kidney damage	toxic to legume crops
Benzene	o		anemia, bone marrow damage	toxic to some fish and aquatic invertebrates
Cadmium		o	suspected causal factor in many human pathologies: tumors, renal dysfunction, hypertension, arteriosclerosis, Itai-itai disease (weakened bones), emphysema	toxic to fish, bioaccumulates significantly in bivalve mollusks.
Carbon tetrachloride	o		kidney and liver damage, heart failure	
Chromium			kidney and gastrointestinal damage, respiratory complications	toxic to some aquatic organisms
Copper			gastrointestinal irritant, liver damage	toxic to juvenile fish & other aquatic organisms
DDT	o	o (minimal)	tremors, convulsions, kidney damage	reproductive failure of birds and fish, bioaccumulates in aquatic organisms, biomagnifies in food chain
Di-n-butyl phthalate			central nervous system damage	eggshell thinning in birds, toxic to some fish
Dioxin	o	o	acute skin rashes, systemic damage, mortality	bioaccumulates, lethal to aquatic organisms birds and mammals
Lead		o	convulsions, anemia, kidney and brain damage	toxic to domestic plants and animals, biomagnifies to some degree in food chain

TABLE 7–165. SELECTED HUMAN HEALTH AND ENVIRONMENTAL EFFECTS FROM TOXIC CHEMICALS (continued)

	Human Health Effects[1]			
Chemical	**Carcin-ogen[2]**	**Tera-togen[2]**	**Others**	**Environmental Effects**
Methyl Mercury		o	irritability, depression, kidney and liver damage, Minamata disease	reproductive failure in fish species, inhibits growth and kills fish; biomagnifies
PCBs	o	o	vomiting, abdominal pain, temporary blindness, liver damage	liver damage in mammals, kidney damage and eggshell thinning in birds, suspected reproductive failure in fish
Phenols			effects on central nervous system, death at high doses	reproductive effects in aquatic organisms toxic to fish
Toxaphene	o	o	pathological changes in kidney & liver; changes in blood chemistry	decreased productivity of phytoplankton communities, birth defects in fish and birds, toxic to fish and invertebrates

[1] In many cases human health effects are based upon the results of animal tests.
[2] If a substance is identified as a carcinogen, there is evidence that it has the potential for causing cancer in humans; if it is identified as a teratogen, it has the potential for causing birth defects in humans.
Source: U.S. Environmental Protection Agency, National Water Quality Inventory, 1984 Report to Congress and The Conservation Foundation, State of the Environment 1982

TABLE 7–166. POTENTIAL EXPOSURE TO LEAD IN TAP WATER

[Percentage of samples taken from kitchen taps exceeding 20 μg/L of lead at different pH levels, by age of house]

		Percent of Samples >20 μg/L	
Age of House	**pH of Water[1]**	**First-Flush**	**Fully-Flushed (2 min)**
0–2 years	≤6.4	93%	51%
	7.0–7.4	83%	5%
	≥8.0	72%	0%
2–5 years	≤6.4	84%	19%
	7.0–7.4	28%	7%
	≥8.0	18%	4%
6+ years	≤6.4	51%	4%
	7.0–7.4	14%	0%
	≤8.0	13%	3%

[1] A measure of the concentration of hydrogenions and potential electrochemical corrosion.
Note: The United States Environmental Protection Agency (EPA) sets drinking water standards and has determined that lead is a health concern at certain levels of exposure. There is currently a standard of 50 parts per billion (μg/L). Based on new health information, EPA is likely to lower this standard significantly. The most common cause of lead entering drinking water is corrosion, a reaction between the water and the lead pipes or the lead-based solder.
When water stands in the pipes of a residence for several hours without use, there is a potential for lead to leach, or dissolve, into the water if a lead source is present.
Soft water (water that makes soap suds easily) can be more corrosive and, therefore, has higher levels of dissolved lead. Some home water treatment devices may also make water more corrosive.
Source: U.S. EPA, 1987, Preliminary Results from "Lead Solder Aging Study"

TABLE 7–167. PESTICIDE RESIDUES IN HUMAN ADIPOSE TISSUE IN THE UNITED STATES

[Concentration levels, 1970–1983 in parts per million (ppm). Data represent geometric means and are based on a sample of measurements of pesticide residues and associated chemicals found in human tissue collected by medical pathologists from selected cities in the conterminous 48 states as part of the National Human Adipose Tissue Monitoring Program]

Pesticide Residues	1970	1971	1972	1973	1974	1975	1976	1977	1978	1979	1980	1981	1983
Sample size	1,386	1,560	1,886	1,092	900	779	682	789	827	796	98	384	407
DDT	7.95	8.06	6.97	5.96	5.15	4.76	4.35	3.14	3.52	3.10	2.82	2.24	1.67
Dieldrin	.16	.22	.18	.17	.14	.12	.09	.09	.09	.08	.10	.05	.06
Oxchlordane	(NA)	(NA)	.10	.12	.12	.11	.11	.10	.11	.10	.12	.09	.09
Heptachlor Epoxide	.09	.09	.07	.09	.08	.08	.08	.07	.07	.07	.08	.09	.09
trans-Nonachlor	(NA)	(NA)	(NA)	(NA)	(NA)	.06	.13	.10	.12	.12	.14	.11	.12
beta-Benzene Hexachloride	.37	.35	.19	.25	.21	.19	.18	.14	.14	.15	.12	.09	.10
Hexachlorobenzene	(NA)	(NA)	(NA)	(NA)	.03	.04	.04	.04	.04	.04	.04	.04	.03

NA = Not available.
Source: U.S. Department of Commerce, Statistical Abstract of the United States 1987

TABLE 7–168. SELECTED SYNTHETIC ORGANIC CHEMICALS DETECTED IN DRINKING WATER WELLS IN THE UNITED STATES

Chemical	Evidence for Carcinogenicity
Benzene	H
alpha-BHC	CA
beta-BHC	NTA
gamma-BHC (Lindane)	CA
Bis (2-ethylhexyl) phthalate	NTA
Bromoform	NTA
Butyl benzyl phthalate	NTA
Carbon tetrachloride	CA
Chloroform	CA
Chloromethane	NTA
Cyclohexane	NTA
Dibromochloropropane (DBCP)	CA
Dibromochloromethane	NTA
1,1-Dichloroethane	SA
1,2-Dichloroethane	CA
1,1-Dichloroethylene	NTA
1,2-Dichloroethylene	NTA
Di-n-butyl phthalate	NTA
Dioxane	CA
Ethylene dibromide (EDB)	CA
Isopropyl benzene	NTA
Methylene chloride	NTA
Parathion	SA
Tetrachloroethylene	CA
Toluene	NTA
1,1,1-Trichloroethane	NA
1,1,2-Trichloroethane	CA
Trichloroethylene (TCE)	CA
Trifluorotrichloroethane	NTA
Vinyl chloride	H, CA
Xylene	NTA

H = Confirmed human carcinogen.
CA = Confirmed animal carcinogen.
SA = Suggested animal carcinogen.
NA = Negative evidence of carcinogenicity from animal bioassay.
NTA = Not tested in animal bioassay.
Source: U.S. Council on Environmental Quality, 1981, Contamination of Ground Water by Toxic Organic Chemicals

TABLE 7–169. SUMMARY OF TOXIC EFFECTS OF SOME ORGANIC AND INORGANIC CHEMICALS KNOWN TO OCCUR IN GROUNDWATER

Contaminant	1	2	3	4	5	6	7	8	9	10	11	12	13	14	15	16	17	18	19	20	21	22	23	24
AROMATIC HYDROCARBONS																								
Alkyl benzene sulfonates	X	X	X											X										
Aniline									X							X								X
Anthracene			X																				X	X
Benzene	X	X	X						X	X	X	X		X										X
Benzidine																								X
Benzyl alcohol		X																						
Chrysene																								X
Dibenz (a.h.) anthracene																								X
4,4-Dinitrosodiphenylamine																								X
Ethylbenzene	X	X		X		X	X	X																
4,4-Methylene-bis-2-chloroaniline (MOCA)	X			X		X	X									X							X	X
Naphthalene	X	X		X							X							X						
o-Nitroaniline																X							X	
Nitrobenzene						X	X		X		X					X			X	X	X			
4-Nitrophenol					X											X								
n-Nitrosodiphenylamine						X																		X
Phenanthrene			X																				X	X
n-Propylbenzene										X														
Pyrene		X			X	X					X			X								X	X	
Styrene (vinyl benzene)	X	X		X		X	X		X					X							X		X	
Toluene	X	X		X		X	X		X				X											
1,2,4-Trimethylbenzene		X			X				X															
Xylenes (m,o,p)	X			X		X	X		X				X											
OXYGENATED HYDROCARBONS																								
Acetic acid	X	X		X	X																			
Acetone	X			X				X																
Benzophenone			X																					
Di-n-butyl phthalate	X			X				X																
Diethyl ether	X	X		X				X																
Diethyl phthalate		X		X																				
1,4-Dioxane	X			X		X	X																	X
Formic acid	X	X		X					X															
Methanol	X					X			X									X						
Phenols (e.g., p-tert-Butyl phenol)		X																						
Tetrahydrofuran		X		X		X	X		X															
HYDROCARBONS WITH SPECIFIC ELEMENTS																								
Aldicarb (sulfoxide and sulfone; Temik)															X									
Aldrin		X				X	X		X											X				X
Atrazine		X			X				X		X										X		X	
Benzoyl chloride		X		X																				
Bromobenzene	X	X			X	X		X															X	
Bromacil						X			X															
Bromochloromethane					X	X	X		X															
Bromoform	X			X		X			X															
Carbofuran						X									X									
Carbon tetrachloride						X	X		X											X				X
Chlordane				X		X			X				X											X
Chlorobenzene	X			X	X	X	X		X											X			X	
Chloroform	X	X				X	X		X					X							X			X
Chloromethane					X	X	X		X															
2-Chloronaphthalene		X				X																		
Chlorpyrifos						X									X									
o-Chlorotoluene	X	X							X															
Dibromochloromethane																							X	
Dibromochloropropane (DBCP)	X	X			X	X	X													X			X	X
Dichlofenthion (DCFT)															X									

TABLE 7–169. SUMMARY OF TOXIC EFFECTS OF SOME ORGANIC AND INORGANIC CHEMICALS KNOWN TO OCCUR IN GROUNDWATER (continued)

Contaminant	1	2	3	4	5	6	7	8	9	10	11	12	13	14	15	16	17	18	19	20	21	22	23	24
HYDROCARBONS WITH SPECIFIC ELEMENTS (Continued)																								
o-Dichlorobenzene	X	X		X	X	X	X		X		X													
p-Dichlorobenzene	X			X	X	X	X		X		X													
Dichlorobenzidine	X			X	X									X										X
Dichlorodiphenyldichloroethane (DDD, TDE)																			X					X
Dichlorodiphenyltrichloroethane (DDT)					X	X			X		X									X			X	X
1,1-Dichloroethane						X	X		X												X			
1,2-Dichloroethane	X	X			X	X	X		X				X	X					X	X			X	X
1,1-Dichloroethylene (vinylidiene chloride)	X	X		X	X	X	X		X				X						X		X		X	X
Dichloromethane (methylene chloride)	X	X				X	X		X												X			
2,4-Dichlorophenoxyacetic acid (2,4-D)									X	X												X	X	
1,2-Dichloropropane						X	X																X	
Dieldrin						X			X											X				X
Dimethyl disulfide																						X		
2,4-Dinotrophenol (Dinoseb, DNBP)	X	X					X		X											X				
Dioxins (e.g., TCDD)						X			X	X							X			X		X	X	X
Dodecyl mercaptan (lauryl mercaptan)	X	X	X						X														X	
Endosulfan						X	X		X												X		X	
Endrin						X			X															
Bis-2-ethylhexylphthalate	X	X				X														X	X	X		X
Fluoroform																							X	
Hexachlorocyclopentadiene	X			X	X	X	X						X						X					
Hexachloroethane						X	X		X												X			X
Kepone						X	X													X	X			X
Methoxychlor							X													X				
Parathion															X					X	X			
Pentachlorophenol (PCP)	X	X		X		X			X												X			
Phorate (Disulfoton)															X									
Polybrominated biphenyls (PBBs)						X						X									X	X		
Polychlorinated biphenyls (PCBs)					X	X						X	X	X						X	X	X	X	X
RDX (Cyclonite)																							X	
Simazine									X															
Tetrachloroethane (1,1,1,2)						X															X		X	
Tetrachloroethane (1,1,2,2)				X		X	X		X					X							X	X	X	X
Tetrachloroethylene (1,1,2,2) (perchloroethylene, PCE)	X			X	X		X		X												X			X
Toxaphene						X	X		X												X		X	X
Trichloroethanes (1,1,1 & 1,1,2)	X	X		X	X	X	X		X				X										X	X
1,1,2-Trichloroethylene (TCE)		X			X	X	X		X		X		X											X
Trichlorofluoromethane (Freon 11)									X				X											
2,4,6-Trichlorophenol						X																	X	X
2,4,5-Trichlorophenoxyacetic acid (2,4,5-T)						X	X		X													X	X	
2,4,5-Trichlorophenoxypropionic acid (2,4,5-TP, or Silvex)						X	X		X															
Vinyl chloride		X				X	X		X		X												X	X
OTHER HYDROCARBONS																								
Alkyl sulfonates	X	X	X																					
Cyclohexane						X	X		X															
Dicyclopentadiene (DCPD)	X	X		X	X		X		X															
Fuel oil		X			X		X		X					X										
Gasoline		X		X					X					X										
Jet fuels						X	X		X	X														
Propane													X											

TABLE 7–169. SUMMARY OF TOXIC EFFECTS OF SOME ORGANIC AND INORGANIC CHEMICALS KNOWN TO OCCUR IN GROUNDWATER (continued)

Contaminant	1	2	3	4	5	6	7	8	9	10	11	12	13	14	15	16	17	18	19	20	21	22	23	24
METALS AND CATIONS																								
Aluminum														X										
Antimony		X			X	X	X				X		X											
Arsenic					X					X			X				X			X		X	X	X
Barium		X		X	X		X			X									X					
Beryllium	X	X	X	X	X	X																		X
Cadmium				X	X		X		X											X	X	X		X
Chromium		X	X	X	X	X															X	X	X	X
Cobalt	X	X	X		X						X		X	X										
Copper						X	X					X												
Iron					X	X								X										
Lead									X		X									X				
Lithium							X		X				X	X										
Magnesium	X								X															
Manganese					X	X			X															
Mercury							X		X									X				X		
Nickel	X		X	X	X				X					X						X				X
Palladium							X				X													
Selenium	X			X	X	X			X					X										
Silver					X		X				X			X			X	X						
Thallium						X	X		X				X	X						X		X		
Vanadium			X		X	X	X		X		X													
Zinc	X	X									X								X	X				
NONMETALS AND ANIONS																								
Ammonia	X	X		X	X	X	X		X															
Cyanides	X	X	X	X					X															
Fluorides							X		X															

Numerical key of toxic effects:

1. Eye irritation
2. Skin irritation
3. Allergic sensitization
4. Upper respiratory tract irritation
5. Lung/respiratory effects
6. Liver damage
7. Kidney damage
8. Pancreatic damage
9. Central nervous system (CNS) effects
10. Peripheral nervous system effects
11. Blood cell disorders
12. Immunological effects
13. Cardiovascular effects
14. Gastrointestinal effects
15. Cholinesterase inhibition
16. Methemoglobinemia
17. Skin damage
18. Visual damage
19. Endocrine effects
20. Reproductive effects
21. Embryotoxicity
22. Teratogenicity
23. Mutagenicity
24. Carcinogenicity

Source: Office of Technology Assessment. Compiled from a partial survey of literature conducted by Environ Corp., 1983

TABLE 7–170. ESTIMATED UPPER STATISTICAL CONFIDENCE LIMITS ON CANCER RISKS FROM LIFETIME CONSUMPTION OF WATER CONTAINING 1 PPB OF A GIVEN CHEMICAL

[A value of 1 x 10^{-6} (one in one million) implies that if a million individuals were exposed at this level (1ppb in the drinking water) over their entire lifetimes, one would develop cancer as a result. A number of assumptions are inherent in these estimates. The table presents upper confidence limits for the human cancer risk of water containing 1 ppb, assuming, as the National Research Council did in 1977, that a person weighs 70 kg (154 pounds) and consumes 2 liters (2 quarts) of water per day. Animal risks were converted to human risks by a surface area conversion procedure. To determine estimates for the upper confidence limit for any other concentration, one need only multiply an entry in the table by the given concentration. From the estimates of carcinogenic risks shown in the table, it is possible to make crude estimates of the carcinogenic risks from drinking water that contains synthetic organic chemicals at various concentrations.]

Chemical	Limits 95%	Limits 99%	Sex, Species	Tumor Type
Benzene	[a]4.4 x 10^{-6}		Human	Leukemia
alpha-BHC	3.5 x 10^{-6}	8.0 x 10^{-6}	M rat	Hepatocellular carcinoma
gamma-BHC	1.3 x 10^{-5}	1.6 x 10^{-5}	M mouse	Hepatocellular carcinoma (pooled controls)
Carbon tetrachloride	1.9 x 10^{-6}	2.2 x 10^{-6}	M mouse	Hepatocellular carcinoma
Chloroform	4.1 x 10^{-6}	4.4 x 10^{-6}	F mouse	Hepatocellular carcinoma
Dibromochloro-propane (DBCP)	2.0 x 10^{-4}	2.2 x 10^{-4}	M rat	Squamous cell carcinoma of the stomach
1,1-Dichloroethane	1.5 x 10^{-4}	1.9 x 10^{-4}	M mouse	All malignant tumors
1,2-Dichloroethane	1.0 x 10^{-6}	1.2 x 10^{-6}	M rat	Hamangiosarcoma of circulatory system
Dioxane	3.9 x 10^{-7}	4.4 x 10^{-7}	M rat	Nasal squamous cell carcinoma
Ethylene dibromide (EDB)	4.8 x 10^{-4}	5.4 x 10^{-4}	M rat	Squamous cell carcinoma of forestomach
Parathion	2.9 x 10^{-5}	3.6 x 10^{-5}	F rat	Adrenal cortical adenoma or carcinoma
Tetrachloroethylene	9.3 x 10^{-7}	1.0 x 10^{-6}	M mouse	Hepatocellular carcinoma
1,1,2-Trichloroethane	1.6 x 10^{-6}	2.0 x 10^{-6}	M mouse	Hepatocellular carcinoma
Trichloroethylene (TCE)	3.0 x 10^{-7}	3.3 x 10^{-7}	M mouse	Hepatocellular carcinoma
Vinyl chloride	4.1 x 10^{-6}	4.6 x 10^{-6}	M, F rat	Liver angiosarcoma

[a] Point estimate.
Source: U.S. Council on Environmental Quality, 1981, Contamination of Ground Water by Toxic Organic Chemicals

CHAPTER 8

Water Resources Management

PAGE

SECTION A. DAMS

TABLE 8–1. HIGH DAMS OF THE WORLD ARRANGED BY COUNTRY

[Height—over 75 meters; Volume—over 1,000,000 cubic meters. Dam types are identified by the following abbreviations: Earth TE; Rockfill ER; Gravity PG; Buttress CB; Arch VA; Multi-Arch MV. Purpose for which reservoir is used is indicated by the following abbreviations: Irrigation I; Hydroelectric H; Flood Control C; Navigation N; Water Supply S; Mine Tailings T; Recreational R. Under year of completion, C is under construction and P is planned.]

Name of Dam	Year of Completion	River	Nearest City	State Province or Country	TYPE	Height Above Lowest Foundation m	Length of Crest m	Volume Content of Dam 10^3m^3	Gross Capacity of Reservoir 10^3m^3	PURPOSE	Maximum Discharge Capacity of Spillways m^3/s
AFGHANISTAN											
Kajakai	1952	Helmand	Kandahar		ER	98	274	3 230	2 680 000	I	9 300
ALBANIA											
Fierze	1978	Drin	B. Curri	Tropoje	ER	167	400	8 000	2 700 000	H	
Koman	C	Drin	Shkoder	Shkoder	ER	133	275	4 500	450 000	H	
ALGERIA											
Bou Hanifia	1948	El Hammam	Bou Hanifia	Mascara	ER	99	464	1530	73 000	IS	5 500
Bou Roumi	C (1984)	Bou Roumi	Bou Medfa	Blida	TE	100	300	4 200	188 000	I	800
Keddara	C (1986)	Boudouaou	Boudouaou	Blida	ER	108	560	4 380	146 000	S	380
Sly	C (1985)	Sly	El Asnam	El Asnam	TE	87	395	3 565	286 000	I	1 700
ARGENTINA											
Las Pirquitas	1961	Del Valle	Piedra Blanca	Catamarca	TE	83	410	2 900	65 000	ICH	1 400
Gral.M. Belgrano											
Cabra Corral res.	1973	Juramento	Coronel Moldes	Salta	TE	112	470	8 000	3 100 000	IH	1 500
Las Maderas (B.L.)	1974	Las Maderas	El Carmen	Jujuy	TE	98	460	4 500	300 000	IH	30
Futaleufu											
Amutui Quimei res.	1976	Futaleufú	Trevelin	Chubut	TE	130	600	6 000	5 600 000	H	3 000
Los Reyunos	1980	Diamante	25 de Mayo	Mendoza	TE	131	266	3 220	260 000	IH	2 300
Alicura	C (1984)	Limay	S.C.de Bariloche	Rio Negro/Neuquén	TE	130	880	13 000	3 215 000	H	3 000
Cerro Pelado	C (1985)	Grande	Amboy	Córdoba	TE	104	410	3 700	370 000	HICR	3 300
AUSTRALIA											
Upper Yarra	1957	Yarra	Melbourne	Victoria	TE/ER	89	610	5 660	207 200	S	2 165
Eucumbene	1958	Eucumbene	Cooma	NSW	TE	116	579	6 735	4 798 400	H	191
South Para	1958	South Para	Adelaide	Sth Aust.	TE	48	284	581	51 190	S	736
Glenbawn	1958	Hunter	Scone	NSW	ER	78	823	7 650	360 000	IC	1 700
Warragamba	1960	Warragamba	Sydney	NSW	PG	137	351	1 233	2 057 000	SH	12 740
Geehi	1966	Geehi	Cooma	NSW	ER	91	265	1 421	21 093	H	1 534
Blowering	1968	Tumut	Tumut	NSW	ER	112	808	8 563	1 628 000	IH	2 350
Corin	1968	Cotter	Canberra	ACT	ER	76	282	1 376	74 970	S	1 190
Talbingo	1971	Tumut	Tumut	N.S.W.	ER	162	701	14 490	921 400	H	4 290
Cethana	1971	Forth	Devonport	Tasmania	ER	110	213	1 376	109 000	H	1 980
Wyangla	1971	Lachlan	Cowra	N.S.W.	ER	85	1 510	3 580	1 220 000	I	14 700
Ord River											
Lake Argyle (res)	1972	Ord	Wyndham	West Aust.	ER	99	341	1 908	5 797 000	I	3 500
Cardinia	1973	Cardinia Creek	Melbourne	Victoria	TE/ER	86	1 542	5 150	288 905	S	13
Copeton	1976	Gwydir	Inverell	N.S.W.	ER	113	1 484	8 333	1 364 000	I	14 800
Dartmouth	1979	Mitta Mitta	Mitta Mitta	Victoria	ER	180	670	14 100	4 000 000	IHS	2 584
Split-Yard Creek	1980	Pryde Creek	Ipswich	Queensland	ER	76	1 140	3 371	28 700	H	570
Winneke	1980	Sugarloaf Creek	Melbourne	Victoria	ER	89	1 000	4 700	100 000	S	7
Mangrove Creek	1982	Mangrove Creek	Wyong	N.S.W.	ER	80	380	1 340	170 000	S	570
Blue Rock	C (1984)	Tanjil	Moe	Victoria	TE	75	640	1 530	200 000	S	1 018
Thomson	C (1984)	Thomson	Moe	Victoria	ER	164	1 180	13 300	1 175 000	SI	1 040
Lower Pieman	C (1985)	Pieman	Queenstown	Tasmania	ER	122	374	2 950	641 000	H	4 714
AUSTRIA											
Gepatsch	1965	Faggenbach	Landeck	Tyrol	ER	153	600	7 100	140 000	H	325
Oscheniksee	1972-1978	tr. Fragant	Obervellach	Carinthia	ER	116	530	2 250	33 000	H	6
Kölnbrein	1977	Malta	Omünd	Carinthia	VA	200	626	1 580	205 000	H	188
Rotlech	1977	tr. Lech	Reutte	Tyrol	PG/ER	32	128	14/11	1 260	H	200
Finstertal	1980	tr. Nederbach	Ötz	Tyrol	ER	150	652	4 500	60 500	H	23
Zillergründl	C (1986)	Ziller	Mayrhofen	Tyrol	VA	186	506	1 355	90 000	H	215
BRAZIL											
Euclides Da Cunha	1960 1977	Pardo	S.José do Rio Pardo	São Paulo	TE	92	312	2 200	13 400	H	2 340
Tres Marias	1960	São Francisco	Tres Marias	Minas Gerais	TE	75	2 700	14 250	19 790 000	CHI	8 700
Furnas	1963	Grande	Passos	Minas Gerais	ER/PG	127	779	9 697	22 950 000	HC	13 000
Estreito	1969	Grande	Pedregulho	M.Gerais/São Paulo	ER/PG	92	715	4 446	1 418 000	H	13 000
Xavantes	1970	Paranapanema	Xavantes	São Paulo	TE/ER	92	500	6 000	8 750 000	H	3 200
Paraitinga	1975	Paraitinga	Paraibuna	São Paulo	TE	105	586	11 051	2 430 000	HC	600
Paraitinga Dike											
Paraitinga Res.	1975		Paraibuna	São Paulo	TE	80	530	3 391	2 430 000		—
Itauba	1978	Jacuí	Julio de Castilhos	Rio Grande do Sul	ER	90	385	3 410	510 000	H	8 130
Paraibuna	1978	Paraibuna	Paraibuna	São Paulo	TE	94	1 285	7 892	2 463 000	HCI	1 500

TABLE 8–1. HIGH DAMS OF THE WORLD ARRANGED BY COUNTRY (continued)

Name of Dam	Year of Com-pletion	River	Nearest City	State Province or Country	TYPE	Height Above Lowest Foun-dation m	Length of Crest m	Volume Content of Dam 10^3m^3	Gross Capacity of Reservoir 10^3m^3	PURPOSE	Maxi-mum Dis-charge Capacity of Spill-ways m^3/s
BRAZIL (continued)											
São Simão	1978	Paranaiba	São Simão	Minas Gerais (Goiás)	TE/ER/PG	120	3 611	27 378	12 540 000	H	24 100
Foz Do Areia	1980	Iguaçu	Bituruna	Paraná	ER	160	850	13 000	6 100 000	H	11 000
Itumbiara	1980	Paranaiba	Itumbiara	Goiás/Minas Gerais	TE/PG	106	6 780	38 820	17 030 000	HC	16 200
Salto Santiago (Main)	1980	Iguaçú	Laranjeiras do Sul	Paraná	ER	80	1 400	9 860	6 750 000	H	24 000
Jequitai	1981	Jequitaí	Jequitaí	Minas Gerais	TE	80	580	2 300	1 200 000	I	1 850
Emborcação	1982	Paranaíba	Araguari	Minas Gerais/Goiás	ER	158	1 607	25 000	17 600 000	H	7 800
Itaipu	1982	Paraná	Foz do Iguaçu	Brasil/Paraguai	ER/PG/TE	196	7 900	29 200	29 000 000	H	61 400
Tucuruí	C (1983)	Tocantis	Tucuruí	Pará	TE/ER/PG	93	10 667	64 300	43 000 000	HN	110 000
Itaparica	C (1986)	São Francisco	Petrolândia	Pernambuco	ER	105	4 150	16 530	10 700 000	IH	28 700
Pedra Do Cavalo	C	Paraguassu	Cachoeira de S. Félix	Bahia	ER	142	510	6 500	5 330 000	HCS	12 000
Segredo	C (1987)	Iguaçu	Pinhão	Paraná	ER	140	700	6 700	3 000 000	H	13 000
BULGARIA											
Belmeken	1976	Kriva	Sestrimo	S	ER	94	760	3 560	145 000	H	
CANADA											
Kenney	1952	Nechako	Prince George	British Columbia	ER	104	457	3 071	23 700 000	H	—
Lajoie	1955	Bridge	Goldbridge	British Columbia	ER	87	1 033	2 860	720 728	H	566
W.A.C. Bennett	1967	Peace	Hudson Hope	British Columbia	TE	183	2 042	43 733	70 308 930	H	10 194
Daniel Johnson (Manic 5)	1968	Manicouagan	Baie Comeau	Quebec	MV	214	1 314	2 255	141 851 350	H	
Outardes 4 No. 1	1968	Outardes	Baie Comeau	Quebec	ER	122	649	7 533		H	
Outardes 4 No. 2	1968	Outardes	Baie Comeau	Quebec	ER	108	726	4 688		H	
Lower Notch	1971	Montreal	North Bay	Ontario	TE/PG	132	1 969	1 817	170 960	H	1 425
Big Horn Abraham Lake Res.	1972	North Sask.	Nordeag	Alberta	TE	150	472	4 330	1 768 000	H	
Mica	1972	Columbia	Revelstoke	British Columbia	TE	242	792	32 111	24 699 800	H	4 248
Manicouagan 3	1975	Manicouagan	Baie Comeau	Quebec	TE	108	366	9 175	10 422 991	H	
Revelstoke	C	Columbia	Revelstoke	British Columbia	PG/ER	153	1 620	13 000	5 180 000	H	7 080
Barrage principal	1978	La Grande Riviere	Radisson	Quebec	ER	168	2 826	23 192	61 715 000	H	16 000
Barrage Nord	1981	La Grande	Radisson	Quebec	ER	93	2 156	13 511	60 020 000	H	9 970
Sud		Riviere	Radisson			93	1 689	8 601			
Barrage	1981	La Grande Riviere	Radisson	Quebec	TE/ER	125	3 780	18 800	19 530	H	7 350
Digue QA-1	1981	R. aux Meandres	Radisson	Quebec	TE	95	1 435	3 000	—	H	—
Digue QA-8	1982	Riviere Stephane	Radisson	Quebec	TE	90	1 940	10 100	—	H	—
CHILE											
Paloma	1967	Grande	Ovalle	IV Región	TE	96	1 000	7 350	740 000	I	6 500
Digua	1968	Cato	Parral	VII Región	TE	89	420	3 650	220 000	I	300
Colihues A	1981	Cauquenes	Rancagua	VI Región	TE	83	1 200	9 400	170 000	T	220
Los Leones -1st stage	1981	Los Leones	Los Andes	V Región	TE	128	330	4 360	42 500	T	10
Colbun	C (1985)	Maule	Linares	VII Región	TE	116	530	13 870	1 490 000	I-H	7 570
CHINA											
Shuifeng	1943	Yalu Jiang	Kuandian	Liaoning	PG	106	899	3 400	14 700 000	H	37 500
Fengman	1955	Songhua Jiang	Jilin	Jilin	PG	91	1 080	1 940	10 778	CIH	10 000
Nangudong	1960	Loushui He	Linxian	Henan	ER	78	164	1 750	64 500	IC	3 285
Sanmenxia	1960	Huang He	Sanmenxia	Henan	PG	106	713	1 630	35 400 000	HI	3 843
Xin'Anjiang	1960	Xin'an Jiang	Hangzhou	Zhejiang	PG	105	462	1 380	21 626	CI	14 000
Songtao	1970	Nandu Jiang	Woxian	Guangdong	TE	80	730	4 915	3 340 000	HIS	6 300
Yunfeng	1970	Yalu Jiang	Jian	Jilin	PG	114	828	2 740	3 911 000	H	24 200
Huairen	1972	Hun Jiang	Huairen	Liaoning	PG	79	593	1 196	3 450 000	HI	23 910
Shanmei	1972	tr.Dong Jiang	Nan'an	Fujian	ER	76	305	3 464	655 000	HIS	8 570
Nanshui	1973	Bei Jiang	Shaoguan	Guangdong	ER	81	215	1 711	1 218 000	CH	400
Zhaikou	1973	Hongnong He	Lingbao	Henan	TE	76	245	4 660	168 000	H	1 950
Danjiangkou	1974	Han Jiang	Xiangfan	Hubei	PG	97	246	2 928	20 900	CIHM	4 700
Bikou	1976	Bailong Jiang	Wenxian	Gansu	ER	101	297	4 241	521 000	IH	2 310
Zhuzhuang	1979	Nanli He	Xingtai	Hebei	VA	95		1 040		I	
Fengtan	1980	Youshui	Yuanling	Hunan	PG	111	488	1 080	1 550 000	IHC	23 300
Wujiangdu	1981	Wu Jiang	Zun'yi	Guizhou	VA	165	368	1 930	2 300 000	H	19 796
Ankang	C	Han Jiang	Ankang	Shaanxi	PG	120	542	2 100	2 580 000	HN	30 900
Baishan	C	Songhua Jiang	Huadian	Jilin	VA	150	670	1 630	6 215 000	H	6 300
Baoquan	C	Luo He	Huixian	Henan	PG	116		2 790	46 000	HIC	
Guxian	C	Luo He	Luoning	Henan	PG	121	315	1 340	1 200 000	HC	
Hunanzhen	C	Wuxi Jiang	Juzhou	Zhejiang	CB	129	440	1 450	2 060 000	HNS	10 600
Longyangxia	C	Huang He	Gonghe	Qinghai	VA	172	342	1 750	24 700 000	HIS	10 500
Lubuge	C	Huangni He	Luoping	Yunnan	ER	97		1 850	110 000	IH	
Shibianyu	C		Changan	Shaanxi	ER	85		2 080	28 000	I	
Shitouhe	C	tr. Wei He	Meixian	Shaanxi	ER	105	600	8 550	147 000	IH	7 150
Wuqiangxi	C	Yuan Shui	Yuanling	Hunan	PG	104		2 670	10 800 000	HI	

TABLE 8–1. HIGH DAMS OF THE WORLD ARRANGED BY COUNTRY (continued)

Name of Dam	Year of Completion	River	Nearest City	State Province or Country	TYPE	Height Above Lowest Foundation m	Length of Crest m	Volume Content of Dam 10^3m^3	Gross Capacity of Reservoir 10^3m^3	PURPOSE	Maximum Discharge Capacity of Spillways m^3/s
COLOMBIA											
Calima I	1965	Calima	Buga	Valle	ER	115	240	2 820	563 000	H	370
Prado	1971	Prado	Ibagué	Tolima	TE	90	260	2 000	1 400 000	HI	1 200
Alto Anchicaya	1974	Anchicayá	Cali	Valle	ER	140	240	2 500	45 000	H	46 000
Chivor(La Esmeralda)	1975	Batá	Guateque	Boyacá	ER	237	280	10 800	815 000	H	10 600
Chuza (Golillas)	1978	Chuza	Bogotá	Cundinamarca	ER	135	106	1 400	257 000	S	545
Punchina	1982	Guatape	Medellín	Antioquía	TE	77	750	5 800	72 000	H	7 500
Betania	C (1984)	Magdalena	Neiva	Huila	ER/PG	90	670	6 300	1 971 000	HICS	19 000
Salvajina	C (1985)	Cauca	Popayan	Cauca	ER	160	360	3 500	904 000	H	3 550
Guavio	C (1987)	Guavio	Gachalá	Cundinamarca	ER	243	390	17 755	1 020 000	H	3 500
CZECHOSLOVAKIA											
Orlík	1963	Vltava	Príbram	Bohême C.	PG	91	550	1 030	703 800	HCS	2 555
Dalesice	1979	Jihlava	Trebíc	Moravie S.	ER	100	330	1 800	127 300	HSRI	442
DOMINICAN REPUBLIC											
Tavera	1974	Yaque del Norte	Tavera	Santiago	TE	82	405	1 850	170 000	IH	6 900
Sabana Yegua	1978	Yaque del Sur	Los Bancos	San Juan	TE	90	1 200	14 700	677 000	IHC	1 885
Tavera-Bao	C (1980)	Bao	Sabana Iglesia	Santiago	TE	112	425	3 050	280 000	IHS	*
ECUADOR											
Daniel Palacios (Amaluza Res.)	C (1983)	Paute	Cuenca	Azuay	PG/VA	167	420	1 200	120 000	H	7 700
Daule-Peripa	C (1990)	Daule	Quevedo	Los Ríos	TE/ER	90	250	3 000	6 000 000	M	3 600
EGYPT											
Aswan High Dam (Sadd-el-Aali)	1970	Daule	Aswan	Egypt	TE/ER	111	3 830	44 300	168 900 000	GR	11 000
EL SALVADOR											
Cerron Grande (Silencio)	1973?	Lempa	Hutyapa		ER	80	900	5 100	1 430 000	H	11 000
FIJI											
Monasavu	1982	Nanuku Ck	Suva	Fiji	ER	85	485	1 738	133 000	H	625
GERMANY (F.R.)											
Rur	1959	Rur	Heimbach	Nordrhein	TE	77	480	2 600	181 800	CNH	450
Frauenau	C (1983)	Kleiner Regen	Zwiesel	Bayern	ER	86	640	2 263	20 300	SNCH	58
FRANCE											
Serre-Poncon	1960	Durance	Gap	Htes Alpes	TE	129	600	14 100	1 270 000	HI	3 430
Mont-Cenis	1968	Cenise	Modane	Savoie	TE/ER	120	1 400	14 850	332 200	H	265
Grand'Maison	C (1985)	Eau d'Olle	Grenoble	Isère	TE/ER	160	550	12 500	140 000	H	65
Pla De Soulcem	C (1983)	Mounicou	Tarascon	Ariège	TE	76	275	1 675	29 300	H	134
GHANA											
Akosombo (Main)	1965	Volta	Accra/Tema	Ghana	ER	134	671	7 991	147 960 000	H	14 160
GREAT BRITAIN											
Scammonden	1970	Black Brook	Huddersfield	West Yorkshire	TE/ER	76	624	4 304	7 873	S	
Llyn Brianne	1972	Towy	Llandovery	Dyfed	ER	91	274	2 085	60 000	S	850
GREECE											
Kremasta	1965	Achelöos	Agrinion	Etolo-Akarnanie	TE	165	460	8 170	4 750 000	H	3 000
Polyphyton	1974	Aliakmon	Kozani	Makedhonia	ER	112	296	3 459	2 244 000	HI	1 375
Mornos	1979	Mornos	Lidhoriki	Phocide	TE	126	815	17 000	780 000	S	1 135
Pournari	1980	Arachthos	Arta	Ipiro	TE	102	574	9 500	730 000	H	6 100
Sfikia	C (1984)	Aliakmon	Veria	Makedhonia	ER	80	230	1 620	99 000	H	1 600
Peghai(Main Dam)	C (1987)	Aöos	Metsovon	Ipiro	TE	78	295	2 800	262 000	H	160
GUATEMALA											
Pueblo Viejo	C (1983)	Chixoy	San Cristobal Verapaz	Alta Verapaz	ER	130	230	3 200	460 000	H	3 850
HONG KONG											
High Island East	1977	Kwun Mun	Sai Kung	New Territories	ER	85	457	3 443	284 375 (both dams)	S	435
High Island West	1977	Kwun Mun	Sai Kung	New Territories	ER	76	762	6 120			
INDIA											
Koyna (Shivaji Sagar)	1961	Koyna	Karad	Maharashtra	PG	103	808	1 555	2 796 500	H	3 823
Rihand (Gobind Ballabh Pant Sagar)	1962	Rihand	Mirzapur	Uttar Pradesh	PG	93	934	1 680	10 600 000	H	13 339

* Bao and Tavera reservoirs are interconnected by a canal and both depend on Tavera spillway.

TABLE 8–1. HIGH DAMS OF THE WORLD ARRANGED BY COUNTRY (continued)

Name of Dam	Year of Completion	River	Nearest City	State Province or Country	TYPE	Height Above Lowest Foundation m	Length of Crest m	Volume Content of Dam 10^3m^3	Gross Capacity of Reservoir 10^3m^3	PURPOSE	Maximum Discharge Capacity of Spillways m^3/s
INDIA (continued)											
Bhakra Dam											
(Gobind Sagar)	1963	Satluj	Nangal Township	Himachal Pradesh	PG	226	518	4 130	9 621 000	IH	8 372
Sholayar	1972	Sholayar	Coimbatore	Tamil Nadu	TE,PG	105	1 282	2 496	160 660	IH	1 474
Ukai Dam	1972	Tapi	Fort Songadh	Gujarat	TE,PG	81	5 065	25 180	8 511 000	IHC	35 960
Nagarjunasagar Dam	1974	Krishna	Hyderabad	Andhra Pradesh	TE,PG	125	4 865	7 960	11 550 000	IH	53 450
Pong Dam											
(Beas Project)	1974	Beas	Mukerian	Himachal Pradesh	TE	133	1 950	35 500	8 570 000	IH	12 374
Cheruthoni	1976	Cheruthoni	Idukki	Kerala	PG	138	650	1 700	1 996 000	H	5 012
Balimela Dam	1977	Sileru	Jeypur	Orissa	TE	75	4 363	19 096	3 610 000	IH	10 930
Ramganga	1978	Ramganga	Dhampur	Uttar Pradesh	TE	128	743	11 013	2 442 600	IH	8 467
Bhatsa	C	Bhatsa	Bombay	Maharashtra	PG	89	938	1 679	915 000	SIH	3 775
Chakra	C	Chakra	Shimoga	Karnataka	TE,ER	84	568	1 715	219 000	H	1 416
Dudhganga	C	Dudhganga	Kolhapur	Maharashtra	TE,PG	89	1 264	1 449	708 260	I	2 247
Hasdeo Project	C	Hasdeo	Bilaspur	Madhya Pradesh	TE,PG	86	2 688	3 219	3 417 000	IH	20 530
Karjan (Lower)	C	Karjan	Rajpipla	Gujarat	PG	100	903	1 937	630 000	I	17 275
Lakhwar	C	Yamuna	Dehradun	Uttar Pradesh	PG	192	440	2 000	580 000	IH	8 000
Sardar Sarovar	C	Narmada	Rajpipla	Gujarat	PG	155	1 210	4 100	9 500 000	IHC	62 269
Srisailam H.E. Project	C	Krishna	Hyderabad	Andhra Pradesh	PG	143	512	1 953	8 722 000	H	37 400
Supa (Kalinadi Project)	C	Kali	Dandeli	Karnataka	PG	101	322	1 150	4 418 000	H	2 830
Tehri Dam	C	Bhagirathi	Tehri	Uttar Pradesh	TE,ER	261	570	22 750	3 539 000	IH	13 150
Thein Dam Ranjit											
Sagar (Res.)	C	Ravi	Pathankot	Punjab	ER,TE	160	565	16 187	3 280 000	IHC	21 890
Warna	C	Warna	Kolhapur	Maharashtra	TE,PG	91	1 580	15 310	964 000	I	3 222
INDONESIA											
IR.H.Juanda											
(Jatiluhur)	1967	Citarum	Purwakarta	W-Java	ER	100	1,225	9,000	3 345 000	IHCSR	3 000
Karangkates	1972	Brantas	Malang	E-Java	ER	100	823	6 156	342 000	IHCR	1 310
Saguling	C	Citarum	Cianjur	W-Java	ER	98	300	2 570	609 000	HCR	2 400
IRAN											
Karoon	1975	Karun	Masjed Soliman	Iran	VA	200	380	1 570	3 005 000	IH	16 200
Ghesh-Lagh	1978	Ghesh-Lagh	Sanandaj	Iran	TE	80	300	2 000	224 000	S	2 600
Lar	1982	Lar	Tehran	Iran	TE	105	1 500	1 300	960 000	IH	1 700
IRAQ											
Derbendikhan	1961	Diyala	Sulaymaniya	Sulaymaniya	ER	128	535	7 480	3 000 000	ICH	11 400
Mosul	1983	Tigris	Mosul	Nienava	ER	131	3 500	23 000	12 500 000	ICH	17 000
ITALY											
Alpe Gera	1964	Cormor	Sondrio	Adige Lombardia	PG	174	528	1 700	68 088	H	59
Place Moulin	1965	Buthier	Aosta	Valle d'Aosta	VA	155	678	1 510	106 000	H	473
JAPAN											
Sakuma	1956	Tenryu	Toyohashi	Aichi	PG	156	294	1 120	327 000	H	7 700
Ogochi	1957	Tama	Ome	Tokyo	PG	149	353	1 680	189 000	SH	1 800
Miboro	1960	Sho	Gifu	Gifu	ER	131	405	7 950	370 000	H	1 800
Tagokura	1960	Tadami	Aizuwakamatsu	Fukushima	PG	145	462	1 990	494 000	H	2 200
Arimine	1961	Joganji	Toyama	Toyama	PG	140	500	1 570	218 000	H	380
Makio	1961	Kiso	Matsumoto	Nagano	ER	105	260	2 616	75 000	IHS	3 200
Okutadami	1961	Tadami	Koide	Fukushima	PG	157	480	1 640	601 000	H	1 500
Kurobe	1964	Kurobe	Omachi	Toyama	VA	186	489	1 360	199 000	H	1 500
Oshirakawa	1964	Oshirakawa	Gifu	Gifu	ER	95	390	1 700	14 200	H	330
Tsuruta	1965	Sendai	Kagoshima	Kagoshima	PG	118	448	1 124	123 000	CH	4 965
Yanase	1965	Nabari	Aki	Kochi	ER	115	202	2 842	105 000	H	1 900
Kuzuryu	1968	Kuzuryu	Ono	Fukui	ER	128	355	6 300	320 000	HC	1 560
Shimokubo	1968	Kanna, tr. Tone	Fujioka	Saitama	PG	129	626	1 190	130 000	IHC	1 600
Kisenyama	1969	Samutani, tr. Yodo	Uji	Kyoto	ER	91	255	2 338	7 230	H	6
Koshibu	1969	Koshibu, tr. Tenryu	Iida	Nagano	VA	105	293	311	58 000	IHC	2 160
Misakubo	1969	Misakubo, tr. Tenryu	Tenryu	Shizuoka	ER	105	258	2 410	30 000	H	900
Shimokotori	1973	Kotori, tr. Jintsu	Takayama	Gifu	ER	119	321	3 530	123 037	H	1 920
Aburatani	1974	Aburatani, tr. Kuma	Yatsushiro	Kumamoto	ER	82	189	1 277	5 420	H	430
Fukuchi	1974	Fukuchi	Nago	Okinawa	ER	92	260	1 622	51 500	ICS	800
Kajigawa	1974	Kaji	Shibata	Niigata	PG	107	286	433	22 500	C	1 920
Kurokawa	1974	Ichi	Himeji	Hyogo	ER	98	325	3 623	33 390	H	175
Matsukawa	1974	Matsu, tr Tenryu	Iida	Nagano	PG	84	165	263	7 400	CIS	900
Miyama	1974	Naka	Kuroiso	Tochigi	ER	76	334	1 967	25 800	IHS	840
Niikappu	1974	Niikappu	Tomakomai	Hokkaido	ER	103	326	3 071	145 000	H	1 500
Sameura	1974	Yoshino	Nankoku	Kochi	PG	106	400	1 200	316 000	IHCS	6 000

TABLE 8–1. HIGH DAMS OF THE WORLD ARRANGED BY COUNTRY (continued)

Name of Dam	Year of Completion	River	Nearest City	State Province or Country	TYPE	Height Above Lowest Foundation m	Length of Crest m	Volume Content of Dam 10^3m^3	Gross Capacity of Reservoir 10^3m^3	PURPOSE	Maximum Discharge Capacity of Spillways m^3/s
JAPAN (continued)											
Taisetsu	1974	Ishikari	Asahikawa	Hokkaido	ER	87	440	3 874	66 000	CHIS	1 000
Hirose	1975	Fuefuki, tr. Fuji	Enzan	Yamanashi	ER	75	255	1 400	14 300	IHCS	1 380
Nabara	1975	Nabara, tr. Ota	Hiroshima	Hiroshima	ER	86	305	2 213	5 658	H	475
Iwaya	1976	Mase, tr. Kiso	Minokamo	Gifu	ER	128	366	5 780	173 500	IHCS	2 950
Kusaki	1976	Watarase, tr. Tone	Kiryu	Gunma	PG	140	405	1 374	60 500	IHCS	4 320
Myojin	1976	Nabara, tr. Ota	Hiroshima	Hiroshima	ER	89	402	3 268	6 145	H	80
Terauchi	1977	Sada, tr. Chikugo	Amagi	Fukuoka	ER	83	420	3 012	18 000	ICS	1 300
Futai	1978	Kiyotsu	Nagaoka	Niigata	ER	87	280	2 350	18 300	H	1 950
Kassa	1978	Kassa	Nagaoka	Niigata	ER	90	487	4 450	13 500	H	114
Miho	1978	Sakawa	Minamiashigara	Kanagawa	ER	95	588	5 816	64 900	CHS	3 100
Nanakura	1978	Takase	Omachi	Nagano	ER	125	340	7 380	32 500	H	1 950
Takase	1978	Takase	Omachi	Nagano	ER	176	362	11 600	76 200	H	1 700
Seto	1978	Setodani	Gojo	Nara	ER	111	343	3 740	16 850	H	230
Tedorigawa	1979	Tedori	Kanazawa	Ishikawa	ER	154	420	10 050	231 000	HCSI	3 500
Terauchi	1980	Sada	Amagi	Fukuoka	ER	83	420	3 000	18 000	CIS	1 300
Urushizawa	1980	Naruse	Furusawa	Miyagi	ER	80	310	2 143	18 000	HCS	1 500
Inamura	1981	Seto	Kochi	Kochi	ER	88	325	3 100	5 800	H	230
Tamahara	1982	Hotchi	Numata	Gunma	ER	116	570	5 435	14 800	H	160
Agigawa	C (1986)	Kiso	Ena	Gifu	ER	102	430	4 400	48 000	CS	2 000
Arakawa	C (1985)	Ara	Kofu	Yamanashi	ER	88	320	3 000	10 800	CS	1 680
Arima	C (1985)	Iruma	Hanno	Saitama	ER	84	260	1 600	7 600	CS	670
Doyo	C (1984)	Doyo	Yonago	Tottori	ER	87	480	2 700	7 680	H	80
Igarashigawa	C (1990)	Igarashi	Sanjo	Niigata	ER	76	360	2 278	21 100	CS	1 420
Jozankei	C (1986)	Ishikari	Sapporo	Hokkaido	PG	113	405	1 150	82 300	CS	675
Kuriyama	C (1985)	Nebesawa	Imaichi	Tochigi	ER	89	340	2 200	6 890	H	52
Kyuragi	C (1985)	Matsuura	Taku	Saga	PG	117	386	1 045	7 400	CSH	1 080
Naramata	C (1986)	Naramata	Numata	Gunma	ER	158	520	12 300	90 000	CIS	1 650
Nitchu	C (1986)	Oshikiri	Kitakata	Fukushima	ER	106	468	5 010	24 600	ICS	1 120
Ogaki	C (1985)	Ukedo	Haramachi	Fukushima	ER	85	262	1 729	19 500	I	1 680
Okawa	C (1984)	Agano	Aizuwakamatsu	Fukushima	PG	78	407	1 000	57 500	CSIH	5 230
Ouchi	C (1985)	Ono	Aizuwakamatsu	Fukushima	ER	102	340	4 400	18 500	H	176
Sagae	C (1985)	Mogami	Sagae	Yamagata	ER	115	510	9 490	109 000	CSIH	2 600
Sagurigawa	C (1988)	Saguri	Ojiya	Niigata	ER	116	420	7 214	27 500	CS	1 690
Takami	C (1984)	Shizunai	Tomakomai	Hokkaido	ER	120	427	5 120	229 000	CH	2 400
Tamagawa	C (1987)	Omono	Omagari	Akita	PG	100	432	1 105	254 000	CISH	3 500
Shichigashuku	C (1988)	Abukuma	Shiroishi	Miyagi	ER	93	565	5 050	109 000	CSI	2 620
Shintsuruko	C (1985)	Nyu	Obanazawa	Yamagata	ER	93	303	2 645	31 500	I	1 100
Shitoki	C (1984)	Shitoki	Iwaki	Fukushima	ER	84	300	2 580	12 100	CSI	1 540
Tokachi	C (1984)	Tokachi	Obihiro	Hokkaido	ER	84	443	3 658	112 000	CH	2 600
Yasaka	C (1987)	Oze	Iwakuni	Yamaguchi	PG	120	540	1 600	48 000	CS	4 050
JORDAN											
King Talal Dam	1977	Zarqa River	Jarash	Salt District	TE/ER	94	330	4 216	52 000	IC	2 950
Wadi Arab Dam	C (1985)	Wadi Arab	North Shuneii	Irbid District	ER	82	482	2 968	20 000	IS	430
LIBYA											
Ghan	1982	Ghan	Ghrian	NE/Ghrian	ER	80	316	1 650	39 500	IC	1 640
MALAYSIA											
Temengor	1978	S. Perak	Grik	Perak	ER	115	128	6 980	570 000	HC	2 720
Kenyir	C (1984)	S. Trengganu	Kuala Brang	Trengganu	ER	150	800	16 500	13 600 000	HC	7 000
Batang Ai	C (1985)	Batang Ai	Lubok Antu	Sarawak	ER	85	810	4 000	2 360 000	H	2 175
MEXICO											
Lazaro Cardenas (El Palmito)	1947	Nazas	C.Lerdo	Durango	TE	95	330	5 300	3 162 000	IC	6 000
Sanalona	1948	Tamazula	Culiacán	Sinaloa	TE	81	1 031	4 900	845 000	IH	6 300
Alvaro Obregon (Oviachic)	1952	Yaqui	Obregón	Sonora	TE	90	1 457	8 773	3 237 000	IHC	11 100
Adolfo Ruiz Cortines (Mocúzari)	1955	Río Mayo	Navojoa	Sonora	TE	81	780	4 196	1 014 700	IHC	8 000
Presidente Aleman (Temascal)	1955	Río Tonto	C.Alemán	Oaxaca	TE	76	830	4 059	6 515 000	ICH	5 500
Miguel Hidalgo (El Mahone)	1956	Río Fuerte	El Fuerte	Sinaloa	TE	86	3 230	10 200	3 290 000	IGH	16 450
Presidente Benito Juarez (El Marques)	1961	Tehuantepec	Tehuantepec	Oaxaca	TE	86	375	3 540	942 000	IC	5 500
El Infiernillo	1963	Balsas	Apatzingan	Michoacán	TE	148	350	5 500	9 340 000	H	10 350
Netzahualcoyotl	1964	Grijalva	Cárdenas	Chiapas	TE	138	478	5 077	8 300 000	HC	21 750
Pte.Adolfo Lopez Mateos (Humaya)	1964	Humaya	Culiacán	Sinaloa	TE	107	820	7 141	3 150 000	IC	5 600

TABLE 8–1. HIGH DAMS OF THE WORLD ARRANGED BY COUNTRY (continued)

Name of Dam	Year of Completion	River	Nearest City	State Province or Country	TYPE	Height Above Lowest Foundation m	Length of Crest m	Volume Content of Dam 10^3m^3	Gross Capacity of Reservoir 10^3m^3	PURPOSE	Maximum Discharge Capacity of Spillways m^3/s
MEXICO (continued)											
Internacional La Amistad	1968	Bravo	Acuña	Coahuila	PG/TE	87	9 760	11 620	4 379 000	IHCS	43 690
La Angostura	1974	Grijalva	Tuxtla Gutierrez	Chiapas	TE	146	323	4 030	9 200 000	H	6 900
Manuel Moreno Torres (Chicoasén)	1980	Grijalva	Tuxtla Gutiérrez	Chiapas	TE	261	485	15 370	1 613 000	H	15 000
Jose Lopez Portillo Pte. (Comedero)	1981	San Lorenzo	Cosalá	Sinaloa	TE	136	400	7 090	2 850 000	IHC	5 000
Gustavo Diaz Ordaz Pdte. (Bacurato)	1982	Sinaloa	Guamuchil	Sinaloa	TE	114	800	9 315	1 800 000	IHC	7 000
El Sabinal	C (1985)	Ocoroni	Guasave	Sinaloa	TE	79	400	2 367	300 000	IC	2 450
Carlos Ramirez Ulloa (Caracol)	C (1985)	Balsas	Iguala	Guerrero	TE	126	347	6 327	782 000	H	17 000
Chilatan	C (1986)	Tepalcatepec	Apatzingan	Jalisco	TE	104	1 150	5 898	600 000	IC	7 000
MOROCCO											
Moulay Youssef	1970	Tessaout	Marrakech	Marrakech	TE	100	725	5 300	200 000	IH	3 000
Hassan Addakhil	1971	Ziz	Er Rachidia	Er Rachidia	TE	85	785	5 800	380 000	I	1 700
Youssf Ben Tachfine	1973	Massa	Tiznit	Tiznit	ER	85	670	3 700	310 000	IS	3 400
Sidi Mohamed Ben Abdellah	1974	Bou Regreg	Rabat	Rabat	ER	100	340	3 000	493 000	SI	5 000
Ait Chouarit	C (1986)	Lakhdar	Demnate	Azilal	TE	145	380	9 500	270 000	IHS	1 820
NEPAL											
Kulekhani Dam	1981	Kulekhani	Kathmandu	Nepal	ER	114	406	4 419	85 300	H	2 540
NEW ZEALAND											
Benmore	1965	Waitaki	Oamaru	Otago	TE	118	957	12 500	2 200 000	H	3 400
Matahina	1965	Rangataiki	Whakatane	South Auckland	TE	86	345	3 500	25 000	H	1 900
Mangatangi	1977	Mangatangi	Manukau	South Auckland	TE	78	340	2 240	39 000	S	510
Patea	C	Patea	Wanganui	Taranaki	TE	82	190	1 100	138 000	H	2 800
NIGERIA											
Shiroro	C (1984)	Kaduna/Dinya	Minna	Niger	ER	125	700	3 457	7 000 000	H	7 500
NORWAY											
Hyttejuvet	1965	Valldalselv	Haugesund	Hordaland	ER	90	350	1 450		H	375
Digea	1970	Sira-Digea	Flekkefjord	Vest-Agder	ER	90	400	2 700		H	45
Svartevatn	1977	Sira	Stavanger	Vest-Agder	ER	129	420	4 715		H	
Sysenvatn	1979	Leiro	Bergen		ER	84	1 140	3 624		H	250
Oddatjorn } Blasjo	C(1987)	Oddeana	Haugesund	Rogaland	ER	140	500	5400	see Storvatn	H	
Storvatn } reservoir	C(1987)	Brattliana	Haugesund	Rogaland	ER	98	1460	9700	3 105 000	H	600
PAKISTAN											
Mangla	1967	Jehlum	Dehlum	Punjab	TE	138	3 139	65 379	7 251 811	HI	31 144
Tarbela	1976	Indus	Taxila	N.W.F.P.	TE/ER	143	2 743	105 570	13 689 644	IH	42 186
Auxiliary-1						105	713	13 770			
Jari	1967	Saddle Dam	Mirpur	A. Kashmir	TE	84	2 073	26 240		I	—
PARAGUAY											
Itaipu	1982	Paraná	Hernandarias	Brazil/Paraguay	ER/PG/TE	190	7 655	33 690	29 000 000	H	62 000
PERU											
Yauliyacu Arriba	1982	Yauliyacu	Casapalca	Lima	TE	75	220	3 000	7 500	M	
Condorma	C (1985)	Colca	Chivay	Arequipa	ER	92	503	4 300	260 000	I	1 300
Gallito Ciego	C (1987)	Jequetepeque	Pacasmayo	Cajamarca	ER	112	750	14 200	400 000	IH	1 830
Chinchan	C (2007)	Yuracocha	Casapalca	Lima	TE	90	230	5 800	13 500	M	
PHILIPPINES											
Ambuklao	1956	Agno	Baguio	Benguet	ER	129	452	6 000	327 170	H	7 300
Binga	1960	Agno	Baguio	Benguet	ER	107	215	2 000	63 000	H	5 200
Angat	1967	Angat	Manila	Bulacan	ER	131	368	7 000	1 099 000	IH	7 500
Pantabangan	1977	Upper Pampanga	Cabanatuan	Nueva Ecija	TE	107	1 615	12 300	2 996 000	IHC	4 200
Magat	1982	Magat	Santiago	Isabela	TE/ER	106	2 925	13 200	1 250 000	IH	30 400
San Roque	C	Agno	Dagupan	Pangasinan	ER	210	1 130	43 150	990 000	IHC	12 600
PORTUGAL											
Paradela	1958	Cávado	Chaves	Vila Real	ER	110	540	2 700	164 500	H	720
Alto Rabagão	1964	Rabagão	Chaves	Vila Real	VA/PG	94	1 897	1 117	569 000	H	500
Santa Clara	1968	Mira	Odemira	Beja	TE	86	428	3 966	485 000	I	208

TABLE 8–1. HIGH DAMS OF THE WORLD ARRANGED BY COUNTRY (continued)

Name of Dam	Year of Com- pletion	River	Nearest City	State Province or Country	TYPE	Height Above Lowest Foun- dation m	Length of Crest m	Volume Content of Dam 10^3m^3	Gross Capacity of Reservoir 10^3m^3	PURPOSE	Maxi- mum Dis- charge Capacity of Spill- ways m^3/s
ROMANIA											
Izvorul Muntelui (Bicaz)	1961	Bistrita	P.Neamt	Neamt	PG	127	430	1 625	1 230 000	H	2 400
Fintinele Somes	1978	Somesul Cald	Huedin	Cluj	ER	92	400	2 320	225 000	HR	700
Cerna Principal	1979	Cerna	Tg.Jiu	Gorj	ER	110	342	2 550	124 000	SHI	1 080
Oasa	1979	Sebes	Sebes	Alba	ER	91	300	1 600	136 000	HCR	264
Colibita	C (1983)	Bistrita	Bistrita	Bistrita Nâsâud	ER	92	250	1 600	90 000	SH	650
Gura Apelor	C (1984)	Rîul Mare	Hateg	Hunedoara	ER	168	450	9 020	225 000	H	1 750
Pecineagu	C (1984)	Dîmbovita	Tîrgoviste	Dîmbovita	TE	105	270	2 400	68 900	SHI	600
Mineciu	C (1985)	Teleajen	Vâlenii de Munte	Prahova	TE	75	720	5 000	60 000	SCHI	1 200
Riusor	C (1985)	R.Tîrgului	Cîmpulung	Arges	ER	120	380	3 500	60 000	SH	620
Siriu	C (1985)	Buzâu	Nehoiu	Buzâu	ER	122	440	8 800	155 000	CISH	3 400
Vija	C (1986)	Bistrita	Tg.Jiu	Gorj	ER	93	270	1 700	29 400	HS	800
Poiana Marului	C (1987)	Bistra Mârului	Otelul Rosu	Caras-Severin	ER	130	400	5 320	96 000	HI	830
SOUTH AFRICA											
P K Le Roux	1977	Orange	Petrusville	Orange Free State	VA	107	853	1 300	3 237 000	IH	20 400
Sterkfontein	1980	Nuwe Jaar Spruit	Harrismith	Orange Free State	TE	93	3 060	19 800	2 656 000	S	-
Goedertrouw	1982	Mhlatuze	Eshowe	Natal	TE	88	660	5 330	321 000	IS	7 000
SOUTH KOREA											
So Yang Gang	1973	Han	Chunchon	Kangwondo	ER	123	530	9 591	2 900 000	IHCR	5 500
Sam Rang Jin Upper part Dam	C	Nakdong	Samrangjin	Kyeongsangnamdo	ER	85	250	1 003	6 140	H	
An Dong	1976	Nakdong	Andong	Kyeongsangnamdo	ER	83	624	4 015	1 248 000	IHCS	5 360
SPAIN											
Mequinenza	1966	Ebro	Mequinenza	Zaragoza	PG	81	451	1 000	1 533 800	H	12 800
Porto De Mouros	1967	Ulla	Arzua	La Coruña	ER	93	460	2 337	297 000	H	1 550
Grado I	1969	Cinca	El Grado	Huesca	PG	130	958	1 225	399 000	IH	3 420
Iznajar	1969	Genil	Rute	Cordoba	PG	122	407	1 450	980 000	IHS	6 550
Almendra	1970	Tormes	Almendra	Salamanca	VA	202	567	2 186	2 649 000	H	3 000
El Atazar	1972	Lozoya	Atazar	Madrid	VA	134	484	1 200	426 000	S	410
Arenos	1979	Mijares	Montanejos	Castellon	ER	108	428	3 014	132 000	I	1 300
Beninar	1983	Grande De Adra	Beninar	Almeria	ER	87	386	3 800	70 000	IS	232
Limonero	1983	Guadalmedina	Malaga	Malaga	ER	93	410	3 188	27 000	S	283
Canales	C (1983)	Benil	Guejar Sierra	Granada	ER	159	340	1 217	71 000	IS	226
Cuevas De Almanzora	C (1983)	Almanzora	Cuevas De Almanzora	Almeria	ER	113	623	6 510	191 000	IH	2 520
Negratin	C (1983)	Guadiana Menor	Freila	Granada	PG/ER	75	439	1 150	546 000	IH	3 440
Sallente	C (1983)	Flamisell	Torre De Capdella	Lerida	ER	89	398	1 100	6 000	H	63
La Viñuela	C (1983)	Guaro	La Viñuela	Malaga	ER/TE	94	460	3 345	25 000	IS	120
Zahara	C (1983)	Guadalete	Zahara	Cadiz	ER	85	500	2 011	212 000	I	1 000
SRI LANKA											
Kotmale	C (1985)	Kotmale Oya	Gampola	CP	ER	87	600	4 159	175 000	H	5 550
Randenigala	C (1986)	Mahaweli	Mahiyangana	UP	ER	94	495	3 700	860 000	HIC	8 085
SWEDEN											
Trangslet	1961	Dalälven	Mora	Kopparberg,M	ER	125	850	7 200	880 000	H	1 000
Holjes	1961	Klarälven	Hagfors	Värmland,M	ER/TE	80	400	1 750	270 000	H	1 280
Messaure	1963	Lule älv	Jokkmokk	Norrbotten,N	TE	101	1 900	10 500	50 000	H	2 300
Letsi	1967	Lule älv	Jokkmokk	Norrbotten,N	ER	85	570	2 300	67 000	H	1 500
Seitevare	1968	Lule älv	Porjus	Norrbotten,N	ER	106	1 450	4 900	1 650 000	H	875
SWITZERLAND											
Marmorera (Castilleto)	1954	Julia	Bivio	Grisons	TE	91	400	2 700	62 600	H	200
Mauvoisin	1957	Drance de Bagnes	Fionnay	Valais	VA	237	520	2 030	181 500	H	100
Goescheneralp	1960	Göschenerreuss	Göschenen	Uri	ER	155	540	9 300	76 000	H	200
Grande Dixence	1961	Dixence	Hérémence	Valais	PG	285	695	6 000	401 000	H	-
Luzzone	1963	Brenno di Luzzone	Olivone	Tessin	VA	208	530	1 330	88 000	H	88
Mattmark	1967	Saaser Vispa	Saas-Fee	Valais	TE	120	780	10 500	101 000	H	150
Emosson	1974	Barberine	Finhaut	Valais	VA	180	555	1 090	227 000	H	60
TAIWAN											
Shihmen	1964	Tahan	Chungli	Taiwan	ER	133	360	7 059	309 120	IHCS	13 400
Tsengwen	1973	Tsengwen	Tainan	Taiwan	TE	133	400	9 296	707 530	ICSH	9 470
THAILAND											
Sirikit	1972	Nan	Uttaradit	N	TE	114	800	9 800	10 550 000	IHC	3 250

TABLE 8–1. HIGH DAMS OF THE WORLD ARRANGED BY COUNTRY (continued)

Name of Dam	Year of Completion	River	Nearest City	State Province or Country	TYPE	Height Above Lowest Foundation m	Length of Crest m	Volume Content of Dam 10^3m^3	Gross Capacity of Reservoir 10^3m^3	PURPOSE	Maximum Discharge Capacity of Spillways m^3/s
THAILAND (continued)											
Bang Lang	1981	Pattani	Yala	S	TE/ER	85	422	2 900	1 360 000	IHC	4 500
Srinagarind	1981	Quae Yai	Kanchanaburi	Central Region	ER	140	610	12 100	17 745 000	IHC	2 420
Khao Laem	C(1984)	Quae Noi	Kanchanaburi	Central Region	ER	90	910	8 000	7 450 000	IHC	3 200
TURKEY											
Seyhan	1956	Seyhan	Adana	South A	TE	77	1955	7 500	1 200 000	ICB	2500
Hirfanli	1959	Kizilirmak	Kirsehir	Inner A	ER	83	364	2 000	5 980 000	ICH	2 300
Demirköprü	1960	Gediz	Manisa	West. A	TE	77	543	4 300	1 320 000	ICH	200+6272
Almus	1966	Yesilirmak	Tokat	NE. A	TE	95	371	3 500	950 000	IH	1550
Kozan	1972	Kilgen	Adana	South.A	TE/ER	83	289	1 195	163 000	I	1 250
Keban	1974	Firat	Elazig	East	PG/ER	207	1126	15585	30 600 000	CH	17 000
Ayvacik	1981	Yesilirmak	Samsun	North	ER	175	405	2 327	10 800 000	CH	11 000
Güzelhisar	1982	Güzelhisar	Izmir	West	ER	89	511	3204	158 000	IS	2 550
Gönen	C	Gönen	Balikesir	West	TE	78	293	2 036	164 000	I	2 785
Doganci	C	Nilüfer	Bursa	West	ER	82	288	2 278	50 000	S	1 978
Çamlidere	C	Bayindir	Ankara	Inner A	TE/ER	106	278	2 487	133 000	IS	662
Aslantas	C	Ceyhan	Adana	South	TE/ER	95	566	8 000	1 150 000	ICH	11 930
Adigüzel	C	B.Menderes	Denizli	West	ER	145	377	5 892	1 188 000	ICH	4 260
Kiliçkaya	C	Kelkit	Sivas	Inner	ER	135	405	6 030	14 000 000	ICH	2 450
Karakaya	C	Firat	Diyarbakir	SE	VA	173	462	2 000	9 580 000	H	17 000
Çatalan	C	Seyhan	Adana	South	TE	95	309	7664	1 629 000	ICH	8 900
Karacaoren	C	Asagiaksu	Burdur	West	TE	95	428	3 500	1 340 000	ICH	4 495
Uluborlu	C	Pupa	Isparta	SE	TE	75	315	1 800	24 000	I	295
Gezende	C	Göksu	Mersin	South	TE	75	171	1 110	66 000	H	4 385
Altinkaya	C	Kizilirmak	Samsun	Nouth	ER	195	604	2 600	5 763 000	ICH	11 800
Menzelet	C	Ceyhan	K.Maras		ER	151	425	8 000	19 500	IH	4 850
Ataturk	C	Firat	Diyarbakir	East	ER	184	746	85 000	48 700	ICH	16 800
UNITED STATES											
Ashokan	1916	Esopus Creek	Olive Bridge	New York	TE	77	1 417	1 950	484 018	S	5 938
Calaveras	1925	Calaveras Creek	Sunol	California	ER	75	366	2 646	123 348	S	702
Dix	1925	Dix	High Bridge	Kentucky	ER	87	311	1 343	222 027	H	1 300
Tieton	1925	Tieton	Naches	Washington	TE	97	280	1 567	244 229	ICR	1 416
Cobble Mountain Reservoir	1931	Little	Westfield	Mass.	TE	80	221	2 294	86 380	S	113
New Exchequer	1926	Merced	Snelling	California	ER	146	378	3 952	1 265 552	H	9 911
Salt Springs	1931	N Fork Mokelumne		California	ER	96	396	2 294	171 947	HS	1 580
El Capitan	1934	San Diego	Lakeside	California	TE/ER	82	357	2 049	88 811	S	4 831
Hoover (Boulder)	1936	Colorado	Boulder	Nevada	VA	221	379	3 364	34 852 028	IHCN	11 327
Fort Peck	1937	Missouri	Frazer	Montana	TE	76	6 534	96 050	22 118 763	CHIN	6 514
Alcova	1938	North Platte	Casper	Wyoming	TE/ER	81	233	1 250	227 577	IHR	1 557
Mathews	1938	Tr Cajalco Creek	Corona	California	TE	84	1 988	7 309	224 494	S	382
Tygart	1938	Tygart	Grafton	W Virginia	PG	76	586	1 055	135 190	NC	8 948
Quabbin Winsor	1939	Swift	Ware	Mass.	TE	85	805	3 058	1 561 256	S	425
San Gabriel No 1	1939	San Gabriel	Azusa	California	ER	123	463	8 104	54 725	CS	7 524
Friant	1942	San Joaquin	Fresno	California	PG	97	1 063	1 632	642 027	ISCR	2 350
Grand Coulee	1942	Columbia	Coulee Dam	Washington	PG	168	1 272	8 093	11 794 553	IHCN	26 986
Marshall Ford Lake Travis (res)	1942	Colorado	Austin	Texas	TE	85	1 230	1 251	1 446 381	HCSR	16 197
Nantahala Lake	1942	Nantahala	Nantahala	N Carolina	TE	76	318	5 532	142 590	H	2 503
Green Mountain	1943	Blue	Hot Sulphur Springs	Colorado	TE/ER	94	351	3 333	190 696	IHR	708
Fontana	1944	Little Tennessee	Fontana Village	N Carolina	PG	146	721	2 734	603 715	H	4 474
Merriman	1945	Roundout Creek	Lackawack	New York	TE	114	732	4 434	189 956	S	5 097
Shasta	1945	Sacramento	Redding	California	PG	183	1 055	6 660	5 614 809	ISHN	5 239
Mud Mountain	1948	White	Buckley	Washington	ER	130	213	1 758	130 698	C	3 892
Watauga	1948	Watauga	Elizabethton	Tennessee	ER	97	274	2 660	398 415	CHNR	1 756
Anderson Ranch	1950	S Fork Boise	Boise	Idaho	TE	139	411	7 380	620 071	ICRH	566
Leroy Anderson	1950	Coyote Creek	San Jose	California	TE	77	421	2 485	112 617	I	1 195
South Holston	1950	S Fork Holston	Bluff City	Tennessee	ER	87	488	4 499	402 115	CHNR	1 756
Bull Shoals	1951	White	Cotter	Arkansas	PG	78	688	1 606	3 759 653	CH	14 158
Center Hill	1951	Caney Fork	Lancaster	Tennessee	TE/PG	76	658	2 736	1 033 658	CHR	12 856
Wolf Creek	1951	Cumberland	Burkesville	Kentucky	TE	79	1 748	8 713	4 927 760	HCR	15 659
Bradbury	1953	Santa Ynez	Santa Barbara	California	TE	85	1 021	5 119	252 864	ISR	4 559
Detroit	1953	N Santiam	Mill City	Oregon	PG	141	482	1 147	561 152	HCRI	4 984
Hungry Horse	1953	S Fork Flathead	Kalispell	Montana	VA	172	645	2 359	4 277 715	IHCN	1 501
Lookout Point	1953	Middle Fork Willamette	Eugene	Oregon	TE	84	968	5 892	562 385	CINH	7 646
Yale	1953	Lewis	Woodland	Washington	TE	98	472	3 211	495 860	HCR	4 899
Lucky Peak	1954	Boise	Boise	Idaho	TE	104	713	4 511	377 445	CRI	2 642
Pine Flat Lake	1954	Kings	Piedra	California	PG	134	561	1 835	1 233 482	CIRH	11 072

TABLE 8–1. HIGH DAMS OF THE WORLD ARRANGED BY COUNTRY (continued)

Name of Dam	Year of Com-pletion	River	Nearest City	State Province or Country	TYPE	Height Above Lowest Foun-dation m	Length of Crest m	Volume Content of Dam 10^3m^3	Gross Capacity of Reservoir 10^3m^3	PURPOSE	Maxi-mum Dis-charge Capacity of Spill-ways m^3/s
UNITED STATES (continued)											
Folsom	1956	American	Sacramento	California	PG	104	3 109	6 866	1 245 817	ISHC	16 056
Beardsley	1957	Mid Fk Stanislaus	Melones	California	TE	85	250	2 294	120 264	H	2 027
Brownlee	1958	Snake	Oxbow Village	Idaho	ER	120	421	4 587	1 759 808	HCR	8 693
Courtwright	1958	Helms Creek	Piedra	California	ER	95	263	1 193	152 088	HS	400
Oahe	1958	Missouri	Pierre	So. Dakota	TE	75	2890	70339	27 432 595	CHIN	2266
Swift	1958	Lewis	Woodland	Washington	TE	186	640	11 774	932 512	HCR	4 106
Wishon	1958	N Fk Kings	Piedra	California	ER	80	1 021	2 829	157 886	HS	1 379
Casitas	1959	Coyote Creek	Santa Barbara	California	TE	102	610	6 967	313 304	ISC	210
Table Rock	1959	White	Branson	Missouri	TE/PG	77	1 958	3 479	3 332 900	CH	15 801
Mammoth Pool	1960	San Joaquin		California	TE	124	250	4 094	151 718	HS	4 757
Arthur R. Bowman (Prineville)	1961	Crooked	Bend	Oregon	TE	75	244	1 089	190 573	IRC	230
Ball Mountain	1961	West	Jamaica	Vermont	TE	83	279	1 767	2 760	CR	4 248
Lewis Smith	1961	Dipsey Fork	Dilworth	Alabama	TE	93	671	3 930	1 714 540	HCR	5 873
Sly Creek	1961	Lost Creek	Oroville	California	TE	83	640	4 000	80 238	H	328
Hills Creek	1962	M Fk Willamette	Oakridge	Oregon	TE	104	703	8 257	439 055	CHIS	4 010
Smith	1962	Smith	Belknap Springs	Oregon	TE	101	351	1 911	18 502	H	255
Trinity	1962	Trinity	Redding	California	TE	164	747	22 486	3 019 563	IHCR	680
Abiquiu	1963	Rio Chama	Abiquiu	New Mexico	TE	99	469	9 017		CR	447
Dillon	1963	Blue	Silverthorne	Colorado	TE	94	1 798	9 140	311 674	S	333
Lemon	1963	Florida	Durango	Colorado	TE	87	415	2 326	49 463	I	272
Navajo	1963	San Juan	Blanco	New Mexico	TE	123	1 112	20 521	2 108 020	IR	963
Union Valley	1963	Silver Crk	Coloma	California	TE	138	549	7 646	334 274	S	1 260
Whiskeytown	1963	Clear Creek	Redding	California	TE	86	1 190	3 412	297 269	IHCR	815
Briones	1964	Bear Creek	El Sobrante	California	TE	87	629	7 578	83285	S	93
Cougar	1964	S Fork Mckenzie	Springfield	Oregon	ER	158	488	9 939	270 463	HCIR	2 152
Homestake	1964	M Fk Homestake	Minturn	Colorado	ER	81	608	2 628	53 780	S	99
Round Butte	1964	Deschutes	Warm Springs	Oregon	TE	134	442	7 340	659 913	HR	1 286
Summersville	1965	Gauley	Swiss	W Virginia	ER	119	695	10 371	236 215	CRS	11 667
Blue Mesa	1966	Gunnison	Montrose	Colorado	TE/ER	119	239	2 355	1 160 706	HCR	954
Glen Canyon	1966	Colorado	Lees Ferry	Arizona	VA	216	475	3 747	33 304 009	HSCR	7 815
John W. Flannagan	1966	Pound	Elkhorn City	Virginia	TE	79	279	1 824	83 261	CR	1 240
Lost Creek	1966	Lost Creek	Devils Slide	Utah	TE	76	329	1 401	27 753	ISRC	70
Lower Hell Hole	1966	Rubicon	Auburn	California	ER	125	472	6 357	257 058	SH	132
Millwood	1966	Little	Ashdown	Arkansas	ER	88	5 350	6 117	189 000	CS	13 403
Yellowtail	1966	Bighorn	Hardin	Montana	VA	160	451	1 182	1 076 830	ICHR	2 605
San Luis	1967	San Luis Creek	Los Banos	California	TE	116	5 639	59 559	2 517 536	ISHR	29
Alamo (res)	1968	Bill Williams	Parker	Arizona	TE	105	297	2 328	539 350	CRI	1 175
Blue River	1968	Blue	Springfield	Oregon	TE	95	381	3 726	110 380	CR	1 501
Oroville	1968	Feather	Oroville	California	TE	230	2 073	59 635	4 297 451	SCHR	4 248
Ruedi	1968	Fryingpan	Basalt	Colorado	TE/ER	98	318	2 863	126 432	IRC	157
International Amistad	1969	Rio Grande	Del Rio	Texas/Mex.	TE/PG	77	9 754	2 652	4 323 847	CIHR	42 673
Lopez	1969	Arroyo Grande Crk	Arroyo Grande	California	TE	85	341	2 705	64 758	S	1 256
New Bullards Bar	1970	North Yuba	Marysville	California	VA	194	671	1 988	1 195 984	SH	4 219
Cedar Springs	1971	W Fk Mojave	Victorville	California	ER	76	725	6 040	96 212	IRS	913
Don Pedro	1971	Tuolumne	La Grange	California	TE	173	549	12 233	2 503 968	H	13 380
Heron	1971	Willow Creek	Tierra Amarillo	New Mexico	TE/ER	84	372	2 317	495 243	ISR	19
Castaic	1973	Castaic Creek	Castaic	California	TE	125	1 585	33 640	431 719	IRS	2 220
Dworshak	1973	N. Fork Clearwater	Ahsahka	Idaho	PG	219	1 002	4 931	4 259 213	HCR	6 258
Jocassee	1973	Keowee		S Carolina	ER	133	549	8 869	1 431 206	H	1 761
Libby	1973	Kootenai	Libby	Montana	PG	129	881	2 875	7 165 296	HCR	4 060
Pyramid	1973	Piru Creek	Piru	California	ER	122	329	5 315	220 793	IRSH	4 248
Soldier Creek	1973	Strawberry	Duchesne	Utah	TE	77	393	2 440	1 365 464	ICR	
Carters	1974	Coosawattee	Carters	Georgia	ER	141	594	11 468	465 146	CHR	520
Cochiti	1975	Rio Grande and Santa Fe	Cochiti Pueblo	New Mexico	TE	77	8 785	50 228	52 726	CIR	4 280
Lost Creek	1976	Rogue	Shady Cove	Oregon	ER	105	1 097	8 257	573 485	CHSR	4 474
Ririe	1976	Willow Creek	Idaho Falls	Idaho	TE	77	326	2 046	123 348	IC	1 133
Gross	1977	South Boulder Crk	Louisville	Colorado	PG	104	332	7 809	50 557	SH	444
Little Blue Run	1977	Little Blue Run of Ohio	East Liverpool	Penn.	TE	122	640	9 939	13 568		48
Gathright	1978	Jackson	Covington	Virginia	ER	78	368	1 988	152 582	RC	173
New Melones	1979	Stanislaus	Modesto	California	ER	191	475	12 233	2 960 356	CIHR	3 171
Bloomington	1981	N Branch of Potomac	Bloomington	Maryland	ER	90	649	7 646	116 793	CSR	5 465
Bath County Upper	C	Little Back Creek	Warm Springs	Virginia	ER/TE	143	731	18 000	43 790	H	509
Warm Springs	C	Dry Creek	Cloverdale	California	TE	97	914	22 920	469 365	CSR	1 056
USSR											
Mingechaur	1953	Kura	Mingechaur	Azerb.SSR	TE	80	1 550	15 600	16 000 000	HNIC	3 600
Bukhtarma	1960	Irtysh	Ust-Kamenogorsk	Kazakh. SSR	PG	90	380	1 170	49 800 000	HN	1 000

TABLE 8–1. HIGH DAMS OF THE WORLD ARRANGED BY COUNTRY (continued)

Name of Dam	Year of Completion	River	Nearest City	State Province or Country	TYPE	Height Above Lowest Foundation m	Length of Crest m	Volume Content of Dam 10^3m^3	Gross Capacity of Reservoir 10^3m^3	PURPOSE	Maximum Discharge Capacity of Spillways m^3/s
USSR (continued)											
Sioni	1963	Iori	Tbilisi	Georg.SSR	TE	86	780	6 300	325 000	IH	596
Bratsk	1964	Angara	Bratsk	Irkutsk	PG	125	1 430	4 415	169 000 000	HNS	7 090
					TE	36	2 987	6 547			
					TE	40	723	2 147			
Serebrianka No 1	1970	Voroniya	Murmansk	Murmansk	ER	78	2 625	5 660	4 170 000	H	675
Sarsang	1976	Terter	Yevlakh	Azerb.SSR	TE	125	590	5 820	560 000	IH	800
Charvak	1977	Chirchik	Tashkent	Uzbek.SSR	ER	168	764	21 600	2 000 000	HI	1 200
Medeo	1977	Malaya Almaatinka	Alma-Ata	Kazakh.SSR	ER	144	530	8 500	-	C	-
Ust-Ilim	1977	Angara	Ust-Ilimsk	Irkutsk	PG	102	1 477	3 800	59 300 000	HN	9 700
					TE/ER	47	2 248	5 066			
Chirkey	1978	Sulak	Makhachkala	Daghest.ASSR	VA	233	333	1 358	2 780 000	HIS	2 870
Toktogul	1978	Naryn	Naryn	Kirgh.SSR	PG	215	293	3 345	19 500 000	HI	2 340
Zeya	1978	Zeya	Blagoveshchensk	Amur.	CB	115	758	2 160	68 400 000	HCN	6 600
Andizhan	1980	Karadarya	Osh	Kirghiz.SSR	CB	115	920	3 700	1 750 000	HI	2 392
Inguri	1980	Inguri	Zugdidi	Georgian SSR	VA	272	680	3 960	1 100 000	HI	2 500
Nurek	1980	Vakhsh	Nurek	Tadjik SSR	TE	300	704	58 000	10 500 000	HI	4 000
Sayano-Shushensk	C	Yenisei	Minusinsk	Krasnoyarsk	VA/PG	245	1 066	9 075	31 300 000	NH	13 600
Bureya	C	Bureya	Blagoveshchensk	Khabarovsk	PG	139	810	3 561	20 900 000	HC	19 100
Irganai	C	Avar Koisu	Makhachkala	Daghest.ASSR	TE/ER	111	312	5 827	705 000	H	2 760
Khudoni	C	Inguri	Dzhavari	Georg.SSR	VA	200	545	1 475	365 000	H	2 030
Kolyma	C	Kolyma	Magadan	Magadan	ER	126	759	12 550	14 600 000	H	17 500
Rogun	C	Vakhsh	Nurek	Tajik SSR	TE/ER	335	660	75 500	13 300 000	HI	3 500
Spandarian	C	Vorotan	Sisian	Armen.SSR	TE	87	317	2 250	277 000	HI	160
Zhinvali	C	Aragvi	Tbilisi	Georg.SSR	TE	102	412	5 200	520 000	HS	2 500
VENEZUELA											
Onia	1978	Onia	El Vigia	Mérida	TE	301	450	1 300	6 600	C	435
Tucupido	C	Tucupido	Guanare	Portuguesa	TE	92	290	3 300		IHC	—
Yacambú	C	Yacambú	Sanare	Lara	TE	158	107	3 000	427 000	ICS	480
Las Palmas	C	Cojedes	Acarigua	Cojedes	TE	77	750	7 000	810 000	IHC	
Las Cuevas	C		San Cristóbal	Tachira	TE	108		7 500	1 400 000	H	
La Vueltosa	C	Caparo	Buena Vista	Mérida	TE	118		15 000	5 300 000	H	
Borde Seco	C	Camburito	San Cristobal	Táchira	TE	120		6 600		H	
Taguaza	C	Taguaza	Sta.Lucia	Miranda	TE	100	300	2 000	212 000	S	430
La Honda	C	Uribante	San.Cristóbal	Táchira	TE	108		7 800	770 000	H	
YUGOSLAVIA											
Kokin Brod	1962	Uvac	Nova Varos	SR Srbija	ER	82	1 220	2 480	250 000	H	1 500
Tikves	1968	Crna Reka	Kavadarci	SR Makedonija	ER	114	338	2 722	475 000	J	2 050
Spilje	1969	Crni Drim	Debar	SR Makedonija	ER	112	330	2 699	520 000	H	2 890
Rama	1969	Rama	Prozor	SR Bosna and Hercegovina	ER	103	230	1 510	487 000	H	400
Turija	1970	Turija	Strumica	SR Makedonija	ER	93	417	1 978	65 000	J	76
Gazivode	1977	Ibar	Titova Mitrovica	SAP Kosovo	ER	108	520	5 000	370 000	H	720
Sjenica	1979	Veliki Uvac	Nova Varos	SR Srbija	ER	106	310	2 430	190 000	H	1 000
Lazici	C (1983)	Beli Rzav	Bajina Basta	SR Srbija	ER	123	120	2 170	150 000	H	180
Zavoj	C (1987)	Visocica	Pirot	SR Srbija	ER	80	250	1 400	16 000	H	1 800
ZAMBIA/ZIMBABWE											
Kariba	1959	Zambezi	Lusaka	Zambia/Zimbabwe	VA	128	579	1 032	160 368	H	9 500

Source: Compiled from World Register of Dams 1984 published by the International Commission on Large Dams, 151 boulevard Haussmann, 75008 Paris

TABLE 8–2. CLASSIFICATION OF DAMS IN THE WORLD BY TYPE AND HEIGHT

	Number of Dams/Height						
	Total	15–30 m	30–60 m	60–100 m	100–150 m	150–200 m	>200 m
Earth (TE) and Rockfill (ER)	28 844	24 567	3 657	477	116	21	6
Gravity (PG)	3 954	2 222	1 294	361	65	8	4
Arch (VA)	1 527	775	428	204	83	24	13
Buttress (CB)	337	175	110	40	12	—	—
Multi-Arch (MV)	136	74	48	13	—	—	1
Total	34 798	27 813	5 537	1 095	276	53	24

Source: International Commission on Large Dams, World Register of Dams 1984

TABLE 8–3. NUMBER OF DAMS IN THE WORLD IN 1950 AND 1982 BY CONTINENT

CONTINENT	1950		1982	
	Dams	%	Dams	%
Europe	1 292	25	3 800	11
Asia	1 541	30	22 701	65
America	2 090	40	7 241	21
Africa	123	2	610	2
Australasia	150	3	446	1
TOTAL	5 196	100	34 798	100
China	8	0.2	18 595	53

Source: International Commission on Large Dams, 1984, World Register of Dams; percentages rounded

TABLE 8–4. COUNTRIES WITH MORE THAN 100 DAMS IN 1982

[Numbers in brackets represent number of dams in service at end of 1977]

China	18 595	(16 500)	Brazil	489	(443)	Germany	184	(158)
USA	5 338	(5 046)	Mexico	487	(433)	Czechoslovakia	142	(131)
Japan	2 142	(2 006)	France	432	(388)	Sweden	134	(132)
India	1 085	(999)	Italy	408	(403)	Switzerland	130	(128)
Spain	690	(630)	Australia	374	(320)	Yugoslavia	114	(93)
Korea	628	(559)	South Africa	342	(316)	Austria	112	(97)
Canada	580	(494)	Norway	219	(205)	Bulgaria	108	(106)
Great Britain	529	(519)				Romania	106	(81)
	29 587			2 751			1 030	

Source: International Commission on Large Dams, 1984, World Register of Dams

TABLE 8–5. RATE OF DAM CONSTRUCTION IN THE WORLD, 1950 TO 1982

Period	Outside of China			China			Ratio China Compared to Outside China
	Number per Period	Total Number	Number per Annum	Number per Period	Total Number	Number per Annum	
Up to 1950	5 188	5 188			8		
1951 to 1974	8 948	14 136	373				
1975 to 1977	752	14 888	251				
1951 to 1977	9 700	14 888	359	16 492	16 500	611	1.70
1978 to 1982	1 315	16 203	263	2 095	18 595	419	1.60
1975 to 1982	2 067	16 203	258				
1951 to 1982	11 015	16 203	344	18 587	18 595	581	1.69

Source: International Commission on Large Dams, 1984, World Register of Dams

TABLE 8–6. WORLD'S HIGHEST DAMS

No.	Height Above Lowest Foundation in m	Type[1]	Name	Country	Year
	335	TE/ER	Rogun	USSR	U/C
1	300	TE	Nurek	USSR	1980
2	285	PG	Grande Dixence	Switzerland	1961
3	272	VA	Inguri	USSR	1980
4	262	VA	Vajont	Italy	1961
5	261	ER	Manuel Moreno Torres (Chicoasén)	Mexico	1980
	261	ER/TE	Tehri	India	U/C
	253	ER/TE	Kishau	India	U/C
	245	VA/PG	Sayano-Shushensk	USSR	U/C
	243	ER	Guavio	Colombia	U/C
6	242	TE	Mica	Canada	1972
7	237	ER	Chivor	Colombia	1975
8	237	VA	Mauvoisin	Switzerland	1957
	234	VA	El Cajón	Honduras	U/C (1984)
9	233	VA	Chirkey	USSR	1978
10	230	TE	Oroville	USA	1968
11	226	PG	Bhakra	India	1963
12	221	VA/PG	Hoover	USA	1936
13	220	VA	Contra	Switzerland	1965
14	220	VA	Mratinje	Yugoslavia	1976
15	219	PG	Dworshak	USA	1973
16	216	VA	Glen Canyon	USA	1966
17	215	PG	Toktogul	USSR	1978
18	214	MV	Daniel Johnson	Canada	1968
19	213	VA	Dez	Iran	1962
	210	ER	San Roque	Philippines	U/C
20	208	VA	Luzzone	Switzerland	1963
21	207	ER/PG	Keban	Turkey	1974
22	202	VA	Almendra	Spain	1970
	201	VA	Khudoni	USSR	U/C
23	200	VA	Karoun	Iran	1975
24	200	VA	Kolnbrein	Austria	1977
25	196	PG/ER/TE	Itaipu	Brazil	1982
	195	ER	Altinkaya	Turkey	U/C
26	194	VA	New Bullard's Bar	USA	1970
	192	PG	Lakhwar	India	U/C
27	191	ER	New Melones	USA	1979
28	186	VA	Kurobe	Japan	1964
29	186	TE	Swift	USA	1958
	186	VA	Zillergründl	Austria	U/C
30	185	VA	Mossyrock	USA	1968
	185	VA	Oymapinar	Turkey	U/C
	184	ER	Atatürk	Turkey	U/C
31	183	PG	Shasta	USA	1945
32	183	TE	W A C Bennett	Canada	1967
33	180	VA	Amir Kabir	Iran	1964
34	180	ER	Dartmouth	Australia	1979
35	180	VA	Emmosson	Switzerland	1974
	180	VA	Tehchi	Taiwan	1974
36	180	VA	Tignes	France	1952
37	176	ER	Takase	Japan	1978
38	175	ER	Ayvacik	Turkey	1981
39	174	PG	Alpe Gera	Italy	1964
40	173	TE	Don Pedro	USA	1971
	173	VA	Karakaya	Turkey	U/C
41	172	VA	Hungry Horse	USA	1953
	172	PG	Longyangxia	China	U/C

TABLE 8–6. WORLD'S HIGHEST DAMS (continued)

No.	Height Above Lowest Foundation in m	Type[1]	Name	Country	Year
41 (cont'd)	171	VA	Cabora Bassa	Mozambique	1974
42	169	VA	Idukki	India	1974
43	168	ER	Charvak	USSR	1977
	168	ER	Gura Apelor	Romania	U/C
44	168	ER	La Grande 2	Canada	1978
45	168	PG	Grand Coulée	USA	1942
46	167	ER	Fierze	Albania	1978
	167	VA/PG	Daniel Palacios	Ecuador	U/C
47	166	VA	Vidraru	Romania	1965
48	165	TE	Kremasta	Greece	1965
49	165	VA	Ross	USA	1949
50	165	PG	Wujiangdu	China	1981
	164	ER	Thomson	Australia	U/C
51	164	TE	Trinity	USA	1962
	162	PG/ER	Guri	Venezuela	U/C
52	162	ER	Talbingo	Australia	1971
53	160	ER	Foz de Areia	Brazil	1980
	160	TE/ER	Grand-Maison	France	U/C
	160	ER	Salvajina	Columbia	U/C
	160	ER/TE	Thein Dam Ranjit	India	U/C
54	160	VA	Yellowtail	USA	1966
	158	ER	Canales	Spain	U/C
	158	TE	Yacambu	Venezuela	U/C
55	158	ER	Cougar	USA	1964
56	158	ER	Emborcacão	Brazil	1982
57	158	VA	Gökcekaya	Turkey	1972
	158	ER	Naramata	Japan	U/C
	157	VA	Dongjiang	China	U/C
58	157	PG	Okutadami	Japan	1961
59	157	VA	Speccheri	Italy	1957
60	156	PG	Sakuma	Japan	1956
61	156	VA-TE	Zeuzier	Switzerland	1957
62	155	ER	Goescheneralp	Switzerland	1960
63	155	VA	Monteynard	France	1962
64	155	VA	Nagawado	Japan	1969
65	155	VA	Place Moulin	Italy	1965
	155	PG	Sadar Sarovar	India	U/C
66	154	VA/PG	Bhumibol	Thailand	1964
67	154	ER	Tedorigawa	Japan	1979
68	153	VA	Curnera	Switzerland	1967
69	153	VA	Flaming Gorge	USA	1964
70	153	ER	Gepatsch	Austria	1965
	153	PG/ER	Revelstoke	Canada	U/C
71	153	VA	Santa Giustina	Italy	1950
	151	PG	Dorna	Spain	U/C
	151	ER	Menzelet	Turkey	U/C
72	151	VA	Zervreila	Switzerland	1957
	150	PG	Baishan	China	U/C
73	150	VA	Canelles	Spain	1960
74	150	ER	Finstertal	Austria	1980
	150	ER	Kenyir	Malaysia	U/C
75	150	VA/CB	Roselend	France	1961
76	150	TE	Big Horn	Canada	1972

U/C Under Construction

[1] TE Earth; ER Rockfill; PG Gravity; CB Buttress; VA Arch; MV Multi-Arch.

Source: International Commission on Large Dams, 1984, World Register of Dams

TABLE 8–7. WORLD'S LARGEST DAMS ACCORDING TO SPILLWAY CAPACITY

No.	Capacity, m^3s	Name	Country	Year
	113 000	Gezhouba	China	U/C
	110 000	Tucurui	Brazil	U/C
1	82 300	Danjiangkou	China	1974
2	64 845	The Dalles	USA	1957
	64 600	Burdekin Falls	Australia	U/C
3	63 713	John Day Lock and Dam	USA	1968
4	62 297	MacNary Lock and Dam	USA	1957
	62 296	Sardar Sarovar	India	U/C
5	61 400	Itaipu	Brazil	1982
	60 000	Oosterscheldekering	Netherlands	U/C
6	59 000	Shuifeng	China/Korea	1943
7	58 500	Saratov	USSR	1967
8	58 400	Gavins Point	USA	1958
9	57 000	Salto Grande	Uruguay/Argentina	1979
	54 862	Jhuj	India	U/C
10	54 400	Panjiakou	China	1979
11	53 450	Nagarjuna Sagar	India	1974
	52 000	Porto Primavera	Brazil	U/C
12	50 000	Jupia	Brazil	1968
	49 800	Wanan	China	U/C
13	49 600	Kadana	India	1978
	47 000	Ankang	China	U/C
	46 970	Gohira	India	U/C
14	46 259	Alvin Wirtz	USA	1950
	45 312	Sri Rama Sagar	India	U/C
15	45 307	Bonneville	USA	1937
	44 752	Bargi	India	U/C
	43 000	Owen Falls	Uganda	1954
16	43 690	International la Amistad	USA/Mexico	1969
17	42 459	Hirakud	India	1957
18	42 186	Tarbela	Pakistan	1976
19	41 280	Xinanjiang	China	1965
20	40 300	Kuibyshev	USSR	1955
21	40 000	Ilha Solteira	Brazil	1973
22	39 644	Priest Rapids	USA	1959
23	39 644	Wanapum	USA	1963
24	39 158	Tom Miller	USA	1939
	37 945	Narayanpur	India	U/C
25	37 500	Shuifeng	China	1943
	37 400	Srisailam HE	India	1943
26	36 200	Fengman	China	1954
27	35 960	Ukai	India	1972
28	35 620	Daheiting	China	1980
	35 000	Guri	Venezuela	U/C
29	33 980	Kentucky	USA	1944
30	33 800	Fuchunjing	China	1968
31	33 600	Chief Joseph	USA	1955
32	33 414	Wells	USA	1967
33	33 131	Conowingo	USA	1928
34	31 400	Xijin	China	1966
35	31 152	Marala	Pakistan	1968
36	31 144	Mangla	Pakistan	1967

U/C Under Construction.
Source: International Commission on Large Dams, 1984, World Register of Dams

TABLE 8–8. CAUSES OF DAM INCIDENTS IN THE UNITED STATES

Cause	Type of Dam								
	Concrete		Embankment		Other[1]		Totals		
	F	A	F	A	F	A	F	A	F&A
Overtopping	6	3	18	7	3		27	10	37
Flow erosion	3		14	17			17	17	34
Slope protection damage				13				13	13
Embankment leakage, piping			23	14			23	14	37
Foundation leakage, piping	5	6	11	43	1		17	49	66
Sliding	2		5	28			7	28	35
Deformation		2	3	29	3		6	31	37
Deterioration		6	2	3			2	9	11
Earthquake instability				3				3	3
Faulty construction	2			3			2	3	5
Gate failures	1	2	1	3			2	5	7
TOTAL	19	19	77	163	7		103	182	285

[1] Steel, masonry-wood, or timber crib.
F = failure.
A = accident (an incident where failure was prevented by remedial work or operating procedures, such as drawing down the pool).
Source: National Academy Press, 1983, Safety of Existing Dams: Evaluation and Improvement. Based on Schnitter, 1979, Lessons from Dam Incidents USA, ASCE/USCOLD, and supplementary data supplied by U.S. Committee on Large Dams for period to 1979

TABLE 8–9. MODES AND CAUSES OF EARTH DAM FAILURES

Form	General Characteristics	Causes	Preventive or Corrective Measures
		Hydraulic Failures (30% of all failures)	
Overtopping	Flow over embankment, washing out dam.	Inadequate spillway capacity. Clogging of spillway with debris. Insufficient freeboard due to settlement, skimpy design.	Spillway designed for maximum flood. Maintenance, trash booms, clean design. Allowance for freeboard and settlement in design; increase crest height or add flood parapet.
Wave erosion	Notching of upstream face by waves, currents.	Lack of riprap, too small riprap.	Properly designed riprap.
Toe erosion	Erosion of toe by outlet discharge.	Spillway too close to dam. Inadequate riprap.	Training walls. Properly designed riprap.
Gullying	Rainfall erosion of dam face.	Lack of sod or poor surface drainage.	Sod, fine riprap; surface drains.
		Seepage Failures (40% of all failures)	
Loss of water	Excessive loss of water from reservoir and/or occasionally increased	Pervious reservoir rim or bottom.	Blanket reservoir with compacted clay or chemical admix; grout seams, cavities.

TABLE 8–9. MODES AND CAUSES OF EARTH DAM FAILURES (continued)

Form	General Characteristics	Causes	Preventive or Corrective Measures
Loss of water (cont.)	seepage or increased groundwater levels near reservoir.	Pervious dam foundation.	Use foundation cutoff; grout; upstream blanket.
		Pervious dam.	Impervious core.
		Leaking conduits.	Watertight joints; waterstops; grouting.
		Settlement cracks in dam.	Remove compressible foundation, avoid sharp changes in abutment slope, compact soils at high moisture.
		Shrinkage cracks in dam.	Use low-plasticity clays for core, adequate compaction.
Seepage erosion or piping	Progressive internal erosion of soil from downstream side of dam or foundation backward toward the upstream side to form an open conduit or "pipe." Often leads to a washout of a section of the dam.	Settlement cracks in dam.	Remove compressible foundation, avoid sharp changes, internal drainage with protective filters.
		Shrinkage cracks in dam.	Low-plasticity soil; adequate compaction; internal drainage with protective filters.
		Pervious seams in foundation.	Foundation relief drain with filter; cutoff.
		Pervious seams, roots, etc., in dam.	Construction control; core; internal drainage with protective filter.
		Concentration of seepage at face.	Toe drain; internal drainage with filter.
		Boundary seepage along conduits, walls.	Stub cutoff walls, collars; good soil compaction.
		Leaking conduits.	Watertight joints; waterstops; materials.
		Animal burrows.	Riprap, wire mesh.
		Structural Failures (30% of all failures)	
Foundation slide	Sliding of entire dam, one face, or both faces in opposite directions, with bulging of foundation in the direction of movement.	Soft or weak foundation.	Flatten slope; employ broad berms; remove weak material; stabilize soil.
		Excess water pressure in confined sand or silt seams.	Drainage by deep drain trenches with protective filters; relief wells.
Upstream slope	Slide in upstream face with little or no bulging in foundation below toe.	Steep slope.	Flatten slope or employ berm at toe.
		Weak embankment soil.	Increased compaction; better soil.
		Sudden drawdown of pond.	Flatten slope, rock berms; operating rules.
Downstream slope	Slide in downstream face.	Steep slope.	Flatten slope or employ berm at toe.
		Weak soil.	Increased compaction; better soil.
		Loss of soil strength by seepage pressure or saturation by seepage or rainfall.	Core; internal drainage with protective filters; surface drainage.
Flow slide	Collapse and flow of soil in either upstream or downstream direction	Loose embankment soil at low cohesion, triggered by shock, vibration, seepage, or foundation movements.	Adequate compaction.

Source: National Academy Press, 1983, Safety of Existing Dams: Evaluation and Improvement. Original source: Sowers, G.F., 1961, The Use and Misuse of Earth Dams, Consulting Engineering, July

SECTION B. RESERVOIRS

TABLE 8–10. NORMAL SURFACE-WATER RESERVOIR CAPACITY IN THE UNITED STATES

[By water-resources region; see Figure 2–6]

Water-Resources Region	Area of Region, in Thousands of Square Miles	Average Renewable Supply, in Billion Gallons per Day[1]	Normal Reservoir Capacity[2] In Million Acre-feet	In Acre-feet of Storage per Square Mile	As a Percentage of Annual Renewable Supply
New England	69	78.4	13.0	188	15
Mid-Atlantic	103	80.7	10.3	100	11
South Atlantic-Gulf	271	233.5	38.7	143	15
Great Lakes	134	74.3	6.9	51	8
Ohio (exclusive of Tennessee Region)	160	139.5	19.6	123	13
Tennessee	43	41.2	11.2	260	24
Upper Mississippi (exclusive of Missouri Region)	181	77.2	12.2	67	14
Mississippi (entire basin)	1,241	464.3	164.8	133	32
Souris-Red Rainy	55	6.5	8.0	145	110
Missouri	511	62.5	84.3	165	120
Arkansas-White-Red	244	68.6	31.8	130	41
Texas-Gulf	178	33.1	24.7	139	67
Rio Grande	137	5.1	10.4	76	182
Upper Colorado	103	14.7	37.7	366	229
Colorado (entire basin)	258	15.6	70.4	273	403
Great Basin	139	9.9	3.3	24	30
Pacific Northwest	271	276.2	60.9	225	20
California	165	70.2	38.8	235	49
Alaska	586	975.5	1.5	3	0.1
Hawaii	6	7.4	0.0	2	0.1
Caribbean	4	5.1	0.3	90	5

[1] Adjusted by adding exports and subtracting imports.
[2] About two-thirds of maximum capacity.
Source: U.S. Geological Survey, 1984, National Water-Summary 1983—Hydrologic Events and Issues, Water-Supply Paper 2250

TABLE 8–11. SUMMARY OF RESERVOIR STORAGE, INCLUDING CONTROLLED NATURAL LAKES, IN THE UNITED STATES AND PUERTO RICO

[Reservoir storage is expressed as normal capacity, which is the total storage space in a reservoir below the normal retention level, including dead storage and inactive storage, and excluding any flood-control or surcharge storage]

Reservoir Storage (Range, in Acre-feet)	Number of Reservoirs	Total Reservoir Storage: Acre-feet	Percentage of Total
Greater than 10,000,000	5	107,655,000	22.4
100,000–10,000,000	569	322,852,000	67.3
50,000–100,000	295	20,557,000	4.3
25,000–50,000	374	13,092,000	2.7
5,000–25,000	1,411	15,632,000	3.3
Total[1]	2,654	479,788,000	100.0

[1] In addition, there are perhaps at least 50,000 reservoirs with capacities ranging from 50 to 5,000 acre-feet, and about 2 million smaller farm ponds used for storage.
Source: U.S. Army Corps of Engineers, 1981

FIGURE 8–1. WATER RESOURCES REGIONS OF THE UNITED STATES

Source: U.S. Geological Survey

TABLE 8–12. RESERVOIR STORAGE REQUIRED TO PRODUCE SELECTED DEPENDABLE FLOWS IN THE UNITED STATES

[Flow is in millions of gallons per day, and storage is in thousands of acre-feet. For regions see Fig. 8–1.]

Region	Storage Required to Produce Indicated Flow 100% of Time									
	Flow	**Storage**	**Flow**	**Storage**	**Flow**	**Storage**	**Flow**	**Storage**	**Flow**	**Storage**
New England	6,300	1,300	9,700	1,900	16,000	4,200	22,000	7,300	39,000	26,000
Delaware and Hudson Rivers	3,200	590	4,800	900	7,800	2,000	11,000	3,800	19,000	11,000
Chesapeake Bay	5,600	960	8,400	1,400	13,000	3,400	18,000	6,000	32,000	20,000
Southeast	21,000	4,900	31,000	7,800	49,000	14,000	71,000	25,000	126,000	78,000
Western Great Lakes	8,400	780	12,000	1,100	16,000	2,800	21,000	5,100	32,000	20,000
Eastern Great Lakes	2,300	720	3,700	1,100	6,500	2,200	9,700	3,700	19,000	11,000
Ohio River	7,400	4,100	9,400	5,200	15,000	7,200	21,000	11,000	46,000	29,000
Cumberland River	1,500	160	2,100	290	3,300	620	4,500	1,100	7,800	3,100
Tennessee River	9,000	400	11,000	600	15,000	1,200	19,000	2,500	28,000	9,600
Upper Mississippi River	7,800	1,100	12,000	2,300	18,000	5,200	25,000	8,700	41,000	26,000
Upper Missouri River	1,200	340	1,800	660	3,200	1,200	4,500	2,100	9,000	5,700
Lower Missouri River	410	660	550	840	1,200	1,300	2,200	2,300	5,800	5,200
Upper Arkansas-Red Rivers	410	250	700	410	1,300	720	2,100	1,200	4,500	3,000
Lower Arkansas-Red and White Rivers	1,000	1,400	2,100	2,500	4,400	5,000	8,400	9,400	20,000	22,000
Lower Mississippi River	2,100	900	3,600	1,800	6,500	3,200	10,000	5,500	21,000	14,000
Upper Rio Grande and Pecos Rivers	[1]	[1]	[1]	[1]	[1]	[1]	[1]	[1]	[1]	[1]
Western Gulf	920	1,300	1,700	1,700	3,400	3,800	5,900	5,500	14,000	12,000
Colorado	210	90	320	120	560	230	830	380	1,700	1,200
Great Basin	300	72	470	100	780	240	1,200	420	2,100	1,000
South Pacific	30	10	40	13	60	26	100	43	180	119
Central Pacific	1,000	880	1,900	1,760	3,800	3,100	6,300	5,300	16,000	11,400
Pacific Northwest	9,700	2,600	21,000	6,600	26,000	9,300	39,000	16,000	76,000	47,000
United States	90,000	24,000	140,000	40,000	210,000	71,000	300,000	120,000	560,000	350,000

[1] Appropriations currently exceed supply.

Source: Select Committee on National Water Resources, U.S. Senate, 1960

TABLE 8–13. ADVANTAGES AND DISADVANTAGES OF SUBSURFACE AND SURFACE RESERVOIRS

Subsurface Reservoirs	Surface Reservoirs
Advantages	Disadvantages
1. Many large-capacity sites available	1. Few new sites available
2. Slight to no evaporation loss	2. High evaporation loss even in humid climate
3. Require little land area	3. Require large land area
4. Slight to no danger of catastrophic structural failure	4. Ever-present danger of catastrophic failure
5. Uniform water temperature	5. Fluctuating water temperature
6. High biological purity	6. Easily contaminated
7. Safe from immediate radioactive fallout	7. Easily contaminated by radioactive material
8. Serve as conveyance systems—canals or pipeline across lands of others unnecessary	8. Water must be conveyed
Disadvantages	Advantages
1. Water must be pumped	1. Water may be available by gravity flow
2. Storage and conveyance use only	2. Multiple use
3. Water may be mineralized	3. Water generally of relatively low mineral content
4. Minor flood control value	4. Maximum flood control value
5. Limited flow at any point	5. Large flows
6. Power head usually not available	6. Power head available
7. Difficult and costly to investigate, evaluate, and manage	7. Relatively easy to evaluate, investigate, and manage
8. Recharge opportunity usually dependent on surplus surface flows	8. Recharge dependent on annual precipitation
9. Recharge water may require expensive treatment	9. No treatment required of recharge water
10. Continuous expensive maintenance of recharge areas or wells	10. Little maintenance required of facilities

Source: U.S. Bureau of Reclamation, 1977, Ground Water Manual, U.S. Department of the Interior

TABLE 8–14. SUMMARY OF STORAGE RESERVOIRS ON UNITED STATES BUREAU OF RECLAMATION PROJECTS

[As of September 30, 1986]

		Active Capacity		Total Capacity	
Category	No. of Reservoirs	Thousand Cubic Meters	Acre-Feet	Thousand Cubic Meters	Acre-Feet
Constructed and operated by Bureau of Reclamation	110	106 709 244	86 510 251	144 632 109	117 254 697
Rehabilitated and operated by Bureau of Reclamation	10	247 069	200 302	248 694	201 618
Constructed by others, operated by Bureau of Reclamation	11	6 337 601	5 137 957	7 009 548	5 682 711
Under construction by Bureau of Reclamation	8	837 118	678 660	5 469 294	4 434 012
Constructed by Bureau of Reclamation, operated by others	109	18 183 188	14 741 292	19 878 391	16 115 611
Rehabilitated by Bureau of Reclamation, operated by others	14	852 561	691 180	865 130	701 370
Constructed and operated by others	71	78 238 538	63 428 765	100 529 146	81 499 984
Constructed or rehabilitated under loan program	22	444 806	360 609	493 310	399 932
TOTAL	355	211 850 129	171 749 017	279 125 626	226 289 936

Source: U.S. Department of the Interior, Bureau of Reclamation, 1986, Statistical Compilation of Engineering Features on Bureau of Reclamation Projects

TABLE 8–15. MAJOR STORAGE RESERVOIRS ON UNITED STATES BUREAU OF RECLAMATION PROJECTS

[As of September 30, 1986; total capacity of 500,000 acre-feet or more]

Project, State / Reservoir Name (Dam Name) / Stream / Operator	Category	Purpose	Active Capacity acre feet	Total Capacity acre feet	Reservoir Area acres	Year Completed
PACIFIC NORTHWEST REGION						
BOISE ID-OR						
CASCADE RESERVOIR	A	I-P-FC	653,000	703,000	28,300	1948
NORTH FORK PAYETTE RIVER						
COLUMBIA BASIN, WA						
BANKS LAKE (NORTH AND DRY FALLS)	A	i	715,000	1,280,000	27,000	1951[2]
OFFSTREAM						
BILLY CLAPP LAKE (PINTO)	A	I-FC	21,200	64,200	1,010	1948
OFFSTREAM						
FRANKLIN D. ROOSEVELT LK (GRAND COULEE)	A	I-P-FC-RR-N	5,190,000	9,390,000	82,300	1942
COLUMBIA RIVER						
POTHOLES RESERVOIR (O SULLIVAN)	A	I-FC	332,000	512,000	27,800	1949
LOWER CRAB CREEK						
HUNGRY HORSE, MT						
HUNGRY HORSE RESERVOIR	A	I-P-FC-N	2,980,000	3,470,000	23,800	1953
SOUTH FORK FLATHEAD RIVER						
MINIDOKA-PALISADES, ID-WY						
AMERICAN FALLS RESERVOIR	B	I-FC-M I	1,670,000	1,670,000	58,100	1978[3]
SNAKE RIVER						
ISLAND PARK RESERVOIR	A	I-FC	127,500	1,280,000	7,794	1938[3]
HENRYS FORK RIVER						
JACKSON LAKE	A	I-FC	624,000	847,000	25,500	1916[3,5]
SNAKE RIVER						
PALISADES RESERVOIR	A	I-P-FC-F W	1,200,000	1,401,000	16,150	1957
SOUTH FORK SNAKE RIVER						
OWYHEE, ID-OR						
LAKE OWYHEE	D	I-FC	715,000	1,120,000	12,700	1932
OWYHEE RIVER						
OWYHEE PROJECT NORTH BOARD OF CONTROL						
YAKIMA, WA						
CLE ELUM LAKE	A	I	437,000	710,000	4,812	1933
CLE ELUM RIVER						
MID-PACIFIC REGION						
CENTRAL VALLEY, CA						
CLAIR ENGLE LAKE (TRINITY)	A	I-P	2,140,000	2,450,000	16,500	1962
TRINITY RIVER						
FOLSOM LAKE	B	I-P-FC	920,000	1,010,000	11,400	1956
AMERICAN RIVER						
LAKE ISABELLA	E	I-FC	567,900	570,000	11,400	1953
KERN RIVER						
U.S. CORPS OF ENGINEERS						
MILLERTON LAKE (FRIANT)	A	I-FC	434,000	521,000	4,900	1942
SAN JOAQUIN RIVER						
NEW HOGAN LAKE	E	I-FC	309,000	324,000	4,410	1964
CALAVERAS RIVER						
U.S. CORPS OF ENGINEERS						
NEW MELONES LAKE	B	I-P-FC	2,100,000	2,400,000	12,500	1979
STANISLAUS RIVER						
O NEILL RESERVOIR	D	I-P	20,800	56,400	2,250	1967
SAN LUIS CREEK						
CALIFORNIA DEPARTMENT OF WATER RESOURCES						
PINE FLAT LAKE	E	I-FC	1,000,000	1,000,000	5,970	1954
KINGS RIVER						
U.S. CORPS OF ENGINEERS						
SAN LUIS RESERVOIR	D	I-P	1,960,000	2,040,000	13,000	1967
SAN LUIS CREEK						
CALIFORNIA DEPARTMENT OF WATER RESOURCES						
SHASTA LAKE	A	I-P-FC-RR-N-M I	3,970,000	4,550,000	29,700	1945
SACRAMENTO RIVER						

TABLE 8–15. MAJOR STORAGE RESERVOIRS ON UNITED STATES BUREAU OF RECLAMATION PROJECTS (continued)

Project, State Reservoir Name (Dam Name) Stream Operator	Category	Purpose	Active Capacity acre feet	Total Capacity acre feet	Reservoir Area acres	Year Com-plet-ed
MID-PACIFIC REGION (continued)						
KLAMATH, CA-OR						
CLEAR LAKE RESERVOIR	A	I-FC	513,000	527,000	25,800	1910
LOST RIVER						
GERBER RESERVOIR	A	I	94,300	94,300	3,830	1925[3]
MILLER CREEK						
UPPER KLAMATH LAKE	E	I-P	465,000	873,000	90,900	1921[1,3]
UPPER KLAMATH LAKE OUTLET						
PACIFIC POWER AND LIGHT						
NEWLANDS, NV						
LAKE TAHOE	D	I-P	732,000	732,000	120,000	1913[3]
TRUCKEE RIVER						
TRUCKEE-CARSON IRRIGATION DISTRICT						
SOLANO, CA						
LAKE BERRYESSA (MONTICELLO)	A	I-FC-M I	1,590,000	1,600,000	20,700	1957
PUTAH CREEK						
LOWER COLORADO REGION						
BOULDER CANYON, AZ-CA-NV						
LAKE MEAD (HOOVER)	A	I-P-FC-RR-N-M I	17,400,000	2,8500,000	163,000	1936
COLORADO RIVER						
COLO R FRONT WORK-LEVEE SYSTEM	A		12,300	13,800	470	1966[2]
SENATOR WASH RESERVOIR						
SENATOR WASH						
PARKER-DAVIS, AZ-CA-NV						
LAKE HAVASU (PARKER)	A	P-FC-M I	180,000	619,400	20,400	1938
COLORADO RIVER						
SALT RIVER, AZ						
THEODORE ROOSEVELT LAKE	D	I-P-M I	1,336,700	1,336,700	17,315	1936[5]
SALT RIVER						
SALT RIVER VALLEY WATER USERS ASSOC.						
UPPER COLORADO REGION						
CENTRAL UTAH, UT						
STRAWBERRY RESERVOIR	A	I-M I-FC	951,000	1,106,500	17,000	1974
STRAWBERRY RIVER						
COLORADO RIVER STORAGE	A	I-P-FC	748,500	940,800	9,180	1966
BLUE MESA RESERVOIR						
GUNNISON RIVER						
FLAMING GORGE RESERVOIR	A	I-P	3,515,700	3,788,700	42,020	1964
GREEN RIVER						
LAKE POWELL (GLEN CANYON)	A	P-R	20,876,000	27,000,000	161,390	1964
COLORADO RIVER						
NAVAJO RESERVOIR	A	I-FC-RR	1,036,100	1,708,600	15,610	1963
SAN JUAN RIVER						
SOUTHWEST REGION						
BRANTLEY, NM						
BRANTLEY RESERVOIR	C	FC-I	42,000	3,485,000	15,320	
CANADIAN RIVER, TX						
LAKE MEREDITH (SANFORD)	D	FC-M I-F W	1,304,554	1,382,478	16,513	1965
CANADIAN RIVER						
CANADIAN RIVER MUNICIPAL WATER AUTHORITY						
COLORADO RIVER, TX						
MARSHALL FORD RESERVOIR	D	I-P-FC-RR-N	1,590,763	1,953,936	18,929	1942
COLORADO RIVER						
LOWER COLORADO RIVER AUTHORITY						
MIDDLE RIO GRANDE, NM						
COCHITI LAKE	E	I-FC-S-F W	676,217	722,000	10,690	1975
RIO GRANDE RIVER						
U.S. CORPS OF ENGINEERS						

TABLE 8–15. MAJOR STORAGE RESERVOIRS ON UNITED STATES BUREAU OF RECLAMATION PROJECTS (continued)

Project, State Reservoir Name (Dam Name) Stream Operator	Category	Purpose	Active Capacity acre feet	Total Capacity acre feet	Reservoir Area acres	Year Com-plet-ed
SOUTHWEST REGION (continued)						
NUECES RIVER, TX						
CHOKE CANYON RESERVOIR	A	M I-F W	689,480	691,130	25,733	1982
FRIO RIVER						
RIO GRANDE, NM-TX						
ELEPHANT BUTTE RESERVOIR	A	I-P	2,060,000	2,110,000	36,521	1916
RIO GRANDE RIVER						
SAN ANGELO, TX						
TWIN BUTTES RESERVOIR	D	I-FC-M I-F W	632,214	640,568	9,079	1963
MIDDLE SOUTH CONCHO RIVER; SPRING CREEK						
SAN ANGELO WATER SUPPLY CORPORATION						
TUCUMCARI, NM						
CONCHAS LAKE	E	I-FC	451,161	528,951	13,664	1940
CANADIAN RIVER						
U. S. CORPS OF ENGINEERS						
MISSOURI BASIN REGION						
COLORADO-BIG THOMPSON, CO						
LAKE GRANBY	A	I-P	466,000	540,000	7,260	1950
COLORADO RIVER						
FORT PECK, MT						
FORT PECK LAKE	E	I-P-FC-N	14,626,000	18,909,000	249,000	1940
MISSOURI RIVER						
U.S. CORPS OF ENGINEERS						
KENDRICK, WY						
SEMINOE RESERVOIR	A	I-P	985,603	1,017,273	20,300	1939
NORTH PLATTE RIVER						
NORTH PLATTE, NE-WY						
PATHFINDER RESERVOIR	A	I-P	985,102	1,016,507	22,000	1909
NORTH PLATTE RIVER						
PICK-SLOAN MISSOURI BASIN PROGRAM						
LAKE FRANCIS CASE (FORT RANDALL)	E	P-FC-N	4,033,000	5,603,000	102,100	1953
MISSOURI RIVER						
U.S. CORPS OF ENGINEERS						
LAKE OAHE	E	I-P-FC	17,886,400	23,337,600	372,800	1962
MISSOURI RIVER						
U.S. CORPS OF ENGINEERS						
LAKE SHARPE (BIG BEND)	E	P	1,882,302	1,884,000	64,000	1964
MISSOURI RIVER						
U.S. CORPS OF ENGINEERS						
LEWIS AND CLARK LAKE (GAVINS POINT)	E	P-FC-N	156,000	504,000	32,000	1955
MISSOURI RIVER						
U.S. CORPS OF ENGINEERS						
BOSTWICK DIVISION, KS-NE						
HARLAN COUNTY LAKE	E	I-FC	691,121	825,782	23,064	1952
REPUBLICAN RIVER						
U.S. CORPS OF ENGINEERS						
BOYSEN DIVISION, WY						
BOYSEN UNIT						
BOYSEN RESERVOIR	A	I-P-FC	802,000	952,400	22,166	1952
WIND RIVER						
GARRISON DIVISION, ND						
AUDUBON LAKE (SNAKE CREEK)	B	I	356,000	500,000	19,400	1965[4]
OFFSTREAM						
LAKE SAKAKAWEA (GARRISON)	E	I-P-FC-N-M I-F W	18,933,500	23,923,000	381,900	1956[1]
MISSOURI RIVER						
U.S. CORPS OF ENGINEERS						

TABLE 8–15. MAJOR STORAGE RESERVOIRS ON UNITED STATES BUREAU OF RECLAMATION PROJECTS (continued)

Project, State / Reservoir Name (Dam Name) / Stream / Operator	Category	Purpose	Active Capacity acre feet	Total Capacity acre feet	Reservoir Area acres	Year Completed
MISSOURI BASIN REGION (continued)						
HELENA-GREAT FALLS DIV, MT						
(CANYON FERRY UNIT)						
CANYON FERRY LAKE	A	I-FC-P-M I-F W	1,605,057	2,050,519	35,181	1954
MISSOURI RIVER						
LOWER BIGHORN DIV, MT-WY						
(YELLOWTAIL UNIT)						
BIGHORN LAKE (YELLOWTAIL DAM)	A	P	834,776	1,328,360	17,298	1966
BIGHORN RIVER						
MARIAS DIVISION, MT						
(LOWER MARIAS UNIT)						
LAKE ELWELL (TIBER)	A	I-F W	790,533	1,368,157	23,152	1956[6]
MARIAS RIVER						
OREGON TRAIL DIV, NE-WY						
(GLENDO UNIT)						
GLENDO RESERVOIR	A	I-P-FC	726,254	789,402	12,400	1958
NORTH PLATTE RIVER						
SOLOMON DIVISION, KS						
(GLEN ELDER UNIT)						
WACONDA LAKE (GLEN ELDER)	A	I-FC-M I	927,104	963,775	12,600	1969
SOLOMON RIVER						

DEFINITIONS

PURPOSE	F W - FISH WILDLIFE FC - FLOOD CONTROL I - IRRIGATION M I - MUNICIPAL INDUSTRIAL N - NAVIGATION P - POWER RR - RIVER REGULATION S - SEDIMENT.
CAPACITY	
ACTIVE -	STORAGE AVAILABLE FOR PROJECT, USUALLY THE STORAGE ABOVE THE LOWEST POINT OF RELEASE.
TOTAL -	STORAGE AT HIGHEST CONTROLLED WATER SURFACE.
RESERVOIR AREA	WATER SURFACE AREA AT ACTIVE CONSERVATION CAPACITY.
YEAR COMPLETED	DATE ORIGINAL CONSTRUCTION COMPLETED EXCEPT AS INDICATED IN FOOTNOTE 5/.
OPERATING AGENCY	AGENCY DIRECTLY AND OFFICIALLY RESPONSIBLE FOR OPERATION AND MAINTENANCE OF FEATURE. DOES NOT INCLUDE AGENCIES THAT MAY ADMINISTER RECREATIONAL OR OTHER SECONDARY FUNCTION AS A SERVICE TO THE PRIMARY OPERATOR.

FOOTNOTES

[1] THE BUREAU OF RECLAMATION HAS RESPONSIBILITY FOR THE PROPER CARE AND MANAGEMENT OF FIVE RESERVOIRS BY SPECIAL AGREEMENT.
[2] REREGULATING RESERVOIR.
[3] DEAD STORAGE NOT EVALUATED.
[4] NOT YET IN OPERATION.
[5] DATE INDICATES BUREAU OF RECLAMATION REHABILITATION WORK.
[6] TIBER DAM MODIFICATION COMPLETED IN 1981.

CATEGORY

A. CONSTRUCTED AND OPERATED BY BUREAU OF RECLAMATION
B. CONSTRUCTED BY OTHERS, OPERATED BY BUREAU OF RECLAMATION
C. UNDER CONSTRUCTION BY BUREAU OF RECLAMATION
D. CONSTRUCTED BY BUREAU OF RECLAMATION, OPERATED BY OTHERS
E. CONSTRUCTED AND OPERATED BY OTHERS

Source: U.S. Department of the Interior Bureau of Reclamation, 1986, Statistical Compilation of Engineering Features on Bureau of Reclamation Projects

TABLE 8–16. FLOOD CONTROL RESERVOIRS IN THE UNITED STATES

[Operable as of September 30, 1986]

Name	River Basin	Stream	Community in Vicinity	Year Placed in Operation	Total Storage (Acre Ft.)	Permanent Pool (Acreage) or No Pool (NPP)	Project Functions[1]	Characteristics of Dam: Type	Height (Feet)	Length (Feet)
ALASKA										
Chena River Lakes	Yukon-Kuskokwim	Chena River	Fairbanks	1981	2,000	NPP	FR	Earth	50	40,200
ARIZONA										
Adobe	Gila	Skunk Creek	Phoenix	1982	18,350	NPP	FR	Earth	109	2,275
Alamo	Colorado	Bill Williams River	Wenden	1968	1,046,310	560	FRWX	Earth	283	975
Cave Buttes	Gila	Cave Creek	Phoenix	1979	46,600	NPP	FRX	Earth	109	2,275
Dreamy Draw	Gila	Dreamy Draw	Phoenix	1973	320	NPP	FRX	Earth	50	448
New River	Gila	New River	Phoenix	1985	43,520	NPP	F	Earth	104	2,320
Painted Rock	Gila	Gila River	Gila Bend	1959	2,491,700	NPP	FRWX	Earth	181	4,780
Tat Momolikot	Gila	Santa Rosa Wash	Casa Grande	1974	198,550	NPP	FWX	Earth	75.5	12,500
Whitlow Ranch	Gila	Queen Creek	Superior	1960	35,500	NPP	FX	Earth	149	837
ARKANSAS										
Blakely Mountain Dam	Ouachita	Ouachita	Hot Springs	1955	2,768,500	20,900	FP	Earth	235	1,100
Blue Mountain	Arkansas	Petit Jean River	Paris	1947	257,900	2,910	FRWX	Earth	115	2,800
DeGray	Ouachita	Caddo	Arkadelphia	1971	881,900	6,400	FPQRS	Earth	243	3,400
DeQueen	Red	Rolling Fork River	DeQueen	1977	136,100	1,680	FSQRW	Earth	160	2,360
Dierks	Red	Saline River	Dierks	1975	96,800	1,360	FSRAW	Earth & Rock	153	2,500
Gillham	Red	Cossatot River	Gillham	1975	221,800	1,370	FSQW	Earth & Rock	160	1,750
Millwood	Red	Little River	Ashdown	1966	1,854,930	29,200	FSW	Earth	88	17,554
Narrows Dam	Ouachita	Little Missouri	Murfreesboro	1949	407,900	2,500	FP	Concrete	175	941
Nimrod	Arkansas	Fourche La Fave River	Plainview	1942	336,010	3,550	FSWX	Concrete	97	1,012
CALIFORNIA										
Black Butte	Sacramento	Stony Creek	Orland	1963	160,000	770	FIRX	Earth	156	2,970
Brea	Santa Ana	Brea Creek	Fullerton	1942	4,010	NPP	FRX	Earth	87	1,765
Buchanan Dam-H.V. Eastman Lake	San Joaquin	Chowchilla River	Chowchilla	1975	150,000	470	FIRW	Earth & Rock	205.5	1,800
Carbon Canyon	Santa Ana	Carbon Canyon River	Brea	1961	6,610	NPP	FRX	Earth	99	2,610
Coyote Valley	Russian	East Fork Russian River	Ukiah	1959	122,500	1,700	FRX	Earth	160	3,500
Dry Creek (Warm Springs) Lake and Channel	Russian	Dry Creek	Healdsburg	1983	381,000	500	FRSW	Earth	319	3,000
Farmington	San Joaquin	Littlejohn Creek	Farmington	1952	52,000	NPP	F	Earth	60	7,800
Fullerton	Santa Ana	East Fullerton Creek	Fullerton	1941	760	NPP	FRX	Earth	46	575
Hansen	Los Angeles	Big Tujunga Wash	Los Angeles	1940	25,450	120	FRWX	Earth	97	10,475
Harry L. Englebright	Sacramento	Yuba River	Marysville	1941	69,000	400	DR	Concrete	280	1,142
Hidden Dam-Hensley Lake	San Joaquin	Fresno River	Madera	1975	90,000	5,000	FIRW	Earth	163	5,730
Isabella	San Joaquin	Kern River	Bakersfield	1953	570,000	1,850	FIRW	Earth	185	4,952
Lopez	Los Angeles	Pacoima Wash	San Fernando	1954	440	NPP	FX	Earth	50	1,300
Martis Creek	Sacramento	Martis Creek	Reno	1971	20,400	71	FSR	Earth	113	2,670
Merced County Stream Group:										
Bear	San Joaquin	Bear Creek	Merced	1954	7,700	NPP	F	Earth	92	1,830
Burns	San Joaquin	Burns Creek	Merced	1950	7,000	NPP	F	Earth	55	4,075
Mariposa	San Joaquin	Mariposa Creek	Merced	1948	15,000	NPP	F	Earth	88	1,330
Owens	San Joaquin	Owens Creek	Merced	1949	3,600	NPP	F	Earth	75	790
Mojave River	Mojave	Mojave	Victorville	1971	89,670	NPP	FRWX	Earth	200	2,200
New Hogan	San Joaquin	Calaveras	Valley Springs	1963	325,000	715	FIRX	Earth & Rock	210	1,960
North Fork	Sacramento	American River	Auburn	1939	14,700	280	DR	Concrete	155	620
Pine Flat	San Joaquin	Kings River	Piedra	1954	1,000,000	NPP	FIRX	Concrete	429	1,820
Prado	Santa Ana	Santa Ana River	Corona	1941	196,240	NPP	FRX	Earth	106	2,280
San Antonio	Santa Ana	San Antonio Creek	Upland	1956	7,700	NPP	FX	Earth	160	3,850
Santa Fe	San Gabriel	San Gabriel River	Duarte	1948	32,110	NPP	FRX	Earth	92	23,800
Sepulveda	Los Angeles	Los Angeles River	Van Nuys	1941	17,430	NPP	FRX	Earth	57	15,444
Success	San Joaquin	Tule River	Porterville	1960	85,000	400	FIRX	Earth	142	3,490
Terminus	San Joaquin	Kaweah River	Visalia	1961	150,000	345	FIRX	Earth	250	2,375
Whittier Narrows	San Gabriel	San Gabriel River and Rio Hondo	El Monte	1957	35,150	NPP	FRWX	Earth	56	16,960
COLORADO										
Bear Creek	Missouri	Bear Creek	Denver	1978	30,810	109	FRX	Earth	180	5,300
Chatfield	Missouri	South Platte River	Denver	1974	231,429	1,412	FRX	Earth	148	12,500

TABLE 8–16. FLOOD CONTROL RESERVOIRS IN THE UNITED STATES (continued)

								Characteristics of Dam		
Name	**River Basin**	**Stream**	**Community in Vicinity**	**Year Placed in Operation**	**Total Storage (Acre Ft.)**	**Permanent Pool (Acreage) or No Pool (NPP)**	**Project Functions[1]**	**Type**	**Height (Feet)**	**Length (Feet)**
COLORADO (continued)										
Cherry Creek	Missouri	Cherry Creek	Denver	1950	93,920	852	FRX	Earth	141	14,300
John Martin	Arkansas	Arkansas River	Lamar	1943	615,500	1,844	FIR	Concrete & Earth	106	13,274
Trinidad	Arkansas	Purgatoire River	Trinidad	1977	123,500	280	FIRX	Earth	200	6,610
CONNECTICUT										
Black Rock	Housatonic	Branch Brook	Thomaston	1970	8,700	20	FR	Earth	154	933
Colebrook River	Connecticut	West Branch, Farmington River	Riverton	1969	97,700	760	FRSX	Earth	223	1,300
Hancock Brook	Housatonic	Hancock Brook	Plymouth	1960	4,030	40	FRW	Earth	57	630
Hop Brook	Housatonic	Hop Brook	Middlebury	1968	6,970	21	FR	Earth	97	520
Mansfield Hollow	Thames	Natchaug River	Willimantic	1952	52,000	450	FRW	Earth	68	12,420
Northfield Brook	Thames	Northfield Brook	Thomaston	1965	2,430	8	FRW	Earth	118	810
Thomaston	Housatonic	Naugatuck River	Thomaston	1960	42,000	NPP	F	Earth	142	2,000
West Thompson	Thames	Quinebaug	Thompson	1965	26,800	200	FRW	Earth	70	2,550
IDAHO										
Lucky Peak	Columbia	Boise River	Boise	1956	306,000	2,850	FIR	Earth	250	1,700
ILLINOIS										
Carlyle	Upper Mississippi	Kaskaskia River	Carlyle	1967	983,000	26,000	FSNRWA	Earth	67	6,570
Farmdale	Upper Mississippi	Farm Creek	East Peoria	1951	15,500	NPP	F	Earth	80	1,275
Fondulac	Upper Mississippi	Fondulac Creek	East Peoria	1951	3,780	NPP	F	Earth	67	1,000
Shelbyville	Upper Mississippi	Kaskaskia River	Shelbyville	1970	684,000	11,100	FSNRW	Earth	108	3,000
Rend Lake	Upper Mississippi	Big Muddy River	Benton	1970	294,000	18,900	FQRSW	Earth	54	10,600
INDIANA										
Brookville	Ohio	East Fork of Whitewater River	Brookville	1974	359,600	2,250	FRSW	Earth & Rock	182	3,000
Cagles Mill	Ohio	Mill Creek	Terre Haute	1952	228,120	1,400	FRX	Earth	150	950
Cecil M. Harden	Ohio	Raccoon Creek	Rockville	1960	132,800	1,100	FRX	Earth	117	1,790
Huntington	Ohio	Wabash River	Huntington	1969	153,100	500	FRW	Earth	91	5,332
Mississinewa	Ohio	Mississinewa	Peru	1967	368,400	1,100	FRW	Earth	137	8,100
Monroe	Ohio	Salt Creek	Harrodsburg	1964	441,000	3,280	FARS	Earth	93	1,400
Patoka	Ohio	Patoka River	Ellsworth	1978	301,600	2,010	FRSQW	Earth & Rock	84	1,550
Salamonie	Ohio	Salamonie	Wabash	1966	263,600	976	FRW	Earth	133	6,100
IOWA										
Coralville	Upper Mississippi	Iowa River	Iowa City	1958	492,000	1,820	FARW	Earth	100	1,400
Rathbun	Missouri	Chariton River	Centerville	1969	552,000	11,000	FNRWXQ	Earth	86	10,600
Red Rock	Upper Mississippi	Des Moines River	Des Moines	1969	1,830,000	8,950	FARWQ	Earth	110	5,676
Saylorville	Upper Mississippi	Des Moines River	Des Moines	1975	602,000	74,000	FARWQ	Earth	125	6,750
KANSAS										
Clinton	Missouri	Wakarusa River	Lawrence	1977	397,200	7,000	FSWAXR	Earth	114	9,250
Council Grove	Arkansas	Grand (Neosho)	Council Grove	1964	112,265	3,235	FSQR	Earth	96	6,500
El Dorado	Arkansas	Walnut River	El Dorado	1981	236,200	8,000	FSQR	Earth	99	20,930
Elk City	Arkansas	Elk River	Independence	1966	284,300	4,450	FSQ	Earth	107	4,840
Fall River	Arkansas	Fall River	Fall River	1949	256,400	2,350	FSX	Earth	94	6,015
Hillsdale	Missouri	Big Bull Creek	Kansas City	1981	160,000	4,580	FSQR	Earth	75	11,600
John Redmond	Arkansas	Grand (Neosho)	Burlington	1964	630,250	9,280	FSQR	Earth	86.5	21,790
Kanopolis	Missouri	Smoky Hill River	Salina	1948	450,000	3,815	FRWX	Earth	131	15,360
Marion	Arkansas	Cottonwood River	Marion	1968	143,850	6,200	FRQS	Earth	67	8,375
Melvern	Missouri	Marias Des Cygnes	Melvern	1972	363,000	6,930	FRQWX	Earth	98	9,700
Milford	Missouri	Republican River	Junction City	1965	1,160,000	15,600	FRSXWQ	Earth & Rock	126	6,300
Pearson Skubitz Big Hill	Arkansas	Big Hill Creek	Cherryvale	1981	40,600	1,240	FSR	Earth	83	3,870
Perry	Missouri	Delaware River	Perry	1969	770,000	12,200	FRSXW	Earth & Rock	95	7,750
Pomona	Missouri	110 Mile Creek	Pomona	1963	230,000	4,000	FRSWXQ	Earth & Rock	85	7,750
Toronto	Arkansas	Verdigris River	Toronto	1960	200,800	2,660	FX	Earth	90	4,712
Tuttle Creek	Missouri	Big Blue River	Manhattan	1962	2,346,000	15,800	FRWXQAN	Earth & Rock	157	7,500
Wilson	Missouri	Saline River	Wilson	1964	776,000	9,000	FIRWXNA	Earth	160	5,600

TABLE 8–16. FLOOD CONTROL RESERVOIRS IN THE UNITED STATES (continued)

Name	River Basin	Stream	Community in Vicinity	Year Placed in Operation	Total Storage (Acre Ft.)	Permanent Pool (Acreage) or No Pool (NPP)	Project Functions[1]	Characteristics of Dam: Type	Height (Feet)	Length (Feet)
KENTUCKY										
Barren River	Ohio	Barren River	Glasgow	1964	815,200	4,340	FARS	Earth	146	3,970
Buckhorn	Ohio	Middle Fork of Kentucky River	Buckhorn	1960	168,000	550	FR	Earth	162	1,020
Carr Fork	Ohio	Carr Fork	Hazard	1976	47,700	530	FQRW	Earth & Rock	130	720
Cave Run	Ohio	Licking River	Farmers	1974	614,100	6,790	FQRW	Earth & Rock	148	2,740
Dewey	Ohio	Johns Creek	Paintsille	1949	93,000	1,100	FARW	Earth	118	913
Fishtrap	Ohio	Levisa Fork, Big Sandy River	Pikeville	1968	164,360	569	FARW	Rock	195	1,100
Grayson	Ohio	Little Sandy	Grayson	1967	118,990	1,050	FQRW	Earth & Rock	120	1,460
Green River	Ohio	Green River	Campbellsville	1969	723,200	5,070	FRSQW	Earth & Rock	142	2,350
Martins Fork	Cumberland	Martins Fork	Harlan	1978	21,000	578	FQ	Concrete	97	574
Paintsville	Ohio	Paint Creek	Paintsville	1983	73,500	261	FQRW	Earth & Rock	160	1,600
Nolin	Ohio	Nolin River	Kyrock	1963	609,400	2,890	FAR	Earth & Rock	174	990
Rough River	Ohio	Rough River	Leitchfield	1958	334,400	2,180	FRX	Earth & Rock	124	1,530
Taylorsville	Ohio	Salt River	Taylorsville	1983	291,670	1,625	FQRW	Earth & Rock	164	1,280
LOUISIANA										
Bayou Bodcau	Red	Bayou Bodcau	Shreveport	1949	357,300	NPP	FRW	Earth	76	12,850
Caddo Lake	Red	Cypress Bayou	Shreveport	1971	175,000	32,700	NFRS	Concrete & Earth		3,600
Wallace Lake	Red	Cypress Bayou	Shreveport	1946	88,300	2,300	FR	Earth	48	4,934
MARYLAND										
Jennings Randolph Lake	Potomac	North Branch	Barnum	1981	130,900	952	FQRS	Earth & Rock	296	2,130
MASSACHUSETTS										
Barre Falls	Connecticut	Ware River	Barre	1958	24,000	NPP	FRW	Earth & Rock	62	885
Birch Hill	Connecticut	Millers River	So.Royalston	1941	49,900	NPP	FRW	Earth & Rock	56	1,400
Buffumville	Thames	Little River	Charlton	1958	12,700	200	FRW	Earth & Rock	66	3,255
Charles River Natural Valley Storage	Charles	Charles River	Millis	1983	35,000	NPP	F	Nonstructural	—	—
Conant Brook	Connecticut	Conant Brook	Monson	1966	3,740	NPP	F	Earth & Rock	85	1,060
East Brimfield	Connecticut	Quinebaug River	Fiskdale	1960	30,000	360	FRW	Earth & Rock	55	520
Hodges Village	Connecticut	French River	Oxford	1959	12,800	NPP	FRW	Earth & Rock	55	2,140
Knightville	Connecticut	Westfield River	Huntington	1941	49,000	NPP	FRW	Earth & Rock	160	1,200
Littleville	Connecticut	Middle Branch, Westfield River	Chester	1965	32,400	275	FRWS	Earth & Rock	1,164	1,360
Tully	Connecticut	Tully River	Fryville	1949	22,000	300	FRW	Earth & Rock	62	1,570
West Hill	Blackstone	West River	Uxbridge	1960	12,350	NPP	FRW	Earth & Rock	51	2,400
Westville	Thames	Quinebaug River	Sturbridge	1961	11,100	23	FRW	Earth & Rock	78	560
MINNESOTA										
Big Stone Lake-Whetstone River	Upper Mississippi	Minnesota River	Ortonville	1973	45,000	12,700	FRW	Earth	25	13,700
Lac Qui Parle Chippewa River	Upper Mississippi	Chippewa River	Montevideo	1950	–	NPP	FRWX	Earth & Rock	23.3	17,975
Lac Qui Parle	Upper Mississippi	Minnesota River	Montevideo	1950	122,800	6,500	FRWX	Earth & Rock	21	4,100
Marsh Lake	Upper Mississippi	Minnesota River	Montevideo	1950	35,000	5,100	FRWX	Earth & Rock	19.5	11,800
Orwell	Red River of the North	Otter Tail River	Fergus Falls	1953	14,100	210	FARS	Earth & Rock	47	1,355
Red Lake	Red River of the North	Red Lake River	Red River	1951	2,680,000	279,000	FARSX	Earth & Rock	15.5	36,500
MISSISSIPPI										
Arkabutla Lake	Lower Mississippi	Coldwater River	Arkabutla	1943	525,300	5,100	F	Earth & Rock	81	11,500
Enid Lake	Lower Mississippi	Yocona River	Enid	1951	660,000	6,100	F	Earth & Rock	99	8,400
Grenada Lake	Lower Mississippi	Yalobusha River	Grenada	1954	1,337,400	9,800	F	Earth & Rock	102	13,900
Okatibbee	Pascagoula	Okatibbee Creek	Meridian	1969	142,400	1,280	FQSR	Earth	67	6,543
Sardis Lake	Lower Mississippi	Little Tallahatchie River	Sardis	1940	1,570,000	10,700	F	Earth & Rock	117	15,300
MISSOURI										
Clearwater	White	Black River	Piedmont	1948	413,700	1,630	FRWX	Earth & Rock	154	4,225
Long Branch	Grande Chariton	Little Chariton	Macon	1980	65,000	2,430	FRSQW	Earth	71	3,800

TABLE 8–16. FLOOD CONTROL RESERVOIRS IN THE UNITED STATES (continued)

Name	River Basin	Stream	Community in Vicinity	Year Placed in Operation	Total Storage (Acre Ft.)	Permanent Pool (Acreage) or No Pool (NPP)	Project Functions[1]	Characteristics of Dam: Type	Height (Feet)	Length (Feet)
MISSOURI (continued)										
Longview	Missouri	Little Blue River	Kansas City	1986	46,900	930	FRWQ	Earth & Rock	120	1,900
Pomme de Terre	Missouri	Pomme de Terre River	Hermitage	1961	650,000	7,820	FRWX	Earth & Rock	155	4,630
Smithville	Missouri	Little Platte River	Smithville	1982	246,500	7,190	FSQRW	Earth	95	4,200
Wappapello	Lower Mississippi	St. Francis River	Wappapello	1941	613,200	4,100	FR	Earth & Rock	109	2,700
NEBRASKA										
Harlan County	Missouri	Republican River	Republican City	1952	850,000	13,600	FIRWX	Earth & Rock	107	11,827
Papillion Creek and Tributaries:										
Glenn Cunningham (Site 11)	Missouri	Knight Creek	Omaha	1975	17,910	392	FQRX	Earth	67	1,940
Standing Bear Lake (Site 16)	Missouri	Trib. of Big Papillion Creek	Omaha	1973	5,220	137	FRX	Earth	70	1,460
Salt Creek & Tributaries:										
Olive Creek (Site 2)	Missouri	S. Trib. Olive Br. Creek	Kramer	1964	5,470	174	FR	Earth	45	3,020
Blue Stem (Site 4)	Missouri	N. Trib. Olive Br. Creek	Sprague	1963	10,260	315	FR	Earth	57	2,760
Wagon Train (Site 8)	Missouri	N. Trib. Hickman Creek	Holland	1963	9,280	303	FR	Earth	52	1,650
Stagecoach (Site 9)	Missouri	S. Trib. Hickman Creek	Hickman	1964	6,640	196	FR	Earth	48	2,250
Yankee Hill (Site 10)	Missouri	Cardwell Creek	Denton	1966	7,560	208	FR	Earth	52	3,100
Conestoga (Site 12)	Missouri	Holmes Creek	Denton	1964	10,640	230	FR	Earth	63	3,000
Twin Lake (Site 13)	Missouri	Middle Creek	Pleasantdale	1966	8,080	255	FR	Earth	58	2,075
Pawnee (Site 14)	Missouri	N. Middle Creek	Emerald	1965	29,520	728	FR	Earth	65	5,000
Holmes Park Lake (Site 17)	Missouri	Antelope Creek	Lincoln	1963	6,510	100	FR	Earth	55	7,700
Branched Oak (Site 18)	Missouri	Oak Creek	Raymond	1968	97,560	1,780	FR	Earth	70	5,200
NEVADA										
Mathews Canyon	Colorado	Mathews Canyon	Caliente	1957	6,270	NPP	FX	Earth	71	800
Pine Canyon	Colorado	Pine Canyon	Caliente	1957	7,750	NPP	FX	Earth	92	884
NEW HAMPSHIRE										
Blackwater	Merrimack	Blackwater River	Webster	1941	46,000	NPP	FRW	Earth	75	1,150
Edward MacDowell	Merrimack	Nubanusit Brook	West Peterborough	1950	12,800	NPP	FRW	Earth	67	1,030
Franklin Falls	Merrimack	Pemigewasset River	Franklin	1943	154,000	NPP	FRW	Earth	140	1,740
Hopkinton-Everett	Merrimack	Contoocook River	West Hopkinton	1962	71,500	200	FRW	Earth	76	790
	Merrimack	Piscataquog River	East Weare	1962	87,500	120	FRW	Earth	115	2,000
Otter Brook	Connecticut	Otter Brook	Keene	1958	18,300	85	FRW	Earth	133	1,288
Surry Mountain	Connecticut	Ashuelot River	Keene	1941	32,500	265	FRW	Earth	86	1,670
NEW MEXICO										
Abiquiu	Rio Grande	Rio Chama	Abiquiu	1963	1,212,000	NPP	FXS	Earth	325	1,540
Cochiti	Rio Grande	Rio Grande	Pena Blanca	1975	596,300	1,200	FRWX	Earth	251	28,300
Conchas	Arkansas	Canadian River	Tucumcari	1939	529,000	3,000	FI	Concrete & Earth	200	19,400
Galisteo	Rio Grande	Galisteo Creek	Albuquerque	1970	89,000	NPP	FX	Earth	158	2,820
Jemez Canyon	Rio Grande	Jemez River	Bernalillo	1953	102,700	NPP	FX	Earth	135	780
Two Rivers:										
Diamond "A" Dam	Rio Grande	Rio Hondo	Roswell	1963	168,000	NPP	FX	Earth	98	4,885
Rocky Dam	Rio Grande	Rocky Arroyo							118	2,940
Santa Rosa Dam Reservoir	Pecos	Pecos	Santa Rosa	1979	447,000	NPP	FIX	Earth	212	1,950
NEW YORK										
Almond	Susquehanna	Canacadea Creek	Hornell	1949	14,600	124	FRW	Earth	90	1,260
Arkport	Susquehanna	Canisteo Creek	Arkport	1940	7,900	NPP	F	Earth	113	1,200
East Sidney	Susquehanna	Ouleout Creek	Franklin	1950	33,550	210	FRW	Concrete & Earth	130	2,010
Mount Morris	Genesee	Genesee River	Mount Morris	1952	337,000	170	FR	Concrete	210	1,028
Whitney Point	Susquehanna	Otselic River	Whitney Point	1942	86,440	1,200	FRW	Earth	95	4,900

TABLE 8–16. FLOOD CONTROL RESERVOIRS IN THE UNITED STATES (continued)

Name	River Basin	Stream	Community in Vicinity	Year Placed in Operation	Total Storage (Acre Ft.)	Permanent Pool (Acreage) or No Pool (NPP)	Project Functions[1]	Characteristics of Dam: Type	Height (Feet)	Length (Feet)
NORTH CAROLINA										
B. Everett Jordan	Cape Fear	New Hope	Durham	1982	753,500	14,300	FQRSWX	Earth	112	1,330
Falls	Neuse	Neuse	Raleigh	1983	335,600	11,300	FQRSWX	Earth	92	1,915
W. Kerr Scott	Yadkin-Pee Dee	Yadkin	Wilkesboro	1963	153,000	1,470	FARSX	Earth	148	1,740
NORTH DAKOTA										
Baldhill	Red River of the North	Sheyenne River	Valley City	1950	70,000	325	FARS	Earth	61	1,650
Bowman-Haley	Missouri	North Fork, Grand River	Haley	1967	92,980	1,750	FSRWX	Earth	79	5,730
Homme	Red River of the North	South Branch of Park River	Park River	1951	3,650	51	FARS	Earth	67	865
Pipestem	James River	Pipestem Creek	Jamestown	1974	146,880	885	FRWX	Earth	108	4,000
OHIO										
Alum Creek	Ohio	Alum Creek	Africa	1975	134,800	348	FRSW	Concrete & Earth	93	10,000
Berlin	Ohio	Mahoning River	Deerfield	1943	91,200	240	FARSWQ	Concrete & Earth	96	5,750
Caesar Creek	Ohio	Caesar Creek	Wilmington	1978	242,200	13,300	FRSQW	Earth & Rock	165	2,750
Clarence J. Brown	Ohio	Buck Creek	Springfield	1974	63,700	1,010	FQRW	Earth & Rock	72	6,620
Deer Creek	Ohio	Deer Creek	New Holland	1968	102,500	727	FRW	Earth	93	3,880
Delaware	Ohio	Olentangy River	Delaware	1951	132,000	950	FARWX	Earth	92	18,600
Dillon	Ohio	Licking River	Zanesville	1961	273,000	1,325	FRWX	Earth	118	1,400
Michael J. Kirwan	Ohio	West Branch, Mahoning River	Newton Falls	1966	78,700	580	FAQRSW	Earth	83	9,900
Mosquito Creek	Ohio	Mosquito Creek	Cortland	1944	104,100	700	FARSWQ	Earth	47	5,650
Muskingum River Reservoirs:										
Atwood	Ohio	Indian Fork	New Cumberland	1937	49,700	1,540	FRX	Earth	65	3,700
Beach City	Ohio	Sugar Creek	Beach City	1937	71,700	420	FRX	Earth	64	5,600
Bolivar	Ohio	Sandy Creek	Bolivar	1938	149,600	NPP	FR	Earth	87	6,300
Charles Mill	Ohio	Black Fork	Mufflin	1936	88,000	1,350	FRX	Earth	48	1,390
Clendening	Ohio	Brushy Fork	Tippecanoe	1937	54,000	1,800	FRX	Earth	64	950
Dover	Ohio	Tuscarawas River	Dover	1938	203,000	350	FRX	Concrete	83	824
Leesville	Ohio	McGuire Creek	Leesville	1937	37,400	1,000	FRX	Earth	74	1,694
Mohawk	Ohio	Walhonding River	Nellie	1937	285,000	NPP	FR	Earth	111	2,330
Mohicanville	Ohio	Lake Fork	Mohicanville	1936	102,000	NPP	FR	Earth	46	1,220
Piedmont	Ohio	Stillwater Creek	Piedmont	1937	65,000	2,270	FRX	Earth	56	1,750
Pleasant Hill	Ohio	Clear Fork	Perrysville	1938	87,700	850	FRX	Earth	113	775
Senecaville	Ohio	Seneca Fork	Senecaville	1937	88,500	3,550	FRSX	Earth	45	2,350
Tappan	Ohio	Little Stillwater Creek	Tappan	1936	61,600	2,350	FRX	Earth	52	1,550
Wills Creek	Ohio	Wills Creek	Conesville	1937	196,000	900	FRX	Earth	87	1,950
North Branch, Kokosing River Lake	Ohio	North Branch of Kokosing River	Fredericktown	1973	14,900	98	FRW	Earth	71	1,400
Paint Creek	Ohio	Paint Creek	New Petersburg	1972	145,000	710	FRSQW	Earth & Rock	118	700
Tom Jenkins	Ohio	East Branch, Sunday Creek	Gloucester	1951	26,900	394	FRSWX	Concrete	84	944
West Fork Mill Creek	Ohio	Mill Creek	Mount Healthy	1952	11,380	200	FRX	Earth	100	1,100
William H. Harsha	Ohio	Little Miami River	Williamsburg	1978	284,500	18,760	FRSQW	Earth	200	1,450
OKLAHOMA										
Birch	Arkansas	Birch Creek	Barnsdall	1977	58,200	1,137	FSQRW	Earth	97	3,193
Canton	Arkansas	North Canadian River	Canton	1948	377,100	7,910	FSI	Earth	73	15,140
Copan	Arkansas	Little Caney River	Copan	1983	227,700	4,850	FSQRW	Earth	70	7,730
Fort Supply	Arkansas	Wolf Creek	Fort Supply	1942	100,700	1,820	FSX	Earth	85	12,225
Great Salt Plains	Arkansas	Salt Fork, Arkansas River	Cherokee	1941	271,400	8,690	FRWX	Earth	68	6,010
Heyburn	Arkansas	Polecat Creek	Sapulpa	1950	55,030	880	FRWXS	Earth	89	2,920
Hugo	Red	Kiamichi River	Hugo	1974	966,700	13,250	FSQRW	Earth	101	10,200
Hulah	Arkansas	Caney River	Caney	1951	289,000	3,570	FSAX	Earth	94	5,200
Kaw	Arkansas	Arkansas River	Ponca City	1976	1,348,000	17,040	FSQRW	Earth	125	9,466
Oologah	Arkansas	Verdigris River	Oologah	1963	1,519,000	29,460	FSN	Earth	137	4,000
Optima	Arkansas	North Canadian River	Hardesty	1978	229,500	5,340	FSRW	Earth	120	15,200
Pine Creek	Red	Little River	Wright City	1969	465,780	3,750	FSQW	Earth	124	7,712

TABLE 8–16. FLOOD CONTROL RESERVOIRS IN THE UNITED STATES (continued)

Name	River Basin	Stream	Community in Vicinity	Year Placed in Operation	Total Storage (Acre Ft.)	Permanent Pool (Acreage) or No Pool (NPP)	Project Functions[1]	Characteristics of Dam: Type	Height (Feet)	Length (Feet)
OKLAHOMA (continued)										
Sardis	Red	Jackfork Creek	Clayton	1983	429,600	14,360	FSRW	Earth	81	14,138
Skiatook	Arkansas	Hominy Creek	Skiatook	1985	500,700	10,190	FSQRW	Earth	143	3,590
Waurika	Red	Beaver Creek	Waurika	1977	343,500	10,100	FISQWR	Earth	106	16,600
Wister	Arkansas	Poteau River	Wister	1949	427,900	5,360	FSAX	Earth	99	5,700
OREGON										
Applegate	Rogue River	Applegate River	Medford	1981	82,000	988	AFIQRSW	Gravel Embankment	242	1,300
Blue River	Columbia	Blue River	Blue River	1968	85,000	975	FINR	Earth	319	1,329
Cottage Grove	Columbia	Coast Fork, Willamette River	Cottage Grove	1942	30,060	1,155	FINR	Concrete & Earth	114	2,110
Dorena	Columbia	Row River	Cottage Grove	1949	70,500	1,885	FINR	Concrete & Earth	145	3,352
Fall Creek	Columbia	Middle Fork, Willamette River	Eugene	1965	115,000	1,865	FINR	Rockfill & Concrete	193	5,100
Fern Ridge	Columbia	Long Tom River	Eugene	1941	110,000	10,305	FINR	Rockfill & Concrete	49	6,624
Willow Creek	Columbia	Willow Creek	Heppner	1983	13,250	96	FRN	Roller Compacted Concrete	160	1,780
PENNSYLVANIA										
Alvin R. Bush	Susquehanna	Kettle Creek	Renovo	1962	75,000	160	FRW	Earth & Rock	165	1,350
Aylesworth Creek	Susquehanna	Aylesworth Creek	Archbald	1970	1,700	NPP	F	Earth & Rock	90	1,270
Beltzville	Delaware	Pohopoco	Lehighton	1971	68,250	947	FQRSW	Earth & Rock	170	4,300
Blue Marsh	Delaware	Tulephocken	Blue Marsh	1978	22,900	963	FAQRS	Earth & Rock	98	1,775
Conemaugh	Ohio	Conemaugh River	Saltsburg	1952	274,000	300	FW	Concrete & Earth	137	1,265
Cowanesque	Susquehanna	Cowanesque River	Lawrenceville	1980	89,000	410	FR	Earth & Rock	151	3,100
Crooked Creek	Ohio	Crooked Creek	Ford City	1940	93,900	350	FRW	Earth	143	1,480
Curwensville	Susquehanna River	West Branch, Susquehanna River	Curwensville	1965	124,200	790	FR	Earth	131	2,850
East Branch, Clarion River	Ohio	East Branch, Clarion River	Wilcox	1952	84,300	90	FARQW	Earth	184	1,725
Foster Joseph Sayers	Susquehanna	Bald Eagle Creek	Blanchard	1969	99,000	1,730	FRW	Earth	100	6,835
Francis E. Walter (Bear Creek)	Delaware	Lehigh River	White Haven	1961	110,000	90	FNRW	Earth & Rock	234	3,000
Gen. Edgar Jadwin	Delaware	Dyberry Creek	Honesdale	1960	24,500	NPP	F	Earth	109	1,225
Indian Rock	Susquehanna	Codorus Creek	York	1942	28,000	NPP	FRW	Earth	83	1,000
Kinsua	Ohio	Allegheny River	Warren	1965	1,180,000	1,900	PFAQRW	Concrete & Earth	177	1,877
Loyalhanna	Ohio	Loyalhanna Creek	Saltsburg	1942	95,300	210	FRW	Concrete & Earth	114	960
Mahoning Creek	Ohio	Mahoning Creek	New Bethlehem	1941	74,200	170	FRW	Concrete	162	926
Prompton	Delaware	Lackawaxen River	Honesdale	1960	52,000	290	FNRW	Earth	140	1,230
Raystown	Susquehanna	Raystown Branch, Juniata River	Huntingdon	1973	762,000	8,300	FRW	Earth & Rock	225	1,700
Shenango	Ohio	Shenanago River	Sharpsville	1966	191,400	1,910	FAQRW	Concrete	68	720
Stillwater	Susquehanna	Lackawanna River	Uniondale	1960	12,000	85	FS	Earth	77	1,700
Tioga-Hammond Lakes	Susquehanna	Tioga River	Tioga	1978	62,000	470	FR	Earth & Rock	140	2,710
Hammond Lakes	Susquehanna	Crooked Creek	Tioga	1978	63,000	680	FR	Earth & Rock	122	6,450
Tionesta	Ohio	Tionesta Creek	Tionesta	1940	133,400	480	FRW	Earth	154	1,050
Union City	Ohio	French Creek	Union City	1970	47,640	NPP	F	Earth	88	1,420
Woodcock Creek	Ohio	French Creek	Meadville	1973	20,000	118	FQRA	Earth	90	4,650
Youghiogheny River	Ohio	Youghiogheny River	Confluence	1943	254,000	450	FARWQ	Earth	184	1,610
SOUTH DAKOTA										
Cold Brook	Missouri	Cold Brook	Hot Springs	1953	7,200	36	FRWX	Earth	127	925
Cottonwood Springs	Missouri	Cottonwood Springs Creek	Hot Springs	1970	8,385	41	FRWX	Earth	123	1,190
Lake Traverse:										
Reservation Control Dam	Red River of the North	Bois de Sioux River	Wheaton	1941	164,500	10,925	FRX	Earth	14	9,100
White Rock	Red River of the North	Bois de Sioux River	Wheaton	1941	85,000	6,500	FRX	Earth	16	14,400

TABLE 8–16. FLOOD CONTROL RESERVOIRS IN THE UNITED STATES (continued)

Name	River Basin	Stream	Community in Vicinity	Year Placed in Operation	Total Storage (Acre Ft.)	Permanent Pool (Acreage) or No Pool (NPP)	Project Functions[1]	Characteristics of Dam: Type	Height (Feet)	Length (Feet)
TEXAS										
Addicks	San Jacinto	South Mayde Creek	Addicks	1948	204,500	NPP	FX	Earth	49	61,166
Aquilla	Brazos	Aquilla Creek	Hillsboro	1983	146,000	3,280	FSX	Earth	104.5	11,890
Bardwell	Trinity	Waxahachie Creek	Ennis	1965	140,000	3,570	FRSX	Earth	82	15,400
Barker	San Jacinto	Buffalo Bayou	Barker	1945	207,000	NPP	FX	Earth	37	72,844
Belton	Brazos	Leon River	Belton	1954	1,097,600	12,300	FIRSX	Earth	192	5,524
Benbrook	Trinity	Clear Fork, Trinity River	Fort Worth	1952	258,600	3,770	FNRXA	Earth	130	9,130
Canyon	Guadalupe	Guadalupe	New Braunfels	1964	740,900	8,240	FRSX	Earth	224	6,830
Ferrells Bridge Dam-Lake O' the Pines	Red	Cypress Creek	Jefferson	1959	842,100	18,700	FRS	Earth	97	10,600
Granger Dam and Lake	Brazos	San Gabriel River	Granger	1980	244,200	4,400	FRSWX	Earth	115	16,320
Grapevine	Trinity	Denton Creek	Grapevine	1952	425,500	7,280	FNRSXA	Earth	137	12,850
Hords Creek	Colorado	Hords Creek	Coleman	1948	25,310	510	FARSX	Earth	91	6,800
Lake Kemp	Red	Wichita River	Wichita Falls	1972	502,900	15,590	FX	Earth & Rock	115	8,890
Lavon	Trinity	East Fork, Trinity River	Fort Worth	1953	748,200	21,400	FRSX	Earth	81	19,483
Lewisville	Trinity	Elm Fork, Trinity River	Lewisville	1954	989,700	23,280	FRSX	Earth	125	32,888
Navarro Mills	Trinity	Richland Creek	Corsicana	1962	212,200	5,070	FRSX	Earth	82	7,570
North San Gabriel Dam, Lake Georgetown	Brazos	North Fork, San Gabriel River	Georgetown	1980	130,800	1,310	FRSWX	Rock	164	6,700
O. C. Fisher	Colorado	North Concho River	San Angelo	1952	396,400	5,440	FRSX	Earth	128	40,885
Pat Mayse	Red	Sanders Creek	Paris	1967	189,100	5,993	FRSX	Earth	96	7,080
Proctor	Brazos	Leon River	Comanche	1963	374,200	4,610	FRSX	Earth	86	13,460
Somerville	Brazos	Yegua Creek	Somerville	1967	507,500	11,460	FRSX	Earth	80	26,175
Stillhouse Hollow	Brazos	Lampasas River	Belton	1968	630,400	6,430	FRSX	Earth	200	15,624
Waco	Brazos	Bosque River	Waco	1965	726,400	7,270	FRSX	Concrete & Earth	140	24,618
Wright Patman	Red	Sulphur River	Texarkana	1957	2,654,300	20,300	FRSX	Earth	100	18,500
VERMONT										
Ball Mountain	Connecticut	West River	Jamaica	1961	54,600	75	FRW	Concrete & Earth	265	915
North Hartland	Connecticut	Ottauguechee River	North Hartland	1960	71,420	220	FRW	Concrete & Earth	185	1,520
North Springfield	Connecticut	Black River	Springfield	1960	51,067	290	FRW	Concrete & Earth	120	2,940
Townshend	Connecticut	West River	Townshend	1961	33,700	100	FRW	Concrete & Earth	133	1,700
Union Village	Connecticut	Ompompanoosuc River	Union Village	1950	38,000	NPP	FRW	Concrete & Earth	170	1,100
VIRGINIA										
John W. Flannagan	Ohio	Pound River	Haysi	1963	145,700	310	FAQR	Concrete & Earth	250	960
Gathright Dam & Lake Moomaw	James	Jackson	Alleghany	1979	123,739	2,532	FQR	Earth & Rock	257	1,172
North Fork of Pound River	Ohio	North Fork, Pound River	Pound	1966	11,293	106	FR	Rock	122	600
WASHINGTON										
Howard A. Hanson	Green	Green River	Kanaskat	1961	106,000	1,600	FAS	Rock	235	675
Mill Creek	Columbia	Mill Creek	Walla Walla	1942	8,300	225	FR	Earth	145	3,200
Mud Mountain	Puyallup	White River	Enumclaw	1953	106,000	NPP	FR	Rock	425	700
Wynoochee	Chehalis	Wynoochee River	Montesano	1972	70,000	1,150	FSARI	Concrete & Earth	177	1,700
WEST VIRGINIA										
Beech Fork	Ohio	Beech Fork	Lavalette	1977	37,540	450	FRW	Earth	86	1,080
Bluestone	Ohio	New River	Hinton	1952	631,000	1,800	FRWX	Concrete	180	2,048

TABLE 8–16. FLOOD CONTROL RESERVOIRS IN THE UNITED STATES (continued)

Name	River Basin	Stream	Community in Vicinity	Year Placed in Operation	Total Storage (Acre Ft.)	Permanent Pool (Acreage) or No Pool (NPP)	Project Functions[1]	Characteristics of Dam: Type	Height (Feet)	Length (Feet)
WEST VIRGINIA (continued)										
Burnsville	Ohio	Little Kanawha	Burnsville	1977	65,400	550	FQRW	Earth & Rock	89	1,400
East Lynn	Ohio	Twelve Pole Creek	East Lynn	1970	82,500	823	FQRW	Earth & Rock	122	650
R. D. Bailey	Ohio	Guynabdot River	Justice	1979	203,700	440	FQRW	Earth & Rock	310	1,397
Summersville Lake	Ohio	Gauley River	Summersville	1965	413,800	407	FANR	Rock	390	2,280
Sutton	Ohio	Elk River	Sutton	1960	265,300	270	FARWX	Concrete	220	1,178
Tygart River	Ohio	Tygart River	Grafton	1938	287,700	620	FNAR	Concrete	230	1,921
WISCONSIN										
Eau Galle	Chippewa	Eau Galle	Spring Valley	1969	43,600	150	FR	Earth	122	1,600

[1] Project Functions:
A Low Flow Augmentation
D Debris Control
F Flood Control
I Irrigation
N Navigation
P Power
Q Water Quality Control
R Public Recreation (Annual Attendance exceeding 5,000)
S Water Supply
W Fish and Wildlife (Federal and State)
X Water Conservation and Sedimentation

Source: Department of the Army Annual Report FY86 of the Secretary of the Army on Civil Works Activities, Washington DC

TABLE 8–17. WORLD'S LARGEST RESERVOIRS IN TERMS OF CAPACITY

No	Capacity, $10^6 m^3$	Name	Country	Year
1	204 800	Owen Falls[1]	Uganda	1954
2	169 270	Bratsk	USSR	1964
3	168 900	High Aswan	Egypt	1970
4	160 368	Kariba	Zimbabwe/Zambia	1959
5	147 960	Akosombo	Ghana	1965
6	141 851	Daniel Johnson	Canada	1968
	135 000	Guri	Venezuela	U/C
7	73 300	Krasnoyarsk	USSR	1967
8	70 309	W A C Bennett	Canada	1967
9	68 400	Zeya	USSR	1978
	63 000	Cabora Bassa	Mozambique	1974
10	61 715	La Grande 2	Canada	1978
11	60 020	La Grande 3	Canada	1981
12	59 300	Ust-Ilim	USSR	1977
13	58 000	Kuibyshev	USSR	1955
14	53 790	Caniapiscau Barrage KA 3	Canada	1980
	50 700	Upper Wainganga	India	U/C

TABLE 8–17. WORLD'S LARGEST RESERVOIRS IN TERMS OF CAPACITY (continued)

No	Capacity, $10^6 m^3$	Name	Country	Year
15	49 800	Bukhtarma	USSR	1960
	48 000	Atatürk	Turkey	U/C
16	46 000	Irkutsk	USSR	1956
	43 000	Tucurui	Brazil	U/C
17	35 900	Vilyui	USSR	1967
18	35 400	Sanmenxia	China	1960
19	34 852	Hoover	USA	1936
20	34 100	Sobradinho	Brazil	1979
21	33 304	Glen Canyon	USA	1966
22	32 203	Skins Lake No 1	Canada	1953
23	31 790	Jenpeg	Canada	1975
24	31 500	Volgograd	USSR	1958
	31 300	Sayano-Shushensk	USSR	U/C
25	30 600	Keban	Turkey	1974
26	29 959	Iroquois	Canada	1958
27	29 000	Itaipu	Brazil	1982
28	28 973	Churchill Falls (GR-1)	Canada	1971
29	28 370	Missi Falls Control	Canada	1976
30	28 100	Kapchagay	USSR	1970
31	29 000	Loma de la Lata	Argentina	1977
32	27 920	Garrison	USA	1953
33	27 675	Kossou	Ivory Coast	1972
34	27 433	Oahe	USA	1958
35	26 000	Razzaza Dyke	Iraq	1970
36	25 400	Rybinsk	USSR	1941
	24 700	Longyangxia	China	U/C
37	24 700	Mica	Canada	1972
38	24 000	Tsimlyansk	USSR	1952
39	23 700	Kenney	Canada	1952
40	23 500	Ust-Khantaika	USSR	1970
41	22 950	Furnas	Brazil	1963
42	22 119	Fort Peck	USA	1937
43	21 626	Xinanjiang	China	1960
44	21 166	Ilha Solteira	Brazil	1973

U/C under construction.

[1] This capacity is not fully obtained by a dam; the major part of it is the natural capacity of a lake; Owen Falls is not the greatest man-made lake.

Source: International Commission on Large Dams, 1984, World Register of Dams

FIGURE 8–2. TENNESSEE VALLEY REGION SYSTEM OF RESERVOIRS

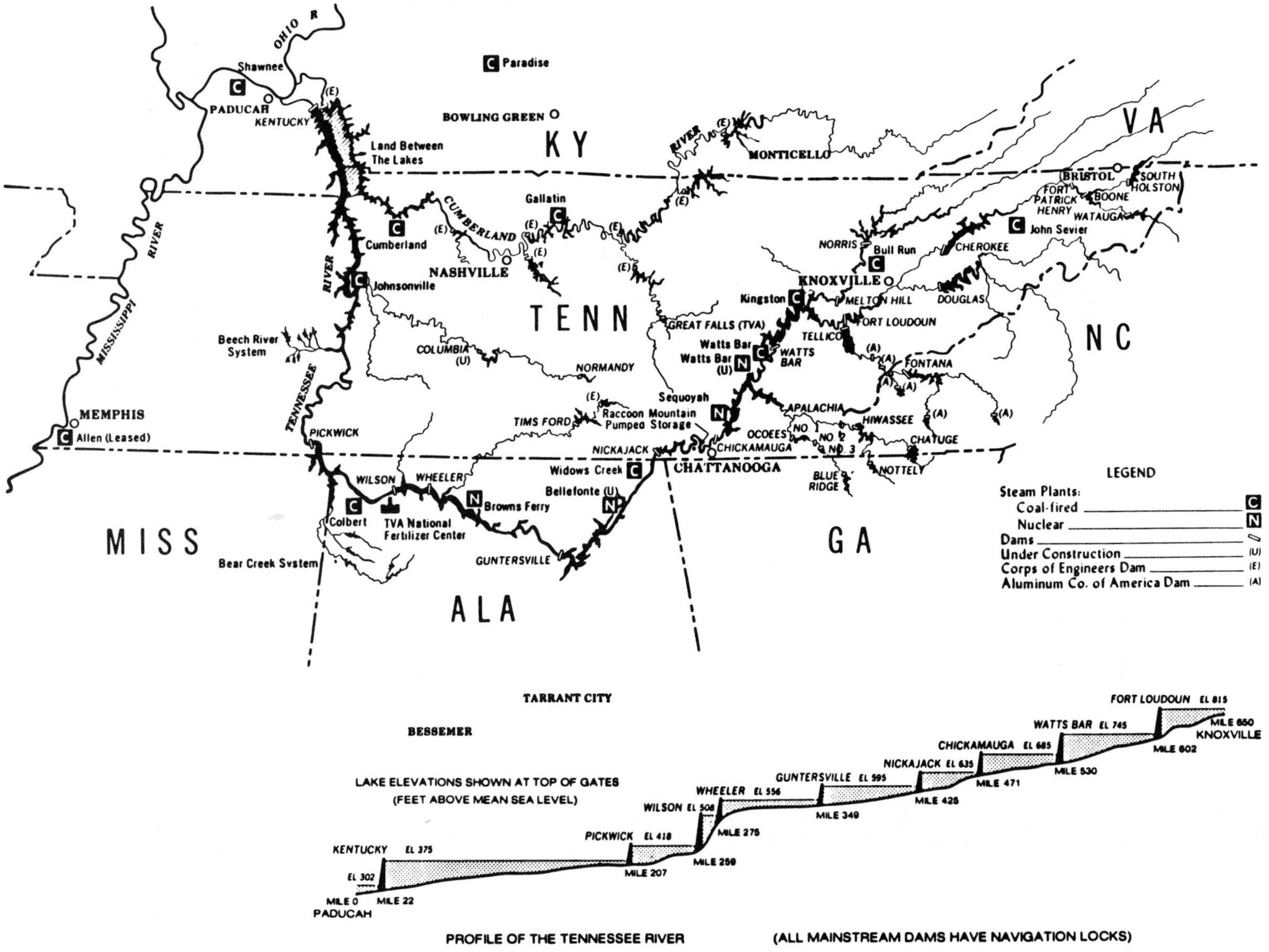

Source: Tennessee Valley Authority, 1988

THE TENNESSEE VALLEY AUTHORITY MULTIPURPOSE RESERVOIR SYSTEM

The TVA reservoir system was designed to provide for navigation, flood control and the production of hydroelectricity. Today the reservoir system is also operated for such other purposes as providing municipal and industrial water supplies, regulating flows to minimize the effects of effluents including thermal discharges, fluctuating water levels for control of mosquitos and troublesome aquatic vegetation, and controlling flows and levels for various recreational uses.

This reservoir system encompasses more than 11,000 miles of shoreline and 600,000 acres of water. The overall benefit of this system to the Nation and the region are incalculable, but some indication is provided by a consideration of the more quantifiable benefits: flood damages prevented, transportation savings, hydroelectric power production, and recreation visits.

There are 39 dams operating in the Tennessee Valley's integrated water control system. They are:

35 TVA dams - Apalachia, Blue Ridge, Boone, Chatuge, Cherokee, Chickamauga, Douglas, Fontana, Fort Loudoun, Fort Patrick Henry, Guntersville, Hiwassee, Kentucky, Melton Hill, Nickajack, Normandy, Norris, Nottely, Ocoee No. 1, Ocoee No. 2, Ocoee No. 3, Pickwick Landing, South Holston, Tellico, Tims Ford, Watauga, Watts Bar, Wheeler, Wilson, Bear Creek, Little Bear Creek, Nolichucky, Upper Bear Creek, Cedar Creek, and Wilbur.

4 Alcoa dams - Calderwood, Cheoah, Chilhowee, and Santeetlah.

Raccoon Mountain Pumped Storage Project stores energy (generated elsewhere) to meet peak power demands.

There are 9 dams in the Cumberland River basin for which TVA distributes power generation. They include one TVA dam, Great Falls, and 8 Corps of Engineer dams - Barkley, Center Hill, Cheatham, Cordell Hull, Dale Hollow, J. Percy Priest, Old Hickory, and Wolf Creek.

The 4 Bear Creek projects (Bear, Little Bear, and Upper Bear and Cedar), the 2 Duck River projects (Normandy and Columbia), and the Tims Ford dam were planned under Tributary Area Development programs but have been counted among those in the Integrated Water Control System because they were all partially justified as having system flood control value. Tims Ford also contributes power to the system.

In the 1960's TVA built two systems of small dams in tributary watersheds:

Beech River Project - Beech, Cedar, Dogwood, Lost Creek, Pine, Pin oak, Redbud, and Sycamore.

Bristol Project - Beaver Creek and Clear.

There are 12 dams owned by Alcoa in The Little Tennessee River Valley which are not included in TVA's integrated water control system.

The Elk River Dam operated by the Air Force, the Burnett Dam operated by the city of Asheville, and the Walters Dam of the Carolina Power and Light Company are other structures in the Tennessee Valley, but are not controlled by TVA.

TABLE 8–18. TENNESSEE VALLEY AUTHORITY RESERVOIRS

Main River Projects	Length of Lake (Miles)	Length of Lake Shoreline (Miles)	Area of Lake (Acres)	Lake Elevation (Feet Above Sea Level)		Lake Volume (Acre-Feet) At Top of Gates	Useful Controlled Storage in Reservoir (Acre-Feet)
				Normal Minimum	Top of Gates		
Kentucky	184.3	2,380	160,300	354	375	6,129,000	4,008,000
Pickwick Landing	52.7	496	43,100	408	418	1,105,000	417,700
Wilson	15.5	154	15,500	504.5	507.88	640,200	53,600
Wheeler	74.1	1,063	67,100	550	556.28	1,069,000	349,000
Guntersville	75.7	949	67,900	593	595.44	1,049,000	162,400
Nickajack	46.3	192	10,370	632	635	251,600	31,500
Chickamauga	58.9	810	35,400	675	685.44	737,300	345,300
Watts Bar	95.5	771	39,000	735	745	1,175,000	379,000
Fort Loudoun	60.8	360	14,600	807	815	393,000	111,000
Raccoon Mtn. (Pumped Storage Project)	—	—	528	1,530	—	37,310	35,110
Tributary Projects							
Columbia	54	236	12,600	603	635	363,000	283,000
Normandy	17	73	3,160	859	880	126,100	60,500
Tims Ford	34.2	246	10,600	865	895	608,000	282,600
Apalachia	9.8	31	1,100	1,272	1,280	57,800	8,650
Hiwassee	22.2	163	6,090	1,450	1,526.5	434,000	306,000
Chatuge	13	132	7,050	1,905	1,928	240,500	123,000
Ocoee No. 1	7.5	47	1,890	811	830.76	83,300	31,030
Ocoee No. 2	—	—	—	—	1,115.2	—	Silted
Ocoee No. 3	7	24	480	1,413	1,435	4,180	3,629
Blue Ridge	11	65	3,290	1,590	1,691	195,900	183,900
Nottely	20.2	106	4,180	1,735	1,780	174,300	117,140
Melton Hill	44	173	5,690	790	796	126,000	31,900
Norris	129	800	34,200	960	1,034	2,552,000	1,922,000
Tellico	33.2	373	15,860	807	815	424,000	120,000
Fontana	29	248	10,640	1,580	1,710	1,443,000	946,000
Douglas	43.1	555	30,400	940	1,002	1,461,000	1,251,000
Cherokee	54	393	30,300	1,020	1,075	1,541,000	1,148,000
Ft. Patrick Henry	10.4	37	872	1,258	1,263	26,900	4,250
Boone	32.7	130	4,310	1,330	1,385	193,400	148,400
S. Holston	23.7	168	7,580	1,675	1,742	764,000	438,300
Watauga	16.3	106	6,430	1,915	1,975	677,000	353,000
Wilbur	1.8	3.6	72	1,645	1,650	715	327
Great Falls	22	120	3,080	780	805.3	50,200	35,700
Nolichucky	—	26	383	1,238.9	1,240.9	2,003	496
Totals[a)]		11,195	641,455			23,771,708	13,408,432

[a)] Does not include the Columbia Dam Project

Source: Tennessee Valley Authority, 1988

FIGURE 8–3. THE MISSOURI RIVER BASIN SYSTEM OF RESERVOIRS

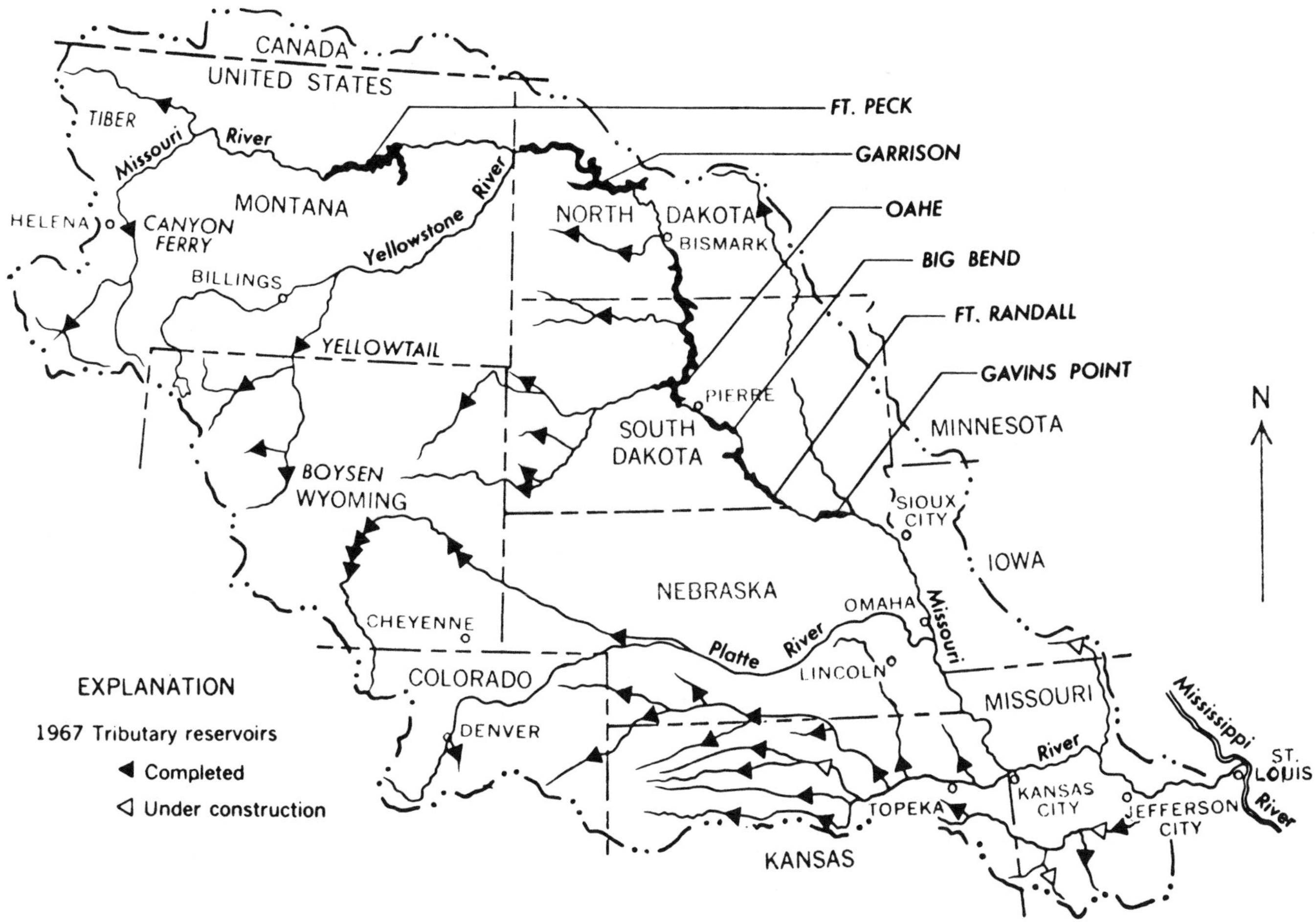

Source: U.S. Water Resources Council, 1968

TABLE 8–19. DEPTH OF RESERVOIRS IN THE UNITED STATES AND LOSS BY EVAPORATION

[For regions see Fig. 8-1]

Region	Median Depth (feet)	Loss in Feet per Acre
New England	19	0.55
Delaware and Hudson	29	.96
Chesapeake Bay	25	.64
Southeast	23	.53
Eastern Great Lakes	16	.68
Western Great Lakes	18	.79
Ohio	24	.47
Cumberland	44	.54
Tennessee	40	.36
Upper Mississippi	7	.83
Lower Mississippi	20	.81
Upper Missouri	25	1.97
Lower Missouri	21	1.11
Upper Arkansas-White-Red	9(15)	3.27
Lower Arkansas-White-Red	10	1.33
Western Gulf	8(15)	2.75
Upper Rio Grande and Pecos	20	4.26
Colorado	16	3.79
Great Basin	15	3.00
Pacific Northwest	25	1.44
Central Pacific	33	2.58
South Pacific	44	4.81

Source: U. S. Geological Survey, 1960

TABLE 8-20. MULTIPURPOSE RESERVOIRS IN THE UNITED STATES

[In operation September 30, 1986]

Project	River	Community in Vicinity	Total Storage Capability (acre-feet)[1]	Flood Control and/or Nav. Feature Placed in Useful Operation	Initial Power In FY	Existing Installation (KW)	Scheduled Installation (KW)	Ultimate Installation (KW)	Project Functions	Type[2]	Height (Feet)	Length (Feet)
Albeni Falls, ID	Pend Oreille	Newport, WA	1,153,000	1952	1955	42,600		42,600	NFPR	C	90	1,055
Allatoona Lake, GA	Etowah	Cartersville, GA	670,000	1950	1950	74,000		74,000	FPRW	C	190	1,250
Barkley Dam & Lake Barkley, KY & TN	Cumberland	Grand Rivers, KY	2,082,000	1964	1966	130,000		130,000	NPFR	CE	157	9,959
Beaver Lake, AR	White	Eureka Springs, AR	1,952,000	1963	1965	112,000		112,000	FPSR	CE	228	2,575
Big Bend Dam (Lake Sharpe), SD	Missouri	Chamberlain, SD	1,883,000	1964	1965	468,000		468,000	FPRIW	E	95	10,570
Blakely Mountain Dam-Lake Ouachita, AR	Ouachita	Mt. Pine, AR	2,768,000	1953	.1956	75,000		75,000	FPRW	E	235	1,100
Bonneville L&D Lake Bonneville, OR & WA	Columbia	Bonneville, OR	537,000	1938	1938	1,076,600		1,076,600	NPR	C	122	2,690
Broken Bow Lake, OK	Mountain Fork	Broken Bow, OK	1,368,230	1968	1970	100,000		100,000	FPWSR	E	225	2,750
Buford Dam Lanier, GA	Chattahoochee	Buford, GA	2,554,000	1956	1957	86,000		86,000	NFPW	E	192	5,400
Bull Shoals Lake, AR & MO	White	Mountain Home, AR	5,408,000	1952	1953	340,000		340,000	FPR	C	258	2,256
Clarence Cannon Dam	Salt	Perry, MO	1,428,000	1983	1985	58,000		58,000	FNPRSW	CE	138	1,700
Carters Dam, GA	Coosawatte	Carters, GA	472,756	1975	1975	500,000		500,000	FPRW	ER	450	1,950
Center Hill Lake, TN	Caney Fork	Lancaster, TN	2,092,000	1948	1951	135,000		135,000	FPR	CE	250	2,160
Cheatham L&D, TN	Cumberland	Ashland City, TN	104,000	1952	1958	36,000		36,000	NPR	C	75	801
Chief Joseph Dam (Rufus Woods Lake), WA	Columbia	Bridgeport, WA	593,100	1955	1956	2,069,000	204,160[3]	2,273,160	PIR	C	220	4,300
Clarks Hill Lake, GA & SC	Savannah	Augusta, GA	2,900,000	1952	1953	280,000		280,000	NFPRSW	CE	200	5,680
Cooper River, Charleston Harbor, SC	Santee	St. Stephen, SC	2,560,000	N/A	1985	84,000		84,000	PW	CE	86	876

TABLE 8–20. MULTIPURPOSE RESERVOIRS IN THE UNITED STATES (continued)

Project	River	Community in Vicinity	Total Storage Capability (acre-feet)[1]	Flood Control and/or Nav. Feature Placed in Useful Operation	Initial Power In FY	Existing Installation (KW)	Scheduled Installation (KW)	Ultimate Installation (KW)	Project Functions	Type[2]	Height (Feet)	Length (Feet)
Cordell Hull L&D, TN	Cumberland	Carthage, TN	310,900	1973	1974	100,000		100,000	NPR	CE	93	1,306
Cougar Lake, OR	S. Fork McKenzie	Blue River, OR	219,000	1963	1964	25,000		64,600	NFPRI	ER	445	1,738
Dale Hollow Lake, TN & KY	Obey	Celina, TN	1,706,000	1943	1949	54,000		54,000	FPR	C	200	1,717
Dardanelle L&D, AR	Arkansas	Dardanelle, AR	486,200	1969	1965	124,000		124,000	NPR	C	68	2,683
DeGray Lake, AR	Caddo	Arkadelphia, AR	831,900	1969	1972	68,000		108,000	FNPRS	E	243	3,400
Denison Dam (Lake Texoma), TX & OK	Red	Denison, TX	5,312,300	1944	1945	70,000		175,000	FPSRN	E	165	17,200
Detroit Lake, OR, including Big Cliff Lake, OR	North Santiam	Mill City, OR	461,000	1953	1954	118,000		118,000	NFPRI	C	382	1,528
Dworshak Dam & Reservoir, ID	N. Fork, Clearwater	Orofino, ID	3,453,000	1972	1973	400,000		1,060,000	PNFR	C	717	3,300
Eufaula Lake, OK	Canadian	Eufaula, OK	3,825,400	1964	1964	90,000		90,000	ENPS	E	114	3,200
Fort Gibson Lake, OK	Grand (Neosho)	Ft. Gibson, OK	1,284,400	1950	1953	45,000		67,500	FP	CE	110	2,990
Fort Peck Lake, MT	Missouri	Glasgow, MT	18,909,000	1938	1944	185,250		185,250	NFPRIW	E	251	21,026
Fort Randal Dam (Lake Francis Case), SD	Missouri	Lake Andes, SD	5,574,000	1953	1954	320,000		320,000	NFPRIW	E	165	10,700
Garrison Dam (Lake Sakakawea), ND	Missouri	Riverdale, ND	24,137,000	1954	1956	430,000		430,000	NFPRIW	E	210	11,300
Gavins Point Dam (Lewis & Clark Lake), SD & NE	Missouri	Yankton, SD	504,000	1956	1957	100,000		100,000	NFPRIW	E	74	8,700
Green Peter Lake, OR, including Foster Lake, OR	Middle Santiam	Sweet Home, OR	491,000	1967	1967	100,000		100,000	PFNIR	C	340	1,380
Greers Ferry Lake, AR	Little Red	Heber Springs, AR	2,844,000	1962	1964	96,000		96,000	FPRS	C	243	1,704
Harry S. Truman Dam & Res.	Osage	Warsaw, MO	5,202,000	1982	1982[4]	160,000		160,000	FPRW	CE	96	5,000
Hartwell Lake, GA & SC	Savannah	Hartwell, GA	2,842,700	1961	1962	344,000		344,000	NFPRS	CE	204	17,852
Hills Creek Lake, OR	Middle Fork Willamette	Oakridge, OR	356,000	1961	1962	30,000		30,000	NFPRI	GE	304	2,150
Ice Harbor L&D (Lake Sacajawea), WA	Snake	Pasco, WA	417,000	1962	1962	603,000		603,000	NPRI	CE	130	2,790
Jim Woodruff Dam (Lake Seminole), FL, GA & AL	Appalachicola	Chattahoochee, FL	367,300	1957	1957	30,000		30,000	NPRW	CE	67	6,150
John Day L&D (Lake Umatilla), OR & WA	Columbia	Rufus, OR	2,500,000	1968	1969	2,160,000		2,700,000	NPRFI	CE	161	5,900
John H. Kerr Dam & Reservoir, NC & VA	Roanoke	Boydton, VA	2,750,300	1952	1953	204,000		204,000	FPRW	CE	144	22,285
Robert F. Henry L&D, AL	Alabama	Benton, AL	234,200	1972	1975	68,000		68,000	NPRW	CE	101	14,962
J. Percy Priest Dam & Reservoir, TN	Stones	Nashville, TN	652,000	1967	1970	28,000		28,000	FPRW	CE	147	2,716
Keystone Lake, OK	Arkansas	Tulsa, OK	1,737,600	1964	1968	70,000		70,000	FNPWS	E	121	4,600
Laurel River Lake, KY	Laurel	London, KY	435,600	1973	1978	61,000		61,000	FPRW	R	282	1,420
Libby Dam, Lake Koocanusa, MT	Kootenai	Libby, MT	5,809,000	1972	1975	525,000		840,000	FPR	C	420	3,055
Little Goose L&D (Lake Bryan), WA	Snake	Starbuck, WA	565,000	1970	1970	810,000		810,000	NPRI	CE	160	2,670
Lookout Point Lake, including Dexter Lake, OR	Willamette Middle Fork, Willamette	Lowell, OR	483,000	1954	1955	135,000		135,000	NFPRI	CE	243	3,381
Lost Creek Lake, OR	Rogue	Trail, OR	465,000	1977	1977	49,000		49,000	DFPISWR	CE	345	3,600
Lower Granite L&D, WA	Snake	Pomeroy, WA	484,000	1975	1975	810,000		810,000	NPRIF	CE	146	3,200
Lower Monumental L&D, WA	Snake	Kahlotus, WA	376,000	1969	1969	810,000		810,000	NPRI	CE	135	3,800

TABLE 8–20. MULTIPURPOSE RESERVOIRS IN THE UNITED STATES (continued)

Project	River	Community in Vicinity	Total Storage Capability (acre-feet)[1]	Flood Control and/or Nav. Feature Placed in Useful Operation	Initial Power In FY	Existing Installation (KW)	Scheduled Installation (KW)	Ultimate Installation (KW)	Project Functions	Type[2]	Height (Feet)	Length (Feet)
McNary L&D—Lake Wallula, OR & WA	Columbia	Umatilla, OR	1,550,000	1953	1954	980,000		1,625,000	NPRI	CE	183	7,300
Millers Ferry L&D, AL	Alabama	Camden, AL	331,800	1969	1970	75,000		75,000	NPRW	CE	90	11,380
Narrows Dam—Lake Greeson, AR	Little Missouri	Murfreesboro, AR	407,900	1950	1950	25,500		25,500	FPRW	C	183.5	941
New Melones Lake, CA[5]	Stanislaus	Oakdale, CA	2,400,000	1978	1979	300,000		300,000	FIPRW	ER	625	1,560
Norfolk Lake, AR & MO	North Fork	Norfolk, AR	1,983,000	1943	1944	80,550		163,000	FPRS	C	216	2,624
Oahe Dam (Lake Oahe), SD & ND	Missouri	Pierre, SD	23,337,000	1959	1962	595,000		595,000	NFPRIW	E	245	9,300
Old Hickory L&D, TN	Cumberland	Hendersonville, TN	545,000	1954	1957	100,000		100,000	NPR	CE	98	3,605
Ozark-Jeta Taylor L&D, AR	Arkansas	Ozark, AR	148,400	1969	1973	100,000		100,000	NPR	C	58	2,480
Philpott Lake, VA	Roanoke	Bassett, VA	318,500	1951	1954	14,000		14,000	FPR	C	220	892
Robert S. Kerr L&D and Reservoir, OK	Arkansas	Sallisaw, OK	525,700	1970	1971	110,000		110,000	NPR	E	75	7,230
Sam Rayburn Dam & Reservoir, TX	Angelina	Jasper, TX	3,997,600	1965	1966	52,000		52,000	FPWR	CE	120	19,430
St. Mary's River, MI	Great Lakes	Sault Ste. Marie, MI	—	1855	1952	18,400		18,400	NP	Control Gate		
Snettisham, AK[6]	Speel	Juneau, AK	352,400		1973	46,700	27,000[7]	73,700	P	C[8]	18	338
Stockton Lake, MO	Sac	Stockton, MO	1,674,000	1969	1973	45,200		45,200	FPRW	CE	128	5,100
Table Rock Lake, AR & MO	White	Branson, MO	3,462,000	1958	1959	200,000		200,000	FPR	CE	252	6,423
Tenkiller Lake, OK	Illinois	Gore, OK	1,230,800	1952	1953	34,000		34,000	FP	E	197	3,000
The Dalles L&D (Lake Celilo), WA & OR	Columbia	The Dalles, OR	53,000	1957	1957	1,806,800		1,806,800	NPR	CR	300	8,875
Walter F. George L&D, GA & FL	Chattahoochee	Fort Gaines, GA	934,000	1963	1963	130,000		130,000	NPRW	CE	114	13,585
Webbers Falls L&D, OK	Arkansas	Webbers Falls, OK	170,100	1970	1973	60,000		60,000	NP	E	84	4,370
West Point Lake, AL & GA	Chattahoochee	West Point, GA	604,500	1975	1975	73,375		108,375	FPRW	CE	97	7,250
Whitney Lake, TX	Brazos	Whitney, TX	1,999,500	1953	1954	30,000		30,000	FPR	CE	159	17,695
Wolf Creek Dam (Lake Cumberland), KY	Cumberland	Jamestown, KY	6,089,000	1950	1952	270,000		270,000	FPR	CE	258	5,736

FY Fiscal Year
[1] Total of all storage functions, including inactive and dead storage to normal full pool level.
[2] G: gravel; R: rock; C: concrete; E: earth.
[3] Chief Joseph Additional Units & Operating Units 1-16.
[4] All units are synchronized-to-line and two units have passed the pumpback test.
However, due to damaging effects to fish, no further pumping will be done for their testing or operation until a solution to the problem is found.
[5] Being operated for the Department of Interior by the Bureau of Reclamation.
[6] Being operated by the Alaska Power Administration.
[7] Crater Lake Unit.
[8] Weir for Long Lake.

Nomenclature for Project Functions

D—Debris Control
F—Flood Control
I—Irrigation
N—Navigation
P—Power
R—Public Recreation (Annual Attendance exceeding 5,000)
S—Water Supply
W—Fish and Wildlife (Federal or State)

Source: Department of the Army Annual Report FY86 of the Secretary of the Army on Civil Works Activities, Washington, DC

SECTION C. HYDROELECTRIC POWER

TABLE 8–21. CONVENTIONAL AND PUMPED STORAGE HYDROELECTRIC PLANTS IN THE UNITED STATES

[Federally owned, developed, under construction, and authorized January 1, 1984]

Plant	Stream	State	Year of Initial Oper.	Gross Static Head Ft	Usable Power Storage 1,000 Ac-Ft[1]	Installed Capacity - Kilowatts[2]		
						Developed	Under Construction	Authorized
Corps of Engineers								
Lincoln School	St John R	ME	—	76	32	—	—	70,000
Tocks Island	Delaware R	NJ	—	107	426	—	—	70,000
Gathright	Jackson R	VA	—	194	61	—	—	49,000
John H Kerr	Roanoke R	VA	1952	90	1029	204,000	—	204,000
Philpott	Smith R	VA	1953	164	111	14,000	—	14,000
St Stephen	Santee & Cooper R	SC	—	70	1420	—	84,000	84,000
Clark Hill	Savannah R	GA	1953	152	1510	280,000	—	280,000
Richard B Russell	Savannah R	GA	—	162	127	—	300,000	300,000
Richard B Russell	Savannah R	GA	—	162	127	—	300,000 R	300,000 R
Hartwell	Savannah R	GA	1962	178	1415	344,000	—	344,000
Jim Woodruff L&D	Apalachicola R	FL	1957	31	37	30,000	—	30,000
Walter F George L&D	Chattahoochee R	GA	1963	92	207	130,000	—	130,000
West Point	Chattahoochee R	GA	1975	78	86	73,375	—	108,375
Buford	Chattahoochee R	GA	1957	155	1049	86,000	—	86,000
Lower Auchumpkee	Flint R	GA	—	84	124	—	—	77,000
Lazer Creek	Flint R	GA	—	123	60	—	—	83,000
Spewrell Bluff	Flint R	GA	—	144	242	—	—	100,000
Spewrell Bluff	Flint R	GA	—	144	242	—	—	50,000 R
Millers Ferry L&D	Alabama R	AL	1970	49	17	75,000	—	75,000
Jones Bluff	Alabama R	AL	1975	45	12	68,000	—	68,000
Carters	Coosawattee R	GA	1975	387	135	250,000	—	250,000
Carters	Coosawattee R	GA	1977	387	135	250,000 R	—	250,000 R
Allatoona	Etowah R	GA	1950	145	284	74,000	—	110,000
St Marys Falls	St Marys R	MI	1951	21	NA	18,400	—	18,400
Barkley	Cumberland R	KY	1966	46	258	130,000	—	130,000
Cheatham L&D	Cumberland R	TN	1958	22	19	36,000	—	36,000
J Percy Priest	Stones R	TN	1970	85	34	28,000	—	28,000
Old Hickory L&D	Cumberland R	TN	1957	46	63	100,000	—	100,000
Center Hill	Caney Fk, Cumberland R	TN	1950	161	492	135,000	—	135,000
Cordell Hull	Cumberland R	TN	1973	49	54	100,000	—	100,000
Dale Hollow	Obey R	TN	1948	135	496	54,000	—	54,000
Celina	Cumberland R	KY	—	60	15	—	—	108,000
Wolf Creek	Cumberland R	KY	1951	163	2142	270,000	—	270,000
Laurel	Laurel R	KY	1977	251	185	61,000	—	61,000
Booneville	S Fk Kentucky R	KY	—	140	258	—	—	7,300
Bluestone	New R	WV	—	125	238	—	—	180,000
Rowlesburg	Cheat R	WV	—	938	8	—	—	350,000 R
Clarence F Cannon	Salt R	MO	—	107	437	—	27,000	27,000
Clarence F Cannon	Salt R	MO	—	107	437	—	31,000 R	31,000 R
Red Rock	Des Moines R	IA	—	53	400	—	—	17,200
Rock Island L&D 15[3]	Sylvan Slough, Miss R	IL	1919	16	NA	2,752	—	2,752
Narrows	Little Missouri R	AR	1950	145	202	25,500	—	25,500
Degray	Caddo R	AR	1971	188	393	40,000	—	80,000
Degray	Caddo R	AR	1971	188	393	28,000 R	—	28,000 R
Blakely Mountain	Ouachita R	AR	1955	181	1286	75,000	—	75,000
Harry S Truman	Osage R	MO	1979	46	90	160,000 R	—	160,000 R
Pomme de Terre	Pomme de Terre R	MO	—	117	165	—	—	16,800
Stockton	Sac R	MO	1973	86	660	45,200	—	45,200
Gavins Point	Missouri R	NE	1956	44	100	100,000	—	100,000
Fort Randall	Missouri R	SD	1954	131	3500	320,000	—	320,000

TABLE 8–21. CONVENTIONAL AND PUMPED STORAGE HYDROELECTRIC PLANTS IN THE UNITED STATES (continued)

Plant	Stream	State	Year of Initial Oper.	Gross Static Head Ft	Usable Power Storage 1,000 Ac-Ft[1]	Installed Capacity - Kilowatts[2]		
						Developed	Under Construction	Authorized
Corps of Engineers								
Big Bend	Missouri R	SD	1964	67	260	468,000	—	468,000
Oahe	Missouri R	SD	1962	189	17000	595,000	—	595,000
Garrison	Missouri R	ND	1956	172	17900	430,000	—	430,000
Fort Peck	Missouri R	MT	1943	213	13800	185,300	—	185,300
Broken Bow	Mountain Fk R, Little R	OK	1970	194	9	100,000	—	100,000
Denison	Red R	TX	1944	108	1730	70,000	—	175,000
Dardanelle	Arkansas R	AR	1965	49	65	124,000	—	124,000
Ozark	Arkansas R	AR	1973	23	19	100,000	—	100,000
Robert S Kerr	Arkansas R	OK	1971	42	NA	110,000	—	110,000
Eufaula	Canadian R	OK	1964	82	1481	90,000	—	90,000
Tenkiller Ferry	Illinois R	OK	1953	143	345	39,100	—	39,100
Webbers Falls	Arkansas R	OK	1973	30	NA	60,000	—	60,000
Fort Gibson	Neosho R	OK	1953	60	NA	45,000	—	67,500
Keystone	Arkansas R	OK	1968	76	351	70,000	—	70,000
Greers Ferry	Little Red R	AR	1964	183	763	96,000	—	96,000
Norfork[4]	North Fork R	AR	1944	174	445	80,550	—	165,550
Bull Shoals	White R	AR	1952	194	952	340,000	—	340,000
Table Rock	White R	MO	1959	204	1098	200,000	—	200,000
Beaver	White R	AR	1965	190	956	112,000	—	112,000
Town Bluff	Neches R	TX	—	24	NA	—	—	3,000
Sam Rayburn	Angelina R	TX	1966	71	1400	52,000	—	52,000
Dam A	Neches R	TX	—	13	NA	—	—	2,700
Rockland	Neches R	TX	—	61	1488	—	—	13,500
Belton	Leon R	TX	—	110	399	—	—	19,000
Whitney	Brazos R	TX	1953	89	133	30,000	—	30,000
Big Cliff RRG	N Santiam R	OR	1954	97	243	18,000	—	18,000
Detroit	N Santiam R	OR	1953	357	40	100,000	—	100,000
Foster RRG	S Santiam R	OR	1968	284	28	20,000	—	20,000
Green Peter	M Fk Santiam R	OR	1967	310	63	80,000	—	80,000
Strube RRG	S Fk McKenzie R	OR	—	64	3	—	—	4,500
Cougar	S Fk McKenzie R	OR	1964	435	10	25,000	—	60,000
Dexter RRG	M Fk Willamette R	OR	1955	48	4	15,000	—	15,000
Lookout Point	M Fk Willamette R	OR	1954	231	12	120,000	—	120,000
Hills Creek	M Fk Willamette R	OR	1962	318	49	30,000	—	30,000
Bonneville	Columbia R	OR	1938	59	87	518,400	—	544,600
Bonneville 2nd Ph	Columbia R	WA	1981	59	87	558,220	—	558,220
The Dalles	Columbia R	WA	1957	83	53	1,806,800	—	1,806,800
John Day	Columbia R	OR	1968	105	300	1,957,500	—	1,957,500
John Day	Columbia R	WA	1968	105	300	202,500	—	742,500
McNary	Columbia R	OR	1953	74	185	980,000	—	980,000
Ice Harbor	Snake R	WA	1961	98	24	603,000	—	603,000
Lower Monumental	Snake R	WA	1969	100	20	810,000	—	810,000
Little Goose	Snake R	WA	1970	98	49	810,000	—	810,000
Lower Granite	Snake R	WA	1975	100	44	810,000	—	810,000
Dworshak	N Fk Clearwater R	ID	1973	630	2000	400,000	—	1,060,000
Chief Joseph	Columbia R	WA	1955	167	NA	2,069,000	—	2,069,000
Albeni Falls	Pend Oreille R	ID	1955	28	1153	42,600	—	42,600
Libby	Kootenai R	MT	1975	341	4934	420,000	420,000	840,000
Lost Creek	Rogue R	OR	1977	321	315	49,000	—	49,000
Marysville	Yuba R	CA	—	210	775	—	—	50,000
Total Conventional						19,011,197	831,000	22,297,897
Total Reversible						438,000 R	331,000 R	1,169,000 R

TABLE 8–21. CONVENTIONAL AND PUMPED STORAGE HYDROELECTRIC PLANTS IN THE UNITED STATES (continued)

Plant	Stream	State	Year of Initial Oper.	Gross Static Head Ft	Usable Power Storage 1,000 Ac-Ft[1]	Installed Capacity - Kilowatts[2] Developed	Under Construction	Autho-rized
Bureau of Reclamation								
Prairie Creek	Prairie Cr, Platte R	NE	—	51	NA	—	—	16,800 R
Flatiron 1 and 2	Rattlesnake Cr, Big Thom	CO	1954	1113	1	74,500[5]	—	74,500
Flatiron 3	Dry Cr, Big Thompson R	CO	1954	292	100	8,500 R	—	8,500 R
Big Thompson	Big Thompson R	CO	1959	184	NA	4,500	—	4,500
Pole Hill	Dry Cr, Big Thompson R	CO	1954	835	NA	33,250	—	33,250
Estes	Big Thompson R	CO	1950	570	NA	45,000	—	45,000
Marys Lake	Fish Cr, Big Thompson R	CO	1951	217	NA	8,100	—	8,100
Guernsey	N Platte R	WY	1927	92	41	4,800	—	4,800
Glendo	N Platte R	WY	1958	133	511	24,000	—	24,000
Alcova	N Platte R	WY	1955	164	1842	36,000	—	36,000
Fremont Canyon	N Platte R	WY	1960	350	1015	48,000	—	48,000
Kortes	N Platte R	WY	1950	204	4	36,000	—	36,000
Seminoe	N Platte R	WY	1939	218	1011	45,000	—	45,000
Yellowtail	Bighorn R	MT	1966	463	613	250,000	—	250,000
Heart Mountain	Shoshone R	WY	1948	275	421	5,000[6]	—	5,000
Boysen	Bighorn R	WY	1952	110	742	15,000	—	15,000
Pilot Butte	Wyoming Cnl (Wind R)	WY	1935	106	NA	1,600	—	1,600
Canyon Ferry	Missouri R	MT	1953	147	1512	50,000	—	50,000
Otero	Arkansas Cnl (Arkansas R)	CO	—	270	93	—	—	11,000
Mt Elbert	Arkansas Cnl (Arkansas R)	CO	1981	464	7	100,000 R	100,000 R	200,000 R
Elephant Butte	Rio Grande	NM	1940	190	1730	24,300	—	24,300
Glen Canyon	Colorado R	AZ	1964	566	20876	1,042,000	—	1,042,000
Flaming Gorge	Green R	UT	1963	435	3516	108,000	—	108,000
Fontenelle	Green R	WY	1968	110	149	10,000	—	10,000
Crystal	Gunnison R	CO	1977	220	18	28,000	—	28,000
Morrow Point	Gunnison R	CO	1970	403	42	120,000	—	120,000
Blue Mesa	Gunnison R	CO	1967	356	774	60,000	—	60,000
Upper Molina	Cottonwood Cr, Plateau Cr	CO	1962	2663	6	8,640	—	8,640
Lower Molina[7]	Plateau Cr, Colorado R	CO	1962	1600	6	4,860	—	4,860
Green Mountain	Blue R	CO	1943	261	146	26,000	—	26,000
Senator Wash	Senator Wash Cr	CA	—	74	9	—	—	7,210 R
Parker	Colorado R	CA	1942	78	218	120,000	—	120,000
Davis	Colorado R	AZ	1951	131	1809	240,000	—	240,000
Hoover	Colorado R	AZ	1936	530	7927	717,000	—	717,000
Hoover	Colorado R	NV	1936	530	7927	717,000	—	717,000
Stampede	L Truckee R	CA	—	183	NA	—	—	3,000
Watasheamu	E Fk Carson R	NV	—	236	90	—	—	8,000
Deer Creek	Provo R	UT	1958	144	149	4,950	—	4,950
Dyne	Diamond Fk Pipeline	UT	—	848	700[8]	—	—	33,000
Sixth Water	Sixth Water Cr, Diamond	UT	—	823	700[8]	—	—	90,000
Syar	Strawberry Offstream	UT	—	431	700	—	—	10,500
Black Canyon	Payette R	ID	1925	94	NA	8,000	—	8,000
Boise	Boise R	ID	1912	40	NA	1,500	—	1,500
Anderson Ranch	S Fk Boise R	ID	1950	300	423[9]	40,000	—	40,000
Minidoka	Snake R	ID	1909	48	95[10]	13,400	—	13,400
Palisades	Snake R	ID	1957	244	1200	118,750	—	118,750
Chandler	Chandler Pwr Cnl (Yakima	WA	1956	121	NA	12,000	—	12,000
Roza (Canal)	Roza Cnl (Yakima R)	WA	1958	160	NA	11,250	—	11,250
Grand Coulee	Columbia R	WA	1941	343	5232[11]	6,180,000[12]	—	6,180,000
Grand Coulee P/G	Columbia R	WA	1974	280	53	150,000 R	150,000 R	300,000 R

TABLE 8–21. CONVENTIONAL AND PUMPED STORAGE HYDROELECTRIC PLANTS IN THE UNITED STATES (continued)

Plant	Stream	State	Year of Initial Oper.	Gross Static Head Ft	Usable Power Storage 1,000 Ac-Ft[1]	Installed Capacity - Kilowatts[2]		
						Developed	Under Construction	Authorized
Bureau of Reclamation (continued)								
Hungry Horse	S Fk Flathead R	MT	1952	477	2982[13]	285,000	—	285,000
Green Springs	Emigrant Cr, Bear Cr	OR	1960	1984	U	16,000	—	16,000
Trinity	Trinity R	CA	1964	469	2285	105,556	—	105,556
Nimbus	American R	CA	1955	43	2	13,500	—	13,500
Folsom	American R	CA	1955	333	920	198,720	—	198,720
Auburn	N Fk American R	CA	—	675	1966	—	—	300,000
Judge Francis Carr	Clear Cr Tnl (Trinity R),	CA	1963	695	2289	160,000	—	160,000
Keswick	Sacramento R	CA	1949	87	23	75,000	—	75,000
Spring Creek	Spring Cr, Sacramento R	CA	1964	625	2517	150,000	—	150,000
Shasta	Sacramento R	CA	1944	482	4050	539,000[15]	—	539,000
New Melones	Stanislaus R	CA	1979	583	2050	300,000[14]	—	300,000
O Neill	Delta Mendota Cnl (San Jose)	CA	1967	56	20	25,200 R	—	25,200 R
San Luis	Calif Aqueduct	CA	1968	327	1961	424,000 R[16]	—	424,000 R
Siphon Drop	Yuma Main Cnl (New R)	CA	—	15	NA	—	—	1,600
Total Conventional						12,139,176		12,596,276
Total Reversible						707,700 R	250,000 R	981,710 R
Tennessee Valley Authority								
Great Falls	Caney Fk, Cumberland R	TN	1916	148	39	31,860	—	31,860
Kentucky	Tennessee R	KY	1944	50	715	175,000	—	175,000
Pickwick landing	Tennessee R	TN	1938	46	236	220,040	—	220,040
Wilson	Tennessee R	AL	1942	88	47	629,840	—	629,840
Wheeler	Tennessee R	AL	1936	48	328	361,800	—	361,800
Tims Ford	Elk R	TN	1972	133	240	45,000	—	45,000
Guntersville	Tennessee R	AL	1939	39	131	115,200	—	115,200
Nickajack	Tennessee R	TN	1968	35	21	103,950	—	103,950
Raccoon MT	Tennessee R	TN	1979	1040	35	1,530,000 R	—	1,530,000 R
Chickamauga	Tennessee R	TN	1940	45	220	120,000	—	120,000
Ocoee 1	Ocoee R	TN	1912	111	33	18,000	—	18,000
Ocoee 2	Ocoee R	TN	1913	245	NA	21,000	—	21,000
Ocoee 3	Ocoee R	TN	1943	308	4	28,800	—	28,800
Blue Ridge	Toccoa R	GA	1931	148	186	20,000	—	20,000
Apalachia	Hiwassee R	TN	1943	440	35	82,800	—	82,800
Hiwassee/	Hiwassee R	NC	1940	246	350	57,600	—	57,600
Hiwassee	Hiwassee R	NC	1956	246	352	59,500 R	—	59,500 R
Nottely	Nottely R	GA	1956	168	171	15,000	—	15,000
Chatuge	Hiwassee R	NC	1954	124	229	10,000	—	10,000
Watts Bar	Tennessee R	TN	1942	54	214	166,500	—	166,500
Melton Hill	Clinch R	TN	1964	58	25	72,000	—	72,000
Norris	Clinch R	TN	1936	193	1761	100,800	—	100,800
Fontana	Little Tennessee R	NC	1945	430	1136	238,500	—	238,500
Fort Loudoun	Tennessee R	TN	1943	70	81	139,140	—	139,140
Douglas	French Broad R	TN	1943	135	1136	120,600	—	120,600
Cherokee	Holston R	TN	1942	147	1411	135,180	—	135,180
Ft Patrick Henry	S Fk Holston R	TN	1953	67	4	36,000	—	36,000
Boone	S Fk Holston R	TN	1953	119	150	75,000	—	75,000
Wilbur	Watauga R	TN	1912	65	U	10,700	—	10,700
Watauga	Watauga R	TN	1949	312	518	57,600	—	57,600

TABLE 8–21. CONVENTIONAL AND PUMPED STORAGE HYDROELECTRIC PLANTS IN THE UNITED STATES (continued)

Plant	Stream	State	Year of Initial Oper.	Gross Static Head ft	Usable Power Storage 1,000 Ac-Ft[1]	Installed Capacity - Kilowatts [2] Developed	Under Construction	Autho-rized
Tennessee Valley Authority (continued)								
South Holston	S Fk Holston R	TN	1951	240	519	35,000	—	35,000
	Total Conventional					3,242,910		3,242,91
	Total Reversible					1,589,500 R	R	1,589,500 R
National Park Service								
Cascades	Merced R	CA	1918	356	NA	2,000	—	2,000
	Total Conventional					2,000		2,000
International Boundary & Water Commission								
Falcon	Rio Grande	TX	1954	115	2767	31,500	—	31,500
Amistad	Rio Grande	TX	1983	213	500	66,000	—	66,000
	Total Conventional					97,500		97,500
Bureau of Indian Affairs								
Coolidge	Gila R	AZ	1929	204	1164	10,000	—	10,000
Drop 2	Yakima R (Offstream)	WA	1942	30	NA	2,500	—	2,500
Drop 3	Yakima R (Offstream)	WA	1932	34	NA	1,700	—	1,700
Big Creek	Big Cr, Flathead R	MT	1916	585	NA	360	—	360
	Total Conventional					14,560		14,560
Alaska Power Administration								
Eklutna[17]	Eklutna R	AK	1955	851	174	30,000	—	30,000
Snettisham[18]	Speel R	AK	1973	823	138	47,160	—	74,160
	Total Conventional					77,160		104,160
	Grand Total Conventional					34,584,503	831,000	38,355,303
	Grand Total Reversible					2,735,200 R	581,000 R	3,740,210 R

[1] U—Less than 1,000 acre-feet.
NA—DATA not available
[2] Includes main generating units only
[3] Operated by Rock Island Army Arsenal. Nameplate rating is 3,440 Kilowatts which cannot be obtained
[4] Storage Capacity for ultimate development is 420,000 acre-feet
[5] All water used for generation must be returned by pumping
[6] Storage in Buffalo Bill Reservoir is used jointly for irrigation and power
[7] Pipeline collects water from Big Creek and Cottonwood Creek, tributaries of Plateau Creek
[8] Water to be stored in the enlarged Strawberry Reservoir which is located upstream
[9] Used jointly for irrigation, power, and flood control
[10] Used jointly for irrigation and power
[11] Used jointly for irrigation, power, flood control, and navigation
[12] Includes two of three 10,000 kilowatt station service units used to supply commercial power
[13] Used jointly for power and flood control
[14] Dam and reservoir were constructed by the Corps of Engineers
[15] Includes one of two 2,000 Kilowatt station service units used to supply commercial power
[16] 222,000 kilowatts are allocated to the state of California under contract
[17] Constructed by the Bureau of Reclamation
[18] Constructed by the Corps of Engineers
R Capacity shown is in reversible equipment
Source: Federal Energy Regulatory Commission, 1984, Hydroelectric Power Resources of the United States—Developed and Undeveloped, FERC-0070, Washington, DC

TABLE 8–22. HYDROELECTRIC POWER WATER USE AND POWER GENERATED IN THE UNITED STATES, 1985

[Figures may not add to totals because of independent rounding. Mgal/d = million gallons per day; GWh = gigawatt hours]

State	Water Use		Power Generated, in GWh
	Mgal/d	Thousand acre-feet per year	
Alabama	114000	127000	7020
Alaska	1480	1650	764
Arizona	36300	40700	13900
Arkansas	59900	67200	4430
California	83800	93900	32000
Colorado	7270	8150	2400
Connecticut	4150	4650	276
Delaware	.0	.0	.0
D.C.	.0	.0	.0
Florida	8040	9010	252
Georgia	40300	45200	3330
Hawaii	283	317	63
Idaho	103000	115000	11700
Illinois	23300	26100	594
Indiana	9620	10800	362
Iowa	17200	19300	918
Kansas	893	1000	7.4
Kentucky	91000	102000	2940
Louisiana	1380	1550	81
Maine	44300	49600	1690
Maryland	17400	19500	1540
Massachusetts	98100	110000	1990
Michigan	13400	15000	635
Minnesota	22700	25400	1050
Mississippi	.0	.0	.0
Missouri	20200	22600	3930
Montana	65500	73500	10200
Nebraska	7080	7940	683
Nevada	8910	9980	4350
New Hampshire	14500	16200	1470
New Jersey	77	86	3.5
New Mexico	717	803	129
New York	515000	577000	31400
North Carolina	53500	60000	6580
North Dakota	12700	14200	2180
Ohio	8290	9300	169
Oklahoma	68800	77200	4010
Oregon	437000	489000	38900
Pennsylvania	60700	68100	1330
Rhode Island	1410	1580	24
South Carolina	42100	47200	2660
South Dakota	60500	67900	6090
Tennessee	118000	132000	8420
Texas	15200	17000	1320
Utah	3340	3740	1010
Vermont	8640	9680	779
Virginia	17700	19900	1320
Washington	628000	704000	76900
West Virginia	16000	18000	907
Wisconsin	63300	71000	1730
Wyoming	6440	7220	1070
Puerto Rico	884	990	216
Virginia Islands	.0	.0	.0
Total	3,050,000	3,420,000	296,000

Source: Solley, W.B., Merk, C.F., and Pierce, R.R., 1988, Estimated Water Use in the United States in 1985, U.S. Geological Survey Circular 1004

TABLE 8–23. SMALL HYDROELECTRIC CAPACITY IN THE UNITED STATES

[Developed, Under Construction, and Projected]

	5,000 kW and Less		15,000 kW and Less		30,000 kW and Less	
Category	**1980**	**1984**	**1980**	**1984**	**1980**	**1984**
Developed						
Number of sites	751	864	946	1,124	1,071	1,252
Capacity (MW)	1,194.6	1,351.1	3,294.7	3,729.7	5,834.6	6,574.0
Generation (GWh)	4.8	16.8	16.8	18.6	28.0	30.8
Under Construction						
Number of sites	16	137	29	153	33	158
Capacity (MW)	23.5	168.6	135.6	325.8	229.9	444.8
Generation (GWh)	0.1	0.7	1.0	1.4	1.3	2.0
Planned (NERC)[1]						
Number of sites	12	25	23	39	25	50
Capacity (MW)	34.2	54.3	136.2	191.4	182.4	436.8
Generation (GWh)	0.1	0.3	0.5	0.9	0.6	1.9
Projected[2]						
Number of sites	157	1,699	227	2,278	279	2,367
Capacity (MW)	317.5	2,587.1	1,241.8	6,669.4	2,317.5	8,572.4
Generation (GWh)	1.2	20.0	4.9	29.4	8.9	37.3
Totals						
Number of sites	936	2,725	1,225	3,594	1,408	3,827
Capacity (MW)	1,569.8	4,160.8	4,808.3	10,916.3	8,564.4	16,028.0
Generation (GWh)	6.2	27.6	23.2	50.3	38.8	72.0

[1] In reports of the Regional Electric Reliability Councils.
[2] Potential developments not under construction or included in NERC reports but which have FERC licensing or exemption status, are authorized or recommended for Federal construction, or have structural provisions for plant additions.
Source: Federal Energy Regulatory Commission, 1984

TABLE 8–24. TRENDS IN PUMPED STORAGE CAPACITY DEVELOPMENT IN THE UNITED STATES[1]

Year as of January 1	Installed Capacity in Reversible Units (millions of kilowatts): Developed			Under Construction		
	Pure	**Combined**	**Total**	**Pure**	**Combined**	**Total**
1960	0	0.1	0.1	0	0.2	0.2
1964	0.4	0.3	0.7	0.7	0.5	1.2
1968	1.6	0.5	2.1	1.2	1.6	2.8
1972	2.6	1.3	3.9	6.0	1.4	7.4
1976	7.3	2.4	9.7	2.7	1.6	4.3
1980	9.3	3.6	12.9	3.2	1.5	4.7
1984	10.1	3.7	13.8	4.9	0.4	5.3

[1] A pure pumped storage project with a large peaking capacity can be developed at a site with two potential reservoirs of reasonable size in close proximity and with a relatively large difference in elevations. Projects are usually more economically developed at sites with high usable heads; consequently, the more favorable sites are normally located in mountainous terrain. However, consideration has been given to the construction of pumped storage projects in areas of level terrain by placing the lower reservoir in an underground cavern or excavated area. For any development, an assured supply of water at least sufficient to replace evaporation, seepage, and other losses is essential.
Source: Federal Energy Regulatory Commission, 1984

TABLE 8–25. HYDROELECTRIC PLANTS HAVING POTENTIAL CONVENTIONAL CAPACITY OVER 1,000,000 KILOWATTS

			Installed Capacity in Conventional Units Kilowatts		
Plant	**River**	**Owner**[1]	**Developed**	**Under Construction**	**Ultimate Authorized**
Grand Coulee	Columbia	Bureau	6,180,000	0	6,180,000
John Day	Columbia	COE	2,160,000	0	2,700,000
Chief Joseph	Columbia	COE	2,069,000	0	2,069,000
R. Moses Niagara	Niagara	PASNY	1,950,000	0	1,950,000
The Dalles	Columbia	COE	1,806,800	0	1,806,800
Hoover	Colorado	Bureau	1,434,000	0	1,434,000
Rocky Reach	Columbia	CC PUD No. 1	1,213,950	0	1,213,950
Wanapum	Columbia	GC PUD No. 2	831,250	0	1,151,250
Priest Rapids	Columbia	GC PUD No. 2	788,500	0	1,108,500
Bonneville	Columbia	COE	1,076,620	0	1,102,820
Dworshak	N. Fork Clearwater	COE	400,000	0	1,060,000
Glen Canyon	Colorado	Bureau	1,042,000	0	1,042,000
Boundary	Pend Oreille	Seattle	634,600	392,000	1,026,600
	TOTAL		21,586,720	392,000	23,834,920

[1] Bureau—Bureau of Reclamation; COE—Corps of Engineers; PASNY—Power Authority, State of New York; GC—Grant County; CC—Chelan County; and Seattle—Seattle Dept. of Lighting.
Source: Federal Energy Regulatory Commission, 1984

TABLE 8–26. FEDERAL HYDROELECTRIC CAPACITY BY OPERATING AGENCY, JANUARY 1, 1984

	Installed Capacity Conventional and Reversible (kilowatts)		
	In Operation	**Under Construction**	**Ultimate Authorized**
Corps of Engineers	19,449,197	1,162,000	23,466,897
Bureau of Reclamation	12,846,876	250,000	13,577,986
Tennessee Valley Authority	4,832,410	—	4,832,410
International Boundary and Water Commission	97,500	—	97,500
Alaska Power Administration	77,160	—	104,160
Bureau of Indian Affairs	14,560	—	14,560
National Park Service	2,000	—	2,000
TOTAL	37,319,703	1,412,000	42,095,513

Source: Federal Energy Regulatory Commission, 1984

TABLE 8–27. CONVENTIONAL HYDROELECTRIC POWER IN THE UNITED STATES DEVELOPED, UNDEVELOPED AND TOTAL POTENTIAL—JANUARY 1, 1984

[By water resources regions (see Figure 8-4)]

Water Resource Region	Developed			Undeveloped			Total Potential		Percent Total Potential Cap. Now Developed
	No. of Plants	Installed Capacity KW	Average Annual Generation 1,000 KWH	No. of Sites	Installed Capacity KW[1]	Average Annual Generation 1,000 KWH	Installed Capacity KW	Average Annual Generation 1,000 KWH	
North Atlantic Region	248	1,538,631	6,359,847	669	4,387,567	13,299,362	5,926,198	19,659,209	25.9
Mid Atlantic Region	122	1,506,299	6,091,861	409	4,786,750	13,732,622	6,293,049	19,824,483	23.9
South Atlantic-Gulf Region	128	6,094,348	16,190,853	261	5,396,194	10,633,130	11,490,542	26,823,983	53.0
Great Lakes Region	209	3,895,123	24,354,228	229	2,243,431	9,852,598	6,138,554	34,206,826	63.4
Ohio Region	36	1,646,672	6,104,170	200	4,313,176	13,339,111	5,959,848	19,443,281	27.6
Tennessee Region	47	3,749,075	16,249,440	33	1,150,985	2,777,781	4,900,060	19,027,221	76.5
Upper Mississippi Region	105	617,237	3,187,639	104	1,267,796	5,459,595	1,885,033	8,,47,234	32.7
Lower Mississippi Region	5	205,800	430,400	29	1,000,075	3,106,275	1,205,875	3,536,675	17.0
Souris-Red Rainy Region	8	13,000	68,000	8	45,880	158,290	58,880	226,290	22.0
Missouri Region	62	3,462,023	15,549,519	234	5,307,568	17,595,761	8,769,591	33,145,280	39.4
Arkansas-White-Red Region	23	1,854,460	5,341,446	128	3,581,693	9,065,790	5,436,153	14,407,236	34.1
Texas-Gulf Region	18	393,970	1,066,359	87	1,481,627	3,032,279	1,875,597	4,098,638	21.0
Rio Grande Region	5	131,408	389,521	25	233,886	632,431	365,294	1,021,952	35.9
Upper Colorado Region	26	1,452,497	5,923,214	137	2,370,246	6,221,787	3,822,743	12,145,001	37.9
Lower Colorado Region	17	1,910,463	6,601,700	33	1,650,983	5,085,441	3,561,446	11,687,141	53.6
Great Basin Region	60	188,422	655,837	112	765,697	1,862,008	954,119	2,517,845	19.7
Pacific Northwest Region	176	29,909,275	143,542,447	1128	22,104,337	67,221,564	52,013,612	210,764,011	57.5
California Region	211	8,033,509	37,644,560	740	10,099,470	25,983,648	18,132,979	63,628,208	44.3
Alaska Region	26	155,737	634,082	96	5,041,107	22,192,656	5,196,844	22,826,738	2.9
Hawaii Region	14	17,652	104,100	15	62,890	310,950	80,542	415,050	21.9
Total United States	1546	66,775,601	296,489,223	4677	77,291,358	231,563,079	144,066,959	528,052,302	46.3

[1] Includes potential capacity additions or subtractions at existing plants.
Source: Federal Energy Regulatory Commission, 1984

TABLE 8–28. CONVENTIONAL HYDROELECTRIC POWER IN THE UNITED STATES DEVELOPED, UNDEVELOPED AND TOTAL POTENTIAL-JANUARY 1, 1984

[By geographic regions (see Figure 8-5)]

Geographic Division	Developed			Undeveloped			Total Potential		Percent Total Potential Cap. Now Developed
	No. of Plants	Installed Capacity KW	Average Annual Generation 1,000 KWH	No. of Sites	Installed Capacity KW[1]	Average Annual Generation 1,000 KWH	Installed Capacity KW	Average Annual Generation 1,000 KWH	
New England	279	1,618,736	6,684,066	744	4,555,434	13,887,230	6,174,170	20,571,296	26.2
Middle Atlantic	157	4,297,372	25,703,831	421	5,370,969	18,152,584	9,668,341	43,856,415	44.4
East North Central	210	1,087,054	5,159,366	208	1,789,913	7,158,041	2,876,967	12,317,407	37.7
West North Central	65	2,808,465	12,187,539	157	3,011,876	9,999,018	5,820,341	22,186,557	48.2
South Atlantic	166	6,013,182	17,216,643	355	8,395,423	17,795,208	14,408,605	35,011,851	41.7
East South Central	57	5,734,292	22,024,618	119	2,842,266	9,337,538	8,576,558	31,362,156	66.8
West South Central	43	2,338,020	6,516,705	214	5,328,495	12,734,953	7,666,515	19,251,658	30.4
Mountain	192	7,869,611	32,866,583	876	18,201,259	53,698,630	26,070,870	86,565,213	30.1
Pacific	337	34,835,480	167,391,690	1472	22,691,726	66,296,271	57,527,206	233,687,961	60.5
Alaska	26	155,737	634,082	96	5,041,107	22,192,656	5,196,844	22,826,738	2.9
Hawaii	14	17,652	104,100	15	62,890	310,950	80,542	415,050	21.9
Total United States	1546	66,775,601	296,489,223	4677	77,291,358	231,563,079	144,066,959	528,052,302	46.3

[1] Includes potential capacity additions or subtractions at existing plants.
Source: Federal Energy Regulatory Commission, 1984

FIGURE 8–4. CONVENTIONAL HYDROELECTRIC POWER DEVELOPED AND UNDEVELOPED, JANUARY 1, 1984

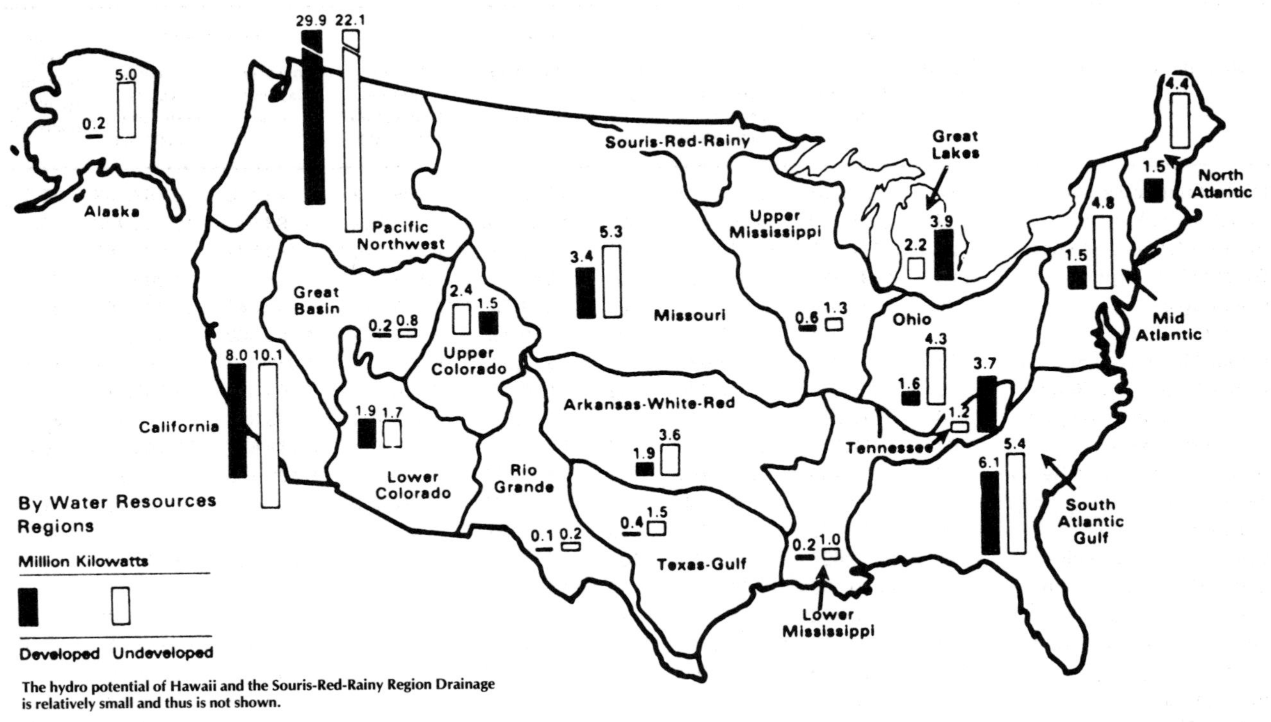

Source: Federal Energy Regulatory Commission, 1984

FIGURE 8–5. CONVENTIONAL HYDROELECTRIC POWER DEVELOPED AND UNDERDEVELOPED, JANUARY 1, 1984

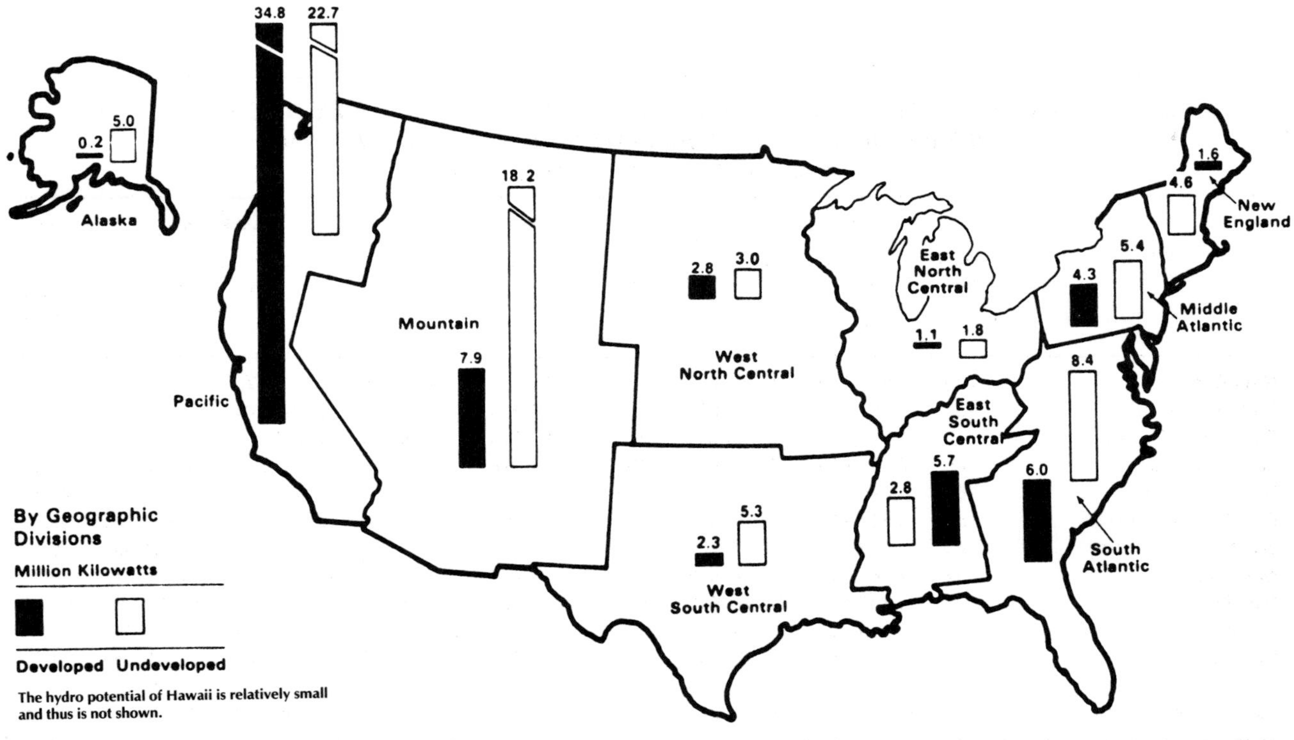

Source: Federal Energy Regulatory Commission, 1984

FIGURE 8–6. RIVER SEGMENTS COVERED BY THE WILD AND SCENIC RIVERS ACT AND SPECIAL ACTS PRECLUDED FROM HYDROELECTRIC DEVELOPMENT

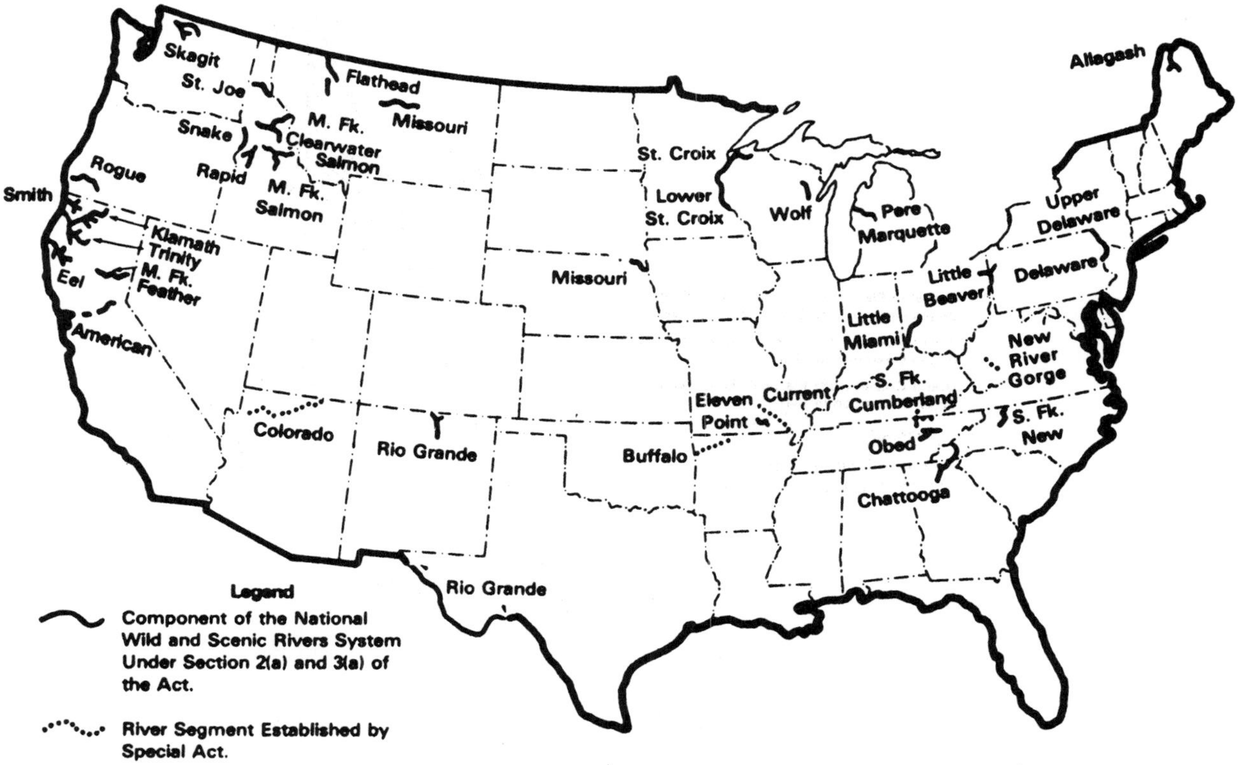

FIGURE 8–7. RIVER SEGMENTS DESIGNATED FOR STUDY UNDER SECTION 5(a) OF THE WILD AND SCENIC RIVERS ACT

Source: Federal Energy Regulation Commission, 1984

FIGURE 8–8. HYDROELECTRIC CAPACITY IN THE UNITED STATES 1882-2000

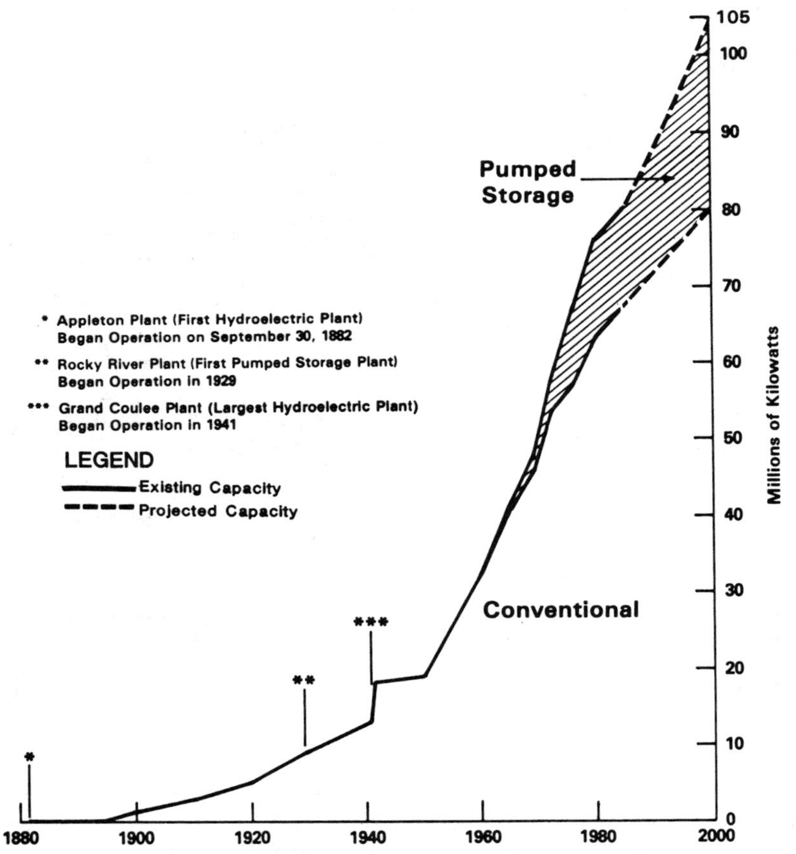

Source: Federal Energy Regulatory Commission, 1984

TABLE 8–29. HYDROELECTRIC POWER CAPACITY IN CANADA, 1984

Provinces	Developed (MW)	Technically Developable Potential (MW)
Northwest Territories	47	6,000
Yukon Territory	78	10,430
British Columbia	10,842	27,119
Alberta	734	12,425
Saskatchewan	575	1,703
Manitoba	3,641	6,100
Ontario	7,126	6,164
Quebec	25,133	31,367
New Brunswick	901	556
Nova Scotia	368[1]	93
Prince Edward Island	—	—
Newfoundland	6,301	6,507
TOTAL	55,746	108,464

[1] Excludes 18 MW of tidal power at Annapolis Royal.

Source: Pearse, P.H., and others, 1985, Currents of Change, Final Report Inquiry on Federal Water Policy, Ottawa, Canada

SECTION D. COSTS OF WATER PROJECTS

TABLE 8–30. TYPICAL COST STRUCTURES FOR WATER-SUPPLY SYSTEMS IN THE UNITED STATES

	Small Systems[1]		Large Systems[2]	
	Percentage of Operating Expenses	Percentage of Operating Expenses Excluding Interest Charges	Percentage of Operating Expenses	Percentage of Operating Expenses Excluding Interest Charges
Acquisition	19	22	15	19
Treatment	15	18	10	13
Distribution and Transmission	36	43	31	38
Support Services	14	17	25	30
Interest Charges	16	—	19	—
Total	100	100	100	100

[1] Serving between 300 and 75,000 people.
[2] Serving over 75,000 people.
Source: U.S. Environmental Protection Agency, 1986, Guidelines for Ground-Water Classification under the EPA Ground-Water Protection Strategy; ACT Systems Inc., 1977, 1979

TABLE 8–31. TYPICAL WATER SYSTEM COSTS IN THE UNITED STATES[1]

[1984 $/million of gallons produced]

	Source	
Population Served by System	Surface Water	Ground Water
1,000 — 3,300	1,085	1,493
3,300 — 10,000	1,063	924
10,000 — 25,000	795	718
25,000 — 75,000	727	710
75,000 — 500,000	596	606
over 500,000	457	574

[1] Operating expenses (including depreciation and capital charges), inflated to 1984 dollars.
Source: U.S. Environmental Protection Agency, 1986, Guidelines for Ground-Water Protection Classification under the EPA Ground-Water Protection Strategy; Survey of Operating and Financial Characteristics of Community Water Systems, Temple, Barker and Sloane Inc., 1982

TABLE 8–32. COSTS OF GROUND-WATER SUPPLY SYSTEMS IN THE UNITED STATES[1]

[By population size category; 1984 $/million gallons produced]

Population Served by System	Annual Cost
25 — 1,000	4,616
1,000 — 3,300	1,493
3,300 — 10,000	924
10,000 — 25,000	718
25,000 — 75,000	710
75,000 — 500,000	606
over 500,000	574

[1] Operating expenses (including depreciation and capital charges, inflated to 1984 dollars.
Source: U.S. Environmental Protection Agency, 1986, Guidelines for Ground-Water Protection Strategy; Survey of Operating and Financial Characteristics of Community Water Systems, Temple, Barker and Sloane, Inc., 1982

FIGURE 8–9. MAP OF PHYSIOGRAPHIC REGIONS IN THE UNITED STATES (FOR RESERVOIR COSTS)

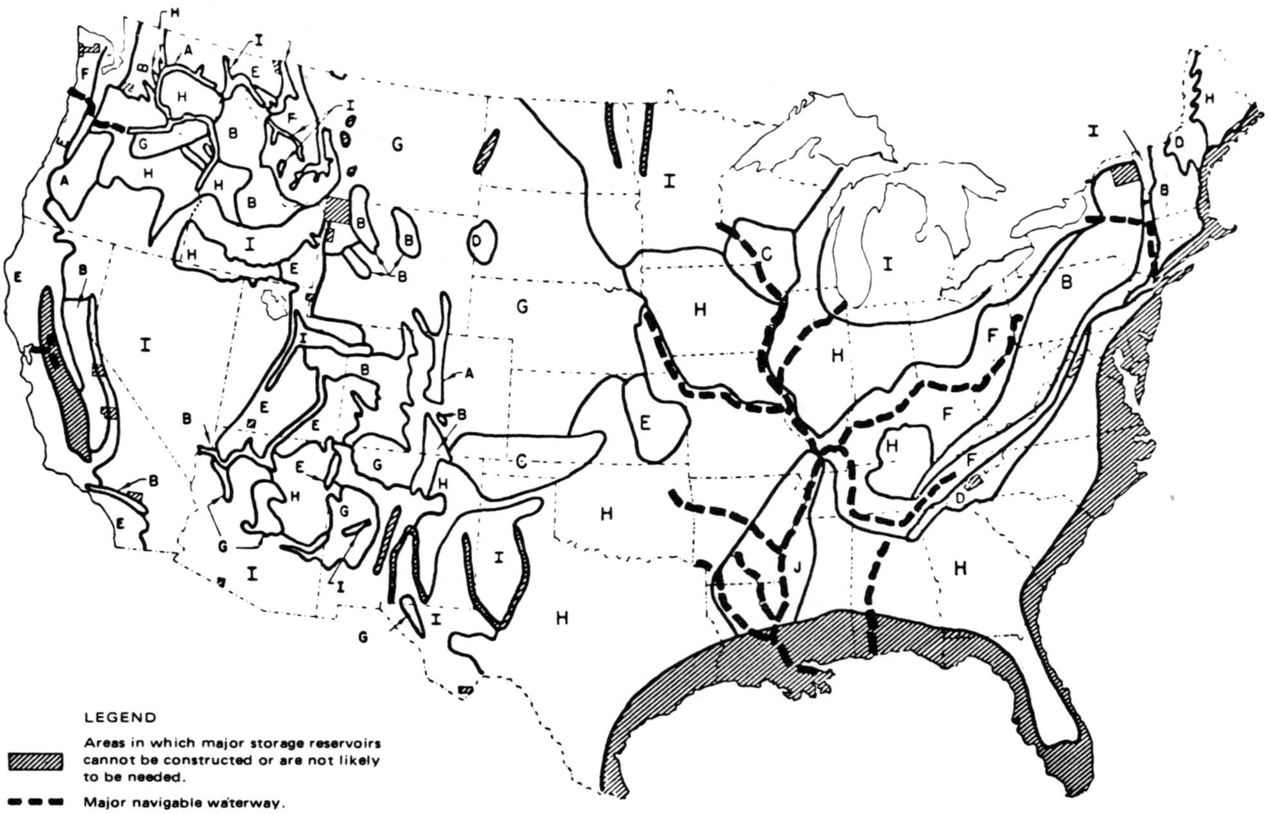

Source: Corps of Engineers, U. S. Army, 1960

TABLE 8–33. COSTS OF RESERVOIRS IN THE PHYSIOGRAPHIC REGIONS OF THE UNITED STATES

[Values are costs in dollars of reservoir storage per acre-feet. Size class in thousand acre-feet. Physiographic regions are shown in Fig. 8-9]

Physio-graphic Region	Size Class										
	10	30	50	80	150	300	700	1,500	3,000	7,000	30,000
A	$230	$190	$175	$162	$145	$130	$113	$100	$88	$77	$60
B	200	160	145	132	115	100	85	75	64	55	40
C	180	135	120	110	95	80	66	56	49	40	28
D	160	120	105	93	78	65	52	42	35	28	18
E	155	115	98	85	73	60	47	38	32	25	16
F	145	106	90	80	65	55	43	34	28	21	14
G	140	103	87	75	60	50	38	30	25	19	12
H	120	86	73	62	50	40	30	24	18	14	10
I	95	65	55	46	38	30	22	18	15	10	8
J	65	43	37	32	25	20	15	12	10	8	6

Source: Corps of Engineers, U.S. Army, 1960

TABLE 8-34. COSTS OF COMMUNITY WATER SUPPLY TECHNOLOGY IN DEVELOPING COUNTRIES

[Capital and recurrent costs for a community of 400 people]

Technology	Low			High		
	Handpumps	Standpipes	Yardtaps	Handpumps	Standpipes	Yardtaps
Capital cost (US$)						
Wells[1]	4,000	2,000	2,500	10,000	5,000	6000
Pumps (hand/motor)	1,300	4,000	4,500	2,500	8,000	9000
Distribution[2]	none	4,500	16,000	none	10,000	30000
Sub-total	5,300	10,500	23,000	12,500	23,000	45000
Cost per capita	13.3	26.3	57.5	31.2	57.5	112.5
Annual cost (US$/year)						
Annualized capital[3]	700	1,500	3,200	1,400	3,000	6000
Maintenance	200	600	1,000	400	1,200	2000
Operation (fuel)	none	150	450	none	300	900
Sub-total (cash)	900	2,250	4,650	1,800	4,700	8900
Haul costs (labor)[4]	1,400	1,100	none	3,000	2,200	none
Total (including labor)	2,300	3,350	4,650	4,800	6,900	8900
Total annualized cost per capita						
Cash only	2.3	5.6	11.6	4.5	11.8	22.3
Cash + labor	5.8	8.4	11.6	12.0	17.3	22.3

[1] Pumping water level assumed to be 20 meters. Two wells assumed for handpump system (200 persons per handpump).
[2] Distribution system includes storage, piping,and taps with soakaway pits.
[3] Capital costs with replacement of mechanical equipment after 10 years annualized at a discount rate of 10% over 20 years.
[4] Labor costs for walking to the water point, queuing, filling the container, and carrying the water back to the house. Time valued at US $0.125/hour.
Source: Arlosoroff, Saul, and others, 1987, Community Water Supply: The Handpump Option, The World Bank, Washington, D.C., Copyright; Reprinted with permission

TABLE 8-35. TYPICAL COSTS OF IRRIGATION DEVELOPMENT IN LATIN AMERICA, AFRICA, ASIA, AND FAR EAST

[1980 prices; US $ per hectare]

Region	$/ha	Gravity Schemes Share in Total %	Pump and Tubewells $/ha	All Schemes Weighted Average Cost $/ha
Latin America	6 000	50	3 000	4 000
Africa (excl. Sudan)	11 000	70	6 000	9 500
Sudan	5 000	50	4 000	4 500
Near East	7 000	70	4 000	6 100
Asia and Far East (excl. South Asia)	4 000	60	2 000	3 200
South Asia (Bangladesh, India, Pakistan)	2 500	40	1 000	1 600
Rehabilitation in all regions except South Asia — $1 760/ha				
Rehabilitation in South Asia — $ 800/ha				

Source: FAO, 1982, The State of Food and Agriculture

SECTION E. PROJECT PLANNING AND ANALYSIS

TABLE 8–36. ELEMENTS COMPRISING A WATER RESOURCES PROJECT INVESTIGATION

A. Purposes of a multiple-purpose project:

1. Irrigation
2. Drainage
3. Domestic or industrial water supply
4. Flood control
5. Hydro-power generation
6. Navigation
7. Fish and wildlife conservation
8. Recreation
9. Water quality control
10. Salinity control
11. Watershed management
12. National defense
13. International relation

B. Land resources:

1. Land classification
2. Land use and capabilities
3. Development
4. Settlement
5. Drainage

C. Water resources:

1. Water supply, surface and ground water and salvage
2. Water quality and treatment
3. Water requirements, all purposes
4. Water rights including international treaties
5. Flood studies
6. Sediment, including transport, erosion and aggradation
7. Project operation studies
8. Forecasting for operation
9. Hydraulic design requirements

D. Engineering and geology:

1. Aerial photography, surveying and mapping
2. Geology, foundation and materials
3. Anticipated construction problems
4. Plans and cost estimates, physical plan formulation
5. Anticipated operation, maintenance and replacement problems and estimates of cost

E. Economics:

1. Existing economy and resource use
2. Future economy without the project
3. Future economy with the project, and regional and national impact
4. Economic criteria for plan formulation
5. Economic justification

F. Financial considerations:

1. Cost allocation to various purposes
2. Repayment of capital investment
3. Payment of annual operation, maintenance and replacement costs

G. Legal considerations:

1. Rights to use of water
2. International agreements and treaties
3. Land acquisition and rights-of-way

H. Public relations:

1. Determination of public interest in contemplated development
2. Dissemination of factual information on progress and objectives of investigation
3. Establishment of government policy and enabling legislation

I. Reports:

1. Reconnaissance reports
   - (a) Basin plan
   - (b) Preliminary project report
2. Special interim or progress reports
3. Feasibility
4. Definite plan

J. Administration:

1. Organizational requirements for supervision of construction and operation of proposed projects
2. Program and budget requirements and control

Source: ECAFE, United Nations, 1964

TABLE 8–37. PURPOSES OF A WATER RESOURCES PROJECT

Purpose	Description	Type of Works and Measures
Flood control	Flood-damage prevention or reduction, protection of economic development, conservation storage, river regulation, recharging of groundwater, water supply, development of power, protection of life	Dams, storage reservoirs, levees, floodwalls, channel improvements, floodways, pumping stations, floodplain zoning, flood forecasting
Irrigation	Agricultural production	Dams, reservoirs, wells, canals, pumps and pumping plants, weed-control and desilting works, distribution systems, drainage facilities, farmland grading
Hydroelectric	Provision of power for economic development and improved living standards	Dams, reservoirs, penstocks, power plants, transmission lines
Navigation	Transportation of goods and passengers	Dams, reservoirs, canals, locks, open-channel improvements, harbor improvements
Domestic and industrial water supply	Provision of water for domestic, industrial, commercial, municipal, and other uses	Dams, reservoirs, wells, conduits, pumping plants, treatment plants, saline-water conversion, distribution systems
Watershed management	Conservation and improvement of the soil, sediment abatement, runoff retardation, forests and grassland improvement, and protection of water supply	Soil-conservation practices, forest and range management practices, debris-detention dams, small reservoirs, and farm ponds
Recreational use of water	Increased well-being and health of the people	Reservoirs, facilities for recreational use, works for pollution control, reservations of scenic and wilderness areas
Fish and wildlife	Improvement of habitat for fish and wildlife, reduction or prevention of fish or wildlife losses associated with man's works, enhancement of sports opportunities, provision for expansion of commercial fishing	Wildlife refuges, fish hatcheries, fish ladders and screens, reservoir storage, regulation of streamflows, stocking of streams and reservoirs with fish, pollution control, and land management
Pollution abatement	Protection or improvement of water supplies for municipal, domestic, industrial, and agricultural use and for aquatic life and recreation	Treatment facilities, reservoir storage for augmenting low flows, sewage-collection systems, legal control measures
Insect control	Public health, protection of recreational values, protection of forests and crops	Proper design and operation of reservoirs and associated works, drainage, and extermination measures
Drainage	Agricultural production, urban development, and protection of public health	Ditches, tile drains, levees, pumping stations, soil treatment
Sediment control	Reduction or control of silt load in streams and protection of reservoirs	Soil conservation, sound forest practices, proper highway and railroad construction, desilting works, channel and revetment works, bank stabilization, special dam construction and reservoir operations
Salinity control	Abatement or prevention of salt-water contamination of agricultural, industrial, and municipal water supplies	Reservoirs for augmenting low streamflow, barriers, groundwater recharge, coastal jetties
Artificial precipitation	Control of precipitation within meteorological limits	Portable cloud-seeding equipment, ground generators
Employment	Stimulation of employment and sources for increased income in depressed areas of unemployment and under-development	Area Redevelopment Act and Area Redevelopment Administration
Public works acceleration	Acceleration of Federal, state, and local constructions of public works on cost-sharing basis	Public Law 87-658
New water-resources policies	New policies to be used by Federal agencies, according to S. Document no. 97 [27], approved by the President May 15, 1962, affecting the economies of project justification as well as project formulation and composition	Senate Document no. 97

Source: Dixon, in Chow, Handbook of Applied Hydrogology, McGraw-Hill, Copyright 1964

TABLE 8–38. ITEMS TO BE CONSIDERED IN PLANNING A MULTIPLE-PURPOSE WATER RESOURCES PROJECT

A. Physical and Related Items

1. Project area
   a. Physical geography: location and size; physiography; climate; soils
   b. Settlement: history; population; cultural background, both rural and urban
   c. Development: industry; transportation; communication; commerce; power; land uses; water uses; minerals; undeveloped resources
   d. Economic conditions: general; relief problems; community needs; national needs
   e. Investigations and reports: previous investigations; history; scope

2. Hydrologic data
   a. Hydrologic records and networks: gaging and observation stations; data-collecting agencies
   b. Hydrometeorological data: precipitation; evaporation and evapotranspiration
   c. Surface water: low flows; normal flows; maximum floods; "design flood"; drought; quality
   d. Groundwater: aquifers; recharge; quality

3. Supply of water
   a. Sources of supply: surface-water supply; groundwater supply; reservoirs
   b. Variation of supply: variability; consumptive use; regulation; diversion requirements; return flow; evapotranspiration losses; seepage losses or gains
   c. Quality of water: physical, chemical, biological and radioactive qualities; quality requirements; pollution
   d. Legal rights: water rights; development of project rights; operation rights

4. General considerations for design and planning
   a. Geology: explorations; geological formations; foundation problems
   b. Design problems: design criteria; methods of analysis; project operation and maintenance
   c. Construction problems: accessibility to project site; rights of way and relocation; construction materials; construction period; flow diversion; manpower; equipment; accessibility
   d. Alternative plans: comparison of alternative plans; supplementary plans; possible alternative plans; relationships to areas to be served
   e. Estimates of costs
   f. Intrastate, interstate, and international problems
   g. Organizations involved: public and/or private; technical, social, and political

5. Flood control
   a. Flood characteristics of the project area: Historical floods; flood magnitude and frequency
   b. Design criteria: project design storms and floods; degree of protection
   c. Damage: survey of flooding areas and things affected by floods, nearby or quite a distance away, including commerce, good will, dates of delivery of goods, etc.
   d. Measure of control: reduction of peak flow by reservoirs; confinement of flow by levees, floodwalls, or a closed conduit; reduction of peak stage by channel improvement; diversion of floodwater through bypasses or floodways; flood-plain zoning and evacuation; floodproofing and flood insurance of specific properties; reduction of flood runoff by watershed management
   e. Existing remedial works

6. Agricultural use of water (irrigation and related drainage)
   a. Factors for land classification: soil texture; depth to sand, gravel, shale, raw soil, or penetrable lime zone; alkalinity; salinity; slopes; surface cover and profile; drainage; water logging
   b. Present and anticipated development: crops; livestock; financial resources; improvements; organizations; development period
   c. Water requirements, if any: total crop requirement; irrigation-water demand; farm-delivery losses; diversion amounts
   d. Available water: sources; quality; quantity; distribution
   e. Irrigation methods: flooding; furrow irrigation; sprinkling; subirrigation; supplemental irrigation
   f. Structural works: storage reservoirs; dams; spillways; diversion works; canals and distribution systems

7. Hydroelectric power
   a. Development: sources; present potential and future capacities
   b. Alternative sources of power: stream; oil; gas; nuclear power; interties
   c. Types of power plants: run-of-river; storage; pumped storage
   d. Structural components: dams; canals; tunnels; penstock; forebay; powerhouse; tailrace
   e. Power problems: load demand and distribution; interties (interconnections with other power transmission systems)
   f. Markets; revenues; costs

TABLE 8–38. ITEMS TO BE CONSIDERED IN PLANNING A MULTIPLE-PURPOSE WATER RESOURCES PROJECT (continued)

8. Navigation
   a. Water traffic: present and future needs and savings in shipping costs, if any, on the basis of which the justifications are primarily judged at the present time
   b. Alternative means of transportation: air; land
   c. Navigation requirements: depth, width, and alignment of channels; locking time; current velocity; terminal facilities
   d. Methods of improving navigation: channel improvement by reservoir regulation; contraction works; bank stabilization, straightening, and snag removal; lock-and-dam construction; canalization; dredging

9. Domestic and industrial water supply
   a. Sources of supply: surface and/or groundwater; location and capacity; desalination
   b. Water demand: climate; population characteristics; industry and commerce; water rates and metering; size of project area; fluctuation
   c. Water requirements: quantity; pressure; quality (tastes, odors, color, turbidity, bacteria content, chemicals, temperature, etc.)
   d. Methods of purification: plain sedimentation; chemical sedimentation or coagulation; filtration; disinfection; aeration; water softening
   e. Treatment plant: location; design; purpose or purposes
   f. Distribution systems: reservoirs; pumping stations; elevated storage; layout and size of pipe systems; location of fire hydrants
   g. Waterworks organizations: maintenance and operation of supply, distribution, and treatment facilities

10. Recreational use of water
    a. Population tributary (population near enough to the project area to use it for recreational purposes)
    b. Facilities: boating; fishing; swimming; etc.
    c. Water requirements: depth of water; area of water surface; sanitation

11. Fish and wildlife
    a. Biological data: species; habits
    b. Facilities: reservoirs; fish ladders
    c. Water requirements: temperature; current velocity; biological qualities

12. Drainage
    a. Existing projects
    b. Drainage conditions: rainfall excess; soil condition; topography; disposal of water
    c. Drainage system: urban; farmland

13. Water-quality control
    a. Problems involved: sources; nature and degree of pollution; sediment; salinity; temperature; oxygen content; radioactive contamination
    b. Hydrologic information and measurement
    c. Methods of control

B. Economic Aspects of Project Formulation

1. Benefits and damages: identification and evaluation

2. Costs: identification and estimation

3. Financial feasibility

4. Allocation of costs

5. Reimbursement requirements and sharing of allocated costs

6. Methods and costs of financing the project, whether federal, state, or local, bringing all benefits and all costs to an annual basis and recognizing interest on the investment not only during construction, but throughout the entire proposed "life of the project"

7. Benefit-cost-ratio analysis: alternative plans

Source: Dixon, in Chow, Handbook of Applied Hydrology, McGraw-Hill, Copyright 1964

TABLE 8–39. SOME EXAMPLES OF WATER MANAGEMENT PURPOSES AND THE NEED FOR PLANNING

Function	Primary Responsibility[1]	Capital Investment	Operating Plans
Irrigation	F,S,L,P	Best construction of systems	Best use of water and money
Municipal water supply	L	Cost effectiveness	Contingencies, best use of facilities
Industrial water supply	P	Cost effectiveness	Contingencies, best use of facilities
Energy cooling water	P	Development of supplies	Best use of facilities, meeting standards
Hydropower	F,P	Development of economic power	Maximization of energy production
Wastewater treatment	L	Cost effectiveness	Meeting standards, reducing costs
Navigation	F	National economic efficiency	Operation of facilities
Flood damage reduction	F,S,L	Optimum facilities	Optimum operation
Urban drainage	L	Plans for economical systems	Maintenance, warning, etc.
Agricultural drainage	L	Plans for systems	Operation of systems
Recreation	F,S,L	Development of facilities	Effective operation
Fish and wildlife	F,S	Preservation and enhancement of species	Effective operation
Watershed management	L	Best plans	Maintenance and operation
Preservation of ecological systems	F,S	Preservation of systems	NA
Preservation of systems of unique value	F,S	Preservation of systems	NA

[1] F = federal; S = state; L = local; P = private.
Source: Grigg, N.S., 1985, Water Resources Planning. Copyright 1985 McGraw-Hill, Inc. Reprinted with permission

TABLE 8–40. STAGES OF PLANNING FOR WATER RESOURCES MANAGEMENT

Policy Planning Examples of Activities	Framework Planning Examples of Activities	General Appraisal Planning Examples of Activities	Implementation Planning Examples of Activities
Assess broad national needs	From viewpoint of broad regionwide totals and on "no-project" basis:	On basis of local projects or measures, and over regional or watershed areas:	For specific projects or measures:
Hypothesize national goals and objectives	Inventory and evaluate available data	Estimate present and future water use and environmental needs	Evaluate specific water use and environmental needs
Identify problems and opportunities in achieving goals	Assess present and future water use and environmental needs	Estimate available water and related land resources	Evaluate available water and related land resources
Identify costs and benefits in achieving goals	Assess available water and related land resources	Make preliminary evaluations of alternative water quality management approaches	Evaluate regulation potential for different degrees of storage
Recommend goals and objectives	Evaluate general regulation potential and identify water quality management approaches	Make preliminary estimates of costs, benefits and consequences of specific alternative projects and measures	Evaluate degree of water quality control with different types of facility
Recommend policy choices	Inventory present status of development	Compare alternative projects and measures	Prepare conceptual designs and cost estimates
Coordinate national priorities	Inventory general means available to satisfy needs	Devise alternative early action and future programs	Make economic analyses of benefits and consequences
Review programs for achievement of goals	Assess general alternatives to meet different goals	Recommend specific early action and alternative future programs, including selection of projects or measures for implementation study	Make financial analyses to demonstrate payout
	Identify problem areas that need priority attention		Compare alternatives on basis of costs, benefits and payout
	Recommend actions that can be taken at present and those that require further study		Recommend an alternative to be selected
			Recommend concerning authorization

Source: National Water Commission, 1972, Water Resources Planning

FIGURE 8–10. BENEFIT CATEGORIES FOR ANALYSIS OF WATER QUALITY PROGRAMS

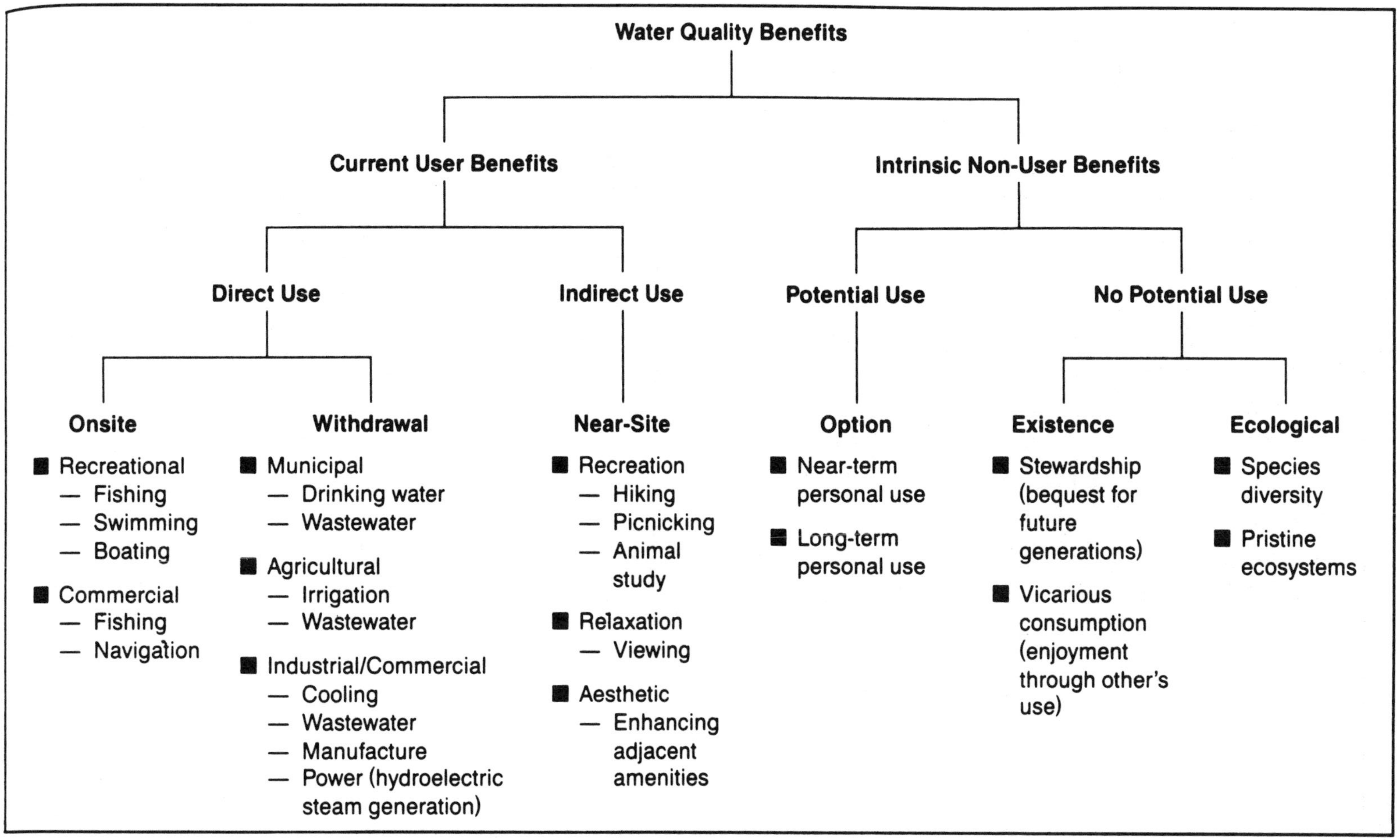

Source: U.S. Environmental Protection Agency, National Water Quality Inventory - 1986 Report to Congress. EPA-440/4-87-008

TABLE 8–41. LOW-COST TECHNICAL APPROACHES TO COMMUNITY WATER QUALITY IMPROVEMENTS IN DEVELOPING COUNTRIES

Technical Approach	Pre-Conditions for Improving the Bacteriological Quality of Water
Sedimentation techniques (jars)	• Source not excessively contaminated initially • Transport in clean covered vessels • Sedimentation for at least 12-24 hours • Dipping from the top with care not to agitate the contents • Addition of an disinfectant (iodide or a hypochlorite) if possible • Frequent cleansing of the jar before the last water is drawn from the bottom
(tanks)	• Lined with concrete or bricks, if possible • Non-use for washing, bathing, urinating, defecating, swimming, or animal watering • Tank equipped with outlets for water drawing • Frequent removal of algae and debris especially after rains including use of duckweed for continuous clearing of water • Frequent and adequate maintenance and repair of lining and outlets
Filtration techniques (in this case infiltration galleries)	• Selection of an adequate sandy or gravelly bed to lay the gallery or other conduit • Placement of an adequate layer of stones, gravel, and sand around the gallery or conduit • Adequate depth of the well • Care taken to line the well and raise the wall to a sufficient height to avoid surface runoff • Adequate depth of sand and gravel layer at the base of the well • Careful attention to maintenance: repair of cracks in well and gallery or conduit; cleansing of the sand and gravel layer • Use of a single bucket for drawing water or a special stand for hanging individual buckets to avoid putting them on the ground
Rainwater catchment	• Use of a foul-flush mechanism to promote runoff of roof debris at the first rain or after storms • Periodic cleaning and repairs of gutters and drains • Covering the cistern, with trap door or spigot mechanism for water-drawing • Single bucket if trap door • Adequate maintenance of cistern: repair of cracks • Adequate height of cistern to avoid surface runoff • Use of a filter box to clear water flowing from gutters • Careful seal around the inlet cap used for cleaning the cistern and around the spigot • Frequent cleaning of the cistern
Spring capping (multiple methods)	• Selection of springs for capping that do not drain likely contaminated areas such as latrines, garbage dumps, or pastures • Adequate gutters dug around the spring site to catch surface runoff and divert it into the drainage channel at the foot of the spring • Use of a single pipe or tube emanating from the spring so as to avoid the need to dip individual recipients into the basin • Complete closure of the basin, if possible • If not, protection of the basin from intrusion by animals and children from the sides, and from dust, leaves, and debris from above. • Avoidance of washing, bathing, swimming, urinating, defecating, and animal-watering at the spring site itself; rather setting aside areas below the water-drawing site for these activities including a fenced off area for animal-watering. • Maintenance of all structures including especially the repair of cracks and leaks
Hand-dug wells: open or with bucket and pulley	• Protection of the first three meters of the well-lining with adequately impermeable material • Lining the entire well if necessary in the case of highly permeable soil • Parapet high enough (0.5-1.0 meter) and well-constructed to avoid entirely any possibility of surface runoff, or of children or animals falling in. • Use of chlorine "pot" to disinfect the well • Construction of an adequately-inclined apron around the well linked to a drainage channel that leads to a soakaway • In cattle-raising areas, construction of animal-watering troughs at a sufficient distance from the well (at least 30 meters) and connected to the well by means of well constructed channels • Construction of a substantial barrier around the well to avoid encroachment by domestic animals on domestic water gathering

TABLE 8–41. LOW-COST TECHNICAL APPROACHES TO COMMUNITY WATER QUALITY IMPROVEMENTS IN DEVELOPING COUNTRIES (continued)

Technical Approach	Pre-Conditions for Improving the Bacteriological Quality of Water
Hand-dug wells: open or with bucket and pulley (continued)	• If the well is left open, application of a concrete or wooden cover with a trap opening to permit water drawing-enforcement of closing during periods of non-use • If a bucket and pulley system, continuous and adequate maintenance • If not, establishment of a rack on which individual buckets can be hung so as to avoid contact with the ground • Avoidance of throwing trash and other debris into the well
Hand-dug wells with handpumps	• Protection of the first three meters of the well lining with an adequately impermeable material • Lining the entire well, if necessary, in the case of impermeable soil • Parapet high enough (0.5 - 1.0 meters) to avoid surface runoff into well • "Shock" chlorination after construction • Construction of an adequately inclined apron around the well linked to a drain leading to a soakaway • In cattle-raising areas, construction of animal-watering troughs at a sufficient distance from the well (at least 30 meters) and connected to the well by well-constructed channels • Construction of a substantial barrier around the well to avoid encroachment by domestic animals on domestic water-gathering • Careful sealing of the well around the slab and the base of the pump • Avoidance of pump-priming with contaminated water • "Shock" chlorination whenever the pump is removed for repairs, or the well repaired
Boreholes with hand pumps	• Same upper well construction features as for hand-dug wells • Careful sealing of both the slab of the well and the attachment of the pump to the tube so as to avoid the entrance of surface runoff
Gravity-fed systems	• Selection of initially nearly pure sources (less than 100 coliforms per liter) • Stringent avoidance of agricultural, grazing, or settlement encroachment on water drainage areas used for systems • Careful bacteriologic surveillance of source to help alert authorities to contamination • Care to bury pipes deeply enough (at least 0.5-1.0 meter) to avoid breakage - care to mark path of pipes to avoid breakage during plowing • Careful sealing of all joints • Careful watertight construction of all reservoirs • Strict enforcement of certain hygienic rules around tapsites: no washing of clothes or dishes, babies' bottoms, etc. • Careful maintenance of the entire system for cracks and leaks
Motorized systems or hydraulic rams	• See notes for wells or springs as appropriate • Careful attention to sealing all connections is most important matter

Source: Isely, R.B., 1985, Low-Cost Water Supply and Sanitation Technologies, Community Participation, and Health and Socio-Economic Outcomes: An Analysis of Interrelationships, U.S. Public Health Service, Research Triangle Park, NC 27709

SECTION F. RESEARCH AND EXPENDITURES

TABLE 8–42. WATER RESOURCES RESEARCH CATEGORIES

I. NATURE OF WATER

Category I deals with fundamental research on the water substance.

A. *Properties of water*—Study of the physical and chemical properties of pure water and its thermodynamic behavior in its various states.

B. *Aqueous solutions and suspensions*—Study of the effects of various solutes on the properties of water; surface interactions; colloidal suspensions.

II. WATER CYCLE

Category II covers generally research on the natural processes involving water. It is an essential supporting effort to applied problems in later categories.

A. *General*—Studies involving two or more phases of the water cycle such as hydrologic models; rainfall-runoff relations; surface and groundwater relationships; watershed studies, etc.

B. *Precipitation*—Investigation of spatial and temporal variations of precipitation; physiographic effects; time trends; extremes; probable maximum precipitation; structure of storms, etc.

C. *Snow, ice, and frost*—Studies of the occurrence and thermodynamics of water in the solid state in nature; spatial variations of snow and frost; formation of ice and frost; breakup of river and lake ice; glaciers, permafrost, etc.

D. *Evaporation and transpiration*—Investigation of the process of evaporation from lakes, soil, and snow and of the transpiration process in plants; methods of estimating actual evapotranspiration; energy balance; etc.

E. *Streamflow*—Mechanics of flow in streams; flood routing; bank storage; space and time variations (includes high and low-flow frequency); droughts; floods, etc.

F. *Groundwater*—Study of the mechanics of groundwater movements; multiphase systems; sources of natural recharge; mechanics of flow to wells and drains; subsidence; properties of aquifers; etc.

G. *Water in soils*—Infiltration, movement and storage of water in the zone of aeration, including soil.

H. *Lakes*—Hydrologic, hydrochemical, and thermal regimes of lakes; water level fluctuations; currents and waves.

I. *Water and plants*—Role of plants in hydrologic cycle; water requirements of plants; interception.

J. *Erosion and sedimentation*—Studies of the erosion process; prediction of sediment yield; sedimentation in lakes and reservoirs; stream erosion; sediment transport, etc.

K. *Chemical processes*—Chemical interactions between water and its natural environment; chemistry of precipitation.

L. *Estuarine problems*—Special problems of the estuarine environment; effect of tides on flow and stage; deposition of sediments; sea water intrusion in estuaries.

III. WATER SUPPLY AUGMENTATION AND CONSERVATION

As water use increases we must pay increasing attention to methods for augmenting and conserving available supplies. Research in Category III is largely applied research devoted to this problem area.

A. *Saline water conversion*—Research and development related to methods of desalting sea water and brackish water.

B. *Water yield improvement*—Increasing streamflow or improving its distribution through land management; water harvesting from impervious areas; phreatophyte control; reservoir evaporation suppression.

C. *Use of water of impaired quality*—Research on methods of agricultural use of water of high salinity; use of poor quality water in industry; crop tolerance to salinity.

D. *Conservation in domestic use*—Methods for reducing domestic water needs without impairment of service.

E. *Conservation in industry*—Reduction in both consumption and diversion requirements for industry.

F. *Conservation in agriculture*—More efficient irrigation practices. Chemical control of evaporation and transpiration; lower water use plants, etc.

IV. WATER QUANTITY MANAGEMENT AND CONTROL

Category IV includes research directed to the management of water, exclusive of conservation, and the effects of related activities on water.

A. *Control of water on the land*—Effects of land management on runoff; land drainage, potholes; etc.

B. *Groundwater management*—Artificial recharge; conjunctive operation; relation to irrigation.

C. *Effects of man's related activities on water*—Impact of urbanization, highways, logging, etc., on water yields and flow rates.

D. *Watershed protection*—Methods of controlling erosion to reduce sediment load of streams and conserve soil.

V. WATER QUALITY MANAGEMENT AND PROTECTION

An increasing population increases the wastes and other pollutants entering our water supplies. Category V deals with methods of identifying, describing and controlling this pollution.

A. *Identification of pollutants*—Techniques of identification of physical, chemical and biological pollutants; rational measures of character and strength of wastes.

B. *Sources and fate of pollution*—Determination of the sources of pollutants in water; the nature of the pollution from various sources; path of pollutant from source to stream of groundwater.

C. *Effects of pollution*—Definition of the effect of pollutants, singly and in combination, on man, aquatic life, agriculture and industry under conditions of sustained use; eutrophication.

TABLE 8–42. WATER RESOURCES RESEARCH CATEGORIES (continued)

D. *Waste treatment processes*—Research to improve conventional treatment methods to gain efficiency or reduce cost; processes to treat new types of waste; advanced treatment methods for more complete removal of pollutants including purification for direct reuse.
E. *Ultimate disposal of wastes*—Disposal of residual material removed from water and sewage during the treatment process; disposal of waste brines.
F. *Water treatment*—Development of more efficient and economical methods of making water suitable for domestic or industrial use.
G. *Water quality control*—Research on methods to control stream and reservoir water quality such as flow augmentation; stream and reservoir aeration; control of natural pollution; control of pollution from pesticides and agricultural chemicals; control of acid mine drainage; etc.

VI. WATER RESOURCES PLANNING

The problems of achieving an optimal plan of water development are becoming increasingly complex. Category VI covers research devoted to determining the best way to plan, the appropriate criteria for planning and the nature of the economic, legal and institutional aspects of the planning process.

A. *Techniques of planning*—Application of systems analysis to project planning; treatment of uncertainty; probability studies.
B. *Evaluation process*—Development of methods, concepts and criteria for evaluating project benefits; discount rate; project life; methods for economic, social and technological projections; reliability of projections; research on the value of water in various uses.
C. *Cost allocation, cost sharing, pricing/repayment*—Research on methods of calculating repayment and establishing prices for vendible products; techniques of cost allocation; cost sharing, pricing and repayment policy.
D. *Water demand*—Research on the water quantity and quality requirements of various uses, both diversion and consumption.
E. *Water law and institutions*—A study of state and Federal water law looking to changes and additions which will encourage greater efficiency in use; investigation of institutional structures and constraints which influence decisions on water at all levels of government.
F. *Nonstructural alternatives*—Exploration of methods to achieve water development aims by nonstructural methods such as flood plain zoning.
G. *Ecologic impact of water development*—Effects of water management operations on overall ecology of the area. Excludes effect of pollution under V-C.

VII. RESOURCES DATA

Planning and management of our water resources require information. Category VII includes research oriented to data needs and the most efficient methods of meeting these needs.

A. *Network design*—Studies of data requirements and of the most effective methods of collecting the data.
B. *Data acquisition*—Research on new and improved instruments and techniques for collection of water resources data; telemetering equipment.
C. *Evaluation, processing and publication*—Studies of effective methods of processing data, form and nature of published data; maps of data.

VIII. ENGINEERING WORKS

To implement water development plans requires engineering works. Category VIII describes research on design, materials and construction which is generally useful to all aspects of water management. Works relevant to a single specific goal, such as water treatment or desalination, are included elsewhere if an appropriate category exists.

A. *Design*—Research leading to improved design of dams, canals, pipelines, locks, fishways, and other works required for water resource development.
B. *Materials*—Research to improve existing structural materials and to develop new materials; subsurface exploration of foundations; corrosion; etc.
C. *Construction and operation*—Research on efficient construction methods, operating systems, and maintenance procedures.

IX. MANPOWER, GRANTS AND FACILITIES

Trained manpower is an essential ingredient of research on water resources and the planning and design of water development projects. Category IX describes plans for support of education and training. It also includes grant and contract programs for which advance distribution to specific categories is impossible.

A. *Education*—extramural—Support of education in water resources at universities (not including research support under other categories).
B. *Education*—in-house—Government employee training programs.
C. *Research facilities*—Laboratories, field stations, etc.
D. *Grants, contracts and research act allotments*—Allotments to University Water Resources Research Institutes under P.L. 88-379; OWRR, HEW, NSF, CSRS, and other grants which cannot be distributed to categories in advance.

Source: Committee on Water Resources Research, Federal Council for Science and Technology, 1966

TABLE 8–43. WATER RESOURCES RESEARCH IN THE UNITED STATES—AREAS OF INTEREST IN 1989

GENERAL AREAS OF INTEREST

1. Aspects of the hydrologic cycle.
2. Supply and demand for water.
3. Demineralization of saline and other impaired waters.
4. Conservation and best use of available supplies of water and methods of increasing such supplies.
5. Water reuse.
6. Depletion and degradation of groundwater supplies.
7. Improvements in the productivity of water when used for agricultural, municipal, and commercial purposes.
8. The economic, legal, engineering, social, recreational, biological, geographic, ecological, and other aspects of water problems.
9. Scientific information-dissemination activities, including identifying, assembling, and interpreting the results of scientific and engineering research on water-resources problems.
10. Providing means for improved communications of research results, having due regard for the varying conditions and needs for the respective States and regions.

SPECIFIC AREAS OF INTEREST

A. Groundwater quality

1. An improved understanding of the transport and fate of contaminants in groundwater and in the unsaturated zone, from both field and modeling studies. This will require attention to the physical, microbial, and chemical interactions encountered by dissolved, miscible, and immiscible substances in groundwater transport.
2. Better equipment and methodology for monitoring and characterizing the chemical, physical, and biological condition of the subsurface environment, including sampling strategies that lead to an approved understanding of groundwater contamination at scales larger than single contamination sources.

B. Water quality management.

1. An improved understanding of hazardous substances in water, including interactions of synthetic and natural organics, interactions of waterborne toxic materials with organisms, the binding of chemical species to waterborne sediments, and microbial alteration of organic compounds in aqueous environments.
2. Improved methods for treatment or control of contaminants in natural water systems, emphasizing reduction, removal, or alteration to prevent further contamination or to increase potable supplies.
3. Applications of biotechnology to water resources, including detection of specific pollutants using DNA probes and genetic engineering of microorganisms for water quality management.
4. Use of waters of impaired quality, including improved physical and chemical methods for treatment of impaired waters and the cost-effective utilization of brackish waters.

C. Institutional change in water resource management

1. Analysis of the social aspects and the distributive consequences of alternative institutional arrangements.
2. Description and evaluation of institutions for consideration/compensation of third-party interests; i.e., those individuals or groups affected by, but external to, reallocation decisions.
3. A description of the market structure that has evolved or is likely to evolve if institutional change involves development of water markets.
4. Representative case studies providing estimates of economic efficiency gains and the distributive consequences from institutional changes.

Source: U.S. Geological Survey, 1988, Announcement 7442, Request for Applications Under the Water Resources Research Act of 1984 (Section 105)

TABLE 8–44. EXPENDITURES FOR WATER MANAGEMENT PROJECTS BY THE UNITED STATES CORPS OF ENGINEERS, 1977 TO 1986

Item	1986	1985	1984	1983	1982	1981	1980	1979	1978	1977
APPROPRIATIONS ($ Millions)										
Navigation	299	304	257	577	552	607	485	395	357	261
Flood control total	673	676	647	882	774	772	873	802	895	859
Flood control										
Mississippi River and tributaries[1]	(227)	(225)	(203)	(291)	(177)	(145)	(211)	(158)	(188)	(173)
Multipurpose, including power	93	112	171	221	247	275	538	358	155	464
Beach erosion control	17	12	25	5	7	12	9	12	12	16
Total new work[2]	1,082	1,104	1,127	1,685	1,580	1,666	1,905	1,567	1,719	1,600
Other work[3]	1,658	1,797	1,561	1,734	1,417	1,331	1,296	1,223	1,070	887
TOTAL	2,740	2,901	2,688	3,419	2,997	2,997	3,201	2,790	2,789	2,487

[1] Included in flood control total.
[2] Advance engineering and design, and construction (including major rehabilitation projects). Savings and slippage applied to projects.
[3] Operation and maintenance, surveys, administration and miscellaneous programs and activities.
Source: Department of the Army Annual Report FY86 of the Secretary of the Army on Civil Works Activities, Washington, DC

TABLE 8–45. UNITED STATES GEOLOGICAL SURVEY BUDGET FOR WATER-RESOURCES INVESTIGATIONS, 1981 TO 1986

[By sources of funds; dollars in thousands]

Budget Activity	1981	1982	1983	1984	1985	1986
Water Resources Investigations	194,016	190,096	199,697	220,390	238,131	248,598
Direct program	115,458	108,637	115,096	129,441	133,408	135,152
Reimbursable program	78,558	81,459	84,601	90,949	104,723	113,446
States, counties, and municipalities	45,138	46,938	48,782	52,113	56,500	56,650
Miscellaneous non-Federal sources	2,088	2,679	3,914	3,600	3,327	2,161
Other Federal agencies	31,332	31,842	31,905	35,236	44,896	54,635

TABLE 8–46. UNITED STATES GEOLOGICAL SURVEY BUDGET FOR WATER-RESOURCES INVESTIGATIONS FOR FISCAL YEAR 1986

[By activity and program element; dollars in thousands]

Activity/Subactivity/Program Element	Fiscal Year 1986[1] Enacted
Water Resources Investigations	$134,802
National Water Resources Research and Information System—Federal Program	72,875
Data Collection and Analysis	19,492
National Water Data and Information Access Program	2,090
Coordination of National Water Data Activities	903
Regional Aquifer System Analysis	13,474
Core Program Hydrologic Research	8,082
Improved Instrumentation	1,888
Water Resources Assessment	1,287
Toxic Substances Hydrology	11,182
Nuclear Waste Hydrology	7,020
Acid Rain	2,998
Scientific and Technical Publications	2,081
National Water-Quality Assessment Program	2,378
National Water Resources Research and Information System—Federal-State Cooperative Program	49,774
Data Collection and Analysis, Areal Appraisals, and Special Hydrological Studies	41,956
Water Use	3,670
Coal Hydrology	4,148
National Water Resources Research and Information System—State Research Institute and Research Grants Program	12,153
State Water Resources Research Institutes	6,159
National Water Resources Research Grants Program	4,767
Program Administration	1,227

[1] Funding shown represents appropriated dollars and does not include reimbursable funding from federal, state, and other non-federal sources.
Source: U.S. Geological Survey Yearbook Fiscal Year 1986

FIGURE 8–11. SOURCE OF FUNDS FOR U.S. GEOLOGICAL SURVEY WATER RESOURCES INVESTIGATIONS IN 1986

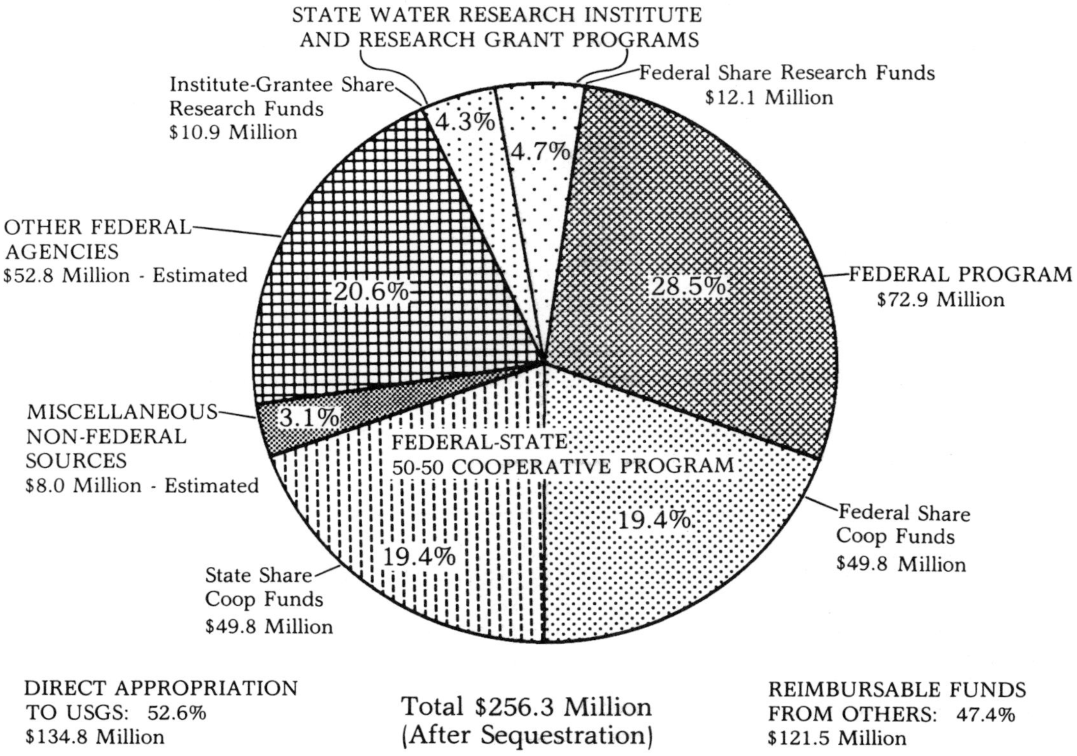

Source: Cardin, G.W., and others, 1986, Water Resources Division in the 1980's, U.S. Geological Survey Circular 1005

TABLE 8–47. WATER POLLUTION ABATEMENT EXPENDITURES IN THE UNITED STATES, 1975 TO 1984

[In millions of dollars]

Year	Total[2]	Industrial		Public Sewer Systems[3]	
		Facilities	Operations[1]	Facilities	Operations[1]
1975	22,840	4,200	2,950	8,997	3,428
1976	24,445	4,625	3,375	9,370	3,713
1977	24,652	4,415	3,706	9,409	4,055
1978	26,631	4,277	3,934	10,090	4,392
1979	26,470	4,013	4,222	9,758	4,583
1980	24,647	3,725	4,081	8,942	4,694
1981	21,984	3,259	4,180	6,882	4,880
1982	21,199	3,080	4,022	6,148	5,156
1983	21,461	2,813	4,402	5,551	5,475
1984, preliminary	23,192	2,902	4,651	6,353	5,773

[1] Operation of facilities.
[2] Includes nonpoint sources not shown separately.
[3] Includes expenditures for private connectors to sewer systems, by owners of animal feedlots, and by government enterprises.
Source: U.S. Department of Commerce, 1987 Statistical Abstract of the United States

TABLE 8-48. PROPOSED COST SHARING FOR NEW WATER PROJECTS AFTER 1983

Project Purpose	Up-front Nonfederal Share of Costs, %
Hydropower	100
Municipal and industrial supply	100
Flood control	35
Recreation	50[1]
Commercial navigation	75[2]
Irrigation	35
Beach erosion	50

[1] Could be repayment instead of up-front payment.
[2] Twenty-five percent of federal financing is reimbursable; the rest must be an up-front cash contribution.
Source: Grigg, N.S., 1985, Water Resources Planning. Copyright 1985 McGraw-Hill, Inc. Reprinted with permission. Original source: S.1031

TABLE 8-49. FEDERAL GROUND-WATER QUALITY RESEARCH AND DEVELOPMENT IN THE UNITED STATES

[The listing is not exhaustive but covers principal programs and activities related to groundwater quality R&D. Examples of other Federal R&D activities omitted here address quantity estimates, use patterns, source inventories, recharge information exchange, socioeconomic effects of alternative supplies, and environmental effect of contamination]

Federal organization	Categories of Groundwater Quality R&D[1]									
	1	2	3	4	5	6	7	8	9	10
National Science Foundation			X		X				X	
Department of Agriculture										
Agricultural Research Services			X			X				
Forest Service			X							
Soil Conservation Service					X	X			X	
Department of Commerce										
National Bureau of Standards	X									
Department of Defense										
Army Corps of Engineers		X			X	X			X	X
Army Medical Bioengineering R&D Laboratory	X									
Army Toxic and Hazardous Materials Agency	X								X	
Department of Energy		X								
Department of Interior										
Bureau of Indian Affairs					X					
Bureau of Land Management					X					
Bureau of Reclamation					X	X				
Fish and Wildlife Service					X					
Geological Survey		X	X	X	X	X	X			
National Park Service				X	X					
Office of Surface Mining					X		X			
Office of Water Policy		X	X					X	X	
Environmental Protection Agency										
Environmental Monitoring Systems Laboratory	X		X							
R.S. Kerr Environmental Research Laboratory			X							
Environmental Research Laboratory			X							
Office of Pesticide Programs			X							
Office of Radiation Programs	X				X					
Office of Research and Development		X	X		X				X	X
Office of Solid Waste					X					
Office of Water								X		
Nuclear Regulatory Commission		X	X							

[1] Key for categories of groundwater research and development:
1 - Standards certification, quality assurance, and water quality criteria.
2 - Hydrogeologic investigations and dynamics of groundwater flow.
3 - Subsurface fate and transport of contaminants.
4 - Background monitoring of groundwater quality.
5 - Detection of groundwater contamination from various sources.
6 - Salt-water intrusion and salinity problems.
7 - Surface water-groundwater interactions.
8 - Control of groundwater contamination from various sources.
9 - Treatment technologies.
10 - Evaluation of alternatives.
Source: Office of Technology Assessment, 1984, Protecting the Nation's Groundwater from Contamination

TABLE 8–50. FEDERAL FUNDS AUTHORIZED FOR IMPLEMENTATION OF THE SAFE DRINKING WATER ACT (AS AMENDED)

	Fiscal Year				
	87	88	89	90	91
Area Funded	Millions of Dollars				
Sole source aquifer demonstration program	$10	$15	$17.5	$17.5	$17.5
Wellhead protection areas	20	20	35	35	35
Grants to alleviate emergency situations	7.65	7.65	8.05	8.05	8.05
Technical assistance and training	35.6	35.6	38.02	38.02	38.02
Technical assistance to small public water systems	10	10	10	10	10
State program grants					
PWS program	37.2	37.2	40.15	40.15	40.15
UIC program	19.7	19.7	20.85	20.85	20.85
Assistance to small systems for monitoring unregulated contaminants	30	—	—	—	—

Source: American Water Resources Association, 1988, New Dimensions in Safe Drinking Water. Copyright AWWA. Reprinted with permission

TABLE 8–51. ANNUAL EXPENDITURES OF THE WATER INDUSTRY IN THE UNITED STATES IN 1984

	Annual Expenditures, U.S. $ Billions
Local water supply operations	12.0
Local wastewater operations	14.0
Local water management, other than above	2.5
Hydropower and thermoelectric cooling	8.5
Industrial pollution control	3.0
Industrial water supply	7.5
Regional water management	8.0
Federal agency operations	3.0
Dams, ports, harbors, and waterways	7.5
Home water treatment	1.0
Bottled water and sparkling water	0.8
Well drilling	3.0
Irrigation	6.0
State government	0.2
Education, research, publishing	0.2
	77.2

Source: Grigg, N.S., 1985, Water Resources Planning. Copyright 1985 McGraw-Hill, Inc. Reprinted with permission

TABLE 8–52. CANADIAN FEDERAL RESOURCES DIRECTED TO WATER-RELATED PROGRAMS BY TYPE OF PROGRAM, FISCAL YEAR 1985–1986

[Expenditures in Canadian dollars]

Program	Agencies	Person Years		Expenditures	
		Number	% of Total	Millions of $	% of Total
1. Research and Information					
Water research	Environment, Fisheries & Oceans	725	20	60	16
Water data collection	Environment, Fisheries & Oceans	750	21	54	14
2. Water Quality					
Water pollution control	Environment, Fisheries & Oceans	630	18	44	12
Drinking water	Agriculture, Indian Affairs and Northern Development, Health & Welfare	10	—	41	11
3. Navigation and Ports					
Canals and channels	Environment, Transport, Public Works	420	12	51	14
Ports and harbors	Fisheries & Oceans, Transport	60	2	36	10
4. Water Supplies					
Irrigation	Agriculture	500	14	38	10
Flood control	Environment, Transport	160	5	18	5
5. International and Interprovincial Agreements	Environment, External Affairs, International Joint Commission	230	6	24	6
6. Administration of Northern Waters	Indian Affairs and Northern Development	55	2	7	2
TOTAL		3,540	100	373	100

Source: Pearse, P.H., and others, 1985, Currents of Change, Final Report on Federal Water Policy, Ottawa, Canada

TABLE 8–53. CANADIAN FEDERAL RESOURCES DIRECTED TO WATER-RELATED PROGRAMS BY DEPARTMENT, FISCAL YEAR 1985–1986

[Expenditures in Canadian dollars]

Agencies	Person Years	Expenditures: Millions of $	Expenditures: % of Total
Environment	1,880	169	45
Fisheries & Oceans	620	62	17
Indian Affairs and Northern Development	60	39	10
Transport and Public Works	270	37	10
Agriculture	545	50	13
Energy, Mines and Resources	80	8	3
External Affairs	25	4	1
International Joint Commission	45	3	1
Other	15	1	—
TOTAL	3,540	373	100

Source: Pearse, P.H. and others, 1985, Currents of Change, Final Report on Federal Water Policy, Ottawa, Canada

TABLE 8–54. WORLD BANK LENDING FOR WATER SUPPLY AND SEWERAGE PROJECTS, 1985–1987

[Millions of United States dollars]

Region	1985		1986		1987		1985–1987	
	$	%	$	%	$	%	$	%
Eastern and Southern Africa	49.00	6.3	9.50	1.6	54.80	5.6	113.30	4.8
Western Africa	101.00	12.9	10.00	1.7	7.00	0.7	118.00	5.0
East Asia & Pacific	175.00	22.4	212.30	35.1	—	—	387.30	16.4
South Asia	—	—	78.00	12.9	284.00	29.3	362.00	15.4
Europe, Middle East & N Africa	292.00	37.4	120.00	19.8	559.60	57.7	971.60	41.3
Latin America & Caribbean	163.80	21.0	175.00	28.9	64.00	6.6	402.80	17.1
Total	780.80	100.0	604.80	100.0	969.40	100.00	2,355.00	100.0

Source: World Water, October 1987, Mediterranean Region Swamps World Bank's 1987 Lending. Thomas Telford Ltd, Box 124 Liverpool L69 2LQ, England

SECTION G. DESALINATION

TABLE 8-55. CLASSIFICATION OF DESALTING PROCESSES BASED ON TYPE OF ENERGY REQUIRED

A. Processes requiring thermal energy
- Multiple effect distillation
- Multiple stage flash distillation
- Solar distillation
- Supercritical distillation

B. Processes requiring mechanical energy
- Vapor compression distillation
- Freeze separation
- Hydrate separation
- Hyperfiltration or reverse osmosis

C. Processes requiring electrical energy
- Electrodialysis

D. Processes requiring chemical energy
- Ion exchange
- Solvent extraction

Source: Howe, University of California, Berkeley, 1968

TABLE 8-56. CLASSIFICATION OF DESALTING PROCESSES BASED ON PROPERTIES OF SOLIDS AND LIQUIDS

A. Processes dependent on phase changes of water

1. Evaporation

   Multiple-effect distillation, in which the latent heat comes from a solid surface.
   Multiple stage flash distillation, in which the latent heat comes from cooling of the liquid being evaporated.
   Supercritical distillation, in which all evaporation occurs above the critical temperature of pure water.
   Solar distillation in which the latent heat is derived from direct solar radiation.
   Vapor compression distillation, in which the latent heat is obtained regeneratively.

2. Crystallization

   Freeze-separation, in which the crystals involved are those of pure water.
   Hydrate-separation, in which the crystals contain molecules of the hydrating agent.

B. Processes dependent on the surface properties of membranes in contact with water.

1. Electrodialysis, in which the unwanted ions are caused to migrate through membranes due to electrical forces.

2. Hyperfiltration or reverse osmosis, in which water is caused to migrate through membranes preferentially to the salt ions, due to pressure.

C. Processes dependent on the properties of solids and liquids in contact with water.

1. Ion exchange, in which unwanted ions are exchanged for less offensive ions loosely bonded to certain double salts in solid form.

2. Solvent extraction, in which certain liquids dissolve water more readily than the salt ions contained in the saline water.

Source: Howe, University of California, Berkeley, 1968

TABLE 8-57. CLASSIFICATION OF WATER DEMINERALIZATION PROCESSES BASED ON THE VARIATION OF ENERGY REQUIREMENT WITH INITIAL SALINITY

Type of Energy	Conversion Process	
Processes in which the energy requirement is essentially independent of initial salinity.	Multiple-effect distillation Multistage flash distillation Vapor compression distillation Supercritical distillation	Vacuum flash distillation Solar distillation Freezing Reverse osmosis
Processes in which the energy requirement depends on initial salinity	Electrodialysis Ion exchange Chemical precipitation	

Source: California Department of Water Resources, 1960

TABLE 8–58. TYPICAL APPLICATIONS OF DESALINATION TECHNIQUES

Technique	Typical Applications			
	Brackish Water			
	0–3,000 ppm	3,000–10,000 ppm	Seawater 35,000 ppm	Higher Salinity Brines
Distillation	T	S	P	P
Electrodialysis	P	S	T	P
Reverse osmosis	P	P	P	S
Ion exchange	P			

KEY: P = Primary application
S = Secondary application
T = Technically possible, but not economic
Source: Office of Technology Assessment, 1987; Office of Technology Assessment, 1988, Using Desalination Technologies for Water Treatment, OTA-BP-0–46

TABLE 8–59. RELATIVE DISTRIBUTION OF DIFFERENT TYPES OF DESALINATION PLANTS WORLDWIDE

Process	Number of Plants	Percent of Total	Capacity (mgd)	Percent of Total
Distillation				
Multi-Stage Flash (MSF)	532	15.1	1,955	64.5
Multiple Effect (ME)	329	9.3	145	4.8
Vapor Compression (VC)	275	7.8	66	2.2
Membrane				
Reverse Osmosis (RO)	1,742	49.4	709	23.4
Electrodialysis (ED)	564	16.0	139	4.6
Other	85	2.4	18	0.6
Total	3,527	100	3,032	100

Source: International Desalination Association desalination plant inventory, 1987; Office of Technology Assessment, 1988, Using Desalination Technologies for Water Treatment, OTA-BP-0–46

FIGURE 8–12. CUMULATIVE WORLDWIDE CAPACITY OF LAND-BASED DESALTING PLANTS
[Plants capable of producing 100 m^3/unit or more of fresh water]

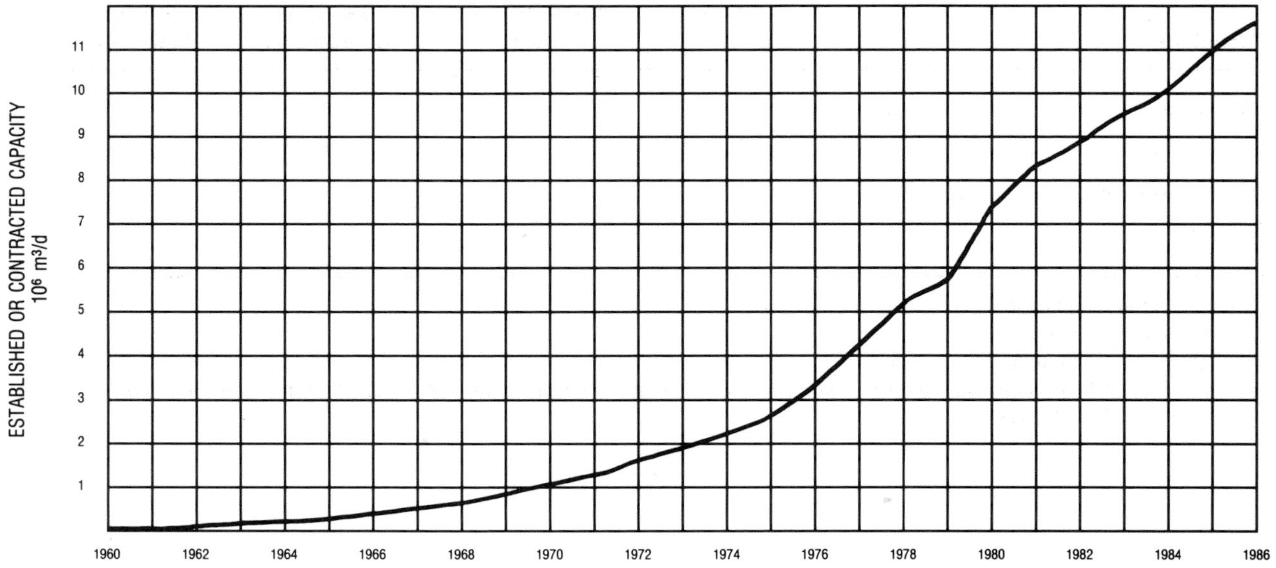

Source: Wangnick, Klaus, 1987, 1986 World Market of Desalting Plants, The IDA Magazine, Vol. 1, no.4. Copyright International Desalting Association. Reproduced with permission

FIGURE 8–13. PROCESSES USED BY LAND-BASED DESALTING PLANTS WORLDWIDE

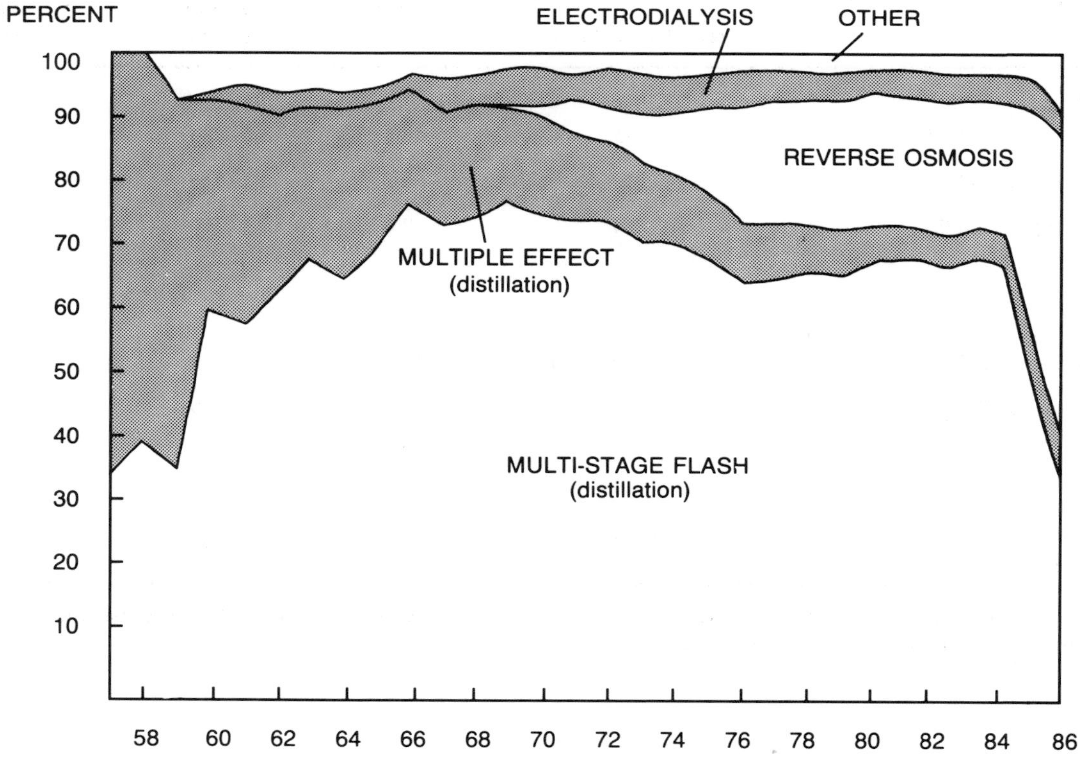

Source: Wangnick, Klaus, 1987, 1986 World Market of Desalting Plants, The IDA Magazine, Vol. 1, no. 4. Copyright International Desalting Association. Reproduced with permission

FIGURE 8–14. WORLDWIDE DISTRIBUTION OF DESALTING PLANT CAPACITY

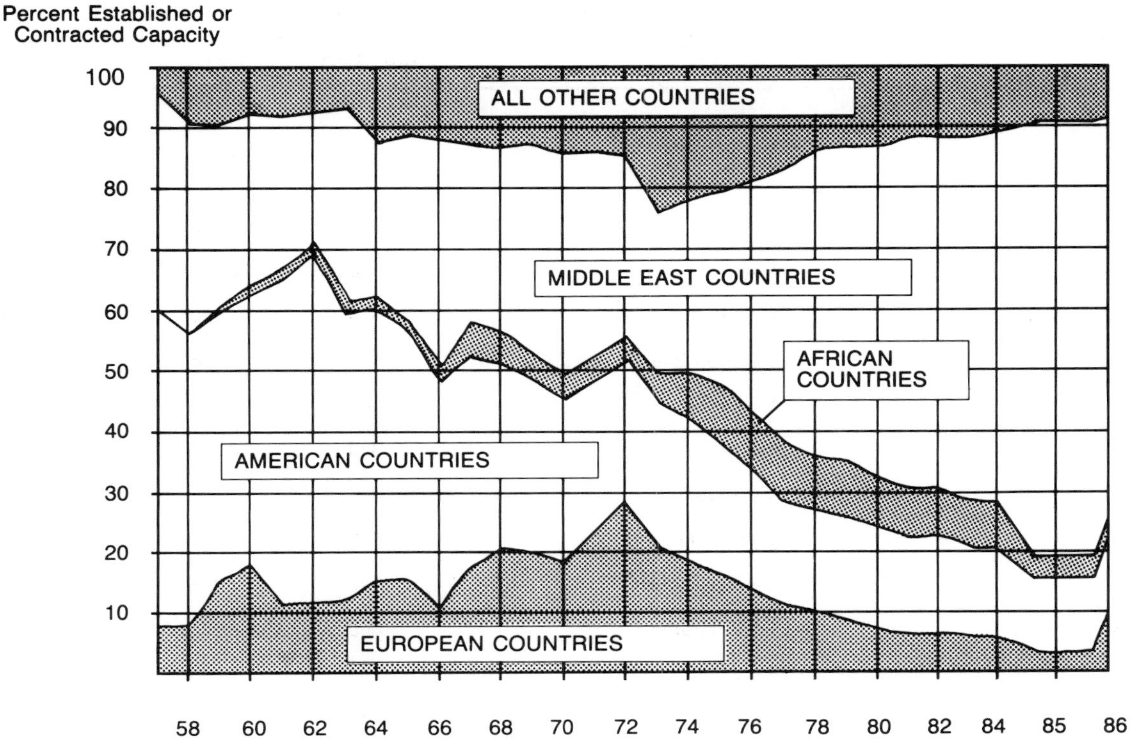

Source: Wangnick, Klaus, 1987, 1986 World Market of Desalting Plants, The IDA Magazine. Vol. 1 , no.4. Copyright International Desalting Association. Reproduced with permission

FIGURE 8–15. INDIVIDUAL CAPACITY OF DESALTING PLANTS WORLDWIDE (in m³/d)

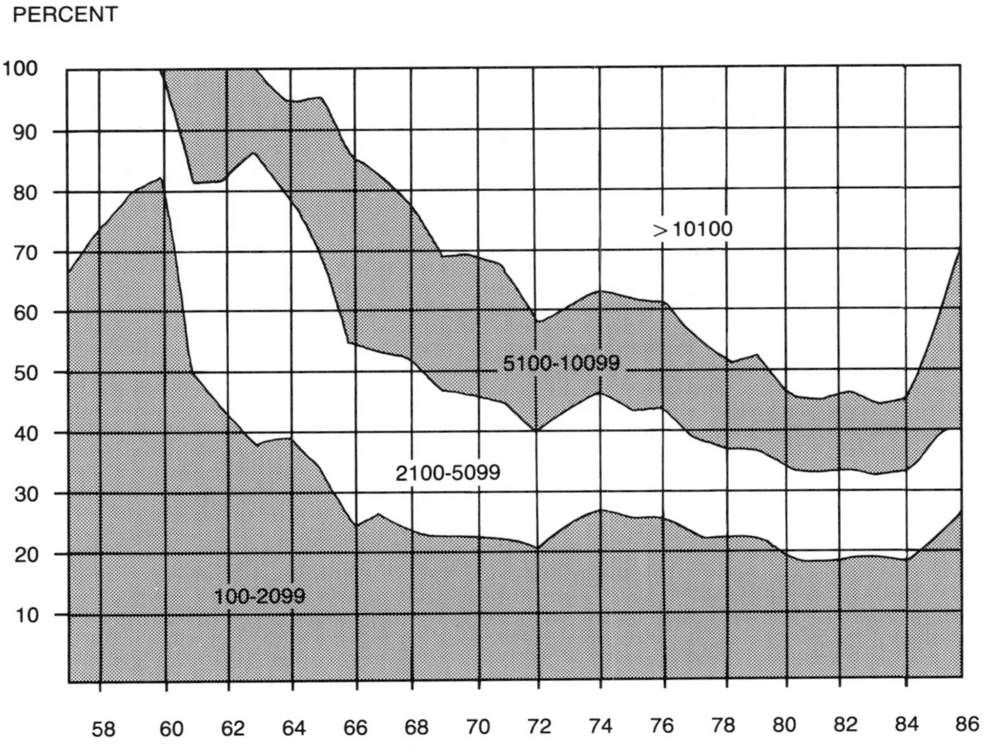

Source: Wangnick, Klaus, 1987, 1986 World Market of Desalting Plants, The IDA Magazine, Vol.1, No. 4. Copyright International Desalting Association. Reproduced with permission

TABLE 8–60. SEAWATER DESALINATION PLANTS USING REVERSE OSMOSIS PERMASEP PERMEATORS

Location	Country	Capacity 1000 GPD	Capacity m³/d	Start-Up
INSTALLED				
Ras Abu-Jarjur[1]	Bahrain	12 000	45 420	84
Ghar Lapsi	Malta	5 300	20 000	83
Ghar Lapsi Expansion	Malta	1 060	4 600	86
Ras Tanajib	Saudi Arabia	3 500	13 248	85
Key West	USA	3 000	11 355	80
Yanbu	Saudi Arabia	1 300	4 920	80
Cadafe I	Venezuela	1 000	3 788	80
Cadafe II	Venezuela	1 000	3 788	83
AlBirk	Saudi Arabia	600	2 271	83
Jeddah Air Defense	Saudi Arabia	600	2 271	83
KISR	Kuwait	264	1 000	84
Paradise Island	Bahamas	240	908	81
Grand Cayman	USA	160	600	86
Cape Verde	Cape Verde Islands	150	568	83
PRASA	Puerto Rico	150	568	82
Jeddah Conference Palace	Saudi Arabia	150	568	84
Magna	Saudi Arabia	150	568	84
Mykonos	Greek Island	132	500	81
Ithaki	Greek Island	132	500	83
Puerto Santo	Madeira Island	132	500	80
Formentera	Spain	132	500	85
Raymond Int.	Ras Al Khaimah, UAE	125	473	78
Dura	Saudi Arabia	100	379	84
Al-Wajh (Coast Guard)	Saudi Arabia	100	379	85
Hakato Cho	Japan	79	300	81
Ras Al Mishab	Saudi Arabia	75	284	80
UNDER CONSTRUCTION				
Al-Dur (SWCC Gift)	Bahrain	12 100	46 000	87/88
Lanzarote	Spain	1 300	5 000	86
Safaniya	Saudi Arabia	1 044	4 000	86
Kutch Lignite[1]	India	1 000	3 788	87

TABLE 8–60. SEAWATER DESALINATION PLANTS USING REVERSE OSMOSIS PERMASEP PERMEATORS (continued)

Location	Country	Capacity 1000 GPD	Capacity m³/d	Start-Up
UNDER CONSTRUCTION (continued)				
St. Maarten	Netherlands	400	1 514	85/86
Codelco	Chile	160	600	86
Mullet Bay, St. Maarten	Netherlands	150	568	85/86
Shipyard	Italy	135	511	85/86
Misurata	Libya	132	500	86
Maho Bay	USA	100	379	86
WED Remote Islands	Abu Dhabi, UAE	4800	18 168	86/87

[1] Highly brackish.
Source: The IDA Magazine, vol.1, no.4, 1987. Copyright International Desalting Association. Reprinted with permission

TABLE 8–61. DESALINATION COSTS IN THE UNITED STATES

	Plant Size (mgd)	Overall Cost (1985 dollars/1,000 gal)
BRACKISH WATER:		
Reverse osmosis	1	1.67
	3	1.41
	5	1.33
	10	1.23
	25	1.21
Electrodialysis (reversing)	1	1.72
	5	1.47
	10	1.37
	25	1.26
SEAWATER:		
Distillation		
Multi-stage flash	1[1]	9.73
	5[1]	6.78
	10[1]	6.50
	25[1]	6.10[2]
Multiple-effect	1	8.31
	5[1]	5.70
	10[1]	5.36
	25[1]	5.36[2]
Reverse osmosis	0.01	13.42
	0.1	9.88
	1	7.40
	3	6.64
	5[1]	6.36
	10[1]	6.03[3]
	25[1]	5.96[3]

[1] Theoretical costs, since no plants of this size are operating in the United States.
[2] Approximated from Reed.
[3] Extrapolated cost
Source: United Nations, "The Use of Nonconventional Water Resources in Developing Countries,"; adopted from Reed, S.A., "Desalting Seawater and Brackish Water: 1981 Cost Update,".

FIGURE 8–16. TRENDS IN COSTS OF DESALINATION

[Desalination costs (including capital and operating costs) for distillation and RO over the last 40 years for plants producing 1 mgd to 5 mgd of "polished" water ready to drink. Costs may be higher than the curves indicate when desalination equipment is not operated efficiently. The increasing distillation costs during the 1970s primarily reflect capital and energy costs]

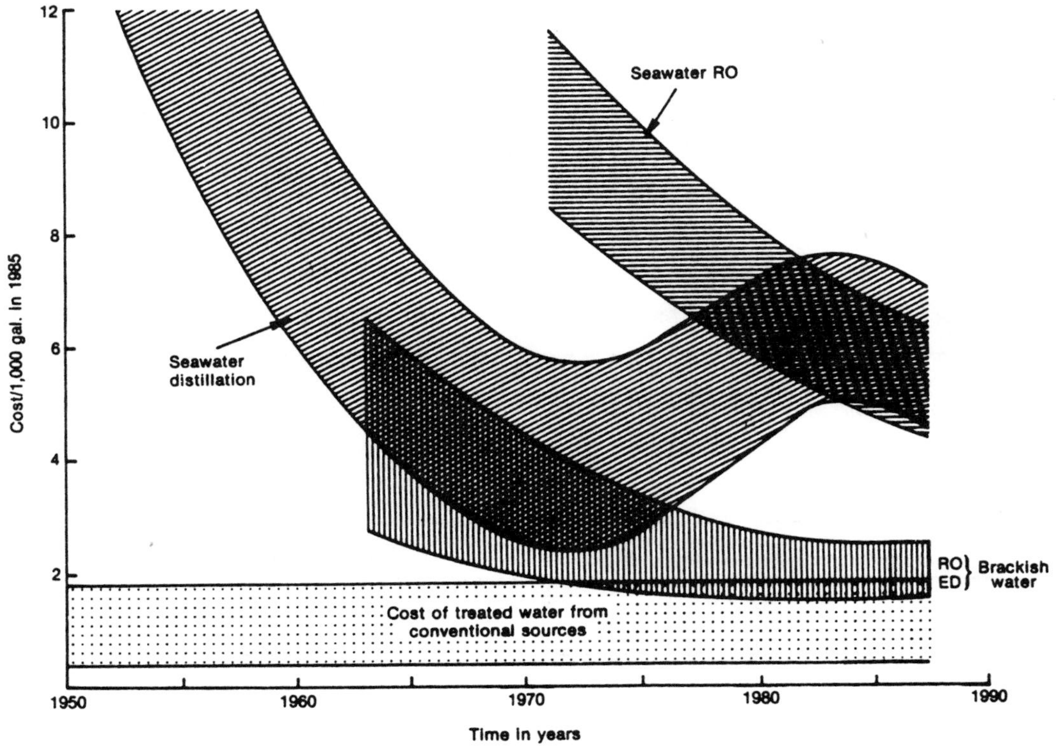

RO Reverse Osmosis
ED Electrodialysis

Source: Office of Technology Assessment, 1988, Using Desalination Technologies for Water Treatment, OTA-BP-O-46

TABLE 8–62. DISCHARGE OF DISSOLVED SOLIDS FROM SURFACE AND GROUND WATER INTO THE OCEANS

Continent	Surface Water; Dissolved Solids Discharge in million t/year	Average Salinity of River Water in g/L	Ground Water; Dissolved Solids Discharge in million t/year	Average Dissolved Solids of Ground Water in g/L
Europe	240	0.077	60	0.4
Asia	850	0.065	296	0.9
Africa	310	0.072	288	1.0
North America	410	0.069	149	0.4
South America	550	0.053	113	0.3
Australia (including Tasmania, New Guinea, New Zealand)	120	0.060	199	0.5
WORLD'S TOTAL	2480	0.063	1045	0.6
Percent of surface water; dissolved solids discharge			42	

Note: Data on surface water discharge of dissolved solids and average river water salinity are according to Lvovich, 1974, "World's Water Resources and Their Future;" data on discharge of dissolved solids in ground water on islands excluded.
Source: Zektser, I.S. and others, 1984, The Effect of Groundwater on the Salt Balance of Seas and Oceans, Water Quality Bulletin, vol. 9, no. 1

SECTION H. WATER TRANSFER

TABLE 8–63. INTERBASIN WATER TRANSFERS IN THE WESTERN UNITED STATES

[Between Water Resources Regions, water years 1978–1982; in acre-feet; for map of Water Resources Regions, see Figure 2–6]

Water Resource Regions Origin	Exported to	1978	1979	1980	1981	1982
Missouri	Upper Mississippi	267	511	754	1,005	1,257
	Arkansas-White-Red	15,980	15,980	17,269	6,327	11,226
	Upper Colorado	NA	NA	NA	NA	NA
	Total	16,247	16,491	18,023	7,332	12,483
Arkansas-White-Red	Missouri	12,669	12,539	9,429	12,519	2,329
Texas-Gulf	Lower Mississippi	—	—	—	32,570	32,430
	Arkansas-White-Red				460	8,940
	Total				33,030	41,370
Rio Grande	Arkansas-White-Red	517	1,460	900	181	1,290
Upper Colorado	Missouri	511,409	354,254	272,346	437,214	437,222
	Arkansas-White-Red	126,114	156,661	128,461	102,357	173,600
	Rio Grande	108,351	167,852	147,439	57,711	132,388
	Great Basin	98,440	116,820	83,330	111,080	77,340
	Total	844,314	795,587	631,576	708,362	820,550
Lower Colorado	Great Basin	3,060	3,420	3,780	4,170	3,420
	California	4,107,730	4,060,690	4,174,130	4,234,260	3,808,170
	Mexico	274,720	253,326	284,510	241,830	255,800
	Total	4,385,510	4,317,436	4,462,420	4,480,260	4,067,390
Great Basin	California	3,130	3,940	3,950	3,740	5,010
Pacific Northwest	Great Basin	900	1,000	1,000	1,300	1,400
California	Pacific Northwest	25,540	21,000	20,900	21,400	23,300
Grand Total		5,288,827	5,169,453	5,148,198	5,268,124	4,975,122

Source: Compiled from Petsch, H. E., Jr., 1985, Inventory of Interbasin Transfers of Water in the Conterminous United States, U.S. Geological Survey Open-File Report 85–166

TABLE 8–64. MAJOR AQUEDUCTS IN CALIFORNIA

	Capacity in Cubic Feet per Second[1]	Length in Miles	Owner	Initial Year of Operation
Los Angeles	710	244	LADWP	1913
Mokelumne River	590	90	EBMUD	1929
Hetch Hetchy	460	152	SF	1934
All American	15,100	80	USBR	1938
Contra Costa	350	48	USBR	1940
Colorado River	1,600	242	MWD	1941
Friant-Kern	4,000	152	USBR	1944
Coachella	2,500	123	USBR	1947
San Diego No. 1	200	71	SD	1947
Delta-Mendota	4,600	116	USBR	1951
Madera	1,000	36	USBR	1952
Putah South	960	35	USBR	1957
Santa Rosa-Sonoma	62	31	SCWA	1959
San Diego No. 2	1,000	93	SD	1960
Corning	500	21	USBR	1960
Petaluma	16	26	SCWA	1961
Tehama-Colusa	2,530	113	USBR	1961[2]
South Bay	360	43	DWR	1965
North Bay	46	27	DWR	1968[3]
California	13,100	444	DWR	1972[4]

TABLE 8–64. MAJOR AQUEDUCTS IN CALIFORNIA (continued)

	Capacity in Cubic Feet per Second[1]	Length in Miles	Owner	Initial Year of Operation
Folsom South	3,500	27	USBR	1973[5]
Cross Valley	740	20	KCWA	1975

[1] Initial reach only for most irrigation canals
[2] Tehama and Glenn Counties
[3] Interim facilities
[4] To Southern California
[5] Reaches 1 and 2
DWR = California Department of Water Resources
EBMUD = East Bay Municipal Utility District
KCWA = Kern County Water Agency
LADWP = Los Angeles Department of Water and Power
MWD = Metropolitan Water District of Southern California
SCWA = Sonoma County Water Agency
SD = City of San Diego
SF = City and County of San Francisco
USBR = U. S. Bureau of Reclamation
Source: California Department of Water Resources, 1987. California Water: Looking to the Future, Bulletin 160–87

TABLE 8–65. APPORTIONMENT OF COLORADO RIVER WATER SUPPLY IN CALIFORNIA

Priority Number	Agency and Description of Service Area	Beneficial Consumptive Use in Acre-Feet per Year: Per California Seven-Party Agreement	Beneficial Consumptive Use in Acre-Feet per Year: After Start of Central Arizona Project
1	Palo Verde Irrigation District		
2	Yuma Project, California portion		
3	Imperial Irrigation District	3,850.000	3,850,000
	Coachella Valley Water District		
	Palo Verde Irrigation District (mesa lands)		
4	Metropolitan Water District	550.000[1]	550,000[1]
5	Metropolitan Water District	662.000	0
6	Imperial Irrigation District		
	Coachella Valley Water District	300,000	0
	Palo Verde Irrigation District (mesa lands)		
	TOTALS	5,362,000[2]	4,400,000[2]

[1] Includes Indian water rights and miscellaneous present perfected rights totalling 58,000 acre-feet that reduce Metropolitan's entitlement to 492,000 acre-feet.
[2] Plus not more than one-half of any excess or surplus water in the lower Colorado River.
Source: California Department of Water Resources, 1987. California Water: Looking to the Future, Bulletin 160–87

TABLE 8–66. MAJOR CANALS ON UNITED STATES BUREAU OF RECLAMATION PROJECTS

[Canals with capacity of 50 m^3/s or greater]

Project, State, and Canal Name	Length, Kilometers	Initial Capacity, m^3s	Initial Reach Width,m	Initial Reach Depth,m	Date Constructed
Boise, ID, Main South					
Side (New York) Canal	64	79.3	21.3	2.4	1906–1908
Columbia Basin, WA					
East Low Canal	132.6	127.4	20.7	4.7	1946–1954
Feeder Canal	2.9	736.3	24.4	7.6	1946–1951
Main Canal	20.8	373.8	15.2	6.3	1946–1951
Potholes East Canal	100.4	110.4	14.6	4.8	1949–1953
West Canal	132.3	144.4	11.6	5.0	1946–1955
Minidoka-Palisades, ID-WY					
Milner-Gooding Canal	112.7	76.5	15.2	3.8	1928–1932
Yakima, WA					
Roza Div. Roza Canal	145.5	62.3	16.8	3.7	–
Central Valley, CA					
Delta Cross Channel	1.9	99.1	64.0	7.9	1950–1951
Folsom South Canal	43.0	99.1	10.4	5.4	1970–1973
Friant-Kern Canal	243.3	141.6	11.0	5.2	1945–1951
O'Neill Forebay Inlet Channel	.8	118.9	24.4	—	1965–1966
San Luis Canal	163.5	371.0	33.5	10.0	1963–1968
Tehama-Colusa Canal	88.8	71.6	7.3	4.8	1965–1980
Klamath, CA-OR					
Lost River Diversion Channel	12.6	85.0	18.3	—	1911–1912
Boulder Canyon, AZ-CA-NV					
All American Canal	128.7	429.1	48.8	6.3	1934–1940
Coachella Canal	199.6	70.8	18.3	3.1	1938–1948
Central Arizona, AZ					
Granite Reef Aqueduct	279.4	85.0	7.3	5.0	1974–1985
Salt Gila Aqueduct	92.4	77.9	7.3	4.8	1981–
Tucson Aqueduct	60.7	63.7	6.7	4.3	1984–
Gila, AZ					
Gila Gravity Main Canal	32.2	62.3	6.7	4.1	1936–1939
Salt River, AZ					
Arizona Canal	62.0	53.8	15.2	2.1	1911–1912
Yuma, AZ-CA					
Yuma Main Canal	7.6	56.6	15.2	2.7	1907–1909
Harlingen Irr. Dist., TX					
Project Canals	47.0	99.1	6.1	1.2	1960
North Platte, NE-WY					
Interstate Canal	152.2	59.5	13.4	3.0	1905–1915
Garrison Division, ND					
McClusky Canal	118.4	55.2	7.6	5.3	1970–1978
Wind Division, WY; Riverton Unit	100.4	62.3	19.8	3.0	1920–1951
Wyoming Canal					

Source: Adapted from U.S. Department of the Interior Bureau of Reclamation, 1986, Statistical Compilation of Engineering Features on Bureau of Reclamation Projects

TABLE 8–67. WATER DIVERSIONS IN AND AFFECTING CANADA, 1985

No.	Jurisdiction	Project	Contributing Basin(s)	Receiving Basin	Average Annual Diversion (cfs)	Uses	Operational Date	Owner
1	British Columbia	Kemano	Nechako (Fraser)	Kemano	4,060	Hydro	1952	Alcan Ltd.
2	British Columbia		Bridge	Seton Lake	3,250	Hydro	(1934)1959	British Columbia Hydro
3	British Columbia		Cheakamus	Squamish	1,300	Hydro	1957	British Columbia Hydro
4	British Columbia		Coquitam Lake	Buntzen Lake	1,000	Hydro	(1902)1912	British Columbia Hydro
5	Saskatchewan		Tazin Lake	Charlot (Lake Athabasca)	1,000	Hydro	1958	Eldorado Nuclear
6	Manitoba	Churchill Diversion	Churchill (Southern Indian Lake)	Rat. Burntwood (Nelson)	27,350	Hydro	1976	Manitoba Hydro
7	Ontario		Lake St. Joseph (Albany)	Root (Winnipeg)	3,020	Hydro	1957	Ontario Hydro
8	Ontario		Ogoki (Albany)	Lake Nipigon (Superior)	3,990	Hydro	1943	Ontario Hydro
9	Ontario		Long Lake (Albany)	Lake Superior	1,480	Hydro/logging	1939	Ontario Hydro
10	Ontario		Little Abitibi (Moose)	Abitibi (Moose)	1,410	Hydro	1963	Ontario Hydro
11	Ontario	Welland	Lake Erie	Lake Ontario	8,830	Hydro navigation	(1829)1951	Government of Canada
12	Quebec	James Bay	Eastmain-Opinaca	La Grande	28,600	Hydro	1980	J.B. Energy Corp.
13	Quebec	James Bay	Fregate	La Grande	1,090	Hydro	1982	J.B. Energy Corp.
14	Quebec	James Bay	Caniapiscau	La Grande		Hydro	1983	J.B. Energy Corp.
15	Newfoundland	Churchill Falls	Julian-Unknown	Churchill	27,400	Hydro	1971	Newfoundland and Labrador Hydro
16	Newfoundland	Churchill Falls	Naskaupi	Churchill	7,060	Hydro	1971	Newfoundland and Labrador Hydro
17	Newfoundland	Churchill Falls	Kanairktok	Churchill	4,590	Hydro	1971	Newfoundland and Labrador Hydro
18	Newfoundland	Bay d'Espoir	Victoria, White Bear, Grey, and Salmon	Northwest Brook (Bay d'Espoir)	6,530	Hydro	1969	Newfoundland and Labrador Hydro
19	Illinois	Chicago diversion	Lake Michigan	Illinois (Mississippi)	3,200 sanitation	Municipal	(1848)1900	Chicago Sanitary Dist.

Source: Quinn, F., Interbasin Water Diversions: A Canadian Perspective. Journal of Soil & Water Conservation, November-December, 1987. Copyright Soil and Water Conservation Society. Reprinted with permission

TABLE 8-68. PROPOSED INTERBASIN WATER TRANSFER SCHEMES FROM CANADA INTO THE UNITED STATES

Proposal (Author)	Year Proposed	Water Source	Annual Diversion (km³)	Estimated Construction Costs (billions of $)
GRAND Canal Plan (Kierans)	1959 to 1983	James Bay diked, water diverted to Great Lakes and United States	347	100
Great Lakes-Pacific Waterways Plan (Decker)	1963	Skeena, Nechako & Fraser of B.C., Peace, Athabasca, Saskatchewan of Prairie Provinces	142	Not Available
North America Water & Power Alliance (NAWAPA) (Parsons)	1964 1964	Primarily the Pacific & Arctic drainage of Alaska, Yukon and B.C.; also tributaries of James Bay	310	100
Magnum Plan (Magnusson)	1965	Peace, Athabasca & N. Saskatchewan in Alberta	31 at border	Not Available
Kuiper Plan (Kuiper)	1967	Peace, Athabasca & N. Saskatchewan in Alberta, Nelson & Churchill in Manitoba	185	50
Central North American Water Project (CeNAWP) (Tinney)	1967	Mackenzie, Peace, Athabasca, N. Saskatchewan, Nelson & Churchill	185	30–50
Western States Water Augmentation Concept (Smith)	1968	Primarily Liard & Mackenzie drainages	49 at border	90
NAWAPA-MUSHEC or Mexican-United States Hydroelectric Commission (Parsons)	1968	NAWAPA sources + lower Mississippi & Sierra Madre Oriental Rivers of Southern Mexico	195 + 159 NAWAPA MUSCHEC	Not Available
North American Waters, A Master Plan (NAWAMP) (Tweed)	1968	Yukon & Mackenzie Rivers, drainage to Hudson Bay	1850	Not Available

Source: Pearse, P.H., Bertrand, F., and MacLaren, J.W., 1985, Currents of Change, Final Report on Federal Water Policy, Ottawa, Canada

TABLE 8-69. COSTS OF SOME WATER TRANSFER PROJECTS

Country	Project	Capacity km³/year	Route Length km	Lift Height m	Approximate Cost billion dollars Total	Per 1 km³	Year of Plan (Construction)
USA, Canada	NAWAPA	100–300	2000	—	100	0.33–1.0	Project of 1964
USA	Beck Plan	12.2	1000	840	3.5	0.29	Project of 1968
USA	Grand Canal	20.0	—	—	5.0	0.25	Project of 1959
USA	Texas Water Plan	20.5	1800	1300	9.0	0.44	Project of 1968
USA	Hudson Institute Plan	41.5	1800	—	12.2	0.29	Project of 1968
USA	California Aqueduct	5.2	300	1100	2.3	0.44	1973
USSR	Irtysh–Central Asia	25	2300	130	19	0.76	Project of 1980
USSR	Sukhona-Volga	4.0	480	6	0.43	0.11	Project of 1980
USSR	Onega R. (upper)-Volga	2.0	390	9	0.22	0.11	Project of 1980
USSR	Pechora-Kama	9.8	300	30	2.3	0.24	Project of 1980
USSR	Onega Bay	38	800	130	8.2	0.22	Project of 1981
India	Rajasthan Canal	16.5	180	—	0.56	0.034	Under construction
India	Sarda Sahayak	15.4	260	—	0.23	0.015	Under construction
Australia	Snowy Mountains	2.4	220	—	0.80	0.33	1974
Israel	Yarkon-Negev	0.23	110	220	0.051	0.22	In operation
Israel	Jordan Waterway	0.32	240	350	0.115	0.36	In operation

Source: Shiklomanov, I.A., 1985, Large Scale Water Transfers. Chapter 12 of Facets of Hydrology II, edited by John C. Rodda.

SECTION I. GROUND WATER

TABLE 8–70. MAJOR COMPONENTS OF AN INFORMATION SYSTEM NEEDED FOR GROUND-WATER MANAGEMENT DECISIONS

- Hydrogeology
  - Soil and unsaturated zone characteristics
  - Aquifer characteristics
    - Depths involved
    - Flow patterns
    - Recharge characteristics
    - Transmissive and storage properties
    - Ambient water quality
    - Interaction with surface water
    - Boundary conditions
    - Mineralogy, including organic content
- Water extraction [withdrawals] and use patterns
  - Locations
  - Amounts
  - Purpose (domestic, industrial, agricultural)
  - Trends
- Potential contamination sources and characteristics
  - Point sources
    - Industrial and mining waste discharges
    - Commercial waste discharges
    - Hazardous material and waste storage
    - Domestic waste discharges
  - Nonpoint sources
    - Agricultural
    - Septic tanks
    - Land applications of waste
    - Urban runoff
    - Transportation spills (can also be considered a point source)
    - Pipelines (energy and waste) (can also be considered a point source)
- Population patterns
  - Demographic
  - Economic trends
  - Land-use patterns

Source: U.S. Geological Survey, National Water Summary 1986; modified from National Research Council, 1986

TABLE 8–71. CRITERIA FOR EFFECTIVE GROUND-WATER PROTECTION PROGRAMS

Criterion	Scope
Goals and objectives	Protection programs should clearly define goals and objectives, reflect understanding of ground-water problems, have adequate legal authority, and have criteria for evaluating program success and the need for modifications.
Information	Programs should be based on information that permits resources and issues to be defined and preventive strategies to be evaluated.
Technical basis	Effective programs require a sound technical basis with which to link actions to results.
Source elimination and control	Long-term program goal should be to eliminate or reduce the sources of ground-water contamination.
Intergovernmental and interagency linkages	Comprehensive protection program must link actions at every level of government into coherent, coordinated action.
Effective implementation and adequate funding	Programs must have adequate legal authority, resources, and stable institutional structures to be effective.
Economic, social, political, and environmental impacts	A preventive program assumes that ground-water protection is the least costly strategy in the long run. Protective actions should be evaluated in terms of their economic, social, political, and environmental effects.
Public support and responsiveness	Programs must be responsive and credible to the public.

Source: U.S. Geological Survey, National Water Summary 1986; modified from National Research Council, 1986

TABLE 8–72. OVERVIEW OF ACTIVITIES INCLUDED IN STATE GROUND-WATER PROTECTION PROGRAMS

Activities	Number of States Initiating	Percent of States Initiating
Ground-Water Mapping/Resource Assessment	51	91
Ground-Water Monitoring Program	46	82
Policy/Strategy Development	38	68
Ground-Water Discharge Permit Program	26	46
Ground-Water Classification System	23	41
Public Information/Education Program	23	41
Ground-Water Standards Development	22	39
Data Management System	14	25
Local Management Plan Development	2	4
Specific Source Control		
Underground Injection Control Program	47	84
Septic Management	38	68
Underground Storage Tank Program	30	54
Hazardous Waste Sites	34	61
Agricultural Contamination	23	41
Landfill Sites	20	36
Surface Impoundments	14	25

Source: U.S. Environmental Protection Agency, National Water Quality Inventory—1986 Report to Congress. EPA–440/4–87–008

FIGURE 8–17. DESIGNATED SOLE-SOURCE AQUIFERS IN THE UNITED STATES

[For explanation and listing of aquifers see Table 8–73]

Source: U.S. Environmental Protection Agency, 1988

TABLE 8–73. DESIGNATED SOLE-SOURCE AQUIFERS IN THE UNITED STATES

[As of July 1988; the Sole Source Aquifer (SSA) program allows individuals and organizations to petition the Environmental Protection Agency (EPA) to designate aquifers as the "sole or principal" source of drinking water for an area. The program was established under Section 1424(e) of the Safe Drinking Water Act (SDWA) of 1974. The primary purpose of the designation is to provide EPA review of Federal financially assisted projects planned for the area to determine their potential for contaminating the aquifer. Based on this review, no commitment of Federal financial assistance may be made for projects "which the Administrator (of EPA) determines may contaminate such (an) aquifer," although Federal funds may be used to modify projects to ensure that they will not contaminate the aquifer; for location of aquifers see Figure 8–17]

USEPA Region	Map Number	Aquifer and/or Location	State	Date Filed	Federal Register Notice	
					Citation	Publication Date
VI	1.	Edwards Aquifer	TX	1/03/75	40 FR 58344	12/16/75
X	2.	Spokane Valley Rathdrum Prairie Aquifer	WA-ID	fall of '76	43 FR 5566	02/09/78
IX	3.	Northern Guam	Guam	11/20/75	43 FR 17868	04/26/78
II	4.	Nassau/Suffolk Counties Long Island	NY	1/21/75	43 FR 26611	06/21/78
IX	5.	Fresno County	CA	8/09/76	44 FR 52751	09/10/79
IV	6.	Biscayne Aquifer	FL	5/08/78	44 FR 58797	10/11/79
II	7.	Buried Valley Aquifer System	NJ	1/15/79	45 FR 30537	05/08/80
III	8.	Maryland Piedmont Aquifer Montgomery, Frederick, Howard, Carroll Counties	MD	09/12/75	45 FR 57165	08/27/80
X	9.	Camano Island Aquifer	WA	4/13/81	47 FR 14779	04/06/82
X	10.	Whidbey Island Aquifer	WA	4/13/81	47 FR 14779	04/06/82
I	11.	Cape Cod Aquifer	MA	3/04/81	47 FR 30282	07/13/82
II	12.	Kings/Queens Counties	NY	6/18/79	49 FR 2950	01/24/84
II	13.	Ridgewood Area	NY/NJ	7/04/79	49 FR 2943	01/24/84
II	14.	Upper Rockaway River Basin Area	NJ	11/30/79	49 FR 2946	01/24/84
IX	15.	Upper Santa Cruz & Avra Altar Basin Aquifers	AZ	6/29/81	49 FR 2948	01/24/84
I	16.	Nantucket Island Aquifer	MA	12/02/82	49 FR 2952	01/24/84
I	17.	Block Island Aquifer	RI	2/18/83	49 FR 2952	01/24/84
II	18.	Schenectady/Niskayuna Schenectady, Saratoga and Albany Counties	NY	8/20/82	50 FR 2022	01/14/85
IX	19.	Santa Marqarita Aquifer Scotts Valley, Santa Cruz County	CA	9/07/77	50 FR 2023	01/14/85
II	20.	Clinton Street-Ballpark Valley, Aquifer System, Broome and Tioga Counties	NY	2/26/81	50 FR2025	01/14/85
III	21.	Seven Valleys Aquifer York County	PA	9/24/81	50 FR 9126	03/06/85

TABLE 8–73. DESIGNATED SOLE-SOURCE AQUIFERS IN THE UNITED STATES (continued)

USEPA Region	Map Number	Aquifer and/or Location	State	Date Filed	Federal Register Notice Citation	Federal Register Notice Publication Date
X	22.	Cross Valley Aquifer	WA	7/29/83	52 FR 18606	05/18/87
III	23.	Prospect Hill Aquifer Clark County	VA	6/27/85	52 FR 21733	06/09/87
V	24.	Pleasant City Aquifer	OH	8/27/84	52 FR 32342	08/27/87
II	25.	Cattaraugus Creek-Sardinia	NY	2/28/85	52 FR 36100	09/25/87
V	26.	Bass Island Aquifer Catawba Island	OH	3/17/86	52 FR 37009	10/02/87
X	27.	Newberg Area Aquifer	WA	1/16/84	52 FR 42474	11/5/87
II	28.	Highlands Aquifer System	NY/NJ	3/14/85	52 FR 37213	10/05/87
X	29.	North Florence Dunal Aquifer	OR	6/02/85	52 FR 37519	10/07/87
IV	30.	Volusia-Florida Aquifer	FL	6/18/82	52 FR 44221	11/18/87
IX	31.	Southern Oahu Basal Aquifer	HI	5/03/83	52 FR 45496	11/30/87
I	32.	Martha's Vineyard Regional Aquifer	MA	6/16/87	53 FR 3451	02/05/88
V	33.	Buried Valley Aquifer System (BVAS)	OH	11/25/87	53 FR 15876	05/04/88
I	34.	Pawcatuck Basin Aquifer System	RI/CT	11/30/87	53 FR 17108	05/13/88
I	35.	Hunt-Annaquatucket-Pettaquamscutt Aquifer	RI	12/30/87	53 FR 19026	05/26/88
VI	36.	Chicot Aquifer	LA	12/05/86	53 FR 20893	06/09/88
VI	37.	Edwards Aquifer Austin Area	TX	08/29/86	53 FR 20897	06/07/88
VIII	38.	Missoula Valley Aquifer	MT	11/23/87	53 FR 20895	06/07/88
II	39.	Cortland-Homer-Preble Aquifer System	NY	9/15/87	53 FR 22045	06/13/88
V	40.	St. Joseph Aquifer System (Elkhart Co)	IN	12/11/87	53 FR 23682	06/23/88
II	41.	N.J. Fifteen Basin Aquifer Systems	NJ/NY	11/18/85	53 FR 23685	06/23/88
II	42.	N.J. Coastal Plain Aquifer System	NJ	12/04/78	53 FR 23791	06/24/88
I	43.	Monhegan Island	ME	5/16/88	53 FR 24496	06/29/88
V	44.	OKI - Miami Buried Valley Aquifer	OH	03/10/88	53 FR 25670	07/08/88
VI	45.	Southern Hills Aquifer System	LA/MS	5/19/80	53 FR 25538	07/07/88
X	46.	Cedar Valley Aquifer	WA	1/88		

Source: U.S. Environmental Protection Agency, 1988

FIGURE 8–18. LOCATIONS OF CURRENT AND POTENTIAL INTERSTATE AND INTERNATIONAL COMPETITION FOR GROUND-WATER RESOURCES.

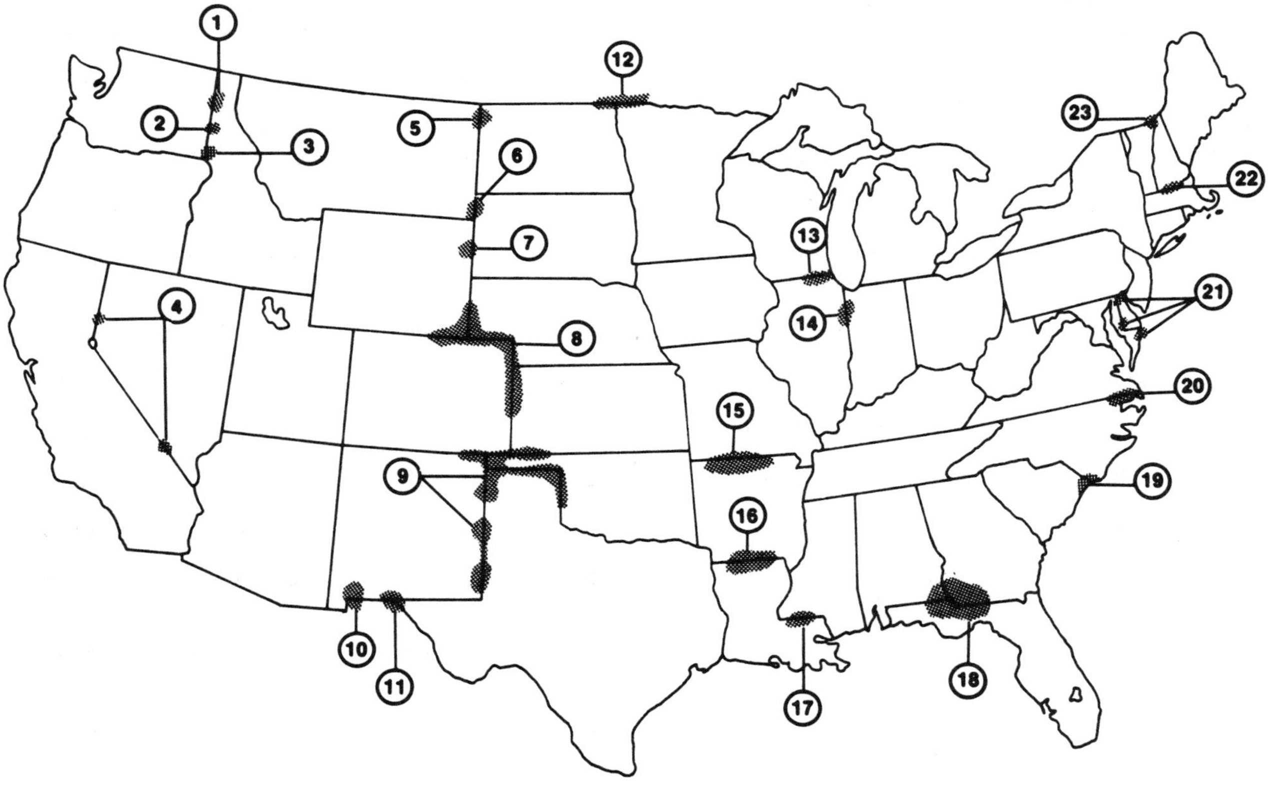

Source: Interstate and International Competition for Water Resources by Zachary A. Smith, Water Resources Bulletin, American Water Resources Association, Oct. 1987

Areas and shared aquifers

1. Rathdrum Prairie-Spokane Valley aquifer.
2. Moscow, ID; University of Idaho; Pullman, WA; Washington State University.
3. Clarkston, WA and Lewiston, ID.
4. Long Valley; Pahrump Valley.
5. Northeast North Dakota.
6. Madison aquifer.
7. Madison and Ogallala aquifers.

8,9. Ogallala aquifer.

10. Mimbres Basin, NM.
11. Mesilla Basin, NM and Hueco Bolson, NM.
12. Several aquifers.
13. Cambro-Ordovician aquifers.
14. Silurian-Devonian aquifers.
15. Gunter-Roubidoux aquifers.
16. Sparta Sand.
17. Southern Hills aquifer.
18. Floridan aquifer.

19, 20. Cretaceous aquifers.

21. Piney Point and Pocomoke-Ocean City-Manokin aquifers.
22. Small valley aquifers.
23. Canaan, VT.

TABLE 8–74. RELATIVE EVALUATION OF THE ELEMENTS OF THE BISCAYNE AQUIFER PROTECTION PLAN (FLORIDA)

Category	Element Number	Description	Ease of Implementation	Effectiveness	Personnel Resources[1]	Cost[2]	Time to Implement[3]	Public Acceptability[4]
Regulation	2	Regulate land use within well field protection zones	Difficult	High	Moderate	Moderate	Moderate	Moderate
Regulation	3	Surveillance over small quantity generators	Moderately difficult	Moderate	Moderate	Low	Short	Moderate
Regulation	4	Regulate activities of small quantity generators	Difficult	High	High	Moderate	Long	Moderate
Regulation	6	Regulate storage tanks	Moderately difficult	High	Moderate	High	Moderate	Moderate
Regulation	7	Control handling/disposal by commercial haulers	Moderately difficult	High	Low	Low	Short	High
Regulation	18	Stormwater/wastewater review	Easy	Moderate	Moderate	Moderate	Short	Moderate
Waste Management	1	Local waste storage/transfer facilities for small generators, individuals	Moderately difficult	High	Low	Moderate	Short	Moderate
Waste Management	5	Public awareness/education program	Easy	High	Moderate	Moderate	Moderate	High
Waste Management	9	Spill prevention, control, and countermeasure program	Moderately difficult	High	Moderate	Moderate	Moderate	High
Waste Management	12	Adopt liability policy	Moderately difficult	High	Low	Low	Moderate	High
Waste Management	13	Emergency spill cleanup program	Difficult	Moderate	Low	High	Moderate	High
Waste Management	15	Control agricultural groundwater pollution	Difficult	Moderate	Moderate	Moderate	Moderate	Moderate
Waste Management	16	Collect/recycle automobile drain oils	Easy	Moderate	Low	Low	Short	High
Construction/Treatment	8	Construct leak-proof sewers	Difficult	Moderate	Low	High	Moderate	High
Construction/Treatment	10	Pretreat commercial/industrial waste	Moderately difficult	Moderate	Moderate	Moderate	Short	High
Construction/Treatment	11	Control sewer exfiltration	Difficult	Moderate	Moderate	Moderate	Long	High
Information Needs	14	Public reporting program on improper disposal	Easy	Moderate	Low	Low	Short	High
Information Needs	17	Tri-county coordinating committee	Easy	Low	Low	Low	Short	High
Information Needs	19	Determine safe soil contamination levels	Difficult	Moderate	Moderate	Moderate	Long	Moderate
Information Needs	20	Monitor groundwater	Moderately difficult	Moderate	Moderate	High	Moderate	High

[1] "Personnel Resources" refers to the manpower needed by each county to develop and implement each program. Low = 1 to 4 full-time persons from each county; Moderate = 5 to 10 full-time persons; High = more than 10 full-time persons.
[2] Includes cost of development and implementation of the program by county agencies as well as costs incurred by the public or the industries.
[3] Short = 0 to 2 years; Moderate = 2 to 5 years; Long = greater than 5 years.
[4] Public acceptability tended to be lower for a specific program if large numbers of people were directly affected.
Source: Sing, U.P., Orban, J.E., and Docal, A.L., 1985, The Biscayne Aquifer Protection Plan, American Water Resources Association Proceedings Tucson AZ Symposium on Groundwater Contamination and Reclamation. Copyright AWWA. Reprinted with permission

FIGURE 8–19. IMPLEMENTATION PHASES OF THE BISCAYNE AQUIFER PROTECTION PLAN

Measure	Implementation phase	Strong program	Complete program
1. Local waste storage/transfer facilities for small generators, individuals	IMMEDIATE IMPLEMENTATION (BASIC PROGRAM)	STRONG PROGRAM	COMPLETE PROGRAM
2. Regulate land use within well field protection zones			
3. Surveillance over small quantity generators			
4. Regulate activities of small quantity generators			
5. Public awareness/education program			
6. Regulate storage tanks			
7. Control handling/disposal by commercial haulers			
8. Construct leak-proof sewers	SHORT-TERM IMPLEMENTATION		
9. Spill prevention, control, and countermeasure program			
10. Pretreat commercial/industrial waste			
11. Control sewer exfiltration			
12. Adopt liability policy			
13. Emergency spill cleanup program			
14. Public reporting program on improper disposal			
15. Control agricultural groundwater pollution	FUTURE IMPLEMENTATION		
16. Collect/recycle automobile drain oils			
17. Tri-county coordinating committee			
18. Stormwater/wastewater review			
19. Determine "safe" soil contamination levels			
20. Monitor groundwater			

Source: Sing, U.P., Orban, J.E., and Docal, A.L., 1985, The Biscayne Aquifer Protection Plan, American Water Resources Association Proceedings, Tucson AZ Symposium on Groundwater Protection and Reclamation. Copyright AWWA. Reprinted with permission

SECTION J. WATER CONSERVATION

TABLE 8–75. WATER CONSERVATION MEASURES

Management	Regulations	Education
Metering	Federal and State Laws and Policies	Direct Mail
Meter Maintenance	Local Codes and Ordinances	pamphlets
Pressure Regulation	plumbing codes for new structures	leaflets
Leak Detection and Repair	retrofitting resolutions	posters
System Rehabilitation	sprinkling ordinances	bill inserts
Pricing (Conservation Oriented)	changes in landscape design	newsletters
marginal cost pricing	reduction in lawn sizes	handbooks
seasonal peak load pricing	increases in impervious area	buttons
uniform unit pricing	planting of low-water using plants	bumper stickers
demand changes	water recycling	News Media
summer surcharge	hook-up moratorium	radio/TV ads
excess-use charge	Restrictions	movie
increasing unit	Rationing by	radio announcements
hook-up fees	fixed allocations	newspaper articles
penalty charges	variable percentage plan	Personal Contact
Economic incentives	per capita use	speaker program
rebates	prior use bases	slide show
tax credits	Determination of priority uses	booths at fairs
subsidies	restrictions on private and public recreational uses	customer assistance
penalties	restrictions on commercial and institutional uses	Special Events
Implementing Water-Saving Devices	banned wasteful practices	school talks
Devices for new construction	car washing	slogan/poster contests
shallow trap toilet	pool filling	posters around town
pressurized tank toilet	landscape irrigation	billboards
vacuum toilet	watering with hand-held hose only	displays
incinerator toilet	scheduled irrigation	reminder items
pressurized flush toilet		decals
wastewater recycling toilet		serving water on request in restaurants
oil flush toilet		county fair exhibit
freeze toilet		
composer toilet		
dual flush toilet		
micropore toilet		
premixed water system		
water recycling system		
compressed air toilet		
Retrofit devices		
water closet inserts		
water dams		
toilet flush adapters		
flush valve toilets		
shower mixing valves		
shower flow-control devices		
air-assisted showerheads		
pressure-reducing valves		
toilet inserts		
faucet aerators		
faucet flow restrictors		
spray taps		
pressure balancing mixing valves		
hot water pipe insulation		
swimming pool covers		
low water-using dishwashers		
low flush toilets		
thermostatic mixing valves		
minimum use showers		
low flow shortheads		
low water-using clothes washers		
Devices for landscape irrigation		
moisture sensors		
hose meters		
sprinkler timers		
Distribution of water conservation kits		
Free distribution and installation of water-saving devices		
Distribution of leak detection kits		
Reuse of water works facility washwater		

Source: U.S. Army Corps of Engineers, Engineer Institute for Water Resources, Analytical Bibliography for Water Supply and Conservation Techniques, Contract Report 82–C07, 1982

TABLE 8–76. WATER CONSERVATION TECHNIQUES

Inspect the plumbing system for leaks.

Install flow control devices in showers.

Turn off all water during vacations or long periods of absence.

Check the frequency with which home water softening equipment regenerates and backwashes. It can use as much as 100 gallons of water each time it does this.

Insulate hot water pipes to avoid having to clear the "hot" line of cold water during use.

Check all faucets, inside and out, for drips. Make repairs promptly. These problems get worse—never better.

Reduce the volume of water in the toilet flush tank with a quart plastic bottle filled with water (bricks lose particles, which can damage the valve).

Never use the toilet as a trash basket for facial tissues, etc. Each flush uses 5 to 7 gallons of water. Items carelessly thrown in could clog the sewage disposal system.

Accumulate a full laundry load before washing, or use a lower water level setting.

Take showers instead of baths.

Turn off shower water while soaping body, lathering hair, and massaging scalp.

Bottle and refrigerate water to avoid running excess water from the lines to get cold water for meals. Shake bottle before serving to incorporate air in the water so that it doesn't taste flat.

To get warm water, turn hot water on first; then add cold water as needed. This is quicker this way and saves water, too.

Wash only full loads of dishes. A dishwasher uses about 9 to 13 gallons to water per cycle.

When washing dishes by hand, use one pan of soapy water for washing and a second pan of hot water for rinsing. Rinsing in a pan requires less water than rinsing under a running faucet.

Use rinse water—"gray water"—saved from bathing or clothes washing to water indoor plants. Do not use soapy water on indoor plants. It could damage them.

Vegetables requiring more water should be grouped together in the garden to make maximum use of water applications.

Mulch shrubs and other plants to retain moisture in the soil longer.
Spread leaves, lawn clippings, chopped bark or cobs, or plastic around the plants. Mulching also controls weeds that complete with garden plants for water.
Mulches should permit water to soak into the soil.

Try "trickle" or "drip" irrigation systems in outdoor gardens. These methods use 25 to 50 percent less water than hose or sprinkler methods. The tube for the trickle system has many tiny holes to water closely spaced plants. The drip system tubing contains holes or openings at strategic places for tomatoes and other plants that are more widely spaced.

Less frequent but heavier lawn watering encourages a deeper root system to withstand dry weather better.

Plan landscaping and gardening to minimize watering requirements.

When building or remodeling, consider:
—Installing smaller than standard bath tubs to save water.
—Locating the water heater near area where hottest water is needed—usually in the kitchen/laundry area.

Source: U.S. Environmental Protection Agency, 1988, The Lake and Reservoir Restoration Guidance Manual, EPA 440/5–88–002

TABLE 8–77. MEANS TO IMPROVE EFFECTIVE USE OF WATER AND ENERGY FOR IRRIGATION

Direct Means	
1. Reduce water applied	More efficient irrigation. More uniform irrigation. Better use of natural rainfall. Reduction of losses. Compromise yield. (Optimize yield ratio)
2. Reduce operating pressures	Use different irrigation devices. Use different irrigation practices.
Indirect Means	
1. Change irrigation practices to allow:	Less tillage Fewer secondary operations Spraying Fertilizing Cultivating
2. Change irrigation practices to prevent:	Need for ripping. Need for land preparation. Loss of fertilizer and nutrients.
3. Substitute irrigation for other more energy consumptive operations.	Environmental control Sprinkling for frost protection vs. wind machines or heaters.
4. Improve yields thereby reducing acreage needed.	Control amount and timing of water application. Irrigation scheduling.
5. Time irrigation to better use limited energy resource.	Peak use avoidance (load management) Irrigation scheduling.

Source: von Bernuth, R.D., 1980, New Concepts in AG Sprinklers, Copyright, The Irrigation Assoc., 1980 Technical Conference Proceedings. Reprinted with permission

TABLE 8–78. HYDROZONES FOR URBAN LANDSCAPE WATER CONSERVATION

[Allocation of different intensities of irrigation to different areas]

	Area with Highest Intensity of Watering	Area with Lowest Intensity of Watering
Level of human contact and activity	Most direct, active, intense	Least direct and intense; most passive
Level of visibility	Highly visible	Least visible
Maximum compatible water use	Most intense water use	Least intense water use
Priority for watering during drought	First priority	Last priority
Typical types of plants	Turf; exotic ornamentals	Native and drought-tolerant trees and shrubs
Examples on golf course	Greens	Rough or fairway
Examples in commercial area	Urban plaza; building entry area	Parking and service screens
Examples in residential community	Play area; building entry area; central common area	Buffer zones; parking screens
Examples in single-family residence	Active areas in rear yard	Side yard; privacy screens

Source: Ferguson, B.K., 1987, Water Conservation Methods in Urban Landscape Irrigation: An Exploratory Overview, Water Resources Bull., v. 23, no. 1. Copyright American Water Resources Assoc. Reprinted with permission

TABLE 8–79. WATER CONSERVING PLANTS, SHRUBS, AND TREES IN THE WESTERN UNITED STATES

Plant	Mature Size	Flower Color	Water Require-ments
FLOWERING PLANTS			
Callistemon citrinus Lemon bottlebrush	25' high 15' wide	O,R	S
Cassia artemisiodes Feathery cassia	3' to 5' high	Y	V
Cistus Rockrose	2' to 6' high 6' wide	V	V
Coreopsis verticillata	2-1/2' high 2' wide	Y	C
Cytisus and Spartium Broom	3' to 8' high 4' to 6' wide	V	V
Escallonia	3' to 15'	V	V
Lantana	1' to 6' high	V	V
Nerium oleander Oleander	8' to 12' high 8' to 12' wide	V	S
Ochna serrulata Mickey Mouse plant	4' to 8' high	Y	V
Plumbago auriculata Cape plumbago	To 6' high 8' to 10' wide	B,W	V
Poinciana gilliesii Bird of paradise bush	10' high and nearly as wide	Y	V
Limonium Perezii Sea Lavender	2' to 4' high	L	V
FOLIAGE PLANTS			
Arbutus unedo Strawberry tree	8' to 35' high and wide	W	V
Dodonaea viscosa Hopseed bush	12' high 6' to 8' wide	—	C
Elaeagnus	Mostly very big shrubs	—	V
Prunus lyoni *P. ilicifolia* *P. caroliniana*	6' to 40' high	W	C
Rhus ovata Sugar bush	2-1/2' to 10' high and wide	P	C
Xylosma congestum	8' to 10' high and often wider	—	V
Yucca	2' to 20' high and more	W	V

Flower Colors:
B-Blue, G-Gray, L-Lavender, O-Orange, P-Pink, R-Red, V-Varies, W-White, Y-Yellow

Plant	Mature Size	Flower Color	Water Require-ments
VINES			
Bougainvillea	Big mound or huge vine	V	V
Campsis Trumpet creeper	Can bury a house or yard	O,R	V
Solanum jasminoides Potato vine	To 30'	W	S
Tecomaria capensis Cape honeysuckle	Bush or fence vine	R	V
Wisteria	100' or more	B,W	V
TREES			
Acacia certain species	Large shrubs small trees	Y	V
Cupressus glabra	20' to 40' high	—	C
Eucalyptus	Shrubs or trees	V	C
Geijera parvifolia	To 30' high 20' high	—	C
Heteromeles arbutifolia Toyon	Shrub to 10' or 25' tree	W	C
Olea europaea Olive	25' to 30'	—	V
Palms	20' to 100'	—	V
Rhus lancea	To 25' spreading	—	V
Robinia (tree forms) Locust	40' to 70' high	P,W	V
GROUND COVER			
Aloe	1' to 18' high	O,R	V
Baccharis pilularis Dwarf coyote bush	8" to 24" mat with 6' spread	—	V
Gazania	6" to 8" high	V	S
Pennisetum setaceum Fountain grass	2' to 4' clumps	G	C
Rosmarinus officinalis Prostrata	Up to 3'	B	V
OTHER			
Ceanothus	Mat, shrubs trees 2' to 14'	B	C
Juniperus Juniper	Ground covers, shrubs, trees	—	S

Water Requirements:
S - Somewhat unthirsty; more than 6 waterings annually
V - Very unthirsty; 4 to 6 waterings annually
C - Completely unthirsty; annual rainfall sufficient

Source: Los Angeles Department of Water and Power, October 1986

TABLE 8–80. OBSERVED HOUSEHOLD WATER SAVINGS

Water Conservation Practice	Observed Water Savings (gal per capita per day)
3-gallon per minute shower head	7.2
0.5-gallon per minute shower head	13.8
3.5-gallon/flush toilet	8.0
0.5-gallon/flush toilet	19.6
Water-efficient dishwasher	1.0
Water-efficient clothes washer	1.7

Note: Savings are compared to conventional nonconserving fixtures and appliances.

Source: U.S. Department of Housing and Urban Development, Office of Policy Development and Research, Building Technology Division, Residential Water Conservation Projects: Summary Report, June 1984

TABLE 8–81. POTENTIAL WATER SAVINGS OF HOUSEHOLD FIXTURES

Water User	Without Conservation	With Conservation
Toilet	5-to-7 gallons per flush	3-to-4 gallons per flush British dual cycle water closets use 2.5 gallons to flush solids and 1.25 gallons to flush liquids. One-gallon toilets have recently become available for about $150. One-half gallon toilets which are aided by compressed air are also available.
Showers	6-to-10 gallons per minute	2-to-3 gallons per minute
Faucets	5 gallons per minute	1.5 gallons per minute
Dishwasher	10-to-14 gallons per load	8.5-to-10 gallons per load
Washing machine	47.5 gallons per load	42 gallons per load

Source: New York State Senate Research Service Task Force on Critical Problems, 1986, Water Conservation, The Hidden Supply, Albany, NY

CHAPTER 9

Water Law and Treaties

TABLE 9–1. POTENTIAL INSTITUTIONAL CONSTRAINTS TO DEVELOPMENT OF A WATER SOURCE

Probably Binding Constraints

- Water is subject to international treaty
- Water is subject to interstate water apportionment compact
- Water is allocated by the U.S. Supreme Court as a result of litigation among states
- Water is subject to Federal or Indian reserved right

Possibly Binding Constraints

- Water is allocated by litigation among persons
- Water is allocated by permit
- Water is allocated by local water district or another local authority

- Amount of water that may be used is limited:
  - by public trust doctrine
  - by instream flow protection requirements
  - by state law
  - by permit
  - by local management authority
  - by prior appropriation(s) that are all for highest beneficial use
  - by Federal navigational servitude
- Place of use of water is limited:
  - by state law
  - by permit
  - by local authority

Constraints Unlikely to be Binding[a]

- Water is subject to prior appropriation (unless for highest beneficial use)
- Water is subject to riparian right
- Physical access to property is restricted:
  - by property rights of other persons limiting rights-of-way for pipes, ditches, conduits, etc.
  - by Federal or State statutes requiring environmental impact assessment or establishing other procedural requirements.

[a] Upon application of simple administrative procedures or market transactions.
Source: U.S. Environmental Protection Agency, 1986, Guidelines for Ground-Water Classification under the EPA Ground-Water Protection Strategy

TABLE 9–2. WATER LAW IN THE UNITED STATES–SUMMARY AND COMPARISON OF THE DOCTRINES OF APPROPRIATION AND RIPARIAN RIGHTS

Appropriation	Riparian
1. Beneficial use, independent of land ownership, is the basis of the water right.	1. Land ownership is the basis of the water right. Water may be used for any reasonable purpose.
2. Priority of use is the basis of allocation between rival claimants. Rights of the appropriators are not equal.	2. Co-sharing equality is the basis of allocation between rival claimants.
3. Rights are to a definite quantity of water.	3. Rights not fixed to a definite quantity of water.
4. Water may be used on nonriparian land.	4. Use of water may be restricted to riparian land.
5. Right may be lost by nonuse or abandonment.	5. Right does not depend on use and is not subject to abandonment.
6. There is no natural flow requirement.	6. There is a qualified right to natural flow in some jurisdictions.

Source: Driscoll, F.G., 1986, Ground Water and Wells. Copyright Johnson Division, Reprinted with permission. Original source: Tank, R.W., 1983, Legal Aspects of Geology. Plenum Press, NY

TABLE 9–3. BASIC LEGAL THEORIES USED TO ALLOCATE RIGHTS TO WITHDRAW PERCOLATING GROUND WATER[a] IN THE UNITED STATES

1. The *English Rule of Capture*, or *Absolute Ownership* rule. Originating in English common-law doctrine, the owners of land overlying a ground-water resource are allowed to withdraw from their wells all the water they wish for whatever purpose they desire. The water withdrawn can be used for any purpose on or off the owner's land. Under this rule, the landowner could even waste the water, thereby injuring a neighbor, but still have no liability under the law.

2. The *American*, or *Reasonable Use, Doctrine*. Under this doctrine, landowners can withdraw ground water to the extent that they must exercise their rights reasonably in relation to the similar rights of others. Furthermore, the owner's use of ground water for offlying land may be unreasonable and, therefore, unlawful if the withdrawals for the offlying land injure a neighbor.

3. The *Restatement of the Law of Torts*. This is a version of the reasonable use doctrine that establishes a process to balance competing uses, whether they are on or off the overlying land. In this interpretation, the landowner who withdraws ground water from the land and uses it for a beneficial purpose is not subject to liability for interference with the use of water by another, unless:

   a. The withdrawal of ground water causes unreasonable harm through lowering of the water table or reduction of confined pressures.

   b. The ground water occurs in a distinct underground stream.

   c. The withdrawal of water has a substantial effect upon a stream, river, or lake.

4. *Correlative Rights*. This interpretation derives from the concept that water users will share the resource during droughts, based on the relative areal extent of the land owned by the competing landowners. If no competition for water exists, then correlative rights are the same as reasonable use.

5. *Appropriation*. In this system, all water is declared to be public and subject to appropriation on the basis of the "first in time, first in right" principle. Control of well use is usually accomplished by permits.

6. *Combination*. Increased ground-water use among competing interests is leading many states to adopt more than one way of handling the legal aspects of resource allocation and protection.

[a] Ground water is treated under two classifications-underground "stream" water and percolating water. Ground water occurring in streams (a rare occurrence except in karstic terrains or lava flows) is treated under the same general rules as those applying to surface streams-the Riparian Doctrine, the Appropriation Doctrine, and a combination of these two.

Source: Adapted from Driscoll, F.G., 1986, Ground Water and Wells. Copyright Johnson Division. Reprinted with permission

TABLE 9–4. GENERAL THEORY OF STATE GROUND-WATER LAW IN THE UNITED STATES

Reasonable Use	Correlative Rights	Absolute Ownership	Appropriation
Alabama	California	Connecticut	Alaska
Arizona		Hawaii	Colorado
Arkansas		Indiana	Florida
Delaware		Louisiana	Idaho
Georgia		Maine	Montana
Illinois		Massachusetts	Nevada
Iowa		Mississippi	New Mexico
Kansas		Ohio	North Dakota
Kentucky		Pennsylvania	Oklahoma
Maryland		Rhode Island	Oregon
Michigan		South Carolina	South Dakota
Minnesota		Texas	Utah
Missouri		Vermont	Washington
Nebraska			Wyoming
New Hampshire			
New Jersey			
New York			
North Carolina			
Tennessee			
Virginia			
West Virginia			
Wisconsin			

This table is subject to change through permit legislation and exemptions.
Source: Driscoll, F.G., 1986, Ground Water and Wells. Copyright Johnson Division. Reprinted with permission. Original source: Smith, R.J., 1980, Current Trends in Ground Water Law; paper presented at the Water Well and Construction Institute, University of Wisconsin Extension, Madison, WI

TABLE 9–5. MAJOR FEDERAL LEGISLATION AUTHORIZING COST SHARING OF WATER RESOURCES PROJECTS IN THE UNITED STATES BY PROJECT PURPOSE

Water Resources Development Purpose	Affected Agency[a]	Authorizing Legislation
Urban flood damage reduction	Corps	Flood Control Act of 1936 (Pub. L. 74–738) Flood Control Act of 1938 (Pub. L. 75–761)
Rural flood damage reduction	SCS	Watershed Protection Act (Pub. L. 83–566)
	Corps	Flood Control Act of 1936 Flood Control Act of 1938 Flood Control Act of 1928 (Pub. L. 70–391)
	Bureau	Small Projects Act (Pub. L. 84–984) Reclamation Projects Act of 1939 (Pub. L. 76–260)
	TVA	TVA Act (Pub. L. 73–017)
Drainage	SCS	Soil Conservation Act (Pub. L. 40–460) Watershed Protection Act
	Corps	Flood Control Act of 1944 (Pub. L. 78–534)
Irrigation	SCS	Soil Conservation Act Watershed Protection Act
	Corps	Flood Control Act of 1944 Reclamation Act of 1902 (Pub. L. 57–161)
	Bureau	Small Projects Act Reclamation Projects Act
Municipal and industrial water supply	SCS	Watershed Protection Act
	Corps	Water Supply Act of 1958 (Pub. L. 85–500)
	Bureau	Small Projects Act Reclamation Projects Act
Streamflow regulation	Corps	Federal Water Pollution Control Act of 1961 (Pub. L. 87–088)
Water quality (point source)	Corps	Federal Water Pollution Control Act of 1972 (Pub. L. 92–500)
Fish and wildlife	SCS	Watershed Protection Act
	Corps	Flood Control Act of 1944 Water Resources Protection Act of 1965 (Pub. L. 89–072) Water Resources Development Act of 1974 (Pub. L. 93–251)
	Bureau	Water Resources Development Act of 1974
Ports and harbors	Corps	Rivers and Harbors Act of 1920 (Pub. L. 66–263)
Inland waterways	Corps	Rivers and Harbors Act of 1920
	TVA	TVA Act
Hydropower	Corps	Flood Control Act of 1944 1937 Bonneville Power Act (Pub. L. 75–329)
	Bureau	Reclamation Projects Act
	TVA	TVA Act
Area redevelopment	Corps	Economic Development Act of 1965 (Pub. L. 89–136)
	TVA	TVA Act

[a] Corps = Corps of Engineers, SCS = Soil Conservation Service, Bureau = Bureau of Reclamation, TVA = Tennessee Valley Authority.

Source: Grigg, N.S., 1985, Water Resources Planning. Copyright 1985 McGraw-Hill, Inc. Reprinted with permission

TABLE 9–6. FEDERAL LAWS RELATED TO THE PROTECTION OF GROUND-WATER QUALITY IN THE UNITED STATES

Statutes
Atomic Energy Act
Clean Water Act
Coastal Zone Management Act
Comprehensive Environmental Response, Compensation, and Liability Act
Federal Insecticide, Fungicide, and Rodenticide Act
Federal Land Policy and Management Act (and associated mining laws)
Hazardous Materials Transportation Act
National Environmental Policy Act
Reclamation Act
Resource Conservation and Recovery Act
Safe Drinking Water Act
Surface Mining Control and Reclamation Act
Toxic Substances Control Act
Uranium Mill Tailings Radiation Control Act
Water Research and Development Act

Source: Office of Technology Assessment, 1984, Protecting the Nation's Groundwater from Contamination

TABLE 9–7. GROUND-WATER MONITORING PROVISIONS OF FEDERAL STATUTES IN THE UNITED STATES

Statute	Monitoring Objectives	Ground-Water Monitoring Provisions
Atomic Energy Act (AEA)	Obtain background water-quality data and evaluate ground-water contamination.	Monitoring for low-level radioactive disposal sites. Facility licenses must specify monitoring requirements for the source.
	Confirm geotechnical and design parameters and ensure that design of geologic repositories accommodates actual field conditions.	Conduct monitoring related to development of geologic repositories.
	Characterize contamination and select and review corrective measures.	DOE may conduct monitoring on remedial actions at storage and disposal facilities for radioactive substances.
Clean Water Act (CWA) — Sections 201 and 405	Evaluate ground-water contamination.	Monitoring for land application of sewage treatment plant by-products.
— Section 208	Characterize contamination and select and review corrective measures.	No requirements established. Some ongoing monitoring of agricultural practices.
Coastal Zone Management Act (CZMA)		No source regulations authorized.
Reclamation Act		No explicit requirements; monitoring may be conducted as part of water-supply development projects.
Resource Conservation and Recovery Act (RCRA)		Monitoring specified for all hazardous wastes land disposal facilities.

TABLE 9–7. GROUND-WATER MONITORING PROVISIONS OF FEDERAL STATUTES IN THE UNITED STATES (continued)

Statute	Monitoring Objectives	Ground-Water Monitoring Provisions
— Subtitle C	Obtain background water-quality data and evaluate ground-water contamination.	*Interim Status* monitoring required until receipt of final permit. Owner/operator can waive monitoring requirements if a qualified geologist/engineer determines low potential for waste migration.
	Obtain background water-quality data, determine ground-water contamination, determine compliance with standards, evaluate corrective action measures.	Fully permitted facilities must have a program for: — detection monitoring — compliance monitoring — corrective action monitoring. Exemptions may be granted in low-risk situations.
— Subtitle D		State solid waste programs may require monitoring. Federal govt. recommends State programs require monitoring.
Safe Drinking Water Act (SDWA) — Part C (Underground Injection Control)	Evaluate ground-water contamination.	Monitoring may be specified for — injection wells used for in-situ or solution mineral mining where injection is into a formation containing < 10,000 mg/L TDS. — wells injecting beneath the deepest underground sources of drinking water.
Surface Mining Control and Reclamation Act (SMCRA)	Obtain background water-quality data and evaluate ground-water contamination.	Monitoring is specified for surface and underground coal mining operations. Monitoring of a particular water-bearing stratum may be waived if it is determined that the stratum is not part of the project's cumulative impact area.
Toxic Substance Control Act (TSCA) — Section 6	Obtain background water-quality data.	Monitoring is required prior to commencement of disposal operations for PCBs.
Uranium Mill Tailings Radiation Control Act (UMTRCA)	Obtain background water-quality data, evaluate ground-water contamination, determine compliance, evaluate corrective action measures.	Requirements, for the most part, are similar to RCRA Subtitle C.
Water Research and Development Act	Obtain background water-quality data, characterize contamination.	Optional monitoring of inactive sites to determine contamination problems and select remedial actions. No provisions for development of source regulations.

Source: Office of Technology Assessment, 1984, Protecting the Nation's Groundwater From Contamination

TABLE 9–8. INTERSTATE COMPACTS ON WATER RESOURCES OF THE UNITED STATES

Compact Reference[a]	Minimum Flow Requirements	Allocation
New England Region Connecticut River Flood Control Compact, 1951, 64 Stat. 45 CT, MA, NH, VT	Flood control	Provides for flood control dams in Massachusetts, Vermont, and New Hampshire, provides for apportionment of tax losses due to Federal acquisition of lands for these dams among the States of Massachusetts, Vermont, New Hampshire, and Connecticut
Merrimack River Flood Control Compact, 1957, 71 Stat. 18 MA, NH	Flood control	Provides for construction of certain dams and reservoirs, for an apportionment if tax and other revenue losses result from U.S. acquisition of land for this purpose, and a procedure whereby other dams may be built by the United States or the States
New England Interstate Water Pollution Control Compact, 1947, 61 Stat. 682 CT, ME, MA, NH, RI, VT	Water quality	Provides for the study and control of pollution of interstate and tidal waters in the States of Connecticut, Maine, Massachusetts, New Hampshire, Rhode Island, and Vermont
New York Harbor (Tri-State) Interstate Sanitation Compact, 1935, 49 Stat. 932 NJ, NY, CT	Water quality	Provides for the regulation and control of water pollution and sewage treatment and discharge in designated areas of Connecticut, New Jersey, and New York
Thames River Flood Control Compact, 1957, 72 Stat. 364 MA, CT	Flood control	Agrees to the construction by the United States of certain dams and reservoirs for flood control in Connecticut and Massachusetts
Mid-Atlantic Region Potomac River Basin Compact, 1939, 54 Stat. 748, as amended, 84 Stat. 856	Water quality	Provides for the formation of the Potomac Valley Conservancy District in the mentioned States and for the study of and the making and recommending of standards for water pollution and sewage disposal within that District
Delaware River Basin Compact, 1961, 75 Stat. 688. Release required determined by adjudication: 283 U.S. 805 (1931) superseded by 347 U.S. 995 (1954) DE, NJ, NY, US	1,750 cfs required at Montague based on adjudication and a basin yield of 0.45 cfs per sq. mi. (from earlier attempts at a compact)	Allocations made by the Commission
Susquehanna River Basin Compact, 1970, 84 Stat. 1509 NY, PA, MD	Releases: by Commission's decision No required releases Correlative rights Possible low flow min. in Oct. of 2,800 cfs to be agreed upon	By permit or by Commission's decision
Great Lakes Region Great Lakes Basin Compact, 1968, 82 Stat. 414 PA, IL, IN, MI, MN, OH, PA, WI	Flood control	In compliance with State release criteria to protect downstream damages and reduce economic losses
Ohio Region Ohio River Valley Water Sanitation Compact, 1939, 54 Stat. 752 IL, IN, KY, NY, OH, PA, TN, WV	Water quality	Provides for the study, control, and regulation of water pollution, sewage treatment, and discharge of Industrial waste into the waters of the Ohio River Valley
Wheeling Creek Watershed Protection and Flood Prevention District Compact, 1967, 81 Stat. 553 PA, WV	Flood control	Creates a watershed protection and flood control district and provides for the construction and maintenance of dams and reservoirs
Tennessee Region Tennessee River Basin Water Pollution Control Compact, 1955, 72 Stat. 823 TN, MS, KY	Water quality	Provides for the study of water pollution problems in the river basin, for the establishment of water quality standards, and for the enforcement of those standards

TABLE 9–8. INTERSTATE COMPACTS ON WATER RESOURCES OF THE UNITED STATES (continued)

Compact Reference[a]	Minimum Flow Requirements	Allocation
Souris-Red-Rainy Region Red River of the North Compact, 1937, 52 Stat. 150 ND, SD, MN	Flood control and water quality	Creates the Tri-State Waters Area and provides for the administration and coordination of all public works on boundary waters in the area
Missouri Region Yellowstone River Compact, 1950, 65 Stat. 663 MT, ND, WY	No minimum	Allocates percentages of flow: Wyoming Above Rock Lake-Clarks Fork 60% Above Yellowstone-Bighorn Rivers 80% Above Yellowstone-Tongue Rivers 40% Above Yellowstone-Powder Rivers 42%
Upper Niobara Basin Compact, 1969, 83 Stat. 86 WY, NB	No minimum	Allocates storage and ground-water development at Anchor Dam
Belle Fourche River Compact, 1943, 58 Stat. 94 SD, WY	No minimum	90% to South Dakota, 10% to Wyoming
South Platte River Compact, 1923, 44 Stat. 195 CO, NE South Platte River Lodgepole Creek	120 cfs at State line, April 1 to October 15	Colorado 100% October 15 to April 1
Big Blue River Compact, 1971, 86 Stat. 193 KS, NE Big Blue River Little Blue River	Minimum at the State-line gauging station: Little Blue / Big Blue May 45 cfs / 45 cfs Jun 45 cfs / 45 cfs Jul 75 cfs / 80 cfs Aug 80 cfs / 90 cfs Sep 60 cfs / 65 cfs	Same as releases
Republican River Compact, 1942, 57 Stat. 86 CO, KS, NE	No minimum Guide check station 132,000 cfs near State line	Detailed allocation by total ac-ft/yr. State totals are: Colorado 54,100 ac-ft/yr Kansas 190,300 ac-ft/yr Nebraska 234,500 ac-ft/yr
Arkansas-White-Red Region Arkansas River Compact, 1948, 63 Stat. 145 CO, KS Arkansas and Purgatoire Rivers	At junction of Arkansas and Purgatoire Rivers below John Martin Dam Summer—April 1 to October 1: To Colorado: 750 cfs (600 when total storage is below 20,000 ac-ft) To Kansas: 500 cfs (400 when total storage is below 20,000 ac-ft) TOTAL: 1,250 cfs (1,000 when total storage is below 20,000 ac-ft) Winter—November 1 to March 31: To Colorado: 100 cfs	Same as releases; when water stored is near exhaustion, Colorado is allotted 100%
Arkansas River Basin Compact, 1965, 80 Stat. 1409 KS, OK Arkansas River, Grand River, Neosho River, Salt Fork River, Verdigris River, Cimarron River	No specified releases—unrestricted use	Allocates reservoir construction Kansas gets 100% of average natural flow plus portion of the new reservoir storage

TABLE 9–8. INTERSTATE COMPACTS ON WATER RESOURCES OF THE UNITED STATES (continued)

Compact Reference[a]	Minimum Flow Requirements	Allocation
Arkansas River Basin Compact, 1970, 87 Stat. 569 AR, OK	No minimum	Arkansas: No more than 50% flow from Spavinaw basin No more than 60% flow from Illinois basin No more than 100% flow from Lee Creek No more than 60% flow from Poteau basin Oklahoma: No more than 60% flow from Arkansas basin
Canadian River Compact, 1950, 66 Stat. 74 NM, OK, TX	No minimum	Allocates reservoir storage. Flow decided by Commission
Texas-Gulf Region Sabine River Compact, 1953, 68 Stat. 690, as amended, 76 Stat. 34 LA, TX	36 cfs at state line	50% to each state
Rio Grande Region Costilla Creek Compact, 1963, 77 Stat. 350 CO, NM	No minimum	Diversions / Release Colorado 50.62 cfs 36.5% New Mexico 89.08 cfs 63.5% New Mexico 13.42 cfs above cost
Rio Grande Compact, 1938, 53 Stat. 785 CO, NM, TX	No minimum; 790,000 ac-ft/yr is given as the "normal release" but seldom has been attained	Allocation by proportion of an upstream index flow
Pecos River Compact, 1948, 63 Stat. 159 NM, TX	Minimum is the "1947 condition"—Use inflow-outflow method to determine delivery to Texas. The method and amounts are presently in dispute. Flow of the Pecos averaged 136,900 ac-ft per year, for 36 years through 1973; but was 43,200 ac-ft in 1972 calendar year. This shows great variability	50% of flood waters to each State, 43% of salvaged water from New Mexico to Texas
Upper Colorado Region Upper Colorado River Basin Compact, 1948, 63 Stat. 31 AZ, CO, NM, UT, WY		Arizona 50,000 ac-ft, and Colorado 51.75%; New Mexico 11.25%; Utah 23.00%; Wyoming 14.00% (after Arizona's allocation and the delivery of the required flow to the lower basin)
Colorado River Compact, 1922, 45 Stat. 1057, 1064 AZ, CA, CO, NV, NM, UT, WY	75,000,000 ac-ft at Lee Ferry, Arizona, in any 10-year period	Nevada, 300,000 ac-ft/yr. Arizona 2,800,000 ac-ft/yr. Upper basin to lower basin, 7,500,000 ac-ft/yr, delivered at Lee Ferry, Arizona. 4,400,000 ac-ft/yr. to California after Central Arizona Project is in use
La Plata River Compact, 1922, 43 Stat. 796 CO, NM	No minimum; releases by index schedule	Unrestricted use when flow is 100 cfs or more at the state line; otherwise, 50% each
Great Basin Region Bear River Compact, 1955, 72 Stat. 38 ID, UT, WY	Over 350 cfs at the Wyoming-Idaho border or a water emergency is declared	Allocation by schedule of percentage of flow when a water emergency has been declared. Storage capacity is allocated
Pacific Northwest Region Snake River Compact, 1949, 64 Stat. 29 ID, WY	No minimum	Idaho 96%; Wyoming 4%
California Region Klamath River Basin Compact, 1957, 71 Stat. 497 OR, CA	No release or diversion requirement	Acreage limitation: California 100,000 acres Oregon 200,000 acres

[a] Compact reference: Water resource region
Compact name, date, and statute reference
States involved; (subregion)
River—Often omitted because of inclusion in compact name

Source: U.S. Water Resources Council, 1978, The Nation's Water Resources 1975–2000, Second National Water Assessment

TABLE 9–9. BOUNDARY WATER TREATIES AND AGREEMENTS BETWEEN THE UNITED STATES AND CANADA

Treaties and Conventions
- Boundary Waters Treaty (1909)
- Lake of the Woods Convention and Protocol (1925)
- Rainy Lake Convention (1940)
- Niagara River Water Diversion Treaty (1950)
- Columbia River Treaty (1961) and Protocol (1964)
- Skagit River Treaty (1984)

Agreements
- St. Lawrence Seaway Project (1952)
- Great Lakes Water Quality Agreement (1972, 1978)

Source: Pearse, P.H., 1985, Currents of Change, Final Report Inquiry on Federal Water Policy, Ottawa, Canada

TABLE 9–10. INTERNATIONAL WATER TREATIES OF THE UNITED STATES

Treaty Reference	Minimum Flow Requirements	Allocation
Great Lakes and Missouri Regions Canadian Boundary Waters Treaty, 1909, 36 Stat. 2448 United States and Canada	No minimum No minimum No minimum Lake Ontario EL 248.0 and 3200 cfs out of Chicago 5100 cfs in at Albany	20,000 cfs N.Y. Niagara River 36,000 cfs Ontario above falls St. Mary River 500 cfs or 75% - Canada Milk River 500 cfs or 75% - U.S. Great Lakes—Level Control Lake St. Francis—Level Control, Water Quality St. Lawrence—Level Control and Regulation of Discharge Lake Memphremagog - Level Control St. Croix River Basin Rainy River - Water Quality
Great Lakes Region Niagara River Water Diversion Treaty, 1950, 1 U.S.T. 695 (TIAS 2130) United States and Canada	Minimum 100,000 cfs over the falls 8AM-10PM April 1 to September 15; 100,000 cfs over the falls 8AM-8PM September 16 to October 31: 50,000 cfs over the falls at any time	Niagara Falls and River All waters above the minimum flows over the falls may be diverted for power production 50% to each country
Souris-Red-Rainy Region Lake of the Woods Convention 1925, 44 Stat. 2108 United States and Canada	Regulates levels of Lake of the Woods between EL 1056 and 1061.25	When level of the lake rises above EL 1061 or below EL 1056 discharge from lake becomes subject to board control No interbasin transfers permitted without agreement
Souris-Red-Rainy Region Rainy Lake Convention 1938, 54 Stat. 1800 United States and Canada	By commission - Kettle Fall Dam, International Falls Dam	By commission
Rio Grande Region Rio Grande Convention, 1906, 34 Stat. 2953 United States and Mexico	Rio Grande at Juarez 60,000 ac-ft/yr at Juarez Jan 0 / Jul 8,180 Feb 1,090 / Aug 4,370 Mar 5,460 / Sep 3,270 Apr 12,000 / Oct 1,090 May 12,000 / Nov 540 Jun 12,000 / Dec 0	Regardless of the amount delivered, U.S. may use all water between Juarez and Ft. Quitman
Rio Grande Convention, 1933, 48 Stat. 1621 United States and Mexico	Flood control	No specific allocation of flow
Rio Grande, Colorado, and Tijuana Treaty, 1944, T.S. 994, 59 Stat. 1219	Rio Grande below Ft. Quitman, Texas Delivery from Mexico to U.S. minimum of 350,000 ac-ft/yr	50% of the unallocated flow of the Rio Grande to each country. Other rivers in the basin are allocated by percentage of flow
Lower Colorado Region	Colorado River at Border Delivery from U.S. to Mexico of minimum of 1,500,000 ac-ft/yr; maximum of 1,700,000 ac-ft/yr	No allocation other than the deliveries as noted
Pacific Northwest Region Columbia River Basin Cooperative Development Treaty, 1961, TIAS 5638 (15 U.S.T. 1555) United States and Canada	Minimum of 2,500 cfs in the Kootenay River at the border (Montana) (after 2020 AD) Mica CK (Canada)—3,000 cfs Arrow Lake (Canada)—5,000 cfs Duncan Lake (Canada)—1,000 cfs.	Allocates storage and hydroelectric power development. Forbids diversions except for consumptive use, except from the Kootenay River
Alaska Region Yukon River Agreement, 1871 United States and Canada	From Canada to Alaska Navigation - Exceptional because of restriction to other uses	Provides Canada with a guarantee to the use of the Yukon River for access to the sea and prohibits construction or actions restricting traffic on the Yukon River (diversions)

Source: U.S. Water Resources Council, The Nation's Water Resources 1975–2000, Second National Water Assessment

CHAPTER 10

Agencies and Organizations

PAGE

SECTION A. FEDERAL AGENCIES

TABLE 10–1. FEDERAL AGENCIES CONCERNED WITH WATER-RELATED MATTERS

FEDERAL INFORMATION CENTERS
Government Services Administration Office of Information Resources Management

Phone numbers of local Federal Information Centers can be found in nearest metropolitan telephone directories. These centers can guide you toward the appropriate federal agency to answer questions.

AGRICULTURAL STABILIZATION AND CONSERVATION SERVICE, P.O. Box 2415, Washington, DC 20013 (202) 447–5237

BUREAU OF RECLAMATION Office of Public Affairs, C and 18th Street, NW, Washington, DC 20240 (202) 343–4662

COAST GUARD, U.S. Coast Guard, Washington, DC 20593–0001 (202) 267–2229

COUNCIL ON ENVIRONMENTAL QUALITY, 722 Jackson Pl., NW, Washington, DC 20503 (202) 395–5750

EXTENSION SERVICE, DEPT. OF AGRICULTURE, Washington, DC 20250 (202) 447–3377

NATIONAL MARINE FISHERIES SERVICE, U.S. Dept. of Commerce, NOAA, Washington, DC 20235 (202) 673–5828

NATIONAL OCEANIC AND ATMOSPHERIC ADMINISTRATION, Rockville, MD 20852 (301) 443–8910

NATIONAL SEA GRANT COLLEGE PROGRAM, 6010 Executive Blvd., Rockville, MD 20852 (301) 443–8923

NATIONAL WEATHER SERVICE, 8060 13th Street, Silver Spring, MD 20910 (301) 427–7622

OCEAN POLLUTION DATA AND INFORMATION NETWORK, National Environmental Satellite, Data and Information Service, NOAA, 1825 Connecticut Ave., NW, Washington, DC 20235

OFFICE OF OCEAN AND COASTAL RESOURCE MANAGEMENT, 1825 Connecticut Ave., NW, Suite 700, Washington, DC 20235 (202) 673–5115

SAINT LAWRENCE SEAWAY DEVELOPMENT CORPORATION, 400 7th St., SW, Washington, DC 20590 (202) 426–3574

SOIL CONSERVATION SERVICE, P.O. Box 2890, Washington, DC 20013 (202) 447–4543

TENNESSEE VALLEY AUTHORITY, 400 West Summit Hill Dr., Knoxville, TN 37902 (615) 632–2101

U.S. ARMY CORPS OF ENGINEERS (See Separate Listing)

U.S. ENVIRONMENTAL PROTECTION AGENCY (See Separate Listing)

UNITED STATES FISH AND WILDLIFE SERVICE, Washington, DC 20240 (202) 343–4131

U.S. GEOLOGICAL SURVEY WATER RESOURCES DIVISION (See Separate Listing)

TABLE 10–2. U.S. ARMY CORPS OF ENGINEERS

Only includes divisions/districts with a civil works program.

The U.S. Army Corps of Engineers is the Federal government's largest water resources development and management agency. The Corps began its water resources (Civil Works) program in 1824 when Congress for the first time appropriated money for improving navigation. Since then, the Corps has been involved in improving river navigation and harbors, reducing flood damage and controlling beach erosion. At projects designed for these missions, the Corps also generates hydropower; supplies water for cities, industry and agriculture; and manages a recreation and natural resources program. The Corps also has a legal mandate to regulate development by others in the Nation's waterways and wetlands, and at sea and lake shores.

Headquarters
United States Army
Corps of Engineers
Washington, D.C. 20314–1000

DIVISIONS AND DISTRICTS:

U.S. ARMY ENGINEER DIVISION, LOWER MISSISSIPPI VALLEY
P.O. Box 80
Vicksburg, Mississippi 39180–0080
U.S. Army Engineer District, **Memphis**
B-202 Clifford Davis Federal Building
Memphis, Tenn. 38103–1894
U.S. Army Engineer District, **New Orleans**
P.O. Box 60267
New Orleans, La. 70160–0267
U.S. Army Engineer District, **St. Louis**
210 Tucker Boulevard, North
St. Louis, Mo. 63101–1986
U.S. Army Engineer District, **Vicksburg**
P.O. Box 60
Vicksburg, Miss. 39180–0060

U.S. ARMY ENGINEER DIVISION, MISSOURI RIVER
P.O. Box 103
Downtown Station
Omaha, Neb. 68101–0103
U.S. Army Engineer District, **Kansas City**
700 Federal Building
601 East 12th Street
Kansas City, Mo. 64106–2896
U.S. Army Engineer District, **Omaha**
Rm. 6014 U.S. Post Office and Court House
215 North 17th Street
Omaha, Neb. 68102–4978

U.S. ARMY ENGINEER DIVISION, NEW ENGLAND
424 Trapelo Road
Waltham, Mass. 02254–9149

U.S. ARMY ENGINEER DIVISION, NORTH ATLANTIC
90 Church Street
New York, N.Y. 10007–9998
U.S. Army Engineer District, **Baltimore**
P.O. Box 1715
Baltimore, Md. 21203–1715
U.S. Army Engineer District, **New York**
26 Federal Plaza
Norfolk, Va. 23510–1096
U.S. Army Engineer District, **Philadelphia**
U.S. Custom House
2nd Chestnut Street
Philadelphia, Penn. 19106–2991

U.S. ARMY ENGINEER DIVISION, NORTH CENTRAL
536 South Clark Street
Chicago, Ill. 60605–1592
U.S. Army Engineer District, **Buffalo**
1776 Niagara Street
Buffalo, N.Y. 14207–3199
U.S. Army Engineer District, **Chicago**
219 South Dearborn Street
Chicago, Ill. 60604–1797
U.S. Army Engineer District, **Detroit**
P.O. Box 1027
Detroit, Mich. 48231–1027
U.S. Army Engineer District, **Rock Island**
Clock Tower Building,
P.O. Box 2004
Rock Island, Ill. 61204–2004
U.S. Army Engineer District, **St. Paul**
1135 U.S. Post Office and Custom House
St. Paul, Minn. 55101–1479

U.S. ARMY ENGINEER DIVISION, NORTH PACIFIC
P.O. Box 2870
Portland, Ore. 97208–2870
U.S. Army Engineer District, **Alaska**
P.O. Box 898
Anchorage, Alaska 99506–0898
U.S. Army Engineer District, **Portland**
P.O. Box 2946
Portland, Ore. 97208–2946
U.S. Army Engineer District, **Seattle**
P.O. Box C-3755
Seattle, Wash. 98124–2255
U.S. Army Engineer District, **Walla Walla**
Building 602
City-County Airport
Walla Walla, Wash. 99362–9265

U.S. ARMY ENGINEER DIVISION, OHIO RIVER
P.O. Box 1159
Cincinnati, Ohio 45201–1159
U.S. Army Engineer District, **Huntington**
502 8th Street
Huntington, W.Va. 25701–2070
U.S. Army Engineer District, **Louisville**
P.O. Box 59
Louisville, Ky. 40201–0059
U.S. Army Engineer District, **Nashville**
P.O. Box 1070
Nashville, Tenn. 37202–1070
U.S. Army Engineer District, **Pittsburgh**
Federal Building
1000 Liberty Avenue
Pittsburgh, Penn. 15222–4186

U.S. ARMY ENGINEER DIVISION, PACIFIC OCEAN
Building 230
Fort Shafter, Hawaii
(Write APO San Francisco 96858–5440)

U.S. ARMY ENGINEER DIVISION, SOUTH ATLANTIC
510 Title Building
30 Pryor Street, S.W.
Atlanta, Ga. 30335–6801
U.S. Army Engineer District, **Charleston**
P.O. Box 919
Charleston, S.C. 29402–0919
U.S. Army Engineer District, **Jacksonville**
P.O. Box 4970
Jacksonville, Fla. 32232–0019
U.S. Army Engineer District, **Mobile**
P.O. Box 2288
Mobile, Ala. 36628–0001
U.S. Army Engineer District, **Savannah**
P.O. Box 889
Savannah, Ga. 31402–0889
U.S. Army Engineer District, **Wilmington**
P.O. Box 1890
Wilmington, N.C. 28402–1890

U.S. ARMY ENGINEER DIVISION, SOUTH PACIFIC
630 Sansome Street
Room 720
San Francisco, Calif. 94111–2206
U.S. Army Engineer District, **Los Angeles**
P.O. Box 2711
Los Angeles, Calif. 90053–2325
U.S. Army Engineer District, **Sacramento**
650 Capitol Mall
Sacramento, Calif. 95814–4794
U.S. Army Engineer District, **San Francisco**
211 Main Street
San Francisco, Calif. 94105–1905

U.S. ARMY ENGINEER DIVISION, SOUTHWESTERN
1114 Commerce Street
Dallas, Texas 75242–0216
U.S. Army Engineer District, **Albuquerque**
P.O. Box 1580
Albuquerque, N.M. 87103–1580
U.S. Army Engineer District, **Galveston**
P.O. Box 1229
Galveston, Texas 77553–1229
U.S. Army Engineer District, **Little Rock**
P.O. Box 867
Little Rock, Ark. 72203–0867
U.S. Army Engineer District, **Tulsa**
P.O. Box 61
Tulsa, Okla. 74121–0061

FIGURE 10–1. ORGANIZATION OF THE U.S. ENVIRONMENTAL PROTECTION AGENCY

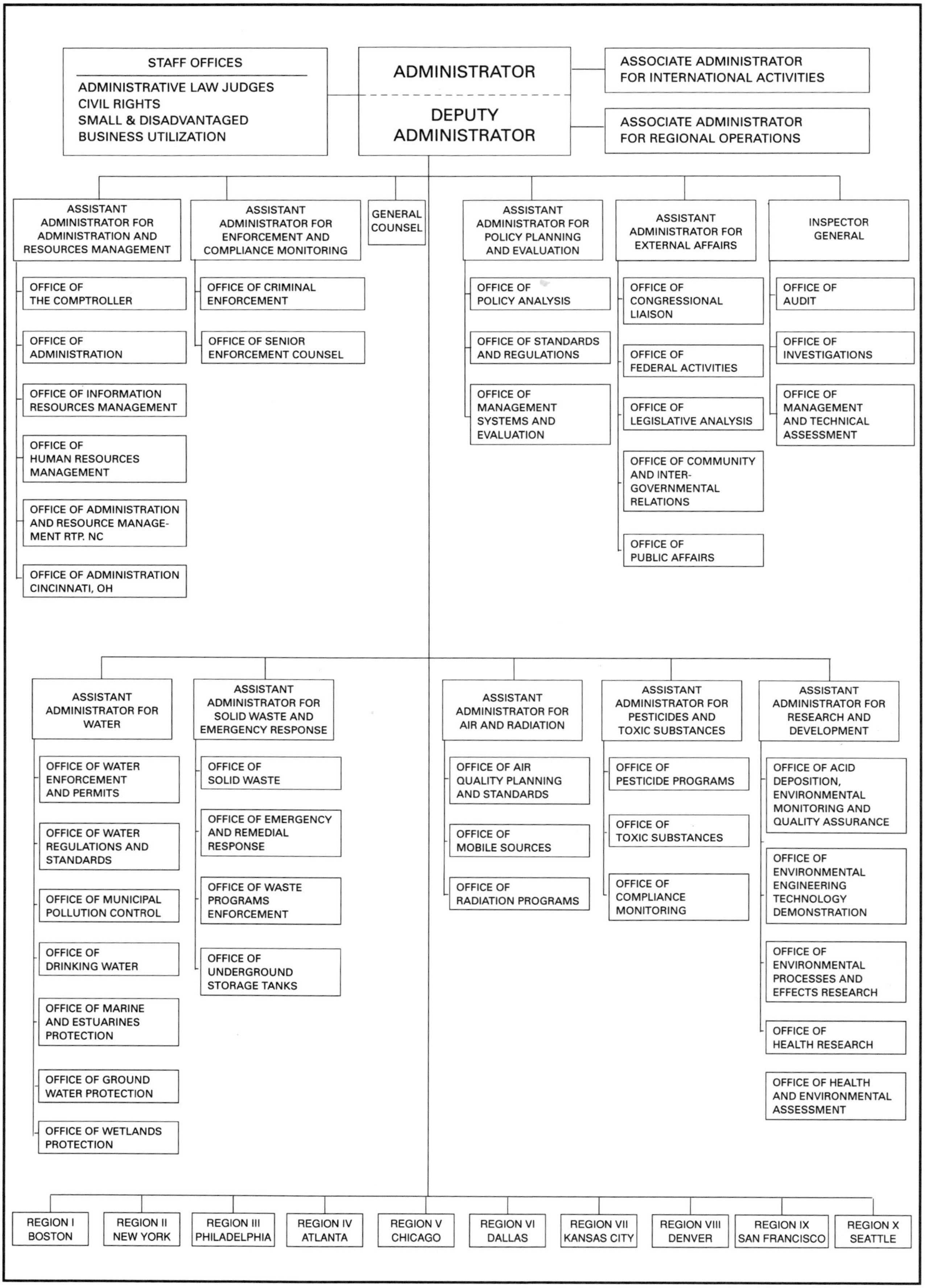

FIGURE 10–2. STATES COVERED BY U.S. ENVIRONMENTAL PROTECTION AGENCY REGIONAL OFFICES

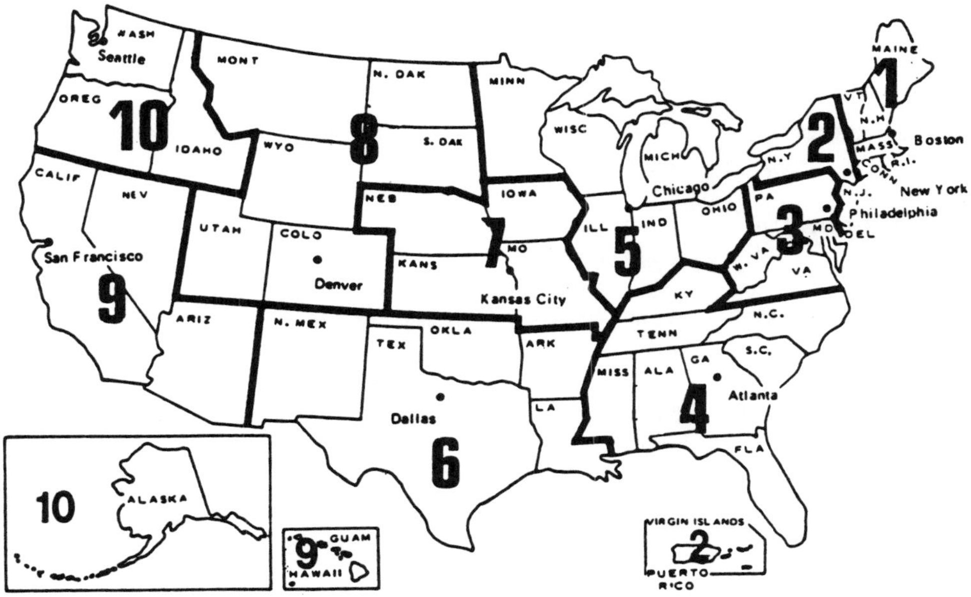

Regions

4 — Alabama
10 — Alaska
9 — Arizona
6 — Arkansas
9 — California
8 — Colorado
1 — Connecticut
3 — Delaware
3 — D.C.
4 — Florida
4 — Georgia
9 — Hawaii
10 — Idaho
5 — Illinois
5 — Indiana
7 — Iowa
7 — Kansas
4 — Kentucky
6 — Louisiana

Regions

1 — Maine
3 — Maryland
1 — Massachusetts
5 — Michigan
5 — Minnesota
4 — Mississippi
7 — Missouri
8 — Montana
7 — Nebraska
9 — Nevada
1 — New Hampshire
2 — New Jersey
6 — New Mexico
2 — New York
4 — North Carolina
8 — North Dakota
5 — Ohio
6 — Oklahoma
10 — Oregon

Regions

3 — Pennsylvania
1 — Rhode Island
4 — South Carolina
8 — South Dakota
4 — Tennessee
6 — Texas
8 — Utah
1 — Vermont
3 — Virginia
10 — Washington
3 — West Virginia
5 — Wisconsin
8 — Wyoming
9 — American Samoa
9 — Guam
2 — Puerto Rico
2 — Virgin Islands

TABLE 10–3. REGIONAL OFFICES OF THE U.S. ENVIRONMENTAL PROTECTION AGENCY

EPA REGION I	John F. Kennedy Federal Bldg., Room 2203, Boston, MA 02203 (617) 835–3424
EPA REGION II	26 Federal Plaza, New York, NY 10278 (212) 264–2515
EPA REGION III	841 Chestnut St., Philadelphia, PA 19107 (215) 597–9370
EPA REGION IV	345 Courtland St., N.E., Atlanta, GA 30365 (404) 257–3004
EPA REGION V	230 South Dearborn, Chicago, IL 60604 (312) 353–2073
EPA REGION VI	1445 Ross Ave., 12th Floor, Suite 1200, Dallas, TX 75202 (214) 255–2200
EPA REGION VII	726 Minnesota Ave., Kansas City, KS 66101 (913) 236–2803
EPA REGION VIII	999 18th St., Suite 500, Denver, CO 80202 (303) 293–1692
EPA REGION IX	215 Fremont St., San Francisco, CA 94105 (415) 974–8071
EPA REGION X	1200 6th Ave., Seattle, WA 98101 (206) 399–1466

TABLE 10–4. EPA LIBRARIES AND INFORMATION CENTERS

The U.S. Environmental Protection Agency (EPA) is comprised of a Headquarters Office in Washington, D.C., 10 Regional offices, and 13 specialized, scientific laboratories located throughout the country. There are 28 EPA network libraries located in Headquarters and all Regional offices and laboratories to support this organizational structure.

The libraries contain a combined collection of over 122,422 books, 5,414 journals, 377,217 hard copy reports, 3,075,443 documents on microfilm and microfiche, 9,000 journal article reprints and 2,000 maps. Most of the EPA library network's holdings are cataloged on OCLC, a national cataloging system.

U.S. EPA
HEADQUARTERS LIBRARY PM-211A
401 M Street, SW
Room 2904 WSM
Washington, DC 20460
(202) 382–5921

U.S. EPA
ENVIRONMENTAL RESEARCH LAB LIBRARY
South Ferry Road
Narragansett, RI 02882
(401) 789–1071

U.S. EPA
REGION 2 FIELD OFFICE LIBRARY
Edison, NJ 08837
(201) 321–6762

U.S. EPA
LAW LIBRARY, LE 130L
401 M Street, SW
Room 2902
Washington, DC 20460
(202) 382–5919

U.S. EPA
OFFICE OF TOXIC SUBSTANCES
Non-Confidential Info Center, TS-793
401 M St., SW, Northeast Mall, Rm B002
Washington, DC 20460
(202) 382–2320

U.S. EPA
LIBRARY SERVICES OFFICE, MD-35
Research Triangle Pk, NC 27711
(919) 541–2777

TABLE 10–4. EPA LIBRARIES AND INFORMATION CENTERS (continued)

U.S. EPA
ENVIRONMENTAL RESEARCH LAB LIBRARY
College Station Rd.
Athens, GA 30613
(404) 546–3324

U.S. EPA
REGION 1 LIBRARY
JFK Building
Boston, MA 02203
(617) 565–3715

U.S. EPA
REGION 2 LIBRARY
Room 402
26 Federal Plaza
New York, NY 10278
(212) 264–2881

U.S. EPA
REGION 3 INFORMATION RESOURCE CENTER
841 Chestnut Street
Philadelphia, PA 19107
(215) 597–0580

U.S. EPA
CENTRAL REGIONAL LABORATORY LIBRARY
839 Bestgate Road
Annapolis, MD 21401
(301) 266–9180

U.S. EPA
REGION 4 LIBRARY
345 Courtland Street, NE
Atlanta, GA 30365–2401
(404) 347–4216

U.S. EPA
ENVIRONMENTAL RESEARCH LAB LIBRARY
Sabine Island
Gulf Breeze, FL 32561
(904) 932–5311

U.S. EPA
OFFICE OF AIR QUALITY PLANNING AND STANDARDS
826 Mutual Plaza, MD-16
Research Triangle Pk, NC 27711
(919) 541–5514

U.S. EPA
ASRL-METEOROLOGY DIVISION LIBRARY (MD-80)
Research Triangle Pk, NC 27711
(919) 541–4536

U.S. EPA
ANDREW W. BRIEDENBACH ENVIRONMENTAL RESEARCH CENTER - LIBRARY
26 W. St. Clair Street
Cincinnati, OH 45268
(513) 569–7707

U.S. EPA
ENVIRONMENTAL RESEARCH LABORATORY LIBRARY
6201 Congdon Blvd.
Duluth, MN 55804
(218) 720–5538

U.S. EPA
ROBERT S. KERR ENVIRONMENTAL RESEARCH LABORATORY LIBRARY
P.O. Box 1198
Ada, OK 74820
(405) 332–8800

U.S. EPA
REGION 8 LIBRARY, 8PM-IML
999 18th Street
Suite 1300
Denver, CO 80202
(303) 293–1444

U.S. EPA
REGION 9 LIBRARY
215 Fremont Street, 6th Floor
San Francisco, CA 94105
(415) 974–8082

U.S. EPA
REGION 10 LIBRARY
1200 Sixth Avenue
Seattle, WA 98101
(206) 442–1289

U.S. EPA
REGION 5 LIBRARY
230 South Dearborn Street
Room 1670
Chicago, IL 60604
(312) 353–2022

U.S. EPA
MOTOR VEHICLE EMISSIONS LABORATORY LIBRARY
2565 Plymouth Road
Ann Arbor, MI 48105
(313) 668–4311

U.S. EPA
REGION 6 LIBRARY
1445 Ross Avenue
Allied Bank Tower
Dallas, TX 75202–2733
(214) 655–6444

U.S. EPA
REGION 7 LIBRARY
726 Minnesota Avenue
Kansas City, KS 66101
(913) 236–2828

U.S. EPA
NAT'L ENFORCEMENT INVESTIGATIONS CENTER
Building 53, Box 25227
Denver Federal Center
Denver, CO 80225
(303) 236–3219

U.S. EPA
ENVIRONMENTAL MONITORING SYSTEMS LABORATORY
944 E. Harmon Avenue
P.O. Box 15027
Las Vegas, NV 89109
(702) 798–2646

U.S. EPA
CORVALLIS ENVIRONMENTAL RESEARCH LABORATORY LIBRARY
200 SW 35th Street
Corvallis, OR 97333
(503) 757–4731

Source: U.S. Environmental Protection Agency, 1987, Guide to EPA Libraries and Information Centers, Report No. EPA/IMSD-87-004

TABLE 10–5. DATABASES USED BY EPA LIBRARIES

BRS (BIBLIOGRAPHIC RETRIEVAL SYSTEM) a commercial vendor of databases, provides access to more than 60 databases in science, technology, business economics, humanities, social sciences and other areas.

CARL (COLORADO ALLIANCE OF RESEARCH LIBRARIES) is an online bibliographic database which contains the holdings of six major Colorado research libraries.

CAS ONLINE, a product of Chemical Abstract Service, contains records for the documents covered since 1967 in the printed version of Chemical Abstracts. Journals, patents, technical reports, books, conference proceedings, and dissertations from all areas of chemistry and chemical engineering are abstracted and the file is updated biweekly.

CDS (COMPLIANCE DATA SYSTEM) is an EPA management information system that stores, sorts, and reports compliance data for approximately 30,000 stationary sources of air pollution that are subject to air quality regulations.

CERCLIS (COMPREHENSIVE ENVIRONMENTAL RESPONSE, COMPENSATION AND LIABILITY INFORMATION SYSTEM), an EPA database, contains an inventory of potential hazardous waste sites and serves as a vehicle for the EPA Regions to report to Headquarters on the status of major stages of cleanup at sites.

CIS (CHEMICAL INFORMATION SYSTEM) is a commercially available collection of scientific and regulatory databases. It contains numeric, textual, and some bibliographic information in the areas of toxicology, environment, regulations, spectroscopy, chemical and physical properties, and nucleotide sequencies.

CONSENT DECREE SYSTEM, an EPA/National Enforcement Investigations Center-operated and maintained system consisting of a hardcopy library of consent decrees to which EPA is a party, a computerized inventory of that library, and computerized summaries of the contents of decrees by facility.

DIALOG INFORMATION RETRIEVAL SERVICE, a commercial vendor of databases, has more than 200 databases available through its system. The type of information varies with the database accessed; database records may be bibliographic citations, abstracts, directory listing or statistical tables. Some of the DIALOG databases commonly used by EPA libraries are Pollution Abstracts, Enviroline, Chemical Regulations and Guidelines, Social SciSearch, and The National Technical Information Service (NTIS).

DUN AND BRADSTREET, a credit-reporting firm, provides business information reports for privately-and publicly-owned companies and government activity reports which list Federal contracts, grants, fines and debarments for specific companies.

FINDS (FACILITIES INDEX SYSTEM), an EPA database, provides descriptive information on facilities/establishments which are of interest to EPA programs, with a cross reference to EPA programmatic information systems which have information on these facilities/establishments.

GROUND WATER ON-LINE, a bibliographic database, contains references on the occurrence and utilization of surface and ground water, and on water well technology. Indexed titles include trade and technical journals and newsletters, books, and government documents. Because EPA established the Center, there is special emphasis in the database on EPA-sponsored reports.

HAZARDLINE, a product of the Occupational Health Service, Inc., contains data on hazardous chemicals, including chemical name, formula, synonyms, and Chemical Abstracts Service Registry Number.

HAZARDOUS WASTE COLLECTION DATABASE, an EPA system, contains references to key materials on hazardous waste in the EPA library network. Bibliographic descriptions, keywords, abstracts, locations and other information are listed for books, EPA reports, Office of Solid Waste and Emergency Response (OSWER) policy and guidance directives, periodicals and commercial databases containing information on hazardous waste. Hard copies of the documents are available in the Headquarters and Regional libraries, the National Enforcement Investigations Center/Denver, CO, and the laboratory libraries in Cincinnati, OH; Edison, NJ; Research Triangle Park, NC; Ada, OK; and Las Vegas, NV. A list of the documents is available at all other laboratory libraries.

HWDMS (HAZARDOUS WASTE DATA MANAGEMENT SYSTEM) is an EPA database which includes information on the status of responses to EPA regulations by the 5,000 facilities that treat, store or dispose of hazardous waste and 60,000 handlers who generate or transport hazardous waste.

IRIS (INTEGRATED RISK INFORMATION SYSTEM) is an E-Mail based system containing EPA data on a chemical by chemical basis. Data include, when available, oral reference doses, carcinogencity assessments, acute health hazard data and risk management summaries.

TABLE 10–5. DATABASES USED BY EPA LIBRARIES (continued)

INFORMATION SYSTEMS INVENTORY (ISI) is an EPA database listing over 500 EPA information systems and environmental models. The database includes a description of each environmental measurement system/model and can be accessed at several points, including system name, acronym, associated Congressional act or law, number of users and responsible personnel. Headquarters Library staff can search this inventory to obtain information for the purpose of referring requestors to the appropriate office within EPA.

JURIS (JUSTICE RETRIEVAL AND INQUIRY SYSTEM), is a computerized full-text legal information retrieval system designed and maintained by the Department of Justice (DOJ) for use by the Federal legal community. The database contains Federal and State case materials; Federal statutes and regulatory materials; administrative decisions; Presidential documents; treaties; DOJ briefs; Shepard's Citations; and special litigation support files.

LEXIS is the largest full-text legal database in existence, containing Federal and State cases, administrative decisions, U.S. Code, CFR, Federal Register, selected State statutes, law reviews, ABA materials, and separate libraries dealing with Federal issues: Admiralty; Banking; Bankruptcy; Energy; Environment; Federal Communications; Federal Securities; Federal Tax; International Trade; Labor; Patent, Trademark and Copyright; Public Contracts; Trade Regulation. Auto-Cite and Shepard's Citations are also available.

NATIONAL LIBRARY OF MEDICINE (NLM) ON-LINE SERVICES database consists of bibliographic citations on biomedical literature, including health care services and administrative, clinical, and policy topics. Special files on population, cancer, and toxic substances are also available on the NLM database.

NATIONAL PESTICIDE INFORMATION RETRIEVAL SYSTEM, developed by Purdue University, contains information that describes the key characteristics of pesticides, including approximately 50,000 products registered by EPA, as well as State registrations.

NEWSNET, is a full-text database of newsletters, press releases, and wire services. Subject areas in NewsNet include the environment, chemicals, government and regulation, public relations, farming and food, research and development, and electronics and computers.

NEXIS is a full-text database that contains general and business news. NEXIS sources include magazines, newspapers, wire services, newsletters, and government documents.

OCLC (ONLINE CATALOGING LIBRARY CENTER) supports resource sharing among more than 2,700 member libraries. Participants use the OCLC Cataloging module to catalog books, serials, and various other library materials and the Interlibrary Loan module to facilitate interlibrary lending.

ORBIT is an online service which provides access to more than 70 databases in science, technology, business, economics, humanities, social sciences and other areas.

PCS (PERMIT COMPLIANCE SYSTEM) is an EPA management information system for tracking permit, compliance, and enforcement status for the National Pollutant Discharge Elimination System program under the Clean Water Act.

SFFAS (SUPERFUND FINANCIAL ASSESSMENT SYSTEM), is an EPA computer application designed to calculate the remedial costs a responsible party can theoretically afford to pay for cleanup of a site.

STORET (STORAGE AND RETRIEVAL OF WATER QUALITY INFORMATION), an EPA database, provides a capability to store, retrieve and analyze water quality information.

TOXLINE, a product of the National Library of Medicine, contains citations and abstracts to the worldwide literature in all areas of toxicology, including chemicals and pharmaceuticals, pesticides, environmental pollutants, and mutagens and teratology.

TOXNET is a computerized system of toxicologically oriented data banks operated by the National Library of Medicine.

VU-TEXT contains the full text of 19 daily newspapers, including nationally recognized papers such as the Boston Globe and Chicago Tribune and regional papers such as the Orlando Sentinel.

WATSTORE (NATIONAL WATER DATA STORAGE AND RETRIEVAL SYSTEM), a product of the U.S. Geological Survey, contains data collected as a result of measuring and quantifying the occurrence and quality of U.S. water resources and the effect of development and utilization on those resources.

WESTLAW is a full-text database containing statutory and case law, citator services, administrative materials and legal literature indexes.

Source: U.S. Environmental Protection Agency, 1987, Guide to EPA Libraries and Information Centers, Report No. EPA/IMSD-87–004

TABLE 10–6. GROUND-WATER ACTIVITIES OF U.S. ENVIRONMENTAL PROTECTION AGENCY OFFICES

Office	Description of Activities
OFFICE OF ENFORCEMENT AND COMPLIANCE MONITORING (OECM)	Enforces provisions of all EPA-administered laws.
OFFICE OF PESTICIDES AND TOXIC SUBSTANCES (OPTS)	
- **OFFICE OF PESTICIDES PROGRAMS (OPP)**	Regulates the use of pesticides to protect the environment and human health. Restricts use of pesticides that have significant potential to leach to ground water. Participates in special surveys such as the one on pesticides in drinking water and leads the development of the agricultural chemicals in the ground-water strategy.
- **OFFICE OF TOXIC SUBSTANCES (OTS)**	Regulates the use in commerce of toxic chemicals. Examines ways to more effectively use provisions of TSCA to protect ground water from toxic substances. Participates, for instance, in the development of the agricultural chemicals in the ground-water strategy.
OFFICE OF POLICY, PLANNING AND EVALUATION (OPPE)	Undertakes special projects to support EPA's policymaking. For example, it looks at comparative risk from various sources of ground-water contamination, attempts to delineate the dimension of the problem from some unaddressed sources, and provides policy and economic expertise to other offices.
OFFICE OF RESEARCH AND DEVELOPMENT (ORD)	Conducts a wide variety of research projects to enhance EPA's understanding of the movement of ground water and contaminants, to improve source/contaminant detection, and to improve source controls.
OFFICE OF SOLID WASTE AND EMERGENCY RESPONSE (OSWER)	
- **OFFICE OF EMERGENCY AND REMEDIAL RESPONSE (OERR)**	Implements the Superfund program which encompasses both emergency response and longer term remedial action to releases or threatened releases of hazardous substances that pose an actual or potential threat to human health or the environment, including ground water.
- **OFFICE OF SOLID WASTE (OSW)**	Regulates a variety of waste management sources of contamination such as landfills, surface impoundments, land application, waste piles under the RCRA. Sets requirements for transporters, generators and treatment, storage and disposal facilities, including ground-water monitoring requirements for facilities.
- **OFFICE OF UNDERGROUND STORAGE TANKS (OUST)**	Conducts the new program to prevent ground-water contamination from leaky underground storage tanks.
- **OFFICE OF WASTE PROGRAMS ENFORCEMENT (OWPE)**	Enforces the Superfund and RCRA programs. Inspects RCRA-regulated facilities for compliance with regulatory requirements including those for ground-water monitoring. Ascertains that private parties are fulfilling administrative orders for Superfund-mandated cleanups.
OFFICE OF WATER (OW)	
- **OFFICE OF DRINKING WATER (ODW)**	Administers the Underground Injection Control Program, which is designed to protect underground sources of drinking water from contamination from unsound injection wells.
- **OFFICE OF DRINKING WATER (ODW)**	Implements the public water supply program under the SDWA, setting and enforcing primary and secondary drinking water standards and water supply monitoring requirements.

TABLE 10–6. GROUND-WATER ACTIVITIES OF U.S. ENVIRONMENTAL PROTECTION AGENCY OFFICES (continued)

Office	Description of Activities
- **OFFICE OF GROUND-WATER PROTECTION (OGWP)**	Carries out the goals of the Ground-Water Protection Strategy, which are: • Building ground-water institutions at the state level • Assessing the problems from unaddressed sources of contamination • Issuing guidelines for EPA decisions affecting ground-water protection and cleanup • Strengthening EPA's organization for ground-water management at the Headquarters and Regional levels and strengthening EPA's cooperation with federal and state governments. Implements two state grant ground-water protection programs established by the SDWAA. These are the Sole Source Aquifer Demonstration Program and the Wellhead Protection Program. Implements grants to states under the Clean Water Act to develop state ground-water protection strategies and adopt measures to control nonpoint sources.
OFFICE OF AIR AND RADIATION (OAR)	
- **OFFICE OF RADIATION PROGRAMS (ORP)**	Participates in investigations of radionuclides and radon in the nation's ground waters and in standard-setting to protect health.

Source: U.S. Environmental Protection Agency, 1987, Improved Protection of Water Resources from Long-Term and Cumulative Pollution: Prevention of Ground-Water Contamination in the United States

FIGURE 10–3. ORGANIZATION OF THE U.S. GEOLOGICAL SURVEY

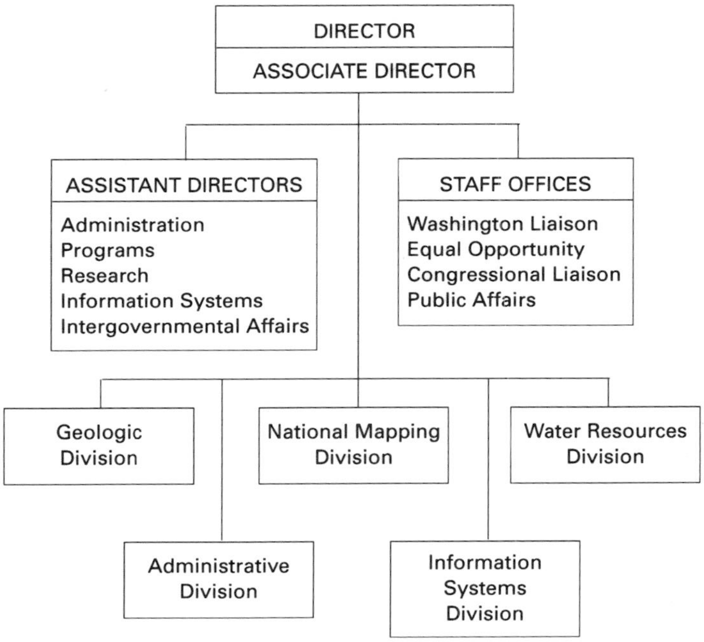

FIGURE 10–4. ORGANIZATION OF THE WATER RESOURCES DIVISION OF THE U.S. GEOLOGICAL SURVEY

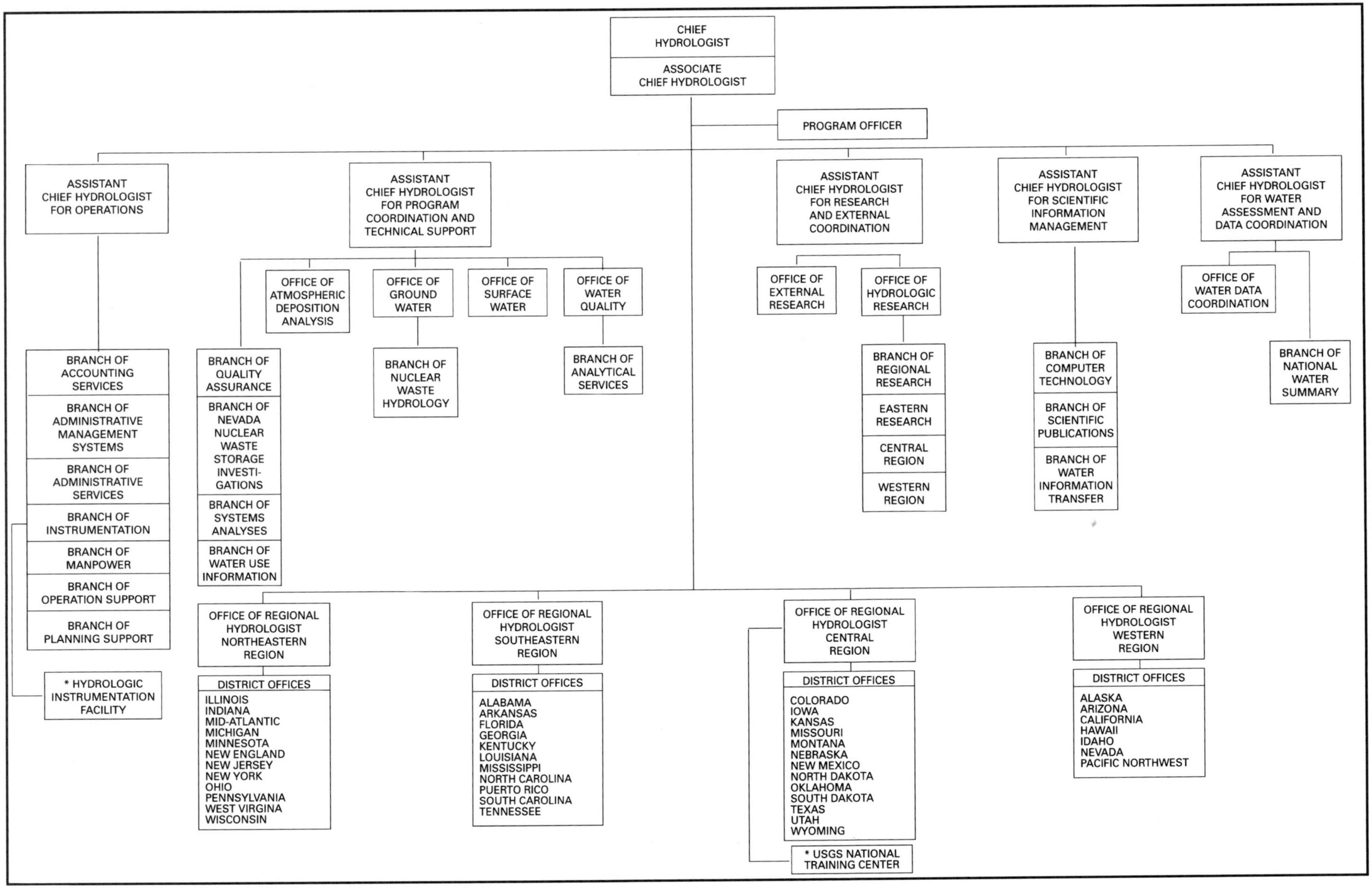

FIGURE 10–5. U.S. GEOLOGICAL SURVEY WATER RESOURCES DIVISION OFFICES

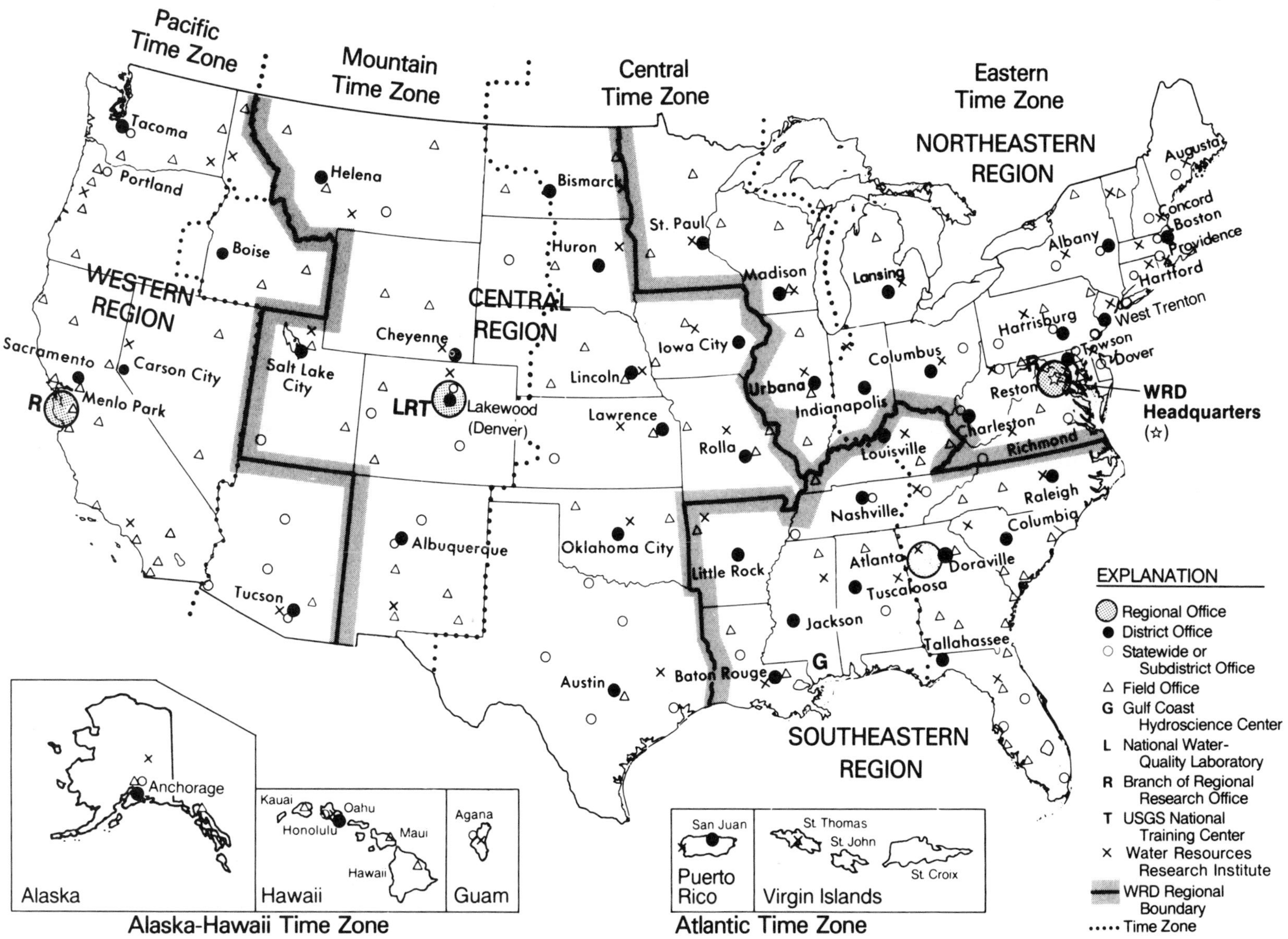

TABLE 10–7. DIRECTORY—U.S. GEOLOGICAL SURVEY WATER RESOURCES DIVISION

HEADQUARTERS

U.S. Geological Survey
(Mail Stop Number) National Center
Reston, Virginia 22092
(703) 648–5215

Position	Mail stop
Chief Hydrologist	409
Associate Chief Hydrologist	408
Program Officer	406A
Assistant Chief Hydrologist for Operations	441
Deputy Assistant Chief Hydrologist for Operations	441
Branch of Accounting Services	443
Branch of Administrative Data Systems	447
Branch of Administrative Services	442
Branch of Instrumentation	460
Branch of Manpower	406
Branch of Operational Support	405
Branch of Planning Support	404
Assistant Chief Hydrologist for Program Coordination and Technical Support	414
Branch of Systems Analysis	410
Branch of Water Use Information	414
Office of Atmospheric Deposition Analysis	416
Office of Ground Water	411
Branch of Nuclear Waste Hydrology	410
Office of Surface Water	415
Office of Water Quality	412
National Water Quality Assessment Program	412
Toxic Substances Hydrology Program	412
Assistant Chief Hydrologist for Research and External Coordination	436
Deputy Assistant Chief Hydrologist for Research and External Coordination	326
Office of External Research	424
Office of Hydrologic Research	436
Assistant Chief Hydrologist for Water Assessment and Data Coordination	407
Branch of National Water Summary	407
Office of Water Data Coordination	417
Assistant Chief Hydrologist for Scientific Information Management	440
Deputy Assistant Chief Hydrologist for Scientific Information Management	444
Branch of Computer Technology	440
WATSTORE Program	437
Branch of Scientific Publications	444
Scientific Publications Program	439
Hydrologic Information Unit	419
Branch of Water Information Transfer	446
Environmental Affairs Program	423
Information Transfer Program	423
NAWDEX Program	421
WRSIC Program	425

TABLE 10–7. DIRECTORY—U.S. GEOLOGICAL SURVEY WATER RESOURCES DIVISION (continued)

NORTHEASTERN REGION

Connecticut, Delaware, Illinois, Indiana, Maine, Maryland, Massachusetts, Michigan, Minnesota, New Hampshire, New Jersey, New York, Ohio, Pennsylvania, Rhode Island, Vermont, Virginia, Washington, D.C., West Virginia, Wisconsin

OFFICE OF THE REGIONAL HYDROLOGIST

Regional Hydrologist
U.S. Geological Survey
433 National Center
Reston, VA 22092
(703) 648–5817

DISTRICT AND STATE OFFICES

CONNECTICUT See also Massachusetts
Hydrologist-in-Charge
Connecticut Office, WRD
U.S. Geological Survey
Abraham A. Ribicoff Federal Bldg., Rm. 525
450 Main Street
Hartford, CT 06103
(203) 240–3060

DELAWARE See also Maryland
Hydrologist-in-Charge
Delaware Office, WRD
U.S. Geological Survey
Federal Bldg., Rm. 1201
300 S. New Street
Dover, DE 19901
(302) 734–2506

DISTRICT OF COLUMBIA See Maryland

ILLINOIS
District Chief, WRD
U.S. Geological Survey
Busey Bank County Plaza
102 E. Main St., 4th floor
Urbana, IL 61801
(217) 398–5353

INDIANA
District Chief, WRD
U.S. Geological Survey
5957 Lakeside Blvd.
Indianapolis, IN 46278
(317) 290–3333

MAINE See also Massachusetts
Hydrologist-in-Charge
Maine Office, WRD
U.S. Geological Survey
26 Ganneston Drive
Augusta, ME 04330
(207) 622–8201

MARYLAND Mid-Atlantic District (Delaware, Maryland, Virginia, Washington, D.C.)
District Chief, WRD
U.S. Geological Survey
208 Carroll Bldg.
8600 La Salle Road
Towson, MD 21204
(301) 828–1535

MASSACHUSETTS New England District (Connecticut, Massachusetts, Maine, New Hampshire, Rhode Island, and Vermont)
District Chief, WRD
U.S. Geological Survey
150 Causeway Street, Suite 1309
Boston, MA 02114
(617) 565–6860

MICHIGAN
District Chief, WRD
U.S. Geological Survey
6520 Mercantile Way, Suite 5
Lansing, MI 48911
(517) 377–1608

MINNESOTA
District Chief, WRD
U.S. Geological Survey
702 Post Office Bldg.
180 E. Kellogg Boulevard
St. Paul, MN 55101
(612) 229–2600

NEW HAMPSHIRE See also Massachusetts
Hydrologist-in-Charge
New Hampshire Office, WRD
U.S. Geological Survey
525 Clinton St., RFD 12
Bow, NH 03301
(603) 225–4681

NEW JERSEY
District Chief, WRD
U.S. Geological Survey
Suite 206, Mountain View Office Park
810 Bear Tavern Rd.
West Trenton, NJ 08628
(609) 771–3900

NEW YORK
Mailing address:
District Chief, WRD
U.S. Geological Survey
P.O. Box 1669
Albany, NY 12201
Office address:
343 U.S. Post Office and Courthouse
Albany, NY 12201
(518) 472–3107

OHIO
District Chief, WRD
U.S. Geological Survey
975 West Third Avenue
Columbus, OH 43212
(614) 469–5553

PENNSYLVANIA
Mailing address:
District Chief, WRD
U.S. Geological Survey
P.O. Box 1107
Harrisburg, PA 17108
Office address:
Federal Bldg., 4th Floor
228 Walnut Street
Harrisburg, PA 17108
(717) 782–4514

RHODE ISLAND See also Massachusetts
Hydrologist-in-Charge
Rhode Island Office, WRD
U.S. Geological Survey
John O. Pastore Federal Bldg., Room 237
Providence, RI 02903
(401) 528–5135

VERMONT See Massachusetts

VIRGINIA See also Maryland
Hydrologist-in-Charge
Virginia Office, WRD
U.S. Geological Survey
3600 West Broad Street, Rm. 606
Richmond, VA 23230
(804) 771–2427

WEST VIRGINIA
District Chief, WRD
U.S. Geological Survey
603 Morris Street
Charleston, WV 25301
(304) 347–5130

WISCONSIN
District Chief, WRD
U.S. Geological Survey
6417 Normandy Lane
Madison, WI 53719–1133
(608) 274–3535

TABLE 10–7. DIRECTORY—U.S. GEOLOGICAL SURVEY WATER RESOURCES DIVISION (continued)

SOUTHEASTERN REGION

Alabama, Arkansas, Florida, Georgia, Kentucky, Louisiana, Mississippi, North Carolina, Puerto Rico, South Carolina, Tennessee, Virgin Islands

OFFICE OF THE REGIONAL HYDROLOGIST

Address:
Regional Hydrologist
U.S. Geological Survey
Richard B. Russell Federal Bldg.
75 Spring Street, SW, Rm. 772
Atlanta, GA 30303
(404) 331–5174

DISTRICT OFFICES

ALABAMA
District Chief, WRD
U.S. Geological Survey
520 19th Avenue
Tuscaloosa, AL 35401
(205) 752–8104

ARKANSAS
District Chief, WRD
U.S. Geological Survey
Federal Office Bldg., Rm. 2301
700 West Capitol Avenue
Little Rock, AR 72201
(501) 378–6391

FLORIDA
District Chief, WRD
U.S. Geological Survey
227 N. Bronough St., Suite 3015
Tallahassee, FL 32301
(904) 681–7620

GEORGIA
District Chief, WRD
U.S. Geological Survey
6481-B Peachtree Industrial Blvd.
Doraville, GA 30360
(404) 331–4858

KENTUCKY
District Chief, WRD
U.S. Geological Survey
2301 Bradley Ave.
Louisville, KY 40217
(502) 582–5241

LOUISIANA
Mailing address:
District Chief, WRD
U.S. Geological Survey
P.O. Box 66492
Baton Rouge, LA 70896
Office address:
6554 Florida Boulevard
Baton Rouge, LA 70806
(504) 389–0281

MISSISSIPPI
District Chief, WRD
U.S. Geological Survey
Federal Office Bldg., Suite 710
100 West Capitol Street
Jackson, MS 39269
(601) 965–4600

NORTH CAROLINA
Mailing address:
District Chief, WRD
U.S. Geological Survey
P.O. Box 2857
Raleigh, NC 27602
Office address:
300 Fayetteville Street Mall
Century Postal Station, Rm. 436
Raleigh, NC 27602
(919) 856–4510

PUERTO RICO Caribbean District (Puerto Rico and U.S. Virgin Islands)
Mailing address:
District Chief, WRD
U.S. Geological Survey
GPO Box 4424
San Juan, PR 00936
Office address:
GSA Center, Building 652
Highway 28, km. 7.2, Pueblo Viejo
San Juan, PR 00936
(809) 783–4660

SOUTH CAROLINA
District Chief, WRD
U.S. Geological Survey
Suite 677 A
1835 Assembly Street
Columbia, SC 29201
(803) 765–5966

TENNESSEE
District Chief, WRD
U.S. Geological Survey
Federal Bldg., and U.S. Court House, Rm. A-413
Nashville, TN 37203
(615) 736–5424

VIRGIN ISLANDS See Puerto Rico

TABLE 10–7. DIRECTORY—U.S. GEOLOGICAL SURVEY WATER RESOURCES DIVISION (continued)

CENTRAL REGION

Colorado, Iowa, Kansas, Missouri, Montana, Nebraska, New Mexico, North Dakota, Oklahoma, South Dakota, Texas, Utah, Wyoming

OFFICE OF THE REGIONAL HYDROLOGIST

Regional Hydrologist
U.S. Geological Survey
Mail Stop 406, Box 25046
Denver Federal Center
Lakewood, CO 80225
(303) 236–5920

DISTRICT OFFICES

COLORADO
District Chief, WRD
U.S. Geological Survey
Mail Stop 415, Box 25046
Denver Federal Center, Bldg. 53
Lakewood, CO 80225
(303) 236–4882

IOWA
Mailing address:
District Chief, WRD
U.S. Geological Survey
P.O. Box 1230
Iowa City, IA 52244–1230
Office address:
Rm. 269, Federal Bldg.
400 S. Clinton St.
Iowa City, IA 52244–1230
(319) 337–4191

KANSAS
District Chief, WRD
U.S. Geological Survey
1950 Constant Ave. - Campus West
Lawrence, KS 66046
(913) 864–4321

MISSOURI
District Chief, WRD
U.S. Geological Survey
1400 Independence Road, Mail Stop 200
Rolla, MO 65401
(314) 341–0824

MONTANA
District Chief, WRD
U.S. Geological Survey
Federal Bldg., Rm. 428
301 South Park Avenue
Helena, MT 59626–0076
(406) 449–5263

NEBRASKA
District Chief, WRD
U.S. Geological Survey
Federal Bldg., Rm. 406
100 Centennial Mall North
Lincoln, NE 68508
(402) 437–5082

NEW MEXICO
District Chief, WRD
U.S. Geological Survey
Pinetree Office Park, Suite 200
4501 Indian School Rd., N.E.
Albuquerque, NM 87110
(505) 262–6630

NORTH DAKOTA
District Chief, WRD
U.S. Geological Survey
821 E. Interstate Avenue
Bismarck, ND 58501
(701) 250–4601

OKLAHOMA
District Chief, WRD
U.S. Geological Survey
215 Dean A. McGee Avenue, Rm. 621
Oklahoma City, OK 73102
(405) 231–4256

SOUTH DAKOTA
District Chief, WRD
U.S. Geological Survey
Federal Bldg., Rm. 317
200 4th Street, SW
Huron, SD 57350–2469
(605) 353–7176

TEXAS
District Chief, WRD
U.S. Geological Survey
Building 1
8011 Cameron Road
Austin, TX 78753
(512) 832–5791

UTAH
District Chief, WRD
U.S. Geological Survey
Administration Bldg., Rm. 1016
1745 West 1700 South
Salt Lake City, UT 84104
(801) 524–5663

WYOMING
Mailing address:
District Chief, WRD
U.S. Geological Survey
P.O. Box 1125
Cheyenne, WY 82003
Office address:
J.C. O'Mahoney Federal Center
Room 4006
2120 Capitol Avenue
Cheyenne, WY 82003
(307) 772–2153

TABLE 10–7. DIRECTORY—U.S. GEOLOGICAL SURVEY WATER RESOURCES DIVISION (continued)

WESTERN REGION

Alaska, Arizona, California, Guam, Hawaii, Idaho, Nevada, Oregon, Washington

OFFICE OF THE REGIONAL HYDROLOGIST
Address:
Regional Hydrologist
U.S. Geological Survey
345 Middlefield Road, Mail Stop 470
Menlo Park, CA 94025

DISTRICT AND STATE OFFICES

ALASKA
District Chief, WRD
U.S. Geological Survey
4230 University Dr., Suite 201
Anchorage, AK 99508–4664

ARIZONA
District Chief, WRD
U.S. Geological Survey
Federal Bldg., FB-44
300 West Congress Street
Tucson, AZ 85701

CALIFORNIA
District Chief, WRD
U.S. Geological Survey
Federal Bldg., Rm. W-2234
2800 Cottage Way
Sacramento, CA 95825

GUAM See also Hawaii
Mailing address:
Hydrologist-in-Charge
Subdistrict Office, WRD
U.S. Geological Survey
P.O. Box 188
FPO San Francisco, CA 96630
Office address:
U.S. Navy Public Works Center, Bldg. 104
Agana, GU 96910

HAWAII Hawaii - Guam District
Mailing address:
District Chief, WRD
U.S. Geological Survey
P.O. Box 50166
Honolulu, HI 96850
Office address:
300 Ala Moana Blvd., Rm. 6110
Honolulu, HI 96850

IDAHO
District Chief, WRD
U.S. Geological Survey
230 Collins Road
Boise, ID 83702

NEVADA
District Chief, WRD
U.S. Geological Survey
Federal Bldg., Rm. 224
705 North Plaza Street
Carson City, NV 89701

OREGON See also Washington
Hydrologist-in-Charge
Oregon Office, WRD
U.S. Geological Survey
847 NE 19th Ave., Suite 300
Portland, OR 97232

WASHINGTON Pacific Northwest District
District Chief, WRD
U.S. Geological Survey
1201 Pacific Avenue, Suite 600
Tacoma, WA 98402

HEADQUARTERS BRANCH FIELD LOCATIONS

OFFICE OF THE ASSISTANT CHIEF HYDROLOGIST FOR PROGRAM COORDINATION AND TECHNICAL SUPPORT
(see previous listing)
(703) 648–5229

BRANCH OF NEVADA NUCLEAR WASTE STORAGE INVESTIGATIONS
U.S. Geological Survey
Mail Stop 421
Denver Federal Center
Lakewood, CO 80225
(303) 236–0516

OFFICE OF WATER QUALITY (see previous listing)

BRANCH OF ANALYTICAL SERVICES
U.S. Geological Survey
Mail Stop 407
5293 Ward Road-B
Arvada, CO 80002
(303) 236–5345

OFFICE OF THE ASSISTANT CHIEF HYDROLOGIST FOR RESEARCH AND EXTERNAL COORDINATION

OFFICE OF HYDROLOGIC RESEARCH
(see previous listing)
(703) 648–5043

BRANCHES OF REGIONAL RESEARCH

EASTERN REGION
Chief, Branch of Regional Research
U.S. Geological Survey
432 National Center
Reston, VA 22092
(703) 648–5833

CENTRAL REGION
Chief, Branch of Regional Research
U.S. Geological Survey
Mail Stop 418, Box 25046
Denver Federal Center
Lakewood, CO 80225
(303) 236–5021

WESTERN REGION
Chief, Branch of Regional Research
U.S. Geological Survey
Mail Stop 472
345 Middlefield Road
Menlo Park, CA 94025
(415) 329–4412

FIGURE 10–6. FEDERAL AGENCIES WITH GROUND-WATER PROTECTION ROLES

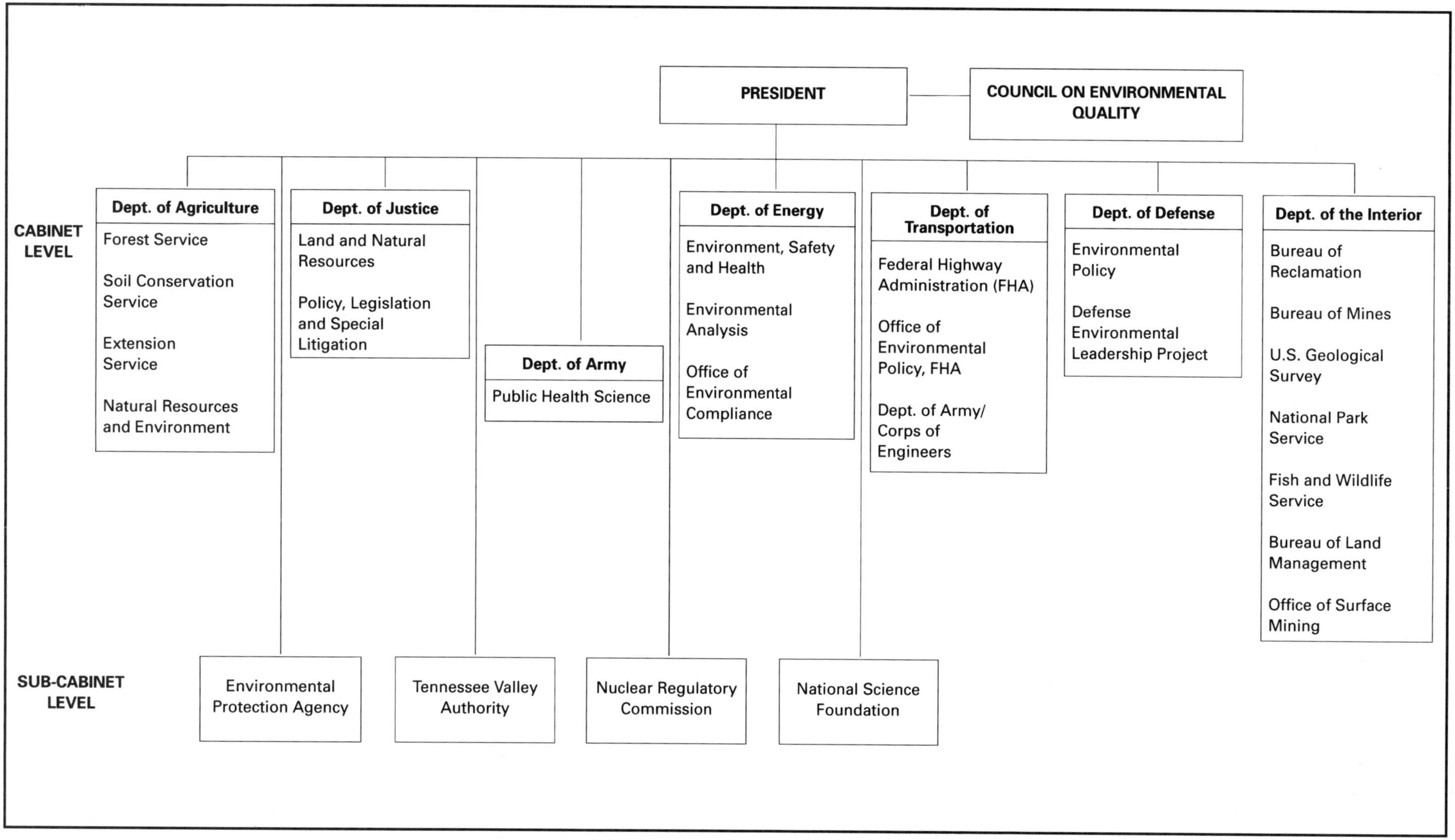

Source: U.S. Environmental Protection Agency, 1987, Improved Protection of Water Resources from Long-Term and Cumulative Pollution: Prevention of Ground-Water Contamination in the United States

TABLE 10–8. GROUND-WATER ROLES OF SELECTED FEDERAL AGENCIES

Agency	Description of Activities
DEPARTMENT OF AGRICULTURE (USDA)	
- **AGRICULTURE RESEARCH SERVICE**	Conducts limited number of research projects related to (1) ground-water recharge, (2) impacts of agricultural activities on ground-water quality.
- **FOREST SERVICE**	Conducts research projects on fate and transport of pesticides (under the National Agricultural Pesticide Impact Assessment Program).
DEPARTMENT OF COMMERCE (DOC)	
- **NATIONAL BUREAU OF STANDARDS**	Responsible for projects on quality assurance standards used by other federal agencies to monitor analytical performance of laboratories.
DEPARTMENT OF DEFENSE (DOD)	Participates in program to identify and evaluate hazardous waste disposal sites on military installations, undertake remedial action (Installation Restoration Program).
	Provides technical support in certain branches for the Installation Restoration Program, conducts program related research.
	Develops water quality criteria for certain munitions compounds.
	With EPA, is working on design, construction, research for CERCLA-designated sites.
ENVIRONMENTAL PROTECTION AGENCY (EPA)	Develops and implements a wide range of source control programs, assesses the quality of drinking water and sets national drinking water standards, conducts research on contaminant movement and health effects as well as treatment technologies.
DEPARTMENT OF ENERGY (DOE)	Operates programs for identifying and decommissioning contaminated nuclear materials storage and processing facilities, conducting site-specific hydrogeologic investigations.
DEPARTMENT OF HOUSING AND URBAN DEVELOPMENT (HUD)	Conducts environmental assessments for housing projects, takes ground water into account.
DEPARTMENT OF THE INTERIOR (DOI)	
- **BUREAU OF LAND MANAGEMENT**	Inventories hazardous waste sites on public lands and manages ground-water resources under these lands.
- **NATIONAL PARK SERVICE**	Conducts ground-water monitoring studies at national parks.
- **U.S. GEOLOGICAL SURVEY**	Collects and analyzes hydrogeologic information, conducts research on hydrogeology, and coordinates federal activities on data gathering.
- **FISH AND WILDLIFE SERVICE**	Inventories hazardous waste sites for all FWS lands and facilities.
- **BUREAU OF INDIAN AFFAIRS**	Inventories hazardous waste sites on or near Indian reservations.
NATIONAL SCIENCE FOUNDATION (NSF)	Supports research projects and diverse hydrogeology projects. Conducts policy-related research.
NUCLEAR REGULATORY COMMISSION (NRC)	Researches fate and transport of radioactive substances.

Source: Office of Technology Assessment, 1984, Protecting the Nation's Groundwater from Contamination.

SECTION B. STATE WATER AGENCIES

TABLE 10–9. A SELECTED LIST OF STATE AGENCY CONTACTS— ENVIRONMENTAL PROTECTION

ENVIRONMENTAL AND FACILITY STANDARDS ADMINISTRATION
Room 328, State Office Building
Montgomery, AL 36130
(205) 261–5004

DEPARTMENT OF ENVIRONMENTAL CONSERVATION
Pouch O
Juneau, AK 99811
(907) 465–2600

DEPARTMENT OF HEALTH SERVICES
2005 North Central, Room 202
Phoenix, AZ 85007
(602) 257–2300

DEPARTMENT OF POLLUTION CONTROL AND ECOLOGY
P.O. Box 9583
Little Rock, AR 72219
(501) 562–7444

DEPARTMENT OF ENVIRONMENTAL AFFAIRS
P.O. Box 2815
Sacramento, CA 95812
(916) 322–5840

DEPARTMENT OF HEALTH
4210 East 11th Avenue
Denver, CO 80220
(303) 320–8333

DEPARTMENT OF ENVIRONMENTAL PROTECTION
165 Capitol Avenue
Hartford, CT 06106
(203) 566–2110

DEPARTMENT OF NATURAL RESOURCES & ENVIRONMENTAL CONTROL
89 Kings Highway, Box 1401
Dover, DE 19903
(302) 736–4403

DEPARTMENT OF CONSUMER & REGULATORY AFFAIRS
614 H Street, NW
Washington, DC 20001
(202) 727–7395

DEPARTMENT OF ENVIRONMENTAL REGULATION
2600 Blairstone Road
Tallahassee, FL 32301
(904) 488–4805

DEPARTMENT OF NATURAL RESOURCES
270 Washington Street, SW
Atlanta, GA 30334
(404) 656–3500

DEPARTMENT OF HEALTH
1250 Punchbowl Street
Honolulu, HI 96801
(808) 548–4139

DEPARTMENT OF HEALTH AND WELFARE
276 State House
Boise, ID 83720
(208) 334–4059

ENVIRONMENTAL PROTECTION AGENCY
2200 Churchill Road
Springfield, IL 62706
(217) 782–3397

ENVIRONMENTAL MANAGEMENT BOARD
1330 West Michigan Street
Indianapolis, IN 46206
(317) 633–0700

WATER, AIR AND WASTE MANAGEMENT COMMISSION
Wallace State Office Building
Des Moines, IA 50319
(515) 281–8854

DEPARTMENT OF HEALTH AND ENVIRONMENT
Forbes Field
Topeka, KS 66620
(913) 862–9360

NATURAL RESOURCES & ENVIRONMENTAL PROTECTION CABINET
Capital Plaza Tower
Frankfort, KY 40601
(502) 564–2150

DEPARTMENT OF ENVIRONMENTAL QUALITY
P.O. Box 44066
Baton Rouge, LA 70804
(504) 342–5007

DEPARTMENT OF ENVIRONMENTAL PROTECTION
State House Station 17
Augusta, ME 04333
(207) 289–2811

DEPARTMENT OF NATURAL RESOURCES
177 Admiral Cochrane Drive
Annapolis, MD 21401
(301) 224–7221

EXECUTIVE OFFICE OF ENVIRONMENTAL AFFAIRS
100 Cambridge Street, 20th Floor
Boston, MA 02202
(617) 727–9800

DEPARTMENT OF NATURAL RESOURCES
P.O. Box 30028
Lansing, MI 48909
(517) 373–7917

POLLUTION CONTROL AGENCY
1935 West Country Road B2
Roseville, MN 55113
(612) 296–7202

DEPARTMENT OF NATURAL RESOURCES
P.O. Box 10385
Jackson, MS 39209
(601) 961–5100

DEPARTMENT OF NATURAL RESOURCES
P.O. Box 1368
Jefferson City, MO 65102
(314) 751–3241

TABLE 10–9. A SELECTED LIST OF STATE AGENCY CONTACTS—ENVIRONMENTAL PROTECTION (continued)

DEPARTMENT OF HEALTH AND ENVIRONMENTAL SCIENCES
Cogswell Building
Helena, MT 59620
(406) 444–3948

DEPARTMENT OF ENVIRONMENTAL CONTROL
301 Centennial Mall South, Box 94877
Lincoln, NE 68509
(402) 471–2186

DEPARTMENT OF CONSERVATION AND NATURAL RESOURCES
201 South Fall Street
Carson City, NV 89710
(702) 885–4670

WATER SUPPLY AND POLLUTION CONTROL COMMISSION
22 Hazen Drive
Concord, NH 03301
(603) 271–3503

DEPARTMENT OF ENVIRONMENTAL PROTECTION
John Fitch Plaza, CN402
Trenton, NJ 08625
(609) 292–2885

DEPARTMENT OF HEALTH AND ENVIRONMENT
P.O. Box 968
Santa Fe, NM 87504
(505) 984–2850

DEPARTMENT OF ENVIRONMENTAL CONSERVATION
50 Wolf Road
Albany, NY 12233
(518) 457–3446

DEPARTMENT OF NATURAL RESOURCES & COMMUNITY DEVELOPMENT
512 North Salisbury Street
Raleigh, NC 27611
(919) 733–4984

DEPARTMENT OF HEALTH
1200 Missouri Avenue
Bismarck, ND 58505
(701) 224–2374

ENVIRONMENTAL PROTECTION AGENCY
361 East Broad Street
Columbus, OH 43266
(614) 466–8318

DEPARTMENT OF POLLUTION CONTROL
100 NE 10th Street,
Oklahoma City, OK 73152
(405) 271–4677

DEPARTMENT OF ENVIRONMENTAL QUALITY
522 SW Fifth Avenue, P.O. Box 1760
Portland, OR 97207
(503) 229–5395

DEPARTMENT OF ENVIRONMENTAL RESOURCES
Fulton Building
Harrisburg, PA 17120
(717) 787–5027

CHAIRMAN, ENVIRONMENTAL QUALITY BOARD
P.O. Box 11488
Santurce, PR 00910
(809) 725–5140

DEPARTMENT OF ENVIRONMENTAL MANAGEMENT
83 Park Street
Providence, RI 02903
(401) 277–2771

HEALTH AND ENVIRONMENTAL CONTROL
2600 Bull Street
Columbia, SC 29201
(803) 758–5631

DEPARTMENT OF WATER AND NATURAL RESOURCES
Foss Building
Pierre, SD 57501
(605) 773–3151

DEPARTMENT OF HEALTH AND ENVIRONMENT
344 Cordell Hull Building
Nashville, TN 37219
(615) 741–3657

DEPARTMENT OF WATER RESOURCES
Box 13087, Capitol Station
Austin, TX 78711
(512) 463–8412

DEPARTMENT OF HEALTH
1460 West 288 North
Salt Lake City, UT 84114
(801) 538–6121

DEPARTMENT OF ENVIRONMENTAL CONSERVATION
State Office Building
Montpelier, VT 05602
(802) 828–2871

COUNCIL ON THE ENVIRONMENT
Ninth Street Office Building
Richmond, VA 23219
(804) 786–4500

DEPARTMENT OF ECOLOGY
Mail Stop PV-11
Olympia, WA 98504
(206) 459–6168

DEPARTMENT OF NATURAL RESOURCES
1800 Washington Street, East, Building 3,
Charleston, WV 25305
(304) 348–2754

DEPARTMENT OF NATURAL RESOURCES
P.O. Box 7921
Madison, WI 53707
(608) 266–1099

DEPARTMENT OF ENVIRONMENTAL QUALITY
Herschler Building
Cheyenne, WY 82002
(307) 777–7937

TABLE 10–10. STATE DRINKING WATER CONTACTS

ALABAMA
Water Supply Branch
Department of Environmental Management
1751 Federal Drive
Montgomery, Alabama 36130
(205) 271–7773

ALASKA
Alaska Drinking Water Program
Water Quality Management
Department of Environmental Conservation
Post Office Box O
Juneau, Alaska 99811
(907) 465–2653

ARIZONA
Compliance Unit
Waste and Water Quality Management
Room 202
2005 North Central Avenue
Phoenix, Arizona 85004
(602) 257–2305

ARKANSAS
Division of Engineering
Arkansas State Department of Health
4815 West Markham Street
Little Rock, Arkansas 72201
(501) 661–2000

CALIFORNIA
Sanitary Engineering Branch
California Department of Health
714 P Street
Sacramento, California 95814
(916) 323–6111

COLORADO
Drinking Water/Ground Water Section
Colorado Department of Health
4210 East 11th Avenue
Denver, Colorado 80220
(303) 320–8333

CONNECTICUT
Water Supplies Section
Connecticut Department of Health
79 Elm Street
Hartford, Connecticut 06115
(203) 566–1251

DELAWARE
Office of Sanitary Engineering
Division of Public Health
Jesse Cooper Memorial Building
Capital Square
Dover, Delaware 19901
(302) 736–4731

DISTRICT OF COLUMBIA
Water Hygiene Branch
Department of Consumer and Regulatory Affairs
5010 Overlook Avenue, SW,
Washington, DC 20032
(202) 767–7370

FLORIDA
Drinking Water Program
Department of Environmental Regulation
Twin Towers Office Building
2600 Blair Stone Road
Tallahassee, Florida 32301–8241
(904) 487–1779

GEORGIA
Water Protection Branch
Environmental Protection Division
Department of Natural Resources
270 Washington Street, SW,
Atlanta, Georgia 30334
(404) 656–3530

HAWAII
Drinking Water Program
Sanitation Branch
Environmental Protection and Health Services
Division
Post Office Box 3378
Honolulu, Hawaii 96801
(808) 548–4682

IDAHO
Bureau of Water Quality
Division of Environment
Idaho Department of Health and Welfare
Statehouse
Boise, Idaho 83720
(208) 334–5867

ILLINOIS
Division of Public Water Supplies
Illinois Environmental Protection Agency
2200 Churchill Road
Springfield, Illinois 62706
(217) 785–8653

INDIANA
Division of Public Water Supply
Indiana State Board of Health
1330 West Michigan Street
Indianapolis, Indiana 46206
(317) 633–0174

IOWA
Water Supply Division
Department of Water, Air, and Waste Management
Wallace State Office Building
900 East Grant Street
Des Moines, Iowa 53019
(515) 281–8998

KANSAS
Support Services Section
Kansas Department of Health and the Environment
Forbes Field
Topeka, Kansas 66605
(913) 862–9360 Ext. 235

KENTUCKY
Division of Water
Department of Environmental Protection
15 Reilly Road, Fort Boone Plaza
Frankfort, Kentucky 40601
(502) 564–3410 Ext. 543

TABLE 10–10. STATE DRINKING WATER CONTACTS (continued)

LOUISIANA
Office of Preventive and Public Health Services
Louisiana Department of Health and Human Resources
Post Office Box 60630
New Orleans, Louisiana 70160
(504) 568–5105

MAINE
Division of Health Engineering
Maine Department of Human Services
State House (STA 10)
Augusta, Maine 04333
(207) 289–5685

MARYLAND
Division of Water Supply
Office of Environmental Programs
201 West Preston Street
Baltimore, Maryland 21201
(301) 225–6361

MASSACHUSETTS
Division of Water Supply
Department of Environmental Quality Engineering
One Winter Street
Boston, Massachusetts 02108
(617) 292–5770

MICHIGAN
Division of Water Supply
Michigan Department of Public Health
P.O. Box 30035
Lansing, Michigan 48909
(517) 335–8318

MINNESOTA
Section of Public Water Supplies
Minnesota Department of Health
717 Delaware Street
Minneapolis, Minnesota 55440
(612) 623–5330

MISSISSIPPI
Division of Water Supply
State Board of Health
Post Office Box 1700
Jackson, Mississippi 39205
(601) 354–6616/490–4211

MISSOURI
Public Drinking Water Program
Division of Environmental Quality
Post Office Box 176
Jefferson City, Missouri 65102
(314) 751–3241

MONTANA
Bureau of Water Quality
Health and Environmental Services
Cogswell Building, Room A206
Helena, Montana 59620
(406) 444–2406

NEBRASKA
Division of Environmental Health and Housing Surveillance
Nebraska Department of Health
301 Centennial Mall South
Lincoln, Nebraska 68509
(402) 471–2674

NEVADA
Public Health Engineering
Nevada Department of Human Resources
Consumer Health Protection Services
505 East King Street, Room 103
Carson City, Nevada 89710
(702) 885–4750

NEW HAMPSHIRE
Water Supply Division
New Hampshire Water Supply and Pollution Control Commission
Post Office Box 95, Hazen Drive
Concord, New Hampshire 03301
(603) 271–3139

NEW JERSEY
Bureau of Safe Drinking Water
Division of Water Resources
New Jersey Department of Environmental Protection
Post Office Box CN-029
Trenton, New Jersey 06825
(609) 984–7945

NEW MEXICO
Drinking Water Section
New Mexico Health and Environment Department
Post Office Box 968
Santa Fe, New Mexico 87504–0968
(505) 827–2778

NEW YORK
Bureau of Public Water Supply Protection
New York Department of Health
Corning Tower Building, Room 478
Nelson A. Rockefeller Empire State Plaza
Albany, New York 12237
(518) 474–5577

NORTH CAROLINA
Water Supply Branch
Division of Health Services
Department of Human Resources
Bath Building
Post Office Box 2091
Raleigh, North Carolina 27602–2091
(919) 733–2321

NORTH DAKOTA
Division of Water Supply and Pollution Control
State Department of Health
1200 Missouri Avenue
Bismarck, North Dakota
(701) 224–2354

OHIO
Office of Public Water Supply
Ohio Environmental Protection Agency
361 East Broad Street
Post Office Box 1049
Columbus, Ohio 43216
(614) 466–8307

TABLE 10–10. STATE DRINKING WATER CONTACTS (continued)

OKLAHOMA
Water Facility Engineering Service
Oklahoma State Department of Health
Post Office Box 53551
Oklahoma City, Oklahoma 73152
(405) 271–5204

OREGON
Drinking Water Program
Health Division
Department of Human Resources
1400 SW, 5th Avenue
Portland, Oregon 97201
(503) 229–6310

PENNSYLVANIA
Division of Water Supplies
Department of Environmental Resources
Post Office Box 2357
Harrisburg, Pennsylvania 17120
(717) 787–9035

PUERTO RICO
Water Supply Supervision Program
Puerto Rico Department of Health
Post Office Box 70184
San Juan, Puerto Rico 00936
(809) 766–1616

RHODE ISLAND
Division of Water Supply
Rhode Island Department of Health
75 Davis Street, Health Building
Providence, Rhode Island 02908
(401) 277–6867

SOUTH CAROLINA
Division of Water Supply
Department of Health and Environmental Control
2600 Bull Street
Columbia, South Carolina 29201
(803) 734–5310

SOUTH DAKOTA
Bureau of Drinking Water
Water and Natural Resources
Joe Foss Building
523 Capital Avenue, East
Pierre, South Dakota 57501
(605) 773–3151

TENNESSEE
Division of Water Supply
Tennessee Department of Health and Environment
150 9th Avenue, North
Nashville, Tennessee 37219–5404
(615) 741–6636

TEXAS
Division of Water Hygiene
Texas Department of Health
1100 West 49th Street
Austin, Texas 78756
(512) 458–7533

UTAH
Bureau of Public Water Supplies
Utah Department of Health
Post Office Box 16690
Salt Lake City, Utah 84116–0690
(801) 538–6159

VERMONT
Environmental Health Division
Vermont Department of Health
60 Main Street
Post Office Box 70
Burlington, Vermont 05401
(802) 863–7220

VIRGIN ISLANDS
Department of Conservation and Cultural Affairs
Government of Virgin Islands
Post Office Box 4340
Charlotte Amalie
St. Thomas, Virgin Islands 00801
(809) 774–6420

VIRGINIA
Bureau of Water Supply Engineering
Virginia Department of Health
James Madison Building
109 Governor Street
Richmond, Virginia 23219
(804) 786–1766

WASHINGTON
Drinking Water Program Section
Department of Social and Health Services
Mail Stop LD-11
Olympia, Washington 98504
(206) 753–5954

WEST VIRGINIA
Drinking Water Division
Office of Environmental Health Services
State Department of Health
1800 Washington Street, East
Charleston, West Virginia 25305
(304) 348–2981

WISCONSIN
Bureau of Water Supply
Public Water Supply Section
Department of Natural Resources
101 S. Webster
Madison, Wisconsin 53707
(608) 267–7651

WYOMING
Water Quality Division
Department of Environmental Quality
401 West 19th Street
Cheyenne, Wyoming 82002
(307) 777–7781

SECTION C. COMMISSIONS

TABLE 10–11. COMMISSIONS, COMPACTS AND JOINT AGENCIES

ATLANTIC STATES MARINE FISHERIES COMMISSION, 1717 Massachusetts Ave., NW, Washington, DC 20036 (202) 387–5330
Established by the Atlantic States Marine Fisheries Compact to promote the better utilization of the fisheries of the 15 Atlantic seaboard states, through the development of a joint program for the promotion and protection of such fisheries, and by the prevention of physical waste of the fisheries. Organized: 1942.

GREAT LAKES COMMISSION, 2200 Bonisteel Blvd., Ann Arbor, MI 48109 (313) 665–9135
Recommendatory and advisory agency for the eight Great Lakes states in regional water resource matters. Great Lakes Compact ratified by Great Lakes states' legislatures 1955; Congressional consent: 1968.

GREAT LAKES FISHERY COMMISSION, 1451 Green Rd., Ann Arbor, MI 48105 (313) 662–3209
The 1955 Canada-U.S. Convention on Great Lakes Fisheries established the commission to advise governments on ways to improve fisheries, to develop and coordinate fishery research programs, and to develop measures and implement programs to manage sea lamprey.

GULF STATES MARINE FISHERIES COMMISSION, P.O. Box 726, Ocean Springs, MS 39564 (601) 875–5912
An interstate compact of the states of Alabama, Florida, Louisiana, Mississippi, and Texas. The compact was signed in July 1949. The purpose of the compact is to promote proper utilization of the fisheries common to the seaboard of the Gulf Coast states.

INTER-AMERICAN TROPICAL TUNA COMMISSION, c/o Scripps Institution of Oceanography, La Jolla, CA 92093 (619) 546–7100
Charged with the investigation and conservation of the tuna and dolphin resources of the eastern Pacific Ocean. Member nations: U.S., France, Japan, Nicaragua, Panama. Established by convention between the U.S. and Costa Rica: 1949.

INTERNATIONAL JOINT COMMISSION, 2001 S St., NW, Second Fl., Washington, DC 20440 (202) 673–6222; 100 Metcalfe St., Ottawa, Ontario K1P 5M1 (613) 995–2984); Regional Office: Suite 800, 100 Ouellette Ave., Windsor, Ontario N9A 6T3 (519) 256–7821/313, 226–2170
Established by the Boundary Waters Treaty of 1909 between the United States and Great Britain for the purpose of preventing disputes regarding the use of the waters on the U.S.-Canadian Boundary. Regional Office monitors, evaluates, and reports on compliance with the Great Lakes Water Quality Agreement of November 22, 1978. Commission functions in quasi-judicial, investigative, and coordination capacities.

INTERNATIONAL NORTH PACIFIC FISHERIES COMMISSION, 6640 NW Marine Dr., Vancouver, British Columbia, Canada V6T 1X2 (604) 228–1128
Established by convention between Canada, Japan, and the U.S. in 1952 for the conservation of the fisheries resources of the North Pacific Ocean.

INTERNATIONAL PACIFIC HALIBUT COMMISSION, P.O. Box 95009, Seattle, WA 98145–2009 (206) 634–1838
Scientific investigation and management of the Pacific halibut resource. Established in 1923 by a convention between Canada and the United States.

INTERNATIONAL WHALING COMMISSION, The Red House, 135 Station Rd., Histon, Cambridge CB4 4NP England 022023 3971
Established under the International Convention for the Regulation of Whaling, 1946, to provide for the conservation, development and optimum utilization of whale resources.

INTERSTATE COMMISSION ON THE POTOMAC RIVER BASIN, 6110 Executive Blvd., Suite 300, Rockville, MD 20852–3903 (301) 984–1908
Established in 1940 by Maryland, Pennsylvania, Virginia, West Virginia, and the District of Columbia. Coordinates tabulates, and summarizes data on condition of streams in Potomac Watershed; promotes uniform legislation; disseminates information; cooperates in studies. Areas of interest are water quality, water supply and land resources.

MARINE MAMMAL COMMISSION, 1625 I St., NW, Washington, DC 20006 (202) 653–6237
Established by the Marine Mammal Protection Act of 1972, P.L. 92–522, in consultation with the Committee of Scientific Advisors on Marine Mammals, to periodically review the status of marine mammal populations; to manage a research program concerned with their conservation; and to develop, review, and make recommendations of federal activities and policies which affect the protection and conservation of marine mammals.

MINNESOTA-WISCONSIN BOUNDARY AREA COMMISSION, 619 2nd St., Hudson, WI 54016 (612) 436–7131
To conduct studies, develop recommendations, and coordinate planning for protection, use, and development of lands, river valleys, and waters that form the boundary between Minnesota and Wisconsin, principally on the St. Croix and Mississippi rivers. Also serves as service center for Upper and Lower St. Croix National Riverways. Founded: 1965 under interstate compact.

NEW ENGLAND INTERSTATE WATER POLLUTION CONTROL COMMISSION, 85 Merrimac St., Boston, MA 02114 (617) 367–8522
Coordinates the work of the member states in the control of pollution of interstate waters; establishes water quality standards and approves and enforces stream classifications; trains waste treatment plant operators, and develops public information and education programs. Compact ratified July 31, 1947.

NORTH-EAST ATLANTIC FISHERIES COMMISSION, Rm. 336, Great Westminster House, Horseferry Rd., London, SW 1P 2AE. England 01, 216–6102
Promotes the conservation and optimum utilization of the fishery resources in the northeast Atlantic, within a framework appropriate to the regimes of contracting parties' own jurisdiction over fisheries in their waters; and encourages international cooperation and consultation. Founded in 1959.

TABLE 10–11. COMMISSIONS, COMPACTS AND JOINT AGENCIES (continued)

OHIO RIVER VALLEY WATER SANITATION COMMISSION, 49 E. Fourth St., Suite 815, Cincinnati, OH 45202 (513) 421–1151
An interstate agency representing Illinois, Indiana, Kentucky, New York, Ohio, Pennsylvania, Virginia, and West Virginia for control of water pollution in the Ohio River Valley Compact District. Established: 1948.

PACIFIC MARINE FISHERIES COMMISSION, 2000 SW 1st Ave., Suite 170, Portland, OR 97201 (503) 294–7025
Organized in 1947, the commission serves the Pacific states of Alaska, California, Idaho, Oregon, and Washington to promote conservation, development, and management of marine and anadromous fisheries through a coordinated regional approach to fisheries research, monitoring, and utilization.

SUSQUEHANNA RIVER BASIN COMMISSION, Interior Bldg., Washington, DC 20240 (202) 343–4091
Conservation and development of water resources and water-related resources in the river basin, comprising parts of Maryland, New York, and Pennsylvania.

UPPER COLORADO RIVER COMMISSION, 355 S. 4th East St., Salt Lake City, UT 84111 (801) 531–1150
An administrative agency, composed of commissioners appointed by the states of the Upper Division of the Colorado River—Colorado, New Mexico, Utah, and Wyoming—and by the President of the U.S. Created: 1949.

Source: Selected Listings from 1988 Conservation Directory, National Wildlife Federation, Washington D.C.

TABLE 10–12. JOINT UNITED STATES AND CANADA WATER BOARDS

Reporting to the International Joint Commission	Reporting Directly to Governments
Boards of Control	
Columbia River	Columbia River Permanent Engineering Board (Treaty)
Kootenay Lake	Lake Memphramagog Board
Lake Champlain	Niagara River Committee (Treaty)
Lake of the Woods	Niagara River Toxics Committee
Lake Superior	Poplar River Bilateral Monitoring Committee
Niagara River	Saint John Water Quality Committee
Osoyoos Lake	
Rainy Lake	
St. Croix River	
St. Lawrence River	
St. Mary-Milk Rivers	
Souris River	
Investigative-Engineering Boards	
Flathead River	
Great Lakes Diversion & Consumptive Uses	
Great Lakes Levels Advisory	
Lake Erie Regulation	
Souris and Red Rivers	
Technical Information Network	
Pollution Advisory Boards	
Rainy River	
Red River	
St. Croix River	
Great Lakes Water Quality Agreement Boards	
Great Lakes Water Quality	
Great Lakes Science Advisory	

Source: Pearse, P.H. and others, 1985, Currents of Change, Final Report Inquiry on Federal Water Policy, Ottawa, Canada

SECTION D. UNIVERSITIES

FIGURE 10–7. STATE WATER RESOURCES RESEARCH INSTITUTES

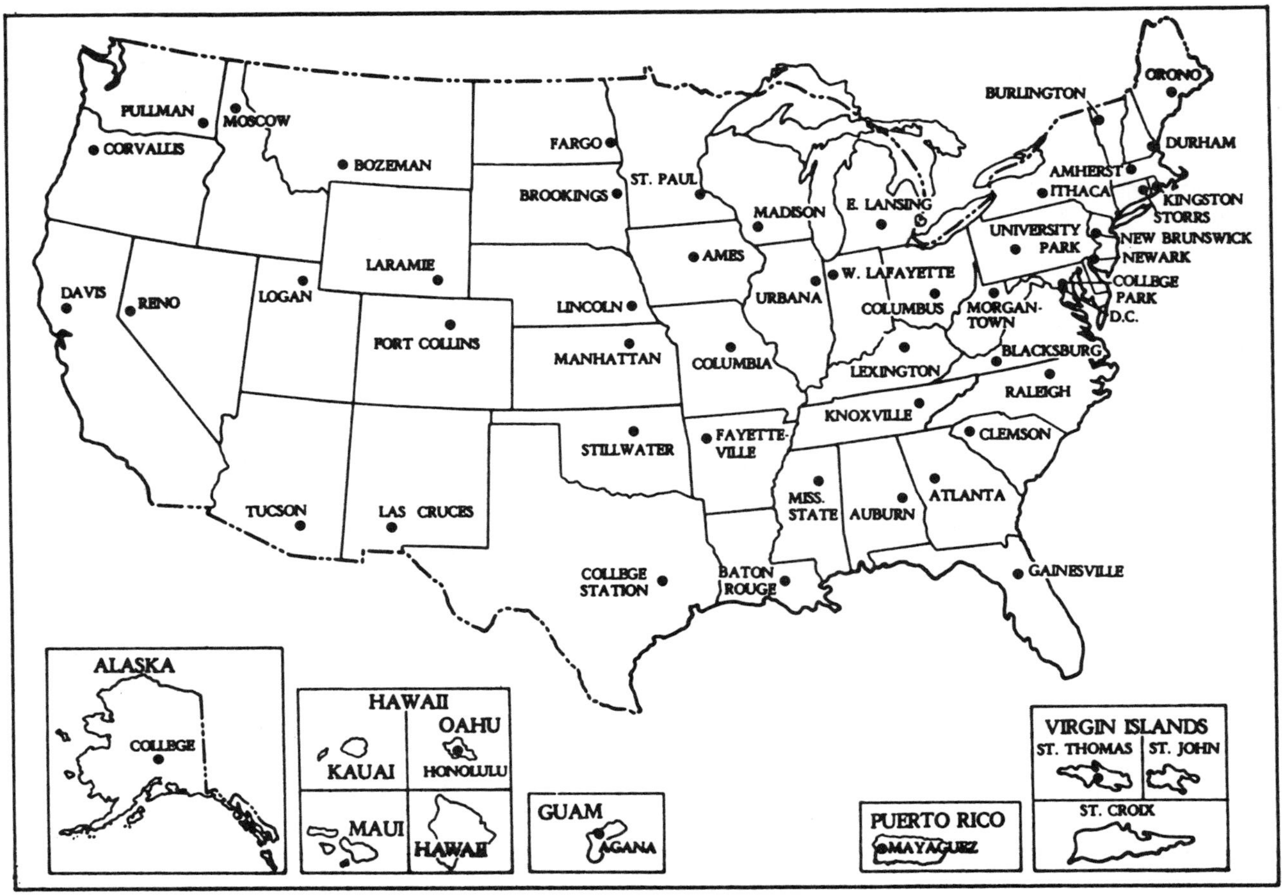

State Water Resources Research Institutes and National Water Resources Research Grants Programs

The Water Resources Research Act of 1984 (Public Law 98–242) directed the Secretary of the Interior to administer programs of grants and contracts for research, technology development, and information transfer that will aid the Nation and the States in solving water resources problems. Responsibility for administration of these programs has been delegated to the Director, U.S. Geological Survey.

Purpose:
- Provide matching grants to finance annually part of the operation of up to 57 State Water Resources Research Institutes.
- Provide matching grants for research concerning any aspect of a water-resource-related problem deemed to be in the national interest.
- Provide grants and contracts for technology development concerning any aspect of water-related technology deemed to be of State, regional, and national importance, including technology associated with improvement of waters of impaired quality.

Activities:
- Evaluate the Institutes at least once in each 4-year period to determine eligibility to receive future grants.
- Define priority areas of research and technology development for funding under the Water Resources Research and Technology Development programs.
- Solicit, evaluate, and select proposals for funding under the programs.
- Monitor progress of research and development projects funded under the Water Resources Research Act.

Source: U.S. Geological Survey, 1986, Water Resources Division in the 1980's, Circular 1005

TABLE 10–13. DIRECTORY—STATE WATER RESOURCES RESEARCH INSTITUTES

ALABAMA
Water Resources Research Institute
202 Hargis Hall
Auburn University, Alabama 36849
(205) 826–5075

ALASKA
Water Research Center
University of Alaska-Fairbanks
Fairbanks, Alaska 99775–1760
(907) 474–7775

ARIZONA
Water Resources Research Center
Geology Building #11
The University of Arizona
Tucson, Arizona 85721
(602) 621–7607

ARKANSAS
Arkansas Water Resources Research Center
113 Ozark Hall
University of Arkansas
Fayetteville, Arkansas 72701
(501) 575–4403

CALIFORNIA
Water Resources Center
Rubidoux Hall
University of California
Riverside, California 92521
(714) 787–4327

COLORADO
Colorado Water Resources Research Institute
410 University Services Center
Colorado State University
Fort Collins, Colorado 80523
(303) 491–6308

CONNECTICUT
Institute of Water Resources
The University of Connecticut
Room 228, Hall Building, U-18
362 Fairfield Road
Storrs, Connecticut 06268
(203) 486–4523

DELAWARE
Water Resources Center
101 Hullihen Hall
University of Delaware
Newark, Delaware 19716
(302) 451–2191

DISTRICT OF COLUMBIA
Water Resources Research Center, MB5004
4200 Connecticut Avenue, N.W. (mail only)
University of the District of Columbia
Washington, DC 20008
(202) 282–7333

FLORIDA
Water Resources Research Center
424 Black Hall
University of Florida
Gainesville, Florida 32611
(904) 392–0840

GEORGIA
Environmental Resources Center
205 Old Civil Engineering Building
Georgia Institute of Technology
Atlanta, Georgia 30332
(404) 894–3776

GUAM
Water and Energy Research
Institute of the Western Pacific
University of Guam
UOG Station
Mangilao, Guam 96913
(Country Code: 671) 734–3132

HAWAII
Water Resources Research Center
University of Hawaii
Holmes Hall 283, 2540 Dole Street
Honolulu, Hawaii 96822
(808) 948–7847

IDAHO
Water Resources Research Institute
Morrill Hall 106
University of Idaho
Moscow, Idaho 83843
(208) 885–6429

ILLINOIS
Water Resources Center
University of Illinois
208 N. Romine Street
Urbana, Illinois 61801
(217) 333–0536

INDIANA
Water Resources Research Center
Purdue University
Lilly Hall
West Lafayette, Indiana 47907
(317) 494–8041

IOWA
Iowa State Water Resources Research Institute
355 Town Engineering Building
Iowa State University
Ames, Iowa 50011
(515) 294–8921

KANSAS
Kansas Water Resources Research Institute
Kansas State University
Waters Hall 144
Manhattan, Kansas 66506
(913) 532–5729

KENTUCKY
Kentucky Water Resources Research Institute
University of Kentucky
219 Anderson Hall
Lexington, Kentucky 40506–0046
(606) 257–1299

LOUISIANA
Louisiana Water Resources Research Institute
Louisiana State University
2401-A CEBA Building
Baton Rouge, Louisiana 70803
(504) 388–8508/8509

TABLE 10–13. DIRECTORY—STATE WATER RESOURCES RESEARCH INSTITUTES (continued)

MAINE
Land and Water Resources Center
University of Maine
11 Coburn Hall
Orono, Maine 04469
(207) 581–1492

MARYLAND
Water Resources Research Center
University of Maryland
0313 Symons Hall
College Park, Maryland 20742
(301) 454–6406

MASSACHUSETTS
Water Resources Research Center
University of Massachusetts
Blaisdell House
Amherst, Massachusetts 01003
(413) 545–2842

MICHIGAN
Institute of Water Research
Michigan State University
East Lansing, Michigan 48824
(517) 353–3742 or 355–3421

MINNESOTA
Water Resources Research Center
University of Minnesota
866 Bioscience Center
1445 Gortner Avenue
St. Paul, Minnesota 55108
(612) 624–9282

MISSISSIPPI
Water Resources Research Institute
Mississippi State University
P. O. Drawer AD
Mississippi State, Mississippi 39762
(601) 325–3620

MISSOURI
Water Resources Research Center
University of Missouri
Sinclair Road - Route 3
Columbia, Missouri 65203
(314) 882–2151

MONTANA
Montana Water Resources Research Center
Montana State University
Suite 309, Montana Hall
Bozeman, Montana 59717
(406) 994–6690

NEBRASKA
Nebraska Water Resources Center
University of Nebraska
113 Nebraska Hall
Lincoln, Nebraska 68588–0517
(402) 472–3305

NEVADA
Water Resources Center
Desert Research Institute
University of Nevada System
P. O. Box 60220
Reno, Nevada 89506
(702) 673–7361

NEW HAMPSHIRE
Water Resources Research Center
224 Science/Engineering Research Bldg.
University of New Hampshire
Durham, New Hampshire 03824
(603) 862–2144

NEW JERSEY
Division of Water Resources
Center for Coastal & Environmental Studies
Rutgers, The State University of New Jersey
Doolittle Hall, Busch Campus
New Brunswick, New Jersey 08903
(201) 932–3596

NEW MEXICO
Water Resources Research Institute
New Mexico State University
P. O. Box 3167
Las Cruces, New Mexico 88003
(505) 646–4337

NEW YORK
Water Resources Institute
468 Hollister Hall
Cornell University
Ithaca, New York 14853
(607) 255–7535

NORTH CAROLINA
Water Resources Research Institute of the University of N.C.
North Carolina State University
Campus Box 7912
Raleigh, North Carolina 27695–7912
(919) 737–2815

NORTH DAKOTA
Water Resources Research Institute
202 Old Main
North Dakota State University
Fargo, North Dakota 58105
(701) 237–7034

OHIO
Water Resources Center
The Ohio State University
1791 Neil Avenue
Columbus, Ohio 43210
(614) 292–2334

OKLAHOMA
Oklahoma Water Research Institute
Oklahoma State University
003 Life Sciences East
Stillwater, Oklahoma 74078–0281
(405) 624–6995

OREGON
Water Resources Research Institute
100 Merryfield Hall
Oregon State University
Corvallis, Oregon 97331
(503) 754–4022

PENNSYLVANIA
Environmental Resources Research Institute
The Pennsylvania State University
103 Land & Water Research Building
University Park, Pennsylvania 16802
(814) 863–0291

TABLE 10–13. DIRECTORY—STATE WATER RESOURCES RESEARCH INSTITUTES (continued)

PUERTO RICO
Water Resources Research Institute
RUM-UPR
P.O. Box 5000
Mayaguez, Puerto Rico 00709–5000
(809) 832–4040

RHODE ISLAND
Water Resources Center
Room 310 Bliss Hall
University of Rhode Island
Kingston, Rhode Island 02881
(401) 792–2297

SOUTH CAROLINA
Water Resources Research Institute
310 Lowry Hall, Clemson University
Clemson, South Carolina 29634–2900
(803) 656–3273

SOUTH DAKOTA
Water Resources Institute
South Dakota State University
P. O. Box 2120
Brookings, South Dakota 57007
(605) 688–4910

TENNESSEE
Water Resources Research Center
The University of Tennessee
428 South Stadium Hall
Knoxville, Tennessee 37996–0750
(615) 974–2151

TEXAS
Texas Water Resources Institute
The Texas A&M University
College Station, Texas 77843
(409) 845–1851

UTAH
Center for Water Resources Research
Utah State University
Logan, Utah 84322–8200
(801) 750–3168

VERMONT
Water Resources Research Center
George D. Aiken Center, Room 320
The University of Vermont
Burlington, Vermont 05405
(802) 656–4057

VIRGINIA
Virginia Water Resources Research Center
Virginia Polytechnic Institute and State University
617 North Main Street
Blacksburg, Virginia 24060
(703) 961–5624

VIRGIN ISLANDS
Virgin Islands Water Resources Research Center
Caribbean Research Institute
University of the Virgin Islands
St. Thomas, Virgin Islands 00802
(809) 776–9200

WASHINGTON
State of Washington Water Research Center
Washington State University
Albrook Lab 202B
Pullman, Washington 99164–3002
(509) 335–5531

WEST VIRGINIA
Water Research Institute
West Virginia University
258 Stewart Street
Morgantown, West Virginia 26506
(304) 293–2757

WISCONSIN
Water Resources Center
The University of Wisconsin
1975 Willow Drive - 2nd Floor
Madison, Wisconsin 53706
(608) 262–3577

WYOMING
Wyoming Water Research Center
The University of Wyoming
P. O. Box 3067, University Station
Laramie, Wyoming 82071
(307) 766–2143

TABLE 10–14. UNIVERSITIES COUNCIL ON WATER RESOURCES (UCOWR)

The Universities Council on Water Resources (UCOWR) is an organization of universities actively engaged in teaching, research, and public service in the broad disciplines of water science, engineering, and management. Its purposes are to: (1) facilitate water-related education at all levels; (2) promote meaningful research and technology transfer on contemporary and emerging water resource issues; (3) compile and disseminate information on water resource problems and solutions; and (4) inform the various publics on water issues with the objective of promoting informed decisions at all levels of society.

The organization was founded in 1962 as the Universities Council on Hydrology (UCOH) amid a flurry of water resource activity by scientific committees, executive agencies, and the U.S. Congress in the late 1950s and early 60s. Sixteen universities were charter members. In 1964 UCOH chose to broaden its scope to include all disciplines in water resources, and the organization was given its present name. Since then the membership has increased to 83 universities.

MEMBERSHIP
THE UNIVERSITIES COUNCIL ON WATER RESOURCES, INC.

Auburn University
University of Alabama
University of Alaska
University of Arizona
University of Arkansas
Baylor University
California Institute of Technology
California State University
University of California, Davis
University of California, Irvine
University of California, LA
University of Southern California
City University of New York
Clemson University
University of Colorado
Colorado State University
University of Connecticut
Cornell University
University of Delaware
Duke University
Florida International University
University of Florida
University of South Florida
Georgia Institute of Technology
University of Georgia
University of Hawaii
University of Idaho
Southern Illinois University
University of Illinois
Iowa State University
University of Iowa
Johns Hopkins University
Kansas State University
University of Kansas
University of Kentucky
Louisiana State University
Louisiana Tech University
University of Maine
University of Massachusetts
Massachusetts Institute of Tech.
Michigan State University
University of Michigan
University of Minnesota
Mississippi State University
University of Missouri
Montana State University
University of Nebraska
University of Nevada at Reno
University of New Hampshire
New Mexico Inst. of Mining & Tech
New Mexico State University
University of New Mexico
North Carolina State University
University of North Carolina
Ohio State University
Oklahoma State University
Oregon State University
Pennsylvania State University
Polytechnic University
Pratt Institute of Technology
University of Puerto Rico
University of Rhode Island
Rutgers University
South Dakota State University
Tarleton State University
University of Tennessee
Tennessee Tech University
Texas A & M University
Texas Tech University
University of Texas
Utah State University
Vanderbilt University
University of Vermont
Virginia Military Institute
Virginia Polytechnic Institute
University of Virginia
U. S. Military Academy
Washington State University
West Virginia University
University of Wisconsin
University of Wyoming

Affiliate Members

University of Alberta
Asian Institute of Technology
Israel Institute of Technology
Monash University
National Taiwan University
University of New South Wales
University of Ottawa
University of Saskatchewan
University of Waterloo

TABLE 10–15. COLLEGES AND UNIVERSITIES OFFERING PROGRAMS IN GROUND WATER

(Note: For more detailed information contact the National Water Well Association Information Resources Dept., 6375 Riverside Drive, Dublin, OH 43017 (614) 761–1711

AUBURN UNIVERSITY
Dept of Civil Eng
238 Harbert Eng Ctr
Auburn University, AL 36849

AUBURN UNIVERSITY
210 Petrie Hall
Geology Dept
Auburn, AL 36849

UNIVERSITY OF NORTH ALABAMA
Physics & Earth Science
Florence, AL 35632–0001

ARIZONA WESTERN COLLEGE
Geology Dept
Yuma, AZ 85364

NO. ALBERTA INST. OF TECHNOLOGY
11762–106 Street
Edmonton Alberta
Canada, AZ T5G 2R1

NORTHERN ARIZONA UNIVERSITY
Geology Dept Box 6030
Flagstaff, AZ 86011

UNIVERSITY OF ARIZONA
Dept of Hydrology
Tucson, AZ 85721

UNIVERSITY OF WINDSOR
Dept of Geology Engr
Windsor, Ontario
Canada, AZ N9B 3P4

UNIVERSITY OF ARKANSAS
College of Arts & Sci/Geol Dep
118 Ozark Hall
Fayetteville, AR 72701

CALIFORNIA STATE UNIVERSITY/NORTHRIDGE
Dept of Geological Sciences
18111 Nordhoff St
Northridge, CA 91330

CALIFORNIA STATE UNIVERSITY/SACRAMENTO
Geology Dept
Sacramento, CA 95814
Dept of Geography
Univ of California
Santa Barbara, CA 93106

LOYOLA MARYMOUNT UNIVERSITY
Civil Engr and Enviro Science
7101 W 80th St
Los Angeles, CA 90045

SAN DIEGO STATE UNIVERSITY
Dept of Geology
San Diego, CA 92182

SONOMA STATE UNIVERSITY
1801 E Cotati Avenue
Geology Dept
Rohnert Park, CA 94928

STANFORD UNIVERSITY
Applied Earth Sciences Dept
Mitchell Bldg
Stanford, CA 94305

UNIVERSITY OF CALIFORNIA BERKELEY
Material Sci & Mineral Engr
210 Hearst Mining Bldg
Berkeley, CA 94720

UNIVERSITY OF CALIFORNIA-IRVINE
Dept of Civil Engineering
Irvine, CA 92717

UNIVERSITY OF CALIF SANTA CRUZ
Board of Earth Sciences
Santa Cruz, CA 95064

UNIVERSITY OF CALIFORNIA
Civil Engineering Dept
4531 C BH UCLA
Los Angeles, CA 90024–1600

UNIVERSITY OF CALIFORNIA
Earth Sciences
Riverside, CA 92521

UNIVERSITY OF CALIFORNIA
760 Davis Hall
Civil Engineering
Berkeley, CA 94720

UNIVERSITY OF THE PACIFIC
Engineering/Civil Engr
UOP Engineering
Stockton, CA 95211

UNIVERSITY OF CALIFORNIA
206 Walker Hall
Dept of Civil Engineering
Davis, CA 95616

COLORADO SCHOOL OF MINES
Dept of Geological Engr
Golden, CO 80401

UNIVERSITY OF CALIFORNIA/DAVIS
Veihmeyer Hall
Davis, CA 95616

UNIVERSITY OF COLORADO
Civil Environ & Architectural Engr
Campus Box 428
Boulder, CO 80309

UNIVERSITY OF COLORADO
Dept of Geological Sciences
Boulder, CO 80302

UNIVERSITY OF CONNECTICUT
Geology & Geophysics
U-45 Beach Hall
Storrs, CT 06268

WESTERN CONNECTICUT STATE UNIVERSITY
Geology Dept
Danbury, CT 06810

UNIVERSITY OF DELAWARE
Civil Engineering
Newark, DE 19716

GEORGE WASHINGTON UNIVERSITY
Dept of Geology
Washington, DC 20052

TABLE 10–15. COLLEGES AND UNIVERSITIES OFFERING PROGRAMS IN GROUND WATER (continued)

UNIVERSITY OF DELAWARE
Dept of Geology
Newark, DE 19716

ECKERD COLLEGE
P O Box 12560
Marine Sciences
St Petersburg, FL 33733

FLORIDA INTERNATIONAL UNIVERSITY
Dept of Civil & Enviro Engr
Tamiami Campus
Miami, FL 33199

FLORIDA STATE UNIVERSITY
Dept of Geology
Tallahassee, FL 32306

UNIVERSITY OF FLORIDA
Hydraulic Lab-College of Engrs
Dept of Civil Engineering
Gainesville, FL 32606

GEORGIA SOUTHERN COLLEGE
Geology and Geography
LB 8149
Statesboro, GA 30460

UNIVERSITY OF HAWAII/MANOA
2540 Dole St Holmes Hall 283
Water Resources Research Ctr
Honolulu, HI 96822

UNIVERSITY OF HAWAII/MANOA
2540 Dole St Holmes 383
Dept of Civil Engineering
Honolulu, HI 96822

UNIVERSITY OF IDAHO
Dept of Geology
Moscow, ID 83843

BRADLEY UNIVERSITY
Civil Engineering & Construction
1501 W Bradley
Peoria, IL 61625

MONMOUTH COLLEGE
Dept of Geology
Monmouth, IL 61462

NORTHEASTERN ILLINOIS UNIVERSITY
Dept of Earth Science
5500 N St Louis Ave
Chicago, IL 60625–4699

NORTHERN ILLINOIS UNIVERSITY
Dept of Geology
Dekalb, IL 60115

NORTHWESTERN UNIVERSITY
Civil Engineering
Evanston, IL 60201

OLIVET NAZARENE UNIVERSITY
Geological Sciences
Kankakee, IL 60901

ROOSEVELT UNIVERSITY
Geography-Geology
430 S Michigan Ave
Chicago, IL 60605

SOUTHERN ILLINOIS UNIVERSITY
Dept of Geology
Carbondale, IL 62901

UNIVERSITY OF ILLINOIS URBANA/CHAMPAIGN
2515 Hydrosystem Lab
208 N Romine St
Urbana, IL 61801

BALL STATE UNIVERSITY
Geology Dept
Muncie, IN 47306

EARLHAM COLLEGE
Geology Dept
Richmond, IN 47374

ROSE-HULMAN INSTITUTE OF TECHNOLOGY
5500 Wabash Avenue
Civil Engineering
Terre Haute, IN 47801

IOWA STATE UNIVERSITY
Civil Engineering
Town Engineering Bldg
Ames, IA 50011

UNIVERSITY OF IOWA
Dept Civil & Env Engineering
Iowa City, IA 52242

KANSAS STATE UNIVERSITY
Dept of Civil Engineering
Seaton Hall
Manhattan, KS 66506

EASTERN KENTUCKY UNIVERSITY
Geology Dept, Roark 9
Richmond, KY 40475

UNIVERSITY OF KENTUCKY
Dept of Geological Sciences
Boman Hall
Lexington, KY 40506–0059

UNIVERSITY OF KENTUCKY
Dept of Geology
Lexington, KY 40506

WESTERN KENTUCKY UNIVERSITY
Dept of Geography & Geology
Bowling Green, KY 42101

LOUISIANA TECH UNIVERSITY
Civil Engineering Dept
Ruston, LA 71272

UNIVERSITY OF MAINE
Geology Div
Math/Science Dept
Farmington, ME 04938

JOHN HOPKINS UNIVERSITY
Dept Earth & Planetary Sci
Baltimore, MD 21218

BOSTON UNIVERSITY
Dept of Geology
675 Commonwealth Ave
Boston, MA 02215

BRIDGEWATER STATE COLLEGE
Earth Science Dept
Bridgewater, MA 02324

GRAND VALLEY STATE UNIVERSITY
College Landing
Allendale, MI 49401

TABLE 10–15. COLLEGES AND UNIVERSITIES OFFERING PROGRAMS IN GROUND WATER (continued)

WAYNE STATE UNIVERSITY
Civil Engineering
Engineering Bldg
Detroit, MI 48202

WESTERN MICHIGAN UNIVERSITY
Geology Dept
Kalamazoo, MI 49008

UNIVERSITY OF MICHIGAN
2340 GG Brown Bldg
Civil Engineering
Ann Arbor, MI 48109

ST CLOUD TECHNICAL INSTITUTE
1540 Northway Drive
St Cloud, MN 56301

MISSISSIPPI STATE UNIVERSITY
P O Box 5174
Mississippi State, MS 39762

NORTHWEST MISSOURI STATE UNIVERSITY
Geology-Geography Dept
Maryville, MO 64468

UNIVERSITY OF MISSOURI/COLUMBIA
Civil Eng
Columbia, MO 65211

UNIVERSITY OF MISSOURI/KANSAS CITY
5100 Rockhill Rd
Dept of Geosciences
Kansas City, MO 64110

UNIVERSITY OF MISSOURI/ROLLA
Dept of Geol & Geophysics
125 McNutt Hall
Rolla, MO 65401

MONTANA STATE UNIVERSITY
Earth Sciences Dept
Bozeman, MT 59717–0001

MONTANA COLLEGE OF MINERAL SCI/TECH
Geological Engineering
West Park Street
Butte, MT 59701

ROCKY MOUNTAIN COLLEGE
Geology Dept
1511 Poly Dr
Billings, MT 59102

UNIVERSITY OF MONTANA
Dept of Geology
Missoula, MT 59812

UNIVERSITY OF NEBRASKA
214 Bessey Hall
Dept of Geology
Lincoln, NE 68508

UNIVERSITY OF NEVADA/RENO
P O Box 60220
Reno, NV 89506–0220

DARTMOUTH COLLEGE
Dept of Earth Sciences
Hanover, NH 03755

UNIVERSITY OF NEW HAMPSHIRE
Civil Engineering
236 Kingsbury Hall
Durham, NH 03824

UNIVERSITY OF NEW HAMPSHIRE
Dept of Earth Science
117 James Hall
Durham, NH 03824

NEW MEXICO INSTITUTE OF MINING AND TECHNOLOGY
Dept of Geoscience
Socorro, NM 87801

PRINCETON UNIVERSITY
Dept Civil Engineering
Princeton, NJ 08544

STOCKTON STATE COLLEGE
Natural Science & Math
Pomona, NJ 08240

ALFRED UNIVERSITY
Dept of Geology
Alfred, NY 14802

CLARKSON UNIVERSITY
Civil & Environmental Dept
Potsdam, NY 13676

HOFSTRA UNIVERSITY
Dept of Geology
Hempstead, NY 11550

LONG ISLAND UNIVERSITY
Southampton Campus
Natural Science Division
Southampton, NY 11968

MANHATTAN COLLEGE
Enviro Engr & Science Programs
3825 Corlear Ave
Bronx, NY 10471

POLYTECHNIC UNIVERSITY
Civil & Environmental Engrg
333 Jay St
Brooklyn, NY 11201

RENSSELAER POLYTECHNIC INSTITUTE
Troy, NY 12181

STATE UNIVERSITY OF NEW YORK
Dept Civil Engr
Buffalo, NY 14260

STATE UNIVERSITY OF NEW YORK
4240 Ridge Lea Rd
Box 47
Buffalo, NY 14226

STATE UNIVERSITY OF NEW YORK
Earth Sciences Water Resource
College at Oneonta
Oneonta, NY 13820

STATE UNIVERSITY OF NEW YORK
Dept of Geology
Cortland, NY 13045

UNION COLLEGE
Civil Engineering Dept
Schenectady, NY 12308

NORTH CAROLINA STATE UNIVERSITY
Box 5068
Dept of Marine & Atomic Sci
Raleigh, NC 27650

TABLE 10–15. COLLEGES AND UNIVERSITIES OFFERING PROGRAMS IN GROUND WATER (continued)

NORTH CAROLINA STATE UNIVERSITY
Civil Engineering Dept
Box 7908
Raleigh, NC 27695

UNIVERSITY OF NORTH CAROLINA/CHAPEL HILL
107 Mitchell Hall 029A
Geology Dept
Chapel Hill, NC 27514

UNIVERSITY OF NORTH CAROLINA/CHAPEL HILL
School Pub Hlth Dept Envro Sci
Rosenau Hall 201 H ESE
Chapel Hill, NC 27514

NORTH DAKOTA STATE UNIVERSITY
College of Engineering & Archi
Dept of Civil Engineering
Fargo, ND 58105

ANTIOCH COLLEGE
Environmental Studies/Geology
Yellow Springs, OH 45387

CASE WESTERN RESERVE UNIVERSITY
Geological Sciences Dept
Cleveland, OH 44106

KENT STATE UNIVERSITY
Geol Dept
Kent, OH 44242

MUSKINGUM AREA TECHNICAL COLLEGE
1555 Newark Rd
Zanesville, OH 43701

OHIO STATE UNIVERSITY
Geology & Mineralogy
106 W Oval Mall
Columbus, OH 43210

OHIO UNIVERSITY
Dept. of Geological Sciences
Porter Hall
Athens, OH 45701

UNIVERSITY OF AKRON
Dept of Geology
Akron, OH 44325

UNIVERSITY OF TOLEDO
2801 W Bancroft St
Dept of Geology
Toledo, OH 43606

OKLAHOMA STATE UNIVERSITY
Dept of Geology
Stillwater, OK 74078

UNIVERSITY OF OKLAHOMA
202 W Boyd Rm 334
Norman, OK 73019

OREGON GRADUATE CENTER
19600 N W Von Neumann Dr
Beaverton, OR 97006

BLOOMSBURG UNIVERSITY
Geography & Earth Science Dept
Bloomsburg, PA 17815

CALIFORNIA UNIVERSITY OF PENNSYLVANIA
Dept of Earth Sciences
California, PA 15419

SLIPPERY ROCK UNIVERSITY
Dept of Geology
Slippery Rock, PA 16057

TEMPLE UNIVERSITY
Computer Science & Architecture
Enviro Engineering Prog
Philadelphia, PA 19122

TEMPLE UNIVERSITY
Dept of Geology
Philadelphia, PA 19122

UNIVERSITY OF PITTSBURGH
Dept of Civil Engineering
949 Beh
Pittsburgh, PA 15261

UNIVERSITY OF PUERTO RICO
Research Center Eng
Mayaguez, PR 00708

WEST CHESTER UNIVERSITY
Geology and Astronomy
West Chester, PA 19383

UNIVERSITY OF RHODE ISLAND
Civil Eng Dept
Kingston, RI 02881

SOUTH DAKOTA SCHOOL OF MINES
Dept of Geology & Geological
501 E St Joseph
Rapid City, SD 57701

UNIVERSITY OF SOUTH DAKOTA
Dept of Earth Science/Physics
Vermillion, SD 57069

AUSTIN PEAY STATE UNIVERSITY
Dept of Geology & Geography
500 College St
Clarksville, TN 37044

BAYLOR UNIVERSITY
Geology Dept
B U Box 7354
Waco, TX 76798–7354

CORPUS CHRISTI STATE UNIVERSITY
6300 Ocean Dr
Corpus Christi, TX 78412

RICE UNIVERSITY
POB 1892
Houston, TX 77251

STEPHEN F AUSTIN STATE UNIVERSITY
Dept of Geology
Nacogdoches, TX 75926

TARLETON STATE UNIVERSITY
Physical Sciences Dept
Box T-69 Tarleton Station
Stephenville, TX 76402

TARLETON STATE UNIVERSITY
Hydrology Division
Box T-69 Tarleton Station
Stephenville, TX 76402

TEXAS A & M UNIVERSITY
Civil Engr Dept
College Station, TX 77843

TABLE 10–15. COLLEGES AND UNIVERSITIES OFFERING PROGRAMS IN GROUND WATER (continued)

TEXAS TECH UNIVERSITY
Civil Engineering
PO Box 4089
Lubbock, TX 79409

UNIVERSITY OF TEXAS
Dept Geological Sciences
Austin, TX 78713

UNIVERSITY OF HOUSTON
Dept Geosciences
University Park
Houston, TX 77004

UNIVERSITY OF HOUSTON
University Part
Dept of Civil Engineering
Houston, TX 77004

UNIVERSITY OF TEXAS AT DALLAS
Programs in Geosciences
UTD FO2.1, PO Box 8306688
Richardson, TX 75083–0688

UNIVERSITY OF TEXAS AT EL PASO
Civil Engineering Dept
El Paso, TX 79968

UNIVERSITY OF TEXAS AT SAN ANTONIO
Earth & Physical Sciences
College of Sciences & Engrg
San Antonio, TX 78285

BRIGHAM YOUNG UNIVERSITY
258 ESC
Geology Department
Provo, UT 84602

BRIGHAM YOUNG UNIVERSITY
36 CB BYU
Civil Engineering
Provo, UT 84602

UNIVERSITY OF UTAH
Dept Civil Eng
Salt Lake City, UT 84112

UTAH STATE UNIVERSITY
Dept of Geology
Logan, UT 84322–0705

UTAH STATE UNIVERSITY
Utah Water Res Lab UMC-82
Logan, UT 84322

MIDDLEBURY COLLEGE
Dept of Geology
Middlebury, VT 05753

NORWICH UNIVERSITY
Dept of Engineering Technology
Northfield, VT 05663

COLLEGE OF WILLIAM & MARY
Dept of Geology
Williamsburg, VA 23185

OLD DOMINION UNIVERSITY
Civil Engineering
Norfolk, VA 23508–8546

UNIVERSITY OF VIRGINIA
Clark Hall
Dept of Environmental Sciences
Charlottesville, VA 22903

VIRGINIA UNIVERSITY
Geology Dept
Box 5801
Radford, VA 24142

NORTH SEATTLE COMMUNITY COLLEGE
9600 College Way N
Engineering Technologies
Seattle, WA 98103

PACIFIC LUTHERAN UNIVERSITY
Mortcedt Lib/Earth Science
Tacoma, WA 98447

UNIVERSITY OF WASHINGTON
Dept of Civil Engineering
Seattle, WA 98195

WASHINGTON STATE UNIVERSITY
Dept of Geologic Engin
Pullman, WA 99163

MARSHALL UNIVERSITY
Dept of Geology
Huntington, WV 25701

UNIVERSITY OF WISCONSIN
1215 W Dayton St
Madison, WI 53706

UNIVERSITY OF WISCONSIN
Dept of Geology
Oshkosh, WI 54901

UNIVERSITY OF WISCONSIN
3409 N Downer Ave Sabin Hall
Rm 123 Dept of Geol & Geophs
Milwaukee, WI 53706

UNIVERSITY OF WISCONSIN/PLATTEVILLE
Civil Engineering
1 University Plaza
Platteville, WI 53818

UNIVERSITY OF WISCONSIN/EAU CLAIRE
Geology Dept
Eau Claire, WI 54701

UNIVERSITY OF WISCONSIN/RIVER FALLS
Plant & Earth Science
River Falls, WI 54022

UNIVERSITY OF WYOMING
Dept of Geology
Laramie, WY 82070

SIR SANFORD FLEMING COLLEGE
Frost Campus LIB-FR-#204
Box 8000 Lindsey ONT
Canada K9V 5E6

UNIVERSITY OF WATERLOO
Dept of Earth Sciences
Waterloo Ontario
Canada N2L 3G1

SECTION E. PROFESSIONAL, TRADE AND ENVIRONMENTAL GROUPS

TABLE 10–16. A SELECTED LIST OF PROFESSIONAL ORGANIZATIONS, TRADE ASSOCIATIONS AND ENVIRONMENTAL GROUPS CONCERNED WITH WATER-RELATED MATTERS

AMERICAN INSTITUTE OF HYDROLOGY
P.O. Box 14251, St. Paul, MN 55114
(612) 379–1030

AMERICAN LITTORAL SOCIETY
Sandy Hook, Highlands, NJ 07732
(201) 291–0055

AMERICAN PUBLIC HEALTH ASSOCIATION
1015 15 St., NW, Washington, DC 20036
(202) 789–5600

AMERICAN PUBLIC WORKS ASSOCIATION
1313 E. 60th St., Chicago, IL 60637
(312) 667–2200

AMERICAN SOCIETY OF AGRONOMY
677 South Segoe Road
Madison, WI 53711
(608) 273–8080

AMERICAN SOCIETY OF CIVIL ENGINEERS
345 East 47th Street, New York, NY 10017
(212) 705–7496

AMERICAN SOCIETY OF ICHTHYOLOGISTS AND HERPETOLOGISTS
Department of Biochemistry, Louisiana State University School of Medicine,
1901 Perdido St., New Orleans, LA 70112
(504) 568–4734

AMERICAN SOCIETY OF LIMNOLOGY AND OCEANOGRAPHY
Monterey Bay Aquarium Research Institute,
P.O. Box 20, Pacific Grove, CA 93950

AMERICAN SOCIETY OF SANITARY ENGINEERS
Box 40326
Bay Village, OH 44140
(216) 835–3040

AMERICAN SOCIETY FOR TESTING MATERIALS
1916 Race St., Philadelphia, PA 19103
(215) 299–5400

AMERICAN WATER RESOURCES ASSOCIATION
5410 Grosvenor Lane, Ste. 220, Bethesda, MD 20814
(301) 493–8600

AMERICAN WATER WORKS ASSOCIATION
6666 West Quincy Ave., Denver, CO 80235
(303) 794–7711

ASSOCIATED PUBLIC WORKS CONTRACTORS
2835 North Mayfair Rd., Milwaukee, WI 53222
(414) 778–1050

ASSOCIATION OF METROPOLITAN SEWERAGE AGENCIES
1015 18 Street, NW, Suite 1002
Washington, DC 20036
(202) 659–9161

ASSOCIATION OF METROPOLITAN WATER AGENCIES
477 H St., NW, Washington, DC 20001
(202) 682–3998.

ASSOCIATION OF STATE DRINKING WATER ADMINISTRATORS
1911 North Fort Meyer Drive, Arlington, Virginia 22209
(703) 524–2428.

ASSOCIATION OF STATE AND INTERSTATE WATER POLLUTION CONTROL ADMINISTRATORS
444 North Capitol Street, NW, Suite 330
Washington, DC 20001
(202) 624–7782

CHEMICAL MANUFACTURERS ASSOCIATION
2501 M St., NW, Washington, DC 20037
(202) 887–1100

CLEAN WATER ACTION PROJECT
317 Pennsylvania Ave., SE, Washington, DC 20003
(202) 547–1196.

COASTAL CONSERVATION ASSOCIATION
4801 Woodway, Suite 220 West, Houston, TX 77056
(713) 626–4222

COASTAL SOCIETY
5410 Grosvenor Ln., Suite 110, Bethesda, MD 20814
(301) 897–8616

CONCERN
1794 Columbia Road, NW, Washington, D.C. 20009
(202) 328–8160

CONSUMER'S UNION
2001 S St., NW, Washington, DC 20009
(202) 462–6262

CONSERVATION FOUNDATION
1250 24th St., NW, Washington, DC 20037
(202) 293–4800

COUSTEAU SOCIETY
930 W. 21st St., Norfolk, VA 23517
(804) 627–1144

DUCKS UNLIMITED
One Waterfowl Way,
Long Grove, IL 60047
(312) 438–4300

ENVIRONMENTAL ACTION FOUNDATION
1525 New Hampshire Ave., NW, Washington, DC 20036
(202) 745–4870

ENVIRONMENTAL DEFENSE FUND
Headquarters, 257 Park Avenue South, New York, NY 10010
(212) 505–2100

ENVIRONMENTAL DEFENSE FUND
1616 P Street, Suite 150, NW, Washington, D.C. 20036
(202) 387–3500

ENVIRONMENTAL LAW INSTITUTE
1616 P St., NW, Suite 200, Washington, DC 20036
(202) 328–5150

ENVIRONMENTAL POLICY INSTITUTE
218 D Street, SE, Washington, DC 20003
(202) 544–2600

ENVIRONMENTAL TASK FORCE
1012 Fourteenth St., NW, 15th Fl., Washington, DC 20005
(202) 842–2222

FOOD AND AGRICULTURE ORGANIZATION OF THE UNITED NATIONS
Via delle Terme di Caracalla, Rome 00100, Italy
(Telephone: 57971)

FRESHWATER FOUNDATION
2500 Shadywood Road, Box 90, Navarre, MN 55392–0090
(612) 471–8407

TABLE 10–16. A SELECTED LIST OF PROFESSIONAL ORGANIZATIONS, TRADE ASSOCIATIONS AND ENVIRONMENTAL GROUPS CONCERNED WITH WATER-RELATED MATTERS (continued)

FRIENDS OF THE EARTH
530 7th St., SE, Washington, DC 20003
(202) 462–1177.

GREENPEACE USA
1611 Connecticut Ave., NW, Washington, DC 20009
(202) 462–1177

GROUNDWATER MANAGEMENT DISTRICTS ASSOCIATION
1125 Maize Road
Colby, KS 67701
(913) 462–3915

INTERNATIONAL BOTTLED WATER ASSOCIATION
113 N. Henry St.
Alexandria, VA 22314
(703) 683–5213

INTERNATIONAL OCEANOGRAPHIC FOUNDATION
3979 Rickenbacker Causeway, Virginia Key, Miami, FL 33149
(305) 361–5786

INTERNATIONAL WATER RESOURCES ASSOCIATION
University of Illinois
205 North Mathews Ave., Urbana, IL 61801
(217) 333–0536

INTERSTATE CONFERENCE ON WATER PROBLEMS
955 L'Enfant Plaza Square, SW, Washington, DC
(202) 466–7287

IZAAK WALTON LEAGUE OF AMERICA
1701 N. Fort Myer Dr., Suite 1100, Arlington, VA 22209
(703) 528–1818

LEAGUE OF WOMEN VOTERS OF THE U.S.
1730 M St., NW, Washington, DC 20036
(202) 429–1965

NATIONAL AUDUBON SOCIETY
950 Third Ave., New York, NY 10022
(212) 832–3200

NATIONAL AUDUBON SOCIETY
645 Pennsylvania Ave., SE, Washington, DC 20003
(202) 547–9009

NATIONAL DEMONSTRATION WATER PROJECT
1111 North 19th St., Arlington, VA 22209
(703) 527–2282

NATIONAL FISH AND WILDLIFE FOUNDATION
18th and C Streets, NW, Rm. 2626, Washington, DC 20240
(202) 343–1040

NATIONAL RURAL WATER ASSOCIATION
2915 South 13th, P.O. Box 1428, Duncan, Oklahoma 73534
(405) 252–0629

NATIONAL SANITATION FOUNDATION
3475 Plymouth Road, P.O. Box 1468, Ann Arbor, Michigan 48106
(313) 769–8010

NATIONAL SOLID WASTES MANAGEMENT ASSOCIATION
1730 Rhode Island Avenue, NW, Suite 1000
Washington, DC 20036
(202) 659–4613

NATIONAL WATER RESOURCES ASSOCIATION
955 L'Enfant Plaza North, SW, Suite 1202
Washington, DC 20024
(202) 488–0610

NATIONAL WATERWAYS CONFERENCE
1130 17th St., NW, Washington, DC 20036
(202) 296–4415

NATIONAL WATER WELL ASSOCIATION
6375 Riverside Drive, Dublin, OH 43017
(614) 761–1711

NATIONAL WILDLIFE FEDERATION
1412 16th Street, NW, Washington, DC 20036
(202) 637–3700

NATURAL RESOURCES DEFENSE COUNCIL
122 E. 42nd St., New York, NY 10168
(212) 949–0049

NATURE CONSERVANCY
Suite 800, 1800 N. Kent St., Arlington, VA 22209
(703) 841–5300

OCEANIC SOCIETY
Executive Offices, P.O. Box 6032, NW, Washington, DC 20005
(202) 328–0098

RESOURCES FOR THE FUTURE
1616 P St., NW, Washington, DC 20036
(202) 328–5000

SIERRA CLUB
730 Polk St., San Francisco, CA 94109
(415) 776–2211

SIERRA CLUB LEGAL DEFENSE FUND
2044 Fillmore St., San Francisco, CA 94115
(415) 567–6100

SOIL AND WATER CONSERVATION SOCIETY
(formerly Soil Conservation Society of America),
7515 NE Ankeny Rd., Ankeny, IA 50021–9764
(515) 289–2331

TROUT UNLIMITED
National Headquarters, 501 Church St., NE
Vienna, VA 22180
(703) 281–1100

UNITED NATIONS ENVIRONMENT PROGRAMME
P.O. Box 30552, Nairobi, Kenya
Rm. DC2-0803 2 UN Plaza, NY 10017

U.S. PUBLIC INTEREST RESEARCH GROUP (U.S.PIRG),
215 Pennsylvania Ave., SE, Washington, DC 20003,
(202) 546–9707

WATER POLLUTION CONTROL FEDERATION
601 Wythe St., Alexandria, VA 22314–1994
(703) 684–2400

WATER QUALITY ASSOCIATION
4151 Naperville Road, Lisle, IL 60532
(312) 369–1600

WATER AND WASTEWATER EQUIPMENT MANUFACTURERS ASSOCIATION
P.O. Box 17402, Dulles Int'l Airport, Washington, DC 20041
(703) 661–8011

WILDERNESS SOCIETY
1400 St., NW, 10th Fl., Washington, DC 20005
(202) 842–3400

WORLD WILDLIFE FUND/CONSERVATION FOUNDATION
1255 23rd Street, NW, Washington, DC 20037
(202) 293–4800

CHAPTER 11

Constants and Conversion Factors

SECTION A. PHYSICAL PROPERTIES OF WATER

TABLE 11–1. VARIATION OF PROPERTIES OF PURE WATER WITH TEMPERATURE

Temperature, °C	Temperature, °F	Density at 1 Atmosphere, (gm-cm^{-3})	Dynamic Viscosity, in Centipoises (10^{-2} dyne-sec cm^{-2})	Kinematic Viscosity, in Centistokes (10^{-2} cm^2 sec^{-1})	Surface Tension Against Air, (dyne cm^{-1})	Vapor Pressure, (mm Hg)
5	41.0	0.999965	1.5188	1.5189	74.92	6.543
6	42.8	.999941	1.4726	1.4727	74.78	7.013
7	44.6	.999902	1.4288	1.4289	74.64	7.513
8	46.4	.999849	1.3872	1.3874	74.50	8.045
9	48.2	.999781	1.3476	1.3479	74.36	8.609
10	50.0	.999700	1.3097	1.3101	74.22	9.209
11	51.8	.999605	1.2735	1.2740	74.07	9.844
12	53.6	.999498	1.2390	1.2396	73.93	10.518
13	55.4	.999377	1.2061	1.2069	73.78	11.231
14	57.2	.999244	1.1748	1.1757	73.64	11.987
15	59.0	.999099	1.1447	1.1457	73.49	12.788
16	60.8	.998943	1.1156	1.1168	73.34	13.634
17	62.6	.998774	1.0875	1.0889	73.19	14.530
18	64.4	.998595	1.0603	1.0618	73.05	15.477
19	66.2	.998405	1.0340	1.0357	72.90	16.477
20	68.0	.998203	1.0087	1.0105	72.75	17.535
21	69.8	.997992	0.9843	0.9863	72.59	18.650
22	71.6	.997770	.9608	.9629	72.44	19.827
23	73.4	.997538	.9380	.9403	72.28	21.068
24	75.2	.997296	.9161	.9186	72.13	22.377
25	77.0	.997044	.8949	.8976	71.97	23.756
26	78.8	.996783	.8746	.8774	71.82	25.209
27	80.6	.996512	.8551	.8581	71.66	26.739
28	82.4	.996232	.8363	.8395	71.50	28.349
29	84.2	.995944	.8181	.8214	71.35	30.043
30	86.0	.995646	.8004	.8039	71.18	31.824
31	87.8	.995340	.7834	.7871	[a] 71.02	33.695
32	89.6	.995025	.7670	.7708	[a] 70.86	35.663
33	91.4	.994702	.7511	.7551	[a] 70.70	37.729
34	93.2	.994371	.7357	.7399	[a] 70.53	39.898
35	95.0	.99403	.7208	.7251	70.38	42.175
36	96.8	.99368	.7064	.7109	[a] 70.21	44.563
37	98.6	.99333	.6925	.6971	[a] 70.05	47.067
38	100.4	.99296	.6791	.6839	[a] 69.88	49.692

[a] Interpolated

Source: U.S. Department of the Interior, Bureau of Reclamation, 1977, Ground Water Manual

TABLE 11–2. DENSITY OF WATER

Temperature, °F	Density, gm/cm³
32	0.99987
35	0.99996
39.2	1.00000
40	0.99999
50	0.99975
60	0.99907
70	0.99802
80	0.99669
90	0.99510
100	0.99318
120	0.98870
140	0.98338
160	0.97729
180	0.97056
200	0.96333
212	0.95865

Density of Sea Water

Approximately 1.025 gm/cm^3 at 15° C.

TABLE 11–3. WEIGHT OF WATER

1 cubic inch	=	0.0360	lb.
1 cubic foot	=	62.4	lb.
1 gallon	=	8.33	lb.
1 Imperial gallon	=	10.0	lb.

Temperature	Weight, lb/ft³
32	62.416
35	62.421
39.2	62.424
40	62.423
50	62.408
60	62.366
70	62.300
80	62.217
90	62.118
100	61.998
120	61.719
140	61.386
160	61.006
180	60.586
200	60.135
212	59.843

Weight of Sea Water

Approximately 63.93 lb/ft^3 at 15° C.

TABLE 11–4. PROPERTIES OF WATER AT VARIOUS ALTITUDES

Altitude Feet	Barometric Pressure Inches Hg	Atmospheric Pressure psia	Atmospheric Pressure Feet Water
-500	30.5	15.0	34.6
0 (Seasonal)	29.9	14.7	33.9
500	29.4	14.5	33.4
1,000	28.9	14.2	32.8
1,500	28.3	13.9	32.1
2,000	27.8	13.7	31.5
4,000	25.8	12.7	29.2
6,000	24.0	11.8	27.2
8,000	22.2	10.9	25.2

Source: Soil Conservation Service, U.S. Department of Agriculture, 1971, Drainage of Agricultural Land

TABLE 11–5. PROPERTIES OF WATER AT VARIOUS TEMPERATURES

Temperature, ° F.	Vapor Pressure psfa	Vapor Pressure Feet Water	Specific Weight pcf	Specific Gravity
32.0	12.7	0.20	62.42	.9999
39.2[a]	16.9	0.27	62.427	1.0000
50.0	25.6	0.41	62.41	.9997
60.0	36.8	0.59	62.37	.9990
70.0	52.3	0.84	62.30	.9980
80.0	73.0	1.17	62.22	.9966
100.0	136.0	2.19	62.00	.9931

[a] Temperature when specific gravity = 1.0000
Source: Soil Conservation Service, U.S. Department of Agriculture, 1971, Drainage of Agricultural Land

TABLE 11–6. VARIATION OF BOILING POINT OF WATER WITH ELEVATION

[U. S. Standard Atmosphere is assumed]

Elevation, ft. above mean sea level	Boiling Point, ° F.
-1000	213.8
0	212.0
1000	210.2
2000	208.4
3000	206.5
4000	204.7
5000	202.9
6000	201.1
7000	199.2
8000	197.4
9000	195.6
10,000	193.7
11,000	191.9
12,000	190.1
13,000	188.2
14,000	186.4

TABLE 11–7. HEAT OF VAPORIZATION, VISCOSITY, VAPOR PRESSURE, SURFACE TENSION, AND ELASTIC MODULUS OF WATER

Temp, °F.	Heat Vaporization, Btu/lb	Viscosity		Vapor Pressure		Surface Tension lb/ft	Elastic Modulus lb/in^2
		Absolute, lb-sec/ft^2	Kinematic, ft^2/sec	Millibars	lb/in.^{2}		
32	1073	0.374×10^{-4}	1.93×10^{-5}	6.11	0.09	0.00518	287,000
40	1066	0.323	1.67	8.36	0.12	.00514	296,000
50	1059	0.273	1.41	12.19	0.18	.00508	305,000
60	1054	0.235	1.21	17.51	0.26	.00504	313,000
70	1049	0.205	1.06	24.79	0.36	.00497	319,000
80	1044	0.180	0.929	34.61	0.51	.00492	325,000
90	1039	0.160	0.828	47.68	0.70	.00486	329,000
100	1033	0.143	0.741	64.88	0.95	.00479	331,000
120	1021	0.117	0.610	....	1.69	.00466	332,000
140	1010	0.0979	0.513	....	2.89		
160	999	0.0835	0.440	....	4.74		
180	988	0.0726	0.385	....	7.51		
200	977	0.0637	0.341	....	11.52		
212	970	0.0593	0.319	....	14.70		

Heat of fusion at 32° F: 143.5 Btu/lb.

TABLE 11–8. THERMAL CONDUCTIVITY OF WATER

Temperature, ° C (° F.)	Cal/sec/cm^2 under 1° C./cm	Btu/sec/ft^2 under 1° F./in.
0 (32)	0.00139	0.00111
4 (39.2)	0.00138	0.00110
15 (59.0)	0.00144	0.00115
20 (68.0)	0.00143	0.00114

TABLE 11–9. VELOCITY OF SOUND IN WATER AND SEA WATER

[Assuming atmospheric pressure and sea water salinity of 35,000 ppm]

Temp., ° C	Pure Water, meters per second	Sea Water, meters per second
0	1,400	1,445
10	1,445	1,485
20	1,480	1,520
30	1,505	1,545

Source: Berry, Bollay, and Beers, Handbook of Meteorology, McGraw-Hill, Copyright 1945

SECTION B. WATER QUALITY

TABLE 11–10. CONSTANTS AND CONVERSION FACTORS FOR WATER QUALITY

For most waters, 1 part per million = 1 milligram per liter. The U.S. Geological Survey has arbitrarily selected 7,000 ppm as the concentration above which a density correction must be made: the determined concentration of each constituent must be divided by the density to give the correct parts per million. The computation is made before rounding the results.

Parts per million are converted to equivalents per million by multiplying ppm by the reciprocals of combining weights of the appropriate ions (see table below). The combining weight is equal to the molecular or ionic weight divided by the ionic charge, or valence.

SPECIFIC CONDUCTANCE RELATIONS
(Adapted from USGS Water Supply Paper 1454)

For most natural waters of a mixed type, the specific conductance, in micromhos,* multiplied by 0.65 approximates the residue on evaporation in parts per million. This does not approach an exact relation, and usually increases as the total dissolved solids exceed 2000 to 3000 ppm. For waters containing appreciable concentrations of free acid, caustic alkalinity, or sodium chloride, the factor may be much less than 0.65. The relationship for any water becomes indefinite as saturation is approached.

With similar limitations, the specific conductance divided by 100 approximates the equivalents per million of cations or anions.

* 1 micromho = 10^{-6} mho. The mho is unit of conductance; thus, a body has a conductance of one mho if one ampere of electric current flows when the potential difference is one volt. The conductance of a body in mhos is the reciprocal of the value of its resistance in ohms.

To convert	To	Multiply by
Grains per Imperial gallon	Parts per million	14.3
Grains per gallon	Parts per million	17.12
Parts per million	Grains per gallon	.05841
Parts per million	Tons per acre-foot	.00136
Parts per million	Tons per day	Second-feet x .0027
Ca^{++}	$CaCO_3$	2.497
$CaCl_2$	$CaCO_3$	.9018
HCO_3^-	$CaCO_3$	.8202
HCO_3^-	CO_3^{--}	.4917
Mg^{++}	$CaCO_3$	4.116
$MgCl_2$	$CaCO_3$	1.051
Na_2CO_3	$CaCO_3$	.9442
Fe^{+++}	H_2SO_4	2.634
NO_3^-	N	.2259
N	NO_3	4.4266

Parts per million to equivalents per million

Ion	Multiply by	Ion	Multiply by
Aluminum (Al^{+++})	0.11119	Iron (Fe^{+++})	0.05372
Barium (Ba^{++})	.01456	Lead (Pb^{++})	.00965
Bicarbonate (HCO_3^-)	.01639	Lithium (Li^+)	.14409
Bromide (Br^-)	.01251	Magnesium (Mg^{++})	.08224
Calcium (Ca^{++})	.04990	Manganese (Mn^{++})	.03640
Carbonate (CO_3^{--})	.03333	Manganese (Mn^{4+})	.07281
Chloride (Cl^-)	.02820	Nitrate (NO_3^-)	.01613
Chromium (Cr^{6+})	.11536	Phosphate (PO_4^{---})	.03159
Copper (Cu^{++})	.03148	Potassium (K^+)	.02558
Fluoride (F^-)	.05263	Sodium (Na^+)	.04350
Hydrogen (H^+)	.99206	Strontium (Sr^{++})	.02282
Hydroxide (OH^-)	.05880	Sulfate (SO_4^{--})	.02082
Iodide (I^-)	.00788	Sulfide (S^-)	.06237
Iron (Fe^{++})	.03581	Zinc (Zn^{++})	.03059

TABLE 11–10. CONSTANTS AND CONVERSION FACTORS FOR WATER QUALITY (continued)

EXPRESSIONS OF HARDNESS

1 grain per gallon	= 1 grain $CaCO_3$ per U.S. Gallon
1 part per million	= 1 part $CaCO_3$ per 1,000,000 parts of water
1 English, or Clark, degree	= 1 grain $CaCO_3$ per Imperial gallon
1 French degree	= 1 part $CaCO_3$ per 100,000 parts of water
1 German degree	= 1 part CaO per 100,000 parts of water

Conversions

1 grain per U.S. gallon	= 17.1 ppm, as $CaCO_3$
1 English degree	= 14.3 ppm, as $CaCO_3$
1 French degree	= 10 ppm, as $CaCO_3$
1 German degree	= 17.9 ppm, as $CaCO_3$

TABLE 11–11. FORMULA WEIGHTS AND EQUIVALENT WEIGHTS OF IONS FOUND IN WATER

[Weights expressed to three significant figures]

Ion	Formula Weight	Equivalent Weight	Ion	Formula Weight	Equivalent Weight
Al^{3+}	27.0	9.00	$Fe^{3}+$	55.8	18.6
Ba^{++}	137.	68.7	Pb^{++}	207.	104.
HCO_3^-	61.0	61.0	Li^{+}	6.94	6.94
Br^-	79.9	79.9	Mg^{++}	24.3	12.2
Ca^{++}	40.1	20.0	Mn^{++}	54.9	27.5
CO_3^{--}	60.0	30.0	Mn^{4+}	54.9	13.7
Cl^-	35.5	35.5	NO_3^-	62.0	62.0
Cr^{6+}	52.0	8.67	PO_4^{3-}	95.0	31.7
Cu^{++}	63.6	31.8	K^+	39.1	39.1
F^-	19.0	19.0	Na^+	23.0	23.0
H^+	1.01	1.01	Sr^{++}	87.6	43.8
OH^-	17.0	17.0	SO_4^{--}	96.1	48.0
I^-	127.	127.	S^{--}	32.1	16.0
Fe^{++}	55.8	27.9	Zn^{++}	65.4	32.7

SECTION C. LENGTH

TABLE 11–12. CONVERSION FACTORS FOR LENGTH

1 mil	= 1×10^{-3} in.
	= 0.0254 mm.
1 inch	= 0.08333 ft.
	= 0.0278 yd.
	= 25.4 mm.
	= 2.54 cm.
	= 0.0254 m.
1 foot	= 12 in.
	= 0.3333 yd.
	= 1.89×10^{-4} mi.
	= 30.48 cm.
	= 0.3048 m.
	= 3.048×10^{-4} km.
1 yard	= 36 in.
	= 3 ft.
	= 5.68×10^{-4} mi.
	= 91.44 cm.
	= 0.9144 m.
	= 9.144×10^{-4} km.
1 mile	= 5280 ft.
	= 1760 yd.
	= 1609.3 m.
	= 1.609 km.
1 micron	= 1×10^{-3} mm.
	= 3.937×10^{-5} in.
1 mm	= 0.03937 in.
	= 0.1 cm.
	= 1×10^{-3} m.
1 centimeter	= 0.3937 in.
	= 0.03281 ft.
	= 0.01094 yd.
	= 10 mm.
	= 0.01 m.
	= 1×10^{-5} km.
1 meter	= 39.37 in.
	= 3.281 ft.
	= 1.094 yd.
	= 6.21×10^{-4} mi.
	= 1000 mm.
	= 100 cm.
	= 1×10^{-3} km.
1 kilometer	= 3280.8 ft.
	= 1093.6 yd.
	= 0.621 mi.
	= 1×10^{5} cm.
	= 1000 m.
1 fathom	= 6 ft.
1 league (land)	= 5280 yd.
	= 3 mi.
	= 4.828 km.
1 international nautical mile	= 6076.1 ft.
	= 1.151 mi.
	= 1.852 km.

10 millimeters	=	1 centimeter
10 centimeters	=	1 decimeter
10 decimeters	=	1 meter
10 meters	=	1 dekameter
10 dekameters	=	1 hectometer
10 hectometers	=	1 kilometer

SECTION D. AREA

TABLE 11–13. CONVERSION FACTORS FOR AREA

1 sq. inch	$= 6.944 \times 10^{-3}$ ft.^{2}
	$= 7.716 \times 10^{-4}$ yd.^{2}
	= 6.452 cm.^{2}
1 sq. foot	= 144 in.^{2}
	= 0.1111 yd.^{2}
	$= 2.296 \times 10^{-5}$ acre
	$= 3.587 \times 10^{-8}$ mi.^{2}
	= 929.0 cm.^{2}
1 sq. yard	= 1296 in.^{2}
	= 9 ft.^{2}
	$= 2.066 \times 10^{-4}$ acres
	$= 3.228 \times 10^{-7}$ mi.^{2}
	= 0.8361 m.^{2}
	$= 8.361 \times 10^{-5}$ hectare
1 acre	= 43,560 ft.^{2}
	= 4840 yd.^{2}
	$= 1.562 \times 10^{-3}$ mi.^{2}
	= 4046.9 m.^{2}
	= 0.4047 hectare
	$= 4.047 \times 10^{-3}$ km.^{2}
1 sq. mile	$= 2.788 \times 10^{7}$ ft.^{2}
	$= 3.098 \times 10^{6}$ yd.^{2}
	= 640 acres
	$= 2.590 \times 10^{6}$ m.^{2}
	= 259.0 hectares
	= 2.590 km.^{2}
	= 1 section (of land)
1 township	= 36 mi.^{2}
	= 36 sections (of land)
1 sq. centimeter	= 0.1550 in.^{2}
	$= 1.076 \times 10^{-3}$ ft.^{2}
	$= 1.196 \times 10^{-4}$ yd.^{2}
	$= 1 \times 10^{-4}$ m.^{2}
1 sq. meter	= 1550.0 in.^{2}
	= 10.76 ft.^{2}
	= 1.196 yd.^{2}
	$= 2.471 \times 10^{-4}$ acre
	$= 1 \times 10^{-4}$ hectare
	$= 1 \times 10^{-6}$ km.^{2}
1 hectare	$= 1.076 \times 10^{5}$ ft.^{2}
	$= 1.196 \times 10^{4}$ yd.^{2}
	= 2.471 acres
	$= 3.861 \times 10^{-3}$ mi.^{2}
	$= 1 \times 10^{4}$ m.^{2}
	= 0.01 km.^{2}
1 sq. kilometer	$= 1.076 \times 10^{7}$ ft.^{2}
	$= 1.196 \times 10^{6}$ yd.^{2}
	= 247.1 acres
	= 0.3861 mi.^{2}
	$= 1 \times 10^{6}$ mi.^{2}
	= 100 hectares

SECTION E. VOLUME

TABLE 11–14. CONVERSION FACTORS FOR VOLUME

1 cubic inch	= 4.329×10^{-3} gal. = 5.787×10^{-4} ft.^{3} = 16.39 cm.^{3} = 0.01639 liter
1 gallon	= 231 in.^{3} = 0.1337 ft.^{3} = 4.951×10^{-3} yd.^{3} = 3.785×10^{-3} cm.^{3} = 3.785 liters = 3.785×10^{-3} m.^{3}
1 cubic foot	= 1728 in.^{3} = 7.481 gal = 0.03704 yd.^{3} = 2.832×10^4 cm.^{3} = 28.32 liters = 0.02832 m.^{3}
1 cubic yard	= 4.666×10^4 in.^{3} = 27 ft.^{3} = 0.7646 m.^{3} = 6.198×10^{-4} acre-foot
1 acre-inch	= 2.715×10^4 gal. = 3630 ft^3
1 acre-foot	= 3.259×10^5 gal. = 43,560 ft.^{3} = 1234 m.^{3}
1 cubic centimeter	= 0.06102 in.^{3} = 2.642×10^{-4} gal. = 3.532×10^{-5} ft.^{3} = 1×10^{-3} liter
1 liter	= 61.02 in.^{3} = 0.2642 gal. = 0.03532 ft.^{3} = 1000 cm.^{3} = 1×10^{-3} m.^{3}
1 cubic meter	= 264.2 gal. = 35.32 ft.^{3} = 1.307 yd.^{3} = 1×10^6 cm^3 = 1000 liters
1 Imperial gallon	= 277.4 in.^{3} = 1.201 gal. = 0.1604 ft.^{3} = 4.546 liters
1 cfs-day	= 1.98 acre-feet = 0.0372 in. -mi.^{2}
1 inch of rain	= 5.610 gal/yd^2 = 2.715×10^4 gal./acre
1 inch-mi.^{2}	= 2.323×10^6 ft.^{3} = 53.3 acre-feet = 26.9 cfs-days
1 barrel (oil)	= 42 gal.
1 million gallons	= 3.069 acre-feet
1 pint	= 28.875 in.^{3} = 0.5 qt. = 0.473 liter
1 quart	= 57.75 in.^{3} = 2 pt. = 0.25 gal. = 0.946 liter

TABLE 11–15. AREAS AND VOLUMES OF VERTICAL CYLINDRICAL TANKS AND WELLS

Diameter	Area, sq ft, or cu ft per ft of depth	U.S. gal per ft of depth	Diameter	Area, sq ft, or cu ft per ft of depth	U.S. gal per ft of depth	Diameter	Area, sq ft, or cu ft per ft of depth	U.S. gal per ft of depth
1'	0.785	5.87	6'	28.27	211.5	28'	615.8	4,606
1' 1"	0.922	6.89	6' 3"	30.68	229.5	28' 6"	637.9	4,772
1' 2"	1.069	8.00	6' 6"	33.18	248.2	29'	660.5	4,941
1' 3"	1.227	9.18	6' 9"	35.78	267.7	29' 6"	683.5	5,113
1' 4"	1.396	10.44	7'	38.48	287.9	30'	706.9	5,288
1' 5"	1.576	11.79	7' 3"	41.28	308.8	31'	754.8	5,646
1' 6"	1.767	13.22	7' 6"	44.18	330.5	32'	804.3	6,016
1' 7"	1.969	14.73	7' 9"	47.17	352.9	33'	855.3	6,398
1' 8"	2.182	16.32	8'	50.27	376.0	34'	907.9	6,792
1' 9"	2.405	17.99	8' 3"	53.46	399.9	35'	962.1	7,197
1' 10"	2.640	19.75	8' 6"	56.75	424.5	36'	1,018	7,616
1' 11"	2.885	21.58	8' 9"	60.13	449.8	37'	1,075	8,043
2'	3.142	23.50	9'	63.62	475.9	38'	1,134	8,483
2' 1"	3.409	25.50	9' 3"	67.20	502.7	39'	1,195	8,940
2' 2"	3.637	27.58	9' 6"	70.88	530.2	40'	1,257	9,404
2' 3"	3.976	29.74	9' 9"	74.66	558.5	41'	1,320	9,876
2' 4"	4.276	31.99	10'	78.54	587.5	42'	1,385	10,360
2' 5"	4.587	34.31	10' 6"	86.59	647.7	43'	1,452	10,860
2' 6"	4.909	36.72	11'	95.03	710.9	44'	1,521	11,370
2' 7"	5.241	39.21	11' 6"	103.9	777.0	45'	1,590	11,900
2' 8"	5.585	41.78	12'	113.1	846.0	46'	1,662	12,430
2' 9"	5.940	44.43	12' 6"	122.7	918.0	47'	1,735	12,980
2' 10"	6.305	47.16	13'	132.7	992.9	48'	1,810	13,540
2' 11"	6.681	49.98	13' 6"	143.1	1,071	49'	1,886	14,110
3'	7.069	52.88	14'	153.9	1,152	50'	1,964	14,690
3' 1"	7.467	55.86	14' 6"	165.1	1,235	52'	2,124	15,890
3' 2"	7.876	58.92	15'	176.7	1,322	54'	2,290	17,130
3' 3"	8.296	62.06	15' 6"	188.7	1,412	56'	2,463	18,420
3' 4"	8.727	65.28	16'	201.1	1,504	58'	2,642	19,760
3' 5"	9.168	68.58	16' 6"	213.8	1,600	60'	2,827	21,150
3' 6"	9.621	71.97	17'	227.0	1,698	62'	3,019	22,580
3' 7"	10.08	75.44	17' 6"	240.5	1,799	64'	3,217	24,060
3' 8"	10.56	78.99	18'	254.5	1,904	66'	3,421	25,590
3' 9"	11.04	82.62	18' 6"	268.8	2,011	68'	3,632	27,170
3' 10"	11.54	86.33	19'	283.5	2,121	70'	3,848	28,790
3' 11"	12.05	90.13	19' 6"	298.6	2,234	72'	4,072	30,450
4'	12.57	94.00	20'	314.2	2,350	74'	4,301	32,170
4' 1"	13.10	97.96	20' 6"	330.1	2,469	76'	4,536	33,930
4' 2"	13.64	102.0	21'	346.4	2,591	78'	4,778	35,740
4' 3"	14.19	106.1	21' 6"	363.1	2,716	80'	5,027	37,600
4' 4"	14.75	110.3	22'	380.1	2,844	82'	5,281	39,500
4' 5"	15.32	114.6	22' 6"	397.6	2,974	84'	5,542	41,450
4' 6"	15.90	119.0	23'	415.5	3,108	86'	5,809	43,450
4' 7"	16.50	123.4	23' 6"	433.7	3,245	88'	6,082	45,490
4' 8"	17.10	128.0	24'	452.4	3,384	90'	6,362	47,590
4' 9"	17.72	132.6	24' 6"	471.4	3,527	92'	6,648	49,720
4' 10"	18.35	137.3	25'	490.9	3,672	94'	6,940	51,920
4' 11'	18.99	142.0	25' 6"	510.7	3,820	96'	7,238	54,140
5'	19.63	146.9	26'	530.9	3,972	98'	7,543	56,420
5' 3"	21.65	161.9	26' 6"	551.5	4,126	100'	7,854	58,750
5' 6"	23.76	177.7	27'	572.6	4,283			
5' 9"	25.97	194.3	27' 6"	594.0	4,443			

TABLE 11–16. VOLUMES OF HORIZONTAL CYLINDRICAL TANKS

[U.S. gallons per foot of length]

	PERCENT FULL									
Diameter	**10**	**20**	**30**	**40**	**50**	**60**	**70**	**80**	**90**	**100**
1 ft.	.3	.8	1.4	2.1	2.9	3.6	4.3	4.9	5.5	5.87
2 ft.	1.2	3.3	5.9	8.8	11.7	14.7	17.5	20.6	22.2	23.50
3 ft.	2.7	7.5	13.6	19.8	26.4	33.0	39.4	45.2	50.1	52.88
4 ft.	4.9	13.4	23.8	35.0	47.0	59.0	70.2	80.5	89.0	94.00
5 ft.	7.6	20.0	37.0	55.0	73.0	92.0	110.0	126.0	139.0	146.9
6 ft.	11.0	30.0	53.0	78.0	106.0	133.0	158.0	182.0	201.0	211.5
7 ft.	15.0	41.0	73.0	107.0	144.0	181.0	215.0	247.0	272.0	287.9
8 ft.	19.0	52.0	96.0	140.0	188.0	235.0	281.0	322.0	356.0	376.0
9 ft.	25.0	67.0	112.0	178.0	238.0	298.0	352.0	408.0	450.0	475.9
10 ft.	30.0	83.0	149.0	219.0	294.0	368.0	440.0	504.0	556.0	587.5
11 ft.	37.0	101.0	179.0	265.0	356.0	445.0	531.0	610.0	672.0	710.9
12 ft.	44.0	120.0	214.0	315.0	423.0	530.0	632.0	714.0	800.0	846.0
13 ft.	51.0	141.0	250.0	370.0	496.0	621.0	740.0	850.0	940.0	992.9
14 ft.	60.0	164.0	291.0	430.0	576.0	722.0	862.0	989.0	1084.0	1152.0
15 ft.	68.0	188.0	334.0	494.0	661.0	829.0	988.0	1134.0	1253.0	1322.0

SECTION F. VELOCITY

TABLE 11–17. CONVERSION FACTORS FOR VELOCITY

1 foot/second	= 0.682 miles/hour
	= 0.3048 meters/second
1 mile/hour	= 1.467 feet/second
	= 0.869 international knots
	= 1.609 kilometers/hour
1 international knot	= 1.151 miles/hour
	= 1.852 kilometers/hour
1 meter/second	= 3.6 kilometers/hour
	= 3.28 feet/second
	= 2.237 miles/hour
	= 1.94 knots
1 kilometer/hour	= 0.621 miles/hour
	= 0.540 international knots

SECTION G. FLOW RATE

TABLE 11–18. CONVERSION FACTORS FOR FLOW RATE

1 gallon per minute	= 1.44×10^{-3} mgd
	= 2.23×10^{-3} cfs
	= 4.42×10^{-3} AF/day
	= 0.0631 l./sec.
	= 6.31×10^{-5} m.^{3}/sec.
	= 5.42 m.^{3}/day
1 million gallons per day	= 694 gpm
	= 1.55 cfs
	= 3.07 AF/day
	= 43.7 l./sec.
	= 0.0437 m.^{3}/sec.
	= 3770 m.^{3}/day
1 cubic foot per second	= 449 gpm
	= 0.646 mgd
	= 1.98 AF/day
	= 28.3 l./sec.
	= 0.0283 m.^{3}/sec.
	= 2450 m.^{3}/day
1 acre-foot per day	= 226 gpm
	= 0.326 mgd
	= 0.505 cfs
	= 14.2 l./sec.
	= 0.0142 m.^{3}/sec.
	= 1230 m.^{3}/day
1 liter per second	= 15.9 gpm
	= 0.0228 mgd
	= 0.0353 cfs
	= 0.0703 AF/day
	= 0.001 m.^{3}/sec.
	= 86.4 m.^{3}/day
1 cubic meter per second	= 1.58×10^4 gpm
	= 22.8 mgd
	= 35.3 cfs
	= 70.0 AF/day
	= 1000 l./sec.
	= 8.64×10^4 m.^{3}/day
1 cubic meter per day	= 0.183 gpm
	= 2.64×10^{-4} mgd
	= 4.09×10^{-4} cfs
	= 8.11×10^{-4} AF/day
	= 0.0116 l./sec.
	= 1.16×10^{-5} m.^{3}/sec.
1 miner's inch	= 0.025 cubic foot per second (in Arizona, California, Montana, and Oregon)
	= 0.02 cubic foot per second (in Idaho, Kansas, Nebraska, New Mexico, North and South Dakota, and Utah)
	= 0.026 cubic foot per second (in Colorado)
	= 0.028 cubic foot per second (in British Columbia)

TABLE 11–19. CONVERSION OF DAILY CONSUMPTIVE USE RATES TO EQUIVALENT CONTINUOUS FLOW RATES PER UNIT AREA

Consumptive Use of Water by Crop		**Equivalent Continuous Flow Rates**			
		Per Hectare		**Per Acre**	
Millimetres per day	*Inches per day*	*Litres per second*	*Cubic metres per day*	*Cubic feet per second*	*U.S. gallons per minute*[a]
2	0.08	0.23	20	0.0033	1.5
3	0.12	0.35	30	0.0050	2.2
4	0.16	0.46	40	0.0067	3.0
5	0.20	0.58	50	0.0083	3.7
6	0.24	0.69	60	0.0100	4.5
7	0.28	0.81	70	0.0117	5.2
8	0.32	0.92	80	0.0133	6.0
9	0.36	1.04	90	0.0150	6.7
10	0.40	1.15	100	0.0167	7.5

[a] To convert to imperial gallons per minute, multiply values by 0.833.
Source: Food and Agriculture Organization of the United Nations, 1974, Surface Irrigation. Copyright FAO. Reprinted with permission.

TABLE 11–20. TIME REQUIRED TO SUPPLY ONE ACRE WITH ONE INCH OF WATER

Flow Rate		**Time**	
gpm	**cfs**	**hr**	**min**
100	0.22	4	32
150	0.33	3	01
200	0.45	2	16
250	0.56	1	48
300	0.67	1	30
350	0.78	1	18
400	0.89	1	08
450	1.00	1	00
500	1.11	0	54
600	1.34	0	46
700	1.56	0	39
800	1.78	0	34
900	2.01	0	30
1000	2.23	0	27
1100	2.45	0	24
1200	2.67	0	22
1300	2.90	0	21
1400	3.12	0	20
1500	3.34	0	18

SECTION H. WEIGHT

TABLE 11–21. CONVERSION FACTORS FOR WEIGHT

1 grain	= 2.286×10^{-3} oz.
	= 1.429×10^{-4} lb.
1 ounce	= 437.5 gr.
	= 0.0625 lb.
	= 28.35 gm.
1 pound	= 7000 gr.
	= 16 oz.
	= 453.6 gm.
	= 0.4536 kg.
1 short ton	= 2000 lb.
	= 907.2 kg.
	= 0.9072 m. ton
1 long ton	= 2240 lb.
	= 1016.0 kg.
	= 1.016 m. ton
1 metric ton	= 2204.6 lb.
	= 1000 kg.
	= 1.102 short ton
	= 0.9842 long ton
1 gram	= 15.43 gr.
	= 0.03527 oz.
	= 2.205×10^{-3} lb.
	= 0.001 kg.
1 kilogram	= 35.27 oz.
	= 2.205 lb.
	= 1000 gm.
	= 0.001 m. ton

SECTION I. PRESSURE

TABLE 11–22. CONVERSION FACTORS FOR PRESSURE

1 inch of water at 62° F	= 0.0361 lb./in.^{2}
	= 5.196 lb./ft.^{2}
	= 0.0735 inch of mercury at 62° F
1 foot of water at 62° F	= 0.433 lb./in.^{2}
	= 62.36 lb./ft.^{2}
	= 0.833 inch of mercury at 62° F
	= 2.950 x 10^{-2} atmosphere
1 atmosphere	= 14.7 lb./in.^{2}
	= 2116 lb./ft.^{2}
	= 33.95 feet of water at 62° F
	= 29.92 inches of mercury at 32° F
	= 1013.2 millibars
	= 760 mm. of mercury at 32° F
	= 1033 gm./cm.^{2}
	= 1.033 kg./cm.^{2}
1 pound per sq. inch	= 2.309 feet of water at 62° F
	= 2.036 inches of mercury at 32° F
	= 0.06804 atmosphere
	= 0.07031 kg./cm.^{2}
1 inch of mercury at 32° F	= 0.4192 lb./in.^{2}
	= 1.133 feet of water at 32° F
1 kilogram per sq. centimeter	= 14.22 lb./in.^{2}
1 bar	= 1 x 10^6 dynes/cm.^{2}
	= 1019.8 gm./cm.^{2}
	= 0.9869 atmosphere
	= 14.50 lb./in.^{2}
	= 1000 millibars

SECTION J. PERMEABILITY

TABLE 11–23. UNITS AND CONVERSION FACTORS FOR PERMEABILITY

Permeability units

Laboratory (standard) coefficient of permeability:

$$1K_S = \frac{\text{1 gal. of water at 60° F/day}}{(\text{1 ft.}^2)\ (\text{1 ft./ft.})}$$

Field coefficient of permeability:

$$1K_f = \frac{\text{1 gal. of water at field temperature/day}}{(\text{1 ft. x 1 mile})\ (\text{1 ft./mile})}$$

Specific (intrinsic) permeability:

$$\text{1 darcy} = \frac{(\text{1 centipoise x 1 cm.}^3\text{/sec.})/\text{1 cm.}^2}{\text{1 atmosphere/1 cm.}}$$

In ground-water investigations, the water-transmitting capability of an aquifer is commonly expressed in terms of a Coefficient of Transmissivity, which is the product of the Coefficient of Permeability of the aquifer material and the thickness of the aquifer.

Conversion factors

Permeability		
	$1K_S$	= 4.72×10^{-5} cm./sec.
		= 0.0408 m./day
		= 0.134 ft./day
		= 1.43×10^{-6} ft./sec.
	1 darcy	= 0.987×10^{-8} cm.^{2}
		= 1.062×10^{-11} ft.^{2}
		= $18.2K_S$ for water at 60° F
		= 0.966×10^{-3} cm./sec. for water at 20° C
	1 millidarcy	= 1.82×10^{-2} K_S for water at 60° F
	1 cm./sec.	= 1.02×10^{-5} cm.^{2} for water at 20° C
	1 m/sec.	= 2.12×10^{6} gpd/ft.^{2}
	1 m/day	= 24.5 gpd/ft.^{2}
Transmissivity		
	1 m^2/sec.	= 6.96×10^{6} gpd/ft.
	1 m^2/day	= 80.5 gpd/ft.

SECTION K. POWER

TABLE 11–24. CONVERSION FACTORS FOR POWER

1 H. P.	= .746 kilowatts or 746 watts
	= 33,000 ft. lbs. per minute
	= 550 ft. lbs. per second
	= 8760 horsepower-hours per year
	= 6535 kilowatt-hours per year
H. P. Input	= Horsepower input to motor
	= 1.34 x kilowatts input to motor
Water H. P.	= H. P. required to lift water at a definite rate to a given distance assuming 100% efficiency
	$= \frac{\text{G. P. M. x total head (in ft.)}}{3960}$
Brake H. P.	= H. P. delivered by motor
	= H. P. required by pump
	= H. P. input motor x efficiency
	= 1.34 x KW input x motor efficiency
	$= \frac{\text{Water H. P}}{\text{Pump efficiency}}$
	$= \frac{\text{G. P. M. x total head (ft.)}}{\text{3960 x pump efficiency}}$
	$= \frac{\text{G. P. H. x total head (lbs. per sq. in.)}}{\text{103,000 x pump efficiency}}$
1 kilowatt	= 1.3405 Horsepower
	= 8760 kilowatt-hours per year
Efficiency	$= \frac{\text{Power output}}{\text{Power input}}$
Motor Efficiency	$= \frac{\text{H. P. output}}{\text{K. W. input x 1.34}}$
Pump Efficiency	$= \frac{\text{G. P. M. x total head (ft.)}}{\text{3960 x B. H. P.}}$
Plant Efficiency	$= \frac{\text{G. P. M. x total head (ft.)}}{\text{5300 x KW input.}}$

TABLE 11–25. HORSEPOWER REQUIRED TO PUMP WATER

[Efficiency of pumping plant 50 percent of theoretical. Use for estimating only.]

Gallons per Minute	Cubic Feet per Second	Horsepower Required for Elevations of 10 ft	20 ft	30 ft	40 ft	50 ft	60 ft	70 ft	80 ft
100	0.22	0.5	1.0	1.5	2.0	2.5	3.0	3.5	4.0
150	0.33	0.8	1.5	2.3	3.0	3.8	4.6	5.3	6.1
200	0.45	1.0	2.0	3.0	4.0	5.0	6.1	7.1	8.1
250	0.56	1.3	2.5	3.8	5.0	6.3	7.6	8.8	10.1
300	0.67	1.5	3.0	4.6	6.1	7.6	9.1	10.6	12.1
350	0.78	1.8	3.5	5.3	7.1	8.8	10.6	12.4	14.1
400	0.89	2.0	4.0	6.1	8.1	10.1	12.1	14.1	16.2
450	1.00	2.3	4.6	6.8	9.1	11.4	13.6	15.9	18.2
500	1.11	2.5	5.0	6.7	10.1	12.6	15.2	17.7	20.2
600	1.34	3.0	6.1	9.1	12.1	15.2	18.2	21.2	24.2
700	1.56	3.5	7.1	10.6	14.1	17.7	21.2	24.8	28.3
800	1.78	4.0	8.1	12.1	16.2	20.2	24.2	28.3	32.3
900	2.01	4.6	9.1	13.6	18.2	22.7	27.3	31.8	36.4
1000	2.23	5.0	10.1	15.2	20.2	25.2	30.3	35.4	40.4
1250	2.78	6.3	12.6	18.9	25.2	31.6	37.9	44.2	50.5
1500	3.34	7.6	15.2	22.7	30.3	37.9	45.4	53.0	60.6

Source: New Mexico Agricultural Experiment Station

TABLE 11–26. THEORETICAL POWER OF AMERICAN MULTIBLADE WINDMILLS OF DIFFERENT DIAMETERS FOR SEVERAL WIND VELOCITIES

Wind velocity (miles per hour)	Horsepower[a] from wheel diameter (in feet) of— 6	8	10	12	14	16	18
6				0.05	0.06	0.08	0.09
8		0.05	0.07	.10	.14	.18	.23
10	0.05	.09	.14	.21	.28	.36	.46
12	.09	.15	.24	.36	.49	.63	.81
15	.18	.31	.48	.68	.94	1.21	1.55
20	.40	.72	1.12	1.62	2.20	2.88	3.65
25	.79	1.41	2.21	3.20	4.34	5.67	7.20

[a] Horsepower = $CD^2 V^3$. Where C, the power coefficient for American multiblade type wheel = 0.45×10^{-6}, when the tip speed is approximately the same as the velocity of the wind; D = diameter of the wheel in feet; and V = indicated velocity of the wind in feet per second (Robinson anemometer).

Source: U. S. Dept. of Agriculture

SECTION L. TEMPERATURE

TABLE 11–27. CONVERSION FACTORS FOR TEMPERATURE

TEMPERATURE CONVERSION Fahrenheit ◄——► Centigrade

°F to °C (RANGE 0° to 279° F)

°F	°C	°F	°C	°F	°C	°F	°C	°F	°C
0	-17.78	56	13.33	112	44.44	168	75.56	224	106.67
1	-17.22	57	13.89	113	45.00	169	76.11	225	107.22
2	-16.67	58	14.44	114	45.56	170	76.67	226	107.78
3	-16.11	59	15.00	115	46.11	171	77.22	227	108.33
4	-15.56	60	15.56	116	46.67	172	77.78	228	108.89
5	-15.00	61	16.11	117	47.22	173	78.33	229	109.44
6	-14.44	62	16.67	118	47.78	174	78.89	230	110.00
7	-13.89	63	17.22	119	48.33	175	79.44	231	110.56
8	-13.33	64	17.78	120	48.89	176	80.00	232	111.11
9	-12.78	65	18.33	121	49.44	177	80.56	233	111.67
10	-12.22	66	18.89	122	50.00	178	81.11	234	112.22
11	-11.67	67	19.44	123	50.56	179	81.67	235	112.78
12	-11.11	68	20.00	124	51.11	180	82.22	236	113.33
13	-10.56	69	20.56	125	51.67	181	82.78	237	113.89
14	-10.00	70	21.11	126	52.22	182	83.33	238	114.44
15	- 9.44	71	21.67	127	52.78	183	83.89	239	115.00
16	- 8.89	72	22.22	128	53.33	184	84.44	240	115.56
17	- 8.33	73	22.78	129	53.89	185	85.00	241	116.11
18	- 7.78	74	23.33	130	54.44	186	85.56	242	116.67
19	- 7.22	75	23.89	131	55.00	187	86.11	243	117.22
20	- 6.67	76	24.44	132	55.56	188	86.67	244	117.78
21	- 6.11	77	25.00	133	56.11	189	87.22	245	118.33
22	- 5.56	78	25.56	134	56.67	190	87.78	246	118.89
23	- 5.00	79	26.11	135	57.22	191	88.33	247	119.44
24	- 4.44	80	26.67	136	57.78	192	88.89	248	120.00
25	- 3.89	81	27.22	137	58.33	193	89.44	249	120.56
26	- 3.33	82	27.78	138	58.89	194	90.00	250	121.11
27	- 2.78	83	28.33	139	59.44	195	90.56	251	121.67
28	- 2.22	84	28.89	140	60.00	196	91.11	252	122.22
29	- 1.67	85	29.44	141	60.56	197	91.67	253	122.78
30	- 1.11	86	30.00	142	61.11	198	92.22	254	123.33
31	- 0.56	87	30.56	143	61.67	199	92.78	255	123.89
32	0.00	88	31.11	144	62.22	200	93.33	256	124.44
33	0.56	89	31.67	145	62.78	201	93.89	257	125.00
34	1.11	90	32.22	146	63.33	202	94.44	258	125.56
35	1.67	91	32.78	147	63.89	203	95.00	259	126.11
36	2.22	92	33.33	148	64.44	204	95.56	260	126.67
37	2.78	93	33.89	149	65.00	205	96.11	261	127.22
38	3.33	94	34.44	150	65.56	206	96.67	262	127.78
39	3.89	95	35.00	151	66.11	207	97.22	263	128.33
40	4.44	96	35.56	152	66.67	208	97.78	264	128.89
41	5.00	97	36.11	153	67.22	209	98.33	265	129.44
42	5.56	98	36.67	154	67.78	210	98.89	266	130.00
43	6.11	99	37.22	155	68.33	211	99.44	267	130.56
44	6.67	100	37.78	156	68.89	212	100.00	268	131.11
45	7.22	101	38.33	157	69.44	213	100.56	269	131.67
46	7.78	102	38.89	158	70.00	214	101.11	270	132.22
47	8.33	103	39.44	159	70.56	215	101.67	271	132.78
48	8.89	104	40.00	160	71.11	216	102.22	272	133.33
49	9.44	105	40.56	161	71.67	217	102.78	273	133.89
50	10.00	106	41.11	162	72.22	218	103.33	274	134.44
51	10.56	107	41.67	163	72.78	219	103.89	275	135.00
52	11.11	108	42.22	164	73.33	220	104.44	276	135.56
53	11.67	109	42.78	165	73.89	221	105.00	277	136.11
54	12.22	110	43.33	166	74.44	222	105.56	278	136.67
55	12.78	111	43.89	167	75.00	223	106.11	279	137.22

TABLE 11–27. CONVERSION FACTORS FOR TEMPERATURE (continued)

° F TO ° C (RANGE 280° to 300° F)		° C TO ° F (RANGE 0° to 150° C)					
° F	° C	° C	° F	° C	° F	° C	° F
280	137.78	0	32.0	56	132.8	112	233.6
281	138.33	1	33.8	57	134.6	113	235.4
282	138.89	2	35.6	58	136.4	114	237.2
283	139.44	3	37.4	59	138.2	115	239.0
284	140.00	4	39.2	60	140.0	116	240.8
285	140.56	5	41.0	61	141.8	117	242.6
286	141.11	6	42.8	62	143.6	118	244.4
287	141.67	7	44.6	63	145.4	119	246.2
288	142.22	8	46.4	64	147.2	120	248.0
289	142.78	9	48.2	65	149.0	121	249.8
290	143.33	10	50.0	66	150.8	122	251.6
291	143.89	11	51.8	67	152.6	123	253.4
292	144.44	12	53.6	68	154.4	124	255.2
293	145.00	13	55.4	69	156.2	125	257.0
294	145.56	14	57.2	70	158.0	126	258.8
295	146.11	15	59.0	71	159.8	127	260.6
296	146.67	16	60.8	72	161.6	128	262.4
297	147.22	17	62.6	73	163.4	129	264.2
298	147.78	18	64.4	74	165.2	130	266.0
299	148.33	19	66.2	75	167.0	131	267.8
300	148.89	20	68.0	76	168.8	132	269.6
		21	69.8	77	170.6	133	271.4
		22	71.6	78	172.4	134	273.2
		23	73.4	79	174.2	135	275.0
		24	75.2	80	176.0	136	276.8
		25	77.0	81	177.8	137	278.6
		26	78.8	82	179.6	138	280.4
		27	80.6	83	181.4	139	282.2
		28	82.4	84	183.2	140	284.0
		29	84.2	85	185.0	141	285.8
		30	86.0	86	186.8	142	287.6
		31	87.8	87	188.6	143	289.4
		32	89.6	88	190.4	144	291.2
		33	91.4	89	192.2	145	293.0
		34	93.2	90	194.0	146	294.8
		35	95.0	91	195.8	147	296.6
		36	96.8	92	197.6	148	298.4
		37	98.6	93	199.4	149	300.2
		38	100.4	94	201.2	150	302.0
		39	102.2	95	203.0		
		40	104.0	96	204.8		
		41	105.8	97	206.6		
		42	107.6	98	208.4		
		43	109.4	99	210.2		
		44	112.2	100	212.0		
		45	113.0	101	213.8		
		46	114.8	102	215.6		
		47	116.6	103	217.4		
		48	118.4	104	219.2		
		49	120.2	105	221.0		
		50	122.0	106	222.8		
		51	123.8	107	224.6		
		52	125.6	108	226.4		
		53	127.4	109	228.2		
		54	129.2	110	230.0		
		55	131.0	111	231.8		

Temperature Conversions:

$° F = 9/5 (° C) + 32$

$° C = 5/9 (° F - 32)$

INDEX